Handbook of Risk and Insurance Strategies for Certified Public Risk Officers and Other Water Professionals

Handbook of Risk and Insurance Strategies for Certified Public Risk Officers and Other Water Professionals

Frank R. Spellman, Lorilee Medders and Paul Fuller

Foreword by

Gordon Graham

Part of the AAWD&M Certified Public Risk Officer-Water/Wastewater
(CPRO-W^2) Designation and Learning Series

CRC Press
Taylor & Francis Group
Boca Raton London New York

CRC Press is an imprint of the
Taylor & Francis Group, an **informa** business

First edition published 2022
by CRC Press
6000 Broken Sound Parkway NW, Suite 300, Boca Raton, FL 33487-2742

and by CRC Press
2 Park Square, Milton Park, Abingdon, Oxon, OX14 4RN

© 2022 American Association of Water Distribution & Management

CRC Press is an imprint of Taylor & Francis Group, LLC

Library of Congress Cataloging-in-Publication Data

Names: Spellman, Frank R., author.
Title: Handbook of risk and insurance strategies for certified public risk
officers and other water professionals / Frank Spellman, PhD, CPRO-W2,
Lorilee Medders, PhD, CPRO-W2, Paul Fuller, MS, CPCU, CPRO-W2; foreword
by Gordon Graham, MS, JD, CPRO-W2.
Description: First edition. | Boca Raton, FL : CRC Press, 2022. | Includes
index.
Identifiers: LCCN 2021015964 (print) | LCCN 2021015965 (ebook) | ISBN
9781032072074 (hbk) | ISBN 9781032074849 (pbk) | ISBN 9781003207146
(ebk)
Subjects: LCSH: Water treatment plants--Risk management. |
Water--Purification--Risk assessment. | Sewage--Purification--Risk
assessment. | Health risk assessment. | Water purification equipment
industry--Insurance.
Classification: LCC TD434 .S636 2022 (print) | LCC TD434 (ebook) | DDC
363.6/10681--dc23
LC record available at https://lccn.loc.gov/2021015964
LC ebook record available at https://lccn.loc.gov/2021015965

ISBN: 978-1-032-07207-4 (hbk)
ISBN: 978-1-032-07484-9 (pbk)
ISBN: 978-1-003-20714-6 (ebk)

DOI: 10.1201/9781003207146

Typeset in Times
by Deanta Global Publishing Services, Chennai, India

Contents

PART I Operations for Public Water Systems

PART II   Operations for Publicly Owned Treatment Works (POTW)

PART III   Practical Risk Management for Public Water/Wastewater Systems

PART IV *Insurance Essentials for Public Water/Wastewater Systems*

Disclaimer

The information contained in the CPRO-W2 Designation and Learning Series is meant to provide the reader with a general understanding of certain aspects of technical operations, risk management and insurance structure for public water/wastewater systems. The information is not to be construed as legal, risk management, or insurance advice and is not meant to be a substitute for legal, risk management, or insurance advice. AAWD&M, its stakeholders, and the authors expressly disclaim any and all liability with respect to any actions taken or not taken based upon the information contained in the CPRO-W2 Designation and Learning Series or with respect to any errors or omissions contained in such inform

Foreword

by Gordon Graham, MS, JD, CPRO-W[2]

Recent events have taken me out of airports and airplanes where I have spent most of the past 20 years as a lecturer in the field of managing risk. As a result, I have spent a considerable amount of time at home getting caught up on things. After finishing Netflix, Prime, Hulu, and several other streaming services I came across a fascinating offering in CNBC – "The Profit." For those of you who have not watched, the show the creator and star is Marcus Lemonis. He is a very successful businessman with a tremendous "story" who now spends time assisting businesses in trouble and making them successful.

Very quickly into watching this program I learned is the approach to success in any business. People, Product, and Process – and he repeated this over and over again and his thinking works. An organization that has good people and a good product and a good process on making it all work is in route to success. These "three pillars" are inextricably intertwined with each other – and if one (or more) is not present you will not be successful.

The above two paragraphs are foundational thoughts for this foreword. You are in the water/wastewater "business." The mere fact that you have this book and are reading this foreword tells me you are "good people." I could digress for hours on how true this observation is, but frankly I am shocked at how few people read about their profession and have the drive and desire to make things better. After more than 60,000 hours of standing in front of audiences around the world I have learned a few things – one of which is that too many people in too many professions consider their work "a job" and they show up and do what they are told to do and that is that. You are commencing the reading of a book about how to improve things in your profession – water/wastewater operations – and thank you for doing so.

Back to Mr. Lemonis and his thinking – you now need a good product – and you certainly have that. What is more important than water? And what do we do with this water once used? Again, I digress but too many Americans have never traveled – and I am not talking about going to other "first world" nations – I am talking about really traveling and learning about the "rest of the world" where water is not taken for granted. Here in North America we nonchalantly go to the faucet and turn it on and water comes out and we use it and it goes into a drain and that is that. I have been to places and have met people who spend a lot of their time getting water from a distant source and hand carrying it back for their daily survival and if they saw what goes on in industrialized nations they would be shocked. Your product – WATER – is absolutely essential for survival and we cannot survive without it.

This brings us to the "third pillar" – the Process component. How does this all work? Where does water come from and how do we transport it and then distribute a high-quality product to the endpoint user? What do we do with the "wastewater" and reconstitute it for use again? What are the processes involved in the generation, transmission, distribution, and water quality?

When I am introduced in my live lectures, what most people hear is "he is a former motorcycle cop who became a lawyer" – and that is true. But what people often miss is that prior to law school I did my graduate work at the University of Southern California's Institute of Safety and Systems Management. During that three-year window, I got hooked on the discipline of "risk management" and the study of tragedies in high-risk operations – and how tragedies can be avoided if organizations have appropriate "systems" in place. A quick look at Webster's will give you this definition of systems. "A set of principles or procedures according to which something is done; an organized framework or method."

Or to make it a simple one-word definition – back to Mr. Lemonis and his "third pillar" – Process! You can have great people who have a great product – but without "well designed process in place" to provide the customer with your "product" – you will fail! And please remember I am not talking about making ice cream or selling commercial real estate or building cars – I am talking about the essential component for the survival of the human race – water! So my question to you is simple. Do you have appropriate processes in place to make this all work? Are these processes properly designed, are they up to date, do they encapsulate the best practices available in your industry? Are they fully understood by your personnel? Are they taken seriously? Do you have a robust auditing component in place to make sure what you say you are doing is in fact being done? Do you have a continuous improvement strategy in place to get better and better at what you do?

When I was asked to write this foreword and started to think about the message I wanted to deliver my brain was all over the place but I settled on this – the importance of the Certified Public Risk Officer – Water/Wastewater – or more simply stated – CPRO-W2. That is what this book is all about – and you need to read and reread this book and use it as a reference throughout your career and share with others who care and understand the critical mission in which you are involved.

The AAWD&M CPRO-W2 designation and the accompanying learning series encapsulates the purpose of *real* risk management (please note the italics – not the lip service stuff I see in so many organizations) in water/wastewater organizations. That's important to me because my life's passion as a motorcop, lawyer, and risk manager has been the study of tragedies and finding ways to avoid tragedies. This inquisitive journey has allowed me to travel throughout our great country and beyond and speak to a variety of professions including public safety, aviation, mining, power generation, ground transportation, chemical operations, and yes – water and wastewater operations.

In each engagement, I share my frustration that risk management isn't extensively taught in vocational, trade, or academic schools. Such an absence means women and men in complex, high-risk professions like yours receive inadequate training in this important field. It also leads to a professional misperception that risk management is simply the safety stuff. That faulty thinking enables problems lying in wait to emerge as tragedies, which subsequently creates a self-repeating cycle of injury and death to personnel, civil liability, organizational embarrassment, internal investigations, and criminal filings. This wheel of tragedy is avoidable when women and men broaden their risk management knowledge.

Real risk management, not just the safety stuff, is best described as a relentless pursuit and infused conviction by high-reliability organizations to do the right thing, the right way every time. Its derivation is predicated on the actuality of human erraticism and, therefore, necessitates the design of resilient systems (i.e., control measures, policies, procedures, processes, rules, and checklists) to protect organizations against this unpredictability. Resilient systems that are continually improved aren't enough to prevent tragedies. They must be paired with an organizational culture that drives employee understanding, adherence, questioning, and enforcement of these systems. Success is conditioned on this interdependent relationship, meaning employees do the right thing, the right way every time because they unequivocally support the underlying rationale of their organizational systems, mission, and purpose. The common link for these attributes is self-regulating employee behavior through discipline, and that notion is the energy keeping *real* risk management in motion. There are many subparts to this philosophy, but its underpinnings are systems and culture. You need both, along with the adhesion of discipline, to counter the unchanging constant of human errors.

The CPRO-W2 designation and learning series delivers an important knowledge platform and goes further than any other profession-specific risk management curriculum I've encountered. The book focuses on the core components of water/wastewater operations while promoting the key tenets of *real* risk management and the necessity of properly designed insurance policies. It's required reading for any executive, manager, and supervisor who has responsibility for implementing risk management and insurance deliverables. The material is purposely cataloged in a manner that's easy to access and will serve as a trusted resource for future review. A significant learning feature for the technical insurance policy material is the inclusion of checklists. I'm a big proponent of checklists because they allow all readers a consistent and methodical means to complete complex and technical activities. The negotiation, structure, and selection of insurance policies is one such activity. These checklists provide directional guidance to properly complete the procurement of organizational insurance policies.

Your decision to embark on this knowledge-based learning excursion will broaden your perspective, sharpen your acuity, and challenge your thinking. Real risk managers are committed to continuous improvement and situational curiosity because they know a well-informed and diverse outlook is necessary to correctly address the technical, administrative, and external challenges confronting water/wastewater organizations. Reading this book will hone your problem-solving skills while concurrently expanding the role of real risk management from a siloed department to an enterprise-wide way of thinking. Moreover, and most importantly, it offers a governing luminesce to do the right thing, the right way every time and a pathway to provide systematic answers to organizational tribulations.

I hope you enjoy the book as much as I did – and please remember how critical your role is in the scheme of things.

Until next time, work safely.

Gordon Graham

Acknowledgment

Our acknowledgment and thanks to the following individual for his contribution in reviewing *Part IV – Insurance Essentials* of the AAWD&M designation and learning series:

W. Kurt Fickling, CRM, CIC, CRIS
Instructor, Risk Management and Insurance
College of Business, Department of Finance and Insurance
East Carolina University
Greenville, NC 27858

Author biographies

Frank R. Spellman, PhD (author of Parts I and II) is a retired full-time adjunct assistant professor of environmental health at Old Dominion University, Norfolk, Virginia, and the author of more than 148 books covering topics ranging from concentrated animal feeding operations (CAFOs) to all areas of environmental science and occupational health. Many of his texts are readily available online at Amazon.com and Barnes and Noble.com, and several have been adopted for classroom use at major universities throughout the United States, Canada, Europe, and Russia; two have been translated into Spanish for South American markets. Dr. Spellman has been cited in more than 850 publications. He serves as a professional expert witness for three law groups and as an incident/accident investigator at wastewater treatment facilities and/or incidents involved with MRSA contraction supposedly due to contact with wastewater and fatalities occurring at treatment plants for the US Department of Justice and a northern Virginia and Nebraska law firm.

In addition, he consults on homeland security vulnerability assessments for critical infrastructures including water/wastewater facilities nationwide and conducts pre-Occupational Safety and Health Administration (OSHA)/Environmental Protection Agency EPA audits throughout the country. Dr. Spellman receives frequent requests to coauthor with well-recognized experts in several scientific fields; for example, he is a contributing author of the prestigious text *The Engineering Handbook,* 2nd ed. (CRC Press). Dr. Spellman lectures on wastewater treatment, water treatment, and homeland security and safety topics throughout the country and teaches water/wastewater operator short courses at Virginia Tech (Blacksburg, Virginia). In 2011–2012, he traced and documented the ancient water distribution system at Machu Pichu, Peru, and surveyed several drinking water resources in Amazonia-Coco, Ecuador. Dr. Spellman also studied and surveyed two separate potable water supplies in the Galapagos Islands; he also studied and researched Darwin's finches while in the Galapagos. He is a Certified Safety Professional (CSP), a Certified Hazardous Materials Manager (CHMM), and Certified Environmental Trainer (Industrial Hygiene). He holds a BA in public administration, a BS in business management, an MBA, and an MS and PhD in environmental engineering.

Lorilee Medders, PhD (author of Part III) joined Appalachian State University's Walker College of Business in the Department of Finance, Banking & Insurance in August 2017, and serves as both the Joseph F. Freeman Distinguished Professor of Insurance and Director of the Honors Program for the Walker College. Prior to joining Appalachian State, she was on faculty at Florida State University (risk and insurance, 2009–2017), Georgia State University (risk and insurance, 1999–2008), and Georgia Southern University (finance, 1994–1998).

Dr. Medders serves on the Board of Directors for the American Association of Water Distribution and Management (AAWD&M) and chairs its Education Committee. She previously served (2009–2017) the State of Florida as a statistics expert member of the Florida Commission on Hurricane Loss Projection Methodology (Commission), and for the last several years of her membership, as chair of the Commission. Lori's time as chair included the period during which the Commission developed its Flood Standards for loss model review. Additionally, as business risk consultant for more than 20 years, she has experience in investment advisory, risk modeling, decision analytics, insurance coverage disputes, and risk management program review.

Dr. Medders has been published extensively in both the academic and industry press on topics related to risk management, insurance, and decision-making, and has received multiple awards for the quality of her work on these fronts. She has particularly keen interests in catastrophe risk management and finance and is active in researching solutions to the challenges of financing, mitigating, and adapting to risks that have disaster potential, not the least of which include sea-level rise and flooding.

Dr. Medders holds a PhD in Business Administration, majoring in risk management and insurance with a concentration in decision sciences, from Georgia State University (1995). She earned a BS in Commerce and Business, *magna cum laude*, from The University of Alabama (1990).

Paul Fuller (author of Part IV) is an insurance professional specializing in public water/wastewater systems. He is a director of the American Association of Water Distribution & Management (AAWD&M), CEO of Allied Public Risk, LLC (a national Managing General Underwriter exclusively focused on public entities and public water/wastewater systems), and insurance administrator for the California Association of Mutual Water Companies Joint Powers Risk & Insurance Management Authority (JPRIMA). Throughout his 25+-year career, Paul has underwritten thousands of public water/wastewater systems and overseen tens of thousands of claims throughout the United States.

Mr. Fuller earned his Bachelor of Business Administration from University of San Diego and Master of Science in Risk Management & Insurance from Georgia State University. His past industry experience includes: president, Alteris, Inc.; Wholesale & Specialty Programs Operations (WSPO) president, Glatfelter Insurance Group; and president, S.N. Potter Insurance Agency, Inc. He holds the Chartered Property Casualty Underwriter (CPCU) designation, a California Water Operator license (D2/T1), a California qualified Claims Manager license, and is nationally licensed as a Property & Casualty Agent and Surplus Lines Broker.

Paul is a frequent speaker on risk management and emerging issues, with presentations to the following associations: American Water Works Association (AWWA); Association of Metropolitan Water Agencies (AMWA); National Association of Clean Water Agencies (NACWA); California Foundation on the Environment and Economy (CFEE); Association of California Water Agencies (ACWA); California Association of Sanitation Agencies (CASA); California Water Association (CWA); California Association of Mutual Water Companies (CalMutuals); and Risk and Insurance Management Society (RIMS). His articles on inverse condemnation have been published in the *Public Law Journal* and *Defense Counsel Journal*.

Additional details on Mr. Fuller can be found on the following links: www.alliedpublicrisk.com; www.aawdm.org; and www.linkedin.com/in/paul-fuller-cpcu-b56a8612.

Part I

Operations for Public Water Systems

Frank R. Spellman

© Las Vegas Valley Water District and used with permission.

DOI: 10.1201/9781003207146-1

Operations for Public Water Systems: Introduction

2019 Photo Courtesy of San Diego County Water Authority.

INTRODUCTION

You will soon read in *Part IV – Insurance Essentials* of this designation and learning series that insurance is an essential risk management technique for controlling many of the observable and imperceptible exposures facing public water systems. This is a wholly accurate statement that goes to the heart of safe work practices and risk management in public water operations. Additionally, those charged with all levels of management within public water operations must not only understand risk but also understand the actual operation of the public water system. And this statement goes to the heart of the functional manager and management. In order to be a functional manager, at any level, you must be credible. To be credible you must possess several different positive traits such as the ability to meet the standards of others, the ability to communicate, to be accountable, have the ability to demonstrate intentions and actions, possess the ability to maintain correct behavior, have the ability to be consistent, and, most importantly, possess the knowledge of what public water system operation is all about. This last point is critical. Why? If you cannot speak and understand the language, the terminology, the jargon of public water system operation, it is analogous to being lost at sea without a sail or paddle or compass or any thought of a successful outcome.

DOI: 10.1201/9781003207146-2

So it is important for the Certified Public Risk Officer to possess knowledge of the operation of public water systems. The required and essential level of knowledge needed does not necessarily mean that managers, at all levels, understand the nuts and bolts of every water operation. Instead, it means that managers involved with risk in public water operations know the nomenclature of and understand water operations.

It is the purpose of *Part I – Operations for Public Water Systems* to provide potential Public Risk Officers with a four-pronged delivery of fundamental knowledge and nomenclature to help ensure credibility. Prong one involves the current issues involved with public water systems. Prong two involves the basic fundamentals of water system operation. Prong three involves important management aspects involved with the proper operation of the public water system. Prong four details the safety and health requirements and precautions needed to safely operate a public water system. The bottom line: *Part I* of this designation and learning series will deliver the information needed to be a credible Certified Public Risk Officer.

AUTHOR'S COMMENTARY

Part I – Operations for Public Water Systems provides an overview of current treatment, distribution, and energy technologies for the water purveyance industry. These collective techniques have not materially changed for decades except for additional filtration options and different types of disinfectants. The industry's renaissance of academic writing occurred from the early 1970s to the late 1990s and was spurred by the Environmental Protection Agency (EPA). Many of my references and citations reflect this timeline. In contrast, academic writings for the wastewater industry are currently in their renaissance period because of reuse. The material included in *Part I – Operations for Public Water Systems* is current and reflective of actual water purveyance operations as of this writing.

1 Public Water System Overview

Perhaps the profession of doing good may be full, but everybody should be kind at least to himself. Take a course in good water …, and in the eternal youth of Nature, you may renew your own. Go quietly, alone; no harm will befall you. Some have strange, morbid fears as soon as they find themselves with Nature even in the kindest and wildest of her solitudes, like very sick children afraid of their mother – as if God were dead and devil were King.

– John Muir (1888)

LEARNING OBJECTIVES

After studying this chapter, you should be able to:

- Discuss how human activity has affected the pathways water takes through the landscape and what results that has on water quality.
- Discuss the natural water cycle.
- Discuss nature's way with water.
- Identify the classification of public water systems.
- Define community water systems.
- Define noncommunity water systems.
- Discuss regulatory compliance and the effect of 1972's Clean Water Act (CWA) on American water supplies.
- Discuss Occupational Safety & Health Administration (OSHA).
- Identify the key locations of water and the percentages each location holds in global water distribution.
- Discuss an organization's safety policy.
- Define the two general sources of water.

DOI: 10.1201/9781003207146-3

- Describe surface water's place in the water cycle and identify the major sources of surface water.
- Discuss surface water's advantages as a freshwater source and its disadvantages.
- Describe the various factors that influence surface water's flow.
- Describe and discuss groundwater's place in the water cycle and the effects soil and rock contribute.
- Discuss groundwater's advantages and its disadvantages as a water source.
- Identify the major users of water in the United States.
- Describe and discuss domestic, commercial and industrial, and public use of water.
- Discuss emerging challenges.
- Describe maintaining infrastructure.
- Define privatization and reengineering.
- Define benchmarking.
- Discuss security needs.
- Discuss technical vs. professional management.
- Discuss energy conservation measures and sustainability.
- Discuss sustainable water infrastructure.

KEY DEFINITIONS

Benchmarking – used to measure or gauge performance based on specific indicators such as cost per unit of measure, productivity per unit of measure, the cycle time of some value per unit of measure, or defects per unit of measure. It is the comparing of your operation to the best in the class.

Best Available Technology (BAT) – defined at CWA section 304 (b)(2). In general, BAT represents the best available economically achievable performance of plants in the industrial subcategory or category.

Biochemical Oxygen Demand (BOD) – the amount of oxygen required by bacteria to stabilize decomposable organic matter under aerobic conditions.

Biodegradable – transformation of a substance into new compounds through biochemical reactions or the actions of microorganisms such as bacteria.

Catchment – an area that serves to catch water.

Clean Water Act (CWA) – federal law dating back to 1972 (with several amendments) with the objective to restore and maintain the chemical, physical, and biological integrity of the nation's waters. Its long-range goal is to eliminate the discharge of pollutants into navigable waters and to make national waters fishable and swimmable.

Community Water System – a public water system that serves at least 15 service connections used by year-round residents or regularly serves at least 25 year-round residents.

Domestic Water – water used for normal household purposes, such as drinking, food preparation, bathing, washing clothes and dishes, flushing toilets, and watering lawns and gardens. The water may be obtained from a public supplier or may be self-supplied. Also called residential water use.

Drainage Area – the drainage area of a stream at a specified location is that area, measured in a horizontal plane, which is enclosed by a drainage divide.

Drainage Basin – the land area drained by a river or stream.

Eutrophication – the process by which water becomes enriched with plant nutrients, most commonly phosphorus and nitrogen.

Evaporation Losses – the process by which water is discharged to the atmosphere as a result of evaporation.

Ground Slope – ground that slopes upward or downward.

Groundwater – the fresh water found under the earth's surface, usually in aquifers. Groundwater is a major source of drinking water and a source of growing concern in areas where leaching agricultural or industrial pollutants or substances from leaking underground storage tanks are contaminating groundwater.

GWUDISW – groundwater under the direct influence of surface water.

Human Influences – various human activities have a definite impact on surface water runoff. Most human activities tend to increase the rate of water flow. For example, canals and ditches are usually constructed to provide steady flow, and agricultural activities generally remove ground cover that would work to retard the runoff rate. On the opposite extreme, man-made dams are generally built to retard the flow of runoff.

Infiltration – the downward movement of water from the atmosphere into the soil or porous rock.

Infrastructure – the basic physical and organizational structures and facilities needed for the operation of an enterprise.

Motor Control Center – motor power panel control center.

NEPA – in response to the need to make a coordinated effort to protect the environment, the National Environmental Policy Act (NEPA) was signed into law on January 1, 1970. In December of that year, a new independent body, the U.S. Environmental Protection Agency (USEPA), was created to bring under one roof all of the pollution-control programs related to air, water, and solid wastes. In 1972, the Water Pollution Control Act Amendments expanded the role of the federal government in water pollution control and significantly increased federal funding for the construction of wastewater treatment plants.

Noncommunity Water System – is composed of transient and non-transient water systems.

- Transient noncommunity water systems provide water to 25 or more people for at least 60 days per year but not to the same people and on a regular basis (e.g., gas stations, campgrounds, etc.).
- Non-transient noncommunity water systems regularly supply to at least 25 of the same people for at least 6 months per year, but not year round (e.g., factories, schools, office buildings, and hospitals which have their own water systems).

Nutrients – any inorganic or organic compound needed to sustain plant life.

Organic Matter – containing carbon but possibly also containing hydrogen, oxygen, chlorine, nitrogen, and other elements.

Pathogens – types of microorganisms that can cause disease.

Potable Water – water that is safe for use and consumption.

Primary Standard – this is a legally enforceable standard that applies to public water systems. Primary standards protect drinking water quality by limiting the levels of specific contaminants that can adversely affect public health and are known or anticipated to occur in water. They take the form of maximum contaminant levels or treatment techniques.

Privatization – means allowing private enterprise to compete with government in providing public services, such as water operations. Privatization is often proposed as one solution to the numerous woes facing public water and wastewater systems, including corruption, inefficiencies (dysfunctional management), and the lack of capital for needed service improvements and infrastructure upgrades and maintenance. Existing management, on the other hand, can accomplish reengineering, internally, or it can be used (and usually is) during the privatization process.

Process Safety Management (PSM) – (29 CFR 1910.119–OSHA). These regulations put the use of elemental chlorine (and other listed hazardous materials) under scrutiny. Moreover, because of these regulations, plant managers throughout the country are forced to choose which side of a double-edged sword cuts their way the most. One edge calls for full compliance with the regulations (analogous to stuffing the regulation through the eye of a needle). The other edge calls for substitution. That is, replacing elemental chlorine with a non-listed chemical (e.g., hypochlorite) or a physical (ultraviolet irradiation, UV) disinfectant; either way, a very costly undertaking.

Public Use – is the water furnished to public buildings and used for public services. This includes water for schools, public buildings, fire protection, and for flushing streets.

Quantity and Quality – these are the Q and Q factors that set the standard for safe and healthy water that is accessible for public use.

Rainfall Duration – the total time rainfall is present in a particular region or area.

Rainfall Intensity – degree of rainfall intensity.

Reengineering – the systematic transformation of an existing system into a new form to realize quality improvements in operation, system capability, functionality, performance, or evolvability at a lower cost, schedule, or risk to the customer.

Risk Management Planning (RMP) – similar to PSM but administered by USEPA and not OSHA.

Safety Policy – an organization's safety rules or guidelines.

SDWA – Safe Drinking Water Act.

Secondary Standard – this is a nonenforceable guideline regarding contaminants that may cause cosmetic effects (such as skin or tooth discoloration) or aesthetic effects (such as taste, odor, or color) in drinking water. USEPA recommends secondary standards to public water systems but does not require systems to comply. However, states may choose to adopt them as enforceable standards. This information focuses on national primary standards.

Soil Composition – the components that make up soil.

Soil Moisture – water occurring in the pore spaces between the soil particles in the unsaturated zone from which water is discharged by the transpiration of plants or by evaporation from the soil.

Surface Water – all water naturally open to the atmosphere, and all springs, wells, or other collectors that are directly influenced by surface water.

Sustainability – the long definition: ensuring that water treatment operations occur indefinitely without negative impact. The short definition: the capacity of water operations to endure. What does sustainability really mean in the real world of water treatment operations?

Total Quality Management (TQM) – giving the employees a voice through empowerment.

Toxic Metals – metal contaminants in water.

Vegetation Cover – runoff is limited by ground cover. Roots of vegetation and pine needles, pinecones, leaves, and branches create a porous layer (a sheet of decaying natural organic substances) above the soil. This porous "organic" sheet (groundcover) readily allows water into the soil. Vegetation and organic waste also act as a cover to protect the soil from hard, driving rains. Hard rains can compact bare soils, close off void spaces, and increase runoff. Vegetation and ground cover work to maintain the soil's infiltration and water-holding capacity. Note that vegetation and groundcover also reduce the evaporation of soil moisture as well.

Water Use – water is used at the present time in the United States at a rate of $1.6 \times 1,012$ liters per day, which amounts to almost a tenfold increase in liters used since the turn of the 1900s. Of the billions of gallons of water available for use in the United States, where is this water used? The National Academy of Sciences (1962) estimates that approximately: (1) 310 billion gallons per day (bgd) are withdrawn; (2) 142 bgd are used for irrigation; (3) 142 bgd are used for industry (principally utility cooling water – 100 bgd); (4) 26 bgd are used in municipal applications; (5) 90 bgd are consumed (principally irrigation, loss to the ground, and evaporation); and (6) 220 bgd are returned to streams.

Watershed – the land area that drains into a river, river system, or other body of water.

OVERVIEW

For those of us residing in the United States, we are fortunate to have one of the safest public drinking water supplies in the world. And that is important, of course, because water is an essential compound in the maintenance of all forms of life on earth. This fact has resulted in the development of direct relationships between the abundance of water and the quality of the water. Remember, having mega-gallons of water in storage, readily available, is of little use if the water is unsuitable for human use and reuse; quality is what it is all about.

DID YOU KNOW?

There are close to 152,000 public water systems in the United States. USEPA classifies public water systems according to the number of people they serve, the source of their water, and whether they serve the same customers year-round or on an occasional basis.

In the pre-Columbian era, on the North American continent, the natural water cycle was able to deliver clean water – clear as dew – to the landscape. Today, the natural water cycle has been changed, tainted in a number of ways. We have dredged, drained, dammed, channeled, tampered with, and sometimes eliminated the ecological niches where water cleans itself. We have simplified the pathways that water takes through the American landscape. As a result, water that is no longer able to clean itself naturally must be regulated and cleaned using various advanced technologies. Because of nature's human-caused limitations to purifying water, we have employed treatment procedures in nature's way by utilizing various treatment procedures. This is accomplished in a variety of ways, but first it is important to understand that public drinking water suppliers are classified as per usage by population.

With regard to the classification of public water systems, the United States Environmental Protection Agency (USEPA) classifies public drinking water systems as community and noncommunity systems, as shown in Figure 1.1.

A community water system (CWS) supplies water to the same population year round. It serves at least 25 people at their primary residences or at least 15 residences that are primary residences (e.g., municipalities, mobile home parks, and subdivisions).

Noncommunity water systems are composed of transient and non-transient water systems. Transient noncommunity water systems provide water to 25 or more people for at least 60 days per year but not to the same people and on a regular basis (e.g., gas stations, campgrounds, etc.). Non-transient noncommunity water systems regularly supply to at least 25 of the same people at least 6 months per year, but not year round (e.g., factories, schools, office buildings, and hospitals which have their own water systems).

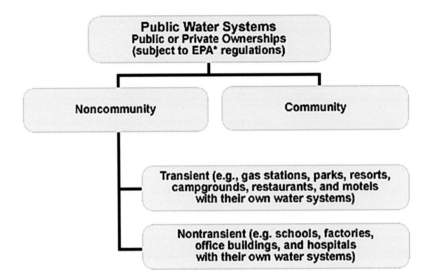

FIGURE 1.1   Classification of public water systems.Source: Adaptation from CDC (2014) Accessed @ www.cdc.gov/healthywater/drinking/public/index.html.

DID YOU KNOW?

Almost 300 million Americans get their tap water from a community water system.

REGULATORY COMPLIANCE

In the United States, in response to a national crisis in drinking water quality, water has been regulated since 1972. The purpose of these water regulations is to restore and maintain the chemical, physical, and biological integrity of the waterways (this was the case because many of the nation's waterways are sources of drinking water). By 1995, the discharge of pollutants into streams, rivers, and lakes was to cease. The goal was to make these freshwater sources not only wholesome sources of drinking water influent for treatment but also to make the waterbodies fishable and swimmable. The means to accomplish this goal was to be provided by technology. Every US city was required to build a wastewater treatment plant with secondary treatment capability. Every industry had to incorporate the *best available technology* (BAT) to decrease the discharge of pollutants into the waterways.

In the years that followed 1972, the stranglehold that pollutant wastes had on the nation's surface waterbodies was eased. However, the task is far from complete. A generation after the CWA was promulgated, about 30% of the stream miles and lake acres in the United States are still polluted (Outwater, 1997). A great deal must still be done. Despite our best legislative efforts, our waterways are still impaired. Simply, more is required. With the explosive growth of populations and the accompanying increased need for more natural resources and habitable land, we will have to work at restoring and maintaining water quality. This can only be accomplished by the correct and judicial use of technology to clean our water. Water must be available for human use and in abundant supply. Available water must have specific characteristics; water quality is defined in terms of those characteristics (Tchobanoglous & Schroeder, 1987).

When we are talking about regulatory compliance, it is just not USEPA that you must comply with but also OSHA; a word to the wise, don't forget OSHA. Section 5 (a) of Public Law 91-596 of December 29, 1970, Occupational Safety and Health Act (OSH Act) requires each employer:

- Shall furnish to each of his employees, employment, and a place of employment which is free from recognized hazards that are causing, or are likely to cause death, or serious physical harm to his employees; and
- Shall comply with occupational safety and health standards, promulgated under this Act.

No matter the number of regulations, standards, and laws that are made to ensure worker safety, and no matter how experienced, and/or motivated the plant's designated safety official is, he/she is powerless without dedicated support from the highest levels of plant management. Simply put, without a strong commitment from upper management, the safety effort is doomed. On the other hand, when organizational management states that it is the company's objective to place "*Safety First*" even before productivity and quality, then the proper atmosphere is present for the safety official to accomplish the intended objective. That is, the safety official will be able to provide and ensure a safe place for all employees to work in.

When assigned the duty as the plant's safety official, the first thing to be done is to meet with upper management and determine what the safety objective is. The newly appointed safety official should not start on the plant's safety program until he/she is quite certain of what is expected of him/her. The pertinent matters that need to be addressed during this initial meeting include the development of a written *safety policy*, *a safety budget*, exactly what *authority the safety official has*, and *to whom does he/she report*. In addition, the organization's safety rules and safety committee structure must be formulated as soon as possible.

THE ORGANIZATION'S SAFETY POLICY

The plant safety official should propose that an organizational safety policy be written, and approved, by the general manager, or other top plant manager. A well-written organizational safety policy should be the cornerstone of an organization's safety program. There are several examples of safety policies used by Fortune 500 companies, and others, from which to model your own plant's safety policy. The key to producing a powerful, tell-it-like-it-is safety policy is to keep it short, to the point, and germane to the overall goal. Many organizational safety policies are well written but are too lengthy, too philosophical. The major point to remember is that the organization's safety policy should be written, not only so that it might be understood by every employee, but also so that all employees will actually read it. An example of a short, to the point, and hard-hitting safety policy that we have used in the past, and found to be effective, is provided in Figure 1.2. The powerful effect of Figure 1.2 on an organization's safety policy is in its brevity. The major point is made in as short of verse as is necessary; thus, this is the type of safety policy that will be read by the employee. More importantly, the safety policy sends the desired message.

WATER RESOURCES

Note: Much of the information contained in the following sections is based on Spellman's *The Science of Water: Concepts and Applications*, 3rd ed. (Boca Raton, FL: CRC Publishing Company, 2013). Where on planet Earth is potable (drinkable) water readily found for human consumption?

First, we must define potable water.

Potable water is water fit for human consumption and domestic use, which is sanitary and normally free of minerals, organic substances, and toxic agents in excess of reasonable amounts for domestic usage in the area served and normally adequate in quantity for the minimum health requirements of the persons served.

–Spellman, 2013

NO JOB IS SO CRITICAL AND NO SERVICE IS SO URGENT THAT
WE CANNOT TAKE THE TIME TO PERFORM OUR WORK SAFELY

While it is true that the major emphasis is on efficient operations, it is also true that this must be accomplished with a minimum of accidents and losses. I cannot overemphasize the importance that the organization places on the health and well-being of each, and every employee. The organization's commitment to occupational health and safety is absolute. The organization's safety goal is to integrate hazard control into all operations, including compliance with applicable standards. I encourage active leadership, direct participation, and enthusiastic support of the entire organization in supporting <u>our</u> safety programs and policies.

_____

General Manager

FIGURE 1.2 Sample organization safety policy.

For a potential potable water supply, the keywords are quality and quantity (the Q & Q factor). If we have a water supply that is unfit for human consumption, then we have a quality problem. If we do not have an adequate supply of quality water, then we have a quantity problem (Spellman, 2003). Look at a map of the world, one that clearly indicates the world's population centers (cities). Take a look at the United States, for example. The first American settlers built their settlements along rivers. Rivers provided the water settlers needed to sustain life, and the principal source of power for early industry and an easy means of transportation.

Most of the earliest settlements in the United States occurred on its east coast. In most cases (the early Jamestown, Virginia, settlement is an exception because potable water quality there was poor), settlers along this eastern seaboard area were lucky. They had settled along river systems of excellent quality. These rivers were ideally suited for paper and textile manufacturing, which were among the earliest industries started. As more settlers arrived in North America, they began to branch out inland (west – where there was more elbow room) from the earliest settlements, and in many cases, they found that finding potable water was not so easy. The further west they traveled, the higher the salinity of rivers and streams, especially with long rivers and streams that flowed through and over areas of relatively soluble rock formations.

In western regions of the United States (especially in desert regions), the US map shows only sparse settlements; those arid areas lack water to support a larger population. These regions are occupied by fewer people and different (more tolerant of water shortage) species than other biomes. A real estate saying you may have heard states: Location! Location! Location!

Is location everything? Though we don't think of it consciously, when you examine the beginnings of human settlements, "Location, Location, Location!" really meant, "Water! Water! Water!" To be suitable as a living place, potable water was essential (Spellman, 1998). On land, the availability of a regular supply of potable water is the most important factor affecting the presence or absence of many life forms. Consider that most people (and other living things) are found in regions of the United States and other parts of the world where potable water is readily available.

MAJOR SOURCES OF DRINKING WATER

Approximately 326 million cubic miles of water cover planet Earth, but only about 3% of this total is freshwater. Most of this freshwater supply is locked up in polar ice caps, in glaciers, in lakes, in flows through soil, and in river or stream systems. Out of this 3%, only 0.027% is available for human consumption.

TABLE 1.1
World Water Distribution

Location and Land Areas	%
Freshwater lakes	0.009
Saline lakes and inland seas	0.008
Rivers (average instantaneous volume)	0.0001
Soil moisture	0.005
Groundwater (above the depth of 4,000 m)	0.61
Ice caps and glaciers	2.14
	2.8
Atmosphere (water vapor)	0.001
Oceans	97.3
Total All Locations (rounded)	**100.0**

Source: Adapted from Peavy et al. (1985), p. 12.

Where is the world's water distributed? The data contained in Table 1.1 should help answer this question. When you review Table 1.1, it should be obvious that the major sources of drinking water are surface water, groundwater, and groundwater that is under the direct influence of surface water (GWUDISW; i.e., a spring or a shallow well).

SURFACE WATER

WORLD WATER DISTRIBUTION

Surface water (water on the earth's surface as opposed to subsurface water – groundwater) is mostly a product of precipitation – rain, snow, sleet, or hail. Surface water is exposed or open to the atmosphere and results from the movement of water on and just under the earth's surface (overland flow). This overland flow is the same thing as surface runoff, which is the amount of rainfall that passes over the earth's surface. Specific sources of surface water include:

- Rivers;
- Streams;
- Lakes;
- Impoundments (man-made lakes made by damming a river or stream);
- Very shallow wells that receive input via precipitation;
- Springs affected by precipitation (their flow or quantity is directly affected by precipitation);
- Rain catchments (drainage basin); and
- Tundra ponds or muskegs (peat bogs) and other wetlands (swamps).

As a source of potable water, surface water does have some advantages. Surface water is usually easy to locate – you do not have to be (or hire) a geologist or hydrologist to find it. Normally, surface water is not tainted with chemicals precipitated from the earth's strata. Surface water also has its disadvantages. The biggest disadvantage of using surface waters as a source of potable water is that they are easily contaminated (polluted) with microorganisms that can cause waterborne diseases and from chemicals that enter the surface waters from surrounding runoff and upstream discharges. Problems can also occur with water rights (i.e., who owns the water?).

Most surface water is the result of surface runoff. The amount and flow rate of surface runoff are highly variable. This variability comes into play for two main reasons: (1) human interference (influences) and (2) natural conditions. In some cases, surface water runs quickly off land. Generally quick runoff is undesirable (from a water resources standpoint) because water needs time to infiltrate into the ground and recharge groundwater aquifers. Other problems associated with quick surface water runoff are erosion and flooding. Probably the only good thing that can be said about surface water that quickly runs off the land is that it does not have enough time (usually) to become contaminated with high mineral content. Surface water running slowly off the land may be expected to have all the opposite effects.

Surface water travels over the land to what amounts to a predetermined destination. What factors influence how surface water moves? Surface water's journey over the face of the earth typically begins at its drainage basin, sometimes referred to as its *drainage area, catchment,* and/or *watershed.* For a groundwater source, this is known as the *recharge area – the area from which precipitation* flows into an underground water source. A surface water drainage basin is usually an area measured in square miles, acres, or sections, and if a city takes water from a surface water source, the size of and what lies within the drainage basin is essential information for water quality assessment.

Water doesn't run uphill on its own. Surface water runoff (like the flow of electricity), follows along the path of least resistance. Generally speaking, water within a drainage basin will naturally be shunted (by the geological formation of the area) toward one primary watercourse (a river, stream, creek, and brook) unless some man-made distribution system diverts the flow. Various factors directly influence the surface water's flow over land.

DID YOU KNOW?

Of the fresh water on earth, much more is stored in the ground than is available in lakes and rivers. More than 2,000,000 m^3 (8,400,000 km^3) of fresh water is stored in the earth, most within one-half mile of the surface. Contrast that with the 60,000 mi^3 (250,000 km^3) of water stored as fresh water in lakes, inland seas, and rivers (USGS 2011).

The principal factors are:

- Rainfall duration;
- Rainfall intensity;
- Soil moisture;
- Soil composition;
- Vegetation cover;
- Ground slope; and
- Human interference.

Rainfall Duration – Rainstorm length affects runoff amount. Even a light, gentle rain eventually saturates the soil if it lasts long enough. Once the saturated soil can absorb no more water, rainfall builds up on the surface and begins to flow as runoff.

Rainfall Intensity – The harder and faster it rains, the more quickly the soil becomes saturated. With hard rains, the surface inches of the soil quickly become inundated, and with short, hard storms, most of the rainfall may end up as surface runoff; the moisture is carried away before significant amounts of water are absorbed into the earth.

Soil Moisture – Obviously, if the soil is already laden with water from previous rains, the saturation point is reached sooner than with dry soil. Frozen soil also inhibits water absorption: up to 100% of snowmelt or rainfall on frozen soil ends as runoff, because the frozen ground is impervious.

Soil Composition – Runoff amount is directly affected by soil composition. Hard rock surfaces shed all rainfall, obviously, but so will soils with heavy clay composition. Clay soils possess small void spaces that swell when wet. When the void spaces close, they form a barrier that does not allow additional absorption or infiltration. On the opposite end of the spectrum course sand allows easy water flow-through, even in a torrential downpour.

Vegetation Cover – Runoff is limited by ground cover. Roots of vegetation and pine needles, pinecones, leaves, and branches create a porous layer (a sheet of decaying natural organic substances) above the soil. This porous "organic" sheet (groundcover) readily allows water into the soil. Vegetation and organic waste also act as a cover to protect the soil from hard, driving rains. Hard rains can compact bare soils, close off void spaces, and increase run-off. Vegetation and ground cover work to maintain the soil's infiltration and water-holding capacity. Note that vegetation and ground cover also reduce the evaporation of soil moisture as well.

Ground Slope – Flat land water flow is usually so slow that large amounts of rainfall can infiltrate the ground. Gravity works against infiltration on steeply sloping ground; up to 80% of rainfall may become surface runoff.

Human Influences – Various human activities have a definite impact on surface water runoff. Most human activities tend to increase the rate of water flow. For example, canals and ditches are usually constructed to provide steady flow, and agricultural activities generally remove ground cover that would work to retard the runoff rate. On the opposite extreme, man-made dams are generally built to retard the flow of runoff. Human habitations, with paved streets, tarmac, paved parking lots, and buildings create surface runoff potential, since so many surfaces are impervious to infiltration. As all these surfaces hasten the flow of water, they also increase the possibility of flooding, often with devastating results. Moreover, water running off of impervious areas, such as roads and parking lots, can contain a lot of contaminants, such as oil and garbage. This runoff often goes directly into streams. Additionally, following summer storms, runoff from heated roads and parking lots causes rapid increases in stream temperatures that can produce thermal shock and death in many fish. Because of urban increases in runoff, a relatively new field (industry) has developed: Stormwater Management. Paving over natural surface acreage has another serious side effect. Without enough area available for water to infiltrate the ground and percolate into the soil to eventually reach and replenish/recharge groundwater sources, those sources may eventually fail, with devastating impact on the local water supply.

GROUNDWATER

Approximately 3 feet of water falls each year on every square foot of land. About 6 inches of this goes back to the sea. Evaporation takes up about 2 feet. What remains, approximately 6 inches, seeps into the ground, entering and filling every interstice, each hollow and cavity, like an absorbent. Although comprised of only one-sixth of the total (1,680,000 miles of it), if it could be ladled up and spread out over the earth's surface, it would blanket all land to a depth of 1,000 feet. This gigantic water source (literally an ocean beneath our feet) forms a reservoir that feeds all the natural fountains and springs of the earth. Eventually, it works its way to the surface. Some come out clean and cool, a liquid blue-green phantom; and some, occupying the deepest recesses, pressurizes and shoots back to the surface in white, foamy, wet chaos, as geysers.

Fortunately, most of the rest lies within easy reach, just beneath the surface. This is groundwater. –

Spellman, 1998, pp. 22–23

Water falling to the ground as precipitation normally follows three courses: (1) some runs off directly to rivers and streams; (2) some infiltrates to ground reservoirs; and (3) the rest evaporates or transpires through vegetation. The water in the ground (groundwater) is "invisible" and may be thought of as a temporary natural reservoir (ASTM, 1969). Almost all groundwater is in constant motion toward rivers or other surface waterbodies.

Groundwater is defined as water below the earth's crust but above a depth of 2,500 feet. Thus, if water is located between the earth's crust and the 2,500-foot level, it is considered usable (potable) fresh water. In the United States, "at least 50% of total available fresh water storage is in underground aquifers" (Kemmer, 1979, p. 17).

Groundwater is usually obtained from wells or from springs that are not influenced by surface water or local hydrologic events. Groundwater, in relation to surface water, has several advantages:

- Unlike surface water, groundwater is not easily contaminated.
- Groundwater sources are usually lower in bacteriological contamination than surface waters.
- Supply usually remains stable throughout the year.
- In the United States, for example, groundwater is available in most locations.

When comparing groundwater with surface water sources, groundwater does have some disadvantages:

- Contamination is usually hidden from view.
- Groundwater is usually loaded with minerals (with an increased level of hardness) because it is in contact longer with minerals; removing contaminants from groundwater supplies is very difficult.
- Because it must be pumped from the ground, operating costs are usually higher.
- Groundwater sources near coastal areas may be subject to salt-water intrusion.

WATER USE

In the United States, rainfall averages approximately $4,250 \times 10^9$ gallons a day; about two-thirds of the rainfall returns to the atmosphere through evaporation from the surface of rivers, streams, and lakes or from transpiration from plant foliage. This leaves approximately $1,250 \times 10^9$ gallons a day to flow across or through the earth to the sea (Kemmer, 1979).

Water is used at the present time in the United States at a rate of 1.6×10^{12} liters per day, which amounts to almost a tenfold increase in liters used since the turn of the 1900s. Of the billions of gallons of water available for use in the United States, where is this water used? The National Academy of Sciences (1962) estimates that approximately: (1) 310 bgd are withdrawn; (2) 142 bgd are used for irrigation; (3) 142 bgd are used for industry (principally utility cooling water – 100 bgd); (4) 26 bgd are used in municipal applications; (5) 90 bgd are consumed (principally irrigation, loss to the ground, and evaporation); and (6) 220 bgd are returned to streams. From the preceding, that much of this increase in use is accounted for by high agricultural and industrial use is evident, each of which accounts for more than 40% of total consumption. Municipal use consumes the remaining 10–12% (Manahan, 1997).

We are primarily concerned with water use for municipal applications (demand). Municipal water demand is usually classified according to the nature of the user. These classifications are:

- *Domestic* – Domestic water is supplied to houses, schools, hospitals, hotels, restaurants, and so forth, for culinary, sanitary, and other purposes. Use varies with the economic level of the consumer, the range being 20–100 gallons per capita per day. Note that these figures include water used for watering gardens, lawns, and washing cars.
- *Commercial and Industrial* – Commercial and industrial water is supplied to stores, offices, and factories. The importance of commercial and industrial demand is based on whether large industries use water supplied from the municipal system, as large industrial facilities can make heavy demands on a municipal system. Large industries

demand a quantity directly related to the number of persons employed, to the actual floor space or area of each establishment, and to the number of units manufactured or produced.

- *Public Use* – Public use water is the water furnished to public buildings and used for public services. This includes water for schools, public buildings, fire protection, and flushing streets.
- *Loss and Waste* – Water that is lost or wasted (unaccounted for) is attributable to leaks in the distribution system, inaccurate meter readings, and unauthorized connections. Loss and waste of water can be expensive. To reduce these costs requires a regular program that includes maintenance of the system and replacement and/or recalibration of meters (McGhee, 1991).
- Manahan (1997) points out that water is not destroyed, but it can be lost for practical use. The three ways in which this may occur include:
- *Evaporative losses* that occur during spray irrigation, and when water is used for evaporative cooling;
- *Infiltration* of water into the ground, often in places and ways that preclude its later use as water source groundwater; and
- *Degradation* from pollutants, such as salts picked up by water used for irrigation (p. 133).

EMERGING CHALLENGES

Problems come and go, shifting from century to century, decade to decade, year to year, and site to site. They range from the problems caused by natural forces (storms, earthquakes, fires, floods, and droughts) to those caused by social forces, currently including terrorism. In general, seven areas are of concern to many water management personnel:

- Complying with regulations, and coping with new and changing regulations;
- Maintaining infrastructure;
- Privatization and/or reengineering;
- Benchmark It!;
- Upgrading security;
- Technical vs. professional management; and
- Energy conservation measures and sustainability.

COMPLIANCE WITH NEW, CHANGING, AND EXISTING REGULATIONS

Adapting the workforce to the challenges of meeting changing regulations and standards for water treatment is a major concern. As mentioned, drinking water standards are regulations that USEPA sets to control the level of contaminants in the nation's drinking water. These standards are part of the SDWA's multiple-barrier approach to drinking water protection. There are two categories of drinking water standards:

- *A National Primary Drinking Water Regulation (Primary Standard)* – This is a legally enforceable standard that applies to public water systems. Primary standards protect drinking water quality by limiting the levels of specific contaminants that can adversely affect public health and are known or anticipated to occur in water. They take the form of Maximum Contaminant Levels or Treatment Techniques.
- *A National Secondary Drinking Water Regulation (Secondary Standard)* – This is a non-enforceable guideline regarding contaminants that may cause cosmetic effects (such as skin or tooth discoloration) or aesthetic effects (such as taste, odor, or color) in drinking

water. USEPA recommends secondary standards to public water systems but does not require systems to comply. However, states may choose to adopt them as enforceable standards. This information focuses on national primary standards.

Drinking water standards apply to public water systems, which provide water for human consumption through at least 15 service connections or regularly serve at least 25 individuals. Public water systems include municipal water companies, mutual water companies, homeowner associations, schools, businesses, campgrounds, and shopping malls. More recent requirements, for example, the Clean Water Act Amendments that went into effect in February 2001, require water treatment plants to meet tougher standards, presenting new problems for treatment facilities to deal with, and offering some possible solutions to the problems of meeting the new standards. These regulations provide for communities to upgrade existing treatment systems, replacing aging and outdated infrastructure with new process systems. Their purpose is to ensure that facilities are able to filter out higher levels of impurities from drinking water, thus reducing the health risk from bacteria, protozoa, and viruses and that they are able to decrease levels of turbidity and reduce concentrations of chlorine by-products in drinking water.

MAINTAINING INFRASTRUCTURE

During the 1950s and 1960s, the US government encouraged the prevention of pollution by providing funds for the construction of municipal wastewater treatment plants, water pollution research, and technical training and assistance. New processes were developed to treat sewage, analyze wastewater, and evaluate the effects of pollution on the environment. In spite of these efforts, however, expanding population and industrial and economic growth caused the pollution and health difficulties to increase. In response to the need to make a coordinated effort to protect the environment, the NEPA was signed into law on January 1, 1970. In December of that year, a new independent body, the USEPA was created to bring under one roof all of the pollution control programs related to air, water, and solid wastes. In 1972, the Water Pollution Control Act Amendments expanded the role of the federal government in water pollution control and significantly increased federal funding for the construction of wastewater treatment plants.

Many of these locally or federally funded treatment plants are aging; based on experience and observation, some can be rated as dinosaurs. The point is many facilities are facing problems caused by aging equipment, facilities, and infrastructure. Complicating the problems associated with natural aging is the increasing pressure on inadequate older systems to meet the demands of increased population and urban growth. Facilities built in the 1960s and 1970s are now 50–60 years old, and not only are they showing signs of wear and tear, they simply were not designed to handle the level of growth that has occurred in many municipalities. Regulations often necessitate a need to upgrade. By matching funds or providing federal money to cover some of the costs, municipalities can take advantage of a window of opportunity to improve their facility at a lower direct cost to the community. Those federal dollars, of course, do come with strings attached; they are to be spent on specific projects in specific areas. On the other hand, many times new regulatory requirements are put in place without the financial assistance needed to implement. When this occurs, either the local community ignores the new requirements (until caught and forced to comply), or they face the situation and implement through local tax hikes to pay the cost of compliance.

An example of how a change in regulations can force the issue is demonstrated by the demands made by OSHA and USEPA in their PSM/RMP regulations (29 CFR 1910.119-OSHA). These regulations put the use of elemental chlorine (and other listed hazardous materials) under scrutiny. Moreover, because of these regulations, plant managers throughout the country are forced to choose which side of a double-edged sword cuts their way the most. One edge calls for full compliance with the regulations (analogous to stuffing the regulation through the eye of a needle). The other edge calls for substitution. That is, replacing elemental chlorine with a non-listed chemical (e.g., hypochlorite) or a physical (ultraviolet irradiation, UV) disinfectant – either way, a very costly undertaking.

PRIVATIZATION AND/OR REENGINEERING

Water treatment operations are undergoing a new paradigm shift. This paradigm shift focuses on the holistic approach to treating water. The shift is, however, more inclusive. It also includes thinking outside the box. In order to remain efficient and therefore competitive in the real world of operations, water treatment facilities have either bought into the new paradigm shift or been forcibly "shifted" to doing other things (often these "other" things have little to do with water operations) (Johnson & Moore, 2002).

Experience has shown that few words conger up more fear among plant managers than "privatization" or "reengineering." *Privatization* means allowing private enterprises to compete with the government in providing public services, such as water operations. Privatization is often proposed as one solution to the numerous woes facing certain public water and wastewater systems, including situations where there is corruption, inefficiencies (dysfunctional management), and the lack of capital for needed service improvements and infrastructure upgrades and maintenance. Existing management, on the other hand, can accomplish reengineering, internally, or it can be used (and usually is) during the privatization process. *Reengineering* is the systematic transformation of an existing system into a new form to realize quality improvements in operation, system capability, functionality, performance, or evolvability at a lower cost, schedule, or risk to the customer.

Some on-site managers in certain communities consider privatization and/or reengineering schemes threatening. In the worst-case scenario, a private contractor could bid the entire staff out of their jobs. In the best case, privatization and/or reengineering are often a very real threat that forces on-site managers into workforce cuts, improving efficiency and cutting costs. While at the same time, on-site managers work to ensure the community receives safe drinking water and the facility meets standards and permits, with fewer workers and without injury to workers, the facility, or the environment.

Local officials take a hard look at privatization and reengineering for a number of reasons:

- *Decaying Infrastructures* – Many water operations include water infrastructures that date back to the early 1900s. The most recent systems were built with federal funds during the 1970s, and even these now need upgrading or replacing. The USEPA recently estimated that the nation's 75,000+ drinking water systems alone would require more than $100 billion in investments over the next 20 years.
- *Mandates* – The federal government has reduced its contributions to local public water and wastewater systems over the past 30 years, while at the same time imposing more stringent water quality and effluent standards under the Clean Water Act and Safe Drinking Water Act. Moreover, as previously mentioned, new unfunded mandated safety regulations, such as OSHA's Process Safety Management and USEPA's Risk Management Planning, are expensive to implement using local sources of revenues or state revolving loan funds.
- *Hidden Function* – Earlier, we stated that much of the work of water/wastewater treatment is a "hidden function." Because of this lack of visibility, it is often difficult for local officials to commit to making the necessary investments in community water and wastewater systems. Simply, the local politicians lack the political will; water pipes and interceptors are not visible and not perceived as immediately critical for adequate funding. Thus, it is easier for some elected officials to ignore them in favor of expenditures of more visible services, such as police and fire. Additionally, in some communities, raising water and sewage rates to cover operations and maintenance is not always affected because it is an unpopular move for elected officials to make. This means that water and sewer rates do not adequately cover the actual cost of providing services in many municipalities.

In many locations throughout the United States, expenditures on water services are the largest facing local governments today. Thus, this area presents a great opportunity for cost savings. Through privatization, some public water systems can take advantage of advanced technology, more flexible management practices, and streamlined procurement and construction practices to lower costs and make critical improvements more quickly. With regard to privatization, the common view taken is that ownership of water resources and treatment plants should be maintained by the public (local government entities) to prevent a Tragedy of the Commons-like event [i.e., free access and unrestricted demand for water (or other natural resource) that ultimately structurally dooms the resource through over-exploitation by private interests]. However, because management is also a "hidden function" of some public service operations (e.g., water and wastewater operations), privatization may be a better alternative in select circumstances and in select communities.

BENCHMARK IT*

Benchmarking is now more than just a buzzword in water operations; it has become a tool for effective management and operations (see Figure 1.3). Cobblers are credited with coining the term *benchmarking*. They used the term to measure people's feet for shoes. They would place the person's foot on a "bench" and mark it out to make the pattern for the shoes. Benchmarking is still used to measure but now specifically gauges performance based on specific indicators such as cost per unit of measure, productivity per unit of measure, the cycle time of some value per unit of measure, or defects per unit of measure.

It is interesting to note that there is no specific benchmarking process that has been universally adopted; this is the case because of its wide appeal and acceptance (and humans are innovative). Accordingly, benchmarking manifests itself via various methodologies. Robert Camp (1989) wrote one of the earliest books on benchmarking and developed a 12-stage approach to benchmarking. Camp's 12 stage methodology consists of:

- Select subject;
- Define the process;
- Identify potential partners;

Select subject;

Define the process;

Identify potential partners;

Identify data sources;

Collected data and select partners;

Determine the gap;

Establish process differences;

Target future performance;

Communicate;

Adjust goal;

Implement; and

Review and recalibrate.

FIGURE 1.3 Benchmarking process.

* Adapted from F. R. Spellman (2013). *Water & Wastewater Infrastructure: Energy Efficiency and Sustainability.* Boca Raton, FL: CRC Press.

- Identify data sources;
- Collect data and select partners;
- Determine the gap;
- Establish process differences;
- Target future performance;
- Communicate;
- Adjust goal;
- Implement; and
- Review and recalibrate.

DID YOU KNOW?

Benchmarking can be useful, but no two public water systems are ever exactly the same. You'll have some characteristics that affect your relative performance and that are beyond a utility's control.

With regard to improving energy efficiency and sustainability in drinking water operations, benchmarking is simply defined as the process of comparing the energy usage of their drinking water treatment operation to similar operations. Local public water systems of similar size and design are excellent points of comparison. Broadening the search, one can find several resources discussing the "typical" energy consumption across the United States for a public water system of a particular size and design. Keep in mind that in drinking water treatment systems (and other utilities and industries), benchmarking is often used by management personnel to increase efficiency and ensure the sustainability of energy resources but is also used, as mentioned earlier, to ensure their own self-preservation (i.e., to retain their lucrative positions). With self-preservation as their motive, benchmarking is used as a tool to compare operations with best in class like facilities or operations to improve performance to avoid the current (and ongoing) trend to privatize water and other public operations.

In energy efficiency and sustainability projects, before the benchmarking tool is used, an energy team should be formed and assigned the task of studying how to implement energy-saving strategies and how to ensure sustainability in the long run. Keep in mind that forming a "team" is not the same as fashioning a silver bullet; the team is only as good as its leadership and its members. Again, benchmarking is a process for rigorously measuring your performance vs. best-in-class operations and using the analysis to meet and exceed the best in class; thus, those involved in the benchmarking process should be the best of the best (Spellman, 2009).

What Is Benchmarking?

- Benchmarking vs. best practices gives water operations a way to evaluate their operations overall.
- How effective?
- How cost-effective?
- Benchmarking shows plants both how well their operations stack up, and how well those operations are implemented.
- Benchmarking is an objective-setting process.
- Benchmarking is a new way of doing business.
- Benchmarking forces an external view to ensure the correctness of objective setting.
- Benchmarking forces internal alignment to achieve plant goals.
- Benchmarking promotes teamwork by directing attention to those practices necessary to remain competitive.

POTENTIAL RESULTS OF BENCHMARKING

Benchmarking may indicate the direction of required change rather than specific metrics: costs must be reduced; customer satisfaction must be increased; return on assets must be increased; improved maintenance; and improved operational practices. Best practices translate into operational units of measure.

BENCHMARKING: THE PROCESS

When forming a benchmarking team, the goal should be to provide a benchmark that evaluates and compares privatized and reengineered water treatment operations to your operation in order to be more efficient and remain competitive and make continual improvements. It is important to point out that benchmarking is more than simply setting a performance reference or comparison; it is a way to facilitate learning for continual improvements. The key to the learning process is looking outside one's own plant to other plants that have discovered better ways of achieving improved performance.

BENCHMARKING STEPS

As shown in Figure 1.3, the benchmarking process consists of five major steps:

1. *Planning* – Managers must select a process (or processes) to be benchmarked. A benchmarking team should be formed. The process of benchmarking must be thoroughly understood and documented. The performance measure for the process should be established (i.e., cost, time, and quality).
2. *Research* – Information on the best-in-class performer must be determined through research. The information can be derived from the industry's network, industry experts, industry and trade associations, publications, public information, and other award-winning operations.
3. *Observation* – The observation step is a study of the benchmarking subject's performance level, processes, and practices that have achieved those levels, and other enabling factors.
4. *Analysis* – In this phase, comparisons in performance levels among facilities are determined. The root causes for the performance gaps are studied. To make accurate and appropriate comparisons, the comparison data must be sorted, controlled for quality, and normalized.
5. *Adaptation* – This phase is putting what is learned throughout the benchmarking process into action. The findings of the benchmarking study must be communicated to gain acceptance, functional goals must be established, and a plan must be developed. Progress should be monitored and, as required, corrections in the process made.

UPGRADING SECURITY

You may say Homeland Security is a Y2K problem that doesn't end Jan. 1 of any given year.–

Governor Tom Ridge

Worldwide conflicts are ongoing and seem never-ending. One of the most important conflicts of our time, such as the ongoing Israeli–Palestinian conflict, is in fact conflict over scarce but vital water resources. This conflict over water, unfortunately, may be a harbinger of things to come.

UNITED STATES SEES INCREASE IN CYBER ATTACKS ON INFRASTRUCTURE

The top United States military official responsible for defending the United States against cyberattacks said Thursday that there had been a 17-fold increase in computer attacks on United States infrastructure between 2009 and 2011, initiated by criminal gangs, hackers and other nations.–

New York Times, 07/27/2012

DID YOU KNOW?

As of 2003, community water systems serve by far the largest proportion of the United States population – 273 million out of a total population of 290 million (USEPA 2004).

According to USEPA (2004), there are approximately 160,000 public water systems (PWSs) in the United States, each of which regularly supplies drinking water to at least 25 persons or 15 service connections. Eighty-four percent of the total US population is served by PWSs, while the remainder is served primarily by private wells. PWSs are divided into community water systems (CWSs) and noncommunity water systems (NCWSs). Examples of CWSs include municipal water systems that serve mobile home parks or residential developments. Examples of NCWSs include schools, factories, churches, commercial campgrounds, hotels, and restaurants.

Because drinking water is consumed directly, health effects associated with contamination have long been major concerns. In addition, interruption or cessation of the drinking water supply can disrupt society, impacting human health and critical activities such as fire protection. Although they have no clue as to its true economic value and to its future worth, the general public correctly perceives drinking water as central to the life of an individual and of society.

Federal and state agencies have long been active in addressing these risks and threats to public water systems through regulations, technical assistance, research, and outreach programs. As a result, an extensive system of regulations governing maximum contaminant levels of 90 conventional contaminants (most established by the USEPA), construction and operating standards (implemented mostly by the states), monitoring, emergency response planning, training, research, and education have been developed to better protect the nation's drinking water supplies.

Since the events of 9/11, EPA has been designated as the sector-specific agency responsible for infrastructure protection activities for the nation's drinking water system. EPA is utilizing its position within the water sector and working with its stakeholders to provide information to help protect the nation's drinking water supply from terrorism or other intentional acts. Ideally, in a perfect world, water infrastructure would be secured in a layered fashion (aka the barrier approach). Layered security systems are vital. Using the protection "in depth" principle, requiring that an adversary defeat several protective barriers or security layers to accomplish its goal, water, and wastewater infrastructure can be made more secure. *Protection in depth* is a term commonly used by the military to describe security measures that reinforce one another, masking the defense mechanisms from the view of intruders, and allowing the defender time to respond to intrusion or attack.

TECHNICAL VS PROFESSIONAL MANAGEMENT

Water treatment operations management is management that is directed toward providing water of the right quality, in the right quantity, at the right place, at the right time, and at the right price to meet various demands. The techniques of management are manifold in water resource management operations. In water treatment operations, for example, management techniques may include (Mather, 1984):

- Storage to detain surplus water available at one time of the year for use later, transportation facilities to move water from one place to another, manipulation of the pricing structure for water to reduce demand, use of changes in legal systems to make better use of the supplies available, introduction to techniques to make more water available through watershed management, cloud seeding desalination of saline or brackish water, or area-wide educational programs to teach *conservation or reuse of water.*
- Managing a waterworks requires a manager must be a well-rounded, highly skilled individual. No one questions the need for incorporation of these highly trained practitioners – well-versed

in the disciplines and practice of sanitary engineering, biology, chemistry, hydrology, environmental science, safety principles, accounting, auditing, technical aspects, energy conservation, security needs, and operations – in both professions. Based on years of experience in the water profession and personal experience dealing with high-level public service managers, however, engineers, biologists, chemists, and others with no formal management training and no proven leadership expertise are often hindered (limited) in their ability to solve the complex management problems currently facing both industries.

So what is dysfunctional management? How is it defined? Well, we have all encountered one or more exposures to dysfunctional managers in our working careers; thus, there is no need to discuss this matter any further here.

ENERGY CONSERVATION MEASURES AND SUSTAINABILITY

There is a number of long-term economic, social, and environmental trends (Elkington's, 1999, so-called Triple Bottom Line) evolving around us. Many of these long-term trends are developing because of us and specifically for us or simply to sustain us. Many of these long-term trends follow general courses and can be described by the jargon of the day; that is, they can be alluded to or specified by a specific buzzword or buzzwords common in usage today. We frequently hear these buzzwords used in general conversation (especially in abbreviated texting form). Buzzwords such as empowerment, outside the box, streamline, wellness, synergy, generation X, face time, exit strategy, LOL, clear goal, and so on and so forth are just part of our daily vernacular.

A popular buzzword that we have become concerned with is *sustainability*, which is often used in business. However, in water treatment practice and operations, sustainability is much more than just a buzzword; it is a way of life (or should be). There are numerous definitions of sustainability that are overwhelming, vague, and/or indistinct. For our purposes, there are long and short definitions of sustainability. The long definition: ensuring that water treatment operations occur indefinitely without negative impact. The short definition: the capacity of water operations to endure. What does sustainability really mean in the real world of water treatment operations?

We defined sustainability in what we call long and short terms. Note however that sustainability in water treatment operations can be characterized in broader or all-encompassing terms than those simple definitions. As mentioned, using the Triple Bottom Line scenario, the environmental aspects, economic aspects, and social aspects of water treatment operations can define today's and tomorrow's needs more specifically. How one chooses to define infrastructure is not important. What is important is to maintain and manage infrastructure in the most efficient and economical manner possible to ensure its sustainability. This is no easy task. Consider, for example, the 2009 Report Card for American Infrastructure produced by the American Society of Civil Engineers shown in Table 1.2.

Not only must water treatment managers maintain and operate aging and often under-funded infrastructure, they must also comply with stringent environmental regulations and they must also keep stakeholders and ratepayers satisfied with operations and with rates. Moreover, in line with these considerations, managers must incorporate economic considerations into every decision. For example, as mentioned, they must meet regulatory standards for the quality of treated drinking water. They must also plan for future upgrades or retrofits that will enable the water facility to meet future water quality and future effluent regulatory standards. Finally, and most importantly, managers must optimize the use of manpower, chemicals, and electricity.

Sustainable Water Infrastructure

(EPA, 2012) points out that sustainable development can be defined as that which meets the needs of the present generation without compromising the ability of future generations to meet their needs.

TABLE 1.2

2009 Report Card for American Infrastructure

Infrastructure	Grade
Bridges	C
Dams	D
Drinking water	D
Energy	D+
Hazardous waste	D
Roads	D–
Schools	D
Wastewater	D
America's infrastructure grade point average (GPA)	**D**

Source: Modified from American Society of Civil Engineers (2012) Report Card for American Infrastructure 2009. Accessed 01/04/2012 @ www.infrastructurereportcard.org/.

The current United States population benefits from the investments that were made over the past several decades to build up the nation's water infrastructure. Practices that encourage public water systems and their customers to address existing needs so that future generations will not be left to address the approaching wave of needs resulting from aging water infrastructure must continuously be promoted by sector professionals.

To be on a sustainable path, investments need to result in efficient infrastructure and infrastructure systems and be at a pace and level that allow the water and wastewater sectors to provide the desired levels of service over the long term. Sounds easy enough: the water manager simply needs to put his/her operation on a sustainable path; moreover, he/she can simply accomplish this by investing, right? Well, investing what? Investing in what? Investing how much? These are questions that require answers, obviously. Before moving on with this discussion it is important first to discuss plant infrastructure basics and second to discuss funding.

THE WATER INFRASTRUCTURE GAP

A 2002 EPA report referenced a water infrastructure gap analysis that compared current spending trends at the nation's drinking water treatment facilities to the expenses they can expect to incur for both capital and operations and maintenance costs. The "gap" is the difference between projected and needed spending and was found to be over $500 billion over a 20-year period. This important 2002 EPA gap analysis study is just as pertinent today as it was back then. Moreover, the author draws upon many of the tenets presented in the EPA analysis in formulating many of the basic points and ideas presented herein.

ENERGY EFFICIENCY: WATER TREATMENT OPERATIONS

Waterworks managers and operators are concerned with water infrastructure. This could mean they are concerned with the pipes, treatment plants, and other critical components that deliver safe drinking water to our taps. Although any component or system that makes up water infrastructure is important, remember that all water-related infrastructure can't function without the aid of some motive force. This motive force (energy source) can be provided by gravitational pull, mechanical means, or electrical energy. We simply can't sustain the operation of water and infrastructure without energy. As a case in point, consider that drinking water systems account for approximately

3–4% of energy use in the United States, adding over 45 million tons of greenhouse gases annually. Further, drinking water plants are typically the largest energy consumers of municipal governments, accounting for 30–40% of total energy consumed. Energy as a percentage of operating costs for drinking water systems can also reach as high as 40% and is expected to increase 20% in the next 15 years due to population growth and tightening drinking water regulations. Not all the news is bad, however. Studies estimate potential savings of 15–30% that are "readily achievable" in water and wastewater treatment plants, with substantial financial returns in the thousands of dollars and within payback periods of only a few months to a few years.

CHAPTER SUMMARY

Going to your kitchen sink and drawing a glass of water to drink or to fill the sink and use to wash dishes is simple enough; it is routine and hardly even or ever thought about. However, those in the know, that is, the water practitioners know that getting safe water to a kitchen tap (or anywhere else, for that matter) is no easy task and certainly should never be taken for granted. This chapter has set the foundation for the chapters to follow that explain the process of providing safe drinking water for consumption or for other uses.

REVIEW QUESTIONS

1. Which of the following actions has had little effect on the natural water cycle?
   a. Dredging
   b. Damming
   c. Damaged ecological niches
   d. Desertification

2. Which public water system supplies water to the same population year round?
   a. Noncommunity
   b. Transient system
   c. Community
   d. Non-transient

3. What type of public water system is typically used in campgrounds?
   a. Transient water system
   b. Community water system
   c. Non-transient water system
   d. County water system

4. Water quality is defined in terms of _____.
   a. Location
   b. Source
   c. Size of water supply
   d. Water characteristics

5. Treatment plant safety depends principally on _____.
   a. Upper management support
   b. Number of hazards present
   c. Attitude of outside contractors
   d. Luck of the draw

6. Potable water must be free of _____.
   a. Rocks
   b. Fish
   c. Non-sanitary properties
   d. Vegetation

7. Which of the following is not a source of drinking water?
    a. Freshwater lakes
    b. Rivers
    c. Glaciers
    d. Moon rocks

8. A drainage basin is naturally shunted toward _____.
    a. Streams
    b. Creeks
    c. Brooks
    d. All of the above

9. Water that is unaccounted for in a distribution system is attributable to _____.
    a. Theft
    b. Evaporation
    c. Transpiration
    d. Inaccurate meter readings

10. Reengineering is _____.
    a. Tearing down the old and rebuilding with the new
    b. Engineering out a problem
    c. Systematic transformation of an existing system to a new form
    d. Unengineering the engineering

ANSWER KEY

1. Which of the following **actions** has had little effect on the natural water cycle?
    a. Dredging
    b. Damming
    c. Damaged ecological niches
    d. **Desertification**

2. Which public water system supplies water to the same population year round?
    a. Noncommunity
    b. Transient system
    c. **Community**
    d. Non-transient

3. What type of public water system is typically used in campgrounds?
    a. **Transient water system**
    b. Community water system
    c. Non-transient water system
    d. County water system

4. Water quality is defined in terms of _____.
    a. Location
    b. Source
    c. Size of water supply
    d. **Water characteristics**

5. Treatment plant safety depends principally on _____.
    a. **Upper management support**
    b. Number of hazards present
    c. Attitude of outside contractors
    d. Luck of the draw

6. Potable water must be free of _____.
   a. Rocks
   b. Fish
   c. **Non-sanitary properties**
   d. Vegetation

7. Which of the following is not a source of drinking water?
   a. Freshwater lakes
   b. Rivers
   c. Glaciers
   d. **Moon rocks**

8. A drainage basin is naturally shunted toward _____.
   a. Streams
   b. Creeks
   c. Brooks
   d. **All of the above**

9. Water that is unaccounted for in a distribution system is attributable to _____.
   a. Theft
   b. Evaporation
   c. Transpiration
   d. **Inaccurate meter readings**

10. Reengineering is _____.
    a. Tearing down the old and rebuilding with the new
    b. Engineering out a problem
    c. **Systematic transformation of an existing system to a new form**
    d. Unengineering the engineering

BIBLIOGRAPHY

Angele, F.J., Sr. 1974. *Cross Connections and Backflow Protection*, 2nd ed. Denver, CO: American Water Association.

ASTM 1969. *Annual Book of ASTM Standards, Section 11, Water and Environmental Technology*. Philadelphia, PA: American Society of Testing Materials.

Camp, R. 1989. *The Search for Industry Best Practices that Lead 2 Superior Performance*. Boca Raton, FL: Productivity Press (CRC Press).

Capra, F. 1982. *The Turning Point: Science, Society and the Rising Culture*. New York: Simon & Schuster, p. 30.

Daly, H.E. 1980. Introduction to the Steady-State Economic. In: *Ecology, Ethics: Essays toward a Steady State Economy*. New York: W.H. Freeman & Company.

De Villiers, M. 2000. *Water: The Fate of Our Most Precious Resource*. Boston, MA: Mariner Books.

DOE 2001. *21 Steps to Improve Cyber Security of SCADA Networks*. Washington, DC: Department of Energy.

Drinan, J.E. 2001. *Water & Wastewater Treatment: A Guide for the Non-Engineering Professional*. Boca Raton, FL: CRC Press.

Drinan, J.E. and Spellman, F.R. 2012. *Water & Wastewater Treatment: A Guide for the Nonengineering Professional*, 2nd ed. Boca Raton, FL: CRC Press.

EIA 2012. The Current and Historical Monthly Retail Sales, Revenues and Average Revenue per Kilowatt Hour by State and by Sector. Accessed 06/09/19 @ http://www.eia.doe.gov/cneaf/electricity/page/sales_revenue.xis.

Elkington, J. 1999. *Cannibals with Forks*. New York: Wiley.

EPA 2002. *The Clean Water and Drinking Water Infrastructure Gap Analysis*. Washington, D.C: United States Environmental Protection Agency - EPA-816-R-02-020.

EPA 2003. EFAB Newsletter volume 3, issue 2: Providing Advice on How to Pay for Environmental Protection: *Diamonds and Water*. Accessed 06/07/19 @ www.epa.gov/efinpage/efab/newslaters/newsletters6.html.

EPA 2012. *Frequently Asked Questions: Water Infrastructure and Sustainability.* Accessed 06/01/2019 @ http://water.ep.gov/infrastrure/sustain/si_faqs.cfm.

Ezell, B.C. 1998. *Risks of Cyber Attack to Supervisory Control and Data Acquisition.* Charlottesville, VA: University of Virginia.

FBI 2000. *Threat to Critical Infrastructure.* Washington, DC: Federal Bureau of Investigation.

FBI 2004. *Ninth Annual Computer Crime and Security Survey.* Washington, DC: FBI, Computer Crime Institute and Federal Bureau of Investigations.

GAO 2003. *Critical Infrastructure Protection: Challenges in Securing Control System.* Washington, DC: United States General Accounting Office.

Garcia, M.L. 2001. *The Design and Evaluation of Physical Protection Systems.* Oxford, UK: Butterworth-Heinemann.

Gellman, B. 2002. Cyber-Attacks by Al Qaeda Feared: Terrorists at Threshold of Using Internet as Tool of Bloodshed, Experts Say. *Washington Post*, June 27, p. A01.

Gleick, P.H. 1998. *The World's Water 1998–1999: The Biennial Report on Freshwater Resources.* Washington, DC: Island Press.

Gleick, P.H. 2000. *The World's Water 2000–2001: The Biennial Report on Freshwater Resources.* Washington, DC: Island Press.

Gleick, P.H. 2004. *The World's Water 2004–2005: The Biennial Report on Freshwater Resources.* Washington, DC: Island Press.

Harper, S. 2007. VA Grants to Fuel Green Research. *The Virginian-Pilot*, Norfolk, VA, June 30.

Henry, K. 2002. New Face of Security. *Government Security.* April, pp. 30–31.

Holyningen-Huene, P. 1993. *Reconstructing Scientific Revolutions.* Chicago, IL: University of Chicago, p. 134.

IBWA 2004. *Bottled Water Safety and Security.* Alexandria, VA: International Bottled Water Association.

Johnson, R. & Moore, A. 2002. Policy Brief 17 Opening the Floodgates: Why Water Privatization Will Continue. Reason Public Institute. www.rppi.org.pbrief17.

Jones, B.D. 1980. *Service Delivery in the City: Citizens Demand and Bureaucratic Rules.* New York: Longman, p.2.

Jones, F.E. 1992. *Evaporation of Water.* Chelsea, MI: Lewis Publishers.

Jones, T. 2006. Water-Wastewater Committee: Program Opportunities in the Municipal Sector: Priorities for 2006. In: presentation to CEE June Program Meeting, June 2006. Boston, MA.

Kemmer, F.N. 1979. *Water: The Universal Solvent*, 2nd ed. Oak Brook, IL: Nalco Chemical Company.

Lewis, S.A. 1996. *The Sierra Club Guide to Safe Drinking Water.* San Francisco, CA: Sierra Club Books.

Manahan, S.E. 1997. *Environmental Science and Technology.* Boca Raton, FL: CRC Press.

Mather, J.R. 1984. *Water Resources: Distribution, Use, and Management.* New York: Wiley.

McCoy, K. 2012. USA Today Analysis: Nation's Water Costs Rushing Higher. Accessed 06/02/19 @ http://www.usatoday.com/money/economy/story/2012-09-27/water-rates-rising/57849626/1.

McGhee, T.J. 1991. *Water Supply and Sewerage*, 6th ed. New York: McGraw-Hill, Inc.

Meyer, W.B. 1996. *Human Impact on Earth.* New York: Cambridge University Press.

Muir, J. 1888. *Picturesque California.* Oakland, CA: Sierra Club.

NIPC 2002. *National Infrastructure Protection Center Report.* Washington, DC: National Infrastructure Protection Center.

Outwater, A. 1997. *Water: A Natural History.* New York: Basic Books.

Peavy, H.S., et al. 1985. *Environmental Engineering.* New York: McGraw-Hill, Inc.

Pielou, E.C. 1998. *Fresh Water.* Chicago, IL: University of Chicago Press.

Powell, J.W. 1904. *Twenty-Second Annual Report of the Bureau of American Ethnology to the Secretary of the Smithsonian Institution, 1900–1901.* Washington, D.C.: Government Printing Office.

Spellman, F.R. 1998. *Composting Biosolids.* Lancaster, PA: Technomic Publishing.

Spellman, F.R. 2003. *Handbook of Water and Wastewater Treatment Plant Operations.* Boca Raton, FL: Lewis Publishers.

Spellman, F.R. 2009. *Handbook of Water and Wastewater Treatment Plant Operations*, 2nd ed. Boca Raton, FL: CRC Press.

Spellman, F.R. 2013. *Water and Wastewater Infrastructure.* Boca Raton, FL: CRC Press.

Stamp, J. et al. 2003. *Common Vulnerabilities in Critical Infrastructure Control Systems*, 2nd ed. Livermore, CA: Sandia National Laboratories.

Tchobanglous, G. and Schroeder, E. 1987. *Water Quality.* Boston, MA: Addison-Wesley Publishing Company.

UN 2010. *Sick Water? The Central Role of Wastewater Management in Sustainable Development.* New York: UNEP.

United States News Online. 2000. *USGS Says Water Supply will be One of Challenges in Coming Century.* Accessed 06/2/19 @ http://www.uswaternews.com/archives/arcsupply/tusgay3.html.

USEPA 2004. *Water Security: Basic Information.* Accessed 06/01/19 @ http://cfpub.epa.gov/safewater/watersecurity/basicinformation.cfm.

USEPA 2005a. EPA Needs to Determine What Barriers Prevent Water Systems from Securing Known SCADA Vulnerabilities. In: Harris, J. *Final Briefing Report.* Washington, DC: USEPA.

USEPA 2005b. *Water and Wastewater Security Product Guide.* Accessed 6/06/19 @ http://cfpub.epa.gov.safewater/watersecurity/guide.

USEPA 2006. *Watersheds.* Accessed 06/06/19 @ http://www.epa.gov/owow/watershed/whatis. html.

USEPA 2008. *Ensuring a Sustainable Future: An Energy Management Guidebook for Wastewater and Water Utilities.* Washington, DC: United States Environmental Protection Agency.

US Fish and Wildlife. 2007. *Nutrient Pollution.* Accessed 06/2/19 @ http://www.fws.gov/chesapeakebay/nutrient.html.

USGS 2004. *Estimated Use of Water in the United States in 2000.* Washington, DC: United States Geological Survey.

USGS 2006. *Water Science in Schools.* Washington, DC: United States Geological Survey.

USGS 2011. *Water Withdrawals and Wastewater Discharges.* Washington, DC: U.S. Geological Survey.

WEF 1997. Energy Conservation in Wastewater Treatment Facilities. In: *Manual of Practice No. MFD-2.* Alexandria, VA: Water Environment Federation.

2 Water Quality, Characteristics, and Treatment

Copyright Las Vegas Valley Water District and used with permission

We forget that the water cycle and the life cycle are one.

– Jacques Yves Cousteau

LEARNING OBJECTIVES

After studying this chapter, you should be able to:

- Describe water quality management.
- Describe water in terms of its characteristics.
- Discuss the water cycle.
- Define physical characteristics of water including solids, turbidity, and describe the factors that affect water clarity.
- Describe the conditions and factors that affect watercolor.
- Discuss consumer expectations of color, taste, and odor for water.
- Describe the organic and inorganic contaminants that affect water's smell and taste.
- Discuss palpability and temperature.
- Describe the effect temperature has on water treatment.
- Discuss the definition of water as the universal solvent.
- Identify the chemical constituents commonly found in water and describe how they affect water quality.
- Define total dissolved solids; discuss their sources and how they are controlled.

DOI: 10.1201/9781003207146-4

- Discuss the problems created by high alkalinity.
- Identify the range and description classifications for water hardness and discuss the advantages and disadvantages of both hard and soft water.
- Discuss fluoride's importance in drinking water and describe how fluoride affects teeth.
- Identify and discuss the metals, both toxic and nontoxic, commonly found in water supplies and describe their effects on water quality.
- Identify the primary effects organic matter has on water quality.
- Discuss biodegradable and nonbiodegradable organic matter in terms of biological oxygen demand (BOD) and chemical oxygen demand (COD).
- Identify the essential nutrients water carries and discuss how excessive or deficient quantities of these nutrients affect the environment.
- Discuss how biological water characteristics can affect human health and well-being, and describe how pathogens are transported by the water system.
- Discuss how the presence or absence of biological organisms works as indicators of water quality to water specialists.
- Identify the ways organisms extract nutrients and energy from their waste environment.
- Describe the role oxygen plays in organism metabolism.
- Define the types of aquatic organisms (bacteria, viruses, protozoa, worms, rotifers, crustaceans, fungi, and algae) that can inhabit aquatic environments and identify the physical, chemical, and biological factors necessary for their existence.
- Describe and define the water treatment process, including pretreatment, aeration, screening, chemical addition, iron and manganese removal, hardness, corrosion control, coagulation, flocculation, sedimentation, filtration, disinfection, and removal of taste and odors.

KEY DEFINITIONS

Aeration – a physical treatment method that promotes biological degradation of organic matter. The process may be passive (when waste is exposed to air) or active (when a mixing or bubbling device introduces the air).

Algae – chlorophyll-bearing nonvascular, primarily aquatic species that have no true roots, stems, or leaves; most algae are microscopic, but some species can be as large as vascular plants.

Alkalinity – imparted to water by bicarbonate, carbonate, or hydroxide components, is a measure of water's ability to absorb hydrogen ions without significant pH change. Stated another way, alkalinity is a measure of the buffering capacity of water. Alkalinity measures the ability of water to neutralize acids and is the sum of all titratable bases down to about pH 4.5.

Anaerobic – pertaining to, taking place in, or caused by the absence of oxygen.

Bacteria – single-celled microscopic organisms.

Biodegradable – (tends to break down) material consists of organics that can be used for nutrients (food) by naturally occurring microorganisms within a reasonable length of time. These materials usually consist of alcohols, acids, starches, fats, proteins, esters, and aldehydes.

Biofouling – is the accumulation of microorganisms, plants, algae, or animals on wetted surfaces.

Biological Oxygen Demand (BOD) – the amount of oxygen required by bacteria to stabilize decomposable organic matter under aerobic conditions.

Biostimulent – a chemical that can stimulate biological growth.

Chemical Oxygen Demand (COD) – the amount of oxygen needed to chemically oxidize a waste; is a more complete and accurate measurement of the total depletion of dissolved oxygen in water.

Coagulation – a chemical water treatment method that causes very small suspended particles to attract one another and form larger particles. This is accomplished by the addition of a coagulant that neutralizes the electrostatic charges that cause particles to repel each other.

Coliform Bacteria – a group of bacteria predominantly inhabiting the intestines of humans or animals, but also occasionally found elsewhere. The presence of the bacteria in water is used as an indication of fecal contamination (contamination by animal or human wastes).

Colloidal Material – consists of solid or liquid particles, each about a few hundred nanometers in size, dispersed in a fluid.

Color – a physical characteristic of water. Color is most commonly tan or brown from oxidized iron, but contaminants may cause other colors, such as green or blue. Color differs from turbidity, which is water's cloudiness.

Demagnetization – chemical precipitation treatments for iron and manganese removal.

Dimerization – chemical precipitation treatments for iron and manganese removal.

Disinfection – a chemical treatment method. The addition of a substance (e.g., chlorine, ozone, or hydrogen peroxide), which destroys or inactivates harmful microorganisms, or inhibits their activity.

Equivalent Weight – of a substance is its atomic or molecular weight divided by n.

Eutrophication – the process by which water becomes enriched with plant nutrients, most commonly phosphorus and nitrogen.

Filtration – a physical treatment method for removing solid (particulate) matter from water by passing the water through porous media such as sand or a man-made filter.

Floc – well-defined aggregation of particles by gentle agitation for a much longer time.

Flocculation – the agglomeration of small particles into well-defined floc by gentle agitation for a much longer time.

Fluoride – in small concentrations is beneficial for controlling dental caries. Water containing the proper amount of fluoride can reduce tooth decay by 65% in children between ages 12 and 15. Adding fluoride to provide a residual of 1.5 to 2.5 mg/L has become common practice for municipal water plants. Concentrations above 5 mg/L are detrimental and limited by drinking water standards.

Fungi – constitute an extremely important and interesting group of aerobic microbes ranging from the unicellular yeasts to the extensively mycelial molds. Fungi are not considered plants but are instead a distinctive life form of great practical and ecological importance. Like bacteria, fungi metabolize dissolved organic matter. As the principal organisms responsible for the decomposition of carbon in the biosphere, fungi are essential. Fungi are unique (in a sense, when compared to bacteria), in that they can grow in low moisture areas and in low pH solutions, which aids in the breakdown of organic matter in aquatic environments.

Hardness – a characteristic of water caused primarily by the salts of calcium and magnesium. It causes deposition of scale in boilers, damage in some industrial processes, and sometimes objectionable taste. It may also decrease soup's effectiveness.

Helminths – worms are important in water quality assessment from a standpoint of human disease. Normally, worms inhabit organic mud and organic slime. They have aerobic requirements but can metabolize solid organic matter not readily degraded by other microorganisms. Water contamination may result from human and animal waste that contains worms. Worms pose hazards primarily to those persons who come into direct contact with untreated water. Thus, swimmers in surface water polluted by sewage or stormwater runoff from cattle feedlots are at particular risk. The Tubifex worm is a common organism used as an indicator of pollution in streams.

Iron and Manganese – commonly found in groundwater, but surface waters, at times, may also contain significant amounts.

Laxative Effect – hardness, especially with the combined presence of magnesium sulfates, can lead to the development of a laxative effect on new consumers.

Metals – iron and manganese are commonly found in groundwater, but surface waters, at times, may also contain significant amounts. Metals in water are classified as either toxic or nontoxic. Only those metals that are harmful in relatively small amounts are labeled toxic; other metals fall into the nontoxic groups. In natural waters, sources of metals include dissolution from natural deposits and discharges of domestic, agricultural, or industrial wastewaters.

Molds – extensive mycelial molds.

Nitrates – an ion consisting of nitrogen and oxygen (NO_3). Nitrate is a plant nutrient and is very mobile in soils.

Nonbiodegradable – substances that do not break down easily in the environment.

Nonpoint Source – a source (of any water-carried material) from a broad area, rather than from discrete points.

Nonvolatile Solids – substances that do not vaporize rapidly.

Nutrients – any inorganic or organic compound needed to sustain plant life.

Organic Matter – containing carbon, but possibly also containing hydrogen, oxygen, chlorine, nitrogen, and other elements.

Oxidation – when a substance either gains oxygen or loses hydrogen or electrons in a chemical reaction. One of the chemical treatment methods.

Pathogens – types of microorganisms that can cause disease.

Pretreatment – any physical, chemical, or mechanical process used before the main water treatment processes. It can include screening, presedimentation, and chemical addition.

Protozoa – are mobile, single-celled, complete, self-contained organisms that can be free-living or parasitic, pathogenic or nonpathogenic, microscopic or macroscopic. Protozoa range in size from two to several hundred microns in length. They are highly adaptable and widely distributed in natural waters, although only a few are parasitic. Most protozoa are harmless. Only a few cause illness in humans.

Screening – removes large debris (leaves, sticks, and fish) that can foul or damage plant equipment.

Sedimentation – removes settleable particles.

Solids – other than gases, all contaminants of water contribute to water's total solids (filterable + non-filterable solids) content. Solids are classified by size and state, chemical characteristics, and size distribution. Solids can be dispersed in water in both suspended and dissolved forms. Solids are also size classified as suspended, settleable, colloidal, or dissolved, depending on their behavioral attributes.

Surface Water – water contained in surface lakes, streams, and rivers.

Taste and Odor – one of the parameters on which water is judged.

Temperature – one of the parameters on which water is judged.

Total Dissolved Solids – dissolved solids present in water.

Toxic Metals – refers to mercury and lead in water.

Trihalomethanes – by-product formed when some organics are exposed to chlorine.

Turbidity – refers to the degree of water clarity.

Universal Solvent – physical trait of water.

Viruses – obligate parasitic particles that require a host to live in. They are the smallest biological structures known and can only be seen with the aid of an electron microscope.

Volatile Solids – those solids in water that are lost when solids are combusted.

Water Cycle – natural cycle where precipitation constantly changes form but returns to cycle again and again.

Water Quality – measure of water's fitness for intended use.

WATER QUALITY MANAGEMENT

Are we to wait until all frogs "croak?"

The earliest chorus of frogs – those high-pitched rhapsodies of spring peepers, those "jug-o-rum" calls of bullfrogs, those banjo-like bass harmonies of green frogs, those long and guttural cadences of leopard frogs, their singing a prelude to the splendid song of birds – beside an otherwise still pond on an early spring evening heralds one of nature's dramatic events: The drama of metamorphosis. This metamorphosis begins with masses of eggs that soon hatch into gill-breathing, herbivorous, fish-like tadpole larvae. As they feed and grow, warmed by the spring sun, almost imperceptibly a remarkable transformation begins. Hind legs appear and gradually lengthen. Tails shorten. Larval teeth vanish, and lungs replace gills. Eyes develop lids. Forelegs emerge. In a matter of weeks, the aquatic, vegetarian tadpole will (should it escape the many perils of the pond) complete its metamorphosis into an adult, carnivorous frog.

This springtime metamorphosis is special. This anticipated event (especially for the frog) marks the end of winter, the rebirth of life, a rekindling of hope (especially for mankind). This yearly miracle of change sums up in a few months each spring what occurred over 3,000 million years ago, when the frog evolved from its ancient predecessor. Today, however, something is different, strange, and wrong with this striking and miraculous event. In the first place, where are all the frogs? Where have they gone? Why has their population decreased so dramatically in recent years? The second problem is that this natural metamorphosis process (perhaps a reenactment of some Paleozoic drama whereby, over countless generations, the first amphibian types equipped themselves for life on land) now demonstrates aberrations of the worst kind, of monstrous proportions and dire results to frog populations in certain areas. For example, reports have surfaced of deformed frogs in certain sections of the United States, specifically Minnesota. Moreover, USEPA has received many similar reports from the United States and Canada as well as parts of Europe.

Most of the deformities have been in the rear legs and appear to be developmental. The question is why? Researchers have noted that neurological abnormalities have also been found. Again, the question is why? Researchers have pointed the finger of blame at parasites, pesticides, and other chemicals, ultraviolet radiation, acid rain, and metals. Something is going on. What is it? We do not know! The next question, then, is what are we going to do about it? Are we to wait until all the frogs croak before we act – before we find the source, the cause, the polluter – before we see this reaction result in other species; maybe in our own? The final question is obvious: When frogs are forced by mutation into something else, is this evolution by gunpoint? Are we holding the gun? (Spellman, 2013).

OVERVIEW

Managing the quality of water for frogs, animals of all kinds, and humans, whether it is used for drinking, human-implemented irrigation, or recreational purposes, is significant for health in both developing and developed countries worldwide. The first problem with water is rather obvious: A source of water must be found. Second, when accessible water is found, it must be suitable for human consumption. Remember, in Chapter 1, we described this as the Q & Q factor (Quality and Quantity). Meeting the water needs of all those living organisms that populate earth is an ongoing challenge. New approaches to meeting these water needs will not be easy to implement: economic and institutional structures still encourage the wasting of water and the destruction of ecosystems (Gleick, 2001). Again, finding a water source is the first problem. Finding a source of water that is safe to drink is the other problem.

Water quality is important; it can have a major impact on health, both through outbreaks of waterborne disease and by contributing to the background rates of disease. Accordingly, water quality standards are important to protect public health. In this chapter, water quality refers to those characteristics or range of characteristics that make water appealing and useful. Keep in mind that useful also means non-harmful or nondisruptive to either ecology or the human condition within the very broad spectrum of possible uses of water. For example, the absences of odor, turbidity, or color are desirable immediate qualities. However, there are imperceptible qualities that are also important; that is, the chemical qualities. The fact is the presence of materials such as toxic metals, (e.g., mercury and lead), excessive nitrogen and phosphorous, or dissolved organic material may not be readily perceived by the senses but may exert substantial negative impacts on the health of a stream and/or on human health. The ultimate impact of these imperceptible qualities of water (chemicals) on the user may be nothing more than the loss of aesthetic values. On the other hand, water-containing chemicals could also lead to a reduction in biological health or to an outright degradation of human health. Simply stated, the importance of water quality cannot be overstated.

With regard to water treatment operations, water quality management begins with a basic understanding of how water moves through the environment, is exposed to pollutants, and transports and deposits pollutants. The simplified hydrologic (water) cycle depicted in Figure 2.1 illustrates the general links between the atmosphere, soil, surface waters, groundwater, and plants.

WATER CYCLE

The water cycle describes how water moves through the environment and identifies the links between groundwater, surface water, and the atmosphere (see Figure 2.1). Components of the water cycle include:

- Atmosphere;
- Condensation;
- Evaporation;
- Evapotranspiration;
- Freshwater lakes and rivers;
- Groundwater flow;
- Groundwater storage;
- Ice and snow;
- Infiltration;
- Oceans;
- Precipitation;
- Snowmelt;
- Springs;
- Stream flow;

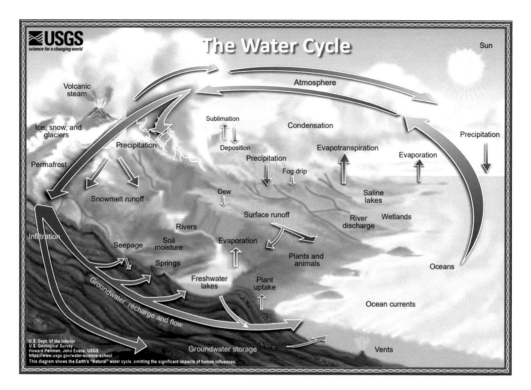

FIGURE 2.1 Hydrologic (water) cycle. Source: USGS (2019). Accessed @ https://www.usgs.gov/special-topic-water-science.

- Sublimation; and
- Surface runoff.

As illustrated, water is taken from the earth's surface to the atmosphere by evaporation from the surface of lakes, rivers, streams, and oceans. This evaporation process occurs when the sun heats water. The sun's heat energizes surface molecules, allowing them to break free of the attractive force binding them together and then evaporate and rise as invisible vapor in the atmosphere. Water vapor is also emitted from plant leaves by a process called transpiration. Every day, an actively growing plant transpires five and up to ten times as much water as it can hold at once. As water vapor rises, it cools and eventually condenses, usually on tiny particles of dust in the air. When it condenses, it becomes a liquid again or turns directly into a solid (ice, hail, or snow).

These water particles then collect and form clouds. The atmospheric water formed in clouds eventually falls to earth as precipitation. The precipitation can contain contaminants from air pollution. The precipitation may fall directly onto surface waters, be intercepted by plants or structures, or fall onto the ground. Most precipitation falls in coastal areas or in high elevations. Some of the water that falls in high elevations becomes runoff water, the water that runs over the ground (sometimes collecting nutrients from the soil) to lower elevations to form streams, lakes, and fertile valleys. The water we see is known as *surface water*. Surface water can be broken down into five categories: (1) oceans; (2) lakes; (3) rivers and streams; (4) estuaries; and (5) wetlands. 170,000 small-scale suppliers provide drinking water to approximately 200+ million Americans by 60,000 community water supply systems and to nonresidential locations, such as schools, factories, and campgrounds. The rest of Americans are served by private wells. The majority of the drinking water used in the United States is supplied from groundwater. Untreated water drawn from groundwater and surface waters, and used as a drinking water supply, can contain contaminants that pose a threat to human health.

Note: USEPA (2002) has reported that American households use approximately 146,000 gallons of freshwater annually and that Americans drink 1 billion glasses of tap water per day.

Obviously, with a limited amount of drinking water available for use, water that is available must be reused, or we will be faced with an inadequate supply to meet the needs of all users. Water use/ reuse is complicated by water pollution. Pollution is relative and hard to define. For example, floods and animals (dead or alive) are polluters, but their effects are local and tend to be temporary. Today, water is polluted in many sources, and pollution exists in many forms. It may appear as excess aquatic weeds, oil slicks, a decline in sport fishing, and an increase in carp, sludge worms, and other forms of life that readily tolerate pollution. Maintaining water quality is important because water pollution is not only detrimental to health but also to recreation, commercial fishing, aesthetics, and private, industrial, and municipal water supplies.

At this point you might be asking: With all the recent publicity about pollution and the enactment of new environmental regulations, hasn't water quality in the United States improved recently? Answer: With the recent pace of achieving fishable and swimmable waters under the Clean Water Act (CWA), one might think so. The 1994 *National Water Quality Inventory Report to Congress* indicated that 63% of the nations' lakes, rivers, and estuaries meet designated uses – only a slight increase over that reported in 1992. The main culprit is *Nonpoint Source Pollution* (NPS) (to be discussed in detail later). NPS is the leading cause of impairment for rivers, lakes, and estuaries. Impaired sources are those that do not fully support designated uses, such as fish consumption, drinking water supply, groundwater recharge, aquatic life support, or recreation. According to Fortner and Schechter (1996), the five leading sources of water quality impairment in rivers are: (1) agriculture; (2) municipal wastewater treatment plants; (3) habitat and hydrologic modification; (4) resource extraction; and (5) urban runoff and storm sewers.

The health of rivers and streams is directly linked to the integrity of habitat along the river corridor and in adjacent wetlands. Stream quality will deteriorate if activities damage vegetation along riverbanks and in nearby wetlands. Trees, shrubs, and grasses filter pollutants from runoff and reduce soil erosion. Removal of vegetation also eliminates shade that moderates stream temperature. Stream temperature, in turn, affects the availability of dissolved oxygen in the water column for fish and other aquatic organisms. Lakes, reservoirs, and ponds may receive water-carrying pollutants from rivers and streams, melting snow, runoff, or groundwater. Lakes may also receive pollution directly from the air. Thus, in attempting to answer the original question: Hasn't water quality in the United States improved recently? The best answer probably is: We are holding our own in controlling water pollution, but we need to make more progress. This understates an important point; that is, when it comes to water quality, we need to make more progress on a continuing basis.

CHARACTERISTICS OF WATER

Note: To gain a fundamental understanding of water's properties it is important to recognize water's characteristics. Water is more than just plain old water; it actually has some very unusual properties. It would be boring if you were told that water is wet, clear, and very necessary. So, instead, first water's major characteristics are described and then a brief USGS quiz on water's properties is provided to test your knowledge about real, non-boring water facts. When attempting to characterize water, it is normal to jump on the obvious features available: appearance, taste, and smell. These physical attributes are important but so are chemical and biological characteristics as well.

The chemical characteristics of water are important because even though water may appear, smell, and even taste okay, it doesn't necessarily mean that some chemical contaminant is not present. Today, with all the pesticides used in agriculture and various other industrial activities, chemical contamination is a very real possibility. The biological characteristics of water are extremely important to anyone who might drink the water. Even before Typhoid Mary, people suspected a relationship between microorganisms and disease. Today, we know with certainty that waterborne diseases are a real threat to human health and life. We also know that contaminated water (whether physically, chemically, or biologically contaminated) does not start out contaminated. Precipitation in the form of rain, snow, hail, or sleet (aside from what we contaminate ourselves) contains very few (if any) impurities. It is true that rain may pick up trace amounts of mineral matter, gases, and other substances as it forms and falls through the earth's atmosphere, but the precipitation has virtually no bacterial content – no waterborne disease (that we know of).

When precipitation reaches the earth's surface, many opportunities for the introduction of mineral and organic substances, microorganisms, and other forms of pollution (contamination) are presented. When water runs over and through the ground surface, it may pick up particles of soil. This attaches a physical property to water that can be readily seen: cloudiness or turbidity. Water, as it courses its inexorable way along the earth's surface, also picks up organic matter and bacteria. As water seeps down into the soil and through the underlying material to the water table, most of the suspended particles are filtered out. This natural filtration may be a two-edged sword: it can be partially effective in removing bacteria and other particulate matter, but it may change the chemical properties of water as it comes in contact with mineral deposits. Substances that alter the quality of water or its characteristics as it moves over or below the surface of the earth are of major concern to water practitioners.

PHYSICAL CHARACTERISTICS OF WATER

What makes water wet? Why is water wet? David Clary (1997), a chemist at University College London, points out that water does not start to behave like a liquid until at least six molecules form a cluster. He found that groups of five water molecules or fewer have planar structures, forming films one molecule thick. However, when a sixth molecule is added, the cluster switches to a three-dimensional cage-like structure, and suddenly it has the properties of water – it becomes wet. Beyond the physical property of wetness, other physical characteristics of interest to us include water's solids, turbidity, color, taste and odor, and temperature – all of which are apparent to our senses of smell, taste, sight, and touch.

Solids

Other than gases, all contaminants of water contribute to water's total solids (filterable + non-filterable solids) content. Solids are classified by size and state, chemical characteristics, and size distribution. Solids can be dispersed in water in both suspended and dissolved forms. Solids are also size classified as suspended, settleable, colloidal, or dissolved, depending on their behavioral attributes. *Colloidal material* in water is sometimes beneficial and sometimes harmful. Beneficial colloids are those that provide a dispersant effect by acting as protective colloids. Some colloids (silica-based types) can be troublesome, forming very hard scale when it deposits on heat transfer surfaces (Kemmer, 1988). Chemically, solids are also characterized as being *volatile* (solids that volatize at 550°C) or *nonvolatile*.

Suspended material present in water is objectionable because it provides adsorption sites for biological and chemical changes. These adsorption sites provide attached microorganisms a protective barrier against the chemical action of chlorine used in the disinfection process. Suspended solids in water may also be objectionable because they may be degraded biologically into unwanted by-products. Obviously, the removal of these solids is of primary concern in the production of clean, safe drinking water. In the water treatment process, the most effective means of removing solids from water is filtration. However, not all solids (colloids and other dissolved solids) can be removed by filtration.

Turbidity

One of the first conditions we notice about water is its clarity which we measure by *turbidity*, an assessment of the extent to which light is either absorbed or scattered by suspended material in water. Both the size and surface characteristics of the suspended material influence absorption and scattering. In surface water, most turbidity results from the erosion of very small colloidal material (rock fragments, silt, clay, and metal oxides from the soil). Microorganisms and vegetable material may also contribute to turbidity. In running waters, turbidity interferes with light penetration and photosynthetic reactions critical to aquatic plants. In water treatment, turbidity is a useful indicator of water quality.

COLOR

Watercolor is a physical characteristic often used to judge water quality. Pure water is colorless. Water takes on color when foreign substances, organic matter from soils, vegetation, minerals, and aquatic organisms, are present. For the most part, color in water is a mixture of colloidal organic compounds that represent breakdown products of high molecular weight substances produced by living cells. Consider, for example, water with a yellowish or tea color. The source of this yellow color is decayed vegetation leached from the watershed by runoff. These organic materials are broadly classified as humic substances (Kemmer, 1988). Color can also be contributed to water by municipal and industrial wastes.

Color in water is classified as either true color or *apparent color*. Water whose color is partly due to dissolved solids that remain after the removal of suspended matter is known as true color. Color contributed by suspended matter is said to have apparent color. In water treatment, true color is the most difficult to remove (Spellman, 1998). Colored water is generally unacceptable to the general public. People tend to prefer clear, uncolored water. Another problem with colored water is the effect it has on manufacturing, textiles, food preparation/processing, papermaking, and laundering. The color of water has a profound effect on its marketability for both domestic and industrial use. In water treatment, color is not usually considered unsafe or unsanitary but is a treatment problem related to exerting chlorine demand, which reduces the effectiveness of chlorine as a disinfectant. Some of the processes used in treating colored water include filtration, softening, oxidation, chlorination, and adsorption.

TASTE AND ODOR

Taste and odor are used jointly in the vernacular of freshwater science. In drinking water, taste and odor are usually not a consideration until the consumer complains. The problem is, of course, that most consumers find any taste and odor in water aesthetically displeasing. Taste and odor do not normally present a health hazard, but they can cause the customer to seek out water that might taste and smell better but may also be unsafe to drink. The fact is consumers expect water to be tasteless and odorless; if it isn't, they consider it substandard. If a consumer can taste or smell the water, he/she automatically associates that with contamination (Spellman, 1998).

Water contaminants are attributable to contact with natural substances (rocks, vegetation, soil, etc.) or human use. Taste and odor are caused by foreign matter, such as organic compounds, inorganic salts, or dissolved gases. Again, these substances may come from domestic, agricultural, or natural sources. Some substances found naturally in groundwater, while not necessarily harmful, may impart a disagreeable taste or undesirable property to the water. Magnesium sulfate, sodium sulfate, and sodium chloride are a few of these (Corbitt, 1990).

In addition, a number of other distinct smells are commonly encountered (see Table 2.1). Objectionable tastes and odors also caused by biological activity include tastes and odors contributed by various algae species. In water treatment, one of the common methods used to remove taste and odor is oxidation (usually with potassium permanganate and chlorine) of the problem material. Another standard treatment method is to feed powdered activated carbon to the flow prior to filtration. The activated carbon has numerous small openings that adsorb the components that cause the odor and taste.

TABLE 2.1

Categories of Offensive Odors Often Encountered in Water

Compound	Descriptive Quality
Amines	Fishy
Ammonia	Ammoniacal
Diamines	Decayed Flesh
Hydrogen Sulfide	Rotten Egg
Mercaptans	Skunk Secretion
Organic Sulfides	Rotten Cabbage
Skatole	Fecal

Source: Adaptation from The Chemical Senses (Moncrief, 1967).

TEMPERATURE

Most consumers prefer drinking waters that are consistently cool and do not have temperature fluctuations of more than a few degrees (groundwater from mountainous areas generally meets these criteria). Water with a temperature between 10°C and 15°C (50°F and 60°F) is generally the most palatable (Corbitt, 1990). Heat is added to surface and groundwater in many ways. Some of these are natural, some artificial. A problem associated with heat or excessive temperature in surface waters is that it affects the solubility of oxygen in water, the rate of bacterial activity, and the rate at which gases are transferred to and from the water.

In the actual examination of water (for its suitability for consumption), temperature is not normally a parameter used in evaluation. However, temperature is one of the most important parameters in natural surface water systems which are subject to great temperature variations. It affects a number of important water quality parameters. Temperature has an effect on the rate at which chemicals dissolve and react. When water is cold, more chemicals are required for efficient coagulation and flocculation to take place. When the water temperature is high, the result may be a higher chlorine demand because of increased reactivity and because of an increased level of algae and other organic matter in raw water. Temperature also has a pronounced effect on the solubility of gases in water.

Ambient temperature (temperature of the surrounding atmosphere) has the most profound and universal effect on the temperature of shallow natural water systems. When water is used by industry to dissipate process waste heat, the discharge points into surface waters may experience dramatic localized temperature changes. Other sources of increased temperatures in running water systems result because of clear-cutting practices in forests (where protective canopies are removed) and also from irrigation flows returned to a body of running water. Many people hold a misconception related to water temperature. Although the temperature of groundwater seems relatively "cool" in summer and warm in winter, its temperature remains nearly constant throughout the year. Human perception of temperature is relative to air temperature; the slight temperature fluctuation is not usually detectable. Contrary to popular belief, colder water is not obtained by drilling deeper wells. Beyond the 100-foot depth mark, the temperature of groundwater actually increases steadily at the rate of about 0.6°C (1°F) for every 100 feet or so of depth. This rate may increase dramatically in volcanic regions.

CHEMICAL CHARACTERISTICS OF WATER

Other parameters used to define water quality are water's chemical characteristics. Water's chemical composition is changed by the nature of the rocks that form the earth's crust. As surface water

TABLE 2.2

Chemical Constituents Commonly Found in Water

Chemical Constituents	
Calcium	Fluorine
Magnesium	Nitrate
Sodium	Silica
Potassium	TDS
Iron	Hardness
Manganese	Color
Bicarbonate	pH
Carbonate	Turbidity
Sulfate Chloride	Temperature

seeps down to the water table, it dissolves and carries portions of the minerals contained in/by soils and rocks. Because of this, groundwater usually is heavier in dissolved mineral content than surface water. Each chemical constituent that water may contain affects water use in some manner, by either restricting or enhancing specific uses (see Table 2.2 for principal constituents). Water, commonly called the universal solvent, is a solvent because of its chemical characteristics. Water analysts test a water supply to determine the supply's chemical characteristics; to determine if other harmful substances are present; to determine if substances are present that will enhance corrosion (of metals in water heaters, for example); and to determine if the chemicals responsible for staining fixtures and clothing are present in the water. The exact analyses to be conducted on a municipal water supply are mandated by the Public Health Service Drinking Water Standards promulgated by the United States Department of Health, Education, and Welfare.

Along with the elements and other constituents listed in Table 2.2 and toxic substances, water-quality managers are concerned with the presence of total dissolved solids (TDS), alkalinity, hardness, fluorides, metals, organics, and nutrients that might be present in a water supply. We discuss these chemical parameters in the following sections.

TOTAL DISSOLVED SOLIDS

TDS come from the minerals dissolved in water from rocks and soil as water passes over and through them. The measurement of TDS constitutes a part of the total solids in water; it is the residue remaining in a water sample after filtration or evaporation and is expressed in mg/L. TDS is an important water quality parameter, commonly used to measure salinity. Roughly, *fresh water* has a TDS of less than 1,500 mg/L (drinking water has a recommended maximum TDS level of 500 mg/L); *brackish* water has a TDS up to 5,000 mg/L; and *saline* water has a TDS above 5,000 mg/L. Seawater contains 30,000–34,000 mg/L TDS (Tchobanoglous & Schroeder, 1987).

Dissolved solids may be organic or inorganic. Water may come into contact with these substances within the soil, on surfaces, and/or in the atmosphere. The organic dissolved constituents of water come from the degradation (decay) of products of vegetation, from organic chemicals, and from organic gases. Dissolved solids can be removed from water by distillation, electrodialysis, reverse osmosis, or ion exchange. Removing these dissolved minerals, gases, and organic constituents is desirable because they may cause physiological effects and produce aesthetically displeasing color, taste, and odor. You might think removing all these dissolved substances from water is desirable, but this is not a prudent move. Pure, distilled water tastes flat.

Also, water has an equilibrium state with respect to dissolved constituents. If water is out of equilibrium or undersaturated, it aggressively dissolves materials it comes into contact with. Because of this particular problem, substances that are readily dissolvable are sometimes added to pure water to reduce its tendency to dissolve plumbing fixtures.

ALKALINITY

Alkalinity imparted to water by bicarbonate, carbonate, or hydroxide components is a measure of water's ability to absorb hydrogen ions without significant pH change. Stated another way, alkalinity is a measure of the buffering capacity of water. Alkalinity measures the ability of water to neutralize acids and is the sum of all titratable bases down to about pH 4.5. The bicarbonate, carbonate, and hydroxide constitutes alkalinity originate from carbon dioxide (from the atmosphere and as a by-product of microbial decomposition of organic material) and from their mineral origin (primarily from chemical compounds dissolved from rocks and soil).

Highly alkaline waters have no known significant impact on human health but are unpalatable. The principal problem with alkaline water is the reactions that occur between alkalinity and certain substances in the water. The resultant precipitate can foul water system appurtenances. Alkalinity levels also affect the efficiency of certain water treatment processes, especially the coagulation process.

HARDNESS

Water *hardness* is familiar to those individuals who have washed their hands with a bar of soap and found that they needed more soap to "get up lather." For this reason, originally hardness referred to the soap-consuming power of water. Hardness is the presence in water of multivalent cations, most notably calcium (Ca) and magnesium ions.

Hardness is classified as *carbonate hardness* and *noncarbonate hardness*. The carbonate that is equivalent to the alkalinity is termed carbonate hardness. Hardness is either temporary or permanent. Carbonate hardness (temporary hardness) can be removed by boiling. Noncarbonate hardness cannot be removed by boiling and is classified as permanent. Hardness values are expressed as an equivalent amount or equivalent weight of calcium carbonate (*equivalent weight* of a substance is its atomic or molecular weight divided by *n*). Water with a hardness of less than 50 ppm is *soft*. Above 200 ppm, domestic supplies are usually blended to reduce the hardness value.

The United States Geological Survey uses the following classification:

RANGE OF HARDNESS [mg/liter (ppm) as CaCo₃]	DESCRIPTIVE CLASSIFICATION
1–50	Soft
51–150	Moderately Hard
151–300	Hard
Above 300	Very Hard

Hardness has an economic impact, on soap consumption as well as problems in tanks and pipes. When using a bar of soap in hard water, you work the soap until the lather is built up. When lathering does occur, the water has been "softened" by the soap. However, the precipitate formed by hardness and soap (soap curd) adheres to just about anything (tubs, sinks, dishwashers) and may stain clothing, dishes, and other items. Hardness also affects people in personal ways: residues of the hardness-soap precipitate may remain in skin pores, causing the skin to feel rough and uncomfortable. Today these problems have been largely reduced by the development of synthetic soaps and detergents that do not react with hardness. However, hardness still leads to other problems such as scaling and laxative effect. *Scaling*, of course, occurs when carbonate hard water is heated and calcium carbonate and magnesium hydroxide are precipitated out of solution, forming a rock-hard scale that clogs hot water pipes and reduces the efficiency of boilers, water heaters, and heat exchangers. Hardness, especially with the combined presence of magnesium sulfates, can lead to the development of a *laxative effect* on new consumers. Rowe & Abdel-Magid (1995) point out that using hard water has some advantages. These include: (1) hard water aids in the growth of teeth and bones; (2) hard water reduces toxicity to man by poisoning with lead oxide from pipelines made of lead; and (3) soft waters are suspected to be associated with cardiovascular diseases.

FLUORIDE

Fluoride, seldom found in appreciable quantities in surface waters, appears in groundwater in only a few geographical regions, though it is sometimes found in a few types of igneous or sedimentary rocks. Fluoride is toxic to humans in large quantities and toxic to some animals. Some plants used for fodder can store and concentrate fluoride. Animals that consume this forage ingest an enormous overdose of fluoride. Their teeth mottle, they lose weight, give less milk, grow bone spurs, and become so crippled they must be destroyed (Koren, 1991). Fluoride in small concentrations is beneficial for controlling dental caries. Water containing the proper amount of fluoride can reduce tooth decay by 65% in children between ages 12 and 15. Adding fluoride to provide a residual of 1.5 to 2.5 mg/L has become common practice for municipal water plants. Concentrations above 5 mg/L are detrimental and limited by drinking water standards. How does the fluoridation of a drinking water supply actually work to reduce tooth decay? Fluoride combines chemically with tooth enamel when permanent teeth are forming. The result, of course, is harder, stronger teeth that are more resistant to decay. Adult teeth are not affected by fluoride. The EPA sets the upper limit for fluoride based on ambient temperatures because people drink more water in warmer climates. Fluoride concentrations should be lower in those areas.

METALS

Iron and manganese are commonly found in groundwater, but surface waters, at times, may also contain significant amounts. Metals in water are classified as either toxic or nontoxic. Only those metals that are harmful in relatively small amounts are labeled toxic; other metals fall into the nontoxic groups. In natural waters, sources of metals include dissolution from natural deposits and discharges of domestic, agricultural, or industrial wastewaters.

Nontoxic metals commonly found in water include hardness ions (calcium and manganese), iron, aluminum, copper, zinc, and sodium. Sodium (abundant in the earth's crust and highly reactive with other elements) is by far the most common nontoxic metal found in natural waters. Sodium salts (in excessive concentrations) produce a bitter taste in water and are a health hazard for kidney and cardiac patients. The usual low sodium diets allow for 20 mg/L of sodium in drinking water. Sodium, in large concentrations, is toxic to plants. Although iron and manganese in natural waters (in very small quantities) may cause color problems, they frequently occur together and present no health hazards at normal concentrations. Some bacteria, however, use manganese compounds as an energy source, and the resulting slime growth may produce taste and odor problems. The recommended limit for iron is 0.3 and 0.05 mg/L for manganese. Very small quantities of other nontoxic metals are found in natural water systems. Most of these metals cause taste problems well before they reach toxic levels.

Toxic metals are present, fortunately, in only minute quantities in most natural water systems. However, even in small quantities, these toxic metals can be especially harmful to humans and other organisms. Arsenic, barium, cadmium, chromium, lead, mercury, and silver are toxic metals that dissolve in water. Arsenic, cadmium, lead, and mercury, all cumulative toxins, are particularly hazardous to human health. Concentrated in organism bodies, these toxins are passed up the food chain and pose the greatest threat to organisms at the top of the chain.

Organic Matter

Organic matter (a broad category) includes both natural and synthetic molecules containing carbon, and usually hydrogen. All living matter is made up of organic molecules. Some organics are extremely soluble in water (alcohol and sugar are good examples) or may be quite insoluble (plastics). Tchobanoglous and Schroeder (1987) point out that the presence of organic matter in water is troublesome for the following reasons: "(1) color formation, (2) taste and odor problems, (3) oxygen depletion in streams, (4) interference with water treatment processes, and (5) the formation of halogenated compounds when chlorine is added to disinfect water" (p. 94). The main source of organic matter in water, although total amounts in water are low, is from decaying vegetation (leaves, weeds, and trees). The general category of "organics" in natural waters includes organic matter whose origins could be from both natural sources and from human activities. Distinguishing natural organic compounds from solely man-made compounds (e.g., pesticides and other synthetic organic compounds) is critical. Those natural water-soluble organic materials are generally limited to contamination of surface water only. Dissolved organics are usually divided into two categories: biodegradable and nonbiodegradable. *Biodegradable* (tends to break down) material consists of organics that can be used for nutrients (food) by naturally occurring microorganisms within a reasonable length of time. These materials usually consist of alcohols, acids, starches, fats, proteins, esters, and aldehydes. They may result from domestic or industrial wastewater discharges, or they may be end products of the initial decomposition of plant or animal tissue. The principal problem associated with biodegradable organics is the effect resulting from the action of microorganisms. Secondary problems include color, taste, and odor problems.

For microorganisms to use dissolved organic material effectively requires *oxidation and reduction* processes. In oxidation, oxygen is added, or hydrogen is deleted from elements of the organic molecule. Reduction occurs when hydrogen is added to or oxygen is deleted from elements of the organic molecule. The oxidation process is by far more efficient and is predominant when oxygen is available. In *aerobic (oxygen-present)* environments, the end products of microbial decomposition of organics are stable and acceptable compounds. *Anaerobic (oxygen-absent)* decomposition results in unstable and objectionable end products. The quantity of oxygen-consuming organics in water is usually determined by measuring the *biological oxygen demand (BOD)*: the amount of dissolved oxygen needed by aerobic decomposers to break down the organic materials in a given volume of water over a five-day incubation period at 20°C (6°F).

Nonbiodegradable organics (resistant to biological degradation and thus considered refractory) include constituents of woody plants (tannin and lignic acids, phenols, and cellulose) and are found in natural water systems. Some polysaccharides with exceptionally strong bonds and benzene with its ringed structure are essentially nonbiodegradable. Some organics are toxic to organisms and thus are nonbiodegradable. These include organic pesticides and compounds that have combined with chlorine. Pesticides and herbicides have found widespread use in agriculture, forestry, and mosquito control. Surface streams are contaminated via runoff and wash-off by rainfall. These toxic substances are harmful to some fish, shellfish, predatory birds, and mammals. Some compounds are toxic to humans. Certain nonbiodegradable chemicals can react with oxygen dissolved in water. The *chemical oxygen demand (COD)*, the amount of oxygen needed to chemically oxidize a waste, is a more complete and accurate measurement of the total depletion of dissolved oxygen in water.

Nutrients

Nutrients are elements (carbon, nitrogen, phosphorous, sulfur, calcium, iron, potassium, manganese, cobalt, and boron) essential to the growth and reproduction of plants and animals. Aquatic species depend on their watery environment to provide their nutrients. In water quality terms, however, nutrients can be considered pollutants when their concentrations are sufficient to encourage the excessive growth of aquatic plants such as algal blooms. The nutrients required in most abundance by aquatic species are carbon, nitrogen, and phosphorous. Plants, in particular, require large amounts of each of these three nutrients, or their growth is *limited.*

Carbon is readily available from a number of natural sources including alkalinity, decaying products of organic matter, and dissolved carbon dioxide from the atmosphere. Since carbon is usually readily available, it is seldom a *limiting nutrient* (the nutrient least available relative to the plant's needs) (Masters, 1991). The limiting nutrient concept is important because it suggests that identifying and reducing the supply of a particular nutrient can control algal growth. In most cases, nitrogen and phosphorous are essential growth factors and are the limiting factors in aquatic plant growth. According to Welch (1980), seawater is most often limited by nitrogen, while freshwater systems are most often limited by phosphorous.

Nitrogen gas (N_2), which is extremely stable, is the primary component of the earth's atmosphere. Major sources of nitrogen in water include runoff from animal feedlots, fertilizer runoff from agricultural fields, from municipal wastewater discharges, and from certain bacteria and blue-green algae that obtain nitrogen directly from the atmosphere. Certain forms of acid rain can also contribute nitrogen to surface waters. In water, nitrogen is frequently found in the form of nitrate (NO_3). Nitrate in drinking water can lead to serious problems, most notably nitrate poisoning. Human and animal infants can be affected by nitrate poisoning, which can cause serious illness and even death, if bacteria commonly found in the infant's intestinal tract converts nitrates to highly toxic nitrites (NO_2). Nitrite can replace oxygen in the bloodstream and result in oxygen starvation that causes a bluish discoloration of the infant ("blue baby" syndrome).

In aquatic surface water systems, nitrogen (and phosphorous in the form of phosphate) is a chemical that can stimulate biological growth and is classified as a *biostimulant.* As biostimulants, nitrogen and phosphates (derived from fertilizers and detergents) are impurities that can result in greatly increased *eutrophication* (or slow death) of the body of water. Eutrophication of water systems, especially in lakes, is usually a natural phenomenon that occurs over time. Increase in biostimulants (nitrogen and especially phosphate, or any other growth-limiting nutrient) speed up eutrophication, affecting the natural process.

UNITED STATES GEOLOGICAL SURVEY (USGS) QUIZ

What are some of the physical and chemical properties of water?

Check your answers with those provided at the end of the quiz.

1. True/False: Water contracts (gets smaller) when it freezes.
2. True/False: Water has high surface tension.
3. True/False: Condensation is water coming out of the air.
4. True/False: More things can be dissolved in sulfuric acid than in water.
5. True/False: Rainwater is the purest form of water.
6. True/False: It takes more energy to heat water at room temperature to 212°F than it does to change 212°F water to steam.
7. True/False: If you evaporate an 8-inch glass full of water from the Great Salt Lake (with a salinity of about 20% by weight), you will end up with about 1 inch of salt.
8. True/False: Seawater is slightly more basic (the pH value is higher) than most natural water.
9. True/False: Raindrops are tear shaped.
10. True/False: Water boils at a lower temperature at Denver, Colorado, than at the beach.

CHAPTER QUIZ: ANSWERS AND EXPLANATIONS

Water Science for Schools (2011). Available @ http://ga2.er.usgs.gov/edunew/sc3action.cfm.

1. Like most liquids, water contracts (gets smaller) when it freezes. **False**
   - *Explanation*: Actually, water expands (gets less dense) when it freezes, which is unusual for liquids. Think of ice – it is one of the few items that float as a solid. If it didn't, then tanks would freeze from the bottom up (that would mean we'd have to wear wet suits when ice skating!), and more lakes way up north would be permanent blocks of ice.

2. Water has a high surface tension. **True**
   - *Explanation*: Water has the highest surface tension among common liquids (mercury is higher). Surface tension is the ability of a substance to stick to itself (cohere). That is why water forms drops, and also why when you look at a glass of water, the water "rises" where it touches the glass (the "meniscus"). Plants are happy that water has a high surface tension because they use capillary action to draw water from the ground up through their roots and stems.

3. Condensation is water coming out of the air. **True**
   - *Explanation*: This is actually true; water that forms on the outside of a cold glass or on the inside of a window in winter is liquid water condensing from water vapor in the air. Air contains water vapor (humidity). In cold air, water vapor condenses faster than it evaporates. So, when the warm air touches the outside of your cold glass, the air next to the glass gets chilled, and some of the water in that air turns from water vapor to tiny liquid water droplets. Clouds in the sky and the "cloud" you see when you exhale on a cold day are condensed water vapor particles. (It is a myth that clouds form because cold air cannot hold as much water vapor as warm air!)

4. More things can be dissolved in sulfuric acid than in water. **False**
   - *Explanation*: Not true. Sulfuric acid might be able to dissolve a car, but water isn't known as the "Universal Solvent" for nothing! It can dissolve more substance than any other liquid. This is lucky for us … what if all the sugar in your soft drink ended up as a pile at the bottom of the glass? The water you see in rivers, lakes, and the ocean may look clear, but it actually contains many dissolved elements and minerals, and because these elements are dissolved, they can easily move with water over the surface of the earth.

5. Rainwater is the purest form of water. **False**
   - *Explanation*: Distilled water is "purer." Rainwater contains small amounts of dissolved minerals that have been blown into the air by winds. Rainwater contains tiny particles of dust and dissolved gases, such as carbon dioxide and sulfur dioxide (acid rain). That doesn't mean rainwater isn't very clean – normally only about 1/100,000th of the weight of rain comes from these substances. In a way, the distillation process is responsible for rainwater. Distilled water comes from water vapor condensing in a closed container (such as a glass jar). Rain is produced by water vapor evaporating from the earth and condensing in the sky. Both the closed jar and the earth (via its atmosphere) are "closed systems," where water is neither added nor lost.

6. It takes more energy to heat cold water to 212°F than it does to change 212°F water to steam. **False**
   - *Explanation*: First, water at boiling temperature (212°F at sea level) is not really the same as boiling water. When water first reaches boiling it has not begun to turn to steam yet. More energy is needed to begin turning the boiling liquid water into gaseous water vapor. The bonds that hold water molecules as a liquid are not easily broken. It takes about seven times as much energy to turn boiling water into steam as it does to heat water at room temperature to the boiling point.

7. If you filled a glass full of water from the Great Salt Lake, when it evaporated there would be 1 inch of salt left. **True**

- *Explanation*: They don't call it the Great Salt Lake for nothing. Water in the Great Salt Lake varies in salinity by both location and time. In this example, we are assuming about a 20% salt concentration. In other words, about one-fifth of the weight of the water comes from salt. And how much saltier is Great Salt Lake water than seawater? Quite a bit. Seawater has a salt concentration of about 3½%.

8. Seawater is slightly more basic (the pH value is higher) than most natural fresh water. **True**
   - *Explanation*: Neutral water (such as distilled water) has a pH of 7, which is in the middle of being acidic and alkaline. Seawater happens to be slightly alkaline (basic), with a pH of about 8. Most natural water has a pH of 6–8, although acid rain can have a pH as low as 4.

9. Raindrops are tear shaped. **False**
   - *Explanation*: When you think of a drop of falling water you probably think it looks like a tear. When a drop of water comes out of a faucet, yes, it does have a tear shape. That is because the back end of the water drop sticks to the water still in the faucet until it can't hold on any more. But, using high-speed cameras, scientists have found that falling raindrops look more like a small hamburger bun! Gravity and surface tension come into play here. As rain falls the air below the drop pushes up from the bottom, causing the drop to flatten out somewhat. The strong surface tension of water holds the drop together, resulting in a bun shape.

10. Water boils at a lower temperature at Denver, Colorado, than at the beach. **True**
    - *Explanation*: The boiling point of water gets lower as you go up in altitude. At beach level, water boils at 212°F (100°C). But at 5,000 feet, about where Denver is located, water boils at about 203°F (95°C). This is because as the altitude gets higher, the air pressure (the weight of all that air above you) becomes less. Boiling occurs when the vapor pressure of water exceeds atmospheric pressure. Since there is less pressure pushing on a pot of water at a higher altitude, it is easier for the water molecules to break their bonds and attraction to each other and, thus, it boils at a lower temperature.

BIOLOGICAL CHARACTERISTICS OF WATER

Along with the physical and chemical parameters of water quality, the water practitioner is also concerned with the *biological* characteristics of water. This concern is well warranted when the health and well-being of the people who receive and use the product from the "end-of-the-pipe" or "at-the-spigot" are at stake. In this context, remember that water may serve as a medium in which thousands of biological species spend part, if not all, of their life cycles. Note that, to some extent, all members of the biological aquatic community serve as parameters of water quality because their presence or absence may indicate in general terms the characteristics of a given body of water.

The biological characteristics of water impact directly water quality. To a lesser degree this impact includes the development of tastes and odors in surface waters and groundwater and the corrosion of and biofouling of heat transfer surfaces in cooling systems and water supply treatment facilities. However, the presence or absence, in particular, of certain biological organisms is of paramount importance to the water specialist. These organisms are, of course, the pathogens. Pathogens are organisms capable of infecting or transmitting diseases to humans and animals. These organisms are not native to aquatic systems and usually require an animal host for growth and reproduction. They can be, and are, transported by natural water systems. Waterborne pathogens include species of bacteria, viruses, protozoa, and parasitic worms (helminthes). In the sections to follow, we provide a brief "basic" description of each of these species, along with a brief description of rotifers, crustaceans, fungi, and algae, which are also microorganisms of concern in water.

The water practitioner who specializes in water quality must be familiar with the nutritional requirements of aquatic organisms. To grow in an aquatic environment, organisms must be able to extract the nutrients they need for cell synthesis and for the generation of energy from their water environment. Some organisms obtain their energy from photosynthetic light. Others obtain their energy from organic or inorganic matter.

Carbon is a critical ingredient for all aquatic microorganisms (actually, carbon is critical to all organisms). Some aquatic organisms (higher plants, algae, and photosynthetic bacteria) get their carbon from carbon dioxide. Others (bacteria, fauna, protozoa, animals) get their carbon from organic matter. In addition to carbon and energy, oxygen plays a critical role in the growth of cells. Many organisms *(aerobes)* require molecular oxygen (O_2) for their metabolism. Other organisms *(anaerobes)* do not require molecular oxygen and derive the oxygen they need for the synthesis of cells from chemical compounds.

Along with the nutritional requirements of microorganisms, also consider the effects of environmental factors covered earlier in the chapter. These factors include chemical composition, pH, temperature, and light. The environmental practitioner must not only be familiar with the types of aquatic organisms that can inhabit water systems but also must understand the physical, chemical, and biological environmental factors required for their existence. To refresh and reinforce the reader's memory on the important biological characteristics of water, they are briefly summarized.

BACTERIA

The word *bacteria* (singular: bacterium) comes from the Greek word meaning "rod" or "staff," a shape characteristic of many bacteria. Bacteria are single-celled microscopic organisms that multiply by splitting in two (binary fission). To multiply, they need carbon from carbon dioxide if they are *autotrophs*, which synthesize organic substances from inorganic molecules by using light or chemical energy, or from organic compounds (dead vegetation, meat, sewage) if they are *heterotrophs*. Their energy comes from sunlight if they are photosynthetic or from chemical reactions if they are chemosynthetic. Bacteria are present in air, water, earth, rotting vegetation, and the intestines of animals. Gastrointestinal disorders are common symptoms of most diseases transmitted by waterborne pathogenic bacteria.

VIRUS

A virus is an infectious particle consisting of a core of nucleic acid (DNA and RNA) enclosed in a protein shell. It is an entity that carries the information needed for its replication but does not possess the machinery to carry out replication. Thus, viruses are *obligate* parasitic particles that require a host in which to live. They are the smallest biological structures known and can only be seen with the aid of an electron microscope. Waterborne viral contaminants that are known to cause poliomyelitis and infectious hepatitis are usually indicated by disorders with the nervous system rather than of the gastrointestinal tract.

PROTOZOA

Protozoa (singular: protozoan) are mobile, single-celled, complete, self-contained organisms that can be free-living or parasitic, pathogenic or nonpathogenic, microscopic or macroscopic. Protozoa range in size from two to several hundred microns in length. They are highly adaptable and widely distributed in natural waters, although only a few are parasitic. Most protozoa are harmless. Only a few cause illnesses in humans; *Entamoeba histolytica* (amebiasis), and *Giardia lamblia* (giardiasis) being two of the exceptions. Giardiasis (typically contracted by drinking surface water contaminated by wild animals or humans) is the most widespread protozoan disease occurring throughout the world (Tchobanoglous & Schroeder, 1987). Unless properly treated, giardiasis can be chronic. Symptoms usually include diarrhea, nausea, indigestion, flatulence, bloating, fatigue, and appetite and weight loss.

WORMS (HELMINTHES)

Worms also are important in water quality assessment from a standpoint of human disease. Normally, worms inhabit organic mud and organic slime. They have aerobic requirements but can metabolize solid organic matter not readily degraded by other microorganisms. Water contamination may result from human and animal waste that contains worms. Worms pose hazards primarily to those persons who come into direct contact with untreated water. Thus swimmers in surface water polluted by sewage or stormwater runoff from cattle feedlots are at particular risk. The *Tubifix* worm is a common organism used as an indicator of pollution in streams.

Fungi (singular fungus) constitute an extremely important and interesting group of aerobic microbes ranging from the unicellular yeasts to the extensively mycelial molds. Fungi are not considered plants but are instead a distinctive life form of great practical and ecological importance. Like bacteria, fungi metabolize dissolved organic matter. As the principal organisms responsible for decomposition of carbon in the biosphere, fungi are essential. Fungi are unique (in a sense, when compared to bacteria), in that they can grow in low moisture areas and in low pH solutions which aids in the breakdown of organic matter in aquatic environments.

Algae are a diverse group of autotrophic, photosynthetic, eukaryotic microorganisms containing chlorophyll. The feature that distinguishes algae from fungi is the algae's chlorophyll. Alga microorganisms impact water quality by shifting the balance between oxygen and carbon dioxide in the water, by affecting pH levels, and by contributing to odor and taste problems.

WATER TREATMENT PROCESS

Municipal water treatment operations and associated treatment unit processes are designed to provide reliable, high-quality water service for customers and to preserve and protect the environment for future generations. Water management officials and treatment plant operators are tasked with exercising responsible financial management, ensuring fair rates and charges, providing responsive customer service, providing a consistent supply of safe potable water for consumption by the user, and promoting environmental responsibility.

While studying this section on water treatment operations, keep in mind the major point within the chapter: Water treatment plant design can be taught in school, but in managing and operating the plants, experience, attention to detail, and common sense are most important. In the past, water quality was described as "wholesome and delightful" and "sparkling to the eye." Today, we describe water quality as "safe and healthy to drink."

Water treatment systems are installed to remove those materials that cause disease and/or create nuisances. At its simplest level, the basic goal of water treatment operations is to protect public health, with a broader goal to provide potable and palatable water. The bottom line: The water treatment process functions to provide water that is safe to drink and is pleasant in appearance, taste, and odor. Water treatment is defined as any unit process that changes/alters the chemical, physical, and/or bacteriological quality of water with the purpose of making it safe for human consumption and/or appealing to the customer.

Treatment also is used to protect the water distribution system components from corrosion. Many water treatment unit processes are commonly used today. Treatment processes used depend upon the evaluation of the nature and quality of the particular water to be treated and the desired quality of the finished water. In water treatment unit processes employed to treat raw water, one thing is certain: as new USEPA regulations take effect, many more processes will come into use in the attempt to produce water that complies with all current regulations, despite source water conditions.

Small public water systems tend to use a smaller number of the wide array of unit treatment processes available, in part because they usually rely on groundwater as the source, and in part because small makes many sophisticated processes impractical (i.e., too expensive to install, too expensive

TABLE 2.3
Basic Water Treatment Processes

Process/Step	Purpose
Screening	Removes large debris (leaves, sticks, fish) that can foul or damage plant equipment
Chemical Pretreatment	Conditions the water for removal of algae and other aquatic nuisances
Presedimentation	Removes gravel, sand, silt, and other gritty materials
Microstraining	Removes algae, aquatic plants, and small debris
Chemical Feed and Rapid Mix	Adds chemicals (coagulants, pH, adjusters, etc.) to water
Coagulation/Flocculation	Converts nonsettleable to settleable particles
Sedimentation	Removes settleable particles
Softening	Removes hardness-causing chemicals from water
Filtration	Removes particles of solid matter which can include biological contamination and turbidity
Disinfection	Kills disease-causing organisms
Adsorption Using Granular Activated Carbon	Removes radon and many organic chemicals, such as pesticides, solvents, and trihalomethanes
Aeration	Removes volatile organic chemicals (VOCs), radon H_2S, and other dissolved gases; oxidizes iron and manganese
Corrosion Control	Prevents scaling and corrosion
Reverse Osmosis, Electrodialysis	Removes nearly all inorganic contaminants
Ion Exchange	Removes some inorganic contaminants including hardness-causing chemicals
Activated Alumina	Removes some inorganic contamination
Oxidation Filtration	Removes some inorganic contaminants (e.g., iron, manganese, radium)

Source: Adapted from AWWA, Introduction to Water Treatment, Vol 2, 1984.

to operate, too sophisticated for limited operating staff). This section concentrates on those individual treatment unit processes usually found in conventional water treatment systems, corrosion control methods, and fluoridation. A summary of basic water treatment processes, many of which are discussed in this chapter, is presented in Table 2.3.

As mentioned, the purpose of water treatment is to condition, modify, and/or remove undesirable impurities, and to provide water that is safe, palatable, and acceptable to users. While this is the obvious and expected purpose of treating water, various regulations also require water treatment. Some regulations state that if the contaminants listed under the various regulations are found in excess of maximum contaminant levels (MCLs), the water must be treated to reduce the levels. If a well or spring source is surface influenced, treatment is required, regardless of the actual presence of contamination. Some impurities affect the aesthetic qualities (taste, odor, color, and hardness) of the water; if they exceed secondary MCLs established by USEPA and the state, the water may need to be treated. If we assume that the water source used to feed a typical water supply system is groundwater (usually the case in the United States), a number of common groundwater problems may require water treatment. Keep in mind that water that must be treated for any one of these problems may also exhibit several other problems.

Among these other problems are:

- Bacteriological contamination;
- Hydrogen sulfide odors;
- Hard water;
- Corrosive water; and
- Iron and manganese.

STAGES OF WATER TREATMENT

The conventional model of water treatment is made up of various stages, unit processes, or basically a train of processes combined to form one treatment system. Note that a given waterworks may contain all the unit processes discussed in the following or any combination of them. One or more of these stages may be used to treat any one or more of the source water problems listed above. In some small systems, water treatment may consist of nothing more than removal of water via pumping from a groundwater source to storage to distribution. In some small water supply operations, disinfection may be added because it is required.

PRETREATMENT

Simply stated, water pretreatment (also called preliminary treatment) is any physical, chemical, or mechanical process used before main water treatment processes. It can include screening, presedimentation, and chemical addition. Pretreatment in water treatment operations usually consists of oxidation or other treatment for the removal of tastes and odors, iron and manganese, trihalomethane precursors, or entrapped gases like hydrogen sulfide. Unit processes may include chlorine, potassium permanganate or ozone oxidation, activated carbon addition, aeration, and presedimentation. Pretreatment of surface water supplies accomplishes the removal of certain constituents and materials that interfere with or place an unnecessary burden on conventional water treatment facilities. According to the Texas Water Utilities Association's *Manual of Water Utility Operations*, 8th edition, typical pretreatment processes include the following:

- Removal of debris from water from rivers and reservoirs that would clog pumping equipment.
- Desertification of reservoirs to prevent anaerobic decomposition that could result in reducing iron and manganese from the soil to a state that would be soluble in water. This can cause subsequent removal problems in the treatment plant. The production of hydrogen sulfide and other taste- and odor-producing compounds also results from stratification.
- Chemical treatment of reservoirs to control the growth of algae and other aquatic growths that could result in taste and odor problems.
- Pre-sedimentation to remove excessively heavy silt loads prior to the treatment processes.
- Aeration to remove dissolved odor-causing gases such as hydrogen sulfide and other dissolved gases or volatile constituents and to aid in the oxidation of iron and manganese, although manganese or high concentrations of iron are not removed in detention provided in conventional aeration units.
- Chemical oxidation of iron and manganese, sulfides, taste and odor producing compounds, and organic precursors that may produce trihalomethanes upon the addition of chlorine.
- Adsorption for removal of tastes and odors.

Note: An important point to keep in mind is that in small public water systems, using groundwater as a source, pretreatment may be the only treatment process used.

Note: Pretreatment may be incorporated as part of the total treatment process or may be located adjacent to the source before the water is sent to the treatment facility.

AERATION

Aeration is commonly used to treat water that contains trapped gases (such as hydrogen sulfide) that can impart an unpleasant taste and odor to the water. Just allowing the water to rest in a vented tank will (sometimes) drive off much of the gas, but usually some form of forced aeration is needed. Aeration works well (about 85% of the sulfides may be removed) whenever the pH of the water is less than 6.5. Aeration may also be useful in oxidizing iron and manganese, oxidizing humic substances that might form trihalomethanes when chlorinated, eliminating other sources of taste and odor, or imparting oxygen to oxygen-deficient water.

SCREENING

Screening is usually the first major step in the water pretreatment process. It is defined as the process whereby relatively large and suspended debris is removed or retained from the water before it enters the plant. River water, for example, typically contains suspended and floating debris varying in size from small rocks to logs. Removing these solids is important, not only because these items have no place in potable water, but also because this river trash may cause damage to downstream equipment (clogging and damaging pumps, etc.), increase chemical requirements, impede hydraulic flow in open channels or pipes, or hinder the treatment process. The most important criterion used in the selection of a particular screening system for water treatment technology is the screen opening size and flow rate; they range in size from microscreens to trash racks. Other important criteria include: costs related to operation and equipment; plant hydraulics; debris handling requirement; and operator qualifications and availability. Large surface water treatment plants may employ a variety of screening devices including rash screens (or trash rakes), traveling water screens, drum screens, bar screens, or passive screens.

CHEMICAL ADDITION

Two of the major chemical pretreatment processes used in treating water for potable use are iron and manganese and hardness removal. Another chemical treatment process that is not necessarily part of the pretreatment process, but is also discussed in this section, is corrosion control. Corrosion prevention is affected by chemical treatment – not only in the treatment process but also in the distribution process. Before discussing each of these treatment methods in detail, however, it is important to describe chemical addition, chemical feeders, and chemical feeder calibration. When chemicals are used in the pretreatment process, they must be the proper ones, fed in the proper concentration, and introduced to the water at the proper locations. Determining the proper amount of chemical to use is accomplished by testing. The operator must test the raw water periodically to determine if the chemical dosage should be adjusted. For surface supplies, checking must be done more frequently than for groundwater (remember, surface water supplies are subject to change on short notice, while groundwater generally remains stable). The operator must be aware of the potential for interactions between various chemicals and how to determine the optimum dosage (e.g., adding both chlorine and activated carbon at the same point will minimize the effectiveness of both processes, as the adsorptive power of the carbon will be used to remove the chlorine from the water). Note: Sometimes using too many chemicals can be worse than not using enough.

Prechlorination (distinguished from chlorination used in disinfection at the end of treatment) is often used as an oxidant to help with the removal of iron and manganese. However, currently, concern for systems that prechlorinate is prevalent because of the potential for the formation of total trihalomethanes (TTHMs), which form as a by-product of the reaction between chlorine and naturally occurring compounds in raw water. USEPA's TTHM standard does not apply to public water systems that serve less than 10,000 people, but operators should be aware of the impact and causes of TTHMs. Chlorine dosage or application point may be changed to reduce problems with TTHMs. Note: TTHMs such as chloroform are known or suspected to be carcinogenic and are limited by water and state regulations. Note: To be effective, pretreatment chemicals must be thoroughly mixed with the water. Short-circuiting or plug flows of chemicals that do not come in contact with most of the water will not result in proper treatment.

All chemicals intended for use in drinking water must meet certain standards. Thus, when ordering water treatment chemicals, the operator must be assured that they meet all appropriate standards for drinking water use. Chemicals are normally fed with dry chemical feeders or solution (metering) pumps. Operators must be familiar with all of the adjustments needed to control the rate at which the chemical is fed to the water (wastewater). Some feeders are manually controlled and must be adjusted by the operator when the raw water quality or the flow rate changes; other feeders are paced by a flow meter to adjust the chemical feed so that it matches the water flow rate. Operators must also be familiar with chemical solution and feeder calibration. As mentioned, a significant part of a waterworks operator's important daily operational functions includes measuring quantities of chemicals and applying them to water at preset rates. Normally accomplished semi-automatically by use of electro-mechanical-chemical feed devices, waterworks operators must still know what chemicals to add, how much to add to the water, and the purpose of the chemical addition.

CHEMICAL FEEDERS

Simply put, a chemical feeder is a mechanical device for measuring a quantity of the chemical and applying it to water at a preset rate. Two types of chemical feeders are commonly used: solution (or liquid) feeders and dry feeders. Liquid feeders apply chemicals in solutions or suspensions, and dry feeders apply chemicals in granular or powdered forms. In a solution feeder, chemical enters feeder and leaves feeder in a liquid state; in a dry feeder, chemical enters and leaves feeder in a dry state.

IRON AND MANGANESE REMOVAL

Iron and manganese are frequently found in groundwater and in some surface waters. They do not cause health-related problems but are objectionable because they may cause aesthetic problems. Severe aesthetic problems may cause consumers to avoid an otherwise safe water supply in favor of one of unknown or of questionable quality or may cause them to incur unnecessary expense for bottled water. Aesthetic problems associated with iron and manganese include the discoloration of water (iron = reddish water, manganese = brown or black water), staining of plumbing fixtures, imparting a bitter taste to the water, and stimulating the growth of microorganisms.

As mentioned, there are no direct health concerns associated with iron and manganese, although the growth of iron bacteria slimes may cause indirect health problems. Economic problems include damage to textiles, dye, paper, and food. Iron residue (or tuberculation) in pipes increases pumping head, decreases carrying capacity, may clog pipes, and may corrode through pipes. Note: Iron and manganese are secondary contaminants. Their secondary maximum contaminant levels (SMCLs) are: iron = 0.3 mg/L; manganese = 0.05 mg/L. Iron and manganese are most likely found in groundwater supplies, industrial waste, and acid mine drainage, and as by-products of pipeline corrosion. They may accumulate in lake and reservoir sediments, causing possible problems during lake/reservoir turnover. They are not usually found in running waters (streams, rivers, etc.).

IRON AND MANGANESE REMOVAL TECHNIQUES

Chemical precipitation treatments for iron and manganese removal are called *deferrization* and *demagnetization*. The usual process is aeration; dissolved oxygen in the chemical causing precipitation; chlorine or potassium permanganate may be required.

HARDNESS TREATMENT

Hardness in water is caused by the presence of certain positively charged metallic ions in the solution in the water. The most common of these hardness-causing ions are calcium and magnesium; others include iron, strontium, and barium. As a general rule, groundwater is harder than surface waters, so hardness is frequently of concern to the small water system operator. This hardness is derived from contact with soil and rock formations such as limestone. Although rainwater itself will not dissolve many solids, the natural carbon dioxide in the soil enters the water and forms carbonic acid (HCO), which is capable of dissolving minerals.

Where soil is thick (contributing more carbon dioxide to the water) and limestone is present, hardness is likely to be a problem. The total amount of hardness in water is expressed as the sum of its calcium carbonate ($CaCO_3$) and its magnesium hardness. However, for practical purposes, hardness is expressed as calcium carbonate. This means that regardless of the amount of the various components that make up hardness, they can be related to a specific amount of calcium carbonate (e.g., hardness is expressed as mg/L as $CaCO_3$ – milligrams per liter as calcium carbonate). Note: The two types of water hardness are temporary hardness and permanent hardness. Temporary hardness is also known as carbonate hardness (hardness that can be removed by boiling); permanent hardness is also known as noncarbonate hardness (hardness that cannot be removed by boiling). Hardness is of concern in domestic water consumption because hard water increases soap consumption, leaves a soapy scum in the sink or tub, can cause water heater electrodes to burn out quickly, can cause discoloration of plumbing fixtures and utensils, and is perceived as less desirable water. In industrial water use, hardness is a concern because it can cause boiler scale and damage to industrial equipment.

CORROSION CONTROL

Water operators add chemicals (e.g., lime or sodium hydroxide) to water at the source or at the waterworks to control corrosion. Using chemicals to achieve slightly alkaline chemical balance prevents the water from corroding distribution pipes and consumers' plumbing. This keeps substances like lead from leaching out of plumbing and into the drinking water. For our purpose, we define *corrosion* as the conversion of a metal to a salt or oxide with a loss of desirable properties such as mechanical strength. Corrosion may occur over an entire exposed surface or may be localized at micro – or macroscopic discontinuities in metal. In all types of corrosion a gradual decomposition of the material occurs, often due to an electrochemical reaction. Corrosion may be caused by: (1) stray current electrolysis; (2) galvanic corrosion caused by dissimilar metals; or (3) differential concentration cells. Corrosion starts at the surface of a material and moves inward.

The adverse effects of corrosion can be categorized according to health, aesthetics, economic effects, and/or other effects. The corrosion of toxic metal pipes made from lead creates a serious *health hazard*. Lead tends to accumulate in the bones of humans and animals. Signs of lead intoxication include gastrointestinal disturbances, fatigue, anemia, and muscular paralysis. Lead is not a natural contaminant in either surface waters or groundwater, and the MCL of 0.005 mg/L in source waters is rarely exceeded. It is a corrosion by-product from high lead solder joints in copper and lead piping. Small dosages of lead can lead to developmental problems in children. The USEPA's Lead and Copper Rule addresses the matter of lead in drinking water exceeding specified action levels.

CORROSION CONTROL TECHNIQUES

One method used to reduce the corrosive nature of water is *chemical addition*. The selection of chemicals depends on the characteristics of the water, where the chemicals can be applied, how they can be applied and mixed with water, and the cost of the chemicals. If the product of the calcium hardness times the alkalinity of the water is less than 100, treatment may be required. Both lime and CO_2 may be required for proper treatment of the water. If the calcium hardness and alkalinity levels are between 100 and 500, either lime or soda ash (Na_2CO_3) will be satisfactory.

The decision regarding which chemical to use depends on the cost of the equipment and chemicals. If the product of the calcium hardness times the alkalinity is greater than 500, either lime or caustic (NaOH) may be used. Soda ash will be ruled out because of the expense. The chemicals chosen for the treatment of public drinking water supplies modify the water characteristics making the water less corrosive to the pipe. Modification of water quality can increase the pH of the water, reducing the hydrogen ions available for galvanic corrosion, as well as reducing the solubility of copper, zinc, iron, lead, and calcium, and increasing the possibility of forming carbonate protective films.

COAGULATION

The primary purpose of surface water treatment is chemical clarification by coagulation and mixing, flocculation, sedimentation, and filtration. These water treatment unit processes, along with disinfection, work to remove particles, naturally occurring organic matter (NOM – i.e., bacteria, algae, zooplankton, and organic compounds), and microbes to produce water that is noncorrosive. Specifically, coagulation/flocculation work to destabilize particles and agglomerate dissolved and particulate matter. Sedimentation removes solids and provides ½-log *Giardia* and 1-log virus removal. Filtration removes solids and provides 2-log *Giardia* and 1 log virus removal. Finally, disinfection provides microbial inactivation and ½ *Giardia* and 2-log virus removal (1 log = 99.1; 2 log = 99.99; 3 log = 99.999).

Following screening and the other pretreatment processes, the next unit process in a conventional water treatment system is a mixer where chemicals are added and mixed in what is known as coagulation. The exception to this unit process configuration occurs in small systems using groundwater when chlorine or other taste and odor control measures are introduced at the intake and are the extent of treatment. Materials present in raw water may vary in size, concentration, and type. Dispersed substances in the water may be classified as suspended, colloidal, or solution. Suspended particles may vary in mass and size and are dependent on the flow of water.

High flows and velocities can carry larger material. As velocities decrease, the suspended particles settle according to size and mass. Other material may be in solution, for example, salt dissolves in water. Matter in the colloidal state does not dissolve, but the particles are so small that they will not settle out of the water. Color (as in tea-colored swamp water) is mainly due to colloids or extremely fine particles of matter in suspension. Colloidal and solute particles in water are electrically charged. Because most of the charges are alike (negative) and repel each other, the particles stay dispersed and remain in the colloidal or soluble state.

Suspended matter will settle without treatment, if the water is still enough to allow it to settle. The rate of settling of particles can be determined as this settling follows certain laws of physics. However, much of the suspended matter may be so slow in settling that the normal settling processes become impractical, and if colloidal particles are present, settling will not occur. Moreover, water drawn from a raw water source often contains many small unstable (unsticky) particles; therefore, sedimentation alone is usually an impractical way to obtain clear water in most locations, and another method of increasing the settling rate must be used: coagulation, which is designed to convert stable (unsticky) particles to unstable (sticky) particles.

The term *coagulation* refers to the series of chemical and mechanical operations by which coagulants are applied and made effective. These operations are comprised of two distinct phases: (1) rapid mixing to disperse coagulant chemicals by violent agitation into the water being treated and (2) flocculation to agglomerate small particles into well-defined floc by gentle agitation for a much longer time. Coagulation results from adding salts of iron or aluminum to the water. The coagulant must be added to the raw water and perfectly distributed into the liquid; such uniformity of chemical treatment is reached through rapid agitation or mixing.

Common coagulants (salts) include:

- Alum – aluminum sulfate;
- Sodium aluminate;
- Ferric sulfate;
- Ferrous sulfate;
- Ferric chloride; and
- Polymers.

Coagulation is the reaction between one of these salts and water. The simplest coagulation process occurs between alum and water. *Alum* or aluminum sulfate is made by a chemical reaction of bauxite ore and sulfuric acid. The normal strength of liquid alum is adjusted to 8.3%, while the strength of dry alum is 17%. When alum is placed in water, a chemical reaction occurs that produces positively charged aluminum ions. The overall result is the reduction of electrical charges and the formation of a sticky substance; the formation of *floc*, which when properly formed, will settle.

These two destabilizing factors are the major contributions that coagulation makes to the removal of turbidity, color, and microorganisms. The formation of floc is the first step of coagulation; for the greatest efficiency, rapid, intimate mixing of the raw water and the coagulant must occur. After mixing, the water should be slowly stirred so that the very small, newly formed particles can attract and enmesh colloidal particles, holding them together to form larger floc. This slow mixing is the second stage of the process (flocculation), covered later.

A number of factors influence the coagulation process: pH, turbidity; temperature; alkalinity; and the use of polymers. The degree to which these factors influence coagulation depends upon the coagulant used. The raw water conditions, optimum pH for coagulation, and other factors must be considered before deciding which chemical is to be fed and at what levels.

MIXING AND FLOCCULATION

Flocculation follows coagulation in the conventional water treatment process. *Flocculation* is the physical process of slowly mixing the coagulated water to increase the probability of particle collision; unstable particles collide and stick together to form fewer larger flocs. Through experience, we see that effective mixing reduces the required amount of chemicals and greatly improves the sedimentation process, which results in longer filter runs and higher quality finished water. The goal of flocculation is to form a uniform, feather-like material similar to snowflakes; a dense, tenacious floc that entraps the fine, suspended, and colloidal particles and carries them down rapidly in the settling basin. Proper flocculation requires from 15 to 45 minutes. The time is based on water chemistry, water temperature, and mixing intensity. Temperature is the key component in determining the amount of time required for floc formation. To increase the speed of floc formation and the strength and weight of the floc, polymers are often added.

SEDIMENTATION

After raw water and chemicals have been mixed and the floc formed, the water containing the floc (because it has a higher specific gravity than water) flows to the sedimentation or settling basin. *Sedimentation* is also called clarification. Sedimentation removes settleable solids by gravity. Water moves slowly through the sedimentation tank/basin with a minimum of turbulence at entry and exit points with minimum short-circuiting. Sludge accumulates at the bottom of the tank/basin. Typical tanks or basins used in sedimentation include conventional rectangular basins, conventional center-feed basins, peripheral-feed basins, and spiral-flow basins. In conventional treatment plants, the amount of detention time required for settling can vary from 2 to 6 hours. Detention time should be based on the total filter capacity when the filters are passing 2 gpm per square foot of superficial sand area.

For plants with higher filter rates, the detention time is based on a filter rate of 3–4 gpm per square foot of sand area. The time requirement is dependent on the weight of the floc, the temperature of the water, and how quiescent (still) the basin is. A number of conditions affect sedimentation: (1) uniformity of flow of water through the basin; (2) stratification of water due to difference in temperature between water entering and water already in the basin; (3) release of gases that may collect in small bubbles on suspended solids, causing them to rise and float as scum rather than settle as sludge; (4) disintegration of previously formed floc; and (5) size and density of the floc.

FILTRATION

In the conventional water treatment process, *filtration* usually follows coagulation, flocculation, and sedimentation. At present, filtration is not always used in small public water systems. However, recent regulatory requirements under USEPA's Interim Enhanced Surface Water Treatment rules may make water filtering necessary at most water supply systems. Water filtration is a physical process of separating suspended and colloidal particles from water by passing water through a granular material. The process of filtration involves straining, settling, and adsorption. As floc passes into the filter, the spaces between the filter grains become clogged, reducing this opening and increasing removal. Some material is removed merely because it settles on a media grain. One of the most important processes is the adsorption of the floc onto the surface of individual filter grains. This helps collect the floc and reduces the size of the openings between the filter media grains. In addition to removing silt and sediment, floc, algae, insect larvae, and any other large elements, filtration also contributes to the removal of bacteria and protozoans such as *Giardia Lamblia* and *Cryptosporidium*. Some filtration processes are also used for iron and manganese removal.

DISINFECTION

In the United States, 99% of surface water systems provide some treatment to their water, with 99% of these treatment systems using disinfection/oxidation as part of the treatment process. Although 45% of groundwater systems provide no treatment, 92% of those groundwater plants that do provide some form of treatment include disinfection/oxidation as part of the treatment process (EPA, 1997). The most commonly used disinfectants/oxidants (in no particular order) are chlorine, chlorine dioxide, chloramines, ozone, and potassium permanganate. Disinfectants are also used to achieve other specific objectives in drinking water treatment. These other objectives include nuisance control (e.g., for zebra mussels and Asiatic clams), oxidation of specific compounds (i.e., taste and odor-causing compounds, iron, and manganese), and use as a coagulant and filtration aid. The goals of this section are to:

- Provide a brief overview of the need for disinfection in water treatment.
- Provide basic information that is common to all disinfectants.
- Discuss other uses for disinfectant chemicals (i.e., as oxidants).
- Describe trends in disinfection by-product (DBP) formation and the health effects of DBPs found in water treatment.
- Discuss microorganisms of concern in water systems, their associated health impact, and the inactivation mechanisms and efficiencies of various disinfectants.
- Summarize current disinfection practices in the United States, including the use of chlorine as a disinfectant and an oxidant.

The effectiveness of disinfection in a drinking water system is measured by testing for the presence or absence of coliform bacteria. Coliform bacteria found in water are generally not pathogenic, though they are good indicators of contamination. Their presence indicates the possibility of contamination, and their absence indicates the possibility that the water is potable; if the source is adequate, the waterworks history is good, and acceptable chlorine residual is present. Desired characteristics of a disinfectant include the following:

- It must be able to deactivate or destroy any type or number of disease-causing microorganisms that may be in a water supply, in a reasonable time, within expected temperature ranges, and despite changes in the character of the water (pH, for example).
- It must be nontoxic.
- It must not add unpleasant taste or odor to the water.
- It must be readily available at a reasonable cost and be safe and easy to handle, transport, store, and apply.

- It must be quick and easy to determine the concentration of the disinfectant in the treated water.
- It should persist within the disinfected water at a high enough concentration to provide residual protection through the distribution.

A major cause for the number of disease outbreaks in potable water is the contamination of the distribution system from cross-connections and back-siphonage with non-potable water. However, outbreaks resulting from distribution system contamination are usually quickly contained and result in relatively few illnesses compared to contamination of the source water or a breakdown in the treatment system, which typically produces many cases of illnesses per incident. When considering the number of cases, the major causes of disease outbreaks are source water contamination and treatment deficiencies (White, 1992). Historically, about 46% of the outbreaks in the public water systems are found to be related to deficiencies in source water and treatment systems with 92% of the causes of illness due to these two particular problems. All natural waters support biological communities. Because some microorganisms can be responsible for public health problems, the biological characteristics of the source water are one of the most important parameters in water treatment. In addition to public health problems, microbiology can also affect the physical and chemical water quality and treatment plant operation.

REMOVAL OF TASTE AND ODORS THROUGH CHEMICAL OXIDATION

Taste and odors in drinking water are caused by several sources, including microorganisms, decaying vegetation, hydrogen sulfide, and specific compounds of municipal, industrial, or agricultural origin. Disinfectants themselves can also create taste and odor problems. In addition to a specific taste- and odor-causing compound, the sanitary impact is often accentuated by a combination of compounds. More recently, significant attention has been given to tastes and odors from specific compounds such as geosmin, 2-methylisoborneol (MIB), and chlorinated inorganic and organic compounds (AWWARF, 1987).

Oxidation is commonly used to remove taste- and odor-causing compounds. Because many of these compounds are very resistant to oxidation, advanced oxidation processes (ozone/hydrogen peroxide, ozone/UV, etc.) and ozone by itself are often used to address taste and odor problems. The effectiveness of various chemicals to control taste and odors can be site specific. Suffet et al. (1986) found that ozone is generally the most effective oxidant for use in taste and odor treatment. They found ozone doses of 2.5–2.7 mg/L and 10 minutes of contact time (residual 0.2 mg/L) significantly reduce levels of taste and odors. Lalezary et al. (1986) used chlorine, chlorine dioxide, ozone, and permanganate to treat earthy-musty-smelling compounds. In that study, chlorine dioxide was found most effective, although none of the oxidants were able to remove geosmin and MIB by more than 40–60%. Potassium permanganate has been used in doses of 0.25–20 mg/L. Prior experiences with taste and odor treatment indicate that oxidant doses are dependent on the source or the water and causative compounds. In general, small doses can be effective for many taste and odor compounds but not for some of the difficult-to-treat compounds. In general, small doses can be effective for many taste and odor compounds, but some of the difficult-to-treat compounds require strong oxidants such as ozone and/or advanced oxidation processes or alternative technologies such as granular activated carbon (GAC) adsorption.

SUMMARY OF METHODS OF DISINFECTION

The methods of disinfection include:

- *Heat* – Possibly the first method of disinfection. Disinfection is accomplished by boiling water for 5 to 10 minutes. Good, obviously, only for household quantities of water when bacteriological quality is questionable;

- *Ultraviolet (UV) light* – While a practical method of treating large quantities, adsorption of UV light is very rapid, so the use of this method is limited to non-turbid waters close to the light source;
- *Metal Ions* – Silver, copper, and mercury;
- Alkalis and acids;
- *pH Adjustment* – To under 3.0 or over 11.0; and
- *Oxidizing Agents* – Bromine, ozone, potassium permanganate, and chlorine.

The vast majority of drinking water systems in the United States use chlorine for disinfection (Spellman, 2007). Along with meeting the desired characteristics listed above, chlorine has the added advantage of a long history of use; it is fairly well understood. Although some small public water systems may use other disinfectants, we focus on chlorine in this chapter and provide only a brief overview of other disinfection alternatives.

Note: One of the recent developments in chlorine disinfection is the use of multiple and interactive disinfectants. In these applications, chlorine is combined with a second disinfectant to achieve improved disinfection efficiency and/or effective DBP control. Note: As described earlier, the 1995 Community Water System Survey indicated that all surface water and groundwater systems in the United States use chlorine for disinfection (Spellman, 2007).

CHAPTER SUMMARY

One of the blessings of modern life that we enjoy (but usually ignore) is readily available, clean, safe, potable water. As anyone who has carried water along for camping or backpacking, or who has traveled or lived in a country without safe drinking water knows, just being able to turn on the tap to get water is an enormous boon that modern technology has given us. We don't have to carry water in pails from the village well or the river – or our own well or spring. We don't have to boil it or chemically treat it to drink or cook with it. We don't have to brush our teeth with bottled water. We take our safe water for granted, even if we buy spring water to drink because we don't like the taste of the treated water from the faucet. We forget, also, that even when our water supplies were relatively unspoiled, unsafe waters killed many because the technology available to test the water was not available. The physical, chemical, and biological characteristics of water and our understanding of the forces at work on groundwater and surface water allow us to control the quality of the water we rely on so absolutely.

REVIEW QUESTIONS

1. What two minerals are primarily responsible for causing "hard water"?
   a. Salt and borax
   b. Hydrogen sulfide and sulfur dioxide
   c. Alum and chlorine
   d. Calcium (Ca) and magnesium

2. The power of a substance to resist pH changes is referred to as a/an _____.
   a. Antidote
   b. Fix
   c. Buffer
   d. Neutral substance

3. Which of the following complicates use or reuse of water?
   a. Sunlight
   b. Rainfall
   c. Recreational use
   d. Pollution

4. Which of the following is a leading cause of impairment to freshwater bodies?
    a. Rivers
    b. Nonpoint pollution
    c. Air pollution
    d. Snowfall

5. Which of the following is not a feature of water?
    a. Appearance
    b. Wetness
    c. Smell
    d. Odor

6. If water is turbid, it is _____.
    a. Smelly
    b. Wet
    c. Cloudy
    d. Clear

7. Solids are classified as _____.
    a. Settable
    b. Dissolved
    c. Colloidal
    d. All of the above

8. Colloidal material is composed of _____.
    a. Leaves
    b. Branches
    c. Silt
    d. Decayed fish

9. Taste and odor in drinking water may be _____.
    a. No hazard
    b. A tooth destroyer
    c. A contributor to kidney stones
    d. A laxative

10. Which of the following is a measure of water's ability to neutralize acids?
    a. pH
    b. Hardness
    c. Fluoride level
    d. Alkalinity

ANSWER KEY

1. What two minerals are primarily responsible for causing "hard water"?
    a. Salt and borax
    b. Hydrogen sulfide and sulfur dioxide
    c. Alum and chlorine
    d. **Calcium (Ca) and magnesium**

2. The power of a substance to resist pH changes is referred to as a/an _____.
    a. Antidote
    b. Fix
    c. **Buffer**
    d. Neutral substance

3. Which of the following complicates use or reuse of water?
   a. Sunlight
   b. Rainfall
   c. Recreational use
   d. **Pollution**

4. Which of the following is a leading cause of impairment to freshwater bodies?
   a. Rivers
   b. Nonpoint pollution
   c. **Air pollution**
   d. Snowfall

5. Which of the following is not a feature of water?
   a. Appearance
   b. **Wetness**
   c. Smell
   d. Odor

6. If water is turbid, it is _____.
   a. Smelly
   b. Wet
   c. **Cloudy**
   d. Clear

7. Solids are classified as _____.
   a. Settable
   b. Dissolved
   c. Colloidal
   d. **All of the above**

8. Colloidal material is composed of _____.
   a. Leaves
   b. Branches
   c. **Silt**
   d. Decayed fish

9. Taste and odor in drinking water may be _____.
   a. **No hazard**
   b. A tooth destroyer
   c. A contributor to kidney stones
   d. A laxative

10. Which of the following is a measure of water's ability to neutralize acids?
    a. pH
    b. Hardness
    c. Fluoride level
    d. **Alkalinity**

BIBLIOGRAPHY

Aptel, P. and Buckley, C.A., 1996. Categories of Membrane Operations. In *Water Treatment Membrane Processes*. New York: McGraw-Hill.

ASTM, 1969. *Manual of Water*. Philadelphia, PA: American Society for Testing and Materials.

AWWARF, 1987. *Water/Wastewater Systems*. Denver, CO: American Water Works Association Research Foundation.

Bangs, R. and Kallen, C., 1985. *Rivergods*. Colorado Springs, CO: Sierra Club.

Boyce, A., 1997. *Introduction to Environmental Technology*. New York: Van Nostrand Reinhold.

Brady, T.J., et al., 1996. Chlorination Effectiveness for Zebra and Quagga Mussels. *J. AWWA* 88(1):107–110.

Britton, J.C., and Morton, B.A., 1982. Dissection Guide, Field and Laboratory Manual for the Introduced Bivalve Corbicula Fluminea. *Malacol. Rev.* 3(1):69–74.

Butterfield, C.T., et al., 1943. Chlorine vs. Hypochlorite. *Public Health Rep.* 58:1837.

Cameron, G.N., Symons, J.M., Spencer, S.R., and Ma, J.Y., 1989. Minimizing THM Formation During Control of the Asiatic Clam: A comparison of Biocides. *J. AWWA* 81(10):53–62.

Cheng, R.C., et al., 1994. Enhanced Coagulation for Arsenic Removal. *J. AWWA* 9:79–90.

Chick, H., 1908. Investigation of the Laws of Disinfection. *J. Hygiene* 8:92.

Clarke, N.A., et al., 1962. Human Enteric Viruses in Water, Source, Survival, and Removability. In International Conference on Water Pollution Research, New York.

Clary, D., 1997. What Makes Water Wet. *Geraghty & Miller Newsletter*, 39,4.

Clifford, D.A., and Lin, C.C., 1985. *Arsenic (arsenite) and Arsenic (arsenate) Removal from Drinking Water in San Ysidro, New Mexico.* Houston, TX: University of Houston.

Clifford, D.A., et al., 1997. *Final Report: Phases 1 & 2 City of Albuquerque Arsenic Study Field studies on Arsenic Removal in Albuquerque, NM using the University of Houston/EPA Mobile Drinking Water Treatment Research Facility.* Houston, TX: University of Houston.

Coakley, P., 1975. Developments in our Knowledge of Sludge Dewatering Behavior. In 8th Public Health Engineering Conference held in the Department of Civil Engineering, Loughborough: University of Technology.

Connell, G.F., 1996. *The Chlorination/Chloramination Handbook.* Denver, CO: American Water Works Association.

Corbitt, R.A., 1990. *Standard Handbook of Environmental Engineering.* New York: McGraw-Hill.

Craun, G.F., 1981. Outbreaks of Waterborne Disease in the United States. *J. AWWA* 73(7):360.

Craun, G.F., and Jakubowski, W., 1996. Status of Waterborne Giardiasis Outbreaks and Monitoring Methods. In American Water Resources Association, Water Related Health Issue Symposium. Atlanta, GA.

Culp, G.L., and Culp, R.L., 1994. Outbreaks of Waterborne Disease in the United States. *J. AWWA* 73(7):360.

Culp, G.L. et al., 1986. *Handbook of Public Water Systems.* New York: Van Nostrand Reinhold.

Demers, L.D. and Renner, R.C., 1992. *Alternative Disinfection Technologies for Small Drinking Water Systems.* Denver, CO: AWWA and AWWART.

Edwards, M.A., 1994. Chemistry of Arsenic Removal during Coagulation and Fe-Mn Oxidation. *J. AWWA*:64–77.

Finch, G.R., et al., 1994. Ozone and Chlorine Inactivation of *Cryptosporidium*. In Conference Proceedings, Water Quality Technology Conference, Part II, San Francisco, CA.

Fortner, B., and Schechter, D., 1996. United States Water Quality shows little Improvement Over 1992 Inventory. *Water Environ. Technol.* 8:2.

Gilcreas, F.W., Sanderson, W.W., and Elmer, R.P., 1953. Two Methods for the Determination of Grease in Sewage. *Sewage Ind. Wastes* 25:1379.

Gleick, P.H., 2001. Freshwater Forum. *United States Water News*, vol. 18, No. 6.

Gordon, G., et al., 1995. *Minimizing Chlorate Ion Formation in Drinking Water when Hypochlorite Ion is the Chlorinating Agent.* Denver, CO: AWWA.-AWWARF.

Gurol, M.D., and Pidatella, M.A., 1983. Study on Ozone-Induced Coagulation. In Conference Proceedings, ASCE Environmental Engineering Division Specialty Conference. Allen Medicine and Michael Anderson (editors), Boulder, CO.

Gyurek, L.L., et al., 1996. Disinfection of *Cryptosporidium Parvum* Using Single and Sequential Application of Ozone and Chlorine Species. In Conference Proceedings, AWWA Water Quality Technology Conference, Boston, MA.

Hass, C.N., and Englebrecht, R.S., 1980. Physiological Alterations of Vegetative Microorganisms Resulting from Aqueous Chlorination. *J. Water Pollut. Control Fed.* 52(7):66–69.

Herbert, P.D.N., et al., 1989. Ecological and Genetic Studies on *Dresissmena polymorpha* (Pallas): A New Mollusc in the Great Lakes. *Can. J. Fish. Aquat. Sci.* 46:187.

Hering, J.G., and Chiu, V.Q., 1998. The Chemistry of Arsenic: Treatment and Implications of Arsenic Speciation and Occurrence. In AWWA Inorganic Contaminants Workshop, San Antonio, TX.

Hoff, J.C., et al., 1984. Disinfection and the Control of Waterborne Giardiasis. In Conference Proceedings, ASCE Specialty Conference. Denver, CO.

IOA, 1997. *Survey of Water Treatment Plants.* Stanford. CT: International Ozone Association.

Kemmer, F.N., 1979. *Water: The Universal Solvent*, 2nd ed. Oak Brook, IL: Nalco Chemical Company.

Kemmer, F.N., 1988. *The Nalco Water Handbook*, 2nd ed., New York: McGraw-Hill.

Klerks, P.L and Fraleigh, P.C. 199. Controlling Adult Zebra Mussels with Oxidants. *J. AWWA* 83(12):92–100.

Koch, B., et al., 1991. Predicting the Formation of DBPs by the Simulate Distribution System. *J. AWWA* 83(10):62–70.

Koren, H., 1991. *Handbook of Environmental Health & Safety: Principles and Practices*. Chelsea, MI: Lewis Publishers.

Krasner, S.W., 1989. The Occurrence of Disinfection Byproducts in US Drinking Water. *J. AWWA* 81(8):41–53.

Laine, J.M., 1993. Influence of Bromide on Low-Pressure Membrane Filtration for controlling DBPs in Surface Waters. *J. AWWA* 85(6):87–99.

Lalezary, S., et al., 1986. Oxidation of Five Earthy-Musty Taste and Odor Compounds. *J. AWWA* 78(3):62.

Lang, C.L., 1994. The Impact of the Freshwater Macrofouling Zebra Mussel (*Dretssena Polymorpha*) on Drinking Water Supplies. In Conference Proceedings, AWWA Water Quality Technology Conference Part II, San Francisco, CA.

Lewis, S.A., 1996. *Safe Drinking Water*. San Francisco, CA: Sierra Club Books.

Liu, O.C., et al., 1971. Relative Resistance of Twenty Human Enteric Viruses to Free Chlorine. Virus and Water A Quality: Occurrence and Control. In Conference Proceedings, Thirteenth Water Quality Conference, Urban-Champaign, IL: University of Illinois.

Manahan, S.E., 1997. *Environmental Science and Technology*, Boca Raton: FL, Lewis Publishers.

Masschelein, W.J., 1992. *Unit Processes in Drinking Water Treatment*. New York, Brussels, Hong Kong: Marcel Decker D.D.

Masters, G.M., 1991. *Introduction to Environmental Engineering and Science*. Englewood Cliffs, NJ: Prentice-Hall.

Matisoff, G., et al., 1996. Toxicity of Chlorine Dioxide to Adult Zebra Mussels. *J. AWWA* 88(8):93–106.

McGhee, T.J., 1991. *Water Supply and Sewerage*. New York: McGraw-Hill.

Metcalf & Eddy, Inc., 2003. *Wastewater Engineering: Treatment, Disposal, Reuse*, 4th ed. New York: McGraw-Hill.

Moncrief, R.W., 1967. *The Chemical Senses*, 3rd ed. London: Leonard Hill.

Montgomery, J.M., 1985. *Water Treatment Principles and Design*. New York: Wiley.

Morrison, A., 1983. In Third World Villages, a Simple Handpump Saves Lives. *Civil Engineering*: 68–72, October.

Muilenberg, T., 1997. Microfiltration Basics: Theory and Practice. In Proceedings Membrane Technology Conference, New Orleans, LA.

Nathanson, J.A., 1997. *Basic Environmental Technology: Water Supply, Waste Management, and Pollution Control*. Upper Saddle River, NJ: Prentice Hall.

Nieminski, E.C., et al., 1993. The Occurrence of DBPs in Utah Drinking Waters. *J. AWWA* 85(9):98–105.

Norse, E.A., 1985. *Animal Extinctions*, Hoage, R.J. (editor).

Oliver, B.G., and Shindler, D.B., 1980. Trihalomethanes for Chlorination of Aquatic Algae. *Environ. Sci. Technol.* 1492:1502.

Outwater, A., 1996. *Water: A Natural History*. New York: Basic Books.

Peavy, H.S., Rowe, D.R., and Tchobanoglous, G., 1985. *Environmental Engineering*. New York: McGraw-Hill.

Prendiville, P.W., 1986. Ozonation at the 900 cfs Los Angeles Water Purification Plant, *Ozone: Sci. Eng.* 8:77.

Reckhow, D.A. and Singer, P.C., 1985. Mechanisms of Organic Halide formation during Fulvic Acid Chlorination and Implications with Respect to Prezonation. In *Water Chlorination: Chemistry, Environmental Impact and Health Effects*, Volume 5. Jolley, R.L., et al. (editors). Chelsea, MI: Lewis Publishers.

Reckhow, D.A., et al., 1986. Ozone as a Coagulant Aid. In Seminar Proceedings, Ozonation, Recent Advances and Research Needs. AWWA Annual Conference, Denver, CO.

Reckhow, D.A., et al., 1990. Chlorination of Humic Materials: Byproduct Formation and Chemical Interpretations. *Environ. Sci. Technol.* 24(11):1655.

Rice, R.G., et al., 1998. Ozone Treatment for Small Water Systems. Presented at First International Symposium on Safe Drinking Water in Small Systems, NSF International, Arlington, VA.

Roberts, R., 1990. Zebra Mussel Invasion threatens United States Waters. *Science* 249:1370.

Rowe, D.R., and Abdel-Magid, I.M., 1995. *Handbook of Wastewater Reclamation and Reuse*, Boca Raton, FL: Lewis Publishers.

Sawyer, C.N., 1994. *Chemistry for Environmental Engineering*. New York: McGraw-Hill, Inc.

Scarpino, P.V., et al., 1972. A Comparative Study of the Inactivation of Viruses in Water by Chlorine. *Water Research* 6:959.

Simms, J., et al, 2000. Arsenic Removal Studies and the Design of a 20,000 m^3 Per Day Plant in U.K. AWWA Inorganic Contaminants Workshop. Albuquerque, NM.

Sinclair, R.M., 1964. Clam Pests in Tennessee Water Supplies. *J AWWA* 56(5):592.

Singer, P.C., 1989. Correlations between Trihalomethanes and Total Organic Halides Formed during Water Treatment. *J. AWWA* 81(8):61–65.

Singer P.C., 1992. Formation and Characterization of Disinfection Byproducts. Presented at the First International Conference on the Safety of Water Disinfection: Balancing Chemical and Microbial Risks. Los Angeles.

Singer, P.C., and Chang, S.D., 1989. Correlations between Trihalomethanes and Total Organic Halides Formed during Water Treatment. *J. AWWA* 81(8):61–65.

Singer, P.C., and Harrington, G.W., 1993. Coagulation of DBP Precursors: Theoretical and Practical Considerations. In Conference Proceedings, AWWA Water Quality Technology Conference, Miami, FL.

Snead, M.C., et al., 1980. *Benefits of Maintaining a Chlorine Residual in Water Supply Systems.* EPA600/2-80-010.

Spellman, F.R., 1996. *Stream Ecology and Self-Purification: An Introduction for Wastewater and Water Specialists.* Lancaster, PA: Technomic Publishing Company.

Spellman, F.R., 1997. *Microbiology for Water/Wastewater Operators,* Lancaster, PA: Technomic Publishing Company.

Spellman, F.R., 1998. *Composting Biosolids.* Lancaster, PA: Technomic Publishing.

Spellman, F.R., 2003. *Handbook of Water and Wastewater Treatment Plant Operations.* Boca Raton, FL: Lewis Publishers.

Spellman, F.R., 2007. *The Science of Water,* 2nd ed. Boca Raton, FL: CRC Press.

Spellman, F.R., 2013. *Water and Wastewater Infrastructure.* Boca Raton, FL: CRC Press.

Sterritt, R.M., and Lester, J.M., 1988. *Microbiology for Environmental and Public Health Engineers.* London: E. and F.N. Spoon.

Stevens, A.A., 1976. Chlorination of Organics in Drinking Water. *J. AWWA* 8(11):615.

Subramanian, K.D. et al., 1997. Manganese Greensand for Removal of Arsenic in Drinking Water. *Water Quality Research Journal Canada.* 32(3):551561.

Suffet, I.H., et al., 1986. Removal of Tastes and Odors by Ozonation. In Conference Proceedings, AWWA Seminar on Ozonation: Recent Advances and Research Needs, Denver, CO.

Tchobanoglous, G., and Schroeder, E.D., 1987. *Water Quality.* Reading, MA: Addison-Wesley.

Thibaud, H., et al., 1988. Effects of Bromide Concentration on the Production of Chloropicrin during Chlorination of Surface Waters: Formation of Brominated Trihalonitromethanes. *Water Res.* 22(3):381.

USDA, 2011. *Water Use Facts.* Accessed 06/06/19 @ http.www.fs.fed.us/r5/publications/water_resources/html/water_uaw:facts.html.

USEPA, 1991. *Manual of Individual and Non-Public Works Supply Systems.* Office of Water, EPA5709-91-004. Washington, DC: United States Environmental Protection Agency

USEPA, 1997. *Community Water System Survey: Volumes I and II; Overview,* EPA 815-R-97-001a. Washington, DC: United States Environmental Protection Agency

USEPA, 1998. *National Primary Drinking Water Regulations: Interim Enhanced Surface Water Treatment Final Rule,* 63 FR 69477. Washington, DC: United States Environmental Protection Agency

USEPA, 1999a. *Guidance Manual: Alternative Disinfectants and Oxidants,* Chapters 1 and 2. Washington, DC: United States Environmental Protection Agency.

USEPA, 1999b. *Lead and Copper Rule Minor Revisions: Fact Sheet,* EPA 815-F-99-010. Washington, DC: United States Environmental Protection Agency.

USEPA, 1999c. *Microbial and Disinfection Byproduct Rules Simultaneous Compliance Guidance Manual,* EPA-815-R-99-015. Washington, DC: United States Environmental Protection Agency.

USEPA, 1999d. *Turbidity Requirements: OESWTR Guidance Manual: Turbidity Provisions.* Washington, DC: United States Environmental Protection Agency.

USEPA, 2000. *Technologies and Costs for the Removal of Arsenic from Drinking Water,* EPA-815-R-00-028. Washington, DC: United States Environmental Protection Agency.

USEPA, 2002. *Onsite Wastewater Treatment System Manual.* Washington, DC: U.S. Environmental Protection Agency.

USEPA, 2007. *Protecting America's Public Health.* Accessed 06/07/19 @ www.epa.gov/safewater/public outreach.html.

USGS, 2011. *Water Science for Schools.* Accessed 06/05/11 @ http://ga.water.usgs.gov/edu/edu/sc#.html.

USGS, 2019. *Wastewater Treatment Use.* Reston, VA: U.S. Geological Survey.

U.S. Watersheds Have Water Quality Problems, 1997. *Environmental Technology.* Nov/Dec., p. 10.

Van Benschoten, J.E., et al., 1995. Zebra Mussel Mortality with Chorine. *J. AWWA* 87(5):101–108.

Vickers, J.C., et al., 1997. Bench Scale Evaluation of Microfiltration for Removal of Particles and Natural Organic Matter. In Proceedings Membrane Technology Conference, New Orleans, LA.

Water Quality, 1995. 2nd ed., Denver, CO: American Water Works Association.

Watson, H.E., 1908. A Note on the Variation of the Rate of disinfection with change in the Concentration of the Disinfectant. *J. Hygiene.* 8:538.

Welch, E.G., 1980. *Ecological Effects of Waste Water.* Cambridge, UK: Cambridge University Press.

White, G.C., 1992. *Handbook of Chlorination and Alternative Disinfectants.* New York: Van Nostrand Reinhold.

Witherell, D.J., et al., 1988. Investigation of Legionella Pneumophila in Drinking Water. *J. AWWA* 80(2):88–93.

3 Water Monitoring

2019© Photo Courtesy of Placer County Water Agency, Placer County California 95604.

An estimated 1.1 billion people worldwide lack clean drinking water, and 2.4 billion lack access to basic sanitation. Targets adopted by the United Nations in September 2000 aim to halve these figures by 2015, but projections suggest those goals, which would require more than 100,000 people every day to be connected to clean water supplies, will not be met.

–Patricia Brett (2005)

LEARNING OBJECTIVES

After studying this chapter, you should be able to:

- Discuss how drinking water monitoring is conducted and its benefits.
- Discuss water quality factors.
- Describe water in terms of "good" and "bad" water.
- Discuss how State water programs are used to protect the public.
- Identify the key elements involved in water sampling.
- Define sampling preparations.
- Describe water sample types.
- Discuss how water samples are collected.
- Describe the various water test methods.

DOI: 10.1201/9781003207146-5

- Describe and discuss dissolved oxygen sampling and testing.
- Discuss biochemical oxygen demand (BOD) sampling and testing.
- Discuss temperature, hardness, and pH sampling.
- Describe turbidity measuring.
- Identify, describe, and discuss the phosphorus cycle.
- Discuss nitrate sampling.
- Define turbidity and describe the factors that affect water clarity.
- Describe the conditions and factors that affect watercolor.
- Discuss consumer expectations of color, taste, and odor for water.
- Describe the organic and inorganic contaminants that affect water's smell and taste.
- Discuss total solids.
- Describe the effect temperature has on water treatment.
- Discuss conductivity testing.
- Identify the fecal bacterium constituents commonly found in water and describe how they affect water quality.
- Define total dissolved solids, discuss their sources, and how they are controlled.
- Discuss the problems created by high alkalinity.
- Identify the range and description classifications for water hardness and discuss the advantages and disadvantages of both hard and soft waters.
- Discuss apparent color.

KEY DEFINITIONS

Apparent Color – attributed to dissolved and suspended matter in water.

Biochemical Oxygen Demand (BOD) – a measure of the amount of oxygen consumed by microorganisms in decomposing organic matter in water.

Composite Sample – consists of a series of individual grab samples collected over a specified period of time in proportion to flow.

Conductivity – is a measure of the ability of water to pass an electrical current.

Corrosivity – water's quality of being corrosive.

Dissolved Oxygen – in water, a parameter indicating amount of oxygen in the water.

E. coli – bacteria specific to fecal contamination of water.

Fecal Bacteria – used as indicators of possible sewage contamination because they are commonly found in human and animal feces.

Feed Pie Diagram – simplified diagram used to determine chemical feed, flow, and dose rates.

Grab Samples – are collected at one time and one location. They are representative only of the conditions at the time of collection.

Hardness – a characteristic of water caused primarily by the salts of calcium and magnesium. It causes deposition of scale in boilers, damage in some industrial processes, and sometimes objectionable taste. It may also decrease soup's effectiveness.

Monitoring – in water, is based on three criteria: (1) water quality monitoring is accomplished to ensure to the extent possible that the water is not a danger to public health; (2) to ensure that the water provided at the tap is as aesthetically pleasing as possible; and (3) to ensure compliance with applicable regulations.

Nitrates – are a form of nitrogen found in several different forms in terrestrial and aquatic ecosystems.

Odor – in water is caused by chemicals that may come from municipal and industrial waste discharges, or natural sources, such as decomposing vegetable matter or microbial activity. Odor affects the acceptability of drinking water, the aesthetics of recreation water, and the taste of aquatic foodstuffs.

pH – is a measure of the concentration of hydrogen ions (the acidity of a solution) determined by the reaction of an indicator that varies in color, depending on hydrogen ion levels in the water.

Phosphorus – in water, it is the nutrient in short supply in most freshwater systems; thus, even a modest increase in phosphorus can (under the right conditions) set off a whole chain of undesirable events in a stream, including accelerated plant growth, algae blooms, low dissolved oxygen, and the death of certain fish, invertebrates, and other aquatic animals.

Secchi Disk – used to measure water clarity/transparency.

Titrant – adding a solution of known strength (the titrant) to a specific volume of a treated sample in the presence of an indicator.

Titrimetric – analyzes based on adding a solution of known strength (the titrant) to a specific volume of a treated sample in the presence of an indicator.

Total Alkalinity – is determined by measuring the amount of acid (e.g., sulfuric acid) needed to bring the sample to a pH of 4.2. At this pH all the alkaline compounds in the sample are "used up."

Total Coliform – the standard test used in water because their presence indicates contamination of a water supply by an outside source.

Total Solids – is a direct measurement of the amount of material suspended and dissolved in water.

Water Quality Monitoring – defined as the sampling and analysis of water constituents and conditions.

Water Sampling – the taking of water samples for laboratory testing and analysis.

DRINKING WATER MONITORING

During the intake of source water for treatment (no matter the source: stream, river, lake, etc.) and during the conveyance of the water from one water treatment unit process to the next within the treatment train, licensed water operators and qualified laboratory personnel monitor and test the quality of the water as required by federal, state, and in many cases, local regulations. Drinking water monitoring is based on three criteria: (1) to ensure to the extent possible that the water is not a danger to public health; (2) to ensure that the water provided at the tap is as aesthetically pleasing as possible; and (3) to ensure compliance with applicable regulations. To meet these goals, all public water systems must monitor water quality factors to some extent (see Figure 3.1).

Before the Ground Water Rule of 2006 was implemented, the degree of monitoring employed was dependent on local needs, requirements, and on the type of public water system – small public water systems using good-quality water from deep wells may only need to provide occasional monitoring, but systems using surface water sources must test water quality frequently (AWWA, 1995). Again, the Ground Water Rule of 2006 modified sampling and monitoring requirements for all public water systems that use groundwater sources at risk of microbial contamination; they are now required to take corrective action to protect consumers from harmful bacteria and viruses. Monitoring is a key element of this risk-targeted approach. Note: Compliance monitoring ensures that systems already providing 99.99% (4-log) inactivation, removal, or a state-approved combination of inactivation and removal of viruses are achieving this level of treatment. Note: "Log" is short for logarithm, which is a power to which a base, such as 10, can be raised to produce a given number. As an example, Log 4 represents 104 10 × 10 × 10 × 10 or 10,000.

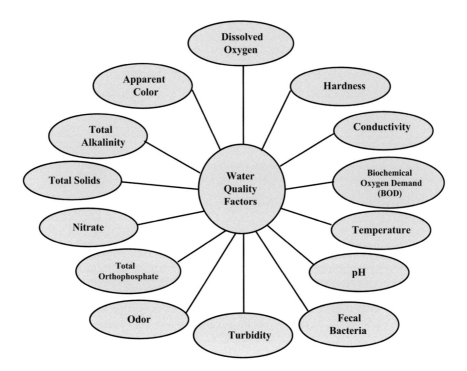

FIGURE 3.1 Water quality factors.

Taking corrective action to protect consumers means treating water to the required safe level for consumption. As explained in Chapter 2, treating the water to a safe level means ensuring that the unit processes are operating correctly. In addition, from monitoring, sampling, and testing of the water throughout the treatment process, water operators are able to determine what adjustments if necessary are needed to be taken. These adjustments, beyond ensuring correct unit process operations, may need to be chemically adjusted; the operator must be able to determine, through testing and/or laboratory analysis, the exact amount of chemical to add to the treatment stream.

DID YOU KNOW?

Drinking water must be monitored to provide adequate control of the entire water drawing/treatment/conveyance system. *Adequate control* is defined as monitoring employed to assess the present level of water quality, so action can be taken to maintain the required level (whatever that might be).

Although this chapter does not cover water operator mathematics in detail, it is important to note that math is a critical element required to properly treat water. For example, when the operator determines from monitoring and testing that the water in the treatment train needs chemical addition, he/she must determine not only the amount of chemical to feed in pounds per day (lbs/day) but also the exact dosage in milligrams per liter (mg/L) to apply and at what flow rate in million gallons per day (MGD). To accomplish this, the operator must perform simple math calculation, such as:

Dose (mg/L) = Feed (lbs/day) ÷ Flow (MGD) ÷ (8.34 lbs/gal)
Flow (MGD) = Feed (lbs/day) ÷ Dose (mg/L) ÷ (8.34 lbs/gal)

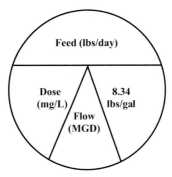

FIGURE 3.2 Feed pie.

Feed (lbs/day) = Dose (mg/L) × Flow (MGD) × (8.34 lbs/gal)*
*The weight of a gallon of water is 8.34 lbs/gal.

For those who have difficulty remembering math formulas and how to perform math operations, it is advisable to train operators on how to use, in the case of applying proper chemical feed, the Feed Pie (shown in Figure 3.2). This simplified diagram, which should be available to all operators, makes calculations for dose, flow, and feed easy. For example, if the amount of chemical feed in lbs/day needs to be determined, the operator simply places a finger of the feed section of the pie. In doing so, he/she can easily see that to find the proper feed amount the math operation to use is dose × flow × 8.34 lbs/gal. For the other parameters shown in the Feed Pie, the operator simply covers the parameter to be determined with his/her finger and then performs the operation.

Water quality monitoring is defined as the sampling and analysis of water constituents and conditions. When we monitor, we collect data. As a monitoring program is developed, deciding the reason for collecting the information is important. The reasons for gathering it are defined by establishing a set of objectives, which include a description of who will collect the information. It may surprise you to know that today the majority of people collecting data are volunteers, not necessarily professional drinking water practitioners. These volunteers have a vested interest in their local stream, lake, or other body of water; in many cases, they are proving they can successfully carry out a water quality monitoring program.

Is the Water Good or Bad?*

To answer the question "Is the water good or bad?" we must factor in two requirements. First, we return to the basic principles of water quality monitoring: sampling and analyzing water constituents and conditions. These constituents may include:

- Introduced pollutants, such as pesticides, metals, and oil; and
- Constituents found naturally in water that can nevertheless be affected by human sources, such as dissolved oxygen, bacteria, and nutrients.

The magnitude of these constituents' effects is influenced by properties such as pH and temperature. Temperature influences the quantity of dissolved oxygen that water is able to contain, and pH affects the toxicity of ammonia, for example.

---

* Much of the information presented in the following sections is from USEPA's 2.841B97003 *Volunteer Stream Monitoring: A Methods Manual*, 1997; and adapted from F.R. Spellman, 2007. *The Science of Water*, 2nd ed. Boca Raton, FL: CRC Press.

Let's get back to the "good" and "bad" of water. The second factor to be considered is that the only valid way to answer this question is to conduct laboratory tests – these tests must be compared to some form of water quality standards. If simply assigning a "good" and "bad" value to each test factor were possible, the meters and measuring devices in water quality test kits would be much easier to design and manufacture. Instead of fine graduations, they could simply have a "good" and a "bad" zone.

When you get right down to it, *water quality*, the difference between "good" and "bad" water, must be interpreted according to the intended use of the water. For example, the perfect balance of water chemistry that assures a sparkling clear, sanitary swimming pool would not be acceptable as drinking water and would be a deadly environment for many biotas. Consider Table 3.1.

In another example, widely different levels of fecal coliform bacteria are considered acceptable, depending on the intended use of the water (Table 3.2).

State and local water quality practitioners, as well as volunteers, have been monitoring water quality conditions for many years. In fact, until the past decade or so (until biological monitoring protocols were developed and began to take hold), water quality monitoring was generally considered the primary way of identifying water pollution problems. Today, professional water quality practitioners and volunteer program coordinators alike are moving toward approaches that combine chemical, physical, and biological monitoring methods to achieve the best picture of water quality conditions.

Water quality monitoring can be used for many purposes:

- To identify whether waters are meeting designated uses. All states have established specific criteria (limits on pollutants) identifying what concentrations of chemical pollutants are allowable in their waters. When chemical pollutants exceed maximum or minimum allowable concentrations, waters may no longer be able to support the beneficial uses such as fishing, swimming, and drinking for which they have been designated (see Table 3.2).
- Designated or intended uses and the specific criteria that protect them (along with anti-degradation statements that say waters should not be allowed to deteriorate below existing or anticipated uses) together form water quality standards. State water quality professionals assess water quality by comparing the concentrations of chemical pollutants found in

TABLE 3.1
Total Residual Chlorine

Total Residual Chlorine (TRC) mg/L	
0.06	Toxic to Striped Bass Larvae
0.31	Toxic to White Perch Larvae
0.5–1.0	Typical Drinking Water Residual
1.0–3.0	Recommended for Swimming Pools

TABLE 3.2
Fecal Coliform Bacteria

Fecal Coliform Bacteria (per 100 mL of Water)		
Desirable	Permissible	Type of Water Use
0	0	Potable and Well Water (for drinking)
<200	<1,000	Primary Contact Water (for swimming)
<1,000	<5,000	Secondary Contact Water (for boating and fishing)

streams to the criteria in the state's standards and so judge whether streams are meeting their designated uses.

- Water quality monitoring, however, might be inadequate for determining whether aquatic life needs are being met in a stream. While some constituents such as dissolved oxygen and temperature are important to maintaining healthy fish and aquatic insect populations, other factors such as the physical structure of the stream and the condition of the habitat play an equal or greater role. Biological monitoring methods are generally better suited to determining whether or not aquatic life is supported.
- To identify specific pollutants and sources of pollution. Water quality monitoring helps link sources of pollution to stream quality problems because it identifies specific problem pollutants. Since certain activities tend to generate certain pollutants (bacteria and nutrients are more likely to come from an animal feedlot than an automotive repair shop), a tentative link on what would warrant further investigation or monitoring can be formed.
- To determine trends. Chemical constituents that are properly monitored (i.e., using consistent time of day and on a regular basis using consistent methods) can be analyzed for trends over time.
- To screen for impairment. Finding excessive levels of one or more chemical constituents can serve as an early warning "screen" for potential pollution problems.

State Water Quality Standards Programs

Each state has a program to set standards for the protection of each body of water within its boundaries. Standards are developed for each body of water that:

- Depend on the water's designated use and other purposes;
- Are based on USEPA national water quality criteria and other scientific research into the effects of specific pollutants on different types of aquatic life and on human health; and
- May include limits based on the biological diversity of the body of water (the presence of food and prey species).

State water quality standards set limits on pollutants and establish water quality levels that must be maintained for each type of water body, based on its designated use. Resources for this type of information include:

- USEPA Water Quality Criteria Program; and
- US Fish and Wildlife Service Habitat Suitability Index Models (for specific species of local interest).

Your own monitoring test results can be plotted against these standards to provide a focused, relevant, required assessment of water quality.

GENERAL SAMPLING

Considerations

The sections that follow detail specific and equipment considerations and analytical procedures for each of the most common water quality parameters.

Preparation of Sampling Containers

Sampling devices should be corrosion resistant, easily cleaned, and capable of collecting desired samples safely and in accordance with test requirements. Whenever possible, assign a sampling device to each sampling point. Sampling equipment must be cleaned on a regular schedule to avoid contamination. Note: Some tests require special equipment to ensure the sample is representative. Dissolved oxygen and fecal bacteria sampling require special equipment and/or procedures to prevent the collection of nonrepresentative samples. Reused sample containers and glassware must be cleaned and rinsed before the first sampling run and after each run.

Sample Types

Two basic types of samples are commonly used for water quality testing: grab samples and composite samples. The type of sample used depends on the specific test, the reason the sample is being collected, and the applicable regulatory requirements.

Grab samples are collected at one time and one location. They are representative only of the conditions at the time of collection. Grab samples must be used to determine pH, total residual chlorine (TRC), dissolved oxygen (DO), and fecal coliform concentrations. Grab samples may also be used for any test which does not specifically prohibit their use. Note: Before collecting samples for any test procedure, reviewing the sampling requirements of the test is best.

Composite samples consist of a series of individual grab samples collected over a specified period of time in proportion to flow. The individual grab samples are mixed together in proportion to the flow rate at the time the sample was collected to form the composite sample. Note: In the following sections, a stream, any stream, is used as the example and source of drinking water. Keep in mind that a lake or river or groundwater could also be used as the source.

Collecting Samples (from a stream)

In general, sample away from the stream bank in the main current. Never sample stagnant water. The outside curve of the stream is often a good place to sample, since the main current tends to hug this bank. In shallow stretches, carefully wade into the center current to collect the sample. A boat is required for deep sites. Try to maneuver the boat into the center of the main current to collect the water sample.

Sample Preservation and Storage

Samples can change very rapidly. However, no single preservation method will serve for all samples and constituents. If analysis must be delayed, follow the instructions for sample preservation and storage listed in *Standard Methods* or those specified by the laboratory that will eventually process the samples (see Table 3.3). In general, handle the sample in a way that prevents changes from biological activity, physical alterations, or chemical reactions. Cool the sample to reduce biological and chemical reactions. Store in darkness to suspend photosynthesis. Fill the sample container completely to prevent the loss of dissolved gases. Metal cations such as iron and lead and suspended particles may adsorb onto container surfaces during storage.

References used for sampling and testing must correspond to those listed in the most current federal regulation. For the majority of tests, to compare the results of either different water quality monitors or the same monitors over the course of time requires some form of standardization of the methods. The American Public Health Association (APHA) recognized this requirement when in 1899 the association appointed a committee to draw up standard procedures for the analysis of water. The report (published in 1905) constituted the first edition of what is now known as *Standard Methods for the Examination of Water and Wastewater* or "*Standard Methods.*" This book is now in its 23rd edition and serves as the primary reference for water testing methods and as the basis for most EPA-approved methods.

Note: Some of the methods used by volunteer monitors are based directly on procedures outlined in the APHA *Standard Methods*. Although many methods used by volunteer monitors are not described in *Standard Methods*, they can be standardized to provide repeatable and comparable data. Instructions, training, and audit procedures all play a role in standardizing the methods used by volunteer monitors.

TABLE 3.3

Recommended Sample Storage and Preservation Techniques

Test Factor	Container Type	Preservation	Max. Storage Time Recommended/Regulatory
Alkalinity	P,G	Refrigerate	24 hr/14 days
BOD	P,G	Refrigerate	6 hr/48 hr
Conductivity	P,G	Refrigerate	28 days/28 days
Hardness	P,G	Lower pH to <2	6 mos/6 mos
Nitrate	P,G	Analyze ASAP	48 hr/48 hr
Nitrite	P,G	Analyze ASAP	none/48 hr
Odor	G	Analyze ASAP	6 hr/NR
Oxygen, Dissolved:	G	Immediately Analyze	0.5 hr/stat
Electrode Winkler	G	"Fix" Immediately	8 hr/8 hr
pH	P,G	Immediately Analyze	2 hr/stat
Phosphate	G(A)	Filter Immediately, Refrigerate	48 hr/NR
Salinity	G, wax seal	Immediately Analyze or Use Wax Seal	6 mos/NR
Temperature	P,G	Immediately Analyze	stat/stat
Turbidity	P,G	Analyze Same Day or Store in Dark up to 24 hr, Refrigerate	24 hr/48 hr

Refrigerate = store at 4°C, in dark
P = plastic (polyethylene or equivalent)
G = glass

G (A) = glass rinsed with 1 + 1 HNO_3
stat = no storage allowed
NR = none recommended

Source: Adapted from Standard Methods (2017), 23rd ed.

TEST METHODS*

Descriptions of general methods to help you understand how each works in specific test kits follow. Always follow the specific instructions included with the equipment and individual test kits. Most water analyses are conducted either by titrimetric analyzes or colorimetric analyzes. Both methods are easy to use and provide accurate results.

Titrimetric: Titrimetric analyses are based on adding a solution of known strength (the titrant) to a specific volume of a treated sample in the presence of an indicator. The indicator produces a color change indicating the reaction is complete. Titrants are generally added by a titrator (microburette) or a precise glass pipette.

Colorimetric: Two basic types of colorimetric tests are commonly used: (1) The pH is a measure of the concentration of hydrogen ions (the acidity of a solution) determined by the reaction of an indicator that varies in color, depending on hydrogen ion levels in the water. (2) Tests which determine a concentration of an element or compound are based on Beer's Law. Simply, this law states that the higher the concentration of a substance, the darker the color produced in the test reaction, and therefore the more light absorbed. Assuming a constant view path, the absorption increases exponentially with concentration.

VISUAL METHODS

The Octet Comparator uses standards that are mounted in a plastic comparator block. It employs eight permanent translucent color standards and built-in filters to eliminate optical distortion. The

---

* Much of the information is adapted from *Standard Methods & The Monitor's Handbook*, LaMotte Company, Chestertown, MD, 1992.

sample is compared using either of two viewing windows. Two devices that can be used with the comparator are the B-color Reader, which neutralizes color or turbidity in water samples, and view path, which intensifies faint colors of low concentrations for easy distinction.

ELECTRONIC METHODS

Although the human eye is capable of differentiating color intensity, interpretation is quite subjective. Electronic colorimeters consist of a light source which passes through a sample and are measured on a photodetector with an analog or digital readout.

Electronic Meters

Besides electronic colorimeters, specific electronic instruments are manufactured for lab and field determination of many water quality factors, including pH, total dissolved solids (TDS)/conductivity, dissolved oxygen, temperature, and turbidity.

DISSOLVED OXYGEN (DO) AND BIOCHEMICAL OXYGEN DEMAND (BOD)*

A stream system (in this case, the hypothetical one that provides the source of water used in our discussion) produces and consumes oxygen. It gains oxygen from the atmosphere and from plants as a result of photosynthesis. Because of running water's churning, it dissolves more oxygen than still water; in a reservoir behind a dam, for example. Respiration by aquatic animals, decomposition, and various chemical reactions consume oxygen; and oxygen is poorly soluble in water. Its solubility is related to pressure and temperature. In water supply systems, *dissolved oxygen* in raw water is considered the necessary element to support the life of many aquatic organisms.

From the drinking water practitioner's point of view, DO is an important indicator of the water treatment process and an important factor in corrosivity. Oxygen is measured in its dissolved form as dissolved oxygen. If more oxygen is consumed than produced, dissolved oxygen levels decline, and some sensitive animals may move away, weaken, or die. DO levels fluctuate over a 24-hour period and from season to season. They vary with water temperature and altitude. Cold water holds more oxygen than warm water (Table 3.4), and water holds less oxygen at higher altitudes. Thermal discharges (such as water used to cool machinery in a manufacturing plant or a power plant) raise the temperature of water and lower its oxygen content. Aquatic animals are most vulnerable to lowered DO levels in the early morning on hot summer days when stream flows are low, water temperatures are high, and aquatic plants have not been producing oxygen since sunset.

SAMPLING AND EQUIPMENT CONSIDERATIONS

In contrast to lakes, where DO levels are most likely to vary vertically in the water column, DO in rivers and streams changes move horizontally along the course of the waterway. This is especially true in smaller, shallow streams. In larger, deeper rivers, some vertical stratification of dissolved oxygen might occur. The DO levels in and below riffle areas, waterfalls, or dam spillways are typically higher than those in pools and slower-moving stretches. If you wanted to measure the effect of a dam, sampling for DO behind the dam, immediately below the spillway, and upstream of the dam would be important. Since DO levels are critical to fish, a good place to sample is in the pools that fish tend to favor or in the spawning areas they use.

---

* Note: In this section and the sections that follow, we discuss several water quality factors that are routinely monitored in drinking water operations. We do not discuss the actual test procedures to analyze each water quality factor; we refer you to the latest edition of *Standard Methods* for the correct procedure to use in conducting these tests.

TABLE 3.4
Maximum Dissolved Oxygen Concentrations vs.
Temperature Variations

Temperature °C	DO (mg/L)	Temperature °C	DO (mg/L)
0	14.60	23	8.56
1	14.19	24	8.40
2	13.81	25	8.24
3	13.44	26	8.09
4	13.09	27	7.95
5	12.75	28	7.81
6	12.43	29	7.67
7	12.12	30	7.54
8	11.83	31	7.41
9	11.55	32	7.28
10	11.27	33	7.16
11	11.01	34	7.05
12	10.76	35	6.93
13	10.52	36	6.82
14	10.29	37	6.71
15	10.07	38	6.61
16	9.85	39	6.51
17	9.65	40	6.41
18	9.45	41	6.31
19	9.26	42	6.22
20	9.07	43	6.13
21	8.90	44	6.04
22	8.72	45	5.95

An hourly time profile of DO levels at a sampling site is a valuable set of data because it shows the change in DO levels from the low point (just before sunrise) to the high point (sometime near midday). However, this might not be practical for a volunteer monitoring program. Note the time of your DO sampling to help judge when in the daily cycle the data were collected.

DO is measured either in milligrams per liter or "percent saturation." Milligrams per liter is the amount of oxygen in a liter of water. Percent saturation is the amount of oxygen in a liter of water relative to the total amount of oxygen that the water can hold at that temperature. DO samples are collected using a special BOD bottle: a glass bottle with a "turtleneck" and a ground glass stopper. You can fill the bottle directly in the stream if the stream is wadable or boatable, or you can use a sampler dropped from a bridge or boat into water deep enough to submerse the sampler. Samplers can be made or purchased. Dissolved oxygen is measured primarily either by using some variation of the Winkler method or by using a meter and probe.

WINKLER METHOD

The *Winkler Method* involves filling a sample bottle completely with water (no air is left to bias the test). The dissolved oxygen is then "fixed" using a series of reagents that form a titrated acid compound. Titration involves the drop-by-drop addition of a reagent that neutralizes the acid compound, causing a change in the color of the solution. The point at which the color changes is the "endpoint" and is equivalent to the amount of oxygen dissolved in the sample. The sample is usually fixed and titrated in the field at the sample site. Preparing the sample in the field and delivering it to a lab for titration is possible.

Dissolved oxygen field kits using the Winkler method are relatively inexpensive, especially compared to a meter and probe. Field kits run from $35 to $200, and each kit comes with enough reagents to run 50–100 DO tests. Replacement reagents are inexpensive, and you can buy them already measured out for each test in plastic pillows. You can also buy the reagents in larger quantities in bottles and measure them out with a volumetric scoop. The pillows' advantage is that they have a longer shelf life and are much less prone to contamination or spillage. Buying larger quantities in bottles has the advantage of considerably lower cost per test. The major factor in the expense of the kits is the method of titration they use; eyedropper, syringe-type titrator, or digital titrator. Eyedropper and syringe-type titration are less precise than digital titration because a larger drop of titrant is allowed to pass through the dropper opening, and on a microscale, the drop size (and thus the volume of titrant) can vary from drop to drop. A digital titrator or a burette (a long glass tube with a tapered tip like a pipette) permits much more precision and uniformity in the amount of titrant it allows to pass.

If a high degree of accuracy and precision in DO results is required, a digital titrator should be used. A kit that uses an eye dropper-type or syringe-type titrator is suitable for most other purposes. The lower cost of this type of DO field kit might be attractive if several teams of samplers and testers at multiple sites at the same time are relied on.

METER AND PROBE

A *dissolved oxygen meter* is an electronic device that converts signals from a probe placed in the water into units of DO in milligrams per liter. Most meters and probes also measure temperature. The probe is filled with a salt solution and has a selectively permeable membrane that allows DO to pass from the stream water into the salt solution. The DO that has diffused into the salt solution changes the electric potential of the salt solution, and this change is sent by electric cable to the meter, which converts the signal to milligrams per liter on a scale that the volunteer can read.

DID YOU KNOW?

Oxygen concentration in air is about 21%, but in water it is only slightly soluble. Oxygen saturation ranges from 7 mg/L in hot water to 15 mg/L in cold water and is 9.2 mg/L at 20°C and atmospheric pressure at sea level.

DO meters are expensive compared to field kits that use the titration method. Meter/probe combinations run from $500 to $1,200, including a long cable to connect the probe to the meter. The advantage of a meter/probe is that DO and temperature can be quickly read at any point where the probe is inserted into the stream. DO levels can be measured at a certain point on a continuous basis. The results are read directly as milligrams per liter, unlike the titration methods, in which the final titration result might have to be converted by an equation to milligrams per liter.

However, DO meters are more fragile than field kits, and repairs to a damaged meter can be costly. The meter/probe must be carefully maintained and calibrated before each sample run, and if many tests are done, between sampling. Because of the expense, a small waterworks might only have one meter/probe, which means that only one team of samplers can sample DO and they must test all the sites. With field kits, on the other hand, several teams can sample simultaneously.

LABORATORY TESTING OF DISSOLVED OXYGEN

If a meter and probe are used, the testing must be done in the field because dissolved oxygen levels in a sample bottle change quickly from the decomposition of organic material by microorganisms or the production of oxygen by algae and other plants in the sample. This lowers the DO reading. If a variation of the Winkler method is used, "fixing" the sample in the field, then delivering it to a lab for titration is possible. This might be preferable if sampling is conducted under adverse conditions or if time spent collecting samples is an issue. Titrating samples in the lab is a little easier, and more quality control is possible because the same person can do all the titrations.

WHAT IS BIOCHEMICAL OXYGEN DEMAND AND WHY IS IT IMPORTANT?

Biochemical oxygen demand (BOD) measures the amount of oxygen consumed by microorganisms in decomposing organic matter in stream water. BOD also measures the chemical oxidation of inorganic matter (the extraction of oxygen from water via chemical reaction). A test is used to measure the amount of oxygen consumed by these organisms during a specified period of time (usually 5 days at 20°C). The rate of oxygen consumption in a stream is affected by a number of variables: temperature, pH, the presence of certain kinds of microorganisms, and the type of organic and inorganic material in the water. BOD directly affects the amount of dissolved oxygen in rivers and streams. The greater the BOD, the more rapidly oxygen is depleted in the stream, leaving less oxygen available to higher forms of aquatic life. The consequences of high BOD are the same as those for low dissolved oxygen: aquatic organisms become stressed, suffocate, and die. Most river waters used as water supplies have a BOD of less than 7 mg/L; therefore, dilution is not necessary. Sources of BOD include leaves and woody debris, dead plants and animals, animal manure, effluents from pulp and paper mills, wastewater treatment plants, feedlots, and food-processing plants, failing septic systems, and urban stormwater runoff. Note: To evaluate raw water's potential for use as a drinking water supply it is usually sampled, analyzed, and tested for biochemical oxygen demand when turbid, polluted water is the only source available.

SAMPLING CONSIDERATIONS

BOD is affected by the same factors that affect DO. Aeration of stream water – by rapids and waterfalls, for example – will accelerate the decomposition of organic and inorganic material. Therefore, BOD levels at a sampling site with slower, deeper waters might be higher for a given column of organic and inorganic material than the levels for a similar site in high aerated waters. Chlorine can also affect BOD measurement by inhibiting or killing the microorganisms that decompose the organic and inorganic matter in a sample. If sampling in chlorinated waters, (such as those below the effluent from a sewage treatment plant) neutralizing the chlorine with sodium thiosulfate is necessary (see *Standard Methods*).

   BOD measurement requires taking two samples at each site. One is tested immediately for dissolved oxygen, and the second is incubated in the dark at 20°C for 5 days, then tested for the amount of dissolved oxygen remaining. The difference in oxygen levels between the first test and the second test (in milligrams per liter is the amount of BOD. This represents the amount of oxygen consumed by microorganisms and used to break down the organic matter present in the sample bottle during the incubation period. Because of the 5-day incubation, the tests are conducted in a laboratory. Sometimes by the end of the 5-day incubation period, the dissolved oxygen level is zero. This is especially true for rivers and streams with a lot of organic pollution. Since knowing when the zero point was reached is not possible, determining the BOD level is also impossible. In this case, diluting the original sample by a factor that results in a final dissolved oxygen level of at least 2 mg/L is necessary. Special dilution water should be used for the dilutions (see *Standard Methods*). Some experimentation is needed to determine the appropriate dilution factor for a particular sampling site. The final result is the difference in dissolved oxygen between the first measurement and the second, after multiplying the second result by the dilution factor. *Standard Methods* prescribes all phases of procedures and calculations for BOD determination. A BOD test is not required for monitoring water supplies.

Useful Conversions

$20°C = 68°F$

$0°C = 32°F$

$100°C = 212°F$

TEMPERATURE

We stated earlier that an ideal water supply should have, at all times, an almost constant temperature or one with minimum variation. Knowing the temperature of the water supply is important because the rates of biological and chemical processes depend on it. Temperature affects the oxygen content of the water (oxygen levels become lower as temperature increases); the rate of photosynthesis by aquatic plants; the metabolic rates of aquatic organisms; and the sensitivity of organisms to toxic wastes, parasites, and diseases. Causes of temperature change include weather, removal of shading stream bank vegetation, impoundments (a body of water confined by a barrier, such as a dam), discharge of cooling water, urban stormwater, and groundwater inflows to the stream.

Temperature Sampling and Equipment Considerations

Temperature in a stream varies with width and depth, and the temperature of well-sunned portions of a stream can be significantly higher than the shaded portion of the water on a sunny day. In a small stream, the temperature will be relatively constant as long as the stream is uniformly in sun or shade. In a large stream, temperature can vary considerably with width and depth, regardless of shade. If safe to do so, temperature measurements should be collected at varying depths and across the surface of the stream to obtain vertical and horizontal temperature profiles. This can be done at each site at least once to determine the necessity of collecting a profile during each sampling visit. Temperature should be measured at the same place every time. Temperature is measured in the stream with a thermometer or a meter. Alcohol-filled thermometers are preferred over mercury-filled ones because they are less hazardous if broken. Armored thermometers for field use can withstand more abuse than unprotected glass thermometers and are worth the additional expense. Meters for other tests [such as pH (acidity) or dissolved oxygen] also measure temperature and can be used instead of a thermometer.

HARDNESS

Water *hardness* refers primarily to the amount of dissolved calcium and magnesium in the water. A more appropriate definition of hardness is the water's effect on scaling, corrosion, and soap. With hard water, it is difficult to produce a soap lather. Hard waters leave spots on glasses, dingy film on laundry and hair, and crusty deposits on bathroom fixtures. The presence of hardness in water supplies contributes to taste, odor, color, or turbidity to the water. Water hardness is purely an aesthetic property and has no health significance (Hauser, 2002). Calcium and magnesium enter water mainly by leaching of rocks. Calcium is the most abundant dissolved cationic constituent of natural fresh waters. Calcium is an important component of aquatic plant cell walls and the shells and bones of many aquatic organisms. Magnesium is an essential nutrient for plants and is a component of the chlorophyll molecule. Hardness test kits express test results in ppm of $CaCO_3$, but these results can be converted directly to calcium or magnesium concentrations: Note: Because of less contact with soil minerals and more contact with rain, surface raw water is usually softer than groundwater. As a general rule of thumb, when hardness is greater than 150 mg/L, softening treatment may be required for public water systems. Hardness determination via testing is required to ensure the efficiency of treatment.

DID YOU KNOW?

Household equipped with ion exchanger water softness may identify hardness in terms of "grains per gallon." The direct conversion: 17.1 mg/L = 1 grain/gal.

Measuring Hardness

A hardness test follows a procedure similar to an alkalinity test, but the reactions involved are different. The sample must be carefully measured; then a buffer is added to the sample to correct pH for the test and an indicator to signal the titration endpoint. The indicator reagent is normally blue in a sample of pure water, but if calcium or magnesium ions are present in the sample, the indicator combines with them to form a red-colored complex. The titrant in this test is EDTA (ethylenediaminetetraacetic acid, used with its salts in this method), the *titration method*, a "chelant" which actually "pulls" the calcium and magnesium ions away from the red-colored complex. The EDTA is added dropwise to the sample until all the calcium and magnesium ions have been "chelated" away from the complex and the indicator returns to its normal blue color. The amount of EDTA required to cause the color change is a direct indication of the amount of calcium and magnesium ions in the sample. Some hardness kits include an additional indicator that is specific for calcium. This type of kit will provide three readings: total hardness, calcium hardness, and magnesium hardness. For interference, precision, and accuracy, consult the latest edition of *Standard Methods*.

pH

pH is a term used to indicate the alkalinity or acidity of a substance as ranked on a scale from 1.0 to 14.0.

Calcium Hardness as ppm $CaCO_3 \times 0.40 =$ ppm Ca
Magnesium Hardness as ppm $CACO_3 \times 0.24 =$ ppm Mg

pH: Analytical and Equipment Considerations

pH can be analyzed in the field or in the lab. If analyzed in the lab, it must be measured within 2 hours of the sample collection because the pH will change from the carbon dioxide from the air as it dissolves in the water, bringing the pH toward 7. If your program requires a high degree of accuracy and precision in pH results, the pH should be measured with a laboratory quality pH meter and electrode. Meters of this quality range in cost from $250 to $1,000. Color comparators and pH "pocket pals" are suitable for most other purposes. The cost of either of these is in the $50 range. The lower cost of the alternatives might be attractive if multiple samplers are used to sample several sites at the same time.

pH Meters

A pH meter measures the electric potential (millivolts) across an electrode when immersed in water. This electric potential is a function of the hydrogen ion activity in the sample; therefore, pH meters can display results in either millivolts (mV) or pH units. A pH meter consists of a *potentiometer*, which measures electric potential where it meets the water sample; a reference electrode, which provides a constant electric potential; and a *temperature compensating device*, which adjusts the readings according to the temperature of the sample (since pH varies with temperature). The reference and glass electrodes are frequently combined into a single probe called a *combination electrode*. A wide variety of meters are available, but the most important part of the pH meter is the electrode. Thus, purchasing a good, reliable electrode and following the manufacturer's instructions for proper maintenance is important. Infrequently used or improperly maintained electrodes are subject to corrosion, which makes them highly inaccurate.

DID YOU KNOW?

Synergy is the process whereby two or more substances combine and produce effects greater than their sum. For example, 2 + 2 = 4 (mathematically). But synergistically, 2 + 2 = more than 4! Synergy is a mathematical impossibility but a chemical reality (Jacobson, 1991).

"Pocket Pals" and Color Comparators

pH "pocket pals" are electronic handheld "pens" that are dipped in the water, providing a digital readout of the pH. They can be calibrated to only one pH buffer (lab meters, on the other hand, can be calibrated to two or more buffer solutions and thus are more accurate over a wide range of pH measurements). Color comparators involve adding a reagent to the sample that colors the sample water. The intensity of the color is proportional to the pH of the sample, then matched against a standard color chart. The color chart equates particular colors to associated pH values, which can be determined by matching the colors from the chart to the color of the sample. For instructions on how to collect and analyze samples, refer to *Standard Methods*.

TURBIDITY

We discussed *turbidity* earlier. Turbidity is a measure of water clarity – how much the material suspended in water decreases the passage of light through the water. Suspended materials include soil particles (clay, silt, and sand), algae, plankton, microbes, and other substances. These materials are typically in the size range of 0.004 mm (clay)–1.0 mm (sand). Turbidity can affect the color of the water. Higher turbidity increases water temperatures, because suspended particles absorb more heat. This in turn reduces the concentration of DO because warm water holds less DO than cold. Higher turbidity also reduces the amount of light penetrating the water, which reduces photosynthesis and the production of DO. Suspended materials can clog fish gills, reducing resistance to disease in fish, lowering growth rates, and affecting egg and larval development. As the particles settle, they can blanket the stream bottom (especially in slower waters) and smother fish eggs and benthic macroinvertebrates.

Sources of turbidity include:

- Soil erosion;
- Waste discharge;
- Urban runoff;
- Eroding stream banks;
- Large numbers of bottom feeders (such as carp), which stir up bottom sediments; and
- Excessive algal growth.

SAMPLING AND EQUIPMENT CONSIDERATIONS

Turbidity can be useful as an indicator of the effects of runoff from construction, agricultural practices, logging activity, discharges, and other sources. Turbidity often increases sharply during a rainfall, especially in developed watersheds, which typically have relatively high proportions of impervious surfaces. The flow of stormwater runoff from impervious surfaces rapidly increases stream velocity, which increases the erosion rates of stream banks and channels. Turbidity can also rise sharply during dry weather if earth-disturbing activities occur in or near a stream without erosion control practices in place. Regular monitoring of turbidity can help detect trends that might indicate increasing erosion in developing watersheds. However, turbidity is closely related to streamflow and velocity and should be correlated with these factors. Comparisons of the change in turbidity over time, therefore, should be made at the same point at the same flow.

Turbidity is not a measurement of the amount of suspended solids present or the rate of sedimentation of a stream since it measures only the amount of light that is scattered by suspended particles. Measurement of total solids is a more direct measurement of the amount of material suspended and dissolved in water. Turbidity is generally measured by using a turbidity meter. Volunteer programs may also take samples to a lab for analysis. Another approach is to measure transparency (an integrated measure of light scattering and absorption) instead of turbidity. Water clarity/transparency can be measured using a *Secchi disk* (see Figure 3.3) or transparency tube. The Secchi disk can only be used in deep, slow-moving rivers; the transparency tube (a comparatively new development) is gaining acceptance in and around the country but is not yet in wide use.

A turbidity meter consists of a light source that illuminates a water sample and a photoelectric cell that measures the intensity of light scattered at 90° angle by the particles in the sample. It measures turbidity in nephelometric turbidity units (NTUs). Meters can measure turbidity over a wide range from 0 to 1,000 NTUs. A clear mountain stream might have a turbidity of around 1 NTU, whereas a large river like the Mississippi might have dry-weather turbidity of 10 NTUs. Because these values can jump into hundreds of NTU during runoff events, the turbidity meter to be used should be reliable over the range in which you will be working. Meters of this quality cost about $800. Many meters in this price range are designed for field or lab use.

Although turbidity meters can be used in the field, samplers might want to collect samples and take them to a central point for turbidity measurements because of the expense of the meter. Most programs can afford only one and would have to pass it along from site to site, complicating logistics and increasing the risk of damage to the meter. Meters also include glass cells that must remain optically clear and free of scratches for operation. Samplers can also take turbidity samples to a lab for meter analysis at a reasonable cost.

USING A SECCHI DISK

A Secchi disk is a black-and-white disk that is lowered by hand into the water to the depth at which it vanishes from sight (see Figure 3.3). The distance to vanishing is then recorded; the clearer the

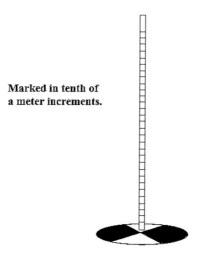

Marked in tenth of a meter increments.

FIGURE 3.3   Using a Secchi disk to measure transparency. Lower the disk into the water until it is no longer visible. That point is the Secchi disk depth.

water, the greater the distance. Secchi disks are simple to use and inexpensive. For river monitoring they have limited use, however, because in most cases the river bottom will be visible, and the disk will not reach a vanishing point. Deeper, slower-moving rivers are the most appropriate places for Secchi disk measurement, although the current might require that the disk be extra-weighted, so it does not sway and make measurement difficult. Secchi disks cost about $50 but can be homemade. The line attached to the Secchi disk must be marked according to units designated by the sampling program, in waterproof ink. Many programs require samplers to measure to the nearest 1/10 meter. Meter intervals can be tagged (e.g., with duct tape) for ease of use.

To measure water clarity with a Secchi disk:

- Check to make sure that the Secchi disk is securely attached to the measured line.
- Lean over the side of the boat and lower the Secchi disk into the water, keeping your back to the sun to block glare.
- Lower the disk until it disappears from view. Lower it one-third of a meter and then slowly raise the disk until it just reappears. Move the disk up and down until you find the exact vanishing point.
- Attach a clothespin to the line at the point where the line enters the water. Record the measurement on your datasheet. Repeating the measurement provides you with a quality control check.

The key to consistent results is to train samplers to follow standard sampling procedures, and if possible, have the same individual take the reading at the same site throughout the season.

ORTHOPHOSPHATE

Both phosphorus and nitrogen are essential nutrients for the plants and animals that make up the aquatic food web. Since phosphorus is the nutrient in short supply in most freshwater systems, even a modest increase in phosphorus can (under the right conditions) set off a whole chain of undesirable events in a stream, including accelerated plant growth, algae blooms, low dissolved oxygen, and the death of certain fish, invertebrates, and other aquatic animals. Phosphorus comes from many sources, both natural and human. These include soil and rocks, wastewater treatment plants, runoff from fertilized lawns and cropland, failing septic systems, runoff from animal manure storage areas, disturbed land areas, drained wetlands, water treatment, and commercial cleaning preparations.

Forms of Phosphorus

Phosphorus has a complicated story. Pure, "elemental" phosphorus (P) is rare. In nature, phosphorus usually exists as part of a phosphate molecule (PO_4). Phosphorus in aquatic systems occurs as organic phosphate and inorganic phosphate. Organic phosphate consists of a phosphate molecule associated with a carbon-based molecule, as in a plant or animal tissue. Phosphate that is not associated with organic material is inorganic, the form required by plants. Animals can use either organic or inorganic phosphate. Both organic and inorganic phosphorus can either be dissolved in the water or suspended (attached to particles in the water column).

The Phosphorus Cycle

Phosphorus cycles through the environment, changing form as it does so (see Figure 3.4). Aquatic plants take in dissolved inorganic phosphorus as it becomes part of their tissues. Animals get the organic phosphorus they need by eating either aquatic plants, other animals, or decomposing plant and animal material. In water bodies, as plants and animals excrete wastes or die, the organic phosphorus they contain sinks to the bottom where bacterial decomposition converts it back to inorganic phosphorus, both dissolved and attached to particles. This inorganic phosphorus gets back into the

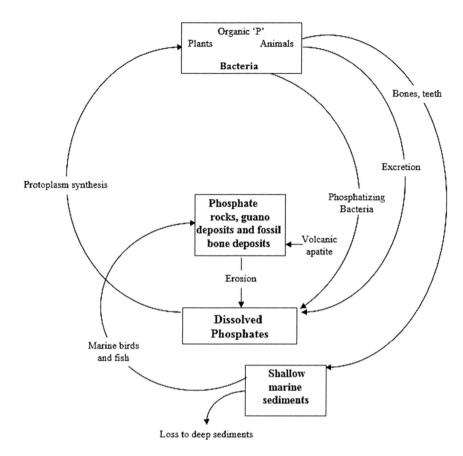

FIGURE 3.4 The phosphorus cycle.

water column when the bottom is stirred up by animals, human activity, chemical interactions, or water currents. Then it is taken up by plants, and the cycle begins again.

In a stream system, the phosphorus cycle tends to move phosphorus downstream as the current carries decomposing plant and animal tissue and dissolved phosphorus. It becomes stationary only when it is taken up by plants or is bound to particles that settle to the bottom of ponds. In the field of water quality chemistry, phosphorus is described in several terms. Some of these terms are chemistry based (referring to chemically based compounds), and others are methods based (they describe what is measured by a particular method).

The term *orthophosphate* is a chemistry-based term that refers to the phosphate molecule all by itself. *Reactive phosphorus* is a corresponding method-based term that describes what is actually being measured when the test for orthophosphate is being performed. Because the lab procedure isn't quite perfect, mostly orthophosphate is obtained along with a small fraction of some other forms. More complex inorganic phosphate compounds are referred to as *condensed phosphates* or *polyphosphates*. The method-based term for these forms is *acid hydrolyzable*.

Monitoring Phosphorus

Monitoring phosphorus is challenging because it involves measuring very low concentrations – down to 0.01 mg/L or even lower. Even such very low concentrations of phosphorus can have a dramatic impact on streams. Less sensitive methods should be used only to identify serious problem areas. While many tests for phosphorus exist, only four are likely to be performed by most monitors:

- The *total orthophosphate* test is largely a measure of orthophosphate. Because the sample is not filtered, the procedure measures both dissolved and suspended orthophosphate. The USEPA approved method for measuring total orthophosphate is known as the ascorbic acid method. Briefly, a reagent (either liquid or powder) containing ascorbic acid and ammonium molybdate reacts with orthophosphate in the sample to form a blue compound. The intensity of the blue color is directly proportional to the amount of orthophosphate in the water.
- The *total phosphorus* test measures all the forms of phosphorus in the sample (orthophosphate, condensed phosphate, and organic phosphate) by first "digesting" (heating and acidifying) the sample to convert all the other forms to orthophosphate; then the orthophosphate is measured by the ascorbic acid method. Because the sample is not filtered, the procedure measures both dissolved and suspended orthophosphate.
- The *dissolved phosphorus* test measures that the fraction of the total phosphorus that is in solution in the water as opposed to being attached to suspended particles. It is determined by first filtering the sample, then analyzing the filtered sample for total phosphorus.
- *Insoluble phosphorus* is calculated by subtracting the dissolved phosphorus result from the total phosphorus result.

All these tests have one thing in common; they all depend on measuring orthophosphate. The total orthophosphate test measures the orthophosphate that is already present in the sample. The others measure that which is already present and that which is formed when the other forms of phosphorus are converted to orthophosphate by digestion.

Phosphorus: Sampling and Equipment Considerations

Monitoring phosphorus involves two basic steps:

- Collecting a water sample; and
- Analyzing it in the field or lab for one of the types of phosphorus described above.

This chapter does not address laboratory methods. Refer to *Standard Methods.*

Sample Containers – Sample containers made of either some form of plastic or Pyrex® glass are acceptable to the USEPA. Because phosphorus molecules have a tendency to "adsorb" (attach) to the inside surface of sample containers, if containers are to be reused, they must be acid washed to remove adsorbed phosphorus. The container must be able to withstand repeated contact with hydrochloric acid. Plastic containers, either high-density polyethylene or polypropylene, might be preferable to glass from a practical standpoint because they are better able to withstand breakage. Some programs use disposable, sterile, plastic Whirl-pak® bags. The size of the container depends on the sample amount needed for the phosphorus analysis method chosen and the amount needed for other analysis to be performed.

Dedicated Labware – All containers that will hold water samples or come into contact with reagents used in this test must be dedicated. They should not be used for other tests to eliminate the possibility that reagents containing phosphorus will contaminate the labware. All labware should be acid washed.

The only form of phosphorus this chapter recommends for field analysis is total orthophosphate, which uses the ascorbic acid method on an untreated sample. Analysis of any of the other forms requires adding potentially hazardous reagents, heating the sample to boiling, and using too much time and too much equipment to be practical. In addition, analysis for other forms of phosphorus is prone to errors and inaccuracies in field situations. Pretreatment and analysis for these other forms should be handled in a laboratory.

Ascorbic Acid Method – In the ascorbic acid method, a combined liquid or prepackaged powder reagent consisting of sulfuric acid, potassium antimonyl tartrate, ammonium molybdate, and ascorbic acid (or comparable compounds) is added to either 50 or 25 mL of the water sample. This colors the sample blue in direct proportion to the amount of orthophosphate in the sample. Absorbance or transmittance is then measured after 10 minutes, but before 30 minutes, using a color comparator with a scale in milligrams per liter that increases with the increase in color hue or an electronic meter that measures the amount of light absorbed or transmitted at a wavelength of 700–880 nanometers (again, depending on manufacturer's directions).

A color comparator may be useful for identifying heavily polluted sites with high concentrations (greater than 0.1 mg/L). However, matching the color of a treated sample to a comparator can be very subjective, especially at low concentrations, and can lead to variable results. A field spectro-photometer or colorimeter with a 2.5-cm light path and an infrared photocell (set for a wavelength of 700–880 nm) is recommended for accurate determination of low concentrations (between 0.2 and 0.02 mg/L). The use of a meter requires that a prepared known standard concentration be analyzed ahead of time to convert the absorbance readings of a stream sample to milligrams per liter or that the meter reads directly in milligrams per liter.

For information on how to prepare standard concentrations and how to collect and analyze samples, refer to *Standard Methods* and USEPA's *Methods for Chemical Analysis of Water and Wastes*, 2nd ed., Method 365.2, US Environmental Protection Agency, Washington, D.C.

NITRATES

As mentioned, *nitrates* are a form of nitrogen found in several different forms in terrestrial and aquatic ecosystems. These forms of nitrogen include ammonia (NH_3), nitrates (NO_3), and nitrites (NO_2). Nitrates are essential plant nutrients, but excess amounts can cause significant water quality problems. Together with phosphorus, nitrates in excess amounts can accelerate eutrophication, causing dramatic increases in aquatic plant growth and changes in the types of plants and animals that live in the stream. This, in turn, affects dissolved oxygen, temperature, and other indicators. Excess nitrates can cause hypoxia (low levels of dissolved oxygen) and can become toxic to warm-blooded animals at higher concentrations (10 mg/L or higher) under certain conditions. The natural level of ammonia or nitrate in the surface water is typically low (less than 1 mg/L); in the effluent of wastewater treatment plants, it can range up to 30 mg/L. Sources of nitrates include wastewater treatment plants, runoff from fertilized lawns and cropland, failing on-site septic systems, runoff from animal manure storage areas, and industrial discharges that contain corrosion inhibitors.

Nitrates: Sampling and Equipment Considerations

Nitrates from land sources end up in rivers and streams more quickly than other nutrients like phosphorus because they dissolve in water more readily than phosphorus, which has an attraction for soil particles. As a result, nitrates serve as a better indicator of the possibility of a source of sewage or manure pollution during dry weather. Water that is polluted with nitrogen-rich organic matter might show low nitrates. Decomposition of the organic matter lowers the dissolved oxygen level, which in turn slows the rate at which ammonia is oxidized to nitrite (NO_2) and then to nitrate (NO_3). Under such circumstances, monitoring for nitrites or ammonia (considerably more toxic to aquatic life than nitrate) might be also necessary. (See *Standard Methods* section 4500-NH_3 and 4500-NH_2 for appropriate nitrite methods.)

Water samples to be tested for nitrate should be collected in glass or polyethylene containers. Most monitoring programs usually use two methods for nitrate testing: the cadmium reduction method and the nitrate electrode. The more commonly used cadmium reduction method produces a color reaction measured either by comparison to a color wheel or by use of a spectrophotometer. A few programs also use a nitrate electrode, which can measure in the range of 0–100 mg/L nitrate. A newer colorimetric immunoassay technique for nitrate screening is also now available.

Cadmium Reduction Method

The *cadmium reduction method* is a colorimetric method that involves contact of the nitrate in the sample with cadmium particles, which cause nitrates to be converted to nitrites. The nitrites then react with another reagent to form a red color in proportional intensity to the original amount of nitrate. The color is measured either by comparison to a color wheel with a scale in milligrams per liter that increases with the increase in color hue or by use of an electronic spectrophotometer that measures the amount of light absorbed by the treated sample at a 543-nanometer wavelength. The absorbance value converts to the equivalent concentration of nitrate against a standard curve. Methods for making standard solutions and standard curves are presented in *Standard Methods.*

This curve should be created by the program advisor before each sampling run. The curve is developed by making a set of standard concentrations of nitrate, reacting them and developing the corresponding color, and then plotting the absorbance value for each concentration against concentration. A standard curve could also be generated for the color wheel. Use of the color wheel is appropriate only if nitrate concentrations are greater than 1 mg/L. For concentrations below 1 mg/L, use a spectrophotometer. Matching the color of a treated sample at low concentrations to a color wheel (or cubes) can be very subjective and can lead to variable results. Color comparators can, however, be effectively used to identify sites with high nitrates.

This method requires that the samples being treated are clear. If a sample is turbid, filter it through a 0.45-micron filter. Be sure to test to make sure the filter is nitrate free. If copper, iron, or other metals are present in concentrations above several mg/L, the reaction with the cadmium will slow down and the reaction time must be increased. The reagents used for this method are often prepackaged for different ranges, depending on the expected concentration of nitrate in the stream. For example, the Hach Company provides reagents for the following ranges: low (0–0.40 mg/L), medium (0–45 mg/L), and high (0–30 mg/L). Determining the appropriate range for the stream being monitored is important.

Nitrate Electrode Method

A nitrate electrode (used with a meter) is similar in function to a dissolved oxygen meter. It consists of a probe with a sensor that measures nitrate activity in the water; this activity affects the electric potential of a solution in the probe. This change is then transmitted to the meter, which converts the electric signal to a scale that is read in millivolts, and the millivolts converted to mg/L of nitrate by plotting them against a standard curve. The accuracy of the electrode can be affected by high concentrations of chloride or bicarbonate ions in the sample water. Fluctuating pH levels can also affect the meter reading. Nitrate electrodes and meters are expensive compared to field kits that employ the cadmium reduction method. (The expense is comparable, however, if a spectrophotometer is used rather than a color wheel.) Meter/probe combinations run from $700 to $1200, including a long cable to connect the probe to the meter. If the program has a pH meter that displays readings in millivolts, it can be used with a nitrate probe and no separate nitrate meter is needed. Results are read directly as milligrams per liter.

Although nitrate electrodes and spectrophotometer can be used in the field, they have certain disadvantages. These devices are more fragile than the color comparators and are therefore more at risk of breaking in the field. They must be carefully maintained and must be calibrated before each sample run, and if many tests are being run, between samplings. This means that samples are best tested in the lab. Note that samples to be tested with a nitrate electrode should be at room temperature, whereas color comparators can be used in the field with samples at any temperature.

TOTAL SOLIDS

Total solids are dissolved solids plus suspended and settleable solids in water. In stream water, dissolved solids consist of calcium, chlorides, nitrate, phosphorus, iron, sulfur, and other ions – particles that will pass through a filter with pores of around 2 microns (0.002 cm) in size. Suspended solids include silt and clay particles, plankton, algae, fine organic debris, and other particulate matter. These are particles that will not pass through a 2-micron filter.

The concentration of total dissolved solids affects the water balance in the cells of aquatic organisms. An organism placed in water with a very low level of solids (distilled water, for example) swells because water tends to move into its cells, which have a higher concentration of solids. An organism placed in water with a high concentration of solids shrinks somewhat because the water in its cells tends to move out. This in turn affects the organism's ability to maintain the proper cell density, making keeping its position in the water column difficult. It might float up or sink down to a depth to which it is not adapted, and it might not survive. Higher concentrations of suspended solids can serve as carriers of toxics, which readily cling to suspended particles. This is particularly a concern where pesticides are being used on irrigated crops. Where solids are high, pesticide concentrations may increase well beyond those of the original application as the irrigation water travels down irrigation ditches. Higher levels of solids can also clog irrigation devices and might become so high that irrigated plant roots will lose water rather than gain it.

A high concentration of total solids will make drinking water unpalatable and might have an adverse effect on people who are not used to drinking such water. Levels of total solids that are too high or too low can also reduce the efficiency of wastewater treatment plants, as well as the operation of industrial processes that use raw water. Total solids affect water clarity. Higher solids decrease the passage of light through water, thereby slowing photosynthesis by aquatic plants. Water heats up more rapidly and holds more heat; this, in turn, might adversely affect aquatic life adapted to a lower temperature regime. Sources of total solids include industrial discharges, sewage, fertilizers, road runoff, and soil erosion. Total solids are measured in milligrams per liter (mg/L).

Total Solids: Sampling and Equipment Considerations

Total solids are important to measure in areas where discharges from sewage treatment plants, industrial plants, or extensive crop irrigation may occur. In particular, streams and rivers in arid regions where water is scarce and evaporation is high tend to have higher concentrations of solids and are more readily affected by the human introduction of solids from landuse activities. Total solids measurements can be useful as an indicator of the effects of runoff from construction, agricultural practices, logging activities, sewage treatment plant discharges, and other sources.

As with turbidity, concentrations often increase sharply during rainfall, especially in developed watersheds. They can also rise sharply during dry weather if earth-disturbing activities occur in or near the stream without erosion control practices in place. Regular monitoring of total solids can help detect trends that might indicate increasing erosion in developing watersheds. Total solids are closely related to stream flow and velocity and should be correlated with these factors. Any change in total solids over time should be measured at the same site at the same flow. Total solids are measured by weighing the amount of solids present in a known volume of sample, and accomplished by weighing a beaker, filling it with a known volume, evaporating the water in an oven and completely drying the residue, then weighing the beaker with the residue. The total solids concentration is equal to the difference between the weight of the beaker with the residue and the weight of the beaker without it. Since the residue is so light in weight, the lab needs a balance that is sensitive to weights in the range of 0.0001 gram. Balances of this type are called analytical or Mettler balances and they are expensive (around $3,000). The technique requires that the beakers be kept in a desiccator, a sealed glass container that contains material that absorbs moisture and ensures that the weighing is not biased by water condensing on the beaker. Some desiccants change color to indicate moisture content. The measurement of total solids cannot be done in the field. Samples must be collected using clean glass or plastic bottles, or Whirl-pak® bags and taken to a laboratory where the test can be run.

Conductivity

Conductivity is a measure of the ability of water to pass an electrical current. Conductivity in water is affected by the presence of inorganic dissolved solids such as chloride, nitrate, sulfate, and phosphate anions (ions that carry a negative charge) or sodium, magnesium, calcium, iron, and aluminum cations (ions that carry a positive charge). Organic compounds like oil, phenol, alcohol, and sugar do not conduct electrical current very well, and therefore have a low conductivity when in water. Conductivity is also affected by temperature: the warmer the water, the higher the conductivity. Conductivity in streams and rivers is affected primarily by the geology of the area through which the water flows. Streams that run through areas with granite bedrock tend to have lower conductivity because granite is composed of more inert materials that do not ionize (dissolve into ionic components) when washed into the water. On the other hand, streams that run through areas with clay soils tend to have higher conductivity because of the presence of materials that ionize when washed into the water. Groundwater inflows can have the same effects, depending on the bedrock they flow through. Discharges to streams can change the conductivity depending on their makeup. A failing sewage system would raise the conductivity because of the presence of chloride, phosphate, and nitrate; an oil spill would lower conductivity.

The basic unit of measurement of conductivity is the mho or siemens. Conductivity is measured in micromhos per centimeter (μmhos/cm) or microsiemens per centimeter (μs/cm). Distilled water has a conductivity in the range of 0.5–3 μmhos/cm. The conductivity of rivers in the United States generally ranges from 50 to 1,500 μmhos/cm. Studies of inland fresh waters indicate that streams supporting good mixed fisheries have a range between 150 and 500 μmhos/cm. Conductivity outside this range could indicate that the water is not suitable for certain species of fish or macroinvertebrates. Industrial waters can range as high as 10,000 μmhos/cm.

Conductivity: Sampling and Equipment Considerations

Conductivity is useful as a general measure of stream water quality. Each stream tends to have a relatively constant range of conductivity that once established can be used as a baseline for comparison with regular conductivity measurements. Significant changes in conductivity could indicate that a discharge or some other source of pollution has entered a stream. Conductivity is measured with a probe and a meter. Voltage is applied between two electrodes in a probe immersed in the sample water. The drop in voltage caused by the resistance of the water is used to calculate the conductivity per centimeter. The meter converts the probe measurement to micromhos per centimeter (μmhos/cm) and displays the result for the user. Note: Some conductivity meters can also be used to test for total dissolved solids and salinity. The total dissolved solids concentration in milligrams per liter (mg/L) can also be calculated by multiplying the conductivity result by a factor between 0.55 and 0.9, which is empirically determined (see *Standard Methods* #2510).

Suitable conductivity meters cost about $350. Meters in this price range should also measure temperature and automatically compensate for temperature in the conductivity reading. Conductivity can be measured in the field or the lab. In most cases, collecting samples in the field and taking them to a lab for testing is probably better. In this way, several teams can collect samples simultaneously. If testing in the field is important, meters designed for field use can be obtained for around the same cost mentioned above. If samples will be collected in the field for later measurement, the sample bottle should be a glass or polyethylene bottle that has been washed in phosphate-free detergent and rinsed thoroughly with both tap and distilled water. Factory-prepared Whirl-pak® bags may be used.

Total Alkalinity

Alkalinity is a measure of the capacity of water to neutralize acids. Alkaline compounds in the water such as bicarbonates (baking soda is one type), carbonates, and hydroxides remove H^+ ions and lower the acidity of the water (which means increased pH). They usually do this by combining with the H^+ ions to make new compounds. Without this acid-neutralizing capacity, any acid added to a stream would cause an immediate change in the pH. Measuring alkalinity is important in determining a stream's ability to neutralize acidic pollution from rainfall or wastewater – one of the best measures of the sensitivity of the stream to acid inputs.

Alkalinity in streams is influenced by rocks and soils, salts, certain plant activities, and certain industrial wastewater discharges. Total alkalinity is determined by measuring the amount of acid (e.g., sulfuric acid) needed to bring the sample to a pH of 4.2. At this pH all the alkaline compounds in the sample are "used up." The result is reported as milligrams per liter of calcium carbonate (mg/L $CaCO_3$).

Total Alkalinity: Analytical and Equipment Considerations

For total alkalinity, a double endpoint titration using a pH meter (or pH "pocket pal") and a digital titrator or burette is recommended. This can be done in the field or in the lab. If alkalinity must be analyzed in the field, a digital titrator should be used instead of a burette because burettes are fragile and more difficult to set up and use in the field. The alkalinity method described below was developed by the Acid Rain Monitoring Project of the University of Massachusetts Water Resources Research Center.*

Burettes, Titrators, and Digital Titrators for Measuring Alkalinity

The total alkalinity analysis involves titration. In this test, titration is the addition of small, precise quantities of sulfuric acid (the reagent) to the sample until the sample reaches a certain pH (known as an endpoint). The amount of acid used corresponds to the total alkalinity of the sample. Alkalinity can be measured using a burette, titrator, or digital titrator described as follows:

- A *burette* is a long, graduated glass tube with a tapered tip like a pipette and a valve that opens to allow the reagent to drip out of the tube. The amount of reagent used is calculated by subtracting the original volume in the burette from the column left after the endpoint has been reached. Alkalinity is calculated based on the amount used.
- *Titrators* forcefully expel the reagent by using a manual or mechanical plunger. The amount of reagent used is calculated by subtracting the original volume in the titrator from the volume left after the endpoint has been reached. Alkalinity is then calculated based on the amount used or is read directly from the titrator.
- *Digital titrators* have counters that display numbers. A plunger is forced into a cartridge containing the reagent by turning a knob on the titrator. As the knob turns, the counter changes in proportion to the amount of reagent used. Alkalinity is then calculated based on the amount used. Digital titrators cost approximately $90.

Digital titrators and burettes allow for much more precision and uniformity in the amount of titrant that is used.

FECAL BACTERIA[†]

Members of two bacteria groups (coliforms and fecal streptococci) are used as indicators of possible sewage contamination because they are commonly found in human and animal feces. Although they are generally not harmful themselves, they indicate the possible presence of pathogenic

* From River Watch Network. Total alkalinity and pH field and laboratory procedures. Based on University of Massachusetts Acid Rain Monitoring Project, July 1, 1992.

[†] Much of the information in this section is from USEPA Test methods for *Escherichia coli* and enterococci in water by the membrane filter procedure (Method #1103.1). EPA 600/4-85-076, 1985 and USEPA Bacteriological ambient water quality criteria for marine and fresh recreational waters. EPA 440/5-84-002. Cincinnati, OH: United States Environmental Protection Agency, Office of Research and Development, 1986.

(disease-causing) bacteria, viruses, and protozoans that also live in human and animal digestive systems. Their presence in streams suggests that pathogenic microorganisms might also be present and that swimming in and/or eating shellfish from the waters might present a health risk. Since testing directly for the presence of a large variety of pathogens is difficult, time-consuming, and expensive, water is usually tested for coliforms and fecal streptococci instead. Sources of fecal contamination to surface waters include wastewater treatment plants, on-site septic systems, domestic and wild animal manure, and storm runoff. In addition to the possible health risk associated with the presence of elevated levels of fecal bacteria, they can also cause cloudy water, unpleasant odors, and an increased oxygen demand.

Indicator Bacteria Types

The most commonly tested fecal bacteria indicators are total coliforms, fecal coliforms, *Escherichia coli*, fecal streptococci, and enterococci. All but *E. coli* are composed of a number of species of bacteria that share common characteristics, including shape, habitat, or behavior; *E. coli* is a single species in the fecal coliform group. Total coliforms are widespread in nature. All members of the total coliform group can occur in human feces, but some can also be present in animal manure, soil, and submerged wood and in other places outside the human body. The usefulness of total coliforms as an indicator of fecal contamination depends on the extent to which the bacteria species found are fecal and human in origin. For recreational waters, total coliforms are no longer recommended as an indicator. For drinking water, total coliforms are still the standard test because their presence indicates contamination of a water supply by an outside source. Fecal coliforms, a subset of total coliform bacteria, are more fecal specific in origin. However, even this group contains a genus, *Klebsiella*, with species that are not necessarily fecal in origin. *Klebsiella* are commonly associated with textile and pulp and paper mill wastes. If these sources discharge to a local stream, consideration should be given to monitoring more fecal and human-specific bacteria. For recreational waters, this group was the primary bacteria indicator until relatively recently when the USEPA began recommending *E. coli* and enterococci as better indicators of health risk from water contact. Fecal coliforms are still being used in many states as indicator bacteria.

E. coli is a species of fecal coliform bacteria specific to fecal material from humans and other warm-blooded animals. The USEPA recommends *E. coli* as the best indicator of health risk from water contact in recreational waters; some states have changed their water quality standards and are monitoring accordingly. Fecal streptococci generally occur in the digestive systems of humans and other warm-blooded animals. In the past, fecal streptococci were monitored together with fecal coliforms, and a ratio of fecal coliforms to streptococci was calculated. This ratio was used to determine whether the concentration was of human or nonhuman origin. However, this is no longer recommended as a reliable test. Enterococci are a subgroup within the fecal streptococcus group. Enterococci are distinguished by their ability to survive in salt water, and in this respect, they more closely mimic many pathogens than do the other indicators. Enterococci are typically more human specific than the larger fecal streptococcus group. The USEPA recommends enterococci as the best indicator of health risk in salt water used for recreation and as a useful indicator in fresh water as well.

Which Bacteria Should Be Monitored?

Which bacteria chosen for testing depends on what is to be determined. Does swimming in the local stream pose a health risk? Does the local stream meet state water quality standards? Studies conducted by the USEPA to determine the correlation between different bacterial indicators and the occurrence of digestive system illness at swimming beaches suggest that the best indicators of health risk from recreational water contact in fresh water are *E. coli* and enterococci. For salt water, enterococci are the best. Interestingly, fecal coliforms as a group were determined to be a poor indicator of the risk of digestive system illness. However, many states continue to use fecal coliforms as their primary health risk indicator. If your state is still using the total of fecal coliforms measurement as the indicator bacteria, and you want to know whether the water meets state water quality standards, you should monitor fecal coliforms. However, if you want to know the health risk from recreational water contact, the results of the USEPA studies suggest that you should consider switching to the *E. coli* or enterococci method for testing fresh water. In any case, consulting with the water quality division of your state's environmental agency is best, especially if you expect to use your data.

Fecal Bacteria: Sampling and Equipment Considerations

Bacteria can be difficult to sample and analyze for many reasons. Natural bacteria levels in streams can vary significantly; bacteria conditions are strongly correlated with rainfall, thus comparing wet and dry-weather bacteria data can be a problem; many analytical methods have a low level of precision yet can be quite complex to accomplish; and absolutely sterile conditions are essential to maintain while collecting and handling samples. The primary equipment decision to make when sampling for bacteria is what type and size of sample container you will use. Once you have made that decision, the same straightforward collection procedure is used, regardless of the type of bacteria being monitored.

When monitoring bacteria, it is critical that all containers and surfaces with which the sample will come into contact are sterile. Containers made of either some form of plastic or Pyrex glass are acceptable to the USEPA. However, if the containers are to be reused, they must be sturdy enough to survive sterilization using heat and pressure. The containers can be sterilized by using an autoclave, a machine that sterilizes with pressurized steam. If using an autoclave, the container material must be able to withstand high temperatures and pressure. Plastic containers – either high-density polyethylene or polypropylene – might be preferable to glass from a practical standpoint because they will better withstand breakage. In any case, be sure to check the manufacturer's specifications to see whether the container can withstand 15 minutes in an autoclave at a temperature of 121°C without melting. (Extreme caution is advised when working with an autoclave.) Disposable, sterile, plastic Whirl-pak® bags are used by a number of programs. The size of the container depends on the sample amount needed for the bacteria analysis method you choose, and the amount needed for other analyses.

The two basic methods for analyzing water samples for bacteria in common use are the membrane filtration and multiple-tube fermentation methods. Given the complexity of the analysis procedures and the equipment required, field analysis of bacteria is not recommended. Bacteria can either be analyzed by the volunteer at a well-equipped lab or sent to a state-certified lab for analysis. If you send a bacteria sample to a private lab, make sure that the lab is certified by the state for bacteria analysis. Consider state water quality labs, university and college labs, private labs, wastewater treatment plant labs, and hospitals. You might need to pay these labs for analysis. This chapter does not address laboratory methods because several bacteria types are commonly monitored, and the methods are different for each type. For more information on laboratory methods, refer to *Standard Methods*. Note: If you decide to analyze your samples in your own lab, be sure to carry out a quality assurance/quality control program.

APPARENT COLOR

> Some aspects of water quality can be judged by its color. Noticeable color is an objectionable characteristic that makes the water psychologically unacceptable to the consumer.
>
> **– De Zuane, 1997**

Pure water is colorless, but water in nature is often colored by foreign substances. As we've said, water with color is partly due to dissolved solids that remain after removal of suspended matter is known as true color. *Apparent color* (the topic of this section) results from dissolved substances and suspended matter and provides useful information about the water's source and content. Simply stated, when turbidity is present, so is apparent color. Natural metallic ions, plankton, algae, industrial pollution, plant pigments from humus and peat may all produce color in water. Pure water absorbs different wavelengths (colors) of light at different rates. Blue and blue-green light are the wavelengths which are best transmitted through water, so a white surface under "colorless" water looks blue (e.g., Caribbean and some South Pacific Island waters above white sand). Over the years, several attempts to standardize the method of describing the "apparent" color of water using comparisons to color standards have been made. *Standard Methods* recognizes the visual comparison method as a reliable method of analyzing water from the distribution system. One of the visual comparison methods is the Forel-Ule Color Scale, consisting of a dozen shades ranging from deep blue to khaki green, typical of offshore and coastal bay waters. By using established color standards, people in different areas can compare test results.

Another visual comparison method is the Borger Color System, which provides an inexpensive, portable color reference for shades typically found in natural waters and can also be used for its original purpose: describing the colors of insects and larvae found in streams or lakes. The Borger Color System also allows the recording of the color of algae and bacteria on stream beds. Note: Do not leave color standard charts and comparators in direct sunlight. Measured levels of color in water can serve as indicators for a number of conditions. For example, transparent water with a low accumulation of dissolved minerals and particulate matter usually appears blue and indicates low productivity. Yellow to brown color normally indicates that the water contains dissolved organic materials, humic substances from soil, peat, or decaying plant material water. Deeper yellow to reddish colors in water indicate some algae and dinoflagellates in the water. Water rich in phytoplankton and other algae appears green. A variety of yellows, reds, browns, and grays are indicative of soil runoff. To ensure reliable and accurate descriptions of apparent color, use a system of color comparison that is reproducible and comparable to the systems used by other groups.

ODOR

Odor in water is caused by chemicals that may come from municipal and industrial waste discharges or natural sources such as decomposing vegetable matter or microbial activity. Odor affects the

TABLE 3.5
Types of Wastewater Odors

Nature of Odor	Description	Such as Odors of:
Aromatic	(Spicy) Camphor, Cloves, Lavender, Lemon	
Balsamic	(Flowery) Geranium, Violet, Vanilla	
Chemical	Industrial Wastes or Treatments	
	Chlorinous	Chlorine
	Hydrocarbon	Oil Refinery Wastes
	Medicinal	Phenol and Iodine
	Sulfur	Hydrogen Sulfide (Rotten Egg)
Disagreeable	Fishy	Dead Algae
	Pigpen	Algae
	Septic	Stale Sewage
Earthy	Damp Earth	
	Peaty	Peat
Grassy	Crushed Grass	
Musty	Decomposing Straw	
	Moldy	Damp Cellar
Vegetable	Root Vegetables	

Source: Adapted from Standard Methods (2017), 23rd ed.

acceptability of drinking water, the aesthetics of recreation water, and the taste of aquatic food-stuffs. A wide variety of smells can be accurately detected by the human nose, which is the best odor-detection and -testing device presently available. To measure odor, collect a sample in a large-mouthed jar. After waving off the air above the water sample with your hand, smell the sample. Use the list of odors provided in *Standard Methods* (a system of qualitative description that helps monitor, describe, and record detected odors; see Table 3.5) to describe the smells. Record all observations.

CHAPTER SUMMARY

All of the elements that comprise the standard practices associated with proper water monitoring provide drinking water practitioners with the technical and scientific data needed to determine the level and types of treatment needed to successfully condition the water obtained from surface and groundwater sources.

REVIEW QUESTIONS

1. Which of the following is a primary reason for ensuring the safety of drinking water?
   a. Reducing cost to customers
   b. Ridding an area of excess water
   c. Ensuring public health
   d. Political considerations

2. 99.999% of anything is equivalent to _____.
   a. 5-log
   b. 6-logM
   c. No log
   d. Excessive water quality

3. Which of the following are water quality factors?
   a. Fecal bacteria
   b. Odor
   c. Hardness
   d. All of the above

4. In chemical addition to water, what is most important?
   a. Using the correct additives
   b. Using the correct amount of additive
   c. Metering additives
   d. Using the best brand of chemical

5. Water pollutants include _____.
   a. Snow
   b. Rain
   c. Oil
   d. Grass seed

6. Which of the following is more likely to be found in groundwater?
   a. Minerals
   b. Salt
   c. *Giardia*
   d. Organic matter

7. Which of the following is the definition of a grab sample?
   a. Represents of flow
   b. A single sample of water collected at a particular time and place which represents the composition of the water at that time and place

    c. Represents dosage

    d. Composite sample

8. What establishes an MCL based on the presence and absence of total coliforms or *E. coli*?
    a. Cryptosporidium rule
    b. Coliform rule
    c. Sampling plan
    d. Sanitary survey

9. What causes water turbidity?
    a. Bicarbonate
    b. Hardness
    c. Suspended material
    d. Dissolved solids

10. Which water sampling parameter must be taken in the field?
    a. Salinity
    b. Fecal coliform
    c. Color
    d. pH and temperature

ANSWER KEY

1. Which of the following is a primary reason for ensuring the safety of drinking water?
    a. Reducing cost to customers
    b. Ridding an area of excess water
    c. **Ensuring public health**
    d. Political considerations

2. 99.999% of anything is equivalent to _____.
    a. **5-log**
    b. 6-log
    c. No log
    d. Excessive water quality

3. Which of the following are water quality factors?
    a. Fecal bacteria
    b. Odor
    c. Hardness
    d. **All of the above**

4. In chemical addition to water, what is most important?
    a. Using the correct additives
    b. **Using the correct amount of additive**
    c. Metering additives
    d. Using the best brand of chemical

5. Water pollutants include _____.
    a. Snow
    b. Rain
    c. **Oil**
    d. Grass seed

6. Which of the following is more likely to be found in groundwater?
    a. **Minerals**
    b. Salt

    c. *Giardia*

    d. Organic matter

7. Which of the following is the definition of a grab sample?

    a. Represents of flow

    b. **A single sample of water collected at a particular time and place which represents the composition of the water at that time and place**

    c. Represents dosage

    d. A composite sample

8. What establishes an MCL based on the presence and absence of total coliforms or *E. coli*?

    a. Cryptosporidium rule

    b. **Coliform rule**

    c. Sampling plan

    d. Sanitary survey

9. What causes water turbidity?

    a. Bicarbonate

    b. Hardness

    c. **Suspended material**

    d. Dissolved solids

10. Which water sampling parameter must be taken in the field?

    a. Salinity

    b. Fecal coliform

    c. Color

    d. **pH and temperature**

REFERENCES

AWWA. *Water Treatment* (2nd ed.). Denver, CO: American Water Works Association, 1995.

De Zuane, J. *Handbook of Drinking Water Quality*. New York: Wiley, 1997.

Hauser, B.A. *Drinking Water Chemistry*. Boca Raton, FL: CRC Press-Lewis Publishers, 2002.

Jacobson, C. *Water, Water Everywhere*. Loveland, CO: Hach CO, 1991.

Standard Methods for Examination of Water and Wastewater. Washington, DC: APHA (American Public Health Association), Water Environment Federation (WEF), and American Water Works Association (AWWA), 2017.

4 Water Hydraulics and Flow Measurement

2019 Photo Courtesy of San Diego County Water Authority.

Water is the best of all things.

–Pindar (c. 522–c. 438 B.C, Olympian Odes)

LEARNING OBJECTIVES

After studying this chapter, you should be able to:

- Describe basic hydraulics.
- Describe water hydraulics.
- Discuss Stevin's Law.
- Define density and specific gravity.
- Describe the properties of water.
- Discuss force and pressure.
- Describe the organic and inorganic contaminants that affect water's smell and taste.
- Discuss hydrostatic pressure.
- Describe head.
- Discuss pressure and head.
- Describe water in motion.
- Define the Law of Continuity.

DOI: 10.1201/9781003207146-6

- Discuss the pressure and velocity connection.
- Discuss Piezometric surface.
- Discuss Bernoulli's Theorem.
- Describe the Conservation of Energy.
- Discuss energy head.
- Discuss head loss.
- Identify hydraulic grade line.
- Discuss well hydraulics.
- Discuss wet well hydraulics.
- Identify friction head loss.
- Describe flow in pipelines.
- Define major head loss.
- Describe slope.
- Discuss minor head loss.
- Discuss basic piping hydraulics.
- Discuss flow measurement.
- Identify flow measuring devices.

KEY DEFINITIONS

Cone of Depression – in unconfined aquifers, there is a flow of water in the aquifer from all directions toward the well during pumping. The free water surface in the aquifer then takes the shape of an inverted cone or curved funnel line.

Density – mass per unit volume.

Discharge – the quantity of water passing a given point in a pipe or channel during a given period.

Drawdown – the difference, or the drop, between the static water level and the pumping water level.

Energy Head – measured in feet of water.

Flow – measured in either gallons (gal) or in cubic feet (ft^3).

Flow Nozzle – a tapered length of tube that causes a falloff pressure head in water flowing through it and from which the flow rate can be calculated.

Flow Rate – measured in gallons per minute (gpm), million gallons per day (MGD), or cubic feet per second (cfs).

Force – the push or pull influence that causes motion.

Friction Head – the equivalent distance of the energy that must be supplied to overcome friction.

Head – the vertical distance the water must be lifted from the supply tank to the discharge.

Hydraulics – concerned with the conveyance of liquid through pipes, especially as a source of mechanical force or control.

Hydrostatic Pressure – pressure exerted by a fluid at equilibrium at a given point within the fluid, due to the force of gravity.

Kinetic Energy – is present when water is in motion.

Law of Continuity – states that the discharge at each point in a pipe or channel is the same as the discharge at any point.

Law of Conservation of Energy – states that energy can neither be created nor destroyed, but it can be converted from one to another.

Major Head Loss – consists of pressure decreases along the length of pipe caused by friction created as water encounters the surfaces of the pipe.

Orifice – in water operations, it is an opening through which water may pass.

Piezometric Surface – an imaginary surface that coincides with the level of the water to which the water in the system would rise in a piezometer.

Positive-Displacement Flow Meter – a hydraulic motor with high volumetric efficiency that absorbs a small amount of energy from the flowing stream.

Potential Energy – results due to water pressure.

Pressure – a force per unit of area.

Pumping Water Level – the water level when the pump is off. When water is pumped out of a well, the water level usually drops below the level in the surrounding aquifer and eventually stabilizes at an equal volume of water – the specific gravity of water is 1.

Slope – head loss per foot.

Specific Capacity – the pumping rate per foot of drawdown (gpm/ft). Specific capacity is one of the most important concepts in well operation and testing.

Specific Gravity – the weight (or density) of a substance compared to the weight (or density) of a reference substance.

Standard Temperature and Pressure (STP) – defined as a temperature of 273 15K (0°C, 32°F) and an absolute pressure of 105 Pa (100 kPa, 1 bar).

Static Head – the actual vertical distance water must be raised.

Static Water Level – the water level in a well when no water is being taken from the groundwater source (i.e., the water level when the pump is switched off). Static water level is normally measured as the distance from the ground surface to the water surface. This is an important parameter because it is used to measure changes in the water table.

Stevin's Law – states pressure at any point in a fluid at rest depends on the distance measured vertically to the free surface and the density of the fluid.

Thrust – the force which water exerts on a pipeline as it rounds a bend.

Velocity Head – the equivalent distance of energy consumed in achieving and maintaining the desired velocity in the system.

Venturi – a short tube with a tapering constriction in the middle that causes an increase in the velocity of the flow of water.

Water Hammer – momentary increase in pressure that occurs when there is a sudden change of direction or velocity of the water.

Well Yield – the rate of water withdrawal that a well can supply over a long period, or, alternatively, the maximum pumping rate that can be achieved without increasing the drawdown.

Zone (or radius) of Influence – the distance between the pump shaft and the outermost area affected by drawdown. The distance depends on the porosity of the soil and other factors. This parameter becomes important in well fields with many pumps. If wells are set too close together, the zones of influence will overlap, increasing the drawdown in all wells. Obviously, pumps should be spaced apart to prevent this from happening.

BASIC WATER HYDRAULICS

Beginning students of water hydraulics and its principles often come to the subject matter with certain misgivings. For example, water operators quickly learn on the job that their primary operational/ maintenance concerns involve a daily routine of monitoring, sampling, laboratory testing, operation, and process maintenance. How does water hydraulics relate to daily operations? The hydraulic functions of the treatment process have already been designed into the plant. Why learn water hydraulics at all? Simply put, while having hydraulic control of the plant is obviously essential to the treatment process, maintaining and ensuring continued hydraulic control is also essential. No water treatment facility (and/or distribution collection system) can operate without proper hydraulic control. The operator must know what hydraulic control is and what it entails in order to know how to ensure proper hydraulic control. Moreover, in order to understand the basics of piping and pumping systems, water maintenance operators must have a fundamental knowledge of basic water hydraulics.

How about water administrators and upper- and mid-level water managers? Do they need to know what the operators know? Maybe. Maybe not. Let's put it this way, water administrators and upper- and mid-level water managers must know a lot. What is a lot? Well, a lot is a holistic view. What is that? What does that mean? For example, water administrators and mid-level water managers do not need to know the micro-minutia of how to overhaul a centrifugal pump, or a positive-displacement pump, or how to repair a chemical feeder, or weld or cement various piping accessories together. They do not need to know the exact valve to open and close in a valve train to properly operate a water system. In the holistic approach to waterworks operations, water administrators and upper- and mid-level managers need to know the "big picture" – they also need enthusiasm, energy, courage, personal maturity, and optimism. They also need to be well grounded in ethics and know how to handle stress. Is that all they need? No. What else? Good question. The what else can be summed up in one word: Credibility, Credibility, Credibility! The waterworks administrator, upper or mid-level managers must possess two types of credibility:

- *Technical Credibility* – Perceived by anyone they interface with as possessing the technical knowledge necessary to direct people and the facility.
- *Administrative Credibility* – Keeping operations on schedule and within costs and regulatory requirements. They must have the ability to ensure that the facility and operators have the tools and equipment they need to maintain the waterworks in optimum operational condition.

Okay, but to obtain, possess, maintain, and demonstrate credibility, what exactly do the administrator and upper- and mid-level managers need to know about water hydraulics? Another good question. Administrators and upper- and mid-level managers need to know water hydraulics nomenclature; they need to speak "waterworks language" related to water hydraulics; they need to be credible when dealing with water operators, customers, and regulators and others. Again, these upper-level personnel do not need to know how to overhaul a centrifugal pump, or a valve, or the exact meaning of Stevin's Law. They do need to know, however, basic information about all these and several other aspects of water hydraulics and hydraulic machines (pumps). And with regard to flow measurement, administrators and managers certainly need to know where the water is, where the water is going, what it costs to treat the water, and keep track of what flows into the plant and what flows out of the plant to the customers. Do the customers receive the water they need when they need it?

This chapter, like all the others, is designed to provide basic information to enhance the administrators' and upper- and mid-level managers' credibility, which in turn helps to make these individuals functional and definitely not dysfunctional (hopefully). Note: The practice and study of water hydraulics is not new. Even in medieval times, water hydraulics was not new. "Medieval Europe had inherited a highly developed range of Roman hydraulic components" (Magnusson, 2001). The basic conveyance technology, based on low-pressure systems of pipe and channels, was already established. In studying "modern" water hydraulics, it is important to remember that the science of water hydraulics is the direct result of two immediate and enduring problems: "The acquisition of freshwater and access to continuous strips of land with a suitable gradient between the source and the destination" (Magnusson, 2001).

WHAT IS WATER HYDRAULICS?

The word "hydraulic" is derived from the Greek words "hydro" (meaning water) and "aulis" (meaning pipe). Originally, the term "hydraulics" referred only to the study of water at rest and in motion (flow of water in pipes or channels). Today it is taken to mean the flow of any "liquid" in a system. What is a liquid? In terms of hydraulics, a liquid can be either oil or water. In fluid power systems used in modern industrial equipment, the hydraulic liquid of choice is oil. Some common examples of hydraulic fluid power systems include automobile braking and power steering systems, hydraulic elevators, and hydraulic jacks or lifts. Probably the most familiar hydraulic fluid power systems in water operations are used on dump trucks, front-end loaders, graders, and earth-moving and excavations equipment. In this chapter, we are concerned with liquid water. Many find the study of water hydraulics difficult and puzzling (especially the licensure examination questions), but we know it is not mysterious or difficult. It is the function or output of practical applications of the basic principles of water physics. Because water treatment is based on the principles of water hydraulics, concise, real-world training is necessary for operators who must operate the plant and for those sitting for state licensure/certification examinations.

BASIC CONCEPTS

$$\text{Air Pressure (@ Sea Level)} = 14.7 \text{ pounds per square inch (psi)}$$

This relationship shown above is important because our study of hydraulics begins with air. A blanket of air, many miles thick surrounds the earth. The weight of this blanket on a given square inch of the earth's surface will vary according to the thickness of the atmospheric blanket above that point. As shown above, at sea level, the pressure exerted is 14.7 pounds per square inch (psi). On a mountaintop, air pressure decreases because the blanket is not as thick.

$$1 \text{ ft}^3 \text{ H}_2\text{O} = 62.4 \text{ lb}$$

The relationship shown above is also important: both cubic feet and pounds are used to describe a volume of water. There is a defined relationship between these two methods of measurement. The specific weight of water is defined relative to a cubic foot. One cubic foot of water weighs 62.4 pounds (see Figure 4.1). This relationship is true only at a temperature of 4°C and at a pressure of one atmosphere, conditions referred to as *standard temperature and pressure (STP)*. Note that 1 atmosphere = 14.7 lb/in² at sea level and 1 ft³ of water contains 7.48 gal.

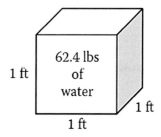

FIGURE 4.1 Hydrostatic pressure.

The weight varies so little that, for practical purposes, this weight is used for temperatures ranging from 0 to 100°C. One cubic inch of water weighs 0.0362 lb. Water 1 foot deep will exert a pressure of 0.43 lb/in² on the bottom area (12 in×0.0362 lb/in³). A column of water 2 feet high exerts 0.86 psi (2 ft×0.43 psi/ft); and one 55 feet high exerts 23.65 psi (55 ft×0.43 psi/ft). A column of water 2.31 feet high will exert 1.0 psi (2.31 ft×0.43 psi/ft). To produce a pressure of 50 psi requires a 115.5-ft water column (50 psi×2.31 ft/psi). Remember the important points being made here:

1 ft³ H₂O=62.4 lb (see Figure 4.1)
A column of water 2.31 ft high will exert 1.0 psi
Another relationship is also important:
1 gal H₂O=8.34 lb

At STP, 1 ft³ of water contains 7.48 gal. With these two relationships, we can determine the weight of one gallon of water. This is accomplished by:

Weight of gal of water=62.4 lb ÷ 7.48 gal=8.34 lb/gal
thus,
1 gal H₂O=8.34 lb

Note: Further, this information allows cubic feet to be converted to gallons by simply multiplying the number of cubic feet by 7.48 gal/ft³. Note: The term head is used to designate water pressure in terms of the height of a column of water in feet. For example, a 10-foot column of water exerts 4.3 psi. This can be called 4.3 psi pressure or 10 feet of head.

Stevin's Law

Stevin's Law deals with water at rest. Specifically, Stevin's Law states: "The pressure at any point in a fluid at rest depends on the distance measured vertically to the free surface and the density of the fluid."

Density and Specific Gravity

Table 4.1 shows the relationship between temperature, specific weight, and density of water. When we say that iron is heavier than aluminum, we say that iron has greater density than aluminum. In

TABLE 4.1

Water Properties (Temperature, Specific Weight, and Density)

Temperature (°F)	Specific Weight (lb/ft³)	Density (slugs/ft³)	Temperature (°F)	Specific Weight (lb/ft³)	Density (slugs/ft³)
32	62.4	1.94	130	61.5	1.91
40	62.4	1.94	140	61.4	1.91
50	62.4	1.94	150	61.2	1.90
60	62.4	1.94	160	61.0	1.90
70	62.3	1.94	170	60.8	1.89
80	62.2	1.93	180	60.6	1.88
90	62.1	1.93	190	60.4	1.88
100	62.0	1.93	200	60.1	1.87
110	61.9	1.92	210	59.8	1.86
120	61.7	1.92			

practice, what we are really saying is that a given volume of iron is heavier than the same volume of aluminum.

Note: What is density? *Density* is the *mass per unit volume* of a substance. Suppose you had a tub of lard and a large box of cold cereal, each having a mass of 600 g. The density of the cereal would be much less than the density of the lard because the cereal occupies a much larger volume than the lard occupies. In water treatment, perhaps the most common measures of density are pounds per cubic foot (lb/ft^3) and pounds per gallon (lb/gal):

$$1 \text{ cu ft of water weighs } 62.4 \text{ lb} - \text{Density} = 62.4 \text{ lb/cu/ft}$$

$$\text{One gallon of water weighs } 8.34 \text{ lb} - \text{Density} = 8.34 \text{ lb/gal}$$

The density of dry material, such as cereal, lime, soda, and sand, is usually expressed in pounds per cubic foot. The density of a liquid, such as liquid alum, liquid chlorine, or water, can be expressed either as pounds per cubic foot or as pounds per gallon. The density of a gas, such as chlorine gas, methane, carbon dioxide, or air, is usually expressed in pounds per cubic foot.

As shown in Table 4.1, the density of a substance like water changes slightly as the temperature of the substance changes. This occurs because substances usually increase in volume (size they expand) as they become warmer. Because of this expansion with warming, the same weight is spread over a larger volume, so the density is lower when a substance is warm than when it is cold. Note: What is specific gravity? Specific gravity is the weight (or density) of a substance compared to the weight (or density) of an equal volume of water. The specific gravity of water is 1.

This relationship is easily seen when a cubic foot of water, which weighs 62.4 lb, is compared to a cubic foot of aluminum, which weighs 178 lb. Aluminum is 2.8 times heavier than water. It is not that difficult to find the specific gravity of a piece of metal. All you have to do is to weigh the metal in the air, then weigh it under water. Its loss of weight is the weight of an equal volume of water. Note: In a calculation of specific gravity, it is essential that the densities be expressed in the same units.

Again, the specific gravity of water is 1, which is the standard, the reference by which all other liquid or solid substances are compared. Specifically, any object that has a specific gravity greater than one will sink in water (rocks, steel, iron, grit, floc, sludge). Substances with a specific gravity of less than 1 will float (wood, scum, and gasoline). Considering the total weight and volume of a ship, its specific gravity is less than 1; therefore, it can float.

The most common use of specific gravity in water treatment operations is in gallons-to-pounds conversions. In many cases, the liquids being handled have a specific gravity of 1.00 or very nearly 1.00 (between 0.98 and 1.02), so 1.00 may be used in the calculations without introducing significant error. However, in calculations involving a liquid with a specific gravity of less than 0.98 or greater than 1.02, the conversions from gallons to pounds must consider specific gravity.

Force and Pressure

Water exerts force and pressure against the walls of its container, whether it is stored in a tank or flowing in a pipeline. Force and pressure are different, although they are closely related. *Force* is the push or pull influence that causes motion. In the English system, force and weight are often used in the same way. The weight of a cubic foot of water is 62.4 lb. The force exerted on the bottom of a one-foot cube is 62.4 lb (see Figure 4.1). If we stack two cubes on top of one another, the force on the bottom will be 124.8 pounds. *Pressure* is a force per unit of area.

Earlier we pointed out that pounds per square inch (lb/in^2 or psi) or pounds per square foot (lb/ft^2) are common expressions of pressure. The pressure on the bottom of the cube is 62.4 lb/ft^2 (see Figure 4.1). It is normal to express pressure in pounds per square inch (psi). This is easily accomplished by determining the weight of 1 in^2 of 1-ft cube. If we have a cube that is 12 inches on each side, the number of square inches on the bottom surface of the cube is $12 \times 12 = 144$ in^2. Dividing the weight by the number of square inches determines the weight on each square inch.

$$\text{psi} = \frac{62.4 \text{ lb/ft}}{144 \text{ in}^2} = 0.433 \text{ psi/ft}$$

This is the weight of a column of water 1 in^2 and 1 foot tall. If the column of water were 2 feet tall, the pressure would be 2 ft$\times$0.433 psi/ft=0.866. Note: 1 foot of water=0.433 psi. With the above information, feet of the head can be converted to psi by multiplying the feet of head times 0.433 psi/ft. Important Point: One of the problems encountered in a hydraulic system is storing the liquid. Unlike air, which is readily compressible and is capable of being stored in large quantities in relatively small containers, a liquid such as water cannot be compressed. Therefore, it is not possible to store a large amount of water in a small tank, as 62.4 lb of water occupies a volume of 1 ft^3, regardless of the pressure applied to it.

Hydrostatic Pressure

Figure 4.2 shows a number of differently shaped, connected, open containers of water. Note that the water level is the same in each container, regardless of the shape or size of the container. This occurs because pressure is developed, within water (or any other liquid), by the weight of the water above. If the water level in any one container were to be momentarily higher than that in any of the other containers, the higher pressure at the bottom of this container would cause some water to flow into the container having the lower liquid level.

In addition, the pressure of the water at any level (such as Line T) is the same in each of the containers. Pressure increases because of the weight of the water. The farther down from the surface, the more pressure is created. This illustrates that the weight, not the volume, of water contained in a vessel determines the pressure at the bottom of the vessel. Nathanson (1997) pointed out some very important principles that always apply for hydrostatic pressure:

- The pressure depends only on the depth of water above the point in question (not on the water surface area).
- The pressure increases in direct proportion to the depth.
- The pressure in a continuous volume of water is the same at all points that are at the same depth.
- The pressure at any point in the water acts in all directions at the same depth.

Effects of Water under Pressure

Hauser (1993) points out that water under pressure and in motion can exert tremendous forces inside a pipeline. One of these forces, called hydraulic shock or *water hammer*, is the momentary increase in pressure that occurs when there is a sudden change of direction or velocity of the water. When

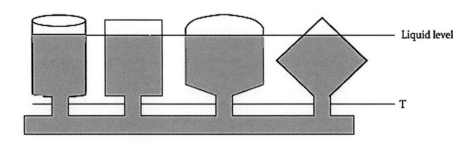

FIGURE 4.2   Hydrostatic pressure.

a rapidly closing valve suddenly stops water flowing in a pipeline, pressure energy is transferred to the valve and pipe wall. Shock waves are set up within the system. Waves of pressure move in horizontal yo-yo fashion – back and forth – against any solid obstacles in the system. Neither the water nor the pipe will compress to absorb the shock, which may result in damage to pipes, valves, and shaking of loose fittings.

Another effect of water under pressure is called thrust. *Thrust* is the force which water exerts on a pipeline as it rounds a bend. As shown in Figure 4.3, thrust usually acts perpendicular (at 90°) to the inside surface it pushes against. It affects not only bends but also reducers, dead ends, and tees. Uncontrolled, the thrust can cause movement in the fitting or pipeline, which will lead to separation of the pipe coupling away from both sections of the pipeline, or at some other nearby coupling upstream or downstream of the fitting.

Two types of devices commonly used to control thrust in larger pipelines are thrust blocks and thrust anchors. A *thrust block* is a mass of concrete cast in place onto the pipe and around the outside bend of the turn. An example is shown in Figure 4.4. These are used for pipes with tees or elbows that turn left or right or slant upward. The thrust is transferred to the soil through the larger bearing surface of the block. A *thrust anchor* is a massive block of concrete, often a cube, cast in place below the fitting to be anchored. As shown in Figure 4.5, embedded steel shackle rods anchor the fitting to the concrete block, effectively resisting upward thrusts. The size and shape of a thrust control device depend on pipe size, type of fitting, water pressure, water hammer, and soil type.

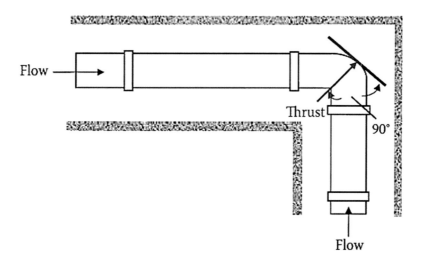

FIGURE 4.3 Shows direction of thrust in a pipe in a trench (viewed from above).

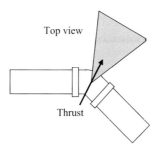

FIGURE 4.4 Thrust block.

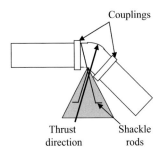

FIGURE 4.5　Thrust anchor.

Head

Head is defined as the vertical distance the water must be lifted from the supply tank to the discharge, or as the height a column of water would rise due to the pressure at its base. A perfect vacuum plus atmospheric pressure of 14.7 psi would lift the water 34 feet. When the top of the sealed tube is opened to the atmosphere and the reservoir is enclosed, the pressure in the reservoir is increased; the water will rise in the tube.

Because atmospheric pressure is essentially universal, we usually ignore the first 14.7 psi of actual pressure measurements and measure only the difference between the water pressure and the atmospheric pressure; we call this *gauge pressure*. For example, water in an open reservoir is subjected to the 14.7 psi of atmospheric pressure; subtracting this 14.7 psi leaves a gauge pressure of 0 psi, indicating that the water would rise 0 feet above the reservoir surface. If the gauge pressure in a water main were 120 psi, the water would rise in a tube connected to the main:

$$120 \text{ psi} \times 2.31 \text{ ft/psi} = 277 \text{ ft (rounded)}$$

The *total head* includes the vertical distance the liquid must be lifted (static head), the loss to friction (friction head), and the energy required to maintain the desired velocity (velocity head).

- *Static head* is the actual vertical distance the liquid must be lifted.
- *Friction head* is the equivalent distance of the energy that must be supplied to overcome friction. Engineering references include tables showing the equivalent vertical distance for various sizes and types of pipes, fittings, and valves. The total friction head is the sum of the equivalent vertical distances for each component.
- *Velocity head* is the equivalent distance of the energy consumed in achieving and maintaining the desired velocity in the system.
- Total Dynamic Head (Total System Head).

$$\text{Total head} = \text{Static head} + \text{friction head} + \text{velocity head}$$

Pressure and Head

The pressure exerted by water is directly proportional to its depth or head in the pipe, tank, or channel. If the pressure is known, the equivalent head can be calculated.

FLOW AND DISCHARGE RATES: WATER IN MOTION

The study of fluid flow is much more complicated than that of fluids at rest, but it is important to have an understanding of these principles because the water in a waterworks and distribution system is nearly always in motion. Discharge (or flow) is the quantity of water passing a given point in a pipe or channel during a given period. Stated another way for open channels, the flow rate through an open channel is directly related to the velocity of the liquid and the cross-sectional area of the liquid in the channel.

$$Q = A \times V$$

where
    Q = Flow – discharge in cubic feet per second (cfs)
    A = Cross-sectional area of the pipe or channel (ft^2)
    V = Water velocity in feet per second (fps or ft/sec)

Discharge or flow can be recorded as gallons/day (gpd), gallons/minute (gpm), or cubic feet per second (cfs). Flows treated by many waterworks are large, and often referred to in million gallons per day (MGD). The discharge or flow rate can be converted from cubic feet per second (cfs) to other units such as gpm or MGD by using appropriate conversion factors. Note: Flow may be *laminar* (streamline, see Figure 4.6) or *turbulent* (see Figure 4.7).

Laminar flow occurs at extremely low velocities. The water moves in straight parallel lines called streamlines or laminae, which slide upon each other as they travel, rather than mixing up. Normal pipe flow is turbulent flow, which occurs because of friction encountered on the inside of the pipe. The outside layers of flow are thrown into the inner layers; the result is that all the layers mix and are moving in different directions and at different velocities. However, the direction of flow is forward. Note: Flow may be steady or unsteady. For our purposes, we consider steady-state flow only.

Area and Velocity

The *law of continuity* states that the discharge at each point in a pipe or channel is the same as the discharge at any other point (if water does not leave or enter the pipe or channel). That is, under the assumption of steady-state flow, the flow that enters the pipe or channel is the same flow that exits the pipe or channel.

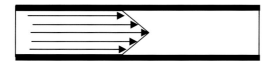

FIGURE 4.6 Laminar (streamline) flow.

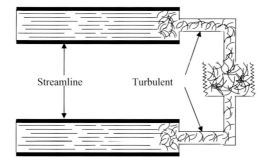

FIGURE 4.7 Turbulent flow.

Pressure and Velocity

In a closed pipe flowing full (under pressure), the pressure is indirectly related to the velocity of the liquid. This principle, when combined with the law of continuity in the previous section, forms the basis for several flow measurement devices (venturi meters and rotameters) as well as the injector used for dissolving chlorine into water.

PIEZOMETRIC SURFACE AND BERNOULLI'S THEOREM

> They will take your hand and lead you to the pearls of the desert, those secret wells swallowed by oyster crags of wadi, underground caverns that bubble rusty salt water you would sell your own mothers to drink.
>
> **– Holman, 1998**

To keep the systems in your plant operating properly and efficiently, you must understand the basics of hydraulics, the laws of force, motion, and others. As stated previously, most applications of hydraulics in water treatment systems involve water in motion in pipes under pressure or in open channels under the force of gravity. The volume of water flowing past any given point in the pipe or channel per unit time is called the flow rate or discharge rate or just flow. The continuity of flow has been discussed. Along with the continuity of flow principle and continuity equation, the law of conservation of energy, piezometric surface, and Bernoulli's theorem (or principle) are also important to our study of water hydraulics.

Conservation of Energy

Many of the principles of physics are important to the study of hydraulics. When applied to problems involving the flow of water, few of the principles of physical science are more important and useful to us than the Law of Conservation of Energy. Simply, the Law of Conservation of Energy states that energy can neither be created nor destroyed, but it can be converted from one form to another. In a given closed system, the total energy is constant.

Energy Head

In hydraulic systems, two types of energy (kinetic and potential) and three forms of mechanical energy (potential energy due to elevation, potential energy due to pressure, and kinetic energy due to velocity) exist. Energy has the units of foot-pounds (ft-lbs). It is convenient to express hydraulic energy in terms of an energy head, in feet of water. This is equivalent to foot-pounds per pound of water (ft-lb/lb = ft).

Piezometric Surface

As mentioned earlier, we have seen that when a vertical tube, open at the top, is installed onto a vessel of water, the water will rise in the tube to the water level in the tank. The water level to which the water rises in a tube is the *piezometric surface*. That is, the piezometric surface is an imaginary surface that coincides with the level of the water to which water in a system would rise in a piezometer (an instrument used to measure pressure). The surface of water that is in contact with the atmosphere is known as the *free water surface*. Many important hydraulic measurements are based on the difference in height between the free water surface and some point in the water system. The piezometric surface is used to locate this free water surface in a vessel, where it cannot be observed directly.

To understand how a piezometer actually measures pressure, consider the following example. If a clear, see-through pipe is connected to the side of a clear glass or plastic vessel, the water will rise in the pipe to indicate the level of the water in the vessel. Such a see-through pipe, the piezometer, allows you to see the level of the top of the water in the pipe; this is the piezometric surface. In practice, a piezometer is connected to the side of a tank or pipeline. If the water-containing vessel is not under pressure (as is the case in Figure 4.8), the piezometric surface will be the same as the free water surface in the vessel, just as it would if a drinking straw (the piezometer) were left standing in a glass of water. When a tank and pipeline system is pressurized, as they often are, the pressure will cause the piezometric surface to rise above the level of the water in the tank. The greater the pressure, the higher the piezometric surface (see Figure 4.9). An increased pressure in a water pipeline system is usually obtained by elevating the water tank. Note: In practice, piezometers are not installed on water towers because water towers are hundreds of feet high, or on pipelines. Instead, pressure gauges are used that record pressure in feet of water or in psi. Water only rises to the water level of the main body of water when it is at rest (static or standing water). The situation is quite different when water is flowing. Consider, for example, an elevated storage tank feeding a distribution system pipeline. When the system is at rest, with all valves closed, all the piezometric surfaces are the same height as the free water surface in storage. On the other hand, when the valves are opened, and the water begins to flow, the piezometric surface changes. This is an important point because as water continues to flow down a pipeline, less and less pressure is exerted. This happens because some pressure is lost (used up) keeping the water moving over the interior surface of the pipe (friction). The pressure that is lost is called head loss.

Head Loss

Head loss is best explained by example. Figure 4.10 shows an elevated storage tank feeding a distribution system pipeline. When the valve is closed (Figure 4.10-A), all the piezometric surfaces

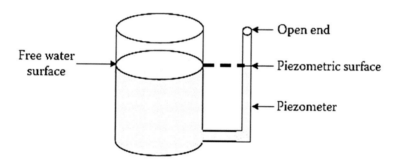

FIGURE 4.8   Container not under pressure where the piezometric surface is the same as the free water surface in the vessel.

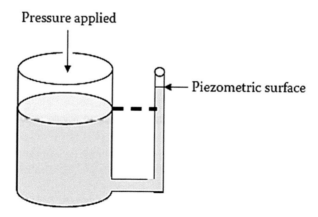

FIGURE 4.9   Container under pressure where the piezometric surface is above the level of the water in the tank.

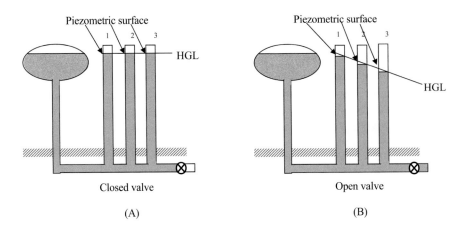

FIGURE 4.10   A&B shows head loss and/or piezometric surface changes when water is flowing.

are the same height as the free water surface in storage. When the valve opens, and water begins to flow (Figure 4.10-B), the piezometric surfaces *drop*. The farther along the pipeline, the lower the piezometric surface, because some of the pressure is used up keeping the water moving over the rough interior surface of the pipe. Thus, pressure is lost and is no longer available to push water up in a piezometer; this is the head loss.

Hydraulic Grade Line (HGL)

When the valve shown in Figure 4.10 is opened, flow begins with a corresponding energy loss due to friction. The pressures along the pipeline can measure this loss. In Figure 4.10-B, the difference in pressure heads between sections 1, 2, and 3 can be seen in the piezometer tubes attached to the pipe. A line connecting the water surface in the tank with the water levels at sections 1, 2, and 3 shows the pattern of continuous pressure loss along the pipeline. This is called the *HGL* or *Hydraulic Gradient* of the system.

Note: It is important to point out that in a static water system, the HGL is always horizontal. The HGL is a very useful graphical aid when analyzing pipe flow problems. Note: During the early design phase of a treatment plant, it is important to establish the HGL across the plant because both the proper selection of the plant site elevation and the suitability of the site depend on this consideration. Typically, most conventional water treatment plants required 16–17 feet of head loss across the plant. Key Point: Changes in the piezometric surface occur when water is flowing.

Bernoulli's Theorem

Nathanson (1997) noted that Swiss physicist and mathematician Samuel Bernoulli developed the calculation for the total energy relationship from point to point in a steady-state fluid system in the 1700s. Before discussing Bernoulli's energy equation, it is important to understand the basic principle behind Bernoulli's equation. Water (and any other hydraulic fluid) in a hydraulic system possesses two types of energy – kinetic and potential. *Kinetic energy* is present when the water is in motion. The faster the water moves; the more kinetic energy is used. *Potential energy* is a result of the water pressure. The *total energy* of the water is the sum of the kinetic and potential energy.

Bernoulli's principle states that the total energy of the water (fluid) always remains constant. Therefore, when the water flow in a system increases, the pressure must decrease. When water starts to flow in a hydraulic system, the pressure drops. When the flow stops, the pressure rises again. The pressure gauges shown in Figure 4.11 indicate this balance more clearly. Note: This discussion of Bernoulli's equation ignores friction losses from point to point in a fluid system employing steady-state flow.

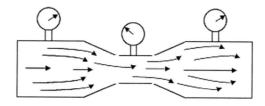

FIGURE 4.11 Demonstrates Bernoulli's principle.

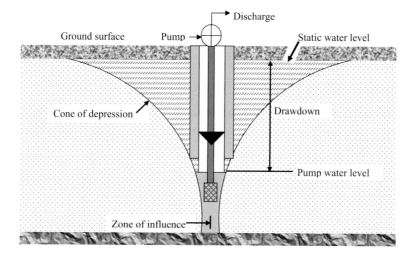

FIGURE 4.12 Hydraulic characteristics of a well.

WELL AND WET WELL HYDRAULICS

When the source of water for a water distribution system is from a groundwater supply, it is important that all involved with the public water system have a basic understanding of well hydraulics. Basic well hydraulics terms are presented and defined, and they are related pictorially (see Figure 4.12). Also discussed are wet wells, which are important, in water operations.

WELL HYDRAULICS

- *Static Water Level* – The water level in a well when no water is being taken from the groundwater source (i.e., the water level when the pump is off; see Figure 4.12). Static water level is normally measured as the distance from the ground surface to the water surface. This is an important parameter because it is used to measure changes in the water table.
- *Pumping Water Level* – The water level when the pump is off. When water is pumped out of a well, the water level usually drops below the level in the surrounding aquifer and eventually stabilizes at a lower level; this is the pumping level (see Figure 4.12).
- *Drawdown* – The difference, or the drop, between the static water level and the pumping water level, measured in feet. Simply, it is the distance the water level drops once pumping begins (see Figure 4.12).
- *Cone of Depression* – In unconfined aquifers, there is a flow of water in the aquifer from all directions toward the well during pumping. The free water surface in the aquifer then takes the shape of an inverted cone or curved funnel line. The curve of the line extends from the Pumping Water Level to the Static Water Level at the outside edge of the Zone (or

Radius) of Influence (see Figure 4.12). Note: The shape and size of the cone of depression are dependent on the relationship between the pumping rate and the rate at which water can move toward the well. If the rate is high, the cone will be shallow, and its growth will stabilize. If the rate is low, the cone will be sharp and continue to grow.

- *Zone (or Radius) of Influence* – The distance between the pump shaft and the outermost area affected by drawdown (see Figure 4.12). The distance depends on the porosity of the soil and other factors. This parameter becomes important in well fields with many pumps. If wells are set too close together, the zones of influence will overlap, increasing the drawdown in all wells. Obviously, pumps should be spaced apart to prevent this from happening.

Two important parameters not shown in Figure 4.12 are well yield and specific capacity.

- *Well yield* is the rate of water withdrawal that a well can supply over a long period, or, alternatively, the maximum pumping rate that can be achieved without increasing the drawdown. The yield of small wells is usually measured in gallons per minute (liters per minute) or gallons per hour (liters per hour). For large wells, it may be measured in cubic feet per second (cubic meters per second).
- *Specific capacity* is the pumping rate per foot of drawdown (gpm/ft). Specific capacity is one of the most important concepts in well operation and testing. The calculation should be made frequently in the monitoring of well operation. A sudden drop in specific capacity indicates problems such as pump malfunction, screen plugging, or other problems that can be serious. Such problems should be identified and corrected as soon as possible.

Wet Well Hydraulics

Water pumped from a wet well by a pump set above the water surface exhibits the same phenomena as the groundwater well. In operation, a slight depression of the water surface forms right at the intake line (drawdown), but in this case it is minimal because there is always free water at the pump entrance (at least there should be). The most important consideration in wet well operations is to ensure that the suction line is submerged far enough below the surface so that air entrained by the active movement of the water at this section is not able to enter the pump.

Because water flow is not always constant or at the same level, variable speed pumps are commonly used in wet well operations, or several pumps are installed for single or combined operation. In many cases, pumping is accomplished in an on/off mode. Control of pump operation is in response to the water level in the well. Level control devices such as mercury switches are used to sense a high and low level in the well and transmit the signal to pumps for action.

Friction Head Loss

Materials or substances capable of flowing cannot flow freely. Nothing flows without encountering some type of resistance. Consider electricity: the flow of free electrons in a conductor. Whatever type of conductor used (i.e., copper, aluminum, silver, etc.) offers some resistance. In hydraulics, the flow of water is analogous to the flow of electricity. Within a pipe or open channel, for instance, flowing water, like electron flow in a conductor, encounters resistance. However, resistance to the flow of water is generally termed friction loss (or more appropriately, head loss).

Flow in Pipelines

The problem of water flow in pipelines – the prediction of flow rate through pipes of given characteristics, the calculation of energy conversions therein, and so forth – is encountered in many applications of water operations and practice. The subject of pipe flow embraces only those problems in which pipes flow completely full (as in water lines). Resistance to flow in pipes is not only the result of long reaches of pipe but also due to pipe fittings, such as bends and valves, which dissipate energy by producing relatively large-scale turbulence.

Major Head Loss

Major head loss consists of pressure decreases along the length of pipe caused by friction created as water encounters the surfaces of the pipe. It typically accounts for most of the pressure drop in a pressurized or dynamic water system. The components that contribute to major head loss: roughness, length, diameter, and velocity:

- *Roughness* – Even when new, the interior surfaces of pipes are rough. The roughness varies, of course, depending on pipe material, corrosion (tuberculation and pitting), and age. Because normal flow in a water pipe is turbulent, the turbulence increases with pipe roughness, which, in turn, causes pressure to drop over the length of the pipe.
- *Pipe Length* – With every foot of pipe length, friction losses occur. The longer the pipe, the more head loss. Friction loss because of pipe length must be factored into head loss calculations.
- *Pipe Diameter* – Generally, small diameter pipes have more head loss than large diameter pipes. This is the case because in large diameter pipes less of the water actually touches the interior surfaces of the pipe (encountering less friction) than in a small diameter pipe.
- *Water Velocity* – Turbulence in a water pipe is directly proportional to the speed (or velocity) of the flow. Thus, the velocity head also contributes to head loss.

Note: For the same diameter pipe, when flow increases, head loss increases.

Slope

Slope is defined as the head loss per foot. In open channels, where the water flows by gravity, slope is the amount of incline of the pipe and is calculated as feet of drop per foot of pipe length (ft/ft). Slope is designed to be just enough to overcome frictional losses so that the velocity remains constant, the water keeps flowing, and solids will not settle in the conduit. In piped systems, where pressure loss for every foot of pipe is experienced, slope is not provided by slanting the pipe but instead by pressure added to overcome friction.

Minor Head Loss

In addition to the head loss caused by friction between the fluid and the pipe wall, losses also are caused by turbulence created by obstructions (i.e., valves and fittings of all types) in the line, changes in direction, and changes in flow area. Note: In practice, if the minor head loss is less than 5% of the total head loss, it is usually ignored.

BASIC PIPING HYDRAULICS

Water, regardless of the source, is conveyed to the waterworks for treatment and distributed to the users. Conveyance from the source to the point of treatment occurs by aqueducts, pipelines, or open channels, but the treated water is normally distributed in pressurized closed conduits. After use, whatever the purpose, the water becomes wastewater, which must be disposed of somehow, but almost always ends up being conveyed back to a treatment facility before being out-falled to some water body, to begin the cycle again. We call this an urban water cycle because it provides a human-generated imitation of the natural water cycle. Unlike the natural water cycle, however, without pipes the cycle would be nonexistent or, at the very least, short-circuited.

For use as water mains in a distribution system, pipes must be strong and durable in order to resist applied forces and corrosion. The pipe is subjected to internal pressure from the water and to external pressure from the weight of the backfill (soil) and vehicles above it. The pipe may also have to withstand a water hammer. Damage due to corrosion or rusting may also occur internally because of the water quality or externally because of the nature of the soil conditions.

PIPING NETWORKS

It would be far less costly and make for more efficient operation if public water systems were built with separate single pipe networks extending from the treatment plant to the user's residence, or from the user's sink or bathtub drain to the local wastewater treatment plant. Unfortunately, this ideal single-pipe scenario is not practical for real-world applications. Instead of a single piping system, a network of pipes is laid under the streets. Each of these piping networks is composed of different materials that vary (sometimes considerably) in diameter, length, and age.

These networks range in complexity to varying degrees, and each of these joined together pipes contribute energy losses to the system. Waterflow networks may consist of pipes arranged in series, parallel, or some complicated combination. In any case, an evaluation of friction losses for the flows is based on energy conservation principles applied to the flow junction points. Note: Demonstrating the procedure for making these complex computations is beyond the scope of this chapter. We only present the operator "need to know" aspects of complex or compound piping systems in this chapter.

When two pipes of different sizes or roughnesses are connected in series (see Figure 4.13), head loss for a given discharge, or discharge for a given head loss, may be calculated by applying the appropriate equation between the bonding points, taking into account all losses in the interval. Thus, head losses are cumulative. Series pipes may be treated as a single pipe of constant diameter to simplify the calculation of friction losses. The approach involves determining an "equivalent length" of a constant diameter pipe that has the same friction loss and discharge characteristics as the actual series pipe system.

In addition, application of the continuity equation to the solution allows the head loss to be expressed in terms of only one pipe size. Note: In addition to the head loss caused by friction between the water and the pipe wall, losses also are caused by minor issues: obstructions in the line, changes in directions, and changes in flow area. In practice, the method of equivalent length is often used to determine these losses. The method of equivalent length uses a table to convert each valve or fitting into an equivalent length of straight pipe.

Two or more pipes connected (as in Figure 4.14) so that flow is first divided among the pipes and is then rejoined comprise a parallel pipe system. A parallel pipe system is a common method for increasing the capacity of an existing line. Determining flows in pipes arranged in parallel are also made by the application of energy conservation principles – specifically, energy losses through all pipes connecting common junction points must be equal. Each leg of the parallel network is treated as a series piping system and converted to a single equivalent length pipe. The friction losses through the equivalent length parallel pipes are then considered equal, and the respective flows are determined by proportional distribution.

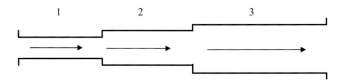

FIGURE 4.13   Pipes in series.

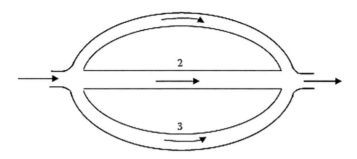

FIGURE 4.14   Pipe in parallel.

FLOW MEASUREMENT

Although it is clear that maintaining water flow is at the heart of any treatment process, clearly, it is the measurement of flow that is essential to ensuring the proper operation of a water treatment system. Few knowledgeable operators would argue with this statement. (Hauser, 1996) asks: "Why measure flow?" Then she explains:

> The most vital activities in the operation of water treatment plants are dependent on a knowledge of how much water is being processed. Certainly, water administrators and managers must keep an accurate account of water flow not only for the cost of operations but also to determine the fees to charge users.

In the statement above, Hauser makes clear that flow measurement is not only important but also routine in water operations. Routine, yes, but also the most important variable measured in a treatment plant. Hauser also pointed out that there are several reasons to measure flow in a treatment plant. (The American Water Works Association, 1995) lists several additional reasons to measure flow:

- The flow rate through the treatment processes needs to be controlled so that it matches distribution system use.
- It is important to determine the proper feed rate of chemicals added in the processes.
- The detention times through the treatment processes must be calculated.
- Flow measurement allows operators to maintain a record of water furnished to the distribution system for periodic comparison with the total water metered to customers. This provides a measure of "water accounted for," or conversely (as pointed out earlier by Hauser), the amount of water wasted, leaked, or otherwise not paid for; that is, lost water.
- Flow measurement allows operators to determine the efficiency of pumps. Pumps that are not delivering their designed flow rate are probably not operating at maximum efficiency and so power is being wasted.
- For well systems, it is very important to maintain records of the volume of water pumped and the hours of operation for each well. The periodic computation of well-pumping rates can identify problems such as worn pump impellers and blocked well screens.
- Reports that must be furnished to the state by most public water systems must include records of raw and finished water pumpage.
- Water generated by a treatment system must also be measured and recorded.
- Individual meters are often required for the proper operation of individual pieces of equipment. For example, the makeup water to a fluoride saturator is always metered to assist in tracking the fluoride feed rate.

Note: Simply put, measurement of flow is essential for operation, process control, and record-keeping in water treatment plants. All of the uses just discussed create the need, obviously, for a number of flow-measuring devices, often with different capabilities. In this section, we discuss many of the major flow-measuring devices currently used in water operations.

FLOW MEASUREMENT: THE OLD FASHIONED WAY

An approximate but very simple method to determine open-channel flow has been used for many years. The procedure involves measuring the velocity of a floating object moving in a straight uniform reach of the channel or stream. If the cross-sectional dimensions of the channel are known and the depth of flow is measured, then flow area can be computed. From the relationship $Q = A \times V$, the discharge Q can be estimated. In preliminary fieldwork, this simple procedure is useful in obtaining a ballpark estimate for the flow rate but is not suitable for routine measurements.

Example

- *Problem:* A floating object is placed on the surface of water flowing in a drainage ditch and is observed to travel a distance of 20 meters downstream in 30 seconds. The ditch is 2 meters wide and the average depth of flow is estimated to be 0.5 meters. Estimate the discharge under these conditions.
- *Solution*: The flow velocity is computed as distance over time or:

$$V = D/T = 20 \text{ m}/30 \text{ s} = 0.67 \text{ m/s}$$

The channel area is $A = 2 \text{ m} \times 0.5 \text{ m} = 1.0 \text{ m}^2$

The discharge $Q = A \times V = 1.0 \text{ m}^2 \times 0.66 \text{ m}^2 = 0.66 \text{ m}^3/\text{sec}$.

BASIS OF TRADITIONAL FLOW MEASUREMENT

Flow measurement can be based on the flow rate or flow amount. Flow rate is measured in gpm, MGD, and cfs. Water operations need flow rate meters to determine process variables within the treatment plant and in potable water distribution. Typically, flow rate meters used are pressure differential meters, magnetic meters, and ultrasonic meters. Flow rate meters are designed for metering flow in a closed pipe or open channel flow.

Flow amount is measured in either gallons (gal) or in cubic feet (cu ft). Typically, a totalizer, which sums up the gallons or cubic feet that pass through the meter, is used. Most service meters are of this type. They are used in private, commercial, and industrial activities where the total amount of flow measured is used in determining customer billing.

FLOW MEASURING DEVICES

In recent decades, flow measurement technology has evolved rapidly from the "old fashioned way" of measuring flow, discussed earlier, to the use of simple practical measuring devices to much more sophisticated devices. Physical phenomena discovered centuries ago have been the starting point for many of the viable flow meter designs used today. Moreover, the recent technology explosion has enabled flow meters to handle many more applications than could have been imagined centuries ago.

Before selecting a particular type of flow measurement device (Kawamura (2000) recommends consideration of several questions:

- Is liquid or gas flow being measured?
- Is the flow occurring in a pipe or in an open channel?

- What is the magnitude of the flow rate?
- What is the range of flow variation?
- Is the liquid being measured clean, or does it contain suspended solids or air bubbles?
- What is the accuracy requirement?
- What is the allowable head loss by the flow meter?
- Is the flow corrosive?
- What types of flow meters are available to the region?
- What types of post-installation service is available to the area?

DIFFERENTIAL PRESSURE FLOW METERS

For many years *differential pressure* flow meters have been the most widely applied flow-measuring device for water flow in pipes that require accurate measurement at a reasonable cost. The differential pressure type of flow meter makes up the largest segment of the total flow measurement devices currently being used. Differential pressure-producing meters currently on the market are the venturi, Dall type, Hershel venturi, universal venturi, and venture inserts.

The differential pressure-producing device has a flow restriction in the line that causes a differential pressure or "head" to be developed between the two measurement locations. Differential pressure flow meters are also known as head meters, and, of all the head meters, the orifice flow meter is the most widely applied device. The advantages of differential pressure flow meters include:

- Simple construction;
- Relatively inexpensive;
- No moving parts;
- Transmitting instruments are external;
- Low maintenance;
- Wide application of flowing fluid; suitable for measuring both gas and liquid flow;
- Ease of instrument and range selection; and
- Extensive product experience and performance database.

Disadvantages include:

- Flow rate is a nonlinear function of the differential pressure; and
- Low flow rate rangeability with normal instrumentation.

Note: Optimum measurement accuracy is maintained when the flow meter is calibrated, the flow meter is installed in accordance with standards and codes of practice, and the transmitting instruments are periodically calibrated.

TYPES OF DIFFERENTIAL PRESSURE FLOW METERS

The most commonly used differential pressure flow meter types used in water treatment are:

- Orifice;
- Venturi;
- Nozzle; and
- Pitot-static tube.

Orifice

The most commonly applied *orifice* is a thin, concentric, and flat metal plate with an opening in the plate, installed perpendicular to the flowing stream in a circular conduit or pipe. Typically, a sharp-edged hole is bored in the center of the orifice plate. As the flowing water passes through the orifice, the restriction causes an increase in velocity. A concurrent decrease in pressure occurs as potential energy (static pressure) is converted into kinetic energy (velocity). As the water leaves the orifice, its velocity decreases and its pressure increases as kinetic energy is converted back into potential energy according to the laws of conservation of energy.

However, there is always some permanent pressure loss due to friction, and the loss is a function of the ratio of the diameter of the orifice bore (d) to the pipe diameter (D). The orifice differential pressure flow meter is the lowest cost differential flow meter, is easy to install, and has no moving parts. However, it also has high permanent head loss (ranging from 40% to 90%), higher pumping costs, and an accuracy of ±2% for a flow range of 4:1, and is affected by wear or damage.

Venturi

A *venturi* is a restriction with a relatively long passage with smooth entry and exit (see Figure 4.15). It has a long life expectancy, simplicity of construction, relatively high-pressure recovery (i.e., produces less permanent pressure loss than a similar-sized orifice) but is more expensive, is not linear with flow rate, and is the largest and heaviest differential pressure flow meter. It is often used in wastewater flows since the smooth entry allows foreign material to be swept through instead of building up as it would in front of an orifice. The accuracy of this type of flow meter is ±1% for a flow range of 10:1. The head loss across a venturi flow meter is relatively small, ranging from 3% to 10% of the differential, depending on the ratio of the throat diameter to the inlet diameter (a.k.a., beta ratio).

Flow Nozzle

Flow nozzles (flow tubes) have a smooth entry and sharp exit. For the same differential pressure, the permanent pressure loss of a nozzle is of the same order as that of an orifice, but it can handle wastewater and abrasive fluids better than an orifice can. Note that, for the same line size and flow rate, the differential pressure at the nozzle is lower (head loss ranges from 10% to 20% of the differential) than the differential pressure for an orifice; hence, the total pressure loss is lower than that of an orifice. Nozzles are primarily used in steam service because of their rigidity, which makes them dimensionally more stable at high temperatures and velocities than orifices.

Note: A useful characteristic of nozzles is that they reach a critical flow condition, that is, a point at which further reduction in downstream pressure does not produce a greater velocity through the nozzle. When operated in this mode, nozzles are very predictable and repeatable.

Pitot Tube

A *pitot tube* is a point velocity-measuring device (see Figure 4.16). It has an impact port; as fluid hits the port, its velocity is reduced to zero and kinetic energy (velocity) is converted to potential energy (pressure head). The pressure at the impact port is the sum of the static pressure and the velocity head. The pressure at the impact port is also known as stagnation pressure or total pressure. The pressure difference between the impact pressure and the static pressure measured at the same point

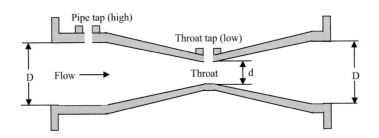

FIGURE 4.15   Venturi tube.

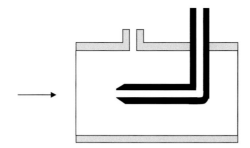

FIGURE 4.16 Pitot tube.

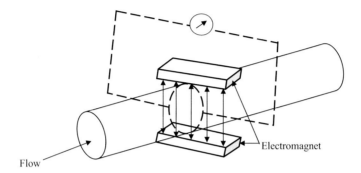

Flow

Electromagnet

FIGURE 4.17 Magnetic flow meter.

is the velocity head. The flow rate is the product of the measured velocity and the cross-sectional area at the point of measurement.

Note that the pitot tube has negligible permanent pressure drop in the line, but the impact port must be located in the pipe where the measured velocity is equal to the average velocity of the flowing water through the cross section.

Magnetic Flow Meters

Magnetic flow meters are relatively new to the water industry (USEPA, 1991). They are volumetric flow devices designed to measure the flow of electrically conductive liquids in a closed pipe. They measure the flow rate based on the voltage created between two electrodes (in accordance with Faraday's Law of Electromagnetic Induction) as the water passes through an electromagnetic field (see Figure 4.17). Induced voltage is proportional to flow rate. Voltage depends on magnetic field strength (constant), the distance between electrodes (constant), and the velocity of flowing water (variable). Properties of the magnetic flow meter include: (1) minimal head loss (no obstruction with line size meter); (2) no effect on flow profile; (3) suitable for size range between 0.1 and 120 inches; (4) have an accuracy rating of form 0.5–2% of flow rate; and (5) it measures forward or reverse flow.

The advantages of magnetic flow meters include:

- Obstruction less flow;
- Minimal head loss;
- Wide range of sizes;
- Bi-directional flow measurement;
- Variations in density, viscosity, pressure, and temperature yield negligible effect; and
- No moving parts.

Disadvantages include:

- The metered liquid must be conductive (but you wouldn't use this type meter on clean fluids anyway); and
- They are bulky, expensive in smaller sizes, and may require periodic calibration to correct drifting of the signal.

The combination of the magnetic flow meter and the transmitter is considered as a system. A typical system, schematically illustrated in Figure 4.18, shows a transmitter-mounted remote from the magnetic flow meter. Some systems are available with transmitters mounted integral to the magnetic flow meter. Each device is individually calibrated during the manufacturing process, and the accuracy statement of the magnetic flow meter includes both pieces of equipment. One is not sold or used without the other. It is also interesting to note that since 1983 almost every manufacturer now offers the microprocessor-based transmitter.

With regard to minimum piping straight-run requirements, magnetic flow meters are quite forgiving of piping configuration. The downstream side of the magnetic flow meter is much less critical than the upstream side. Essentially, all that is required of the downstream side is that sufficient backpressure is provided to keep the magnetic flow meter full of liquid during flow measurement. Two diameters downstream should be acceptable (Mills, 1991).

Note: Magnetic flow meters are designed to measure conductive liquids only. If air or gas is mixed with the liquid, the output becomes unpredictable.

ULTRASONIC FLOW METERS

Ultrasonic flow meters use an electronic transducer to send a beam of ultrasonic sound waves through the water to another transducer on the opposite side of the unit. The velocity of the sound beam varies with the liquid flow rate, so the beam can be electronically translated to indicate flow volume. The accuracy is $\pm 1\%$ for a flow velocity ranging from 1 to 25 ft/s, but the meter reading is greatly affected by a change in the fluid composition. Two types of ultrasonic flow meters are in general use for closed pipe flow measurements. The first (time of flight or transit time) usually uses pulse transmission and is for clean liquids, while the second (Doppler) usually uses continuous wave transmission and is for dirty liquids.

VELOCITY FLOW METERS

Velocity or turbine flow meters use a propeller or turbine to measure the velocity of the flow passing the device. The velocity is then translated into a volumetric amount by the meter register. Sizes exist from a variety of manufacturers to cover the flow range from 0.001 gpm to over 25,000 gpm for liquid service. End connections are available to meet the various piping systems. The flow meters

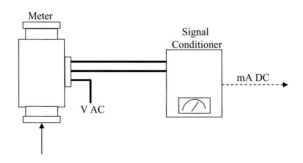

FIGURE 4.18   Magnetic flow meter system.

are typically manufactured of stainless steel but are also available in a wide variety of materials, including plastic. Velocity meters are applicable to all clean fluids. Velocity meters are particularly well suited for measuring intermediate flow rates on clean water (Oliver, 1991). The advantages of the velocity meter include:

- Accuracy;
- Composed of corrosion-resistant materials;
- Long-term stability;
- Liquid or gas operation;
- Wide operating range;
- Low pressure drop;
- Wide temperature and pressure limits;
- High shock capability; and
- Wide variety of electronics available.

As shown in Figures 4.19 and 4.20, a turbine flow meter consists of a rotor mounted on a bearing and shaft in a housing. The fluid to be measured is passed through the housing, causing the rotor to spin with a rotational speed proportional to the velocity of the flowing fluid within the meter. A device to measure the speed of the rotor is employed to make the actual flow measurement. The sensor can be a mechanically gear-driven shaft to a meter or an electronic sensor that detects the passage of each rotor blade generating a pulse. The rotational speed of the sensor shaft and the frequency of the pulse are proportional to the volumetric flow rate through the meter.

POSITIVE-DISPLACEMENT FLOW METERS

Positive-displacement flow meters are most commonly used for customer metering; they have long been used to measure liquid products. These meters are very reliable and accurate for low flow rates because they measure the exact quantity of water passing through them. Positive-displacement flow

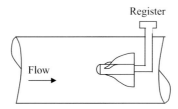

FIGURE 4.19 Propeller meter.

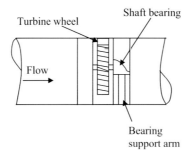

FIGURE 4.20 Turbine meter.

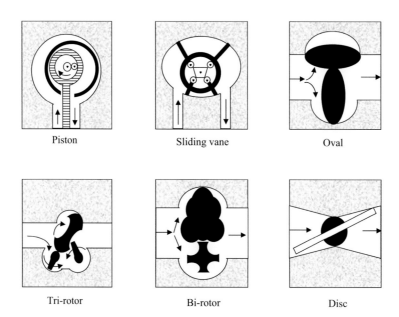

Piston   Sliding vane   Oval

Tri-rotor   Bi-rotor   Disc

FIGURE 4.21   Six common positive-displacement meter principles.

meters are frequently used for measuring small flows in a treatment plant because of their accuracy. Repair or replacement is easy since they are so common in the distribution system (Barnes, 1991).

In essence, a positive-displacement flow meter is a hydraulic motor with high volumetric efficiency that absorbs a small amount of energy from the flowing stream. This energy is used to overcome internal friction in driving the flow meter and its accessories and is reflected as a pressure drop across the flow meter. Pressure drop is regarded as unavoidable that must be minimized. It is the pressure drop across the internals of a positive-displacement flow meter that actually creates a hydraulically unbalanced rotor, which causes rotation.

A positive-displacement flow meter continuously divides the flowing stream into known volumetric segments, isolates the segments momentarily, and returns them to the flowing stream while counting the number of displacements. A positive-displacement flow meter can be broken down into three basic components: the external housing, the measuring unit, and the counter drive train. The external housing is the pressure vessel that contains the product being measured. The measuring unit is a precision metering element and is made up of the measuring chamber and the displacement mechanism. The most common displacement mechanisms include the oscillating piston, sliding vane, oval gear, tri-rotor, bi-rotor, and nutating disc types (see Figure 4.21). The counter drive train is used to transmit the internal motion of the measuring unit into a usable output signal. Many positive-displacement flow meters use a mechanical gear train that requires a rotary shaft seal or packing gland where the shaft penetrates the external housing. The positive-displacement flow meter can offer excellent accuracy, repeatability, and reliability in many applications. The positive-displacement flow meter has satisfied many needs in the past and should play a vital role in serving future needs as required.

CHAPTER SUMMARY

Water hydraulics and water measurement are important in water treatment and distribution. Simply put, the mission of many public and private water treatment operations is to manage and conserve existing water supplies and to treat water to safe levels for consumption and use and to distribute the treated, monitored, and tested product for human use. In order to accomplish this, water administrators and managers must make sound technical and economic decisions concerning new and existing water needs, while respecting the environment by sustaining or restoring the water supply ecosystem which may be affected. One key to better administrative and management practices is reliable and accurate water measurement.

REVIEW QUESTIONS

1. Bernoulli's principle states that the total energy of a hydraulic fluid is _____.
   a. Flexible
   b. Adjustable
   c. Flat
   d. Constant

2. The difference, or the drop, between static water level and the pumping water head is known as _____.
   a. Drawdown
   b. Draw up
   c. Slope
   d. Potential energy

3. What causes water motion?
   a. Inertia
   b. Energy
   c. Force
   d. Momentum

4. This is present when water is in motion _____.
   a. Potential energy
   b. Force
   c. Kinetic energy
   d. Inertia

5. Friction created as water encounters the surface of pipe causes _____.
   a. Major head loss
   b. No loss
   c. Minor head loss
   d. Heat

6. Force per unit area is _____.
   a. Force
   b. Head
   c. Hydrostatic charge
   d. Pressure

7. What is head loss measured in feet commonly called?
   a. Slope
   b. Head gain
   c. Head
   d. Static system

8. What is the force as it rounds a bend in a pipe called?
   a. Inertia
   b. Dynamic load
   c. Thrust
   d. Cavitation

9. What is the action that a sudden change in water pressure and direction in a pipe called?
   a. Pressure loss
   b. Water hammer
   c. Pressure increase
   d. Dynamic load

10. What does a gallon of water weigh in pounds?
    a. 50 lbs
    b. 62.14 lbs
    c. 8.34 lbs
    d. It is weightless

ANSWER KEY

1. Bernoulli's principle states that the total energy of a hydraulic fluid is _____.
   a. Flexible
   b. Adjustable
   c. Flat
   d. **Constant**

2. The difference, or the drop, between static water level and the pumping water head is known as _____.
   a. **Drawdown**
   b. Draw up
   c. Slope
   d. Potential energy

3. What causes water motion?
   a. Inertia
   b. Energy
   c. **Force**
   d. Momentum

4. This is present when water is in motion _____.
   a. Potential energy
   b. Force
   c. **Kinetic energy**
   d. Inertia

5. Friction created as water encounters the surface of pipe causes _____.
   a. Major head loss
   b. No loss
   c. Minor head loss
   d. **Head**

6. Force per unit area is _____.
   a. Force
   b. Head
   c. Hydrostatic charge
   d. **Pressure**

7. What is head loss measured in feet commonly called?
   a. **Slope**
   b. Head gain
   c. Head
   d. Static system

8. What is the force as it rounds a bend in a pipe called?
    a. Inertia
    b. Dynamic load
    c. **Thrust**
    d. Cavitation

9. What is the action that a sudden change in water pressure and direction in a pipe called?
    a. Pressure loss
    b. **Water hammer**
    c. Pressure increase
    d. Dynamic load

10. What does a gallon of water weigh in pounds?
    a. 50 lbs
    b. 62.14 lbs
    c. **8.34 lbs**
    d. It is weightless

REFERENCES

AWWA. 1995. *Basic Science Concepts and Applications: Principles and Practices of Water Supply Operations*, 2 ed. Denver, CO: American Water Works Association.

Barnes, R.G. 1991. Positive Displacement Flow Meters for Liquid Measurement, *in Flow Measurement*, Spitzer, D.W. (ed.). Research Triangle Park, NC: Instrument Society of America.

Hauser, B.A. 1993. *Hydraulics for Operators*. Boca Raton, FL: Lewis Publishers.

Hauser, B.A. 1996. *Practical Hydraulics Handbook*, 2nd ed. Boca Raton, FL: Lewis Publishers.

Holman, S. 1998. *A Stolen Tongue*. New York: Anchor Press, Doubleday.

Kawamura, S. 2000. *Integrated Design and Operation of Water Treatment Facilities*, 2nd ed. New York: Wiley.

Magnusson, R. J. 2001. *Water Technology in the Middle Ages*. Baltimore: The John Hopkins University Press.

Mills, R.C. 1991. Magnetic Flow meters, *in Flow Measurement*, Spitzer, D.W. (ed.). Research Triangle Park, NC: Instrument Society of America.

Nathanson, J.A. 1997. *Basic Environmental Technology: Water Supply Waste Management, and Pollution Control*, 2nd ed. Upper Saddle River, NJ: Prentice Hall.

Oliver, P.D. 1991. Turbine Flow Meters, *in Flow Measurement*, Spitzer, D.W. (ed.). Research Triangle Park, NC: Instrument Society of America.

USEPA 1991. *Flow Instrumentation: A Practical Workshop on Making Them Work*. Sacramento, CA: Water & Wastewater Instrumentation Testing Association.

5 Pumping Water

Copyright Las Vegas Valley Water District and used with permission.

Pump operations usually control only one variable: flow, pressure, or level. All pump control systems have a measuring device that compares a measured value with a desired one.

This information relays to a control element that makes the changes.

The user may obtain control with manually operated valves or sophisticated microprocessors. Economics dictate the accuracy and complication of a control system.

– Wahren (1997)

LEARNING OBJECTIVES

After studying this chapter, you should be able to:

- Explain why it is important for public water system administrators, upper level and lower level non-water operators to know, recognize, and understand pumping nomenclature and basic pumping operations.
- Discuss pumping water.
- Describe hydraulic machines and their operation in water treatment.
- Discuss the history of the modern Archimedean screw pump.
- Describe why math operations are important in water pumping.
- Describe head, static head, static discharge head, friction head, velocity head, and total head.

DOI: 10.1201/9781003207146-7

- Discuss pump horsepower.
- Describe affinity laws.
- Discuss specific speed.
- Describe net positive suction head (NPSH).
- Discuss centrifugal pumps and applications.
- Identify centrifugal pump modifications.
- Define and discuss pumping theory.
- Discuss pump characteristics.
- Identify the advantages and disadvantages of centrifugal pumps for water pumping.
- Discuss pumping controls.
- Identify and discuss the importance and function of motor controllers.
- Describe submersible pumps and their uses.
- Discuss purpose and use of vortex pumps.
- Discuss purpose and use of turbine pumps.
- Describe and discuss positive displacement pumps.

KEY DEFINITIONS

Brake Horsepower (bhp) – is the horsepower applied to the pump.

Centrifugal Pump (and its modifications) – is the most widely used type of pumping equipment in water operations. This type of pump is capable of moving high volumes of water (and other liquids) in a relatively efficient manner. The centrifugal pump is very dependable, has relatively low maintenance requirements, and can be constructed out of a wide variety of construction materials. It is considered one of the most dependable systems available for water transfer.

Efficiency – is the power produced by the unit, divided by the power used in operating the unit.

Float Control System – is the simplest of the centrifugal pump controls.

Friction Head – is the amount of energy used to overcome resistance to the flow of liquids through the system.

Head – gives the water energy and causes it to flow.

Hydraulic Machine or Pump – is a device that raises, compresses, or transfers fluids.

Motor Horsepower (hp) – is the horsepower applied to the motor.

NPSH – the total suction head in feet of liquid absolute determined at the suction nozzle and referred to as datum less the vapor pressure of the liquid in feet absolute.

Positive Displacement Pumps – force or displace water through the pumping mechanism. Most have a reciprocating element that draws water into the pump chamber on one stroke and pushes it out on the other.

Pump Impeller – imparts (transfers) energy to the water.

Packing – is the material that is placed around the pump shaft to seal the shaft opening in the casing and prevent air leakage into the casing.

Peristaltic Pumps (sometimes called tubing pumps) – use a series of rollers to compress plastic tubing to move the liquid through the tubing. A rotary gear turns the rollers at a constant speed to meter the flow. Peristaltic pumps are mainly used as chemical feed pumps.

Piston or Reciprocating Pump – is one type of positive displacement pump. This pump works just like the piston in an automobile engine – on the intake stroke, the intake valve opens, filling the cylinder with liquid.

Pneumatic Control Systems (also called a bubbler tube control system) – are relatively simple systems that can be used to control one or more pumps.

Specific Speed – is the quantitative index of optimization; that is, the higher the specific speed of a pump, the higher its efficiency.

Static Head – is the vertical distance the liquid travels from the supply tank to the discharge point.

Static Suction Head – refers to when the supply is located above the pump datum (i.e., an item of data).

Static Suction Lift – refers to when the supply is located below the pump datum.

Submersible Pump – is placed directly in the wet well or groundwater well. It uses a waterproof electric motor located below the static level of the wet well/well to drive a series of impellers. In some cases, only the pump is submerged; while in other cases, the entire pump-motor assembly is submerged.

Suction Specific Speed (n_{ss}) – also an impeller design characteristic, is an index of the suction characteristics of the impeller (i.e., the suction capacities of the pump).

Turbine Pump – consists of a motor, drive shaft, a discharge pipe of varying lengths, and one or more impeller-bowl assemblies. It is normally a vertical assembly where water enters at the bottom, passes axially through the impeller-bowl assembly where the energy transfer occurs, then moves upward through additional impeller-bowl assemblies to the discharge pipe.

Velocity Head – is the amount of head or energy required to maintain a stated velocity in the suction and discharge lines.

Water Horsepower (whp) – is the power necessary to lift the water to the required height.

OVERVIEW

The first four chapters of *Part I – Operations for Public Water Systems* have provided grounding and foundational materials for public water system administrators and upper- and mid-level managers. Chapter 5 continues this trend by discussing the workhorse of water operations and their critical importance in water operations; that is, the hydraulic machine (i.e., pumps and pumping). Again, those who perform administrative functions and support for public water systems (e.g., procurement personnel and others) do not need to know how to overhaul a pump. They do need to know, however, pumping basics; that is, what pumps do and the kinds or types of pumps. The need for credibility can't be overstated. Thus, fundamental knowledge about the pumping in water treatment operations is needed. At the very minimum, familiarity with pumping terminology and nomenclature is needed.

HYDRAULIC MACHINES: PUMPS

A hydraulic machine, or pump, is a device that raises, compresses, or transfers fluids. Garay (1990) points out that "few engineered artifacts are as essential as pumps in the development of the culture which our western civilization enjoys." This statement is germane to any discussion about pumps simply because humans have always needed to move water from one place to another against the forces of nature. As the need for potable water increases, the need to pump the water from distant locations to where it is most needed is also increasing. Initially, humans relied on one of the primary forces of nature – gravity – to assist them in moving water from one place to another. Gravity only works, of course, if the water is moved downhill on a sloping grade. Humans soon discovered that if they accumulated water behind the water source (e.g., behind a barricade, levy, or dam), pressure moved the water further. But when pressure is dissipated by various losses (e.g., friction loss), when water in low-lying areas is needed in higher areas, the energy needed to move that water must be created. Simply, some type of pump is needed.

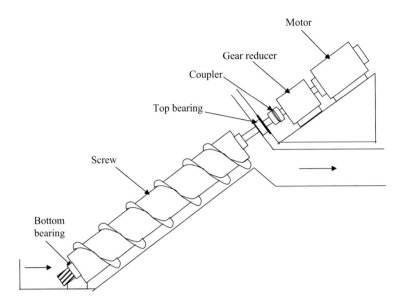

FIGURE 5.1　Modern Archimedean screw pump.

In 287 B.C., Archimedes (Greek mathematician, physicist, and engineer) invented the screw pump (see Figure 5.1). The Roman emperor, Nero, around 100 A.D., is often given credit for the development of the piston pump. In operation, the piston pump displaces volume after volume of water with each stroke. The piston pump has two basic problems: (1) its size limits capacity; and (2) it is a high energy consumer. It was not until the 19th century before pumping technology took a leap forward from its rudimentary beginnings. The first fully functional centrifugal pumps were developed in the 1800s. Centrifugal pumps can move great quantities of water with much smaller units than the pumps previously in use.

The pump is a type of hydraulic machine. Pumps convert mechanical energy into fluid energy. Whether it is taken from groundwater or a surface water body, or from one water unit treatment process to another, or to the storage tank for eventual final delivery through various sizes and types of pipes to the customer, *pumps* are the usual source of energy necessary for the conveyance of water. Again, the only exception may be, of course, where the source of energy is supplied entirely by gravity. Waterworks maintenance operators must therefore be familiar with pumps, pump characteristics, pump operation, and properly trained to perform needed maintenance. There are three general requirements of pump and motor combinations.

These requirements are: (1) reliability; (2) adequacy; and (3) economy. *Reliability* is generally obtained by installing in duplicate the very best equipment available and by the use of an auxiliary power source. *Adequacy* is obtained by securing liberal sizes of pumping equipment. *Economics* can be achieved by taking into account the life and depreciation, first cost, standby charges, interest, and operating costs (Texas Utilities Association, 1988).

Over the past several years, it has become more evident that many waterworks facilities have been unable to meet their optimum supply and/or treatment requirements for one of three reasons:

- Untrained operations and maintenance staff;
- Poor plant maintenance; and
- Improper plant design.

Note: It is important to point out that water and maintenance operators are expected to perform various mathematical operations related to pumps and pumping operations.

Calculations, calculations, calculations, and more calculations! Indeed, the licensed operator and/or trainee seeking licensure must be able to perform fundamental math operations. They simply can't get away from them – not in water treatment and distribution operations or licensure certification examinations. Basic calculations are a fact of life that the water maintenance operator soon learns, and hopefully learns well enough to use as required to operate a water facility correctly. This designation and learning series, however, is not about obtaining operator licensure; thus, typical math operations required to be performed by water operators and maintenance personnel are only briefly mentioned and not practiced in this discussion. Again, fundamental knowledge of waterworks operations is what this and the other chapters are all about; therefore, we only list those calculations that they may be required to perform.

Important Note: Before discussing pumping operations and nomenclature, it can be simply pointed out that a pump is a device for the transportation of water from one location to another, for lifting water to a higher elevation, or for increasing the water pressure.

WHERE PUMPING MATH OPERATIONS ARE USED

Flow Velocity

The speed or velocity of a fluid flowing through a channel or pipeline is related to the cross-sectional area of the pipeline and the quantity of water moving through the line; for example, if the diameter of a pipeline is reduced, then the velocity of the water in the line must increase to allow the same amount of water to pass through the line. If the cross-sectional area is decreased, the velocity of the flow must be increased. Mathematically we can say that the velocity and cross-sectional area are inversely proportional when the amount of flow (Q) is constant.

Note: The concept just explained is extremely important in the operation of a centrifugal pump and will be discussed later.

Pressure–Velocity Relationship

A relationship similar to that of velocity and cross-sectional area exists for velocity and pressure. As the velocity of flow in a full pipe increases, the pressure of the liquid decreases. Again, this is another important hydraulics principle that is very important to the operation of a centrifugal pump.

DID YOU KNOW?

Think about static head as the vertical distance between a reference point to the water surface when water is not moving – water not moving (static) is the key point.

Static Head

Pressure at a given point originates from the height or depth of water above it. It is this pressure, or *head*, which gives the water energy and causes it to flow. By definition, *static head* is the vertical distance the liquid travels from the supply tank to the discharge point. In many cases, it is desirable to separate the static head into two separate parts: (1) the portion that occurs before the pump (suction head or suction lift); and (2) the portion that occurs after the pump (discharge head). When this is done, the center (or datum) of the pump becomes the reference point.

Static Suction Head

Static suction head refers to when the supply is located above the pump datum (i.e., an item of data).

Static Suction Lift

Static suction lift refers to when the supply is located below the pump datum.

Static Discharge Head – If the total static head is to be determined after calculating the static suction head or lift and static discharge head, individually, two separate calculations can be used, depending on whether there is a suction head or a suction lift.

Friction Head

Various formulae calculate friction losses. Hazen-Williams wrote one of the most common for smooth steel pipe. Usually, we do not need to calculate the friction losses because handbooks such as the *Hydraulic Institute Pipe Friction Manual* tabulated these long ago. This important manual also shows velocities in different pipe diameters at varying flows, as well as the resistance coefficient (K) for valves and fittings (Wahren, 1997). Friction head (in feet) is the amount of energy used to overcome resistance to the flow of liquids through the system. It is affected by the length and diameter of the pipe, the roughness of the pipe, and the velocity head. It is also affected by the physical construction of the piping system.

The number of and types of elbows, valves, T's, etc., will greatly influence the friction head for the system. These must be converted to their equivalent length of pipe and included in the calculation. The *roughness factor (f)* varies with length and diameter as well as the condition of the pipe and the material from which it is constructed; it normally ranges from 0.01 to 0.04. Important Point: For centrifugal pumps, good engineering practice is to try to keep velocities in the suction pipe to 3 feet/sec or less. Discharge velocities higher than 11 feet/sec may cause turbulent flow and/or erosion in the pump casing.

It is also possible to compute friction head using tables. Friction head can also be determined on both the suction side of the pump and the discharge side of the pump. In each case, it is necessary to determine:

- The length of pipe;
- The diameter of the pipe;
- Velocity; and
- Pipe equivalent of valves, elbows, T's etc.

Velocity Head

Velocity head is the amount of head or energy required to maintain a stated velocity in the suction and discharge lines. The design of most pumps makes the total velocity head for the pumping system zero. Note: Velocity head only changes from one point to another on a pipeline if the diameter of the pipe changes. Note: There is no velocity head in a static system. The water is not moving.

Total Head

Total head is the sum of the static, friction, and velocity head.

CONVERSION OF PRESSURE HEAD

Pressure is directly related to the head. If liquid in a container subjected to a given pressure is released into a vertical tube, the water will rise 2.31 feet for every pound per square inch of pressure. This calculation can be very useful in cases where liquid is moved through another line that is under pressure. Since the liquid must overcome the pressure in the line it is entering, the pump must supply this additional head.

Horsepower

The unit of work is foot pound; the amount of work required to lift a 1-lb object 1 foot off the ground (ft/lb). For practical purposes, in water treatment operations we consider the amount of work being done. It is more valuable, obviously, to be able to work faster; that is, for economic reasons we consider the rate at which work is being done (i.e., power or foot pound/second). At some point, the horse was determined to be the ideal work animal; it could move 550 pounds 1 foot, in 1 sec, considered to be equivalent to 1 horsepower.

$$550 \text{ ft/lb/sec} = 1 \text{ horsepower (hp)}$$

or

$$33,000 \text{ ft/lb/min} = 1 \text{ horsepower (hp)}$$

A pump performs work while it pushes a certain amount of water at a given pressure. The two basic terms for horsepower are: (1) *hydraulic horsepower;* and (2) *brake horsepower.*

Hydraulic (Water) Horsepower (whp) – A pump has power because it does work. A pump lifts water (which has weight) a given distance in a specific amount of time (feet/lb/min). One whp provides the necessary power to lift the water to the required height.

Brake Horsepower (bhp) – A water pump does not operate alone. It is driven by the motor, and electrical energy drives the motor. Brake horsepower is the horsepower applied to the pump. A pump's bhp equals its hydraulic horsepower divided by the pump's efficiency.

Note that neither the pump nor its prime mover (motor) is 100% efficient. There are friction losses within both these units, and it will take **more** horsepower applied to the pump to get the required amount of horsepower to move the water, and even more horsepower applied to the motor to get the job done (Hauser, 1993).

Important Points – (1) *Water horsepower* is the power necessary to lift the water to the required height; (2) *Brake horsepower* is the horsepower applied to the pump; (3) *Motor horsepower* (hp) is the horsepower applied to the motor; and (4) *Efficiency* is the power produced by the unit, divided by the power used in operating the unit.

Specific Speed

The capacity of flow rate of a centrifugal pump is governed by the impeller thickness (Lindeburg, 1986). For a given impeller diameter, the deeper the vanes, the greater the capacity of the pump. Each desired flow rate or a desired discharge head will have one optimum impeller design. The impeller that is best for developing a high discharge pressure will have different proportions from an impeller designed to produce a high flow rate. The quantitative index of this optimization is called *specific speed* (n_s). The higher the specific speed of a pump, the higher its efficiency.

Note: The specific speed of an impeller is its speed when pumping 1 gpm of water at a differential head of 1 ft.

Pump specific speeds vary between pumps. Although no absolute rule sets the specific speed for different kinds of centrifugal pumps, the following rule of thumb for N_s can be used:

$$\text{Volute, diffuser, and vertical turbine} = 500 - 5000$$

$$\text{Mixed flow} = 5000 - 10,000$$

$$\text{Propeller pumps} = 9000 - 15,000$$

Suction Specific Speed

Suction specific speed (n_{ss}), also an impeller design characteristic, is an index of the suction characteristics of the impeller (i.e., the suction capacities of the pump) (Wahren, 1997). For practical purposes, n_{ss} ranges from about 3,000 to 15,000. The limit for the use of n_{ss} impellers in water is approximately 11,000. Ideally, n_{ss} should be approximately 7,900 for single suction pumps and 11,200 for double suction pumps.

Affinity Laws – Centrifugal Pumps

Most parameters (impeller diameter, speed, and flow rate) determining a pump's performance can vary.

Net Positive Suction Head (NPSH)

NPSH is different from both suction head and suction pressure. This important point tends to be confusing to those first introduced to the term, and to pumping technology in general. When an impeller in a centrifugal pump spins, the motion creates a partial vacuum in the impeller eye. The NPSH is the height of the column of liquid that will fill this partial vacuum without allowing the liquid's vapor pressure to drop below its flash point; that is, the NPSH required (NPSHR) for the pump to function properly.

The Hydraulic Institute (1990) defines NPSH as "the total suction head in feet of liquid absolute determined at the suction nozzle and referred to datum less the vapor pressure of the liquid in feet absolute." This defines the NPSH *available* (NPSHA) for the pump.

(Note that NPSHA is the actual water energy at the inlet.) The important point is a pump will run satisfactorily if the NPSHA equals or exceeds the NPSHR. Most authorities recommend the NPSHA be at least 2 feet absolute or 10% larger than the NPSHR, whichever number is larger. Note: With regard to NPSHR, contrary to popular belief, water is not sucked into a pump. A positive head (normally atmospheric pressure) must push the water into the impeller (i.e., flood the impeller). NPSHR is the minimum water energy required at the inlet by the pump for satisfactory operation. The pump manufacturer usually specifies NPSHR.

It is important to point out that if NPSHA is less than NPSHR the water will cavitate. *Cavitation* is the vaporization of fluid within the casing or suction line. If the water pressure is less than the vapor pressure, pockets of vapor will form. As vapor pockets reach the surface of the impeller, the local high pressure will collapse them, causing noise, vibration, and possible structural damage to the pump.

Pumps in Series and Parallel

Parallel operation is obtained by having two pumps discharging into a common header. This type of connection is advantageous when the system demand varies greatly. An advantage of operating pumps in parallel is that when two pumps are online, one can be shut down during low demand. This allows the remaining pump to operate close to its optimum efficiency. *Series* operation is achieved by having one pump discharge into the suction of the next. This arrangement is used primarily to increase the discharge head, although a small increase in capacity also results.

CENTRIFUGAL PUMPS

The *centrifugal pump* (and its modifications) is the most widely used type of pumping equipment in water operations. This type of pump is capable of moving high volumes of water (and other liquids) in a relatively efficient manner. The centrifugal pump is very dependable, has relatively low maintenance requirements, and can be constructed out of a wide variety of construction materials. It is considered one of the most dependable systems available for water transfer.

The centrifugal pump consists of a rotating element (*impeller*) sealed in a casing (*volute*). The rotating element is connected to a drive unit (*motor/engine*) which supplies the energy to spin the rotating element. As the impeller spins inside the volute casing, an area of low pressure is created in the center of the impeller. This low pressure allows the atmospheric pressure on the liquid in the supply tank to force the liquid up to the impeller. Because the pump will not operate if there is no low-pressure zone created at the center of the impeller, it is important that the casing be sealed to prevent air from entering the casing. To ensure the casing is airtight, the pump employs some type of seal (*mechanical* or *conventional packing*) assembly at the point where the shaft enters the casing. This seal also includes lubrication, provided by water, grease, or oil, to prevent excessive wear.

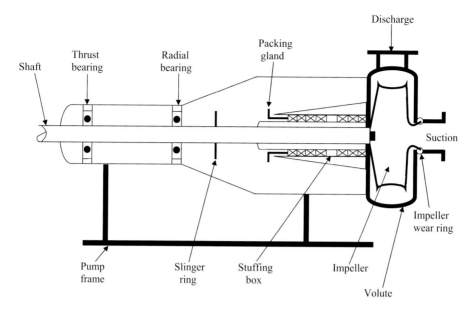

FIGURE 5.2 Centrifugal pump – major components.

From a hydraulic standpoint, note the energy changes that occur in the moving water. As water enters the casing, the spinning action of the impeller imparts (transfers) energy to the water. This energy is transferred to the water in the form of increased speed or velocity. The liquid is thrown outward by the impeller into the volute casing where the design of the casing allows the velocity of the liquid to be reduced, which, in turn, converts the velocity energy (*velocity head*) to pressure energy (*pressure head*).

The process by which this change occurs is described later. The liquid then travels out of the pump through the pump discharge. The major components of the centrifugal pump are shown in Figure 5.2.

Key Point: A centrifugal pump is a pumping mechanism whose rapidly spinning impeller imparts a high velocity to the water that enters, then converts that velocity to pressure upon exit.

TERMINOLOGY

To understand centrifugal pumps and their operation, we must understand the terminology associated with centrifugal pumps:

Base Plate – The foundation under a pump. It usually extends far enough to support the drive unit. The base plate is often referred to as the pump frame.

Bearings – Devices used to reduce friction and to allow the shaft to rotate easily. Bearings may be sleeve, roller, or ball.

Thrust Bearing – In a single suction pump, it is the bearing located nearest the motor, farthest from the impeller. It takes up the major thrust of the shaft, which is opposite from the discharge direction.

Radial (Line) Bearing – In a single suction pump, it is the one closest to the pump. It rides free in its own section and takes up and down stresses.

Note: In most cases, where pump and motor are constructed on a common shaft (no coupling), the bearings will be part of the motor assembly.

DID YOU KNOW?

Velocity pumps function as propeller meters.

Casing – The housing surrounding the rotating element of the pump. In the majority of centrifugal pumps, this casing can also be called the *volute*.

Split Casing – A pump casing that is manufactured in two pieces fastened together by means of bolts. Split casing pumps may be vertically (perpendicular to the shaft direction) split, or horizontally (parallel to the shaft direction) split.

Coupling – Device to join the pump shaft to the motor shaft. If pump and motor are constructed on a common shaft it is called a close-coupled arrangement.

Extended Shaft – For a pump constructed on one shaft that must be connected to the motor by a coupling.

Frame – The housing that supports the pump bearing assemblies. In an end suction pump, it may also be the support for the pump casing and the rotating element.

Impeller – The rotating element in the pump that actually transfers the energy from the drive unit to the liquid. Depending on the pump application, the impeller may be open, semi-open, or closed. It may also be single or double suction.

Impeller Eye – The center of the impeller, the area that is subject to lower pressures due to the rapid movement of the liquid to the outer edge of the casing.

Priming – Filling the casing and impeller with liquid. If this area is not completely full of liquid, the centrifugal pump will not pump efficiently.

Seals – Devices used to stop the leakage of air into the inside of the casing around the shaft.

Packing – This is the material that is placed around the pump shaft to seal the shaft opening in the casing and prevent air leakage into the casing. Packing and mechanical seals also prevent pump water leakage.

Stuffing Box – The assembly located around the shaft at the rear of the casing. It holds the packing and lantern ring. If excessive water leaks from the stuffing box, the packing gland needs to be tightened.

Lantern Ring – Also known as the seal cage, it is positioned between the rings of packing in the stuffing box to allow the introduction of a lubricant (water, oil, or grease) onto the surface of the shaft to reduce the friction between the packing and the rotating shaft.

Gland – Also known as the packing gland, it is a metal assembly that is designed to apply even pressure to the packing to compress it tightly around the shaft.

Mechanical Seal – A device consisting of a stationary element, a rotating element, and a spring to supply force to hold the two elements together. Mechanical seals may be either single or double units.

Shaft – The rigid steel rod that transmits the energy from the motor to the pump impeller. Shafts may be either vertical or horizontal.

Shaft Sleeve – A piece of metal tubing placed over the shaft to protect the shaft as it passes through the packing or seal area. In some cases, the sleeve may also help to position the impeller on the shaft.

Shut-Off Head – The head or pressure at which the centrifugal pump will stop discharging. It is also the pressure developed by the pump when it is operated against a closed discharge valve. This is also known as a cut-off head.

Shroud – The metal plate that is used to either support the impeller vanes (open or semi-open impeller) or to enclose the vanes of the impeller (closed impeller).

Slinger Ring – A device to prevent pumped liquids from traveling along the shaft and entering the bearing assembly. A slinger ring is also called a deflector.

Wear Rings – Devices that are installed on stationary or moving parts within the pump casing to protect the casing and the impeller from wear due to the movement of liquid through points of small clearances. Restricts leakage from the impeller discharge.

DID YOU KNOW?

The main purpose of seal water is to cool pump packing.

Impeller Ring – A wearing ring installed directly on the impeller.

Casing Ring – A wearing ring installed in the casing of the pump. A casing ring is also known as the suction head ring.

Stuffing Box Cover Ring – A wearing ring installed at the impeller in an end suction pump to maintain the impeller clearances and to prevent casing wear.

Pump Theory

The *volute-cased centrifugal pump* (see Figure 5.3) provides the pumping action necessary to transfer liquids from one point to another. First, the drive unit (usually an electric motor) supplies energy to the pump impeller to make it a spin. This energy is then transferred to the water by the impeller. The vanes of the impeller spin the liquid toward the outer edge of the impeller at a high rate of speed or velocity. This action is very similar to that which would occur when a bucket full of water with a small hole in the bottom is attached to a rope and spun. When sitting still, the water in the bucket will drain out slowly. However, when the bucket is spinning, the water will be forced through the hole at a much higher rate of speed.

Centrifugal pumps may be single stage, having a single impeller, or they may be multiple stage, having several impellers through which the fluid flows in series. Each impeller in the series increases the pressure of the fluid at the pump discharge. Pumps may have thirty or more stages in extreme cases. In centrifugal pumps, a correlation of pump capacity, head, and speed at optimum efficiency is used to classify the pump impellers with respect to their specific geometry. This correlation is called *specific speed* and is an important parameter for analyzing pump performance (Garay, 1990).

The volute of the pump is designed to convert velocity energy to pressure energy. As a given volume of water moves from one cross-sectional area to another with the volute casing, the velocity or speed of the water changes proportionately. The volute casing has a cross-sectional area, which

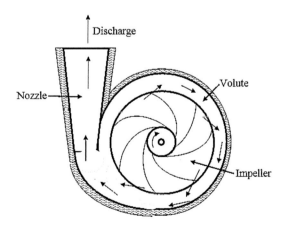

FIGURE 5.3 Cross-sectional diagram showing the features of a centrifugal pump.

is extremely small at the point in the case that is farthest from the discharge (see Figure 5.3). This area increases continuously to the discharge. As this area increases, the velocity of the water passing through it decreases as it moves around the volute casing to the discharge point. As the velocity of the water decreases the velocity head decreases and the energy is converted to pressure head. There is a direct relationship between the velocity of the water and the pressure it exerts. Therefore, as the velocity of the water decreases, the excess energy is converted to additional pressure (pressure head). This pressure head supplies the energy to move the water through the discharge piping.

PUMP CHARACTERISTICS

The centrifugal pump operates on the principle of an energy transfer and, therefore, has certain definite characteristics that make it unique. The type and size of the impeller limit the amount of energy that can be transferred to the water, the characteristics of the material being pumped, and the total head of the system through which the water is moving. For any one centrifugal pump, there is a definite relationship between these factors along with head (capacity), efficiency, and brake horsepower.

Head (Capacity)

As might be expected, the capacity of a centrifugal pump is directly related to the total head of the system. If the total head on the system is increased, the volume of the discharge will be reduced proportionately. As the head of the system increases, the capacity of the pump will decrease proportionately until the discharge stops. The head at which the discharge no longer occurs is known as the *cut-off head*.

As pointed out earlier, the total head includes a certain amount of energy to overcome the friction of the system. This friction head can be greatly affected by the size and configuration of the piping and the condition of the system's valving. If the control valves on the system are closed partially, the friction head can increase dramatically. When this happens, the total head increases, and the capacity or volume discharged by the pump decreases. In many cases, this method is employed to reduce the discharge of a centrifugal pump. It should be noted, however, that this does increase the load on the pump and drive system causing additional energy requirements and additional wear.

The total closure of the discharge control valve increases the friction head to the point where all the energy supplied by the pump is consumed in the friction head and is not converted to pressure head. Consequently, the pump exceeds its cut-off head, and the pump discharge is reduced to zero. Again, it is important to note that although the operation of a centrifugal pump against a closed discharge may not be hazardous (as with other types of pumps), it should be avoided because of the excessive load placed on the drive unit and pump. Our experience has shown that on occasion the pump can produce pressure higher than the pump discharge piping can withstand. Whenever this occurs, the discharge piping may be severely damaged by the operation of the pump against a closed or plugged discharge.

Efficiency

Every centrifugal pump will operate with varying degrees of efficiency over its entire capacity and head ranges. The important factor in selecting a centrifugal pump is to select a unit that will perform near its maximum efficiency in the expected application.

Brake Horsepower Requirements

In addition to the head capacity and efficiency factors, most pump literature includes a graph showing the amount of energy in horsepower that must be supplied to the pump to obtain optimal performance.

Advantages and Disadvantages of the Centrifugal Pump

The primary reason centrifugal pumps have become one of the most widely used types of pumps is the several advantages this type offers, including:

- *Construction* – The pump consists of a single rotating element and simple casing, which can be constructed using a wide assortment of materials. If the fluids to be pumped are highly corrosive, the pump parts that are exposed to the fluid can be constructed of lead or other material that is not likely to corrode. If the fluid being pumped is highly abrasive, the internal parts can be made of abrasion-resistant material or coated with a protective material. Also, the simple design of a centrifugal pump allows the pump to be constructed in a variety of sizes and configurations. No other pump currently available has the range of capacities or applications available through the use of the centrifugal pump.

DID YOU KNOW?

A foot valve is installed on the suction pipe of a pump to prevent water from draining out of the pump.

- *Operation* – "Simple and quiet" best describes the operation of a centrifugal pump. An operator-in-training with a minimum amount of experience may be capable of operating facilities that use centrifugal-type pumps. Even when improperly operated, the centrifugal pump's rugged construction allows it to operate (in most cases) without major damage.
- *Maintenance* – The amount of wear on a centrifugal pump's moving part is reduced and its operating life is extended because its moving parts are not required to be constructed to very close tolerances.
- *Pressure Is Self-Limited* – Because of the nature of its pumping action, the centrifugal pump will not exceed a predetermined maximum pressure. Thus, if the discharge valve is suddenly closed, the pump cannot generate additional pressure that might result in damage to the system or could potentially result in a hazardous working condition. The power supplied to the impeller will only generate a specified amount of head (pressure). If a major portion of this head or pressure is consumed in overcoming friction or is lost as heat energy, the pump will have a decreased capacity.
- *Adaptable to High-Speed Drive Systems* – Centrifugal pumps can make use of high-speed, high-efficiency motors. In situations where the pump is selected to match a specific operating condition, which remains relatively constant, the pump drive unit can be used without the need for expensive speed reducers.
- *Small Space Requirements* – For most pumping capacities, the amount of space required for installation of the centrifugal-type pump is much less than that of any other type of pump.
- *Fewer Moving Parts* – Rotary rather than reciprocating motion employed in centrifugal pumps reduces space and maintenance requirements due to the fewer number of moving parts required.

Although the centrifugal pump is one of the most widely used pumps, it does have a few disadvantages:

- *Additional Equipment Needed for Priming* – The centrifugal pump can be installed in a manner that will make it self-priming, but it is not capable of drawing water to the pump impeller unless the pump casing and impeller are filled with water. This can cause problems because if the water in the casing drains out, the pump would cease pumping until it is refilled. Therefore, it is normally necessary to start a centrifugal pump with the discharge valve closed. The valve is then gradually opened to its proper operating level. Starting the pump against a closed discharge valve is not hazardous provided the valve is not left closed for extended periods.

- *Air Leaks Affect Pump Performance* – Air leaks on the suction side of the pump can cause reduced pumping capacity in several ways. If the leak is not serious enough to result in a total loss of prime, the pump may operate at a reduced head or capacity due to air mixing with the water. This causes the water to be lighter than normal and reduces the efficiency of the energy transfer process.
- *Narrow Range of Efficiency* – Centrifugal pump efficiency is directly related to the head capacity of the pump. The highest performance efficiency is available for only a very small section of the head-capacity range. When the pump is operated outside of this optimum range, the efficiency may be greatly reduced.
- *Pump May Run Backward* – If a centrifugal pump is stopped without closing the discharge line, it may run backward because the pump does not have any built-in mechanism to prevent flow from moving through the pump in the opposite direction (i.e., from discharge side to suction). If the discharge valve is not closed or the system does not contain the proper check valves, the flow that was pumped from the supply tank to the discharge point will immediately flow back to the supply tank when the pump shuts off. This results in increased power consumption due to the frequent startup of the pump to transfer the same liquid from supply to discharge. Note: It is sometimes difficult to tell whether a centrifugal pump is running forward or backward because it appears and sounds like it is operating normally when operating in reverse.
- *Pump Speed Is Difficult to Adjust* – Centrifugal pump speed cannot usually be adjusted without the use of additional equipment, such as speed-reducing or speed-increasing gears or special drive units. Because the speed of the pump is directly related to the discharge capacity of the pump, the primary method available to adjust the output of the pump other than a valve on the discharge line is to adjust the speed of the impeller. Unlike some other types of pumps, the delivery of the centrifugal pump cannot be adjusted by changing some operating parameter of the pump.

CENTRIFUGAL PUMP APPLICATIONS

The centrifugal pump is probably the most widely used pump available at this time because of its simplicity of design and wide-ranging diversity (it can be adjusted to suit a multitude of applications). Proper selection of the pump components (impeller, casing, etc.) and construction materials can produce a centrifugal pump capable of transporting not only water but also other materials ranging from material/chemical slurries to air (centrifugal blowers). To attempt to list all of the various applications for the centrifugal pump would exceed the limitations of this chapter. Therefore, our discussion of pump applications is limited to those that frequently occur in water/wastewater operations.

- *Large Volume Pumping* – In water operations, the primary use of centrifugal pumps is large volume pumping. In large volume pumping, generally low speed, moderate head, vertically shafted pumps are used. Centrifugal pumps are well suited for water/wastewater system operations because they can be used in conditions where high volumes are required and a change in flow is not a problem. As the discharge pressure on a centrifugal pump is increased, the quantity of water/wastewater pumped is reduced. Also, centrifugal pumps can be operated for short periods with the discharge valve closed.
- *Non-clog Pumping* – These specifically designed centrifugal pumps use closed impellers with, at most, two to three vanes. It is usually designed to pass solids or trash up to 3 inches in diameter.

- *Dry Pit Pump* – Depending on the application, may be either a large volume pump or a non-clog pump. It is located in a dry pit that shares a common wall with the wet well. This pump is normally placed in such a position to ensure that the liquid level in the wet well is sufficient to maintain the pump's prime.
- *Wet Pit or Submersible Pump* – This type of pump is usually a non-clog type pump that can be submerged, with its motor, directly in the wet well. In a few instances, the pump may be submerged in the wet well while the motor remains above the water level. In these cases, the pump is connected to the motor by an extended shaft.
- *Underground Pump Stations* – When utilizing a wet well/dry well design, the pumps are located in an underground facility. Wastes are collected in a separate wet well, then pumped upward and discharged into another collector line or manhole. This system normally uses a non-clog type pump and is designed to add sufficient head to water flow to allow gravity to move the flow to the plant or the next pump station.
- *Recycle or Recirculation Pumps* – Because the liquids being transferred by the recycle or recirculation pump normally do not contain any large solids, the use of the non-clog type centrifugal pump is not always required. A standard centrifugal pump may be used to recycle trickling filter effluent, return activated sludge, or digester supernatant.
- *Service Water Pumps* – The wastewater plant effluent may be used for many purposes, such as, to clean tanks, water lawns, provide the water to operate the chlorination system, and to backwash filters. Because the plant effluent used for these purposes is normally clean, the centrifugal pumps used closely parallel the units used for potable water. In many cases, the double suction, closed impeller, or turbine type pump will be used.

PUMP CONTROL SYSTEMS

Most centrifugal pumps require some form of a pump control system. The only exception to this practice is when the plant pumping facilities are designed to operate continuously at a constant rate of discharge. The typical pump control system includes a sensor to determine when the pump should be turned on or off and the electrical/electronic controls to actually start and stop the pump. The control systems currently available for the centrifugal pump range from a very simple on-off float control to an extremely complex system capable of controlling several pumps in sequence. In the following sections, we briefly describe the operation of various types of control devices/systems used with centrifugal pumps.

Float Control

Currently, the *float control system* is the simplest of the centrifugal pump controls (see Figure 5.4). In the float control system, the float rides on the water's surface in the well, storage tank, or clear well, attached to the pump controls by a rod with two collars. One collar activates the pump when the liquid level in the well or tank reaches a preset level and a second collar shuts the pump off when the level in the well reaches a minimum level. This type of control system is simple to operate and relatively inexpensive to install and maintain. The system has several disadvantages. The system operates at one discharge rate. This can result in: (1) extreme variations in the hydraulic loading on succeeding units; and (2) long periods of non-operation due to low flow periods or maintenance activities.

Pneumatic Controls

*Pneumatic control sy*stems (also called a bubbler tube control system) are relatively simple systems that can be used to control one or more pumps. The system consists of an air compressor, a tube extending into the well, clear well, or storage tank/basin, and pressure-sensitive switches with varying on/off set points and a pressure relief valve (see Figure 5.5). The system works on the basic

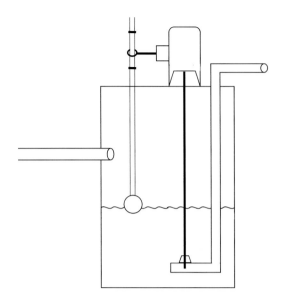

FIGURE 5.4   Float system for pump motor control.

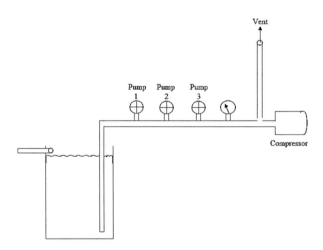

FIGURE 5.5   Pneumatic system for pump motor control.

principle which measures the depth of the water in the well or tank by determining the air pressure that is necessary to just release a bubble from the bottom of the tube (see Figure 5.5), hence, the name bubbler tube. The air pressure required to force a bubble out of the tube is determined by the liquid pressure, which is directly related to the depth of the liquid (1 psi = 2.31 feet). By installing a pressure switch on the airline to activate the pump starter at a given pressure, the level of the water can be controlled by activating one or more pumps.

Installation of additional pressure switches with slightly different pressure settings allows several pumps to be activated in sequence. For example, the first pressure switch can be adjusted to activate a pump when the level in the wet/tank is 3.8 feet (1.6 psi) and shut off at 1.7 feet (0.74 psi). If the flow into the pump well/tank varies greatly, and additional pumps are available to ensure that the level in the well/tank does not exceed the design capacity, additional pressure switches may be installed. These additional pressure switches are set to activate a second pump when the level in the well/tank reaches a preset level (i.e., 4.5 feet or 1.95 psi) and cut off when the well/tank level is reduced to a preset level (i.e., 2.7 feet or 1.2 psi). If the first pump's capacity is less than the rate of flow into the well/tank, the level of the well/tank continues to rise. Upon reaching the preset level (4-feet level), it will activate the second pump. If necessary, a third pump can be added to the system set to activate at a third preset well/tank depth (4.6 feet or 1.99 psi) and cut off a preset depth (3.0 feet or 1.3 psi).

The pneumatic control system is relatively simple with minimal operation and maintenance requirements. The major operational problem involved with this control system is the clogging of the bubbler tube. If, for some reason, the tube becomes clogged, the pressure on the system can increase and may activate all pumps to run even when the well/tank is low. This can result in excessive power consumption, which in turn, may damage the pumps.

ELECTRODE CONTROL SYSTEMS

The electrode control system uses a probe or electrode to control the pump on and off cycle. A relatively simple control system, it consists of two electrodes extended into the clear well, storage tank, or basin. One electrode is designed to activate the pump starter when it is submerged in the water; the second electrode extends deeper into the well/tank and is designed to open the pump circuit when the water drops below the electrode (see Figure 5.6). The major maintenance requirement of this system is keeping the electrodes clean. *Important Point*: Because the electrode control system uses two separate electrodes, the unit may be locked into an on-cycle or off-cycle depending on which electrode is involved.

Other Control Systems

Several other systems that use electrical energy are available for control of the centrifugal pump. These include a *tube-like device* that has several electrical contacts mounted inside (see Figure 5.7). As the water level rises in the clear well, storage tank, or basin, the water rises in the tube contacting the electrical contacts and activates the motor starter. Again, this system can be used to activate several pumps in series by the installation of several sets of contact points. As the water level drops in the well/tank, the level in the tube drops below a second contact that deactivates the motor and stops the pumping. Another control system uses a *mercury switch* (or a similar type of switch) enclosed in a protective capsule. Again, two units are required per pump. One switch activates the

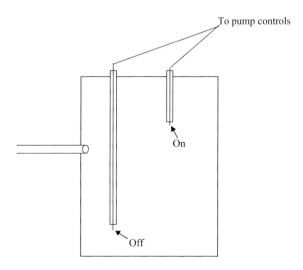

FIGURE 5.6   Electrode system for pump motor control.

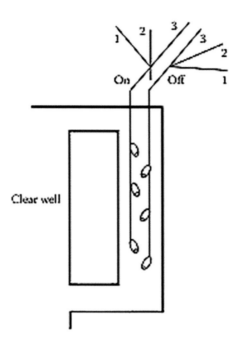

FIGURE 5.7   Electrical contacts for pump motor control.

pump when the liquid level rises, and the second switch shuts the pump off when the level reaches the desired minimum depth.

ELECTRICAL CONTROL SYSTEMS

Several centrifugal pump control systems are available that use electronic systems for control of pump operation. A brief description of some of these systems is provided in the sections that follow.

Flow Equalization System – In any multiple pump operation, the flow delivered by each pump will vary due to the basic hydraulic design of the system. To obtain equal loads on each pump when two or more are in operation, the flow equalization system electronically monitors the delivery of each pump and adjusts the speed of the pumps to obtain similar discharge rates for each pump.

Sonar or Other Transmission Type of Controllers

A *sonar* or *low-level radiation system* can be used to control centrifugal pumps. This type of system uses a transmitter and receiver to locate the level of the water in a tank, clear well, or basin. When the level reaches a predetermined set point, the pump is activated and when the level is reduced to a predetermined set point, the pump is shut off. Basically, the system is very similar to a radar unit. The transmitter sends out a beam that travels to the liquid, bounces off the surface, and returns to the receiver. The time required for this is directly proportional to the distance from the liquid to the instrument. The electronic components of the system can be adjusted to activate the pump when the time interval corresponds to a specific depth in the well or tank. The electronic system can also be set to shut off the pump when the time interval corresponds to a preset minimum depth.

Motor Controllers

Several types of controllers are available that start and stop motors. They also protect the motor from overloads and also from short-circuit conditions. Many motor controllers also function to adjust motor speed to increase or decrease the discharge rate for a centrifugal pump. This type of control may use one of the previously described controls to start and stop the pump, and, in some cases, adjust the speed of the unit. As the depth of the water in a well or tank increases, the sensor automatically increases the speed of the motor in predetermined steps to the maximum design speed. If the level continues to increase, the sensor may be designed to activate an additional pump.

Protective Instrumentation

Protective instrumentation of some type is normally employed in pump or motor installation. Note that the information provided in this section applies to the centrifugal pump as well as to many other types of pumps. Protective instrumentation for centrifugal pumps (or most other types of pumps) is dependent on pump size, application, and the amount of operator supervision. That is, pumps under 500 hp often only come with pressure gauges and temperature indicators. These gauges or transducers may be mounted locally (on the pump itself) or remotely (in suction and discharge lines immediately upstream and downstream of the suction and discharge nozzles). If transducers are employed, readings are typically displayed and taken (or automatically recorded) at a remote operating panel or control center.

Temperature Detectors

Resistance temperature devices (RTDs) and *thermocouples* (see Figure 5.8) (Grimes, 1976) are commonly used as temperature detectors on the pump prime movers (motors) to indicate temperature problems. In some cases, dial thermometers, armored glass-stem thermometers, or bimetallic-actuated temperature indicators are used. Whichever device is employed, it typically monitors temperature variances that may indicate a possible source of trouble. On electric motors greater than 250 hp, RTD elements are used to monitor temperatures in stator winding coils. Two RTDs per phase is standard. One RTD element is usually installed in the shoe of the loaded area employed on journal bearings in pumps and motors. Normally, tilted-pad thrust bearings have an RTD element in the active, as well as the inactive side. RTD elements are used when remote indication, recording, or automatic logging of temperature readings is required. Because of their smaller size, RTDs provide more flexibility in locating the measuring device near the measuring point. When dial thermometers are installed, they monitor oil thrown from bearings. Sometimes temperature detectors also monitor bearings with water-cooled jackets to warn against water supply failure. Pumps with heavy wall casing may also have casing temperature monitors.

Vibration Monitors

Vibration sensors are available to measure either bearing vibration or shaft vibration direction directly. Direct measurement of shaft vibration is desirable for machines with stiff bearing supports where bearing-cap measurements will be only a fraction of the shaft vibration. Wahren (1997) noted that pumps and motors 1,000 hp and larger may have the following vibration monitoring equipment:

- Seismic pickup with double set points installed on the pump outboard housing;
- Proximators with X-Y vibration probes complete with interconnecting co-axial cables at each radial and thrust journal bearing; and
- Key phasor with proximator and interconnecting co-axial cables.

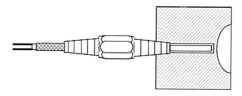

FIGURE 5.8   Thermocouple installation in journal bearing.

Supervisory Instrumentation

Supervisory instruments are used to monitor the routine operation of pumps, their prime movers, and their accessories in order to sustain a desired level of reliability and performance. Generally, these instruments are not used for accurate performance tests or for automatic control, although they may share connections or functions. Supervisory instruments consist of annunciators and alarms that provide operators with warnings of abnormal conditions that, unless corrected, will cause pump failure. Annunciators used for both alarm and pre-alarm have both visible and audible signals.

CENTRIFUGAL PUMP MODIFICATIONS

The centrifugal pump can be modified to meet the needs of several different applications. If there is a need to produce higher discharge heads, the pump may be modified to include several additional impellers. If the material being pumped contains a large amount of material that could clog the pump, the pump construction may be modified to remove a major portion of the impeller from direct contact with the material being pumped. Although there are numerous modifications of the centrifugal pump available, the scope of this discussion covers only those that have found wide application in the water distribution and treatment fields. Modifications to be presented in this section include:

- Submersible pumps;
- Recessed impeller or vortex pumps; and
- Turbine pumps.

SUBMERSIBLE PUMPS

The *submersible pump* is, as the name suggests, placed directly in the wet well or groundwater well. It uses a waterproof electric motor located below the static level of the wet well/well to drive a series of impellers. In some cases, only the pump is submerged; while in other cases, the entire pump-motor assembly is submerged. Figure 5.9 illustrates this system.

Description

The submersible pump may be either a close-coupled centrifugal pump or an extended shaft centrifugal pump. If the system is a close-coupled system, then both motor and pump are submerged in the liquid being pumped. Seals prevent water and wastewater from entering the inside of the motor protecting the electric motor in a close-coupled pump from shorts and motor burnout. In the extended shaft system, the pump is submerged while the motor is mounted above the pump wet well. In this situation, an extended shaft assembly must connect the pump and motor.

Applications

The submersible pump has wide applications in the water/wastewater treatment industry. It generally can be substituted in any application of other types of centrifugal pumps. However, it has found its widest application in the distribution or collector system pump stations.

Advantages

In addition to the advantages discussed earlier for a conventional centrifugal pump, the submersible pump has additional advantages:

- Because it is located below the surface of the liquid, there is less chance that the pump will lose its prime, develop air leaks on the suction side of the pump, or require initial priming.

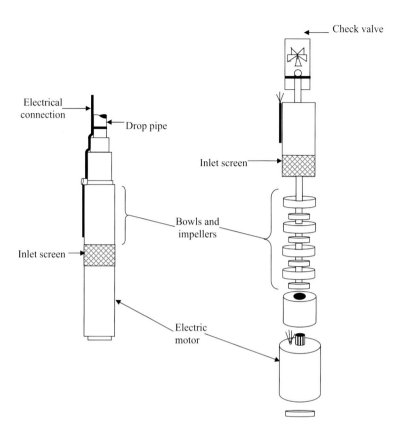

FIGURE 5.9   Submersible pump.

- The pump or the entire assembly is located in the well/wet well; there is less cost associated with the construction and operation of this system. It is not necessary to construct a dry well or a large structure to hold the pumping equipment and necessary controls.

Disadvantages

The major disadvantage associated with the submersible pump is the lack of access to the pump or pump and motor. The performance of any maintenance requires either drainage of the wet well or extensive lift equipment to remove the equipment from the wet well or both. This may be a major factor in determining if a pump receives the attention it requires.

   Also, in most cases, all major maintenance on close-coupled submersible pumps must be performed by outside contractors due to the need to re-seal the motor to prevent leakage.

Recessed Impeller or Vortex Pumps

The *recessed impeller* or *vortex pump* uses an impeller that is either partially or wholly recessed into the rear of the casing (see Figure 5.10). The spinning action of the impeller creates a vortex or whirlpool. This whirlpool increases the velocity of the material being pumped. As in other centrifugal pumps, this increased velocity is then converted to increased pressure or head.

Applications

The recessed impeller or vortex pump is used widely in applications where the liquid being pumped contains large amounts of solids or debris and slurries that could clog or damage the pump's impeller.

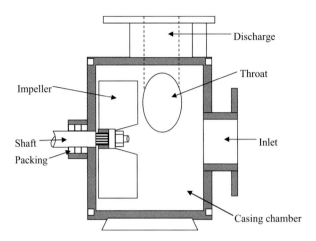

FIGURE 5.10 Schematic of a recessed impeller or vortex pump.

It has found increasing use as a sludge pump in facilities that withdraw sludge continuously from their primary clarifiers.

Advantages

The major advantage of this modification is the increased ability to handle materials that would normally clog or damage the pump impeller. Because the majority of the flow does not come in direct contact with the impeller, there is much less potential for problems.

Disadvantages

Because there is less direct contact between the liquid and the impeller, the energy transfer is less efficient. This results in somewhat higher power costs and limits the pump's application to low to moderate capacities. Objects that might have clogged a conventional type centrifugal pump are able to pass through the pump. Although this is very beneficial in reducing pump maintenance requirements, it has, in some situations, allowed material to be passed into a less accessible location before becoming an obstruction. To be effective, the piping and valving must be designed to pass objects of a size equal to that which the pump will discharge.

Turbine Pumps

The turbine pump consists of a motor, drive shaft, a discharge pipe of varying lengths and one or more impeller-bowl assemblies (see Figure 5.11). It is normally a vertical assembly where water enters at the bottom, passes axial through the impeller-bowl assembly where the energy transfer occurs, then moves upward through additional impeller-bowl assemblies to the discharge pipe. The length of this discharge pipe will vary with the distance from the wet well to the desired point of discharge.

Applications

Due to the construction of the turbine pump, the major applications have traditionally been for the pumping of relatively clean water. The line shaft turbine pump has been used extensively for drinking water pumping, especially in those situations where water is withdrawn from deep wells.

Advantages

The turbine pump has a major advantage in the amount of heat it can produce. By installing additional impeller-bowl assemblies, the pump is capable of even greater production. Moreover, the

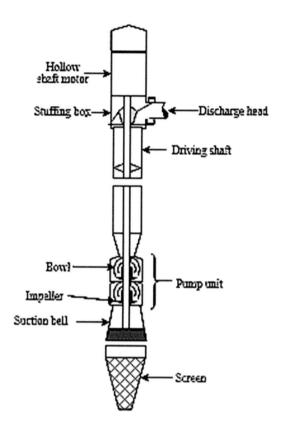

FIGURE 5.11   Vertical turbine pump.

turbine pump has a simple construction, a low noise level, and is adaptable to several drive types: motor, engine, or turbine.

Disadvantages

High initial cost and high repair costs are two of the major disadvantages of turbine pumps. In addition, the presence of large amounts of solids within the liquid being pumped can seriously increase the amount of maintenance the pump requires; consequently, the unit has not found widespread use in any situation other than service water pumping.

POSITIVE DISPLACEMENT PUMPS

Positive displacement pumps force or displace water through the pumping mechanism. Most have a reciprocating element that draws water into the pump chamber on one stroke and pushes it out on the other. Unlike centrifugal pumps that are meant for low pressure, high flow applications, positive displacement pumps can achieve greater pressures but are slower-moving, low-flow pumps. Other positive displacement pumps include the piston pump, diaphragm pump, and peristaltic pumps, which are the focus of our discussion. In the drinking water industry, positive displacement pumps are most often found as chemical feed pumps. It is important to remember that positive displacement pumps cannot be operated against a closed discharge valve. As the name indicates, something must be displaced with each stroke of the pump. Closing the discharge valve can cause rupturing of the discharge pipe, the pump head, the valve, or some other component.

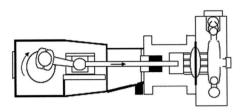

FIGURE 5.12   Diaphragm pump.

PISTON PUMP OR RECIPROCATING PUMP

The *piston or reciprocating pump* is one type of positive displacement pump. This pump works just like the piston in an automobile engine on the intake stroke; the intake valve opens, filling the cylinder with liquid. As the piston reverses direction, the intake valve is pushed closed and the discharge valve is pushed open; the liquid is pushed into the discharge pipe. With the next reversal of the piston, the discharge valve is pulled closed and the intake valve pulled open, and the cycle repeats. A piston pump is usually equipped with an electric motor and a gear and cam system that drives a plunger connected to the piston. Just like an automobile engine piston, the piston must have packing rings to prevent leakage and must be lubricated to reduce friction. Because the piston is in contact with the liquid being pumped, only good grade lubricants can be used when pumping materials that will be added to drinking water. The valves must be replaced periodically as well.

DIAPHRAGM PUMP

A *diaphragm pump* is composed of a chamber used to pump the fluid; a diaphragm that is operated by either electric or mechanical means; and two valve assemblies – a suction and a discharge valve assembly (see Figure 5.12). A diaphragm pump is a variation of the piston pump in which the plunger is isolated from the liquid being pumped by a rubber or synthetic diaphragm. As the diaphragm is moved back-and-forth by the plunger, liquid is pulled into and pushed out of the pump. This arrangement provides better protection against leakage of the liquid being pumped and allows the use of lubricants that otherwise would not be permitted. Care must be taken to assure that diaphragms are replaced before they rupture. Diaphragm pumps are appropriate for discharge pressures up to about 125 psi but do not work well if they must lift liquids more than about four feet. Diaphragm pumps are frequently used for chemical feed pumps. By adjusting the frequency of the plunger motion and the length of the stroke, extremely accurate flow rates can be metered. The pump may be driven hydraulically by an electric motor or by an electronic driver in which the plunger is operated by a solenoid. Electronically driven metering pumps are extremely reliable (few moving parts) and inexpensive.

PERISTALTIC PUMPS

Peristaltic pumps (sometimes called tubing pumps) use a series of rollers to compress plastic tubing to move the liquid through the tubing. A rotary gear turns the rollers at a constant speed to meter the flow. Peristaltic pumps are mainly used as chemical feed pumps. The flow rate is adjusted by changing the speed the roller-gear rotates (to push the waves faster) or by changing the size of the tubing (so there is more liquid in each wave). As long as the right type of tubing is used, peristaltic pumps can operate at discharge pressures up to 100 psi.

Note that the tubing must be resistant to deterioration from the chemical being pumped. The principal item of maintenance is the periodic replacement of the tubing in the pump head. There are no check valves or diaphragms in this type of pump.

CHAPTER SUMMARY

This chapter pointed out the importance of water delivery to the waterworks for treatment and water conveyance for customer use. In order to achieve water delivery the water must flow. In order for water to flow it must be provided the motive force of gravity or flow by hydraulic machines. The hydraulic machines are pumps. Because the centrifugal pump is the workhorse of water flow and delivery, emphasis in this chapter is focused on centrifugal pumps and their various applications.

REVIEW QUESTIONS

1. What is a static head?
   a. Horizontal distance between a reference point to the water surface
   b. The output of a centrifugal pump
   c. Friction loss
   d. The vertical distance between a reference point to the water surface, when water is not moving

2. What is the purpose of packing and mechanical seal?
   a. Increases head
   b. Supports the shaft assembly
   c. Prevents pump water leakage
   d. Prevents bearing grease leakage

3. If excessive water leaks from the stuffing box, _____.
   a. Impeller has crumbled
   b. Broken check valve
   c. Packing gland needs to be tightened
   d. Bearing failure

4. _____ is installed on the suction pipe of a pump to prevent water from draining out of the pump.
   a. Foot valve
   b. Slinger ring
   c. Pressure valve
   d. Emergency valve

5. What is used to restrict leakage from the impeller discharge?
   a. Slinger ring
   b. Wear rings
   c. Lantern ring
   d. Shaft sleeve

6. Which of the following is a positive displacement pump?
   a. Feeder pump
   b. Sub pump
   c. Plston pump
   d. Well pump

7. What is the main purpose of seal water in pumps?
   a. Prevent water hammer
   b. Protect bearings
   c. Cool packing
   d. Protect the slinger ring

8. The velocity pump is what type of pump?
   a. Piston pump
   b. Air pump
   c. Centrifugal pump
   d. Positive displacement pump

9. What is the function of a velocity pump?
   a. Piston
   b. Propeller meter
   c. Motor controller
   d. Relief valve

10. The _____ pump has a piston inside its casing.
    a. Centrifugal
    b. Venturi
    c. Reciprocating
    d. Peristaltic

ANSWER KEY

1. What is a static head?
   a. Horizontal distance between a reference point to the water surface
   b. The output of a centrifugal pump
   c. Friction loss
   d. **The vertical distance between a reference point to the water surface, when water is not moving**

2. What is the purpose of packing and mechanical seal?
   a. Increases head
   b. Supports the shaft assembly
   c. **Prevents pump water leakage**
   d. Prevents bearing grease leakage

3. If excessive water leaks from the stuffing box, _____.
   a. Impeller has crumbled
   b. Broken check valve
   c. **Packing gland needs to be tightened**
   d. Bearing failure

4. _____ is installed on the suction pipe of a pump to prevent water from draining out of the pump.
   a. **Foot valve**
   b. Slinger ring
   c. Pressure valve
   d. Emergency valve

5. What is used to restrict leakage from the impeller discharge?
   a. Slinger ring
   b. **Wear rings**
   c. Lantern ring
   d. Shaft sleeve

6. Which of the following is a positive displacement pump?
   a. Feeder pump
   b. Sub pump

c. **Piston pump**

d. Well pump

7. What is the main purpose of seal water in pumps?
   a. Prevent water hammer
   b. Protect bearings
   c. **Cool packing**
   d. Protect the slinger ring

8. The velocity pump is what type of pump?
   a. Piston pump
   b. Air pump
   c. **Centrifugal pump**
   d. Positive displacement pump

9. What is the function of a velocity pump?
   a. Piston
   b. **Propeller meter**
   c. Motor controller
   d. Relief valve

10. The _____ pump has a piston inside its casing.
    a. Centrifugal
    b. Venturi
    c. **Reciprocating**
    d. Peristaltic

REFERENCES

Garay, P.N. *Pump Application Desk Book*. Lilburn, GA: The Fairmont Press, Inc., p. 1, 1990.

Grimes, A.S. Supervisory and Monitoring Instrumentation. In *Pump Handbook*, Karassik, I.J. et al. (eds.). New York: McGraw-Hill Book Company, 1976.

Hauser, B.A. *Hydraulics for Operators*. Boca Raton, FL: Lewis Publishers, 1993.

Hydraulic Institute *The Hydraulic Institute Engineering Data Book*, 2nd ed. Cleveland, OH: Hydraulic Institute, 1990.

Hydraulic Institute *Hydraulic Institute Complete Pump Standards*, 4th ed. Cleveland, OH: Hydraulic Institute, 1994.

Lindeburg, M.R. *Civil Engineering Reference Manual*, 4th ed. San Carlos, CA: Professional Publications, Inc., 1986.

Texas Manual. *Manual of Water Utility Operations*, 8th ed. Texas Utilities Association, 1988.

Wahren, U. *Practical Introduction to Pumping Technology*. Houston, TX: Gulf Publishing Company, 1997.

6 Drinking Water Conveyance and Distribution

2019 Photo Courtesy of California Water Service Group.

When the well's dry, we know the worth of water.

– Benjamin Franklin

LEARNING OBJECTIVES

After studying this chapter, you should be able to:

- Describe the six functional components of a water distribution system.
- Describe surface water/groundwater intake systems.
- Discuss surface distribution.
- Describe distribution line networks.
- Describe the conditions and factors that affect watercolor.
- Describe distribution storage.
- Describe pipes used in water distribution systems.

DOI: 10.1201/9781003207146-8

- Discuss energy losses in pipe networks.
- Describe the difference between fluids and liquids.
- Identify piping color codes.
- Discuss maintaining fluid flow.
- Define scaling.
- Discuss pumping system maintenance.
- Identify piping system accessories.
- Discuss flow control.
- Identify and discuss valves, types of valves and piping system protective devices.
- Identify water piping ancillaries.

KEY DEFINITIONS

Absolute Pressure – gauge pressure plus atmospheric pressure.

Air Gaps – vertical air gaps are used to prevent back-siphonage and back pressure.

Asbestos – fibrous mineral form of magnesium silicate.

Automatic Check Valve – required to prevent back pressure.

Backflow – a term in plumbing for an unwanted flow of water in the reverse direction. It can be a serious health risk for the contamination of potable water supplies with foul water.

Backflow Preventers – they are intended to prevent backflow of sewage on the sanitary sewer line during a flood or sewer blockage and have no connection with potable water.

Back-Siphonage – a condition in which the pressure in the distribution system is less than atmospheric pressure allowing contamination to enter a water system through a cross-connection.

Bimetallic – made of two different types of metal.

Bourbon Tube – a semicircular tube of elliptical cross section, used to sense pressure changes.

Brazing – soldering with a nonferrous alloy that melts at a lower temperature than that of the metals being joined, also known as hard soldering.

Butterfly Valve – a valve in which a disk rotates on a shaft as the valve opens and closes. In the fully open position, the disk is parallel to the axis of the pipe.

C Factor – a representation of the roughness of pipe. High C factor means the pipe has a smooth interior.

Check Valve – a valve designed to open in the direction of normal flow and close with reversal of flow. An approved check valve has substantial construction and suitable materials, is positive in closing, and permits no leakage in a direction opposite to normal flow.

Critical Flow – flow at the critical depth and velocity. Critical flow minimizes the specific energy and maximizes discharge.

Diaphragm Valve – a valve in which the closing element is a thin, flexible disk often used in low-pressure systems.

Differential Pressure – the difference between the inlet and outlet pressures in a piping system.

Equivalent Pipe Theory – states that two pipes are equivalent if the head loss generated by the water velocity is the same in both pipes.

Filter – an accessory fitting used to remove solids from a fluid stream.

Fluids – any substance that flows.

Gate Valve – a valve in which the closing element consists of a disk that slides across an opening to stop the flow of water.

Gauge Pressure – the amount by which the total absolute pressure exceeds the ambient atmospheric pressure.

Globe Valve – a valve having a round, ball-like shell and horizontal disk.

Joint – a connection between two lengths of pipe or between a length of pipe and a fitting.

Laminar Flow – ideal water flow; water particles move along straight, parallel paths, in layers or streamlines. There is no turbulence in the water and no friction loss.

Nominal Pipe Size – the thickness given in the product material specifications or standard to which manufacturing tolerances are applied.

Parallel Pipe Systems – two or more pipes of different ages, materials, sizes or lengths, laid side by side, with the flow splitting among them.

Piping Systems – a complete network of pipes, valves, and other components.

Pressure-Regulating Valve – a valve with a horizontal disk for automatically reducing water pressure in a main to a preset value.

Series Pipe Systems – two or more pipes of different ages, materials, or sizes laid end to end.

Strainer – an accessory fitting used to remove large particles of foreign matter from a fluid.

Throttle – controlling flow through a valve by means of intermediate steps between fully open and fully closed.

Trap – an accessory fitting used to remove condensate from steam lines.

Turbulent Flow – normal for a water system; water particles move in a haphazard fashion and continually cross each other in all directions.

Uniform Flow – occurs when the magnitude and direction of velocity do not change from point to point.

Vacuum Breaker – a mechanical device that allows air into the piping system thereby preventing backflow that could otherwise be caused by the siphoning action created by a partial vacuum.

Varied Flow – flow that has a changing depth along the water course, with respect to location, not time.

Viscosity – the thickness or resistance to flow of a liquid.

Water Hammer – the concussion of moving water against the sides of pipe, caused by a sudden change in the rate of flow or stoppage of flow in the line.

OVERVIEW

A typical public water-supply system consists of six functional elements: (1) a source or sources of supply; (2) storage facilities (impoundment reservoirs, for example); (3) transmission facilities used for transporting water from the point of storage to the treatment plant; (4) treatment facilities for altering water quality; (5) transmission and storage facilities for transporting water to intermediate points (such as water towers or standpipes); and (6) distribution facilities for bringing water to individual users (see Figure 6.1).

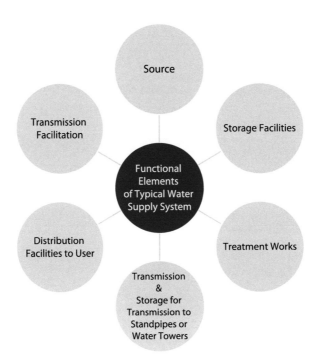

FIGURE 6.1   Elements of a typical water supply system.

DID YOU KNOW?

Recent archaeological work and personal verification has uncovered and/or collaborated an elaborate water distribution network at Machu Picchu, Peru. Amazingly, by 1450 A.D., Incan engineers had devised a spring collection system that fed 16 fountains – at an altitude of more than 8,000 feet.

When precipitation falls on a watershed or catchment area, it either flows as runoff above ground to streams and rivers or soaks into the ground to reappear in springs or to where it can be drawn from wells. A water supply can come from a catchment area that may contain several thousands of acres (or hectares) of land, draining to streams whose flow is retained in impoundment reservoirs. If a water supply is drawn from a large river or lake, the catchment area is the entire area upstream from the point of intake.

Obviously, the amount of water that enters a water supply system depends on the amount of precipitation and the volume of the runoff. The annual average precipitation in the United States is about 30 inches, of which two-thirds is lost to the atmosphere by evaporation and transpiration. The remaining water becomes runoff into rivers and lakes, or through infiltration replenishes groundwater. Precipitation and runoff vary greatly with geography and season. Drinking water comes from surface water and/or groundwater. Large-scale water supply systems tend to rely on surface water resources; smaller public water systems tend to use groundwater. If surface water is the source of supply for a particular drinking water supply system, the water is obtained from lakes, streams, rivers, or ponds. Storage reservoirs (artificial lakes created by constructing dams across stream valleys) can hold back higher-than-average flows and release them when greater flows are needed. Water supplies may be taken directly from reservoirs, or from locations downstream of the dams. Reservoirs may serve other purposes in addition to water supply, including flood mitigation, hydroelectric power, and water-based recreation.

Groundwater is pumped from natural springs, from wells, and from infiltration galleries, basins, or cribs. Most small and some large US public water systems use groundwater as their source of supply. Groundwater may be drawn from the pores of alluvial, glacial, or eolian deposits of granular unconsolidated material (such as sand and gravel); from the solution passages, caverns, and cleavage planes of sedimentary rocks (such as limestone, slate, and shale); and from combinations of these geologic formations. Groundwater sources may have intake or recharge areas that are miles away from points of withdrawal (water-bearing stratum or aquifer). Water quality in aquifers (geologic formations that contain water) and water produced by wells depend on the nature of the rock, sand, or soil in the aquifer where the well withdraws water. Drinking water wells may be shallow (50 feet or less) or deep (more than 1,000 feet). Including the approximately 23 million Americans who use groundwater as private drinking water sources, slightly more than half of the population receives its drinking water from groundwater sources.

DRINKING WATER CONVEYANCE AND DISTRIBUTION

Water as shown in Figures 6.1 and 6.2, regardless of the source, is conveyed to the waterworks for treatment and distributed to the users. Conveyance from the source to the point of treatment occurs by aqueducts, pipelines, or open channels, but the treated water is normally distributed in pressurized closed conduits. After use, whatever the purpose, the water becomes wastewater, which must be disposed of somehow, but almost always ends up being conveyed back to a treatment facility before being out-falled to some waterbody to begin the cycle again.

We can call the portrayals depicted in Figures 6.1 and 6.2 simplified presentations of the urban water cycle because they provide a human-generated imitation of the natural water cycle. Unlike the natural water cycle, however, without pipes the cycle would be non-existent or, at the very least, short-circuited. With regard to open-channel flow, water is transported over long distances through aqueducts to locations where it is to be used and/or treated. Selection of an aqueduct type rests on such factors as topography, head availability, climate, construction practices, economics, and water quality protection. Along with pipes and tunnels, aqueducts may also include or be solely composed of open channels.*

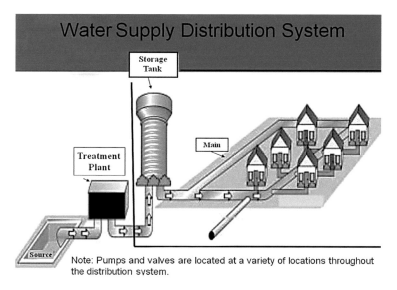

FIGURE 6.2 Simplified drinking water distribution system.Source: USEPA (2019). Accessed 7/19 @ https:/ /www.epa.gov/dwsixyearreviw/drinking-water-distribution-systems.

---

* Viessman, W., Jr., & Hammer, M.J., *Water Supply and Pollution Control*, 6th ed. Menlo Park, CA: Addison-Wesley, p. 119, 1998.

Public water system administrators and administrative managers need to possess a fundamental knowledge of the water conveyance and distribution system. Again, this knowledge does not necessarily need to be of the nuts and bolts variety but instead needs to be an overall fundamental understanding of how the public water system delivers water to its customers. For example, administrators and associated managers need to know that water distribution systems consist of an interconnected series of pipes, storage facilities, and other components that convey drinking water. Water distribution needs to meet fire protection needs for cities, homes, schools, hospitals, business, industries, and other facilities. Public water systems depend on distribution systems to provide an uninterrupted supply of pressurized safe drinking water to all consumers.

Distribution systems span almost one million miles in the United States. They represent the vast majority of physical infrastructure for water supplies. Distribution system wear and tear can pose intermittent or persistent health risks and result in large costs and accompanying lawsuits. The purpose of this chapter is to provide the fundamental knowledge needed by public water system nontreatment operating personnel on all of these functions and issues.

SURFACE WATER/GROUNDWATER DISTRIBUTION SYSTEMS

Major water supply systems can generally be divided into two categories based on the source of water they use. The water source, in turn, impacts the design, construction, and operation of the water distribution systems. The types of systems classified by source (see Figures 6.3 and 6.4) are:

- Surface water supply systems; and
- Groundwater supply systems.

Surface water (acquired from rivers, lakes, or reservoirs) flows through an intake structure into the transmission system (Figure 6.3). From a groundwater source, flow moves through an intake pipe from the groundwater source and is then pumped through a transmission conduit that conveys the water to a distribution system (Figure 6.4). Groundwater is generally available in most of the United States; however, the amount available for withdrawal at any particular location is usually limited. Surface water and groundwater supply systems may contain canals, pipes, and other conveyances; pumping plants; and distribution reservoirs or tanks to assist in balancing supplies and demands for water and to control pressures, other appurtenances, and treatment works.

Note: To illustrate drinking water supply conveyance and distribution in its most basic form, we concentrate on the major components of a typical "surface" water supply and distribution system.

In a typical community water supply system, water is transported under pressure through a distribution network of buried pipes. Smaller pipes (house service lines) attached to the main water

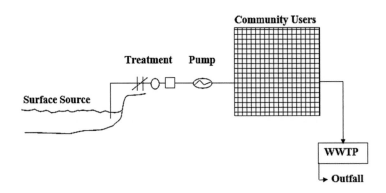

FIGURE 6.3 Example of surface water supply system.

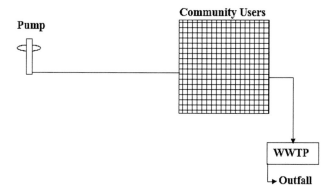

FIGURE 6.4   Example of groundwater supply system.

lines bring water from the distribution network to households. In many community water supply systems, pumping water up into storage tanks that store water at higher elevations than the households they serve provides water pressure. The force of gravity then "pushes" the water into homes when household taps open. Households on private supplies usually get their water from private wells. A pump brings the water out of the ground and into a small tank within the home, where the water is stored under pressure.

In cities, while the distribution system generally follows street patterns, it is also affected by topography and by the types of residential, commercial, and industrial development, as well as the location of treatment facilities and storage works. A distribution system is often divided into zones that correspond to different ground elevations and service pressures. The water pipes (mains) are generally enclosed loops so that supply to any point can be provided from at least two directions. Street mains usually have a minimum diameter of 6–8 inches to provide adequate flows for buildings and for fighting fires. The pipes connected to buildings may range down to as small as 1 inch for small residences.

Surface Water Intake

Withdrawing water from a lake, reservoir, or river requires an intake structure. Because surface sources of water are subject to wide variations in flow, quality, and temperature, intake structures must be designed so that the required flow can be withdrawn despite these natural fluctuations. Surface water intakes consist of screened openings and conduits that convey the flow to a sump from which it may be pumped to the treatment works. Typical intakes are towers, submerged ports, and shoreline structures.

Intakes function primarily to supply the highest quality water from the source and to protect downstream piping, equipment, and unit processes from damage or clogging as a result of floating and submerged debris, flooding, and wave action. To facilitate this, intakes should be located to consider the effects of anticipated variations in water level, navigation requirements, local currents and patterns of sediment deposition and scour, spatial and temporal variations in water quality, and the quantity of floating debris.

For *lakes* and *impounding reservoirs* where fluctuating water levels and variations in water quality with depth are common, intake structures that permit withdrawal over a wide range of elevations are typically used. Towers are commonly used for reservoirs and lakes. A tower water intake provides ports located at several depths, avoiding the problems of water quality that stem from locating a single inlet at the bottom, since the water quality varies with both time and depth. With the exception of brief periods in spring and fall when overturns may occur, water quality is usually best close to the surface; thus, intake ports located at several depths (see Figure 6.5) permit selection of

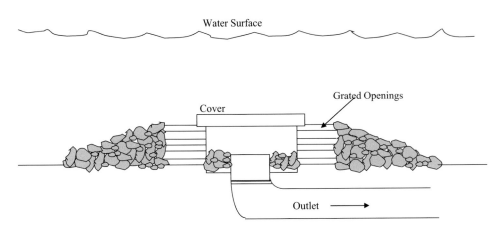

FIGURE 6.5   Submerged intake used for both lake and river sources.

the most desirable water quality in any season of the year. Submerged ports also have the advantage of remaining free from ice and floating debris. Selection of port levels must be related to characteristics of the waterbody (Spellman, 2014).

Other considerations affect lake intake location, including: (1) locate as far as possible from any source of pollution; (2) factor in wind and current effects on the motion of contaminants; and (3) provide sufficient water depth (typically 20–30 feet) necessary to prevent blocking of the intake by ice jams that may fill shallower lake areas to the bottom (McGhee, 1991). River intakes are typically designed to withdraw water from slightly below the surface to avoid both sediment in suspension at lower levels and floating debris, and if necessary, at levels low enough to meet navigation requirements. Generally, river intakes are submerged (see Figure 6.5) or screened shore intakes. Because of low costs, the submerged type is widely used for small river and lake intakes. Note: If they need repair, submerged intakes are not readily accessible, hampering any repair or maintenance activity. If used in lakes or reservoirs, another distinct disadvantage is the lack of alternate withdrawal levels to choose the highest water quality available throughout the year.

SURFACE WATER DISTRIBUTION

Along with providing potable water to the household tap, water distribution systems are ordinarily designed to adequately satisfy the water requirements for a combination of domestic, commercial, industrial, and fire-fighting purposes. The system should be capable of meeting the demands placed on it at all times and at satisfactory pressures. Pipe systems, pumping stations, storage facilities, fire hydrants, house service connections, meters, and other appurtenances are the main elements of the system (Cesario, 1995). Water is normally distributed using one of three different means: gravity distribution, pumping without storage, or pumping with storage. *Gravity distribution* is possible only when the source of supply is located substantially above the level of the community. *Pumping without storage* (the least desirable method because it provides no reserve flow and pressures fluctuate substantially) uses sophisticated control systems to match an unpredictable demand. *Pumping with storage* is the most common method of distribution (McGhee, 1991).

DISTRIBUTION LINE NETWORK

Distribution systems may be generally classified as grid systems, branching systems, a combination of these, or dead-end systems. The branching system is not the preferred distribution network because it does not furnish supply to any point from at least two directions and because it includes several terminals or dead ends. Normally, grid systems are the best arrangement for distributing water. All of the arterials and secondary mains are looped and interconnected, eliminating dead ends and permitting water circulation in such a way that a heavy discharge from one main allows drawing water from other pipes. Newly constructed distribution systems avoid the antiquated dead-end system, and such systems are often retrofitted later by incorporating proper looping.

Service Connection to Household Tap

A typical service connection consists of a pipe from the distribution system to a turnoff valve located near the property line.

Distribution Storage

Distribution reservoirs and other storage facilities/vessels are in place to provide a sufficient amount of water to average or equalize daily demands on the water supply system. Storage also serves to increase operating convenience; to level out pumping requirements; to decrease power costs; to provide water during power source or pump failure; to provide large quantities of water to meet fire demands; to provide surge relief; to increase detention time; and to blend water sources. Generally, six different types of storage units are employed in storing potable water: (1) clear wells; (2) elevated tanks; (3) stand pipes; (4) ground-level reservoirs; (5) hydro pneumatic or pressure tanks; and (6) surge tanks.

- *Clear Wells* – Used to store filtered water from treatment works, and as chlorine contact tanks.
- *Elevated Tanks* – Located above the service zone and used primarily to maintain an adequate and fairly uniform pressure to the service zone (see Figure 6.6).
- *Stand Pipes* – Tanks that stand on the ground, with a height greater than their diameter.
- *Ground-Level Reservoirs* – Located above service area to maintain the required pressures.

FIGURE 6.6   Elevated Water Tower – Norfolk, VA. Photo by F.R. Spellman.

- *Hydro Pneumatic or Pressure Tanks* – Usually used on small public water systems such as with a well or booster pump. The tank is used to maintain water pressures in the system and to control the operation of the well pump or booster pump.
- *Surge Tanks* – Not necessarily storage facilities, but used mainly to control water hammer, or to regulate water flow.

PIPES

For use as water mains in a distribution system, pipes must be strong and durable in order to resist applied forces and corrosion. The pipe is subjected to internal pressure from the water and to external pressure from the weight of the backfill (soil) and vehicles above it. The pipe may also have to withstand water hammer. Damage due to corrosion or rusting may also occur internally because of the water quality or externally because of the nature of the soil conditions. Pipes used in a water system must be strong and durable to resist the abrasive and corrosive properties of the waste. Moreover, water pipes must also be able to withstand stresses caused by the soil backfill material and the effect of vehicles passing above the pipeline. Joints between water pipe sections should be flexible, but tight enough to prevent excessive leakage, either of water out of the pipe, or groundwater or contaminants into the pipe. Of course, pipes must be constructed to withstand the expected conditions of exposure, and pipe configuration systems for water distribution must be properly designed and installed in terms of water hydraulics.

DID YOU KNOW?

Fluids travel through a piping system at various pressures, temperature, and speeds.

PIPING NETWORKS

It would be far less costly and make for more efficient operation if municipal water systems were built with separate single pipe networks extending from treatment plant to user's residence, or from user's sink or bathtub drain to the local wastewater treatment plant. Unfortunately, this ideal single-pipe scenario is not practical for real world applications. Instead of a single piping system, a network of pipes is laid under the streets. Each of these piping networks is composed of different materials that vary (sometimes considerably) in diameter, length, and age. These networks range in complexity to varying degrees, and each of these joined-together pipes contributes energy losses to the system.

ENERGY LOSSES IN PIPE NETWORKS

Water flow networks may consist of pipes arranged in series, parallel, or some complicated combination. In any case, an evaluation of friction losses for the flows is based on energy conservation principles applied to the flow junction points. Methods of computation depend on the particular piping configuration. In general, however, they involve establishing a sufficient number of simultaneous equations or employing a friction loss formula where the friction coefficient depends only on the roughness of the pipe.

Note: Excessive air bubbles in a pipe can cause air pockets to form at the high points in pipes.

CONVEYANCE SYSTEMS

With regard to early conveyance systems, the prevailing practice in medieval England was the use of closed pipes. This practice was contrary to the Romans who generally employed open channels in their long-distance aqueducts and used pipes mainly to distribute water within cities. The English preferred to lay long runs of pipes from the water source to the final destination. The Italians, on the other hand, where antique aqueduct arches were still visible, seem to have had more of a tendency to follow the Roman tradition of long-distance channel conduits. At least some of the channel aqueducts seem to have fed local distribution systems of lead or earthenware pipes (Magnusson, 2001).

With today's water conveyance, not that much has changed from the past. Our goal today remains the same: Convey water from source to treatment facility to user. In water operations, the term *conveyance* or *piping system* refers to a complete network of pipes, valves, and other components. For water operations in particular, the piping system is all-inclusive; it includes both the network of pipes, valves, and other components that bring the flow (water or wastewater) to the treatment facility, as well as piping, valves, and other components that distribute treated water to the end user. In short, all piping systems are designed to perform a specific function and can be composed of several different components (see Figure 6.7). Figure 6.7 shows a single-line diagram that is similar to an electrical schematic. It uses symbols for all the diagram components. A double-line diagram (not shown here) is a pictorial view of the pipe, joints, valves, and other major components similar to an electrical wiring diagram vs. an electrical schematic. To read piping schematics correctly, maintenance operators must identify and understand the symbols used. It is not necessary to memorize them, but maintenance operators should keep a table of basic symbols handy and refer to it whenever the need arises. In the following sections, many of the common symbols used in general piping systems are described.

Note: To connect a new water main to an existing water main a coupling or Dresser is used.

IMPORTANCE OF CONVEYANCE SYSTEMS

Probably the best way to illustrate the importance of a piping system is to describe many of its applications used in water operations. In the modern water treatment, plant piping systems are critical to successful operation. In water operations, fluids and gases are used extensively in processing

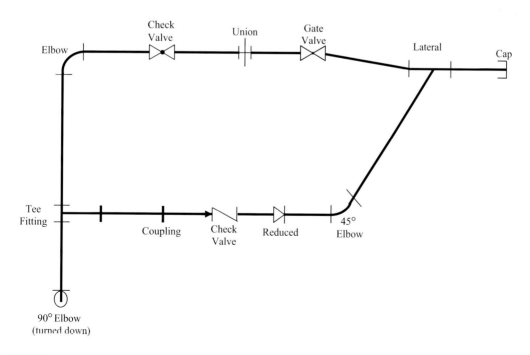

FIGURE 6.7 Shows various components in a single-line piping diagram.

operations; they usually are conveyed through pipes. Piping carries water into plants for treatment, fuel oil to heating units, steam to steam services, lubricants to machinery, compressed air to pneumatic service outlets for air-powered tools, etc., and chemicals to unit processes.

For water treatment alone, as Kawamura (1999) pointed out, there are

> six basic piping systems: (1) raw water and finished waste distribution mains; (2) plant yard piping that connects the unit processes; (3) plant utility, including the fire hydrant lines; (4) chemical lines; (5) sewer lines; and (6) miscellaneous piping, such as drainage and irrigation lines.

Any waterworks has many piping systems, not just the systems that convey water. Along with those mentioned earlier, keep in mind that plant piping systems also include those that provide hot and cold water for plant personnel use. Another system heats the plant, while still another may be used for air conditioning.

Water managers, operators, and maintenance operators have many responsibilities and basic skills. The typical plant operator is skilled in HVAC (heating, ventilation, and air-conditioning) systems, chemical feed systems, mechanical equipment operation and repair, and in piping system maintenance activities. However, only the fluid transfer systems themselves are important to us in this chapter.

FLUIDS VERSUS LIQUIDS

Fluids describe the substance(s) being conveyed through various piping systems from one part of the plant to another. We normally think of pipes conveying some type of liquid substance, which most of us take to have the same meaning as fluid; however, there is a subtle difference between the two terms. The dictionary's definition of *fluid* is any substance that flows – which can mean a liquid or gas (air, oxygen, nitrogen, etc.). Some fluids carried by piping systems include thick viscous mixtures such as sludge (biosolids) in a semi-fluid state. Although sludge (biosolids) and other such materials might seem more solid (at times) than liquid, they do flow, and are considered fluids. In addition to carrying liquids such as oil, hydraulic fluids, and chemicals, piping systems carry compressed air and steam, which also are considered fluids because they flow.

SIDEBAR: THE PIPING MAZE — COMMENTARY ON A REAL-WORLD EXAMPLE

It is rather simple and straightforward to say that public water systems use pipes, a lot of pipes and miles of pipes, to carry various fluids from place to place throughout the system. However, it could be an entirely different matter, a difficult matter: befuddling, head-scratching, muddling, and/or urbanely defined as a "*mazing matter*" to guess what a particular water piping system or single pipe is actually conveying. To make the point about how confusing a piping maze, a massive gallery of above ground (usually ceiling hung) pipes can be to the uninformed, uninitiated rookie water person or a student can be, the following example is used:

In teaching water treatment principles to undergrad and grad students in my college classes, eventually the topic turns to water conveyance systems (piping and valves). Before the class presentation on piping and valves lectures, I have all my students meet me behind the college gymnasium so that we can walk across the street and into a fairly large (60 MGD) wastewater treatment plant located on the banks of the Elizabeth River in Norfolk, Virginia. I am quite familiar with this plant and know exactly where to lead the students to the location that will make the point that I am trying to get across to them. We enter into a large tunnel system, the kind you can drive a pickup truck-type vehicle into and through with room to spare on all sides. The tunnel system is located beneath several bar screening devices and a dozen massive primary clarifiers. It is into one of the corner entrance areas beneath the bar screens and clarifiers that I lead them. Once gathered inside, I tell them to take their time to look around and to mentally note what they see. Eventually, I have all of them turn their gazes upward to the massive pipe gallery that runs the entire length of the tunnel system. Then I make my point by asking: "What do you think those pipes are conveying?" No answer … none … well, none I can repeat here. Then I ask: "Do you think it is important to know what is inside those pipes?" Some of the students shake their heads sideways to mean no. Some of the students shake their heads up and down to mean yes. Then I give them a real-world situation that occurs often within the bowels of any treatment plant, water or wastewater. "Suppose you are a maintenance person who has been directed to go to this exact location to cut into a certain pipe to install a new isolation valve."

You know it is fascinating to me to see the looks on some of the faces of those students who anticipate where I am leading them before the full explanation. Anyway, I go on. "Now, as that maintenance person arrives here and looks up at the same maze of pipes above our heads, how do you suppose he/she is able to find the correct pipe to cut into?" Then I see that look on a few faces that indicates that they *may* know what I know. So, I ask them, "Do any of you know how he/she will pick the right pipe to cut into?" It does not take long to get a response … usually a few students respond in unison. "The pipes are colored … that is, they are painted … in different colors … they probably use a color-code system." Ah, when I recognize that many of these students are smart enough to figure out the key to solving the what-pipe-is-which-puzzle, I just want to state, "Right on, grasshoppers." So, I do; I say it. Before leaving the plant tunnel and heading for the classroom, I know that someone, anyone, sometimes a whole bunch of students at the same time will make my Eureka Moment … and they do; they always do. Someone, anyone, or the entire bunch mentions (I do not have dumb students, never have, never will have) that it would be impossible, maybe, for any maintenance person, even one knowing the coatings color scheme to pick out the correct pipe to work on. Why? Because as the student(s) pointed out there are many pipes of the exact same color above our heads in that piping gallery. That is the Eureka Moment (for me, anyway).

After this comment I move on down tunnel; my students follow until about the 20- or 25-foot mark and then I tell them to look up again. There, clear as marking allows, are stenciled markings in black or white in color to contrast with the pipe background color as appropriate; assorted letter sizes are dependent on the size of the pipes. The students also notice that the colored and stenciled pipes are also numbered. The numbering is the key to determining which pipe is which. The maintenance person would simply have to refer to the piping diagrams to determine which pipe the supervisor has directed him/her to cut into and install the valve. In the tunnel, before we leave, I make one final point about the piping color identification scheme and marking: If the plant's piping system were not color-coded and stenciled and the maintenance person had to pick willy-nilly which pipe to use his or her cutting torch on, what would happen if the maintenance person picked the wrong pipe and instead put his or her torch to a pipe containing explosive methane? All points having been made it was time to get back to the classroom.

Note: If a new pipe has significant tuberculation inside the pipe it indicates that the water is corrosive.

PIPING AND EQUIPMENT COLOR IDENTIFICATION CODE

As made clear in the sidebar, the objective of a piping system's Piping and Equipment Color Identification Code is to provide easy and rapid identification of the various unit process equipment and piping in a treatment plant. The large amount of equipment and piping in a water treatment plant requires that a simple system be used to identify and distinguish the equipment and piping.

DID YOU KNOW?

If piping is stainless steel and/or PVC and/or there are other viable reasons not to paint (piping is already painted and may not be economical to repaint), then the following is recommended or should be considered: All piping should be distinctly and frequently marked with color-coded bands (as listed in Table 6.2) from around the pipe and labeled.

PIPING COLORING CODING AND MARKING REQUIREMENTS

The Color Shade Code should be used to provide a uniform shade of color when using various manufacturers and types of coatings (see Table 6.1). The color should be based on the US Government Federal Register 595 Paint Color Number.

The Color Selection Code should be used to select the color for all equipment and piping in the treatment plant. The selected color for individual equipment and piping would be based on the type of flow involved. As stated, the Equipment and Piping Color Identification Code should be used to identify all equipment in the treatment plant. The identification for all equipment should include *both* the proper coating *color* and the proper stenciled *markings*.

Equipment Markings

- Stenciled markings should be placed on multiple units of equipment to indicate the unit number, i.e., No. 1, No. 2, etc.
- Stenciled markings should be black or white in color to contrast with the equipment background color as appropriate.
- Stenciled markings should be 2.5 inches in size.

The Equipment and Piping Color Identification Code should be used to identify all piping in the treatment plant. The identification for all piping should include *both* the proper coating *color* and the proper stenciled *markings*.

Piping Markings

Stenciled markings should include both the proper unit process flow abbreviations and a direction arrow indicating the proper flow direction as appropriate. Stenciled markings should be placed wherever pipes pass through walls, at the influent and effluent of pumps or valves, at intersections, and at regular intervals along pipe runs as appropriate. Stenciled markings should be black or white

TABLE 6.1
Color Shade Code

Color Shade	US Government Federal Register 595 Paint Color Number
Aluminum	17178
Black	17038
Blue	15050
Brown	10091
Gray	16473
Green, Dark	14062
Green, Light	14533
Orange	12246
Red	11105
White	17875
Yellow	13655

TABLE 6.2
Water Treatment Plant Color Coding

Type of Pipe	Use of Pipe	Color of Pipe
Water Lines	Raw Water	Olive Green
	Settled or Clarified Water	Aqua
	Finished or Portable Water	Dark Blue
Chemical Lines	Alum or Primary Coagulant	Orange
	Ammonia	White
	Carbon Slurry	Black
	Caustic	Yellow w/ Green Band
	Chlorine Gas or Solution	Yellow
	Fluoride	Light Blue w/ Red Band
	Lime Slurry	Light Green
	Ozone	Yellow w/ Orange Band
	Phosphate Compounds	Light Green w/ Red Band
	Polymers or Coagulant Aids	Orange w/ Green Band
	Potassium Permanganate	Violet
	Soda Ash	Light Green w/ Orange Band
	Sulfuric Acid	Yellow w/ Red Band
	Sulfur Dioxide	Light Green w/ Yellow Band
Waste Lines	Backwash Waste	Light Brown
	Sludge	Dark Brown
	Sewer (sanitary or other)	Dark Grey
Other Lines	Compressed Air	Dark Green
	Gas	Red
	Other Pipes	Light Gray

---

in color to contrast with the pipe background color as appropriate. Stenciled markings for pipes should be all capital letters, sized as follows (see Table 6.2):

Pipe Size	Letter Size
Up to 1.5 inch	.5 inch
2–6 inches	1.25 inch
8 inches and up	2.5 inches

MAINTAINING FLUID FLOW IN PIPING SYSTEMS

The primary purpose of any piping system is to maintain a free and smooth flow of fluids through the system. Another purpose is to ensure that the fluids being conveyed are kept in good condition (i.e., free of contamination). Piping systems are purposely designed to ensure a free and smooth flow of fluids throughout the system, but additional system components are often included to ensure that fluid quality is maintained. Piping system filters are one example, and strainers and traps are two others. It is extremely important to maintain free and smooth flow and fluid quality in piping systems, especially those that feed vital pieces of equipment/machinery.

Consider the internal combustion engine, for example. Impurities such as dirt and metal particles can damage internal components and cause excessive wear and eventual breakdown. To help prevent such wear, the oil is run continuously through a filter designed to trap and filter out the impurities.

Other piping systems need the same type of protection that the internal combustion engine does, which is why most piping systems include filters, strainers, and traps. These filtering components may prevent damage to valves, fittings, the pipe itself, and downstream equipment/machinery. Chemicals, various types of waste products, paint, and pressurized steam are good examples of potentially damaging fluids. Filters and strainers play an important role in piping systems, protecting both the piping system and the equipment that the piping system serves.

Scaling

Because sodium and calcium hypochlorite are widely used in water treatment operations, problems in piping systems feeding this chemical are of special concern. In this section, we discuss *scaling* problems that can occur in piping systems that convey hypochlorite solution. To maintain the chlorine in solution (used primarily as a disinfectant), sodium hydroxide (caustic) is used to raise the pH of the hypochlorite; the excess caustic raises the shelf life. A high-pH caustic solution raises the pH of the dilution water to greater than pH 9.0 after it is diluted. The calcium in the dilution water reacts with dissolved CO_2 and forms calcium carbonate.

Experience has shown that 2-inch pipes have turned into 3/4-inch pipes due to scale buildup. The scale deposition is greatest in areas of turbulence such as pumps, valves, rotameters, and backpressure devices. If lime (calcium oxide) is added (for alkalinity), plant water used as dilution water will have higher calcium levels and will generate more scale. While it is true that softened water will not generate scale, it is also true that it is expensive in large quantities. Many facilities use softened water on hypochlorite mist odor scrubbers only.

Scaling also often occurs in solution rotameters, making flow readings impossible and freezing the flow indicator in place. Various valves can freeze up and pressure-sustaining valves freeze and become plugged. Various small diffuser holes fill with scale. To slow the rate of scaling, many facilities purchase water from local suppliers to dilute hypochlorite for miscellaneous uses. Some facilities have experimented with this system by not adding lime to it. When they did this, manganese dioxide (black deposits) developed on the rotameters glass, making viewing the float impossible. In many instances, moving the point of hypochlorite addition to downstream of the rotameter seemed to solve the problem.

If remedial steps are not taken, scaling from hypochlorite solutions can cause problems. For example, scale buildup can reduce the inside diameter of pipe so much that the actual supply of hypochlorite solution required to properly disinfect water was reduced. As a result, the water sent to the customer may not be properly disinfected. Because of the scale buildup, the treatment system itself will not function as designed and could result in a hazardous situation in which the reduced pipe size increases the pressure level to the point of catastrophic failure. Scaling, corrosion, or other clogging problems in certain piping systems are far from an ideal situation.

DID YOU KNOW?

A basic principle in fluid mechanics states that fluid flowing through a pipe is affected by friction – the greater the friction, the greater the loss of pressure. Another principle or rule states that the amount of friction increases as the square of the velocity. (Note that speed and velocity are not the same, but common practice refers to the "velocity" of a fluid). In short, if the velocity of the fluid doubles, the friction is increased four times what it was before. If the velocity is multiplied by five, the friction is multiplied by 25, and so on.

Piping System Maintenance

Maintaining a piping system can be an involved process. However, good maintenance practices can extend the life of piping system components, and rehabilitation can further prolong their life. The performance of a piping system depends on the ability of the pipe to resist unfavorable conditions and to operate at or near the capacity and efficiency that it was designed for. This performance can be checked in several ways: flow measurement; fire flow tests; loss-of-head tests; pressure tests; simultaneous flow and pressure tests; tests for leakage; and chemical and bacteriological water tests. These tests are an important part of system maintenance. They should be scheduled as part of the regular operation of the system (AWWA, 2003).

Most piping systems are designed with various protective features included that are designed to minimize wear and catastrophic failure and, therefore, the amount of maintenance required. Such protective features include pressure relief valves, blow-off valves, and clean-out plugs.

- *Pressure Relief Valves* – A valve that opens automatically when the fluid pressure reaches a *preset* limit to relieve the stress on a piping system.
- *Blow-Off Valve* – A valve that can be opened to blow out any foreign material in a pipe.
- *Clean-Out Plug* – A threaded plug that can be removed to allow access to the inside of the pipe for cleaning.

Important Point: Use caution when removing a clean-out plug from a piping system. Before removing the plug, pressure must be cut off and the system bled of residual pressure. Many piping systems (including water distribution networks and wastewater lines and interceptors) can be cleaned either by running chemical solvents through the lines or by using mechanical clean-out devices. Note: C factor is a representation of the roughness of the pipe. High C factor means the pipe has a smooth interior.

Piping System Accessories

Along with valves, piping systems typically include accessories such as pressure and temperature gauges, filters, strainers, and pipe hangers and supports.

- *Pressure gauges* indicate the pressure in the piping system.
- *Temperature gauges* indicate the temperature in the piping system.
- *Filters and strainers* are installed in piping systems to help keep fluids clean and free from impurities.
- *Pipe hangers and supports* support piping to keep the lines straight and prevent sagging, especially in long runs. Various types of pipe hangers and supports are shown in Figure 6.8.

CONVEYANCE FLOW CONTROL DEVICES

Any water operation will have many valves that require attention. Simply as a matter of routine, a maintenance operator must be able to identify and locate different valves in order to inspect them, to adjust them, and to repair or replace them. For this reason, the operator should be familiar with all valves, especially those that are vital parts of a piping system. Note: To maintain an acceptable water pressure downstream of a high-pressure zone in a water distribution system, a regulator station (steps down pressure like a transformer steps down current flow) is used.

Valves: Definition and Function

A valve is defined as any device by which the flow of fluid may be started, stopped, or regulated by a movable part that opens or obstructs passage. As applied in fluid power systems, valves are used for controlling the flow, the pressure, and the direction of the fluid flow through a piping system. The fluid may be a liquid, a gas, or some loose material in bulk (like biosolids slurry). Designs of valves vary, but all valves have two features in common: a passageway through which fluid can flow and some kind of movable (usually machined) part that opens and closes the passageway.

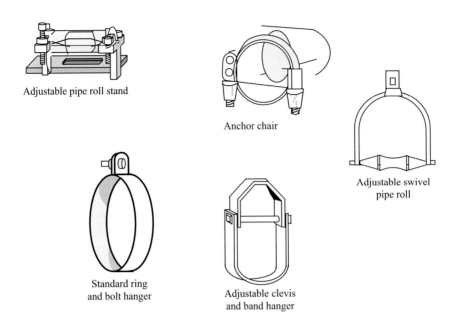

Adjustable pipe roll stand

Anchor chair

Adjustable swivel
pipe roll

Standard ring
and bolt hanger

Adjustable clevis
and band hanger

FIGURE 6.8   Pipe hangers and supports.

Important Point: It is all but impossible, obviously, to operate a practical fluid power system without some means of controlling the volume and pressure of the fluid and directing the flow of fluid to the operating units. This is accomplished by the incorporation of different types of valves. Whatever type of valve is used in a system, it must be accurate in the control of fluid flow and pressure and the sequence of operation. Leakage between the valve element and the valve seat is reduced to a negligible quantity by precision-machined surfaces, resulting in carefully controlled clearances. This is, of course, one of the very important reasons for minimizing contamination in fluid power systems.

Contamination causes valves to stick, plugs small orifices, and causes abrasions of the valve seating surfaces, which results in leakage between the valve element and valve seat when the valve is in the closed position. Any of these can result in inefficient operation or complete stoppage of the equipment. Valves may be controlled manually, electrically, pneumatically, mechanically, hydraulically, or by combinations of two or more of these methods. Factors that determine the method of control include the purpose of the valve, the design and purpose of the system, the location of the valve within the system, and the availability of the source of power. Valves are made from bronze, cast iron, steel, Monel®, stainless steel, and other metals. They are also made from plastic and glass (see Table 6.3). Special valve trim is used where seating and sealing materials are different from the basic material of construction. (*Valve trim* usually means those internal parts of a valve controlling the flow and in physical contact with the line fluid.) Valves are made in a full range of sizes, which match pipe and tubing sizes. Actual valve size is based upon the internationally agreed-upon definition of nominal size. *Nominal size (DN)* is a numerical designation of the size that is common to all components in a piping system other than components designated by outside diameters. It is a convenient number for reference purposes and is only loosely related to manufacturing dimensions. Valves are made for service at the same pressures and temperatures that piping and tubing are subject to. Valve pressures are based upon the internationally agreed upon definition of nominal pressure. *Nominal pressure (PN)* is a pressure that is conventionally accepted or used for reference purposes. All equipment of the same nominal size *(DN)* designated by the same nominal pressure *(PN)* number must have the same mating dimensions appropriate to the type of end connections. The permissible working pressure depends upon materials, design, and working temperature, and should be selected from the (relevant) pressure/temperature tables.

TABLE 6.3

Valves – Materials of Construction

Materials of Construction	
Cast Iron	Grey cast iron, also referred to as flake graphite iron
Ductile Iron	May be malleable iron, or spheroidal graphite (nodular) cast iron
Carbon Steel	May be as steel forgings, or steel castings, according to the method of manufacture; carbon steel valves may also be manufactured by fabrication using wrought steels
Stainless Steel	May also be in the form of forgings, castings, or wrought steels for fabrication
Copper Alloy	May be gunmetal, bronze, or brass; aluminum bronze may also be used
High Duty Alloys	Are usually those nickel or nickel–molybdenum alloys manufactured under various trade names
Other Metals	Are those pure metals having extreme corrosion resistance such as titanium or aluminum
Non-metals	Are typically the plastics materials such as PVC or polypropylene

The pressure rating of many valves is designated under the American (ANSI) class system. The equivalent class rating to PN ratings is based upon international agreement. Usually, valve end connections are classified as flanged, threaded, or other. Valves are also covered by various codes and standards, as are the other components of piping and tubing systems. Many valve manufacturers offer valves with special features. Table 6.3 lists a few of these special features; however, this is not an exhaustive list and for more details on other features, the manufacturer should be consulted. The various types of valves used in fluid power systems, their classifications, and their applications are discussed in this section. Depending on the complexity of the piping system, the number of valves included in a system can range from no more than one in a small, simple system to a large number in very complex systems such as water distribution systems. Again, valves are necessary for both the operation of a piping system and for control of the system and system components. In water/wastewater treatment, this control function is used to control various unit processes, pumps, and other equipment. Valves also function as protective devices. For example, valves used to protect a piping system may be designed to open automatically to vent fluid out of the pipe when the pressure in the lines becomes too high. In lines that carry liquids, relief valves preset to open at a given pressure are commonly used. Important Point: Not all valves function as safety valves. For example, hand-operated gate and globe valves function primarily as control valves. The size and type of valve are selected depending on its intended use. Most valves require periodic inspection to ensure that they are operating properly.

Valve Construction

In a common valve (see Figure 6.9), fluid flows into the valve through the inlet. The fluid flows through passages in the body and past the opened element that closes the valve. It then flows out of the valve through the outlet or discharge. If the closing element is in the closed position, the passageway is blocked. Fluid flow is stopped at that point. The closing element keeps the flow blocked until the valve is opened again. Some valves are opened automatically, and manually operated hand wheels control others. Other valves, such as check valves, operate in response to pressure or the direction of flow. To prevent leakage whenever the closing element is positioned in the closed position, a seal is used.

Types of Valves

The types of valves covered in this chapter include:

- Ball valves;
- Gate valves;

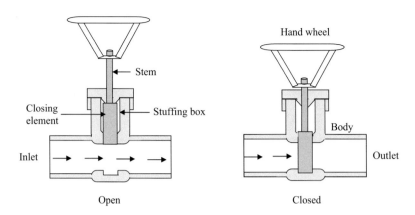

FIGURE 6.9   Basic valve operation.

- Globe valves;
- Needle valves;
- Butterfly valves;
- Plug valves;
- Check valves;
- Quick-opening valves;
- Diaphragm valves;
- Regulating valves;
- Relief valves; and
- Reducing valves.

Each of these valves is designed to perform either control of the flow, the pressure, and the direction of fluid flow, or some other special application. With a few exceptions, these valves take their names from the type of internal element that controls the passageway. The exceptions are the check valve, quick-opening valve, regulating valve, relief valve, and reducing valves.

Ball Valves

Ball valves, as the name implies, are stop valves that use a ball to stop or start fluid flow. The ball performs the same function as the disk in other valves. As the valve handle is turned to open the valve, the ball rotates to a point where part or all of the hole through the ball is in line with the valve body inlet and outlet, allowing fluid to flow through the valve; when the ball is rotated so the hole is perpendicular to the flow openings of the valve body, flow of fluid stops. Most ball valves are the quick-acting type. They require only a 90° turn to either completely open or close the valve. However, many are operated by planetary gears. This type of gearing allows the use of a relatively small hand wheel and operating force to operate a large valve. The gearing does, however, increase the operating time for the valve. Some ball valves also contain a swing check located within the ball to give the valve a check valve feature. The two main advantages of using ball valves are: (1) the fluid can flow through it in either direction, as desired; and (2) when closed, pressure in the line helps to keep it closed.

Gate Valve

Gate valves are used when a straight-line flow of fluid and minimum flow restriction is needed; they are the most common type of valve found in a water distribution system. Gate valves are so-named because the part that either stops or allows flow through the valve acts somewhat like a gate. The gate is usually wedge-shaped. When the valve is wide open, the gate is fully drawn up into the valve bonnet. This leaves an opening for flow through the valve the same size as the pipe in which the valve is installed.

For these reasons, the pressure loss (pressure drop) through these types of valves is about equal to the loss in a piece of pipe of the same length. Gate valves are not suitable for throttling (which means to control the flow as desired, by means of intermediate steps between fully open and fully closed) purposes. The control of flow is difficult because of the valve's design, and the flow of fluid slapping against a partially open gate can cause extensive damage to the valve.

Globe Valves

Probably the most common valve type in existence, the globe valve is commonly used for water faucets and other household plumbing and is usually used to control flow and pressure. As shown in Figure 6.10, the valves have a circular disk (the globe) that presses against the valve seat to close the valve. The disk is the part of the globe valve that controls the flow. The disk is attached to the valve stem. As shown in Figure 6.10, fluid flow through a globe valve is at right angles to the direction of flow in the conduits. Globe valves seat very tightly and can be adjusted with fewer turns of the wheel than gate valves; thus, they are preferred for applications that call for frequent opening and closing. On the other hand, globe valves create high head loss when fully open; thus, they are not suited in systems where head loss is critical.

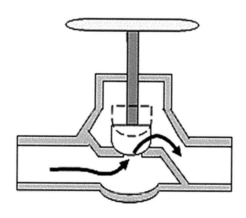

FIGURE 6.10 Globe valve.

Needle Valves

Although similar in design and operation to the globe valve, the *needle* valve (a variation of globe valves) has a closing element in the shape of a long tapered point, which is at the end of the valve stem. Figure 6.11 shows a cross-sectional view of a needle valve. The long taper of the valve-closing element permits a much smaller seating surface area than that of the globe valve; accordingly, the needle valve is more suitable as a throttle valve. In fact, needle valves are used for very accurate throttling.

Butterfly Valves

Figure 6.12 shows a cross-sectional view of a butterfly valve. The valve itself consists of a body in which a disk (butterfly) rotates on a shaft to open or close the valve. Butterfly valves may be flanged or wafer design, the latter intended for fitting directly between pipeline flanges. In the fully open position, the disk is parallel to the axis of the pipe and the flow of fluid. In the closed position, the disk seals against a rubber gasket-type material bonded either on the valve seat of the body or on the edge of the disk. Because the disk of a butterfly valve stays in the fluid path in the open position, the valve creates more turbulence (higher resistance to flow equaling a higher-pressure loss) than a gate valve. On the other hand, butterfly valves are compact. They can also be used to control flow in either direction. This feature is useful in water treatment plants that periodically backwash to clean filter systems.

Plug Valves

A *Plug* valve (also known as a *cock*, or *petcock*) is similar to a ball valve. Plug valves:

- Offer high capacity operation, 1/4 turn operation;
- Use either a cylindrical or conical plug as the closing member;
- Are directional;

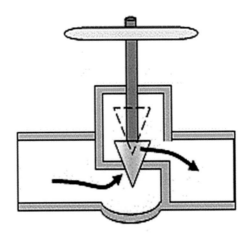

FIGURE 6.11   Common needle valve.

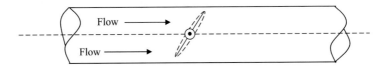

FIGURE 6.12   Cross section of butterfly valve.

- Offer moderate vacuum service;
- Allow flow throttling with interim positioning;
- Are of simple construction; O-ring seal;
- Are not necessarily full on and off;
- Are easily adapted to automatic control; and
- Can safely handle gases and liquids.

Check Valves

Check valves are usually self-acting and designed to allow the flow of fluid in one direction only. They are commonly used at the discharge of a pump to prevent backflow when the power is turned off. When the direction of flow is moving in the proper direction, the valve remains open. When the direction of flow reverses, the valve closes automatically from the fluid pressure against it.

There are several types of check valves used in water/wastewater operations, including the following:

- Slanting disk check valves;
- Cushioned swing check valves;
- Rubber flapper swing check valves;
- Double door check valves;
- Ball check valves;
- Foot valves; and
- Backflow prevention devices.

In each case, pressure from the flow in the proper direction pushes the valve element to an open position. Flow in the reverse direction pushes the valve element to a closed position. Important Point: Check valves are also commonly referred to as *non-return* or *reflux* valves.

Quick-Opening Valves

Quick-opening valves are nothing more than adaptations of some of the valves already described. Modified to provide a quick on/off action, they use a lever device in place of the usual threaded stem and control handle to operate the valve. This type of valve is commonly used in water operations where deluge showers and emergency eyewash stations are installed in work areas where chemicals are loaded, transferred, and/or where chemical systems are maintained. They also control the air supply for some emergency alarm horns around chlorine storage areas, for example. Moreover, they are usually used to cut off the flow of gas to a main or to individual outlets.

Diaphragm Valves

Diaphragm valves are glandless valves that use a flexible elastomeric diaphragm (a flexible disk) as the closing member and, in addition, create an external seal. They are well suited to serve in applications where tight, accurate closure is important. The tight seal is effective whether the fluid is a gas or a liquid. This tight closure feature makes these valves useful in vacuum applications. Diaphragm valves operate similar to globe valves and are usually multi-turn in operation; they are available as weir type and full bore. A common application of diaphragm valves in water operations is to control fluid to an elevated tank.

Regulating Valves

As their name implies, *regulating* valves regulate either pressure or temperature in a fluid line, keeping them very close to a preset level. If the demands and conditions of a fluid line remained steady at all times, no regulator valve would be needed. In the real world, however, ideal conditions do not occur.

Pressure-regulating valves regulate fluid pressure levels to meet flow demand variations. Flow variations vary with the number of pieces of equipment in operation and the change in demand as pumps and other machines operate. In such fluid line systems, demands are constantly changing. Probably the best example of this situation is seen in the operation of the plant's low-pressure air supply system. For shop use, no more than 30 psi air is usually required (depending on required usage, of course). This air is supplied by the plant's air compressor, which normally operates long enough to fill an accumulator with pressurized air at a set pressure level. When shop air is required, for whatever reason, compressed air is drawn from the connection point in the shop. The shop connection point is usually connected via a pressure reducer (sets the pressure at the desired usage level) that, in turn, is fed from the accumulator, where the compressed air is stored. If the user draws a large enough quantity of compressed air from the system (from the accumulator), a sensing device within the accumulator will send a signal to the air compressor to start, which will produce compressed air to recharge the accumulator. In addition to providing service in air lines, pressure-regulating valves are also used in liquid lines. The operating principle is much the same for both types of service. Simply, the valve is set to monitor the line and to make needed adjustments in response to a signal from a sensing device.

Temperature-regulating valves (also referred to as *thermostatic control* valves) are closely related to pressure-regulating valves (see Figure 6.13). Their purpose is to monitor the temperature in a line or process solution tank and to regulate it – to raise or lower the temperature as required. In water and wastewater operations, probably the most familiar application whereby temperature-regulating valves (see Figure 6.14) are used is in heat exchangers. A *heat exchanger-type* water system utilizes a water-to-coolant heat exchanger for heat dissipation. This is an efficient and effective method to dispose of unwanted heat. Heat exchangers are equipped with temperature-regulating valves that automatically modulate the shop process water, limiting usage to just what is required to achieve the desired coolant temperature.

Relief Valves

Some fluid power systems, even when operating normally, may temporarily develop excessive pressure. For example, whenever an unusually strong work resistance is encountered, dangerously high pressure may develop. *Relief* valves are used to control this excess pressure. Such valves are automatic valves; they start to open at a preset pressure but require a 20% overpressure to open wide.

As the pressure increases, the valve continues to open farther until it has reached its maximum travel. As the pressure drops, it starts to close, and finally shuts off at about the set pressure. Main

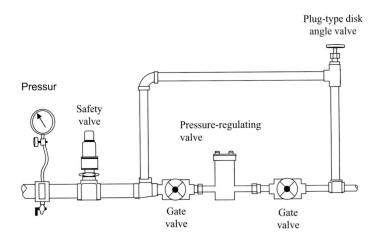

FIGURE 6.13   Pressure-regulating valve system.

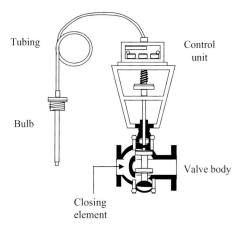

FIGURE 6.14   Temperature-regulating valve assembly.

system relief valves are generally installed between the pump or pressure source and the first system isolation valve. The valve must be large enough to allow the full output of the hydraulic pump to be delivered back to the reservoir. Important Point: Relief valves do not maintain flow or pressure at a given amount but prevent pressure from rising above a specific level when the system is temporarily overloaded.

Reducing Valves

Pressure-reducing valves provide a steady pressure into a system that operates at a lower pressure than the supply system. In practice, they are very much like pressure-regulating valves. A pressure-reducing valve reduces pressure by throttling the fluid flow. A reducing valve can normally be set for any desired downstream pressure within the design limits of the valve. Once the valve is set, the reduced pressure will be maintained regardless of changes in supply pressure (as long as the supply pressure is at least as high as the reduced pressure desired) and regardless of the system load, provided the load does not exceed the design capacity of the reducer.

VALVE OPERATORS

In many modern water operations, devices called operators or actuators mechanically operate many valves. These devices may be operated by air, electricity, or fluid; that is, pneumatic, hydraulic, and magnetic operators.

Valve Maintenance

As with any other mechanical device, effective valve maintenance begins with its correct operation. As an example of incorrect operation, consider the standard household water faucet. As the faucet washers age, they harden and deteriorate. The valve becomes more difficult to operate properly – eventually, the valve begins to leak. A common practice is simply to apply as much force as possible to the faucet handle. Doing so, however, damages the valve stem and the body of the valve body. Good maintenance includes preventive maintenance, which, in turn, includes inspection of valves, correct lubrication of all moving parts, and the replacement of seals or stem packing.

PIPING SYSTEM: PROTECTIVE DEVICES

Piping systems must be protected from the harmful effects of undesirable impurities (solid particles) entering the fluid stream. Because of the considerable variety of materials carried by piping systems, there is an equal range of choices in protective devices. Such protective devices include strainers, filters, and traps. In this section, we describe the design and function of strainers, filters, and traps.

PIPING PROTECTIVE DEVICES: APPLICATIONS

Filters, strainers, and traps are normally thought of in terms of specific components used in specific systems. However, it is important to keep in mind that the basic principles apply in many systems. While the examples used in this chapter include applications found in water treatment and distribution systems, the applications are also found in almost every plant: hot- and cold-water lines, lubricating lines, pneumatic and hydraulic lines, and steam lines. With regard to steam lines, it is important to note that traps are primarily used in steam systems where they remove unwanted air and condensate from lines.

Other system applications of piping protective devices include conveyance of hot and chilled water for heating and air conditioning and lines that convey fluids for various processes. Any foreign contamination in any of these lines can cause potential trouble. Piping systems can become clogged, thereby causing greatly increased friction and lower line pressure. Foreign contaminants (dirt and other particles) can also damage valves, seals, and pumping components.

Strainers

Strainers, usually wire mesh screens, are used in piping systems to protect equipment sensitive to contamination that may be carried by the fluid. Strainers can be used in pipelines conveying air, gas, oil, steam, water, and nearly any other fluid conveyed by pipes. Generally, strainers are installed ahead of valves, pumps, regulators, and traps in order to protect them against the damaging effects of corrosion products that may become dislodged and conveyed throughout the piping system (Geiger, 2000). A common strainer is shown in Figure 6.15.

This type of strainer is generally used upstream of traps, control valves, and instruments. This strainer resembles a lateral branch fitting with the strainer element installed in the branch. The end of the lateral branch is removable, to permit servicing of the strainer. In operation, the fluid passes through the strainer screen, which catches most of the contaminants. Then the fluid passes back into the line. Contaminants in the fluid are caught in two ways. Either they do not make it through the strainer screen, or they do not make the sharp turn that the fluid must take as it leaves the unit. The bottom of the unit serves as a sump where the solids collect. A blowout connection may be provided in the end cap to flush the strainer. The blowout plug can be removed, and the pressure in the line can be used to blow the fixture clean (Geiger, 2000).

Filters

The purpose of any filter is to reduce or remove impurities or contaminants from a fluid (liquid or gas) to an acceptable or predetermined level. This is accomplished by passing the fluid through some

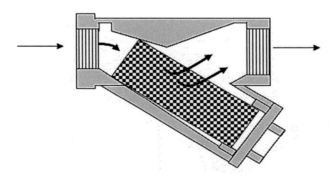

FIGURE 6.15   A common strainer.

kind of porous barrier. Filter cartridges have replaceable elements made of paper, wire cloth, nylon cloth, or fine-mesh nylon cloth between layers of coarse wire. These materials filter out unwanted contaminants, which collect on the entry side of the filter element. When clogged, the element is replaced.

Traps

Traps, used in steam processes, are automatic valves that release condensate (condensed steam) from a steam space while preventing the loss of live steam. Condensate is undesirable because water produces rust and water plus steam leads to water hammer. In addition, steam traps remove air and non-condensate from the steam space. Note: Water hammer can be used by incorporating surge tanks in the system. The operation of a trap depends on what is called *differential pressure* (or delta-P; ΔP) (measured in psi). Differential pressure is the difference between the inlet and outlet pressures. A trap will not operate correctly at a differential pressure higher than the one at which it was designed.

Thermostatic traps have a corrugated bellows-operating element that is filled with an alcohol mixture that has a boiling point lower than that of water (see Figure 6.16). The bellows contracts when in contact with condensate and expands when stream is present. If a heavy condensate load occurs, the bellows will remain in the contracted state, allowing condensate to flow continuously. As steam builds up, the bellows closes. Thus, at times the trap acts as a "continuous flow" type, while at other times it acts intermittently as it opens and closes to condensate and stream, or it may remain totally closed (Bandes & Gorelick, 2000).

PIPING ANCILLARIES

Earlier, we described various devices associated with process piping systems designed to protect the system. In this section, we discuss some of the most widely used ancillaries (or accessories)

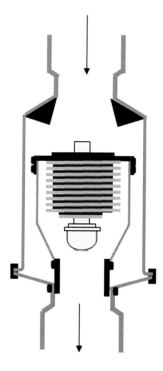

FIGURE 6.16   A thermostatic trap (shown in the open position).

designed to improve the operation and control of the system. These include pressure and tempera-ture gauges, vacuum breakers, accumulators, receivers, and heat exchangers.

GAUGES

In order to properly operate a system, any system, the operator must know certain things. For example, to operate a plant air compressor, the operator needs to know: (1) how to operate it; (2) how to maintain it; (3) how to monitor its operation; and, in many cases, (4) how to repair it. In short, the operator must know system parameters and how to monitor them. Simply, operating parameters refer to those physical indications of system operation. The term *parameter* refers to a system's limits or restrictions. For example, let's consider, again, the plant's air compressor. Obviously, it is important to know how the air compressor operates, or at least how to start and place the compres-sor on line properly. However, it is also just as important to determine if the compressor is operating as per design. Gauges are the main devices that provide us with parameter indications that we need to determine equipment or system operation. With regard to the air compressor, the parameter we are most concerned about now is air pressure (gauge pressure). Not only is correct pressure gen-eration by the compressor important, but also correct pressure in system pipes, tubes, and hoses is essential. Keeping air pressure at the proper level is necessary for four main reasons:

- Safe operation;
- Efficient, economic conveyance of air through the entire system, without waste of energy;
- Delivery of compressed air to all outlet points in the system (the places where the air is to be used) at the required pressure; and
- Prevention of too much or too little pressure (either condition can damage the system and become hazardous to personnel).

DID YOU KNOW?

Pressure in any fluid pushes equally in all directions. The total force on any surface is the psi multiplied by the area in square inches. For example, a fluid under a pressure of 10 psi, pushing against an area of 5 in^2, produces a total force against that surface of 50 lb (10 × 5).

We pointed out that before starting the air compressor, certain pre-start checks must be made. This is important for all machinery, equipment, and systems. In the case of our air compressor example, we want to ensure that proper lubricating oil pressure is maintained. This is important, of course, because pressure failure in the lubricating line that serves the compressor can mean inad-equate lubrication of bearings, and, in turn, expensive mechanical repairs.

PRESSURE GAUGES

As mentioned, many pressure-measuring instruments are called *gauges*. Generally, pressure gauges are located at key points in piping systems. Usually expressed in pounds per square inch (psi), there is a difference between *gauge pressure* (psig) and *absolute pressure* (psia). Simply, "gauge pressure" refers to the pressure level indicated by the gauge. However, even when the gauge reads zero, it is subject to ambient atmospheric pressure (i.e., 14.7 psi at sea level). When a gauge reads 50 psi, that is 50 pounds *gauge pressure* (psig). The true pressure is the 50 pounds shown plus the 14.7 pounds of atmospheric pressure acting on the gauge. The total of "actual" pressure is called the *absolute pressure*: gauge pressure plus atmospheric pressure (50 psi + 14.7 psi = 64.7). It is written 64.7 psia.

Spring-Operated Pressure Gauges

Pressure, by definition, must operate against a surface. Thus, the most common method of measuring pressure in a piping system is to have the fluid press against some type of surface – a flexible surface that moves slightly. This movable surface, in turn, is linked mechanically to a gear-lever mechanism that moves the indicator arrow to indicate the pressure on the dial (i.e., a pressure gauge).

Bourdon Tube Gauges

Many pressure gauges in use today use a coiled tube as a measuring element called a *Bourdon tube*. (The gauge is named for its inventor, Eugene Bourdon, a French engineer). The Bourdon tube is a device that senses pressure and converts the pressure to displacement. Under pressure, the fluid fills the tube (see Figure 6.17a and 6.17b). Since the Bourdon tube displacement is a function of the pressure applied, it may be mechanically amplified and indicated by a pointer. Thus, the pointer position indirectly indicates pressure.

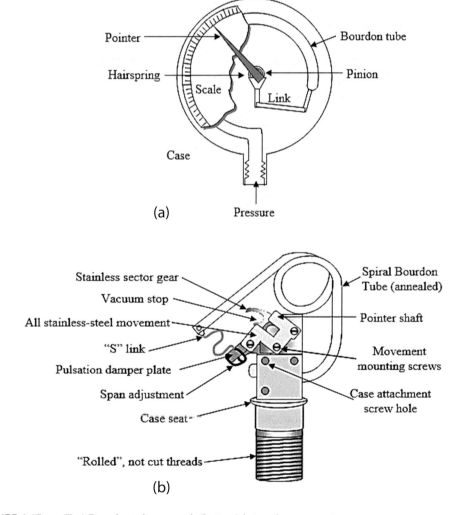

FIGURE 6.17   a (Top) Bourdon tube gauge; b (bottom) internal components.

Bellows Gauge

Figure 6.18 shows how a simplified *bellows* gauge works. The bellows itself is a convoluted unit that expands and contracts axially with changes in pressure. The pressure to be measured can be applied to either the outside or the inside of the bellows; in practice, most bellows measuring devices have the pressure applied to the outside of the bellows. When pressure is released, the spring returns the bellows and the pointer to the zero position.

Plunger Gauge

Most of us are familiar with the simple tire-pressure gauge. This device is a type of plunger gauge. Figure 6.19 shows a *plunger* gauge used in industrial hydraulic systems. The bellows gauge is a spring-loaded gauge, where pressure from the line acts on the bottom of a cylindrical plunger in the center of the gauge and moves it upward. At full pressure, the plunger extends above the gauge, indicating the measured pressure. As the pressure drops, the spring contracts to pull the plunger downward, back into the body (the zero-reading indication).

TEMPERATURE GAUGES

As mentioned, ensuring that system pressures are properly maintained in equipment and piping systems is critical to safe and proper operation. Likewise, ensuring that the temperature of fluids in

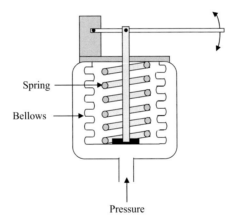

FIGURE 6.18   Bellows gauge.

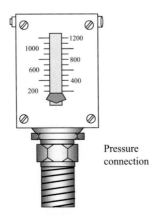

FIGURE 6.19   Plunger gauge.

industrial equipment and piping systems is correct is just as critical. For measuring the temperature of fluids in industrial systems, various temperature-measuring devices are available.

Temperature has been defined in a variety of ways. One example defines temperature as the measure of heat (thermal energy) associated with the movement (kinetic energy) of the molecules of a substance. This definition is based on the theory that molecules of all matter are in continuous motion that is sensed as heat. For our purposes, we define temperature as the degree of hotness or coldness of a substance measured on a definite scale. Temperature is measured when a measuring instrument is brought into contact with the medium being measured (e.g., a thermometer). All temperature-measuring instruments use some change in a material to indicate temperature. Some of the effects that are used to indicate temperature are changes in physical properties and altered physical dimensions (e.g., the change in the length of a material in the form of expansion and contraction). Figure 6.20 shows an *industrial-type thermometer* that is commonly used for measuring the temperature of fluids in industrial piping systems. This type of measuring instrument is nothing more than a rugged version of the familiar mercury thermometer. The bulb and capillary tube are contained inside a protective metal tube called a *well*. The thermometer is attached to the piping system (vat, tank, or other components) by a union fitting.

Another common type of temperature gauge is the *bimetallic* gauge shown in Figure 6.21. Bimetallic means that if two materials with different linear coefficients of expansion (i.e., how much a material expands with heat) are bonded together, as the temperature changes their rate of expansion will be different. This will cause the entire assembly to bend in an arc. When the temperature

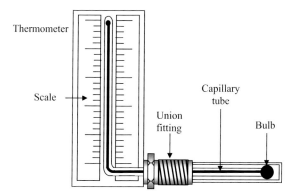

FIGURE 6.20 Industrial thermometer.

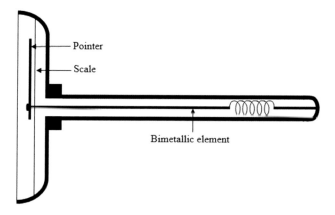

FIGURE 6.21 Bimetallic gauge.

is raised, an arc is formed around the material with the smaller expansion coefficient. The amount of arc is reflected in the movement of the pointer on the gauge. Because two dissimilar materials form the assembly, it is known as a bimetallic element, which is also commonly used in thermostats.

VACUUM BREAKERS

Another common ancillary device found in pipelines is a *vacuum breaker* (components shown in Figure 6.22). Simply, a vacuum breaker is a mechanical device that allows air into the piping system, thereby preventing backflow that could otherwise be caused by the siphoning action created by a partial vacuum. In other words, a vacuum breaker is designed to admit air into the line whenever a vacuum develops. A vacuum, obviously, is the absence of air. Vacuum in a pipeline can be a serious problem. For example, it can cause fluids to run in the wrong direction, possibly mixing contaminants with purer solutions. In public water systems, back-siphonage can occur when a partial vacuum pulls non-potable liquids back into the supply lines (AWWA, 2003). In addition, it can cause the collapse of tubing or equipment.

As illustrated in Figure 6.22, this particular type of vacuum breaker uses a ball that usually is held against a seat by a spring. The ball is contained in a retainer tube mounted inside the piping system or inside the component being protected. If a vacuum develops, the ball is forced (sucked) down into the retainer tube, where it works against the spring. Air flows into the system to fill the vacuum. In public water systems, when air enters the line between a cross-connection and the source of the vacuum, then the vacuum will be broken and back-siphonage is prevented (AWWA, 2003). The spring then returns the ball to its usual position, which acts to seal the system again.

ACCUMULATORS

As mentioned, in a plant compressed air system, a means of storing and delivering air as needed is usually provided. In a hydraulic system, an accumulator provides the functions provided by an air receiver for an air system. That is, the accumulator (usually a dome-shaped or cylindrical chamber or tank attached to a hydraulic line) in a hydraulic system works to help store and deliver energy as required. Moreover, accumulators work to help keep pressure in the line smoothed out. For example, if pressure in the line rises suddenly, the accumulator absorbs the rise, preventing shock to the piping. If pressure in the line drops, the accumulator acts to bring it up to normal.

AIR RECEIVERS

As shown in Figure 6.23, an air receiver is a tank- or cylindrical-type vessel used for a number of purposes. Most important is their ability to store compressed air. Much like accumulators, they cushion shock from sudden pressure rises in an airline. That is, the air receiver serves to absorb the shock of valve closure and load starts, stops, and reversals. There is no liquid in an air receiver. The air compresses as pressure rises. As pressure drops, the air expands to maintain pressure in the line.

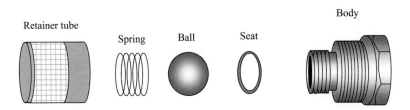

FIGURE 6.22   Vacuum breaker components.

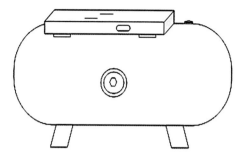

FIGURE 6.23   Air receiver.

HEAT EXCHANGERS

Operating on the principle that heat flows from a warmer body to a cooler one, *heat exchangers* are devices used for adding or removing heat and cold from a liquid or gas. The purpose may be to cool one body, or to warm the other; nonetheless, whether used to warm or to cool, the principle remains the same. Various designs are used in heat exchangers. The simplest form consists of a tube, or possibly a large coil of tubing, placed inside a larger cylinder. In an oil lubrication system, the purpose of a heat exchanger is to cool the hot oil. However, a heat exchanger system can also be used to heat up a process fluid circulating through part of the heat exchanger while steam circulates through its other section.

CHAPTER SUMMARY

Whether a community water supply is taken from surface or groundwater sources, the needs of individual community members are similar, whether household or industrial. Consumers need consumable water, wash water, irrigation water, and waste conveyance water. While the average consumer may give little thought as to how the water reaches their tap, these systems provide the mechanics of potable water provision to the customer; an essential and valuable service, as anyone who has had to transport potable water during an emergency will tell you. In addition, we have discussed the basic but major water distribution and conveyance systems and components. Moreover, ancillary or accessory equipment used in many piping systems was described. It is important to point out that there are other accessories commonly used in piping systems (e.g., rotary pressure joints, actuators, intensifiers, pneumatic pressure line accessories, and so forth); however, discussion of these accessories is beyond the scope of this chapter.

REVIEW QUESTIONS

1. _____ control fluid flow through piping systems.
   a. Robots
   b. Valves
   c. Accumulators
   d. Receivers

2. What does a high C factor for a pipe indicate?
   a. No corrosion protection
   b. Easy to bend
   c. Smooth pipe interior
   d. Large pipe

3. When air pockets form at the high points in a pipeline, the cause is usually _____.
   a. Flow rate too high
   b. Excessive air bubbles
   c. Oil contamination
   d. Pump impeller cracked

4. Surge tanks are used to control _____.
   a. Leaks
   b. Air bubbles
   c. Water hammer
   d. Corrosion

5. A coupling (a Dresser) is used to _____.
   a. Repair clamps
   b. Saddle a pipe
   c. Install a volute
   d. Connect a new water main to an existing water main

6. When a new pipe suffers tuberculation, what does this indicate?
   a. Scaling
   b. Low pH
   c. Water is corrosive
   d. Neutral pH

7. The valve type commonly used to control pressure and flow is a/an _____.
   a. Emergency relief valve
   b. Check valve
   c. Foot valve
   d. Globe valve

8. Deposition of calcium carbonate in a water main is known as _____.
   a. Overloading
   b. Scaling
   c. Pigging
   d. Corrosion

9. What is a regulator station used for?
   a. Feeding chlorine
   b. Preventing corrosion
   c. Maintaining an acceptable water pressure within a piping system
   d. Releasing air within the pipe

10. What type of valve is a corporation stop?
    a. Relief valve
    b. Check valve
    c. Foot valve
    d. Ball valve or plug valve

ANSWER KEY

1. _____ control fluid flow through piping systems.
   a. Robots
   b. **Valves**
   c. Accumulators
   d. Receivers

2. What does a high C factor for a pipe indicate?
    a. No corrosion protection
    b. Easy to bend
    c. **Smooth pipe interior**
    d. Large pipe

3. When air pockets form at the high points in a pipeline, the cause is usually _____.
    a. Flow rate too high
    b. **Excessive air bubbles**
    c. Oil contamination
    d. Pump impeller cracked

4. Surge tanks are used to control _____.
    a. Leaks
    b. Air bubbles
    c. **Water hammer**
    d. Corrosion

5. A coupling (a Dresser) is used to _____.
    a. Repair clamps
    b. Saddle a pipe
    c. Install a volute
    d. **Connect a new water main to an existing water main**

6. When a new pipe suffers tuberculation, what does this indicate?
    a. Scaling
    b. Low pH
    c. **Water is corrosive**
    d. Neutral pH

7. The valve type commonly used to control pressure and flow is a/an _____.
    a. Emergency relief valve
    b. Check valve
    c. Foot valve
    d. **Globe valve**

8. Deposition of calcium carbonate in a water main is known as _____.
    a. Overloading
    b. **Scaling**
    c. Pigging
    d. Corrosion

9. What is a regulator station used for?
    a. Feeding chlorine
    b. Preventing corrosion
    c. **Maintaining an acceptable water pressure within a piping system**
    d. Releasing air within the pipe

10. What type of valve is a corporation stop?
    a. Relief valve
    b. Check valve
    c. Foot valve
    d. **Ball valve or plug valve**

BIBLIOGRAPHY

ACPA, 1987. *Concrete Pipe Design Manual*. Vienna, VA: American Concrete Pipe Association.

ASME, 1996. American Society of Mechanical Engineers. In *ASME b 36.10M. Welded and Seamless Wrought Steel Pipe*. New York: ASME.

AWWA, 2003. *Water Transmission and Distribution*, 3rd ed. Denver, CO: American Water Works Association.

Babcock & Wilcox, 1972. *Steam: Its Generation and Use*. Cambridge, ON: The Babcock & Wilcox Company.

Bales, R.C., New Kirk, D.D., and Hayward, S.B., 1984. Chrysol Tile Asbestos in California Surface Waters from Upstream Rivers through Water Treatment. *J. Am. Water Works Assoc.* 76, 66.

Bandes, A., and Gorelick, B., 2000. *Inspect Steam Traps for Efficient System*. Terre Haute, IN: TWI Press.

Basavaraju, C., 2000. Pipe properties, Geiger, E.L., 2000. Tube Properties. In *Piping Handbook*, 7th ed., Nayyar, M.L., Ed. New York: McGraw-Hill.

Casada, D., 2000. *Valve Replacement Savings*. Oak Ridge, TN: Oak Ridge Laboratories.

Cesario, L., 1995. *Water Distribution Systems*. Denver, CO: American Water Works Association.

Coastal Video Communications Corp., 1994. Virginia Beach, VA: Asbestos Awareness.

Crocker, S., Jr., 2000. Hierarchy of Design Documents. In *Piping Handbook*, 7th ed., Nayyar, M.L., Ed. New York: McGraw-Hill.

Gagliardi, M.G., and Liberatore, L.J., 2000. Water Piping Systems. In *Piping Handbook*, 7th ed., Nayyar, M.L., Ed. New York: McGraw-Hill.

Geiger, E.L., 2000. *Tube Properties*. New York: McGraw-Hill.

Geiger, E.L., 2000. Piping Components. In *Piping Handbook*, 7th ed., Nayyar, M.L., Ed. New York: McGraw-Hill.

Giachino, J.W., and Weeks, W., 1985. *Welding Skills*. Homewood, IL: American Technical Publishers.

Globe Valves, 1998. Integrated Publishing's Official Web Page, http/tpub.com/fluid/ch2c.htm.

Kawamura, S., 1999. *Integrated Design and Operation of Water Treatment Facilities*, 2nd ed. New York: Wiley.

Lindeburg, M.R., 2008. *Civil Engineering Reference Manual*, 4th ed. San Carlos, CA: Professional Publications, Inc.

Lohmeir, A., and Avery, D.R., 2000. Manufacture of Metallic Pipe. In *Piping Handbook*, 7th ed., Nayyar, M.L., Ed. New York: McGraw-Hill.

Magnusson, R.J., 2001. *Technology in the Middle Ages*. Baltimore, MD: Hopkins University.

Marine, C.S., 1999. Hydraulic Transient Design for Pipeline Systems. In *Water Distribution Systems Handbook*, Mays, L.W., Ed. New York: McGraw-Hill.

McGhee, T.J., 1991. *Water Supply and Sewerage*, 6th ed. New York: McGraw-Hill.

Miller, R., and Miller, M.R., 2004. *Pumps & Hydraulics*, 6th ed. New York: Wiley.

Nathanson, J.A., 1997. *Basic Environmental Technology: Water Supply, Waste Management, and Pollution Control*, 2nd ed. Upper Saddle River, NJ: Prentice Hall.

Nayyar, M.L., 2000. Introduction to piping. In *Piping Handbook*, 7th ed., Nayyar, M.L., Ed. New York: McGraw-Hill.

OSHA, 1978. *Drain on Air Receivers*. 29 CFR 1910.169. Washington, DC: OSHA.

Snoek, P.E., and Carney, J.C., 1981. Pipeline Material Selection for Transport of Abrasive tailings. In Proceedings of the 6th International Technical Conference on Slurry Transportation. Las Vegas, NV.

Spellman, F.R., 2014. *Handbook of Water and Wastewater Treatment Plant Operations*, 3rd ed. Boca Raton, FL: CRC Press.

USEPA, 2006. *Emerging Technologies for Conveyance Systems*. Washington, DC: EPA, 832-R-06-004.

Viessman, W., Jr., and Hammer, M.J., 1998. *Water Supply and Pollution Control*, 6th ed. Menlo Park, CA: Addison-Wesley.

Webber, J.S., Covey, J.R., and King, M.V. 1989. Asbestos in Drinking Water Supplied Through Grossly Deteriorated Pipe. *J. Am. Water Works Assoc.* 81, 80.

7 Public Service Management, Water Treatment Infrastructure, and Energy Conservation

For it matters not how small the beginning may seem to be; what is once well done is well done forever.

– Henry David Thoreau

LEARNING OBJECTIVES

After studying this chapter, you should be able to:

- Describe water quality management.
- Describe water in terms of its characteristics.
- Discuss the water cycle.
- Define physical characteristics of water including solids, turbidity, and describe the factors that affect water clarity.
- Describe the conditions and factors that affect watercolor.
- Discuss consumer expectations of color, taste, and odor for water.
- Describe the organic and inorganic contaminants that affect water's smell and taste.

DOI: 10.1201/9781003207146-9

- Discuss palpability and temperature.
- Describe the effect temperature has on water treatment.
- Discuss the definition of water as the universal solvent.
- Identify the chemical constituents commonly found in water and describe how they affect water quality.
- Define total dissolved solids, discuss their sources, and how they are controlled.
- Discuss the problems created by high alkalinity.
- Identify the range and description classifications for water hardness and discuss the advantages and disadvantages of both hard and soft waters.
- Discuss fluoride's importance in drinking water and describe how fluoride affects teeth.
- Identify and discuss the metals, both toxic and nontoxic, commonly found in water supplies, and describe their effects on water quality.
- Identify the primary effects that organic matter has on water quality.
- Discuss biodegradable and nonbiodegradable organic matter in terms of biological oxygen demand (BOD) and chemical oxygen demand (COD).
- Identify the essential nutrients water carries and discuss how excessive or deficient quantities of these nutrients affect the environment.
- Discuss how biological water characteristics can affect human health and well-being and describe how pathogens are transported by the water system.
- Discuss how the presence or absence of biological organisms works as indicators of water quality to water specialists.
- Identify the ways organisms extract nutrients and energy from their waste environment.
- Describe the role oxygen plays in organism metabolism.
- Define the types of aquatic organisms (bacteria, viruses, protozoa, worms, rotifers, crustaceans, fungi, and algae) that can inhabit aquatic environments, and identify the physical, chemical, and biological factors necessary for their existence.
- Describe and define the water treatment process including pretreatment, aeration, screening, chemical addition, iron and manganese removal, hardness, corrosion control, coagulation, flocculation, sedimentation, filtration, disinfection, and removal of taste and odors.

KEY DEFINITIONS

Absorb – to take in. Many things absorb water.

Acre-Feet (acre-foot) – an expression of water quantity. One acre-foot will cover one acre of ground one foot deep. An acre-foot contains 43,560 cubic feet, 1,233 cubic meters, or 325,829 gallons (United States). Also abbreviated as ac-ft.

Activated Carbon – derived from vegetable or animal materials by roasting in a vacuum furnace. Its porous nature gives it a very high surface area per unit mass – as much as 1,000 square meters per gram, which is 10 million times the surface area of 1 gram of water in an open container. Used in adsorption (see definition), activated carbon adsorbs substances that are not (or are only slightly) adsorbed by other methods.

Adsorption – the adhesion of a substance to the surface of a solid or liquid. Adsorption is often used to extract pollutants by causing them to attach to such adsorbents as activated carbon or silica gel. Hydrophobic (water-repulsing adsorbents) are used to extract oil from waterways in oil spills.

Aerobic – conditions in which free, elemental oxygen is present. Also used to describe organisms, biological activity, or treatment processes that require free oxygen.

Agglomeration – floc particles colliding and gathering into a larger settleable mass.

Air Gap – the air space between the free-flowing discharge end of a supply pipe and an unpressurized receiving vessel.

Ambient – the expected natural conditions that occur in water unaffected or uninfluenced by human activities.

Anaerobic–conditions in which no oxygen (free or combined) is available. Also used to describe organisms, biological activity, or treatment processes that function in the absence of oxygen.

Aquifer – a water-bearing stratum of permeable rock, sand, or gravel.

Aquifer System – a heterogeneous body of introduced permeable and less permeable material that acts as a water-yielding hydraulic unit of regional extent.

Average Monthly Discharge Limitation – the highest allowable discharge over a calendar month.

Average Weekly Discharge Limitation – the highest allowable discharge over a calendar week.

Backflow – reversal of flow when pressure in a service connection exceeds the pressure in the distribution main.

Backwash – fluidizing filter media with water, air, or a combination of the two so that individual grains can be cleaned of the material that has accumulated during the filter run.

Base – a substance that has a pH value from 7 to 14.

Basin – a groundwater reservoir defined by the overlying land surface and underlying aquifers that contain water stored in the reservoir.

Beneficial Use of Water – the use of water for any beneficial purpose. Such uses include domestic use, irrigation, recreation, fish and wildlife, fire protection, navigation, power, industrial use, etc. The benefit varies from one location to another and by custom. What constitutes beneficial use is often defined by statute or court decisions.

Biochemical Oxygen Demand (BOD^5) – the oxygen used in meeting the metabolic needs of aerobic microorganisms in water rich in organic matter.

Biosolids – from *Merriam-Webster's Collegiate Dictionary*, Tenth Ed. (1998): biosolids *n.* (1977) solid organic matter recovered from a sewage treatment process and used especially as fertilizer [or soil amendment], usually used in plural. Note: In this chapter, *biosolids* is used in many places to replace the standard term *sludge*. The authors view the term *sludge* as an ugly four-letter word inappropriate to use to describe biosolids. Biosolid is a product that can be reused; it has some value. Because biosolids have value, it certainly should not be classified as a "waste" product – and when biosolids for beneficial reuse is addressed, it is made clear that it is not.

Biota – all the species of plants and animals indigenous to a certain area.

Boiling Point – the temperature at which a liquid boils. The temperature at which the vapor pressure of a liquid equals the pressure on its surface. If the pressure of the liquid varies, the actual boiling point varies. The boiling point of water is 212° Fahrenheit or 100° Celsius.

Breakpoint – point at which chlorine dosage satisfies chlorine demand.

Breakthrough – in filtering, when unwanted materials start to pass through the filter.

Carbonaceous Biochemical Oxygen Demand, CBOD^5 – the amount of biochemical oxygen demand that can be attributed to carbonaceous material.

Chemical Oxygen Demand (COD) – the amount of chemically oxidizable materials present in the wastewater.

Chlorination – disinfection of water using chlorine as the oxidizing agent.

Clarifier – a device designed to permit solids to settle or rise and be separated from the flow. Also known as a settling tank or sedimentation basin.

Consumptive Use – (1) the quantity of water absorbed by crops and transpired or used directly in the building of plant tissue, together with the water evaporated from the cropped area; (2) the quantity of water transpired and evaporated from a cropped area or the normal loss of water from the soil by evaporation and plant transpiration; and (3) the quantity of water discharged to the atmosphere or incorporated in the products of the process in connection with vegetative growth, food processing, or an industrial process.

Contamination (Water) – damage to the quality of water sources by sewage, industrial waste, or other material.

Cross-Connection – a connection between a storm drain system and a sanitary collection system; a connection between two sections of a collection system to handle anticipated overloads of one system; or a connection between drinking (potable) water and an unsafe water supply or sanitary collection system.

Daily Discharge – the discharge of a pollutant measured during a calendar day or any 24-hour period that reasonably represents a calendar day for the purposes of sampling.

Disinfection – water treatment process that kills pathogenic organisms.

Dissolved Oxygen (DO) – the amount of oxygen dissolved in water or sewage. Concentrations of less than five parts per million (ppm) can limit aquatic life or cause offensive odors. Excessive organic matter present in water because of inadequate waste treatment and runoff from agricultural or urban land generally causes low DO.

Domestic Consumption (use) – water used for household purposes such as washing, food preparation, and showers. The quantity (or quantity per capita) of water consumed in a municipality or district for domestic uses or purposes during a given period. It sometimes encompasses all uses, including the quantity wasted, lost, or otherwise unaccounted for.

Drinking Water Standards – established by state agencies, United States Public Health Service, and Environmental Protection Agency (EPA) for drinking water in the United States.

Effluent – something that flows out, usually a polluting gas or liquid discharge.

Effluent Limitation – any restriction imposed by the regulatory agency on quantities, discharge rates, or concentrations of pollutants discharged from point sources into state waters.

Energy – in scientific terms, the ability or capacity of doing work. Various forms of energy include kinetic, potential, thermal, nuclear, rotational, and electromagnetic. One form of energy may be changed to another, as when coal is burned to produce steam to drive a turbine, which produces electric energy.

Evaporation – the process by which water becomes a vapor at a temperature below the boiling point.

Filtration – the mechanical process that removes particulate matter by separating water from solid material, usually by passing it through sand.

Flocculation – slow mixing process in which particles are brought into contact, with the intent of promoting their agglomeration.

Force Main – a pipe that carries water under pressure from the discharge side of a pump to a point of gravity flow downstream.

Head Loss – amount of energy used by water in moving from one point to another.

Holding Pond – a small basin or pond designed to hold sediment laden or contaminated water until it can be treated to meet water quality standards or used in some other way.

Hydraulic Cleaning – cleaning pipe with water under enough pressure to produce high water velocities.

Hydraulic Gradient – a measure of the change in groundwater head over a given distance.

Hydraulic Head – the height above a specific datum (generally sea level) that water will rise in a well.

Hydrologic Cycle (water cycle) – the cycle of water movement from the atmosphere to the earth and back to the atmosphere through various processes. These processes include: precipitation, infiltration, percolation, storage, evaporation, transpiration, and condensation.

Hydrology – the science dealing with the properties, distribution, and circulation of water.

Infiltration – the gradual downward flow of water from the surface into soil material.

Influent – water entering a tank, channel, or treatment process.

Langelier Saturation Index (LI) – a numerical index that indicates whether calcium carbonate will be deposited or dissolved in a distribution system.

License – a certificate issued by the State Board of Waterworks/Wastewater Works Operators authorizing the holder to perform the duties of a wastewater treatment plant operator.

Maximum Contaminant Level (MCL) – an enforceable standard for protection of human health.

Mechanical Cleaning – clearing pipe by using equipment (bucket machines, power rodders, or hand rods) that scrapes, cuts, pulls, or pushes the material out of the pipe.

Metering Pump – a chemical solution feed pump that adds a measured amount of solution with each stroke or rotation of the pump.

Milligrams/Liter (mg/L) – a measure of concentration equivalent to parts per million (ppm).

Nephelometric Turbidity Unit (NTU) – indicates amount of turbidity in a water sample.

Nonpoint Source (NSP) Pollution – forms of pollution caused by sediment, nutrients, organic, and toxic substances originating from land use activities that are carried to lakes and streams by surface runoff. Nonpoint source pollution occurs when the rate of materials entering these waterbodies exceeds natural levels.

NPDES Permit – National Pollutant Discharge Elimination System permit, which authorizes the discharge of treated wastes and specifies the conditions that must be met for discharge.

Nutrients – substances required to support living organisms. Usually refers to nitrogen, phosphorus, iron, and other trace metals.

Organic Chemicals/Compounds – animal or plant-produced substances containing mainly carbon, hydrogen, and oxygen, such as benzene and toluene.

Parts per Million (ppm) – the number of parts by weight of a substance per million parts of water. This unit is commonly used to represent pollutant concentrations. Large concentrations are expressed in percentages.

pH – a way of expressing both acidity and alkalinity on a scale of 0–14, with 7 representing neutrality; numbers less than 7 indicate increasing acidity, and numbers greater than 7 indicate increasing alkalinity.

Point Source Pollution – a type of water pollution resulting from discharges into receiving waters from easily identifiable points. Common point sources of pollution are discharges from factories and municipal sewage treatment plants.

Pollution – the alteration of the physical, thermal, chemical, or biological quality of, or the contamination of, any water in the state that renders the water harmful, detrimental, or injurious to humans, animal life, vegetation, property, or to public health, safety, or welfare, or impairs the usefulness or the public enjoyment of the water for any lawful or reasonable purpose.

Potable Water – water satisfactorily safe for drinking purposes from the standpoint of its chemical, physical, and biological characteristics.

Precipitate – a deposit on the earth of hail, rain, mist, sleet, or snow. The common process by which atmospheric water becomes surface or subsurface water; the term *precipitation* is also commonly used to designate the quantity of water precipitated.

Preventive Maintenance (PM) – regularly scheduled servicing of machinery or other equipment using appropriate tools, tests, and lubricants. This type of maintenance can prolong the useful life of equipment and machinery and increase its efficiency by detecting and correcting problems before they cause a breakdown of the equipment.

Purveyor – an agency or person that supplies potable water.

Recharge – the addition of water into a groundwater system.

Reservoir – a pond, lake, tank, or basin (natural or human made) where water is collected and used for storage. Large bodies of groundwater are called groundwater reservoirs; water behind a dam is also called a reservoir of water.

Sedimentation – a process that reduces the velocity of water in basins so that suspended material can settle out by gravity.

Septic Tanks – used to hold domestic wastes when a sewer line is not available to carry them to a treatment plant. The wastes are piped to underground tanks directly from a home or homes. Bacteria in the wastes decompose some of the organic matter, the sludge settles on the bottom of the tank, and the effluent flows out of the tank into the ground through drains.

Settleability – a process control test used to evaluate the settling characteristics of the activated sludge. Readings taken at 30–60 minutes are used to calculate the settled sludge volume (SSV) and the sludge volume index (SVI).

Sludge – the mixture of settleable solids and water removed from the bottom of the settling tank.

Stormwater – runoff resulting from rainfall and snowmelt.

Stream – a general term for a body of flowing water. In hydrology, the term is generally applied to the water flowing in a natural channel as distinct from a canal. More generally, it is applied to the water flowing in any channel, natural or artificial. Some types of streams: (1) Ephemeral: a stream that flows only in direct response to precipitation, and whose channel is always above the water table. (2) Intermittent or Seasonal: a stream that flows only at certain times of the year when it receives water from springs, rainfall, or from surface sources such as melting snow. (3) Perennial: a stream that flows continuously. (4) Gaining: a stream or reach of a stream that receives water from the zone of saturation. An effluent stream. (5) Insulated: a stream or reach of a stream that neither contributes water to the zone of saturation nor receives water from it. It is separated from the zones of saturation by an impermeable bed. (6) Losing: a stream or reach of a stream that contributes water to the zone of saturation. An influent stream. (7) Perched: a perched stream is either a losing stream or an insulated stream that is separated from the underlying groundwater by a zone of aeration.

Surface Water – lakes, bays, ponds, impounding reservoirs, springs, rivers, streams, creeks, estuaries, wetlands, marshes, inlets, canals, gulfs inside the territorial limits of the state, and all other bodies of surface water, natural or artificial, inland or coastal, fresh or salt, navigable or non-navigable, and including the beds and banks of all watercourses and bodies of surface water, that are wholly or partially inside or bordering the state or subject to the jurisdiction of the state. Waters in treatment systems that are authorized by state or federal law, regulation, or permit and which are created for the purpose of water treatment are not considered to be waters in the state.

Thermal Pollution – the degradation of water quality by the introduction of a heated effluent. Primarily the result of the discharge of cooling waters from industrial processes (particularly from electrical power generation); waste heat eventually results from virtually every energy conversion.

Toxicity – the occurrence of lethal or sublethal adverse effects on representative sensitive organisms due to exposure to toxic materials. Adverse effects caused by conditions of temperature, dissolved oxygen, or nontoxic dissolved substances are excluded from the definition of toxicity.

Vaporization – the change of a substance from a liquid or solid state to a gaseous state.

Wastewater – the water supply of a community after it has been spoiled by use.

Water Cycle – the process by which water travels in a sequence from the air (condensation) to the earth (precipitation) and returns to the atmosphere (evaporation). It is also referred to as the hydrologic cycle.

Water Quality – a term used to describe the chemical, physical, and biological characteristics of water with respect to its suitability for a particular use.

Water Supply – any quantity of available water.

Waterborne Disease – a disease caused by a microorganism that is carried from one person or animal to another by water.

Watershed – the area of land that contributes surface runoff to a given point in a drainage system.

PUBLIC SECTOR MANAGEMENT: THE HIDDEN FUNCTION

Communities are finding that there is a gap between how their public sector utilities function and how the interests of the community want them to function, both in the present and the future. A common complaint is a lack of dedication to the underlying values of public service and the interests of the citizens served. The typical response seems to be promoting a certain type of manager. The most important role of a manager is to solve the problems and challenges within a specific environment. In this chapter, the specific environment discussed is public sector utility management with specific emphasis placed on modern public water systems.

The current economic environment in public sector utility management is one of uncertainty and change. The US economy has shown signs of limited recovery since mid-1999 but seems to be expanding at the present time. Even with economic expansion, local government, operating under ever-increasing fiscal constraints and diminishing resources, is forced to reduce expenditures and limit expansion. Good management is a critical component of good public service. When a public-sector utility worker, as with almost any other worker in the general workforce, is promoted to any level of management, certain benefits accrue with the promotion. In many cases, these promotions include increased pay, increased benefits, increased pride and self-satisfaction, and increased responsibilities. It is this last "benefit," increased responsibility, that is often seen as more of an obligation than a benefit. For many newly promoted managers that added responsibility is most difficult to accept and to effect.

Let's state this same opinion in a different manner, by asking a question: When a public sector utility worker is promoted to any level of management, is the new manager (no matter how well motivated) properly equipped to assume the responsibility incumbent with the position? In many respects this question is relative, almost nebulous, because it depends on the selection process and the qualifications of the worker. This is a universal problem, of course, faced by all those who make new worker-to-management selections. The key part of this question is "how well equipped" is the new manager? The answer, of course, will vary. However, if the new manager is not properly prepared, the transition can be analogous to running a minefield while blindfolded.

To be effective and successful at the rank-and-file (worker) level, public sector workers need only follow directions from the supervisor and perform work output at a level of effectiveness approved by the supervisor. The worker's focus is on performing to a narrow standard of accomplishment, which, in turn, can result in job security. Thus, the typical public sector worker is less likely to give much thought to current and pressing issues facing the overall industry that are not part and parcel of his/her daily work routine; that is, their view is narrow, not holistic. This is the case even though many of these current and pressing issues may directly or indirectly impact the worker's employment future in the public sector job. There is another public-sector utility management level that needs to be addressed: mid-level management. Mid-level manager jobs in utilities are usually filled by college graduates who are specialists (based on educational achievements) in narrow fields. For example, in water treatment it is not unusual and actually quite normal to hire biologists, environmental scientists, chemists, and engineers to fill various technical positions. In electrical and natural gas utility work, engineering graduates are commonly hired. Engineers and/or hydrologists usually fill stormwater management mid-level management positions. Experience has shown that all of these technical specialists are needed in public sector utility work.

The problem arises when technical specialists who are not experienced in management are expected to manage. This is not to say that a biologist or chemist or engineer cannot manage. Instead, it may be difficult (and sometimes disastrous) if an engineer or biologist or other technical expert is required to perform the dual function of technical expert and manager, especially when the person is not properly trained or suited to manage. Based on well-documented examples, this problem is more often the case than the exception. One thing is certain: Graduation from college does not automatically qualify one to manage. In order to manage effectively, the manager must not only have the ability and experience required to manage, to lead, but also must be cognizant of the problems facing public sector utilities. These management problems are many and varied. Consider, for example, one area of concern. During the 1980s and 1990s, a wave of new buzzwords, reengineering, privatizing, and rightsizing swept the developed and developing world, affecting the many public sector operations partially or fully privatized/reengineered/rightsized in the transition economies. The transition from the old way to the new way of doing public sector utility business is still evolving, but at a rapid rate.

The question is why. Why this apparent rush to change, especially in developed countries? Public-owned and -operated utilities (i.e., water, wastewater, electric, natural gas, stormwater management, and to some degree public transportation, telecommunications, and others) have been around for a long time (the exception being stormwater management and telecommunications, which are relatively recent). Many utilities have operated well and accomplished their intended missions with little or no public controversy. In most cases, when the public demands water at the tap, a waste-emptied toilet bowl after the flush, electricity at a throw of the on-switch, likewise for natural gas, and controlled stormwater runoff, these services have been readily available. At least this is one view residing on one edge of a double-edged sword.

The other edge? The public, always a demanding customer, wants more and better quality from public sector utilities, which many perceive to be wasteful, ineffective, and in many cases just plain unresponsive to its needs. It is within this atmosphere that we have witnessed the rise of accountability and competition for public sector programs. Programs that are unable to justify their value are being downsized or eliminated altogether. Again, what is the problem? Why change something that apparently works – to a degree at least? We all know that change is controversial, difficult, costly, troublesome, and often cumbersome. So why not leave well enough alone? Sorting through the current arguments for and against change in public utility management and operations evokes an emotional firestorm that would test Solomon. Almost inevitably, the eye of the beholder colors conclusions. But a few facts seem clear. In the United States, for example, state and local governments now face a series of unprecedented challenges: budget deficits, bloated workforces, decaying infrastructures, suffocating regulatory compliance, shrinking tax bases, citizen opposition to new taxes and facilities, taxpayer (ratepayer) imposed tax (rates), and spending limitations. These problems are often compounded by the perception of ratepayers and politicians who view rate payments or budget expenditures to public sector utilities as nothing more than throwing good money at bad money; money that never seems to make a dent in filling the black hole.

For the typical water operator, stormwater manager, and/or electric and natural gas worker (i.e., public utility rank-and-file or public utility-person) and some mid-level managers, budget deficits, bloated workforces, decaying infrastructures, shrinking tax bases, and the other associated problems are as foreign as the man in the moon. These problems are not in their domain; they're someone else's problem, typically the managers' problem. Unfortunately, such problems, while not necessarily germane to public utility-workers, only remain floating in the rarefied atmosphere occupied by those holding upper management positions. Accordingly, public utility-persons may be aware of obvious utility shortcomings (a plant unit process that belongs in the boneyard along with the other dinosaurs) but have no power or inclination to effect change.

The worker's situation or view does change, of course, whenever he/she joins the upper echelons of management through promotion. These "new managers" are, unfortunately, often unsuited to take over the responsibilities incumbent with their new leadership status. Their new position and subsequent outlook are similar to the sideline cheerleader who is suddenly suited out in full football garb, thrown onto the playing field, and expected to perform in a positive way in a team effort to win the game. The problems facing utilities currently require a new breed of public-sector manager. This new breed, inspired by the successful streamlining of American business, is trying to meet these challenges, not by increasing taxes or government spending, but by fundamentally transforming government through a process called privatization, reengineering, or rightsizing. Privatizing, reengineering, and/or rightsizing means establishing clear priorities and asking questions that successful utilities regularly ask, such as: If we were not doing this already, would we start? Is this activity central to our mission? What is our mission? Do we have the need for a mission? Does a central all-controlling authority better make decisions or are they better made by consensus management or by empowering the employees to share input on decision-making? If we were to design this organization from scratch, given what we now know about modern technology, what would it look like?

Any current seasoned or newly advanced manager can ask these questions; the problem arises in attempting to answer them. Knowledge, awareness, thinking outside the box, learned management skill, and common sense all play a crucial role in answering these questions. Coming up with the "correct" answer(s) is even more challenging. This dilemma points to the need to pretrain prospective managers before they are promoted. Unfortunately, this process is usually more of an afterthought than common practice. Experience has shown that promotions within the public sector often are based on seniority, the next available worker, and not on potential management skill.

DID YOU KNOW?

The term *consumptive use of water* defines or describes: (1) the quantity of water absorbed by crops and transpired or used directly in the building of plant tissue, together with the water evaporated from the cropped area; (2) the quantity of water transpired and evaporated from a cropped area or the normal loss of water from the soil by evaporation and plant transpiration; and (3) the quantity of water discharged to the atmosphere or incorporated in the products of the process in connection with vegetative growth, food processing, or an industrial process.

Today's public-sector utility manager faced with privatization, reengineering, or rightsizing problems is also faced with many additional challenges. One of these major challenges struck home with the 9/11 terrorist attack. Many managers found out the brutal way that security is no longer one of those backburner concerns, just a cumbersome necessity. Security is not only on the front burner now and not only a lifestyle-sustaining necessity but also a huge challenge for any manager, no matter how seasoned. Complicating matters are even more managerial challenges. For example, complying with current, new, and changing regulations is not exactly a cakewalk in a pristine high alpine meadow on a sunny spring day. Because all the challenges facing today's utility managers are too long to list, in this chapter we aim our focus on five areas of pressing concern to most water public sector managers. It should be noted, however, that though intentionally directed toward public water systems, the information provided in this chapter applies to all utilities:

- Complying with regulations, and coping with new and changing regulations;
- Maintaining infrastructure;
- Privatization, reengineering, and rightsizing;
- Benchmarking; and
- Upgrading security.

This chapter does not provide all the solutions to the problems currently facing managers in the public sector. However, the chapter does provide information for constructing a roadmap for successful accomplishment and/or compliance with the listed items above. This is accomplished by describing each challenge and by pointing out the many pitfalls that should be avoided in taking on the challenges. Written primarily for new managers, this chapter also is designed to provide insight and information to managers that have been managing for some time.

The focus of this chapter is not on the primary and traditional management skills of coordinating, organizing, and directing the workforce; many excellent texts are available on developing and enhancing these skills. Instead, the focus is on the five areas listed earlier as they relate to waterworks infrastructure and energy conservation, all of which are germane to the current or new manager's challenges in managing utility operations in a competitive, performance-based manner critical to the current culture. At present, many available management texts ignore, avoid, or pay cursory lip service coverage to these important areas.

Though directed at public utility sector managers and prospective managers in water operations, this chapter will serve the needs of students, teachers, consulting engineers, and technical personnel in city, state, and federal public sector employment. Moreover, in order to maximize the usefulness of the material contained in the chapter, it is written in straightforward, user-friendly, plain English – a characteristic of the author's style; that is, a failure to communicate is not allowed or even given a first, second, or third thought. To assure correlation to modern practice and design, illustrative problems and case studies are presented throughout the chapter in terms of commonly used managerial parameters.

This chapter and its content are accessible to those who have no experience with public sector utility management. If you work through the chapter systematically, an understanding of and skill in management techniques can be acquired, adding a critical component to your professional knowledge. The bottom line: Public sector work is a hidden function; many of its appurtenances are buried beneath the soil until the customer complains. Moreover, leadership within the public sector work enterprise is a hidden function until the manager is proven to be dysfunctional in the eyes of the customer(s).

THE INFRASTRUCTURE DILEMMA

The United States economy, because it is so energy wasteful, is much less efficient than either the European or Japanese economies. It takes us twice as much energy to produce a unit of GDP as it does in Europe and Japan. So, we're fundamentally less efficient and therefore less competitive, and the sooner we begin to tighten up, the better it will be for our economy and society.

– Hazel Henderson, on ENN Radio

SETTING THE STAGE

There are a number of long-term economic, social, and environmental trends [Elkington's (1999) so-called Triple Bottom Line] evolving around us. Many of these long-term trends are developing because of us, specifically for us, or simply to sustain us. Many of these long-term trends follow general courses and can be described by the jargon of the day; that is, they can be alluded to or specified by a specific buzzword or buzzwords common in usage today. We frequently hear these buzzwords used in general conversation especially in abbreviated texting form. Buzzwords such as empowerment, outside the box, streamline, wellness, synergy, generation X, face time, exit strategy, clear goal, and so on and so forth are just part of our daily vernacular.

In this chapter, the popular buzzword that we are concerned with, sustainability, is often used in business. However, in water treatment, sustainability is much more than just a buzzword; it is a way of life or should be. There are numerous definitions of sustainability that are overwhelming, vague, and/or indistinct. For our purposes, there is a long definition and a short definition of sustainability. The long definition: ensuring that water treatment operations occur indefinitely without negative impact. The short definition: the capacity of water operations to endure. Whether we define long or short fashion, what does sustainability really mean in the real world of water treatment operations? We defined sustainability in what we call long and short terms. Note: However, that sustainability in water treatment operations can be characterized in broader or all-encompassing terms than those simple definitions. As mentioned, using the Triple Bottom Line scenario in regard to sustainability, the environmental aspects, economic aspects, and social aspects of water treatment operations can define today's and tomorrow's needs more specifically.

Infrastructure is another term used in this chapter; it can be used to describe water operations in the whole or can identify several individual or separate elements of water treatment operations. For example, as individual water treatment infrastructure components, fundamental systems, or unit processes, we could list source water intake, pretreatment, screening, coagulation and mixing, flocculation, settling and biosolids processing, filtering, disinfection, storage, and distribution systems. Otherwise we could simply describe water treatment plant operations as the infrastructure. How one chooses to define infrastructure is not important. What is important is to maintain and manage infrastructure in the most efficient and economical manner possible to ensure its sustainability. This is no easy task. Consider, for example, the 2009 Report Card for American Infrastructure produced by the American Society of Civil Engineers shown in Table 7.1.

Not only must water treatment managers maintain and operate aging and often underfunded infrastructure, they must also comply with stringent environmental regulations, and they must also keep stakeholders and ratepayers satisfied with operations and with rates. Moreover, in line with these considerations, managers must incorporate economic considerations into every decision. For example, as mentioned, they must meet regulatory standards for the quality of treated drinking water. They must also plan for future upgrades or retrofits that will enable the water treatment facility to meet future water quality and future effluent regulatory standards. Finally, and most importantly, managers must optimize the use of manpower, chemicals, and electricity.

TABLE 7.1

2009 Report Card for American Infrastructure

Infrastructure	Grade
Bridges	C
Dams	D
Drinking Water	D–
Energy	D+
Hazardous Waste	D
Rail	C–
Roads	D–
Schools	D
Wastewater	D–

America's Infrastructure GPA: D

Source: Modified from American Society of Civil Engineers (2012). Report Card for American Infrastructure 2009. Accessed 06/04/2019 @ http://www.infrastructurereportcard.org/.

SUSTAINABLE WATER INFRASTRUCTURE

EPA (2012) points out that *sustainable development* can be defined as that which meets the needs of the present generation without compromising the ability of future generations to meet their needs. The current US population benefits from the investments that were made over the past several decades to build our nation's water/wastewater infrastructure. Sector professionals must continuously promote practices that encourage water sector utilities and their customers to address existing needs so that future generations will not be left to address the approaching wave of needs resulting from aging water infrastructure.

To be on a sustainable path, investments need to result in efficient infrastructure and infrastructure systems and be at a pace and level that allow the water sector to provide the desired levels of service over the long term. Sounds easy enough: The water manager simply needs to put his or her operation on a sustainable path; moreover, he/she can simply accomplish this by investing, right? Well, investing what? Investing in what? Investing how much? These are questions that require answers, obviously. Before moving on with this discussion it is important first to discuss plant infrastructure basics.

Maintaining Sustainable Infrastructure

During the 1950s and 1960s, the US government encouraged the prevention of pollution by providing funds for the construction of municipal water treatment plants, water-pollution research, and technical training and assistance. New processes were developed to analyze and evaluate the effects of pollution on the water environment. In spite of these efforts, however, expanding population and industrial and economic growth caused the pollution and health difficulties to increase.

In response to the need to make a coordinated effort to protect the environment, the National Environmental Policy Act (NEPA) was signed into law on January 1, 1970. In December of that year, a new independent body, the USEPA was created to bring under one roof all of the pollution-control programs related to air, water, and solid wastes. In 1972, the Water Pollution Control Act Amendments expanded the role of the federal government in water pollution control and significantly increased federal funding for the construction of wastewater treatment plants.

Many of the water and wastewater treatment plants in operation today are the result of federal grants made over the years. For example, because of the 1977 Clean Water Act Amendment to the Federal Water Pollution Control Act of 1972 and the 1987 Clean Water Act reauthorization bill, funding for water and wastewater treatment plants was provided. Many large sanitation districts, with their multiple plant operations, and even a larger number of single plant operations in smaller communities in operation today are a result of these early environmental laws. Because of these laws, the federal government provided grants of several hundred million dollars to finance the construction of water and wastewater treatment facilities throughout the country.

DID YOU KNOW?

Looking at water distribution piping only, EPA's 2000 survey on community water systems found that in systems that serve more than 100,000 people, about 40% of drinking water pipes are more than 40 years old. However, it is important to note that age, in and of itself, does not necessarily point to problems. If a system is well maintained, it can operate over a long period (EPA, 2012).

Many of these locally or federally funded treatment plants are aging; based on experience, we can rate some as dinosaurs. The point is, many facilities are facing problems caused by aging equipment, facilities, and infrastructure. Complicating the problems associated with natural aging is the increasing pressure on inadequate older systems to meet the demands of increased population and urban growth. Facilities built in the 1960s and 1970s are now 40–50+ years old, and not only are they showing signs of wear and tear, they simply were not designed to handle the level of growth that has occurred in many municipalities.

Regulations often necessitate a need to upgrade. By receiving matching funds or federal money to cover some of the costs, municipalities can take advantage of a window of opportunity to improve their facility at a lower direct cost to the community. Those federal dollars, of course, do come with strings attached; they are to be spent on specific projects in specific areas. On the other hand, many times new regulatory requirements are put in place without the financial assistance needed to implement them. When this occurs, either the local community ignores the new requirements until caught and forced to comply, or they face the situation and implement through local tax hikes or ratepayer hikes to pay the cost of compliance.

An example of how a change in regulations can force the issue is demonstrated by the demands made by OSHA and USEPA in their Process Safety Management (PSM)/Risk Management Planning (RMP) regulations. These regulations put the use of elemental chlorine (and other listed hazardous materials) under close scrutiny. Moreover, because of these regulations, plant managers throughout the country are forced to choose which side of a double-edged sword cuts their way the most. One edge calls for full compliance with the regulations (analogous to stuffing the regulation through the eye of a needle). The other edge calls for substitution. That is, replacing elemental chlorine (probably the EPA's motive in the first place; see the note) with a non-listed hazardous chemical (e.g., hypochlorite) or a physical (ultraviolet irradiation, UV) disinfectant; either way, a very costly undertaking (Spellman, 2008). Note: Many of us who have worked in water and wastewater treatment for years characterize PSM and RMP as the elemental chlorine "killer." You have probably heard the old saying, "If you can't do away with something in one way, then regulate it to death."

Note: Changes resulting because of regulatory pressure sometimes mean replacing or changing existing equipment, increased chemical costs (e.g., substituting hypochlorite for chlorine typically increases costs threefold), and could easily involve increased energy and personnel costs. Equipment condition, new technology, and financial concerns are all considerations when upgrades or new processes are chosen. In addition, the safety of the process must be considered, of course, because of the demands made by USEPA and OSHA. The potential of harm to workers, the community, and the environment are all under study, as are the possible long-term effects of chlorination on the human population.

Important Point: Remember that because drinking water is distributed and conveyed by pipeline networks to customers, elemental chlorine is typically used to ensure safe and healthy water is sent through the system; some of the systems are extensive and residual chlorine is important in delivering a safe product. Note that water treatment plants typically have a useful life of 20–50 years before they require expansion or rehabilitation. Various unit processes (e.g., filtering systems) and distribution pipes have life cycles that can range from 15 to 100 years, depending on the type of material and where they are laid. Long-term corrosion reduces a pipe's carrying capacity, requiring increasing investments in power and pumping. When water pipes age to the point of failure, the result can be contamination of drinking water and its release into our surface waters or basements, and high costs both to replace the pipes and repair any resulting damage. With pipes, the material used and proper installation of the pipe can be a greater indicator of failure than age.

DID YOU KNOW?

More than 50% of Americans drink bottled water occasionally or as their major source of drinking water – an astounding fact given the high quality and low cost of US tap water.

Is a Drinking Water System a Cash Cow or Cash Dog?

Maintaining the sustainable operations of water and wastewater treatment facilities is expensive. If funding is not available from federal, state, or local governmental entities, then the facilities must be funded by ratepayers. Water and wastewater treatment plants are usually owned, operated, and managed by the community (the municipality) where they are located. While many of these facilities are privately owned, the majority of water treatment plants (WTPs) and wastewater treatment plants (WWTPs) are publicly owned treatment works (POTW) (i.e., owned by local government agencies). These publicly owned facilities are managed and operated on-site by professionals in the field. On-site management, however, is usually controlled by a board of elected, appointed, or hired directors/commissioners, who set policy, determine budget, plan for expansion or upgrading, hold decision-making power for large purchases, set rates for ratepayers, and in general control the overall direction of the operation. Final decisions on matters that affect plant performance are sometimes in the hands of a board of directors that comprises elected and appointed city officials. Their knowledge of the science, engineering, and the hands-on problems at the treatment facility may be very different from those who work on-site. Matters that are of critical importance to those in on-site management may mean little to those on the board. The board of directors may indeed also be responsible for other city services and have an agenda that encompasses more than just the water facility. Thus, decisions that affect on-site management can be affected by political and financial concerns that have little to do with the successful operation of a WTP or POTW.

Finances and funding are always of concern, no matter how small or large, or how well supported or underfunded the municipality. Publicly owned treatment works are generally funded from a combination of sources. These include local taxes, state and federal monies (including grants and matching funds for upgrades), as well as usage fees for water customers. In smaller communities, in fact, the water plants may be the only city service that actually generates income. This is especially true in water treatment and delivery, which is commonly looked upon as the cash cow of city services. As a cash cow, the water treatment works generate cash in excess of the amount of cash needed to maintain the treatment works.

These treatment works are "milked" continuously with as little investment as possible. Funds generated by the facility do not always stay with the facility. Funds can be reassigned to support other city services – "the cow is milked dry" – and when facility upgrade time comes, funding for renovations can be problematic. On the other end of the spectrum, spent water (wastewater), treated in a POTW, is often looked upon as one of the cash dogs of city services. Typically, these units make only enough money to sustain operations. This is the case, of course, because managers and oversight boards or commissions are fearful, for political reasons, of charging ratepayers too much for treatment services. Some progress has been made, however, in marketing and selling treated wastewater for reuse in industrial cooling applications and some irrigation projects and even advancements to treated drinking water standards and pipe-to-tap connections.

Planning is essential for funding, for controlling expenses, and for ensuring water infrastructure sustainability. The infrastructure we build today will be with us for a long time and, therefore, must be efficient to operate, offer the best solution in meeting the needs of a community, and be coordinated with infrastructure investments in other sectors such as transportation and housing. It is both important and challenging to ensure that a plan is in place to renew and replace it at the right time, which may be years away. Replacing an infrastructure asset too soon means not benefiting from the remaining useful life of that asset. Replacing an asset too late can lead to expensive, emergency repairs that are significantly more expensive than those which are planned (EPA, 2012). Additionally, making retrofits to newly constructed infrastructure that was not designed or constructed correctly is expensive. Doing the job correctly the first time requires planning and a certain amount of competence.

DID YOU KNOW?

Electricity prices generally reflect the costs to build, finance, maintain, manage, and operate power plants and the electricity grid (the complex system of power transmission and distribution lines) and to operate and administer the utilities that supply electricity to consumers. Some utilities are for-profit, and their prices include a return for the owners and shareholders.

ENERGY-EFFICIENT OPERATING STRATEGIES

The California Energy Commission (2012) points out that electrical energy consumption at water treatment plants is increasing because of more stringent regulations and customer concerns about water quality. As a result, more facility managers are turning to energy management to reduce operating costs. Along with the cost of labor (licensed operators and environmental scientists and others), the cost of chemicals, the cost of maintenance and repair parts, the use of electricity to power plant operations ranks right up there with the other costs, and if not properly managed and conserved electrical costs can be even more costly. Reducing energy consumption by managing your electrical load, however, is only part of the equation. Operating strategies to reduce energy usage also include proper chemical usage and management, operation and maintenance practices, and inflow and infiltration control; these operating strategies are discussed in this chapter.[*]

Electric Load Management

By choosing when and where to use electricity, drinking water facilities can often save as much (or more) money as they could by reducing energy consumption. Note that electricity is typically billed in two ways: by the quantity of energy used over a period, measured in kilowatt-hours (kWh); and by demand, the rate of flow of energy, measured in kilowatts.

---

[*] Based on information from EPA 2012 Water & Energy Efficiency in Water and Wastewater Facilities. Accessed 04/21/12 @ http://www.epa.gov/region09/waterinfrastructure/technology.html.

Rate Schedules

Electric utilities often structure rates to encourage customers to minimize demand during peak periods because it is costly to provide generating capacity for use during periods of peak demand. That's why drinking water treatment plants should investigate the variety of rate schedules offered by electric utilities. They may achieve substantial savings simply by selecting a rate schedule that better fits their pattern of electricity use.

Time-of-Use Rates – In most areas of the country, time-of-use rates, which favor off-peak electrical use, are available. Under the time-of-use rates, energy and demand charges vary during different block periods of the day. For example, energy charges in the summer may be only five cents per kWh with no demand charge between 9:30 p.m. and 8:30 a.m. but increase to nine cents per kWh demand charge of $10 per kilowatt between noon and 6:00 p.m. The monthly demand charge is often based on the highest 15-minute average demand for the month.

Interruptible Rates – Offer users discounts in exchange for a user commitment to reduce demand on request. On the rare occasions when a plant receives such a request, it can run standby power generators.

Power Factor Charges – As mentioned earlier, power factor, also known as reactive power or kVAR, reflects the extent that the current and voltage cycle in phase. Low power factor such as that caused by a partly loaded motor results in excessive current flow. Many electric utilities charge extra for low power factor because of the cost of providing the extra current.

Future Pricing Options – As the electrical industry is deregulated, many new pricing options will be offered. *Real-time pricing*, where pricing varies continuously based on regional demand, and *block power*, or electricity priced in low-cost, constant-load increments, are only two of the many rate structures that may be available. Facilities that know how and when they use energy and have identified flexible electric loads can select a rate structure that offers the highest economy while meeting their energy needs.

Energy Demand Management

Energy demand management, also known as demand side management (DSM), is the modification of consumer (utility) demand for energy through various methods. DSM programs consist of the planning, implementing, and monitoring activities of electric utilities that are designed to encourage consumers to modify their level and pattern of electricity usage.

Energy Management Strategies

- *Conduct an Energy Survey* – The first step to an effective energy management program for your facility is to learn how and when each piece of equipment uses energy. Calculate the demand and monthly energy consumption for the largest motors in your plant. Staff may be surprised at the results; a 100-horsepower motor may cost over $4,500 per month if run continuously. The rate at which energy is used will vary throughout the day, depending upon factors such as the demand from the distribution system and reservoir and well levels for public water systems or influent flows and biological oxygen demand loading for wastewater systems. Plot daily electrical load as a function of time for different plant loading conditions and note which large equipment can be operated off-peak. Examine all available rate schedules to determine which can provide the lowest cost in conjunction with appropriate operational changes.
- *Reduce Peak Demand* – Look for opportunities to improve the efficiency of equipment that must run during the peak period, such as improving pump efficiency or upgrading a wastewater plant's aeration system. During on-peak periods, avoid using large equipment simultaneously: Two 25-kilowatt pumps that run only two hours each day can contribute 50-kilowatts to demand if run at the same time.
- *Shift Load to Off-Peak* – Many large loads can be scheduled for off-peak operation. For example, plants can use system storage to ride out periods of highest load, rather than operating pumps. Avoid running large intermittent pumps when operating the main pumps.

- *Improve Power Factor* – Low power factor is frequently caused by motors that run less than fully loaded. This also wastes energy because motor efficiency drops off below full load. Examine motor systems to determine if the motor should be resized or if a smaller motor can be added to handle lower loads. Power factors can also be corrected by installing a capacitor in parallel with the offending equipment.

ELECTRICAL LOAD MANAGEMENT SUCCESS STORIES[*]

Moulton Niguel Water District

- Service Area: 37.5 square miles.
- Potable Water System Capacity: 48 million gallons per day (MGD).
- Wastewater System Capacity: 17 MGD.
- Wastewater Treatment Type: Tertiary.
- Secondary Treatment Method: Activated sludge.
- Annual System-wide Purchased Electricity: $1,310,000 (15.5 million kWh).
- Annual Savings Attributed to Energy Efficient Strategies: $332,000.

For over a decade, automation and instrumentation have helped Southern California's Moulton Niguel Water District supply water and treat wastewater economically and efficiently. Facing a major rise in energy costs, the agency explored other methods to increase energy efficiency. Working closely with Southern California Edison and San Diego Gas & Electric to identify optimal rate schedules and energy-efficiency strategies, the district implemented a program in 1992 that has yielded substantial savings in the reservoir-fed branches of their distribution system. Additionally, the district plans to investigate potential improvements to their potable water systems requiring full-speed pumping to maintain system pressure.

Key Improvements
Moulton Niguel Water District staff implemented changes in the following areas:

- Install programmable logic controllers to benefit from lower off-peak utility rates – Moulton Niguel uses automated controls and programmable logic controllers to enable 77 district pumping stations to benefit from lower off-peak utility rates. The controls activate pumps during off-peak hours, bringing reservoirs to satisfactory levels. On-peak, pumping is halted, allowing reservoir levels to fall. Stations employing this strategy are on "reservoir duty" – meaning the system is pressurized by the static head of the reservoir. All stations previously operated in "closed grid mode," running pumps 24 hours a day to maintain system pressure.
- The programmable logic controllers' sophisticated internal clock and calendar automatically adjusts for seasonal changes, consistently keeping equipment running off-peak. This strategy has decreased pumping costs (saving nearly $320,000 annually) and allowing reservoir levels to fall has improved water quality.
- Regulate lift station wastewater levels using proportional, integral, and derivative controls to automatically transmit data to a central computer – Moulton Niguel previously cycled constant-speed wastewater pump drives on and off to distribute wastewater. As a result, drive control was limited; pump motors were subject to starting surges, and the system shutdown left sewage sitting in pipes, producing offensive odors.

---

[*] From California Energy Commission 2012. Success Story Encina Wastewater Authority. Accessed 04.22/12 @ www.energy.ca.gov/process/pubs/encina.pdf.

- They decided to replace their standard motor drives with variable-frequency drives linked to proportional, integral, and derivative controls. This system regulates wastewater levels by sending a signal to modulate wastewater flow to the controllers of variable-frequency drives. The new proportional, integral, and derivative/variable-frequency drives system provides a continuous, modulating flow that uses less energy, reduces motor wear and high energy demands from motor starting surges. This ensures that sewage does not remain stagnant in pipes, thereby reducing odor problems and reducing energy costs by about 4%.
- Install variable-frequency drives on the wastewater system to control pump speed in coordination with the proportional, integral, and derivative system, reducing costs.
- Specify that all motors used in new construction be 95–97% efficient; replace standard-efficiency motors with energy-efficient motors (ongoing).

OPERATION AND MAINTENANCE: ENERGY-/COST-SAVINGS PROCEDURES[*]

From 2008 to 2010, EPA developed and conducted energy workshops for over 500 public water and wastewater system associates in a handful of western states. The following subsections identify the systems that participated in the program and provide a summary of their results. Many of the projects required no additional resources outside of existing staff time and minor equipment purchases made within existing expense accounts. Some facilities focused on collecting and using renewable energy (specifically energy generated from the force of water dropping in elevation while traveling through pipes).

Other sites concentrated on reducing energy consumption by increasing energy efficiency. Some public water systems reduced energy use during the day when energy costs more and increased energy use during times of the day when energy costs less. At several of the facilities, optimizing operations resulted in significant savings without requiring a large capital outlay. One step that each of the ten facilities profiled below developed and implemented was an Energy Improvement Management Plan. More details on the specific projects can be found in the case studies developed by the public water systems and described in the subsections that follow:

Real-World Case Studies

Chandler Municipal Utilities, Arizona (February 21, 2012)

- Facility Profile – Chandler Municipal Utilities is located in the City of Chandler, Arizona, within Maricopa County. The Municipal Utilities Department oversees wastewater treatment, reclaimed water, and the drinking water supply for the city. The utility selected their potable water system for the Energy Management Initiative. The Chandler potable water system serves 255,000 customers. The system treats an average of 52 MGD of groundwater and surface water at two treatment plants. The use of groundwater from 31 wells requires more energy than surface water from the Salt River Project and Central AZ Project. Additional energy is needed to bring groundwater to the surface for treatment and distribution.
- Baseline Data – Chandler spent $2.9 million on electricity last year treating potable water. The annual electricity used at the water treatment plant was 33,880,000 kWh. In 2010 the plant's energy consumption generated 22,268 metric tons of carbon dioxide equivalent (MTCO$_2$) of greenhouse gas (GHG) emissions.
- Energy Improvement Management Plan – Chandler Municipal Utilities chose to reduce energy consumption by optimizing the potable water system. They did this in two days.

---

[*] From EPA 2011 United States EPA Region 9 Energy Management Initiatives for Public Wastewater and Drinking Water Utilities Facilitating Utilities toward Sustainable Energy Management. Washington, DC: United States Environmental Protection Agency.

First, they revised tank management practices based on hydraulic modeling and master planning to find the best configuration and operating program to reduce groundwater pumping. Second, the utility staff upgraded pumps and revised pressure zones to operate more efficiently under the new operating program. Based on this energy management approach, Chandler's goal was to reduce the number of kilowatt-hours used to produce and distribute 1 million gallons of potable water by 5% from 2010 levels. Chandler chose to develop a strong team as an area of focus for the Energy Management System.

- Challenges – One of the biggest challenges Chandler faced was changing staff attitudes and long-established habits associated with operating a small groundwater-based system to the practicality of operating a large surface water-dominated system. A series of small victories led to a staff-driven team approach to system optimization.

The City of Chandler's potable water production and distribution system expanded rapidly to meet the growth of the 1990s–early 2000s. Wells and mains were added so the system was able to meet all of its demands. The recent economic slowdown gave Chandler staff the opportunity to analyze the system as a whole rather than a collection of separate parts. Complicating factors included a lack of consistent historic design philosophy and evolution of the system from a small groundwater-based utility to a surface water-dominated system serving a population of 250,000 as well as several major industrial and commercial customers.

Surface water is the most cost-effective source of water for Chandler, but due to a lack of dedicated transmission infrastructure, staff had a difficult time filling tanks with surface water. Facing budget constraints, staff revisited all aspects of the system. The result was (1) an expanded second pressure zone; (2) consistent hydraulic grade lines for the pressure zones; (3) focused rehabilitation of key facilities; and (4) a new tank management strategy.

During this time the programmable logic controllers the system used were no longer being supported by the manufacturer, so they were replaced by controllers with much better information collection capabilities. The new technology gave the operators much better information and control of the system. Given the new tools, the operators developed and tested new operation strategies which have resulted in a more robust system that produces better quality water while using fewer resources.

Accomplishments

- Chandler fell a bit short of the 5% goal but was able to achieve a reduction of 4.2%. They also produced 4.7% more potable water in 2011 than they did in 2010. Had Chandler not reduced the energy necessary to produce and distribute 1 million gallons, they would have used an additional 1,445,000 more kilowatt-hours. Using the average cost per kilowatt-hour Chandler paid for power in 2011, this amounts to an energy savings of almost $130,000 and avoids the generation of 950 $MTCO_2$ of GHG emissions.
- One aspect of Chandler's potable water system optimization approach involved using a higher percentage of surface water than in previous years. This resulted in substantial savings in water resource costs and a significant reduction in chlorine use.
- Chandler adopted a team approach to optimization that resulted in a high level of understanding of system dynamics throughout the organization, an involved staff that continually identifies ways to improve the efficiency and operation of the system, improvements in data acquisition and management, and multiple open channels of communication.
- Annual energy savings – 1,445,000 kWh.
- Annual cost savings – $130,000.
- Annual GHG reductions – 950 $MTCO_2$, equal to the removal of 195 passenger vehicles from the road.
- Project cost – No additional funds required.
- Payback period – Immediate.

Future Steps – Staff is continuing to evaluate system performance and seek additional opportunities for optimization. Some lighting has been upgraded; more lighting upgrades are planned in the future. Staff is investigating the feasibility of on-site power generation using solar panels and in-pipe hydraulic power generation.

Airport Water Reclamation Facility, Prescott, AZ (February 21, 2012)

- Facility Profile – The Airport Water Reclamation Facility (AWRF) is one of two facilities owned and operated by the City of Prescott, within Yavapai County. Prescott is positioned close to the center of AZ between Phoenix and Flagstaff, just outside the Prescott National Forest. The original wastewater treatment plant was built in 1978 and received a major facility upgrade in 1999. The next major upgrade began in 2012. The City of Prescott also operates another wastewater treatment plant called Sundog.
- Baseline Data – AWRF treats 1.1 million gallons of wastewater per day for approximately 18,000 residents. The facility spends $160,000 annually on electricity costs and uses 1.8 million kWh. GHG emissions for the SWTF are 1,023 metric tons of carbon dioxide ($MTCO_2$).
- Energy Improvement Management Plan – The City of Prescott elected to construct a hydro-turbine electric generation unit as part of the Energy Improvement Management Plan at the Airport Water Reclamation Facility to conserve energy and better manage resources. The turbine will be placed at the discharge point of the recharge water pipeline to convert the potential energy in the flowing water to electricity. This hydro turbine has the potential to produce 125,000 kWh per year, which would save the City approximately $12,000 per year.
- Challenges – The greatest challenge so far has been the time commitment required to accomplish the project concurrent with the design of the facility expansion. Initially, there was also some difficulty connecting with the appropriate staff at the power company; however, that has since been resolved.

Accomplishments

- Energy and cost savings will result with the new hydro-turbine electric generation unit installation at the Airport Water Reclamation Facility. This project is estimated to produce 125,000 kWh and save $12,000 in electrical costs per year. The Energy Management Program has promoted an awareness of energy uses and potential savings associated with minor and major changes to operations. The program also highlighted many other programs/projects that improved the City's knowledge of energy-saving considerations. The City was successful in developing and adopting an Energy Conservation Policy.
- Annual projected GHG reductions – 86 $MTCO_2$, equal to the removal of 17 passenger vehicles from the road.
- Project cost – $25,000.
- Payback period – 25 months.

Future Steps – Going out to bid with the hydro turbine project with a major upgrade project ready to bid as well.

Somerton Municipal Water, Arizona (February 21, 2012)

- Facility Profile – Somerton Municipal Water is located in the City of Somerton, Arizona, within Yuma County. Yuma County is situated in the southwest corner of Arizona close to the California and Mexico borders. The water treatment plant was built in 1985. In 1998 the plant carried out a major facility upgrade by installing a new 1.2-million-gallon storage tank and a new 100 horsepower booster pump. The drinking water treatment plant is called the Somerton Municipal Water System (System), and it sources water from 3–300-feet deep wells.

- Baseline Data – SMW serves a population of 14,267 residents and treats 2 million gallons of water per day (MGD). Approximately 907,000 kWh is used to run the system and it costs an average of $85,000 per year for electricity. The plant generated 512.79 $MTCO_2$ of GHG emissions.
- Energy Improvement Management Plan – Water treatment plant staff started the energy improvement process by first evaluating the energy efficiency of existing wells and pumps. Upon inspection, repairs were made to two wells and three pumps. By fixing the wells and pumps, Somerton Municipal Eater System will save $27,000 each year on their electricity bill. Moreover, they will save an average 250,000 kWh annually which will result in a reduction in CO_2 emissions by an estimated 172 $MTCO_2$.
- Challenges – It was difficult to find the time to participate in the Energy Management sessions. No additional staff resources were available to complete the project.

Accomplishments

- Somerton staff gained a better understanding of how projects are selected and increased their effectiveness in working with management.
- By upgrading wells and booster pumps, they were able to save approximately $27,000/year per pump on energy costs.
- Annual projected GHG reductions – 172 $MTCO_2$, equal to the removal of 34 passenger vehicles from the road.
- Project cost – $131,203, with $33,500 covered by incentives.
- Payback period – 4 years.

Future Steps – Complete; begin a solar project that will produce 1.5 million kWh.

Hawaii County Department of Water Supply (February 21, 2012)

- Facility Profile – The Hawaii County Department of Water Supply (DWS) is one of 21 departments within the county of Hawaii. The county encompasses the entire island of Hawaii, and the administrative offices are located in Hilo, Hawaii. The DWS is the public potable water distribution utility; however, the Island of Hawaii also has several other public water systems that are not owned and operated by DWS. Hawaii, the largest island in the Hawaiian chain, is 93 miles long and 76 miles wide with a land area of approximately 4,030 square miles. The DWS operates and maintains 67 water sources and almost 2,000 miles of water distribution pipeline. Over 90% of the water served is from a groundwater source that requires minimal treatment with chlorine for disinfection.
- Baseline Data – The DWS serves 41,507 customers and produces 31.1 MGD. Because of the vast area, mountainous terrain of the island, and the many separate water sources, the energy needed to pump water to the surface is significant, and plant employees drive a significant distance to operate the water distribution system. In 2010, the plant spent $16.5 million on energy costs, used 54,781,373 kWh for electricity and 95,100 gallons of gas and diesel. Hawaii County DWS emits an estimated 31,784.35 $MTCO_2$ of GHG emissions.
- Energy Improvement Management Plan – Hawaii County DWS energy reduction strategy's main goal is to reduce gas and diesel consumption. The performance target was to reduce fuel purchases by 200 gallons per month or 2,400 gallons annually. The target was met by establishing more efficient operators' routes around Hilo, using an automated SCADA system, and installing GPS equipment in 30 vehicles. A new automated SCADA system

replaced the manual system. By reducing fuel consumption and increasing efficiency, DWS will reduce CO_2 emissions by an estimated 17.3 TCO_2 per year. This project was fully implemented on December 31, 2011. DWS also chose to develop an Energy Policy.

- Challenges – Introducing a new automated SCADA system (to replace the old manual system) required programming time and duplicate systems until the new system was proven. The data being collected was all manual until the new vehicle equipment was installed. DWS decided to purchase vehicle GPS units so a new vehicle policy was needed. The pilot project modified the city of Hilo's operators' routes. The pilot project covers about one-third of the island. Implementing route changes met resistance because operators lost overtime. Establishing an Energy Management Team was unsuccessful, so the project was implemented by one person, but there were many moving parts.

Accomplishments

- Created an energy policy.
- Annual projected energy savings – 1,965 gallons of gasoline.
- Annual overtime savings – $9,780.
- Annual projected cost savings – $17,670.
- Annual projected GHG reductions – 18 $MTCO_2$, equal to the removal of 3.4 passenger vehicles from the road.
- Project cost – $25,300.
- Payback period – 1.5 years.

Future Steps

- Complete purchase of vehicle GPS systems.
- Fully implement project throughout the island, which will include 100 vehicles.
- Begin wind power project to generate renewable energy.

Eastern Municipal Water District, California (February 21, 2012)

- Facility Profile – Eastern Municipal Water District (EMWD) has over 250 operating facilities that have been constructed over the past 60 years. EMWD selected a drinking water filtration plant for the Energy Management Initiative. Perris Water Filtration Plant is located in Perris, California, within Riverside County. Perris is situated southeast of Los Angeles along the Escondido Freeway. The plant treats water pumped from the Colorado River and/or from the CA State Water Project.
- Baseline Data – Perris Water Filtration Plant.
- Water Treatment Design Capacity – 24 MGD.
- Annual Energy Consumption – 5.6 million kWh.
- Annual Cost of Energy – $695,000.
- GHG Emissions: 3,862 $MTCO_2$.
- Design Average Flow (FY 2011) – 10.2 MGD.
- Energy Improvement Management Plan – EMWD chose to reduce their energy consumption and GHG emissions by producing renewable energy on-site. The municipality will install a Renewable Power Generator (In-Conduit Hydro Generation) at the Perris Water Filtration Plant. The renewable power generator will produce up to 290,000 kWh of energy each year. The project will take one year to complete once funding is secure and will cost approximately $350,000.
- Challenges – Challenges associated with the proposed project include identifying hydro-generation technology capable of meeting the low head pressure and varying flow

conditions existing at the Perris Water Filtration Plant. These were technical challenges that were eventually overcome. Also, a grant application for funding from the US Bureau of Reclamation was not funded, but EMWD obtained feedback on the proposal and will fund the project without a grant.

Accomplishments

- Raw water supply to EMWD's Perris Valley Water Filtration plant is controlled through an existing valve which requires frequent, and costly, replacement. The benefits of this project are that it will eliminate the need for this replacement and provide the necessary flow control capabilities combined with energy generation. EMWD has contracted for a feasibility study which shows the viability of the proposed project. The project, which is capable of producing nearly 300,000 kWh of electricity, has now completed the design phase. Included in the design are revised cost estimates and inclusion of hydro turbine technology that meets the unique challenges of this application (low head, with the highly variable flow).
- EMWD increased awareness of systematic processes for analyzing overall energy management efforts and provided structure and a strategic approach to energy management as a whole.
- Annual projected energy savings – 290,000 kWh.
- Annual projected cost savings – $36,000.
- Annual projected GHG reductions – 200 $MTCO_2$, equal to the removal of 39 passenger vehicles from the road.
- Project cost – $350,000
- Payback period – 5 years (factoring in the need to replace an existing, non-energy generating valve); 10 years (if valve didn't need to be replaced).

Future Steps – Board approval of funding, go out to bid.

Port Drive Water Treatment Plant, Lake Havasu, Arizona (February 21, 2012)

- Facility Profile – Port Drive Water Treatment plant (PDWTP) is located in Mohave County, Arizona, and serves the majority of the population of Lake Havasu City. The city is located along the Colorado River on the eastern shores of Lake Havasu. The water treatment plant was built between 2002 and 2004 and has never received a major facility upgrade. Currently, an estimated 86% of the water treated is sourced from groundwater wells and a small percentage comes directly from Lake Havasu.
- Baseline Data – PDWTP uses a natural biological process to remove iron and manganese from 11 million gallons of water per day. The treated drinking water is then distributed to 50,000 customers. Annually, the plant uses 6,636,960 kWh of energy at a cost of $612,749 to the City each year. This equates to 4,577 $MTCO_2$ of GHG emissions.
- Energy Improvement Management Plan – Lake Havasu City Water Treatment Plant staff, encouraged by the EPA, completed a test "Change of Operations" of the North and South High Service Pump Station.
- Challenges – Overall, since the 2009 recession, one major challenge to any noncore activity, including the effort, has been a lack of personnel resources. Despite these challenges, the City was able to implement small incremental changes. A major benefit in the City was the realization that those changes can and will be valuable in the future. This was seen in the demonstration project described in more detail in the accomplishments section. The lack of staff hours prevented the City from moving forward with an Energy Improvement Plan. This also delayed the implementation of a pump study until November 2011.

Accomplishments

- The City changed how much and when water would be released at each lead pump. The lead pump for each system was changed to operate at full water flow until it reached the turn-off level set point. Prior to this, the variable-frequency drive (VFD) controller was programmed to slow the lead pump at a nearly full tank level to provide continuous flow through the WTP process. This change resulted in an average 8.7% energy reduction in pump stations for the months of March, April, and May.
- For the months of June, July, and August, the average energy reduction was 6% for the North Pumps, and 6.7% for the South, compared to the same months in 2010. There was no change in the water quality or ability to supply water on demand with these changes.
- All light fixtures were replaced.
- In the last 10–12 years, an increase in electric costs has not been passed on to users. Last year there was a 24% rate hike, but Lake Havasu reduced energy use by 30%.
- In addition to the test project, Lake Havasu City qualified for an energy audit which was conducted for the North Regional Wastewater Treatment Plant by a consultant funded through the EPA. This resulted in a draft report submitted to Lake Havasu City in September 2011. The report identified seven potential projects that may be scheduled in the future. Many of the suggestions in their report may be relevant to Lake Havasu City's other treatment plants. Approximately 4 MGD of wastewater is treated by these facilities.
- The anticipated energy savings of the Lake Havasu City water treatment plant should equate to approximately 130,000 kWh annually.
- Annual projected GHG reductions – 90 $MTCO_2$, equal to the removal of 18 passenger vehicles from the road.
- Project cost – Zero.
- Payback period – Zero.

Future Steps – Test savings of 6–8.7% during the next 6 months; implement energy audit recommendations. All parts of the plant are being examined for energy-saving opportunities.

Truckee Meadows Water Authority, Reno, Nevada (February 21, 2012)

- Facility Profile – Truckee Meadows Water Authority (TMWA) has chosen its drinking water facility, Chalk Bluff Water Treatment Plant, for the Energy Management Initiative. The plant was built in 1994 and serves more than 330,000 customers throughout 110 square miles within Washoe County, Nevada. The Chalk Bluff Plant treats water from the Truckee River, which flows from Lake Tahoe and the Sierra Nevada mountain range.
- Baseline Data – TMWA serves 93,000 customer connections. In 2009 the water authority spent just under $7 million on electricity and $88,000 on natural gas. The Chalk Bluff Water Treatment Plant uses 13.5 gigawatt hours (GWh) and 74,452 therms per year and spends $1.3 million for electricity. The total energy use results in estimated annual GHG emissions of 9,309 $MTCO_2$.
- Energy Improvement Management Plan – While TMWA relies on gravity as much as possible in a mountainous community, pumping water is a reality. The Chalk Bluff Plant is TMWA's largest water producer and the highest energy use facility. The high energy use is due to the pumping of water uphill from the river into the plant. To reduce energy consumption at Chalk Bluff, the implementation plan consists of two parts: (1) optimizing the time-of-use operating procedures and (2) water supply capital improvements.
  - The first strategy is to optimize time-of-use operating procedures by creating a mass flow/electric cost model of the treatment and effluent pumping processes. The model will be used to predict how changes to the operating procedure will affect electricity cost. In 2010, TMWA spent $938,000 on 7.8 GWh for non-water supply processes of the plant. This project is intended to reduce non-supply electric costs by 15% or $141,000.

- The second project involves water supply improvements to the Highland Canal which transports 90% of Chalk Bluff's water directly to the plant using gravity. The improvement plan will allow 100% of the water to be brought to the plant using the Highland Canal and meets multiple objectives. Improvements will be made during the winter months when customer water demand is lowest to reduce the water supply pumping costs during construction. Currently, TMWA spends $60,000 on 0.5 GWh for water supply pumping at the Chalk Bluff Plant. Energy use will be zero when the project is complete. The design life of the new infrastructures is over 100 years and it will require no energy to operate.
- Challenges – Originally scheduled to begin construction during the all of 2011, delays in obtaining highway encroachment permits resulted in postponements. To minimize water supply pumping costs during construction and therefore continue to reduce energy costs, this project was delayed until the fall of 2012.

TMWA attempted to use a mass balance/electric cost model to optimize time-of-use operating procedures. However, the mass balance/electrical cost model was not capable of making sophisticated decisions like those of experienced water plant operators. Therefore, the purpose of the model shifted from generating decisions to being one of several techniques useful for improving time-of-use energy optimization at the Chalk Bluff Plant.

Accomplishments

- TMWA began setting and tracking time-of-use electricity goals for the Chalk Bluff Plant in November of 2010. The goals depend on the time of day (e.g., 200 kW On-Peak, 400 kW Mid-Peak, and 950 kW Off-Peak) and vary with season, based on the electric utility's tariffs. Water plant operators have the ability to be innovative in order to meet electricity use goals and system demands. The mass balance/electric cost modeling effort was valuable to establish baseline energy usage by (1) formally inventorying energy-intensive unit processes, (2) establishing kW draw of equipment, (3) establishing and ranking historic kWh usage of equipment, and (4) suggesting starting point kW targets for further optimization by operators.
- TMWA considers the time-of-use optimization project a great success due to its ability to save energy costs and will continue to optimize and track the project's results. For the 12 months from November 2010 to October 2011, the time-of-use optimization has saved more than $225,000 (24.4%) compared to the same period the previous year. During this time, electric energy usage was reduced by only 0.45 GWh (5.8%), indicating the savings were primarily due to improved time-of-use cost management.
- Going through the process of identifying the energy needs of each process was eye opening. Talking to the operators and getting them to work toward the time-of-use goals was educational, engaging, and served to strengthen the team.
- Annual projected GHG reductions – 310 $MTCO_2$, equal to the removal of 61 passenger vehicles from the road.
- Project cost – zero.
- Payback period – zero
- Design was substantially complete for the water supply improvement project, and highway encroachment permits were expected in time for the project to proceed in the fall of 2012.
- Annual projected GHG reductions – 345 $MTCO_2$, equal to the removal of 68 passenger vehicles form the road.
- Project cost: – $3,000,000.
- Payback period – 50 years.

Future Steps – Complete; continue to optimize and track project results and get permits; construction.

Tucson Water, Arizona (February 21, 2012)

- Facility Profile – Tucson Water provides clean drinking water to a 330 square-mile service area in Tucson, Arizona. Located in Pima country, Tucson is positioned along Highway 10 about 70 miles north of the United States/Mexico border. The potable water system serves roughly 85% of the Tucson metropolitan area, serving 228,000 customers. The potable system includes 212 production wells, 65 water storage facilities, and over 100 distribution pumps. A separate reclaimed water system serves parks, golf courses, and other turf irrigation. Tucson Water sources drinking water through groundwater and the Colorado River, by recharging and recovering river water delivered through the Central AZ Project. Tucson Water uses recycled water for its reclaimed water system.
- Baseline Data – Tucson Water spends on average $12.5 million on the energy to operate its portable system and produces approximately 110 MGD of potable water per year. Annually, Tucson Water uses approximately 115,000,000 kWh and 5,000,000 therms of energy to run the system. This energy use results in an estimated 83,367.43 $MTCO_2$ of GHG emissions.
- Energy Improvement Management Plan – Tucson Water is partnering with Tucson Electric Power (TEP), a privately held, regulated electric utility, to reduce peak demand. TEP contracted EnerNOC (develops and provides energy management applications and services for commercial, institution, and industrial customers, as well as electric power grid operators and utilities) to facilitate a new demand management program to reduce the energy load during peak hours. EnerNOC will pay customers to shed the load. Tucson Water will participate by identifying sites that are appropriate for load shedding. It has been estimated that the program will save Tucson Water 24,000 kWh during peak energy use and created $9,000 in offsetting revenue to be put toward energy cost.
- As part of the City of Tucson's award under the Department of Energy's Energy Efficiency and Conservation Block Grant (funded by the American Recovery and Reinvestment Act), Tucson Water is implementing a Water System Distribution Pump Efficiency Project. The project is designed to establish baseline data and data management tools for system booster pumps, provide energy-savings recommendations for the distribution system, and implement prioritized energy-savings upgrades. In addition, training will be provided and results from the project will provide actionable information on the cost effectiveness of continuing a program without grant funding. Projected energy and cost savings for the project are 350,000 kWh and $30,000 (year one, past project). The project is scheduled to be completed in the early fall of 2012.
- Challenges – It was difficult to complete the project within one year. A two-year program would have given more time to implement the project in our Energy Management Plan.

Accomplishments

- In addition to the projected $9,000 cost savings of the EnerNOC program, energy data will inform any plans to expand real-time energy monitoring. At the end of the grant-funded booster pump project, the utility will realize energy and cost savings and have the information necessary to scope a more permanent pump efficiency program.
- Annul projected energy savings – 24,000 kWh during peak energy use periods.
- Annual projected cost savings – $9,000.
- Annual projected GHG reductions – None.
- Project cost – Zero.
- Payback period – Zero.

Future Steps – The peak demand project is complete. Tucson will continue implementing an ARRA-funded pump efficiency project that is estimated to save $30,000 and 350,000 kWh per year.

Chino Water Production Facility, Prescott, Arizona (February 21, 2012)

- Facility Profile – The Prescott-China Water Production Facility (Facility) is located within Yavapai County in the town of Chino Valley, Arizona. Prescott is positioned close to the center of Arizona between Phoenix and Flagstaff just outside the Prescott National Forest. The Facility consists of a production well field, reservoir, and booster pump facility. The Facility was built in 1947 and received its last major facility upgrade in 2004.
- Baseline Data – The Facility supplies 50,000 residents with drinking water. During the winter the plant treats 4.5 million gallons of water per day and peaks at 12 million gallons during the summer. The cost of electricity for the plant is $1,600,000 for 11,000,000 kWh per year. Annual GHG emissions are 7,581 $MTCO_2$.
- Energy Improvement Management Plan – The Facility has wells that are 15 miles north and lower in elevation than the treatment plant. The challenge was to reduce the $2,000/month demand charge. The Energy Management Plan for the Facility includes replacing the existing step voltage starts on three wells with soft-start units. The soft-start units will reduce the instantaneous demand on the power supply which will reduce the demand charge on the utility bill. It will also help to reduce the power surge on the power distribution system. In addition, softer starters help to extend the life of the well motors. Overall, the project saves money and electricity through reduced demand charges, energy use, and maintenance. The estimated immediate cost savings associated with the reduced demand is $1,260/month, for a total savings of $15,120/year. The savings toward the maintenance and reduced strain on the electrical distribution system will result in long-term progressive savings.
- Challenges – The main challenge has been the loss of two members of the City's Energy Management Team, which reduced management support and buy-in from staff. This delayed the ultimate implementation and construction of the soft-start project.

Accomplishments

- As the City neared the end of the program, the project gained acceptance. The City has completed a specification package and will soon advertise for construction. The Energy Management Program has promoted an awareness of the City's energy use and potential savings associated with minor or major changes to operations. It has provided a good networking opportunity to learn, expand concepts, and consider new options. This project will reduce direct electrical costs, long-term maintenance costs, and result in unseen benefits like improved safety due to reduced instantaneous electrical demands. The Energy Management Program also highlighted many other programs/projects that improved the City's awareness of energy savings. The City has implemented a new Energy Conservation Policy.
- Annual projected GHG reductions – None.
- Project cost – $42,000.
- Payback period – 34 months.

Future Steps – Advertise for construction; and begin 2-megawatt solar power generation project to offset 30% of water booster costs.

CAPITAL STOCK AND IMPACT ON OPERATIONS AND MAINTENANCE[*]

The different components of capital stock (total physical capital) that make up our nation's wastewater and drinking water systems vary in complexity, materials, and the degree to which they

---

[*] From USEPA 2002. The Clean Water and Drinking Water Infrastructure Gap Analysis.

are subjected to wear and tear. The expenditures that public water systems must make to address the maintenance of systems are largely driven by the condition and age of the components of infrastructure.

USEFUL LIFE OF ASSETS

The life of an asset can be estimated based on the material, but many other factors related to environment and maintenance can affect the useful life of a component of the infrastructure. It is not feasible to conduct a condition assessment of all drinking water infrastructure systems throughout the United States. However, approximation tools can be used to estimate the useful life of these infrastructure systems.

One approximation tool that can be used is the useful life matrix. This matrix can serve as a tool for developing initial cost estimates and for long-range planning and evaluating programmatic scenarios. Table 7.2 shows an example of a matrix developed as an industry guide in Australia. Although the useful life of a component will vary according to the materials, environment, and maintenance, matrices such as that shown in Table 7.2 can be used at the local level as a starting point for repair and replacement, strategic planning, and cost projects. The United States as well as other industrialized countries have engineering and design manuals that instruct professional designers to the accepted standards of practice for design life considerations. The United States Army Corps of Engineers, the American Society for Testing Materials, the Water Environment Federation (WEF), the American Society of Civil Engineers, and several associates maintain data that provide guidance on design and construction of conduits, culverts, and pipes and related design procedures.

Most of the assets of both wastewater and drinking water treatment systems are comprised of the pipe. The useful life of pipe varies considerably based on a number of factors. Some of these factors

---

TABLE 7.2
Useful Life Matrix

Years	Component
	Wastewater
80–100	Collections
50	Treatment Plants – Concrete Structures
15–25	Treatment Plants – Mechanical and Electrical
25	Force Mains
50	Pumping Stations – Concrete Structures
15	Pumping Stations – Mechanical and Electrical
90–100	Interceptors
	Drinking Water
50–80	Reservoirs and Dams
60–70	Treatment Plants – Concrete Structures
15–25	Treatment Plants – Mechanical and Electrical
65–95	Trunk Mains
60–70	Pumping Stations – Concrete Structures
25	Pumping Stations – Mechanical and Electrical
65–95	Distribution

Source: *Adaption from* The International Infrastructure Management Manual, Version 1.0. *Australia/New Zealand Edition, 2000.*

include the material of which the pipe is made, the conditions of the soil in which it is buried, and the character of the water or wastewater flowing through it. In addition, pipes do not deteriorate at a constant rate. During the initial period following installation, the deterioration rate is likely to be slow and repair and upkeep expenses low. For pipe, this initial period may last several decades. Later in the life cycle, pipe will deteriorate more rapidly. The best way to determine the remaining useful life of a system is to conduct periodic condition assessments. At the local level, it is essential for local service providers to complete periodic condition assessments in order to make the best life-cycle decisions regarding maintenance and replacement.

OPERATION AND MAINTENANCE (O&M) CAPITAL STOCK

Since 1970, spending in constant dollars on operations and maintenance (O&M) for waste-water treatment operation and drinking water treatment operations has grown significantly. In 1994, for example, 63% of the total spending for wastewater treatment operation was for O&M, and 70% of the total spending for drinking water operations was for O&M (CBO, 1999). Likely explanations for the increase in wastewater and drinking water O&M costs include the following:

- Expansion and improvement of services, which translated into an increase in capital stock and a related increase in operations and maintenance costs.
- Aging infrastructure, which requires increasing repairs and increasing maintenance costs.

Additionally, increases in water operations and maintenance have been driven, in large part, by a large number of solids handling (biosolids) facilities coming on-line. The installation of these facilities has increased O&M costs since the mid-1980s. Over the next 20 years, O&M expenses are likely to increase in response to the aging of the capital stock: that is, as infrastructure begins to deteriorate, the costs of maintaining and operating the equipment will increase. An American Water Works Association (AWWA) study found that projected expenditures for deteriorating infrastructure would increase steadily over the next 30 years (AWWA, 2001). The projected increase in O&M costs finds support in the historical spending data, which indicate an upward trend for O&M.

Increasing O&M needs will present a significant challenge to the financial resources of water and drinking water systems. As the nation's water infrastructure ages, systems should expect to spend more on O&M. Some systems might even postpone capital investments to meet the rising costs of O&M, assuming that their total level of spending remains constant. The majority of systems likely would increase spending to ensure that both capital and O&M needs are fulfilled, and thus total spending would increase significantly. Many systems would recognize that delaying new capital investments would only increase expenditures on O&M, as old and deteriorated infrastructure would need to be maintained at increasingly higher costs.

DRINKING WATER CAPITAL STOCK

The capital stock of an individual drinking water system can be broken down into four principal components: source, treatment, storage, and transmission and distribution mains. Each of these components fulfills an important function in delivering safe drinking water to the public. While there is no study available that directly addresses the capital make-up of our nation's drinking water systems, a general picture can be obtained from the 2007 EPA Drinking Water Infrastructure Needs Survey. The survey found that the total nationwide infrastructure need is $334.8 billion for the 20-year period from January 2007 through December 2026. Although it is the least visible component of a public water system, the buried pipes of a transmission and distribution network generally comprise most of a system's capital value.

DID YOU KNOW?

Pipes are expensive but invisible. Most people do not realize the huge magnitude of the capital investment that has been made to develop the vast network of distribution mains and pipes – the infrastructure – that makes clean and safe water available at the turn of a tap. Water is by far the most capital intensive of all utility services, mostly due to the cost of these pipes, water infrastructure that is literally a buried treasure beneath our streets (AWWA, 2001).

Transmission and distribution needs accounted for 60% of the total need reported in the 2007 survey. Treatment facilities that are needed to address contaminants with acute and chronic health effects represented the second-largest category with 22% of the total need. Storage projects needed to construct or rehabilitate finished water storage tanks represented 11% of the total need. Projects needed to address sources of water accounted for 6%. The source category included needs for constructing or rehabilitating surface water intakes, raw water pumping facilities, drilled wells, and spring collectors. Neither the storage nor source categories considered needs associated with the construction or rehabilitation of raw water reservoirs or dams (USEPA, 2007).

DID YOU KNOW?

On average, the replacement cost value of water mains is about $6,300 per household in today's dollars (2001) in the relatively large public water systems studied. If water treatment plants, pumps, etc., are included, the replacement cost value rises to just under $10,000 per household, on average (AWWA, 2001).

The need to replace aging transmission and distribution components is a critical part of any drinking water system's capital improvement plan. A recent AWWA 2001 report, Dawn of the Replacement Era, surveyed the inventory of pipe and the year in which the pipe was installed for 20 cities in an effort to predict when replacement of the pipe would be needed. While the 20 cities in the sample were not selected at random, the cities likely represent a broad range of systems of various ages and sizes from across the country.

More importantly, the study provides the only available data on the age of pipe from a reasonably large number of systems. Age is one factor that affects the life expectancy of pipe. A simple aging model, therefore, was developed to predict when pipes for these 20 cities would need to be replaced. It was assumed that pipes installed before 1910 last an average of 120 years. Pipe installed 1911–1945 is assumed to last an average of 100 years.

Pipe installed after 1945 is assumed to last an average of 75 years. In estimating when the current inventory of pipe will be replaced, the model assumes that the actual life span of the pipe will be distributed normally around its expected average life; that is, pipe expected to last 75 years will last 50–100 years, pipe expected to last 100 years will last from 66 to 133 years, and pipe expected to last 120 years will last 80–160 years (AWWA, 2001).

This assumption greatly simplifies reality, as the deterioration rates of pipe will vary considerably as a function not only of age, but also of climatic conditions, pipe material, and soil properties. Pipe of the same material, for example, can last from 15 years to over 200 years depending on the soil characteristics alone. In the absence of data that would allow for the development of a model to estimate pipeline (i.e., accounting for local variability of pipe deterioration), the application of a normal distribution to an average life expectancy may provide a reasonable approximation of replacement rates.

This model also does not account for other factors, most notably inadequate capacity, that may have equal or greater importance than deterioration in determining pipe replacement rates. Applying this simple aging model to the historical inventory of pipe for the 20 cities reveals that most of the projected replacement needs for those cities will occur beyond the 20-year period of the analysis with peak annual replacement occurring in 2040. This conclusion makes sense considering that most of the nation's drinking water lines were installed after the 1940s. Moreover, we need to remember that pipes are hearty but ultimately mortal (AWWA, 2001).

Costs of Providing Service

Although many purveyors of water services obtain funds from the federal government to finance the costs of capital improvements, most of the funds that systems use for both capital and operations and maintenance come from revenues derived from user fees. As public water systems look to address future capital needs and increasing O&M costs, they need to increase fees to obtain the funding needed for these activities. While there is no complete source of national data on how rates have changed through time, the state of Ohio has information that can serve as an example for the purposes of a simple discussion. For more than 15 years, the state has conducted an annual survey of water and sewer rates for communities in the state. Data from communities that reported rates for both 1989 and 1999 reveal that there has been an upward shift in the number of communities paying higher annual fees with time.

Higher user rates that may be required to meet the increasing costs of providing service have the potential to negatively impact low-income families. Data from the Census Bureau show that from 1980 to 1998, incomes at the lower range (as a percentage share of aggregate income for households) declined or stagnated (US Census Bureau, 2000). If rates increase to fund increasing needs, public water systems may be challenged to develop rate structures that will minimize impacts on the less affluent segments of society.

CHAPTER SUMMARY

Water treatment and distribution is a business. To be successful, a public water system must be run and managed like a business – a successful business, that is. Good managers help. Smart managers help. Trained managers help. Natural-born managers not only help but are priceless! Public water system administrators, mid- and upper-level managers are critical because the present and future issues are serious, significant, and very important. We are not only talking about ensuring the delivery of a safe and healthy product to the ratepayer, but many facilities accomplish this via their aging and jerry-rigged infrastructure. Moreover, costs continue to rise, almost daily. Thus, the administrator and manager must operate the treatment plant at the optimum level but at the same time conserve financial resources; all of these responsibilities can certainly be classified as head-scratchers.

REVIEW QUESTIONS

1. What is the best procedure for ensuring equipment proper operation?
   a. All new equipment
   b. Preventive maintenance
   c. 24-hour operator surveillance
   d. Good luck

2. What causes the water to look cloudy?
   a. High pH
   b. Wastewater
   c. Hardness
   d. Suspended solids

3. What is a plant vulnerability study used for?
    a. Ensure security
    b. Evaluate potential threats
    c. Find out who is vulnerable
    d. Prepare for OSHA inspection

4. Why is a chlorine residual important?
    a. Oxidizes iron and manganese
    b. Provides chlorine residual to save on costs
    c. Oxidizes nitrate
    d. Continues treatment after treatment

5. One of the most prominent industrial wastes found in surface water is _____.
    a. Milk
    b. Caustic
    c. Organic solvents
    d. None of the above

6. Who is responsible for proper operation and maintenance and plant equipment?
    a. Top manager
    b. CEO
    c. Custodian
    d. All plant employees

7. What is the main disadvantage of performing preventive maintenance on plant equipment?
    a. Okay equipment could be damaged
    b. Too expensive
    c. Someone might be injured
    d. Might shutdown plant operations

8. Energy is expensive. What is best way to save on energy costs?
    a. Turn out the lights when not in use
    b. Turn off equipment not needed
    c. Purchase only energy-efficient equipment
    d. All of the above

9. Which of the following is a recommended way to protect and conserve plant infrastructure?
    a. Purchase good insurance policy
    b. Proper operating procedures
    c. If it runs, leave it alone mantra
    d. Increase funding

10. Why is/are public water systems "invisible" for many people?
    a. Piping is buried underground, and treatment plants are located in remote locations
    b. Many people have no idea where their drinking water comes from
    c. Many people do not care where their water comes from
    d. Water is not often thought about until the tap is dry or they are dying of thirst

ANSWER KEY

1. What is the best procedure for ensuring equipment proper operation?
    a. All new equipment
    b. **Preventive maintenance**
    c. 24-hour operator surveillance
    d. Good luck

2. What causes the water to look cloudy?
   a. High pH
   b. Wastewater
   c. Hardness
   d. **Suspended solids**

3. What is a plant vulnerability study used for?
   a. Ensure security
   b. **Evaluate potential threats**
   c. Find out who is vulnerable
   d. Prepare for OSHA inspection

4. Why is a chlorine residual important?
   a. Oxidizes iron and manganese
   b. Provides chlorine residual to save on costs
   c. Oxidizes nitrate
   d. **Continues treatment after treatment**

5. One of the most prominent industrial wastes found in surface water is _____.
   a. Milk
   b. Caustic
   c. **Organic solvents**
   d. None of the above

6. Who is responsible for proper operation and maintenance and plant equipment?
   a. Top manager
   b. CEO
   c. Custodian
   d. **All plant employees**

7. What is the main disadvantage of performing preventive maintenance on plant equipment?
   a. **Okay equipment could be damaged**
   b. Too expensive
   c. Someone might be injured
   d. Might shutdown plant operations

8. Energy is expensive. What is best way to save on energy costs?
   a. Turn out the lights when not in use
   b. Turn off equipment not needed
   c. Purchase only energy-efficient equipment
   d. **All of the above**

9. Which of the following is a recommended way to protect and conserve plant infrastructure?
   a. Purchase good insurance policy
   b. **Proper operating procedures**
   c. If it runs, leave it alone mantra
   d. Increase funding

10. Why is/are public water systems "invisible" for many people?
    a. **Piping is buried underground, and treatment plants are located in remote locations**
    b. Many people have no idea where their drinking water comes from
    c. Many people do not care where their water comes from
    d. Water is not often thought about until the tap is dry or they are dying of thirst

REFERENCES

AWWA 2001. *Dawn of the Replacement Era: Reinvesting in Drinking Water Infrastructure.* Denver, CO: American Water Works Association.

California Energy Commission 2012. Managing Your Electrical Load. In *EPA Water & Energy Efficiency in Water and Wastewater Facilities.* Accessed 04/21/19 @ http://www.epa.gov/reion09/waterinfrastructure/technology.html.

CBO 1999. *Trends in Public Infrastructure.* Washington, DC: Congressional Budget Office.

Elkington, J. 1999. *Enter the Triple Bottom Line.* Accessed 6/6/20 @ https//johnelkington.com/archive/TBL

EPA 2000. *Water Fact Sheet Package Plants.* Washington, DC: U.S. Environmental Protection Agency.

EPA 2012. *Water & Energy Efficiency in Water and Wastewater Facilities.* Accessed 04/28/12 @ http://www.epa.gov/region09/waterinfrastructure/technology.html.

United States Census Bureau 2000. *The Changing Shape of the Nation's Income Distribution.* United States Census P60–204, Current Population Reports Series. Washington, DC: United States Census Bureau.

USEPA 2002. *The Clean Water and Drinking Water Infrastructure Gap Analysis.* Washington, DC: United States Environmental Protection Agency, Document EPA-816-R-02-020.

USEPA 2007. *EPA's 2007 Drinking Water Infrastructure Needs Survey and Assessment.* Washington, DC: United States Environmental Protection Agency.

8 Plant Safety

Las Virgenes Municipal Water District

The safety of the people shall be the highest law.

– Marcus Tullius Cicero, Roman philosopher born in 106 BC

LEARNING OBJECTIVES

After studying this chapter, you should be able to:

- Describe waterworks common safety hazards, safety policy, high injury rate, duties of the plant safety person, applicable regulations, safety programs required, the safety budget, safety person's authority, and accident investigation.
- Describe the importance of safety rules, safety committees, and worker input.
- Discuss hazard communication.
- Discuss lockout/tagout.
- Discuss confined space entry.

DOI: 10.1201/9781003207146-10

- Discuss respiratory protection.
- Discuss noise control.
- Discuss personal protection equipment.
- Discuss first aid.
- Discuss thermal hazards.
- Discuss electrical safety.
- Discuss fire safety.
- Discuss welding and hot work safety.
- Discuss laboratory safety.
- Discuss machine guarding.
- Discuss blood-borne pathogens.
- Discuss ergonomics.

KEY DEFINITIONS

Acceptable Entry Conditions – the conditions that must exist in a permit space to allow entry and to ensure that employees involved with a permit-required confined space entry can safely enter and work within the space.

Aerosol – a suspension of solid particles or liquid droplets in a gaseous medium.

Asbestos – a broad mineralogical term applied to numerous fibrous silicates composed of silicon, oxygen, hydrogen, and metallic ions, like sodium, magnesium, calcium, and iron. At least six forms of asbestos occur naturally.

Attendant – an individual stationed outside one or more permit spaces who monitors the authorized entrants and who performs all attendant's duties designated by the employer's permit space program.

Attenuate – to reduce the amplitude of sound pressure (*noise*).

Audible Range – the frequency range over which normal ears hear: approximately 20 Hz through 20,000 Hz.

Audiogram – a chart, graph, or table resulting from an audiometric test showing an individual's hearing threshold levels as a function of frequency.

Authorized Entrant – an employee who is authorized by the employer to enter a permit space.

Authorized Employee – a person who locks out or tags out machines or equipment to perform servicing or maintenance on that machine or equipment. An affected employee becomes an authorized employee when that employee's duties include performing servicing or maintenance covered under the company's lockout/tagout program.

Background Noise – noise coming from sources other than the *particular noise* sources being monitored.

Banana Oil – a liquid which has a strong smell of bananas, used to check for general sealing of a respirator during fit-testing.

Baseline Audiogram – the audiogram against which future audiograms are compared.

Breathing Resistance – the resistance that can build up in a chemical respirator cartridge that has become clogged by particulates.

Capable of Being Locked Out – an energy isolating device is capable of being locked out if it has a hasp or other means of attachment to which (or through which) a lock can be affixed, or it has a locking mechanism built into it. Other energy isolating devices are capable of being locked out if lockout can be achieved without the need to dismantle, rebuild, or replace the energy isolating device or permanently alter its energy control capability.

Chemical – any element, compound, or mixture of elements and/or compounds.

Chemical Hazard – any chemical that has the capacity to produce injury or illness when taken into the body.

Cleaning Respirators – cleaning respirators involves washing with mild detergent and rinsing with potable water.

Combustible Liquid – any liquid having a flashpoint at or above 100°F (37.8°C) but below 200°F (93.3°C).

Confined Space – a space that is large enough, and so configured that an employee can bodily enter and perform the assigned work and has limited or restricted means for entry or exit (e.g., tanks, vessels, silos, storage bins, hoppers, vaults, and pits are spaces that may have limited means of entry); and is not designed for continuous employee occupancy.

Container – any bag, barrel, bottle, box, can, cylinder, drum, reaction vessel, storage tank, or the like that contains a hazardous chemical.

Decibel (dB) – unit of measurement of sound level.

Double Hearing Protection – a combination of both ear plug and ear muff-type hearing protection devices. These devices are required for employees who have demonstrated Temporary Threshold Shift during audiometric examination and for those who have been advised to wear double protection, by a medical doctor, in work areas that exceed 104 dBA.

Emergency – any occurrence or event (*including any failure of hazard control or monitoring equipment*) internal or external to the permit space that could endanger entrants.

Energized – connected to an energy source or containing residual or stored energy.

Energy Isolating Device – a mechanical device that physically prevents the transmission or release of energy, including (but not limited to) the following: a manually operated electrical circuit breaker, a disconnect switch, a manually operated switch by which the conductors of a circuit can be disconnected from all ungrounded supply conductors, and, in addition, in which no pole can be operated independently; a line valve; a block; and any similar device used to block or isolate energy. Push buttons, selector switches, and other control circuit-type devices are not energy isolating devices.

Energy Source – any source of electrical, mechanical, hydraulic, pneumatic, chemical, thermal, or other energy.

Entry – the action by which a person passes through an opening into a permit-required confined space. Entry includes ensuing work activities in that space and is considered to have occurred, as soon as any part of the entrant's body breaks the plane of an opening into the space.

Entry Permit (permit) – the written or printed document provided by the employer to allow and control entry into a permit space and that contains the information shown in an approved entry permit.

Entry Supervisor – the person (such as the employer, foreperson, or crew chief) responsible for determining whether acceptable entry conditions are present at a permit space where entry is planned for authorizing entry and overseeing entry operations, and for terminating entry, as required by the Confined Space Entry Standard.

Explosive – a chemical that causes a sudden almost instantaneous release of pressure, gas, and heat when subjected to sudden shock, pressure, or high temperature.

Exposure – the actual or potential subjection of an employee to a hazardous chemical through any route of entry during employment.

Fit-Testing – an evaluation of the ability of a respiratory device to interface with the wearer in such a manner as to prevent the workplace atmosphere from entering the worker's respiratory system.

Flammable Aerosol – an aerosol that, when tested by the method described in 16 CFR 1500.45, yields a flame projection exceeding 18 inches at full valve opening or a flashback (*flame extending back to the valve*) at any degree of valve opening.

Flammable Gas – a gas that at ambient temperature and pressure forms a flammable mixture with air at a concentration of 13% by volume or less, or a gas that at ambient temperature and pressure forms a range of flammable mixtures with air wider than 12% by volume regardless of the lower limit.

Flammable Liquid – a liquid having a flashpoint of 100°F (37.8°C).

Flashpoint – the minimum temperature at which a liquid gives off a vapor in sufficient concentration to ignite.

Hazard Warning – any words, pictures, symbols, or combination thereof appearing on a label or other appropriate form of warning which conveys the hazards of the chemical(s) in the container.

Hazardous Chemical – any chemical which is a health or physical hazard.

Health Hazard – a chemical for which there is statistically considerable evidence based on at least one study conducted in accordance with established scientific principles that acute or chronic health effects may occur in exposed employees.

Hot Work Permit – the employer's written authorization to perform operations capable of providing a source of ignition (*e.g., riveting, welding, cutting, brazing, burning, and heating*).

Immediately Dangerous to Life or Health (IDLH) – any condition that poses an immediate or delayed threat to life and would cause irreversible adverse health effects or would interfere with an individual's ability to escape unaided from a permit space.

Inerting – the displacement of the atmosphere in a permit space by a noncombustible gas (such as nitrogen) to such an extent that the resulting atmosphere is noncombustible. Note: This procedure produces an IDLH oxygen-deficient atmosphere.

Isolation – the process by which a permit space is removed from service and completely protected against the release of energy and material into the space by such means as: blanking or blinding; realigning or removing sections of lines, pipes, or ducts; a double block and bleed system; lockout or tagout of all sources of energy; or blocking or disconnecting all mechanical linkages.

Label – any written, printed, or graphic material displayed on, or affixed to, containers or hazardous chemicals.

Line Breaking – the intentional opening of a pipe, line, or duct that is or has been carrying flammable, corrosive, or toxic material, an inert gas, or any fluid at a volume, pressure, or temperature capable of causing injury.

Lockout – the placement of a lockout device on an energy isolating device, in accordance with an established procedure, ensuring that the energy isolating device and the equipment being controlled cannot be operated until the lockout device is removed.

Lockout Device – a device that utilizes a positive means (such as a lock, either key or combination type) to hold an energy isolating device in the safe position and prevent the energizing of a machine or equipment. Included are blank flanges and bolted slip blinds.

NFPA Hazardous Chemical Label – a color-code labeling system developed by the National Fire Protection Association (NFPA) which rates the severity of the health hazard, fire hazard, reactivity hazard, and special hazard of the chemical.

Hearing Conservation Record – employee's audiometric record. Includes name, age, job classification, TWA exposure, date of audiogram, and name of audiometric technician. Each employee's record is to be retained for duration of employment for OSHA and to be kept indefinitely for Workers' Compensation.

NIOSH – National Institute of Occupational Safety & Health.

Noise Dose – the ratio, expressed as a percentage, of (1) the time integral, over a stated time or event, of the 0.6 power of the measured SLOW exponential time-averaged, squared A-weighted sound pressure, and (2) the product of the criterion duration (8 hours) and the 0.6 power of the squared sound pressure, corresponding to the criterion sound level (90 dB).

Noise Dosimeter – an instrument that integrates a function of sound pressure over a period of time, to directly indicate a noise dose.

Noise Hazard Area – any area where noise levels are equal to or exceed 85 dBA. OSHA requires employers to designate work areas and to post warning signs which warn employees when work practices exceed 90 dBA as a "Noise Hazard Area." Hearing protection must be worn whenever 90 dBA is reached or exceeded.

Noise Hazard Work Practice – performing or observing work where 90 dBA is equaled or exceeded. Some work practices will be specified, however, as a "Rule of Thumb," whenever attempting to hold a normal conversation with someone who is one foot away, and shouting must be used to be heard, it can be assumed that a 90-dBA noise level, or greater exists, and hearing protection is required. Typical examples of work practices where hearing protection is required are jackhammering, heavy grinding, heavy equipment operations, and similar activities.

Noise Level Measurement – total sound level within an area. Includes workplace measurements indicating the combined sound levels of tool noise (from ventilation systems, cooling compressors, circulation pumps, etc.).

Noise Reduction Ratio – the number of decibels of sound reduction achieved by a particular hearing protection device.

Non-permit Confined Space – a confined space that does not contain (or with respect to atmospheric hazards) or have the potential to contain any hazard capable of causing death or serious physical harm.

Oxygen Deficient Atmosphere – an atmosphere containing less than 19.5% oxygen by volume.

Oxygen Enriched Atmosphere – an atmosphere containing more than 23.5% oxygen by volume.

Permissible Exposure Limit (PEL) – the maximum time-weighted, average concentration of a substance in air that a person can be exposed to during an 8-hour shift.

Permit-Required Confined Space (permit space) – a confined space that has one or more of the following characteristics:

- Contains or has a potential to contain a hazardous atmosphere;
- Contains a material that has the potential for engulfing an entrant;
- Has a configuration such that an entrant could be trapped or asphyxiated by inwardly converging walls, or by a floor which slopes downward and tapers to a smaller cross section; or
- Contains any other recognized serious safety or health hazard.

Personal Protective Device – items such as earplugs or earmuffs used as protection against hazardous noise.

Physical Hazard – a chemical for which there is scientifically valid evidence that it is a combustible liquid, a compressed gas explosive, flammable, an organic peroxide, an oxidizer, pyrophoric, unstable, (reactive) or water reactive.

Portable Container – a storage vessel which is mobile such as a drum, side-mounted tank, tank truck, or vehicle fuel tank.

Primary Route of Entry – the primary means (by inhalation, ingestion, skin contact, etc.) whereby an employee is subjected to a hazardous chemical.

Rescue Service – the personnel designated to rescue employees from permit spaces.

Respirator – a face mask which filters out harmful gases and particles from air, enabling a person to breathe and work safely.

Respiratory Hazard – any hazard that enters the human body by inhalation.

Retrieval System – the equipment (including a retrieval line, chest or full-body harness, wristlets [if appropriate] and a lifting device, or anchor-usually a tripod, and winch assembly) used for non-entry rescue of persons from permit spaces.

"Right to Know" Work Station – provides employees with a central information work station where they can have access to site SDS sheets, Hazardous Chemicals Inventory List, and Company's written Hazard Communication Program.

"Right to Know" Station Binder – a station binder located in the "right to know" work station that contains Company's Hazard Communication Program, the Hazardous Chemicals Inventory List and corresponding SDS, and the Hazard Communication Program Review, and Signature Form.

Safety Data Sheet – the written or printed material concerning a hazardous chemical, developed in accordance with 29 CFR 1910.

Signal Word – a word used on the label to indicate the relative level of severity of a hazard and to alert the reader to a potential hazard. The signal words used in this section are "Danger" and "Warning." "Danger" is used for the more severe hazards, while "Warning" is used for the less severe.

Sound Level – ten times the common logarithm of the ratio of the square of the measured A-weighted sound pressure, to the square of the standard reference pressure of 20 micro pascals unit: decibels (dB).

Sound Level Meter – an instrument for the measurement of sound level.

Stationary Container – a permanently mounted chemical storage tank.

Tagout – the placement of a tagout device on an energy isolating device, in accordance with an established procedure, to indicate that the energy isolating device and the equipment being controlled may not be operated until the tagout device is removed.

Tagout Device – a prominent warning device, such as a tag and a means of attachment, which can be securely fastened to an energy isolating device in accordance with an established procedure to indicate that the energy isolating device and the equipment being controlled may not be operated until the tagout device is removed.

Temporary Threshold Shift (TTS) – temporary loss of normal hearing level brought on by brief exposure to high-level sound. TTS is greatest immediately after exposure to excessive noise and progressively diminishes with increasing rest time.

Time-Weighted Average Sound Level – the amount of sound level which if constant over an 8-hour exposure would result in the same noise dose as is measured.

Unstable (reactive chemical) – a chemical which in its pure state, or when produced or transported, will vigorously polymerize, decompose, condense, or will become self-reactive under changing conditions of shock, pressure, or temperature.

Water Reactive (chemical) – a chemical that reacts with water to release a gas that is either flammable or presents a health hazard.

Work Center – any convenient or logical grouping of designated unit processes or related maintenance actions.

SETTING THE STAGE

Several statistical reports have presented historical evidence showing that the water treatment industries are an extremely unsafe occupational field. This less than stellar historical safety performance has continued to deteriorate, even in the age of the Occupational Safety and Health Act (OSH Act). The question is: "Why is the water treatment industry's on-the-job injury rate so high?" There are several reasons that help to explain this high injury rate. In the first place, all the major classifications of hazards exist at a water treatment plant:

- Oxygen deficiency;
- Physical injuries;
- Toxic gases and vapors;
- Infections;
- Fire;
- Explosion;
- Electrocution; and
- Workplace violence.

Along with having all the major classifications of hazards, other factors serve to cause the high incidence of injury in the water treatment industries. Some of these can be attributed to the following:

- Complex treatment systems;
- Shift work;
- New employees;
- Liberal workers compensation laws;
- Absence of safety laws; and
- Absence of written and enforced safe work practices and safety programs.

Experience has shown that a lack of well-managed safety programs and safe work practices are major factors which cause the water treatment industries to belong to an industrial group which ranks near the top of the National Safety Council's (NSC) worst industries regarding worker safety. Because of these findings, this critical issue, the absence of safe work practices and safety programs, is the last item that this chapter is designed to address. One might ask: "If the water treatment industries have such a high incidence of on-the-job injury occurrence, how will well-managed safety programs and safe work practices make a difference?"

To begin with, workers involved with water treatment and distribution work have a high incidence of injury because of the diversity that is required of them in performing their assigned duties. The average water worker is not only required to be a licensed operator but must also be a "Jack or Jill" of all trades. For example, operating the plant is one thing. That is, taking samples, operating, monitoring, and determining settings for chemical feed systems and high-pressure pumps, along with performing laboratory tests, and then recording the results in the plant's daily operating log, are routine functions performed by most water operators. It is the "nonroutine" functions that cause the problems.

For example, the typical water treatment plant operator must not only perform the functions stated above, but he/she must also (in many plants) make emergency repairs to systems (e.g., welding a broken machine part to keep the equipment on-line), perform material handling operations, make chemical additions to process flow, respond to hazardous materials emergencies, perform site landscaping duties, and perform several other functions that are not usually part of the operator's job classification but are required to maintain satisfactory plant operation and site appearance. Remember, the plant operator's job is to keep the plant running in accordance with permit(s); keeping the plant running at 3:00 a.m. during a snowstorm and icy conditions may require the operator to perform mechanical tasks that he/she is not trained to do.

Because the water treatment operator is expected to be a diverse, multitalented, extremely capable individual who can do whatever is required to maintain smooth plant operation, several safety considerations come into play during a water operator's normal plant shift. Thus, there is a need for a wide variety of safety programs and safe work practices in order to cover a wide variety of diverse job functions. It logically follows that diverse job functions expose the worker to a myriad of hazards.

Let us now examine the safety organization within a typical water treatment plant. In a water treatment facility, it is not unusual for the personnel manager, or other designated organizational representative, to be thrust into the all-encompassing world of safety. This is especially the case in smaller organizations. As a case in point, consider the example of the small water treatment facility that employs less than 50 full-time workers. Along with a chief operator or plant superintendent, the small water treatment facility generally employs someone in the capacity of personnel manager/payroll clerk/timekeeper and safety person. Yes, more often than not an "all in one" position.

Performing personnel functions, including timekeeping and payroll accounting, are tasks that are difficult enough in themselves. Add safety to the mixture, and the ingredients do not always easily blend. Indeed, safety is not only a science in its own right but also an endeavor that requires full-time attention. The average person who might be thrown into the above described predicament may have no knowledge, or at the most, a very limited knowledge of safety. As a matter of fact, this same individual may have difficulty in properly explaining the term OSHA. (This shortcoming is resolved quickly, however, whenever the organization is cited by OSHA.)

In the first place, the primary lesson the water treatment facility "safety person" must learn to be successful is to be an advocate for safe work conditions in his/her facility and not just a regulator of safe work conditions. Secondly, when the uninitiated person is thrown into the position as the "safety person," he/she must quickly come to grips with the fact that on-the-job injuries are very real and can be frequent occurrences. On the other hand, it can take a much longer time for the "rookie safety person" to realize that Grimaldi and Simonds (1989) put it right when they stated that "many (injurious) events, almost 9 out of 10 that occur in work places... can be predicted" (p. 3). The point Grimaldi and Simonds make is that knowledge exists not only on how to predict injuries but also on how to prevent their occurrence.

"*Where do I start?*" This is a natural question for the new "safety person" to ask. Typically, (as previously stated) someone is assigned the additional duty of a safety officer for their plant as a collateral duty. It is not unusual to find senior plant operators or chief operators who have been assigned this collateral duty. It would be difficult to find a more challenging or more mind-boggling, collateral duty assignment than that of the "safety person" job.

This statement may seem strange to those managers who view safety as a duty that only requires "someone" to keep track of accident statistics, to conduct plant safety meetings, and perhaps to place a safety notice or safety poster on the plant bulletin boards. The fact of the matter is in this day and age of highly technical safety standards and government regulations, the safety person has

FIGURE 8.1   General Duties of "the Safety Person's Job." Illustration by Kat Welsh and F. Spellman.

much more to do than place posters on the bulletin board and also has much more responsibility. Figure 8.1 will give you an indication of the varied duties of the "safety person" job.

Beyond answering the question "Where do I start?", this chapter provides a guide on how to maintain the safety effort after it is on-line. Moreover, this chapter will point out the pitfalls and the failures that have been experienced and the lessons learned through years of safety experience. This is not to say that any one person has all the answers; no one does. Safety knowledge is something that must be gained through a common-sense approach, blended with years of experience. Experience can only be gained through doing.

The "safety person" must be a **DOER**. A key element that can aid the DOER in doing what is correct is a dogged determination to make his/her workplace the safest one possible.

The other major elements of the safety profession can be learned by reading several of the outstanding texts that are available on the subject. The safety person can never stop learning. Simply stated, one cannot learn all there is to know about safety. One can only try to learn the main factors involved with preventing injuries. When you get right down to it, isn't preventing injuries what safety is all about? So, for the personnel manager, water operator, chief operator, maintenance operator, or other person who suddenly finds him/herself assigned to the prestigious, but absolutely demanding collateral or full-time duty as plant safety official, this chapter is designed for you.

Additionally, the good news is this chapter is designed to explain a technique-procedure, a paradigm, or model that works; it has been tested. Because the methodology described in this chapter has been used successfully in water treatment, it will provide answers to several questions. For example:

- What types of safety programs are needed at a water treatment facility?
- What are the health and safety concerns that are unique to the water industries?
- What are the applicable regulations?
- Which safe work practices should be used in the water industry?

- When one determines which plant safety programs to implement, how does he/she maintain these safety programs?
- Secondarily, how does he/she measure the results of his/her efforts?

There is one question this chapter does not answer; that is, how do you know when you are finished or when the job has been completed? An effective safety effort is never finished; it is never completed. It never stops. It continues to flow like the water in a large stream that eventually crests at a waterfall, and then the turmoil settles, and smooth water prevails ... for a while ... if you are vigilant, and dedicated, and steadfast. Again, sometimes the stream flow is interrupted by obstacles in the way. Sometimes the flow is contaminated by unsafe acts.

Sometimes the flow comes to a complete stop because of a dam placed in its way by some unenlightened plant official. This is where the designated safety official must step in to free up the flow. To free the flow, the safety official must be armed with facts – facts are weapons for progress. This chapter is designed to provide the designated safety person not only with the facts but also the knowledge from lessons learned and gained through years of making good judgments and by making some not-so-good judgments.

Moreover, although it is true that there are several outstanding safety texts available, it is also true that many of these texts are targeted for use by officials who have some background in safety. This chapter, on the other hand, is user-friendly - it is written to be used by that clear majority of uninitiated individuals, who might find themselves assigned the dubious task of plant safety official. In other words, this chapter is designed to provide the answer for those individuals who might ask the question: "Where do I start?"

SAFETY STARTS AT THE TOP

Section 5(a) of Public Law 91-596 of December 29, 1970, OSH Act requires each employer:

- Shall furnish to each of his employees, employment, and a place of employment which is free from recognized hazards that are causing, or are likely to cause death, or serious physical harm to his employees;
- Shall comply with occupational safety and health standards, promulgated under this Act.

No matter the number of regulations, standards, and laws that are made to ensure worker safety, and no matter how experienced and/or motivated the plant's designated safety official is, he/she is powerless without dedicated support from the highest levels of plant management. Simply put, without a strong commitment from upper management, the safety effort is doomed. On the other hand, when organizational management states that it is the company objective to place **"Safety First"** even before productivity and quality, then the proper atmosphere is present for the safety official to accomplish the intended objective. That is, the safety official and managers will be able to provide a safe place for all employees to work in.

When assigned the duty as the plant's safety official, the first thing to be done is to meet with upper management and determine what the safety objective is. The newly appointed safety official should not start on the plant's safety program until he/she is quite certain of what is expected of him/her. The pertinent matters that need to be addressed during this initial meeting include the development of a written safety policy, a safety budget, exactly what authority the safety official has, and to whom does he/she report. In addition, the organization's safety rules and safety committee structure must be formulated as soon as possible.

THE ORGANIZATION'S SAFETY POLICY

The plant safety official should propose that an organizational safety policy be written and approved by the general manager or another top plant manager. A well-written organizational safety policy

---

**NO JOB IS SO CRITICAL AND NO SERVICE
IS SO URGENT -- THAT WE CANNOT
TAKE THE TIME TO PERFORM
OUR WORK SAFELY.**

While it is true that the major emphasis is on efficient operations, it is also true that this must be accomplished with a minimum of accidents, and losses. I cannot overemphasize the importance that the Organization places on the health, and well-being of each, and every employee. The Organization's commitment to occupational health and safety is absolute. The Organization's safety goal is to integrate hazard control into all operations, including compliance with applicable standards. I encourage active leadership, direct participation, and enthusiastic support of the entire organization in supporting <u>our</u> safety programs and policies.

_____

General Manager

FIGURE 8.2   Sample plant safety policy.

should be the cornerstone of an organization's safety program. There are several examples of safety policies used by Fortune 500 companies, and others, from which to model your own plant's safety policy. The key to producing a powerful, tell-it-like-it-is safety policy is to keep it short, to the point, and germane to the overall goal. Many organizational safety policies are well written but are too lengthy, too philosophical.

The major point to remember is that the organization's safety policy should be written not only so that it might be understood by every employee, but also so that all employees will read it. An example of a short, to the point, and hard-hitting safety policy that we have used in the past, and found to be effective, is provided in Figure 8.2. The powerful effect of Figure 8.2 on an organization's safety policy is in its brevity. The major point is made in as short of verse as is necessary; thus, this is the type of Safety Policy that will be read by the employee. More importantly, the Safety Policy sends the desired message.

THE SAFETY BUDGET

The safety budget is critical. No one who knows the requirements of an effective safety program ever said that safety is inexpensive; it is not. On the contrary, it is not unusual for safety divisions to expend six-figure budgets per year on safety and health programs and equipment for water organizations with large workforces. On the other hand, in smaller water treatment facilities where money for safety is either hard to find or is nonexistent, the total safety budget might be limited to a few hundred dollars per year. However, seasoned safety professionals, and others, recognize that a few hundred dollars spent to keep workers safe is much cheaper than lawsuits filed by injured workers due to unsafe conditions.

As an example of budgeting woes, consider water treatment facilities in rural and sparsely populated counties. It is not unusual in these small Publicly Owned Treatment Works (POTWs) for water to have a total budget of not more than $200 per year to use for the plant's safety programs. Funding the plant's safe operation sometimes takes a back seat to other more urgent county or state requirements. More importantly, when you consider that an environmental air monitor can cost $2,500 or more, it is not difficult to understand why many POTWs do not possess even one of these critical air monitoring devices. This has been found to be the case, even though air monitoring is required to meet various regulations and for confined space entry operations. When it comes to budgeting, management is concerned with the bottom line. It is often difficult for management to discern the value of safety in terms of the cost–benefit relationship (until expensive lawsuits are experienced). It is the safety official's job to enlighten management on the significance of a sound safety program and how it relates to the organization's bottom line. This is not an easy undertaking.

Additionally, making an argument for funding that does not exist can be extremely frustrating. It is true, however, that sometimes the plant safety official can convince those who control the finances to budget more money for safety; that is, if he/she can present a compelling case or argument for the additional funding. For example, the safety official must make the point to management that it is less expensive to incorporate safety into the organization than it is to pay for the loss of life, severe injury, hazardous materials incidents that affect the public, medical expenses, destruction of property, employees' lost time, workers' compensation expenses, a possible violation of the plant's operating permit, and citations issued by OSHA or other regulators.

The plant safety official who is attempting to increase funding for safety must understand from the very beginning that when organizational money is spent, upper management wants results. It wants to see what its money has bought. For the designated safety official, this is a critical area. Some would call it "blowing your own horn." In reality it should be called communicating success to the extreme. For example, when OSHA inspects one of the organization's facilities and can find very little wrong and nothing that can be cited, then this information must be passed on to upper management because you just saved the organization from embarrassment and possible costly fines; remember, a publicly funded organization that are not in compliance with regulators is not good business, to say the least. Upper management must get the message that its commitment to spending money on safety has paid off; it has saved money; it has prevented fines and embarrassment. More importantly, a strong commitment to an effective safety program can and will prevent injuries and fatalities.

When talking about the organization's bottom line, the safety official must convince upper management that the organization's real bottom line is the health and well-being of its employees. If upper management's bottom line is putting financial gain before protecting the employee, then the organization does not need a safety official and the appropriate safety funding, but instead it needs very deep pockets to pay for very expert legal counsel.

SAFETY OFFICIAL'S AUTHORITY

The degree and extent of the safety official's authority are important. For example, the safety official must have the authority to conduct in-house audits. These safety audits are designed to reveal unsafe conditions and/or practices. And, more importantly, these audits must be followed up. In other words, the safety official must have the authority, and the backing of upper management, to ensure that supervisors correct any deficiencies that are found during the audit. The safety official must have the authority to shut down work in progress, on the spot, immediately, whenever unsafe work practices or unsafe conditions are discovered. Although this type of authority is important, it should also be stressed that this is latent authority, authority that is reserved and only to be used with great discretion. Remember, the safety official must take on the facade of being "Good Neighbor Sam," not that of a Gestapo Agent. The safety official must have the authority to convene plant-wide mandatory training sessions. Training is at the heart of safety. Employees cannot be expected to abide by safe work practices unless they have been properly trained on what is required of them.

Safety training is more effective whenever the training is provided by those who are "expert" in the subject matter and by those who take on a personal approach. The organization's safety official must be viewed by supervisors and employees as part of the organization. This can be accomplished if the safety official takes an active role in learning the operation of the plant. In large organizations where there may be several employees working at several separate locations, it is important to train supervisors so that they will be able to recognize safety hazards and to take the correct remedial actions. Additionally, well-trained supervisors should augment the safety official's effort in providing employee safety training. Training provided by the worker's immediate supervisor, who is competent in the subject matter, is often more effective than training provided by other officials.

The importance of using supervisors in safety training cannot be overstated. The supervisor's importance in this vital area can be seen when one considers that in order to train workers on the proper and safe performance of assigned duties, input from the technical expert (the supervisor) for each job function is critical. Moreover, when supervisors are asked for their input and advice in formulating safe work practices, they generally buy into the overall safety program. Thus, these same supervisors often become valuable allies of the safety official and staunch supporters of the organization's safety effort.

ACCIDENT INVESTIGATIONS

The plant safety official must have the authority to properly perform other functions. Accident investigations can turn up causal factors that point to a disregard of established safety rules and/or safe work practices, or disobedience of direct orders. All the actions just mentioned are going to lay the finger of blame on someone. Caution is advised here: The safety official must tread a fine line in this area; his/her intention should be to determine the cause, recommend remedial action, and then follow up to ensure that corrective procedures have been put in place. The safety official should never act or perform investigations in a gestapo-type manner. The safety official must be professional, tactful, efficient, observant, and thoughtful. The safety official should never target individuals for blame. Remember, Dragnet's Joe Friday said it best: "Just the facts, ma'am." The organizational safety official should stick to and report only the facts.

SAFETY RULES

One of the first items on the newly assigned safety official's agenda should be the generation and incorporation of the organization's safety rules. Figure 8.3 gives a clear and concise illustration of the safety official's main duties. However, before submitting a list of safety rules to a higher authority for approval, the safety official should think through what he/she is proposing. Rules are everywhere. All through our lives we have functioned according to some set of rules. Workers generally do not like rules. This is especially the case when the rules are unclear, cumbersome, and arbitrary. In putting the plant's safety rules together, it is wise for the plant safety official to abide by the old acronym and saying: KISS - Keep It Simple Stupid. Note: As shown in Figure 8.3, it is not uncommon to call the organization's safety official the Safety Engineer. No matter the title, the job description should be clear and the mission undaunted.

Safety rules should be straightforward, easily understood, and limited to as few as possible. Concocting volumes of complicated safety rules will result in much-wasted effort and another "dust collector" for the shelf. Employees will not follow nor abide by rules that occupy voluminous manuals. Additionally, supervisors will have difficulty enforcing too many rules. The best safety rules are those that can be read and understood in short order. Rules can be effective for two major reasons. First, the safety rules are limited in number and easily understood. Secondly, the rules are printed on cards designed for billfold occupancy. Thus, they are a ready reference for the employee. Additionally, a copy of these rules is placed at strategic locations within each work center. The work

FIGURE 8.3   Safety official's duties.

FIGURE 8.4   Some equipment required for employee work safety.

center safety rule posters are printed in larger, bolder print on 20 × 16-inch rigid poster board. It is best to provide enforcement in the form of both punishment and praise. Punishment should be provided, depending upon the severity of the infraction, to those employees who disregard or disobey safety rules. Moreover, when a good safety effort is observed, praise should be provided in the form of letters of commendation and appropriate notations on the worker's performance record. To ensure the best possible chance for safety in the organization's operation, it is obvious that having the proper safety equipment on hand is one of the most crucial factors. Figure 8.4 is an illustration of some of the most important safety equipment to have available.

Safety Committee/Council

When working with upper management in the formulation of the organization's safety program, it is important to set up a safety committee or council. The safety committee can provide valuable assistance to the plant safety official. The safety committee should be composed of a cross section of the organization's workforce. Additionally, the safety committee should consist of a combination of senior managers, as well as employees, at mid-grade supervisory levels. If the organization is unionized, then a designated union representative must be assigned.

Worker Input

Safety officials sometimes overlook a valuable resource that is always present in any organization: WORKERS. Some would argue that workers not only make up the organization but also that workers are the organization. Safety officials are hired primarily to protect workers from injury. Safety officials sometimes forget their mission; that is, they forget that they are primarily tasked with ensuring that workers have a safe place to work. One of the areas of importance is to mandate that workers wear personal safety gear. A sampling of the type and kind of safety equipment that the employer should provide to water treatment employees is shown in the examples shown in Figure 8.5.

Workers also have a role in their own safety. When workers hear this statement, they might be surprised. At the beginning of this chapter, part of the OSH Act was stated; the part dealing with the employer's responsibilities under the Act. It may surprise many people to know (it almost always surprises the workers) that the OSH Act mandates the following:

Section 5(b) mandates that each employee shall comply with occupational safety, and health standards, and all rules, regulations, and orders issued pursuant to this Act, which are applicable to his own actions, and conduct.

Several organizational safety programs, policies, manuals, or directives specifically define who is responsible for safety. When the reader reviews such programs, policies, manuals, or directives, it usually is clear to him/her whom has been made responsible for safety – but, on the other hand, when personnel are "designated" as being responsible for safety, the reader might wonder about the rest of the organization's personnel; that is, those personnel who are not "designated" as being responsible for safety. Shouldn't everyone share this responsibility? Providing a safe place in which to work can be better accomplished when all organizational personnel have input, especially the

FIGURE 8.5   Examples of personal safety items.

workers – the rank and file. Input can be in the form of a discussion between the worker and the supervisor. Input can also be in the form of a discussion between the worker and the work center's safety committee member. On occasion, workers provide input directly to the safety official.

Whatever form of safety input is used is not the most principal factor. The important thing is to get the input. An organizational safety program should encourage worker empowerment. Worker empowerment provides for worker input. For example, input received from workers during an accident investigation in which a severe injury occurred due to a workplace hazard is important to the investigating official in formulating his/her final report and recommending remedial action. On the other hand, if this same worker input had been provided to the work center supervisor or safety official prior to the mishap's occurrence, then proper remedial action might have been affected in removing the hazard and thus preventing the mishap from having occurred in the first place. The question is, "How does the worker provide workplace hazard information to the safety official?" There are several ways to accomplish this. Input to the organizational safety official can be made through the work center's safety committee, for example.

Accident Reporting

Experience shows that we do not know how many on-the-job injuries we could have prevented if only workforce personnel had informed us of near misses, and/or conditions, situations, or equipment operations that bothered them. Possibly because they were near-miss victims, or otherwise felt unsafe while near, or involved in operations, or situations that they did not feel comfortable in. It can't be overstressed for the need for workplace supervisors to prepare and maintain records for reporting worker injuries and accident investigations. In water industries, exposure to various microorganisms (some that are harmful to humans) which exist in some raw water is a routine occurrence. When an employee is exposed to these organisms, even the most minor scratch, cut, or abrasion should immediately be reported to the supervisor. Some would say that it is burdensome for the supervisor and the safety official to process paperwork that details minor, on-the-job injuries. This might be the case in other industries, but in water treatment plants, where exposure to tainted water might be a daily occurrence, it is important to require employees to report all injuries, no matter how minor.

Safety Audits

Safety audits or inspections can be valuable tools to use in detecting worksite hazards that may lead to worker injury. The obvious purpose of safety audits is to identify and correct workplace hazards. Not surprisingly, a newly assigned safety official is sometimes apprehensive about conducting safety audits on their own of the new organization's facilities. This apprehension stems from fear of antagonizing the site supervisor. Moreover, unless the "rookie" safety official has previous safety inspection experience, he/she often feels that he/she lacks knowledge and expertise in conducting safety inspections.

Water treatment plants possess most of the industrial hazards that are present in other industrial settings. The following is a concise list of some of the typical hazards that are present at water treatment plants:

- Machinery;
- Flammable-combustible materials;
- Walking-working surfaces;
- Welding, cutting, and brazing operations;
- Electrical equipment and appliances;
- Ladders and scaffolds;
- Compressed gases;

- Materials handling and storage;
- Hand and portable powered tools; and
- Process-generated hazardous and toxic substances.

Even with our many years of experience and the thousands of inspections we have conducted, there is no way we are able to remember every item at every plant that we should look at or inspect; thus, we have found that a valuable tool to use when conducting the safety audit is a safety inspection checklist. If your organization does not have a safety inspection checklist, one should be generated as soon as possible.

Once the organization's safety checklist is generated, the safety official should consider this document as a "living" document that will continue to grow with additional checklist items as time passes. Many discrepancies found during safety audits can be remedied right then and there, on the spot. It should be remembered that after having discovered a hazard or hazards, it is the prompt and correct remedial action that is the goal of the safety audit.

SAFETY PROGRAMS

HAZARD COMMUNICATION

Under its Hazard Communication Standard (more commonly known as "HazCom" or the "Right to Know Law"), OSHA requires employers, who use or produce chemicals on the worksite, to inform all employees of the hazards that might be involved with those chemicals. HazCom says that employees have the right to know what chemicals they are handling or may be exposed to. HazCom's intent is to make the workplace safer. Under the HazCom Standard, the employer is required to fully evaluate all chemicals on the worksite for possible physical and health hazards. All information relating to these hazards must be made available to the employee 24 hours a day. The standard is written in a performance manner, meaning that the specifics are left to the employer to develop.

HazCom also requires the employer to ensure proper labeling of each chemical, including chemicals that might be produced by a process (process hazards). Labels must be designed to be clearly understood by all workers. Employers are required to provide both training and written materials to make workers aware of what they are working with and what hazards they might be exposed to. Employers are also required to make Safety Data Sheets (SDS) available to all employees. An SDS is a fact sheet for a chemical which poses a physical or health hazard in the workplace. An SDS must be in English and contain the following information:

- Identity of the chemical (label name);
- Physical hazards;
- Control measures;
- Health hazards;
- Whether it is a carcinogen;
- Emergency and first aid procedures;
- Date of preparation of the latest revision; and
- Name, address, and telephone number of manufacturer, importer, or other responsible parties.

Your facility must have an SDS for each hazardous chemical it uses. Copies must be made available to other companies working on your worksite (outside contractors, for example), and they must do the same for you. The facility hazard communication program must be in writing and, along with an SDS, be made available to all workers 24 hours each day/each shift. The plant manager and operators must take personal interest in ensuring that the facility is in full compliance with the Hazard

Communication Standard for three major reasons: (1) it is the law; (2) it is consistently the number one cause of citations issued by OSHA for noncompliance; and (3) compliance with the standard goes a long way toward protecting workers.

The employer's (thus, the safety and health person's) responsibilities include: signs, placards, process sheets, batch tickets, operating procedures, or other such written materials, in lieu of affixing labels to individual stationary process containers – if the alternative method identifies the containers to which it is applicable and conveys the information required on the label. The written materials must be readily accessible to the employees in their work area throughout each shift.

LOCKOUT/TAGOUT

OSHA's Lockout/Tagout

OSHA's 29 CFR 1910.147 states: Employers are required to develop, document, and utilize an energy control procedures program to control potentially hazardous energy.

The energy control procedures must specifically outline:

- The scope, purpose, authorization, rules, and techniques to be utilized for the control of hazardous energy; and
- The means to enforce compliance including, but not limited to, the following:
- A specific statement of the intended use of the procedure;
- Specific procedural steps for shutting down, isolating, blocking, and securing machines and equipment to control hazardous energy;
- Specific procedural steps for the placement, removal, and transfer of lockout devices or tagout devices and the responsibility for them; and
- Specific requirements for testing a machine or equipment to determine and verify the effectiveness of lockout devices, tagout devices, and other energy control measures.

Removing the hazard is always the best way to protect entrants; however, removing all the hazards is, in many cases, impossible. Thus, OSHA requires the control of hazardous energy using isolation, blanking, or blinding, disconnection, and/or lockout/tagout procedures. According to (Caruey, 1991), OSHA estimates that full compliance with the lockout/tagout standard will prevent 120 accidental deaths, 29,000 serious injuries, and 32,000 minor injuries every year. Experience has shown that many workers mistake the results of atmospheric testing that show no hazard exists in a confined space as meaning that the space is totally safe for entry. Indeed, this might be the case; however, many other dangers inherent to confined spaces make entry into them hazardous. For example, if the confined space has some type of open liquid stream flowing through it, the chance for engulfment exists. If the space has electrical devices and circuitry inside, an electrocution hazard exists. If hazardous chemicals are stored and taken into the space, the potential for a hazardous atmosphere exists. Many confined spaces contain physical hazards, including piping and other obstructions – for example, rotating machinery is often housed within confined spaces. The control of hazardous energies by locking or tagging out also applies to most work involved in servicing, adjusting, or maintenance activities involving machines and/or processes that place personnel at elevated risk. In addition to the sources of machine energy mentioned in the opening (electrical, pneumatic, steam, and so forth), an added concern is inadvertent activation when personnel are in contact with the hazards. The elements of the Lockout/Tagout Program and sample Lockout/Tagout Programs are illustrated in Figures 8.6 and 8.7.

CONFINED SPACE ENTRY

Confined spaces can be very unforgiving; entering confined spaces without the proper equipment and training amplifies and acerbates the inherent danger. With regard to confined space rescue, the tendency to leap into a confined space with total disregard for one's own safety is what the author

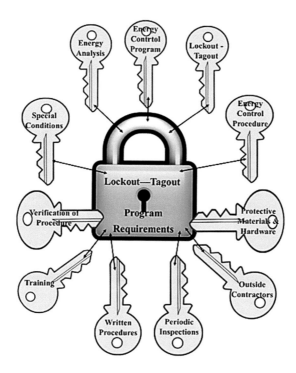

FIGURE 8.6   Elements required for compliance with OSHA's Lockout/Tagout.

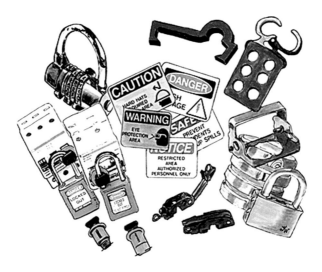

FIGURE 8.7   Lockout devices and tags.

calls the "John Wayne" syndrome. John Wayne, who frequently played larger-than-life heroes, rushed into dangerous situations to rescue victims in movie after movie, with no regard for his own safety or well-being. Workers often disregard their own safety in attempts to rescue fellow workers. Should workers who disregard their own safety by attempting to rescue fellow workers in jeopardy be considered heroes? The point we make here is that serious problems exist with posthumous heroism and that workers should never be placed in the position where such life-threatening decisions are made.

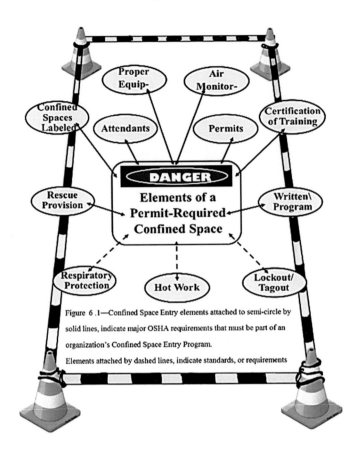

FIGURE 8.8   Elements of confined space program.

Without question, when an organization's safety engineer's duties include compliance with OSHA's 29 CFR 1910.146 Confined Space Entry Standard, his/her hands are full-full on a full-time, continuous basis. There is no effortless way out when it comes to ensuring full compliance with this vital requirement. Full compliance is completely possible, but it requires exceptional attention to regulatory compliance, to detail, and to the ongoing management of the program as it should be managed; the task is to lead, and not to mislead, workers. Figure 8.8 lists the elements needed to ensure full compliance with OSHA's Confined Space Entry requirements. From Figure 8.8, you see that the main program elements are attached to the semi-circle by solid lines, while ancillary or interfacing OSHA standards (Lockout/Tagout and Respiratory Protection) and the Hot Work Permitting requirement are attached separately below by dashed lines. Lockout/Tagout, Respiratory Protection, and Hot Work Permitting are essential in protecting workers from hazards that are sometimes present in confined spaces – however, not all confined spaces present such risks.

Once the hazards have been identified and evaluated, the identity and hazard(s) of each site's confined space must be listed in the organization's written confined space entry program. Obviously, it is important that employees are made well aware of all the hazards (Figure 8.9).

The next step is to develop written procedures and practices for those personnel who are required to enter, for any reason, permit-required confined spaces. The procedures and practices used for permit-required confined space entry must be *in writing*, and at the very least, must include:

- Specifying acceptable entry conditions;
- Isolating the permit space;

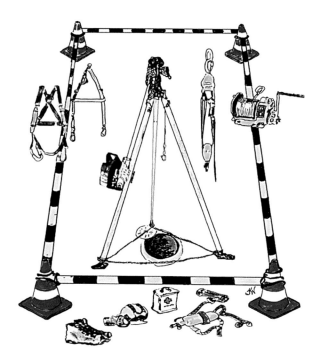

FIGURE 8.9   Confined equipment. Ventilation system is not shown, but typically required. (Red-Devil Blower).

- Purging, inerting, flushing, or ventilating the permit space, as necessary to protect entrants from external hazards;
- Providing pedestrian, vehicle, or other barriers as necessary to protect entrants from external hazards; and
- Verifying that conditions in the permit space are acceptable for entry throughout the duration of an authorized entry.

Under OSHA's program, the employer must also provide specified equipment to employees involved in confined space entry (see Figure 8.10).

Permit System

A permit system for permit-required confined space entry is required by the Confined Space Standard. An entry supervisor (qualified or competent person) must authorize entry, prepare and sign written permits, order corrective measures if necessary, and cancel permits when work is completed. Permits must be available to all permit space entrants at the time of entry and should extend only for the duration of the task. They must be retained for a year to facilitate review of the confined space program (1910.146 (e)(f), The Office of the Federal Register, 1995a,b).

Confined Space Training

The employer shall provide training so that all employees, whose work is regulated by this [standard], acquire the understanding, knowledge, and skills necessary for the safe performance of the duties assigned (1910.146 (g), The Office of the Federal Register, 1995). Any work requirement is easier to perform if the person doing the task is fully trained on the proper way to accomplish it. Training offers another advantage as well – increased safety. In accomplishing any work task safely, proper training is critical.

FIGURE 8.10   Elements of written confined space program.

Confined space entry operations are extremely dangerous undertakings. We stated earlier that confined spaces are very unforgiving – this is the case, even for those workers who have been well trained. However, training helps to reduce the severity of any incident. When something goes wrong, (as if often the case), it is better to have fully trained personnel standing by than to have people standing by who are not trained, who do not know how to properly rescue an entrant, let alone how to rescue themselves. When you get right down to it, having fully trained workers for any job just makes good common sense.

RESPIRATORY PROTECTION

It may not always be obvious to workers just when a respirator is required. Whenever workers must enter a confined space or other vessel for maintenance, entry should not be made until the atmosphere (air) within the confined space or vessel is tested for flammable agents, oxygen content, and for toxic agents (e.g., hydrogen sulfide, methane, carbon monoxide, and other toxic agents). Lack of oxygen is the most common cause of deaths in confined space entry. As an example of what can happen, consider that to prevent fire or explosion, fuel storage tanks that have contained flammable materials are frequently inerted or purged with various gases such as nitrogen prior to allowing personnel to enter the space to perform required maintenance. Purging the tanks will prevent fires from welding or spark-making activities, but if a worker should enter this space without a proper respirator, he/she will be quickly overcome by lack of oxygen.

OSHA requires employers in occupational settings to establish and administer an effective written respiratory protection program. This requirement is vital when you take into consideration that the most common route of entry of chemicals and toxic substances into the body is by inhalation. Respirators are frequently used in water treatment and distribution. If your facility requires the use of respirators, then your facility must have a written respiratory protection program and abide by the OSHA requirements for respirator use.

FIGURE 8.11   Elements of respiratory program.

Respirators

The basic purpose of any respirator is, simply, to protect the respiratory system from inhalation of hazardous atmospheres. Respirators provide protection, either by removing contaminants from the air before it is inhaled or by supplying an independent source of respirable air. The principal classifications of respirator types are based on these categories (NIOSH Guide to Industrial Respiratory Protection, p. 3, 1987). OSHA mandates that written procedures shall be prepared covering safe use of respirators in dangerous atmospheres that might be encountered in normal operations or in emergencies. Personnel shall be familiar with these procedures and with the available respirators [OSHA 29 CFR 1910.134 (c)].

Respirators are devices that can allow workers to safely breathe without inhaling particles or toxic gases. Two basic types are (1) air-purifying, which filter dangerous substances from the air; and (2) air-supplying, which deliver a supply of safe breathing air from a tank (SCBA) or group of tanks (cascade system), or an uncontaminated area nearby via a hose or airline to your mask. A well-planned, well-written respiratory protection program must include the eleven elements shown in Figure 8.11.

In this section, we discuss these eleven elements and explain what they require. This information will enable the safety and health professionals to implement a respiratory protection program that complies with OSHA requirements. A written (it must be in writing) "Respiratory Protection Program" to comply with OSHA regulations (as set forth in 29 CFR 1910.134) is designed to do all that is possible to protect those employees who are filling a job classification that requires respirator use in the performance of their duties.

NOISE CONTROL

OSHA's Hearing Conservation Requirements

In 1983, OSHA adopted a Hearing Conservation Amendment to OSHA 29 CFR 1910.95 that requires employers to implement *hearing* conservation programs in any work setting where employees are

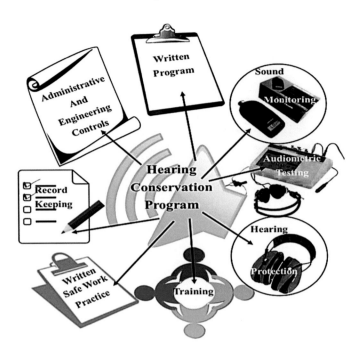

FIGURE 8.12   Elements of a hearing conservation program.

exposed to an 8-hour, time-weighted average of 85 dBA and above (LaBar, 1989). Employers are required to implement hearing conservation procedures in settings where the noise level exceeds a time-weighted average of 90 dBA. They are also required to provide personal protective equipment for any employee who shows evidence of hearing loss, regardless of the noise level at his or her worksite. Figure 8.12 shows the elements of a Hearing Conservation Program and examples of hearing protection devices.

The basic requirement of OSHA's Hearing Conservation Standard includes:

- *Monitoring Noise Levels* – Noise levels should be monitored on a regular basis. Whenever a new process is added, an existing process is altered, or new equipment is purchased, special monitoring should be undertaken immediately.
- *Medical Surveillance* – The medical surveillance component of the regulation specifies that employees who will be exposed to high noise levels be tested when they are hired and at least annually while employed.
- *Noise Controls* – The regulation requires that steps be taken to control noise at the source. Noise controls are required in situations where the noise level exceeds 90 dBA. Administrative controls are sufficient until noise levels exceed 100 dBA. Beyond 100 dBA, engineering controls must be used.
- *Personal Protection* – Personal protective devices are specified as the next level of protection when administrative and engineering controls do not reduce noise hazards to acceptable levels. They are to be used in addition to, rather than instead of, administrative and engineering controls.
- *Education and Training* – The regulation requires the provision of education and training to do the following: to ensure that employees understand: (1) how the ear works; (2) how to interpret the results of audiometric tests; (3) how to select personal protective devices that will protect them against the types of noise hazards to which they will be exposed; and (4) how to properly use the personal protective devices (LaBar, 1989).

In the written Hearing Conservation Program also list or designate the responsibilities and the person responsible; that is, *who* is responsible for managing and enforcing the various components to ensure compliance with the Program. According to OSHA, when information indicates that any employee's exposure equals or exceeds an 8-hour time-weighted average of 85 decibels, the employer must develop and implement a monitoring program. The responsibility for noise monitoring is typically assigned to the organization's safety and health professional. *Audiometric testing* is an essential element of the Hearing Conservation Program for two reasons: (1) it helps to determine the effectiveness of hearing protection and administrative and/or engineering controls; and (2) audiometric surveillance also helps to detect hearing loss before it noticeably affects the employee and before the loss becomes legally compensable under workers' compensation.

Audiometric examinations are usually done by an outside contractor but can be done in-house, with the proper equipment. Wherever they are done, they require properly calibrated equipment used by a trained and certified audiometric technician. The hearing protection element of the Hearing Conservation Program provides hearing protection devices for employees and training in how to wear them effectively whenever hazardous noise levels exist in the workplace. Hearing protection comes in numerous sizes, shapes, and materials, and the cost of this equipment can vary dramatically. Two general types of hearing protection are used widely in industry: (1) the cup muff (commonly called Mickey Mouse Ears); and (2) the plug insert type. Because feasible engineering noise controls have not been developed for many types of industrial equipment, hearing protection devices are the best option for preventing noise-induced hearing loss in these situations. As with the other elements of the Hearing Conservation Program, the hearing protective device element must be in writing and included in the Hearing Conservation Program.

PPE, First Aid, and Thermal Hazards

A hazard, any hazard, if possible, should be "engineered out" of the system or process. Determining when, and how, to engineer out a hazard is one of the safety and health professional's primary functions. However, the safety and health professional can much more effectively accomplish this if he/she is included in the earliest stages of design. Remember, it does little good (and is often very expensive) to attempt to engineer out any hazard once the hazard is in place. While the goal of the safety and health professional is certainly to engineer out all workplace hazards, we realize that this goal is virtually impossible to achieve. Even in this day of robotics, computers, and other automated equipment and processes, the man-machine-process interface still exists. When people are included in the work equation, the opportunity for their exposure to hazards is very real and as injury statistics make clear, it happens.

This is extremely important for two reasons: First, the safety and health professional's primary goal is to engineer out the problem. If this is not possible, the second alternative is to implement administrative controls. When neither is possible, personal protective equipment (PPE) becomes the final choice. The key words here are "the final choice." Secondly, PPE is sometimes incorrectly perceived by both the supervisor and/or the worker as being their first line of defense against all hazards.

This, of course, is incorrect and dangerous. The worker must be made to understand (by means of enforced company rules, policies, and training) that PPE affords only minimal protection against most hazards: It Does Not Eliminate the Hazard. When some workers put on their PPE, they also don a "super person" mentality. What does this mean? Often, when workers use eye, hand, foot, head, hearing, or respiratory protection, they also adopt an "I can't be touched" attitude. They feel safe, as if the PPE somehow magically protects them from the hazard, so they act as if they are protected, are invincible, and are beyond injury. They feel, however illogically, that they are well out of harm's way. Nothing could be further from the truth.

OSHA's PPE Standard

In the past, many OSHA standards have included PPE requirements, ranging from very general to very specific. It may surprise you to know, however, that not until recently (1993–1994) did OSHA incorporate a stand-alone PPE Standard into its 29 CFR 1910/1926 Guidelines. This relatively new Personal Protective Equipment standard is covered (General Industry) under 1910.132-138, but you can find PPE requirements elsewhere in the General Industry Standards. For example, 29 CFR 1910.156, OSHA's Fire Brigades Standard has requirements for firefighting gear. In addition, 29 CFR 1926.95-106 covers the construction industry. The PPE standard focuses on head, feet, eye, hand, respiratory, and hearing protection.

Common PPE classifications and examples include:

- Head protection (hard hats, welding helmets);
- Eye protection (safety glasses, goggles);
- Face protection (face shields);
- Respiratory protection (respirators);
- Arm protection (protective sleeves);
- Hearing protection (ear plugs, muffs);
- Hand protection (gloves);
- Finger protection (cots);
- Torso protection (aprons);
- Leg protection (chaps);
- Knee protection (kneeling pads);
- Ankle protection (boots);
- Foot protection (boots, metatarsal shields);
- Toe protection (safety shoes); and
- Body protection (coveralls, chemical suits).

First Aid in the Workplace

Subpart K of OSHA's 1910 standard addresses issues directly for eye-flushing capabilities in the workplace and indirectly for medical personnel to be readily available. Readily available can mean that there is a clinic or hospital nearby (*response time to worksite is less than 4 minutes*). If such a facility is not located nearby, employers must have a person on site that has had first-aid training. Because of these OSHA requirements, the organization's safety and health professional must, as with all other regulatory requirements, ensure that the organization is in full compliance. First aid awareness/training in the workplace usually involves providing lectures, interactive video presentations, discussions, and hands-on training to teach participants how to:

- Locate all workplace first aid kits and emergency eyewash stations;
- Recognize emergency situations;
- Check the scene and call for help;
- Avoid blood-borne pathogen exposure;
- Care for wounds, bone, and soft-tissue injuries, head, and spinal injuries, burns, and heat, and cold emergencies;
- Manage sudden illnesses, stroke, seizure, bites, and poisoning; and
- Minimize stroke.

First aid services in the workplace typically include training and certification of selected individuals to perform CPR on workers when necessary. This training usually combines lectures, video demonstrations, and hands-on mannequin training. This training teaches participants how to:

- Call and work with EMS;
- Recognize breathing and cardiac emergencies that call for CPR;
- Perform CPR and care for breathing and cardiac emergencies;

- Avoid blood-borne pathogen exposure;
- Know the role of AEDs in the cardiac chain of survival;
- Call and work with EMS;
- Care for conscious and unconscious choking victims;
- Perform rescue breathing and CPR; and
- Use an AED safely on a victim of sudden cardiac arrest.

Thermal Hazards

Exposure to heat or cold can lead to serious illness. Factors such as physical activity, clothing, wind, humidity, working and living conditions, age, and health all influence whether a person will get ill. There are several ways to lessen the chances of succumbing to exposure. Battling the elements safely includes protecting skin from excessive exposure to subfreezing temperatures and from excessive exposure to the sun (Cyr & Johnson, 2002).

Appropriately controlling the temperature, humidity, and air distribution in work areas is an important part of providing a safe and healthy workplace. A work environment in which the temperature is not properly controlled can be uncomfortable. Extremes of either heat or cold can be more than uncomfortable; they can be dangerous. Heat stress and cold stress are major concerns of modern safety and health professionals. This section provides the information they need to know to overcome the hazards associated with extreme temperatures. There are several causal factors involved in heat stress. These include:

- Age, weight, degree of physical fitness, degree of acclimatization, metabolism, use of alcohol or drugs, and a variety of medical conditions such as hypertension all affect a person's sensitivity to heat. However, even the type of clothing worn must be considered. Prior heat injury predisposes an individual to additional injury. It is difficult to predict just who will be affected and when because individual susceptibility varies. In addition, environmental factors include more than the ambient air temperature. Radiant heat, air movement, conduction, and relative humidity all affect an individual's response to heat (OSHA, 2003).
- According to OSHA (2003), heat stress can manifest itself in several ways, depending on the level of stress. The most common types of heat stress are heat stroke, heat exhaustion, heat cramps, heat rash, transient heat fatigue, and chronic heat fatigue. These diverse types of heat stress can cause numerous undesirable bodily reactions, including prickly heat, inadequate venous return to the heart, inadequate blood flow to vital body parts, circulatory shock, cramps, thirst, and fatigue.

ELECTRICAL SAFETY

The normal use of electricity and electrical equipment and appliances has resulted in failure of most persons to appreciate the hazards involved. These hazards can be divided into five principal categories: (1) shock to personnel, (2) ignition of combustible (or explosive) materials; (3) overheating, and damage to equipment; (4) electrical explosions; and (5) inadvertent activation of equipment (Hammer, 1989).

Water workers seem to have a healthy respect for electricity, which is well warranted. For example, data from the Bureau of Labor Statistics indicated that there were nearly 6,000 deaths as the direct result of electrocutions at work in the United States, between 1992 and 2013. What makes these deaths more tragic is that for the most part they could have easily been avoided. In a way, it seems somewhat ironic that most workers have a deep respect for electricity, yet seem to ignore, or abuse, electrical safe work practices. Perhaps the answer lies in the fact that electricity, as a source of power, has become accepted without much thought to the hazards encountered. Because it has become such a familiar part of our surroundings, it often is not treated with the level of respect and/ or fear it deserves (Figure 8.13).

> Water treatment and distribution workers are exposed to electrical equipment and their inherent hazards on a daily basis. For this reason, the water safety official must pay attention to this important safety topic. When seeking information and guidance on electrical hazards and their control, various sources of information are available. For example, OSHA has devoted Subpart S of its 1910 manual to rules governing electrical work. OSHA requires the employer to train all workers in safe work practices for working with electrical equipment. OSHA's training rules distinguish between workers who do not work on or near exposed energized components and those who do. Even if the worker is not qualified to work on or around energized electrical equipment, he/she is required to know the specific safety practices which apply to their jobs.

> **– Frank R. Spellman (2016)**

If you were to take a look at the annual on-the-job injury statistics for all employers in the United States, you would quickly notice that many of these injuries are typically the result of electrical shock, injuries received during electrical fires, and/or injuries received when some electrical component fails due to faulty installation, faulty maintenance conducted on electrical equipment, or equipment malfunction caused by manufacturers' errors. Figure 8.14 illustrates a much too common occurrence when one is working with electricity.

While normally true that most workers fear electricity and its power or at least have a healthy respect for electricity, it is also true that on-the-job electrocutions do occur, and that the number one cause of fire in the workplace is due to electrical causes. For the safety and health professional, electrical safety in water treatment plants is not only an important priority, but also requires constant vigilance on his or her part and on the part of all supervisors and workers to ensure that safe work practices are followed when working with or around electrical circuits and components. All workers must be diligent about maintaining the integrity of all electrical equipment and systems and realize this requires constant vigilance. This includes an organization's "standing order" that any discovered electrical discrepancy is to be reported to responsible parties, immediately.

Another essential element in any electrical safety program is employee awareness. This is accomplished through training and written safe work practices and policies (see Figure 8.15). Employees should be routinely trained on the hazards of electricity and on what to look for and what to do if electrical discrepancies are discovered. Safe work practices are required for those employees required to work with or on electrical circuits and components. The safety and health professional must include a close look at all electrical installations during his or her organizational

FIGURE 8.13   Electricity can be dangerous. Illustration by F. Spellman.

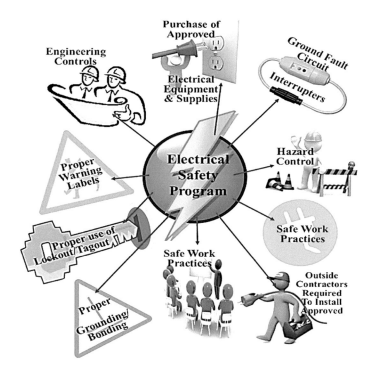

FIGURE 8.14   Elements of electrical safety program.

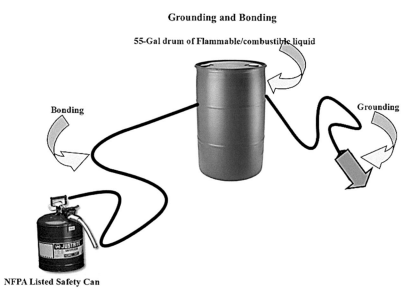

FIGURE 8.15   Grounding/bonding 55 gal drums.

safety inspection (audit). Facility occupational safety and health professionals must also insist that outside contractors hired to install new equipment, renovations, and upgrades accomplish their construction projects in accordance with OSHA, National Electrical Code (NEC), and all local code requirements. The safety and health professional must also ensure that any planned electrical equipment is suitable for installation in the proposed installation areas. For example, if a new electrical

FIGURE 8.16 What is necessary to have an effective ergonomics program.

motor and controller are to be installed in an area that contains explosive vapors, the proper class of electrical motor and control equipment must be installed in such a space to prevent the possibility of explosion, based on National Fire Protection Association (NFPA) recommendations.

OSHA's standards relating to electricity are found in 29 CFR 1910 (Subpart S). They are extracted from the National Electrical Code. Subpart S is divided into the following two categories of standards: (1) Design of Electrical Systems; and (2) Safety-Related Work Practices. Note: Grounding and bonding control the electrical potential between two bodies (see Figure 8.16). If there is a difference of potential between two bodies, a conductor between them will allow charge or current to flow. That flow may be dangerous, particularly as a source of ignition. Lee in *Electrical Grounding: Safe, or Hazardous* (1969, p. 162) and W. Hammer, in *Occupational Safety Management and Engineering* (1989, pp. 362–363) provide valuable information on grounding and bonding.

Equipment grounds may be used on the metal parts of a wiring system, such as the conduit, armor, switch boxes, and connected apparatus, other than the wire, cable, or other circuit components. They may also be provided for equipment, such as metal tables and cabinets that might come in contact with an energized circuit or source of electrical charges. Equipment on which undesirable charges may be induced or generated should also be grounded.

FIRE, WELDING, AND HOT WORK SAFETY

Water workers, supervisors, and safety and health professionals must be prepared for fire and its consequences. The plant must maintain a fire prevention strategy that will ensure that work areas are clean and clutter-free (to ensure fire lane access). Employees must know how to handle and properly store flammable or combustible chemicals and materials, what they are expected to do in case of a fire emergency, and how and whom to call when a fire occurs. If required to use fire extinguishers to fight small workplace fires, employees must know how to properly and safely operate the extinguishers.

Fire Safety

Industrial facilities are not immune to fire and the terrible consequences. Each year fire-related losses in the United States are considerable. According to conservative figures reported by (Brauer, 1994), about 1 million fires involving structures and about 8,000 deaths occur each year. The total annual property loss is more than $7 billion. Okay, fast forward to the present: (NFPA, 2013) reports that United States fire departments responded to an estimated 1,240,000 fires in 2013. These fires resulted in 3,240 civilian fire fatalities, 15,925 civilian fire-related injuries, and an estimated $11.5 billion in direct property loss. Complicating the fire problem is the point that (Cote & Bugbee, 1991) made earlier - the unpredictability of fire. Fortunately, facility safety engineers are aided in their efforts in fire prevention and control by the authoritative and professional guidance readily available from the National Fire Protection Association, the National Safety Council, Fire Code Agencies, local Fire Authorities, and OSHA regulations.

Along with providing fire prevention guidance, OSHA regulates several aspects of fire prevention and emergency response in the workplace. Emergency response, evacuation, and fire prevention plans are required under OSHA's 1910.38. The requirement for fire extinguishers and worker training is addressed in 1910.157. Along with state and municipal authorities, OSHA has listed several fire safety requirements for the general industry. All the advisory and regulatory authorities approach fire safety in much the same manner. For example, they all agree that electrical short circuits or malfunctions usually start fires in the workplace. Other leading causes of workplace fires are friction heat, welding, and cutting of metals, improperly stored flammable/combustible materials, open flames, and cigarette smoking. To gain a better perspective of the chemical reaction known as fire, remember that the combustion reaction normally occurs in the gas phase; generally, the oxidizer is air. If a flammable gas is mixed with air, there is a minimum gas concentration, below which ignition will not occur. That concentration is known as the Lower Flammable Limit (LFL). When trying to visualize LFL and its counterpart UFL (Upper Flammable Limit), it helps to use an example that most people are familiar with the combustion process that occurs in the automobile engine. When an automobile engine has a gas/air mixture that is below the LFL, the engine will not start because the mixture is too lean. When the same engine has a gas/air mixture that is above the UFL, it will not start because the mixture is too rich (the engine is flooded). However, when the gas/air mixture is between the LFL and UFL levels, the engine should start (Spellman, 1996).

The best way in which to try to prevent and control fires in the workplace is to institute a facility Fire Safety Program. Safety experts agree that the best way to reduce the possibility of fire in the workplace is prevention. For the facility safety engineer this begins with developing a Fire Prevention Plan, which must be in writing and must list fire hazards, fire controls, specify control jobs and responsible personnel, and emergency actions to be taken. In the event of a fire emergency, all employees need to know what to do; they need a plan to follow. The Fire Emergency Plan normally is the protocol to follow for fire emergency response and evacuation.

Welding Safety Program

Welding is typically thought of as the electric arc and gas (fuel gas/oxygen) welding process. However, welding can involve many types of processes. Some of these other processes include inductive welding, thermite welding, flash welding, percussive welding, plasma welding, and others. McElroy (1980) points out that the most common type of electric arc welding also has many variants, including gas shielded welding, metal arc welding, gas-metal arc welding, gas-tungsten arc welding, and flux cored arc welding. Welding, cutting, and brazing are widely used processes. OSHA's Subpart Q contains the standards relating to these processes in all their various forms. The primary health and safety concerns are fire protection, employee personal protection, and ventilation. The standards contained in this subpart are as follows:

- General requirements;
- Oxygen-fuel gas welding and cutting;
- Arc welding and cutting;
- Resistance welding;
- Sources of standards; and
- Standards organization.

In looking back on an OSHA study (reported in Professional Safety, February 1989) on deaths related to welding/cutting incidents, it is striking to note that of 200 such deaths over an 11-year period, 80% were caused by failure to practice safe work procedures. Surprisingly, only 11% of deaths involved malfunctioning or failed equipment, and only 4% were related to environmental factors. The implications of this study should be obvious: Equipment malfunctions or failures are not the primary causal factor of hazards presented to workers. Instead, the safety engineer's emphasis should be on establishing and ensuring safe work practices for welding tasks. In this section, we discuss these safe work practices.

Cutting Safety

Whenever torch-cutting operations are conducted, the possibility of fire is very real because proper precautions are often not taken. Torch cutting is particularly dangerous because sparks and slag can travel several feet and can pass through cracks out of sight of the operator. The safety engineer must ensure the persons responsible for supervising or performing cutting of any kind follow accepted safe work practices. Accepted safe work practices for torch-cutting operations typically include:

- Use of a cutting torch where sparks will be a hazard is prohibited.
- If cutting is to be over a wooden floor, the floor must be swept clean and wet down before starting the cutting.
- A fire extinguisher must be kept in reach any time torch-cutting operations are conducted.
- Cutting operations should be performed in wide-open areas, so sparks and slag will not become lodged in crevices or cracks.
- In areas where flammable materials are stored and cannot be removed, suitable fire-resistant guards, partitions, or screens must be used.
- Sparks and flame must be kept away from oxygen cylinders and hoses.
- Never perform cutting near ventilators.
- Firewatchers with fire extinguishers should be used.
- Never use oxygen to dust off clothing or work.
- Never substitute oxygen for compressed air.

Fire Watch Requirements

As stated earlier and as shown on the Hot Work Permit, a Fire Watch must be assigned whenever hot work operations are being performed around hazardous materials in confined spaces and other times when there is the danger of fire and/or explosion from such work. OSHA has specific requirements regarding Fire Watch Duties.

Laboratory Safety

OSHA'S Laboratory Standard

In the not too distant past, it was a widespread practice for academic, industrial, and other laboratories that used hazardous chemicals and materials to follow "homemade" or generic safety standards. These generic safety standards or procedures were usually developed by someone who was designated as being responsible for safety in the lab. From laboratory to laboratory, the content of these safety procedures varied. When OSHA began to publish its various safety standards, some of these new regulations were implemented for use in laboratory operations. Just a few years ago, probably the most commonly used OSHA standard was its 29 CFR 1910.1200 Hazard Communication (HazCom) Standard. HazCom requires facilities where hazardous chemicals and materials are used, handled, or produced to implement a HazCom Program. Under the HazCom Program, the affected facility is required to have a written program, provide employee training, use an easily recognizable labeling system for all hazardous chemicals and materials, and make it available to all employees 24-hours per day Safety Data Sheets for each hazardous chemical and material used in the workplace. HazCom has been an extremely beneficial program that has worked to enhance the safety of workers in the laboratory (it was the foundational document that made up the laboratory safety program).

Eventually, OSHA concluded that laboratories using hazardous chemicals and materials were somewhat different (by being more complex) than other types of work centers that also used these materials. Further, laboratory workers are sometimes exposed to a wider variety of hazardous materials and work activities than the average water and industrial worker. Therefore, in an effort to establish consistent safety guidelines for use in laboratory environments only, OSHA, in 1992, promulgated 29 CFR 1910.1450, "Occupational Exposures to Hazardous Chemicals in the Laboratory." This important standard is now commonly known as the Laboratory Standard.

OSHA's Laboratory Standard is a performance standard. Performance standards indicate the desired result to be achieved but not the methodology by which it is to be accomplished. In plain English, what a performance standard does is give the employer a certain amount of "license" to achieve compliance with the requirements of the standard. The primary emphasis of a performance standard is on the implementation of administrative controls, which are designed to protect workers from overexposure to hazardous chemicals and materials in the laboratory. Two important requirements of OSHA's Laboratory Standard are presented as follows:

1. Laboratories that use hazardous chemicals in their spaces, where workers might be exposed to chemicals, must follow the guidelines provided in the Laboratory Standard. Keep in mind that "workers" can include such personnel as those who work in the lab office, maintenance, and custodial personnel, and others who regularly spend a significant amount of their time within the lab environment.
2. The Lab Standard requires laboratories that use hazardous chemicals to develop and implement *Chemical Hygiene Plans* (CHP). The CHP must outline the specific procedures, safe work practices, and personal protective equipment that are required to be used in the laboratory to control occupational exposures. It is important to point out that all personnel associated with laboratory operation must share chemical hygiene responsibilities. More will be said about the Chemical Hygiene Plan and its requirements later.

Machine Guarding

Safety, and health on the job, begins with sound engineering and design. The engineer and designer should be familiar with most of the common hazards to be dealt with in the design phase. For the senior manager, however, highlighting the most common hazards found in equipment and the ones requiring particular alertness is called for here. The most common sources of mechanical hazards are unguarded shafting, shaft ends, belt drives, gear trains, and projections on rotating parts. Where a moving part passes a stationary part or another moving part, there can be a scissor-like effect on anything caught between the parts. A machine component which moves rapidly with power or a point of operation where the machine performs its work is also typical hazard sources. There are probably over 2 million metalworking machines, and half of the woodworking machines in use are at least 10 years old. Most are poorly guarded, if at all. Even the newer ones may have substandard guards, despite the OSHA requirements. The basic objective of machine guarding is to prevent personnel from coming into contact with revolving or moving parts such as belts, chains, pulleys, gears, flywheels, shafts, spindles, and any working part that creates a shearing or crushing action or that may entangle the worker.

The basic purpose of machine guarding is to prevent contact of the human body with dangerous parts of machines. Moving machine parts have the potential for causing severe workplace injuries, such as crushed fingers or hands, amputations, burns, blindness, just to name a few. Machine guards are essential to protect workers from these needless and preventable injuries. Any machine part, function, or process which may cause injury must be safeguarded. When the operation of a machine or accidental contact with it can injure the operator or others in the vicinity, the hazards must be either eliminated or controlled (OSHA, 2003). Our experience has clearly (and much too frequently) demonstrated that when arms, fingers, hair, or any body part enters or makes contact with moving machinery, the results can be not only gory, bloody, and disastrous but also sometimes fatal. Safety and health professionals have several safeguarding methods to consider when he/she determines that machine guarding is needed. The type of operation, the size, or shape of stock, the method of handling, the physical layout of the work area, the type of material, as well as production requirements, or limitations will help to determine the appropriate safeguarding method for the individual machine. OSHA points out that, as a rule, any power transmission apparatus is best protected by fixed guards that enclose the danger areas. For hazards at the point of operation where moving parts are performing work on the stock, several kinds of safeguarding may be possible. The safety and health professional must always choose the most effective and practical means available. Safeguards include guards, devices, automatic and semiautomatic feeding and ejecting methods, location and distance, and miscellaneous safeguarding accessories.

Blood-Borne Pathogens

Bloodborne Pathogen Standard 29 CFR 1910.1030

Although scientific research has determined (at present) that HIV and other blood-borne pathogens are not found in the water or some other environmental streams, (except in strictly controlled laboratory conditions), the Centers for Disease Control (CDC) does warn that persons who provide emergency First aid could become contaminated. Thus, if your facility requires employees trained in first aid to render medical assistance as part of his/her job activities, your facility is covered under the 29 CFR 1919.1030 OSHA Standard. The major point to get across to all workers is that if they render any kind of first aid assistance whereby the rescuer can or could be exposed to another person's body fluids, then care and caution must be exercised. It is an excellent idea to equip all environmental facilities with first aid kits and other emergency medical kits that are designed to protect against blood-borne pathogens. That is, these kits are equipped with the following:

- Rescue barrier mask – prevents mouth-to-mouth contact;
- Alcohol cleansing wipes – for cleanup;
- Latex gloves – prevents hand contact with body fluids;
- Safety goggles – prevents body fluids from entering eyes; and
- Biohazard bag – for disposal of cleanup materials.

Training should be provided to each worker on the dangers of blood-borne pathogens. Careful attention to personal hygiene habits should be stressed. Workers should be informed that handwashing is one of their best defenses against spreading infection, including HIV. Ensuring worker awareness is the key to complying with this standard.

Ergonomics

What is ergonomics? We state this question right up front (*and subsequently answer it*) because we have found that few people understand the meaning of the word and even fewer understand what ergonomics is all about. Until recently, the term *ergonomics* was used primarily in Europe and the rest of the world to describe Human Factors Engineering (a synonymous term most commonly used in the United States). At present, a widespread practice in the United States and in general safety practice now uses the term *ergonomics*. So again, what is ergonomics? Let's break the word down a bit and see if the Greek it is derived from (*ergon* and *nomos*) will help us in defining the term. *Ergon* means work, and *nomos* means law. Thus, ergonomics means the laws of work. What does that mean? Let's further define ergonomics by pointing out that it relates to the interface between people and a variety of elements: equipment, environments, facilities, vehicles, printed materials, and so forth (Brauer, 1994). Grimaldi and Simonds (1989) define ergonomics as the measurement of work. Ergonomics could be defined as how human physical considerations affect work.

So, we end up with ergonomics meaning the "laws of work," the interface between people and a variety of elements, and/or the measurement of work; however, the definition does not seem complete. To solve this problem, let's sum it all up by using the best definition we have been able to find to date. This term is actually derived, though slightly modified for our purposes, from Chapanis (1985): Ergonomics discovers, and applies information about human behavior, abilities, limitations, and other characteristics to the design of tools, machines, systems, tasks, jobs, and environments for productive, safe, comfortable, and effective human use. Okay, Chapanis' definition for ergonomics seems logical, (and it is the definition we use to describe ergonomics in this chapter), but what is the goal of ergonomics? Stated simply, ergonomics is about instituting measures to prevent awkward postures, repetitive motions, forceful exertions, pressure points, vibration, and temperature extremes so as to protect the worker, to minimize worker error, and to maximize worker efficiency. Figure 8.16 displays what is necessary to have an effective ergonomics program.

CHAPTER SUMMARY

The primary duty and/or obligation of water work operators and public water system managers are to deliver the safest product possible to ratepayers/customers as in possible. It is also an obligation and/or duty for public water system managers to ensure the safety and health of all waterworks personnel. A common statement made in industry is that Safety First! Well, safety first can only be accomplished in a safe work environment where personnel perform their routine functions in a safe manner. This chapter described what is required and/or is necessary to ensure both a safe workplace and safe workers so that they can perform their important functions in the safest manner possible.

REVIEW QUESTIONS

1. Why does the water industry have a high incidence of injury?
    a. Almost all types of hazards are present
    b. Workplace violence
    c. Too many off hours
    d. Very liberal on-the-job injury reporting system

2. Absence of plant safety rules can lead to _____.
    a. Increased injury rate
    b. Lawsuits
    c. Abuse of workers' compensation
    d. All of the above

3. Safety starts at the top. What is the main factor that brings this statement to reality?
    a. Unlimited funds in safety budget
    b. Great insurance coverage
    c. Top management supports safety
    d. Employees buy into safety programs

4. When are Safety Data Sheets required to be available to workers?
    a. Once per year
    b. Whenever supervisor provides them
    c. 24/7
    d. When asked for

5. Confined spaces can be killers due to _____.
   a. Wild animals within
   b. Lack of oxygen
   c. Piping galleries
   d. One way in and out

6. Who is authorized to provide first used in the workplace?
   a. Anyone
   b. Outside responders only
   c. Trained and certified first aid personnel
   d. Human resources specialists

7. Who is authorized to remove a lockout/tagout device?
   a. Supervisor
   b. Anyone
   c. General manager
   d. None of the above

8. The Hazard Communication Standard requires employers to _____.
   a. Label chemical hazards
   b. Train employees
   c. Provide safety data sheets
   d. All of the above

9. During hot work operations, why is a fire watch required?
   a. To prevent fire
   b. To comply with OSHA
   c. To create extra jobs
   d. To satisfy insurance requirements

10. Torch cutting is dangerous because it _____.
   a. Can cause fire
   b. Can burn operators
   c. Can seep into cracks, out of sight of operator
   d. All of the above

ANSWER KEY

1. Why does the water industry have a high incidence of injury?
   a. **Almost all types of hazards are present**
   b. Workplace violence
   c. Too many off hours
   d. Very liberal on-the-job injury reporting system

2. Absence of plant safety rules can lead to _____.
   a. Increased injury rate
   b. Lawsuits
   c. Abuse of workers' compensation
   d. **All of the above**

3. Safety starts at the top. What is the main factor that brings this statement to reality?
   a. Unlimited funds in safety budget
   b. Great insurance coverage
   c. **Top management supports safety**
   d. Employees buy into safety programs

4. When are Safety Data Sheets required to be available to workers?
   a. Once per year
   b. Whenever supervisor provides them
   c. **24/7**
   d. When asked for

5. Confined spaces can be killers due to _____.
   a. Wild animals within
   b. **Lack of oxygen**
   c. Piping galleries
   d. One way in and out

6. Who is authorized to provide first used in the workplace?
   a. Anyone
   b. Outside responders only
   c. **Trained and certified first aid personnel**
   d. Human resources specialists

7. Who is authorized to remove a lockout/tagout device?
   a. Supervisor
   b. Anyone
   c. General manager
   d. **None of the above**

8. The Hazard Communication Standard requires employers to _____.
   a. Label chemical hazards
   b. Train employees
   c. Provide safety data sheets
   d. **All of the above**

9. During hot work operations, why is a fire watch required?
   a. **To prevent fire**
   b. To comply with OSHA
   c. To create extra jobs
   d. To satisfy insurance requirements

10. Torch cutting is dangerous because it _____.
    a. Can cause fire
    b. Can burn operators
    c. Can seep into cracks, out of sight of operator
    d. **All of the above**

BIBLIOGRAPHY

29 CFR 1910.95, OSHA.
29 CFR 1910.38, OSHA.
29 CFR 1910.108, OSHA.
29 CFR 1910.157, OSHA.
29 CFR 1910.251, OSHA.
29 CFR 1910.252, OSHA.
29 CFR 1910.253, OSHA.
29 CFR 1910.254, OSHA.
29 CFR 1910.255, OSHA.
29 CFR 1910.256, OSHA.
29 CFR 1910.257, OSHA.
29 CFR 1910.147, OSHA.

AARP, (1998). Older Workers Survey. *Workplace Visions.* Sept/Oct.

Accident Prevention Manual for Industrial Operations, (1988). 9th ed. Chicago, IL: National Safety Council.

Alpaugh, E.L., (1988). Particulates. In *Fundamental of Industrial Hygiene*, 3rd ed. Ed. Barbara A. Plog. Chicago, IL: National Safety Council.

Alpaugh, E.L. (Revised by T.J. Hogan), (2016). *Fundamentals of Industrial Hygiene*, 3rd ed. Chicago: National Safety Council, pp. 259–260.

American Chemical Society (ACS), (2002). *Principles of Environment Sampling and Analysis - Two Decades Later.* Boston, MA: American Chemical Society Meeting.

American Industrial Hygiene Association, (1984). *Ergonomics Guides, including Kroemer, K.H.E. Ergonomics of VDT Workplaces.* Akron, OH: American Industrial Hygiene Association.

American National Standards Institute (ANSI), (1994). *Calibration Laboratories and Measuring and Texting Equipment General Requirements.* Washington, DC: ANSI.

Anderson, C.K. & Catteral, M.J., (1987). A Simple Redesign Strategy for Storage of Heavy Objects. *Professional Safety* (Nov): 35–38.

ANSI A13.1, (1981). New York: American National Standards Institute.

ANSI B7.1, (1988). *The Use, Care, and Protection of Abrasive Wheels.* New York: American National Standards Institute.

ANSI Z244.1, (n.d.). *Minimum Safety Requirements for A Lockout/Tagout of Energy Sources* New York: ANSI.

ANSI Z535.1, (1991). New York: American National Standards Institute.

ANSI/AAMI, (n.d.). *ES1, Safe Current Limits for Electromechanical Apparatus.*

ANSI/UL 817, (n.d.). *Cord Sets and Power: Supply Cords.*

ANSI/UL 859, (n.d.). *Electrical Personal Grooming Appliances.*

American Conference of Governmental Industrial Hygienists (ACGIH), (1992). *1992–1993 Threshold Limit Values for Chemical Substances and Physical Agents and Biological Exposure Indices.* Cincinnati, OH: American Conference of Governmental Industrial Hygienists.

American Nation Standards Institute, (1984). *American National Standard for Respirator Protection - Respirator Use - Physical Qualifications for Personnel*, ANSI Z88.6. New York: ANSI.

Ayoub, M. & Mital, A., (1989). *Manual Materials Handling.* London: Taylor & Francis.

Baumeister, T (ed.), (1987). *Standard Handbook for Mechanical Engineers*, 9th ed. New York: McGraw-Hill.

Blackman, W.C., (1993). *Basic Hazardous Waste Management.* Boca Raton, FL: Lewis Publishers.

Blum, T.C. et al., (1992). Workplace Drug Testing Programs: A Review of Research and a Survey of Worksites. *Journal of Employee Assistance Research*: 315–349.

Blundell, J.K., (1987). *Safety Engineering: Machine Guarding Accidents.* Del Mar, CA: Hanrow Press.

Brauer, R.L., (1994). *Safety and Health Engineers.* New York: Van Nostrand Reinhold.

Bretherick, L., (1991). Overview of Objectives in Design of Safe Laboratories. In *Safe Laboratories: Principals, and Practices for Design, and Remodeling.* Eds. Ashbrook, P.C.

Breysse, P.N. & P.S.J. Lees, (1999). Analysis of Gases and Vapors. In *The Occupational Environment: Its Evaluation and Control.* Ed. S.R. DiNardi. Fairfax, VA: American Industrial Hygiene Association.

British Standards Institute (BSI), (1975). *Safeguarding of Machinery*, BS 5304. London: British Standards Institute.

Brubaker, S.A., (December, 1997). Optimum Machine Guarding. *Occupational Health & Safety.*

Caillet, R., (1981). *Low Back Pain Syndrome.* Philadelphia: F. A. Davis.

Carson, R., (April 1994). Stand by Your Job. *Occupational Health & Safety*: 38.

Caruey, A., (May 1991). Lock Out the Chance for Injury. *Safety & Health* 143(5): 46.

CDC, (2017). Confined Space Fatality at a Wastewater Treatment Plant in Indiana. Accessed 22, April 2017 @ https://www.cdc.gov/niosh/face/in-house/full8746.html.

Chapanis, A., (1985). Some Reflections on Progress. In Proceedings of the Human Factors Society 29th Annual Meeting. Santa Monica, CA: Human Factors Society.

Coastal Video, (1993). *Confined Space Rescue Booklet.* Virginia Beach, VA: Coastal Video Communication Corporation.

Cote, A. & Bugbee, P., (1991). *Principles of Fire Protection.* Batterymarch Park, MA: National Fire Protection Association.

CoVan, J., (1995). *Safety Engineering.* New York: Wiley.

Cyr, D.L. & Johnson, S.B., (2002). *Battling the Elements Safely.* Accessed at http://www.cdc.gov/nasd/docs/d 000901-d001000/d000925/d000925.html.

daRoza, R.A. & Weaver W., (1985). *Is It Safe to Wear Contact Lenses with a Full-Facepiece Respirator?* Livermore, CA: Lawrence Livermore National Laboratory. manuscript UCRL-53653.

DiBeradinis, L.J., Baum, J.S., First, M.W., Gatwood, G.T., Groden, E., & Seth, A.K., (1993). *Guidelines for Laboratory Design*, 2nd ed. New York: Wiley.

Douglas, B.L., (2000). Know the Company You Keep: Co-Morbidities at Work. *Journal of Workers Compensation* 9(4): 101–107.

NFPA, (1987). *Electrical Standard for Metalworking Machine Tools*, NFPA 79. Quincy, MA: National Fire Protection Association.

Ferry, T., (1990). *Safety and Health Management Planning*. New York: Van Nostrand Reinhold.

Fire Protection Guide on Hazardous Materials, (1986). 9th ed. Quincy, MA: National Fire Protection Association.

Fire Protection Handbook, (2004). latest ed. Quincy, MA: National Fire Protection Association (current edition).

Galvin, D., (1999). Workplace Managed Care: Collaboration for Substance Abuse Prevention. *Journal of Behavioral Health Services and Research*.

Gasaway, D.C., (1985). *Hearing Conservation: A Practical Manual and Guide*. Englewood Cliffs, NJ: Prentice-Hall.

General Industry Safety and Health Standards. Title 29 Code of Federal Regulations 1910.212-1910.222.

Giachino, J. & Weeks, W., (1985). *Welding Skills*. Homewood, IL: American Technical Publications.

Goetsch, D.L., (1996a). *Occupational Safety and Health*, 2nd ed. Englewood Cliffs, NJ: Prentice Hall.

Goetsch, D.L., (1996b). *Occupational Safety and Health: In the Age of High Technology for Technologists, Engineers, and Managers*. Englewood Cliffs, NJ: Prentice Hall.

Gould, G.B., (1997). *First Aid in the Workplace*. New York: Simon & Shuster.

Grantham, D., (1992). Dusts in the Workplace. In *Occupational Health and Hygiene Guidebook for the WHSO*. Brisbane: D.L. Grantham.

Greaney, P.P., (2000). Ensuring Employee Safety in Cold Weather Working Environments. Work Care. Accessed at www.workcare.com/Archive/News_Art_2000_Dec14.htm.

Grimaldi, J.V. & Simonds, R.H., (1989). *Safety Management*. Homewood, IL: Irwin Press.

Hall, S.K., (1994). *Chemical Safety in the Laboratory*. Boca Raton, FL: CRC Press.

Hammer, W., (1989). *Occupational Safety Management and Engineering*. Englewood Cliffs, NJ: Prentice-Hall.

Harrington, D., (1939). *United States Bureau of Minds Information Circular Number 7072*. Washington, DC: U.S. Bureau of mines.

HEHS-97-163, (1997). *Worker Protection: Private Sector Ergonomics Programs Yield Positive Results*. Washington, DC: National Bureau of Standards.

Hermack, F.L., (Dec. 1955). *Static Electricity in Fibrous Materials*. National Bureau of Standards Report 4455.

Hoover, R.L., Hancock, R.L., Hylton, K.L., Dickerson, O.B., & Harris, G.E., (1989). *Health, Safety and Environmental Control*. New York: Van Nostrand Reinhold.

Industrial Ventilation, A Manual of Recommended Practice, (1988). 20th ed. Lansing, MI: American Conference of Governmental Industrial Hygienists, Edward Brothers.

Jacobson, J., (1998). The Supervisor's Tough Job: Dealing with Drug and Alcohol Abusers. In *Supervisors' Safety Update*, 97. Raynam, MA: Eagle Insurance Group, Inc.

Janpuntich, D.A., (1984). Respiratory Particulate Filtration. *Journal of Internatinal Society for Respiratory Protection* 2(1): 137–169.

Kaliokin, A., (October 1988). Six Steps Can Help Prevent Back Injuries and Reduce Compensation Costs. *Safety & Health* 138(4): 50.

Kavianian, H.R. & Wentz, C.A., (1990). *Occupational and Environmental Safety Engineering and Management*. New York: Van Nostrand Reinhold.

Keyserling, W.M., (1986). Occupational Ergonomics: Designing the Job to Match the Worker. *Annual Review of Public Health* 7:77–104.

Knutson, G.W., (1991). Principles of Ventilation in Chemistry Labs. *Safe Laboratories: Principles, and Practices for Design, and Remodeling*. (eds.) Ashbrook, P.C. & Renfrew, M.M. Chelsea, MI: Lewis Publishers.

LaBar, G., (1989). Sound Policies for Protecting Workers' Hearing. *Occupational Hazards*, July 1989, p. 46.

Lawrence, E. et al., (1996). Coerced Treatment for Substance Abuse Problems Detected through Workplace Urine Surveillance: Is It Effective? *Journal of Substance Abuse* 8(1):115–128.

Lee, R.H., (July, 1969). Electrical Grounding: Safe or Hazardous? *Chemical Engineering*.

LePree, J., (1998). *Complying with OSHA's Guarding Standards Can Avert Accidents*. New York, NY: Cahners Business Information.

Levy, B.S. & Wegman, D.H. (eds.), (1988). *Occupational Health: Recognizing and Preventing Work-Related Disease*, 2nd ed. Boston, MA: Little, Brown and Company.

Longman, P., (March 1999). The World Turns Gray. *United States News and World Report.*

Mansdorf, S.Z., (1993). *Complete Manual of Industrial Safety.* Englewood Cliffs, NJ: Prentice-Hall.

Margentino, M.R. & Malinowski, K., (1992). *Operation of Farm Machinery.* Document FS620. New Brunswick, NJ: Rutgers, State University of New Jersey.

Maturity Works, (2003). Accessed at www.maturityworks.co.uk/content/news.asp.

McElroy, (1980). *Electric Arc Welding.* Dublin, TX: McElroy Welding Service

Morris, B.K. (ed.), (August 8, 1990). Ergonomic Problems at Paper Mill Prompt Fine. *Occupational Health & Safety Letter* 20(16): 127.

National Academy of Sciences, (1983). *Prudent Practices for Disposal of Chemicals from Laboratories.* Washington, DC: National Academy Press.

National Academy Press, (1989). *Biosafety in the Laboratory: Prudent Practices for Handling, and Disposal of Infectious Materials.* Washington, DC: National Academy Press.

National Coalition on Ergonomics, (2003). *Home Page.* Washington, DC: National Coalition on Ergonomics. Accessed August 11, 2003 at http://www.necergor.org.

National Institute for Safety and Health (NIOSH), (1987). *NIOSH Guide to Industrial Respiratory Protection,* NIOSH Publication No. 87–116, Cincinnati, OH: National Institute for Safety and Health.

National Research Council, (1981). *Prudent Practices for Handling Hazardous Chemicals in Laboratories.* Washington, DC: National Academy Press.

National Safety Council, (1987). *Guards: Safeguarding Concepts Illustrated,* 5th ed. Chicago, IL: National Safety Council.

National Safety Council, (1988). *Fundamentals of Industrial Hygiene,* 3rd ed. Chicago, IL: National Safety Council.

National Safety Council, (1992). *Accident Prevention Manual for Business & Industry: Engineering & Technology,* 10th ed. Chicago, IL: National Safety Council.

Newton, J., (1987). *A Practical Guide to Emergency Response Planning.* Northbrook, IL: Pudvan Publishing Co.

NFPA, (2012). *Worker Casualties Involving Wastewater, Sewers or Sewage Treatment Plants and Fire Incidents at Water or Sanitation Utilities.* Batterymarch, MA: National Fire Protection Association. National Fire Protection Association Standards: 70 National Electrical Code.

NFPA, (2013). *Fire Protection Handbook.* Batterymarch, MA: National Fire Protection Association.

NIOSH, (1981). *Work Practices Guide for Manual Lifting.* Cincinnati, OH: National Institute for Occupational Safety and Health.

NIOSH, (1987). *Guide to Industrial Respiratory Protection.* Cincinnati, OH: National Institute for Occupational Safety & Health.

NIOSH, (2005). Direct Reading Instruments for Determining Concentrations of Aerosols, Gases, and Vapors. By Robert G. Keenan. www.cidc.gov/niosh/pdgs/74-177-h.pdf (accessed October 4, 2016).

North, C., (April 1991). Heat Stress. In *Safety & Health* 141(4): 55.

Occupational Health Guidelines for Chemical Hazards, (1981). NIOSH Pub. Number 81–123. Washington, DC: Government Printing Office.

Office of the Federal Register, (1986). Notices: Environmental Protection Agency (EPA), Environmental Auditing Policy Statement. *Federal Register,* July 9.

Office of the Federal Register, (January 26, 1989). Guidelines on Workplace Safety, and Health Program Management. *Federal Register* 54: 3904.

Office of the Federal Register & National Archives and Records Administration, (July 1, 1992). Code of Federal Regulations, Title 29, Labor, Part 1910.1450. In *Occupational Exposures to Hazardous Chemicals in Laboratories.* Washington, DC: United States Government Printing Office.

Olishifski, J.B. & McElroy, F.E., (eds.) (1988). *Fundamentals of Industrial Hygiene.* Chicago, IL: National Safety Council.

Orfinger, B., (2002). *Saving More Lives: Red Cross Adds AED Training to CPR Course.* Washington, DC: American Red Cross.

OSHA, (1992). *Managing Worker Safety & Health.* Washington, DC: Office of Consultation Programs, United States Department of Labor, June.

OSHA, (1995). *Occupational Safety and Health Standards for General Industry.* CFR 29 1910.147. Washington, DC: United States Department of Labor.

OSHA, (1996). *OSHA's Hazard Communication Standard.* Rockville, MD: Government Institutes.

OSHA, (1998). *The Cold Stress Equation.* Washington, DC: United States Department of Labor. Occupational Safety and Health Administration (OSHA3156).

OSHA, (2003). *Machine Guarding.* Accessed at http://www.osha-sle.gov/SLTC/machineguarding/August 7, 2003 (Revised).

OSHA, (2011). *OSHA's Bloodborne Pathogens Standard.* Washington, DC: Department of Labor Occupational Safety and Health Administration.

OSHA 29 CFR 1910.144.

OSHA 29 CFR 1910.212–.219.

OSHA 3067, (1992). *Concepts and Techniques of Machine Safeguarding.* Washington, DC: U.S. Dept of Labor.

Owen, B.D., (1986). Exercise Can Help Prevent Low Back Injuries. *Occupational Health & Safety* (June): 33–37.

Paques, J.J., Masse, S., & Belanger, R., (1989). Accidents Related to Lockout in Quebec Sawmills. *Professional Safety* (Sept): 17–20.

Phifer, L.H., (1991). Safe Laboratory Design in the Small Business. In *Safe Laboratories: Principles & Practice for Design, and Remodeling.* (eds.) Ashbrook, P.C. & Renfrew, M.M. Chelsea, MI: Lewis Publishers.

Phiffer, R.W. & McTigue, W.R., (1988). *Handbook of Hazardous Waste Management for Small Quantity Generators.* Chelsea, MI: Lewis Publishers.

Powell, C.H. & Hosey, A.D., (eds.), (n.d.). *The Industrial Environmental: Its Evaluation and Control,* 2nd ed. United States PHS Publication No. 614. Washington, DC: United States Public Health Service.

Power Press Safety Manual, (1988). 3rd ed. Chicago, IL: National Safety Council.

Puddicombe, R., (March 1998). A Guide to Using Safety Interlock Switches. *Occupational Hazards* xii: 45–50.

Putnam, A., (October 1988). How to Reduce the Cost of Back Injuries. *Safety & Health* 138(4): 48–49.

Ramsey, J.D., Buford, C.L., Bestir, M.Y., & Jensen, R.C., (1983). Effects of Workplace Thermal Conditions, or Safe Work Behavior. *Journal of Safety Research* 14: 105–114.

Renfrew, M.M., & Chelsea, MI: Lewis Publishers, (1990). Compressed Gas Association. In *Handbook of Compressed Gases,* 3rd ed. New York: Van Nostrand Reinhold.

Roberts, V.L., (1980). *Machine Guarding: Historical Perspective.* Durham, NC: Institute for Product Safety.

Roister, J.D. & Roister, L.H., (1990). *Hearing Conservation Programs: Practical Guidelines for Success.* Chelsea, MI: Lewis Publishers.

Rowe, M.L., (1982). Are Routine Spine Films on Workers in Industry Cost: Or Risk-Benefit Effective? *Journal of Occupational Medicine* 17(Jan): 41–43.

Ryan, Michael, (1998). How a Little Headwork Saves a Lot of Children. *Parade Magazine,* 5/24/98. New York: 4–5.

Safety Compliance Alert, (1998). Malvern, PA: Progressive Review. June 18.

Saunders, G.T., (1993). *Laboratory Fume Hoods: A User's Manual.* New York: Wiley.

Saunders, M.S. & McCormick E.J., (1993). *Human Factors in Engineering, and Design,* 7th ed. New York: McGraw-Hill.

Sayers, D.L. & Walsh, J.L., (1998). *Thrones, Dominations.* New York: St. Martin's Press, p. 263–264.

Schaefer, W.P., (1991). Safe Laboratory Design: A Users Contribution. In *Safe Laboratories: Principles and Practices for Design and Remodeling.* (eds.), Ashbrook, P.C. & Renfrew, M.M. Chelsea, MI: Lewis Publishers.

Sicker, M., (2002). *The Political Economy of Work in the 21st Century: Implications for American Workforce.* Westport, CT: Greenwood Publishing Group.

Snook S.H., (1988). Approaches to the Control of Back Pain in Industry: Job Design, Job Placement, and Education/Training. *Professional Safety* Aug. 23–31.

Sonnenstuhl, W. & Trice, H., (1986). The social construction of alcohol problems in a union's peer counseling program. *Journal of Drug Issues* 17(3): 223–254.

Sorock, G., (1981). A Review of Back Injury Prevention, and Rehabilitation Research: Suggestions for New Programs (for National Safety Council Back Injury Committee), Chicago, IL: National Safety Council. May 2.

Spellman, F.R., (1996). *Safe Work Practices for Wastewater Operators.* Lancaster, PA: Technomic Publishers.

Spellman, F.R., (1999). *Confined Space Entry: A Guide to Compliance.* Lancaster, PA: Technomic Publishing Company.

Spellman, F.R., (2015a). *Occupational Safety and Health Simplified for the Industrial Workplace.* Lanham, MD: Bernan Press.

Spellman, F.R., (2015b). *Safety and Environmental Management,* 3rd ed. Lanham, MD: Bernan Press.

Stiricoff, R.S. & Walters, D.B., (1995). *Handbook of Laboratory Health & Safety,* 2nd ed. New York: Wiley.

The Bureau of National Affairs, (1971). *The Job Safety and Health Act of 1970,* 1st ed. Washington, DC: Bureau of National Affairs.

The Office of the Federal Register, (1995a). *Code of Federal Regulations Title 29 Parts 1900–1910*. Washington, DC: Office of Federal Register.

The Office of the Federal Register, (1995b). *Code of Federal Regulations Title 29 Parts 1900–1910 (.134)*. Washington, DC: Office of Federal Register.

The Office of the Federal Register, (1995c). *Code of Federal Regulations Title 29 Pars 1900–1910 (.147)*. Washington, DC: Office of Federal Registers.

Timbrell, V., (1965). The Inhalation of Fibrous Dusts, biological Effects of Asbestos. *Annals of the New York Academy of Sciences* 132, 255–273. New York.

United States Bureau of Labor Statistics, (2003). Accessed at http://www.bls.gov/images/curve-wh-lfet-gi/.

United Steelworkers of America, (2001). OSHA Ergonomics Standard Assassinated. Accessed August 11, 2003 at http://www.uswa.org/services/erog040301a.html.

USDL News Release 88-223, April 29, 1988.

USDOL, (1996). *Working Partners: Substance Abuse in the Workplace. The Hospitality Industry*. Washington, DC: The United States Department of Labor.

USPHS, (1973). *The Industrial Environment: Its Evaluation & Control*. Washington, DC: United States Public Health Service.

Walsh et al., (1993). Research and Prevention Alcohol Problems at Work: Toward an Integrative Mode. *American Journal of Health Promotion* 7(4):289–295.

Water Environment Federation, (1994). *Confined Space Entry*. Alexandria, VA: Water Environment Federation.

Wegman, D.H., (2000). Older Workers. In *Occupational Health: Recognizing and Preventing Work: Related Disease and Injury*, Levy, B.S. & Wegman, D.H. (eds.). Philadelphia: Lippincott Williams & Wilkins.

White, W.T. & Kibbe, R.R. (eds.), (1991). Machine Tool Practice. New York: Prentice Hall.

Zacharisson, R., (2011). *Electrical Safety*. Boston: Cengage Learning.

Zenz, C., (1988). *Occupational Medicine: Principles and Practical Applications*, 2nd ed. St. Louis, MO: Mosby Year Book, Inc.

Ziska, M., (2013). *Machine Guarding*. Boston: Precision Safety Consulting.

Operations for Public Water Systems: Conclusion

2019 Photo Courtesy of San Diego County Water Authority.

CONCLUSION

As stated earlier in *Part I – Operations for Public Water Systems*, the credibility of a Certified Public Risk Officer necessitates an understanding of the nomenclature and jargon used in water operations. In order to be a functional manager, at any level, you must be credible. The purpose of *Part I* is not to make you a water professional and/or an operator of a waterworks but simply to provide you with a basic overview of water operations. The point is: If you can speak and understand the language used in water operations, it will go a long way in allowing you to be a much more effective Certified Public Risk Officer.

SAMPLE EXAM QUESTIONS

1. What is one of the first questions that should be answered before planning entry into a permit-required confined space?
    a. Who is going to enter?
    b. Can this task be accomplished without entering the permit space?
    c. Why is permit entry required?
    d. Who is the attendant?

2. What are raw water intakes designed to remove?
    a. pH
    b. THCM
    c. Debris and fish
    d. Hardness

3. Which of the following is more likely to be found in groundwater?
    a. Fish
    b. Microorganisms
    c. Branches and other woody debris
    d. Minerals

4. To knock down algae growth in reservoirs what is commonly used?
    a. Soap
    b. Hydrogen peroxide
    c. Copper sulfate
    d. Ferric oxide

5. OSHA addresses confined space hazards in two specific, *comprehensive* standards. One of the standards covers General Industry, and the other covers:
    a. Agriculture
    b. Long shoring
    c. Shipyards
    d. Space travel

6. How are moving parts in a centrifugal pump cooled?
    a. Bearing grease
    b. Air flow
    c. Leakage within the pump
    d. Installed cooling package

7. A single sample of water collected at a particular time and place is called a:
    a. Composite sample
    b. Grab sample
    c. Total sample
    d. Glass of water

8. Which of the following will increase the pH of water?
    a. More water
    b. Sodium hypochlorite
    c. Gravel
    d. Food coloring

9. The Groundwater Rule is designed to:
    a. Protect the groundwater from microorganisms
    b. Set rules on chlorine addition
    c. Decrease organic contamination
    d. Decrease inorganic contamination

10. What type of pump has a piston within the casing?
    a. Venturi
    b. Centrifugal
    c. Turbine
    d. Reciprocating

11. When a new chemical is introduced into the workplace what must accompany it?
    a. Chemical dipper
    b. Safety data sheet
    c. Container
    d. First aid kit

12. OSHA's definition of confined spaces in general industry, includes:
    a. The space being more than 4 feet deep
    b. Limited or restricted means for entry and exit
    c. The space being designed for short-term occupancy
    d. Having only natural ventilation

13. The depression of a water table is known as:
    a. Cone of depression
    b. Radius of influence
    c. Static water level during a drought
    d. Well depression

14. Which of the following would *not* constitute a hazardous atmosphere, under the permit-required, confined space standard?
    a. Less than 19.5% oxygen
    b. More than the IDLH of hydrogen sulfide
    c. Enough combustible dust that obscures vision at a distance of 5 feet
    d. 5% of LEL

15. OSHA's review of accident data indicates that most confined space deaths and injuries are caused by the following three hazards:
    a. Electrical, falls, toxics
    b. Asphyxiates, flammables, toxics
    c. Drowning, flammables, entrapment
    d. Asphyxiates, explosions, engulfment

16. What parameter is the best way to monitor filter performance?
    a. Turbidity
    b. Dissolved solids
    c. Contact time
    d. Hardness

17. What causes water to look dirty or cloudy?
    a. Suspended material
    b. Bicarbonates
    c. Hardness
    d. Dissolved solids

18. Ammonia is sprayed near a chlorine gas leak to produce what?
    a. Neutral gas
    b. White smoke
    c. Black smoke
    d. Green smoke

19. A ball valve or plug valve is a type of:
    a. Foot valve
    b. Control valve
    c. Corporation stop
    d. Needle valve

20. Toxic gases in confined space can result from:
    a. Products stored in the space, and the manufacturing processes
    b. Work being performed inside the space, or in adjacent areas
    c. Desorption from porous walls, and decomposing organic matter

21. Oxygen deficiency in confined spaces does *not* occur by:
    a. Consumption by chemical reactions, and combustion
    b. Absorption by porous surfaces, such as activated charcoal
    c. Leakage around valves, fittings, couplings, and hoses of oxy-fuel gas welding equipment
    d. Displacement by other gases

22. What is installed in a water pipeline to minimize the number of customers affected by a line break?
    a. An air gap
    b. Back flow preventers
    c. A sufficient number of valves
    d. A control valve

23. The best way to measure your plant's performance against the best system in operation is to:
    a. Work at both plants
    b. Read the literature
    c. Benchmark
    d. Ask for opinions

24. Which of the following actions has had little effect on the natural water cycle?
    a. Dredging
    b. Damming
    c. Damaged ecological niches
    d. Desertification

25. Which public water system supplies water to the same population year-round?
    a. Non-community
    b. Transient system
    c. Community
    d. Non-transient

26. What reading (in % O_2) would you expect to see on an oxygen meter after an influx of 10% nitrogen into a permit space?
    a. 5.0%
    b. 11.1%
    c. 18.9%
    d. 90.0%

27. What type of public water system is typically used in campgrounds?
    a. Transient water system
    b. Community water system
    c. Nontransient water system
    d. County water system

28. An attendant is which of the following?
    a. A person who makes a food run to the local 7–11 store for refreshments for the crew inside the confined space.

b. A person who often enters a confined space, while other personnel are within the same space.

c. A person who watches over a confined space, while other employees are in it, and only leaves if he/she must use the restroom.

d. A person with no other duties assigned, other than to remain immediately outside the entrance to the confined space, and who may render assistance, as needed to personnel inside the space. The attendant never enters the confined space and never leaves the space unattended, while personnel are within the space.

29. Water quality is defined in terms of:
   a. Location
   b. Source
   c. Size of water supply
   d. Water characteristics

30. Treatment plant safety depends principally on:
   a. Upper management support
   b. Number of hazards present
   c. Attitude of outside contractors
   d. Luck of the draw

31. Potable water must be free of:
   a. Rocks
   b. Fish
   c. Non-sanitary properties
   d. Vegetation

32. USEPA classifies public water systems; which of the following systems is not included in their classifications?
   a. Nontransient
   b. Non-community
   c. Private commune supplies
   d. Community

33. Per 1910.146, an atmosphere that contains a substance at a concentration exceeding a permissible exposure limit, intended solely to prevent long-term, (chronic) adverse health effects, is *not* considered to be a hazardous atmosphere, on that basis alone.
   a. True
   b. False

34. In spite of regulations and cleanup efforts our main waterways are still polluted.
   a. True
   b. False

35. Which of the following is not a source of drinking water?
   a. Freshwater lakes
   b. Rivers
   c. Glaciers
   d. Moon rocks

36. A drainage basin is naturally shunted toward:
   a. Streams
   b. Creeks
   c. Brooks
   d. All of the above

37. With regard to the water supply factor Q&Q refers to:
   a. Quick and quality
   b. Quality and quick

   c. Quantity and Quality

   d. None of the above

38. Water that is unaccounted for in a distribution system is attributable to:
   a. Theft
   b. Leaks
   c. Transpiration
   d. Inaccurate meter readings

39. Of the following chemical substances, which one is a simple asphyxiate *and* flammable:
   a. Carbon monoxide (CO)
   b. Methane (CH_4)
   c. Hydrogen sulfide (H_2S)
   d. Carbon dioxide (CO_2)

40. Glaciers and ice caps hold the largest amount of freshwater.
   a. True
   b. False

41. Reengineering is:
   a. Tearing down the old and rebuilding with the new
   b. Engineering out a problem
   c. Systematic transformation of an existing system to a new form
   d. Unengineering the engineering

42. Entry into a permit-required confined space is considered to have occurred:
   a. When an entrant reaches into a space too small to enter
   b. As soon as any part of the body breaks the plane of an opening into the space
   c. Only when there is clear intent to fully enter the space (therefore, reaching into a permit space would not be considered entry)
   d. When the entrant says, "I'm going in now"

43. Do human activities slow or speed up water flow in canals and ditches?
   a. Increase the rate of water flow
   b. Have no effect
   c. Slow water flow
   d. Stop the flow of water

44. If the LEL of a flammable vapor is 1% by volume, how many parts per million is 10% of the LEL?
   a. 10 ppm
   b. 100 ppm
   c. 1,000 ppm
   d. 10,000 ppm

45. Groundwater in relationship to surface water has the following advantages:
   a. Is usually clean
   b. Easy to get to
   c. Flow naturally
   d. Is available everywhere

46. The principal operation of most combustible gas meters used for permit entry testing is:
   a. Electric arc
   b. Double displacement
   c. Electrochemical
   d. Catalytic combustion

47. Groundwater sources have some disadvantages when compared to surface water sources, which of the following are disadvantages?
    a. Contamination is usually hidden from view.
    b. Operating costs are usually higher.
    c. Groundwater may be subject to salt water intrusion.
    d. All of the above.

48. Water can't be destroyed or lost for practical use.
    a. True
    b. False

49. The proper testing sequence for confined spaces is the following:
    a. Toxics, flammables, oxygen
    b. Oxygen, flammables, toxics
    c. Oxygen, toxics, flammables
    d. Flammables, toxics, oxygen

50. The main reason public utility officials take a hard look at privatizing includes:
    a. To save money
    b. To keep OSHA from knocking on the door
    c. To avoid unionization
    d. It's a way of raising rates without ratepayers knowing

51. Circle the following true statement(s).
    a. Employers must document that they have evaluated their workplace, to determine if any spaces are permit-required confined spaces.
    b. If employers decide that their employees will enter permit spaces, they shall develop and implement a written permit space program.
    c. Employers do not have to comply with any of 1910.146 if they have identified the permit spaces, and have told their employees not to enter those spaces.
    d. The employer must identify permit-confined spaces by posting signage.

52. ROAD Gangers are those who:
    a. Work on road gangs
    b. Lay pipe in roadways
    c. Are retired on active duty
    d. Paint stripes on roadways

53. The benchmarking includes the following steps:
    a. Start, stop, adapt
    b. Research, observe, adapt
    c. Adapt, start, stop
    d. Observe, stop, start

54. Is it true that benchmarking is an objective-setting process?
    a. True
    b. False

55. Is it true that the public has no clue as to water's true economic value?
    a. True
    b. False

56. Circle the following true statement(s).
    a. Under paragraph (c)(5) (i.e., alternate procedures), continuous monitoring can be used in lieu of continuous forced air ventilation, if no hazardous atmosphere is detected.
    b. Continuous forced air ventilation eliminates atmospheric hazards.

    c. Continuous atmospheric monitoring is required if employees are entering permit spaces using alternate procedures, under paragraph (c)(5).

    d. Periodic atmospheric monitoring is required when making entries using alternate procedures under paragraph (c)(5).

57. Sustainable development can be defined as that which meets the needs of future generations without compromising the ability of present generations to meet their needs.
    a. True
    b. False

58. OSHA's position allows employers the option of making a space *eligible* for the application of alternate procedures for entering permit spaces, paragraph (c)(5), by first temporarily "eliminating" all non-atmospheric hazards, and then controlling atmospheric hazards by continuous forced air ventilation.
    a. True
    b. False

59. Water contracts (gets smaller) when it freezes.
    a. True
    b. False

60. Respirators allowed for entry into, and escape from, immediately dangerous to life, or health (IDLH) atmospheres are _____.
    a. Airline
    b. Self-contained breathing apparatus (SCBA)
    c. Gas mask
    d. Air purifying
    e. A and B

61. Which of the following is a measure of water's ability to neutralize acids?
    a. pH
    b. Hardness
    c. Fluoride level
    d. Alkalinity

62. What two minerals are primarily responsible for causing "hard water"?
    a. Salt and borax
    b. Hydrogen sulfide and sulfur dioxide
    c. Alum and chlorine
    d. Calcium (Ca) and magnesium

63. Water has high surface tension.
    a. True
    b. False

64. The power of a substance to resist pH changes is referred to as a(n):
    a. Antidote
    b. Fix
    c. Buffer
    d. Neutral substance

65. Which of the following complicates the use or reuse of water?
    a. Sunlight
    b. Rainfall
    c. Recreational use
    d. Pollution

66. Condensation is water coming out of the air.
    a. True
    b. False

67. Which of the following is a leading cause of impairment to freshwater bodies?
    a. Rivers
    b. Nonpoint pollution
    c. Air pollution
    d. Snowfall

68. Which of the following is not a feature of water?
    a. Appearance
    b. Wetness
    c. Smell
    d. Odor

69. Circle the following *false* statement(s):
    a. If all hazards within a permit space are eliminated without entry into the space, the permit space may be reclassified as a non-permit confined space, under paragraph (c)(7).
    b. Minimizing the amounts of regulation that apply to spaces whose hazards have been eliminated encourages employers to remove all hazards from permit spaces.
    c. A certification containing only the date, location of the space, and the signature of the person making the determination that all hazards have been eliminated shall be made available to each employee entering a space that has been reclassified under paragraph (c)(7).
    d. An example of eliminating an engulfment hazard is requiring an entrant to wear a full-body harness, attached directly to a retrieval system.

70. If water is turbid it is
    a. Smelly
    b. Wet
    c. Cloudy
    d. Clear

71. In order for an employee to be approved for respirator use on the job what is required?
    a. Training
    b. Medical approval
    c. Fit-testing
    d. All of the above

72. Solids are classified as
    a. Settable
    b. Dissolved
    c. Colloidal
    d. All of the above

73. Colloidal material is composed of
    a. Leaves
    b. Branches
    c. Silt
    d. Decayed fish

74. Taste and odor in drinking water may be
    a. No hazard
    b. A tooth destroyer
    c. A contributor to kidney stones
    d. A laxative

75. More things can be dissolved in sulfuric acid than in water.
    a. True
    b. False

76. Rainwater is the purest form of water.
    a. True
    b. False

77. Circle the following *false* statement(s):
    a. Compliance with OSHA's Lockout/Tagout Standard is considered to eliminate electrome-chanical hazards.
    b. Compliance with the requirements of the Lockout/Tagout Standard is not considered to eliminate hazards created by flowable materials, such as steam, natural gas, and other substances that can cause hazardous atmospheres or engulfment hazards in a confined space.
    c. Techniques used in isolation are blanking, blinding, misaligning, or removing sections of line soil pipes and a double and bleed system.
    d. Water is considered to be an atmospheric hazard.

78. It takes more energy to heat water at room temperature to 212°F than it does to change 212°F water to steam.
    a. True
    b. False

79. Circle the following *false* statement(s):
    a. Compliance with OSHA's Lockout/Tagout Standard is considered to eliminate electrome-chanical hazards.
    b. Compliance with the requirements of the Lockout/Tagout Standard is not considered to eliminate hazards created by flowable materials, such as steam, natural gas, and other substances that can cause hazardous atmospheres or engulfment hazards in a confined space.
    c. Techniques used in isolation are blanking, blinding, misaligning, or removing sections of line soil pipes, and a double and bleed system.

80. If you evaporate an 8-inch glass full of water from the Great Salt Lake (with a salinity of about 20% by weight), you will end up with about 1 inch of salt.
    a. True
    b. False

81. Pretreatment of water is used to oxidize which of the following:
    a. Iron and manganese
    b. Entrapped gases
    c. Removal of tastes and odors
    d. All of the above

82. Sea water is slightly more basic (the pH value is higher) than most natural water.
    a. True
    b. False

83. Circle the following true statement(s):
    a. "Alarm only" devices which do not provide numerical readings, are considered acceptable direct reading instruments, for initial (pre-entry) or periodic (assurance) testing.
    b. Continuous atmospheric testing must be conducted during permit space entry.
    c. Under alternate procedures, OSHA will accept a minimal "safe for entry" level, as 50% of the level of flammable or toxic substances that would constitute a hazardous atmosphere.
    d. The results of air sampling required by 1910.146, which show the composition of an atmosphere to which an employee is exposed, are not exposure records under 1910.1020.

84. Example(s) of simple asphyxiates are:
    a. Nitrogen (N_2)
    b. Carbon monoxide (CO)
    c. Carbon dioxide (CO_2)
    d. a and c

85. Which statement(s) is/are true, about combustible gas meters (CGMs)?
    a. CGMs can measure all types of gases.
    b. The percent of oxygen will affect the operation of CGMs.
    c. Most CGMs can measure only pure gases.
    d. CGMs will indicate the lower explosive limit for explosive dusts.

86. Solids in water may be classified by:
    a. Where they come from
    b. Size and state
    c. Whether they dissolve or not
    d. None of the above

87. Colloidal material is never beneficial in water.
    a. True
    b. False

88. Why are the adsorption sites on suspended material present in water objectionable?
    a. They can provide protective barriers against the chemical treatment of microorganisms.
    b. They enhance bad smell problems.
    c. They make treatment by screening impossible.
    d. There is no problem.

89. What can cause water to turn the color of tea?
    a. Iron contamination
    b. Air contamination
    c. Decayed vegetation
    d. Salt water contamination

90. Taste and odor of water are usually not an issue until:
    a. The water boils on its own.
    b. The water is viscous.
    c. The customer complains.
    d. None of the above.

91. Circle the following true statement(s):
    a. An off-site rescue service should have a permit space program, before performing confined space rescues.
    b. The only respirator that a rescuer can wear into an IDLH atmosphere, is a self-contained breathing apparatus.
    c. Only members of in-house rescue teams shall practice making permit space rescues, at least once every 12 months.
    d. Each member of the rescue team shall be trained in basic first aid and CPR.
    e. To facilitate non-entry rescue, with no exceptions, retrieval systems shall be used whenever an authorized entrant enters a permit space.

92. Hydrogen sulfide gives off a characteristic odor of:
    a. Rotten egg odor
    b. Sweet odor
    c. Flowery odor
    d. No odor

93. The chemical that gives a rotten cabbage odor is:
    a. Hydrogen sulfide
    b. Calcium oxide
    c. Organic sulfides
    d. Alum

94. The chemical composition of groundwater is changed by:
    a. Atmospheric pollution
    b. The rocks the comes into contact with
    c. Pollution
    d. None of the above

95. The Permit-Required Confined Space standard requires the employer to initially:
    a. Train employees to recognize confined spaces
    b. Measure the levels of air contaminants in all confined spaces
    c. Evaluate the workplace to determine if there are any confined spaces
    d. Develop an effective confined space program

96. Which of the following would increase alkalinity in the water?
    a. Dirt
    b. Sulfur dioxide
    c. Manganese
    d. $CaCO_3$

97. What should be done with a water sample that can't be analyzed right away?
    a. Throw it away
    b. Freeze it
    c. Add chemicals
    d. Refrigerate it

98. The water table is:
    a. Aquifer water
    b. Atmospheric pressure
    c. Water under the table
    d. None of the above

99. In a pump, what is the purpose of the shaft sleeve?
    a. Makes for a tighter connection of the impeller
    b. Acts as a water slinger ring
    c. Protects shaft from wear
    d. Prevents contamination inside the pump

100. If an employer decides that he/she will contract out all confined space work, then the employer:
    a. Has no further requirement under the standard
    b. Must label all spaces with a keep out sign
    c. Must train workers on how to rescue people from confined spaces
    d. Must effectively prevent all employees from entering confined spaces

101. Which of the following is a primary reason for ensuring the safety of drinking water?
    a. Reducing cost to customers
    b. Ridding an area of excess water
    c. Ensuring public health
    d. Political considerations

102. 99.999% of anything is equivalent to:
    a. 5-log
    b. 6-log

    c. No log

    d. Excessive water quality

103. Which of the following are water quality factors?

    a. Fecal bacteria

    b. Odor

    c. Hardness

    d. All of the above

104. In chemical addition to water, what is most important?

    a. Using the correct additives

    b. Using the correct amount of additive

    c. Metering additives

    d. Using the best brand of chemical

105. Water pollutants include:

    a. Snow

    b. Rain

    c. Oil

    d. Grass seed

106. Not required on a permit for confined space entry is:

    a. Names of all entrants

    b. Name(s) of entry supervisors(s)

    c. The date of entry

    d. The ventilation requirements of the space

107. Which of the following is more likely to be found in groundwater?

    a. Minerals

    b. Salt

    c. Giardia

    d. Organic matter

108. Which of the following is the definition of a grab sample?

    a. Represents flow

    b. A single sample of water collected at a particular time and place, which represents the composition of the water at that time and place

    c. Represents dosage

    d. A composite sample

109. Beer's law has to do with what?

    a. A sample's color based on its concentration

    b. Taste

    c. Odor

    d. pH

110. What causes water turbidity?

    a. Bicarbonate

    b. Hardness

    c. Suspended material

    d. Dissolved solids

111. Which water sampling parameter must be taken in the field?

    a. Salinity

    b. Fecal coliform

    c. Color

    d. pH and temperature

112. Circle the following training requirement that is identical for entrant, attendant, and entry supervisor.
    a. Know the hazards that may be faced during entry
    b. The means of summoning rescue personnel
    c. The schematic of the space, to ensure all can get around in the space
    d. The proper procedure for putting on, and using, a self-contained breathing apparatus

113. Which of the following describes the trickle of water from the land surface toward an aquifer?
    a. Conveyance
    b. Suction
    c. Evapotranspiration
    d. Percolation

114. Which of the following must always be done before repairing a machine?
    a. Notify people affected
    b. Lockout/Tagout equipment
    c. Notify supervisor
    d. Wipe down equipment

115. What reading must be taken in the field when obtaining a sample?
    a. pH
    b. Color
    c. Chlorine residual
    d. Hardness

116. Attendants can:
    a. Perform other activities when the entrant is on break inside the confined space.
    b. Summon rescue services, as long as he/she does not exceed a 200-ft radius around the confined space.
    c. Enter the space to rescue a worker, but only when wearing an SCBA, and connected to a lifeline.
    d. Order evacuation if a prohibited condition occurs.

117. An oxygen-enriched atmosphere is considered by 1910.146 to be:
    a. Greater than 22% oxygen
    b. Greater than 23.5% oxygen
    c. Greater than 20.9% oxygen
    d. Greater than 25% oxygen when the nitrogen concentration is greater than 75%

118. The following confined space that would be permit-required is:
    a. A grain silo with inward sloping walls
    b. A 10-gallon methylene chlorine reactor vessel
    c. An overhead crane cab which moves over a steel blast furnace
    d. All the above

119. Performing laboratory work safely requires:
    a. Adding acid to water
    b. Mix acid with a pH of 3 with a base of pH 12.5
    c. Add water to acid
    d. Titrate water to acid

120. What are colloidal particles?
    a. Sand
    b. Large particles
    c. Total solids
    d. Very small particles that do not dissolve

121. A written permit space program requires:
    a. That the employer purchase SCBAs and lifelines, but the employees purchase safety shoes and corrective lens safety glasses
    b. That the employer test all permit-required confined spaces at least once per year, or before entry, whichever is most stringent
    c. That the employer provides one attendant for each entrant up to five, and one for each two entrants, when there are more than five
    d. That the employer develops a system to prepare, issue, and cancel entry permits.

122. Bernoulli's principle states that the total energy of a hydraulic fluid is:
    a. Flexible
    b. Adjustable
    c. Flat
    d. Constant

123. The difference, or the drop, between static water level and the pumping water head is known as:
    a. Drawdown
    b. Draw up
    c. Slope
    d. Potential energy

124. What causes water motion?
    a. Inertia
    b. Energy
    c. Force
    d. Momentum

125. Of the following, which is *not* a duty of the entrant?
    a. Properly use all assigned equipment
    b. Communicate with the attendant
    c. Exit when told to
    d. Continually test the level of toxic chemicals in the space

126. This is present when water is in motion:
    a. Potential energy
    b. Force
    c. Kinetic energy
    d. Inertia

127. Friction created as water encounters the surface of pipe causes:
    a. Major head loss
    b. No loss
    c. Minor head loss
    d. Heat

128. Iron and manganese are most likely to reside in which of the following water sources?
    a. River
    b. Groundwater
    c. Stream
    d. Lake

129. Alkalinity is:
    a. pH of 7
    b. The capacity of water to neutralize acid
    c. pH of 5
    d. The property of neutralizing a base

130. Force per unit area is:
    a. Force
    b. Head
    c. Hydrostatic charge
    d. Pressure

131. Of the following, which is *not* a duty of the entry supervisor?
    a. Summon rescue services
    b. Terminate entry
    c. Remove unauthorized persons
    d. Endorse the entry permit

132. What is head loss measured in feet commonly called?
    a. Slope
    b. Head gain
    c. Head
    d. Static system

133. A drawback of pre-chlorination is:
    a. Mudballs
    b. Sandballs
    c. Scaling
    d. THMs

134. Why do we have secondary MCLs?
    a. For health reasons
    b. Just for the heck of it
    c. For aesthetic reasons
    d. Because of carcinogens

135. An aquifer is typically composed of:
    a. Rocks
    b. Melted rocks
    c. Sand and gravel
    d. Silt

136. What is the force as it rounds a bend in a pipe called?
    a. Inertia
    b. Dynamic load
    c. Thrust
    d. Cavitations

137. When designing ventilation systems for permit space entry:
    a. The air should be blowing into the space.
    b. The air should always be exhausting out of the space.
    c. The configuration, contents, and tasks determine the type of ventilation methods used.
    d. Larger ducts and bigger blowers are better.

138. What is the action that a sudden change in water pressure and direction in a pipe called?
    a. Pressure loss
    b. Water hammer
    c. Pressure increase
    d. Dynamic load

139. Which one of following is used to lower the pH in water?
    a. Sand
    b. Carbon dioxide

    c. Ammonia

    d. Sulfur

140. What does a gallon of water weigh in pounds?

    a. 50 lbs

    b. 62.14 lbs

    c. 8.34 lbs

    d. It is weightless

141. Of the following which is *not* a duty of the attendant:

    a. Know accurately how many entrants are in the space

    b. Communicate with entrants

    c. Summon rescue services when necessary

    d. Continually test the level of toxic chemicals in the space

142. Circle the following true statement(s):

    a. Carbon monoxide gas should be ventilated from the bottom.

    b. The mass of air going into a space equals the amount leaving.

    c. Methane gas should be ventilated from the bottom.

    d. Gases flow by the inverse law of proportion.

143. Turbidity is described as:

    a. Cloudiness or haziness of water

    b. Dissolved solids

    c. Total solids

    d. Mud

144. A disease-causing microorganism is called:

    a. Algae

    b. Fungi

    c. A pathogen

    d. PCP

145. Which of the following is used to mix coagulants?

    a. Diffuser

    b. Mechanical mixing

    c. Hydraulic mixing

    d. All of the above

146. Hot work is going to be performed in a solvent reactor vessel that is 10 feet high and 6 feet in diameter. Which of the following is the *preferred* way to do this?

    a. Use submerged arc-welding equipment

    b. Inert the vessel with nitrogen and provide a combination airline, with an auxiliary SCBA respirator, for the welder

    c. Fill the tank with water, and use underwater welding procedures

    d. Pump all the solvent out, ventilate for 24 hours, and use non-sparking welding sticks

    e. Clean the reactor vessel then weld per 1910.252

147. Your pump is making a pinging noise, what is the most likely cause?

    a. Water hammer

    b. Downstream valves closed

    c. Frequent cycling

    d. Cavitation

148. The efficiency of removing suspended solids in the sedimentation process is controlled by:

    a. Flowrate

    b. Amount of sediment

    c. Chemicals added

    d. Temperature

149. Bacteria is colloidal.

    a. True

    b. False

150. Chlorine is not considered free residual chlorine.

    a. True

    b. False

151. The certification of training required for attendants, entrants, and entry supervisors must contain (circle all that apply):

    a. The title of each person trained

    b. The signature or initials of each person trained

    c. The signature or initials of the trainer

    d. The topics covered by the training

152. What is a static head?

    a. Horizontal distance between a reference point to the water surface

    b. The output of a centrifugal pump

    c. Friction loss

    d. The vertical distance between a reference point to the water surface, when water is not moving

153. What is the purpose of packing and mechanical seal?

    a. Increases head

    b. Supports the shaft assembly

    c. Prevents pump water leakage

    d. Prevents bearing grease leakage

154. Coliform is used as an indicator organism because:

    a. Saves time, money, and cost of analysis

    b. Why not?

    c. Easily detected

    d. Provides a wealth of information

155. A rest stop is an example of a community water system.

    a. True

    b. False

156. Hardness is least likely to impact the formation of flocs during coagulation and flocculation.

    a. True

    b. False

157. If excessive water leaks from the stuffing box:

    a. Impeller has crumbled

    b. Broken check valve

    c. Packing gland needs to be tightened

    d. Bearing failure

158. _____ is installed on the suction pipe of a pump to prevent water from draining out of the pump.

    a. Foot valve

    b. Slinger ring

    c. Pressure valve

    d. Emergency valve

159. What is used to restrict leakage from the impeller discharge?
    a. Slinger ring
    b. Wear rings
    c. Lantern ring
    d. Shaft sleeve

160. A thermocline in a reservoir is a thin layer of water in which temperature changes more rapidly with depth than it does in the layers above or below.
    a. True
    b. False

161. What agency is responsible for protecting worker safety and health?
    a. EPA
    b. NRA
    c. AARP
    d. OSHA

162. Which of the following is a zone of a clarifier?
    a. Top
    b. Sludge
    c. Bottom
    d. Aerator

163. Which of the following is a positive displacement pump?
    a. Feeder pump
    b. Sub pump
    c. Piston pump
    d. Well pump

164. What is the main purpose of seal water in pumps?
    a. Prevent water hammer
    b. Protect bearings
    c. Cool packing
    d. Protect the slinger ring

165. Which of the following impacts the color of water?
    a. Salt
    b. Sulfur dioxide
    c. Vegetation
    d. Carbon dioxide

166. Which of the following would it be likely to find coliform in?
    a. Sand
    b. Tap water
    c. Lake
    d. Deep well

167. What is an example of natural organic matter?
    a. Vegetation
    b. Salt
    c. Dirt
    d. Iron

168. Federal Paragraph 1910.146 (g) requires that training of all employees whose work is regulated by the permit-required confined space standard shall be provided:
    a. On an annual basis
    b. When the employer believes that there are inadequacies in the employee's knowledge of the company's confined space procedures

    c. When the union demands it

    d. All the above

169. What is the function of a velocity pump?

    a. Piston

    b. Propeller meter

    c. Motor controller

    d. Relief valve

170. Short circuiting in a clarifier is shorter settling time.

    a. True

    b. False

171. Adsorption is substances sticking to the surface of a media.

    a. True

    b. False

172. What do you think is the most common water quality complaint from customers?

    a. The well runs dry

    b. The water tastes nasty

    c. The water is rust-filled

    d. Taste and odor issues

173. Which of the following causes hardness in water?

    a. Iron

    b. Calcium

    c. Dirt

    d. Leaves

174. Which of the following range of pH is considered to be basic?

    a. 3–5

    b. 7

    c. 12–14

    d. 0–6

175. Blackish water may be due to precipitated manganese.

    a. True

    b. False

176. What is the drawback of using UV for disinfection?

    a. Not expensive

    b. Electrocution

    c. Longer contact time

    d. High turbidity water

177. The _____ pump has a piston inside its casing.

    a. Centrifugal

    b. Venturi

    c. Reciprocating

    d. Peristaltic

178. Scaling is the common problem with hard water:

    a. True

    b. False

179. The problem with soft water is that it causes:

    a. Scaling

    b. Blockage

    c. Water hammer

    d. Corrosion

180. Chlorine gas is brown in color.
     a. True
     b. False

181. The _____ is responsible for enforcing the drinking water standards.
     a. EPA
     b. State
     c. FDA
     d. USDA

182. _____ control fluid flow through piping systems.
     a. Robots
     b. Valves
     c. Accumulators
     d. Receivers

183. Which is the smallest pathogen?
     a. Cyst
     b. Protozoa
     c. Virus
     d. Bacteria

184. What does a high C factor for a pipe indicate?
     a. No corrosion protection
     b. Easy to bend
     c. Smooth pipe interior
     d. Large pipe

185. Decaying vegetation in water is an example of:
     a. Suspended solids
     b. Dirt
     c. Soluble solids
     d. Colloids

186. When air pockets form at the high points in a pipeline, the cause is usually _____.
     a. Flow rate too high
     b. Excessive air bubbles
     c. Oil contamination
     d. Pump impeller cracked

187. _____ happens during the course of a filter run time.
     a. Headloss increases
     b. No headless
     c. Maximum headloss
     d. Headloss decreases

188. Surge tanks are used to control _____.
     a. Leaks
     b. Air bubbles
     c. Water hammer
     d. Corrosion

189. A coupling (a Dresser) is used to:
     a. Repair clamps
     b. Saddle a pipe
     c. Install a volute
     d. Connect a new water main to an existing water main

190. A metallic flat surface used to minimize leakage around the pump shaft is known as a:
    a. Mechanical seal
    b. Air gap
    c. Bearing ring
    d. Slinger ring

191. When a new pipe suffers tuberculation, what does this indicate?
    a. Scaling
    b. Low pH
    c. Water is corrosive
    d. Neutral pH

192. The valve type commonly used to control pressure and flow is:
    a. Emergency relief valve
    b. Check valve
    c. Foot valve
    d. Globe valve

193. Deposition of calcium carbonate in a water main is known as _____.
    a. Overloading
    b. Scaling
    c. Pigging
    d. Corrosion

194. What is a regulator station used for?
    a. Feed chlorine
    b. Prevent corrosion
    c. Maintain an acceptable water pressure within a piping system
    d. Release air within the pipe

195. A turbine pump is best described as a:
    a. Piston pump
    b. Air pump
    c. Centrifugal pump
    d. Jet pump

196. A chain of flight collectors is commonly used to:
    a. To remove soluble materials from water
    b. To remove sludge from clarifier bottom
    c. Has no function
    d. Protects water from contamination

197. When scheduling preventive maintenance, how often should it be accomplished?
    a. Refer to manufacturer's recommendation
    b. Never
    c. Annually
    d. Every day

198. The purpose of a check valve is to:
    a. Throttle flow
    b. Stop flow
    c. Speed up flow
    d. Allow water to flow in one direction only

199. If you add lime to water what happens to pH?
    a. Goes acid
    b. Increases pH
    c. Decreases turbidity
    d. Increases turbidity

200. What type of valve is a corporation stop?
    a. Relief valve
    b. Check valve
    c. Foot valve
    d. Ball valve or plug valve

201. What is the best procedure for ensuring equipment proper operation?
    a. All new equipment
    b. Preventive maintenance
    c. 24-hour operator surveillance
    d. Good luck

202. A _____ describes groundwater that flows naturally from the ground.
    a. Stream
    b. Well
    c. Spring
    d. Waterfall

203. What is the most critical problem that may be encountered when entering a confined space?
    a. Black widow spiders
    b. Lack of oxygen
    c. Trip hazards
    d. Low temperature

204. Which of the following are commonly used coagulants?
    a. Ferric sulfate
    b. Alum
    c. Polymer
    d. All of the above

205. What causes the water to look cloudy?
    a. High pH
    b. Wastewater
    c. Hardness
    d. Suspended solids

206. If you are exposed to high noise levels for only an hour a day, it will not lessen your hearing acuity.
    a. True
    b. False

207. What is a plant vulnerability study used for?
    a. Ensure security
    b. Evaluate potential threats
    c. Find out who is vulnerable
    d. Prepare for OSHA inspection

208. Why is a chlorine residual important?
    a. Oxidizes iron and manganese
    b. Provides chlorine residual to save on costs
    c. Oxidizes nitrate
    d. Continues treatment after treatment

209. One of the most prominent industrial wastes found in the surface water is:
    a. Milk
    b. Caustic
    c. Organic solvents
    d. None of the above

210. Who is responsible for proper operation and maintenance and plant equipment?
    a. Top manager
    b. CEO
    c. Custodian
    d. All plant employees

211. When the flow rate in the sedimentation process is increased, efficiency is lowered.
    a. True
    b. False

212. Which of the following mechanisms occurs in filter media?
    a. Sedimentation
    b. Adsorption
    c. Straining
    d. All of the above

213. What is the main disadvantage of performing preventive maintenance on plant equipment?
    a. Okay equipment could be damaged
    b. Too expensive
    c. Someone might be injured
    d. Might shutdown plant operations

214. Why does the water industry have a high incidence of injury?
    a. Almost all types of hazards are present
    b. Workplace violence
    c. Too many off hours
    d. Very liberal on-the-job-injury reporting system

215. Turbidity measurement is used to evaluate filtration efficiency.
    a. True
    b. False

216. Absence of plant safety rules can lead to:
    a. Increased injury rate
    b. Lawsuits
    c. Abuse of workers' compensation
    d. All of the above

217. The _____ process does not remove pathogens.
    a. Filtration
    b. Disinfection
    c. Screening
    d. Sedimentation

218. Rupture in pipe(s) is the most common source of waterborne disease.
    a. True
    b. False

219. Humans can't detect chlorine gas contamination at 3 ppm.
    a. True
    b. False

220. Safety starts at the top. What is the main factor that brings this statement to reality?
    a. Unlimited funds in safety budget
    b. Great insurance coverage
    c. Top management supports safety
    d. Employees buy into safety programs

221. Energy is expensive. What is the best way to save on energy costs?
    a. Turn out the lights when not in use.
    b. Turn off equipment not needed.
    c. Purchase only energy-efficient equipment.
    d. All of the above

222. _____ has a rotten egg odor.
    a. Dirt
    b. Sulfur dioxide
    c. Hydrogen sulfide
    d. Alum

223. Which of the following is a recommended way to protect and conserve plant infrastructure?
    a. Purchase a good insurance policy
    b. Proper operating procedures
    c. If it runs, leave it alone mantra
    d. Increase funding

224. Why is/are public water systems "invisible" for many people?
    a. Piping is buried underground and treatment plants are located in remote locations.
    b. Many people have no idea where their drinking water comes from.
    c. Many people do not care where their water comes from.
    d. Water is not often thought about until the tap is dry or they are dying of thirst.

225. When are Safety Data Sheets required to be available to workers?
    a. Once per year
    b. Whenever supervisor provides them
    c. 24/7
    d. When asked for

226. Confined spaces can be killers due to:
    a. Wild animals within
    b. Lack of oxygen
    c. Piping galleries
    d. One way in and out

227. What does a pH of 3 indicate?
    a. Strong acid
    b. Weak acid
    c. Neutral
    d. Very strong base

228. Total dissolved solids in water give the water a bad taste.
    a. True
    b. False

229. Who is authorized to provide first aid used in the workplace?
    a. Anyone
    b. Outside responders only
    c. Trained and certified first aid personnel
    d. Human resources specialists

230. pH of 7.5 is ideal for disinfecting bacteria.
    a. True
    b. False

231. Who is authorized to remove a lockout/tagout device?
    a. Supervisor
    b. Anyone

    c. General manager

    d. None of the above

232. The centrifugal pump is not a positive displacement pump.

    a. True

    b. False

233. Which of the following imparts velocity to water?

    a. Pump housing

    b. Slinger ring

    c. Impeller

    d. Bearings

234. Algal blooms increase the pH of water in a reservoir.

    a. True

    b. False

235. Coagulants have a positive charge.

    a. True

    b. False

236. The Hazard Communication Standard requires employers to:

    a. Label chemical hazards

    b. Train employees

    c. Provide safety data sheets

    d. All of the above

237. High-velocity backwashing of filter systems can result in _____.

    a. Breakdown of filter

    b. Excessive spent media

    c. No problem

    d. Failure to remove contaminants

238. _____ neutralizes the negative charges on colloids during water treatment.

    a. pH

    b. Coagulation and flocculation

    c. Ferric chloride

    d. Alum

239. _____ inactivates many of the pathogens in water.

    a. Alum

    b. Sodium hydroxide

    c. Disinfection

    d. Filtering

240. What is a dial indicator used for?

    a. Feeding chemicals

    b. Changing bearings

    c. Checking pump vibration

    d. Measuring alignment

241. _____ is formed when chlorine is added to water.

    a. Hypochlorous acid

    b. Sludge

    c. Alum

    d. Hypochlorite

242. During hot work operations, why is a fire watch required?
    a. To prevent fire
    b. To comply with OSHA
    c. To create extra jobs
    d. To satisfy insurance requirements

243. After a filter is backwashed, turbidity momentarily _____ in the filter effluent.
    a. Remains the same
    b. Causes headloss to decrease
    c. Increases
    d. Decreases

244. Five log removal is what?
    a. 1015
    b. 10%
    c. 99.999%
    d. 5%

245. Torch cutting is dangerous because:
    a. Can cause fire
    b. Can burn operators
    c. Can seep into cracks, out of sight of the operator
    d. All of the above

246. Heterotrophic use sunlight for energy.
    a. True
    b. False

247. Taste and odor problems in raw surface water are usually caused by _____.
    a. High pH
    b. Algae
    c. C dirt
    d. Sand

248. _____ is used to remove turbidity.
    a. Alum
    b. Carbon
    c. Sand filter
    d. Hypochlorite

249. Those who are required to wear full face respirators in the performance of their daily duties should be clean shaven.
    a. True
    b. False

250. The typical source of lead in drinking water is _____.
    a. Water supply
    b. Groundwater supply
    c. Artesian water
    d. Corrosion of pipe

ANSWER KEY

1. b
2. c
3. d
4. c
5. c
6. c
7. b
8. b
9. a
10. d
11. b
12. b
13. a
14. d
15. b
16. a
17. a
18. b
19. c
20. d
21. c
22. c
23. c
24. d
25. c
26. c
27. a
28. d
29. d
30. a
31. c
32. c
33. a
34. a
35. d
36. d
37. c
38. d
39. b
40. a
41. c
42. b
43. a
44. c
45. a
46. d
47. d
48. b
49. B

50. a
51. b
52. c
53. b
54. a
55. a
56. d
57. b
58. b
59. b
60. e
61. d
62. d
63. a
64. c
65. d
66. a
67. c
68. b
69. d
70. c
71. d
72. d
73. c
74. a
75. b
76. b
77. d
78. b
79. a
80. a
81. d
82. a
83. c
84. d
85. b
86. b
87. b
88. a
89. c
90. c
91. d
92. a
93. c
94. b
95. c
96. d
97. d
98. d
99. c
100. d

101. c
102. d
103. d
104. a
105. c
106. d
107. a
108. b
109. a
110. c
111. d
112. a
113. d
114. b
115. c
116. d
117. b
118. a
119. a
120. d
121. d
122. d
123. a
124. c
125. d
126. c
127. a
128. b
129. b
130. d
131. a
132. a
133. d
134. c
135. c
136. c
137. c
138. b
139. b
140. c
141. d
142. b
143. a
144. c
145. d
146. e
147. d
148. a
149. a
150. b
151. c

152. d
153. c
154. a
155. b
156. a
157. c
158. b
159. b
160. a
161. d
162. b
163. c
164. c
165. c
166. c
167. a
168. b
169. b
170. a
171. a
172. d
173. b
174. c
175. a
176. d
177. c
178. a
179. d
180. b
181. b
182. b
183. c
184. c
185. b
186. b
187. a
188. c
189. d
190. a
191. c
192. d
193. b
194. c
195. c
196. b
197. a
198. d
199. b
200. d
201. b
202. c

203. b
204. d
205. d
206. b
207. b
208. d
209. c
210. d
211. a
212. d
213. a
214. a
215. a
216. d
217. c
218. b
219. b
220. c
221. d
222. c
223. b
224. a
225. c
226. b
227. c
228. a
229. c
230. a
231. d
232. a
233. c
234. a
235. a
236. d
237. b
238. b
239. c
240. d
241. a
242. a
243. c
244. c
245. d
246. b
247. b
248. a
249. a
250. d

Part II

Operations for Publicly Owned Treatment Works (POTW)

Frank R. Spellman

Photo Courtesy of David McNeil

DOI: 10.1201/9781003207146-12

Operations for Publicly Owned Treatment Works (POTW): Introduction

In *Part I* of this designation and learning series, *Operations for Public Water Systems*, fundamental water operations were presented so that the Certified Public Risk Officer–Water and Wastewater (CPRO-W^2) designee would be able to speak the language with a basic understanding of the unit processes involved in water operations. As stated in *Part I* and reasserted in *Part II – Operations for Publicly Owned Treatment Works* (POTW), you must be credible about water and wastewater treatment to be a functional manager at any level. Again, as stated in *Part I*, to be credible you must possess several different positive traits such as: the ability to meet the standards of others, the ability to communicate, to be accountable, have the ability to demonstrate intentions, and, importantly, possess the ability to maintain correct behavior, have the ability to be consistent, and, most importantly, possess the knowledge of what publicly owned treatment works (POTW) system operation is all about. This last point is critical. Why? If you cannot speak and understand the language, the terminology, the jargon of publicly owned treatment works (POTW) system operation, it is analogous to jumping from an airplane without a parachute or any thought of a successful outcome.

So it is important for the Certified Public Risk Officer to possess knowledge of the operation of POTW. Keep in mind that the required and essential level of knowledge does not necessarily mean managers, at all levels, understand the operation of every valve and control device in the operation of the wastewater treatment process or its unit operations. Instead, it means that managers involved with risk in POTW operations know nomenclature and understand process operations. It is my purpose in *Part II – Operations for Publicly Owned Treatment Works* (POTW), to provide potential Public Risk Officers with a fundamental knowledge and understanding of nomenclature in order to ensure credibility. This fundamental knowledge includes pointing out the areas of operation that crossover from the information presented in *Part I*, such as current issues, water hydraulics, and water conveyance, that is germane to the material presented in *Part II*. It is important to remember that the main similarity between water operations and POTW operations is that the influent to be treated is water.

The main difference is that water needs to be treated for public consumption or wastewater (or what I correctly call "used water") that then needs to be treated for safe outfall into various receiving bodies (lakes, rivers, canals, or oceans). Also, keep in mind that with advancing technology, used water (wastewater) is being eyed for more reuse applications. If the wastewater industry can get the consumer to get over his or her yuck factor about drinking treated toilet water, wastewater made safe and healthy to consume will soon become a reality and a very pressing reality.

Although there are several current issues affecting POTW and their management and operations, many of these issues were discussed in *Part I* of this designation and learning series and because of the crossover between water and wastewater treatment operations, will not be discussed in full herein. However, although the management aspects of water treatment and wastewater treatment coincide and are alike, several important management aspects of POTW operation are discussed in this part. In addition, important plant security aspects, both physical and digital, that affect both water and wastewater systems are included in *Part II*.

Be advised that water and wastewater systems are lucrative terrorist targets. Physical targets include the poisoning of the water; however, in wastewater operations the goal is not to poison

DOI: 10.1201/9781003207146-13

or blow up a sewage storage tank but instead to attack the chemical storage areas (e.g., especially chlorine storage where several tons of deadly chlorine may be stored). Also, digital targets include SCADA systems and other operations – the diversion of raw wastewater to contaminate drinking water sources is not hypothetical – it is a terrorist's dream; it is real; it can happen; it has already happened! Finally, *Part II* of this designation and learning series will deliver the information needed to be a credible Certified Public Risk Officer.

AUTHOR'S COMMENTARY

Part II – Operations for Publicly Owned Treatment Works (POTW) provides a review of current treatment, collection, and security technologies (both physical and digital security) for the waste-water treatment industry. As pointed out in Part I, the collective technologies on water treatment operations have not changed dramatically from the processes used in the 1970s to the present time. However, with quality drinking water becoming more difficult to find, technologies have recently been developed for treating wastewater to drinking water quality and standards. Some of the material presented herein is dated but still current and new technologies such as Sustainable Water for Tomorrow (SWIFT) are now being used to augment freshwater aquifers such as the Potomac aquifer in Southeastern Virginia and the Chesapeake Bay area.

The water and wastewater industry's renaissance for academic writing occurred from the early 1970s to the later 1990s and was spurred by the Environmental Protection Agency (EPA). Many of my references and citations reflect this timeline. However, note that because I have been on the cutting edge, aboard the flagship for wastewater reuse and plant security, and have studied, researched, and authored several potential and developing reuse technologies, I am able to present the latest research on wastewater treatment techniques and plant security for both water and wastewater. The material included in *Part II – Operations for Publicly Owned Treatment Works* (POWT) is current and reflective of actual wastewater treatment operations as of this writing.

9 Operations for Publicly Owned Treatment Works (POTW)

Photo courtesy of Las Virgenes Municipal Water District

It will take proactive management programs [to] help protect the integrity of wastewater treatment systems and to avoid future capacity issues. It places the cost of fixing the problem on those who are experiencing it, rather than the entire region, and helps keep the overall cost to a minimum.

– Jason Willet (2019)

KEY OBJECTIVES

After studying this chapter, you should be able to:

- Discuss wastewater.
- Discuss used water.

DOI: 10.1201/9781003207146-14

- Discuss I & I (sewer inflow and infiltration).
- Identify the characteristics of wastewater.
- Identify wastewater sources.
- Define sick water.
- Discuss the difference between blackwater and greywater.
- Discuss forever chemicals.
- Identify PFAS and where they are found and their importance.
- Define industrial wastewater.
- Discuss the general sources of industrial wastes.
- Describe the composition of wastewater.
- Discuss endocrine disruptors.
- Discuss oxygen content of natural waste sources.
- Describe and discuss various types of wastewater treatment systems.
- Discuss wastewater disposal.
- Discuss wastewater reuse.

KEY DEFINITIONS

Benchmarking – used to measure or gauge performance based on specific indicators such as cost/unit of measure, productivity/unit of measure, cycle time of some value/unit of measure, or defects/unit of measure. It is the comparing of your operation to the best in the class.

Biochemical Oxygen Demand (BOD) – the amount of oxygen required by bacteria to stabilize decomposable organic matter under aerobic conditions.

Biodegradable – transformation of a substance into new compounds through biochemical reactions or the actions of microorganisms such as bacteria.

Blackwater – water containing toilet paper, human feces, urine, blood, and other bodily fluids flushed from toilets.

Domestic Water – water used for normal household purposes, such as drinking, food preparation, bathing, washing clothes and dishes, flushing toilets, and watering lawns and gardens. The water may be obtained from a public supplier or may be self-supplied. Also called residential water use.

Endocrine Disruptors – exogenous agents that interfere with the production, release, transport metabolism binding, action, or elimination of natural hormones in the body responsible for the maintenance of homeostasis and the regulation of developmental processes.

Forever Chemicals – are PFAS that never break down once released, and they build up in our bodies.

Greywater – also known as sullage, all waste streams that do not include human waste.

Groundwater – the fresh water found under the earth's surface, usually in aquifers. Groundwater is a major source of drinking water and a source of growing concern in areas where leaching agricultural or industrial pollutants or substances from leaking underground storage tanks are contaminating groundwater.

Human Influences – various human activities have a definite impact on surface water runoff. Most human activities tend to increase the rate of water flow. For example, canals and ditches are usually constructed to provide steady flow, and agricultural activities generally remove ground cover that would work to retard the runoff rate. On the opposite extreme, man-made dams are generally built to retard the flow of runoff.

Hydraulic Fracking – horizontal drilling for oil and natural gas.

Industrial Wastewater – this is water containing contaminates used and produced by various industries.

Infiltration and Inflow (I & I) – this includes water that seeps into pipes and systems from a variety of sources.

Nutrient – any inorganic or organic compound needed to sustain plant life.

Organic Matter – containing carbon, but possibly also containing hydrogen, oxygen, chlorine, nitrogen, and other elements.

Pathogens – types of microorganisms that can cause disease.

PFAS – these are a forever chemical group of per- and polyfluoroalkyl substances that includes many other chemicals.

Potable Water – water that is safe for use and consumption.

PPCPs – pharmaceuticals and personal care products.

Produced Water – this is water that is produced via the production of oil and natural gas.

Sick Water – water flushed down toilets along with unused vitamins, illicit drugs, and unused medicines and other assorted waste products such as body oils, shampoo, and conditioner.

Surface Water – all water naturally open to the atmosphere, and all springs, wells, or other collectors that are directly influenced by surface water.

Toxic Metals – metal contaminants in water.

Wastewater – is any water that has been contaminated by human use; it is also known as used water.

Wastewater Disposal – after treatment, wastewater is disposed of by outfalling to various waters or is reused in industry.

Wastewater Reuse – treated wastewater that is often used in industry as a coolant for machinery; with recent improvements in treatment processes wastewater reuse is being used in some water-starved locations to augment drinking water supplies.

Wastewater Treatment – includes various unit processes that work in sequence to clean the water.

OVERVIEW

We pointed out in *Part I: Operations for Public Water Systems* that those of us residing in the United States are fortunate to have one of the safest public drinking water supplies in the world. This is important because water is an essential compound in the maintenance of all forms of life on earth. Well, we can also state that those residing in the United States are fortunate to have state-of-the-art wastewater treatment systems. In the United States, 75% of wastewater is treated by some type of publicly owned treatment works (POTW); the remainder of the US population is served by private or decentralized septic systems.

Anyone from the US who has traveled the world and needed to use a toilet may have recognized that in some locations, even in the most modern, expensive, and exclusive hotels, (e.g., in Lima, Peru) the users are warned not to flush used toilet paper down the toilet, but instead to deposit the used paper in a container which is generally located within easy reach of the toilet.

In other locations, it is common to use bidets where toilet paper is not provided. In some third-world countries, outhouses or the great outdoors become the depository of human waste. The bottom line: Those of us residing in the US are fortunate to have one of the safest sanitary wastewater systems in the world *(Spellman, 2020a)*.

What Is Wastewater?

Simply, wastewater (aka used water, blackwater, greywater, or "sick water") is any water that has been contaminated by human use. Wastewater is used water from any combination of domestic, industrial, commercial, or agricultural activities, illicit drug disposal, stormwater runoff, and sewer inflow or sewer infiltration (aka I & I) (Tilley et al., 2014; Spellman, 2010). So wastewater is a by-product of domestic, industrial, commercial, or agriculture activities.

The characteristics of wastewater vary depending on the source. Types of wastewater include domestic wastewater from households, sewage from communities (aka municipal wastewater), and industrial wastewater. Wastewater can contain biological, chemical, and physical contaminants.

Households may produce wastewater from flush toilets, dishwashers, washing machines, bathtubs, sinks, and showers. Households that use dry toilets (i.e., composting, or biological toilets) produce less wastewater than those that use flush toilets.

DID YOU KNOW?

A dry (or composting or biological) toilet system contains and processes excrement, toilet paper, carbon additive, and even some food waste. Unlike a septic system, a composting toilet system relies on unsaturated conditions where aerobic bacteria break down waste. Similar to a yard waste composter, if sized and maintained properly, a composting toilet breaks down waste to 10% or 30% of its original volume. The end result product is a soil-like humus that must be either buried or removed by a licensed septage hauler in accordance with local and state regulations (USEPA, 1998; Spellman, 2013).

Wastewater Sources

The Difference between Blackwater and Greywater

The sources of wastewater include the following household or domestic activities:

- Mixed with toilet paper and/or wet wipes is human excreta (feces, urine, blood, and other bodily fluids); when these components are flushed from toilets, they are known as *blackwater.*
- Personal hygiene water (washing water, water for cleaning clothes, cleaning floors, dishes, cars, etc.) is known as sullage (all waste streams that do not include human waste) or commonly referred to as *greywater.*
- Surplus manufactured liquids from domestic sources FOG (fats, oil, and grease), drinks, cooking oil, pesticides, lubricating oil, paint, cleaning detergents, etc.

Note of Caution: Do not dump FOG into your toilets or other drains – if you do so you overload local wastewater treatment systems, and might plug your household drainage system, which may require the need of an expensive plumber to rectify.

Sick water – this is water that is flushed down the toilet along with unused vitamins, illicit drugs, unused medicines, and other assorted waste products such as body oils, lipsticks, perfumes, shampoo and conditioners, sunless tanning lotions, etc. – basically used beauty products washed off during showering or thrown into the toilet and flushed.

The Difference between Blackwater and Greywater

The difference between blackwater and greywater is mainly what makes it dirty. However, how they are being treated to be reused is also a differentiating characteristic – blackwater must be treated differently than other wastewaters because of bacteria content, while greywater is not as harmful and can be easily treated.

What Is Sick Water?

Importance of PFAS

The term "sick water" was coined by the United Nations in a 2010 press release addressing the need to arrest the global tide of sick water. The gist of the UN's report pointed out that transforming waste from a major health and environmental hazard into a clean, safe, and economically attractive resource is emerging as a key challenge in the 21st century. As a practitioner of environmental health, I certainly support the UN's view on this important topic. However, when I discuss sick water, in the context of this text and in many others I have authored or co-authored on the topic, I go a few steps further than the UN in describing the real essence and tragic implications of water that makes people or animals sick (or worse), or at least can be classified as sick; again, in my opinion.

Water that is sick is a filthy medium; wastewater is a cocktail of fertilizer runoff and sewage disposal alongside animal, industrial, agricultural, and other wastes. In addition to these listed wastes of concern, other wastes are beginning to garner widespread attention; they certainly have garnered our attention in our research on the potential problems related to these so-called other wastes.

What are these other wastes? Any waste or product we dispose of in our waters; that we flush down the toilet, pour down the sink, bathtub, or down the drain of a worksite deep sink. Consider the following example of "pollutants" we discharge to our wastewater treatment plants or septic tanks that we do not often consider as waste products but are waste products.

Consider the following, for example:

Each morning a family of four wakes up and prepares for the workday for the two parents and school for the two teenagers. Fortunately, this family has three upstairs bathrooms to accommodate their needs to prepare as in: the morning natural waste disposal, shower and soap usage, cosmetic application, hair treatments, vitamins, sunscreen, fragrances, and prescribed medications. In addition, the overnight deposit of cat and dog waste is routinely picked up and flushed down the toilet.

Let us fashion a short inventory list of what this family of four has disposed of or has applied to themselves as they prepare for their day outside the home.

- Toilet-flushed animal wastes
- Prescription and over-the-counter therapeutic drugs
- Veterinary drugs
- Fragrances (perfumes, etc.)
- Soap
- Shampoo, conditioner, other hair treatment products
- Body lotion, deodorant, body powder
- Cosmetics
- Sunscreen products
- Diagnostic agents
- Nutraceuticals (e.g., vitamins, medical foods, functional foods, etc.)

Even though these bioactive substances have been around for decades, today we group all of them (the exception being animal wastes) under the title of pharmaceuticals and personal care products, called "PPCPs" (see Figure 9.1). I pointed to the human activities of the family of four in contributing PPCPs to the environment, but there are other sources of PPCPs that should also be recognized. For example, residues from pharmaceutical manufacturing, residues from the hospital, illicit drug disposal [i.e., police knock on the door and the frightened user flushes the illicit drugs down the toilet (along with $100 bills, weapons, self-aborted fetuses, dealers' phone numbers, etc.) and into the wastewater stream], veterinary drugs, (especially antibiotics and steroids), and agribusiness are all contributors of PPCPs in the environment.

Regarding personal deposits of PPCPs to the environment and to the local wastewater supply, let us return to the family of four. After having applied or taken in the various substances mentioned

FIGURE 9.1   Origins and fate of PPCPs in the environment. *Source: EPA. Accessed 04/09/2020 @ http://epagov/nerlesd1/chemistry/pharma/.*

earlier, the four individuals involved add these products, PPCPs, to the environment through excretion (the elimination of waste material from the body), through bathing later in the day, and then possibly through disposing of any unwanted medications to sewers and trash. How many of us have found old medical prescriptions in the family medicine chest and decided they were no longer needed? How many of us have grabbed up such unwanted medications and simply disposed of them with a single toilet flush? Many of these medications, for example antibiotics, are not normally found in the environment, and then there are those who dump the deep fat fryer grease down the toilet – please, do not do that!

Earlier I stated that wastewater is a cocktail of fertilizer runoff and sewage disposal alongside animal, industrial, agricultural, and other wastes. When we factor in the addition of PPCPs to this cocktail, we can state analogously that we are simply adding a mixer to the mix. The questions about our mixed waste cocktail are obvious: Does the disposal of antibiotics and/or other medications into the local wastewater treatment system cause problems for anyone or anything else downstream, so to speak? When we ingest locally treated water, are we also ingesting flushed-down-the-toilet or flushed-down-the-drain antibiotics, other medications, illicit drugs, animal excretions, cosmetics, vitamins, vaginal cleaning products, sunscreen products, diagnostic agents, crankcase oil, grease, oil, fats, and veterinary drugs each time we drink a glass of tap water? Well, Katy, bar the door – we certainly hope not. However, hope is not always a fact.

The jury is still out on these questions. Simply, we do not know what we do not know about the fate of PPCPs or their impact on the environment once they enter our wastewater treatment systems, the water cycle, and eventually our drinking water supply systems. This is the case even though some PPCPs are easily broken down and processed by the human body or degraded quickly in the environment.

This question has been around since the mythical hero Hercules (arguably the world's first environmental engineer) was ordered to perform his fifth labor by Eurystheus to clean up King Augeas' stables. Hercules, faced literally with a mountain of horse and cattle waste piled high in the stable area, had to devise some method to dispose of the waste; so, he did. He diverted a couple of river streams to the inside of the stable area so that all the animal waste could simply be deposited into the river: Out of sight out of mind. The waste simply flowed downstream – someone else's problem. Hercules understood the principal point in pollution control technology that is pertinent to this very day; that is, dilution is the solution to pollution.

As applied to today, the fly in the ointment in Hercules' dilution is the solution to pollution approach is today's modern PPCPs. Although he was able to dispose of animal waste into a running water system where eventually the water's self-purification process would clean the stream, he did not have to deal with today's pharmaceuticals and hormones that are given to many types of livestock to enhance their health and growth. The truth is that studies have shown that pharmaceuticals are present in our nation's waterbodies. Further research suggests that certain drugs may cause ecological harm. The EPA and other research agencies are committed to investigating this topic and developing strategies to help protect the health of both the environment and the public. To date, scientists have found no evidence of adverse human health effects from PPCPs in the environment.

Moreover, others might argue that even if PPCPs were present today or in ancient (and mythical) times, the amount present in local public water systems and others would represent only a small fraction (ppt – parts/trillion, 10^{-12}) of the total volume of water, and the critics would be quick to point out that when we are speaking of parts/trillion (ppt) we are speaking of a proportion equivalent to one-twentieth of a drop of water diluted into an Olympic-size swimming pool. I remember when I had one student in my environmental health classes who stated that he did not think the water should be termed "sick water" because it was evident to him that if the water contained so many medications, for example, how could it be sick? Instead, it might be termed "getting well water – making anyone who drinks it well, cured, no longer sick, etc."

It is important to point out that the term "sick water" can be applied to not only PPCP-contaminated water but also to any filthy, dirty, contaminated, vomit-filled, polluted, pathogen-filled drinking water source. The fact is dirty or sick water means more people now die from contaminated and polluted water than from all forms of violence including wars (UN, 2010). The United Nations also points out that dirty or sick water is a key factor in the rise of deoxygenated dead zones that have been emerging in seas and oceans across the globe (Spellman, 2010; UN, 2010).

Are we finished with our discussion of sick water? Oh no … not even close. Consider, for example, the so-called forever contaminants.

Forever chemicals? Yes.

Forever chemicals are human-made PFAS that never break down once released and they build up in our bodies. So what are PFAS? They are a group of per- and polyfluoroalkyl substances that includes PFOA, PFOS, GenX, and many other chemicals (see Figure 9.2 A–B).

So, the question becomes "what is the difference between PFOA, PFOS and GenX and other replacement PFAS?" PFOA and PFOS are the most studied PFAS chemicals and have been voluntarily phased out by industry, though they are still present in the environment. There are other PFAS, including GenX chemicals and PFBS, in use throughout our economy. GenX is a trade name for a technology that is used to make high-performance fluoropolymers (e.g., some nonstick coatings) without the use of perfluorooctanoic acid (PFOA). HFPO dimer acid and its ammonium salt are the major chemicals associated with the GenX technology. GenX chemicals have been found in surface water, groundwater, finished drinking water, rainwater, and air emissions in some areas.

PFAS have been manufactured and used in a variety of industries around the globe, including in the United States. Since the 1940s, PFOA and PFOS have been the most extensively produced and

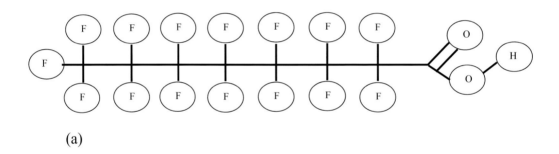

(a)

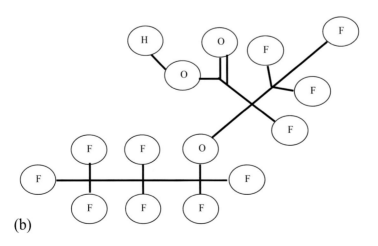

(b)

FIGURE 9.2   (A) PFOA and PFOS. (B) GenX chemicals. *Source: USEPA, 2015.* Accessed 04/08/2020 @ www.epagov/pfas.

studied of these chemicals. Both chemicals are very persistent in the environment and in the human body – meaning they do not break down and they can accumulate over time. There is evidence that exposure to PFAS can lead to adverse human health effects *(USEPA, 2015)*.

PFAS can be found in:

- Food – grown in PFAS-contaminated soil or water, packaged in PFAS-containing materials, or processed with equipment that used PFAS.
- Living organisms – including fish, animals, and humans, where PFAS can build up and persist over time.
- Drinking water – typically localized and associated with a specific facility (e.g., manufacturer, landfill, wastewater treatment plant, firefighter-training facility).
- Workplace – including production facilities or industries (e.g., chrome plating, electronics manufacturing, or oil recovery) that use PFAS.
- Commercial household products – including stain- and water-repellent fabrics, nonstick products (e.g., Teflon), polishes, waxes, cleaning products, and firefighting foams (a major source of groundwater contamination at airports and military bases where firefighting training occurs).

Note: Certain PFAS chemicals are no longer manufactured in the United States as a result of phase outs in which eight major chemical manufacturers agreed to eliminate the use of PFOA- and PROA-related chemicals in their products and as emissions from their facilities. However, PFOA and PFOS are still produced internationally and can be imported into the United States in consumer goods such as carpet, leather and apparel, textiles, paper and packaging, coatings, rubber, and plastics *(2018)*.

DID YOU KNOW?

Recent advances in drilling technologies – including horizontal drilling and hydraulic fracturing (fracking) – have made massive supplies of natural gas economically recoverable in the US. Fracking is not a new technology; it has been around for years. Indeed, hydraulic fracturing, often called fracking (the preferred usage), or well stimulation as many in the industry like to call it, is a well-known practice; it was developed by the US corporation Halliburton 60 years ago.

The use of hydraulic fracturing is a double-edged sword, a Dr. Jekyll and Mr. Hyde operation (i.e., both a benefit and a liability, with the good comes the bad, etc.). One edge or side of the fracturing process, known as a frack job, involves the pressurized injection of fluids commonly made up of large quantities of water and chemical additives into a geologic formation (e.g., gas-bearing shale).

The pressure exceeds the rock strength and the fluid opens or enlarges fractures in the rock. As the formation is fractured, a "propping agent," such as ceramic or sand beads (even peanut and walnut shells have been used) is pumped into the fractures to keep them from closing as the pumping pressure is released. The fracturing fluids (water and chemical additives) are then returned to the surface. On the good edge of the double-edged sword is the oil and natural gas that flows from pores and fractures in the rock into the well for subsequent extraction (Spellman, 2017).

Importance of PFAS

PFAS are found in a wide range of consumer products that people use daily such as pizza boxes, cookware, and stain repellents. The fact is that most people have been exposed to PFAS. Certain PFAS are persistent; they can accumulate and stay in the human body for long periods. The most studied PFAS chemicals are PFOA and PFOS. Studies indicate that PFOA and PFOS can cause reproductive and developmental, liver and kidney, and immunological effects in laboratory animals. Both chemicals have caused tumors in animals. The most consistent findings are increased cholesterol levels among exposed populations, with more limited findings related to:

- Cancer (for PFOA);
- Low infant birth weights;
- Effects on the immune system; and
- Thyroid hormone disruption (for PFOS).

You might be wondering how PFAS can be removed from our drinking water supplies?

Conventional water treatment processes remove many contaminants from the treated water. However, some contaminants including PFAS are not removed by water and wastewater conventional treatment plants. At the present time, it is reverse osmosis that appears to be the major treatment process that removes PFAS (Spellman, 2020b).

INDUSTRIAL WASTEWATER

Activities that produce industrial wastewater include:

- industrial cooling waters (biocides, heat, slimes, silt);
- industrial processing water;
- industrial site drainage (chemical residues, silt, oil, alkali, sand);
- toxic waste from metal plating, cyanide production, pesticide manufacturing, etc.;
- produced water from oil and natural gas production – oil wells sometimes produce large volumes of water with the oil (Spellman, 2017);
- solids and emulsions from paper mills, factories producing lubricants or hydraulic oils, foodstuffs, etc.;
- organic and biodegradable waste including waste from hospitals, abattoirs (aka animal slaughterhouses), creameries, and food factories;
- extreme pH waste from acid and alkali manufacturing;
- organic or nonbiodegradable waste that is difficult to treat from pesticide and pharmaceutical manufacturing;
- urban runoff from highways, roads, railway tracks, car parking lots, roofs, pavement;
- direct and diffuse agricultural pollution; and
- water used in hydraulic fracturing.

COMPOSITION OF WASTEWATER

The actual composition of wastewater varies widely. Chemical, physical, and biological pollutants that may be contained in wastewater include:

- heavy metals, including mercury, lead, and chromium;
- gases such as hydrogen sulfide (w/ its characteristic rotten egg odor), methane, carbon dioxide, etc.;
- thermal pollution from power stations and industrial manufacturers;
- microplastics such as polyethylene and polypropylene beads, polyester, and polyamide (Gatidou et al., 2019);
- toxins such as pesticides, poisons, herbicides, etc.;
- organic particles such as feces, food waste, vomit, paper fibers, plant material, humus, etc.;
- emulsions such as paints, adhesives, mayonnaise, hair colorants, emulsified oils, etc.;
- micro-solids such as sanitary napkins, nappies/diapers, condoms, needles, children's toys, dead animals, or plants, etc.;

- soluble organic material such as urea, fruit sugars, soluble proteins, drugs, pharmaceuticals, etc.;
- soluble inorganic material such as ammonia, road-salt, sea-salt, cyanide, etc.;
- inorganic particles such as grit, sand, rubber residues, ceramics, metal particles, etc.;
- pharmaceuticals, endocrine-disrupting compounds, hormones, fluorinated compounds, illicit drugs of abuse and other hazardous substances, etc. (Spellman, 2020b; Arvaniti & Stasinakis, 2015; Bletsou et al., 2013; Gatidou et al., 2016).

SIDEBAR 1.1 – THE 411 ON ENDOCRINE DISRUPTORS

A growing body of evidence suggests that humans and wildlife species have suffered adverse health effects after exposure to inadvertent consumption of endocrine-disrupting chemicals (aka environmental endocrine disruptors) found in various drinking water sources. Environmental endocrine disruptors are defined as exogenous agents that interfere with the production, release, transport, metabolism binding, action, or elimination of natural hormones in the body responsible for the maintenance of homeostasis and the regulation of developmental processes.

The definition reflects a growing awareness that the issue of endocrine disruptors in the environment extends considerably beyond that of exogenous estrogens and includes antiandrogens and agents that act on other components of the endocrine system such as the thyroid and pituitary glands *(Spellman 2020b).*

Disruption of the endocrine system can occur in various ways. Some chemicals can mimic a natural hormone, fooling the body into over-responding to the stimulus (e.g., a growth hormone that results in increased muscle mass) or responding at inappropriate times (e.g., producing insulin when it is not needed). Other endocrine-disrupting chemicals can block the effects of a hormone from certain receptors. Still others can directly stimulate or inhibit the endocrine system, causing overproduction or underproduction of hormones. Certain drugs are used to intentionally cause some of these effects, such as birth control pills. In many situations involving environmental chemicals, an endocrine effect may not be desirable.

In recent years, some scientists have proposed that chemicals might inadvertently be disrupting the endocrine system of humans and wildlife. Reported adverse effects include declines in populations, increases in cancers, and reduced reproductive function. To date, these health problems have been identified primarily in domestic or wildlife species with relatively high exposures to organochlorine compounds, including DDT and its metabolites, polychlorinated biphenyls (PCBs) and dioxides, or to naturally occurring plant estrogens (phytoestrogens). However, the relationship of human diseases of the endocrine system and exposure to environmental contaminants is poorly understood and scientifically controversial.

Wastewater that contains human feces may contain pathogens of the four types listed below (Spellman, 2020a; WHO, 2006; Anderson et al., 2016):

1. Bacteria
2. Viruses
3. Protozoa
4. Parasites

Note: It can also contain nonpathogenic bacteria and animals such as small fish, arthropods, and insects.

OXYGEN CONTENT OF NATURAL WATER SOURCES

The oxygen content of natural water sources has a direct impact on the quality of all-natural waterways (many of which serve as sources of drinking water). Oxygen content of these water sources is directly impacted by pollution. For example, almost any waste compound introduced into such water sources will initiate biochemical reactions. Those biochemical reactions create what is measured in the laboratory or the *biochemical oxygen demand* (BOD).

Biochemical oxygen demand (BOD): the amount of dissolved oxygen demanded by bacteria to break down the organic materials during the stabilization action of the decomposable organic matter under aerobic conditions over a 5-day incubation period at 20°C (68°F). In the laboratory, *biochemical oxygen demand-5* (BOD_5) measures the amount of organic matter that can be biologically oxidized under controlled conditions (5 days @ 20°C in the dark). Several criteria are used when selecting which BOD_5 dilutions to be used for calculating test results. Consult a laboratory testing reference manual (such as *Standard Methods*) for this information.

WASTEWATER TREATMENT

United Nations World Water Assessment Programme (WWAP, 2017) reports that at a global level around 80% of wastewater produced is discharged into the environment untreated, causing widespread water pollution. For those countries that employ wastewater treatment systems, there are numerous processes that can be used to clean up wastewaters depending on the type and extent of contamination. The availability of numerous processes is an understatement.

To make this point clear, consider the partial listing of wastewater treatment technologies that are currently available:

- Reverse osmosis
- Rotating biological contactor (RBC)
- Sand filter
- Belt filter
- Anaerobic lagoon
- Carbon filtering
- Extended aeration
- Distillation
- Reed bed
- Cesspit
- Dark fermentation
- Membrane bioreactor
- Trickling filter
- Wet oxidation
- Forward osmosis
- Septic tank
- Stabilization pond
- Ion exchange
- Screen filter
- Conventional wastewater treatment (activated sludge process)

Note: It is this last listed process, conventional wastewater treatment (activated sludge process – an aerobic process), that is the focus of *Part II*. This process is based on the maintenance and recirculation of a complex biomass composed of microorganisms able to absorb and adsorb the organic matter carried in the wastewater. Although the focus herein is on conventional wastewater treatment, keep in mind that anaerobic wastewater treatment processes are also widely applied in the stream of industrial wastewaters and biological sludge (aka biosolids). Some wastewater may be highly treated and reused as reclaimed water. Constructed wetlands are also commonly used.

Wastewater Disposal

In many locations, municipal wastewater is carried together with stormwater, in a combined sewer system, to a wastewater treatment plant. All the manholes you might see here and there are important and germane to wastewater practices and methodologies. This can be seen in some urban areas where municipal wastewater is carried separately in sanitary sewers, and runoff from streets is carried in storm drains. Access to these systems, for maintenance purposes, is typically through a manhole. During high precipitation events, a combined sewer system may experience a combined sewage overflow event, which forces untreated sewage to flow directly to receiving waters. This can pose a serious threat to public health and the surrounding environment.

Minimal or no treatment of wastewater is also often the case and is the typical practice in less developed (and some rural) locations where raw sewage is dumped or drained directly into major waterways and watersheds. This usually has serious impacts on the quality of the environment and human health. Pathogens can cause a variety of illnesses. Some chemicals pose risks even at very low concentrations and can remain a threat for long periods because of bioaccumulation of animal or human tissue. Wastewater from industrial activities such as factories and power plants is extensively regulated in developed countries, and treatment is required before discharge to surface waters. Some facilities, such as oil and gas wells, may be permitted to pump their wastewater underground through injection wells. However, keep in mind that wastewater injection has been linked to induced seismicity – increased incidence of earthquakes (Van der Baan & Calixto, 2017).

Wastewater Reuse

Wastewater, if treated correctly, can be used to supply drinking water systems. Treated wastewater is already commonly used in industry (e.g., in cooling towers), in artificial recharge of aquifers (the Potomac Aquifer in Hampton Roads is a good example), in agriculture, and to the rehabilitation of natural ecosystems (wetlands). It is also used to augment drinking water supplies. In doing so, the facility or enterprise doing the augmenting has to overcome the so-called yuck factor involved with drinking toilet water – no easy task, for sure.

Now there are several technologies used to treat wastewater for reuse. In combination, these treatment technologies can be used to treat wastewater to regulatory standards – to safe for consumption standards. Making sure the drinking water is hygienically safe, meaning free from bacteria and viruses, is the goal, the requirement. Some of these technologies include ozonation, ultrafiltration, aerobic treatment (membrane bioreactor), forward osmosis, reverse osmosis, and advanced oxidation.

It is important to point out that some water-demanding activities do not require wastewater treated to drinking water quality and standards. Wastewater can be used with little or no treatment when toilets can be flushed using greywater from baths and showers. Recycled wastewater can also be used in irrigation, and this can be especially beneficial whenever the wastewater contains nutrients, such as nitrogen, phosphorus, and potassium. In developing countries, agriculture is using untreated wastewater for irrigation; this is not wise or safe. There can be significant health hazards related to using untreated wastewater in agriculture.

CHAPTER SUMMARY

Whenever we flush the toilet, wash dishes in the sink, or wash clothes in the washer, we probably give very little thought to the product we have created via our use of the water. What we have done is create wastewater, or what is now commonly called used water. This chapter has set the foundation for the chapters to follow that explain the process of providing proper treatment, disposal, and/or reuse of used water.

REVIEW QUESTIONS

1. Which of the following should never be flushed down in the toilet?
   a. Toilet paper
   b. Urine
   c. Feces
   d. Crankcase oil

2. Why are illicit drugs sometimes flushed down the toilet?
   a. When the user wants to spread the wealth
   b. Whenever users want to give wastewater operators a tax-free windfall
   c. Whenever the police are knocking on the door
   d. To prevent anyone else in the house from using the drugs

3. Why are dry toilets feasible in some locations?
   a. Area is short on water
   b. They are inexpensive
   c. They are part of the new green deal
   d. They provide humus for plant life

4. Why is FOG dumping in toilets not recommended?
   a. Clouds the water
   b. Stinks
   c. Can plug sewer lines and confound sewage treatment processes
   d. Not easy to flush

5. Why is treated blackwater not connected pipe to pipe for household use?
   a. Can be unhealthy for consumption
   b. Too difficult to treat to drinking water quality
   c. In most locations it is illegal to do so
   d. All of the above

6. Can sick water be thoroughly treated by conventional treatment?
   a. Only if the most sophisticated treatment processes are available for treatment
   b. Yes, if the personal care products are biodegradable
   c. Not a problem
   d. The water is not sick

7. PFAS is not found in which of the following?
   a. Living organisms
   b. Chrome plating operations
   c. Air
   d. Peanut butter

8. Which of the following could be a result of drinking water contaminated with PFAS?
   a. Convulsions
   b. Cancer
   c. High infant birth weights
   d. Stunted growth

9. Produced water results from _____
   a. Production of cheese
   b. Bakery waste
   c. Oil and natural gas production
   d. Gold mining

10. Biological oxygen demand increases whenever _____
    a. Drought occurs
    b. Whenever animal waste or other organics are dumped into a stream
    c. Whenever the stream is overcrowded with fish
    d. Whenever the stream is dammed

ANSWER KEY

1. Which of the following should never be flushed down in the toilet?
   a. Toilet paper
   b. Urine
   c. Feces
   d. **Crankcase oil**

2. Why are illicit drugs sometimes flushed down the toilet?
   a. When the user wants to spread the wealth
   b. Whenever users want to give wastewater operators a tax-free windfall
   c. **Whenever the police are knocking on the door**
   d. To prevent anyone else in the house from using the drugs

3. Why are dry toilets feasible in some locations?
   a. **Area is short on water**
   b. They are inexpensive
   c. They are part of the new green deal
   d. They provide humus for plant life

4. Why is FOG dumping in toilets not recommended?
   a. Clouds the water
   b. Stinks
   c. **Can plug sewer lines and confound sewage treatment processes**
   d. Not easy to flush

5. Why is treated blackwater not connected pipe to pipe for household use?
   a. Can be unhealthy for consumption
   b. Too difficult to treat to drinking water quality
   c. In most locations it is illegal to do so
   d. **All of the above**

6. Can sick water be thoroughly treated by conventional treatment?
   a. **Only if the most sophisticated treatment processes are available for treatment**
   b. Yes, if the personal care products are biodegradable
   c. Not a problem
   d. The water is not sick

7. PFAS is not found in which of the following?
   a. Living organisms
   b. Chrome plating operations
   c. **Air**
   d. Peanut butter

8. Which of the following could be a result of drinking water contaminated with PFAS?
   a. Convulsions
   b. **Cancer**
   c. High infant birth weights
   d. Stunted growth

9. Produced water results from _____
   a. Production of cheese
   b. Bakery waste
   c. **Oil and natural gas production**
   d. Gold mining

10. Biological oxygen demand increases whenever _____
    a. Drought occurs
    b. **Whenever animal waste or other organics are dumped into a stream**
    c. Whenever the stream is overcrowded with fish
    d. Whenever the stream is dammed

REFERENCES

Anderson, K., Rosemarin, A., Lamizana, B., Kvamstrom, E., McConville, J., Seidu, R., Dickin, S. and Trimmer, C. (2016). *Sanitation, Wastewater Management and Sustainability.* New York, NY: UNEP.

Arvaniti, O.S. and Stasinakis, A.S. (2015). Review on the occurrence, fate and removal of/fluorinated compounds during wastewater treatment. *Science of the Total Environment* 524–525(August), p.81–92.

Bletsou, A.A. et al. (2013). Mass loading and fate of linear and cyclic Siloxanes in a wastewater treatment plant in Greece. *Environmental Science and Technology* 47(January), p.1824–1832.

Gatidou, G. et al. (2016). Drugs of abuse and alcohol consumption among different groups of population on Greek island of Lesvos through sewage-based epidemiology. *Science of the Total Environment* 563–564(September), p.633–640.

Gatidou, G. et al. (2019). Review on the occurrence and fate of microplastics in Sewage Treatment Plants. *Journal of Hazardous Materials* 367(April), p.504–512.

Spellman, F.R. (2013). *Spellman's Standard Handbook for Water and Wastewater Operators.* Boca Raton, FL: CRC Press.

Spellman, F.R. (2017). *Hydraulic Fracturing Wastewater: Treatment, Reuse, and Disposal.* Boca Raton, FL: CRC Press.

Spellman, F.R. (2020a). *Handbook of Water and Wastewater Treatment Operations*, 4th ed. Boca Raton, FL: CRC Press.

Spellman, F.R. (2020b). *The Science of Water*, 4th ed. Boca Raton, FL: CRC Press.

Spellman, F.R. (2010). *Water and Wastewater Infrastructure.* Boca Raton, FL: CRC Press.

Tilley, E., Ulrich, L., Luthi, C., Reymond, P., Zubrugg, C. (2014). *Compendium on Sanitation Systems and Technology*, 2nd Revised Edition. Accessed 0408/2020 @ http://www.eawag.ch/department/sandec/publications.

UN (2010). *Water Global Annual Assessment.* New York: UN.

United Nations World Water Assessment Programme (WWAP) (2017). *The United Nations World Water Development Report 2017 Wastewater: The Untapped Resource.*

USEPA (1998). *IESWTR Interim Enhanced Surface Water Rule.* Washington, DC: USEPA.

USEPA (2015). *Condition Report of Underground Pipe.* Washington, DC: USEPA.

Van der Baan M., Calixto, F.J. (2017). Human induced seismicity and large-scale hydrocarbon production in the United States and Canada. *Geochemistry Geophysics, Geosystems* 18(7), p.2467–2485.

World Health Organization (2006). *Guidelines for the Safe Use Wastewater, Excreta, and Greywater.* Geneva, Switzerland: World Health Organization. p.31.

10 Wastewater Operators, References, Models, and Terminology

Perhaps the most important person at a wastewater treatment plant is the plant operator because that person is responsible for treating the water to meet or exceed the Federal Safe and Clean Water Act standards and public expectations.

– F.R. Spellman (2013)

Living things depend on water but water does not depend on living things. It has a life of its own.

– E.C. Pielou (1998)

KEY OBJECTIVES

After studying this chapter, you should be able to:

- Define key terms used in wastewater treatment and operations.
- Identify and define key acronyms used in wastewater treatment operations.

DOI: 10.1201/9781003207146-15

- Identify abbreviations commonly used in wastewater treatment operations.
- Identify what wastewater operators do.
- Identify the qualifications to become a wastewater operator.
- Discuss the high injury rate of wastewater operators.
- Define wastewater.
- Explain why wastewater operators are Jacks and Jills of all trades.
- Explain the digital interface with wastewater operations.
- Identify the computer literate wastewater operator's work.
- Discuss the need for emergency response preparedness at wastewater plants.
- Explain operator duties.
- Identify operators' working conditions.
- Discuss wastewater operator certification and licensure.
- Identify the specialized topics wastewater operators must be familiar with.

KEY DEFINITIONS AND ACRONYMS

To learn wastewater treatment operations (or any other technology for that matter), you must master the language associated with the technology. Each technology has its own terms and acronyms and abbreviations with their own accompanying definitions. Many of the terms and acronyms used in wastewater treatment are unique; others combine words from many different technologies and professions. One thing is certain: Wastewater operators without a clear understanding of the terms and acronyms related to their profession are ill equipped to perform their duties in the manner required.

Experience has shown that an early introduction to keywords and definitions is beneficial to readers. Those terms not defined in this section are defined as they appear in the applicable chapters. A short quiz on many of the following terms follows the end of this chapter.

DID YOU KNOW?

It takes between 250 and 600 gallons of water to grow a pound of rice. That is more water than many households use in a week for just a bag of rice (Pearce, 2006).

KEY DEFINITIONS

Absorb – to take in. Many things absorb water.

Acid Rain – the acidic rainfall that results when rain combines with sulfur oxides emissions from combustion of fossil fuels (coal, for example).

Acre-Feet (acre-foot) – an expression of water quantity. One acre-foot will cover one acre of ground one foot deep. An acre-foot contains 43,560 cubic feet, 1,233 cubic meters, or 325,829 gallons (US). Also abbreviated as ac-ft.

Activated carbon – derived from vegetable or animal materials by roasting in a vacuum furnace. Its porous nature gives it a very high surface area per unit mass – as much as 1,000 square meters per gram, which is 10 million times the surface area of 1 gram of water in an open container. Used in adsorption (see definition), activated carbon adsorbs substances that are not or are only slightly adsorbed by other methods.

Activated sludge – the solids formed when microorganisms are used to treat wastewater using the activated sludge treatment process. It includes organisms, accumulated food materials, and waste products from the aerobic decomposition process.

Advanced wastewater treatment – treatment technology to produce an extremely high-quality discharge.

Adsorption – the adhesion of a substance to the surface of a solid or liquid. Adsorption is often used to extract pollutants by causing them to attach to such adsorbents as activated carbon or silica gel.

Hydrophobic (water-repulsing adsorbents) – are used to extract oil from waterways in oil spills.

Aeration – the process of bubbling air through a solution, sometimes cleaning water of impurities by exposure to air.

Aerobic – conditions in which free, elemental oxygen is present. Also used to describe organisms, biological activity, or treatment processes that require free oxygen.

Agglomeration – floc particles colliding and gathering into a larger settleable mass.

Air gap – the air space between the free-flowing discharge end of a supply pipe and an unpressurized receiving vessel.

Algae bloom – a phenomenon whereby excessive nutrients within a river, stream, or lake cause an explosion of plant life that results in the depletion of the oxygen in the water needed by fish and other aquatic life. Algae bloom is usually the result of urban runoff (of lawn fertilizers, etc.). The potential tragedy is that of a "fish kill" where the stream life dies in one mass execution.

Alum – aluminum sulfate; a standard coagulant used in water treatment.

Ambient – the expected natural conditions that occur in water unaffected or uninfluenced by human activities.

Anaerobic – conditions in which no oxygen (free or combined) is available. Also used to describe organisms, biological activity, or treatment processes that function in the absence of oxygen.

Anoxic – conditions in which no free, elemental oxygen is present. The only source of oxygen is combined oxygen, such as that found in nitrate compounds. Also used to describe biological activity of treatment processes that function only in the presence of combined oxygen.

Aquifer – a water-bearing stratum of permeable rock, sand, or gravel.

Aquifer system – a heterogeneous body of introduced permeable and less permeable material that acts as a water-yielding hydraulic unit of regional extent

Artesian water – a well tapping a confined or artesian aquifer in which the static water level stands above the top of the aquifer. The term is sometimes used to include all wells tapping confined water. Wells with water level above the water table are said to have positive artesian head (pressure) and those with water level below the water table, negative artesian head.

Average monthly discharge limitation – the highest allowable discharge over a calendar month.

Average weekly discharge limitation – the highest allowable discharge over a calendar week.

Backflow – reversal of flow when pressure in a service connection exceeds the pressure in the distribution main.

Backwash – fluidizing filter media with water, air, or a combination of the two so that individual grains can be cleaned of the material that has accumulated during the filter run.

Bacteria – any of several one-celled organisms, some of which cause disease.

Bar screen – a series of bars formed into a grid used to screen out large debris from influent flow.

Base – a substance that has a pH value between 7 and 14.

Basin – a groundwater reservoir defined by the overlying land surface and underlying aquifers that contain water stored in the reservoir.

Beneficial use of water – the use of water for any beneficial purpose. Such uses include domestic use, irrigation, recreation, fish and wildlife, fire protection, navigation, power, industrial use, etc. The benefit varies from one location to another and by custom. What constitutes beneficial use is often defined by statute or court decisions.

Biochemical oxygen demand (BOD_5) – the oxygen used in meeting the metabolic needs of aerobic microorganisms in water rich in organic matter.

Biosolids – from *Merriam-Webster's Collegiate Dictionary*, 10th ed. (1998): biosolids *n.* solid organic matter recovered from a sewage treatment process and used especially as fertilizer (or soil amendment) – usually used in plural.

Note: In this text, *biosolids* is used in many places (activated sludge being the exception) to replace the standard term *sludge*. The authors view the term *sludge* as an ugly four-letter word inappropriate to use to describe biosolids. Biosolids is a product that can be reused; it has some value. Because biosolids have value, it certainly should not be classified as a "waste" product – and when biosolids for beneficial reuse is addressed, it is made clear that it is not.

Biota – all the species of plants and animals indigenous to a certain area.

Boiling point – the temperature at which a liquid boils. The temperature at which the vapor pressure of a liquid equals the pressure on its surface. If the pressure of the liquid varies, the actual boiling point varies. The boiling point of water is 212° Fahrenheit or 100° Celsius.

Breakpoint – point at which chlorine dosage satisfies chlorine demand.

Breakthrough – in filtering when unwanted materials start to pass through the filter.

Buffer – a substance or solution that resists changes in pH.

Calcium carbonate – compound principally responsible for hardness.

Calcium hardness – portion of total hardness caused by calcium compounds.

Carbonaceous biochemical oxygen demand ($CBOD_5$) – the amount of biochemical oxygen demand that can be attributed to carbonaceous material.

Carbonate hardness – caused primarily by compounds containing carbonate.

Chemical oxygen demand (COD) – the amount of chemically oxidizable materials present in the wastewater.

Chlorination – disinfection of water using chlorine as the oxidizing agent.

Clarifier – a device designed to permit solids to settle or rise and be separated from the flow. Also known as a settling tank or sedimentation basin.

Coagulation – the neutralization of the charges of colloidal matter.

Coliform – a type of bacteria used to indicate possible human or animal contamination of water.

Combined sewer – a collection system that carries both wastewater and stormwater flows.

Comminution – a process to shred solids into smaller, less harmful particles.

Composite sample – a combination of individual samples taken in proportion to flow.

Connate water – pressurized water trapped in the pore spaces of sedimentary rock at the time it was deposited. It is usually highly mineralized.

Consumptive use – (1) the quantity of water absorbed by crops and transpired or used directly in the building of plant tissue, together with the water evaporated from the cropped area; (2) the quantity of water transpired and evaporated from a cropped area or the normal loss of water from the soil by evaporation and plant transpiration; (3) the quantity of water discharged to the atmosphere or incorporated in the products in connection with vegetative growth, food processing, or an industrial process.

Contamination (water) – damage to the quality of water sources by sewage, industrial waste, or other material.

Cross-connection – a connection between a storm drain system and a sanitary collection system; a connection between two sections of a collection system to handle anticipated overloads of one system; or a connection between drinking (potable) water and an unsafe water supply or sanitary collection system.

Daily discharge – the discharge of a pollutant measured during a calendar day or any 24-hour period that reasonably represents a calendar day for the purposes of sampling. Limitations expressed as weight are total mass (weight) discharged over the day. Limitations expressed in other units are average measurement of the day.

Daily maximum discharge – the highest allowable values for a daily discharge.

Darcy's Law – an equation for the computation of the quantity of water flowing through porous media. Darcy's Law assumes that the flow is laminar and that inertia can be neglected. The law states that the rate of viscous flow of homogenous fluids through isotropic porous media is proportional to, and in the direction of, the hydraulic gradient.

Detention time – the theoretical time water remains in a tank at a given flow rate.

De-watering – the removal or separation of a portion of water present in a sludge or slurry.

Diffusion – the process by which both ionic and molecular species dissolved in water move from areas of higher concentration to areas of lower concentration.

Discharge monitoring report (DMR) – the monthly report required by the treatment plant's National Pollutant Discharge Elimination System (NPDES) discharge permit.

Disinfection – water treatment process that kills pathogenic organisms.

Disinfection by-products (DBPs) – chemical compounds formed by the reaction of disinfectant with organic compounds in water.

Dissolved oxygen (DO) – the amount of oxygen dissolved in water or sewage. Concentrations of less than 5 parts per million (ppm) can limit aquatic life or cause offensive odors. Excessive organic matter present in water because of inadequate waste treatment and runoff from agricultural or urban land generally causes low DO.

Dissolved solid – the total amount of dissolved inorganic material contained in water or wastes. Excessive dissolved solids make water unsuitable for drinking or industrial uses.

Domestic consumption (use) – water used for household purposes such as washing, food preparation, and showers. The quantity (or quantity per capita) of water consumed in a municipality or district for domestic uses or purposes during a given period; it sometimes encompasses all uses, including the quantity wasted, lost, or otherwise unaccounted for.

Drawdown – lowering the water level by pumping. It is measured in feet for a given quantity of water pumped during a specified period or after the pumping level has become constant.

Drinking water standard – established by state agencies, U.S. Public Health Service, and Environmental Protection Agency (EPA) for drinking water in the US.

Effluent – something that flows out, usually a polluting gas or liquid discharge.

Effluent limitation – any restriction imposed by the regulatory agency on quantities, discharge rates, or concentrations of pollutants discharged from point sources into state waters.

Energy – in scientific terms, the ability or capacity of doing work. Various forms of energy include kinetic, potential, thermal, nuclear, rotational, and electromagnetic. One form of energy may be changed to another, as when coal is burned to produce steam to drive a turbine, which produces electric energy.

Erosion – the wearing away of the land surface by wind, water, ice, or other geologic agents. Erosion occurs naturally from weather or runoff but is often intensified by human land use practices.

Eutrophication – the process of enrichment of water bodies by nutrients. Eutrophication of a lake normally contributes to its slow evolution into a bog or marsh and ultimately to dry land. Eutrophication may be accelerated by human activities, thereby speeding up the aging process.

Evaporation – the process by which water becomes a vapor at a temperature below the boiling point.

Facultative – organisms that can survive and function in the presence or absence of free, elemental oxygen.

Fecal coliform – the portion of the coliform bacteria group that is present in the intestinal tracts and feces of warm-blooded animals.

Field capacity – the capacity of soil to hold water. It is measured as the ratio of the weight of water retained by the soil to the weight of the dry soil.

Filtration – the mechanical process that removes particulate matter by separating water from solid material, usually by passing it through sand.

Floc – solids that join to form larger particles that will settle better.

Flocculation – slow mixing process in which particles are brought into contact, with the intent of promoting their agglomeration.

Flume – a flow rate measurement device.

Fluoridation – chemical addition to water to reduce incidence of dental cavities in children.

Food-to-microorganisms ratio (F/M) – an activated sludge process control calculation based upon the amount of food (BOD_5 or COD) available per pound of mixed liquor volatile suspended solids.

Force main – a pipe that carries wastewater under pressure from the discharge side of a pump to a point of gravity flow downstream.

Grab sample – an individual sample collected at a randomly selected time.

Graywater – water that has been used for showering, clothes washing, and faucet uses. Kitchen sink and toilet water is excluded. This water has excellent potential for reuse as irrigation for yards.

Grit – heavy inorganic solids, such as sand, gravel, eggshells, or metal filings.

Groundwater – the supply of fresh water found beneath the earth's surface (usually in aquifers) often used for supplying wells and springs. Because groundwater is a major source of drinking water, concern is growing over areas where leaching agricultural or industrial pollutants or substances from leaking underground storage tanks (USTs) are contaminating groundwater.

Groundwater hydrology – the branch of hydrology that deals with groundwater; its occurrence and movements, its replenishment and depletion, the properties of rocks that control groundwater movement and storage, and the methods of investigation and use of groundwater.

Groundwater recharge – the inflow to a groundwater reservoir.

Groundwater runoff – a portion of runoff that has passed into the ground, has become groundwater, and has been discharged into a stream channel as spring or seepage water.

Hardness – the concentration of calcium and magnesium salts in water.

Headloss – amount of energy used by water in moving from one point to another.

Heavy metal – metallic elements with high atomic weights, e.g., mercury, chromium, cadmium, arsenic, and lead. They can damage living things even at low concentrations and tend to accumulate in the food chain.

Holding pond – a small basin or pond designed to hold sediment-laden or contaminated water until it can be treated to meet water quality standards or used in some other way.

Hydraulic cleaning – cleaning pipe with water under enough pressure to produce high water velocities.

Hydraulic gradient – a measure of the change in groundwater head over a given distance.

Hydraulic head – the height above a specific datum (generally sea level) that water will rise in a well.

Hydrologic cycle (water cycle) – the cycle of water movement from the atmosphere to the earth and back to the atmosphere through various processes. These processes include precipitation, infiltration, percolation, storage, evaporation, transpiration, and condensation.

Hydrology – the science dealing with the properties, distribution, and circulation of water.

Impoundment – a body of water such as a pond, confined by a dam, dike, floodgate, or other barrier, and used to collect and store water for future use.

Industrial wastewater – wastes associated with industrial manufacturing processes.

Infiltration – the gradual downward flow of water from the surface into soil material.

Infiltration/inflow – extraneous flows in sewers; simply, inflow is water discharged into sewer pipes or service connections from such sources as foundation drains, roof leaders, cellar and yard area drains, cooling water from air conditioners, and other clean-water discharges from commercial and industrial establishments.

Defined by Metcalf & Eddy (2003) as follows:

Infiltration – water entering the collection system through cracks, joints, or breaks.

Steady inflow – water discharged from cellar and foundation drains, cooling water discharges, and drains from springs and swampy areas. This type of inflow is steady and is identified and measured along with infiltration.

Direct flow – those types of inflow that have a direct stormwater runoff connection to the sanitary sewer and cause an almost immediate increase in wastewater flows. Possible sources are roof leaders, yard and areaway drains, manhole covers, cross connections from storm drains and catch basins, and combined sewers.

Total inflow – the sum of the direct inflow at any point in the system plus any flow discharged from the system upstream through overflows, pumping station bypasses, and the like.

Delayed inflow – stormwater that may require several days or more to drain through the sewer system. This category can include the discharge of sump pumps from cellar drainage as well as the slowed entry of surface water through manholes in ponded areas.

Influent – wastewater entering a tank, channel, or treatment process.

Inorganic chemical/compound – chemical substances of mineral origin, not of carbon structure. These include metals such as lead, iron (ferric chloride), and cadmium.

Ion exchange process – used to remove hardness from water.

Jar Test – laboratory procedure used to estimate proper coagulant dosage.

Langelier saturation *index* (L.I.) – a numerical index that indicates whether calcium carbonate will be deposited or dissolved in a distribution system.

Leaching – the process by which soluble materials in the soil such as nutrients, pesticide chemicals or contaminants are washed into a lower layer of soil or are dissolved and carried away by water.

License – a certificate issued by the State Board of Waterworks/Wastewater Works Operators authorizing the holder to perform the duties of a wastewater treatment plant operator.

Lift station – a wastewater pumping station designed to "lift" the wastewater to a higher elevation. A lift station normally employs pumps or other mechanical devices to pump the wastewater and discharges into a pressure pipe called a force main.

Maximum contaminant level (MCL) – an enforceable standard for protection of human health.

Mean cell residence time (MCRT) – the average length of time a mixed liquor suspended solids particle remains in the activated sludge process. May also be known as sludge retention time.

Mechanical cleaning – clearing pipe by using equipment (bucket machines, power rodders, or hand rods) that scrapes, cuts, pulls, or pushes the material out of the pipe.

Membrane process – a process that draws a measured volume of water through a filter membrane with small enough openings to take out contaminants.

Metering pump – a chemical solution feed pump that adds a measured amount of solution with each stroke or rotation of the pump.

Mixed liquor – the suspended solids concentration of the mixed liquor.

Mixed liquor volatile suspended solids (MLVSS) – the concentration of organic matter in the mixed liquor suspended solids.

Milligrams/liter (mg/L) – a measure of concentration equivalent to parts per million (ppm).

Nephelometric turbidity unit (NTU) – indicates amount of turbidity in a water sample.

Nitrogenous oxygen demand (NOD) – a measure of the amount of oxygen required to biologically oxidize nitrogen compounds under specified conditions of time and temperature.

Nonpoint Source (NPS) *pollution* – forms of pollution caused by sediment, nutrients, organic and toxic substances originating from land use activities that are carried to lakes and streams by surface runoff. Nonpoint source pollution occurs when the rate of materials entering these waterbodies exceeds natural levels.

NPDES permit – National Pollutant Discharge Elimination System permit, which authorizes the discharge of treated wastes and specifies the conditions that must be met for discharge.

Nutrients – substances required to support living organisms. Usually refers to nitrogen, phosphorus, iron, and other trace metals.

Organic chemicals/compound – animal or plant-produced substances containing mainly carbon, hydrogen, and oxygen, such as benzene and toluene.

Parts per million (ppm) – the number of parts by weight of a substance per million parts of water. This unit is commonly used to represent pollutant concentrations. Large concentrations are expressed in percentages.

Pathogenic – disease causing. A pathogenic organism can cause illness.

Percolation – the movement of water through the subsurface soil layers, usually continuing downward to the groundwater or water table reservoirs.

pH – a way of expressing both acidity and alkalinity on a scale of 0–14, with 7 representing neutrality; numbers less than 7 indicate increasing acidity and numbers greater than 7 indicate increasing alkalinity.

Photosynthesis – a process in green plants in which water, carbon dioxide, and sunlight combine to form sugar.

Piezometric surface – an imaginary surface that coincides with the hydrostatic pressure level of water in an aquifer.

Point source pollution – a type of water pollution resulting from discharges into receiving waters from easily identifiable points. Common point sources of pollution are discharges from factories and municipal sewage treatment plants.

Pollution – the alteration of the physical, thermal, chemical or biological quality of, or the contamination of, any water in the state that renders the water harmful, detrimental, or injurious to humans, animal life, vegetation, property or to public health, safety, or welfare, or impairs the usefulness or the public enjoyment of the water for any lawful or reasonable purpose.

Porosity – that part of a rock that contains pore spaces without regard to size, shape, interconnection, or arrangement of openings. It is expressed as percentage of total volume occupied by spaces.

Potable water – water satisfactorily safe for drinking purposes from the standpoint of its chemical, physical, and biological characteristics.

Precipitate – a deposit on the earth of hail, rain, mist, sleet, or snow. The common process by which atmospheric water becomes surface or subsurface water, the term *precipitation* is also commonly used to designate the quantity of water precipitated.

Preventive maintenance (PM) – regularly scheduled servicing of machinery or other equipment using appropriate tools, tests, and lubricants. This type of maintenance can prolong the useful life of equipment and machinery and increase its efficiency by detecting and correcting problems before they cause a breakdown of the equipment.

Purveyor – an agency or person that supplies potable water.

Radon – a radioactive, colorless, odorless gas that occurs naturally in the earth. When trapped in buildings, concentrations build up and can cause health hazards such as lung cancer.

Recharge – the addition of water into a groundwater system.

Reservoir – a pond, lake, tank, or basin (natural or human made) where water is collected and used for storage. Large bodies of groundwater are called groundwater reservoirs; water behind a dam is also called a reservoir of water.

Reverse osmosis – process in which almost pure water is passed through a semipermeable membrane.

Return activated sludge solids (RASS) – the concentration of suspended solids in the sludge flow being returned from the settling tank to the head of the aeration tank.

River basin – a term used to designate the area drained by a river and its tributaries.

Sanitary wastewater – wastes discharged from residences and from commercial, institutional, and similar facilities that include both sewage and industrial wastes.

Schmutzdecke – layer of solids and biological growth that forms on top of a slow sand filter, allowing the filter to remove turbidity effectively without chemical coagulation.

Scum – the mixture of floatable solids and water removed from the surface of the settling tank.

Sediment – transported and deposited particles derived from rocks, soil, or biological material.

Sedimentation – a process that reduces the velocity of water in basins so that suspended material can settle out by gravity.

Seepage – the appearance and disappearance of water at the ground surface. Seepage designates movement of water in saturated material. It differs from percolation, which is predominantly the movement of water in unsaturated material.

Septic tank – used to hold domestic wastes when a sewer line is not available to carry them to a treatment plant. The wastes are piped to underground tanks directly from a home or homes. Bacteria in the wastes decompose some of the organic matter, the sludge settles on the bottom of the tank, and the effluent flows out of the tank into the ground through drains.

Settleability – a process control test used to evaluate the settling characteristics of the activated sludge. Readings taken at 30–60 minutes are used to calculate the settled sludge volume (SSV) and the sludge volume index (SVI).

Settled sludge volume (SSV) – the volume (in percent) occupied by an activated sludge sample after 30–60 minutes of settling. Normally written as SSV with a subscript to indicate the time of the reading used for calculation (SSV_{60} or SSV_{30}).

Sludge –the mixture of settleable solids and water removed from the bottom of the settling tank.

Sludge retention time (SRT) – see mean cell residence time.

Sludge volume index (SVI) – a process control calculation used to evaluate the settling quality of the activated sludge. Requires the SSV_{30} and mixed liquor suspended solids test results to calculate.

Soil moisture (soil water) – water diffused in the soil. It is found in the upper part of the zone of aeration from which water is discharged by transpiration from plants or by soil evaporation.

Specific heat – the heat capacity of a material per unit mass. The amount of heat (in calories) required to raise the temperature of one gram of a substance 1°C; the specific heat of water is 1 calorie.

Storm sewer – a collection system designed to carry only stormwater runoff.

Stormwater – runoff resulting from rainfall and snowmelt.

Stream – a general term for a body of flowing water. In hydrology, the term is generally applied to the water flowing in a natural channel as distinct from a canal. More generally, it is applied to the water flowing in any channel, natural or artificial. Some types of streams:

- Ephemeral: a stream that flows only in direct response to precipitation and whose channel is always above the water table.
- Intermittent or Seasonal: a stream that flows only at certain times of the year when it receives water from springs, rainfall, or from surface sources such as melting snow.
- Perennial: a stream that flows continuously.
- Gaining: a stream or reach of a stream that receives water from the zone of saturation. An effluent stream.

- Insulated: a stream or reach of a stream that neither contributes water to the zone of saturation nor receives water from it. It is separated from the zones of saturation by an impermeable bed.
- Losing: a stream or reach of a stream that contributes water to the zone of saturation. An influent stream.
- Perched: a perched stream is either a losing stream or an insulated stream that is separated from the underlying groundwater by a zone of aeration.

Supernatant – the liquid standing above a sediment or precipitate.

Surface water – lakes, bays, ponds, impounding reservoirs, springs, rivers, streams, creeks, estuaries, wetlands, marshes, inlets, canals, gulfs inside the territorial limits of the state, and all other bodies of surface water, natural or artificial, inland or coastal, fresh or salt, navigable or non-navigable, and including the beds and banks of all watercourses and bodies of surface water that are wholly or partially inside or bordering the state or subject to the jurisdiction of the state; except that waters in treatment systems which are authorized by state or federal law, regulation, or permit, and which are created for the purpose of water treatment are not considered to be waters in the state.

Surface tension – the free energy produced in a liquid surface by the unbalanced inward pull exerted by molecules underlying the layer of surface molecules.

Thermal pollution – the degradation of water quality by the introduction of a heated effluent. Primarily the result of the discharge of cooling waters from industrial processes (particularly from electrical power generation); waste heat eventually results from virtually every energy conversion.

Titrant – a solution of known strength of concentration; used in titration.

Titration – a process whereby a solution of known strength (titrant) is added to a certain volume of treated sample containing an indicator. A color change shows when the reaction is complete.

Titrator – an instrument [usually a calibrated cylinder (tube-form)] used in titration to measure the amount of titrant being added to the sample.

Total dissolved solid – the amount of material (inorganic salts and small amounts of organic material) dissolved in water and commonly expressed as a concentration in terms of milligrams per liter.

Total suspended solids (TSS) – total suspended solids in water, commonly expressed as a concentration in terms of milligrams per liter.

Toxicity – the occurrence of lethal or sublethal adverse effects on representative sensitive organisms due to exposure to toxic materials. Adverse effects caused by conditions of temperature, dissolved oxygen, or nontoxic dissolved substances are excluded from the definition of toxicity.

Transpiration – the process by which water vapor escapes from the living plant – principally the leaves – and enters the atmosphere.

Vaporization – the change of a substance from a liquid or solid state to a gaseous state.

VOC (volatile organic compound) – any organic compound that participates in atmospheric photochemical reactions except for those designated by the USEPA Administrator as having negligible photochemical reactivity.

Wastewater – the water supply of a community after it has been soiled by use.

Waste activated sludge solids (WASS) – the concentration of suspended solids in the sludge being removed from the activated sludge process.

Water cycle – the process by which water travels in a sequence from the air (condensation) to the earth (precipitation) and returns to the atmosphere (evaporation). It is also referred to as the hydrologic cycle.

Water quality standard – a plan for water quality management containing four major elements: water use, criteria to protect uses, implementation plans, and enforcement plans. An anti-degradation statement is sometimes prepared to protect existing high-quality waters.

Water quality – a term used to describe the chemical, physical, and biological characteristics of water with respect to its suitability for a particular use.

Water supply – any quantity of available water.

Waterborne disease – a disease caused by a microorganism that is carried from one person or animal to another by water.

Watershed – the area of land that contributes surface runoff to a given point in a drainage system.

Weir – a device used to measure wastewater flow.

Zone of aeration – a region in the earth above the water table. Water in the zone of aeration is under atmospheric pressure and would not flow into a well.

Zoogloeal slime – the biological slime that forms on fixed film treatment devices. It contains a wide variety of organisms essential to the treatment process.

ACRONYMS AND ABBREVIATIONS

°C	Degrees Centigrade or Celsius
°F	Degrees Fahrenheit
μ	Micron
μg	Microgram
μm	Micrometer
A-C	Alternating Current
ACEEE	American Council for an Energy Efficient Economy
Al^3	Aluminum Sulfate (or Alum)
Amp	Amperes
ANAMMOX	Anaerobic Ammonia Oxidation
APPA	American Public Power Association
ASCE	American Society of Civil Engineers
A/O	Anoxic/Oxic
A^2/O	Anaerobic/Anaerobic/Oxic
AS	Activated Sludge
ATM	Atmosphere
AT 3	Aeration Tank 3 process
ASE	Alliance to Save Energy
AWWA	American Water Works Association
BABE	Bio-Augmentation Batch Enhanced
BAF	Biological Aerated Filter
BAR	Bioaugmentation Reaeration
BASIN	Biofilm Activated Sludge Innovative Nitrification
BEP	Best Efficiency Point
bhp	Brake Horsepower
BNR	Biological Nutrient Removal
BOD	Biochemical Oxygen Demand
BOD-to-TKN	Biochemical Oxygen Demand-to-Total Kjeldahl Nitrogen Ratio
BOD-to-TP	Biochemical Oxygen Demand-to-Total Phosphorus Ratio
BPR	Biological Phosphorus Removal

CANON	Completely Autotrophic Nitrogen Removal over Nitrate
CAS	Cyclic Activated Sludge
CBOD	Carbonaceous Biochemical Oxygen Demand
CCCSD	Central Contra Costa Sanitary District
CEC	California Energy Commission
CEE	Consortium for Energy Efficiency
CFO	Cost Flow Opportunity
CFM	Cubic Feet per Minute
CFS	Cubic Feet per Second
CHP	Combined Heat and Power
Ci	Curie
COD	Chemical Oxygen Demand
COV	Coefficient of Variation
CP	Central Plant
CWSRF	Clean Water State Revolving Fund
DAF	Dissolved-Air Flotation Unit
DCS	Distributed Control System
DO	Dissolved Oxygen
DOE	Department of Energy
DON	Dissolved Organic Nitrogen
DSIRE	Database of State Incentives for Renewables and Efficiency
EBPR	Enhanced Biological Phosphorus Removal
ECM	Energy Conservation Measure
ENR	Engineering News-Record
EPA	Environmental Protection Agency
EPACT	Energy Policy Act
EPC	Energy Performance Contracting
EPRI	Electric Power Research Institute
ESCO	Energy Services Company
FeCl$_3$	Ferric Chloride
FFS	Fixed Film System
GAO	Glycogen Accumulating Organism
GBMSD	Green Bay (Wisconsin) Metropolitan Sewerage District
GPD	Gallons per Day
GPM	Gallons per minute
HCO$_3$	Bicarbonate
H$_2$CO$_3$	Carbonic acid
HDWK	Headworks
HP	Horsepower
HRT	Hydraulic Retention Time
Hz	Hertz
I&I	Inflow and Infiltration
IFAS	Integrated Fixed-Film Activated Sludge
I&C	Instrumentation and Control
IOA	International Ozone Association
IUVA	International Ultraviolet Association
kW	Kilowatt
kWh/year	Kilowatt-Hours per Year
LPHO	Low Pressure High Output
MBR	Membrane Bioreactor
M	Mega

M	Million
MG	Million Gallons
MGD	Million Gallons per Day
Mg/L	Milligrams per Liter (equivalent to parts per million)
MLE	Modified Ludzack-Ettinger process
MLSS	Mixed Liquor Suspended Solids
MPN	Most Probable Number
MW	Molecular Weight
N	Nitrogen
NAESCO	National Association of Energy Service Companies
NEMA	National Electrical Manufacturers Association
NH$_4$	Ammonium
NH$_4$-N	Ammonia Nitrogen
NL	No Limit
NPDES	National Pollutant Discharge Elimination System
NYSERDA	New York State Research and Development Authority
ORP	Oxidation-reduction Potential
O&M	Operation and Maintenance
Pa	Pascal
PAO	Phosphate Accumulating Organisms
PG&E	Pacific Gas and Electric
PID	Phased Isolation Ditch
PLC	Programmable Logic Controller
PO$_4^3$	Phosphate
POTW	Publicly Owned Treatment Works
PSAT	Pump System Assessment Tool
psi	Pounds per Square Inch
psig	Pounds per Square Inch Gauge
RAS	Return Activated Sludge
rpm	Revolutions per Minute
SCADA	Supervisory Control and Data Acquisition
SCFM	Standard Cubic Feet per Minute
SBR	Sequencing Batch Reactor
SRT	Solids Retention Time
TDH	Total Dynamic Head
TKL	Total Kjeldahl Nitrogen
TMDL	Total Maximum Daily Load
TN	Total Nitrogen
TP	Total Phosphorus
TSS	Total Suspended Solids
TVA	Tennessee Valley Authority
UV	Ultraviolet Light
UVT	UV transmittance
VFD	Variable Frequency Drive
VSS	Volatile Suspended Solids
W	Watt
WAS	Waste Activated Sludge
WEF	Water Environment Federation
WEFTEC	Water Environment Federation Technical Exhibition & Conference
WERF	Water Environment Research Foundation
WMARSS	Waco Metropolitan Area Regional Sewer System

WPCP   Water Pollution Control Plant
WRF   Water Research Foundation
WSU   Washington State University
WWTP   Wastewater Treatment Plant

WASTEWATER OPERATORS

To begin our discussion of water and wastewater operators, it is important that we point out a few significant factors.

- Employment as a wastewater operator is concentrated in local government and sanitary services companies.
- Postsecondary training is increasingly an asset as the number of regulated contaminants grows and treatment unit processes become more complex.
- Operators must pass examinations certifying that they can oversee various treatment processes.
- Plants operate 24/7; therefore, plant operators must be willing to work shifts.
- Operators have a relatively high incidence of on-the-job injuries.

DID YOU KNOW?

Water and wastewater treatment plant and system operators held about 94,000 jobs in 2004. Almost four in five operators worked for local governments. Others worked primarily for private water, sewage, and other systems utilities and for private waste treatment and disposal and waste management services companies. Private firms are increasingly providing operation and management services to local governments on a contract basis (BLS, 2006).

Properly operating a wastewater treatment and collection system usually requires a team of highly skilled personnel filling a variety of job classifications. Typical positions include plant manager/plant superintendent, plant engineer, chief operator, lead operator, operator, maintenance operator, interceptor system technicians, assistant operators, laboratory professionals, electricians, instrument technicians, recycling manager, safety and health manager, accountants, and clerical personnel, to list just a few.

Beyond the distinct job classification titles, over the years those operating wastewater treatment plants have been called by a variety of various titles. These include water jockey, practitioner of wastewater, sewer rat, or just plain wastewater operator. Based on our experience we have come up with a title that perhaps more closely characterizes what the wastewater operator really is: a Jack or Jill of all trades. This characterization seems only fitting when you consider the knowledge and skills required of operators to properly perform their assigned duties. Moreover, operating the plant or distribution/collection system is one thing; taking samples, operating equipment, monitoring conditions, and determining settings for chemical feed systems and high-pressure pumps, along with performing laboratory tests, recording the results in the plant daily operating log, is another.

It is, however, the nontypical functions, the diverse functions, and the off-the-wall functions that cause us to describe operators as Jacks or Jills of all trades. For example, in addition to their normal, routine, daily operating duties, operators may be called upon to make emergency repairs to systems (e.g., making a welding repair to a vital piece of machinery to keep the plant or unit process on line); perform material handling operations; make chemical additions to process flow; respond to hazardous materials emergencies; make confined space entries; perform site landscaping duties; and carry out several other assorted functions. Remember, the plant operator's job is to keep the plant running and to make permit. Keeping the plant running, the flow flowing, and making permit – no matter what – requires not only talent but also the performance of a wide range of functions – many of which are not called for in written job descriptions.

SETTING THE RECORD STRAIGHT

Based on experience, I have found that most people either have a preconceived notion as to what wastewater operations are all about, or they have nary a clue. Most of us understand that clean water is essential for everyday life. Moreover, we have at least a vague concept that wastewater treatment and system operations are all about the image of toilets flushed and a sewer system managed and run by a bunch of sewer rats. Others give wastewater and its treatment and the folks who treat it no thought at all (that is, unless they are irate ratepayers upset at a back-flushed toilet or the cost of wastewater service). Typically, the average person has other misconceptions about wastewater operations. For example, the average person is clueless as to the fate of wastewater. Once the toilet is flushed, it is out of sight, out of mind, and that is that – the old one and done routine. Beyond the few functions, we have pointed out to this point, what exactly is it those wastewater operators, the 50,000+ Jacks or Jills of all trades in the US do?

Operators in wastewater treatment systems control unit processes and equipment to remove or destroy harmful materials, chemical compounds, pathogens, and microorganisms from the water. They also control pumps, valves, and other processing equipment (including a wide array of computerized systems) to convey the wastewater through the various treatment processes (unit processes) and dispose of (or reuse) the removed solids (waste materials: sludge or biosolids). Operators also read, interpret, and adjust meters and gauges to make sure plant equipment and processes are working properly. They operate chemical-feeding devices, take samples of the wastewater, perform chemical and biological laboratory analyses, and adjust the amount of chemicals, such as chlorine, in the waste stream. They use a variety of instruments to sample and measure water quality, and common hand and power tools to make repairs and adjustments. Operators also make minor repairs to valves, pumps, basic electrical apparatus, and other equipment.

As mentioned, wastewater system operators increasingly rely on computers to help monitor equipment, store sampling results, make process-control decisions, schedule and record maintenance activities, and produce reports. Computer-operated automatic sampling devices are beginning to gain widespread acceptance and use, especially at larger facilities. When a system malfunction occurs, operators may use system computers to determine the cause and the solution to the problem.

THE DIGITAL WORLD

At many modern wastewater treatment plants, operators are required to perform skilled treatment plant operations work and to monitor, operate, adjust, and regulate a computer-based treatment process. In addition, the operator is also required to operate and monitor electrical, mechanical, and electronic processing, and security equipment through central and remote terminal locations. In those treatment facilities that are not completely or are partially automated or computer controlled, computers are used in other applications, such as in clerical applications and in a computerized maintenance management system (CMMS). The operator must be qualified to operate and navigate such computer systems.

Typical examples of the computer-literate operator's work (for illustrative purposes only) are provided as follows:

- Monitors, adjusts, starts, and stops automated wastewater treatment processes and emergency response systems to maintain a safe and efficient wastewater treatment operation; monitors treatment plant processing equipment and systems to identify malfunctions and their probable cause following prescribed procedures; places equipment in or out of service or redirects processes around failed equipment; following prescribed procedures, monitors and starts process-related equipment, such as boilers, to maintain process and permit objectives; refers difficult equipment maintenance problems and malfunctions to supervisor; monitors the system through a process integrated control terminal or remote station terminal to assure control devices are making proper treatment adjustments; operates the central control terminal keyboard to perform backup adjustments to such treatment processes as influent and effluent pumping, chemical feed, sedimentation, and disinfection; monitors specific treatment processes and security systems at assigned remote plant stations; observes and reviews terminal screen display of graphs, grids, charts, and digital readouts to determine process efficiency; responds to visual and audible alarms and indicators that indicate deviations from normal treatment processes and chemical hazards; identifies false alarms and other indicators that do not require immediate response; alerts remote control locations to respond to alarms indicating trouble in that area; performs alarm investigations.

- Switches over to semiautomatic or manual control when the computer control system is not properly controlling the treatment process; off-scans a malfunctioning field sensor point and inserts data obtained from field in order to maintain computer control; controls automated mechanical and electrical treatment processes through the computer keyboard when computer programs have failed; performs field tours to take readings when problems cannot be corrected through the computer keyboard; makes regular field tours of the plant to observe physical conditions; manually controls processes when necessary.
- Determines and changes the amount of chemicals to be added for the amount of wastewater or biosolids to be treated; takes periodic samples of treated residuals, biosolids processing products and byproducts;cleans water or wastewater for laboratory analysis; receives, stores, handles, and applies chemicals and other supplies needed for operation of assigned station; maintains inventory records of suppliers on hand and quantities used; prepares and submits daily shift operational reports; records daily activities in plant operation log, computer database, or from a computer terminal; changes chemical feed tanks, chlorine cylinders, and feed systems; flushes clogged feed and sampling lines.
- Notes any malfunctioning equipment; makes minor adjustments when required; reports major malfunctions to higher-level operator; enters maintenance and related task information into a computerized maintenance management system; processes work requests for skilled maintenance personnel.
- Performs routine mechanical maintenance such as packing valves, adjusting belts, and replacing shear pins and air filters; lubricates equipment by applying grease and adding oil; changes and cleans strainers; drains condensate from pressure vessels, gearboxes, and drip traps; performs minor electrical maintenance such as replacing bulbs and resetting low-voltage circuit switches; prepares equipment for maintenance crews by unblocking pipelines, pumps and isolating and draining tanks; checks equipment as part of a preventive and predictive maintenance program; reports more complex mechanical–electrical problems to supervisors.
- Responds, in a safe manner, to chlorine leaks and chemical spills in compliance with OSHA's HAZWOPER (29 CFR 1910.120) requirements and with plant-specific emergency response procedures; participates in chlorine and other chemical emergency response drills.
- Prepares operational and maintenance reports as required, including flow and treatment information; changes charts and maintains recording equipment; utilizes system and other software packages to generate reports and charts and graphs of flow and treatment status and trends; maintains workplace housekeeping.

PLANT OPERATORS AS EMERGENCY RESPONDERS

As mentioned, occasionally operators must work under emergency conditions. Sometimes these emergency conditions are operational and not necessarily life threatening. A good example occurs during a rain event when there may be a temporary loss of electrical power and large amounts of liquid waste flow into sewers, exceeding a plant's treatment capacity. Emergencies can also be caused by conditions inside a plant, such as oxygen deficiency within a confined space or exposure to toxic and/or explosive off-gases such as hydrogen sulfide and methane. To handle these conditions, operators are trained to make an emergency management response and use special safety equipment and procedures to protect co-workers, public health, the facility, and the environment. During emergencies, operators may work under extreme pressure to correct problems as quickly as possible. These periods may create dangerous working conditions; thus, operators must be extremely careful and cautious.

Operators who must aggressively respond to hazardous chemical leaks or spills (e.g., enter a chlorine gas-filled room and install chlorine repair kit B on a damaged 1-ton cylinder to stop the leak) must possess a HAZMAT Emergency Response Technician 24-hour certification. Additionally, many facilities, where elemental chlorine is used for disinfection, odor control, or other process applications, require operators to possess an appropriate certified pesticide applicator training completion certificate. Because of OSHA's specific confined space requirement whereby a standby rescue team for entrants must be available, many plants require operators to hold and maintain CPR First Aid Certification.

Note: It is important to point out that many wastewater facilities have substituted elemental chlorine with sodium or calcium hypochlorite, ozone, or ultraviolet irradiation because of the stringent requirements of OSHA's Process Safety Management Standard (29 CFR 1910.119) and USEPA's Risk Management Program.

OPERATOR DUTIES AND WORKING CONDITIONS

The specific duties of plant operators depend on the type and size of plant. In smaller plants, one operator may control all machinery, perform sampling and lab analyses, keep records, handle customer complaints, and troubleshoot and make repairs, or perform routine maintenance. In some locations, operators may handle both water treatment and wastewater treatment operations. On the other hand, in larger plants with many employees, operators may be more specialized and only monitor one-unit process (e.g., a solids handling operator who operates and monitors an incinerator). Along with treatment operators, plant staffing may include environmentalists, biologists, chemists, engineers, laboratory technicians, maintenance operators, supervisors, clerical help, and various assistants. In the US, notwithstanding a certain amount of downsizing brought on by privatization activities, employment opportunities for wastewater operators have increased in number. The number of operators has increased because of the ongoing construction of new wastewater and solids handling facilities. In addition, operator jobs have increased because of water pollution standards that have become increasingly more stringent since adoption of two major federal environmental regulations: The Clean Water Act of 1972 (and subsequent amendments), which implemented a national system of regulation on the discharge of pollutants, and the Safe Drinking Water Act of 1974.

Operators are often hired in industrial facilities to monitor or pretreat wastes before discharge to municipal treatment plants. These wastes must meet certain minimum standards to ensure that they have been adequately pretreated and will not damage municipal treatment facilities. This often means that additional qualified staff members must be hired to monitor and treat/remove specific contaminants. Complicating the problem is the fact that the list of contaminants regulated by these regulations has grown over time. Operators must be familiar with the guidelines established by federal regulations and how they affect their plant. In addition to federal regulations, operators must be aware of any guidelines imposed by the state or locality in which the treatment process operates.

Another unique factor related to wastewater operators is their working conditions. Wastewater treatment plant operators work indoors and outdoors in all kinds of weather. Operators' work is physically demanding and often is performed in unclean locations (hence, the emanation of the descriptive but inappropriate title, "sewer rat"). They are exposed to slippery walkways, vapors, odors, heat, dust, and noise from motors, pumps, engines, and generators. They work with hazardous chemicals. In wastewater plants, operators may be exposed to many bacterial and viral conditions. As mentioned, dangerous gases such as methane and hydrogen sulfide could be present, so they need to use proper safety gear.

Operators generally work a 5-day, 40-hour week. However, many treatment plants are in operation 24/7, and operators may have to work nights, weekends, holidays, or rotating shifts. Some overtime is occasionally required in emergencies. Over the years, statistical reports have related historical evidence showing that the wastewater industry is an extremely unsafe occupational field. This less-than-stellar safety performance has continued to deteriorate even in the age of the Occupational Safety and Health Act (OHS Act, 1970).

The question is why is the wastewater treatment industry on-the-job injury rate so high? Several reasons help to explain this high injury rate. First, all the major classifications of hazards exist at wastewater treatment plants (typical exception radioactivity): (1) oxygen deficiency, (2) physical injuries, (3) toxic gases and vapors, (4) infections, (5) fire, (6) explosion, and (7) electrocution. Along with all the major classifications of hazards, other factors as well cause the high incidence of injury in the wastewater industry. Some of these injuries can be attributed to: (1) complex treatment systems, (2) shift work; (3) new employees, (4) liberal workers' compensation laws, (5) absence of safety laws, and (6) absence of safe work practices and safety programs.

Experience has shown that a lack of well-managed safety programs and safe work practices are major factors causing the wastewater industry's high incidence of on-the-job injuries (Spellman, 2001).

OPERATOR CERTIFICATION AND LICENSURE

A high school diploma or its equivalency usually is required as the entry-level credential to become a wastewater treatment plant operator-in-training. Operators need mechanical aptitude and should be competent in basic mathematics, chemistry, and biology. They must have the ability to apply data to formulas of treatment requirements, flow levels, and concentration levels. Some basic familiarity with computers also is necessary because of the present trend toward computer-controlled equipment and more sophisticated instrumentation. Certain operator positions, particularly in larger cities, are covered by civil service regulations. Applicants for these positions may be required to pass a written examination testing mathematics skills, mechanical aptitude, and general intelligence.

Because treatment operations are becoming more complex, completion of an associate degree or one-year certificate program in wastewater treatment technology is highly recommended. These credentials increase an applicant's chances for both employment and promotion. Advanced training programs are offered throughout the country. They provide a good general and thorough advanced training on wastewater treatment processes, as well as basic preparation for becoming a licensed operator. They also offer a wide range of computer training courses.

New wastewater operators-in-training typically start out as attendants or assistants and learn the practical aspects of their job under the direction of an experienced operator. They learn by observing, show-and-tell, and doing routine tasks such as recording meter readings; taking samples of liquid waste and sludge; and performing simple maintenance and repair work on pumps, electrical motors, valves, and other plant or system equipment.

Larger treatment plants generally combine this on-the-job training (OJT) with formal classroom or self-paced study programs. Some large sanitation districts operate their own 3–4-year apprenticeship schools. In some of these programs, each year of apprenticeship school completed not only prepares the operator for the next level of certification or licensure but also satisfies a requirement for advancement to the next higher pay grade.

The Safe Drinking Water Act (SDWA) Amendments of 1996, enforced by the USEPA, specify national minimum standards for certification (licensure) and recertification of operators of wastewater systems. As a result, operators must pass an examination to certify that they are capable of overseeing wastewater treatment operations. There are different levels of certification depending on the operator's experience and training. Higher certification levels qualify the operator for a wider variety of treatment processes. Certification requirements vary by state and by size of treatment plants. Although relocation may mean having to become certified in a new location, many states accept other states' certifications.

To ensure currency of training and qualifications and to improve operators' skills and knowledge, most state water pollution control agencies offer ongoing training courses. These courses cover principles of treatment processes and process control methods, laboratory practices, maintenance procedures, management skills, collection system operation, general safe work practices, chlorination procedures, sedimentation, biological treatment, sludge/biosolids treatment, biosolids land application and disposal, and flow measurements. Correspondence courses covering wastewater operations and preparation for state licensure examinations are provided by various state and local agencies. Many employers provide tuition-assistance for formal college training. Whether received from formal or informal sources, training provided for or obtained by wastewater operators must include coverage of very specific subject/topic areas. Though much of their training is similar or the same, Table 10.1 lists many of the specific specialized topics wastewater operators are expected to have a fundamental knowledge of.

Note: It is important that wastewater operators must have fundamental knowledge of basic science and math operations. For many wastewater operators, crossover training or overlapping training is common practice.

TABLE 10.1
Specialized Topics for Wastewater Operators

Wastewater Math	Fecal Coliform Testing
Troubleshooting Techniques	Recordkeeping
Preliminary Treatment	Flow Measurement
Sedimentation	Sludge Dewatering
Ponds	Drying Beds
Trickling Filters	Centrifuges
Rotating Biological Contactors	Vacuum Filtration
Activated Sludge	Pressure Filtration
Chemical Treatment	Sludge Incineration
Disinfection	Land Application of Biosolids
Solids Thickening	Laboratory Procedures
Solids Stabilization	General Safety

Photo courtesy of Las Virgenes Municipal Water District

Part II of this designation and learning series is a compilation or summary of information available in many expert sources. While I have attempted to cover the major aspects of wastewater treatment system operation, let me point out that no one single discussion has *all* the information or all the answers. Moreover, because of the physical limits of any written text, some topics are only given cursory exposure and limited coverage. For those individuals seeking a more in-depth treatment of specific topics germane to wastewater treatment system operations, we recommend consulting one or more of the references listed in Table 10.2, and any of the many other outstanding references referred to throughout this text.

Note: Technomic Publishing Company originally published many of the texts listed in Table 10.2. Technomic is now a part of CRC/Lewis Publishers; the listed Technomic texts are available from CRC Press.

Treatment Process Models

Figure 10.1 shows a basic schematic or model of a wastewater treatment process that provides primary and secondary treatment using the activated sludge process. In secondary treatment (which provides BOD removal beyond what is achievable by simple sedimentation), three approaches are commonly used: trickling filter, activated sludge, and oxidation ponds. We discuss these systems in detail later in the text. We also discuss BNR (biological nutrient removal), and standard tertiary or advanced wastewater treatment. The purpose of the model shown in Figure 10.1 is to allow readers to visually follow the wastewater treatment process step by step as they are presented. The figure helps the reader understand how all the various unit processes sequentially follow and tie into each other. This format simply provides a pictorial presentation along with pertinent written information, enhancing the learning process.

Additional Wastewater Treatment Models

To demonstrate that there are more wastewater treatment models available and used in the industry than just the conventional unit processes shown in Figure 10.1, four different models are described and discussed in this section.

TABLE 10.2
Recommended Reference Material

1   *Small Water System O and M*, Kerri, K. et al. California State University, Sacramento, CA.

2   *Water Distribution System O and M*, Kerri, K. et al. California State University, Sacramento, CA.

3   *Water Treatment Plant Operation*, Vol. 1 and 2. Kerri, K. et al. California State University, Sacramento, CA.

4   *Basic Mathematics*, #3014-g. Atlanta: Centers for Disease Control.

5   *Waterborne Disease Control*. Atlanta: Centers for Disease Control.

6   *Water Fluoridation*, #3017-G. Atlanta: Centers for Disease Control.

7   *Introduction to Water Sources and Transmission* – Volume 1. Denver: American Water Works Association.

8   *Introduction to Water Treatment* – Volume 2. Denver: American Water Works Association.

9   *Introduction to Water Distribution* – Volume 3. Denver: American Water Works Association.

10 *Introduction to Water Quality Analysis* – Volume 4. Denver: American Water Works Association.

11 *Reference Handbook: Basic Science Concepts and Applications*. Denver: American Water Works Association.

12 *Handbook of Water Analysis*, 2nd ed., HACH Chemical Company, PO Box 389, Loveland, CO., 1992.

13 *Methods for Chemical Analysis of Water and Wastes*, US Environmental Protection Agency, Environmental Monitoring Systems Laboratory-Cincinnati (ESSL-CL), EPA-6000/4-79-020, Revised March 1983 and 1979 (where applicable).

14 *Standard Methods for the Examination of Water and Wastewater*, current edition. Washington, D.C.: American Public Health Association.

15 *Basic Math Concepts: For Water and Wastewater Plant Operators*. Price, J.K. Lancaster, PA: Technomic Publishing Company, 1991.

16 *Spellman's Standard Handbook for Wastewater Operators, Vol. 1, 2, & 3*. Spellman, F.R. Lancaster, PA: Technomic Publishing Company, 1994-2000.

17 *The Handbook for Waterworks Operator Certification, Vols. 1, 2, & 3*. Spellman, F. R., & Drinan, J. Lancaster, PA: Technomic Publishing Company, 2001.

18 *Fundamentals for the Water & Wastewater Maintenance Operator Series: Electricity, Electronics, Pumping, Water Hydraulics, Piping and Valves, & Blueprint Reading*. Spellman, F.R., & Drinan, J. Lancaster, PA: Technomic Publishing Company (Distributed by CRC Press, Boca Raton, FL), 2000-2002.

19 *Wastewater Treatment Plants: Planning, Design, and Operation*, 2nd ed. Qasim, S.R. Lancaster, PA: Technomic Publishing Company, Inc., 1999.

20 *Simplified Wastewater Treatment Plant Operations*. Haller, E. Lancaster, PA: Technomic Publishing Company, 1999.

21 *Operation of Wastewater Treatment Plants: A Field Study Program*, Vol. I, 4th ed. Kerri, K., et al. Sacramento, CA: California State University.

22 *Operation of Wastewater Treatment Plants: A Field Study Program*, Vol. II, 4th ed. Kerri, K., et al. Sacramento, CA: California State University.

- **Green Bay Metropolitan Sewerage District**

The Green Bay (Wisconsin) Metropolitan Sewerage District's (GBMSD) De Pere Wastewater Treatment Facility (WWTF) is an 8.0-mgd (average daily flow) two-stage activated sludge plant with biological phosphorus removal and tertiary effluent filtration. It serves the City of De Pere, portions of the village of Ashwaubenon, and portions of the towns of Lawrence, Belleview, and Hobart. GBMSD acquired ownership of the De Pere WWTF from the City of De Pere on January 1, 2008.

The original circa mid-1930's plant (a primary treatment facility with biosolids digestion) was upgraded in 1964 to an activated biosolids process, with chlorination for disinfection. In the later 1970s, there was a major upgrade to the facility (which represents the current operational scheme), including a two-stage activated biosolids process with biological phosphorus removal, tertiary filtration (gravity sand filters), solids dewatering with incineration, and liquid chlorine disinfection. Influent data for the De Pere WWTF are presented in Table 10.3.

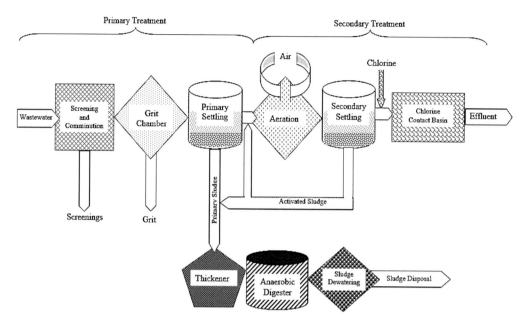

FIGURE 10.1 Schematic of a conventional wastewater treatment facility providing primary and secondary treatment using the activated sludge process.

TABLE 10.3

Profile of De Pere WWTF Influent Data (Y2009)

Parameter	Daily Average
Flow (mgd)	8
BOD (lb/day)	29,070
TSS (lb/day)	18,587
Ammonia-N (lb/day)	Not monitored
Phosphorus (lb/day)	307.5

In 1997, additional upgrades to the facility were initiated, beginning with UV disinfection replacing the liquid chlorine system. The chlorine disinfection system is currently maintained as a back-up system. Several other major upgrades included: replacement of the coarse influent screen with fine screens (1998–1999), renovation of the multimedia tertiary filtration system to a signal media US Filter Multiwash® air scour system (1999–2000), and a solids handling upgrade which included installation of two gravity belt thickeners (replacing dissolved air floatation) and the addition of two filter presses (2001–2002). Figure 10.2 presents the process flow diagram for the GBMSD – De Pere WWTF, a two-stage activated biosolids treatment plant (on line 1978 to present) with tertiary filtration and design flows as follows:

- Average Dry – 8.5 MGD
- Design Flow – 14.2 MGD
- Maximum Hourly Dry – 23.8 MGD
- Maximum Hourly Wet – 30.0 MGD

Influent to the plant undergoes fine screening and is subsequently pumped to preliminary treatment (grit removal followed by grease removal, utilizing two 50 feet × 50 feet clarifiers with grease/scum

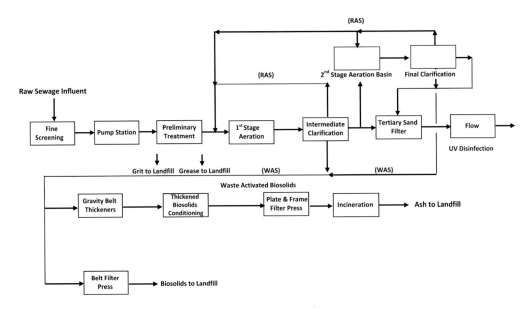

FIGURE 10.2 De Pere WWT Process Flow Diagram (Biological Treatment). Source: EPA 832-R-10-00 (2010). Evaluation of Energy Conservation Measures. Washington, D.C.: Environmental Protection Agency.

collection). The influent pump station consists of four 150 hp, 10 mgd pumps. Screenings are disposed in a landfill. Grit, oil, and grease removed in preliminary treatment units arc also disposed in a landfill. Biological treatment is conducted in two serial states, each with a 1.1-mgd anaerobic zone (for phosphorus removal) followed by a 2.2-mgd aeration zone. Approximately 100% of the mixed liquor suspended solids from the aeration zone is recycled to the anoxic zone. Aeration is provided by six (each), 6,000 standard cubic feet per minute (scfm), 330 hp turbo blowers for the first stage aeration process and thereafter (each), 4,000 scfm, 250 hp multistage centrifugal blowers for the second stage aeration process.

The first stage biological treatment is followed by clarification [two each, 100-foot diameter, 13.7-foot side water depth clarifiers (one online for each aeration basin)]. Clarifier effluent from the first stage biological treatment process can be further treated in the second stage treatment process. However, all wastewater is currently treated only in the first stage biological process. The second stage of biological treatment is not currently utilized since it is not required to achieve discharge compliance. Biological treatment is followed by their 125-foot diameter, 10.9-foot side water depth clarifiers. Clarifier effluent is polished by tertiary sand filtration and disinfected using UV prior to discharge. During periods of high flow, UV disinfection is supplemented by disinfection with liquid chlorine.

Clarifier underflow (WAS – waste activated sludge) from biological treatment undergoes one of two dewatering processes. Approximately 75% of the WAS undergoes thickening (two each, 2-m gravity belt thickeners), chemical conditioning (lime and ferric chloride), dewatering (two each, 1.5 m × 2 m plate and frame filter press), and incineration (18.75-foot diameter, 7 hearth, 7,500 lb/hr multiple hearth incinerator). The incinerator ash is disposed of in a landfill. The balance of the WAS is chemically conditioned with polymer and dewatered in two each, 2-m belt filter presses. The dewatered sludge is disposed of in a landfill. Filtrate from sludge thickening and dewatering operations is returned to first stage biological treatment. The most recent upgrade (2003–2004) replaced the facility's first stage treatment centrifugal blowers with high-speed, magnetic turbo blowers, the first installation of this new, energy-efficient technology in the country. Because the second stage aeration process is currently not utilized, only the first stage process blowers were replaced under the ECM (Energy Conservation Measure) project.

Reference: Shumaker, G. (2007). High Speed Technology Brings Low Costs, Water and Wastes Digest, August.

- **Sheboygan Regional Wastewater Treatment Plant**

The Sheboygan Regional Wastewater Treatment Plant in Sheboygan, MI, is an 11.8-mgd (average daily flow) activated biosolids plant with biological phosphorus removal. The Sheboygan Regional Wastewater Treatment Plant (WWTP) serves approximately 68,000 residential customers in the cities of Sheboygan and Sheboygan Falls, the Village of Kohler, and the towns of Sheboygan, Sheboygan Falls, and Wilson. The plant was originally constructed in 1982 as a conventional activated biosolids plant using turbine aerators with sparger rings. In 1990, the plant was upgraded to include a fine bubble diffused air system with positive displacement blowers. From 1997 to 1999, additional improvements were made to the facility to implement biological nutrient removal and to the bar screens, grit removal facilities, biosolids storage tanks, and the primary and secondary clarifiers. The plant currently operates as an 18.4 MGD biological nutrient removal plant with fine screens, grit removal, primary clarification, biological nutrient removal, activated biosolids thickening, and liquid (6% solids) biosolids storage. Table 10.4 provides average daily influent data for the plant. Figure 10.3 provides a process flow diagram of the plant treatment scheme.

Influent to the plant goes through two automatic self-cleaning fine screens. A 20-feet-diameter cyclone separator removes grit before the wastewater enters primary clarification. Primary clarification is provided by four primary clarifiers. Secondary biological treatment is conducted in six basins. The first two basins are anaerobic to provide phosphorus removal. They are configured with baffles in an "N" pattern. The remaining four basins are currently aerated using two Turblex blowers. Following aeration, secondary clarification is provided by four clarifiers. Return activated sludge (RAS) from the clarifiers is sent to the anaerobic zone. A portion of the RAS is conveyed upstream of the primary clarifier. Plant effluent is disinfected with chlorine and is then dechlorinated before discharge into Lake Michigan. The combined primary and secondary biosolids underflow from the primary clarifier (waste sludge – biosolids) is sent to three primary anaerobic digesters. From the primary digesters the biosolids flows to a single secondary anaerobic digester. Methane from the digesters is used to provide heat to the digesters as well as fuel for ten 30-kW microturbines that provide electricity to the plant. Two belt thickeners (one at 2 meters and one at 3 meters) increase the solids content of the digested biosolids from 2.5% to 6% solids. Digested, thickened biosolids is held in two storage tanks before being land applied.

- **Big Gulch Wastewater Treatment Plant**

The Big Gulch WWTP, owned and operated by the Mukilteo Water and Wastewater District (Washington), is a 1.5-mgd (average daily flow) oxidation ditch plant operating two parallel oxidation ditches. Ditch A treats approximately 40% of the plant flow and Ditch B treats approximately

TABLE 10.4

Profile of Sheboygan WWTP Influent Data (Y2009)

Parameter	Daily Average
Flow (mgd)	11.78
BOD (lb/day)	175
TSS (lb/day)	203
Ammonia-N (lb/day)	Not monitored
Phosphorus (lb/day)	5.7

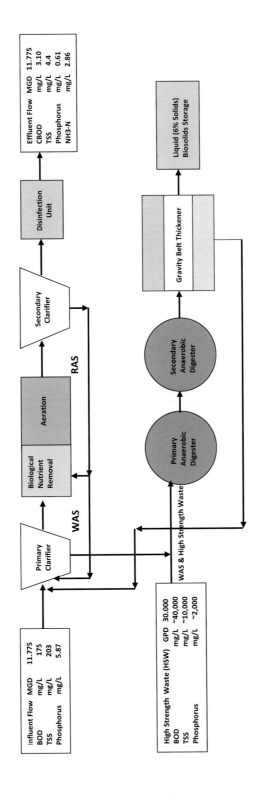

FIGURE 10.3 Sheboygan Wastewater Treatment Plant Process Flow Diagram. Source: EPA, 2010. Adapted from Evaluation of Energy Conservation Measures. EPA 832-R-10-005. Washington, D.C.: Environmental Protection Agency.

60% of the flow. The Big Gulch WWTP provides wastewater treatment service for 22,455 people residing in portions of the City of Mukilteo and Snohomish County (Washington). Originally constructed in 1970, the WWTP consisted of a coarse bar screen and single oxidation ditch using brush rotor aerators, followed by a secondary clarifier and chlorine disinfection.

Between 1989 and 1991, the Big Gulch WWTP underwent significant upgrades including the following:

- New headworks with a girt removal channel
- Influent screw pumps
- Selector tank
- Second oxidation ditch
- Third secondary clarifier
- Aerobic biosolids holding tanks within a rotary drum thickener
- Biosolids return piping
- Scum and waste active biosolids pumps
- Biosolids pumps
- Biosolids dewatering belt filter press
- Chlorine contact chamber

After the 1991 facility upgrade, the following upgrades to the treatment plant were implemented:

- Influent screening (perforated-plate fine screens)
- Submersible mixers (in the oxidation ditches)
- UV disinfection (replacing chlorine disinfection)

To address a need for additional oxidation ditch aeration capacity to handle intermittent increases in BOD loading, the aeration system in both ditches was upgraded with fine bubble diffusers and automatically controlled turbo blowers. Influent data for the Big Gulch WWTP are presented in Table 10.5.

Figure 10.4 presents the process flow diagram for the Big Gulch WWTP, an activated biosolids treatment plant with UV disinfection. Influent to the plant passes through a perforated-plate mechanical fine screen (rate capacity of 6.5 mgd) into a gravity grit channel. Effluent from the grit removal system is returned to the headworks, and grit is sent to the dumpsters. Degritted influent, combined with return active sludge (RAS) from the secondary clarifiers and filtrate from the sludge-dewatering belt filter press, is lifted to the selector mixing basin using the two influent screw lift pumps (3.83 MGD capacity, each). Selector mixing basin effluent is conveyed to the oxidation ditches via overflow channels equipped with adjustable weir gates to distribute the flow to the ditches (40% to Oxidation Ditch A and 60% to Oxidation Ditch B). The two oxidation ditches, operating in parallel and providing a combined 18-hour hydraulic residence time, are followed by three secondary clarifiers. Effluent from the secondary clarifiers is conveyed to the UV disinfection system. The UV system consists of 96 lamps and provides 35 mJ/cm^2 (energy) at a peak flow of

TABLE 10.5
Profile of Big Gulch WWTP Influent Data (Y2004 to Y2010)

Parameter	Average	Minimum	Maximum
Flow (mgd)	1.68	1.21	2.40
CBOD (mg/L)	217	116	462
TSS (mg/L)	255	131	398

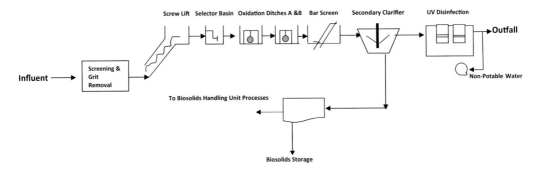

FIGURE 10.4   Simplified schematic flow diagram of Big Gulch WWTP.

TABLE 10.6
Profile of the City of Bartlett WWTP #1 Influent Data (Y2009)

Parameter	Daily Average
Flow (mg/L)	1.0
BOD (mg/L)	130
TSS (mg/L)	180
Ammonia-N (mg/L)	Not monitored
TKN (mg/L)	41
Phosphorus (mg/L)	6

8.7 MGD (based on 60% UV transmittance). The UV disinfection system produces an effluent with fecal coliform counts below the facility's permit limit of 200 colonies/100 ml (monthly average).

Waste activated biosolids and scum from the secondary clarifiers are transferred via a rotary lobe pump to a pair of two-cell aerobic biosolids holding tanks for aerobic digestion, producing Class-B biosolids. In 2006, the aerobic biosolids digestion system was upgraded with fine-bubble air diffusers and positive displacement blowers. Biosolids is thickened through either settling in the aerobic biosolids holding tanks or through rotary drum thickening. In 2007, the rotary drum biosolids thickener was installed to increase digestion capacity. Digested biosolids is dewatered using a gravity belt dewatering press, and the dewatered biosolids is transported for land application.

References: Bridges, T.G. (2010). Phone Conversation-Plant Manager, Big Gulch WWTP. Mukilteo, Washington. As reported in EPA (2010) Evaluation of Energy Conservation Measures, EPA 832-R-10-005. Washington, D.C.: Environmental Protection Agency.

Gray & Osborne, Inc. (2008). Wastewater Treatment Plant Capacity Study and Engineering Report. Seattle, WA.

- **City of Bartlett Wastewater Treatment Plant**

The City of Bartlett, TN, WWTP is a 1.0 mgd (average daily flow) secondary facility utilizing two mechanically aerated oxidation ditches to provide secondary treatment. Each of the aeration basins is equipped with three rotor aerators. The City of Bartlett's Wastewater Treatment Plant (WWTP) #1, located in West Tennessee near Memphis, serves approximately 24,000 residential customers and one school. One hundred percent of the plant influent is domestic wastewater. The facility was originally commissioned in 1994 as a 0.5 mgd aerated lagoon and has undergone three major expansions (in 1994, 2003, 2005) to meet the city's growing population. In 1993, the facility was upgraded to a secondary treatment facility (one oxidation ditch and secondary clarification). In 2003, the facility was upgraded with solids handling (aerobic digester and belt filter press). In 2005, a second oxidation ditch was added. Influent data for the City of Bartlett WWTP #1 is presented in Table 10.6.

Plant influent undergoes mechanical screening followed by biological treatment in two mechanically aerated oxidation ditches. Each oxidation ditch is equipped with three 60 HP rotor aerators. Oxidation ditch effluent undergoes secondary clarification followed by UV disinfection prior to discharge to the Loosahatchie River. Waste biosolids from the secondary clarifiers undergoes aerobic digestion. Digested biosolids is dewatered in a belt filter press and is then land applied as an agricultural soil amendment and fertilizer.

REVIEW QUESTIONS

Matching: Match the definitions in Part A with the terms in Part B.

Part A

1. A nonchemical turbidity removal layer in a slow sand filter. _____
2. Region in earth (soil) above the water table. _____
3. Compound associated with photochemical reaction. _____
4. Oxygen used in water rich inorganic matter. _____
5. A stream that receives water from the zone of saturation. _____
6. The addition of water into a groundwater system. _____
7. The natural water cycle. _____
8. Present in intestinal tracts and feces of animals and humans. _____
9. Discharge from a factory or municipal sewage treatment plant. _____
10. Common to fixed film treatment devices. _____
11. Identified water that is safe to drink. _____
12. The capacity of soil to hold water. _____
13. Used to measure acidity and alkalinity. _____
14. Rain mixed with sulfur oxides. _____
15. Enrichment of water bodies by nutrients. _____
16. A solution of known strength of concentration. ___
17. Water lost by foliage. _____
18. Another name for a wastewater pumping station. _____
19. Plants and animals indigenous to an area. _____
20. The amount of oxygen dissolved in water. _____
21. A stream that flows continuously. _____
22. A result of excessive nutrients within a water body. _____
23. Change in groundwater head over a given distance. _____
24. Water trapped in sedimentary rocks. _____
25. Heat capacity of a material per unit mass. _____ __
26. A compound derived from material that once lived. _____

Part B

a. pH

b. algae bloom

c. zone of aeration

d. hydrological cycle

e. point source pollution

f. perennial

g. organic

h. connate water

i. fecal coliform

j. BOD

k. field capacity

l. transpiration

m. biota

n. specific heat

o. Schmutzdecke

p. recharge

q. Zoogloeal slime

r. reutrophication

s. gaining

t. VOC

u. potable

v. acid rain

w. titrant

x. lift station

y. DO

z. hydraulic gradient

Part C

27. Which of the following is a factor in the high injury rates in wastewater treatment operations?
    a. Video games
    b. Exposure to varying weather conditions
    c. Bad luck
    d. Plane crashes

28. Why are treatment plants upgrading to computerized operations?
    a. Saves money
    b. Right thing to do
    c. Keeps regulators at bay
    d. Makes the workers happy

29. What is the benefit of a computerized maintenance management system?
    a. Ease of scheduling maintenance
    b. Looks professional

    c. Keeps regulators happy

    d. Makes radioing easier

30. Why do some treatment plants use and call 911 in case of HAZMAT incidents?

    a. Replaces the need to train plant personnel as HAZMAT responders

    b. Saves money

    c. Reduces need for more employees

    d. All of the above

ANSWER KEY

PART A: ANSWERS

1. A nonchemical turbidity removal layer in a slow sand filter. **o**

2. Region in earth (soil) above the water table. **c**

3. Compound associated with photochemical reaction. **t**

4. Oxygen used in water rich inorganic matter. **j**

5. A stream that receives water from the zone of saturation. **s**

6. The addition of water into a groundwater system. **p**

7. The natural water cycle. **d**

8. Present in intestinal tracts and feces of animals and humans. **i**

9. Discharge from a factory or municipal sewage treatment plant. **e**

10. Common to fixed film treatment devices. **q**

11. Identified water that is safe to drink. **u**

12. The capacity of soil to hold water. **k**

13. Used to measure acidity and alkalinity. **a**

14. Rain mixed with sulfur oxides. **v**

15. Enrichment of water bodies by nutrients. **r**

16. A solution of known strength of concentration. **w**

17. Water lost by foliage. **l**

18. Another name for a wastewater pumping station. **x**

19. Plants and animals indigenous to an area. **m**

20. The amount of oxygen dissolved in water. **y**

21. A stream that flows continuously. **f**

22. A result of excessive nutrients within a water body. **b**

23. Change in groundwater head over a given distance. **z**

24. Water trapped in sedimentary rocks. **h**

25. Heat capacity of a material per unit mass. **n**

26. A compound derived from material that once lived. **g**

PART C: ANSWERS

27. Which of the following is a factor in the high injury rates in wastewater treatment operations?

    a. Video games

    b. **Exposure to varying weather conditions**

   c. Bad luck

   d. Plane crashes

28. Why are treatment plants upgrading to computerized operations?

   a. **Saves money**

   b. Right thing to do

   c. Keeps regulators at bay

   d. Makes the workers happy

29. What is the benefit of a computerized maintenance management system?

   a. **Ease of scheduling maintenance**

   b. Looks professional

   c. Keeps regulators happy

   d. Makes radioing easier

30. Why do some treatment plants use and call 911 in case of HAZMAT incidents?

   a. Replaces the need to train plant personnel as HAZMAT responders

   b. Saves money

   c. Reduces need for more employees

   d. **All of the above**

BIBLIOGRAPHY

BLS. (2006). *Water and Liquid Waste Treatment Plant and System Operators*. Accessed 09/30/17 @ http://www.bls.gov.oco/229.htm.

Brown, C. (1983). Seabces in Service. In Proceedings, Conference on Coastal Structures '83. American Society of Civil Engineers. Arlington, VA, March 11, 1983.

Brunner, C.R. (1980). *Design of Sewage Incinerator Systems*. Park Ridge, NJ: Noyes Data Corporation.

Brunner, C.R. (1984). *Incinerator Systems: Selection and Design*. New York: Van Nostrand Reinhold Company, Inc.

Coker, C.S., Walden, R., Shea, T.G., & Brinker, M. (1991). Dewatering Municipal Wastewater Sludges for Incineration. *Water Environment & Technology* March, 1991, pp 63–67.

Ekster, A. (2009). Optimization of Pump Station Operation Saves Energy and Reduces Carbon Footprint. In Proceedings of 82nd WEFTEC, Conference and Exposition, Orlando.

Haller, E.J. (1995). *Simplified Wastewater Treatment Plant Operations*. Lancaster, PA: Technomic Publishers, Inc.

Hardaway, C.S., Thomas, G.R., Unger, M.A., Greaves, J., & Rice, G. (1991). Seabees Monitoring Project James River Estuary, VA. *Contract Report to Center of Innovative Technology*. Herndon, Virginia Beach, VA: HRSD.

Hardaway, C.S., Thomas, G.R., Unger, M.A., & Smith, C.L. (1994). Seabee Monitoring Project James River Estuary, Virginia. In *Contract Update Report for Center for Innovative Technology*. Herndon, Virginia Beach, VA: HRSD.

Lemma, I. T., Colby, S., & Herrington, T. (2009). Pulse Aeration of Secondary Aeration Tanks Holds Energy saving Potential without Compromising Effluent Quality. In Proceedings of 82nd WEFTEC, Conference and Exposition, Orlando.

Lester, F.N. (1992). Sewage and Sewage Sludge Treatment. In R. Harrison (Ed.) *Pollution: Causes, Effects, & Control*, pp. 33–62. London: Royal Society of Chemistry.

Lewis, F.M., Haug, R.T., & Lundberg, L.A. (1988). Control of Organics, Particulates and Acid Gas Emissions from Multiple Hearth and Fluidized Bed Sludge Incinerators. Paper presented at 61st Annual Conference and Exposition, WPCF, Dallas, TX.

Metcalf & Eddy, Inc. (2002). *Wastewater Engineering: Treatment, Disposal & Reuse*, 4th ed. New York: McGraw-Hill.

Metcalf & Eddy, Inc. (2003). *Wastewater Engineering: Treatment, Disposal, Reuse*, 5th ed. New York: McGraw-Hill.

OSH Act (1970). *Occupational Safety and Health Act*. Washington, DC: OSHA. U.S. Dept of Labor.

Outwater, A.B. (1994). *Reuse of Sludge and Minor Wastewater Residuals*. Boca Raton, FL: CRC Press.

Pearce, F. (2006). *When the Rivers Run Dry*. Boston, MA: Beacon Press.

Peavy, S., Rowe, D.R., & Tchobanglous, G. (1985). *Environmental Engineering*. New York: McGraw-Hill.

Pielou, E.C. (1998). *Fresh Water.* Chicago: University of Chicago.

Sarai, D.S. (2006). *Water Treatment Made Simple for Operators.* New York: Wiley.

Sinaki, M., Yerrapotu, B., Colby, S., & Lemma, I., (2009). Permit Safe, Energy Smart: Greening Wastewater Treatment Plant Operations. In Proceedings of 82 WEFTEC, Conference and Exposition, Orlando.

Spellman, F.R. (2001). *Safe Work Practices for Wastewater Treatment Plants.* Boca Raton, FL: CRC Press.

Spellman, F.R. (2007). *The Science of Water*, 2nd ed. Boca Raton, FL: CRC Press.

Spellman, F.R. (2013). *Spellman's Standard Handbook for Water and Wastewater Operators.* Boca Raton, FL: CRC Press.

Unger, M.A., & Smith, C.L. (1994). Seabees Monitoring Project James River Estuary, Virginia. In *An Evaluation of the Long Term.* Little Creek, VA: United States Navy.

USEPA (1978). *Operations Manual: Sludge Handling & Conditioning.*

USEPA (1979). *Common Environmental Terms: A Glossary.*

USEPA (1982a). *Dewatering Municipal Wastewater Sludges.* EPA-625/1.82-014. Cincinnati, Ohio: Center for Environment Research Information.

USEPA (1982b). *Resource Conservation & Recovery Act.* Washington, DC: USEPA.

USEPA (1990). *National Sewage Sludge Survey.* Washington, DC: USEPA.

USEPA (1993a). *40 CFR Part 503 Regulation.* Washington, DC: USEPA.

USEPA (1993b). *Standards for Use or Disposal of Sewage Sludge*, pp 9248–9415. Washington, DC: USEPA.

Vesilind, P.A. (1980). *Treatment and Disposal of Wastewater Sludges.* Ann Arbor, MI: Ann Arbor Science.

11 Basic Wastewater Microbiology

© Las Vegas Valley Water District and used with permission.

Scientists picture the primordial Earth as a planet washed by a hot sea and bathed in an atmosphere containing water vapor, ammonia, methane, and hydrogen. Testing this theory, Stanley Miller at the University of Chicago duplicated these conditions in the laboratory. He distilled seawater in a special apparatus, passed the vapor with ammonia, methane, and hydrogen through an electrical discharge at frequent intervals, and condensed the "rain" to return to the boiling seawater. Within a week, the seawater had turned red. Analysis showed that it contained amino acids, which are the building blocks of protein substances.

Whether this is what really happened early in the Earth's history is not important; the experiment demonstrated that the basic ingredients of life could have been made in some such fashion, setting the stage for life to come into existence in the sea The saline fluids in most living things may be an inheritance from such early beginnings.

– Kemmer (1979)

LEARNING OBJECTIVES

After studying this chapter, you should be able to:

- Discuss how microorganisms in wastewater can cause disease.
- Discuss the importance of microorganisms in biological treatment.
- Define microbiology.
- Discuss waterborne pathogenic organisms.
- Describe algae.

DOI: 10.1201/9781003207146-16

- Discuss and describe bacteria.
- Define the biogeological cycles.
- Discuss coliforms.
- Describe fungi.
- Describe human cells and their organelles.
- Describe virus.
- Discuss microorganism classification.
- Discuss microorganism differentiation.
- Describe the structure of the human cell.
- Discuss bacteria growth factors.
- Describe the destruction of bacteria.
- Describe waterborne bacteria.
- Identify protozoans contained in wastewater.
- Describe *Giardia lamblia* and giardiasis.
- Describe *Cryptosporidium* contained in wastewater.
- Describe nematodes contained in wastewater.
- Describe Cyclospora.
- Describe aerobic process.
- Describe anaerobic process.
- Describe anoxic treatment process.
- Describe biogeochemical cycles.
- Describe carbon cycle.
- Describe sulfur cycle.
- Describe nitrogen cycle.
- Describe phosphorus cycle.

KEY DEFINITIONS

Algae – simple plants, many microscopic, containing chlorophyll. Freshwater algae are diverse in shape, color, size, and habitat. They are the basic link in the conversion of inorganic constituents in water into organic constituents.

Algal bloom – sudden spurts of algal growth, which can affect water quality adversely and indicate potentially hazardous changes in local water chemistry.

Anaerobic – able to live and grow in the absence of free oxygen.

Autotrophic organisms – produce food from inorganic substances.

Bacteria – single-cell, microscopic living organisms (single-celled microorganisms) that possess rigid cell walls. They may be aerobic, anaerobic, or facultative; they can cause disease and some are important in pollution control.

Biogeochemical cycle – the chemical interactions between the atmosphere, hydrosphere, and biosphere.

Coliform organism – microorganisms found in the intestinal tract of humans and animals. Their presence in water indicates fecal pollution and potentially adverse contamination by pathogens.

Denitrification – the anaerobic biological reduction of nitrate to nitrogen gas.

Fungi – simple plants lacking in the ability to produce energy through photosynthesis.

Heterotrophic organism – organisms that are dependent on organic matter for foods.

Prokaryotic cell – the simple cell type, characterized by the lack of a nuclear membrane and the absence of mitochondria.

Virus – the smallest form of microorganisms capable of causing disease.

OVERVIEW

Microorganisms are significant in wastewater because of their role in disease transmission and because they are the primary agents of biological treatment. Thus, wastewater practitioners must have considerable knowledge of the microbiological characteristics of wastewater. Simply put, wastewater operators cannot fully comprehend the principles of effective wastewater treatment without knowing the fundamental factors concerning microorganisms and their relationships to one another, their effect on the treatment process, and their impact on consumers, animals, and the environment.

Wastewater operators must know the principal groups of microorganisms found in water supplies (surface and groundwater) and wastewater as well as those that must be treated (pathogenic organisms) and/or removed or controlled for biological treatment processes; they must be able to identify the organisms used as indicators of pollution and know their significance; and they must know the methods used to enumerate the indicator organisms. This chapter provides microbiology fundamentals specifically targeting the needs of water and wastewater specialists.

Note: To have microbiological activity, the body of wastewater must possess the appropriate environmental conditions. Most wastewater treatment processes, for example, are designed to operate using an aerobic process.

The conditions required for aerobic operation are:

- Sufficient free, elemental oxygen;
- Sufficient organic matter (food);
- Sufficient water;
- Enough nitrogen and phosphorus (nutrients) to permit oxidation of the available carbon materials;
- Proper pH (6.5–9.0); and
- Lack of toxic materials.

MICROBIOLOGY: WHAT IS IT?

Biology is generally defined as the study of living organisms (i.e., the study of life). *Microbiology* is a branch of biology that deals with the study of microorganisms so small that they must be studied under a microscope. Microorganisms of interest to the water and wastewater operator include bacteria, protozoa, viruses, algae, and others.

Note: The science and study of bacteria is known as bacteriology (discussed later).

Waterworks operators' primary concern is how to control microorganisms that cause waterborne diseases – waterborne pathogens – to protect the consumer (human and animal). Wastewater operators have the same microbiological concerns as water operators, but instead of directly purifying water for consumer consumption, their focus is on removing harmful pathogens from the waste stream before outfalling it to the environment. To summarize, water and wastewater operators' reliance on knowledge of microbiological principles is described in the following (Spellman, 1996).

- Water operators are concerned with water supply and water purification through a treatment process. In treating water, the primary concern is producing potable water that is safe to drink (free of pathogens) with no accompanying offensive characteristics such as foul taste and odor. The treatment operator must possess a wide range of knowledge in order to correctly examine water for pathogenic microorganisms and to determine the type of treatment necessary to ensure the water quality of the product, potable water, meets regulatory requirements.

• Wastewater operators are also concerned with water quality. However, they are not as concerned as water specialists with the total removal or reduction of most microorganisms. The wastewater treatment process benefits from microorganisms that act to degrade organic compounds and, thus, stabilize organic matter in the waste stream. Thus, wastewater operators must be trained to operate the treatment process in a manner that controls the growth of microorganisms and puts them to work. Moreover, to fully understand wastewater treatment, it is necessary to determine which microorganisms are present and how they function to break down components in the wastewater stream. Then, of course, the operator must ensure before outfalling or dumping treated effluent into the receiving body that the microorganisms that worked so hard to degrade organic waste products (especially the pathogenic microorganisms) are not sent from the plant with effluent as viable organisms.

WASTEWATER MICROORGANISMS

As mentioned, microorganisms of interest to wastewater operators include bacteria, protozoa, rotifers, viruses, algae, fungi, and nematodes. These organisms are the most diverse group of living organisms on earth and occupy important niches in the ecosystem. Their simplicity and minimal survival requirements allow them to exist in diverse situations. Because they are a major health concern, wastewater treatment specialists are mostly concerned about how to control microorganisms that cause *waterborne diseases*. These waterborne diseases are carried by *waterborne pathogens* (i.e., bacteria, virus, protozoa, etc.).

The focus of wastewater operators, on the other hand, is on the millions of organisms that arrive at the plant with the influent. Most of these organisms are nonpathogenic and beneficial to plant operations. From a microbiological standpoint, the predominant species of microorganisms depend on the characteristics of the influent, environmental conditions, process design, and the mode of plant operation. There are, however, pathogenic organisms that may be present. These include the organisms responsible for diseases, such as typhoid, tetanus, hepatitis, dysentery, gastroenteritis, and others.

To understand how to minimize or maximize growth of microorganisms and control pathogens, one must study the structure and characteristics of the microorganisms. In the sections that follow, we will look at each of the major groups of microorganisms (those important to wastewater operators) in relation to their size, shape, types, nutritional needs, and control.

Note: Koren (1991) pointed out that, in a water environment, water is not a medium for the growth of microorganisms but is instead a means of transmission (that is, it serves as a conduit; hence, the name *waterborne*) of the pathogen to the place where an individual is able to consume it and there start the outbreak of disease. This is contrary to the view taken by the average person. That is, when the topic of waterborne disease is brought up, we might mistakenly assume that waterborne diseases are at home in water. Nothing could be further from the truth.

A water-filled ambience is not the environment in which the pathogenic organism would choose to live, that is, if it has such a choice. The point is that microorganisms do not normally grow, reproduce, languish, and thrive in watery surroundings. Pathogenic microorganisms temporarily residing in water are simply biding their time, going with the flow, waiting for their opportunity to meet up with their unsuspecting host or hosts. To a degree, when the pathogenic microorganism finds its host or hosts, it is finally home or may have found its final resting place (Spellman, 1997).

MICROORGANISM CLASSIFICATION AND DIFFERENTIATION

The microorganisms we are concerned with are tiny organisms that make up a large and diverse group of free-living forms; they exist either as single cells, cell bunches, or clusters. Found in abundance almost anywhere on earth, most microorganisms are not harmful. Many microorganisms, or microbes, occur as single cells (unicellular); others are multicellular; and still others, viruses, do not have a true cellular appearance. A single microbial cell, for the most part, exhibits the characteristic features common to other biological systems, such as metabolism, reproduction, and growth.

CLASSIFICATION

Greek scholar and philosopher Aristotle classified animals based on fly, swim, and walk/crawl/run. For centuries thereafter, scientists simply classified the forms of life visible to the naked eye as either animal or plant. When we started to have trouble differentiating microorganisms, this classification had to be changed. In 1735, the Swedish naturalist Carolus Linnaeus organized much of the current knowledge about living things.

The importance of organizing or classifying organisms cannot be overstated, for without a classification scheme, it would be difficult to establish a criterion for identifying organisms and to arrange similar organisms into groups. Probably the most important reason for classifying organisms is to make things less confusing (Wistriech & Lechtman, 1980). Linnaeus was quite innovative in the classification of organisms. One of his innovations is still with us today: the binomial system of nomenclature. Under the binomial system, all organisms are generally described by a two-word scientific name, the genus and species. Genus and species are groups that are part of a hierarchy of groups of increasing size, based on their taxonomy.

This hierarchy follows:

Kingdom
Phylum
Class
Order
Family
Genus
Species

Using this system, a fruit fly might be classified as:

Animalia
Arthropoda
Insecta
Diptera
Drosophilidae
Drosophila
melanogaster

This means that this organism is the species melanogaster in the genus *Drosophila* in the Family Drosophilidae in the Order Diptera in the Class Insecta in the Phylum Arthropoda in the Kingdom Animalia.

To further illustrate how the hierarchical system is exemplified by the classification system, the standard classification of the mayfly is provided below:

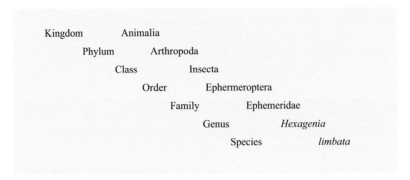

Kingdom   Animalia

Phylum   Arthropoda

Class   Insecta

Order   Ephermeroptera

Family   Ephemeridae

Genus   *Hexagenia*

Species   *limbata*

Utilizing this hierarchy and Linnaeus's binomial system of nomenclature, the scientific name of any organism (as stated previously) includes both the generic and specific names. In the above instances, to uniquely name the species it is necessary to supply the genus and the species, *Drosophila melanogaster* (i.e., the fruit fly) and *Hexagenia limbota* (mayfly).

As shown, the first letter of the generic name is usually capitalized, hence, for example, *E. coli* indicates that coli is the species and Escherichia (abbreviated to E.) is the genus. The largest, most inclusive category – the kingdom – is plant. The names are always in Latin, so they are usually printed in italics or underlined. Some organisms also have English common names.

Microbe names of interest in wastewater treatment include:

- *Escherichia coli* – a coliform bacterium
- *Salmonella typhi* – the typhoid bacillus
- *Giardia lamblia* – a protozoan
- *Shigella* spp.
- *Vibrio cholerae*
- *Campylobacter*
- *Leptospira* spp.
- *Entamoeba histolytica*
- *Crytosporidia*

Note: *Escherichia coli* is commonly known as simply *E. coli*, while *Giardia lamblia* is usually referred to by only its genus name, *Giardia*.

Generally, we use a simplified system of microorganism classification in water science, breaking down classification into the kingdoms of animal, plant, and Protista. As a rule, the animal and plant kingdoms contain all multicellular organisms, and the protists contain all single-cell organisms. Along with microorganism classification based on the animal, plant, and Protista kingdoms, microorganisms can be further classified as being *eukaryotic* or *prokaryotic* (see Table 11.1).

DIFFERENTIATION

Differentiation among the higher forms of life is based almost entirely upon morphological (form or structure) differences. However, differentiation (even among the higher forms) is not as easily accomplished as you might expect because normal variations among individuals of the same species occur frequently. Because of this variation even within a species, securing accurate classification when dealing with single-celled microscopic forms that present virtually no visible structural differences becomes extremely difficult. Under these circumstances, considering physiological,

TABLE 11.1

Simplified Classification of Microorganisms

Kingdom	Members	Cell Classification
Animal	Rotifers	
	Crustaceans	
	Worms and larvae	Eukaryotic
Plant	Ferns	
	Mosses	
Protista	Protozoa	
	Algae	
	Fungi	
	Bacteria	Prokaryotic
	Lower algae forms	

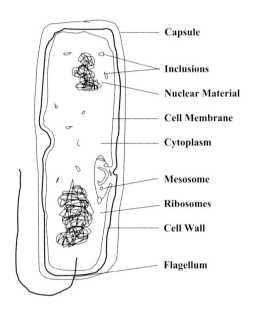

FIGURE 11.1 Bacterial cell.

cultural, and chemical differences are necessary, as well as structure and form. Differentiation among the smaller groups of bacteria is based almost wholly upon chemical differences.

THE CELL

The structural and fundamental unit of both plants and animals, no matter how complex, is the cell. Since the 19th century, scientists have known that all living things, whether animal or plant, are made up of cells. A typical cell is an entity, isolated from other cells by a membrane or cell wall. The cell membrane contains protoplasm and the nucleus (see Figure 11.1). The protoplasm within the cell is a living mass of viscous, transparent material. Within the protoplasm is a dense spherical mass called the nucleus or nuclear material. In a typical mature plant cell, the cell wall is rigid and is composed of nonliving material, while in the typical animal cell, the wall is an elastic living membrane. Cells exist in a very great variety of sizes and shapes, as well as functions. Their average size ranges from bacteria too small to be seen with the light microscope to the largest known single cell, the ostrich egg. Microbial cells also have an extensive size range, some being larger than human cells (Kordon, 1992).

Note: The nucleus cannot always be observed in bacteria.

STRUCTURE OF THE BACTERIAL CELL

The structural form and various components of the bacterial cell are probably best understood by referring to the simplified diagram of a rod-form bacterium shown in Figure 11.1. When studying Figure 11.1, keep in mind that cells of different species may differ greatly, both in structure and chemical composition; for this reason, no typical bacterium exists. Figure 11.1 shows a generalized bacterium used for the discussion that follows. Not all bacteria have all the features shown in the figure, and some bacteria have structures not shown in the figure.

Capsules

Bacterial *capsules* (see Figure 11.1) are organized accumulations of gelatinous materials on cell walls, in contrast to *slime layers* (a water secretion that adheres loosely to the cell wall and commonly diffuses into the cell), which are unorganized accumulations of similar material. The capsule is usually thick enough to be seen under the ordinary light microscope (macro-capsule), while thinner capsules (microcapsules) can be detected only by electron microscopy (Singleton & Sainsbury, 1994). The production of capsules is determined largely by genetics as well as environmental conditions and depends on the presence or absence of capsule-degrading enzymes and other growth factors. Varying in composition, capsules are mainly composed of water; the organic contents are made up of complex polysaccharides, nitrogen-containing substance, and polypeptides. Capsules confer several advantages when bacteria grow in their normal habitat.

For example, they help to:

- Prevent desiccation;
- Resist phagocytosis by host phagocytic cells;
- Prevent infection by bacteriophages; and
- Aid bacterial attachment to tissue surfaces in plant and animal hosts or to surfaces of solids objects in aquatic environments. Capsule formation often correlates with pathogenicity.

Flagella

Many bacteria are motile, and this ability to move independently is usually attributed to a special structure, the *flagella* (singular: flagellum). Depending on species, a cell may have a single flagellum (see Figure 11.1) (*monotrichous* bacteria; *trichous* means "hair"); one flagellum at each end (*amphitrichous* bacteria; *amphi* means "on both sides"); a tuft of flagella at one or both ends (*lophotrichous* bacteria; *lopho* means "tuft"); or flagella that arise all over the cell surface (*peritrichous* bacteria; *peri* means "around"). A flagellum is a threadlike appendage extending outward from the plasma membrane and cell wall. Flagella are slender, rigid, locomotor structures, about 20 μm across and up to 15–20 μm long. Flagellation patterns are very useful in identifying bacteria and can be seen by light microscopy but only after being stained with special techniques designed to increase their thickness. The detailed structure of flagella can be seen only in the electron microscope. Bacterial cells benefit from flagella in several ways. They can increase the concentration of nutrients or decrease the concentration of toxic materials near the bacterial surfaces by causing a change in the flow rate of fluids. They can also disperse flagellated organisms to areas where colony formation can take place. The main benefit of flagella to organisms is their increased ability to flee from areas that might be harmful.

Cell Wall

The main structural component of most prokaryotes is the rigid *cell wall* (see Figure 11.1). Functions of the cell wall include:

- Providing protection for the delicate protoplast from osmotic lysis (bursting);
- Determining a cell's shape;
- Acting as a permeability layer that excludes large molecules and various antibiotics and plays an active role in regulating the cell's intake of ions; and
- Providing a solid support for flagella.

Cell walls of different species may differ greatly in structure, thickness, and composition. The cell wall accounts for about 20–40% of the dry weight of a bacterium.

Plasma Membrane (Cytoplasmic Membrane)

Surrounded externally by the cell wall and composed of a lipoprotein complex, the *plasma membrane* or cell membrane is the critical barrier, separating the inside from the outside of the cell (see Figure 11.1). About 7–8 μm thick and comprising 10–20% of a bacterium's dry weight, the plasma membrane controls the passage of all material into and out of the cell. The inner and outer faces of the plasma membrane are embedded with water-loving (hydrophilic) lips, whereas the interior is hydrophobic. Control of material into the cell is accomplished by screening, as well as by electric charge. The plasma membrane is the site of the surface charge of the bacteria.

In addition to serving as an osmotic barrier that passively regulates the passage of material into and out of the cell, the plasma membrane participates in the entire active transport of various substances into the bacterial cell. Inside the membrane, many highly reactive chemical groups guide the incoming material to the proper points for further reaction. This active transport system provides bacteria with certain advantages, including the ability to maintain a constant intercellular ionic state in the presence of varying external ionic concentrations. In addition to participating in the uptake of nutrients, the cell membrane transport system participates in waste excretion and protein secretions.

Cytoplasm

Within a cell and bound by the cell membrane is a complicated mixture of substances and structures called the *cytoplasm* (see Figure 11.1). The cytoplasm is a water-based fluid containing ribosomes, ions, enzymes, nutrients, storage granules (under certain circumstances), waste products, and various molecules involved in synthesis, energy metabolism, and cell maintenance.

Mesosome

A common intracellular structure found in the bacterial cytoplasm is the Mesosome (see Figure 11.1). *Mesosomes* are invaginations of the plasma membrane in the shape of tubules, vesicles, or lamellae. Their exact function is unknown. Currently many bacteriologists believe that Mesosomes are artifacts generated during the fixation of bacteria for electron microscopy.

Nucleoid (Nuclear Body or Region)

The *nuclear region* of the prokaryotic cell is primitive and a striking contrast to that of the eucaryotic cell (see Figure 11.1). Prokaryotic cells lack a distinct nucleus, the function of the nucleus being carried out by a single, long, double strand of DNA that is efficiently packaged to fit within the nucleoid. The nucleoid is attached to the plasma membrane. A cell can have more than one Nucleoid when cell division occurs after the genetic material has been duplicated.

Ribosomes

The bacterial cytoplasm is often packed with ribosomes (see Figure 11.1). *Ribosomes* are minute, rounded bodies made of RNA and are loosely attached to the plasma membrane. Ribosomes are estimated to account for about 40% of a bacterium's dry weight; a single cell may have as many as 10,000 ribosomes. Ribosomes are the site of protein synthesis and are part of the translation process.

Inclusions

Inclusions (or storage granules) are often seen within bacterial cells (see Figure 11.1). Some inclusion bodies are not bound by a membrane and lie free in the cytoplasm. A single-layered membrane about 2–4 μm thick encloses other inclusion bodies. Many bacteria produce polymers that are stored as granules in the cytoplasm.

BACTERIA

The simplest wholly contained life systems are *bacteria* or *prokaryotes*, which are the most diverse group of microorganisms. As mentioned, they are among the most common microorganisms in water, are primitive, unicellular (single-celled) organisms, possessing no well-defined nucleus, that present a variety of shapes and nutritional needs. Bacteria contain about 85% water and 15% ash or mineral matter. The ash is largely composed of sulfur, potassium, sodium, calcium, and chlorides, with small amounts of iron, silicon, and magnesium. Bacteria reproduce by binary fission.

Note: Binary fission occurs when one organism splits or divides into two or more new organisms.

Bacteria, once called the smallest living organisms (now it is known that smaller forms of matter exhibit many of the characteristics of life), range in size from 0.5 to 2 μm in diameter and about 1–10 μm long.

Note: A *micron* is a metric unit of measurement equal to one-thousandth of a millimeter. To visualize the size of bacteria, consider that about 1,000 bacteria lying side by side would reach across the head of a straight pin.

Note: A eukaryotic organism is characterized by a cellular organization that includes a well-defined nuclear membrane. The prokaryotes have a structural organization that sets them off from all other organisms. They are simple cells characterized by a nucleus *lacking* a limiting membrane, an endoplasmic reticulum, chloroplasts, and mitochondria. They are remarkably adaptable, existing abundantly in the soil, the sea, and freshwater.

Bacteria are categorized into three general groups based on their physical form or shape (though almost every variation has been found; see Table 11.2). The simplest form is the sphere. Spherical-shaped bacteria are called *cocci* (meaning "berries"). They are not necessarily perfectly round, but may be somewhat elongated, flattened on one side, or oval. Rod-shaped bacteria are called *bacilli*. Spiral-shaped bacteria (called Spirilla), which have one or more twists and are never straight, make up the third group (see Figure 11.2). Such formations are usually characteristic of a genus or species. Within these three groups are many different arrangements. Some exist as single cells; others

TABLE 11.2

Forms of Bacteria

Form	Technical Name		Example
	Singular	Plural	
Sphere	Coccus	Cocci	*Streptococcus*
Rod	Bacillus	Bacilli	*Bacillus typhosis*
Curved or spiral	Spirillum	Spirilla	*Spirillum cholera*

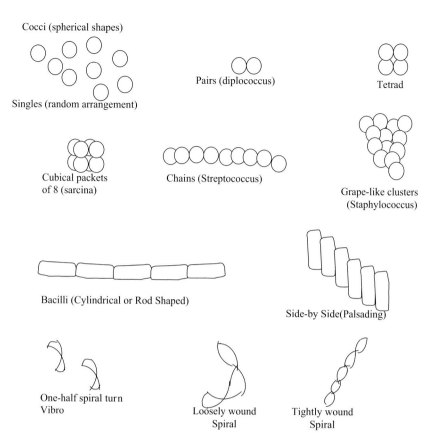

FIGURE 11.2   Bacterial shapes and arrangements.

as pairs, as packets of four or eight, as chain, and as clumps. Most bacteria require organic food to survive and multiply. Plant and animal material that gets into the water provides the food source for bacteria. Bacteria convert the food to energy and use the energy to make new cells. Some bacteria can use inorganics (e.g., minerals such as iron) as an energy source and exist and multiply even when organics (pollution) are not available.

BACTERIAL GROWTH FACTORS

Several factors affect the rate at which bacteria grow, including temperature, pH, and oxygen levels. The warmer the environment, the faster the rate of growth. Generally, for each increase of 10°C, the growth rate doubles. Heat can also be used to kill bacteria. Most bacteria grow best at neutral pH. Extreme acidic or basic conditions generally inhibit growth, though some bacteria may require acidic and some require alkaline conditions for growth.

Bacteria are aerobic, anaerobic, or facultative. If *aerobic*, they require free oxygen in the aquatic environment. *Anaerobic* bacteria exist and multiply in environments that lack dissolved oxygen. *Facultative* bacteria (e.g., iron bacteria) can switch from an aerobic to anaerobic growth or grow in an anaerobic or aerobic environment. Under optimum conditions, bacteria grow and reproduce very rapidly. As stated previously, bacteria reproduce by *binary fission*. An important point to consider in connection with bacterial reproduction is the rate at which the process can take place. The total time required for an organism to reproduce and the offspring to reach maturity is called *generation time*. Bacteria growing under optimal conditions can double their number about every 20–30 minutes. Obviously, this generation time is very short compared with that of higher plants and animals.

Bacteria continue to grow at this rapid rate if nutrients hold out – even the smallest contamination can result in a sizable growth in a very short time.

Note: Even though wastewater can contain bacteria counts in the millions per mL, in wastewater treatment, under controlled conditions, bacteria can help to destroy and to identify pollutants. In such a process, bacteria stabilize organic matter (e.g., activated sludge processes), and thereby assist the treatment process in producing effluent that does not impose an excessive oxygen demand on the receiving body. Coliform bacteria can be used as an indicator of pollution by human or animal wastes.

DESTRUCTION OF BACTERIA

In water and wastewater treatment, the destruction of bacteria is usually called *disinfection*.

Disinfection does not mean that all microbial forms are killed. That would be *sterilization*. However, disinfection does reduce the number of disease-causing organisms to an acceptable number. Growing bacteria are easy to control by disinfection. Some bacteria, however, form spores – survival structures – which are much more difficult to destroy.

Note: Inhibiting the growth of microorganisms is termed "antisepsis," while destroying them is called *disinfection*.

WATERBORNE BACTERIA

All surface waters contain bacteria. Waterborne bacteria, as we have said, are responsible for infectious epidemic diseases. Bacterial numbers increase significantly during storm events when streams are high. Heavy rainstorms increase stream contamination by washing material from the ground surface into the stream. After the initial washing occurs, few impurities are left to be washed into the stream, which may then carry relatively "clean" water. A river of good quality shows its highest bacterial numbers during rainy periods; however, a much-polluted stream may show the highest numbers during low flows because of the constant influx of pollutants. Water and wastewater operators are primarily concerned with bacterial pathogens responsible for the disease. These pathogens enter potential drinking water supplies through fecal contamination and are ingested by humans if the water is not properly treated and disinfected.

Note: Regulations require that owners of all public water supplies collect water samples and deliver them to a certified laboratory for bacteriological examination at least monthly. The number of samples required is usually in accordance with Federal Standards, which generally require that one sample per month be collected for every 1,000 persons served by the waterworks.

PROTOZOA

© Las Vegas Valley Water District and used with permission

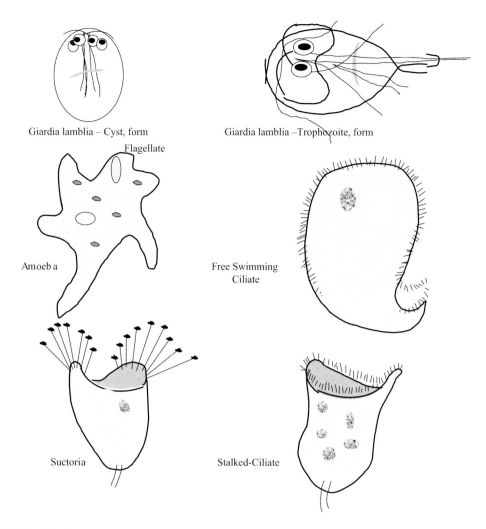

Giardia lamblia – Cyst, form

Giardia lamblia –Trophozoite, form

Flagellate

Amoeba

Free Swimming
Ciliate

Suctoria

Stalked-Ciliate

FIGURE 11.3 Protozoa

Protozoans (or "first animals") are a large group of eucaryotic organisms of more than 50,000 known species belonging to the Kingdom Protista that have adapted a form of cell to serve as the entire body. In fact, protozoans are one-celled animal-like organisms with complex cellular structures. In the microbial world, protozoans are giants, many times larger than bacteria. They range in size from 4 microns to 500 microns. The largest ones can almost be seen by the naked eye. They can exist as solitary or independent organisms [e.g., the stalked ciliates (see Figure 11.3) such as *Vorticella* sp.], or they can colonize like the sedentary *Carchesium* sp. Protozoa get their name because they employ the same type of feeding strategy as animals. That is, they are heterotrophic, meaning they obtain cellular energy from organic substances such as proteins. Most are harmless, but some are parasitic. Some forms have two life stages: *active trophozoites* (capable of feeding) and *dormant cysts.*

The major groups of protozoans are based on their method of locomotion (motility). For example, the *Mastigophora* are motile by means of one or more *flagella* (the whip-like projection that propels the free-swimming organisms – *Giardia lamblia* is a flagellated protozoan); the *Ciliophora* by means of shortened modified flagella called *cilia* (short hair-like structures that beat rapidly and propel them through the water); the *Sarcodina* by means of amoeboid movement (streaming or gliding action – the shape of amoebae change as they stretch, then contract, from place to place); and the *Sporozoa*, which are nonmotile; they are simply swept along, riding the current of the water.

Protozoa consume organics to survive; their favorite food is bacteria. Protozoa are mostly aerobic or facultative regarding oxygen requirements. Toxic materials, pH, and temperature affect protozoan rates of growth in the same way as they affect bacteria. Most protozoan life cycles alternate between an active growth phase (*trophozoites*) and a resting stage (*cysts*). Cysts are extremely resistant structures that protect the organism from destruction when it encounters harsh environmental conditions – including chlorination.

Note: Those protozoans not completely resistant to chlorination require higher disinfectant concentrations and longer contact time for disinfection than normally used in water treatment.

The protozoa and the waterborne diseases associated with them that are of most concern to waterworks operator are:

- *Entamoeba histolytica* – Amoebic dysentery;
- *Giardia lamblia* – Giardiasis;
- *Cryptosporidium* – Cryptosporidiosis.

In wastewater treatment, protozoa are a critical part of the purification process and can be used to indicate the condition of treatment processes. Protozoa normally associated with wastewater include amoeba, flagellates, free-swimming ciliates, and stalked ciliates.

Amoebae are associated with poor wastewater treatment of a young biosolids mass (see Figure 11.3). They move through wastewater by a streaming or gliding motion. Moving the liquids stored within the cell wall effects this movement. They are normally associated with an effluent high in biochemical oxygen demand (BODs) and suspended solids.

Flagellates (flagellated protozoa) have a single, long hair-like or whip-like projection (flagella) that is used to propel the free-swimming organisms through wastewater and to attract food (see Figure 11.3). Flagellated protozoans are normally associated with poor treatment and young biosolids. When the free-swimming ciliated protozoan is the predominant organism, the plant effluent will contain large amounts of BODs and suspended solids.

The *free-swimming ciliated protozoan* uses its tiny, hair-like projections (cilia) to move itself through the wastewater and to attract food (see Figure 11.3). The free-swimming ciliated protozoan is normally associated with a moderate biosolids age and effluent quality. When the free-swimming ciliated protozoan is the predominant organism, the plant effluent will normally be turbid and contain a high number of suspended solids.

The *stalked ciliated protozoan* attaches itself to the wastewater solids and uses its cilia to attract food (see Figure 11.3). The stalked ciliated protozoan is normally associated with a plant effluent that is very clear and contains low amounts of both BODs and suspended solids.

Rotifers make up a well-defined group of the smallest, simplest multicellular microorganisms and are found in nearly all aquatic habitats (see Figure 11.4). Rotifers are a higher life form associated with cleaner waters. Normally found in well-operated wastewater treatment plants, they can be used to indicate the performance of certain types of treatment processes.

MICROSCOPIC CRUSTACEANS

Because they are important members of freshwater zooplankton, microscopic *crustaceans* are of interest to water and wastewater operators. These microscopic organisms are characterized by a rigid shell structure. They are multicellular animals that are strict aerobes, and as primary producers, they feed on bacteria and algae. They are important as a source of food for fish. Additionally, microscopic crustaceans have been used to clarify algae-laden effluents from oxidations ponds. *Cyclops* and *Daphnia* are two microscopic crustaceans of interest to water and wastewater operators.

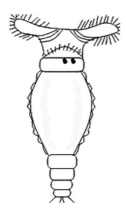

FIGURE 11.4   Physodine, a common rotifer.

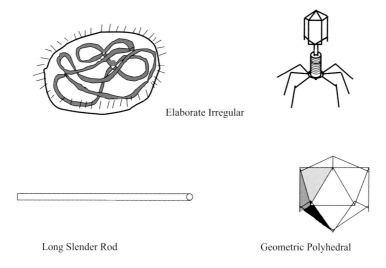

Elaborate Irregular

Long Slender Rod   Geometric Polyhedral

FIGURE 11.5   Virus shapes.

VIRUSES

Viruses are very different from the other microorganisms. Consider their size relationship, for example. Relative to size, if protozoans are the Goliaths of microorganisms, then viruses are the Davids. Stated more specifically and accurately, viruses are intercellular parasitic particles that are the smallest living infectious materials known – the midgets of the microbial world. Viruses are very simple life forms consisting of a central molecule of genetic material surrounded by a protein shell called a *capsid* and sometimes by a second layer called an *envelope*. They contain no mechanisms by which to obtain energy or reproduce on their own; thus, to live, viruses must have a host. After they invade the cells of their specific host (animal, plant, insect, fish, or even bacteria), they take over the host's cellular machinery and force it to make more viruses. In the process, the host cell is destroyed, and hundreds of new viruses are released into the environment. Viruses occur in many shapes, including long slender rods, elaborate irregular shapes, and geometric polyhedrals (see Figure 11.5).

The viruses of most concern to the waterworks operator are the pathogens that cause hepatitis, viral gastroenteritis, and poliomyelitis. Smaller and different from bacteria, viruses are prevalent in water contaminated with sewage. Detecting viruses in water supplies is a major problem because of the complexity of the non-routine procedures involved, although experience has shown that the normal coliform index can be used as a rough guide for viruses and for bacteria. More attention must be paid to viruses, however, whenever surface water supplies have been used for sewage disposal. Viruses are difficult to destroy by normal disinfection practices, requiring increased disinfectant concentration and contact time for effective destruction.

Note: Viruses that infect bacterial cells cannot infect and replicate within cells of other organisms. It is possible to utilize the specificity to identify bacteria, a procedure called *phage typing*.

ALGAE

You do not have to be a water and wastewater operator to understand that algae can be a nuisance. Many ponds and lakes in the United States are currently undergoing *eutrophication*, the enrichment of an environment with inorganic substances (e.g., phosphorus and nitrogen), causing excessive algae growth and premature aging of the water body. When eutrophication occurs, especially when filamentous algae-like *Caldophora* breaks loose in a pond or lake and washes ashore, algae makes its stinking, noxious presence known.

Algae are a form of aquatic plants and are classified by color (e.g., green algae, blue-green algae, golden-brown algae, etc.). Algae come in many shapes and sizes (see Figure 11.6). Although they are not pathogenic, algae do cause problems with water and wastewater treatment plant operations. They grow easily on the walls of troughs and basins, and heavy growth can plug intakes and screens. Additionally, some algae release chemicals that give off undesirable tastes and odors. Although algae are usually classified by their color, they are also commonly classified based on their cellular properties or characteristics.

Several characteristics are used to classify algae including:

- Cellular organization and cell wall structure;
- The nature of the chlorophyll(s);
- The type of motility, if any;
- The carbon polymers that are produced and stored; and
- The reproductive structures and methods.

Many algae (in mass) are easily seen by the naked eye – others are microscopic. They occur in fresh and polluted water, as well as in salt water. Since they are plants, they can use energy from the sun in photosynthesis. They usually grow near the surface of the water because light cannot penetrate very far through the water. Algae are controlled in raw waters with chlorine and potassium permanganate. Algae blooms in raw water reservoirs are often controlled with copper sulfate.

Note: By producing oxygen, which is utilized by other organisms including animals, algae play an important role in the balance of nature.

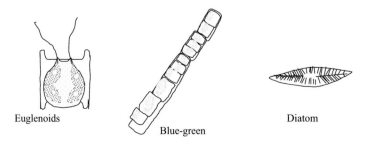

FIGURE 11.6   Algae.

FUNGI

Fungi are of relatively minor importance in water and wastewater operations (except for biosolids composting, where they are critical). Fungi, like bacteria, are also extremely diverse. They are multicellular, autotrophic, photosynthetic protists. They grow as filamentous, mold-like forms or as yeast-like (single-celled) organisms. They feed on organic material.

Note: Aquatic fungi grow as parasites on living plants or animals and as saprophytes on those that are dead.

NEMATODES AND FLATWORMS (WORMS)

Along with inhabiting organic mud, worms also inhabit biological slimes; they have been found in activated sludge and in trickling filter slimes (wastewater treatment processes). Microscopic in size, they range in length from 0.5 to 3 mm and in diameter from 0.01 to 0.05 mm. Most species have a similar appearance. They have a body that is covered by cuticles, are cylindrical, nonsegmented, and taper at both ends. These organisms continuously enter wastewater treatment systems, primarily through attachment to soils that reach the plant through inflow and infiltration (I&I). They are present in large, often highly variable numbers, but as strict aerobes, they are found only in aerobic treatment processes where they metabolize solid organic matter.

When nematodes are firmly established in the treatment process, they can promote microfloral activity and decomposition. They crop bacteria in both the activated sludge and trickling filter systems. Their activities in these systems enhance oxygen penetration by tunneling through floc particles and biofilm. In activated sludge processes, they are present in relatively small numbers because the liquefied environment is not a suitable habitat for crawling, which they prefer over the free-swimming mode. In trickling filters where the fine stationary substratum is suitable to permit crawling and mating, nematodes are quite abundant.

Along with preferring the trickling filter habitat, nematodes play a beneficial role in this habitat; for example, they break loose portions of the biological slime coating the filter bed. This action prevents excessive slime growth and filter clogging. They also aid in keeping slime porous and accessible to oxygen by tunneling through slime. In the activated sludge process, the nematodes play important roles as agents of better oxygen diffusion. They accomplish this by tunneling through floc particles. They also act as parameters of operational conditions in the process, such as low dissolved oxygen levels (anoxic conditions) and the presence of toxic wastes.

Environmental conditions have an impact on the growth of nematodes. For example, in anoxic conditions, their swimming and growth are impaired. The most important condition they indicate is when the wastewater strength and composition have changed. Temperature fluctuations directly affect their growth and survival; population decreases when temperatures increase.

Aquatic flatworms (improperly named because they are not all flat) feed primarily on algae. Because of their aversion to light, they are found in the lower depths of pools. Two varieties of flatworms are seen in wastewater treatment processes: *microtubellarians* are more round than flat and average about 0.5–5 mm in size, and *macrotubellarians* (planarians) are more flat than round and average about 5–20 mm in body size. Flatworms are very hardy and can survive in wide variations in humidity and temperature. As inhabitants of sewage sludge, they play an important part in sludge stabilization and as bioindicators or parameters of process problems. For example, their inactivity or sluggishness might indicate a low dissolved oxygen level or the presence of toxic wastes.

Surface waters grossly polluted with organic matter (especially domestic sewage) have a fauna that can thrive in very low concentrations of oxygen. A few species of tubificid worms dominate this environment. Pennak reported (1989) that the bottoms of severely polluted streams can be literally covered with a "writhing" mass of these tubificids.

The *Tubifex* (commonly known as sludge worms) are small, slender, reddish worms that normally range in length from 25 to about 50 mm. They are burrowers; their posterior end protrudes to obtain nutrients (see Figure 11.7). When found in streams, *Tubifex* are indicators of pollution.

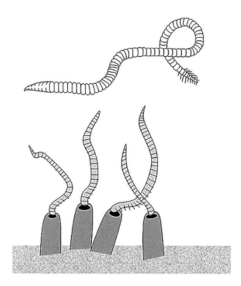

FIGURE 11.7   Tubificid worms.

PATHOGENIC PROTOZOA

Certain types of protozoans can cause disease. Of interest to the wastewater practitioner are the *Entamoeba histolytica* (amebic dysentery and amebic hepatitis), *Giardia lamblia* (Giardiasis), *Cryptosporidium* (Cryptosporidiosis), and the emerging *Cyclospora* (Cyclosporasis). Sewage contamination transports eggs, cysts, and oocysts of parasitic protozoa and helminthes (tapeworms, hookworms, etc.) into raw water supplies, leaving water treatment (filtration) and disinfection as the means by which to diminish the danger of contaminated water for the consumer.

To prevent the occurrence of *Giardia* and *Cryptosporidium* spp. in surface water supplies and to address increasing problems with waterborne diseases, United States Environmental Protection Agency (USEPA) implemented its Surface Water Treatment Rule (SWTR) in 1989. The rule requires both filtration and disinfection of all surface water supplies as a means of primarily controlling *Giardia* spp. and enteric viruses. Since the implementation of its Surface Water Treatment Rule, USEPA has also recognized that *Cryptosporidium* spp. is an agent of waterborne disease. In 1996, in its next series of surface water regulations, the USEPA included *Cryptosporidium*.

To test the need for and the effectiveness of USEPA's Surface Water Treatment Rule, LeChevallier et al. (1991) conducted a study on the occurrence and distribution of *Giardia* and *Cryptosporidium* organisms in raw water supplies to 66 surface water filter plants. These plants were in fourteen states and one Canadian province. A combined immunofluorescence test indicated that cysts and oocysts were widely dispersed in the aquatic environment. *Giardia* spp. was detected in more than 80% of the samples. *Cryptosporidium* spp. was found in 85% of the sample locations. Considering several variables, *Giardia* or *Cryptosporidium* spp. were detected in 97% of the raw water samples. After evaluating their data, the researchers concluded that the Surface Water Treatment Rule might have to be upgraded (subsequently, it has been) to require additional treatment.

Giardia

Giardia (gee-ar-dee-ah) *lamblia* (also known as hiker'/traveler's scourge or disease) is a microscopic parasite that can infect warm-blooded animals and humans. Although *Giardia* was discovered in the 19th century, not until 1981 did the World Health Organization (WHO) classify *Giardia* as a pathogen. *An outer shell called a cyst that allows it to survive outside the body for long periods protects giardia.* If viable cysts are ingested, *Giardia* can cause the illness known as *Giardiasis*, an intestinal illness that can cause nausea, anorexia, fever, and severe diarrhea. The symptoms last only for several days, and the body can naturally rid itself of the parasite in one to two months. However, for individuals with weakened immune systems, the body often cannot rid itself of the parasite without medical treatment. In the United States, *Giardia* is the most identified pathogen in waterborne disease outbreaks. Contamination of a water supply by *Giardia* can occur in two ways:

1. By the activity of animals in the watershed area of the water supply; or
2. By the introduction of sewage into the water supply. Wild and domestic animals are major contributors to contaminating water supplies.

Studies have also shown that, unlike many other pathogens, *Giardia* is not host-specific. In short, *Giardia* cysts excreted by animals can infect and cause illness in humans. Additionally, in several major outbreaks of waterborne diseases, the *Giardia* cyst source was sewage-contaminated water supplies. Treating the water supply, however, can effectively control waterborne Giardia. Chlorine and ozone are examples of two disinfectants known to effectively kill *Giardia* cysts. Filtration of the water can also effectively trap and remove the parasite from the water supply.

   The combination of disinfection and filtration is the most effective water treatment process available today for the prevention of *Giardia* contamination. In drinking water, *Giardia* is regulated under the SWTR. Although the SWTR does not establish a Maximum Contaminant Level (MCL) for *Giardia,* it does specify treatment requirements to achieve at least 99.9% (3-log) removal and/or inactivation of *Giardia.* This regulation requires that all drinking water systems using surface water or groundwater under the influence of surface water must disinfect and filter the water. The Enhanced Surface Water Treatment Rule (ESWTR), which includes Cryptosporidium and further regulates *Giardia*, was established in December 1996.

Giardiasis

Giardiasis is recognized as one of the most frequently occurring waterborne diseases in the United States. *Giardia lamblia* cysts have been discovered in the United States in places as far apart as Estes Park, Colorado (near the Continental Divide); Missoula, Montana; Wilkes-Barre, Scranton, and Hazleton, Pennsylvania; and Pittsfield and Lawrence, Massachusetts, just to name a few (CDC, 1995). Giardiasis is characterized by intestinal symptoms that usually last one 1 week or more and may be accompanied by one or more of the following: diarrhea, abdominal cramps, bloating, flatulence, fatigue, and weight loss. Although vomiting and fever are commonly listed as relatively frequent symptoms, people involved in waterborne outbreaks in the United States have not commonly reported them. While most *Giardia* infections persist only for one or two months, some people undergo a more chronic phase, which can follow the acute phase or may become manifest without an antecedent acute illness. Loose stools and increased abdominal gassiness with cramping, flatulence, and burping characterize the chronic phase. Fever is not common, but malaise, fatigue, and depression may ensue.

   For a small number of people, the persistence of infection is associated with the development of marked malabsorption and weight loss (Weller, 1985). Similarly, lactose (milk) intolerance can be a problem for some people. This can develop coincidentally with the infection or be aggravated by it, causing an increase in intestinal symptoms after ingestion of milk products. Some people may have several of these symptoms without evidence of diarrhea or have only sporadic episodes of diarrhea every three or four days. Still others may not have any symptoms at all. Therefore, the problem may not be whether you are infected with the parasite or not, but how harmoniously you both can live together, or how to get rid of the parasite (either spontaneously or by treatment) when the harmony does not exist or is lost.

Note: Three prescription drugs are available in the United States to treat giardiasis: quinacrine, metronidazole, and furazolidone. In a recent review of drug trials in which the efficacies of these drugs were compared, quinacrine produced a cure in 93% of patients, metronidazole cured 92%, and furazolidone cured about 84% of patients (Davidson, 1984).

Giardiasis occurs worldwide. In the United States, *Giardia* is the parasite most identified in stool specimens submitted to state laboratories for parasitological examination. During a three-year period, approximately 4% of one million stool specimens submitted to state laboratories tested positive for *Giardia* (CDC, 1979). Other surveys have demonstrated *Giardia* prevalence rates ranging from 1% to 20%, depending on the location and ages of persons studied. Giardiasis ranks among the top 20 infectious diseases that cause the greatest morbidity in Africa, Asia, and Latin America; it has been estimated that about two million infections occur per year in these regions (Walsh, 1981).

People who are at the highest risk for acquiring *Giardia* infection in the United States may be placed into five major categories:

1. People in cities whose drinking water originates from streams or rivers, and whose water treatment process does not include filtration, or where filtration is ineffective because of malfunctioning equipment
2. Hikers/campers/outdoor people
3. International travelers
4. Children who attend day-care centers, day-care center staff, and parents and siblings of children infected in day-care centers
5. Homosexual men

People in categories 1, 2, and 3 have in common the same general source of infection, that is, they acquire *Giardia* from fecally contaminated drinking water. The city resident usually becomes infected because the municipal water treatment process does not include the filter necessary to physically remove the parasite from the water. The number of people in the United States at risk (i.e., the number who receive municipal drinking water from unfiltered surface water) is estimated to be 20 million. International travelers may also acquire the parasite from improperly treated municipal waters in cities or villages in other parts of the world, particularly in developing countries. In Eurasia, only travelers to Leningrad (now renamed St. Petersburg) appear to be at increased risk. In prospective studies, 88% of US and 35% of Finnish travelers to Leningrad who had negative stool tests for *Giardia* on departure to the Soviet Union developed symptoms of giardiasis and had a positive test for *Giardia* after they returned home (Brodsky et al., 1974). Except for visitors to Leningrad, however, *Giardia* has not been implicated as a major cause of traveler's diarrhea – it has been detected in fewer than 2% of travelers who develop diarrhea. However, hikers and campers risk infection every time they drink untreated raw water from a stream or river. Persons in categories 4 and 5 become exposed through more direct contact with feces or an infected person by exposure to soiled diapers of an infected child (day-care center-associated cases), or through direct or indirect anal–oral sexual practices in the case of homosexual men.

Although community waterborne outbreaks of giardiasis have received the greatest publicity in the United States during the past decade, about half of the *Giardia* cases discussed with the staff of the Centers for Disease Control (CDC) over a three-year period had a day-care exposure as the most likely source of infection. Numerous outbreaks of *Giardia* in day-care centers have been reported in recent years. Infection rates for children in day-care center outbreaks range from 21% to 44% in the United States and from 8% to 27% in Canada (Black et al., 1981). The highest infection rates are usually observed in children who wear diapers (1–3 years of age). In a study of 18 randomly selected day-care centers in Atlanta, 10% of diapered children were found infected (CDC Unpublished). Transmission from this age group to older children, day-care staff, and household contacts is also common. About 20% of parents caring for an infected child become infected.

Local health officials and managers of wastewater systems need to realize that sources of *Giardia* infection other than municipal drinking water exist. Armed with this knowledge, they are less likely to make a quick (and sometimes wrong) assumption that a cluster of recently diagnosed cases in a city is related to municipal drinking water. Of course, drinking water must not be ruled out as a source of infection when a larger than expected number of cases is present in a community, but the possibility that the cases are associated with a day-care center outbreak, drinking untreated stream water, or international travel should also be entertained.

To understand the finer aspects of *Giardia* transmission and strategies for control, the drinking water practitioner must become familiar with several aspects of the parasite's biology. Two forms of the parasite exist: a *trophozoite* and a *cyst*, both of which are much larger than bacteria (see Figure 11.8). Trophozoites live in the upper small intestine where they attach to the intestinal wall by means of a disc-shaped suction pad on their ventral surface. Trophozoites actively feed and reproduce at this location. At some time during the trophozoite's life, it releases its hold on the bowel wall and floats in the fecal stream through the intestine.

As it makes this journey, it undergoes a morphologic transformation into an egg-like structure called a cyst. The cyst (about 6–9 nanometers in diameter × 8–12 micrometers - 1/100 millimeter - in length) has a thick exterior wall that protects the parasite against the harsh elements that it will encounter outside the body. This cyst form of the parasite is infectious to other people or animals. Most people become infected either directly (by hand-to-mouth transfer of cysts from the feces of an infected individual) or indirectly (by drinking feces-contaminated water). Less common modes of transmission include ingestion of fecally contaminated food and hand-to-mouth transfer of cysts after touching a fecally contaminated surface. After the cyst is swallowed, the trophozoite is liberated through the action of stomach acid and digestive enzymes and becomes established in the small intestine.

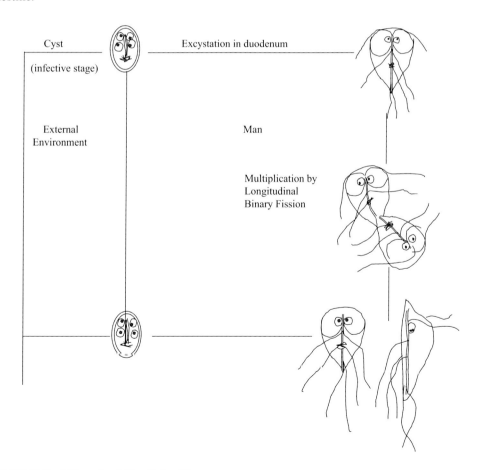

FIGURE 11.8   Life cycle of Giardia lamblia

Although infection after ingestion of only one *Giardia* cyst is theoretically possible, the minimum number of cysts shown to infect a human under experimental conditions is ten (Rendtorff, 1954). Trophozoites divide by binary fission about every 12 hours. What this means in practical terms is that if a person swallowed only a single cyst, reproduction at this rate would result in more than one million parasites 10 days later, and one billion parasites by day 15. The exact mechanism by which *Giardia* causes illness is not yet well understood but is not necessarily related to the number of organisms present. Nearly all the symptoms, however, are related to dysfunction of the gastrointestinal tract. The parasite rarely invades other parts of the body, such as the gall bladder or pancreatic ducts. Intestinal infection does not result in permanent damage.

Note: *Giardia* has an incubation period of 1–8 weeks.

Data reported by the CDC indicate that *Giardia* is the most frequently identified cause of diarrheal outbreaks associated with drinking water in the United States. The remainder of this section is devoted specifically to waterborne transmissions of *Giardia*. *Giardia* cysts have been detected in 16% of potable water supplies (lakes, reservoirs, rivers, springs, groundwater) in the United States at an average concentration of 3 cysts per 100L (Rose et al., 1983). Waterborne epidemics of giardiasis are a relatively frequent occurrence. In 1983, for example, *Giardia* was identified as the cause of diarrhea in 68% of waterborne outbreaks in which the causal agent was identified. From 1965 to 1982, more than 50 waterborne outbreaks were reported (CDC, 1983). In 1984, about 250,000 people in Pennsylvania were advised to boil drinking water for six months because of *Giardia*-contaminated water.

Many of the municipal waterborne outbreaks of *Giardia* have been subjected to intense study to determine their cause. Several general conclusions can be made from data obtained in those studies. Waterborne transmission of *Giardia* in the United States usually occurs in mountainous regions where community drinking water obtained from clear running streams is chlorinated but not filtered before distribution. Although mountain streams appear to be clean, fecal contamination upstream by human residents or visitors, as well as by *Giardia*-infected animals such as beavers, has been well documented. Water obtained from deep wells is an unlikely source of *Giardia* because of the natural filtration of water as it percolates through the soil to reach underground cisterns. Well-waste sources that pose the greatest risk of fecal contamination are poorly constructed or improperly located ones. A few outbreaks have occurred in towns that included filtration in the water treatment process, where the filtration was not effective in removing *Giardia* cysts because of defects in filter construction, poor maintenance of the filter media, or inadequate pretreatment of the water before filtration. Occasional outbreaks have also occurred because of accidental cross-connections between water and sewage systems.

Important Point: From these data, we conclude that two major ingredients are necessary for waterborne outbreak. *Giardia* cysts must be present in untreated source water, and the water purification process must either fail to kill or to remove *Giardia* cysts from the water.

Although beavers are often blamed for contaminating water with *Giardia* cysts, that they are responsible for introducing the parasite into new areas seems unlikely. Far more likely is that they are also victims: *Giardia* cysts may be carried in untreated human sewage discharged into the water by small-town sewage disposal plants or originate from cabin toilets that drain directly into streams and rivers. Backpackers, campers, and sports enthusiasts may also deposit *Giardia*-contaminated feces in the environment, which are subsequently washed into streams by rain. In support of this concept is a growing amount of data that indicate a higher *Giardia* infection rate in beavers living downstream from U.S. National Forest campgrounds when compared with beavers living in more remote areas that have a near-zero rate of infection.

Although beavers may be unwitting victims of the *Giardia* story, they still play an important part in the contamination scheme because they can (and probably do) serve as amplifying hosts. An *amplifying host* is one that is easy to infect, serves as a good habitat for the parasite to reproduce, and in the case of Giardia, returns millions of cysts to the water for every one ingested. Beavers are especially important in this regard, because they tend to defecate in or very near the water, which ensures that most of the *Giardia* cysts excreted are returned to the water.

The microbial quality of water resources and the management of the microbially laden wastes generated by the burgeoning animal agriculture industry are critical local, regional, and national problems. Animal waste from cattle, hogs, sheep, horses, poultry, and other livestock and commercial animals can contain high concentrations of microorganisms, such as *Giardia,* that are pathogenic to humans. The contribution of other animals to waterborne outbreaks of *Giardia* is less clear. Muskrats (another semiaquatic animal) have been found to have high infection rates (30–40%) in several parts of the United States (Frost et al., 1984). Recent studies have shown that muskrats can be infected with *Giardia* cysts from humans and beavers. Occasional *Giardia* infections have been reported in coyotes, deer, elk, cattle, dogs, and cats (but not in horses and sheep) encountered in mountainous regions of the United States. Naturally occurring *Giardia* infections have not been found in most other wild animals (bear, nutria, rabbit, squirrel, badger, marmot, skunk, ferret, porcupine, mink, raccoon, river otter, bobcat, lynx, moose, bighorn sheep) (Frost et al., 1984).

Scientific knowledge about what is required to kill or remove *Giardia* cysts from a contaminated water supply has increased considerably. For example, we know that cysts can survive in cold water (4°C) for at least two months, and they are killed instantaneously by boiling water (100°C) (Frost et al., 1984). We do not know how long the cysts will remain viable at other water temperatures (e.g., at 0°C or in a canteen at 15–20°C), nor do we know how long the parasite will survive on various environment surfaces, e.g., under a pine tree, in the sun, on a diaper-changing table, or in carpets in a day-care center.

The effect of chemical disinfection (chlorination, for example) on the viability of *Giardia* cysts is an even more complex issue. The number of waterborne outbreaks of *Giardia* that have occurred in communities where chlorination was employed as a disinfectant process demonstrates that the amount of chlorine used routinely for municipal water treatment is not effective against *Giardia* cysts. These observations have been confirmed in the laboratory under experimental conditions (Jarroll et al., 1979). This does not mean that chlorine does not work at all. It does work under certain favorable conditions.

Without getting too technical, gaining some appreciation of the problem can be achieved by understanding a few of the variables that influence the efficacy of chlorine as a disinfectant:

- *Water pH* – At pH values above 7.5, the disinfectant capability of chlorine is greatly reduced.
- *Water temperature* – The warmer the water, the higher the efficacy. Chlorine does not work in ice-cold water from mountain streams.
- *Organic content of the water* – Mud, decayed vegetation, or other suspended organic debris in water chemically combines with chlorine, making it unavailable as a disinfectant.
- *Chlorine contact time* – The longer Giardia cysts are exposed to chlorine, the more likely the chemical will kill them.
- *Chlorine concentration* – The higher the chlorine concentration, the more likely chlorine will kill Giardia cysts. Most water treatment facilities try to add enough chlorine to give a free (unbound) chlorine residual at the customer tap of 0.5 mg/L of water.

These five variables are so closely interrelated that improving one can often compensate for another; for example, if chlorine efficacy is expected to be low because water is obtained from an icy stream, the chlorine contact time, chlorine concentration, or both could be increased. In the case of *Giardia*-contaminated water, producing safe drinking water with a chlorine concentration of 1 mg/L and contact time as short as 10 minutes might be possible – if all the other variables were optimal (i.e., pH of 7.0, water temperature of 25°C, and a total organic content of the water close to 0). On the other hand, if all these variables were unfavorable (i.e., pH of 7.9, water temperature of 5°C, and high organic content), chlorine concentrations in excess of 8 mg/L with several hours of contact time may not be consistently effective.

Because water conditions and water treatment plant operations (especially those related to water retention time, and therefore, to chlorine contact time) vary considerably in different parts of the United States, neither the USEPA nor the CDC has been able to identify a chlorine concentration that would be safe yet effective against *Giardia* cysts under all water conditions. Therefore, the use of chlorine as a preventive measure against waterborne giardiasis generally has been used under outbreak conditions when the amount of chlorine and contact time have been tailored to fit specific water conditions and the existing operatonal design of the water system.

In an outbreak, for example, the local health department and water system may issue an advisory to boil water, may increase the chlorine residual at the consumer's tap from 0.5 mg/L to 1 or 2 mg/L, and if the physical layout and operation of the water treatment facility permit, increase the chlorine contact time. These are emergency procedures intended to reduce the risk of transmission until a filtration device can be installed or repaired or until an alternative source of safe water (a well, for example) can be made operational.

The long-term solution to the problem of municipal waterborne outbreaks of giardiasis involves improvements in and more widespread use of filters in the municipal water treatment process. The sand filters most used in municipal water treatment today cost millions of dollars to install, which makes them unattractive for many small communities. The pore sizes in these filters are not sufficiently small to remove a *Giardia* (6–9 micrometers × 8–12 micrometers). For the sand filter to remove *Giardia* cysts from the water effectively, the water must receive some additional treatment before it reaches the filter. The flow of water through the filter bed must also be carefully regulated.

An ideal prefilter treatment for muddy water would include sedimentation (a holding pond where large suspended particles are allowed to settle out by the action of gravity) followed by flocculation or coagulation (the addition of chemicals such as alum or ammonium to cause microscopic particles to clump together). The sand filter easily removes the large particles resulting from the flocculation/coagulation process, including Giardia cysts bound to other microparticulates. Chlorine is then added to kill the bacteria and viruses that may escape the filtration process. If the water comes from a relatively clear source, chlorine may be added to the water before it reaches the filter.

The successful operation of a complete waterworks operation is a complex process that requires considerable training. Troubleshooting breakdowns or recognizing the potential problems in the system before they occur often requires the skills of an engineer. Unfortunately, most small public water systems with water treatment facilities that include filtration cannot afford the services of a full-time engineer. Filter operation or maintenance problems in such systems may not be detected until a *Giardia* outbreak is recognized in the community. The bottom line is that although filtration is the best that water treatment technology has to offer for municipal water systems against waterborne giardiasis, it is not infallible. For municipal water filtration facilities to work properly, they must be properly constructed, operated, and maintained.

Whenever possible, persons out of doors should carry drinking water of known purity with them. When this is not practical, when water from streams, lakes, ponds, and other outdoor sources must be used, time should be taken to properly disinfect the water before drinking it.

CRYPTOSPORIDIUM

Ernest E. Tyzzer first described the protozoan parasite *Cryptosporidium* in 1907. Tyzzer frequently found a parasite in the gastric glands of laboratory mice. Tyzzer identified the parasite as a sporozoan, but of uncertain taxonomic status; he named it *Cryptosporidium muris*. Later, in 1910, after more detailed study, he proposed *Cryptosporidium* as a new genus and *muris* as the type of species. Amazingly, except for developmental stages, Tyzzer's original description of the life cycle (see Figure 11.9) was later confirmed by electron microscopy. Later, in 1912, Tyzzer described a new species, *Cryptosporidium parvum* (Tyzzer, 1912). For almost 50 years, Tyzzer's discovery of the

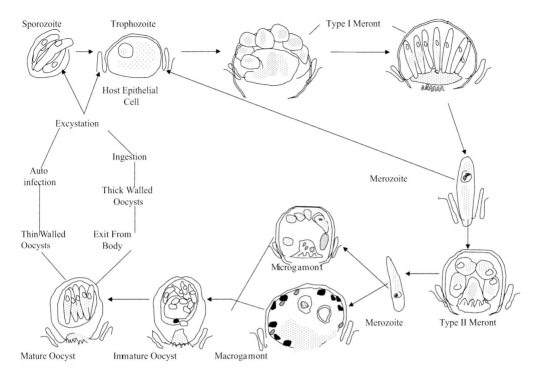

FIGURE 11.9   Life cycle of Cryptosporidium parvum.

genus *Cryptosporidium* remained (like himself) relatively obscure because it appeared to be of no medical or economic importance. Slight rumblings of the genus' importance were felt in the medical community when Slavin (1955) wrote about a new species, *Cryptosporidium melagridis*, associated with illness and death in turkeys. Interest remained slight even when *Cryptosporidium* was found to be associated with bovine diarrhea (Panciera et al., 1971). Not until 1982 did worldwide interest focus in on the study of organisms in the genus *Cryptosporidium*. During this period, the medical community and other interested parties were beginning to attempt a full-scale, frantic effort to find out as much as possible about Acquired Immune Deficiency Syndrome (AIDS). The CDC reported that 21 AIDS-infected males from six large cities in the United States had severe protracted diarrhea caused by *Cryptosporidium*. It was in 1993, though, that when the "bug – the pernicious parasite *Cryptosporidium* – made [itself and] Milwaukee famous" (Mayo Foundation, 1996).

Note: The *Cryptosporidium* outbreak in Milwaukee caused the deaths of 100 people – the largest episode of waterborne disease in the United States in the 70 years since health officials began tracking such outbreaks.

The massive waterborne outbreak in Milwaukee (more than 400,000 persons developed acute and often prolonged diarrhea or other gastrointestinal symptoms) increased interest in *Cryptosporidium* at an exponential level. The Milwaukee Incident spurred both public interest and the interest of public health agencies, agricultural agencies and groups, environmental agencies and groups, and suppliers of drinking water. This increase in interest level and concern has spurred on new studies of *Cryptosporidium* with emphasis on developing methods for recovery, detection, prevention, and treatment (Fayer et al., 1997).

The USEPA has become particularly interested in this "new" pathogen. For example, in the reexamination of regulations on water treatment and disinfection, the USEPA issued MCLG and CCL for *Cryptosporidium*. The similarity to *Giardia lamblia* and the necessity to provide an efficient conventional water treatment capable of eliminating viruses at the same time forced the USEPA to regulate the surface water supplies. The proposed "Enhanced Surface Water Treatment Rule" included regulations from watershed protection to specialized operation of treatment plants (certification of operators and state overview) and effective chlorination. Protection against *Cryptosporidium* included control of waterborne pathogens such as *Giardia* and viruses (DeZuane, 1997).

The Basics of *Cryptosporidium*

Cryptosporidium (crip-toe-spor-id-ee-um) is one of several single-celled protozoan genera in the phylum Apircomplexa (all referred to as coccidian). *Cryptosporidium* along with other genera in the phylum Apircomplexa develop in the gastrointestinal tract of vertebrates through all their life cycle – in short, they live in the intestines of animals and people. This microscopic pathogen causes a disease called *Cryptosporidiosis* (crip-toe-spor-id-ee-O-sis). The dormant (inactive) form of *Cryptosporidium* called an oocyst (O-o-sist) is excreted in the feces (stool) of infected humans and animals. The tough-walled oocysts survive under a wide range of environmental conditions.

Several species of *Cryptosporidium* were incorrectly named after the host in which they were found; subsequent studies have invalidated many species. Now, eight valid species of Cryptosporidium (see Table 11.3) have been named. Upton (1997) reports that *muris* infects the gastric glands of laboratory rodents and several other mammalian species, but (even though several texts state otherwise) is not known to infect humans. However, *parvum* infects the small intestine of an unusually wide range of mammals, including humans, and is the zoonotic species responsible for human Cryptosporidiosis. In most mammals *parvum* is predominately a parasite of neonate (new-born) animals. He points out that even though exceptions occur, older animals generally develop poor infections, even when unexposed previously to the parasite. Humans are the one host that can be seriously infected at any time in their lives, and only previous exposure to the parasite results in either full or partial immunity to challenge infections.

Oocysts are present in most surface bodies of water across the United States, many of which supply public drinking water. Oocysts are more prevalent in surface waters when heavy rains increase runoff of wild and domestic animal wastes from the land or when sewage treatment plants are overloaded or break down. Only laboratories with specialized capabilities can detect the presence of Cryptosporidium oocysts in water. Unfortunately, present sampling and detection methods are unreliable. Recovering oocysts trapped on the material used to filter water samples is difficult. Once a sample is obtained, however, determining whether the oocyst is alive or whether it is the species parvum that can infect humans is easily accomplished by looking at the sample under a microscope.

The number of oocysts detected in raw (untreated) water varies with location, sampling time, and laboratory methods. Water treatment plants remove most, but not always all, oocysts. Low numbers of oocysts are enough to cause Cryptosporidiosis, but the low numbers of oocysts sometimes present in drinking water are not considered cause for alarm in the public. Protecting water supplies from *Cryptosporidium* demands multiple barriers. Why? Because *Cryptosporidium* oocysts have

---

TABLE 11.3

Valid Named Species of Cryptosporidium (Fayer et al., 1997)

Species	Host
C. baileyi	Chicken
C. felis	Domestic cat
C. meleagridis	Turkey
C. murishouse	House mouse
C. nasorium	Fish
C. parvum	House mouse
C. serpentis	Corn snake
C. wrairi	Guinea pig

tough walls that can withstand many environmental stresses and are resistant to the chemical disinfectants such as chlorine that are traditionally used in municipal drinking water systems.

Physical removal of particles (including oocysts) from water by filtration is an important step in the water treatment process. Typically, water pumped from rivers or lakes into a treatment plant is mixed with coagulants, which help settle out particles suspended in the water. If sand filtration is used, even more particles are removed. Finally, the clarified water is disinfected and piped to customers. Filtration is the only conventional method now in use in the United States for controlling *Cryptosporidium*.

Ozone is a strong disinfectant that kills protozoa if enough doses and contact times are used, but ozone leaves no residual for killing microorganisms in the distribution system, as does chlorine. The high costs of new filtration or ozone treatment plants must be weighed against the benefits of additional treatment. Even well operated water treatment plants cannot ensure that drinking water will be completely free of *Cryptosporidium* oocysts. Water treatment methods alone cannot solve the problem; watershed protection and monitoring of water quality are critical. As mentioned, watershed protection is another barrier to *Cryptosporidium* in drinking water. Land use controls such as septic systems regulations and best management practices to control runoff can help keep human and animal wastes out of water.

Under the Surface Water Treatment Rule of 1989, public water systems must filter surface water sources unless water quality and disinfection requirements are met and a watershed control program is maintained. This rule, however, did not address *Cryptosporidium*. The USEPA has now set standards for turbidity (cloudiness) and coliform bacteria (which indicate that pathogens are probably present) in drinking water. Frequent monitoring must occur to provide officials with early warning of potential problems to enable them to take steps to protect public health. Unfortunately, no water quality indicators can reliably predict the occurrence of Cryptosporidiosis. More accurate and rapid assays of oocysts will make it possible to notify residents promptly if their water supply is contaminated with *Cryptosporidium* and thus avert outbreaks.

The bottom line: The collaborative efforts of public water systems, government agencies, health care providers, and individuals are needed to prevent outbreaks of Cryptosporidiosis.

Cryptosporidiosis

Cryptosporidium parvum is an important emerging pathogen in the U.S. and a cause of severe, life-threatening disease in patients with AIDS. No safe and effective form of specific treatment for Cryptosporidiosis has been identified to date. The parasite is transmitted by ingestion of oocysts excreted in the feces of infected humans or animals. The infection can therefore be transmitted from person-to-person, through ingestion of contaminated water (drinking water and water used for recreational purposes) or food, from animal to person, or by contact with fecally contaminated environmental surfaces. Outbreaks associated with all these modes of transmission have been documented. Patients with human immunodeficiency virus infection should be made more aware of the many ways that Cryptosporidium species are transmitted, and they should be given guidance on how to reduce their risk of exposure. Since the Milwaukee outbreak, concern about the safety of drinking water in the U.S. has increased, and new attention has been focused on determining and reducing the risk of Cryptosporidiosis from community and municipal water supplies. Cryptosporidiosis is spread by putting something in the mouth that has been contaminated with the stool of an infected person or animal. In this way, people swallow the Cryptosporidium parasite. As previously mentioned, a person can become infected by drinking contaminated water or eating raw or undercooked food contaminated with Cryptosporidium oocysts; direct contact with the droppings of infected animals or stools of infected humans; or hand-to-mouth transfer of oocysts from surfaces that may have become contaminated with microscopic amounts of stool from an infected person or animal. The symptoms may appear two-ten days after infection by the parasite. Although some persons may not have symptoms, others have watery diarrhea, headache, abdominal cramps, nausea, vomiting, and low-grade fever. These symptoms may lead to weight loss and dehydration. In otherwise healthy persons, these symptoms usually last one- two weeks, at which time the immune system can stop the infection. In persons with suppressed immune systems, such as persons who have AIDS or who recently have had an organ or bone marrow transplant, the infection may continue and become life threatening. Currently, no safe and effective cure for Cryptosporidiosis exists. People with normal immune systems improve without taking antibiotic or antiparasitic medications. The treatment recommended for this diarrheal illness is to drink plenty of fluids and to get extra rest. Physicians may prescribe medication to slow the diarrhea during recovery. The best way to prevent Cryptosporidiosis is to:

- Avoid water or food that may be contaminated.
- Wash hands after using the toilet and before handling food.

If you work in a childcare center where you change diapers, be sure to wash your hands thoroughly with plenty of soap and warm water after every diaper change, even if you wear gloves. During community-wide outbreaks caused by contaminated drinking water, drinking water practitioners should inform the public to boil drinking water for one minute to kill the Cryptosporidium parasite.

Source: Juranek (1995) Clinical Infectious Diseases:

CYCLOSPORA

Cyclospora organisms, which until recently were considered blue-green algae, were discovered at the turn of the century. The first human cases of *Cyclospora* infection were reported in the 1970s. In the early 1980s, *Cyclospora* was recognized as a pathogen in patients with AIDS. We now know that *Cyclospora* is endemic in many parts of the world and appears to be an important cause of traveler's diarrhea. *Cyclospora* are two to three times larger than *Cryptosporidium*, but otherwise have similar features. *Cyclospora* diarrheal illness in patients with healthy immune systems can be cured with a week of therapy with timethoprim-sulfamethoxazole (TMP-SMX). So, exactly what is *Cyclospora?* In 1998, the CDC described *Cyclospora cayetanensis* as a unicellular parasite previously known as a cyanobacterium-like (blue-green algae-like) or coccidian-like body (CLB). The disease is known as Cyclosporasis. *Cyclospora* infects the small intestine and causes an illness characterized by diarrhea with frequent stools. Other symptoms can include loss of appetite, bloating, gas, stomach cramps, nausea, vomiting, fatigue, muscle ache, and fever. Some individuals infected with Cyclospora may not show symptoms. Since the first known cases of illness caused by *Cyclospora* infection were reported in the medical journals in the 1970s, cases have been reported with increased frequency from various countries since the mid-1980s (in part because of the availability of better techniques for detecting the parasite in stool specimens).

Huang et al. (1995) detailed what they believe is the first known outbreak of diarrheal illness associated with *Cyclospora* in the United States. The outbreak, which occurred in 1990, consisted of 21 cases of illness among physicians and others working at a Chicago hospital. Contaminated tap water from a physicians' dormitory at the hospital was the probable source of the organisms. The tap water probably picked up the organism while in a storage tank at the top of the dormitory after the failure of a water pump. The transmission of *Cyclospora* is not a straightforward process. When infected persons excrete the oocyst state of *Cyclospora* in their feces, the oocysts are not infectious and may require from days to weeks to become so (i.e., to sporulate). Therefore, transmission of *Cyclospora* directly from an infected person to someone else is unlikely. However, indirect transmission can occur if an infected person contaminates the environment and oocysts have enough time, under appropriate conditions, to become infectious. For example, *Cyclospora* may be transmitted by ingestion of water or food contaminated with oocysts. Outbreaks linked to contaminated water, as well as outbreaks linked to various types of fresh produce, have been reported in recent years (Herwaldt et al., 1997). How common the various modes of transmission and sources of infection is not yet known, nor is it known whether animals can be infected and serve as sources of infection for humans.

Note: Cyclospora organisms have not yet been grown in tissue cultures or laboratory animal models.

Persons of all ages are at risk for infection. Persons living or traveling in developing countries may be at increased risk; but infection can be acquired worldwide, including in the United States. In some countries of the world, infection appears to be seasonal. Based on currently available information, avoiding water or food that may be contaminated with stool is the best way to prevent infection. Reinfection can occur.

Note: De Zuane (1997) points out that pathogenic parasites are not easily removed or eliminated by conventional treatment and disinfection unit processes. This is particularly true for *Giardia lamblia*, *Cryptosporidium*, and *Cyclospora*. Filtration facilities can be adjusted in depth, prechlorination, filtration rate, and backwashing to become more effective in the removal of cysts. The pretreatment of protected water shed raw water is a major factor in the elimination of pathogenic protozoa.

HELMINTHS

Helminths are parasitic worms that grow and multiply in sewage (biological slimes) and wet soil (mud). They multiply in wastewater treatment plants; strict aerobes, they have been found in activated sludge and particularly in trickling filters, and therefore appear in large concentrations in treated domestic liquid waste. The worm, in one of its many lifecycle phases, enters the skin or is ingested. Generally, they are not a problem in drinking water supplies in the United States because both their egg and larval forms are large enough to be trapped during conventional water treatment. In addition, most Helminths are not waterborne, so chances of infection are minimized (WHO, 1990).

WASTEWATER TREATMENT AND BIOLOGICAL PROCESSES

Uncontrolled bacteria in industrial public water systems produce an endless variety of problems including disease, equipment damage, and product damage. Unlike the microbiological problems that can occur in public water systems, in wastewater treatment, microbiology can be applied as a beneficial science for the destruction of pollutants in wastewater (Kemmer, 1979). It should be noted that all the biological processes used for the treatment of wastewater (in particular) are derived or modeled from processes occurring naturally in nature. The processes discussed in the following are typical examples. It also should be noted that "by controlling the environment of microorganisms, the decomposition of wastes is speeded up. Regardless of the type of waste, the biological treatment process consists of controlling the environment required for optimum growth of the microorganism involved" (Metcalf & Eddy, 2003).

AEROBIC PROCESS

In *aerobic treatment processes*, organisms use free, elemental oxygen and organic matter together with nutrients (nitrogen, phosphorus) and trace metals (iron, etc.) to produce more organisms and stable dissolved and suspended solids and carbon dioxide (see Figure 11.10).

Oxygen		More bacteria
Bacteria	$\Rightarrow$	Stable solids
Organic matter		Settleable solids
Nutrients		Carbon dioxide

FIGURE 11.10   Aerobic decomposition.

ANAEROBIC PROCESS

The *anaerobic treatment process* consists of two steps, occurs completely in the absence of oxygen, and produces a useable by-product, methane gas. In the first step of the process, facultative microorganisms use the organic matter as food to produce more organisms, volatile (organic) acids, carbon dioxide, hydrogen sulfide, and other gases and some stable solids (see Figure 11.11).

In the second step, anaerobic microorganisms use volatile acids as their food source. The process produces more organisms, stable solids, and methane gas that can be used to provide energy for various treatment system components (see Figure 11.12).

ANOXIC PROCESS

In the *anoxic treatment process* (anoxic means without oxygen), microorganisms use the fixed oxygen in nitrate compounds as a source of energy. This process produces more organisms and removes nitrogen from the wastewater by converting it to nitrogen gas that is released into the air (see Figure 11.13).

PHOTOSYNTHESIS

Green algae use carbon dioxide and nutrients in the presence of sunlight and chlorophyll to produce more algae and oxygen (see Figure 11.14).

GROWTH CYCLES

All organisms follow a basic growth cycle that can be shown as a growth curve. This curve occurs when the environmental conditions required for the organism are reached. It is the environmental conditions (i.e., oxygen availability, pH, temperature, presence or absence of nutrients, presence or absence of toxic materials) that determine when a group of organisms will predominate. Obviously, this information can be very useful in operating a biological treatment process (see Figure 11.15).

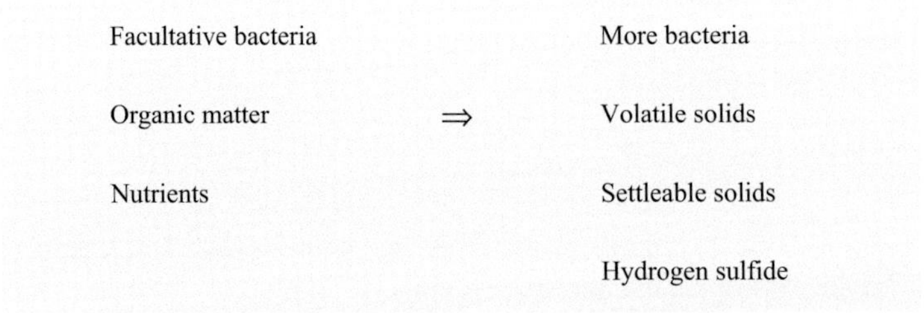

Facultative bacteria		More bacteria
Organic matter	$\Rightarrow$	Volatile solids
Nutrients		Settleable solids
		Hydrogen sulfide

FIGURE 11.11   Anaerobic decomposition – ffirst step.

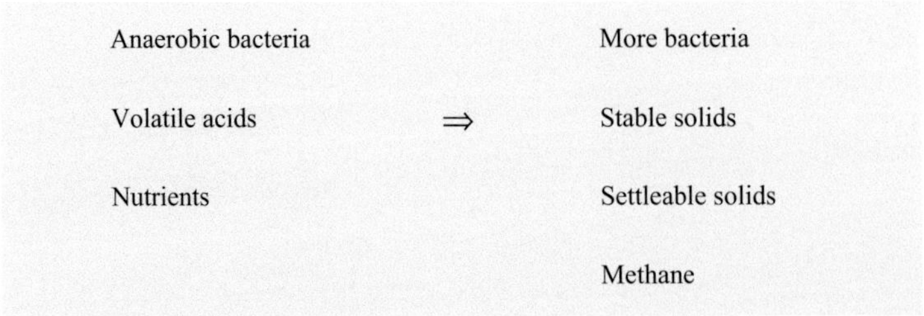

Anaerobic bacteria		More bacteria
Volatile acids	$\Rightarrow$	Stable solids
Nutrients		Settleable solids
		Methane

FIGURE 11.12   Anaerobic decomposition – second step.

FIGURE 11.13   Anoxic decomposition.

FIGURE 11.14   Photosynthesis.

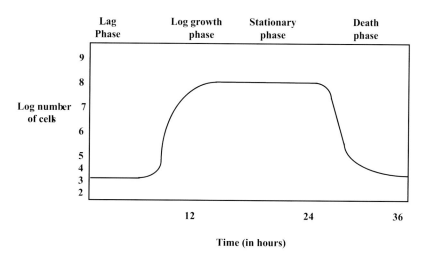

FIGURE 11.15   Microorganism growth curve.

BIOGEOCHEMICAL CYCLES

Several chemicals are essential to life and follow predictable cycles through nature. In these natural cycles or *biogeochemical cycles*, the chemicals are converted from one form to another as they progress through the environment. The water and wastewater operator should be aware of those cycles dealing with the nutrients (e.g., carbon, nitrogen, and sulfur) because they have a major impact on the performance of the plant and may require changes in operation at various times of the year to keep them functioning properly; this is especially the case in wastewater treatment. The microbiology of each cycle deals with the biotransformation and subsequent biological removal of these nutrients in wastewater treatment plants.

Note: Smith categorizes biogeochemical cycles into two types, the *gaseous* and the *sedimentary*. Gaseous cycles include the carbon and nitrogen cycles. The main sink of nutrients in the gaseous cycle is the atmosphere and the ocean. Sedimentary cycles include the sulfur cycle. The main sink for sedimentary cycles is soil and rocks of the earth's crust (Smith, 1974).

CARBON CYCLE

Carbon, which is an essential ingredient of all living things, is the basic building block of the large organic molecules necessary for life. Carbon is cycled into food chains from the atmosphere, as shown in Figure 11.16. From Figure 11.16 it can be seen that green plants obtain carbon dioxide (CO_2) from the air and, through photosynthesis, described by Asimov (1989) as the "most important chemical process on Earth," it produces the food and oxygen that all organisms live on. Part of the carbon produced remains in the living matter; the other part is released as CO_2 in cellular respiration. Miller (1988) points out that the carbon dioxide released by cellular respiration in all living organisms is returned to the atmosphere.

Some carbon is contained in a buried dead animal and plant materials. Much of these buried animal and plant materials were transformed into fossil fuels. Fossil fuels, coal, oil, and natural gas contain large amounts of carbon. When fossil fuels are burned, stored carbon combines with oxygen in the air to form carbon dioxide, which enters the atmosphere. In the atmosphere, carbon dioxide acts as a beneficial heat screen as it does not allow the radiation of the earth's heat into space. This balance is important. The problem is that as more carbon dioxide from burning is released into the atmosphere, the balance can and is being altered.

Odum (1983) warns that the recent increase in consumption of fossil fuels "coupled with the decrease in 'removal capacity' of the green belt is beginning to exceed the delicate balance." Massive increases of carbon dioxide into the atmosphere tend to increase the possibility of global warming. The consequences of global warming "would be catastrophic ... and the resulting climatic change would be irreversible" (Abrahamson, 1988).

NITROGEN CYCLE

Nitrogen is an essential element that all organisms need. In animals, nitrogen is a component of crucial organic molecules such as proteins and DNA and constitutes 1–3% dry weight of cells. Our atmosphere contains 78% by volume of nitrogen, yet it is not a common element on earth. Although nitrogen is an essential ingredient for plant growth, it is chemically very inactive, and before most of the biomass can incorporate it, it must be *fixed*. Special nitrogen-fixing bacteria found in soil and water fix nitrogen. Thus, microorganisms play a major role in nitrogen cycling in the environment.

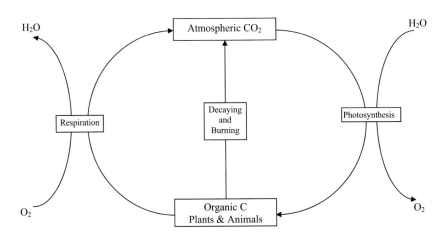

FIGURE 11.16   Carbon cycle.

These microorganisms (bacteria) can take nitrogen gas from the air and convert it to nitrate. This is called *nitrogen fixation*. Some of these bacteria occur as free-living organisms in the soil. Others live in a *symbiotic relationship* (a close relationship between two organisms of different species and one where both partners benefit from the association) with plants.

An example of a symbiotic relationship, related to nitrogen, can be seen, for example, in the roots of peas. These roots have small swellings along their length. These contain millions of symbiotic bacteria, which can take nitrogen gas from the atmosphere and convert it to nitrates that can be used by the plant. Then the plant is plowed into the soil after the growing season to improve the nitrogen content. Price (1984) describes the nitrogen cycle as an example "of a largely complete chemical cycle in ecosystems with little leaching out of the system." Simply, the nitrogen cycle provides various bridges between the atmospheric reservoirs and the biological communities (see Figure 11.17). Atmospheric nitrogen is fixed either by natural or industrial means. For example, nitrogen is fixed by lightning or by soil bacteria that convert it to ammonia, then to nitrite, and finally to nitrates, which plants can use. Nitrifying bacteria make nitrogen from animal wastes. Denitrifying bacteria convert nitrates back to nitrogen and release them as nitrogen gas.

The logical question now is "What does all of this have to do with water?" The best way to answer this question is to ask another question. Have you ever dived into a slow-moving stream and had the noxious misfortune to surface right in the middle of an algal bloom? When this happens to you, the first thought that runs through your mind is, "Where is my nose plug?" Why? Because of the horrendous stench, disablement of the olfactory sense is a necessity. If too much nitrate, for example, enters the water supply – as runoff from fertilizers – it produces an overabundance of algae, called algal bloom. If this runoff from fertilizer gets into a body of water, algae may grow so profusely that they form a blanket over the surface. This usually happens in summer, when the light levels and warm temperatures favor rapid growth.

In their voluminous and authoritative text, *Wastewater Engineering: Treatment, Disposal, & Reuse* (Metcalf & Eddy, 2003), it is noted that nitrogen is found in wastewater in the form of urea. During wastewater treatment, the urea is transformed into ammonia nitrogen. Since ammonia exerts a BOD and chlorine demand, high quantities of ammonia in wastewater effluents are undesirable. The process of nitrification is utilized to convert ammonia to nitrates. *Nitrification* is a biological process that involves the addition of oxygen to the wastewater. If further treatment is necessary,

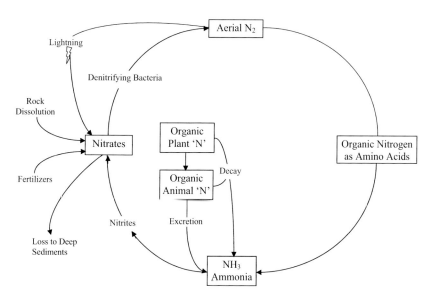

FIGURE 11.17   Nitrogen cycle.

another biological process called *denitrification* is used. In this process, nitrate is converted into nitrogen gas, which is lost to the atmosphere, as can be seen in Figure 11.17. From the wastewater operator's point of view, nitrogen and phosphorus are both considered limiting factors for productivity. Phosphorus discharged into streams contributes to pollution. Of the two, nitrogen is harder to control but is found in smaller quantities in wastewater.

Sulfur Cycle

Sulfur, like nitrogen, is characteristic of organic compounds. The *sulfur cycle* is both sedimentary and gaseous (see Figure 11.18). The principal forms of sulfur that are of special significance in water quality management are organic sulfur, hydrogen sulfide, elemental sulfur, and sulfate (Tchobanoglous & Schroeder, 1985). Bacteria play a major role in the conversion of sulfur from one form to another. In an anaerobic environment, bacteria break down organic matter thereby producing hydrogen sulfide with its characteristic rotten-egg odor. A bacterium called *Beggiatoa* converts hydrogen sulfide into elemental sulfur into sulfates. Other sulfates are contributed by the dissolving of rocks and some sulfur dioxide. Sulfur is incorporated by plants into proteins. Organisms then consume some of these plants. Many heterotrophic anaerobic bacteria liberate sulfur from proteins as hydrogen sulfide.

Phosphorus Cycle

Phosphorus is another chemical element that is common in the structure of living organisms (see Figure 11.19). However, the phosphorus cycle is different from the hydrologic, carbon, and nitrogen cycles because phosphorus is found in sedimentary rock. These massive deposits are gradually eroding to provide phosphorus to ecosystems. A large amount of eroded phosphorus ends up in deep sediments in the oceans and in lesser amounts in shallow sediments. Part of the phosphorus comes to land when marine animals surface.

Decomposing plant or animal tissue and animal droppings return organic forms of phosphorus to the water and soil. For example, fish-eating birds play a role in the recovery of phosphorus. The guano deposit, bird excreta, of the Peruvian coast is an example. Humans have hastened the rate of phosphorus loss through mining and the production of fertilizers, which is washed away and lost.

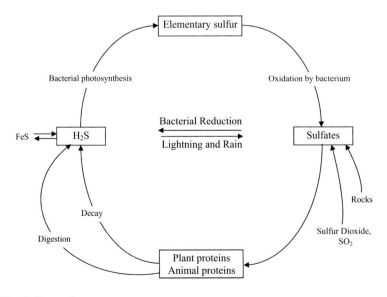

FIGURE 11.18   Sulfur cycle.

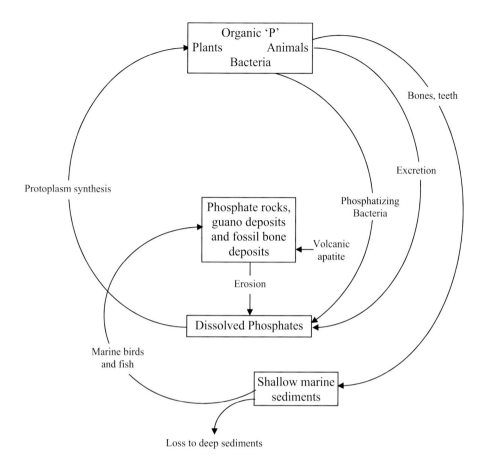

FIGURE 11.19 Phosphorus cycle.

Odum (1983) suggested, however, that there is no immediate cause for concern since the known reserves of phosphate are quite large.

Phosphorus has become very important in water quality studies because it is often found to be a limiting factor. Metcalf & Eddy Inc. (2003) reported that control of phosphorus compounds that enter surface waters and contribute to the growth of algal blooms is of considerable interest and has generated much study. Phosphorus acts as fertilizer upon entering a stream, promoting the growth of undesirable algae populations or alga blooms. As the organic matter decays, DO levels decrease and fish and other aquatic species die.

Although it is true that phosphorus discharged into streams is a contributing factor of stream pollution, it is also true that phosphorus is not the lone factor. Odum (1975) warns against what he calls the one-factor control hypothesis, (i.e., one-problem/one-solution syndrome). He notes that environmentalists in the past have focused on one of two items, like phosphorus contamination, and have failed to understand that the strategy for pollution control must involve reducing the input of all enriching and toxic materials.

CHAPTER SUMMARY

Although wastewater operators and others involved with the industry do not need to be microbiologists to perform many of their routine duties, it is important for them to be aware of the substance they are working with and the components that make up wastewater. Knowledge about and awareness of the microorganisms that may be present in the wastewater is important not only to ensure their removal from the waste stream but also important for operators' safety and health.

REVIEW QUESTIONS

1. The three major groups of microorganisms that cause disease in water are:
    a. Bacteria
    b. Viruses
    c. Protozoa
    d. All the above

2. When does a river of good quality show its highest bacterial numbers?
    a. During nights
    b. Summer
    c. Winter
    d. During rainstorms

3. Are coliform organisms pathogenic?
    a. Yes
    b. No
    c. On occasion
    d. Could be

4. How do bacteria reproduce?
    a. Nuclear fusion
    b. By mating
    c. Binary fission
    d. Singular fission

5. The three common shapes of bacteria are:
    a. Spheres, rods, spirals
    b. Spheres, rods, squares
    c. Squares, rods, spheres
    d. Rods, squares, spirals

6. Three waterborne diseases caused by bacteria are:
    a. Hoof and mouth, dysentery, cholera
    b. Cholera, dysentery, ameobis
    c. Gastroenteritis, cholera, typhoid
    d. Cancer

7. Two protozoa-caused waterborne diseases are:
    a. Diabetes and cancer
    b. Amoebic dysentery, Giardiasis
    c. Host typhoid and cancer
    d. Cholera and Giardiasis

8. When a protozoon is in a resting phase, it is called a _____.
    a. Cyst
    b. Sleep
    c. Dormant
    d. Dud

9. For a virus to live it must have a _____.
    a. Nest
    b. Home

    c. Victim

    d. Host

10. What problems do algae cause in wastewater?

    a. Heartburn

    b. Cancer

    c. Loss of electricity

    d. Plug screens and machinery

ANSWER KEY

1. The three major groups of microorganisms that cause disease in water are:

    a. Bacteria

    b. Viruses

    c. Protozoa

    d. **All the above**

2. When does a river of good quality show its highest bacterial numbers?

    a. During nights

    b. Summer

    c. Winter

    d. **During rainstorms**

3. Are coliform organisms pathogenic?

    a. Yes

    b. **No**

    c. On occasion

    d. Could be

4. How do bacteria reproduce?

    a. Nuclear fusion

    b. By mating

    c. **Binary fission**

    d. Singular fission

5. The three common shapes of bacteria are:

    a. **Spheres, rods, spirals**

    b. Spheres, rods, squares

    c. Squares, rods, spheres

    d. Rods, squares, spirals

6. Three waterborne diseases caused by bacteria are:

    a. Hoof and mouth, dysentery, cholera

    b. Cholera, dysentery, ameobis

    c. **Gastroenteritis, cholera, typhoid**

    d. Cancer

7. Two protozoa-caused waterborne diseases are:

    a. Diabetes and cancer

    b. **Amoebic dysentery, Giardiasis**

    c. Host typhoid and cancer

    d. Cholera and Giardiasis

8. When a protozoon is in a resting phase, it is called a _____.

    a. **Cyst**

    b. Sleep

    c. Dormant

    d. Dud

9. For a virus to live it must have a _____.

    a. Nest

    b. Home

    c. Victim

    d. **Host**

10. What problems do algae cause in wastewater?

    a. Heartburn

    b. Cancer

    c. Loss of electricity

    d. **Plug screens and machinery**

REFERENCES

Abrahamson, D.E., Ed. 1988. *The Challenge of Global Warming.* Washington, DC: Island Press.

Asimov, I. 1989. *How Did We Find out about Photosynthesis?* New York: Walker & Company.

Black, R.E., Dykes, A.C., Anderson, K.E., Wells, J.G., Sinclair, S.P., Gary, G.W., Hatch, M.H., and Ginagaros, E.J. 1981. Handwashing to prevent diarrhea in day-care centers, *Am. J. Epidmilo.* 113, 445–451.

Brodsky, R.E., Spencer, H.C., and Schultz, M.G. 1974. Giardiasis in American travelers to the Soviet Union, *J. Infect. Dis.* 130, 319–323.

CDC. 1979. *Intestinal Parasite Surveillance, Annual Summary 1978.* Atlanta, Georgia: Centers for Disease Control.

CDC. 1983. *Water-Related Disease Outbreaks Surveillance, Annual Summary 1983.* Atlanta, Georgia: Centers for Disease Control.

CDC. 1995. *Giardiasis.* Juranek, D.D., Ed. Atlanta, Georgia: Centers for Disease Control.

Davidson, R.A. 1984. Treating Giardiasis. Accessed 12/06/20 @ https://pubmed.nelos.niha.gov.

De Zuane, J. 1997. *Handbook of Drinking Water Quality.* New York: Wiley.

Fayer, R., Speer, C.A., and Dudley, J.P. 1997. *The General Biology Cryptosporidium in Cryptosporidium and Cryptosporidiosis.* Fayer, R., Ed. Boca Raton, FL: CRC Press.

Frost, F., Plan, B., and Liechty, B. 1984. Giardia prevalence in commercially trapped mammals, *J. Environ. Health* 42, 245–249.

Herwaldt, F.L., et al. 1997. An outbreaking 1996 of Cyclosporasis associated with imported raspberries, *N. Engl. J. Med.* 336, 1548–1556.

Huang, P., Weber, J.T. Sosin, D.M., et al. 1995. Cyclospora, *Ann. Intern. Med.* 123, 401–414.

Jarroll, E.L., Jr., Gingham, A.K, and Meyer, E.A. 1979. Giardia cyst destruction: Effectiveness of six small-quantity water disinfection methods, *Am J. Trop. Med. Hygiene* 29, 8–11.

Juranek, D.D. 1995. Cryptosporidium parvum, *Clinical Infect. Dis.* 21, S37–61.

Kemmer, F.N. 1979. *Water: The Universal Solvent*, 2nd ed. Oak Brook, IL: Nalco Chemical Company.

Kordon, C. 1992. *The Language of the Cell.* New York: McGraw-Hill.

Koren, H. 1991. *Handbook of Environmental Health and Safety: Principles and Practices.* Chelsea, MI: Lewis Publishers.

LeChevallier, M.W., Norton, W.D., and Less, R.G. 1991. Occurrences of Giardia and Cryptosporidium spp. In surface water supplies, *Applied and Environ. Microbiol.* 57, 2610–2616.

Mayo Foundation. 1996. *The "Bug" That made Milwaukee Famous.* Rochester, MN: Mayo Foundation.

Metcalf & Eddy, Inc. 2003. *Wastewater Engineering: Treatment, Disposal and Reuse.* New York: McGraw-Hill, Inc.

Miller, G.T. 1988. *Environmental Science: An Introduction.* Belmont, CA: Wadsworth Publishing Company.

Odum, E.P. 1975. *Ecology: The Link between the Natural and the Social Sciences.* New York: Holt, Rinehart and Winston, Inc.

Odum, E.P. 1983. *Basic Ecology.* Philadelphia, PA: Saunders College Publishing.

Panciera, R.J., Thomasson, R.W., and Garner, R.M. 1971. Cryptosporidium infection in a calf, *Vet Pathol.* 8, 479.

Pennak, R.W. 1989. *Fresh-Water Invertebrates of the United States*, 3rd ed. New York: Wiley.

Price, P.W. 1984. *Insect Ecology.* New York: Wiley.

Rendtorff, R.C. 1954. The experimental transmission of human intestinal protozoan parasites II *Giardia lamblia* cysts given in capsules, *Am. J. Hygiene* 59, 209–220.

Rose, J.B. 1983. Survey of potable water supplies for protozoans. *Environ. Sci. Technol.* 25, 1393–1399.

Rose, J.B., Gerb, C.P., and Jakubowski, W. 1985 Survey of potable water supplies for Cryptosporidium and Giardia, *Environm. Sci. Technol.* 25, 1393–1399.

Singleton, P. and Sainsbury, D. 1994. *Dictionary of Microbiology and Molecular Biology*, 2nd ed. New York: Wiley.

Slavin, D. 1955. Cryptosporidium melagridis, *J. Comp. Pathol.* 65, 262.

Smith, R.L. 1974 *Ecology and Field Biology.* New York: Harper & Row.

Spellman, F.R. 1996. *Stream Ecology and Self-Purification: An Introduction for Wastewater and Water Specialists.* Boca Raton, FL: CRC Press.

Spellman, F.R. 1997. *Microbiology for Water and Wastewater Operators.* Boca Raton, FL: CRC Press.

Tchobanoglous, G. and Schroeder, E.D. 1985. *Water Quality.* Reading, MA: Addison-Wesley Publishing Company.

Tyzzer, E.E. 1912. Cryptosporidium parvum sp.: A Coccidium found in the small intestine of the common mouse, *Arch. Protistenkd.* 26, 394 m.

Upton, S.J. 1997. *Basic Biology of Cryptosporidium*, Manhattan, KS: Kansas State University.

Walsh, J.D., and Warren, K.S. 1979. Selective primary health care: An interim strategy for disease control in developing countries, *N. Engl. J. Med.* 301, 974–976.

Weller, P.F. 1985. Intestinal protozoa: Giardiasis. *Scientific American Medicine.* Accessed 12/6/20 @ https://medicine.nhmedical.com

Welsh, J.D. 1981. Selective primary health care. *N. Engl. J. Med.* 301, 974–976.

Wistriech, G.A., and Lechtman, M.D. 1980. *Microbiology*, 3rd ed. New York: Macmillan Publishing Company.

World Health Organization (WHO). 1990. *Guidelines for Drinking Water Quality*, 2nd ed., Vol. 2. Austria: World Health Organization.

12 Basic Wastewater Ecology

Photo Courtesy of David McNeil

Have you ever peered into a microscope at a drop of pond or stream water and realized that there is more life around all of us than just that which is visible to us each day?

– Frank R, Spellman (1996)

Streams are arteries of earth, beginning in capillary creeks, brooks, and rivulets. No matter the source, they move in only one direction – downhill – the heavy hand of gravity tugs and drags the stream toward the sea. During its inexorable flow downward, now and then there is an abrupt change in geology. Boulders are mowed down by "slumping" (gravity) from their in-place points high up on canyon walls. As stream flow grinds, chisels, and sculpts the landscape, the effort is increased by momentum, augmented by turbulence provided by rapids, cataracts, and waterfalls. These falling waters always hypnotize us just as fire gazing or wave-watching does. Before emptying into the sea, streams often pause, forming lakes. When one stares into a healthy lake, its phantom blue-green eye stares right back. Only for a moment – relatively speaking, of course, because all lakes are ephemeral, doomed. Eventually the phantom blue-green eye is close lidded by the moist verdant green of landfill. For water that escapes the temporary bounds of a lake, most of it evaporates or moves on to the gigantic sink – the sea – where the cycle continues, forever more … it is hoped. The bottom line: It is important for wastewater professionals to have knowledge of stream ecology and a stream's ability to self-purify for one basic but important reason: Conventional wastewater treatment is nothing more than a stream in a box.

–Frank R. Spellman, 2019

DOI: 10.1201/9781003207146-17

LEARNING OBJECTIVES

After studying this chapter, you should be able to:

- Define ecology.
- Discuss why ecology is important to wastewater professionals.
- Discuss leaf processing in streams.
- Discuss levels of organization.
- Define an ecosystem.
- Discuss energy flow in an ecosystem.
- Identify aquatic food chain.
- Discuss food chain efficiency.
- Describe ecological pyramids.
- Define productivity.
- Discuss productivity.
- Define population ecology.
- Discuss patterns of distribution.
- Discuss bare-rock-succession.
- Describe stream genesis.
- Describe stream structure.
- Discuss water flow in a stream.
- Discuss steam water discharge.
- Discuss transport of material in a stream.
- Describe characteristics of stream channels.
- Describe stream profiles.
- Describe stream sinuosity.
- Describe stream riffles, pools, and bars.
- Define a floodplain.
- Describe organism adaptations to stream current.
- Describe benthic life.
- Describe benthic macroinvertebrates.
- Describe non-insect macroinvertebrates.

KEY DEFINITIONS

Abiotic factor – the nonliving part of the environment composed of sunlight, soil, mineral elements, moisture, temperature, topography, minerals, humidity, tide, wave action, wind, and elevation.

Important Note: Every community is influenced by a set of abiotic factors. While it is true that the abiotic factors affect the community members, it is also true that the living (biotic factors) may influence the abiotic factors. For example, the amount of water lost through the leaves of plants may add to the moisture content of the air. Also, the foliage of a forest reduces the amount of sunlight that penetrates the lower regions of the forest. The air temperature is therefore much lower than in non-shaded areas (Tomera, 1989).

Autotroph (green plants) – fixes energy of the sun and manufactures food from simple, inorganic substances.

Biogeochemical cycles – are cyclic mechanisms in all ecosystems by which biotic and abiotic materials are constantly exchanged.

Biotic factor (community) – the living part of the environment composed of organisms that share the same area, are mutually sustaining, interdependent, and constantly fixing, utilizing, and dissipating energy.

Community – in an ecological sense, community includes all the populations occupying a given area.

Consumers and decomposers – dissipate energy fixed by the producers through food chains or webs. The available energy decreases by 80%–90% during transfer from one trophic level to another.

Ecology – is the study of the interrelationship of an organism or a group of organisms and their environment.

Ecosystem – is the community and the nonliving environment functioning together as an ecological system.

Environment – is everything that is important to an organism in its surroundings.

Heterotrophs – (animals) use food stored by the autotroph, rearrange it, and finally decompose complex materials into simple inorganic compounds. Heterotrophs may be carnivorous (meat eaters), herbivorous (plant-eaters), or omnivorous (plant- and meat-eaters).

Homeostasis – is a natural occurrence during which an individual population or an entire ecosystem regulates itself against negative factors and maintains an overall stable condition.

Niche – is the role that an organism plays in its natural ecosystem, including its activities, resource use, and interaction with other organisms.

Pollution – is an adverse alteration to the environment by a pollutant.

OVERVIEW

The "control of nature" is a phrase conceived in arrogance, born of the Neanderthal age of biology and the convenience of man.

– Rachel Carson (1962)

What is ecology? Why is ecology important? Why study ecology? These are all simple, straightforward questions; however, providing simple, straightforward answers is not that easy. Notwithstanding, the inherent difficulty with explaining any complex science in simple, straightforward terms, that is the purpose, the goal, the mission of this chapter. In short, the task of this chapter is to outline basic information that explains the functions and values of ecology and its interrelationships with other sciences, including ecology's direct impact on our lives. In doing so the author not only hopes to dispel the common misconception that ecology is too difficult for the average person to understand but also to instill the concept of ecology as an asset that can not only be learned but also cherished.

WHAT IS ECOLOGY?

Ecology can be defined in various and numerous ways. For example, ecology, or ecological science, is commonly defined in the literature as the scientific study of the distribution and abundance of living organisms and how the distribution and abundance are affected by interactions between the organisms and their environment. The term ecology was coined in 1866 by the German biologist Haeckel, and it loosely means "the study of the household [of nature]." Odum (1983) explains that the word "ecology" is derived from the Greek oikos, meaning home. Ecology then means the study of organisms at home. It means the study of an organism at its home. Ecology is the study of the relation of an organism or a group of organisms to their environment. In a broader sense, ecology is the study of the relation of organisms or groups to their environment.

Important Point: No ecosystem can be studied in isolation. If we were to describe ourselves, our histories, and what made us the way we are, we could not leave the world around us out of our description! So, it is with streams: they are directly tied in with the world around them. They take their chemistry from the rocks and dirt beneath them as well as for a great distance around them (Spellman, 1996).

Charles Darwin (1998) explained ecology in a famous passage in the *Origin*, a passage that helped establish the science of ecology. According to Darwin, a "web of complex relations" binds all living things in any region, Darwin writes. Adding or subtracting even a single species causes waves of change that race through the web, "onwards in ever-increasing circles of complexity." The simple act of adding cats to an English village would reduce the number of field mice. Killing mice would benefit the bumblebees, whose nests and honeycombs the mice often devour. Increasing the number of bumblebees would benefit the heartsease and red clover, which are fertilized almost exclusively by bumblebees. So, adding cats to the village could end by adding flowers. For Darwin, the whole of the Galapagos archipelago argues this fundamental lesson. The island volcanoes are much more diverse in their ecology than their biology. The contrast suggests that in the struggle for existence, species are shaped at least as much by the local flora and fauna as by the local soil and climate. "Why else would the plants and animals differ radically among islands that have the same geological nature, the same height, and climate" (Darwin, 1998).

Probably the best way to understand ecology – to get a good "feel" for it or to get to the heart of what ecology is all about is to read the following by (Carson, 1962):

> We poison the caddis flies in a stream and the salmon runs dwindle and die. We poison the gnats in a lake and the poison travels from link to link of the food chain and soon the birds of the lake margins become victims. We spray our elms and the following springs are silent of robin song, not because we sprayed the robins directly but because the poison traveled, step by step, through the now familiar elm leaf-earthworm-robin cycle. These are matters of record, observable, part of the visible world around us. They reflect the web of life – or death – that scientists know as ecology.

As Carson pointed out, what we do to any part of our environment has an impact upon other parts. In other words, there is an interrelationship between the parts that make up our environment. Probably the best way to state this interrelationship is to define ecology definitively – that is, to define it as it is used in this text: "Ecology is the science that deals with the specific interactions that exist between organisms and their living and nonliving environment" (Tomera, 1989). When environment was mentioned in the proceeding and as it is discussed throughout this text, it (the environment) includes everything important to the organism in its surroundings. The organism's environment can be divided into four parts:

1. Habitat and distribution – its place to live
2. Other organisms – whether friendly or hostile
3. Food
4. Weather-light, moisture, temperature, soil, etc.

There are four major subdivisions of ecology:

1. Behavioral Ecology
2. Population Ecology (Autecology)
3. Community Ecology (Synecology)
4. Ecosystem Ecology

Behavioral ecology is the study of the ecological and evolutionary basis for animal behavior. *Population ecology* (or autecology) is the study of the individual organism or a species, emphasizing life history, adaptations, and behavior. It is the study of communities, ecosystems, and biosphere. An example of autecology would be when biologists spend their entire lifetime studying the ecology of the salmon. *Community ecology* (or synecology), on the other hand, is the study of groups of organisms associated together as a unit and deals with the environmental problems caused by mankind. For example, the effect of discharging phosphorous-laden effluent into a stream involves several organisms. The activities of human beings have become a major component of many natural areas.

As a result, it is important to realize that the study of ecology must involve people. *Ecosystem ecology* is the study of how energy flow and matter interact with biotic elements of ecosystems (Odum, 1971).

Important Point: Ecology is generally categorized according to complexity; the primary kinds of organism under study (plant, animal, insect ecology); the biomes principally studied (forest, desert, benthic, grassland, etc.); the climatic or geographic area (e.g., artic or tropics) and/or the spatial scale (macro or micro) under consideration.

WHY IS ECOLOGY IMPORTANT?

Ecology, in its true sense, is a holistic discipline that does not dictate what is right or wrong. Instead, ecology is important to life on earth simply because it makes us aware, to a certain degree, of what life on earth is all about. Ecology shows us that each living organism has an ongoing and continual relationship with every other element that makes up our environment. Simply, ecology is all about interrelationships, intraspecific and interspecific, and on how important it is to maintain these relationships to ensure our very survival. At this point in this discussion there are literally countless examples that could be used to point out the importance of ecology and interrelationships. However, to demonstrate not only the importance of ecology but also to point out that an ecological principle can be a double-edged sword, depending on point of view (ecological problems along with pollution can be a judgment call – that is, they are a matter of opinion). A famous parable (*The Keeper of the Spring*) adapted from Peter Marshall's *Mr. Jones: Meet the Master* is used here to demonstrate that many misconceptions exist in ecology.

The Keeper of the Spring

This is the story of the keeper of the spring. He lived high in the Alps above an Austrian town and had been hired by the town council to clear debris from the mountain springs that fed the stream that flowed through the town. The man did his work well and the village prospered. Graceful swans floated in the stream. The surrounding countryside was irrigated. Several mills used the water for power. Restaurants flourished for townspeople and for a growing number of tourists. Years went by. One evening at the town council meeting someone questioned the money being paid to the keeper of the spring. No one seemed to know who he was or even if he was still on the job high up in the mountains. Before the evening was over, the council decided to dispense with the old man's services. Weeks went by and nothing seemed to change. Then autumn came. The trees began to shed their leaves. Branches broke and fell into the pools high up in the mountains. Down below, the villagers noticed the water becoming darker. A foul odor appeared. The swans disappeared. Also, the tourists. Soon disease spread through the town. When the town council reassembled, they realized that they had made a costly error. They found the old keeper of the spring and hired him back again. Within a few weeks, the stream cleared up and life returned to the village as they had known it before.

(Marshall, 1950)

After reviewing Marshall's parable and the restoration of the spring, the average person might say to him/herself, "Gee, all is well with the town again." For swans, irrigation, hydropower, and pretty views, residents seem to be pleased that the stream was restored to its "normal" state. The trained ecologist, however, would take a different view of this same stream.

The ecologist would go beyond the hype (as portrayed in the popular media, including literature) about what a healthy stream is. For example, the trained ecologist would know that a perfectly clean stream, clear of all terrestrial plant debris (woody debris and leaves) would not be conducive to ensuring diverse, productive invertebrates and fish, would not preserve natural sediment and water regimes, and would not ensure overall stream health (Dolloff & Webster, 2000).

WHY STUDY ECOLOGY?

Does anyone really need to be an ecologist or a student of ecology to appreciate the following words of Will Carleton (1845–1912) from his classic poem "Autumn Days"?

Sweet and smiling are thy ways, Beauteous, gold Autumn days.

Moreover, does anyone need to study ecology to observe and to relish and to feel and/or to sense the real thing: Nature's annual color palette in full kaleidoscope display – where those "yellow, mellow, ripened days are sheltered in a golden coating?" It is those clear and sunny days and cool and crisp nights of autumn that provide an almost irresistible lure to those of us (ecologist and nonecologist alike) who enjoy the outdoors. To take in the splendor and delight of autumn's color display, many head for the hills, the mountains, countryside, lakes, streams, and recreation areas of the National Forests. The more adventurous horseback ride or backpack through Nature's glory and solitude on trails winding deep into forest tranquility – just being out-of-doors in those golden days rivals any thrill in life. Even those of us fettered to the chains of city life are often exposed to city streets with those columns of life ablaze in color.

No, one need not study ecology to witness, appreciate, and/or understand the enchantment of autumn's annual color display – summer extinguished in a blaze of color. It is a different story, however, for those involved in trying to understand all of the complicated actions – and even more complicated interactions – involving pigments, sunlight, moisture, chemicals, temperatures, site, hormones, length of daylight, genetic traits, and so on that make for a perfect autumn color display (USDA, 1999). This is the work of the ecologist – to probe deeper and deeper into the basics of Nature, constantly seeking answers. To find the answers, the ecologist must be a synthesis scientist; that is, he/she must be a generalist well versed in botany, zoology, physiology, genetics, and other disciplines like geology, physics, and chemistry.

Earlier, we used Marshall's parable to make the point that a clean stream and other downstream water bodies can be a good thing, depending on one's point of view – pollution is a judgment call. It was also pointed out that this view may not be shared by a trained ecologist, especially a stream ecologist. The ecologist knows, for example, that terrestrial plant debris is not only a good thing but that it is necessary. Why? Consider the following explanation (Spellman, 1996).

A stream has two possible sources of primary energy:

1. instream photosynthesis by algae, mosses, and higher aquatic plants, and
2. imported organic matter from streamside vegetation (e.g., leaves and other parts of vegetation). Simply put, a significant portion of the food that is eaten by aquatic organisms grows right in the stream, like algae, diatoms, nymphs and larvae, and fish. This food that originates from within the stream is called autochthonous (Benfield, 1996).

Most food in a stream, however, comes from outside the stream. This is especially the case in small, heavily wooded streams, where there is normally insufficient light to support substantial instream photosynthesis, so energy pathways are supported largely by imported energy. A large portion of this imported energy is provided by leaves. Worms drown in floods and get washed in. Leafhoppers and caterpillars fall from trees. Adult mayflies and other insects mate above the stream, lay their eggs in it, and then die in it. All this food from outside the stream is called allochthonous.

Little brook, sing a song of leaf that sailed along.

Down the golden braided center of your current swift and strong.

James Whitcomb Riley ("The Brook Song")

Leaf Processing in Streams

Autumn leaves entering streams are nutrition poor because trees absorb most of the sugars and amino acids (nutrients) that were present in the green leaves (Suberkoop et al., 1976). Leaves falling into streams may be transported short distances but usually are caught by structures in the stream-bed to form leaf packs. These leaf packs are then processed in place by components of the stream communities in a series of well-documented steps (Figure 12.1) (Peterson & Cummins, 1974).

Within 24–48 hours of entering a stream, many of the remaining nutrients in leaves leach into the water. After leaching, leaves are composed mostly of structural materials like non-digestible cellulose and lignin. Within a few days, fungi (especially hyphomycetes), protozoa, and bacteria process the leaves by microbial processing (see Figure 12.1) (Barlocher & Kendrick, 1975). Two weeks later, microbial conditioning leads to structural softening of the leaf and, among some species, fragmentation. Reduction in particle size from whole leaves (coarse particulate organic matter, CPOM) to fine particulate organic matter (FPOM) is accomplished mainly through the feeding activities of a variety of aquatic invertebrates collectively known as "shredders" (Cummins, 1974; Cummins & Klug, 1979). Shredders (stoneflies, for example) help to produce fragments shredded from leaves but not ingested and fecal pellets, which reduce the particle size of organic matter. The particles then are collected (by mayflies, for example) and serve as a food resource for a variety of micro- and macroconsumers. Collectors eat what they want and send even smaller fragments downstream. These tiny fragments may be filtered out of the water by a true fly larva (i.e., a filterer). Leaves may also be fragmented by a combination of microbial activity and physical factors such as current and abrasion (Benfield et al., 1977; Paul et al., 1978).

Leaf-pack processing by all the elements mentioned above (i.e., leaf species, microbial activity, physical and chemical features of the stream) is important. However, the most important point is that these integrated ecosystem processes convert whole leaves into fine particles which are then distributed downstream and used as an energy source by various consumers.

The bottom line on allochthonous material in a stream: Insects that have fallen into a stream are ready to eat, and may join leaves, exuviae (castoff coverings, as crab shells or the skins of snakes, etc.), copepods, dead and dying animals, rotifers, bacteria, and dislodged algae and immature insects in their float (drift) downstream to a waiting hungry mouth.

Another important reason to study and learn ecology can be garnered from another simple stream ecology example. Consider the following:

Family picnic hosts insect intruders.

On one of their late August holiday outings, a family of 18 picnickers from a couple of small rural towns visited a local stream that coursed its way alongside and/or through one of the towns. This annual outing was looked upon with great anticipation for it was that one time each year when aunts, uncles, and cousins came together as one big family. The streamside setting was perfect for such an outing, but historically, until quite recently, the stream had been posted *"DANGER – NO SWIMMING – CAMPING or FISHING!"*

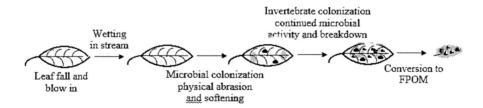

FIGURE 12.1   Leaf processing in streams.

Because the picnic area was such a popular location for picnickers, swimmers, and fishermen over the years, several complaints about the polluted stream were filed with the County Health Department. The Health Department finally took action to restore the stream to a relatively clean condition: sanitation workers removed debris and old tires and plugged or diverted end-of-pipe industrial outfalls upstream of the picnic area. After two years of continuous worker-aided stream clean-up and the stream's natural self-purification process, the stream was given a clean bill of health by the Health Department. The danger postings were removed.

When the stream had been declared clean, fit for use by swimmer and fisher, with postings removed, it did not take long for word to get out. Local folks and others alike made certain, at first opportunity, that they flocked to the restored picnic, swimming, and fishing site alongside the stream.

During most visits to the restored picnic-stream area, visitors, campers, fishermen, and others were pleased with their cleaned-up surroundings. However, during late summer, when the family of 18 and several others visited the restored picnic-stream area, they found themselves swarmed by thousands of speedy dragonflies and damselflies, especially near the bank of the stream. Soon they found the insects too much to deal with, so they stayed clear of the stream. To themselves and to anyone that would listen the same complaint was heard repeatedly: "What happened to our nice clean stream? With all those nasty bugs the stream is polluted again." So, when August arrived with its hordes of dragonfly-type insects, the picnickers, campers, swimmers, and fishermen avoided the place until the insects departed; until the human visitors thought the stream was clean again.

One local family does not avoid the stream-picnic area in August; on the contrary, August is one of their favorite times to visit, camp, swim, take in nature, and fish – and they usually have the site to themselves. The family is led by a local university professor of ecology, and she knows the truth about the picnic-stream area and the dragonflies and other insects. She knows that dragonflies and damselflies are macroinvertebrates indicator organisms; they only inhabit, grow, and thrive in and around streams that are clean and healthy – when dragonflies and damselflies are around, they indicate nonpolluted water. Further, the ecology professor knows that dragonflies are valued as predators, friends, and allies in waging war against flies and controlling populations of harmful insects, such as mosquitoes. Regarding mosquitoes, dragonflies take the wrigglers in the water, and the adults, on swiftest wings (25–35 mph) that are hovering over streams and ponds laying their eggs.

The ecology professor's husband, an amateur poet, also understood the significance of the presence of the indicator insects and had no problem sharing the same area with them. He also viewed the winged insects differently with the eye of a poet. He knew that the poets have been lavish in their attention to the dragonflies and have paid them delightful tributes. James Witcomb Riley (1849–1916) wrote:

> Till the dragon fly, in light gauzy armor burnished bright,
> Came tilting down the waters in a wild, bewildered flight.

LEVELS OF ORGANIZATION

Odum (1983) suggested that the best way to delimit modern ecology is to consider the concept of *levels of organization*. Levels of organization can be simplified as shown in Figure 12.2. In this relationship, organs form an organism; organisms of a species form a population; populations occupying an area form a community. Communities, interacting with nonliving or abiotic factors, separate in a natural unit to create a stable system known as the *ecosystem (the major ecological unit)*; and the part of earth in which ecosystem operates is known as the *biosphere*. Tomera points out "every community is influenced by a particular set of abiotic factors" (Tomera, 1989). Inorganic substances, such as oxygen, carbon dioxide, several other inorganic substances, and some organic substances represent the abiotic part of the ecosystem.

Organs ⮀ Organisms ⮀ Population ⮀ Communities ⮀ Ecosystem ⮀ Biosphere

FIGURE 12.2   Levels of organization.

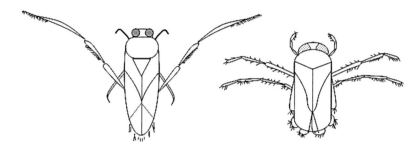

FIGURE 12.3   Notonecta (left) and Corixa (right). (Adapted from Odum, E.P., Basic Ecology, Saunders, Philadelphia, PA, 1983, p. 42.)

The physical and biological environment in which an organism lives is referred to as its *habitat*. For example, the habitat of two common aquatic insects, the "backswimmer" (Notonecta) and the "water boatman" (Corixa) is the littoral zone of ponds and lakes (shallow, vegetation-choked areas) (see Figure 12.3) (Odum, 1983). Within each level of organization of a habitat, each organism has a special role. The role the organism plays in the environment is referred to as its *niche*. A niche might be that the organism is food for some other organism or is a predator of other organisms. Odum refers to an organism's niche as its "profession" (Odum, 1971). In other words, each organism has a job or role to fulfill in its environment. Although two different species might occupy the same habitat, "niche separation based on food habits" differentiates between two species (Odum, 1983). Comparing the niches of the water backswimmer and the water boatman can see such niche separation. The backswimmer is an active predator, while the water boatman feeds largely on decaying vegetation (McCafferty, 1981).

ECOSYSTEMS

An *ecosystem* denotes an area that includes all organisms therein and their physical environment. The ecosystem is the major ecological unit in nature. Living organisms and their nonliving environment are inseparably interrelated and interact upon each other to create a self-regulating and self-maintaining system. To create a self-regulating and self-maintaining system, ecosystems are homeostatic, i.e., they resist any change through natural controls. These natural controls are important in ecology. This is especially the case because it is people, through their complex activities, who tend to disrupt natural controls.

As stated earlier, the ecosystem encompasses both the living and nonliving factors in an environment. The living or biotic part of the ecosystem is formed by two components: *autotrophic* and *heterotrophic*. The autotrophic (self-nourishing) component does not require food from its environment but can manufacture food from inorganic substances. For example, some autotrophic components (plants) manufacture needed energy through photosynthesis. Heterotrophic components, on the other hand, depend upon autotrophic components for food.

The nonliving or abiotic part of the ecosystem is formed by three components: inorganic substances, organic compounds (link biotic and abiotic parts), and climate regime. Figure 12.4 is a simplified diagram showing a few of the living and nonliving components of an ecosystem found in a freshwater pond. An ecosystem is a cyclic mechanism in which biotic and abiotic materials are constantly exchanged through biogeochemical cycles. Biogeochemical cycles are defined as follows: bio refers to living organisms and geo to water, air, rocks, or solids. Chemical is concerned

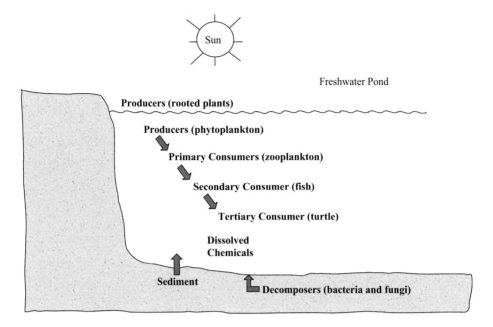

FIGURE 12.4   Major components of a freshwater pond ecosystem.

with the chemical composition of the earth. Biogeochemical cycles are driven by energy, directly or indirectly from the sun.

Figure 12.4 depicts an ecosystem where biotic and abiotic materials are constantly exchanged. Producers construct organic substances through photosynthesis and chemosynthesis. Consumers and decomposers use organic matter as their food and convert it into abiotic components. That is, they dissipate energy fixed by producers through food chains. The abiotic part of the pond in Figure 12.4 is formed of inorganic and organic compounds dissolved and in sediments such as carbon, oxygen, nitrogen, sulfur, calcium, hydrogen, and humic acids. Producers such as rooted plants and phytoplankton represent the biotic part. Fish, crustaceans, and insect larvae make up the consumers. Mayfly nymphs represent detritivores, which feed on organic detritus. Decomposers make up the final biotic part. They include aquatic bacteria and fungi, which are distributed throughout the pond.

As stated earlier, an ecosystem is a cyclic mechanism. From a functional viewpoint, an ecosystem can be analyzed in terms of several factors. The factors important in this study include the biogeochemical cycles and energy and food chains.

ENERGY FLOW IN THE ECOSYSTEM

Simply defined, *energy* is the ability or capacity to do work. For an ecosystem to exist, it must have energy. All activities of living organisms involve work, which is the expenditure of energy. This means the degradation of a higher state of energy to a lower state. Two laws govern the flow of energy through an ecosystem: *The First and Second Laws of Thermodynamics.* The first law, sometimes called the *conservation law*, states that energy may not be created or destroyed. The second law states that no energy transformation is 100% efficient. That is, in every energy transformation, some energy is dissipated as heat. The term *entropy* is used as a measure of the non-availability of energy to a system. Entropy increases with an increase in dissipation. Because of entropy, input of energy in any system is higher than the output or work done; thus, the resultant efficiency, is less than 100%. The interaction of energy and materials in the ecosystem is important. Energy drives the biogeochemical cycles.

Note: Energy does not cycle as nutrients do in biogeochemical cycles. For example, when food passes from one organism to another, energy contained in the food is reduced systematically until all the energy in the system is dissipated as heat. Price (1984) refers to this process as "an *unidirectional flow* of energy through the system, with no possibility for recycling of energy." When water or nutrients are recycled, energy is required. The energy expended in this recycling is not recyclable.

As mentioned, the principal source of energy for any ecosystem is sunlight. Green plants, through the process of photosynthesis, transform the sun's energy into carbohydrates, which are consumed by animals. This transfer of energy, again, is unidirectional – from producers to consumers. Often this transfer of energy to different organisms is called a *food chain*. Figure 12.5 shows a simple aquatic food chain.

All organisms, alive or dead, are potential sources of food for other organisms. All organisms that share the same general type of food in a food chain are said to be at the same *trophic level* (nourishment or feeding level). Since green plants use sunlight to produce food for animals, they are called the *producers*, or the first trophic level. The herbivores, which eat plants directly, are called the second trophic level or the *primary consumers*. The carnivores are flesh-eating consumers; they include several trophic levels from the third on up. At each transfer, a large amount of energy (about 80%–90%) is lost as heat and wastes. Thus, nature normally limits food chains to four or five links. In aquatic ecosystems, however, food chains are commonly longer than those on land. The aquatic food chain is longer because several predatory fish may be feeding on the plant consumers. Even so, the built-in inefficiency of the energy transfer process prevents development of extremely long food chains.

Only a few simple food chains are found in nature. Most simple food chains are interlocked. This interlocking of food chains forms a *food web*. Most ecosystems support a complex food web. A food web involves animals that do not feed on one trophic level. For example, humans feed on both plants and animals. An organism in a food web may occupy one or more trophic levels. Trophic level is determined by an organism's role in its community, not by its species. Food chains and webs help to explain how energy moves through an ecosystem.

An important trophic level of the food web is comprised of the *decomposers*. The decomposers feed on dead plants or animals and play an important role in recycling nutrients in the ecosystem. Simply, there is no waste in ecosystems. All organisms, dead or alive, are potential sources of food for other organisms. An example of an aquatic food web is shown in Figure 12.6.

Algae ➲ Zooplankton ➲ Perch ➲ Bass

FIGURE 12.5 Aquatic food chain.

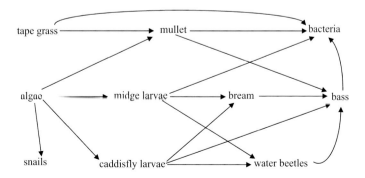

FIGURE 12.6 Aquatic food web.

FOOD CHAIN EFFICIENCY

Earlier, we pointed out that energy from the sun is captured (via photosynthesis) by green plants and used to make food. Most of this energy is used to carry on the plant's life activities. The rest of the energy is passed on as food to the next level of the food chain. Nature limits the amount of energy that is accessible to organisms within each food chain. Not all food energy is transferred from one trophic level to the next. Only about 10% (10% rule) of the amount of energy is transferred through a food chain. For example, if we apply the 10% rule to the diatoms-copepods-minnows-medium fish-large fish food chain shown in Figure 12.7, we can predict that 1,000 grams of diatoms produce 100 grams of copepods, which will produce 10 grams of minnows, which will produce 1 gram of medium fish, which, in turn, will produce 0.1 gram of large fish. Thus, only about 10% of the chemical energy available at each trophic level is transferred and stored in usable form at the next level. The other 90% are lost to the environment as low-quality heat in accordance with the second law of thermodynamics.

ECOLOGICAL PYRAMIDS

In the food chain, from the producer to the final consumer, a community in nature often consists of several small organisms associated with a smaller and smaller number of larger organisms. A grassy field, for example, has a larger number of grasses and other small plants, a smaller number of herbivores like rabbits, and an even smaller number of carnivores like foxes. The practical significance of this is that we must have several more producers than consumers.

This pound-for-pound relationship, where it takes more producers than consumers, can be demonstrated graphically by building an *ecological pyramid*. In an ecological pyramid, separate levels represent the number of organisms at various trophic levels in a food chain or bars placed one above the other with a base formed by producers and the apex formed by the final consumer. The pyramid shape is formed due to a great amount of energy loss at each trophic level. The same is true if the corresponding biomass or energy substitutes numbers. Ecologists generally use three types of ecological pyramids: *pyramids of number, biomass,* and *energy.* Obviously, there will be differences among them. Some generalizations:

- Energy pyramids must always be larger at the base than at the top (because of the second law of thermodynamics and has to do with dissipation of energy as it moves from one trophic level to another).
- Likewise, biomass pyramids (in which biomass is used as an indicator of production) are usually pyramid shaped. This is particularly true of terrestrial systems and aquatic ones dominated by large plants (marshes) in which consumption by heterotrophs is low and organic matter accumulates with time.

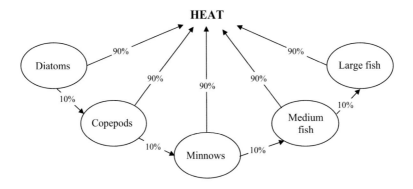

FIGURE 12.7   Simple food chain.

However, biomass pyramids can sometimes be inverted. This is common in aquatic ecosystems in which the primary producers are microscopic planktonic organisms that multiply very rapidly, have very short life spans, and heavy grazing by herbivores. At any single point in time, the amount of biomass in primary producers is less than that in larger, long-lived animals that consumer primary producers. Numbers pyramids can have various shapes (and not be pyramids at all) depending on the sizes of the organisms that make up the trophic levels. In forests, the primary producers are large trees and the herbivore level usually consists of insects, so the base of the pyramid is smaller than the herbivore level above it. In grasslands, the number of primary producers (grasses) is much larger than that of the herbivores above (large grazing animals) (Ecosystems Topics, 2000).

PRODUCTIVITY

As mentioned, the flow of energy through an ecosystem starts with the fixation of sunlight by plants through photosynthesis. In evaluating an ecosystem, the measurement of photosynthesis is important. Ecosystems may be classified into highly productive or less productive. Therefore, the study of ecosystems must involve some measure of the productivity of that ecosystem. Primary production is the rate at which the ecosystem's primary producers capture and store a given amount of energy in a specified time interval. In simpler terms, primary productivity is a measure of the rate at which photosynthesis occurs. Four successive steps in the production process are:

1. Gross primary productivity – the total rate of photosynthesis in an ecosystem during a specified interval.
2. Net primary productivity – the rate of energy storage in plant tissues in excess of the rate of aerobic respiration by primary producers.
3. Net community productivity – the rate of storage of organic matter not used.
4. Secondary productivity – the rate of energy storage at consumer levels.

When attempting to comprehend the significance of the term *productivity* as it relates to ecosystems, it is wise to consider an example. Consider the productivity of an agricultural ecosystem such as a wheat field. Often its productivity is expressed as the number of bushels produced per acre. This is an example of the harvest method for measuring productivity. For a natural ecosystem, several-one-square meter plots are marked off, and the entire area is harvested and weighed to give an estimate of productivity as grams of biomass per square meter per given time interval. From this method, a measure of net primary production (net yield) can be measured.

Productivity, both in the natural and cultured ecosystem, may vary considerably, not only between types of ecosystems but also within the same ecosystem. Several factors influence year-to-year productivity within an ecosystem. Such factors as temperature, availability of nutrients, fire, animal grazing, and human cultivation activities are directly or indirectly related to the productivity of an ecosystem.

Productivity can be measured in several different ways in the aquatic ecosystem. For example, the production of oxygen may be used to determine productivity. Oxygen content may be measured in several ways. One way is to measure it in the water every few hours for a period of 24 hours During daylight, when photosynthesis is occurring, the oxygen concentration should rise. At night, the oxygen level should drop. The oxygen level can be measured by using a simple x-y graph. The oxygen level can be plotted on the y-axis with time plotted on the x-axis, as shown in Figure 12.8.

Another method of measuring oxygen production in aquatic ecosystems is to use light and dark bottles. Biochemical oxygen demand (BOD) bottles (300 mL) are filled with water to a height. One of the bottles is tested for the initial dissolved oxygen (DO), and then the other two bottles (one clear, one dark) are suspended in the water at the depth they were taken from. After a 12-hour period, the bottles are collected, and the DO values for each bottle are recorded. Once the oxygen

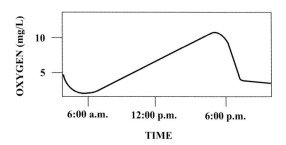

FIGURE 12.8 Diurnal oxygen curve for an aquatic ecosystem.

production is known, the productivity in terms of grams/m/day can be calculated. In the aquatic ecosystem, pollution can have a profound impact upon the system's productivity.

POPULATION ECOLOGY

Webster's Third New International Dictionary defines population as "the total number or amount of things especially within a given area; the organisms inhabiting a particular area or biotype; and a group of interbreeding biotypes that represents the level of organization at which speciation begins." The concept of population is interpreted differently in various sciences. In *human demography* a population is a set of humans in each area. In *genetics*, a population is a group of interbreeding individuals of the same species, which is isolated from other groups. In *population ecology*, a population is a group of individuals of the same species inhabiting the same area.

If we wanted to study the organisms in a slow-moving stream or stream pond, we would have two options. We could study each fish, aquatic plant, crustacean, and insect one by one. In that case, we would be studying individuals. It would be easier to do this if the subject were trout, but it would be difficult to separate and study each aquatic plant. The second option would be to study all the trout, all the insects of each specific kind, all of a certain aquatic plant type in the stream or pond at the time of the study. When ecologists study a group of the same kind of individuals in each location at a given time, they are investigating a *population*. When attempting to determine the population of a species, it is important to remember that time is a factor. Time is important because populations change, whether it is at various times during the day, during the different seasons, or from year to year.

Population density may change dramatically. For example, if a dam is closed off in a river midway through spawning season, with no provision allowed for fish movement upstream (a fish ladder); it would drastically decrease the density of spawning salmon upstream. Along with the swift and sometimes unpredictable consequences of change, it can be difficult to draw exact boundaries between various populations. The population density or level of a species depends on *natality, mortality, immigration,* and *emigration*. The birth rate of a population is called *natality* and the death rate *mortality*. Changes in population density are the result of both births and deaths. In aquatic populations, two factors besides natality and mortality can affect density. For example, in a run of returning salmon to their spawning grounds, the density could vary as more salmon migrated in or as others left the run for their own spawning grounds. The arrival of new salmon to a population from other places is termed *immigration* (ingress). The departure of salmon from a population is called *emigration* (egress). Thus, natality and immigration increase population density, whereas mortality and emigration decrease it. The net increase in population is the difference between these two sets of factors.

Each organism occupies only those areas that can provide for its requirements, resulting in an irregular distribution. How a population is distributed within a given area has considerable influence on density. As shown in Figure 12.9, organisms in nature may be distributed in three ways. In a

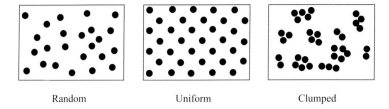

Random   Uniform   Clumped

FIGURE 12.9 Basic patterns of distribution. (Adapted from Odum, E.P., Fundamentals of Ecology, Saunders, Philadelphia, PA 1971, p. 205.)

random distribution, there is an equal probability of an organism occupying any point in space, and "each individual is independent of the others." In a regular or uniform distribution, in turn, organisms are spaced more evenly; they are not distributed by chance.

Animals compete and effectively defend a specific territory, excluding other individuals of the same species. In regular or uniform distribution, the competition between individuals can be quite severe and antagonistic to the point where spacing generated is quite even. The most common distribution is the contagious or clumped distribution where organisms are found in groups; this may reflect the heterogeneity of the habitat. Organisms that exhibit a contagious or clumped distribution may develop social hierarchies in order to live together more effectively. Animals within the same species have evolved many symbolic aggressive displays that carry meanings that are not only mutually understood but also prevent injury or death within the same species.

The size of animal populations is constantly changing due to natality, mortality, emigration, and immigration. As mentioned, the population size will increase if the natality and immigration rates are high. On the other hand, it will decrease if the mortality and emigration rates are high. Each population has an upper limit on size, often called the *carrying capacity*. Carrying capacity is the optimum number of species' individuals that can survive in a specific area over time. Stated differently, the carrying capacity is the maximum number of species that can be supported in a bioregion. A pond may be able to support only a dozen frogs depending on the food resources for the frogs in the pond. If there were 30 frogs in the same pond, at least half of them would probably die because the pond environment would not have enough food for them to live. Carrying capacity is based on the quantity of food supplies, the physical space available, the degree of predation, and several other environmental factors.

The carrying capacity is of two types: ultimate and environmental. *Ultimate carrying capacity* is the theoretical maximum density; that is, it is the maximum number of individuals of a species in a place that can support itself without rendering the place uninhabitable. The *environmental carrying capacity* is the actual maximum population density that a species maintains in an area. Ultimate carrying capacity is always higher than environmental. Ecologists have concluded that a major factor that affects population stability or persistence is *species diversity*. Species diversity is a measure of the number of species and their relative abundance. If the stress on an ecosystem is small, the ecosystem can usually adapt quite easily. Moreover, even when severe stress occurs, ecosystems have a way of adapting. Severe environmental change to an ecosystem can result from such natural occurrences as fires, earthquakes, and floods and from people-induced changes such as land clearing, surface mining, and pollution. One of the most important applications of species diversity is in the evaluation of pollution. Stress of any kind will reduce the species diversity of an ecosystem to a significant degree. In the case of domestic sewage pollution, for example, the stress is caused by a lack of dissolved oxygen for aquatic organisms.

Ecosystems can and do change; for example, if a fire devastates a forest, it will grow back, eventually because of *ecological succession*. Ecological succession is the observed process of change (a normal occurrence in nature) in the species structure of an ecological community over time. Succession usually occurs in an orderly, predictable manner. It involves the entire system.

The science of ecology has developed to such a point that ecologists are now able to predict several years in advance what will occur in each ecosystem. For example, scientists know that if a burned-out forest region receives light, water, nutrients, and an influx or immigration of animals and seeds, it will eventually develop into another forest through a sequence of steps or stages. Ecologists recognize two types of ecological succession: primary and secondary. The type that takes place depends on the condition at a site at the beginning of the process. *Primary succession*, sometimes called *bare-rock succession*, occurs on surfaces such as hardened volcanic lava, bare rock, and sand dunes, where no soil exists, and where nothing has ever grown before (see Figure 12.10).

Obviously, in order to grow, plants need soil. Thus, soil must form on the bare rock before succession can begin. Usually this soil formation process results from weathering. Atmospheric exposure – weathering, wind, rain, and frost – forms tiny cracks and holes in rock surfaces. Water collects in the rock fissures and slowly dissolves the minerals out of the rock's surface. A pioneer soil layer is formed from the dissolved minerals and supports such plants as lichens. Lichens gradually cover the rock surface and secrete carbonic acid, which dissolves additional minerals from the rock. Eventually, mosses replace the lichens. Organisms called *decomposers* move in and feed on dead lichen and moss. A few small animals such as mites and spiders arrive next. The result is what

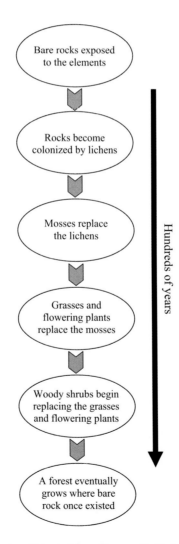

FIGURE 12.10   Bare-rock succession. (Adapted from Tomera, N., Understanding Basic Ecological Concepts, J. Weston Walch, Publisher, Portland, ME, 1989, p. 67.)

is known as a *pioneer community*. The pioneer community is defined as the first successful integration of plants, animals, and decomposers into a bare-rock community.

After several years, the pioneer community builds up enough organic matter in its soil to be able to support rooted plants like herbs and shrubs. Eventually, the pioneer community is crowded out and is replaced by a different environment. This, in turn, works to thicken the upper soil layers. The progression continues through several other stages until a mature or climax ecosystem is developed several decades later. In bare-rock succession, each stage in the complex succession pattern dooms the stage that existed before it.

Secondary succession is the most common type of succession. Secondary succession occurs in an area where the natural vegetation has been removed or destroyed but the soil is not destroyed. For example, succession that occurs in abandoned farm fields, known as *old-field succession*, illustrates secondary succession. An example of secondary succession can be seen in the Piedmont region of North Carolina. Early settlers of the area cleared away the native oak-hickory forests and cultivated the land. In the ensuing years, the soil became depleted of nutrients, reducing the soil's fertility. As a result, farming ceased in the region a few generations later, and the fields were abandoned. Some 150–200 years after abandonment, the climax oak-hickory forest was restored.

In a stream ecosystem, growth is enhanced by biotic and abiotic factors, including:

- Ability to produce offspring;
- Ability to adapt to new environments;
- Ability to migrate to new territories;
- Ability to compete with species for food and space to live;
- Ability to blend into the environment so as not to be eaten;
- Ability to find food;
- Ability to defend itself from enemies;
- Favorable light;
- Favorable temperature;
- Favorable dissolved oxygen content;
- Enough water level.

The biotic and abiotic factors in an aquatic ecosystem that reduce growth include:

- Predators;
- Disease;
- Parasites;
- Pollution;
- Competition for space and food;
- Unfavorable stream conditions (i.e., low water levels);
- Lack of food.

Regarding stability in a freshwater ecosystem, the higher the species diversity the greater the inertia and resilience of the ecosystem are. At the same time, when the species diversity is high within a stream ecosystem, a population within the stream can be out of control because of an imbalance between growth and reduction factors, with the ecosystem at the same time remaining stable. Regarding instability in a freshwater ecosystem, recall that imbalance occurs when growth and reduction factors are out of balance. For example, when sewage is accidentally dumped into a stream, the stream ecosystem via the self-purification process (discussed later) responds and returns to normal. This process can be described as follows:

- Raw sewage is dumped into the stream.
- Decreases the oxygen available as the detritus food chain breaks down the sewage.
- Some fish die at the pollution site and down stream.

- Sewage is broken down and washes out to sea and is finally broken down in the ocean.
- Oxygen levels return to normal.
- Fish populations that were depleted are restored as fish around the spill reproduce and the young occupy the real estate formerly occupied by the dead fish.
- Populations all return to "normal."

A shift in balance in a stream's ecosystem (or in any ecosystem) like the one just described is a common occurrence. In this case, the stream responded (on its own) to the imbalance the sewage caused and through the self-purification process returned to normal. Recall that succession is the method by which an ecosystem either forms itself or heals itself. Thus, we can say that a type of succession has occurred in the polluted stream described above because in the end it healed itself. More importantly, this healing process is a good thing; otherwise, long ago there would have been few streams on earth suitable for much more than the dumping of garbage.

In summary, through research and observation, ecologists have found that the succession patterns in different ecosystems usually display common characteristics. First, succession brings about changes in the plant and animal members present. Second, organic matter increases from stage to stage. Finally, as each stage progresses, there is a tendency toward greater stability or persistence. Remember, succession is usually predictable – this is the case, unless humans interfere.

STREAM GENESIS AND STRUCTURE

Consider the following: Early in the spring on a snow and ice-covered high alpine meadow, the water cycle continues. The cycle's main component, water, has been held in reserve – literally frozen, for the long dark winter months, but with longer, warmer spring days, the sun is higher, more direct, and of longer duration, and the frozen masses of water respond to the increased warmth. The melt begins with a single drop, then two, then increasingly. As the snow and ice melts, the drops join a chorus that continues unending; they fall from their ice-bound lip to the bare rock and soil terrain below.

The terrain on which the snowmelt falls is not like glacial till, which is an unconsolidated, heterogeneous mixture of clay, sand, gravel, and boulders, dug-out, ground-out, and exposed by the force of a huge, slow, and inexorably moving glacier. Instead, this soil and rock ground is exposed to the falling drops of snowmelt because of a combination of wind and tiny, enduring force exerted by drops of water as over seasons after season they collide with the thin soil cover, exposing the intimate bones of the earth. Gradually, the single drops increase to a small rush – they join to form a splashing, rebounding, helter-skelter cascade, many separate rivulets that trickle then run their way down the face of the granite mountain. At an indented ledge halfway down the mountain slope, a pool forms whose beauty, clarity, and sweet iciness provides the visitor with an incomprehensible, incomparable gift – a blessing from earth.

The mountain pool fills slowly, tranquil under the blue sky, reflecting the pines, snow, and sky around and above it, an open invitation to lie down and drink and to peer into the glass-clear, deep phantom blue-green eye, so clear that it seems possible to reach down over 50 feet and touch the very bowels of the mountain. The pool has no transition from shallow margin to depth; it is simply deep and pure. As the pool fills with more meltwater, we wish to freeze time, to hold this place and this pool in its perfect state forever, it is such a rarity to us in our modern world. However, this cannot be – Mother Nature calls, prodding, urging – and for a brief instant, the water laps in the breeze against the outermost edge of the ridge, then a trickle flows over the rim. The giant hand of gravity reaches out and tips the overflowing melt onward and it continues the downward journey, following the path of least resistance to its next destination, several thousand feet below.

When the overflow, still high in altitude, but with its rock-strewn bed bent downward, toward the sea, meets the angled, broken rocks below, it bounces, bursts, and mists its way against steep, V-shaped walls that form a small valley, carved out over time by water and the forces of the earth.

Within the valley confines, the meltwater has grown from drops to rivulets to a small mass of flowing water. It flows through what is at first a narrow opening, gaining strength, speed, and power as the V-shaped valley widens to form a U-shape. The journey continues as the water mass picks up speed and tumbles over massive boulders and then slows again.

At a larger but shallower pool, waters from higher elevations have joined the main body – from the hillsides, crevices, springs, rills, mountain creeks. At the influent poolside, all appears peaceful, quiet, and restful, but not far away, at the effluent end of the pool, gravity takes control again. The overflow is flung over the jagged lip, and cascades downward several hundred feet where the waterfall again brings its load to a violent, mist-filled meeting. The water separates and joins repeatedly, forming a deep, furious, wild stream that calms gradually as it continues to flow over lands that are less steep. The waters widen into pools overhung by vegetation, surrounded by tall trees. The pure, crystalline waters have become progressively discolored on their downward journey, stained brown black with humic acid and literally filled with suspended sediments; the once-pure stream is now muddy.

The mass divides and flows in different directions, over different landscapes. Small streams divert and flow into open country. Different soils work to retain or speed the waters, and in some places the waters spread out into shallow swamps, bogs, marshes, fens, or mires. Other streams pause long enough to fill deep depressions in the land and form lakes. For a time, the water remains and pauses in its journey to the sea. However, this is only a short-term pause because lakes are only a short-term resting place in the water cycle. The water will eventually move on by evaporation or seepage into groundwater. Other portions of the water mass stay with the main flow, and the speed of flow changes to form a river, which braids its way through the landscape, heading for the sea. As it changes speed and slows, the river bottom changes from rock and stone to silt and clay. Plants begin to grow, stems thicken, and leaves broaden. The river is now full of life and the nutrients needed to sustain life. However, the river courses onward, its destiny met when the flowing rich mass slows its last and finally spills into the sea.

Freshwater systems are divided into two broad categories, running waters (lotic systems) and standing waters (lentic systems). We concentrate on lotic systems, although many of the principles described herein apply to other freshwater surface bodies as well, which are known by common names. Some examples include seeps, springs, brooks, branches, creeks, streams, and rivers. Again, because it is the best term to use in freshwater ecology, it is the stream we are concerned with here. Although there is no standard scientific definition of a stream, it is usually distinguished subjectively as follows: A stream is of intermediate size that can be waded from one side to the other.

Physical processes involved in the formation of a stream are important to the ecology of the stream because stream channel and flow characteristics directly influence the functioning of the stream's ecosystem and the biota found therein. Thus, in this section, we discuss the pathways of water flow contributing to streamflow; namely, we discuss precipitation inputs as they contribute to flow. We also discuss streamflow discharge, transport of material, characteristics of stream channels, stream profile, sinuosity, the floodplain, pool-riffle sequences, and depositional features – all of which directly or indirectly impact the ecology of the stream.

Water Flow in a Stream

Most elementary students learn early in their education process that water on earth flows downhill – from land to the sea. However, they may or may not be told that water flows downhill toward the sea by various routes. The route (or pathway) that we are primarily concerned with is the surface water route taken by surface water runoff. Surface runoff is dependent on various factors. For example, climate, vegetation, topography, geology, soil characteristics, and land use determine how much surface runoff occurs compared with other pathways.

The primary source (input) of water to total surface runoff, of course, is precipitation. This is the case even though a substantial portion of all precipitation input returns directly to the atmosphere by evapotranspiration. Evapotranspiration is a combination process, as the name suggests whereby water in plant tissue and in the soil evaporates and transpires to water vapor in the atmosphere. A substantial portion of precipitation input returns directly to the atmosphere by evapotranspiration. It is also important to point out that when precipitation occurs, some rainwater is intercepted by vegetation where it evaporates, never reaching the ground or being absorbed by plants. A large portion of the rainwater that reaches the surface on ground, in lakes and streams also evaporates directly back to the atmosphere. Although plants display a special adaptation to minimize transpiration, plants still lose water to the atmosphere during the exchange of gases necessary for photosynthesis. Notwithstanding the large percentage of precipitation that evaporates, rain- or meltwater that reach the ground surface follows several pathways in reaching a stream channel or groundwater.

Soil can absorb rainfall to its *infiltration capacity* (i.e., to its maximum rate). During a rain event, this capacity decreases. Any rainfall in excess of infiltration capacity accumulates on the surface. When this surface water exceeds the depression storage capacity of the surface, it moves as an irregular sheet of overland flow. In arid areas, overland flow is likely because of the low permeability of the soil.

Overland flow is also likely when the surface is frozen and/or when human activities have rendered the land surface less permeable. In humid areas, where infiltration capacities are high, overland flow is rare. In rain events, where the infiltration capacity of the soil is not exceeded, rain penetrates the soil and eventually reaches the groundwater – from which it discharges to the stream slowly and over a long period. This phenomenon helps to explain why streamflow through a dry weather region remains constant; the flow is continuously augmented by groundwater. This type of stream is known as a *perennial stream*, as opposed to an *intermittent* one, because the flow continues during periods of no rainfall. When a stream courses through a humid region, it is fed water via the water table, which slopes toward the stream channel. Discharge from the water table into the stream accounts for flow during periods without precipitation and explains why this flow increases, even without tributary input, as one proceeds downstream. Such streams are called *gaining* or *effluent*, opposed to *losing* or *influent streams* that lose water into the ground (see Figure 12.11). The same stream can shift between gaining and losing conditions along its course because of changes in underlying strata and local climate.

Stream Water Discharge

The current velocity (speed) of water (driven by gravitational energy) in a channel varies considerably within a stream's cross section owing to friction with the bottom and sides, with sediment, and the atmosphere, and to sinuosity (bending or curving) and obstructions. Highest velocities, obviously, are found where friction is least, generally at or near the surface and near the center of the channel. In deeper streams, current velocity is greatest just below the surface due to the friction with the atmosphere; in shallower streams, current velocity is greatest at the surface due to friction with the bed. Velocity decreases as a function of depth, approaching zero at the substrate surface.

Transport of Material

Water flowing in a channel may exhibit *laminar flow* (parallel layers of water shear over one another vertically), or *turbulent flow* (complex mixing). In streams, laminar flow is uncommon, except at boundaries where flow is very low and in groundwater. Thus, the flow in streams generally is turbulent. Turbulence exerts a shearing force that causes particles to move along the streambed by pushing, rolling, and skipping referred to as *bed load*. This same shear causes turbulent eddies that entrain particles in suspension (called the *suspended* load – particles size under 0.06 mm).

Entrainment is the incorporation of particles when stream velocity exceeds the *entraining velocity* for a particle size. The entrained particles in suspension (suspended load) also include fine sediment, primarily clays, silts, and fine sands that require only low velocities and minor turbulence to

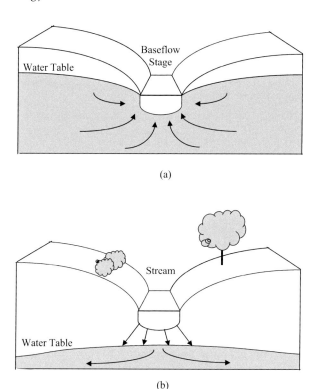

FIGURE 12.11   (a) Cross-section of a gaining stream. (b) cross-section of a losing stream.

remain in suspension. These are referred to as *wash* load (under 0.002 mm). Thus, the suspended load includes the wash load and coarser materials (at lower flows). Together, the suspended load and bed load constitute the *solid load*. It is important to note that in bedrock streams the bed load will be a lower fraction than in alluvial streams where channels are composed of easily transported material. A substantial amount of material is also transported as the *dissolved load*. Solutes are generally derived from chemical weathering of bedrock and soils, and their contribution is greatest in sub-surface flows, and in regions of limestone geology. The relative amount of material transported as solute rather than solid load depends on basin characteristics, lithology (i.e., the physical character of rock), and hydrologic pathways. In areas of very high runoff, the contribution of solutes approaches or exceeds sediment load, whereas in dry regions, sediments make up as much as 90% of the total load.

Deposition occurs when *stream competence* (i.e., the largest particle that can be moved as bed load, and the critical erosion – competent – velocity is the lowest velocity at which a particle resting on the streambed will move) falls below a given velocity. Simply stated: The size of the particle that can be eroded and transported is a function of current velocity.

Sand particles are the most easily eroded. The greater the mass of larger particles (e.g., coarse gravel) the higher the initial current velocities must be for movement. However, smaller particles (silts and clays) require even greater initial velocities because of their cohesiveness and because they present smaller, streamlined surfaces to the flow. Once in transport, particles will continue in motion at somewhat slower velocities than initially required to initiate movement and will settle at still lower velocities. Particle movement is determined by size, flow conditions, and mode of entrainment. Particles over 0.02 mm (medium to coarse sand size) tend to move by rolling or sliding along the channel bed as *traction load*. When sand particles fall out of the flow, they move by *saltation* or repeated bouncing. Particles under 0.06 mm (silt) move as *suspended* load and particles under 0.002 (clay), indefinitely, as *wash* load. Unless the supply of sediments becomes depleted the concentration and amount of transported solids increase. However, discharge is usually too low, throughout most of the year, to scrape or scour, shape channels, or move significant quantities of sediment in all but sand-bed streams, which can experience change more rapidly. During extreme events, the greatest scour occurs, and the amount of material removed increases dramatically. Sediment inflow into streams can both be increased and decreased because of human activities. For example, poor agricultural practices and deforestation greatly increase erosion. Fabricated structures such as dams and channel diversions can, on the other hand, greatly reduce sediment inflow.

CHARACTERISTICS OF STREAM CHANNELS

Flowing waters (rivers and streams) determine their own channels, and these channels exhibit relationships attesting to the operation of physical laws – laws that are not, yet, fully understood. The development of stream channels and entire drainage networks and the existence of various regular patterns in the shape of channels indicate that streams are in a state of dynamic equilibrium between erosion (sediment loading) and deposition (sediment deposit), and governed by common hydraulic processes. However, because channel geometry is four-dimensional with a long profile, cross section, depth, and slope profile, and because these mutually adjust over a time scale as short as years, and if centuries or more, cause-and-effect relationships are difficult to establish. Other variables that are presumed to interact as the stream achieves its graded state include width and depth, velocity, size of sediment load, bed roughness, and the degree of braiding (sinuosity).

STREAM PROFILES

Mainly because of gravity, most streams exhibit a downstream decrease in gradient along their length. Beginning at the headwaters, the steep gradient becomes less so as one proceeds downstream, resulting in a concave longitudinal profile. Though diverse geography provides for almost unlimited variation, a lengthy stream that originates in a mountainous area typically comes into existence as a series of springs and rivulets; these coalesce into a fast-flowing, turbulent mountain stream, and the addition of tributaries results in a large and smoothly flowing river that winds through the lowlands to the sea. When studying a stream system of any length, it becomes readily apparent (almost from the start) that what we are studying is a body of flowing water that varies considerably from place to place along its length. For example, a common variable – the results of which can be readily seen – is whenever discharge increases, it causes corresponding changes in the stream's width, depth, and velocity. In addition to physical changes that occur from location to location along a stream's course, there are a legion of biological variables that correlate with stream size and distance downstream. The most apparent and striking changes are in steepness of slope and in the transition from a shallow stream with large boulders and a stony substrate to a deep stream with a sandy substrate. The particle size of bed material at various locations is also variable along the stream's course. The particle size usually shifts from an abundance of coarser material upstream to mainly finer material in downstream areas.

SINUOSITY

Unless forced by man in the form of heavily regulated and channelized streams, straight channels are uncommon. Streamflow creates distinctive landforms composed of straight (usually in appearance only), meandering, and braided channels, channel networks, and flood plains. Simply put, flowing water will follow a sinuous course. The most used measure is the *sinuosity index* (SI). Sinuosity equals 1 in straight channels and more than 1 in sinuous channels. *Meandering* is the natural tendency for alluvial channels and is usually defined as an arbitrarily extreme level of sinuosity, typically a SI greater than 1.5. Many variables affect the degree of sinuosity. Even in many natural channel sections of a stream course that appear straight, meandering occurs in the line of maximum water or channel depth (known as the *thalweg*). Keep in mind that a stream must meander; that is how they renew themselves. By meandering, they wash plants and soil from the land into their waters, and these serve as nutrients for the plants in the rivers. If rivers are not allowed to meander, if they are *channelized*, the amount of life they can support will gradually decrease. That means less fish, ultimately – and fewer bald eagles, herons, and other fishing birds (Spellman, 1996).

Meander flow follows predictable pattern and causes regular regions of erosion and deposition (see Figure 12.12). The streamlines of maximum velocity and the deepest part of the channel lie close to the outer side of each bend and cross over near the point of inflection between the banks (see Figure 12.12). A huge elevation of water at the outside of a bend causes a helical flow of water toward the opposite bank. In addition, a separation of surface flow causes a back eddy. The result is zones of erosion and deposition and explains why point bars develop in a downstream direction in depositional zones.

Bars, Riffles, and Pools

Implicit in the morphology and formation of meanders are *bars*, *riffles*, and *pools*. Bars develop by deposition in slower, less competent flow on either side of the sinuous mainstream. Onward moving water, depleted of bed load, regains competence and shears a pool in the meander – reloading the stream for the next bar. Alternating bars migrate to form riffles (see Figure 12.13). As streamflow continues along its course, a pool-riffle sequence is formed. The riffle is a mound or hillock, and the pool is a depression.

The Floodplain

A stream channel influences the shape of the valley floor through which it courses. The self-formed, self-adjusted flat area near to the stream is the *flood plain*, which loosely describes the valley floor prone to periodic inundation during over-bank discharges. What is not commonly known is that valley flooding is a regular and natural behavior of the stream. A stream's aquatic community has several unique characteristics. The aquatic community operates under the same ecologic principles as terrestrial ecosystems, but the physical structure of the community is more isolated and exhibits limiting factors that are very different than the limiting factors of a terrestrial ecosystem. Certain materials and conditions are necessary for the growth and reproduction of organisms.

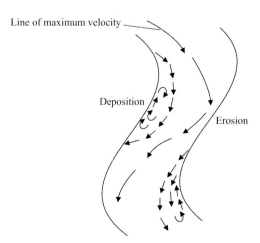

FIGURE 12.12   A meandering reach.

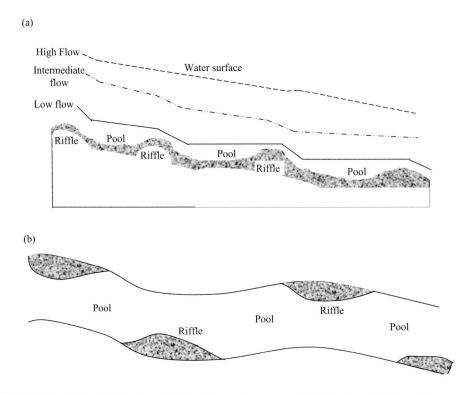

FIGURE 12.13   (a) Longitudinal profile of a riffle-pool sequence; (b) plain view of riffle-pool sequence.

If, for instance, a farmer plants wheat in a field containing too little nitrogen, it will stop growing when it has used up the available nitrogen, even if the wheat's requirements for oxygen, water, potassium, and other nutrients are met. In this case, nitrogen is said to be the limiting factor. A *limiting factor* is a condition or a substance (the resource in shortest supply), which limits the presence and success of an organism or a group of organisms in an area.

Even the smallest mountain stream provides an astonishing number of different places for aquatic organisms to live, or *habitats*. If it is a rocky stream, every rock of the substrate provides several different habitats. On the side facing upriver, organisms with special adaptations, that are very good at clinging to rock, do well here. On the side that faces downriver, a certain degree of shelter is provided from current but still allows organisms to hunt for food. The top of a rock, if it contacts air, is a good place for organisms that cannot breathe underwater and need to surface now and then. Underneath the rock is a popular place for organisms that hide to prevent predation. Normal stream life can be compared to that of a "balanced aquarium" (ASTM, 1969); that is, nature continuously strives to provide clean, healthy, normal streams. This is accomplished by maintaining the stream's flora and fauna in a balanced state. Nature balances stream life by maintaining both the number and the type of species present in any one part of the stream. Such balance ensures that there is never an overabundance of one species compared to another. Nature structures the stream environment so that both plant and animal life is dependent upon the existence of others within the stream. As mentioned, lotic (washed) habitats are characterized by continuously running water of current flow. These running waterbodies, have typically three zones: riffle, run, and pool. The *riffle zone* contains faster flowing, well-oxygenated water, with coarse sediments. In the riffle zone, the velocity of current is great enough to keep the bottom clear of silt and sludge, thus providing a firm bottom for organisms. This zone contains specialized organisms, which are adapted to live in running water. For example, organisms adapted to live in fast streams or rapids (trout) have streamlined bodies, which aid in their respiration and in obtaining food (Smith, 1996).

Stream organisms that live under rocks to avoid the strong current have flat or streamlined bodies. Others have hooks or suckers to cling or attach to a firm substrate to avoid the washing-away effect of the strong current. The *run* zone (or intermediate zone) is the slow-moving, relatively shallow part of the stream with moderately low velocities and little or no surface turbulence. The *pool zone* of the stream is usually a deeper water region where velocity of water is reduced and silt and other settling solids provide a soft bottom (more homogeneous sediments), which is unfavorable for sensitive bottom-dwellers.

Decomposition of some of these solids causes a lower amount of dissolved oxygen. Some stream organisms spend some of their time in the rapids part of the stream and other times can be found in the pool zone (trout, for example). Trout typically spend about the same amount of time in the rapid zone pursuing food as they do in the pool zone pursuing shelter. Organisms are sometimes classified based on their mode of life:

- *Benthos* (Mud Dwellers): the term originates from the Greek word for bottom and broadly includes aquatic organisms living on the bottom or on submerged vegetation. They live under and on rocks and in the sediments. A shallow sandy bottom has sponges, snails, earthworms, and some insects. A deep, muddy bottom will support clams, crayfish, and nymphs of damselflies, dragonflies, and mayflies. A firm, shallow, rocky bottom has nymphs of mayflies, stoneflies, and larvae of water beetles.
- *Periphytons* or *Aufwuchs*: The first term usually refers to microfloral growth upon substrata (i.e., benthic-attached algae). The second term, aufwuchs (pronounce: OWF-vooks; German: "growth upon"), refers to the fuzzy, sort of furry-looking, slimy green coating that attach or cling to stems and leaves of rooted plants or other objects projecting above the bottom without penetrating the surface. It consists not only of algae like Chlorophyta, but also diatoms, protozoans, bacteria, and fungi.
     Planktons (Drifters): They are small, mostly microscopic plants and animals that are suspended in the water column; movement depends on water currents. They mostly float in the direction of the current. There are two types of planktons: (a) phytoplankton are assemblages of small plants (algae) and have limited locomotion abilities; they are subject to movement and distribution by water movements, (b) zooplankton are animals that are suspended in water and have limited means of locomotion. Examples of zooplanktons include crustaceans, protozoans, and rotifers.
- *Nektons or Pelagic Organisms* (capable of living in open waters): They are distinct from other planktons in that they are capable of swimming independently of turbulence. They are swimmers, which can navigate against the current. Examples of nektons include fish, snakes, diving beetles, newts, turtles, birds, and large crayfish.
- *Neustons*: They are organisms that float or rest on the surface of the water (never break water tension). Some varieties can spread out their legs so that the surface tension of the water is not broken; for example, water striders (see Figure 12.14).
- *Madricoles*: Organisms that live on rock faces in waterfalls or seepages.

In a stream, the rocky substrate is the home for many organisms. Thus, we need to know something about the particles that make up the substrate. Namely, we need to know how to measure the particles so we can classify them by size.

Substrate particles are measured with a metric ruler, in centimeters (cm). Because rocks can be long and narrow, we measure them twice: first the width, then the length. By adding the width to the length and dividing by two, we obtain the average size of the rock.

It is important to randomly select the rocks we wish to measure. Otherwise, we would tend to select larger rocks, or more colorful rocks, or those with unusual shapes. Instead, we should just reach down and collect those rocks in front of us and within easy reach. Then measure each rock. Upon completion of measurement, each rock should be classified. Ecologists have developed a

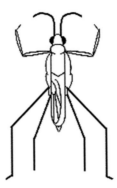

FIGURE 12.14 Water strider. (Source: Adapted from Standard Methods, 15th Edition. Copyright © 1981 by the American Public Health Association, the American Water Works Association, and the Water Pollution Control Federation.)

standard scale (Wentworth scale) for size categories of substrate rock and other mineral materials, along with the different sizes. These are:

Boulder	> 256mm
Cobble	64–256 mm
Pebble	16–64 mm
Gravel	2–16 mm
Sand	0.0625–2 mm
Silt	0.0039–0.0625 mm
Clay	< 0.0039 mm

Organisms that live in/on/under rocks or small spaces occupy what is known as a *microhabitat*. Some organisms make their own microhabitats: many of the caddisflies build a case about themselves and use it for their shelter. Rocks are not the only physical features of streams where aquatic organisms can be found. For example, fallen logs and branches (commonly referred to as Large Woody Debris, or LWD), provide an excellent place for some aquatic organisms to burrow into and surfaces for others to attach themselves, as they might to a rock. They also create areas where small detritus such as leaf litter can pile up underwater. These piles of leaf litter are excellent shelters for many organisms, including large, fiercely predaceous larvae of dobsonflies.

Another important aquatic organism habitat is found in the matter, or *drift*, that floats along downstream. Drift is important because it is the main source of food for many fish. It may include insects such as mayflies (*Ephemeroptera*), some true flies (*Diptera*) and some stoneflies (*Plecoptera*) and caddisflies (*Trichoptera*). In addition, dead or dying insects and other small organisms, terrestrial insects that fall from the trees, leaves, and other matter, are common components of drift. Among the crustaceans, amphipods (small crustaceans) and isopods (small crustaceans including sow bugs and gribbles) also have been reported in the drift.

Adaptations to Stream Current

Current in streams is the outstanding feature of streams and the major factor limiting the distribution of organisms. The current is determined by the steepness of the bottom gradient, the roughness of the streambed, and the depth and width of the streambed. The current in streams has promoted many special adaptations by stream organisms. Odum (1971) listed these adaptations as follows (see Figure 12.15):

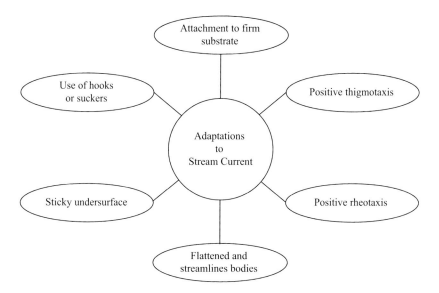

FIGURE 12.15   Adaptations to stream current.

- Attachment to a firm substrate: Attachment is to stones, logs, leaves, and other underwater objects such as discarded tires, bottles, pipes, etc. Organisms in this group are primarily composed of the primary producer plants and animals, such as green algae, diatoms, aquatic mosses, caddisfly larvae, and freshwater sponges.
- The use of hooks and suckers: These organisms have the unusual ability to remain attached and withstand even the strongest rapids. Two Diptera larvae, *Simulium* and *Blepharocera*, are examples.
- A sticky undersurface: Snails and flatworms are examples of organisms that can use their sticky undersurfaces to adhere to underwater surfaces.
- Flattened and streamlined bodies: All macroconsumers have streamlined bodies, i.e., the body is broad in front and tapers posteriorly to offer minimum resistance to the current. All nektons such as fish, amphibians, and insect larvae exhibit this adaptation. Some organisms have flattened bodies, which enable them to stay under rocks and in narrow places. Examples are water penny, a beetle larva, mayfly, and stonefly nymphs.
- Positive chemotaxis (rhea: current; taxis: arrangement): An inherent behavioral trait of stream animals (especially those capable of swimming) is to orient themselves upstream and swim against the current.
- Positive thigmotaxis (thigmo: touch, contact): Another inherent behavior pattern for many stream animals is to cling close to a surface or keep the body in close contact with the surface. This is the reason that stonefly nymphs (when removed from one environment and placed into another) will attempt to cling to just about anything, including each other.

It would take an entire text to describe the great number of adaptations made by aquatic organisms to their surroundings in streams. For our purposes, instead, we cover those special adaptations that are germane to this discussion. The important thing to remember is that an aquatic organism can adapt to its environment in several basic ways.

TYPES OF ADAPTIVE CHANGES

Adaptive changes are classed as genotypic, phenotypic, behavioral, or ontogenic:

- Genotypic changes: Tend to be great enough to separate closely related animals into species, such as mutations or recombination of genes. A salmonid is an example that has evolved a subterminal mouth (i.e., below the snout) in order to eat from the benthos.
- Phenotypic changes: Are the changes that an organism might make during its lifetime to better utilize its environment (e.g., a fish that changes sex from female to male because of an absence of males).
- Behavioral changes: Have little to do with body structure or type: a fish might spend more time under an overhang to hide from predators.
- Ontogenetic change: That which takes place as an organism grows and matures (e.g., a Coho salmon that inhabits streams when young, and migrates to the sea when older, changing its body chemistry to allow it to tolerate salt water).

SPECIFIC ADAPTATIONS

Specific adaptations observed in aquatic organisms include mouths, shape, color, aestivation, and schooling.

- *Mouths:* Aquatic organisms such as fish change mouth shape (morphology) depending on the food the fish eats. The arrangement of the jawbones and even other head bones, the length and width of gill rakers, the number, shape, and location of teeth, and barbels all change to allow fish to eat just about anything found in a stream.
- *Shape:* changes to allow fish to do different things in the water. Some organisms have body shapes that push them down in the water, against the substrate, and allow them to hold their place against even strong current (e.g., chubs, catfish, dace, and sculpins). Other organisms, especially predators, have evolved an arrangement and shape of fins that allow them to lurk without moving then lunge suddenly to catch their prey (e.g., bass, perch, pike, trout, and sunfish).
- *Color:* May change within hours, to camouflage, or within days, or may be genetically predetermined. Fish tend to turn dark in clear water, and pale in muddy water.
- *Aestivation:* Helps fishes survive in arid desert climates, where streams may dry up from time to time. Aestivation refers to the ability of some fishes to burrow into the mud and wait out the dry period.
- *Schooling:* Serves as protection for many fish, particularly those that are subject to predation.

BENTHIC LIFE

The benthic habitat is found in the streambed, or benthos. As mentioned, the streambed is comprised of various physical and organic materials where erosion and/or deposition are a continuous characteristic. Erosion and deposition may occur simultaneously and alternately at different locations in the same streambed. Where channels are exceptionally deep and taper slowly to meet the relatively flattened streambed, habitats may form on the slopes of the channel. These habitats are referred to as littoral habitats. Shallow channels may dry up periodically in accordance with weather changes. The streambed is then exposed to open air and may take on the characteristics of a wetland.

Silt and organic materials settle and accumulate in the streambed of slowly flowing streams. These materials decay and become the primary food resource for the invertebrates inhabiting the streambed. Productivity in this habitat depends upon the breakdown of these organic materials by herbivores. Bottom-dwelling organisms use not all-organic materials; a substantial amount becomes part of the streambed in the form of peat.

In faster moving streams, organic materials do not accumulate so easily. Primary production occurs in a different type of habitat found in the riffle regions where there are shoals and rocky regions for organisms to adhere to. Therefore, plants that can root themselves into the streambed dominate these regions. By plants, we are referring mostly to forms of algae, often microscopic and filamentous, that can cover rocks and debris that have settled into the streambed during summer months.

Note: If you have ever stepped into a stream, the green, slippery slime on the rocks in the streambed is representative of this type of algae.

Although the filamentous algae seem well anchored, strong currents can easily lift it from the streambed and carry it downstream where it becomes a food resource for low-level consumers. One factor that greatly influences the productivity of a stream is the width of the channel; a direct relationship exists between stream width and richness of bottom organisms. Bottom-dwelling organisms are very important to the ecosystem as they provide food for other, larger benthic organisms through consuming detritus.

BENTHIC PLANTS AND ANIMALS

Vegetation is not common in the streambed of slow-moving streams; however, they may anchor themselves along the banks. Alga (mainly green and blue green) as well as common types of water moss attaches themselves to rocks in fast-moving streams. Mosses and liverworts often climb up the sides of the channel onto the banks as well. Some plants like the reeds of wetlands with long stems and narrow leaves can maintain roots and withstand the current. Aquatic insects and invertebrates dominate slow-moving streams. Most aquatic insects are in their larval and nymph forms such as the blackfly, caddisfly, and stonefly. Adult water beetles and water bugs are also abundant. Insect larvae and nymphs provide the primary food source for many fish species, including American eel and brown bullhead catfish. Representatives of crustaceans, rotifers, and nematodes (flatworms) are sometimes present. Leech, worm, and mollusk (especially freshwater mussels) abundance varies with stream conditions but generally favors low phosphate conditions. Larger animals found in slow-moving streams and rivers include newts, tadpoles, and frogs. As mentioned, the important characteristic of all life in streams is adaptability to withstand currents.

BENTHIC MACROINVERTEBRATES

The emphasis on aquatic insect studies, which has expanded exponentially in the last three decades, has been largely ecological. Freshwater macroinvertebrates are ubiquitous; even polluted waters contain some representative of this diverse and ecologically important group of organisms. Benthic macroinvertebrates are aquatic organisms without backbones that spend at least a part of their life cycle on the stream bottom. Examples include aquatic insects – such as stoneflies, mayflies, caddisflies, midges, and beetles – as well as crayfish, worms, clams, and snails. Most hatch from eggs and mature from larvae to adults. Most of the insects spend their larval phase on the river bottom and, after a few weeks to several years, emerge as winged adults. The aquatic beetles, true bugs, and other groups remain in the water as adults. Macroinvertebrates typically collected from the stream substrate are either aquatic larvae or adults.

In practice, stream ecologists observe indicator organisms and their responses to determine the quality of the stream environment. There are several methods for determining water quality based on biologic characteristics. A wide variety of indicator organisms (biotic groups) are used for biomonitoring. These most often include algae, bacteria, fish, and macroinvertebrates. Notwithstanding their popularity, in this text, we use benthic macroinvertebrates for several other reasons.

Simply, they offer several advantages:

- They are ubiquitous, so they are affected by perturbations in many different habitats.
- They are species rich, so the large number of species produces a range of responses.
- They are sedentary, so they stay put, which allows determination of the spatial extent of a perturbation.

- They are long-lived, which allows temporal changes in abundance and age structure to be followed.
- They integrate conditions temporally, so like any biotic group, they provide evidence of conditions over long periods.

In addition, benthic macroinvertebrates are preferred as bioindicators because they are easily collected and handled by samplers; they require no special culture protocols. They are visible to the naked eye, and samplers easily distinguish their characteristics. They have a variety of fascinating adaptations to stream life. Certain benthic macroinvertebrates have very special tolerances and thus are excellent specific indicators of water quality. Useful benthic macroinvertebrate data are easy to collect without expensive equipment. The data obtained by macroinvertebrate sampling can serve to indicate the need for additional data collection, possibly including water analysis and fish sampling.

In short, we base the focus of this discussion on benthic macroinvertebrates (regarding water quality in streams and lakes) simply because some cannot survive in polluted water, while others can survive or even thrive in polluted water. In a healthy stream, the benthic community includes a variety of pollution-sensitive macroinvertebrates. In an unhealthy stream or lake, there may be only a few types of nonsensitive macroinvertebrates present. Thus, the presence or absence of certain benthic macroinvertebrates is an excellent indicator of water quality. Moreover, it may be difficult to identify stream or lake pollution with water analysis, which can only provide information for the time of sampling (a snapshot of time). Even the presence of fish may not provide information about a polluted stream because fish can move away to avoid polluted water and then return when conditions improve. However, most benthic macroinvertebrates cannot move to avoid pollution. Thus, a macroinvertebrate sample may provide information about pollution that is not present at the time of sample collection.

Obviously, before we can use benthic macroinvertebrates to gauge water quality in a stream (or for any other reason), we must be familiar with the macroinvertebrates that are commonly used as bioindicators. Samplers need to be aware of basic insect structures before they can classify the macroinvertebrates they collect. Structures, which need to be stressed, include head, eyes (compound and simple), antennae, mouth (no emphasis on parts), segments, thorax, legs and leg parts, gills, abdomen, etc. Samplers also need to be familiar with insect metamorphosis – both complete and incomplete – as most of the macroinvertebrates collected are larval or nymph stages.

Note: Information on basic insect structures is beyond the scope of this text. Thus, we highly recommend "the" standard guide to aquatic insects of North America: *An Introduction to the Aquatic Insects of North America*, 3rd ed., Merritt, R.W., & Cummins, K.W. (eds.). Dubuque, IA: Kendall/Hunt Publishing Company, 1996.

Identification of Benthic Macroinvertebrates

Before identifying and describing the key benthic macroinvertebrates significant to water and wastewater operators, it is important first to provide foundational information. We characterize benthic macroinvertebrates using two important descriptive classifications: trophic groups and mode of existence. In addition, we discuss their relationship in the food web; that is, what, or whom, they eat.

Trophic groups: Of the trophic groups (i.e., feeding groups) that Merritt and Cummins have identified for aquatic insects, only five are likely to be found in a stream using typical collection and sorting methods (Merritt & Cummins, 1996):

Shredders – These have strong, sharp mouthparts that allow them to shred and chew coarse organic material such as leaves, algae, and rooted aquatic plants. These organisms play an important role in breaking down leaves or larger pieces of organic material to a size that can be used by other macroinvertebrates. Shredders include certain stonefly and caddisfly larvae, sow bugs, scuds, and others.

Collectors – These gather the very finest suspended matter in the water. To do this, they often sieve the water through rows of tiny hairs. These sieves of hairs may be displayed in fans on their heads (blackfly larvae) or on their forelegs (some mayflies). Some caddisflies and midges spin nets and catch their food in them as the water flows through.

Scrapers – These scrape the algae and diatoms off surfaces of rocks and debris, using their mouthparts. Many of these organisms are flattened to hold onto surfaces while feeding. Scrapers include water pennies, limpets and snails, net winged midge larvae, certain mayfly larvae, and others.

Piercers – These herbivores pierce plant tissues or cells and suck the fluids out. Some caddisflies do this.

Predators – Predators eat other living creatures. Some of these are engulfers; that is, they eat their prey completely or in parts. This is very common in stoneflies and dragonflies, as well as caddisflies. Others are piercers, which are like the herbivorous piercers except that they are eating live animal tissues.

Mode of Existence (habit, locomotion, attachment, concealment):

Skaters – Adapted for "skating" on the surface where they feed as scavengers on organisms trapped in the surface film (example: water striders).

Planktonic – Inhabiting the open water limnetic zone of standing waters (lentic; lakes, bogs, ponds). Representatives may float and swim about in the open water but usually exhibit a diurnal vertical migration pattern (example: phantom midges) or float at the surface to obtain oxygen and food, diving when alarmed (example: mosquitoes).

Divers – Adapted for swimming by "rowing" with the hind legs in lentic habitats and lotic pools. Representatives come to the surface to obtain oxygen, dive and swim when feeding or alarmed; may cling to or crawl on submerged objects such as vascular plants (examples: water boatmen; predaceous diving beetle).

Swimmers – Adapted for "fishlike" swimming in lotic or lentic habitats. Individuals usually cling to submerged objects, such as rocks (lotic riffles) or vascular plants (lentic) between short bursts of swimming (example: mayflies).

Clingers – Representatives have behavioral (e.g., fixed retreat construction) and morphological (e.g., long, curved tarsal claws, dorsoventral flattening, and ventral gills arranged as a sucker) adaptations for attachment to surfaces in stream riffles and wave-swept rocky littoral zones of lakes (examples: mayflies and caddisflies).

Sprawlers – Inhabiting the surface of floating leaves of vascular hydrophytes or fine sediments, usually with modifications for staying on top of the substrate and maintaining the respiratory surfaces free of silt (examples: mayflies, dobsonflies, and damselflies).

Climbers – adapted for living on vascular hydrophytes or detrital debris (e.g., overhanging branches, roots and vegetation along streams, and submerged brush in lakes) with modifications for moving vertically on stem-type surfaces (examples: dragonflies and damselflies).

Burrowers – Inhabiting the fine sediments of streams (pools) and lakes. Some construct discrete burrows, which may have sand grain tubes extending above the surface of the substrate, or the individuals may ingest their way through the sediments (examples: mayflies and midges).

Macroinvertebrates and the Food Web

In a stream or lake, the two possible sources of primary energy are:

- photosynthesis by algae, mosses, and higher aquatic plants, and
- imported organic matter from streamside/lakeside vegetation (e.g., leaves and other parts of vegetation).

Simply put, a significant portion of the food that is eaten grows right in the stream or lake, like algae, diatoms, nymphs and larvae, and fish. A food that originates from within the stream is called *autochthonous*. Most food in a stream, however, comes from outside the stream – especially the case in small, heavily wooded streams, where there is normally insufficient light to support substantial instream photosynthesis, so energy pathways are supported largely by imported energy. Leaves provide a large portion of this imported energy. Worms drown in floods and are washed in. Leafhoppers and caterpillars fall from trees. Adult mayflies and other insects mate above the stream, lay their eggs in it, and then die in it. All this food from outside the stream is called *allochthonous*.

Units of Organization

Macroinvertebrates, like all other organisms, are classified and named. Macroinvertebrates are classified and named using a *taxonomic hierarchy*. The taxonomic hierarchy for the caddisfly (a macroinvertebrate insect commonly found in streams) is shown below.

TYPICAL BENTHIC MACROINVERTEBRATES IN RUNNING WATERS

As mentioned, the macroinvertebrates are the best-studied and most diverse animals in streams. We therefore devote our discussion to the various macroinvertebrate groups. While it is true that non-insect macroinvertebrates, such as Oligochaeta (worms), Hirudinea (leeches), and Acari (water mites), are frequently encountered groups in lotic environments, the insects are among the most conspicuous inhabitants of streams. In most cases, it is the larval stages of these insects that are aquatic, whereas the adults are terrestrial.

Kingdom	Animalia (animals)
Phylum	Arthropoda ("jointed legs")
Class	Insecta (insect)
Order	Trichoptera (caddisfly)
Family	Hydropsychidae (net-spinning caddis)
Genus species	*Hydropsyche morosa*

Typically, the larval stage is much extended, while the adult lifespan is short. Lotic insects are found among many different orders, and brief accounts of their biology are presented in the following sections. First, however, we present a brief glossary that includes select terminology used below, and it also includes other important terms mentioned earlier often associated with the description of aquatic macroinvertebrates. The source of most of these definitions is Spellman's (1996) *Stream Ecology and Self Purification* (Boca Raton, FL: CRC Press; Merritt & Cummins, 1996).

Macroinvertebrate Glossary

Abdomen – The third main division of the body; behind the head and thorax.

Anterior – In front (before).

Apical – Near or pertaining to the end of any structure, part of the structure that is farthest from the body.

Basal – Pertaining to the end of any structure that is nearest to the body.

Burrower – Animal that uses a variety of structures designed for moving and burrowing into sand and silt or building tubes within loose substrate.

Carapace – The hardened part of some arthropods that spreads like a shield over several segments of the head and thorax.

Caudal filament – Threadlike projection at the end of the abdomen, like a tail.

Clinger – Animal that uses claws or hooks to cling to the surfaces or rocks, plants or other hard surfaces and often moves slowly along these surfaces.

Concentric – A growth pattern on the opercula of some gastropods, marked by a series of circles that lie entirely within each other; compare multi-spiral and pauci-spiral.

Crawler – An animal, whose main means of locomotion is moving slowly along the bottom, usually has some type of hooks, claws, or specially designed feet to help hold them to surfaces.

Detritus – Disintegrated or broken up mineral or organic material.

Dextral – The curvature of a gastropod shell where the opening is visible on the right when the spire is pointing up.

Distal – Near or toward the free end of any appendage; that part farthest from the body.

Dorsal – Pertaining to, or situated on the back or top, especially of the thorax and abdomen.

Elytra – Hardened shell-like mesothoracic wings of adult beetles (Coleoptera).

Femur – The leg section between the tibia and coxa of Arthropoda, comparable to an upper arm or thigh.

Flagellum – A small fingerlike or whip-like projection.

Gill – Any structure especially adapted for the exchange of dissolved gases between animal and a surrounding liquid.

Glossae – A lobe of lobes front and center on the labium; in Plecoptera, the lobes are between the paraglossae.

Hemimetabolism – Incomplete metamorphosis.

Holometabolism – Complete metamorphosis.

Labium – Lower mouthpart of an arthropod, like a jaw or lip.

Labrum – Upper mouthpart of an arthropod consisting of a single usually hinged plate above the mandibles.

Lateral – Feature or marking located on the side of a body or other structure.

Ligula – Forming the ventral wall of an arthropod's oral cavity.

Lobe – A rounded projection or protuberance.

Mandibles – The first pair of jaws in insects.

Maxillae – The second pair of jaws in insects.

Multispiral – The growth pattern on the opercula of some gastropods marked by several turns from the center to the edge.

Operculum – A lid or covering structure, like a door to an opening.

Palpal lobes – The grasping pinchers at the end of the Odonata lower jaw.

Pauci-spiral – A growth pattern on the opercula of some gastropods marked by a few turns from the center to the edge.

Periphyton – Algae and associated organisms that live attached to underwater surfaces.

Posterior – Behind; opposite of anterior.

Proleg – Any projection appendage that serves for support locomotion or attachment.

Prothorax – The first thoracic segment closest to the head.

Rostrum – A beak or beak-like mouthpart.

Sclerite – A hardened area of an insect body wall, usually surrounded by softer membranes.

Seta – (pl. setae) – Hair-like projection.

Sinistral – The curvature of a gastropod shell where the opening is seen on the left when the spire is pointing up.

INSECT MACROINVERTEBRATES

The most important insect groups in streams are Ephemeroptera (mayflies), Plecoptera (stoneflies), Trichoptera (caddisflies), Diptera (true flies), Coleoptera (beetles), Hemiptera (bugs), Megaloptera (alderflies and dobsonflies), and Odonata (dragonflies and damselflies). The identification of these different orders is usually easy and there are many keys and specialized references (e.g., Merritt and Cummins, *An Introduction to the Aquatic Insects of North America*, 1996) available to help in the identification of species. In contrast, some genera and species specialist taxonomists can often only diagnose particularly the Diptera. As mentioned, insect macroinvertebrates are ubiquitous in streams and are often represented by many species. Although the numbers refer to aquatic species, a majority is to be found in streams.

Mayflies (Order: Ephemeroptera)

Description: Wing pads may be present on the thorax; three pairs of segmented legs attach to the thorax; one claw occurs on the end of the segmented legs; gills occur on the abdominal segments and are attached mainly to the sides of the abdomen, but sometimes extend over the top and bottom of the abdomen; gills consist of either flat plates or filaments; there long thin caudal (tails filaments) usually occur at the end of the abdomen, but there may only be two in some kinds.

Streams and rivers are generally inhabited by many species of mayflies, and, in fact, most species are restricted to streams. For the experienced freshwater ecologist who looks upon a mayfly nymph, recognition is obtained through trained observation: abdomen with leaf-like or feather-like gills, legs with a single tarsal claw, generally (but not always) with three cerci (three "tails"; two cerci, and between them usually a terminal filament; see Figure 12.16). The experienced ecologist knows that mayflies are hemimetabolous insects (i.e., where larvae or nymphs resemble wingless adults) that go through many postembryonic molts, often in the range between 20 and 30. For some species, body length increases about 15% for each instar.

Mayfly nymphs are mainly grazers or collector-gatherers feeding on algae and fine detritus, although a few genera are predatory. Some members filter particles from the water using hair-fringed

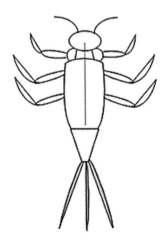

FIGURE 12.16 Mayfly (Ephemeroptera order).

legs or maxillary palps. Shredders are rare among mayflies. In general, mayfly nymphs tend to live mostly in unpolluted streams where with densities of up to 10,000/sq. meter they contribute substantially to secondary producers.

Adult mayflies resemble nymphs but usually possess two pairs of long, lacy wings folded upright; adults usually have only two cerci. The adult lifespan is short, ranging from a few hours to a few days, rarely up to two weeks, and the adults do not feed. Mayflies are unique among insects in having two winged stages, the subimago and the imago. The emergence of adults tends to be synchronous, thus ensuring the survival of enough adults to continue the species.

Stoneflies (Order: Plecoptera)

Description: Long thin antenna project in front of the head; wing pads usually present on the thorax but may only be visible in older larvae; three pairs of segmented legs attach to the thorax; two claws are located at the end of the segmented legs; gills occur on the thorax region, usually on the legs or bottom the thorax, or there may be no visible gills (usually there are none or very few gills on the abdomen); gills are either single or branched filaments; two long thin tails project from the rear of the abdomen. Stoneflies have very low tolerance to many insults; however, several families are tolerant of slightly acidic conditions. Although many freshwater ecologists would maintain that the stonefly is a well-studied group of insects, this is not exactly the case. Despite their importance, less than 5%–10% of stonefly species are well known with respect to life history, trophic interactions, growth, development, spatial distribution, and nymphal behavior. Notwithstanding our lacking extensive knowledge in regard to stoneflies, enough is known to provide an accurate characterization of these aquatic insects.

We know, for example, that stonefly larvae are characteristic inhabitants of cool, clean streams (i.e., most nymphs occur under stones in well-aerated streams). While they are sensitive to organic pollution, or more precisely to low oxygen concentrations accompanying organic breakdown processes, stoneflies seem rather tolerant to acidic conditions. Lack of extensive gills at least partly explains their relative intolerance of low oxygen levels.

Stoneflies are drab-colored, small- to medium-sized 1/6–2¼ inches (4–60 mm), rather flattened insects. Stoneflies have long, slender, many-segmented antennae and two long narrow antenna-like structures (cerci) on the tip of the abdomen (see Figure 12.17). The cerci may be long or short. At rest, the wings are held flat over the abdomen, giving a "square-shouldered" look compared to the roof-like position of most caddisflies and vertical position of the mayflies. Stoneflies have two pairs of wings. The hind wings are slightly shorter than the forewings and much wider, having a large anal lobe that is folded fanwise when the wings are at rest. This fanlike folding of the wings gives the order its name: "*pleco*" (folded or plaited) and "*-ptera*" (wings). The aquatic nymphs

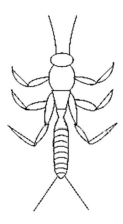

FIGURE 12.17 Stonefly (Plecoptera order).

are generally very similar to mayfly nymphs except that they have only two cerci at the tip of the abdomen. The stoneflies have chewing mouthparts. They may be found anywhere in a nonpolluted stream where food is available. Many adults, however, do not feed and have reduced or vestigial mouthparts. Stoneflies have a specific niche in high-quality streams where they are very important as a fish food source at specific times of the year (winter to spring, especially) and of the day. They complement other important food sources, such as caddisflies, mayflies, and midges.

Caddisflies (Order: Trichoptera)

Description: Head has a thick hardened skin; antennae are very short, usually not visible; no wing pads occur on the thorax; top of the first thorax always has a hardened plate and in several families the second and third section of the thorax have a hardened plate; three pairs of segmented legs attach to the thorax; abdomen has a thin soft skin; single or branched gills on the abdomen in many families, but some have no visible gills; pair of prologs with one claw on each, is situated at the end of the abdomen; most families construct various kinds of retreats consisting of a wide variety of materials collected from the streambed.

Trichoptera (Greek: *trichos*, a hair; *ptera*, wing), is one of the most diverse insect orders living in the stream environment, and caddisflies have nearly a worldwide distribution (the exception: Antarctica). Caddisflies may be categorized broadly into free-living (roving and net spinning) and case-building species.

Caddisflies are described as medium-sized insects with bristle-like and often long antennae. They have membranous hairy wings (which explains the Latin name "Trichos"), which are held tent-like over the body when at rest; most are weak fliers. They have greatly reduced mouthparts and five tarsi. The larvae are mostly caterpillar-like and have a strongly sclerotized (hardened) head with very short antennae and biting mouthparts. They have well-developed legs with a single tarsus. The abdomen is usually ten segmented; in case-bearing species the first segment bears three papillae, one dorsally and the other two laterally which help hold the insect centrally in its case, allowing a good flow of water to pass the cuticle and gills; the last or anal segment bears a pair of grappling hooks.

In addition to being aquatic insects, caddisflies are superb architects. Most caddisfly larvae (see Figure 12.18) live in self-designed, self-built houses, called *cases*. They spin out silk, and either live in silk nets or use the silk to stick together bits of whatever is lying on the stream bottom. These houses are so specialized, that you can usually identify a caddisfly larva to genus if you can see its house (case). With nearly 1,400 species of caddisfly species in North America (north of Mexico), this is a good thing!

Caddisflies are closely related to butterflies and moths (Order: Lepidoptera). They live in most stream habitats and that is why they are so diverse (have so many species). Each species has special adaptations that allow it to live in the environment it is found in. Mostly herbivorous, most caddisflies feed on decaying plant tissue and algae. Their favorite algae are diatoms, which they scrape off rocks. Some of them, though, are predacious.

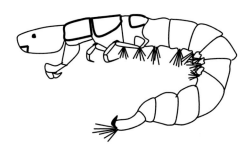

FIGURE 12.18   Caddis (Hydropsyche) larvae.

Caddisfly larvae can take a year or two to change into adults. They then change into *pupae* (the inactive stage in the metamorphosis of many insects, following the larval stage and preceding the adult form) while still inside their cases for their metamorphosis. It is interesting to note that caddisflies, unlike stoneflies and mayflies, go through a "complete" metamorphosis. Caddisflies remain as pupae for 2–3 weeks, and then emerge as adults. When they leave their pupae, splitting their case, they must swim to the surface of the water to escape it. The winged adults fly evening and night, and some are known to feed on plant nectar. Most of them will live less than a month: like many other winged stream insects, their adult lives are brief compared to the time they spend in the water as larvae.

Caddisflies are sometimes grouped into five main groups by the kinds of cases they make: free-living forms that do not make cases, saddle-case makers, purse-case makers, net-spinners and retreat-makers, and tube-case makers. Caddisflies demonstrate their architectural talents in the cases they design and make. For example, a caddisfly might make a perfect, four-sided box case of bits of leaves and bark, or tiny bits of twigs. It may make a clumsy dome of large pebbles. Others make rounded tubes out of twigs or very small pebbles.

In our experience in gathering caddisflies, we have come to appreciate not only their architectural ability but also their flare in the selection of construction materials. For example, we have found many caddisfly cases constructed of silk, emitted through an opening at the tip of the labium, used together with bits of ordinary rock mixed with sparkling quartz and red garnet, green peridot, and bright fool's gold.

In addition to the protection their cases provide them, the cases provide another advantage. The cases help caddisflies breathe. They move their bodies up and down, back and forth inside their cases, and this makes a current that brings them fresh oxygen. The less oxygen there is in the water, the faster they must move. It has been seen that caddisflies inside their cases get more oxygen than those that are outside of their cases – and this is why stream ecologists think that caddisflies can often be found even in still waters, where dissolved oxygen is low, in contrast to stoneflies and mayflies.

True Flies (Order: Diptera)

Description: Head may be a capsule-like structure with thick hard skin; head may be partially reduced so that it appears to be part of the thorax, or it may be greatly reduced with only the mouthparts visible; no wing pads occur on the thorax; false-legs (pseudo-legs) may extend from various sections of the thorax and abdomen composed of entirely soft skin, but some families have hardened plates scattered on various body features. The larval states do not have segmented leg features.

True or two- (*Di*-) winged (*ptera*) flies not only include the flies that we are most familiar with, like fruit flies and houseflies, they also include midges (see Figure 12.19), mosquitoes, crane flies (see Figure 12.20), and others. Houseflies and fruit flies live only on land, and we do not concern ourselves with them. Some, however, spend nearly their whole lives in water; they contribute to the

FIGURE 12.19 Midge larvae.

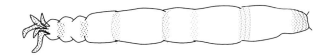

FIGURE 12.20 Crane fly larvae.

ecology of streams. True flies are in the order Diptera, and are one of the most diverse orders of the class Insecta, with about 120,000 species worldwide. Dipteran larvae occur almost everywhere except Antarctica and deserts where there is no running water. They may live in a variety of places within a stream: buried in sediments, attached to rocks, beneath stones, in saturated wood or moss, or in silken tubes, attached to the stream bottom. Some even live below the stream bottom.

True fly larvae may eat almost anything, depending on their species. Those with brushes on their heads use them to strain food out of the water that passes through. Others may eat algae, detritus, plants, and even other fly larvae. The longest part of the true fly's life cycle, like that of mayflies, stoneflies, and caddisflies, is the larval stage. It may remain an underwater larva anywhere from a few hours to five years. The colder the environment, the longer it takes to mature. It pupates and emerges, then, as a winged adult. The adult may live four months – or it may only live for a few days. While reproducing, it will often eat plant nectar for the energy it needs to make its eggs. Mating sometimes takes place in aerial swarms. The eggs are deposited back in the stream; some females will crawl along the stream bottom, losing their wings, to search for the perfect place to put their eggs. Once they lay them, they die.

Diptera serve an important role in cleaning water and breaking down decaying material, and they are a vital food source (i.e., they play pivotal roles in the processing of food energy) for many of the animals living in and around streams. However, the true flies most familiar to us are the midges, mosquitoes, and the crane flies because they are pests. Some midge flies and mosquitoes bite; the crane fly, however, does not bite but looks like a giant mosquito.

Like mayflies, stoneflies, and caddisflies, true flies are mostly in larval form. Like caddisflies, you can also find their pupae because they are holometabolous insects (go through complete metamorphosis). Most of them are free living; that is, they can travel around. Although none of the true fly larvae have the six, jointed legs we see on the other insects in the stream, they sometimes have strange little almost-legs (prolegs) to move around with. Others may move somewhat like worms do, and some – the ones who live in waterfalls and rapids – have a row of six suction discs that they use to move much like a caterpillar does. Many use silk pads and hooks at the ends of their abdomens to hold them fast to smooth rock surfaces.

Beetles (Order: Coleoptera)

Description: Head has thick hardened skin; thorax and abdomen of most adult families have moderately hardened skin, several larvae have a soft-skinned abdomen; no wing pads on the thorax in most larvae, but wing pads are usually visible on adults; three pairs of segmented legs attach to the thorax; no structures. Projections extend from the sides of the abdomen in most adult families, but some larval stages have flat plates or filaments; no prolegs or long tapering filaments at the end of the abdomen. Beetles are one of the most diverse insect groups but are not as common in aquatic environments.

Of the more than 1 million described species of insect, at least one-third are beetles, making the Coleoptera not only the largest order of insects but also the most diverse order of living organisms. Even though the most speciose order of terrestrial insects, surprisingly their diversity is not so apparent in running waters. Coleoptera belongs to the infraclass Neoptera, division Endpterygota. Members of this order have an anterior pair of wings (the *elytra*) that are hard and leathery and not used in flight; the membranous hindwings, which are used for flight, are concealed under the elytra when the organisms are at rest. Only 10% of the 350,000 described species of beetles are aquatic.

Beetles are holometabolous. Eggs of aquatic coleopterans hatch in 1–2 weeks, with diapause occurring rarely. Larvae undergo from three to eight molts. The pupal phase of all coleopternas is technically terrestrial; making this life stage of beetles the only one that has not successfully invaded the aquatic habitat. A few species have diapausing prepupae, but most complete transformation to adults in two to three weeks. Terrestrial adults of aquatic beetles are typically short lived and sometimes nonfeeding, like those of the other orders of aquatic insects. The larvae of Coleoptera are morphologically and behaviorally different from the adults, and their diversity is high.

Aquatic species occur in two major suborders, the Adephaga and the Polyphaga. Both larvae and adults of six beetle families are aquatid. Dytiscidae (predaceous diving beetles), Elmidae (riffle beetles), Gyrinidae (whirligig beetles), Halipidae (crawling water beetles), Hydrophilidae (water scavenger beetles), and Noteridae (burrowing water beetles). Five families, Chrysomelidae (leaf beetles), Limnichidae (marsh-loving beetles),

Psephenidae (water pennies), Ptilodactylidae (toe-winged beetles), and Sciuridae (marsh beetles) have aquatic larvae and terrestrial adults, as do most of the other orders of aquatic insects; adult limnichids, however, readily submerge when disturbed. Three families have species that are terrestrial as larvae and aquatic as adults, Curculionidae (weevils), Dryopidae (long-toed water beetles), and Hydraenidae (moss beetles), a highly unusual combination among insects. Because they provide a greater understanding of a freshwater body's condition (i.e., they are useful indicators of water quality), we focus our discussion on the riffle beetle, water penny, and whirligig beetle.

Riffle beetle larvae (most found in running waters, hence the name Riffle Beetle) are up to ¾-inch long (see Figure 12.21). Their body is not only long but also hard, stiff, and segmented. They have six long segmented legs on upper middle section of body; back end has two tiny hooks and short hairs. Larvae may take three years to mature before they leave the water to form a pupa; adults return to the stream.

Riffle beetle adults are considered better indicators of water quality than larvae because they have been subjected to water quality conditions over a longer period. They walk very slowly under the water (on stream bottom), and do not swim on the surface. They have small oval-shaped bodies (see Figure 12.22) and are typically about ¼ inch in length. Both adults and larvae of most species feed on fine detritus with associated microorganisms that are scraped from the substrate, although others may be xylophagous, that is, wood eating (e.g., *Lara*, Elmidae). Predators do not seem to include riffle beetles in their diet, except perhaps for eggs, which are sometimes attacked by flatworms.

The adult *water penny* is inconspicuous and often found clinging tightly in a sucker-like fashion to the undersides of submerged rocks, where they feed on attached algae. The body is broad, slightly oval, and flat in shape, ranging from 4 to 6 mm (¼ inch) in length. The body is covered with segmented plates and looks like a tiny round leaf (see Figure 12.23). It has six tiny jointed legs (underneath). The color ranges from light brown to almost black. There are 14 water penny species in the United States. They live predominately in clean, fast-moving streams. Aquatic larvae live one year or more (they are aquatic); adults (they are terrestrial) live on land for only a few days. They scrape algae and plants from surfaces.

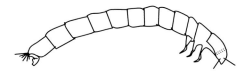

FIGURE 12.21   Riffle beetle larvae.

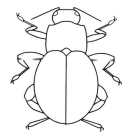

FIGURE 12.22   Riffle beetle adult.

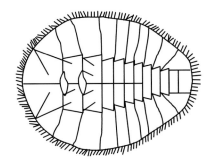

FIGURE 12.23   Water penny larvae.

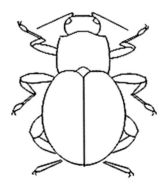

FIGURE 12.24   Whirligig beetle larvae.

Whirligig beetles are common inhabitants of streams and normally are found on the surface of quiet pools. The body has pincher-like mouthparts. Six segmented legs on the middle of the body; the legs end in tiny claws. Many filaments extend from the sides of the abdomen. They have four hooks at the end of the body and no tail (see Figure 12.24).

Note: When disturbed, whirligig beetles swim erratically or dive while emitting defensive secretions.

As larvae, they are benthic predators, whereas the adults live on the water surface, attacking dead and living organisms trapped in the surface film. They occur on the surface in aggregations of up to thousands of individuals. Unlike the mating swarms of mayflies, these aggregations serve primarily to confuse predators. Whirligig beetles have other interesting defensive adaptations. For example, the Johnston's organ at the base of the antennae enables them to echolocate using surface wave signals; their compound eyes are divided into two pairs, one above and one below the water surface, enabling them to detect both aerial and aquatic predators; and they produce noxious chemicals that are highly effective at deterring predatory fish.

Water Strider ("Jesus bugs") (Order: Hemiptera)

Description: The most distinguishing characteristic of the order is the mouthparts that are modified into an elongated, sucking beak. Most adults have hemelytra, which are modified leathery forewings. Some adults and all larvae lack wings; but most mature larvae possess wing pads. Both adults and larvae have three pairs of segmented legs with two tarsal claws at the end of each leg. Many families can also utilize atmospheric oxygen. This order is generally not suited for the biological assessment of flowing waters, due to their ability to use atmospheric oxygen.

It is fascinating to sit on a log at the edge of a stream pool and watch the drama that unfolds among the small water animals. Among the star performers in small streams are the Water Bugs. These are aquatic members of that large group of insects called the "true bugs," most of which live on land. Moreover, unlike many other types of water insects, they do not have gills but get their oxygen directly from the air. Most conspicuous and commonly known are the Water Striders or Water Skaters. These ride the top of the water, with only their feet making dimples in the surface film.

Like all insects, the water striders have a three-part body (head, thorax, and abdomen), six jointed legs, and two antennae. It has a long, dark, narrow body (see Figure 12.25).

The underside of the body is covered with water-repellent hair. Some water striders have wings, others do not. Most water striders are over 5 mm (0.2 inch) long. Water striders eat small insects that fall on the water's surface and larvae. Water striders are very sensitive to motion and vibrations on the water's surface. It uses this ability in order to locate prey. It pushes its mouth into its prey, paralyzes it, and sucks the insect dry. Predators of the water strider, like birds, fish, water beetles, backswimmers, dragonflies, and spiders, take advantage of the fact that water striders cannot detect motion above or below the surface of the water.

Alderflies and Dobsonflies (Order: Megaloptera)

Description: Head and thorax have thick hardened skin, while the abdomen has thin soft skin; prominent chewing mouthparts project in front of the head; no wing pads on the thorax; three pairs of segmented legs attached to the thorax; seven or eight pairs of stout tapering filaments extend from the abdomen; end of the abdomen has either a pair of prologs with two claws on each proleg, or a single long tapering filament with no prologs.

Larvae of all species of Megaloptera ("large wing") are aquatic and attain the largest size of all aquatic insects. Megaloptera is a medium-sized order with less than 5,000 species worldwide. Most species are terrestrial; in North America 64 aquatic species occur. In running waters, alderflies (Family: Sialidae) and dobsonflies (Family: Corydalidae, sometimes called hellgrammites or toe biters) are particularly important, as they are voracious predators, having large mandibles with sharp teeth.

Alderfly brownish-colored larvae possess a single tail filament with distinct hairs. The body is thick-skinned with six to eight filaments on each side of the abdomen; gills are located near the base of each filament. Mature body size: 0–1.25 inches (see Figure 12.26). Larvae are aggressive predators, feeding on other adult aquatic macroinvertebrates (they swallow their prey without chewing); as secondary consumers, other larger predators eat them. Female alderflies deposit eggs on vegetation that overhangs water; larvae hatch and fall directly into water (i.e., into quiet but moving water). Adult alderflies are dark with long wings folded back over the body; they only live a few days.

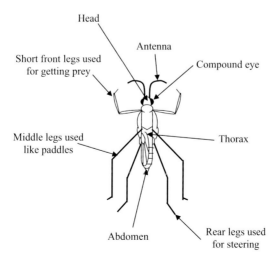

FIGURE 12.25 Water strider.

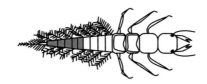

FIGURE 12.26   Alderfly larvae

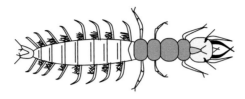

FIGURE 12.27   Dobsonfly larvae

Dobsonfly larvae are extremely ugly (thus, they are rather easy to identify) and can be rather large, anywhere from 25 to 90 mm (1–3.5 inches) in length. The body is stout, with eight pairs of appendages on the abdomen. Brush-like gills at base of each appendage look like "hairy armpits" (see Figure 12.27). The elongated body has spiracles (spines) and has three pairs of walking legs near the upper body and one pair of hooked legs at the rear. The head bears four segmented antennae, small compound eyes, and strong mouthparts (large chewing pinchers).

Coloration varies from yellowish, brown, gray, and black, often mottled. Dobsonfly larvae, commonly known as hellgrammites, are customarily found along stream banks under and between stones. As indicated by the mouthparts, they are predators and feed on all kinds of aquatic organisms.

Dragonflies and Damselflies (Order: Odonata)

Description: Dragonflies: Lower lip (labium) is long and elbowed to fold back against the head when not feeding, thus concealing other mouthparts; wing pads are present on the thorax; three pairs of segmented legs attach to the thorax; no gills on the sides of the abdomen. Dragonflies have three pointed structures that may occur at the end of the abdomen forming a pyramid-shaped opening; bodies are long and stout or somewhat oval. Damselflies have three flat gills at the end of the abdomen forming a tail-like structure, and their bodies are long and slender.

The Odonata (dragonflies, suborder Anisoptera; and damselflies, suborder Zygoptera) is a small order of conspicuous, hemimetabolous insects (lack a pupal stage) of about 5,000 named species and 23 families worldwide. Odonata is a Greek word meaning toothed one. It refers to the serrated teeth located on the insect's chewing mouthparts (mandibles).

Characteristics of dragonfly and damselfly larvae include:

- Large eyes;
- Three pairs of long segmented legs on upper middle section (thorax) of body;
- Large scoop-like lower lip that covers bottom of mouth;
- No gills on sides or underneath of abdomen.

Note: Dragonflies and damselflies are unable to fold their four elongated wings back over the abdomen when at rest.

Dragonflies and damselflies are medium to large insects with two pairs of long equal-sized wings. The body is long and slender, with short antennae. Immature stages are aquatic, and development occurs in three stages (egg, nymph, adult).

Dragonflies are also known as darning needles. Myths about dragonflies warned children to keep quiet or less the dragonfly's "darning needles" would sew the child's mouth shut. The nymphal stage of dragonflies is grotesque creatures, robust and stoutly elongated. They do not have long "tails" (see Figure 12.28). They are commonly gray, greenish, or brown to black in color. They are medium to large aquatic insects, ranging in size from 15 to 45 mm the legs are short and used for perching. They are often found on submerged vegetation and at the bottom of streams in the shallows. They are rarely found in polluted waters. Food consists of other aquatic insects, annelids, small crustacea, and mollusks.

Transformation occurs when the nymph crawls out of the water, usually onto vegetation. There it splits its skin and emerges prepared for flight. The adult dragonfly is a strong flier, capable of great speed (> 60 mph) and maneuverability (fly backward, stop on a dime, zip 20 feet straight up, and slip sideways in the blink of an eye!). When at rest the wings remain open and out to the sides of the body.

A dragonfly's freely movable head has large, hemispherical eyes (nearly 30,000 facets each), which the insects use to locate prey with their excellent vision. Dragonflies eat small insects, mainly mosquitoes (large numbers of mosquitoes), while in flight. Depending on the species, dragonflies lay hundreds of eggs by dropping them into the water and leaving them to hatch or by inserting eggs singly into a slit in the stem of a submerged plant. The incomplete metamorphosis (egg, nymph, mature nymph, and adult) can take 2–3 years. Nymphs are often covered by algal growth.

Note: Adult dragonflies are sometimes called "mosquito hawks" because they eat so many mosquitoes that they catch while they are flying.

Damselflies are smaller and slenderer than dragonflies. They have three long, oar-shaped feathery tails, which are gills, and long slender legs (see Figure 12.29). They are gray, greenish, or brown to black in color. Their habits are like those of dragonfly nymphs, and they emerge from the water as adults in the same manner. The adult damselflies are slow and seem uncertain in flight. Wings are commonly black or clear, and body is often brilliantly colored. When at rest, they perch on vegetation with their wings closed upright. Damselflies mature in one to four years. Adults live for a few

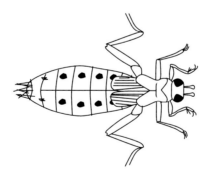

FIGURE 12.28   Dragonfly nymph.

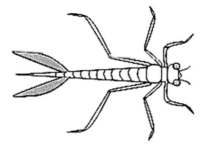

FIGURE 12.29   Damselfly nymph.

weeks or months. Unlike the dragonflies, adult damselflies rest with their wings held vertically over their backs. They mostly feed on live insect larvae.

Note: Relatives of the dragonflies and damselflies are some of the most ancient of the flying insects. Fossils have been found of giant dragonflies with wingspans up to 720 mm (28.4 inches) that lived long before the dinosaurs!

NON-INSECT MACROINVERTEBRATES

Non-insect macroinvertebrates are important to our discussion of stream and freshwater ecology because many of them are used as bioindicators of stream quality. Three frequently encountered groups in running water systems are Oligochaeta (worms), Hirudinea (leeches), and Gastropoda (lung-breathing snails). They are by no means restricted to running water conditions, and the great majority of them occupy slow-flowing marginal habitats where the sedimentation of fine organic materials takes place.

Oligochaeta (Family Tuificidae, Genus: *Tubifex*)

Tubifex worms (commonly known as sludge worms) are unique in the fact that they build tubes. Sometimes there are as many as 8,000 individuals per square meter. They attach themselves within the tube and wave their posterior end in the water to circulate the water and make more oxygen available to their body surface. These worms are commonly red, since their blood contains hemoglobin. *Tubifex* worms may be very abundant in situations when other macroinvertebrates are absent; they can survive in very low oxygen levels and can live with no oxygen at all for short periods. They are commonly found in polluted streams, and feed on sewage or detritus.

Hirudinea (Leeches)

Despite the many different families of leeches, they all have common characteristics. They are soft-bodied worm-like creatures that are flattened when extended. Their bodies are dull in color, ranging from black to brown and reddish to yellow, often with a brilliant pattern of stripes or diamonds on the upper body. Their size varies within species but generally ranges from 5 mm to 45 cm when extended. Leeches are very good swimmers, but they typically move in an inchworm fashion. They are carnivorous and feed on other organisms ranging from snails to warm-blooded animals. Leeches are found in warm protected shallows under rocks and other debris.

Gastropoda (Lung-Breathing Snail)

Lung-breathing snails (pulmonates) may be found in streams that are clean. However, their dominance may indicate that dissolved oxygen levels are low. These snails are different from *right-handed snails* because they do not breathe underwater by use of gills but instead have a lung-like sac called a pulmonary cavity, which they fill with air at the surface of the water. When the snail takes in air from the surface, it makes a clicking sound. The air taken in can enable the snail to breathe underwater for long periods, sometimes hours.

Lung-breathing snails have two characteristics that help us to identify them. First, it has no operculum or hard cover over the opening to its body cavity. Second, snails are either "right handed" or "left handed" and the lung-breathing snails are "left-handed." We can tell the difference by holding the shell so that its tip is upward and the opening toward us. If the opening is to the *left* of the axis of the shell, the snail is termed sinistral – that is it is left handed. If the opening is to the *right* of the axis of the shell, the snail is termed dextral – that is it is right handed, and it breathes with gills. Snails are animals of the substrate and are often found creeping along on all types of submerged surfaces in water from 10 cm to 2 m deep.

Before the Industrial Revolution of the 1800s, metropolitan areas were small and sparsely populated. Thus, river and stream systems within or next to early communities received insignificant quantities of discarded waste. Early on, these river and stream systems were able to compensate for the small amount of wastes they received; when wounded (polluted), nature has a way of fighting back.

In the case of rivers and streams, nature provides their flowing waters with the ability to restore themselves through their own self-purification process. It was only when humans gathered in great numbers to form great cities that the stream systems were not always able to recover from having received great quantities of refuse and other wastes. What exactly is it that man does to rivers and streams? What man does to rivers and streams is to upset the delicate balance between pollution and the purification process. That is, we tend to unbalance the aquarium.

CHAPTER SUMMARY

Whenever I mention the need for wastewater managers and operators to have a fundamental understanding of ecology, they respond with confused looks on their faces – at least to start with. When I ask them, what do they think is a close mimic to the wastewater treatment process, the concerns on their faces grow. However, when I ask them do you think that conventional wastewater treatment is similar to self-purifying processes in running water, streams, and rivers, many of the frowns disappear and they light up like the proverbial light bulb – understanding is powerful. Stream ecology and self-purification in streams is Nature's way of taking care of polluted running waters, at least to a point. Only when a river or stream is overloaded with pollutants is their innate ability to purify quenched, throttled, and hindered – in some cases totally thwarted. Also, a clear understanding that streams and rivers, even polluted ones, are full of life or contain some lower forms of life. The biology of a stream helps us to understand biological treatment of wastewater.

REVIEW QUESTIONS

1. The major ecological unit is _____.
   a. Productivity
   b. Bare-rock succession
   c. Ecosystem
   d. Niche

2. Those organisms residing within or on the bottom sediment are _____.
   a. Fruit flies
   b. Salmon
   c. Benthos
   d. All of the above

3. Organisms attached to plants or rocks are referred to as _____.
   a. Periphyton
   b. Snails
   c. Moss
   d. Algae

4. Small plants and animals that move about with the current are _____.
   a. Spiders
   b. Plankton
   c. Virus
   d. Flowers

5. Free-swimming organisms belong to which group of aquatic organisms?
   a. Snails
   b. Fish
   c. Pelagic
   d. All of the above

6. Organisms that live on the surface of the water are _____.
   a. Leaves
   b. Snails
   c. Fish
   d. Neuston

7. Movement of new individuals into a natural area is referred to as _____.
   a. Entry
   b. Immigration
   c. Assimilation
   d. A natural occurrence

8. Fixes energy of the sun and makes food from simple inorganic substances _____.
   a. Autotrophs
   b. Heterotrophs
   c. Trophs
   d. Salmon

9. The freshwater habitat that is characterized by normally calm water is _____.
   a. Lotic
   b. Lentic
   c. Surface layer
   d. Benthos

10. The amount of oxygen dissolved in water and available for organisms is the _____.
    a. BOD
    b. Acid
    c. Dissolved oxygen solubility
    d. Salt

ANSWER KEY

1. The major ecological unit is _____.
   a. Productivity
   b. Bare-rock succession
   c. **Ecosystem**
   d. Niche

2. Those organisms residing within or on the bottom sediment are _____.
   a. Fruit flies
   b. Salmon
   c. **Benthos**
   d. All of the above

3. Organisms attached to plants or rocks are referred to as _____.
   a. **Periphyton**
   b. Snails
   c. Moss
   d. Algae

4. Small plants and animals that move about with the current are _____.
   a. Spiders
   b. **Plankton**
   c. Virus
   d. Flowers

5. Free-swimming organisms belong to which group of aquatic organisms?
   a. Snails
   b. Fish
   c. **Pelagic**
   d. All of the above

6. Organism that live on the surface of the water are _____.
   a. Leaves
   b. Snails
   c. Fish
   d. **Neuston**

7. Movement of new individuals into a natural area is referred to as _____.
   a. Entry
   b. **Immigration**
   c. Assimilation
   d. A natural occurrence

8. Fixes energy of the sun and makes food from simple inorganic substances _____.
   a. **Autotrophs**
   b. Heterotrophs
   c. Trophs
   d. Salmon

9. The freshwater habitat that is characterized by normally calm water is _____.
   a. Lotic
   b. **Lentic**
   c. Surface layer
   d. Benthos

10. The amount of oxygen dissolved in water and available for organisms is the _____.
    a. BOD
    b. Acid
    c. **Dissolved oxygen solubility**
    d. Salt

BIBLIOGRAPHY

ASTM. 1969. *Manual on Water.* Philadelphia, PA: American Society for Testing and Materials.

Barlocher, R., & Kendrick, L. 1975. Leaf conditioning by microorganisms. *Oecologia* 20: 359–362.

Benfield, E.F. 1996. Leaf breakdown in streams ecosystems. In *Methods in Stream Ecology*, Hauer, F.R., & Lambertic, G.A., Eds. San Diego: Academic Press, pp. 579–590.

Benfield, E.F., Jones, D.R., & Patterson, M.F. 1977. Leaf pack processing in a pastureland stream. *Oikos* 29: 99–103.

Benjamin, C.L., Garman, G.R., & Funston, J.H. 1997. *Human Biology.* New York: McGraw-Hill Co., Inc.

Carson, R. 1962. *Silent Spring.* Boston: Houghton Mifflin.

Clements, E.S. 1960. *Adventures in Ecology.* New York: Pageant Press.

Crossley, D.A. , Jr., G.J. House, R.M. Snider, R.J. Snider, & B.R. Stinner 1984. The positive interactions in agroecosystems. In *Agricultural Ecosystems*, R. Lowrance, B.R. Stinner, & G.J. House, Eds. New York: Wiley.

Cummins, K.W. 1974. Structure and function of stream ecosystems. *Bioscience* 24: 631–641

Cummins, K.W., & Klug, M.J. 1979. Feeding ecology of stream invertebrates. *Annual Review of Ecology and Systematics* 10: 631–641.

Darwin, C. 1998. *The Origin of Species.* Suriano, G. (ed.) New York: Grammercy.

Dolloff, C.A., & Webster, J.R. 2000. Particulate organic contributions from forests to streams: Debris isn't so bad. In *Riparian Management in Forests of the Continental Eastern United States*, E.S. Verry, J.W. Hornbeck, and C.A. Dolloff (eds.) Washington, DC: USDA Department of Agriculture.

Ecology. 2007. *Ecology.* Accessed 02/19/19 @ http://www.newworldencyclopedia.org/preview/ Ecology.

Ecosystem. 2007. *Ecosystem.* Accessed 02/11/19 @ http://en.wikipedia.org/wiki/Ecosystem.

Ecosystem Topics. 2000. Accessed 11/05/19 @http://www.so.gun.edu.Irockwoo/Ecosystems.

Evans, F.C. 1956. Ecosystem as the basic unit in ecology. *Science* 23:1127–1128.

Jessup, B.K., Markowitz, A. and Stribling, J.B. 2002. *Family-Level Key to Stream Invertebrates of Maryland and Surround Areas.* Pasadena, CA: Tetra Tech, Inc.

Krebs, C.H. 1972. *Ecology. The Experimental Analysis of Distribution and Abundance.* New York: Harper and Row.

Lindeman, R.L. 1942. The trophic-dynamic aspect of ecology. *Ecology* 23: 399–418.

Margulis, L., & Sagan, D. 1997. *Microcosmos: Four Billion Years of Evolution from our Microbial Ancestors.* Berkeley, CA: University of California Press.

Marshall, P. 1950. *Mr. Jones, Meet the Master.* Grand Rapids, MI: Fleming H. Revel Company.

McCafferty, P.W. 1981. *Aquatic Entomology.* Boston, MA: Jones and Bartlett Publishers, Inc.

Merrit, R.W., & Cummins, K.W. 1996. *An Introduction to the Aquatic Insects of North America,* 3rd ed. Dubuque, IA: Kendall/Hunt Publishing.

Odum, E.P. 1952. *Fundamentals of Ecology,* 1st ed. Philadelphia, PA: W.B. Saunders Co.

Odum, E.P. 1971. *Fundamentals of Ecology,* 3rd ed. Philadelphia, PA: Saunders.

Odum, E.P. 1983. *Basic Ecology.* Philadelphia, PA: Saunders College Publishing.

Odum, E.P. 1984. Properties of agroecosystems. In *Agricultural Ecosystems,* R. Lowrance, B.R. Stinner, and G.J. House, Eds. New York: Wiley.

Odum, E.P., & Barrett, G.W. 2005. *Fundamentals of Ecology,* 5th ed. Belmont, CA: Thomson Brooks/Cole.

Paul, R.W., Jr., Benfield, E.F., & Cairns, J., Jr. 1978. Effects of thermal discharge on leaf decomposition in a river ecosystem. *Verhandlugen der Internationalen Vereinigung fur Thoeretsche and Angewandte Limnologie* 20: 1759–1766.

Peterson, R.C. & Cummins, K.W. 1974. Leaf processing in woodland streams. *Freshwater Biology* 4:345–368.

Porteous, a. 1992. *Dictionary of Environmental Science and Technology.* New York: Wiley.

Price, P.W. 1984. *Insect Ecology.* New York: Wiley.

Ramalay, F. 1940. *The Growth of a Science.* Colorado Springs, CO: University of Colorado Study. Vol. 26, pp. 3–14.

Smith, C.H. 2007. *Karl Ludwig Willdenow.* Accessed 02/09/19 @ http://www.wku.edu/~smithch/chronob/WILL1765.htm.

Smith, R.L. 1996. *Ecology and Field biology.* New York: Harper Collins College Publishers.

Smith, T.M., & Smith, R.L. 2006. *Elements of Ecology,* 6th ed. San Francisco, CA: Pearson, Benjamin Cummings.

Spellman, F.R. 1996. *Stream Ecology and Self-Purification.* Lancaster, PA: Technomic Publishing Company.

Suberkoop, K., Godshalk, G.L., & Klug, M.J. 1976. Changes in the chemical composition of leaves during processing in a woodland stream. *Ecology* 57: 720–727.

Tansley, A.G. 1935. The use and abuse of vegetational concepts and terms. *Ecology* 16: 284–307.

Tomera, A.N. 1989. *Understanding Basic Ecological Concepts.* Portland, ME: J. Weston Walch, Publisher.

USDA. 1982. *Agricultural Statistics 1982.* Washington, D.C.: U.S. Government Printing Office.

USDA. 1999. *Autumn Colors: How Leaves Change Color.* Accessed 02/08/19 @ http://www.na.fs.fed.us/spfo/pubs/misc/autumn/autumn_colors.htm.

USDA. 2007. *Agricultural Ecosystems and Agricultural Ecology.* Accessed 02/11/19 @ http://nrcs.usda.gov/technical/ECS/agecol/ecosystem.html.

USFWS. 2007. *Ecosystem Conservation.* U.S. Fish & Wildlife Service. Accessed 02/11/19 @ http://www.fws.gov/ecosystems/.

13 Plant Security

© Las Vegas Valley Water District and used with permission.

You may say Homeland Security is a Y2K problem that doesn't end Jan. 1 of any given year.

– Governor Tom Ridge

KEY OBJECTIVES

After studying this chapter, you should be able to:

- Discuss various alarm systems.
- Discuss detection devices (aka sensors).
- Describe arming stations.
- Discuss control panels.
- Describe an annunciator.
- Discuss local alarms.
- Identify perimeter and interior intrusion sensors.
- Discuss fire detection and fire alarm systems.
- Describe backflow prevention devices.
- Define backsiphonage.
- Discuss active security barriers.
- Define wedge barriers.

DOI: 10.1201/9781003207146-18

- Discuss crash beam barriers.
- Discuss security gates.
- Describe bollards.
- Describe portable removable barriers.
- Discuss biometrics.
- Discuss iris recognition.
- Discuss card readers.
- Describe frame grabber.
- Describe fences.
- Discuss hatch security.
- Describe intrusion sensors and buried exterior intrusion sensors.
- Define a ladder access control system.
- Describe security locks and manhole locks.
- Describe manhole intrusion sensors.
- Describe radiation detection instrumentation.
- Describe reservoir covers.
- Discuss vent security.
- Describe visual surveillance.
- Discuss water quality monitoring.
- Discuss arsenic monitoring.
- Describe portable cyanide analyzers.
- Describe portable field monitors to measure VOC.
- Describe monitoring and toxicity meters.
- Discuss electronic controllers.
- Discuss two-way radios.
- Describe wireless data communications.
- Discuss SCADA.

KEY DEFINITIONS

Active security barriers – also known as crash barriers, are large structures that are placed in roadways at entrance and exit points to protected facilities to control vehicle access to these areas.

Alarm system – is a type of electronic monitoring system that is used to detect and respond to specific types of events – such as unauthorized access to an asset, or a possible fire.

Arming station – is the main user interface with the security system, allows the user to arm (turn on), disarm (turn off), and communicate with the system.

Arsenic monitoring – the SDWA requires arsenic monitoring for public water systems. The Arsenic Rule indicates that surface water systems must collect one sample annually; groundwater systems must collect one sample in each compliance period (once every three years). Samples are collected at entry points to the distribution system, and analysis is done in the lab using one of several USEPA-approved methods, including Inductively Coupled Plasma Mass Spectroscopy (ICP-MS, USEPA 200.8) and several atomic absorption (AA) methods.

Backflow prevention devices – designed to prevent backflow, which is the reversal of the normal and intended direction of water flow in a water system.

Backsiphonage – the reverse from normal flow direction within a piping system that is caused by negative pressure in the supply piping (i.e., the reversal of normal flow in a system caused by a vacuum or partial vacuum within the water supply piping).

Biometric – involves measuring the unique physical characteristics or traits of the human body.

Bollard – are vertical barriers at least 3 feet tall and 1 to 2 feet in diameter that are typically set 4–5 feet apart from each other so that they block vehicles from passing between them.

Buried exterior intrusion sensor – electronic devices that are designed to detect potential intruders. The sensors are buried along the perimeters of sensitive assets and can detect intruder activity both above- and belowground. Some of these systems are composed of individual, stand-alone sensor units, while other sensors consist of buried cables.

Card reader – this is a type of electronic identification system that is used to identify a card and then perform an action associated with that card.

Control panel – receives information from the sensors and sends it to an appropriate location, such as to a central operations station or to a 24-hour monitoring facility.

Crash beam barrier – consists of aluminum beams that can be opened or closed across the roadway.

Detection devices (also called *sensors*) – are designed to detect a specific type of event (such as smoke, intrusion, etc.).

Electronic control – a piece of electronic equipment that receives incoming electric signals uses preprogrammed logic to generate electronic output signals based on the incoming signals.

Fence – a physical barrier that can be set up around the perimeter of an asset. Fences often consist of individual pieces (such as individual pickets in a wooden fence or individual sections of a wrought iron fence) that are fastened together.

Fire detection/fire alarm system – consists of different types of fire detection devices and fire alarm systems available.

Frame grabber – device that provides high-resolution image of the iris; it is captured or extracted from the video.

Gates – are often integrated units of a perimeter fence or wall around a facility. Gates are basically movable pieces of fencing that can be opened and closed across a road.

Hatch security – a hatch is basically a door installed on a horizontal plane (such as in a floor, a paved lot, or a ceiling), instead of on a vertical plane (such as in a building wall).

Hydrant lock – a physical security device designed to prevent unauthorized access to the water supply through a hydrant.

Interior intrusion sensors – are designed to detect an intruder who has already accessed the protected asset (i.e., interior intrusion sensors are used to detect intruders once they are already within a protected room or building).

Intrusion sensor – an exterior intrusion sensor is a detection device that is used in an outdoor environment to detect intrusions into a protected area. These devices are designed to detect an intruder and then communicate an alarm signal to an alarm system. Intrusion sensors consist of two main categories: perimeter sensors and interior (space) sensors.

Iris recognition – the iris is the colored or pigmented area of the eye surrounded by the sclera (the white portion of the eye), is a muscular membrane that controls the amount of light entering the eye by contracting or expanding the pupil (the dark center of the eye). The dense, unique patterns of connective tissue in the human iris were first noted in 1936, but it was not until 1994, when algorithms for iris recognition were created and patented, that commercial applications using biometric iris recognition began to be used extensively.

Ladder access control system – consists of some type of cover that is locked or secured over the ladder.

Local alarm – emits a signal at the location of the event (typically using a bell or siren).

Lock – a type of physical security device that can be used to delay or prevent a door, a window, a manhole, a filing cabinet drawer, or some other physical feature from being opened, moved, or operated.

Manhole intrusion sensor – a physical security device designed to detect unauthorized access to the system through a manhole.

Manhole Lock – a physical security device designed to delay unauthorized access to the system through a manhole.

Perimeter intrusion sensor – are typically applied on fences, doors, walls, windows, etc., and are designed to detect an intruder before he/she accesses a protected asset (i.e., perimeter intrusion sensors are used to detect intruders attempting to enter through a door, window, etc.).

Portable cyanide analyzer – detection systems designed to be used in the field to evaluate for potential cyanide contamination of a water asset.

Portable field monitors to measure VOC – volatile organic compounds (VOCs) are a group of highly utilized chemicals that have widespread applications, including use as fuel components, as solvents, and as cleaning and liquefying agents in degreasers, polishes, and dry-cleaning solutions. VOCs are also used in herbicides and insecticides for agriculture applications.

Portable/removable barrier – can include removable crash beams and wedge barriers, are mobile obstacles that can be moved in and out of position on a roadway.

Radiation detection instrumentation – protects against radioactive materials being brought on-site; a facility may set up monitoring sites outfitted with radiation detection instrumentation at entrances to the facility.

Reservoir cover – a structure installed on or over the surface of the reservoir to minimize water quality degradation.

SCADA or Supervisory Control and Data Acquisition System (also sometimes referred to as Digital Control Systems or Process Control Systems) – plays an important role in computer-based control systems. Many water and wastewater systems use computer-based systems to remotely control sensitive processes and system equipment previously controlled manually. These systems (commonly known as SCADA) allow a water and wastewater system to collect data from sensors and control equipment located at remote sites. Common water and wastewater system sensors measure elements such as fluid level, temperature, pressure, water purity, water clarity, and pipeline flow rates. Common water and wastewater system equipment include valves, pumps, and mixers for mixing chemicals in the water supply.

Toxicity monitoring/toxicity meters – devices used to measure general toxicity to biological organisms, and detection of toxicity in any water and wastewater asset can indicate a potential threat, either to the treatment process (in the case of influent toxicity), to human health (in the case of drinking water toxicity) or to the environment (in the case of effluent toxicity).

Two-way radio – is limited to a direct unit-to-unit radio communication, either via single unit-to-unit transmission and reception or via multiple handheld units to a base station radio contact and distribution system. Radio frequency spectrum limitations apply to all handheld units and are directed by the FCC. This also distinguishes a handheld unit from a base station or base station unit [such as those used by an amateur (ham) radio operator], which operate under different wavelength parameters.

Vent security – making the vents tamper resistant or by adding other security features, such as security screens or security covers, can enhance the security of the entire water system.

Visual surveillance – used to detect threats through continuous observation of important or vulnerable areas of an asset.

Water quality monitoring – sensor equipment used to monitor key elements of water or wastewater treatment processes (such as influent water quality, treatment processes, or effluent water quality) to identify anomalies that may indicate threats to the system.

Wedge barriers – are plated, rectangular steel buttresses approximately 2–3 feet high that can be raised and lowered from the roadway.

Wireless data communications – consists of two components: a "Wireless Access Point" (WAP), and a "Wireless Network Interface Card" (sometimes also referred to as a "Client"), which work together to complete the communications link. These wireless systems can link electronic devices, computers, and computer systems together using radio waves, thus eliminating the need for these individual components to be directly connected through physical wires.

Protocol analysis – is the process of capturing, decoding, and interpreting electronic traffic. The protocol analysis method of network intrusion detection involves the analysis of data captured during transactions between two or more systems or devices and the evaluation of these data to identify unusual activity and potential problems. Once a problem is isolated and recorded, problems or potential threats can be linked to pieces of hardware or software. Sophisticated protocol analysis will also provide statistics and trend information on the captured traffic.

Traffic anomaly detection – Traffic anomaly detection identifies potential threatening activity by comparing incoming traffic to "normal" traffic patterns and identifying deviations. It does this by comparing user characteristics against thresholds and triggers defined by the network administrator. This method is designed to detect attacks that span several connections, rather than a single session.

Network honeypot – This method establishes nonexistent services in order to identify potential hackers. A network honeypot impersonates services that don't exist by sending fake information to people scanning the network. It identifies the attackers when they attempt to connect to the service. There is no reason for legitimate traffic to access these resources because they don't exist; therefore, any attempt to access them constitutes an attack.

Anti-intrusion detection system evasion techniques – these methods are designed to identify attackers who may be trying to evade intrusion detection system scanning. They include methods called IP defragmentation, TCP streams reassembly, and deobfuscation.

OVERVIEW

According to USEPA (2004), there are approximately 160,000 public water systems (PWSs) in the United States, each of which regularly supplies drinking water to at least 25 persons or 15 service connections. Eighty-four percent of the total US population is served by PWSs, while the remainder is served primarily by private wells. PWSs are divided into community water systems (CWSs) and noncommunity water systems (NCWSs). Examples of CWSs include municipal water systems that serve mobile home parks of residential developments. Examples, of NCWSs include schools, factories, and churches, commercial campgrounds, hotels, and restaurants.

Wastewater treatment is important for preventing disease and protecting the environment. Wastewater is treated by publicly owned treatment works (POTW) and by private facilities such as industrial plants. There are approximately 2.3 million miles of distribution system pipes and approximately 16,255 POTW in the United States. Seventy-five percent of the total US population is served by POTW with existing flows of less than 1 MGD and are considered small; they number approximately 13,057 systems. For purpose of determining population served, 1 MGD equals approximately 10,000 persons served.

DID YOU KNOW?

As of 2003, community water systems serve by far the largest proportion of the US population – 273 million out of a total population of 290 million (USEPA, 2004).

Disruption of a wastewater treatment system or service can cause loss of life, economic impacts, and severe public health incidents. If structural damage occurs, wastewater systems can become vulnerable to inadequate treatment. The public is much less sensitive to wastewater as an area of vulnerability than it is to drinking water; however wastewater systems do provide opportunities for terrorist threats.

Federal and state agencies have long been active in addressing these risks and threats to water and wastewater systems through regulations, technical assistance, research, and outreach programs. As a result, an extensive system of regulations governing maximum contaminant levels of 90 conventional contaminants (most established by EPA), construction and operating standards (implemented mostly by the states), monitoring, emergency response planning, training, research, and education have been developed to better protect the nation's drinking water supply and receiving waters.

Since the events of 9/11, EPA has been designated as the sector-specific agency responsible for infrastructure protection activities for the nation's drinking water and wastewater system. EPA is utilizing its position within the water sector and working with its stakeholders to provide information to help protect the nation's drinking water supply from terrorism or other intentional acts.

Note: In this chapter, we discuss security provisions and devices that apply to both water and wastewater operations and to water sector infrastructure which includes waterworks and wastewater treatment plants.

SECURITY HARDWARE/DEVICES

Keep in mind that when it comes to making "anything" absolutely secure from intrusion or attack, there is inherently, or otherwise, no silver bullet. However, careful preplanning and installation of security hardware and/or devices-products can significantly affect the plant's ability to weather the storm, so to speak. USEPA (2005) groups the water and wastewater infrastructure security devices or products described below into four general categories:

1. Physical asset monitoring and control devices
2. Water monitoring devices
3. Communication/integration
4. Cyber protection devices

Physical Asset Monitoring and Control Devices

Aboveground, Outdoor Equipment Enclosures

Water and wastewater systems consist of multiple components spread over a wide area and typically include a centralized treatment plant, as well as distribution or collection system components that are typically distributed at multiple locations throughout the community. However, in recent years, distribution and collection system designers have favored placing critical equipment – especially assets that require regular use and maintenance – aboveground.

One of the primary reasons for doing so is that locating this equipment aboveground eliminates the safety risks associated with confined space entry, which is often required for the maintenance of equipment located belowground. In addition, space restrictions often limit the amount of equipment that can be located inside, and there are concerns that some types of equipment (such as backflow prevention devices) can, under certain circumstances, discharge water that could flood pits, vaults, or equipment rooms. Therefore, many pieces of critical equipment are located outdoors and aboveground. Many different system components can be installed outdoors and aboveground.

Examples of these types of components could include:

- Backflow prevention devices;
- Air release and control valves;
- Pressure vacuum breakers;
- Pumps and motors;
- Chemical storage and feed equipment;
- Meters;
- Sampling equipment;
- Instrumentation – Much of this equipment is installed in remote locations and/or in areas where the public cannot access it. One of the most effective security measures for protecting aboveground equipment is to place it inside a building. When/where this is not possible, enclosing the equipment or parts of the equipment using some sort of commercial or homemade add-on structure may help to prevent tampering with the equipment.

Equipment enclosures can generally be categorized into one of four main configurations, which include:

1. One-piece, drop over enclosures
2. Hinged or removable top enclosures
3. Sectional enclosures
4. Shelters with access locks

Other security features that can be implemented on aboveground, outdoor equipment enclosures include locks, mounting brackets, tamper-resistant doors, and exterior lighting.

Alarms

An *alarm system* is a type of electronic monitoring system that is used to detect and respond to specific types of events – such as unauthorized access to an asset or a possible fire. In water and wastewater systems, alarms are also used to alert operators when process operating or monitoring conditions go out of preset parameters (i.e., process alarms). These types of alarms are primarily integrated with process monitoring and reporting systems (i.e., SCADA systems).

Note: This discussion does not focus on alarm systems that are not related to a system's processes.

Alarm systems can be integrated with fire detection systems, intrusion detection systems (IDSs), access control systems, or closed-circuit television (CCTV) systems, such that these systems automatically respond when the alarm is triggered. For example, a smoke detector alarm can be set up to automatically notify the fire department when smoke is detected, or an intrusion alarm can automatically trigger cameras to turn on in a remote location so that personnel can monitor that location. An alarm system consists of sensors that detect different types of events; an arming station that is used to turn the system on and off; a control panel that receives information, processes it, and transmits the alarm; and an annunciator that generates a visual and/or audible response to the alarm. When a sensor is tripped it sends a signal to a control panel, which triggers a visual or audible alarm and/or notifies a central monitoring station. A more complete description of each of the components of an alarm system is provided below.

Detection devices (also called sensors) are designed to detect a specific type of event (such as smoke, intrusion, etc.). Depending on the type of event they are designed to detect, sensors can be located inside or outside of the facility or other asset. When an event is detected, the sensors use some type of communication method (such as wireless radio transmitters, conductors, or cables) to send signals to the control panel to generate the alarm. For example, a smoke detector sends a signal to a control panel when it detects smoke.

An *arming station*, which is the main user interface with the security system, allows the user to arm (turn on), disarm (turn off), and communicate with the system. How a specific system is armed will depend on how it is used. For example, while IDSs can be armed for continuous operation (24 hours/day), they are usually armed and disarmed according to the work schedule at a specific location so that personnel going about their daily activities do not set off the alarms. In contrast, fire protection systems are typically armed 24 hours/day.

The control panel receives information from the sensors and sends it to an appropriate location, such as to a central operations station or to a 24-hour monitoring facility. Once the alarm signal is received at the central monitoring location, personnel monitoring for alarms can respond (such as by sending security teams to investigate or by dispatching the fire department).

The annunciator responds to the detection of an event by emitting a signal. This signal may be visual, audible, electronic, or a combination of these three. For example, fire alarm signals will always be connected to audible annunciators, whereas intrusion alarms may not be.

Alarms can be reported locally, remotely, or both locally and remotely. A local alarm emits a signal at the location of the event (typically using a bell or siren). A "local only" alarm emits a signal at the location of the event but does not transmit the alarm signal to any other location (i.e., it does not transmit the alarm to a central monitoring location). Typically, the purpose of a "local only" alarm is to frighten away intruders and possibly to attract the attention of someone who might notify the proper authorities. Because no signal is sent to a central monitoring location, personnel can only respond to a local alarm if they are in the area and can hear and/or see the alarm signal.

Fire alarm systems must have local alarms, including both audible and visual signals. Most fire alarm signal and response requirements are codified in the National Fire Alarm Code, National Fire Protection Association (NFPA) 72. NFPA 72 discusses the application, installation, performance, and maintenance of protective signaling systems and their components. In contrast to fire alarms, which require a local signal when fire is detected, many IDSs do not have a local alert device because monitoring personnel do not wish to inform potential intruders that they have been detected. Instead, these types of systems silently alert monitoring personnel that an intrusion has been detected, thus allowing monitoring personnel to respond.

In contrast to systems that are set up to transmit "local only" alarms when the sensors are triggered, systems can also be set up to transmit signals to a central location, such as to a control room or guard post at the system or to a police or fire station. Most fire/smoke alarms are set up to signal both at the location of the event and at a fire station or central monitoring station. Many insurance companies require that facilities install certified systems that include alarm communication to a central station. For example, systems certified by the Underwriters Laboratory (UL) require that the alarm be reported to a central monitoring station.

The main differences between alarm systems lie in the types of event detection devices used in different systems. *Intrusion sensors*, for example, consist of two main categories: perimeter sensors and interior (space) sensors. *Perimeter intrusion sensors* are typically applied on fences, doors, walls, windows, etc., and are designed to detect an intruder before he/she accesses a protected asset (i.e., perimeter intrusion sensors are used to detect intruders attempting to enter through a door, window, etc.). In contrast, *interior intrusion sensors* are designed to detect an intruder who has already accessed the protected asset (i.e., interior intrusion sensors are used to detect intruders once they are already within a protected room or building). These two types of detection devices can be complementary, and they are often used together to enhance security for an asset. For example, a typical intrusion alarm system might employ a perimeter glass-break detector that protects against intruders accessing a room through a window, as well as an ultrasonic interior sensor that detects intruders that have gotten into the room without using the window.

Fire Detection/Fire Alarm Systems consist of different types of fire detection devices and fire alarm systems available. These systems may detect fire, heat, smoke, or a combination of any of these. For example, a typical fire alarm system might consist of heat sensors, which are located throughout a facility and which detect high temperatures or a certain change in temperature over a fixed time period. A different system might be outfitted with both smoke and heat detection devices.

When a sensor in an alarm system detects an event, it must communicate an alarm signal. The two basic types of alarm communication systems are hardwired and wireless. Hardwired systems rely on wire that is run from the control panel to each of the detection devices and annunciators. Wireless systems transmit signals from a transmitter to a receiver through the air – primarily using radio or other waves. Hardwired systems are usually lower cost, more reliable (they are not affected by terrain or environmental factors), and significantly easier to troubleshoot than are wireless systems. However, a major disadvantage of hardwired systems is that it may not be possible to hardwire all locations (e.g., it may be difficult to hardwire remote locations). In addition, running wires to their required locations can be both time consuming and costly. The major advantage to using wireless systems is that they can often be installed in areas where hardwired systems are not feasible. However, wireless components can be much more expensive when compared to hardwired systems. In addition, in the past, it has been difficult to perform self-diagnostics on wireless systems to confirm that they are communicating properly with the controller. Presently, the majority of wireless systems incorporate supervising circuitry, which allows the subscriber to know immediately if there is a problem with the system (such as a broken detection device or a low battery) or if a protected door or window has been left open.

Backflow Prevention Devices

As their name suggests, backflow prevention devices are designed to prevent backflow, which is the reversal of the normal and intended direction of water flow in a water system. Backflow is a potential problem in a water system because it can spread contaminated water back through a distribution system. For example, backflow at uncontrolled cross-connections (cross-connections are any actual or potential connection between the public water supply and a source of contamination) or pollution can allow pollutants or contaminants to enter the potable water system. More specifically, backflow from private plumbing systems, industrial areas, hospitals, and other hazardous contaminant-containing systems, into public water mains and wells poses serious public health risks and security problems. Cross-contamination from private plumbing systems can contain biological hazards (such as bacteria or viruses) or toxic substances that can contaminate and sicken an entire population in the event of backflow. Most historical incidences of backflow have been accidental, but growing concern that contaminants could be intentionally back-fed into a system is prompting increased awareness for private homes, businesses, industries, and areas most vulnerable to intentional strikes. Therefore, backflow prevention is a major tool for the protection of public water systems.

Backflow may occur under two types of conditions: backpressure and backsiphonage. Backpressure is the reverse from normal flow direction within a piping system that is the result of the downstream pressure being higher than the supply pressure. These reductions in the supply pressure occur whenever the amount of water being used exceeds the amount of water supplied, such as during water line flushing, firefighting, or breaks in water mains. Backsiphonage is the reverse from normal flow direction within a piping system that is caused by negative pressure in the supply piping (i.e., the reversal of normal flow in a system caused by a vacuum or partial vacuum within the water supply piping). Backsiphonage can occur where there is a high velocity in a pipeline; when there is a line repair or break that is lower than a service point; or when there is lowered main pressure due to high water withdrawal rate, such as during firefighting or water main flushing. To prevent backflow, various types of backflow preventers are appropriate for use.

The primary types of backflow preventers are as follows:

- Air Gap Drains
- Double Check Valves
- Reduced Pressure Principle Assemblies
- Pressure Vacuum Breakers

Barriers

Active security barriers (also known as *crash barriers*) are large structures that are placed in road-ways at entrance and exit points to protected facilities to control vehicle access to these areas. These barriers are placed perpendicular to traffic to block the roadway so that the only way that traffic can pass the barrier is for the barrier to be moved out of the roadway. These types of barriers are typically constructed from sturdy materials, such as concrete or steel, such that vehicles cannot penetrate through them. They are also designed at a certain height off the roadway so that vehicles cannot go over them.

The key difference between active security barriers, which include wedges, crash beams, gates, retractable bollards, and portable barricades; and passive security barriers, which include non-moveable bollards, jersey barriers, and planters, is that active security barriers are designed so that they can be raised and lowered or moved out of the roadway easily to allow authorized vehicles to pass them. Many of these types of barriers are designed so that they can be opened and closed automatically (i.e., mechanized gates, hydraulic wedge barriers), while others are easy to open and close manually (swing crash beams, manual gates).

In contrast to active barriers, passive barriers are permanent, non-movable barriers, and thus they are typically used to protect the perimeter of a protected facility, such as sidewalks and other areas that do not require vehicular traffic to pass them. Several of the major types of active security barriers such as wedge barriers, crash beams, gates, bollards, and portable/removable barricades are described below.

Wedge barriers are plated, rectangular steel buttresses approximately 2–3 feet high that can be raised and lowered from the roadway. When they are in the open position, they are flush with the roadway and vehicles can pass over them. However, when they are in the closed (armed) position, they project up from the road at a 45° angle, with the upper end pointing toward the oncoming vehicle and the base of the barrier away from the vehicle. Generally, wedge barriers are constructed from heavy-gauge steel, or concrete that contains an impact-dampening iron rebar core that is strong and resistant to breaking or cracking, thereby allowing them to withstand the impact from a vehicle attempting to crash through them. In addition, both of these materials help to transfer the energy of the impact over the barrier's entire volume, thus helping to prevent the barrier from being sheared off its base. In addition, because the barrier is angled away from traffic, the force of any vehicle impacting the barrier is distributed over the entire surface of the barrier and is not concentrated at the base, which helps prevent the barrier from breaking off at the base. Finally, the angle of the barrier helps hang up any vehicles attempting to drive over it.

Wedge barriers can be fixed or portable. Fixed wedge barriers can be mounted on the surface of the roadway ("surface-mounted wedges") or in a shallow mount on the road's surface, or they can be installed completely below the road surface. Surface-mounted wedge barricades operate by rising from a flat position on the surface of the roadway, while shallow-mount wedge barriers rise from their resting position just below the road surface. In contrast, below-surface wedge barriers operate by rising from beneath the road surface. Both the shallow-mounted and surface-mounted barriers require little or no excavation, and thus do not interfere with buried utilities. All three barrier mounting types project above the road surface and block traffic when they are raised into the armed position. Once they are disarmed and lowered, they are flush with the road, thereby allowing traffic to pass. Portable wedge barriers are moved into place on wheels that are removed after the barrier has been set into place.

Installing rising wedge barriers requires preparation of the road surface. Installing surface-mounted wedges does not require that the road be excavated; however, the road surface must be intact and strong enough to allow the bolts anchoring the wedge to the road surface to attach properly. Shallow-mount and below-surface wedge barricades require excavation of a pit that is large enough to accommodate the wedge structure, as well as any arming/disarming mechanisms. Generally, the bottom of the excavation pit is lined with gravel to allow for drainage. Areas not sheltered from rain or surface runoff can install a gravity drain or self-priming pump.

Crash beam barriers consist of aluminum beams that can be opened or closed across the roadway. While there are several different crash beam designs, every crash beam system consists of an aluminum beam that is supported on each side by a solid footing or buttress, which is typically constructed from concrete, steel, or some other strong material. Beams typically contain an interior steel cable (typically at least one inch in diameter) to give the beam added strength and rigidity. The beam is connected by a heavy-duty hinge or other mechanisms to one of the footings so that it can swing or rotate out of the roadway when it is open and can swing back across the road when it is in the closed (armed) position, blocking the road and inhibiting access by unauthorized vehicles. The non-hinged end of the beam can be locked into its footing, thus providing anchoring for the beam on both sides of the road and increasing the beam's resistance to any vehicles attempting to penetrate it. In addition, if the crash beam is hit by a vehicle, the aluminum beam transfers the impact energy to the interior cable, which in turn transfers the impact energy through the footings and into their foundation, thereby minimizing the chance that the impact will snap the beam and allow the intruding vehicle to pass through.

Crash beam barriers can employ drop-arm, cantilever, or swing beam designs. Drop-arm crash beams operate by raising and lowering the beam vertically across the road. Cantilever crash beams project structures that are opened and closed by extending the beam from the hinge buttress to the receiving buttress located on the opposite side of the road. In the swing beam design, the beam is hinged to the buttress such that it swings horizontally across the road. Generally, swing beam and cantilever designs are used at locations where a vertical lift beam is impractical. For example, the swing beam or cantilever designs are utilized at entrances and exits with overhangs, trees, or buildings that would physically block the operation of the drop-arm beam design. Installing any of these crash beam barriers involves the excavation of a pit approximately 8 inches deep for both the hinge and the receiver footings. Due to the depth of excavation, the site should be inspected for underground utilities before digging begins.

In contrast to wedge barriers and crash beams, which are typically installed separately from a fence line, *gates* are often integrated units of a perimeter fence or wall around a facility. Gates are basically movable pieces of fencing that can be opened and closed across a road When the gate is in the closed (armed) position, the leaves of the gate lock into steel buttresses that are embedded in concrete foundation located on both sides of the roadway, thereby blocking access to the roadway. Generally, gate barricades are constructed from a combination of heavy-gauge steel and aluminum that can absorb an impact from vehicles attempting to ram through them. Any remaining impact energy not absorbed by the gate material is transferred to the steel buttresses and their concrete foundation.

Gates can utilize a cantilever, linear, or swing design. Cantilever gates are projecting structures that operate by extending the gate from the hinge footing across the roadway to the receiver footing. A linear gate is designed to slide across the road on tracks via a rack and pinion drive mechanism. Swing gates are hinged so that they can swing horizontally across the road. Installation of the cantilever, linear, or swing gate designs described above involve the excavation of a pit approximately 48 inches deep for both the hinge and receiver footings to which the gates are attached. Due to the depth of excavation, the site should be inspected for underground utilities before digging begins.

Bollards are vertical barriers at least 3 feet tall and 1–2 feet in diameter that are typically set 4–5 feet apart from each other so that they block vehicles from passing between them. Bollards can either be fixed in place, removable, or retractable. Fixed and removable bollards are passive barriers that are typically used along building perimeters or on sidewalks to prevent vehicles from them while allowing pedestrians to pass them. In contrast to passive bollards, retractable bollards are active security barriers that can easily be raised and lowered to allow vehicles to pass between them. Thus, they can be used in driveways or on roads to control vehicular access. When the bollards are raised, they protect above the road surface and block the roadway; when they are lowered, they sit flush with the road surface, and thus allow traffic to pass over them. Retractable bollards are typically constructed from steel or other materials that have a low weight-to-volume ratio so that they require low power to raise and lower. Steel is also more resistant to breaking than is a more brittle material, such as concrete, and is better able to withstand direct vehicular impact without breaking apart.

Retractable bollards are installed in a trench dug across a roadway – typically at an entrance or gate. Installing retractable bollards requires preparing the road surface. Depending on the vendor, bollards can be installed either in a continuous slab of concrete or in individual excavations with concrete poured in place. The required excavation for a bollard is typically slightly wider and slightly deeper than the bollard height when extended aboveground. The bottom of the excavation is typically lined with gravel to allow drainage. The bollards are then connected to a control panel which controls the raising and lowering of the bollards. Installation typically requires mechanical, electrical, and concrete work; if system personnel with these skills are available, then the system can install the bollards themselves.

Portable/Removable barriers, which can include removable crash beams and wedge barriers, are mobile obstacles that can be moved in and out of position on a roadway. For example, a crash beam may be completely removed and stored off-site when it is not needed. An additional example would be wedge barriers that are equipped with wheels that can be removed after the barricade is towed into place. When portable barricades are needed, they can be moved into position rapidly. To provide them with added strength and stability, they are typically anchored to buttress boxes that are located on either side of the road. These buttress boxes, which may or may not be permanent, are usually filled with sand, water, cement, gravel, or concrete to make them heavy and aid in stabilizing the portable barrier. In addition, these buttresses can help dissipate any impact energy from vehicles crashing into the barrier itself.

Because these barriers are not anchored into the roadway, they do not require excavation or other related construction for installation. In contrast, they can be assembled and made operational in a short period of time. The primary shortcoming to this type of design is that these barriers may move if they are hit by vehicles. Therefore, it is important to carefully assess the placement and anchoring of these types of barriers to ensure that they can withstand the types of impacts that may be anticipated at that location.

Because the primary threat to active security barriers is that vehicles will attempt to crash through them, their most important attributes are their size, strength, and crash resistance. Other important features for an active security barrier are the mechanisms by which the barrier is raised and lowered to allow authorized vehicle entry, and other factors, such as weather resistance and safety features.

Passive Security Barriers

One of the most basic threats facing any facility is from intruders accessing the facility with the intention of causing damage to its assets. These threats may include intruders actually entering the facility, as well as intruders attacking the facility from outside without actually entering it (i.e., detonating a bomb near enough to the facility to cause damage within its boundaries).

Security barriers are one of the most effective ways to counter the threat of intruders accessing a facility or the facility perimeter. Security barriers are large, heavy structures that are used to control access through a perimeter by either vehicles or personnel. They can be used in many different ways depending on how/where they are located at the facility. For example, security barriers can be used on or along driveways or roads to direct traffic to a checkpoint (i.e., a facility may install jersey barriers in a road to direct traffic in certain direction).

Other types of security barriers (crash beams, gates) can be installed at the checkpoint so that guards can regulate which vehicles can access the facility. Finally, other security barriers (i.e., bollards or security planters) can be used along the facility perimeter to establish a protective buffer area between the facility and approaching vehicles. Establishing such a protective buffer can help in mitigating the effects of the type of bomb blast described above, both by potentially absorbing some of the blast, and also by increasing the "stand-off" distance between the blast and the facility (the force of an explosion is reduced as the shock wave travels further from the source, and thus the further the explosion is from the target, the less effective it will be in damaging the target).

Security barriers can be either "active" or "passive." "Active" barriers, which include gates, retractable bollards, wedge barriers, and crash barriers, are readily movable, and thus they are typically used in areas where they must be moved often to allow vehicles to pass – such as in roadways at entrances and exits to a facility. In contrast to active security barriers, "passive" security barriers, which include jersey barriers, bollards, and security planters, are not designed to be moved on a regular basis, and thus they are typically used in areas where access is not required or allowed – such as along building perimeters or in traffic control areas. Passive security barriers are typically large, heavy structures that are usually several feet high, and they are designed so that even heavy-duty vehicles cannot go over or through them. Therefore, they can be placed in a roadway parallel to the flow of traffic so that they direct traffic in a certain direction (such as to a guardhouse, a gate, or some other sort of checkpoint), or perpendicular to traffic such that they prevent a vehicle from using a road or approaching a building or area.

Biometric Security Systems

Biometrics involves measuring the unique physical characteristics or traits of the human body. Any aspect of the body that is measurably different from person to person – for example fingerprints or eye characteristics – can serve as a unique biometric identifier for that individual. Biometric systems recognizing fingerprints, palm shape, eyes, face, voice, and signature comprise the bulk of the current biometric systems that recognize other biological features.

Biometric security systems use biometric technology combined with some type of locking mechanisms to control access to specific assets. In order to access an asset controlled by a biometric security system, an individual's biometric trait must be matched with an existing profile stored in a database. If there is a match between the two, the locking mechanisms (which could be a physical lock, such as at a doorway, an electronic lock, such as at a computer terminal, or some other type of lock) are disengaged, and the individual is given access to the asset.

A biometric security system typically comprises the following components:

- A sensor, which measures/records a biometric characteristic or trait
- A control panel, which serves as the connection point between various system components. The control panel communicates information back and forth between the sensor and the host computer and controls access to the asset by engaging or disengaging the system lock based on internal logic and information from the host computer
- A host computer, which processes and stores the biometric trait in a database
- Specialized software, which compares an individual image taken by the sensor with a stored profile or profiles
- A locking mechanism which is controlled by the biometric system
- A power source to power the system

Biometric Hand and Finger Geometry Recognition

Hand and finger geometry recognition is the process of identifying an individual through the unique "geometry" (shape, thickness, length, width, etc.) of that individual's hand or fingers. Hand geometry recognition has been employed since the early 1980s and is among the most widely used biometric technologies for controlling access to important assets. It is easy to install and use and is appropriate for use in any location requiring use of two-finger highly accurate, nonintrusion biometric security. For example, it is currently used in numerous workplaces, day-care facilities, hospitals, universities, airports, and power plants.

A newer option within hand geometry recognition technology is finger geometry recognition (not to be confused with fingerprint recognition). Finger geometry recognition relies on the same scanning methods and technologies as does hand geometry recognition, but the scanner only scans two of the user's fingers, as opposed to his entire hand. Finger geometry recognition has been in commercial use since the mid-1990s and is mainly used in time and attendance applications (i.e., to track when individuals have entered and exited a location). To date the only large-scale commercial use of two-finger geometry for controlling access is at Disney World, where season pass holders use the geometry of their index and middle finger to gain access to the facilities.

Hand and finger geometry recognition systems can be used in several different types of applications, including access control and time and attendance tracking. While time and attendance tracking can be used for security, it is primarily used for operations and payroll purposes (i.e., clocking in and clocking out). In contrast, access control applications are more likely to be security related. Biometric systems are widely used for access control and can be used on various types of assets, including entryways, computers, vehicles, etc. However, because of their size, hand/finger recognition systems are primarily used in entryway access control applications.

Iris Recognition

The iris, which is the colored or pigmented area of the eye surrounded by the sclera (the white portion of the eye), is a muscular membrane that controls the amount of light entering the eye by contracting or expanding the pupil (the dark center of the eye). The dense, unique patterns of connective tissue in the human iris were first noted in 1936, but it was not until 1994, when algorithms for iris recognition were created and patented, that commercial applications using biometric iris recognition began to be used extensively. There are now two vendors producing iris recognition technology: both the original developer of these algorithms, as well as a second company, which has developed and patented a different set of algorithms for iris recognition.

The iris is an ideal characteristic for identifying individuals because it is formed *in utero*, and its unique patterns stabilize around eight months after birth. No two irises are alike; neither an individual's right or left irises, nor the irises of identical twins. The iris is protected by the cornea (the clear covering over the eye), and therefore it is not subject to the aging or physical changes (and potential variation) that are common to some other biometric measures, such as the hand, fingerprints, and the face.

Although some limited changes can occur naturally over time, these changes generally occur in the iris' melanin and therefore affect only the eye's color, and not its unique patterns. (In addition, because iris scanning uses only black and white images, color changes would not affect the scan anyway.) Thus, barring specific injuries or certain rare surgeries directly affecting the iris, the iris' unique patterns remain relatively unchanged over an individual's lifetime.

Iris recognition systems employ a monochromatic, or black and white, video camera that uses both visible and near infrared light to take video of an individual's iris. Video is used rather than still photography as an extra security procedure. The video is used to confirm the normal continuous fluctuations of the pupil as the eye focuses, which ensures that the scan is of a living human being and not a photograph or some other attempted hoax. A high-resolution image of the iris is then captured or extracted from the video, using a device often referred to as a *frame grabber*. The unique characteristics identified in this image are then converted into a numeric code, which is stored as a template for that user.

Card Identification/Access/Tracking Systems

A card reader system is a type of electronic identification system that is used to identify a card and then perform an action associated with that card. Depending on the system, the card may identify where a person is or where they were at a certain time, or it may authorize another action, such as disengaging a lock. For example, a security guard may use his card at card readers located throughout a facility to indicate that he has checked a certain location at a certain time. The reader will store the information and/or send it to a central location, where it can be checked later to ensure that the guard has patrolled the area. Other card reader systems can be associated with a lock so that the cardholder must have their card read and accepted by the reader before the lock disengages.

A complete card reader system typically consists of the following components:

- Access cards that are carried by the user
- Card readers, which read the card signals and send the information to control units
- Control units, which control the response of the card reader to the card
- A power source

Numerous card reader systems are available. The three primary differences between card reader systems are the way that data is encoded on the cards, the way these data are transferred between the card and the card reader, and the types of applications for which they are best suited. However, all card systems are similar in the way that the card reader and control unit interact to respond to the card. While card readers are similar in the way that the card reader and control unit interact to control access, they are different in the way data is encoded on the cards and the way these data are transferred between the card and the card reader. There are several types of technologies available for card reader systems.

These include:

- Proximity
- Wiegand
- Smartcard
- Magnetic Stripe
- Bar Code
- Infrared
- Barium Ferrite
- Hollerith
- Mixed Technologies

The determination for the level of security rate (low, moderate, or high) is based on the level of technology a given card reader system has and how simple it is to duplicate that technology and thus bypass the security. Vulnerability ratings are based on whether the card reader can be damaged easily due to frequent use or difficult working conditions (i.e., weather conditions if the reader is located outside). Often this is influenced by the number of moving parts in the system – the more moving parts, the greater the system's potential susceptibility to damage. The life cycle rating is based on the durability of a given card reader system over its entire operational period. Systems requiring frequent physical contact between the reader and the card often have a shorter life cycle due to the wear and tear to which the equipment is exposed. For many card reader systems, the vulnerability rating and life cycle ratings have a reciprocal relationship. For instance, if a given system has a high vulnerability rating it will almost always have a shorter life cycle. Card reader technology can be implemented for facilities of any size, and with any number of users. However, because individual systems vary in the complexity of their technology and in the level of security they can provide a facility, individual users must determine the appropriate system for their needs.

Some important features to consider when selecting a card reader system include:

- What level of technological sophistication and security does the card system have?
- How large is the facility, and what are its security needs?
- How frequently will the card system be used? For systems that will experience a high frequency of use it is important to consider a system that has a longer life cycle and lower vulnerability rating, thus making it more cost effective to implement.
- Under what conditions will the system will be used? (Will it be installed on the interior or exterior of buildings? Does it require light or humidity controls?) Most card reader systems can operate under normal environmental conditions, and therefore this would be a mitigating factor only in extreme conditions.
- What are the system costs?

Fences

A fence is a physical barrier that can be set up around the perimeter of an asset. Fences often consist of individual pieces (such as individual pickets in a wooden fence or individual sections of a wrought iron fence) that are fastened together. Individual sections of the fence are fastened together using posts, which are sunk into the ground to provide stability and strength for the sections of the fence hung between them. Gates are installed between individual sections of the fence to allow access inside the fenced area. Fences are often used as decorative architectural features to separate physical spaces from each other. They may also be used to physically mark the location of a boundary (such as a fence installed along a property line). However, a fence can also serve as an effective means of physically delaying intruders from gaining access to a water or wastewater asset. For example, many utilities install fences around their primary facilities, around remote pump stations, or around hazardous materials storage areas or sensitive areas within a facility. Access to the area can be controlled through security at gates or doors through the fence (e.g., by posting a guard at the gate or by locking it). In order to gain access to the asset, unauthorized persons could either have to go around or through the fence.

Fences are often compared with walls when determining the appropriate system for perimeter security. While both fences and walls can provide adequate perimeter security, fences are often easier and less expensive to install than walls. However, they do not usually provide the same physical strength that walls do. In addition, many types of fences have gaps between the individual pieces that make up the fence (i.e., the spaces between chain links in a chain link fence or the space between pickets in a picket fence). Thus, many types of fences allow the interior of the fenced area to be seen. This may allow intruders to gather important information about the locations or defenses of vulnerable areas within the facility.

Numerous types of materials are used to construct fences, including chain link iron, aluminum, wood, or wire. Some types of fences, such as split rails or pickets, may not be appropriate for security purposes because they are traditionally low fences, and are not physically strong. Potential intruders may be able to easily defeat these fences either by jumping or climbing over them or by breaking through them. For example, the rails in a split fence may be able to be broken easily.

Important security attributes of a fence include the height to which it can be constructed, the strength of the material comprising the fence, the method and strength of attaching the individual sections of the fence together at the posts, and the fence's ability to restrict the view of the assets inside the fence. Additional considerations should include the ease of installing the fence and the ease of removing and reusing sections of the fence. Some fences can include additional measures to delay, or even detect, potential intruders. Such measures may include the addition of barbed wire, razor wire, or other deterrents at the top of the fence. Barbed wire is sometimes employed at the base of fences as well.

This can impede a would be intruder's progress in even reaching the fence. Fences may also be fitted with security cameras to provide visual surveillance of the perimeter. Finally, some facilities have installed motion sensors along their fences to detect movement on the fence. Several manufacturers have combined these multiple perimeter security features into one product and offer alarms and other security features. The correct implementation of a fence can make it a much more effective security measure.

Security experts recommend the following when a facility constructs a fence:

- The fence should be at least 7–9 feet high.
- Any outriggers, such as barbed wire, that are affixed on top of the fence should be angled out and away from the facility, and not in toward the facility. This will make climbing the fence more difficult and will prevent ladders from being placed against the fence.
- Other types of hardware can increase the security of the fence. This can include installing concertina wire along the fence (this can be done in front of the fence or at the top of the fence) or adding intrusion sensors, cameras, or other hardware to the fence.

- All undergrowth should be cleared for several feet (typically 6 feet) on both sides of the fence. This will allow for a clearer view of the fence by any patrols in the area.
- Any trees with limbs or branches hanging over the fence should be trimmed so that intruders cannot use them to go over the fence. Also, it should be noted that fallen trees can damage fences, and so management of trees around the fence can be important. This can be especially important in areas where a fence goes through a remote area.
- Fences that do not block the view from outside the fence to inside the fence allow patrols to see inside the fence without having to enter the facility.
- "No Trespassing" signs posted along fence can be a valuable tool in prosecuting any intruders who claim that the fence was broken and that they did not enter through the fence illegally. Adding signs that highlight the local ordinances against trespassing can further persuade simple troublemakers from illegally jumping/climbing the fence.

Films for Glass Shatter Protection

Most water and wastewater systems have numerous windows on the outside of buildings, in doors, and in interior offices. In addition, many facilities have glass doors or other glass structures, such as glass walls or display cases. These glass objects are potentially vulnerable to shattering when heavy objects are thrown or launched at them, when explosions occur near them, or when there are high winds (for exterior glass). If the glass is shattered, intruders may potentially enter an area. In addition, shattered glass projected into a room from an explosion or from an object being thrown through a door or window can injure and potentially incapacitate personnel in the room. Materials that prevent glass from shattering can help to maintain the integrity of the door, window, or other glass object and can delay an intruder from gaining access. These materials can also prevent flying glass and thus reduce potential injuries.

Materials designed to prevent glass from shattering include specialized films and coatings. These materials can be applied to existing glass objects to improve their strength and their ability to resist shattering. These films have been tested against many scenarios that could result in glass breakage, including penetration by blunt objects, bullets, high winds, and simulated explosions. Thus, the films are tested against both simulated weather scenarios (which could include both the high winds themselves and the force of objects blown into the glass), as well as more criminal/terrorist scenarios where the glass is subject to explosives or bullets. Many vendors provide information on the results of these types of tests, and thus potential users can compare different product lines to determine which products best suit their needs.

The primary attributes of films for shatter protection are:

- The materials from which the film is made;
- The adhesive that bonds the film to the glass surface;
- The thickness of the film.

Fire Hydrant Locks

Fire hydrants are installed at strategic locations throughout a community's water distribution, and they are often located in residential neighborhoods, industrial districts, and other areas where they cannot be easily observed and/or guarded. As a result, they are potentially vulnerable to unauthorized access. Many municipalities, states, and EPA regions have recognized this potential vulnerability and have instituted programs to lock hydrants. For example, EPA Region 1 has included locking hydrants as number 7 on its "Drinking Water Security and Emergency Preparedness" Top Ten List for small groundwater suppliers.

A "hydrant lock" is a physical security device designed to prevent unauthorized access to the water supply through a hydrant. They can also ensure water and water pressure availability to firefighters and prevent water theft and associated lost water revenue. These locks have been successfully used in numerous municipalities and in various climates and weather conditions.

Fire hydrant locks are basically steel covers or caps that are locked in place over the operating nut of a fire hydrant. The lock prevents unauthorized persons from accessing the operating nut and opening the fire hydrant valve. The lock also makes it more difficult to remove the bolts from the hydrant and access the system that way. Finally, hydrant locks shield the valve from being broken off. Should a vandal attempt to breach the hydrant lock by force and succeed in breaking the hydrant lock, the vandal will only succeed in bending the operating valve. If the hydrant's operating valve is bent, the hydrant will not be operational, but the water asset remains protected and inaccessible to vandals. However, the entire hydrant will need to be replaced. Hydrant locks are designed so that the hydrants can be operated by special "key wrenches" without removing the lock. These specialized wrenches are generally distributed to the fire department, public works department, and other authorized persons so that they can access the hydrants as needed. An inventory of wrenches and their serial numbers is generally kept by a municipality so that the location of all wrenches is known. These operating key wrenches may only be purchased by registered lock owners. The most important features of hydrants are their strength and the security of their locking systems. The locks must be strong so that they cannot be broken off. Hydrant locks are constructed from stainless or alloyed steel. Stainless steel locks are stronger and are ideal for all climates; however, they are more expensive than alloy locks. The locking mechanisms for each fire hydrant locking system ensure that the hydrant can only be operated by authorized personnel who have the specialized key to work the hydrant.

Hatch Security

A hatch is basically a door installed on a horizontal plane (such as in a floor, a paved lot, or a ceiling), instead of on a vertical plane (such as in a building wall). Hatches are usually used to provide access to assets that are either located underground (such as hatches to basements or underground storage areas), or to assets located above ceilings (such as emergency roof exits). At water and wastewater facilities, hatches are typically used to provide access to underground vaults containing pumps, valves, or piping, or to the interior of water tanks or covered reservoirs. Securing a hatch by locking it or upgrading materials to give the hatch added strength can help to delay unauthorized access to any asset behind the hatch. Like all doors, a hatch consists of a frame anchored to the horizontal structure, a door or doors, hinges connecting the door/doors to the frame, and a latching or locking mechanism that keeps the hatch door/doors closed. It should be noted that improving hatch security is straightforward and that hatches with upgraded security features can be installed new or can be retrofit for existing applications. Many municipalities already have specifications for hatch security at their water and wastewater system assets.

Depending on the application, the primary security-related attributes of a hatch are the strength of the door and frame, its resistance to the elements and corrosion, its ability to be sealed against water or gas, and its locking features. Hatches must be both strong and lightweight so that they can withstand typical static loads (such as people or vehicles walking or driving over them) while still being easy to open. In addition, because hatches are typically installed at outdoor locations, they are usually designed from corrosion-resistant metal that can withstand the elements. Therefore, hatches are typically constructed from high gauge steel or lightweight aluminum. The hatch locking mechanism is perhaps the most important part of hatch security.

There are a number of locks that can be implemented for hatches, including:

- Slam locks (internal locks that are located within the hatch frame);
- Recessed cylinder locks;
- Bolt locks;
- Padlocks.

Intrusion Sensors

An exterior intrusion sensor is a detection device that is used in an outdoor environment to detect intrusions into a protected area. These devices are designed to detect an intruder and then communicate an alarm signal to an alarm system. The alarm system can respond to the intrusion in many different ways, such as by triggering an audible or visual alarm signal or by sending an electronic signal to a central monitoring location that notifies security personnel of the intrusion. Intrusion sensors can be used to protect many kinds of assets. Intrusion sensors that protect physical space are classified according to whether they protect indoor, or "interior" space (i.e., an entire building or room within a building) or outdoor, or "exterior" space (i.e., a fence line or perimeter). Interior intrusion sensors are designed to protect the interior space of a facility by detecting an intruder who is attempting to enter or who has already entered a room or building. In contrast, exterior intrusion sensors are designed to detect an intrusion into a protected outdoor/exterior area. Exterior protected areas are typically arranged as zones or exclusion areas placed so that the intruder is detected early in the intrusion attempt before the intruder can gain access to more valuable assets (e.g., into a building located within the protected area). Early detection creates additional time for security forces to respond to the alarm.

Buried Exterior Intrusion Sensors

Buried sensors are electronic devices that are designed to detect potential intruders. The sensors are buried along the perimeters of sensitive assets and can detect intruder activity both above- and belowground. Some of these systems are composed of individual, stand-alone sensor units, while other sensors consist of buried cables.

Ladder Access Control

Water and wastewater systems have a number of assets that are raised above ground level, including raised water tanks, raised chemical tanks, raised piping systems, and roof access points into buildings. In addition, communications equipment, antennae, or other electronic devices may be located on the top of these raised assets. Typically, these assets are reached by ladders that are permanently anchored to the asset. For example, raised water tanks typically are accessed by ladders that are bolted to one of the legs of the tank. Controlling access to these raised assets by controlling access to the ladder can increase security at a water or wastewater system.

A typical ladder access control system consists of some type of cover that is locked or secured over the ladder. The cover can be a casing that surrounds most of the ladder or a door or shield that covers only part of the ladder. In either case, several rungs of the ladder (the number of rungs depends on the size of the cover) are made inaccessible by the cover, and these rungs can only be accessed by opening or removing the cover. The cover is locked so that only authorized personnel can open or remove it and use the ladder. Ladder access controls are usually installed at several feet above ground level, and they usually extend several feet up the ladder so that they cannot be circumvented by someone accessing the ladder above the control system. The important features of ladder access control are the size and strength of the cover and its ability to lock or otherwise be secured from unauthorized access.

The covers are constructed from aluminum or some type of steel. This should provide adequate protection from being pierced or cut through. The metals are corrosion resistant so that they will not corrode or become fragile in extreme weather conditions in outdoor applications. The bolts used to install each of these systems are galvanized steel. In addition, the bolts for each cover are installed on the inside of the unit so they cannot be removed from the outside.

Locks

A lock is a type of physical security device that can be used to delay or prevent a door, a window, a manhole, a filing cabinet drawer, or some other physical feature from being opened, moved, or operated. Locks typically operate by connecting two pieces together – such as by connecting a door to a door jamb or a manhole to its casement. Every lock has two modes – engaged (or "locked"), and disengaged (or "opened"). When a lock is disengaged, the asset on which the lock is installed can be accessed by anyone, but when the lock is engaged, only authorized personnel have access to the locked asset.

Locks are excellent security features because they have been designed to function in many ways and to work on many different types of assets. Locks can also provide different levels of security depending on how they are designed and implemented. The security provided by a lock is dependent on several factors, including its ability to withstand physical damage (i.e., can it be cut off, broken, or otherwise physically disabled) as well as its requirements for supervision or operation (i.e., combinations may need to be changed frequently so that they are not compromised and the locks remain secure). While there is no single definition of the "security" of a lock, locks are often described as minimum, medium, or maximum security. Minimum security locks are those that can be easily disengaged (or "picked") without the correct key or code or those that can be disabled easily (such as small padlocks that can be cut with bolt cutters). Higher security locks are more complex and thus are more difficult to pick, or are sturdier and more resistant to physical damage. Many locks (such as door locks) only need to be unlocked from one side. For example, most door locks need a key to be unlocked only from the outside. A person opens such devices, called single-cylinder locks, from the inside by pushing a button or by turning a knob or handle. Double-cylinder locks require a key to be locked or unlocked from both sides.

Manhole Intrusion Sensors

Manholes are located at strategic locations throughout most municipal water, wastewater, and other underground system systems. Manholes are designed to provide access to the underground utilities, and therefore they are potential entry points to a system. For example, manholes in water or wastewater systems may provide access to sewer lines or vaults containing on/off or pressure reducing water valves. Because many utilities run under other infrastructure (roads, buildings), manholes also provide potential access points to critical infrastructure as well as water and wastewater assets. In addition, because the portion of the system to which manholes provide entry is primarily located underground, access to a system through a manhole increases the chance that an intruder will not be seen. Therefore, protecting manholes can be a critical component of guarding an entire community.

The various methods for protecting manholes are designed to prevent unauthorized personnel from physically accessing the manhole and detecting attempts at unauthorized access to the manhole. A manhole intrusion sensor is a physical security device designed to detect unauthorized access to the system through a manhole. Monitoring a manhole that provides access to a water or wastewater system can mitigate two distinct types of threats. First, monitoring a manhole may detect access of unauthorized personnel to water or wastewater systems or assets through the manhole. Second, monitoring manholes may also allow the detection of the introduction of hazardous substances into the water system. Several different technologies have been used to develop manhole intrusion sensors, including mechanical systems, magnetic systems, and fiber-optic and infrared sensors. Some of these intrusion sensors have been specifically designed for manholes, while others consist of standard, off-the-shelf intrusion sensors that have been implemented in a system specifically designed for application in a manhole.

Manhole Locks

A manhole lock is a physical security device designed to delay unauthorized access to the system through a manhole. Locking a manhole that provides access to a water or wastewater system can mitigate two distinct types of threats. First, locking a manhole may delay access of unauthorized personnel to water or wastewater systems through the manhole. Second, locking manholes may also prevent the introduction of hazardous substances into the wastewater or stormwater system.

Radiation Detection Equipment for Monitoring Personnel and Packages

A major potential threat facing water and wastewater facilities is contamination by radioactive substances. Radioactive substances brought on-site at a facility could be used to contaminate the facility, thereby preventing workers from safely entering the facility to perform necessary water treatment tasks. In addition, radioactive substances brought on-site at a water treatment plant could be discharged into the water source or the distribution system, contaminating the downstream water supply. Therefore, detection of radioactive substances being brought on-site can be an important security enhancement. Various radionuclides have unique properties, and different equipment is required to detect different types of radiation. However, it is impractical and potentially unnecessary to monitor for specific radionuclides being brought on-site. Instead, for security purposes, it may be more useful to monitor for gross radiation as an indicator of unsafe substances.

In order to protect against these radioactive materials being brought on-site, a facility may set up monitoring sites outfitted with radiation detection instrumentation at entrances to the facility. Depending on the specific types of equipment chosen, this equipment would detect radiation emitted from people, packages, or other objects being brought through an entrance. One of the primary differences between the different types of detection equipment is the means by which the equipment reads the radiation. Radiation may either be detected by direct measurement or through sampling. Direct radiation measurement involves measuring radiation through an external probe on the detection instrumentation. Some direct measurement equipment detects radiation emitted into the air around the monitored object. Because this equipment detects radiation in the air, it does not require that the monitoring equipment make physical contact with the monitored object. Direct means for detecting radiation include using a walk-through portal-type monitor that would detect elevated radiation levels on a person or in a package or by using a handheld detector, which would be moved or swept over individual objects to locate a radioactive source.

Some types of radiation, such as alpha or low energy beta radiation, have a short range and are easily shielded by various materials. These types of radiation cannot be measured through direct measurement. Instead, they must be measured through sampling. Sampling involves wiping the surface to be tested with a special filter cloth and then reading the cloth in a special counter. For example, specialized smear counters measure alpha and low energy beta radiation.

Reservoir Covers

Reservoirs are used to store raw or untreated water. They can be located underground (buried), at ground level, or on an elevated surface. Reservoirs can vary significantly in size; small reservoirs can hold as little as 1,000 gallons, while larger reservoirs may hold many millions of gallons. Reservoirs can be either natural or man made. Natural reservoirs can include lakes or other contained water bodies, while man-made reservoirs usually consist of some sort of engineered structure, such as a tank or other impoundment structure. In addition to the water containment structure itself, reservoir systems may also include associated water treatment and distribution equipment, including intakes, pumps, pump houses, piping systems, chemical treatment and chemical storage areas. Drinking water reservoirs are of particular concern because they are potentially vulnerable to contamination of the stored water, either through direct contamination of the storage area, or through infiltration of the equipment, piping, or chemicals associated with the reservoir. For example, because many drinking water reservoirs are designed as aboveground, open-air structures, they are potentially vulnerable to airborne deposition, bird and animal wastes, human activities, and dissipation of chlorine or other treatment chemicals.

However, one of the most serious potential threats to the system is direct contamination of the stored water through dumping contaminants into the reservoir. Utilities have taken various measures to mitigate this type of threat, including fencing off the reservoir, installing cameras to monitor for intruders, and monitoring for changes in water quality. Another option for enhancing security is covering the reservoir using some type of manufactured cover to prevent intruders from gaining physical access to the stored water. Implementing a reservoir cover may or may not be practical depending on the size of the reservoir (e,g,, covers are not typically used on natural reservoirs because they are too large for the cover to be technically feasible and cost effective).

This section will focus on drinking water reservoir covers, where and how they are typically implemented, and how they can be used to reduce the threat of contamination of the stored water. While covers can enhance the reservoir's security, it should be noted that covering a reservoir typically changes the reservoir's operational requirements. For example, vents must be installed in the cover to ensure gas exchange between the stored water and the atmosphere. A reservoir cover is a structure installed on or over the surface of the reservoir to minimize water quality degradation.

The three basic design types for reservoir covers are:

- Floating
- Fixed
- Air-supported

A variety of materials are used when manufacturing a cover, including reinforced concrete, steel, aluminum, polypropylene, chlorosulfonated polyethylene, or ethylene interpolymer alloys. There are several factors that affect a reservoir cover's effectiveness and thus its ability to protect the stored water. These factors include:

- The location, size, and shape of the reservoir;
- The ability to lay/support a foundation (e.g., footing, soil, and geotechnical support conditions);
- The length of time the reservoir can be removed from service for cover installation or maintenance;
- Aesthetic considerations; and
- Economic factors, such as capital and maintenance costs.

It may not be practical, for example, to install a fixed cover over a reservoir if the reservoir is too large or if the local soil conditions cannot support a foundation. A floating or air-supported cover may be more appropriate for these types of applications.

In addition to the practical considerations for installation of these types of covers, there are a number of operations and maintenance (O&M) concerns that affect the system of a cover for specific applications, including how different cover materials will withstand local climatic conditions, what types of cleaning and maintenance will be required for each particular type of cover, and how these factors will affect the cover's lifespan and its ability to be repaired when it is damaged.

The primary feature affecting the security of a reservoir cover is its ability to maintain its integrity. Any type of cover, no matter what its construction material, will provide good protection from contamination by rainwater or atmospheric deposition, as well as from intruders attempting to access the stored water with the intent of causing intentional contamination. The covers are large and heavy, and it is difficult to circumvent them to get into the reservoir. At the very least, it would take a determined intruder, as opposed to a vandal, to defeat the cover.

Side-Hinged Door Security

Doorways are the main access points to a facility or to rooms within a building. They are used on the exterior or in the interior of buildings to provide privacy and security for the areas behind them. Different types of doorway security systems may be installed depending on the needs or requirements of the buildings or rooms. For example, exterior doorways tend to have heavier doors to withstand the elements and to provide some security to the entrance of the building. Interior doorways in office areas may have lighter doors that may be primarily designed to provide privacy rather than security. Therefore, these doors may be made of glass or lightweight wood. Doorways in industrial areas may have sturdier doors than do other interior doorways and may be designed to provide protection or security for areas behind the doorway. For example, fireproof doors may be installed in chemical storage areas or in other areas where there is a danger of fire. Because they are the main entries into a facility or a room, doorways are often prime targets for unauthorized entry into a facility or an asset. Therefore, securing doorways may be a major step in providing security at a facility.

A doorway includes four main components:

1. The door, which blocks the entrance. The primary threat to the actual door is breaking or piercing through the door. Therefore, the primary security features of doors are their strength and resistance to various physical threats, such as fire or explosions.
2. The door frame, which connects the door to the wall. The primary threat to a door frame is that the door can be pried away from the frame. Therefore, the primary security feature of a door frame is its resistance to prying.
3. The hinges, which connect the door to the door frame. The primary threat to door hinges is that they can be removed or broken, which will allow intruders to remove the entire door. Therefore, security hinges are designed to be resistant to breaking. They may also be designed to minimize the threat of removal from the door.
4. The lock, which connects the door to the door frame. Use of the lock is controlled through various security features, such as keys, combinations, etc., such that only authorized personnel can open the lock and go through the door. Locks may also incorporate other security features, such as software or other systems to track overall use of the door or to track individuals using the door, etc.

Each of these components is integral in providing security for a doorway, and upgrading the security of only one of these components while leaving the other components unprotected may not increase the overall security of the doorway. For example, many facilities upgrade door locks as a basic step in increasing the security of a facility. However, if the facilities do not also focus on increasing security for the door hinges or the door frame, the door may remain vulnerable to being removed from its frame, thereby defeating the increased security of the door lock. The primary attribute for the security of a door is its strength. Many security doors are 4–20 gauge hollow metal doors consisting of steel plates over a hollow cavity reinforced with steel stiffeners to give the door extra stiffness and rigidity. This increases resistance to blunt force used to try to penetrate through the door. The space between the stiffeners may be filled with specialized materials to provide fire, blast, or bullet resistance to the door. The Window and Door Manufacturers Association has developed a series of performance attributes for doors.

These include:

- Structural Resistance;
- Forced Entry Resistance;
- Hinge Style Screw Resistance;
- Split Resistance;
- Hinge Resistance;
- Security Rating;
- Fire Resistance;
- Bullet Resistance;
- Blast Resistance.

The first five bullets provide information on a door's resistance to standard physical breaking and prying attacks. These tests are used to evaluate the strength of the door and the resistance of the hinges and the frame in a standardized way. For example, the Rack Load Test simulates a prying attack on a corner of the door. A test panel is restrained at one end, and a third corner is supported. Loads are applied and measured at the fourth corner. The Door Impact Test simulates a battering attack on a door and frame using impacts of 200 foot pounds by a steel pendulum. The door must remain fully operable after the test. It should be noted that door glazing is also rated for resistance to shattering, etc. Manufacturers will be able to provide security ratings for these features of a door as well.

Door frames are an integral part of doorway security because they anchor the door to the wall. Door frames are typically constructed from wood or steel, and they are installed such that they extend for several inches over the doorway that has been cut into the wall. For added security, frames can be designed to have varying degrees of overlap with, or wrapping over, the underlying wall.

This can make prying the frame from the wall more difficult. A frame formed from a continuous piece of metal (as opposed to a frame constructed from individual metal pieces) will prevent prying between pieces of the frame. Many security doors can be retrofit into existing frames; however, many security door installations include replacing the door frame as well as the door itself. For example, bullet resistance per Underwriter's Laboratory (UL) 752 requires resistance of the door and frame assembly, and thus replacing the door only would not meet UL 752 requirements.

Valve Lockout Devices

Valves are utilized as control elements in water and wastewater process piping networks. They regulate the flow of both liquids and gases by opening, closing, or obstructing a flow passageway. Valves are typically located where flow control is necessary. They can be located in-line or at pipeline and tank entrance and exit points.

They can serve multiple purposes in a process pipe network, including:

- Redirecting and throttling flow;
- Preventing backflow;
- Shutting off flow to a pipeline or tank (for isolation purposes);
- Releasing pressure;
- Draining extraneous liquid from pipelines or tanks;
- Introducing chemicals into the process network; and
- As access points for sampling process water.

Valves are located at critical junctures throughout water and wastewater systems, both on-site at treatment facilities and off-site within water distribution and wastewater collection systems. They may be located either aboveground or belowground. Because many valves are located within the community, it is critical to provide protection against valve tampering. For example, tampering with a pressure relief valve could result in a pressure buildup and potential explosion in the piping network. On a larger scale, addition of a pathogen or chemical to the water distribution system through an unprotected valve could result in the release of that contaminant to the general population.

Various security products are available to protect aboveground vs. belowground valves. For example, valve lockout devices can be purchased to protect valves and valve controls located aboveground. Vaults containing underground valves can be locked to prevent access. Valve-specific lockout devices are available in a variety of colors, which can be useful in distinguishing different valves. For example, different colored lockouts can be used to distinguish the type of liquid passing through the valve (i.e., treated, untreated, potable, chemical) or to identify the party responsible for maintaining the lockout. Implementing a system of different colored locks on operating valves can increase system security by reducing the likelihood of an operator inadvertently opening the wrong valve and causing a problem in the system.

Vent Security

Vents are installed in aboveground, covered water reservoirs, and in underground reservoirs to allow ventilation of the stored water. Specifically, vents permit the passage of air that is being displaced from, or drawn into, the reservoir as the water level in the reservoir rises and falls due to system demands. Small reservoirs may require only one vent, whereas larger reservoirs may have multiple vents throughout the system. The specific vent design for any given application will vary depending on the design of the reservoir, but every vent consists of an open air connection between the reservoir and the outside environment. Although these air exchange vents are an integral part of covered or underground reservoirs, they also represent a potential security threat. Improving vent security by making the vents tamper resistant or by adding other security features, such as security screens or security covers, can enhance the security of the entire water system. Many municipalities already have specifications for vent security at their water assets. These specifications typically include the following requirements:

- Vent openings are to be angled down or shielded to minimize the entrance of surface and/or rainwater into the vent through the opening.
- Vent designs are to include features to exclude insects, birds, animals, and dust.
- Corrosion-resistant materials are to be used to construct the vents.

Some states have adopted more specific requirements for added vent security at their water system assets. For example, the State of Utah's Department of Environmental Quality, Division of Drinking Water, Division of Administrative Rules (DAR), provides specific requirements for public drinking water storage tanks. The rules for drinking water storage tanks as they apply to venting are set forth in Utah-R309-545-15: "Venting" includes the following requirements:

- Drinking water storage tank vents must have an open discharge on buried structures.
- The vents must be located 24–36 inches above the earthen covering.
- The vents must be located and sized to avoid blockage during winter conditions.

In a second example, Washington State's "Drinking Water Tech Tips: Sanitary Protection of Reservoirs" document states that vents must be protected to prevent the water supply from being contaminated. The document indicates that non-corrodible No. 24 mesh may be used to screen vents on elevated tanks. The document continues to state that the vent opening for storage facilities located underground or at ground level should be 24–36 inches above the roof or ground and that it must be protected with a No. 24-inch mesh non-corrodible screen. New Mexico's Administrative Code also specifies that vents must be covered with No. 24 mesh (NMAC Title 20, Chapter 7, Subpart I, 208.E). Washington and New Mexico, as well as many other municipalities, require vents to be screened using a non-corrodible mesh to minimize the entry of insects, other animals, and rain-borne contamination into the vents. When selecting the appropriate mesh size, it is important to identify the smallest mesh size that meets both the strength and durability requirements for that application.

Visual Surveillance Monitoring

Visual surveillance is used to detect threats through continuous observation of important or vulnerable areas of an asset. The observations can also be recorded for later review or use (e.g., in court proceedings). Visual surveillance systems can be used to monitor various parts of collection, distribution, or treatment systems, including the perimeter of a facility, outlying pumping stations, or entry or access points into specific buildings.

These systems are also useful in recording individuals who enter or leave a facility, thereby helping to identify unauthorized access. Images can be transmitted live to a monitoring station, where they can be monitored in real time, or they can be recorded and reviewed later. Many facilities have found that a combination of electronic surveillance and security guards provides an effective means of facility security. Visual surveillance is provided through a closed-circuit television (CCTV) system, in which the capture, transmission, and reception of an image are localized within a closed "circuit." This is different than other broadcast images, such as over-the-air television, which is broadcast over the air to any receiver within range.

At a minimum, a CCTV system consists of:

- One or more cameras;
- A monitor for viewing the images; and
- A system for transmitting the images from the camera to the monitor.

WATER MONITORING DEVICES

Note: Adapted from Spellman, F.R., *Water Infrastructure Protection and Homeland Security*, Government Institutes Press, Lanham, MD, 2007.

Proper security preparation really comes down to a three-legged approach: detect, delay, respond. The third leg of security, to detect, is discussed in this section. Specifically, this section deals with the monitoring of water samples to detect toxicity and/or contamination.

Many of the major monitoring tools that can be used to identify anomalies in process streams or finished water that may represent potential threats are discussed, including:

- Sensors for monitoring chemical, biological, and radiological contamination;
- Chemical sensor – Arsenic measurement system;
- Chemical sensor for toxicity [adapted biochemical oxygen demand(BOD) analyzer];
- Chemical sensor – total organic carbon analyzer;
- Chemical sensor – Chlorine measurement system;
- Chemical sensor-portable cyanide analyzer;
- Portable field monitors to measure VOCs;
- Radiation detection equipment;
- Radiation detection equipment for monitoring water assets; and
- Toxicity monitoring/toxicity meters.

Water quality monitoring sensor equipment may be used to monitor key elements of water or wastewater treatment processes (such as influent water quality, treatment processes, or effluent water quality) to identify anomalies that may indicate threats to the system. Some sensors, such as sensors for biological organisms or radiological contaminants, measure potential contamination directly, while others, particularly some chemical monitoring systems, measure "surrogate" parameters that may indicate problems in the system but do not identify sources of contamination directly. In addition, sensors can provide more accurate control of critical components in water and wastewater systems and may provide a means of early warning so that the potential effects of certain types of attacks can be mitigated. One advantage of using chemical and biological sensors to monitor for potential threats to water and wastewater systems is that many utilities already employ sensors to monitor potable water (raw or finished) or influent/effluent for Safe Drinking Water Act (SDWA) or Clean Water Act (CWA) water quality compliance or process control.

Chemical sensors that can be used to identify potential threats to water and wastewater systems include inorganic monitors (e.g., chlorine analyzer), organic monitors (e.g., total organic carbon analyzer), and toxicity meters. Radiological meters can be used to measure concentrations of several different radioactive species. Monitors that use biological species can be used as sentinels for the presence of contaminants of concern, such as toxins. At the present time, biological monitors are not in widespread use and very few bio-monitors are used by drinking public water systems in the United States.

Monitoring can be conducted using either portable or fixed-location sensors. Fixed-location sensors are usually used as part of a continuous, online monitoring system. Continuous monitoring has the advantage of enabling immediate notification when there is an upset. However, the sampling points are fixed, and only certain points in the system can be monitored. In addition, the number of monitoring locations needed to capture the physical, chemical, and biological complexity of a system can be prohibitive. The use of portable sensors can overcome this problem of monitoring many points in the system. Portable sensors can be used to analyze grab samples at any point in the system but have the disadvantage that they provide measurements only at one point in time.

Sensors for Monitoring Chemical, Biological, and Radiological Contamination

Toxicity tests measure water toxicity by monitoring adverse biological effects on test organisms. Toxicity tests have traditionally been used to monitor wastewater effluent streams for National Pollutant Discharge Elimination System (NPDES) to permit compliance or to test water samples for toxicity. However, this technology can also be used to monitor drinking water distribution systems or other water and wastewater streams for toxicity. Currently, several types of bio-sensors and toxicity tests are being adapted for use in the water and wastewater security field. The keys to using bio-monitoring or bio-sensors for drinking water or other water and wastewater asset security are rapid response and the ability to use the monitor at critical locations in the system, such as in water distribution systems downstream of pump stations or prior to the biological process in a wastewater treatment plant.

While there are several different organisms that can be used to monitor for toxicity (including bacteria, invertebrates, and fish), bacteria-based bio-sensors are ideal for use as early warning screening tools for drinking water security because bacteria usually respond to toxics in a matter of minutes. In contrast to methods using bacteria, toxicity screening methods that use higher level organisms such as fish may take several days to produce a measurable result. Bacteria-based bio-sensors have recently been incorporated into portable instruments, making rapid response and field testing practical. These portable meters detect decreases in biological activity (e.g., decreases in bacterial luminescence), which are highly correlated with increased levels of toxicity.

At the present time, few utilities are using biologically based toxicity monitors to monitor water and wastewater assets for toxicity, and very few products are now commercially available. Several new approaches to the rapid monitoring of microorganisms for security purposes (e.g., microbial source tracking) have been identified. However, most of these methods are still in the research and development phase.

Chemical Sensors: Arsenic Measurement System

Arsenic is an inorganic toxin that occurs naturally in soils. It can enter water supplies from many sources, including erosion of natural deposits, runoff from orchards, runoff from glass and electronics production wastes, or leaching from products treated with arsenic, such as wood. Synthetic organic arsenic is also used in fertilizer. Arsenic toxicity is primarily associated with inorganic arsenic ingestion and has been linked to cancerous health effects, including cancer of the bladder, lungs, skin, kidney, nasal passages, liver, and prostate. Arsenic ingestion has also been linked to noncancerous cardiovascular, pulmonary, immunological, neurological, and endocrine problems. According to USEPA's Safe Drinking Water Act (SDWA) Arsenic Rule, inorganic arsenic can exert toxic effects after acute (short-term) or chronic (long-term) exposure.

Toxicological data for acute exposure, which is typically given as an LD50 value (the dose that would be lethal to 50% of the test subjects in a given test), suggests that the LD50 of arsenic ranges 1–4 milligrams arsenic per kilogram (mg/kg) of body weight. This dose would correspond to a lethal dose range of 70–280 mg for 50% of adults weighing 70 kg. At nonlethal, but high, acute doses, inorganic arsenic can cause gastroenterological effects, shock, neuritis (continuous pain), and vascular effects in humans. USEPA has set a maximum contaminant level goal of zero for arsenic in drinking water; the current enforceable maximum contaminant level (MCL) is 0.050 mg/L. As of January 23, 2006, the enforceable MCL for arsenic will be 0.010 mg/L.

The SDWA requires arsenic monitoring for public water systems. The Arsenic Rule indicates that surface water systems must collect one sample annually; groundwater systems must collect one sample in each compliance period (once every three years). Samples are collected at entry points to the distribution system, and analysis is done in the lab using one of several USEPA-approved methods, including Inductively Coupled Plasma Mass Spectroscopy (ICP-MS, USEPA 200.8) and several atomic absorption (AA) methods. However, several different technologies, including colorimetric test kits and portable chemical sensors, are currently available for monitoring inorganic arsenic concentrations in the field. These technologies can provide a quick estimate of arsenic concentrations in a water sample. Thus, these technologies may be useful for spot-checking different parts of a drinking water system (e.g., reservoirs, isolated areas of distribution systems) to ensure that the water is not contaminated with arsenic.

Chemical Sensor: Adapted BOD Analyzer

One manufacturer has adapted a BOD analyzer to measure oxygen consumption as a surrogate for general toxicity. The critical element in the analyzer is the bioreactor, which is used to continuously measure the respiration of the biomass under stable conditions. As the toxicity of the sample increases, the oxygen consumption in the sample decreases. An alarm can be programmed to sound if oxygen reaches a minimum concentration (i.e., if the sample is strongly toxic). The operator must then interpret the results into a measure of toxicity.

Note :The current time, it is difficult to directly define the sensitivity and/or the detection limit of toxicity measurement devices because limited data is available regarding specific correlation of decreased oxygen consumption and increased toxicity of the sample.

Chemical Sensor: Total Organic Carbon Analyzer

Total organic carbon (TOC) analysis is a well defined and commonly used methodology that measures the carbon content of dissolved and particulate organic matter present in water. Many public water systems monitor TOC to determine raw water quality or to evaluate the effectiveness of processes designed to remove organic carbon. Some wastewater systems also employ TOC analysis to monitor the efficiency of the treatment process. In addition to these uses for TOC monitoring, measuring changes in TOC concentrations can be an effective "surrogate" for detecting contamination from organic compounds (e.g., petrochemicals, solvents, pesticides). Thus, while TOC analysis does not give specific information about the nature of the threat, identifying changes in TOC can be a good indicator of potential threats to a system. TOC analysis includes inorganic carbon removal oxidation of the organic carbon into CO_2, and quantification of the CO_2. The primary differences between different online TOC analyzers are in the methods used for oxidation and CO_2 quantification.

The oxidation step can be high or low temperature. The determination of the appropriate analytical method (and thus the appropriate analyzer) is based on the expected characteristics of the wastewater sample (TOC concentrations and the individual components making up the TOC fraction). In general, high-temperature (combustion) analyzers achieve more complete oxidation of the carbon fraction than do low-temperature (wet chemistry/UV) analyzers. This can be important both in distinguishing different fractions of the organics in a sample and in achieving a precise measurement of the organic content of the sample. Three different methods are also available for detection and quantification of carbon dioxide produced in the oxidation step of a TOC analyzer.

These are:

- Nondispersive infrared (NDIR) detector;
- Colorimetric methods;
- Aqueous conductivity methods.

The most common detector that online TOC analyzers use for source water and drinking water analysis is the nondispersive infrared detector. Although the differences in analytical methods employed by different TOC analyzers may be important in compliance or process monitoring, high levels of precision and the ability to distinguish specific organic fractions from a sample may not be required for detection of a potential chemical threat. Instead, gross deviations from normal TOC concentrations may be the best indication of a chemical threat to the system.

The detection limit for organic carbon depends on the measurement technique used (high or low temperature) and the type of analyzer. Because TOC concentrations are simply surrogates that can indicate potential problems in a system, gross changes in these concentrations are the best indicators of potential threats. Therefore, high-sensitivity probes may not be required for security purposes.

However, the following detection limits can be expected:

- High-temperature method (between 680°C and 950°C or higher in a few special cases, best possible oxidation): = 1 mg/L carbon
- Low-temperature method (below 100°C, limited oxidation potential): = 0.2 mg/L carbon

The response time of a TOC analyzer may vary depending on the manufacturer's specifications, but it usually takes from 5 to 15 minutes to get a stable, accurate reading.

Chemical Sensors: Chlorine Measurement System

Residual chlorine is one of the most sensitive and useful indicator parameters in water distribution system monitoring. All water distribution systems monitor for residual chlorine concentrations as part of their SDWA requirements, and procedures for monitoring chlorine concentrations are well established and accurate. Chlorine monitoring assures proper residual at all points in the system, helps pace rechlorination when needed, and quickly and reliably signals any unexpected increase in disinfectant demand significant decline or loss of residual chlorine could be an indication of potential threats to the system. Several key points regarding residual chlorine monitoring for security purposes are:

- Chlorine residuals can be measured using continuous online monitors at fixed points in the system or by taking grab samples at any point in the system and using chlorine test kits or portable sensors to determine chlorine concentrations.
- Correct placement of residual chlorine monitoring points within a system is crucial to early detection of potential threats. For example, while dead ends and low-pressure zones are common trouble spots that can show low residual chlorine concentrations, these zones are generally not of great concern for water security purposes because system hydraulics will limit the circulation of any contaminants present in these areas of the system.
- Monitoring point and monitoring procedures for SDWA compliance vs. system security purposes may be different, and utilities must determine the best use of online, fixed monitoring systems vs. portable sensors/test kits to balance their SDA compliance and security needs.

Various portable and online chlorine monitors are commercially available. These range from sophisticated online chlorine monitoring systems to portable electrode sensors to colorimetric test kits. Online systems can be equipped with control, signal, and alarm systems that notify the operator of low chlorine concentrations, and some may be tied into feedback loops that automatically adjust chlorine concentrations in the system. In contrast, use of portable sensors or colorimetric test kits requires technicians to take a sample and read the results. The technician then initiates required actions based on the results of the test.

Several measurement methods are currently available to measure chlorine in water samples, including:

- N, N-diethyl-p-phenylenediamine (DPD) colorimetric method;
- Iodometric method;
- Amperometric electrodes;
- Polarographic membrane sensors.

It should be noted that there can be differences in the specific type of analyte, the range, and the accuracy of these different measurement methods. In addition, these different methods have different operations and maintenance requirements. For example, DPD systems require periodic replenishment of buffers, whereas polarographic systems do not. Users may want to consider these requirements when choosing the appropriate sensor for their system.

Chemical Sensors: Portable Cyanide Analyzer

Portable cyanide detection systems are designed to be used in the field to evaluate for potential cyanide contamination of a water asset. These detection systems use one of two distinct analytical methods – either a colorimetric method or an ion-selective method – to provide a quick, accurate cyanide measurement that does not require laboratory evaluation. Aqueous cyanide chemistry can be complex. Various factors, including the water asset's pH and redox potential, can affect the toxicity of cyanide in that asset. While personnel using these cyanide detection devices do not need to have advanced knowledge of cyanide chemistry to successfully screen a water asset for cyanide, understanding aqueous cyanide chemistry can help users to interpret whether the asset's cyanide concentration represents a potential threat. Therefore, a short summary of aqueous cyanide chemistry, including a discussion of cyanide toxicity, is provided below. For more information, the reader is referred to Greenberg et al. (1999). Cyanide (CN^-) is a toxic carbon-nitrogen organic compound that is the functional portion of the lethal gas hydrogen cyanide (HCN). The toxicity of aqueous cyanide varies depending on its form. At near-neutral pH, "free cyanide" (which is commonly designated as "CN" although it is actually defined as the total of HCN and CN) is the predominant cyanide form in water.

Free cyanide is potentially toxic in its aqueous form, although the primary concern regarding aqueous cyanide is that it could volatilize. Free cyanide is not highly volatile (it is less volatile than most VOCs, but its volatility increases as the pH decrease below 8). However, when free cyanide does volatilize, it volatilizes in its highly toxic gaseous form (gaseous HCN). As a general rule, metal-cyanide complexes are much less toxic than free cyanide because they do not volatilize unless the pH is low. Analyses for cyanide in public water systems are often conducted in certified labs using various USEPA-approved methods, such as the preliminary distillation procedure with subsequent analysis by a colorimetric, ion-selective electrode, or flow injection methods. Lab analyses using these methods require careful sample preservation and pretreatment procedures and are generally expensive and time consuming.

Using these methods, several cyanide fractions are typically defined:

Total Cyanide – includes free cyanide (CN + HCN) and all metal-completed cyanide.

Weak Acid Dissociable (WAD) Cyanide – includes free cyanide (CN + HCN) and weak cyanide complexes that could be potentially toxic by hydrolysis to free cyanide in the pH range 4.5–6.0.

Amendable Cyanide – includes free cyanide (CN + HCN) and weak cyanide complexes that can release free cyanide at high pH (11–12) (this fraction gets its name because it includes measurement of cyanide from complexes that are "amendable" to oxidation by chlorine at high pH). To measure "Amendable Cyanide," the sample is split into two fractions. One of the fractions is analyzed for "Total Cyanide" as above. The other fraction is treated with high levels of chlorine for approximately one hour, dechlorinated, and distilled per the above "Total Cyanide" method. "Amendable Cyanide" is determined by the difference in the cyanide concentrations in these two fractions.

Soluble Cyanide – measures only soluble cyanide. Soluble cyanide is measured by using the preliminary filtration step, followed by "Total Cyanide" analysis described above.

As discussed above, these different methods yield various cyanide measurements which may or may not give a complete picture of that sample's potential toxicity. For example, the "Total Cyanide" method includes cyanide complexed with metals, some of which will not contribute to cyanide toxicity unless the pH is out of the normal range. In contrast, the "WAD Cyanide" measurement includes metal-complexed cyanide that could become free cyanide at low pH, and "Amendable Cyanide" measurements include metal-complexed cyanide that could become free cyanide at high pH. Personnel using these kits should therefore be aware of the potential differences in actual cyanide toxicity versus the cyanide potential differences in actual cyanide toxicity, versus the cyanide measured in the sample under different environmental conditions.

Ingestion of aqueous cyanide can result in numerous adverse health effects and may be lethal. USEPA's Maximum Contamination Level (MCL) for cyanide in drinking water is 0.2 µg/L (0.2 parts per million, or ppm). This MCL is based on free cyanide analysis per the "Amendable Cyanide" method described above (USEPA has recognized that very stable metal-cyanide complexes such as iron-cyanide complex are non-toxic [unless exposed to significant UV radiation], and these fractions are therefore not considered when defining cyanide toxicity). Ingestion of free cyanide at concentrations in excess of this MCL causes both acute effects (e.g., rapid breathing, tremors, and neurological symptoms) and chronic effects (e.g., weight loss, thyroid effects, and nerve damage). Under the current primary drinking water standards, public water systems are required to monitor their systems to minimize public exposure to cyanide levels in excess of the MCL.

Hydrogen cyanide gas is also toxic, and the Office of Safety and Health Administration (OSHA) has set a permissible exposure limit (PEL) of 10 ppmv for HCN inhalation. HCN also has a strong, bitter, almond-like smell and an odor threshold of approximately 1 ppmv. Considering the fact that HCN is relatively nonvolatile (see above), a slight cyanide odor emanating from a water sample suggests very high aqueous cyanide concentrations – greater than 10–50 mg/L, which is in the range of a lethal or near-lethal dose with the ingestion of one pint of water.

Portable Field Monitors to Measure VOCs

Volatile organic compounds (VOCs) are a group of highly utilized chemicals that have widespread applications, including use as fuel components, as solvents, and as cleaning and liquefying agents in degreasers, polishes, and dry-cleaning solutions. VOCs are also used in herbicides and insecticides for agriculture applications. Laboratory-based methods for analyzing VOCs are well established; however, analyzing VOCs in the lab is time consuming – obtaining a result may require several hours to several weeks depending on the specific method. Faster, commercially available methods for analyzing VOCs quickly in the field include use of portable gas chromatographs (GC), mass spectrometer (MS), or gas chromatographs/mass spectrometers (GC/MS), all of which can be used to obtain VOC concentration results within minutes. These instruments can be useful in rapid confirmation of the presence of VOCs in an asset or for monitoring an asset on a regular basis. In addition, portable VOC analyzers can analyze for a wide range of VOCs, such as toxic industrial chemicals (TICs), chemical warfare agents (CWAs), drugs, explosives, and aromatic compounds. There are several easy-to-use, portable VOC analyzers currently on the market that are effective in evaluating VOC concentrations in the field. These instruments utilize gas chromatography, mass spectroscopy, or a combination of both methods to provide near laboratory-quality analysis for VOCs.

Radiation Detection Equipment

Radioactive substances (radionuclides) are known health hazards that emit energetic waves and/or particles that can cause both carcinogenic and non-carcinogenic health effects. Radionuclides pose unique threats to source water supplies and water treatment, storage, or distribution systems because radiation emitted from radionuclides in water systems can affect individuals through several pathways – by direct contact with, ingestion, or inhalation of or external exposure to, the contaminated water. While radiation can occur naturally in some cases due to the decay of some minerals, intentional and non-intentional releases of man-made radionuclides into water systems are also a realistic threat.

Threats to water and wastewater facilities from radioactive contamination could involve two major scenarios. First, the facility or its assets could be contaminated, preventing workers from accessing and operating the facility/assets. Second, at drinking water facilities, the water supply could be contaminated, and tainted water could be distributed to users downstream. These two scenarios require different threat reduction strategies. The first scenario requires that facilities monitor for radioactive substances being brought on-site; the second requires that water assets be monitored for radioactive contamination. While the effects of radioactive contamination are basically the same under both threat types, each of these threats requires different types of radiation monitoring and different types of equipment.

Radiation Detection Equipment for Monitoring Water Assets

Most water systems are required to monitor for radioactivity and certain radionuclides, and to meet maximum contaminant levels for these contaminants, to comply with the Safe Drinking Water Act. Currently, USEPA requires drinking water to meet MCLs for beta/photon emitters (includes gamma radiation), alpha particles, combined radium 226/228, and uranium. However, this monitoring is required only at entry points into the system. In addition, after the initial sampling requirements, only one sample is required every 3–9 years, depending on the contaminant type and the initial concentrations.

While this is adequate to monitor for long-term protection from overall radioactivity and specific radionuclides in drinking water, it may not be adequate to identify short-term spikes in radioactivity, such as from spills, accidents, or intentional releases. In addition, compliance with the SDWA requires analyzing water samples in a laboratory, which results in a delay in receiving results. In contrast, security monitoring is more effective when results can be obtained quickly in the field. In addition, monitoring for security purposes does not necessarily require that the specific radionuclides causing the contamination be identified. Thus, for security purposes, it may be more appropriate to monitor for non-radionuclide-specific radiation using either portable field meters, which can be used as necessary to evaluate grab samples, or online systems, which can provide continuous monitoring of a system.

Ideally, measuring radioactivity in water assets in the field would involve minimal sampling and sample preparation. However, the physical properties of specific types of radiation combined with the physical properties of water make evaluating radioactivity in water assets in the field somewhat difficult. For example, alpha particles can only travel short distances, and they cannot penetrate through most physical objects. Therefore, instruments designed to evaluate alpha emissions must be specially designed to capture emissions at a short distance from the source, and they must not block alpha emissions from entering the detector. Gamma radiation does not have the same types of physical properties, and thus it can be measured using different detectors.

Measuring different types of radiation is further complicated by the relationship between the radiation's intrinsic properties and the medium in which the radiation is being measured. For example, gas-flow proportional counters are typically used to evaluate gross alpha and beta radiation from smooth, solid surfaces, but due to the fact that water is not a smooth surface and because alpha and beta emissions are relatively short range and can be attenuated within the water, these types of counters are not appropriate for measuring alpha and beta activity in water. An appropriate method for measuring alpha and beta radiation in water is by using a liquid scintillation counter. However, this requires mixing an aliquot of water with a liquid scintillation "cocktail." The liquid scintillation counter is a large, sensitive piece of equipment, so it is not appropriate for field use. Therefore, measurements for alpha and beta radiation from water assets are not typically made in the field.

Unlike the problems associated with measuring alpha and beta activity in water in the field, the properties of gamma radiation allow it to be measured relatively well in water samples in the field.

The standard instrumentation used to measure gamma radiation from water samples in the field is a sodium iodide (NaI) scintillator.

Although the devices outlined above are the most used for evaluating total alpha, beta, and gamma radiation, other methods and other devices can be used. In addition, local conditions (i.e., temperature, humidity) or the properties of the specific radionuclides emitting the radiation may make other types of devices or other methods more optimal to achieve the goals of the survey than the devices noted above. Experts or individual vendors should then be consulted to determine the appropriate measurement device for any specific application. An additional factor to consider when developing a program to monitor for radioactive contamination in water assets is whether to take regular grab samples or sample continuously.

For example, portable sensors can be used to analyze grab samples at any point in the system but have the disadvantage that they provide measurements only at one point in time. On the other hand, fixed-location sensors are usually used as part of a continuous, online monitoring system. These systems continuously monitor a water asset and could be outfitted with some type of alarm system that would alert operators if radiation increased above a certain threshold. However, the sampling points are fixed and only certain points in the system can be monitored. In addition, the number of monitoring locations needed to capture the physical and radioactive complexity of a system can be prohibitive.

Toxicity Monitoring/Toxicity Meters

Toxicity measurement devices measure general toxicity to biological organisms, and detection of toxicity in any water and wastewater asset can indicate a potential threat, either to the treatment process (in the case of influent toxicity), to human health (in the case of drinking water toxicity) or to the environment (in the case of effluent toxicity). Currently, whole effluent toxicity tests (WET tests), in which effluent samples are tested against test organisms, are required of many National Pollutant Discharge Elimination System (NPDES) discharge permits. The WET tests are used as a complement to the effluent limits on physical and chemical parameters to assess the overall effects of the discharge on living organisms or aquatic biota. Toxicity tests may also be used to monitor wastewater influent streams for potential hazardous contamination, such as organic heavy metals (arsenic, mercury, lead, chromium, and copper) that might upset the treatment process. The ability to get feedback on sample toxicity from short-term toxicity tests or toxicity "meters" can be valuable in estimating the overall toxicity of a sample. Online real-time toxicity monitoring is still under active research and development. However, there are several portable toxicity measurement devices commercially available. They can generally be divided into two categories based on the different ways they measure toxicity:

* Meters measuring direct biological activity (e.g., luminescent bacteria) and correlating decreases in this direct biological activity with increased toxicity; and
* Meters measuring oxygen consumption and correlating decrease in oxygen consumption with increased toxicity.

COMMUNICATION AND INTEGRATION

This section discusses those devices necessary for communication and integration of water and wastewater system operations, such as electronic controllers, two-way radios, and wireless data communications. Electronic controllers are used to automatically activate equipment (such as lights, surveillance cameras, audible alarms, or locks) when they are triggered. Triggering could be in response to a variety of scenarios, including tripping of an alarm or a motion sensor; breaking of a window or a glass door; variation in vibration sensor readings; or simply through input from a timer. Two-way wireless radios allow two or more users that have their radios tuned to the same frequency to communicate instantaneously with each other without the radios being physically lined together with wires or cables. Wireless data communications devices are used to enable transmission of data between computer systems and/or between a SCADA server and its sensing devices, without individual components being physically linked together via wires or cables. In water and wastewater systems, these devices are often used to link remote monitoring stations (i.e., SCADA components) or portable computers (i.e., laptops) to computer networks without using physical wiring connections.

Electronic Controllers

An electronic controller is a piece of electronic equipment that receives incoming electric signals and uses preprogrammed logic to generate electronic output signals based on the incoming signals. While electronic controllers can be implemented for any application that involves inputs and outputs (e.g., control of a piece of machinery in a factory) in a security application, these controllers essentially act as the system's "brain," and can respond to specific security-related inputs with preprogrammed output response. These systems combine the control of electronic circuitry with a logic function such that circuits are opened and closed (and thus equipment is turned on and off) through some preprogrammed logic. The basic principle behind the operation of an electrical controller is that it receives electronic inputs from sensors or any device generating an electrical signal (e.g., electrical signals from motion sensors), and then uses its preprogrammed logic to produce electrical outputs (e.g., these outputs could turn on power to a surveillance camera or to an audible alarm). Thus, these systems automatically generate a preprogrammed, logical response to a preprogrammed input scenario.

The three major types of electronic controllers are timers, electromechanical relays, and programmable logic controllers (PLCs), which are often called "digital relays." Each of these types of controller is discussed in more detail below. Timers use internal signal/inputs (in contrast to externally – generated inputs) and generate electronic output signals at certain times. More specifically, timers control electric current flow to any application to which they are connected and can turn the current on or off on a schedule pre-specified by the user. Typical timer range (amount of time that can be programmed to elapse before the timer activates linked equipment) is from 0.2 seconds to 10 hours, although some of the more advanced timers have ranges of up to 60 hours. Timers are useful in fixed applications that don't require frequent schedule changes. For example, a timer can be used to turn on the lights in a room or building at a certain time every day. Timers are usually connected to their own power supply (usually 120–240 V).

In contrast to timers, which have internal triggers based on a regular schedule, electromechanical relays and PLCs have both external inputs and external outputs. However, PLCs are more flexible and more powerful than are electromechanical relays, and thus this section focuses primarily on PLCs as the predominant technology for security-related electronic control applications. Electromechanical relays are simple devices that use a magnetic field to control a switch. Voltage applied to the relay's input coil creates a magnetic field, which attracts an internal metal switch. This causes the relay's contacts to touch, closing the switch and completing the electrical circuit. This activates any linked equipment. These types of systems are often used for high voltage applications, such as in some automotive and other manufacturing processes.

Two-Way Radios

Two-way radios, as discussed here, are limited to a direct unit-to-unit radio communication, either via single unit-to-unit transmission and reception or via multiple handheld units to a base station radio contact and distribution system. Radio frequency spectrum limitations apply to all handheld units as directed by the FCC. This also distinguishes a handheld unit from a base station or base station unit (such as those used by an amateur ham radio operator), which operate under different wavelength parameters.

Two-way radios allow a user to contact another user or group of users instantly on the same frequency and to transmit voice or data without the need for wires. They use "half-duplex" communications; or communication that can be only transmitted or received; it cannot transmit and receive simultaneously. In other words, only one person may talk, while other personnel with radio(s) can only listen. To talk, the user depresses the talk button and speaks into the radio. The audio then transmits the voice wirelessly to the receiving radios. When the speaker has finished speaking and the channel has cleared, users on any of the receiving radios can transmit; either to answer the first transmission or to begin a new conversation. In addition to carrying voice data, many types of wireless radios also allow the transmission of digital data, and these radios may be interfaced with computer networks that can use or track these data. For example, some two-way radios can send information such as global positioning system (GPS) data or the ID of the radio. Some two-way radios can also send data through a SCADA system.

Wireless radios broadcast these voice or data communications over the airwaves from the transmitter to the receiver. While this can be an advantage in that the signal emanates in all directions and does not need a direct physical connection to be received at the receiver, it can also make the communications vulnerable to being blocked, intercepted, or otherwise altered. However, security features are available to ensure that the communications are not tampered with.

Wireless Data Communications

A wireless data communication system consists of two components: a "Wireless Access Point" (WAP), and a "Wireless Network Interface Card" (sometimes also referred to as a "Client"), which work together to complete the communications link. These wireless systems can link electronic devices, computers, and computer systems together using radio waves, thus eliminating the need for these individual components to be directly connected through physical wires. While wireless data communications have widespread applications in water and wastewater systems, they also have limitations. First, wireless data connections are limited by the distance between components (radio waves scatter over a long distance and cannot be received efficiently, unless special directional antennae are used). Second, these devices only function if the individual components are in a direct line of sight with each other since radio waves are affected by interference from physical obstructions. However, in some cases, repeater units can be used to amplify and retransmit wireless signals to circumvent these problems. The two components of wireless devices are discussed in more detail next.

The wireless access point provides the wireless data communication service. It usually consists of a housing (which is constructed from plastic or metal depending on the environment it will be used in) containing a circuit board; flash memory that holds software; one of two external ports to connect to existing wired networks; a wireless radio transmitter/receiver; and one or more antenna connections. Typically, the WAP requires a one-time user configuration to allow the device to interact with the local area network (LAN). This configuration is usually done via a web-driven software application which is accessed via a computer.

Wireless network interface card or client is a piece of hardware that is plugged into a computer and enables that computer to make a wireless network connection. The card consists of a transmitter, functional circuitry, and a receiver for the wireless signal, all of which work together to enable communication between the computer, its wireless transmitter/receiver, and its antenna connection. Wireless cards are installed in a computer through a variety of connections, including USB Adapters, or Laptop CardBus (PCMCIA), or Desktop Peripheral (PCI) cards. As with the WAP, software is loaded onto the user's computer, allowing configuration of the card so that it may operate over the wireless network

Two of the primary applications for wireless data communications systems are to enable mobile or remote connections to a LAN and to establish wireless communications links between SCADA remote telemetry units (RTUs) and sensors in the field. Wireless car connections are usually used for LAN access from mobile computers. Wireless cards can also be incorporated into RTUs to allow them to communicate with sensing devices that are located remotely.

CYBER PROTECTION DEVICES

Various cyber protection devices are currently available for use in protecting system computer systems. These protection devices include anti-virus and pest eradication software, firewalls, and network intrusion hardware/software. These products are discussed in this section.

Anti-Virus and Pest Eradication Software

Anti-virus programs are designed to detect, delay, and respond to programs or pieces of code that are specifically designed to harm computers. These programs are known as "malware.". Malware can include computer viruses, worms, and Trojan horse programs (programs that appear to be benign but which have hidden harmful effects). Pest eradication tools are designed to detect, delay, and respond to "spyware" (strategies that websites use to track user behavior, such as by sending "cookies" to the user's computer) and hacker tools that track keystrokes (keystroke loggers) or passwords (password crackers).

Viruses and pests can enter a computer system through the Internet or through infected floppy discs or CDs. They can also be placed onto a system by insiders. Some of these programs, such as viruses and worms, then move within a computer's drives and files, o between computers if the computers are networked to each other. This malware can deliberately damage files, utilize memory and network capacity, crash application programs, and initiate transmissions of sensitive information from a PC. While the specific mechanisms of these programs differ, they can infect files and even the basic operating program of the computer firmware/hardware.

The most important features of an anti-virus program are its abilities to identify potential malware and to alert a user before infection occurs, as well as its ability to respond to a virus already resident on a system. Most of these programs provide a log so that the user can see what viruses have been detected and where they were detected. After detecting a virus, the anti-virus software may delete the virus automatically, or it may prompt the user to delete the virus. Some programs will also fix files or programs damaged by the virus. Various sources of information are available to inform the general public and computer system operators about new viruses being detected. Since anti-virus programs use signatures (or snippets of code or data) to detect the presence of a virus, periodic updates are required to identify new threats. Many anti-virus software providers offer free upgrades that are able to detect and respond to the latest viruses.

Firewalls

A firewall is an electronic barrier designed to keep computer hackers, intruders, or insiders from accessing specific data files and information on a system's computer network or other electronic/computer systems. Firewalls operate by evaluating and then filtering information coming through a public network (such as the Internet) into the system's computer or other electronic systems. This evaluation can include identifying the source or destination addresses and ports and allowing or denying access based on this identification. Two methods are used by firewalls to limit access to the system's computers or other electronic systems from the public network:

- The firewall may deny all traffic unless it meets certain criteria.
- The firewall may allow all traffic through unless it meets certain criteria.

A simple example of the first method is to screen requests to ensure that they come from an acceptable (i.e., previously identified) domain name and Internet protocol address. Firewalls may also use more complex rules that analyze the application data to determine if the traffic should be allowed through. For example, the firewall may require user authentication (i.e., use of a password) to access the system. How a firewall determines what traffic to let through depends on which network layer it operates at and how it is configured. Firewalls may be a piece of hardware, a software program, or an appliance card that contains both. Advanced features that can be incorporated into firewalls allow for the tracking of attempts to log on to the local area network system. For example, a report of successful and unsuccessful log-in attempts may be generated for the computer specialist to analyze. For systems with mobile users, firewalls allow remote access into the private network using secure log-on procedures and authentication certificates.

Most firewalls have a graphical user interface for managing the firewall. In addition, new Ethernet firewall cards that fit in the slot of an individual computer bundle provide additional layers of defense (like encryption and permit/deny) for individual computer transmissions to the network interface function. The cost of these new cards is only slightly higher than traditional network interface cards.

Network Intrusion Hardware and Software

Network intrusion detection and prevention systems are software- and hardware-based programs designed to detect unauthorized attacks on a computer network system. Whereas other applications such as firewalls and anti-virus software share similar objectives with network intrusion systems, network intrusion systems provide a deeper layer of protection beyond the capabilities of these other systems because they evaluate patterns of computer activity rather than specific files.

It is worth noting that attacks may come from either outside or within the system (i.e., from an insider) and that network intrusion detection systems may be more applicable for detecting patterns of suspicious activity from inside a facility (i.e., accessing sensitive data, etc.) than are other information technology solutions. Network intrusion detection systems employ a variety of mechanisms to evaluate potential threats.

The types of search and detection mechanisms are dependent upon the level of sophistication of the system. Some of the available detection methods include:

- *Protocol analysis* – Protocol analysis is the process of capturing, decoding, and interpreting electronic traffic. The protocol analysis method of network intrusion detection involves the analysis of data captured during transactions between two or more systems or devices and the evaluation of these data to identify unusual activity and potential problems. Once a problem is isolated and recorded, problems or potential threats can be linked to pieces of hardware or software. Sophisticated protocol analysis will also provide statistics and trend information on the captured traffic.
- *Traffic anomaly detection* – Traffic anomaly detection identifies potential threatening activity by comparing incoming traffic to "normal" traffic patterns and identifying deviations. It does this by comparing user characteristics against thresholds and triggers defined by the network administrator. This method is designed to detect attacks that span several connections, rather than a single session.
- *Network honeypot* – This method establishes nonexistent services in order to identify potential hackers. A network honeypot impersonates services that don't exist by sending fake information to people scanning the network. It identifies the attacker when they attempt to connect to the service. There is no reason for legitimate traffic to access these resources because they don't exist; therefore, any attempt to access them constitutes an attack.
- *Anti-intrusion detection system evasion techniques* – These methods are designed for attackers who may be trying to evade intrusion detection system scanning. They include methods called IP defragmentation, TCP stream reassembly, and deobfuscation.

These detection systems are automated, but they can only indicate patterns of activity, and a computer administrator or other experienced individual must interpret activities to determine whether they are potentially harmful. Monitoring the logs generated by these systems can be time consuming, and there may be a learning curve to determine a baseline of "normal" traffic patterns from which to distinguish potential suspicious activity.

SCADA

In Queensland, Australia, on April 23, 2000, police stopped a car on the road and found a stolen computer and radio inside. Using commercially available technology, a disgruntled former employee had turned his vehicle into a pirate command center of sewage treatment along Australia's Sunshine Coast. The former employee's arrest solved a mystery that had troubled the Maroochy Shire wastewater system for two months. Somehow the system was leaking hundreds of thousands of gallons of putrid sewage into parks, rivers, and the manicured grounds of a Hyatt Regency hotel – marine life died, the creek water turned black, and the stench was unbearable for residents. Until the former employee's capture – during his 46th successful intrusion – the system's managers did not know why.

Specialists study this case of cyber-terrorism because it is the only one known in which someone used a digital control system deliberately to cause harm. The former employee's intrusion shows how easy it is to break in – and how restrained he was with his power. To sabotage the system, the former employee set the software on his laptop to identify itself as a pumping station and then suppressed all alarms. The former employee was the "central control station" during his intrusions, with unlimited command of 300 SCADA nodes governing sewage and drinking water alike. Gellman (2002). The bottom line: As serious as the former employee's intrusions were, they pale in comparison with what he could have done to the freshwater system – he could have done anything he liked. In 2000, the Federal Bureau of Investigation (FBI) identified and listed threats to critical infrastructure. These threats are listed and described in Table 13.1. In the past few years, especially

TABLE 13.1
Threats to Critical Infrastructure Observed by the FBI

Threat	Description
Criminal group	There is an increased use of cyber intrusions by criminal groups who attack systems for purposes of monetary gain.
Foreign intelligence services	Foreign intelligence services use cyber tools as part of their information gathering and espionage activities.
Hackers	Hackers sometimes crack into networks for the thrill of the challenge or for bragging rights in the hacker community. While remote cracking once required a fair amount of skill or computer knowledge, hackers can now download attack scripts and protocols from the Internet and launch them against victim sites. Thus, while attack tools have become more sophisticated, they have also become easier to use.
Hacktivists	Hacktivism refers to politically motivated attacks on publicly accessible web pages or e-mail servers. These groups and individuals overload e-mail servers and hack into websites to use.
Information warfare	Several nations are aggressively working to develop information warfare doctrine, programs, and capabilities. Such capabilities enable a single entity to have a significant and serious impact by disrupting the supply, communications, and economic infrastructures that support military power-impacts that, according to the Director of Central Intelligence, can affect the daily lives of Americans across the country.
Insider threat	The disgruntled organization insider is a principal source of computer crimes. Insiders may not need a great deal of knowledge about computer intrusions because their knowledge of a victim system often allows them to gain unrestricted access to cause damage to the system or to steal system data. The insider threat also includes outsourcing vendors.
Virus writers	Virus writers are posing an increasingly serious threat. Several destructive computer viruses and "worms" have harmed files and hard drives, including the Melissa Macro Virus, ExploreZip worm, CIH (Chernobyl) Virus, Nimda, and Code Red.

Source: FBI, 2000.

since 9/11, it has been somewhat routine for us to pick up a newspaper, magazine, or view a television news program where a major topic of discussion is cybersecurity or the lack thereof. Many of the cyber intrusion incidents we read or hear about have added new terms or new uses for old terms to our vocabulary. For example, old terms such as Trojan horse, worms, and viruses have taken on new connotations in regards to cybersecurity issues. Relatively new terms such as scanners, Windows NT hacking tools, ICQ hacking tools, mail bombs, sniffer, logic bomb, nukers, dots, backdoor Trojan, key loggers, hackers' Swiss knife, password crackers, and BIOS crackers are now commonly encountered.

Not all relatively new and universally recognizable cyber terms have sinister connotations or meanings, of course. Consider, for example, the following digital terms: backup, binary, bit byte, CD-ROM, CPU, database, e-mail, HTML, icon, memory, cyberspace, modem, monitor, network, RAM, Wi-Fi (wireless fidelity), record, software, World Wide Web – none of these terms normally generate thoughts of terrorism in most of us. There is, however, one digital term, SCADA, that most people have not heard of. This is not the case, however, with those who work with the nation's critical infrastructure, including water and wastewater. SCADA, or **S**upervisory Control and Data Acquisition System (also sometimes referred to as Digital Control Systems or Process Control Systems), plays an important role in computer-based control systems. Many water and wastewater systems use computer-based systems to remotely control sensitive processes and system equipment previously controlled manually. These systems (commonly known as SCADA) allow a water and wastewater system to collect data from sensors and control equipment located at remote sites. Common water and wastewater system sensors measure elements such as fluid level, temperature, pressure, water purity, water clarity, and pipeline flow rates. Common water and wastewater system equipment includes valves, pumps, and mixers for mixing chemicals in the water supply.

What Is SCADA?

Simply, SCADA is a computer-based system that remotely controls processes previously controlled manually. SCADA allows an operator using a central computer to supervise (control and monitor) multiple networked computers at remote locations. Each remote computer can control mechanical processes (pumps, valves, etc.) and collect data from sensors at its remote location. Thus the phrase: Supervisory Control and Data Acquisition, of SCADA The central computer is called the Master Terminal Unit, or MTU. The operator interfaces with the MTU using software called Human Machine Interface, or HMI. The remote computer is called Program Logic Controller (PLC) or Remote Terminal Unit. The RTU activates a relay (or switch) that turns mechanical equipment "on" and "off." The RTU also collects data from sensors.

Initially, utilities ran wires, also known as hardwires or landlines, from the central computer (MTU) to the remote computers (RTUs). Because remote locations can be located hundreds of miles from the central location, utilities have begun to use public phone lines and modems, and leased telephone company lines, and radio and microwave communication. More recently, they have also begun to use satellite links, Internet, and newly developed wireless technologies. Because the SCADA systems' sensors provided valuable information, many utilities established "connections" between their SCADA systems and their business system. This allowed System management and other staff access to valuable statistics, such as water usage. When utilities later connected their systems to the Internet, they were able to provide stakeholders with water and wastewater statistics on the System web pages.

SCADA Applications in Water and Wastewater Systems

As stated above, SCADA systems can be designed to measure a variety of equipment operating conditions and parameters or volumes and flow rates or water quality parameters and to respond to change in those parameters either by alerting operators or by modifying system operation through a feedback loop system without having personnel physically visit each process or piece of equipment on a daily basis to check it and/or ensure that it is functioning properly. SCADA systems can also be used to automate certain functions so that they can be performed without needing to be initiated by an operator (e.g., injecting chlorine in response to periodic low chlorine levels in a distribution system or turning on a pump in response to low water levels in a storage tank).

As described above, in addition to process equipment, SCADA systems can also integrate specific security alarms and equipment, such as cameras, motion sensors, lights, data from card reading systems, etc. thereby providing a clear picture of what is happening at areas throughout a facility. Finally, SCADA systems also provide constant, real-time data on processes, equipment, location access, etc., which allows for the necessary response to be made quickly. This can be extremely useful during emergency conditions, such as when distribution mains break or when potentially disruptive BOD spikes appear in wastewater influent. Because these systems can monitor multiple processes, equipment, and infrastructure and then provide quick notification of, or response to, problems or upsets, SCADA systems typically provide the first line of detection for atypical or abnormal conditions. For example, a SCADA system connected to sensors that measure specific water quality parameters are measured outside of a specific range. A real-time customized operator interface screen could display and control critical systems monitoring parameters. The system could transmit warning signals back to the operators, such as by initiating a call to a personal pager. This might allow the operators to initiate actions to prevent contamination and disruption of the water supply. Further automation of the system could ensure that the system initiated measures to rectify the problem. Preprogrammed control functions (e.g., shutting a valve, controlling flow, increasing chlorination, or adding other chemicals) can be triggered and operated based on SCADA system.

SCADA VULNERABILITIES

According to USEPA (2005), SCADA networks were developed with little attention paid to security, making the security of these systems often weak. Studies have found that, while technological advancements introduced vulnerabilities, many water and wastewater systems have spent little time securing their SCADA networks. As a result, many SCADA networks may be susceptible to attacks and misuse. Remote monitoring and supervisory control of processes began to develop in the early 1960s and adopted many technological advancements. The advent of minicomputers made it possible to automate a vast number of once manually operated switches. Advancements in radio technology reduced the communication costs associated with installing and maintaining buried cable in remote areas. SCADA systems continued to adopt new communication methods including satellite and cellular. As the price of computers and communications dropped, it became economically feasible to distribute operations and to expand SCADA networks to include even smaller facilities. Advances in information technology and the necessity of improved efficiency have resulted in increasingly automated and interlinked infrastructures and created new vulnerabilities due to equipment failure, human error, weather and other natural causes, and physical and cyberattacks. Some areas and examples of possible SCADA vulnerabilities include:

- *Human* – People can be tricked or corrupted, and may commit errors.
- *Communications* – Messages can be fabricated, intercepted, changed, deleted, or blocked.
- *Hardware* – Security features are not easily adapted to small self-contained units with limited power supplies.
- *Physical* – Intruders can break into a facility to steal or damage SCADA equipment.
- *Natural* – Tornadoes, floods, earthquakes, and other natural disasters can damage equipment and connections.
- *Software* – Programs can be poorly written.

A survey found that many public water systems were doing little to secure their SCADA network vulnerabilities (Ezell, 1998); for example, many respondents reported that they had remote access, which can allow an unauthorized person to access the system without being physically present. More than 60% of the respondents believed that their systems were not safe from unauthorized access and use. Twenty percent of the respondents even reported known attempts, successful unauthorized access, or use of their system. Yet 22 of 43 respondents reported that they do not spend any time ensuring their network is safe, and 18 of 43 respondents reported that they spend less than 10% ensuring network safety.

SCADA system computers and their connections are susceptible to different types of information system attacks and misuse such as system penetration and unauthorized access to information. The Computer Security Institute and Federal Bureau of Investigation conduct an annual Computer Crime and Security Survey (FBI, 2004). The survey reported on ten types of attacks or misuse and reported that virus and denial of service had the greatest negative economic impact. The same study also found that 15% of the respondents reported abuse of wireless networks, which can be a SCADA component. On average, respondents from all sectors did not believe that their organization invested enough in security awareness. Utilities as a group reported a lower average computer security expenditure/investment per employee than many other sectors such as transportation, telecommunications, and financial.

Sandia National Laboratories' *Common Vulnerabilities in Critical Infrastructure Control Systems* described some of the common problems it has identified in the following five categories (Stamp et al., 2003):

- *System Data* – Important data attributes for security include availability, authenticity, integrity, and confidentiality. Data should be categorized according to its sensitivity, and ownership and responsibility must be assigned. However, SCADA data is often not classified at all, making it difficult to identify where security precautions are appropriate.
- *Security Administration* – Vulnerabilities emerge because many systems lack a properly structured security policy, equipment and system implementation guides, configuration management, training, and enforcement and compliance auditing.
- *Architecture* – Many common practices negatively affect SCADA security. For example, while it is convenient to use SCADA capabilities for other purposes such as fire and security systems, these practices create single points of failure. Also, the connection of SCADA networks to other automation systems and business networks introduces multiple entry points for potential adversaries.
- *Network* (including communication links) – Legacy systems' hardware and software have very limited security capabilities, and the vulnerabilities of contemporary systems (based on modern information technology) are publicized. Wireless and shared links are susceptible to eavesdropping and data manipulation.
- *Platforms* – Many platform vulnerabilities exist, including default configurations retained, poor password practices, shared accounts, inadequate protection for hardware, and non-existent security monitoring controls. In most cases, important security patches are not installed, often due to concern about negatively impacting system operation; in some cases technicians are contractually forbidden from updating systems by their vendor agreements.

The following incident helps to illustrate some of the risks associated with SCADA vulnerabilities.

During the course of conducting a vulnerability assessment, a contractor stated that personnel from his company penetrated the information system of a system within minutes. Contractor personnel drove to a remote substation and noticed a wireless network antenna. Without leaving their vehicle, they plugged in their wireless radios and connected to the network within 5 minutes. Within 20 minutes they had mapped the network, including SCADA equipment, and accessed the business network and data.

This illustrates what a cybersecurity advisor from Sandia National Laboratories specialized in SCADA stated – that utilities are moving to wireless communication without understanding the added risks.

THE INCREASING RISK

According to GAO (2003), historically, security concerns about control systems (SCADA included) were related primarily to protecting against physical attack and misuse of refining and processing sites or distribution and holding facilities. However, more recently there has been a growing recognition that control systems are now vulnerable to cyberattacks from numerous sources, including hostile governments, terrorist groups, disgruntled employees, and other malicious intruders. In addition to control system vulnerabilities mentioned earlier, several factors have contributed to the escalation of risk to control systems, including:

- The adoption of standardized technologies with known vulnerabilities;
- The connectivity of control systems to other networks;
- Constraints on the implementation of existing security technologies and practices;
- Insecure remote connections; and
- The widespread availability of technical information about control systems.

ADOPTION OF TECHNOLOGIES WITH KNOWN VULNERABILITIES

When a technology is not well known, not widely used, not understood, or publicized, it is difficult to penetrate it and thus disable it. Historically, proprietary hardware, software, and network protocols made it difficult to understand how control systems operated – and therefore how to hack into them. Today, however, to reduce costs and improve performance, organizations have been transitioning from proprietary systems to less expensive, standardized technologies such as Microsoft's Windows and Unix-like operating systems and the common networking protocols used by the Internet. These widely used standardized technologies have commonly known vulnerabilities, and sophisticated and effective exploitation tools are widely available and relatively easy to use. As a consequence, both the number of people with the knowledge to wage attacks and the number of systems subject to attack have increased. Also, common communication protocols and the emerging use of Extensible Markup Language (commonly referred to as XML) can make it easier for a hacker to interpret the content of communications among the components of a control system.

Control systems are often connected to other networks – enterprises often integrate their control system with their enterprise networks. This increased connectivity has significant advantages, including providing decision makers with access to real-time information and allowing engineers to monitor and control the process control system from different points on the enterprise network. In addition, the enterprise networks are often connected to the networks of strategic partners and to the Internet. Further, control systems are increasingly using wide area networks and the Internet to transmit data to their remote or local stations and individual devices. This convergence of control networks with public and enterprise networks potentially exposes the control systems to additional security vulnerabilities. Unless appropriate security controls are deployed in the enterprise network and the control system network, breaches in enterprise security can affect the operation of control systems. According to industry experts, the use of existing security technologies, as well as strong user authentication and patch management practices, are generally not implemented in control systems because control systems operate in real time, typically are not designed with cybersecurity in mind, and usually have limited processing capabilities.

Existing security technologies such as authorization, authentication, encryption, intrusion detection, and filtering of network traffic and communications require more bandwidth, processing power, and memory than control system components typically have. Because controller stations are generally designed to do specific tasks, they use low-cost, resource-constrained microprocessors. In fact, some devices in the electrical industry still use the Intel 8088 processor, introduced in 1978. Consequently, it is difficult to install existing security technologies without seriously degrading the performance of the control system. Further, complex passwords and other strong password practices are not always used to prevent unauthorized access to control systems, in part because this could hinder a rapid response to safety procedures during an emergency. As a result, according to experts weak passwords that are easy to guess, shared, and infrequently changed are reportedly common in control systems, including the use of default passwords or even no password at all.

In addition, although modern control systems are based on standard operating systems, they are typically customized to support control system applications. Consequently, vendor-provided software patches are generally either incompatible or cannot be implemented without compromising service shutting down "always-on" systems or affecting interdependent operations.

Potential vulnerabilities in control systems are exacerbated by insecure connections. Organizations often leave access links – such as dial-up modems to equipment and control information – open for remote diagnostics, maintenance, and examination of system status. Such links may not be protected with authentication of encryption, which increases the risk that hackers could use these insecure connections to break into remotely controlled systems. Also, control systems often use wireless communications systems, which are especially vulnerable to attack, or leased lines that pass through commercial telecommunications facilities. Without encryption to protect data as it flows through these insecure connections or authentication mechanisms to limit access, there is limited protection for the integrity of the information being transmitted.

Public information about infrastructures and control systems is available to potential hackers and intruders. The availability of this infrastructure and vulnerability data was demonstrated by a university graduate student, whose dissertation reportedly mapped every business and industrial sector in the American economy to the fiber-optic network that connects them – using material that was available publicly on the Internet, none of which was classified. Many of the electric system officials who were interviewed for the National Security Telecommunications Advisory Committee's Information Assurance Task Force's Electric Power Risk Assessment expressed concern over the amount of information about their infrastructure that is readily available to the public.

In the electric power industry, open sources of information – such as product data and educational videotapes from engineering associations can be used to understand the basics of the electrical grid. Other publicly available information – including filings of the Federal Energy Regulatory Commission (FERC), industry publications, maps, and material available on the Internet – is sufficient to allow someone to identify the most heavily loaded transmission lines and the most critical substations in the power grid.

In addition, significant information on control systems is publicly available – including design and maintenance documents, technical standards for the interconnection of control systems and RTUs, and standards for communication – all of which could assist hackers in understanding the systems and how to attack them. Moreover, there are numerous former employees, vendor, support contractors, and other end users of the same equipment worldwide with inside knowledge of the operation of control systems.

CYBER THREATS TO CONTROL SYSTEMS

There is a general consensus – and increasing concern – among government officials and experts on control systems about potential cyber threats to the control systems that govern our critical infrastructures. As components of control systems increasingly make critical decisions that were once made by humans, the potential effect of a cyber threat becomes more devastating. Such cyber threats could come from numerous sources, ranging from hostile governments and terrorist groups to disgruntled employees and other malicious intruders. Based on interviews and discussions with representatives throughout the electric power industry, the Information Assurance Task Force of the

National Security Telecommunications Advisory Committee concluded that an organization with sufficient resources, such as a foreign intelligence service or a well-supported terrorist group, could conduct a structured attack on the electric power grid electronically, with a high degree of anonymity and without having to set foot in the target nation.

In July 2002, National Infrastructure Protection Center (NIPC) reported that the potential for compound cyber and physical attacks, referred to as "swarming attacks," is an emerging threat to the US critical infrastructure. As NIPC reports, the effects of a swarming attack include slowing or complicating the response to a physical attack. For instance, a cyberattack that disabled the water supply or the electrical system in conjunction with a physical attack could deny emergency services the necessary resources to manage the consequences – such as controlling fires, coordinating actions, and generating light. Control systems, such as SCADA, can be vulnerable to cyberattacks. Entities or individuals with malicious intent might take one or more of the following actions to successfully attack control systems:

- Disrupt the operation of control systems by delaying or blocking the flow of information through control networks, thereby denying availability of the networks to control system operations;
- Make unauthorized changes to programmed instructions in PLCs, RTUs, or DCS controllers, change alarm thresholds, or issue unauthorized commands to control equipment, which could potentially result in damage to equipment (if tolerances are exceeded), premature shutdown of processes (such as prematurely shutting down transmission lines), or even disabling of control equipment;
- Send false information to control system operators either to disguise unauthorized changes or to initiate inappropriate actions by system operators;
- Modify the control system software, producing unpredictable results; and
- Interfere with the operation of safety systems.

In addition, in control systems that cover a wide geographic area, the remote sites are often unstaffed and may not be physically monitored. If such remote systems are physically breached, the attackers could establish a cyber connection to the control network.

SECURING CONTROL SYSTEMS

Several challenges must be addressed to effectively secure control systems against cyber threats. These challenges include:

- The limitations of current security technologies in securing control systems;
- The perception that securing control systems may not be economically justifiable; and
- The conflicting priorities within organizations regarding the security of control systems.

A significant challenge in effectively securing control systems is the lack of specialized security technologies for these systems. The computing resources in control systems that are needed to perform security functions tend to be quite limited, making it very difficult to use security technologies within control system networks without severely hindering performance. Securing control systems may not be perceived as economically justifiable.

Experts and industry representatives have indicated that organizations may be reluctant to spend more money to secure control systems. Hardening the security of control systems would require industries to expend more resources, including acquiring more personnel, providing training for personnel, and potentially prematurely replacing current systems that typically have a lifespan of about 20 years. Finally, several experts and industry representatives indicated that the responsibility for securing control systems typically includes two separate groups: IT security personnel and control system engineers and operators. IT security personnel tend to focus on securing enterprise systems, while control system engineers and operators tend to be more concerned with the reliable performance of their control systems. Further, they indicate that, as a result, those two groups do not always fully understand each other's requirements and collaborate to implement secure control systems.

STEPS TO IMPROVE SCADA SECURITY

The President's Critical Infrastructure Protection Board and the Department of Energy (DOE) have developed the steps outlined below to help organizations improve the security of their SCADA networks. DOE (2001) points out that these steps are not meant to be prescriptive or all-inclusive. However, they do address essential actions to be taken to improve the protection of SCADA networks. The steps are divided into two categories: specific actions to improve implementation, and actions to establish essential underlying management processes and policies (DOE, 2001).

21 Steps to Increase SCADA Security

The following steps focus on specific actions to be taken to increase the security of SCADA networks:

1. Identify all connections to SCADA networks. Conduct a thorough risk analysis to assess the risk and necessity of each connection to the SCADA network.
   - Develop a comprehensive understanding of all connections to the SCADA network and how well those connections are protected. Identify and evaluate the following types of connections:
   - Internal local area and wide area networks, including business networks
   - The Internet
   - Wireless network devices, including satellite uplinks
   - Modem or dial-up connections
   - Connections to business partners, vendors, or regulatory agencies
2. Disconnect unnecessary connections to the SCADA network.
3. To ensure the highest degree of security of SCADA systems, isolate the SCADA network from other network connections to as great a degree as possible.
   - Any connection to another network introduces security risks, particularly if the connection creates a pathway from or to the Internet. Although direct connections with other networks may allow important information to be passed efficiently and conveniently, insecure connections are simply not worth the risk; isolation of the SCADA network must be a primary goal to provide needed protection. Strategies such as utilization of "demilitarized zones" (DMZs) and data warehousing can facilitate the secure transfer of data from the SCADA network to business networks. However, they must be designed and implemented properly to avoid introduction of additional risk through improper configuration.
4. Evaluate and strengthen the security of any remaining connections to the SCADA networks.
   - Conduct penetration testing or vulnerability analysis of any remaining connections to the SCADA network to evaluate the protection posture associated with these pathways. Use this information in conjunction with risk management processes to develop a robust protection strategy for any pathways to the SCADA network. Since the SCADA network is only as secure as its weakest connecting point, it is essential to implement firewalls, intrusion detection systems (IDSs), and other appropriate security measures at each point of entry. Configure firewall rules to prohibit access from and to the SCADA network, and be as specific as possible when permitting approved connections. For example, an Independent System Operator (ISO) should not be granted

"blanket" network access simply because there is a need for a connection to certain components of the SCADA system. Strategically place IDSs at each entry point to alert security personnel of potential breaches of network security. Organization management must understand and accept responsibility or risks associated with any connection to the SCADA network.

5. Harden SCADA networks by removing or disabling unnecessary services.

- SCADA control servers built on commercial or open-source operating systems can be exposed to attack default network services. To the greatest degree possible, remove or disable unused services and network demons to reduce the risk of direct attack. This is particularly important when SCADA networks are interconnected with other networks. Do not permit a service or feature on a SCADA network unless a thorough risk assessment of the consequences of allowing the service/feature shows that the benefits of the service/feature far outweigh the potential for vulnerability exploitation. Examples of services to remove from SCADA networks include automated meter reading/remote billing systems, e-mail services, and Internet access. An example of a feature to disable is remote maintenance. Numerous secure configurations exist, such as the National Security Agency's series of security guides. Additionally, work closely with SCADA vendors to identify secure configurations and coordinate any and all changes to operational systems to ensure that removing or disabling services does not cause downtime, interruption of service, or loss of support.

6. Do not rely on proprietary protocols to protect your system.

- Some SCADA systems are unique, proprietary protocols for communications between field devices and servers. Often the security of SCADA systems is based solely on the secrecy of these protocols. Unfortunately, obscure protocols provide very little "real" security. Do not rely on proprietary protocols or factor default configuration setting to protect your system. Additionally, demand that vendors disclose any backdoor or vendor interfaces to your SCADA systems, and expect them to provide systems that are capable of being secure.

7. Implement the security features provided by device and system vendors.

- Older SCADA systems (most systems in use) have no security features whatsoever. SCADA system owners must insist that their system vendor implement security features in the form of product patches or upgrades. Some newer SCADA devices are shipped with basic security features, but these are usually disabled to ensure ease of installation. Analyze each SCADA device to determine whether security features are present. Additionally, factory default security settings (such as in computer network firewalls) are often set to provide maximum usability, but minimal security. Set all security features to provide the maximum security only after a thorough risk assessment of the consequences of reducing the security level. Establish strong controls over any medium that is used as a backdoor into the SCADA network. Where backdoors or vendor connections do exist in SCADA systems, strong authentication must be implemented to ensure secure communications. Modems, wireless, and wired networks used for communications and maintenance represent a significant vulnerability to the SCADA network and remote sites. Successful "war dialing" or "war driving" attacks could allow an attacker to bypass all of the other controls and have direct access to the SCADA network or resources. To minimize the risk of such attacks, disable inbound access and replace it with some type of callback system.

8. Implement internal and external intrusion detection systems and establish 24-hour-a-day incident monitoring.

- To be able to effectively respond to cyberattacks, establish an intrusion detection strategy that includes alerting network administrators of malicious network activity originating from internal or external sources. Intrusion detection system monitoring is essential 24 hours a day; this capability can be easily set up through a pager.

Additionally, incident response procedures must be in place to allow an effective response to any attack. To complement network monitoring, enable logging on all systems and audit system logs daily to detect suspicious activity as soon as possible.

9. Perform technical audits of SCADA devices and networks and any other connected networks to identify security concerns.
   - Technical audits of SCADA devices and networks are critical to ongoing security effectiveness. Many commercial and open-sourced security tools are available that allow system administrators to conduct audits of their systems/networks to identify active services, patch level, and common vulnerabilities. The use of these tools will not solve systemic problems but will eliminate the "paths of least resistance" that an attacker could exploit. Analyze identified vulnerabilities to determine their significance, and take corrective actions as appropriate. Track corrective actions and analyze this information to identify trends. Additionally, retest systems after corrective actions have been taken to ensure that vulnerabilities were actually eliminated. Scan non-production environments actively to identify and address potential problems.

10. Conduct physical security surveys and assess all remote sites connected to the SCADA network to evaluate their security.
    - Any location that has a connection to the SCADA network is a target, especially unmanned or unguarded remote sites. Conduct a physical security survey and inventory access points at each facility that has a connection to the SCADA system. Identify and assess any source of information including remote telephone/computer network/ fiber-optic cables that could be tapped; radio and microwave links that are exploitable, computer terminals that could be accessed; and wireless local area network access points. Identify and eliminate single points of failure. The security of the site must be adequate to detect or prevent unauthorized access. Do not allow "live" network access points at remote, unguarded sites simply for convenience.

11. Establish SCADA "Red Teams" to identify and evaluate possible attack scenarios.
    - Establish a "Red Team" to identify potential attack scenarios and evaluate potential system vulnerabilities. Use a variety of people who can provide insight into weaknesses of the overall network, SCADA system, physical systems, and security controls. People who work on the system every day have great insight into the vulnerabilities of your SCADA network and should be consulted when identifying potential attack scenarios and possible consequences. Also, ensure that the risk from a malicious insider is fully evaluated, given that this represents one of the greatest threats to an organization. Feed information resulting from the "Red Team" evaluation into risk management processes to assess the information and establish appropriate protection strategies.

The following steps focus on management actions to establish an effective cybersecurity program:

12. Clearly define cybersecurity roles, responsibilities, and authorities for managers, system administrators, and users.
    - Organization personnel need to understand the specific expectations associated with protecting information technology resources through the definition of clear and logical roles and responsibilities. In addition, key personnel need to be given sufficient authority to carry out their assigned responsibilities. Too often, good cybersecurity is left up to the initiative of the individual, which usually leads to inconsistent implementations and ineffective security. Establish a cybersecurity organizational structure that defines roles and responsibilities and clearly identifies how cybersecurity issues are escalated and who is notified in an emergency.

13. Document network architecture and identify systems that serve critical functions or contain sensitive information that require additional levels of protection.

- Develop and document robust information security architecture as part of a process to establish an effective protection strategy. It is essential that organizations design their network with security in mind and continue to have a strong understanding of their network architecture throughout its lifecycle. Of particular importance, an in-depth understanding of the functions that the systems perform and the sensitivity of the stored information is required. Without this understanding, risk cannot be properly assessed, and protection strategies may not be sufficient. Documenting the information security architecture and its components is critical to understanding the overall protection strategy and identifying single points of failure.

14. Establish a rigorous, ongoing risk management process.
    - A thorough understanding of the risks to network computing resources from denial-of-service attacks and the vulnerability of sensitive information to compromise is essential to an effective cybersecurity program. Risk assessments from the technical basis of this understanding are critical to formulating effective strategies to mitigate vulnerabilities and preserve the integrity of computing resources. Initially, perform a baseline risk analysis based on current threat assessment to use for developing a network protection strategy. Due to rapidly changing technology and the emergence of new threats on a daily basis, an ongoing risk assessment process is also needed so that routine changes can be made to the protection strategy to ensure it remains effective. Fundamental to risk management is identification of residual risk with a network protection strategy in place and acceptance of that risk by management.

15. Establish a network protection strategy based on the principle of defense-in-depth.
    - A fundamental principle that must be part of any network protection strategy is defense-in-depth. Defense-in-depth must be considered early in the design phase of the development process and must be an integral consideration in all technical decision-making associated with the network. Utilize technical and administrative controls to mitigate threats from identified risks to as great a degree as possible at all levels of the network. Single points of failure must be avoided, and cybersecurity defense must be layered to limit and contain the impact of any security incidents. Additionally, each layer must be protected against other systems at the same layer. For example, to protect against the inside threat, restrict users to access only those resources necessary to perform their job functions.

16. Clearly identify cybersecurity requirements.
    - Organizations and companies need structured security programs with mandated requirements to establish expectations and allow personnel to be held accountable. Formalized policies and procedures are typically used to establish and institutionalize a cybersecurity program. A formal program is essential in establishing a consistent, standards-based approach to cybersecurity throughout an organization and eliminates sole dependence on individual initiative. Policies and procedures also inform employees of their specific cybersecurity responsibilities and the consequences of failing to meet those responsibilities. They also provide guidance regarding actions to be taken during a cybersecurity incident and promote efficient and effective actions during a time of crisis. As part of identifying cybersecurity requirements, include user agreements and notification and warning banners. Establish requirements to minimize the threat from malicious insiders, including the need for conducting background checks and limiting network privileges to those absolutely necessary.

17. Establish effective configuration management processes.
    - A fundamental management process needed to maintain a secure network is configuration management. Configuration management needs to cover both hardware configurations and software configurations. Changes to hardware or software can easily introduce vulnerabilities that undermine network security. Processes are required

to evaluate and control any change to ensure that the network remains secure. Configuration management begins with well-tested and documented security baselines for your various systems.

18. Conduct routine self-assessments.

Robust performance evaluation processes are needed to provide organizations with feedback on the effectiveness of cybersecurity policy and technical implementation. A sign of a mature organization is one that is able to self-identify issues, conduct root cause analyzes, and implement effective corrective actions that address individual and systemic problems. Self-assessment processes that are normally part of an effective cybersecurity program include routine scanning for vulnerabilities, automated auditing of the network, and self-assessments of organizational and individual performance.

19. Establish system backups and disaster recovery plans.

Establish a disaster recovery plan that allows for rapid recovery from any emergency (including a cyberattack). System backups are an essential part of any plan and allow rapid reconstruction of the network. Routinely exercise disaster recovery plans to ensure that they work and that personnel are familiar with them. Make appropriate changes to disaster recovery plans based on lessons learned from exercises.

20. Senior organizational leadership should establish expectations for cybersecurity performance and hold individuals accountable for their performance.

Effective cybersecurity performance requires commitment and leadership from senior managers in the organization. It is essential that senior management establish an expectation for strong cybersecurity and communicate this to their subordinate managers throughout the organization. It is also essential that senior organizational leadership establish a structure for implementation of a cybersecurity program. This structure will promote consistent implementation and the ability to sustain a strong cybersecurity program. It is then important for individuals to be held accountable for their performance as it relates to cybersecurity. This includes managers, system administrators, technicians, and users/operators.

21. Establish policies and conduct training to minimize the likelihood that organizational personnel will inadvertently disclose sensitive information regarding SCADA system design, operations, or security controls.

Release data related to the SCADA network only on a strict, need-to-know basis and only to persons explicitly authorized to receive such information. "Social engineering," the gathering of information about a computer or computer network via questions to naïve users, is often the 1st step in a malicious attack on computer networks. The more information revealed about a computer or computer network, the more vulnerable the computer/network is. Never divulge data revealed to a SCADA network, including the names and contact information about the system operators/administrators, computer operating systems, and/or physical and logical locations of computers and network systems over telephones or to personnel unless they are explicitly authorized to receive such information. Any requests for information by unknown persons need to be sent to a central network security location for verification and fulfillment. People can be a weak link in an otherwise secure network. Conduct training and information awareness campaigns to ensure that personnel remain diligent in guarding sensitive network information, particularly their passwords.

CHAPTER SUMMARY

Again, when it comes to the security of our nation and even of water and wastewater treatment facilities, few have summed it up better than Governor Ridge (Henry, 2002). Now, obviously, the further removed we get from 9/11, I think the natural tendency is to let down our guard. Unfortunately, we cannot do that. The government will continue to do everything we can to find and stop those who seek to harm us, and I believe we owe it to the American people to remind them that they must be vigilant, as well.

REVIEW QUESTIONS

1. General categories of security devices include _____.
   a. Water monitoring devices
   b. Control devices
   c. Cyber protection devices
   d. All of the above

2. Responds to specific types of events _____.
   a. Alarms
   b. Flags
   c. Local wildlife
   d. None of the above

3. A sensor is also called a _____.
   a. Flag
   b. Detection devices
   c. Hydrant
   d. Fence

4. A perimeter sensor is a type of _____.
   a. Alarm
   b. Fire annunciator
   c. Intrusion
   d. Flags

5. Backflow can occur under two types of conditions _____.
   a. Back press and backsiphonage
   b. Flag activation
   c. Triggered sprinkler systems
   d. None of the above

6. To install a wedge barrier the _____ must be prepared.
   a. Flag
   b. Alarm panel
   c. Surface area
   d. All of the above

7. A mobile obstacle that can be moved in and out of position on roadway _____.
   a. Alarms
   b. Flags
   c. Removable crash beams
   d. All of the above

8. A biometric system includes the following components _____.
   a. Hand recognition
   b. Host computer
   c. Power source
   d. All of the above

9. A card reader system consists of _____.
   a. Card reader
   b. Power source

    c. Control unit

    d. All of the above

10. When considering installing a fence a comparison can be made with (a) _____.

    a. Drawbridge

    b. Walls

    c. Barricades

    d. Flag poles

ANSWER KEY

1. General categories of security devices include _____.

    a. Water monitoring devices

    b. Control devices

    c. Cyber protection devices

    d. **All of the above**

2. Responds to specific types of events _____.

    a. **Alarms**

    b. Flags

    c. Local wildlife

    d. None of above

3. A sensor is also called a _____.

    a. Flag

    b. **Detection devices**

    c. Hydrant

    d. Fence

4. A perimeter sensor is a type of _____.

    a. Alarm

    b. Fire annunciator

    c. **Intrusion**

    d. Flags

5. Backflow can occur under two types of conditions _____.

    a. **Back press and backsiphonage**

    b. Flag activation

    c. Triggered sprinkler systems

    d. None of the above

6. To install a wedge barrier the _____ must be prepared.

    a. Flag

    b. Alarm panel

    c. **Surface area**

    d. All of the above

7. A mobile obstacle that can be moved in and out of position on roadway _____.

    a. Alarms

    b. Flags

    c. **Removable crash beams**

    d. All of the above

8. A biometric system includes the following components _____.

    a. Hand recognition

    b. Host computer

    c. Power source

    d. **All of the above**

9. A card reader system consists of _____.

    a. Card reader

    b. Power source

    c. Control unit

    d. **All of the above**

10. When considering installing a fence a comparison can be made with (a) _____.

    a. Drawbridge

    b. **Walls**

    c. Barricades

    d. Flag poles

REFERENCES

DOE, 2001. *21 Steps to Improve Cyber Security of SCADA Networks.* Washington, DC: Department of Energy.

Ezell, B.C., 1998. *Risks of Cyber Attack to Supervisory Control and Data Acquisition.* Charlottesville, VA: University of VA.

FBI, 2000. *Threat to Critical Infrastructure.* Washington, DC: Federal Bureau of Investigation.

FBI, 2004. *Ninth Annual Computer Crime and Security Survey.* Washington, DC: FBI: Computer Crime Institute and Federal Bureau of Investigations.

GAO, 2003. *Critical Infrastructure Protection: Challenges in Securing Control System.* Washington, DC: United States General Accounting Office.

Gellman, B., 2002. Cyber-Attacks by Al Qaeda Feared: Terrorists at Threshold of Using Internet as Tool of Bloodshed, Experts Say. *Washington Post*, June 27; p. A01.

Greenberg, A.E. et al., 1999. *Standard Methods.* Washington, DC: American Public Health Association.

Henry, K. 2002. New Face of Security. *Government Security*, April; pp. 30–31.

NIPC, (2002). *National Infrastructure Protection Center Report.* Washington, DC: National Infrastructure Protection Center.

Stamp, J. et al., 2003. *Common Vulnerabilities in Critical Infrastructure Control Systems*, 2nd ed. Livermore, CA: Sandia National Laboratories.

USEPA. 2004. Water Security: Basic Information. Accessed 09/30/2020 @ http://cfpub.epa.gov/ safewater/ watersecurity/basicinformation.cfm.

USEPA. 2005. EPA Needs to Determine What Barriers Prevent Water Systems from Securing Known SCADA Vulnerabilities. In: Harris, J. (ed.), *Final Briefing Report.* Washington, DC: USEPA.

14 Wastewater Treatment

The Code of Federal Regulations (CFR) 40 CFR Part 403, regulations were established in the late 1970s and early 1980s to help Publicly Owned Treatment Works (POTW) control industrial discharges to sewers. These regulations were designed to prevent pass-through and interference at the treatment plants and interference in the collection and transmission systems.

Pass-through occurs when pollutants literally "pass through" a POTW without being properly treated and cause the POTW to have an effluent violation or increase the magnitude or duration of a violation.

Interference occurs when a pollutant discharge causes a POTW to violate its permit by inhibiting or disrupting treatment processes, treatment operations, or processes related to sludge use or disposal.

KEY OBJECTIVES

After studying this chapter, you should be able to:

- Describe conventional wastewater treatment and model.
- Identify unit processes making up conventional treatment.
- Describe duties and qualification of wastewater operators.
- Discuss how plant performance is measured.
- Discuss unit process efficiency.
- Identify wastewater sources and characteristics.
- Discuss classification of wastewater.

DOI: 10.1201/9781003207146-19

- Define a gravity collection system.
- Describe force main collection.
- Discuss pumping stations.
- Identify and discuss preliminary treatment unit processes.
- Identify and discuss primary treatment unit processes.
- Identify and discuss secondary treatment unit processes.
- Describe activated sludge processes.
- Describe disinfection of wastewater.
- Describe advanced wastewater treatment processes.
- Discuss sludge and biosolids handling.
- Discuss permits, records, and reports.

KEY DEFINITIONS

Activated sludge – the solids formed when microorganisms are used to treat wastewater using the activated sludge treatment process. It includes organisms, accumulated food materials, and waste products from the aerobic decomposition process.

Advanced waste treatment – technology to produce an extremely high-quality discharge.

Aerobic – conditions in which free, elemental oxygen is present. Also used to describe organisms, biological activity, or treatment processes, which require free oxygen.

Anaerobic – conditions in which no oxygen (free or combined) is available. Also used to describe organisms, biological activity, or treatment processes which function in the absence of oxygen.

Anoxic – conditions in which no free, elemental oxygen is present, the only source of oxygen is combined oxygen such as that found in nitrate compounds. Also used to describe biological activity or treatment processes, which function only in the presence of combined oxygen.

Average monthly discharge limitation – the highest allowable discharge over a calendar month.

Average weekly discharge limitation – the highest allowable discharge over a calendar week.

Biochemical oxygen demand, BOD_5 – the amount of organic matter which can be biologically oxidized under controlled conditions (5 days @ 20°C in the dark).

Biosolid – From *Merriam-Webster's Collegiate Dictionary, Tenth Edition (1998)*: biosolid *n.* – solid organic matter recovered from a sewage treatment process and used especially as fertilizer – usually used in plural.

Note: In *Part II*, *biosolids* is used in many places (activated sludge being the exception) to replace the standard term *sludge*. The author (along with others in the field) views the term *sludge* as an ugly four-letter word that is inappropriate to use in describing biosolids. Biosolid is a product that can be reused; it has some value. Because biosolid has some value, it should not be classified as a "waste" product and when biosolids for beneficial reuse are addressed, they are not.

Buffer – a substance or solution which resists changes in pH.

Carbonaceous biochemical oxygen demand, $CBOD_5$ – the amount of biochemical oxygen demand which can be attributed to carbonaceous material.

Chemical oxygen demand (COD) – the amount of chemically oxidizable materials present in the wastewater.

Clarifier – a device designed to permit solids to settle or rise and be separated from the flow. Also known as a settling tank or sedimentation basin.

Coliform – a type of bacteria used to indicate possible human or animal contamination of water.

Combined sewer – a collection system which carries both wastewater and stormwater flows.

Comminution – a process to shred solids into smaller, less harmful particles.

Composite sample – a combination of individual samples taken in proportion to flow.

Daily discharge – the discharge of a pollutant measured during a calendar day or any 24-hour period that reasonably represents a calendar day for the purposes of sampling. Limitations expressed as weight is the total mass (weight) discharged over the day. Limitations expressed in other units are average measurements of the day.

Daily maximum discharge – the highest allowable values for a daily discharge.

Detention time – the theoretical time water remains in a tank at a given flow rate.

Dewatering – the removal or separation of a portion of water present in a sludge or slurry.

Discharge Monitoring Report (DMR) – the monthly report required by the treatment plant's NPDES discharge permit.

Dissolved oxygen (DO) – free or elemental oxygen, which is dissolved in water.

Effluent – the flow leaving a tank, channel, or treatment process.

Effluent limitation – any restriction imposed by the regulatory agency on quantities, discharge rates, or concentrations of pollutants which are discharged from point sources into state waters.

Facultative – organisms that can survive and function in the presence or absence of free, elemental oxygen.

Fecal coliform – a type of bacteria found in the bodily discharges of warm-blooded animals. Used as an indicator organism.

Floc – solids which join to form larger particles which will settle better.

Flume – a flow rate measurement device.

Food-to-microorganism ratio (F/M) – an activated sludge process control calculation based upon the amount of food (BOD_5 or COD) available per pound of mixed liquor volatile suspended solids.

Grab sample – an individual sample collected at a randomly selected time.

Grit – heavy inorganic solids such as sand, gravel, eggshells, or metal filings.

Industrial wastewater – wastes associated with industrial manufacturing processes.

Infiltration/inflow – extraneous flows in sewers; defined by Metcalf & Eddy in *Wastewater Engineering: Treatment, Disposal, Reuse*, 5th. ed., New York: McGraw-Hill, 2013, as follows:

Infiltration – water entering the collection system through cracks, joints, or breaks.

Steady inflow – water discharged from cellar and foundation drains, cooling water discharges, and drains from springs and swampy areas. This type of inflow is steady and is identified and measured along with infiltration.

Direct flow – those types of inflow that have a direct stormwater runoff connection to the sanitary sewer and cause an almost immediate increase in wastewater flows. Possible sources are roof leaders, yard and areaway drains, manhole covers, cross connections from storm drains and catch basins, and combined sewers.

Total inflow – the sum of the direct inflow at any point in the system plus any flow discharged from the system upstream through overflows, pumping station bypasses, and the like.

Delayed inflow – stormwater that may require several days or more to drain through the sewer system. This category can include the discharge of sump pumps from cellar drainage as well as the slowed entry of surface water through manholes in ponded areas.

Influent – the wastewater entering a tank, channel, or treatment process.

Inorganic – mineral materials such as salt, ferric chloride, iron, sand, gravel, etc.

License – a certificate issued by the State Board of Waterworks/Wastewater Works Operators authorizing the holder to perform the duties of a wastewater treatment plant operator.

Mean Cell Residence Time (MCRT) – the average length of time a mixed liquor suspended solids particle remains in the activated sludge process. May also be known as the sludge retention time.

Mixed liquor – the combination of return activated sludge and wastewater in the aeration tank.

Mixed Liquor Suspended Solids (MLSS) – the suspended solids concentration of the mixed liquor.

Mixed Liquor Volatile Suspended Solids (MLVSS) – the concentration of organic matter in the mixed liquor suspended solids.

Milligrams/Liter (mg/L) – a measure of concentration. It is equivalent to parts per million (ppm).

Nitrogenous Oxygen Demand (NOD) – a measure of the amount of oxygen required to biologically oxidize nitrogen compounds under specified conditions of time and temperature.

NPDES Permit – National Pollutant Discharge Elimination System permit that authorizes the discharge of treated wastes and specifies the condition, which must be met for discharge.

Nutrient – substances required to support living organisms. Usually refers to nitrogen, phosphorus, iron, and other trace metals.

Organic – materials which consist of carbon, hydrogen, oxygen, sulfur, and nitrogen. Many organics are biologically degradable. All organic compounds can be converted to carbon dioxide and water when subjected to high temperatures.

Pathogenic – disease causing. A pathogenic organism can cause illness.

Point source – any discernible, defined, and discrete conveyance from which pollutants are or may be discharged.

Part per million – an alternative (but numerically equivalent) unit used in chemistry is milligrams per liter (mg/L). As an analogy, think of a ppm as being equivalent to a full shot glass in a swimming pool.

Return Activated Sludge Solids (RASS) – the concentration of suspended solids in the sludge flow being returned from the settling tank to the head of the aeration tank.

Sanitary wastewater – wastes discharged from residences and from commercial, institutional, and similar facilities, which include both sewage and industrial wastes.

Scum – the mixture of floatable solids and water, which is removed from the surface of the settling tank.

Septic – a wastewater which has no dissolved oxygen present. Generally characterized by black color and rotten egg (hydrogen sulfide) odors.

Settleability – a process control test used to evaluate the settling characteristics of the activated sludge. Readings taken at 30–60 minutes are used to calculate the settled sludge volume (SSV) and the sludge volume index (SVI).

Settled sludge volume – the volume in percent occupied by an activated sludge sample after 30–60 minutes of settling. Normally written as SSV with a subscript to indicate the time of the reading used for calculation (SSV_{60}) or (SSV_{30}).

Sewage – wastewater containing human wastes.

Sludge – the mixture of settleable solids and water, which is removed from the bottom of the settling tank.

Sludge Retention Time (SRT) – See Mean Cell Residence Time.

Sludge Volume Index (SVI) – a process control calculation, which is used to evaluate the settling quality of the activated sludge. Requires the SSV_{30} and mixed liquor suspended solids test results to calculate.

Storm sewer – a collection system designed to carry only stormwater runoff.

Storm water – runoff resulting from rainfall and snowmelt.

Supernatant – in a digester it is the amber-colored liquid above the sludge.

Wastewater – the water supply of the community after it has been soiled by use.

Waste Activated Sludge Solids (WASS) – the concentration of suspended solids in the sludge, which is being removed from the activated sludge process.

Weir – a device used to measure wastewater flow.

Zoogleal slime – the biological slime which forms on fixed film treatment devices. It contains a wide variety of organisms essential to the treatment process.

OVERVIEW

In wastewater treatment, unit operations (unit processes) are the components linked together to form a process train as shown in Figure 14.1. Keep in mind that the caboose attached to this train is treated and cleaned wastewater, outfalled to a receiving water body, and is usually cleaner than the water in the receiving body. These unit treatment processes are commonly divided based on the fundamental mechanisms acting within them (i.e., physical, chemical, and biochemical). Physical operations are those, such as sedimentation, that are governed by the laws of physics (gravity).

Chemical operations are those in which strictly chemical reactions occur, such as precipitation. Biochemical operations are those that use living microorganisms to destroy or transform pollutants through enzymatically catalyzed chemical reactions (Grady et al., 2011).

WASTEWATER OPERATORS

Like waterworks operators, wastewater operators are highly trained and artful practitioners and technicians of their trade. Moreover, wastewater operators, again, like waterworks operators are required by the states to be licensed or certified to operate a wastewater treatment plant.

When learning wastewater operator skills, there are several excellent texts available to aid in the training process. Many of these texts are listed in Table 14.1.

THE WASTEWATER TREATMENT PROCESS: THE MODEL

Figure 14.1 shows a basic schematic of an example wastewater treatment process providing primary and secondary treatment using the *Activated sludge process*. This is the model, the prototype, the paradigm used in this chapter. Though it is true that in secondary treatment (which provides BOD removal beyond what is achievable by simple sedimentation) there are actually three commonly

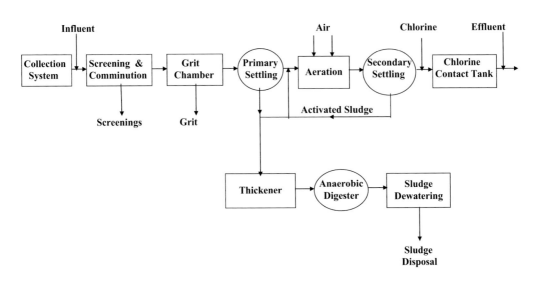

FIGURE 14.1 Schematic of an example wastewater treatment process providing primary and secondary treatment using activated sludge process.

used approaches – trickling filter, activated sludge, and oxidation ponds – we focus, for instructive and illustrative purposes, on the activated sludge process throughout *Part II*.

The purpose of Figure 14.1 is to allow the reader to follow the treatment process step by step as it is presented (and as it is configured in the real world) and to assist understanding of how all the various unit processes sequentially follow and tie into each other. Therefore, we begin certain sections (which discuss unit processes) with frequent reference to Figure 14.1. It is important to begin these sections in this manner because wastewater treatment is a series of individual steps (unit processes) that treat the waste stream as it makes its way through the entire process. Thus, it logically follows that a pictorial presentation along with pertinent written information enhances the learning process. It should also be pointed out, however, that even though the model shown in Figure 14.1 does not include all unit processes currently used in wastewater treatment we do not ignore the other major processes: trickling filters, rotating biological contactors (RBCs), and oxidation ponds.

MEASURING PLANT PERFORMANCE

To evaluate how well a plant or treatment unit process is operating, *performance efficiency* or *percent (%) removal* is used. The results can be compared with those listed in the plant's operation and maintenance manual (O&M) to determine if the facility is performing as expected. In this chapter, sample calculations often used to measure plant performance/efficiency are presented.

PLANT PERFORMANCE/EFFICIENCY

Example 14.1

Problem: The influent BOD_5 is 247 mg/L and the plant effluent BOD is 17 mg/L.

What is the percent removal?

TABLE 14.1

Recommended Reference/Study Materials

Advanced Waste Treatment, A Field Study Program, 2nd ed., Kerri, K., et al. California State University, Sacramento, CA.

Aerobic Biological Wastewater Treatment Facilities, Environmental Protection Agency, EPA 430/9-77-006, Washington, DC, 1977.

Anaerobic Sludge Digestion, Environmental Protection Agency, EPA-430/9-76-001, Washington, DC, 1977.

Annual Book of ASTM Standards, Section 11, "Water and Environmental Technology," American Society for Testing Materials (ASTM), Philadelphia, PA.

Guidelines Establishing Test Procedures for the Analysis of Pollutants. Federal Register (40 CFR 136), April 4, 1995 Volume 60, No. 64, Page 17160.

Handbook of Water Analysis, 2nd ed., HACH Chemical Company, PO Box 389, Loveland, CO, 1992.

Industrial Waste Treatment, A Field Study Program, Volume I, Kerri, K. et al. California State University, Sacramento, CA.

Industrial Waste Treatment, A Field Study Program, Volume 2, Kerri, K. et al. California State University, Sacramento, CA.

Methods for Chemical Analysis of Water and Wastes, U.S. Environmental Protection Agency, Environmental Monitoring Systems Laboratory-Cincinnati (EMSL-CL), EPA-6000/4-79-020, Revised March 1983 and 1979 (where applicable).

O&M of Trickling Filters, RBC and Related Processes, Manual of Practice OM-10, Water Pollution Control Federation (now called Water Environment Federation), Alexandria, VA, 1988.

Operation of Wastewater Treatment Plants, A Field Study Program, Volume I, 4th ed., Kerri, K., et al. California State University, Sacramento, CA.

Operation of Wastewater Treatment Plants, A Field Study Program, Volume II, 4th ed., Kerri, K., et al. California State University, Sacramento, CA.

Standard Methods for the Examination of Water and Wastewater, 18th ed., American Public Health Association, American Water Works Association-Water Environment Federation, Washington, DC, 1992.

Treatment of Metal Waste Streams, K. Kerri, et. al., California State University, Sacramento, CA.

Basic Math Concepts: For Water and Wastewater Plant Operators. Joanne K. Price, Lancaster, PA: Technomic Publishing Company, 1991.

Simplified Wastewater Treatment Plant Operations. Edward J. Haller, Lancaster, PA: Technomic Publishing Company, 1995.

Wastewater Treatment Plants: Planning, Design, and Operation. Syed R. Qasim, Lancaster, PA: Technomic Publishing Company, 1994.

The Science of Water, 4th ed. F.R. Spellman. Boca Raton, FL: CRC Press, 2020.

Water and Wastewater Infrastructure: Energy Efficiency and Sustainability. F.R. Spellman. Boca Raton, CRC Press, 2013.

SOLUTION:

$$\%\,Removal = \frac{(Influent\ Concentration - Effluent\ Concentration)}{Influent\ Concentration} \times 100$$

$$\%\,Removal = \frac{(247\ mg/L - 17\ mg/L) \times 100}{247\ mg/L} = 93\%$$

The calculation used for determining the performance (percent removal) for a digester is different from that used for performance (percent removal) for other processes. Care must be taken to select the right formula.

Example 14.2

Problem: The influent BOD is 235 mg/L and the plant effluent BOD is 169 mg/L.

What is the percent removal?

SOLUTION:

$$\%\,Removal = \frac{(Influent\ Concentration - Effluent\ Concentration)}{Influent\ Concentration} \times 100$$

$$\%\,Removal = \frac{(235\ mg/L - 169\ mg/L) \times 100}{235\ mg/L} = 28\%$$

Unit Process Performance and Efficiency

The concentration entering the unit and the concentration leaving the unit (i.e., primary, secondary, etc.) are used to determine the unit performance.

Hydraulic Detention Time

The term *detention time* or *hydraulic detention time* (HDT) refers to the average length of time (theoretical time) a drop of water, wastewater, or suspended particles remains in a tank or channel. It is calculated by dividing the water and wastewater in the tank by the flow rate through the tank. The units of flow rate used in the calculation are dependent on whether the detention time is to be calculated in seconds, minutes, hours, or days.

Detention time is used in conjunction with various treatment processes, including sedimentation and coagulation-flocculation. Generally, in practice, detention time is associated with the amount of time required for a tank to empty. The range of detention time varies with the process. For example, in a tank used for sedimentation, detention time is commonly measured in minutes.

WASTEWATER SOURCES AND CHARACTERISTICS

Wastewater treatment is designed to use the natural purification processes (self-purification processes of streams and rivers) to the maximum level possible. It is also designed to complete these processes in a controlled environment rather than over many miles of stream or river. Moreover, the treatment plant is also designed to remove other contaminants, which are not normally subjected to natural processes, as well as treating the solids which are generated through the treatment unit steps. The typical wastewater treatment plant is designed to achieve many different purposes:

- Protect public health;
- Protect public water supplies;
- Protect aquatic life;
- Preserve the best uses of the waters; and
- Protect adjacent lands.

Wastewater treatment is a series of steps. Each of the steps can be accomplished using one or more treatment processes or types of equipment. The major categories of treatment steps are:

- *Preliminary Treatment* – removes materials that could damage plant equipment or would occupy treatment capacity without being treated.
- *Primary Treatment* – removes settleable and floatable solids (may not be present in all treatment plants).
- *Secondary Treatment* – removes BODs and dissolved and colloidal suspended organic matter by biological action; organics are converted to stable solids, carbon dioxide, and more organics.

- *Advanced Waste Treatment* – uses physical, chemical, and biological processes to remove additional BOD5, solids, and nutrients (not present in all treatment plants).
- *Disinfection* – removes microorganisms to eliminate or reduce the possibility of disease when the flow is discharged.
- *Sludge Treatment* – stabilizes the solids removed from wastewater during treatment, inactivates pathogenic organisms, and/or reduces the volume of the sludge by removing water.

The various treatment processes described above are discussed in detail later.

WASTEWATER SOURCES

The principal sources of domestic wastewater in a community are the residential areas and commercial districts. Other important sources include institutional and recreational facilities and stormwater (runoff) and groundwater (infiltration). Each source produces wastewater with specific characteristics. In this section, wastewater sources and the specific characteristics of wastewater are described.

Generation of Wastewater

Wastewater is generated by five major sources: human and animal wastes, household wastes, industrial wastes, storm water runoff, and groundwater infiltration.

- *Human and animal wastes* – contain the solid and liquid discharges of humans and animals and are considered by many to be the most dangerous from a human health viewpoint. The primary health hazard is presented by the millions of bacteria, viruses, and other microorganisms (some of which may be pathogenic) present in the waste stream.
- *Household wastes* – are wastes, other than human and animal wastes, discharged from the home. Household wastes usually contain paper, household cleaners, detergents, trash, garbage, and other substances the homeowner discharges into the sewer system.
- *Industrial wastes* – includes industry-specific materials, which can be discharged from industrial processes into the collection system. Typically contains chemicals, dyes, acids, alkalis, grit, detergents, and highly toxic materials.
- *Storm water runoff* – many collection systems are designed to carry both the wastes of the community and stormwater runoff. In this type of system when a storm event occurs, the waste stream can contain large amounts of sand, gravel, and other grit as well as excessive amounts of water.
- *Groundwater infiltration* – groundwater will enter older improperly sealed collection systems through cracks or unsealed pipe joints. Not only can this add large amounts of water to wastewater flows but also additional grit.

Classification of Wastewater

Wastewater can be classified according to the sources of flows: domestic, sanitary, industrial, combined, and stormwater:

- *Domestic (Sewage) Wastewater* – mainly contains human and animal wastes, household waste, small amounts of groundwater infiltration, and small amounts of industrial waste.
- *Sanitary Wastewater* – consists of domestic waste and significant amounts of industrial waste. In many cases, the industrial waste can be treated without special precautions. However, in some cases the industrial waste will require special precautions or a pretreatment program to ensure the waste does not cause compliance problems for the wastewater treatment plant.
- *Industrial Wastewater* – industrial waste only. Often the industry will determine that it is safer and more economical to treat its waste independent of domestic waste.

- *Combined Wastewater* – is the combination of sanitary wastewater and stormwater runoff. All the wastewater and storm water of the community is transported through one system to the treatment plant.
- *Storm Water* – a separate collection system (no sanitary waste) that carries stormwater runoff including street debris, road salt, and grit.

WASTEWATER CHARACTERISTICS

Wastewater contains many different substances which can be used to characterize it. The specific substances and amounts or concentrations of each will vary, depending on the source. Thus, it is difficult to "precisely" characterize wastewater. Instead, wastewater characterization is usually based on and applied to average domestic wastewater. Wastewater is characterized in terms of its physical, chemical, and biological characteristics.

Note: Keep in mind that other sources and types of wastewater can dramatically change the characteristics.

Physical Characteristics

The *physical characteristics* of wastewater are based on color, odor, temperature, and flow.

- *Color* – fresh wastewater is usually a light brownish-gray color. However, typical wastewater is gray and has a cloudy appearance. The color of the wastewater will change significantly if allowed to go septic (if travel time in the collection system increases). Typical septic wastewater will have a black color.
- Odor – odors in domestic wastewater usually are caused by gases produced by the decomposition of organic matter or by other substances added to the wastewater. Fresh domestic wastewater has a musty odor. If the wastewater can go septic, this odor will change significantly – to a rotten egg odor associated with the production of hydrogen sulfide (H_2S).
- Temperature – the temperature of wastewater is commonly higher than that of the water supply because of the addition of warm water from households and industrial plants. However, significant amounts of infiltration or stormwater flow can cause major temperature fluctuations.
- Flow – the actual volume of wastewater is commonly used as a physical characterization of wastewater and is normally expressed in terms of gallons per person per day. Most treatment plants are designed using an expected flow of 100–200 gallons per person per day. This figure may have to be revised to reflect the degree of infiltration or stormwater flow the plant receives. Flow rates will vary throughout the day. This variation, which can be as much as 50–200% of the average daily flow, is known as the diurnal flow variation.

Note: *Diurnal* means occurs in a day or each day; daily.

Chemical Characteristics

When describing the chemical characteristics of wastewater, the discussion generally includes topics such as organic matter, the measurement of organic matter, inorganic matter, and gases. For the sake of simplicity, in *Part II* we specifically describe chemical characteristics in terms of alkalinity, biochemical oxygen demand (BOD), chemical oxygen demand (COD), dissolved gases, nitrogen compounds, pH, phosphorus, solids (organic, inorganic, suspended, and dissolved solids), and water.

- *Alkalinity* – is a measure of the wastewater's capability to neutralize acids. It is measured in terms of bicarbonate, carbonate, and hydroxide alkalinity. Alkalinity is essential to buffer (hold the neutral pH) of the wastewater during the biological treatment processes.

- *Biochemical Oxygen Demand* – a measure of the amount of biodegradable matter in the wastewater. Normally measured by a 5-day test conducted at 20°F, the BOD_5 domestic waste is normally in the range of 100–300 mg/L.
- *Chemical Oxygen Demand* (COD) – a measure of the amount of oxidizable matter present in the sample. The COD is normally in the range of 200–500 mg/L. The presence of industrial wastes can increase this significantly.
- *Dissolved gases* – gases that are dissolved in wastewater. The specific gases and normal concentrations are based upon the composition of the wastewater. Typical domestic wastewater contains oxygen in relatively low concentrations, carbon dioxide, and hydrogen sulfide (if septic conditions exist).
- *Nitrogen compounds* – the type and amount of nitrogen present will vary from the raw wastewater to the treated effluent. Nitrogen follows a cycle of oxidation and reduction. Most of the nitrogen in untreated wastewater will be in the forms of organic nitrogen and ammonia nitrogen. Laboratory tests exist for the determination of both forms. The sum of these two forms of nitrogen is also measured and is known as Total Kjeldahl Nitrogen (TKN). Wastewater will normally contain between 20 and 85 mg/L of nitrogen. Organic nitrogen will normally be in the range of 8–35 mg/L and ammonia nitrogen will be in the range of 12–50 mg/L.
- *pH* – a method of expressing the acid condition of the wastewater. pH is expressed on a scale of 1–14. For proper treatment, wastewater pH should normally be in the range of 6.5–9.0 (ideal 6.5–8.0).
- *Phosphorus* – essential to biological activity and must be present in at least minimum quantities, or secondary treatment processes will not perform. Excessive amounts can cause stream damage and excessive algal growth. Phosphorus will normally be in the range of 6–20 mg/L. The removal of phosphate compounds from detergents has had a significant impact on the amounts of phosphorus in wastewater.
- *Solids* – most pollutants found in wastewater can be classified as solids. Wastewater treatment is generally designed to remove solids or to convert solids to a form which is more stable or can be removed. Solids can be classified by their chemical composition (organic or inorganic) or by their physical characteristics (settleable, floatable, and colloidal). The concentration of total solids in wastewater is normally in the range of 350–1,200 mg/L.
- *Organic solids* – consist of carbon, hydrogen, oxygen, nitrogen and can be converted to carbon dioxide and water by ignition at 550°. Also known as fixed solids or loss on ignition.
- *Inorganic solids* – mineral solids which are unaffected by ignition. Also known as fixed solids or ash.
- *Suspended solid* – will not pass through a glass fiber filter pad. Can be further classified as total suspended solids (TSS), volatile suspended solids, and/or fixed suspended solids. Can also be separated into three components based on settling characteristics: settleable solids, floatable solids, and colloidal solids. TSS in wastewater are normally in the range of 100–350 mg/L.
- *Dissolved solid* – will pass through a glass fiber filter pad. Can also be classified as total dissolved solids (TDS), volatile dissolved solids, and fixed dissolved solids. TDS are normally in the range of 250–850 mg/L.
- *Water* – always the major constituent of wastewater. In most cases water makes up 99.5–99.9% of the wastewater. Even in the strongest wastewater, the total amount of contamination present is less than 0.5% of the total, and in average strength wastes it is usually less than 0.1%.

Biological Characteristics and Processes

After undergoing physical aspects of treatment (i.e., screening, grit removal, and sedimentation) in preliminary and primary treatment, wastewater still contains some suspended solids and other

TABLE 14.2
Typical Domestic Wastewater Characteristics

Characteristic	Typical Characteristic
Color	Gray
Odor	Musty
Dissolved Oxygen	>1.0 mg/L
pH	6.5–9.0
TSS	100–350 mg/L
BOD$_5$	100–300 mg/L
COD	200–500 mg/L
Flow	100–200 gallons/person/day
Total Nitrogen	20–85 mg/L
Total Phosphorus	6–20 mg/L
Fecal Coliform	500,0000-3,000,000 MPN/100 ml

solids that are dissolved in the water. In a natural stream, such substances are a source of food for protozoa, fungi, algae, and several varieties of bacteria.

In secondary wastewater treatment, these same microscopic organisms (which are one of the main reasons for treating wastewater) work as fast as they can to biologically convert the dissolved solids to suspended solids which will physically settle out at the end of secondary treatment.

Raw wastewater influent typically contains millions of organisms. Most of these organisms are nonpathogenic; however, several pathogenic organisms may also be present (these may include the organisms responsible for diseases such as typhoid, tetanus, hepatitis, dysentery, gastroenteritis, and others). Many of the organisms found in wastewater are microscopic (microorganisms); they include algae, bacteria, protozoans (such as amoeba, flagellates, free-swimming ciliates, and stalked ciliates), rotifers, and virus. Table 14.2 is a summary of typical domestic wastewater characteristics.

WASTEWATER COLLECTION SYSTEMS

Wastewater collection systems collect and convey wastewater to the treatment plant. The complexity of the system depends on the size of the community and the type of system selected. Methods of collection and conveyance of wastewater include gravity systems, force main systems, vacuum systems, and combinations of all three types of systems.

GRAVITY COLLECTION SYSTEM

In a *gravity collection system*, the collection lines are sloped to permit the flow to move through the system with as little pumping as possible. The slope of the lines must keep the wastewater moving at a velocity (speed) of 2–4 feet per second. Otherwise, at lower velocities, solids will settle out causing clogged lines, overflows, and offensive odors. To keep collection systems lines at a reasonable depth, wastewater must be lifted (pumped) periodically so that it can continue flowing downhill to the treatment plant. Pump stations are installed at selected points within the system for this purpose.

FORCE MAIN COLLECTION SYSTEM

In a typical *force main collection system*, wastewater is collected to central points and pumped under pressure to the treatment plant. The system is normally used for conveying wastewater long distances. The use of the force main system allows the wastewater to flow to the treatment plant at the desired velocity without using sloped lines. It should be noted that the pump station discharge lines in a gravity system are force mains since the content of the lines is under pressure.

Note: Extra care must be taken when performing maintenance on force main systems since the content of the collection system is under pressure.

VACUUM SYSTEM

In a *vacuum collection system*, wastewaters are collected to central points and then drawn toward the treatment plant under vacuum. The system consists of a large amount of mechanical equipment and requires a large amount of maintenance to perform properly. Generally, the vacuum-type collection systems are not economically feasible.

PUMPING STATIONS

Pumping stations provide the motive force (energy) to keep the wastewater moving at the desired velocity. They are used in both the force main and gravity systems. They are designed in several different configurations and may use different sources of energy to move the wastewater (i.e., pumps, air pressure, or vacuum). One of the more commonly used types of pumping station designs is the wet well/dry well design.

Wet-Well/Dry-Well Pumping Stations

The wet-well/dry-well pumping station consists of two separate spaces or sections separated by a common wall. Wastewater is collected in one section (wet well section), and the pumping equipment (and in many cases, the motors and controllers) is in a second section known as the dry well. There are many different designs for this type of system, but in most cases the pumps selected for this system are of a centrifugal design. There are a couple of major considerations in selecting centrifugal design: (1) allows for the separation of mechanical equipment (pumps, motors, controllers, wiring, etc.) from the potentially corrosive atmosphere (sulfides) of the wastewater; and (2) this type of design is usually safer for workers because they can monitor, maintain, operate, and repair equipment without entering the pumping station wet well.

Note: Most pumping station wet wells are confined spaces. To ensure safe entry into such spaces, compliance with OSHA's 29 CFR 1910.146 (Confined Space Entry Standard) is required.

Wet-Well Pumping Stations

Another type of pumping station design is the *wet well* type. This type consists of a single compartment, which collects the wastewater flow. The pump is submerged in the wastewater with motor controls located in that space or has a weatherproof motor housing located above the wet well. In this type of station, a submersible centrifugal pump is normally used.

Pneumatic Pumping Stations

The *pneumatic pumping station* consists of a wet well and a control system which controls the inlet and outlet value operations and provides pressurized air to force or "push" the wastewater through the system. The exact method of operation depends on the system design. When operating, wastewater in the wet well reaches a predetermined level and activates an automatic valve, which closes the influent line. The tank (wet well) is then pressurized to a predetermined level. When the pressure reaches the predetermined level, the effluent line valve is opened, and the pressure pushes the waste stream out the discharge line.

PRELIMINARY TREATMENT

The initial stage in the wastewater treatment process (following collection and influent pumping) is *preliminary treatment*. Raw influent entering the treatment plant may contain many kinds of materials (trash). The purpose of preliminary treatment is to protect plant equipment by removing these materials which could cause clogs, jams, or excessive wear to plant machinery. In addition, the removal of various materials at the beginning of the treatment process saves valuable space within the treatment plant.

Preliminary treatment may include many different processes; each designed to remove a specific type of material, which is a potential problem for the treatment process. Processes include wastewater collection – influent pumping, screening, shredding, grit removal, flow measurement, preaeration, chemical addition, and flow equalization – the major processes are shown in Figure 14.1. In this section, we describe and discuss each of these processes and their importance in the treatment process.

Note: As mentioned, not all treatment plants will include all the processes shown in Figure 14.1. Specific processes have been included to facilitate discussion of major potential problems with each process and its operation; this is information that may be important to the wastewater operator.

Screening

The purpose of screening is to remove large solids such as rags, cans, rocks, branches, leaves, roots, etc., from the flow before the flow moves on to downstream processes.

Note: Typically, a treatment plant will remove anywhere from 0.5 to 12 ft³ of screenings for each million gallon of influent received.

A bar screen traps debris as wastewater influent passes through. Typically, a bar screen consists of a series of parallel, evenly spaced bars or a perforated screen placed in a channel (see Figure 14.2). The waste stream passes through the screen and the large solids (screenings) are trapped on the bars for removal.

Note: The screenings must be removed frequently enough to prevent accumulation which will block the screen and cause the water level in front of the screen to build up.

The bar screen may be coarse (2–4 in. openings) or fine (0.75–2.0 in. openings). The bar screen may be manually cleaned (bars or screens are placed at an angle of 30° for easier solids removal – see Figure 14.2) or mechanically cleaned (bars are placed at 45°–60° angle to improve mechanical cleaner operation).

The screening method employed depends on the design of the plant, the amount of solids expected, and whether the screen is for constant or emergency use only.

Manually Cleaned Screens

Manually cleaned screens are cleaned at least once per shift (or often enough to prevent buildup which may cause reduced flow into the plant) using a long tooth rake. Solids are manually pulled to the drain platform and allowed to drain before storage in a covered container. The area around the screen should be cleaned frequently to prevent a buildup of grease or other materials, which can

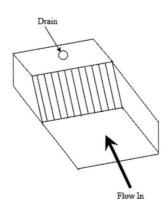

Drain

Flow In

FIGURE 14.2 Basic bar screen.

cause odors, slippery conditions, and insect and rodent problems. Because screenings may contain organic matter as well as large amounts of grease, they should be stored in a covered container. Screenings can be disposed of by burial in approved landfills or by incineration. Some treatment facilities grind the screenings into small particles, which are then returned to the wastewater flow for further processing and removal later in the process.

Mechanically Cleaned Screens

Mechanically cleaned screens use a mechanized rake assembly to collect the solids and move them (carry them) out of the wastewater flow for discharge to a storage hopper. The screen may be continuously cleaned or cleaned on a time- or flow-controlled cycle. As with the manually cleaned screen, the area surrounding the mechanically operated screen must be cleaned frequently to prevent buildup of materials, which can cause unsafe conditions. As with all mechanical equipment, operator vigilance is required to ensure proper operation and that proper maintenance is performed. Maintenance includes lubricating equipment and maintaining it in accordance with the manufacturer's recommendations or the plant's O&M manual (operations and maintenance manual). Screenings from mechanically operated bar screens are disposed of in the same manner as screenings from manually operated screens: landfill disposal, incineration, or ground into smaller particles for return to the wastewater flow.

Screening Safety

The screening area is the first location where the operator is exposed to the wastewater flow. Any toxic, flammable, or explosive gases present in the wastewater can be released at this point. Operators who frequent enclosed bar screen areas should be equipped with personal air monitors. Adequate ventilation must be provided. It is also important to remember that due to the grease attached to the screenings, this area of the plant can be extremely slippery. Routine cleaning is required to minimize this problem.

Note: Never override safety devices on mechanical equipment. Overrides can result in dangerous conditions, injuries, and major mechanical failure.

Shredding

As an alternative to screening, *shredding* can be used to reduce solids to a size which can enter the plant without causing mechanical problems or clogging. Shredding processes include comminution (comminute means to cut up) and barminution devices.

Comminution

The *comminutor* is the most common shredding device used in wastewater treatment. In this device, all the wastewater flow passes through the grinder assembly. The grinder consists of a screen or slotted basket, a rotating or oscillating cutter, and a stationary cutter. Solids pass through the screen and are chopped or shredded between the two cutters. The comminutor will not remove solids which are too large to fit through the slots, and it will not remove floating objects. These materials must be removed manually. Maintenance requirements for comminutors include aligning, sharpening, and replacing cutters, and corrective and preventive maintenance performed in accordance with plant O&M manual.

Barminution

In barminution, the *barminutor* uses a bar screen to collect solids, which are then shredded and passed through the bar screen for removal at a later process. Device cutter alignment and sharpness are critical factors in effective operation. Cutters must be sharpened or replaced, and alignment must be checked in accordance with manufacturer's recommendations. Solids which are not shredded must be removed daily, stored in closed containers, and disposed of by burial or incineration.

Barminutor operational problems are like those listed above for comminutors. Preventive and corrective maintenance as well as lubrication must be performed by qualified personnel and in accordance with the plant's O&M manual. Because of higher maintenance requirements, the barminutor is less frequently used.

GRIT REMOVAL

The purpose of *grit removal* is to remove the heavy inorganic solids, which could cause excessive mechanical wear. Grit is heavier than inorganic solids and includes sand, gravel, clay, eggshells, coffee grounds, metal filings, seeds, and other similar materials. There are several processes or devices used for grit removal. Since grit is heavier than the organic solids, it should be kept in suspension for treatment in the following processes. Grit removal may be accomplished in grit chambers or by the centrifugal separation of sludge. Processes use gravity/velocity, aeration, or centrifugal force to separate the solids from the wastewater.

Gravity/Velocity-Controlled Grit Removal

Gravity/velocity-controlled grit removal is normally accomplished in a channel or tank where the speed or the velocity of the wastewater is controlled to about 1 foot per second (ideal), so that grit will settle while the organic matter remains suspended. If the velocity is controlled in the range of 0.7–1.4 feet per second (fps), the grit removal will remain effective.

Velocity is controlled by the amount of water flowing through the channel, the depth of the water in the channel, by the width of the channel, or by cumulative width of channels in service.

Cleaning

Gravity-type systems may be manually or mechanically cleaned. Manual cleaning normally requires that the channel be taken out of service, drained, and manually cleaned. Mechanical cleaning systems are operated continuously or on a time cycle. Removal should be frequent enough to prevent grit carry-over into the rest of the plant.

Note: Before and during cleaning activities always ventilate the area thoroughly.

AERATION

Aerated grit removal systems use aeration to keep the lighter organic solids in suspension while allowing the heavier grit articles to settle out. Aerated grit removal may be manually or mechanically cleaned; however, most of the systems are mechanically cleaned. During normal operation, adjusting the aeration rate produces the desired separation. This requires observation of mixing and aeration and sampling of fixed suspended solids. Actual grit removal is controlled by the rate of aeration. If the rate is too high, all the solids remain in suspension. If the rate is too low, both grit and organics will settle out. The operator observes the same kinds of conditions as those listed for the gravity/velocity-controlled system but must also pay close attention to the air distribution system to ensure proper operation.

Centrifugal Force

The *cyclone degritter* uses a rapid spinning motion (centrifugal force) to separate the heavy inorganic solids or grit from the light organic solids. This unit process is normally used on primary sludge rather than the entire wastewater flow. The critical control factor for the process is the inlet pressure. If the pressure exceeds the recommendations of the manufacturer, the unit will flood, and grit will carry through with the flow. Grit is separated from flow, washed, and discharged directly to a storage container. Grit removal performance is determined by calculating the percent removal for inorganic (fixed) suspended solids. The operator observes the same kinds of conditions listed for the gravity/velocity-controlled and aerated grit removal systems, except for the air distribution system.

Typical problems associated with grit removal include mechanical malfunctions and rotten egg odor in the grit chamber (hydrogen sulfide formation), which can lead to metal and concrete corrosion problems. The low recovery rate of grit is another typical problem. Bottom scour, over aeration, or not enough detention time normally causes this. When these problems occur, the operator must make the required adjustments or repairs to correct the problem.

PREAERATION

In the *preaeration process* (diffused or mechanical), we aerate wastewater to achieve and maintain an aerobic state (to freshen septic wastes), strip off hydrogen sulfide (to reduce odors and corrosion), agitate solids (to release trapped gases and improve solids separation and settling), and to reduce BOD_5. All of this can be accomplished by aerating the wastewater for 10–30 minutes. To reduce BOD_5, preaeration must be conducted for 45–60 minutes.

CHEMICAL ADDITION

Chemical addition to the waste stream is done (either via dry chemical metering or solution feed metering) to improve settling, reduce odors, neutralize acids or bases, reduce corrosion, reduce BOD_5, improve solids and grease removal, reduce loading on the plant, add or remove nutrients, add organisms, and/or aid subsequent downstream processes. The chemical and amount used depends on the desired result. Chemicals must be added at a point where enough mixing will occur to obtain maximum benefit. Chemicals typically used in wastewater treatment include chlorine, peroxide, acids and bases, miner salts (ferric chloride, alum, etc.), bioadditives, and enzymes.

EQUALIZATION

The purpose of *flow equalization* (whether by surge, diurnal, or complete methods) is to reduce or remove the wide swings in flow rates normally associated with wastewater treatment plant loading; it minimizes the impact of storm flows. The process can be designed to prevent flows above maximum plant design hydraulic capacity; to reduce the magnitude of diurnal flow variations; and to eliminate flow variations. Flow equalization is accomplished using mixing or aeration equipment, pumps, and flow measurement. Normal operation depends on the purpose and requirements of the flow equalization system. Equalized flows allow the plant to perform at optimum levels by providing stable hydraulic and organic loading. The downside to flow equalization is in additional costs associated with the construction and operation of the flow equalization facilities.

AERATED SYSTEMS

Aerated grit removal systems use aeration to keep the lighter organic solids in suspension while allowing the heavier grit particles to settle out. Aerated grit may be manually or mechanically cleaned; however, most of the systems are mechanically cleaned. In normal operation, the aeration rate is adjusted to produce the desired separation, which requires observation of mixing and aeration and sampling of fixed suspended solids. Actual grit removal is controlled by the rate of aeration. If the rate is too high, all the solids remain in suspension. If the rate is too low, both the grit and the organics will settle out.

Cyclone Degritter

The *cyclone degritter* uses a rapid spinning motion (centrifugal force) to separate the heavy inorganic solids or grit from the light organic solids. This unit process is normally used on primary sludge rather than on the entire wastewater flow. The critical control factor for the process is the inlet pressure. If the pressure exceeds the recommendations of the manufacturer, the unit will flood, and grit will pass through with the flow. Grit is separated from the flow and discharged directly to a storage container. Grit removal performance is determined by calculating the percent removal for inorganic (fixed) suspended solids.

PRIMARY TREATMENT (SEDIMENTATION)

The purpose of primary treatment (primary sedimentation or primary clarification) is to remove settleable organic and floatable solids. Normally, each primary clarification unit can be expected to remove 90–95% settleable solids, 40–60% TSS, and 25–35% BOD_5.

Note: Performance expectations for settling devices used in other areas of plant operation are normally expressed as overall unit performance rather than settling unit performance.

Sedimentation may be used throughout the plant to remove settleable and floatable solids. It is used in primary treatment, secondary treatment, and advanced wastewater treatment processes. In this section, we focus on primary treatment or primary clarification, which uses large basins in which primary settling is achieved under relatively quiescent conditions (see Figure 14.1). Within these basins, mechanical scrapers collect the primary settled solids into a hopper, from which they are pumped to a sludge-processing area. Oil, grease, and other floating materials (scum) are skimmed from the surface. The effluent is discharged over weirs into a collection trough.

PROCESS DESCRIPTION

In primary sedimentation, wastewater enters a settling tank or basin where velocity is reduced to approximately 1 ft/min.

Note: Notice that the velocity is based on minutes instead of seconds, as was the case in the grit channels. A grit channel velocity of 1 ft/sec would be 60 ft/min.

Solids which are heavier than water settle to the bottom, while solids which are lighter than water float to the top. Settled solids are removed as sludge and floating solids are removed as scum. Wastewater leaves the sedimentation tank over an effluent weir and on to the next step in treatment. Detention time, temperature, tank design, and condition of the equipment control the efficiency of the process.

OVERVIEW OF PRIMARY TREATMENT

Primary treatment reduces the organic loading on downstream treatment processes by removing a large amount of settleable, suspended, and floatable materials.

- Primary treatment reduces the velocity of the wastewater through a clarifier to approximately 1-2 ft/min so that settling and floatation can take place. Slowing the flow enhances the removal of suspended solids in wastewater.
- Primary settling tanks remove floated grease and scum, remove the settled sludge solids, and collect them for pumped transfer to disposal or further treatment.
- Clarifiers used may be rectangular or circular. In rectangular clarifiers, wastewater flows from one end to the other, and the settled sludge is moved to a hopper at the one end, either by flights set on parallel chains or by a single bottom scraper set on a traveling bridge. Floating material (mostly grease and oil) is collected by a surface skimmer.
- In circular tanks, the wastewater usually enters at the middle and flows outward. Settled sludge is pushed to a hopper in the middle of the tank bottom, and a surface skimmer removes floating material.
- Factors affecting primary clarifier performance include:
- Rate of flow through the clarifier

- Wastewater characteristics (strength; temperature; amount and type of industrial waste; and the density, size, and shapes of particles)
- Performance of pretreatment processes
- Nature and amount of any wastes recycled to the primary clarifier

TYPES OF SEDIMENTATION TANKS

Sedimentation equipment includes septic tanks, two story tanks and plain settling tanks or clarifiers. All three devices may be used for primary treatment while plain settling tanks are normally used for secondary or advanced wastewater treatment processes.

Septic Tanks

Septic tanks are prefabricated tanks that serve as a combined settling and skimming tank and as an unheated-unmixed anaerobic digester. Septic tanks provide long settling times (6–8 hr or more) but do not separate decomposing solids from the wastewater flow. When the tank becomes full, solids will be discharged with the flow. The process is suitable for small facilities (i.e., schools, motels, homes, etc.), but due to the long detention times and lack of control, it is not suitable for larger applications.

Two-Story (Imhoff) Tank

The *two-story* or *Imhoff tank*, named after German engineer Karl Imhoff (1876–1965), is like a septic tank in the removal of settleable solids and the anaerobic digestion of solids. The difference is that the two-story tank consists of a settling compartment where sedimentation is accomplished, a lower compartment where settled solids and digestion takes place, and gas vents. Solids removed from the wastewater by settling pass from the settling compartment into the digestion compartment through a slot in the bottom of the settling compartment.

The design of the slot prevents solids from returning to the settling compartment. Solids decompose anaerobically in the digestion section. Gases produced as a result of the solids decomposition are released through the gas vents running along each side of the settling compartment.

Plain Settling Tanks (Clarifiers)

The plain settling tank or clarifier optimizes the settling process. Sludge is removed from the tank for processing in other downstream treatment units. Flow enters the tank, is slowed and distributed evenly across the width and depth of the unit, passes through the unit, and leaves over the effluent weir. Detention time within the primary settling tank is from 1 to 3 hours (2-hour average). Sludge removal is accomplished frequently on either a continuous or intermittent basis. Continuous removal requires additional sludge treatment processes to remove the excess water resulting from the removal of sludge which contains less than 2–3% solids. Intermittent sludge removal requires the sludge be pumped from the tank on a schedule frequent enough to prevent large clumps of solids rising to the surface but infrequent enough to obtain 4–8% solids in the sludge withdrawn.

Scum must be removed from the surface of the settling tank frequently. This is normally a mechanical process but may require manual start-up. The system should be operated frequently enough to prevent excessive buildup and scum carryover but not so frequent as to cause hydraulic overloading of the scum removal system. Settling tanks require housekeeping and maintenance. Baffles (prevent floatable solids, scum from leaving the tank), scum troughs, scum collectors, effluent troughs, and effluent weirs require frequent cleaning to prevent heavy biological growths and solids accumulations. Mechanical equipment must be lubricated and maintained as specified in the manufacturer's recommendations or in accordance with procedures listed in the plant O&M manual.

Process control sampling and testing is used to evaluate the performance of the settling process. Settleable solids, dissolved oxygen, pH, temperature, total suspended solids and BOD_5, as well as sludge solids and volatile matter testing are routinely carried out.

Primary Clarification: Normal Operation

In primary clarification, wastewater enters a settling tank or basin.

Velocity reduces to approximately 1 ft/min.

Note: Notice that the velocity is based on minutes instead of seconds, as was the case in the grit channels. A grit channel velocity of 1 ft/sec would be 60 ft/min.

Primary Clarification: Operational Parameters (Normal Observations)

- *Flow Distribution* – Normal flow distribution is indicated by flow to each in-service unit being equal and uniform. There is no indication of short-circuiting. The surface-loading rate is within design specifications.
- *Weir Condition* – Weirs are level; flow over the weir is uniform; and weir overflow rate is within design specifications.
- *Scum Removal* – The surface is free of scum accumulations; the scum removal does not operate continuously.
- *Sludge Removal* – No large clumps of sludge appear on the surface; system operates as designed; pumping rate is controlled to prevent coning or buildup; and sludge blanket depth is within desired levels.
- *Performance* – The unit is removing expected levels of BOD_5, TSS, and settleable solids.
- *Unit Maintenance* – Mechanical equipment is maintained in accordance with planned schedules; equipment is available for service as required.

To assist the operator in judging primary treatment operation, several process control tests can be used for process evaluation and control. These tests include the following:

- pH – normal range: 6.5–9.0
- Dissolved oxygen – normal range: <1.0 mg/L
- Temperature – varies with climate and season
- Settleable solids, ml/L – influent: 5–15 ml/L

effluent: 0.3–5 ml/L

- BOD_5, mg/L – influent: 150–400 mg/L

effluent: 50–150 mg/L

- % solids – 4–8%
- % volatile matter – 40–70%
- Heavy metals – as required
- Jar tests – as required

Note: Testing frequency should be determined based on the process influent and effluent variability and the available resources. All should be performed periodically to provide reference information for evaluation of performance.

EFFLUENT FROM SETTLING TANKS

Upon completion of screening, degritting, and settling in sedimentation basins, large debris, grit, and many settleable materials have been removed from the waste stream. What is left is referred to as *primary effluent*. Usually cloudy and frequently gray in color, primary effluent still contains large amounts of dissolved food and other chemicals (nutrients). These nutrients are treated in the next step in the treatment process (secondary treatment), which is discussed in the next section.

Note: Two of the most important nutrients left to remove are phosphorus and ammonia. While we want to remove these two nutrients from the waste stream, we do not want to remove too much. Carbonaceous microorganisms in secondary treatment (biological treatment) need both phosphorus and ammonia.

SECONDARY TREATMENT

The main purpose of *secondary treatment* (sometimes referred to as biological treatment) is to provide BOD removal beyond what is achievable by primary treatment. There are three commonly used approaches, all of which take advantage of the ability of microorganisms to convert organic wastes (via biological treatment) into stabilized, low-energy compounds.

Two of these approaches, the *trickling filter* [and/or its variation, the *rotating biological contactor (RBC)*] and the *activated sludge* process, sequentially follow normal primary treatment. The third, *ponds* (oxidation ponds or lagoons), however, can provide equivalent results without preliminary treatment. In this section, we present a brief overview of the secondary treatment process followed by a detailed discussion of wastewater treatment ponds (used primarily in smaller treatment plants), trickling filters, and RBCs.

We then shift focus to the activated sludge process, which is used primarily in large installations and is the focus of *Part II*.

Secondary treatment refers to those treatment processes which use biological processes to convert dissolved, suspended, and colloidal organic wastes to more stable solids which can either be removed by settling or discharged to the environment without causing harm. Exactly what is the secondary treatment?

As defined by the Clean Water Act (CWA), the secondary treatment produces an effluent with no more than 30 mg/L BOD_5 and 30 mg/L TSS.

Note: The CWA also states that ponds and trickling filters will be included in the definition of secondary treatment even if they do not meet the effluent quality requirements continuously.

Most secondary treatment processes decompose solids aerobically producing carbon dioxide, stable solids, and more organisms. Since solids are produced, all the biological processes must include some form of solids removal (settling tank, filter, etc.). Secondary treatment processes can be separated into two large categories: fixed film systems and suspended growth systems.

Fixed film systems are processes which use a biological growth (biomass or slime). When the wastewater and slime are in contact, the organisms remove and oxidize the organic solids. The media may be stone, redwood, synthetic materials, or any other substance that is durable (capable of withstanding weather conditions for many years) and provides a large area for slime growth while providing open space for ventilation and is not toxic to the organisms in the biomass. Fixed film devices include trickling filters and rotating biological contactors. *Suspended growth systems* are processes that use a biological growth, which is mixed with the wastewater. Typical suspended growth systems consist of various modifications of the activated sludge process.

TREATMENT PONDS

Wastewater treatment can be accomplished using *ponds* (aka, *lagoons*). Ponds are relatively easy to build and manage, they accommodate large fluctuations in flow, and can also provide treatment that approaches conventional systems (producing a highly purified effluent) at a much lower cost. It is the cost (the economics) that drives many managers to decide on the pond option. The actual degree of treatment provided depends on the type and number of ponds used. Ponds can be used as the sole type of treatment, or they can be used in conjunction with other forms of wastewater treatment; that is, other treatment processes followed by a pond, or a pond followed by other treatment processes.

Stabilization ponds (aka treatment ponds) have been used for the treatment of wastewater for over 3,000 years. The first recorded construction of a pond system in the US was in San Antonio, Texas, in 1901. Today, over 8,000 wastewater treatment ponds are in place, involving more than 50% of the wastewater treatment facilities in the US (CWNS, 2000). Facultative ponds account for 62%, aerated ponds 25%, anaerobic 0.04%, and total containment 12% of the pond treatment systems. They treat a variety of wastewaters from domestic wastewater to complex industrial wastes, and they function under a wide range of weather conditions, from tropical to arctic. Ponds can be used alone or in combination with other wastewater treatment processes. As our understanding of pond operating mechanisms has increased, different types of ponds have been developed for application in specific types of wastewater under local environmental conditions. *Part II* focuses on municipal wastewater treatment pond systems.

DID YOU KNOW?

A pond can be judged as an "Attractive Nuisance." The term "Attractive Nuisance" is a legal expression that implies the pond could be attractive to potential users, such as, duck hunters, fishermen, or playing children. Since ponds have fairly steep slopes, the potential for someone falling in and drowning is a significant legal problem that must be a concern. It is important that adequate fencing and signing be provided.

While the tendency in the US has been for smaller communities to build ponds, in other parts of the world, including Australia, New Zealand, Mexico, Latin America, Asia, and Africa, treatment ponds have been built for large cities. As a result, our understanding of the biological, biochemical, physical, and climatic factors that interact to transform the organic compounds, nutrients, and pathogenic organisms found in sewage into less harmful chemicals and unviable organisms (i.e., dead or sterile) has grown since 1983. A wealth of experience has been built up as civil, sanitary, or environmental engineers, operators, public works managers, and public health and environmental agencies have gained more experience with these systems. While some of this information makes its way into technical journals and textbooks, there is a need for a less formal presentation of the subject for those working in the field every day (USEPA, 2011).

Ponds are designed to enhance the growth of natural ecosystems that are either anaerobic (providing conditions for bacteria that grow in the absence of oxygen (O_2) environments), aerobic (promoting the growth of O_2 producing and/or requiring organisms, such as algae and bacteria), or facultative, which is a combination of the two. Ponds are managed to reduce concentrations of BOD, TSS, and coliform numbers (fecal or total) to meet water quality requirements.

Types of Ponds

Ponds can be classified based on their location in the system, by the type of wastes they receive, and by the main biological process occurring in the pond. First, we look at the types of ponds according to their location and the type of wastes they receive: raw sewage stabilization ponds (see Figure 14.3), oxidation ponds, and polishing ponds.

Then, in the following section, we look at ponds classified by the type of processes occurring within the pond:

- Aerobic ponds;
- Anaerobic ponds;
- Facultative ponds; and
- Aerated ponds.

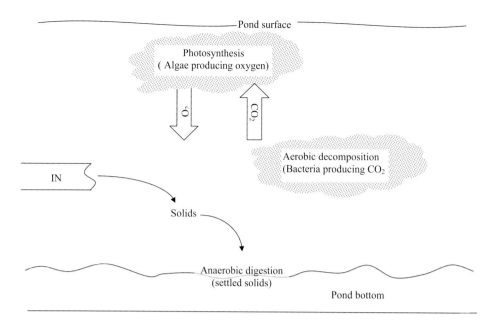

FIGURE 14.3 Stabilization pond processes.

Ponds Based on Location and Types of Wastes They Receive

Raw Sewage Stabilization Pond

The raw sewage stabilization pond is the most common type of pond (see Figure 14.3). Except for screening and shredding, this type of pond receives no prior treatment. Generally, raw sewage stabilization ponds are designed to provide a minimum of 45 days' detention time and to receive no more than 30 pounds of BOD_5 per day per acre. The quality of the discharge is dependent on the time of the year. Summer months produce high BOD_5 removal but excellent suspended solids removal. The pond consists of an influent structure, pond berm or walls, and an effluent structure designed to permit the selection of the best quality effluent. Normal operating depth of the pond is 3–5 feet. The process occurring in the pond involves bacteria decomposing the organics in the wastewater (aerobically and anaerobically) and algae using the products of the bacterial action to produce oxygen (photosynthesis). Because this type of pond is the most used in wastewater treatment, the process that occurs within the pond is described in greater detail in the following text.

When wastewater enters the stabilization pond, several processes begin to occur. These include settling, aerobic decomposition, anaerobic decomposition, and photosynthesis (see Figure 14.3). Solids in the wastewater will settle to the bottom of the pond. In addition to the solids in the wastewater entering the pond, solids which are produced by the biological activity will also settle to the bottom. Eventually this will reduce the detention time and the performance of the pond. When this occurs (20–30 years normal) the pond will have to be replaced or cleaned.

Bacteria and other microorganisms use the organic matter as a food source. They use oxygen (aerobic decomposition), organic matter, and nutrients to produce carbon dioxide, water, stable solids (which may settle out), and more organisms. The carbon dioxide is an essential component of the photosynthesis process occurring near the surface of the pond. Organisms also use the solids that settled out as food material; however, the oxygen levels at the bottom of the pond are extremely low so the process used is anaerobic decomposition. The organisms use the organic matter to produce gases (hydrogen sulfide, methane, etc.) which are dissolved in the water, stable solids, and more organisms. Near the surface of the pond a population of green algae will develop which can use the carbon dioxide produced by the bacterial population, nutrients, and sunlight to produce more algae and oxygen which is dissolved into the water. The dissolved oxygen is then used by organisms in the aerobic decomposition process.

When compared with other wastewater treatment systems involving biological treatment, a stabilization pond treatment system is the simplest to operate and maintain. Operation and maintenance activities include collecting and testing samples for dissolved oxygen (DO) and pH, removing weeds and other debris (scum) from the pond, mowing the berms, repairing erosion, and removing burrowing animals.

Note: Dissolved oxygen and pH levels in the pond will vary throughout the day. Normal operation will result in very high DO and pH levels due to the natural processes occurring.

Note: When operating properly, the stabilization pond will exhibit a wide variation in both dissolved oxygen and pH. This is due to the photosynthesis occurring in the system.

Oxidation Pond

An oxidation pond, which is normally designed using the same criteria as the stabilization pond, receives flows that have passed through a stabilization pond or primary settling tank. This type of pond provides biological treatment, additional settling, and some reduction in the number of the fecal coliform present.

Polishing Pond

A polishing pond, which uses the same equipment as a stabilization pond, receives flow from an oxidation pond or from other secondary treatment systems. Polishing ponds remove additional BOD_5, solids and fecal coliform, and some nutrients. They are designed to provide 1–3 days' detention time and normally operate at a depth of 5–10 feet. Excessive detention time or too shallow a depth will result in algae growth, which increases influent, suspended solids and concentrations.

Ponds Based on the Type of Processes Occurring Within

The type of processes occurring within the pond may also classify ponds. These include the aerobic, anaerobic, facultative, and aerated processes.

Aerobic Ponds

In aerobic ponds, also known as oxidation ponds or high-rate aerobic ponds (which are not widely used), oxygen is present throughout the pond. All biological activity is aerobic decomposition. They are usually 30–45 cm deep, which allows light to penetrate throughout the pond. Mixing is often provided, keeping algae at the surface to maintain maximum rates of photosynthesis and O_2 production and to prevent algae from settling and producing an anaerobic bottom layer. The rate of photosynthetic production of O_2 may be enhanced by surface re-aeration; O_2 and aerobic bacteria biochemically stabilize the waste. Detention time is typically 2–6 days.

These ponds are appropriate for treatment in warm, sunny climates. They are used where a high degree of BOD_5 removal is desired, but land area is limited. The chief advantage of these ponds is that they produce a stable effluent during short detention times with low land and energy requirements. However, their operation is somewhat more complex than that of facultative ponds, and unless the algae are removed the effluent will contain high TSS. While the shallow depths allow penetration of ultraviolet (UV) light that may reduce pathogens, shorter detention times work against effective coliform and parasite die-off. Since they are shallow, bottom paving or veering is usually necessary to prevent aquatic plants from colonizing the ponds. The Advanced Integrated Wastewater Pond System® (AIWPS®) uses the high-rate pond to maximize the growth of microalgae using a low-energy paddle-wheel (USEPA, 2011).

Anaerobic Ponds

Anaerobic ponds are normally used to treat high strength industrial wastes; that is, they receive heavy organic loading, so much so that there is no aerobic zone – no oxygen is present, and all biological activity is anaerobic decomposition. They are usually 2.5–4.5 m in depth and have detention times of 5–50 days. The predominant biological treatment reactions are bacterial acid formation and methane fermentation. Anaerobic ponds are usually used for treatment of strong industrial and agricultural (food processing) wastes, as a pretreatment step in municipal systems, or where an industry is a significant contributor to a municipal system. The biochemical reactions in an anaerobic pond produce hydrogen sulfide (H_2S) and other odorous compounds. To reduce odors, the common practice is to recirculate water from a downstream facultative or aerated pond. This provides a thin aerobic layer at the surface of the anaerobic pond, which prevents odors from escaping into the air. A cover may also be used to contain odors. The effluent from anaerobic ponds usually requires further treatment prior to discharge (USEPA, 2011).

Facultative Pond

The facultative pond, which may also be called an oxidation or photosynthetic pond, is the most common type pond (based on processes occurring). Oxygen is present in the upper portions of the pond, and aerobic processes are occurring. No oxygen is present in the lower levels of the pond where processes occurring are anoxic and anaerobic. Facultative ponds are usually 0.9–2.4 m deep or deeper, with an aerobic layer overlying an anaerobic layer. Recommended detention times vary from 5 to 50 days in warm climates and 90–180 days in colder climates (NEIWPCC, 1998). Aerobic treatment processes in the upper layer provide odor control, nutrient and BOD removal. Anaerobic fermentation processes, such as sludge digestion, denitrification and some BOD removal, occur in the lower layer. The key to successful operation of this type of pond is O_2 production by photosynthetic algae and/or re-aeration at the surface. Facultative ponds are used to treat raw municipal wastewater in small communities and for primary or secondary effluent treatment for small or large cities. They are also used in industrial applications, usually in the process line after aerated or anaerobic ponds, to provide additional treatment prior to discharge. Commonly achieved effluent BOD values, as measured in the BOD_5 test, range from 20 to -60 mg/L, and TSS levels may range from 30 to 150 mg/L. The size of the pond needed to treat BOD loadings depends on specific conditions and regulatory requirements.

Aerated Pond

Facultative ponds overloaded due to unplanned additional sewage volume or higher strength influent from a new industrial connection may be modified by the addition of mechanical aeration. Ponds originally designed for mechanical aeration are generally 2–6 m deep with detention times of 3–10 days. For colder climates, 20–40 days is recommended. Mechanically aerated ponds require less land area but have greater energy requirements. When aeration is used, the depth of the pond and/or the acceptable loading levels may increase. Mechanical or diffused aeration is often used to supplement natural oxygen production or to replace it.

Elements of Pond Processes

The Organisms

Although our understanding of wastewater pond ecology is far from complete, general observations about the interactions of macro- and microorganisms in these biologically driven systems support our ability to design, operate and maintain them.

Bacteria

DID YOU KNOW?

A pond should be drawn down in fall after the first frost and when the algae concentration drops off, the BOD is still low, and when the receiving stream temperature is low with accompanying high dissolved oxygen. A pond should be drawn down in spring before algae concentration increases, when the BOD level is acceptable, and when the receiving stream flows are high (low temperature with high dissolved oxygen helps). During the actual discharge, the effluent must be sampled for BOD, suspended solids, and pH at a frequency specified in the discharge permit.

In this section, we discuss other types of bacteria found in the pond; these organisms help to decompose complex, organic constituents in the influent simple, non-toxic compounds. Certain pathogen bacteria and other microbial organisms (viruses, protozoa) associated with human waste enter in that system with the influent; the wastewater treatment process is designed so that their numbers will be reduced adequately to meet public health standards.

- *Aerobic bacteria* are found in the aerobic zone of a wastewater pond and are primarily the same type as those found in an activated sludge process or in the zoogleal mass of a trickling filter.
- The most frequently isolated bacteria include Beggiatoa alba, Sphaerotilus natans, Achromobacter, Alcaligenes, Flavobacterium, Pseudomonas and Zoogoea spp. (Spellman, 2000; Lynch & Poole, 1979; Pearson, 2005).
- These organisms decompose the organic materials present in the aerobic zone into oxidized end products.
- *Anaerobic bacteria* are hydrolytic bacteria that convert complex organic material into simple alcohols and acids, primarily amino acids, glucose, fatty acid and glycerols (Spellman, 2000; Brockett, 1976; Pearson, 2005; Paterson & Curtis, 2005).
- Acidogenic bacteria convert the sugars and amino acids into acetate, ammonia (NH_3), hydrogen (H), and carbon dioxide (CO_2).
- Methanogenic bacteria break down these products further to methane (CH_4) and CO_2 (Gallert & Winter, 2005).
- Cyanobacteria, formerly classified as blue-green algae, are autotrophic organisms that can synthesize organic compounds using CO_2 as the major carbon source. Cyanobacteria produce O_2 as a by-product of photosynthesis, providing an O_2 source for other organisms in the ponds. They are found in very large numbers as blooms when environmental conditions are suitable (Gaudy & Gaudy, 1980). Commonly encountered cyanobacteria include *Oscillatoria, Arthrospira, Spirulina*, and *Microcystis* (Vasconcelos & Pereira, 2001; Spellman, 2000).
- *Purple Sulfur Bacteria* (Chromatiaceae) may grow in any aquatic environment to which light of the required wavelength penetrates if CO_2, nitrogen (N), and a reduced form of sulfur (S) or H are available. Purple surf bacteria occupy the anaerobic layer below the algae, cyanobacteria, and other aerobic bacteria in a pond. They are commonly found at a specific depth, in a thin layer where light and nutrient conditions are at an optimum (Gaudy & Gaudy, 1980; Pearson, 2005). Their biochemical conversion of odorous sulfide compounds to elemental S or sulfate (SO_4) helps to control odor in facultative and anaerobic ponds.

Algae

Algae constitute a group of aquatic organisms that may be unicellular or multicellular, motile or immotile, and, depending on the phylogenetic family, have different combinations of photosynthetic pigments. As autotrophs, algae need only inorganic nutrients, such as N, phosphorus (P), and a suite of microelements, to fix CO_2 and grow in the presence of sunlight. Algae do not fix atmospheric N; they require an external source of inorganic N in the form of nitrate (NO_3) or NH_3. Some algal species can use amino acids and other organic N compounds. Oxygen is a by-product of these reactions. Algae are generally divided into three major groups, based on the color reflected from the cells by the chlorophyll and other pigments involved in photosynthesis. Green and brown algae are common to wastewater ponds; red algae occur infrequently. The algal species that is dominant at any time is thought to be primarily a function of temperature, although the effects of predation, nutrient availability, and toxins are also important.

- *Green algae* (Chlorophyta) include unicellular, filamentous, and colonial forms. Some green algal genera commonly found in facultative and aerobic ponds are Euglena, Phacus, Chlamydomonas, Ankistrodesmus, Chlorella, Micractinium, Scenedesmus, Selenastrum, Dictyosphaerium, and Volvox.
- *Chrysophytes*, or brown algae, are unicellular and may be flagellated and include the diatoms. Certain brown algae are responsible for toxic red blooms. Brown algae found in wastewater ponds include the diatoms *Navicula* and *Cyclotella*.
- *Red algae (Rhodophyta)* include a few unicellular forms but are primarily filamentous (Gaudy & Gaudy, 1980; Pearson, 2005).

Importance of Interactions between Bacteria and Algae

It is generally accepted that the presence of both algae and bacteria is essential for the proper functioning of a treatment pond. Bacteria break down the complex organic waste components found in anaerobic and aerobic pond environments into simple compounds, which are then available for uptake by the algae. Algae, in turn, produce the O_2 necessary for the survival of aerobic bacteria. In the process of pond reactions of biodegradation and mineralization of waste material by bacteria and the synthesis of new organic compounds in the form of algal cells, a pond effluent might contain a higher than acceptable TSS. Although this form of TSS does not contain the same constituents as the influent TSS, it does contribute to turbidity and needs to be removed before the effluent is discharged. Once concentrated and removed, depending on regulatory requirements, algal TSS may be used as a nutrient for use in agriculture or as a feed supplement (Grolund, 2002).

DID YOU KNOW?

The variation in pH in a facultative pond normally occurs in the upper aerobic zone, while the anaerobic and facultative zones will be relatively constant. This variation happens due to the changes that occur in the concentration of dissolved carbon dioxide. When carbon dioxide is dissolved in water, it forms a weak carbonic acid which would tend to lower pH. The relationship between algae and bacteria affect the carbon dioxide levels.

During intense photosynthesis, algae use carbon dioxide and produce oxygen to be used by bacteria to assimilate organic wastes. The algae use much of the carbon dioxide and the pH can rise significantly (pH in the 11–12 range is not uncommon).

During the night or during cloudy weather, the algae respire and active photosynthesis does not occur. The bacteria continue to use up oxygen and produce carbon dioxide. This can cause a significant drop in the pond pH, especially if the influent wastewater has low alkalinity. This same pH swing can occur in natural ponds, lakes, and stream impoundments.

During peak summer algae activity, the dissolved oxygen of stream impoundments has varied from dawn levels of less than 1 mg/L, to later afternoon values of 13–15 mg/L (supersaturation).

Invertebrates

Although bacteria and algae are the primary organisms through which waste stabilization is accomplished, predator life forms do play a role in wastewater pond ecology. It has been suggested that the planktonic invertebrate *Cladocera* spp. and the benthic invertebrate family Chironomidae are the most significant fauna in the pond community in terms of stabilizing organic material.

The cladocerans feed on the algae and promote flocculation and settling of particulate matter. This in turn results in better light penetration and algal growth at greater depths. Settled matter is further broken down and stabilized by the benthic feeding Chironomidae. Predators, such as rotifers, often control the population levels of certain of the smaller life forms in the pond, thereby influencing the succession of species throughout the seasons.

Mosquitoes can present a problem in some ponds. Aside from their nuisance characteristics, certain mosquitoes are also vectors for such diseases as encephalitis, malaria, and yellow fever and constitute a hazard to public health which must be controlled. *Gambusia*, commonly called mosquito fish, have been introduced to eliminate mosquito problems in some ponds in warm climates (Ullrich, 1967; Pipes, 1961; Pearson, 2005), but their introduction has been problematic as they can out-compete native fish that also feed on mosquito larvae.

There are also biochemical controls, such as the larvicides *Bacillus thuringiensis* israelensis (Bti), and Abate®, which may be effective if the product is applied directly to the area containing mosquito larvae. The most effective means of control of mosquitoes in ponds is the control of emergent vegetation (USEPA, 2011).

Biochemistry in a Pond

Photosynthesis

Photosynthesis is the process whereby organisms use solar energy to fix CO_2 and obtain the reducing power to convert it to organic compounds. In wastewater ponds, the dominant photosynthetic organisms include algae, cyanobacteria, and purple sulfur bacteria (Pipes, 1961; Pearson, 2005).

Photosynthesis may be classified as oxygenic or anoxygenic, depending on the source of reducing power used by an organism. In oxygenic photosynthesis, water serves as the source of reducing power, with O_2 as a by-product.

Oxygenic photosynthetic algae and cyanobacteria convert CO_2 to organic compounds, which serve as the major source of chemical energy for other aerobic organisms. Aerobic bacteria need the O_2 produced to function in their role as primary consumers in degrading complex organic waste material.

Anoxygenic photosynthesis does not produce O_2 and, in fact, occurs in the complete absence of O_2. The bacteria involved in anoxygenic photosynthesis are largely strict anaerobes, unable to function in the presence of O_2. They obtain energy by reducing inorganic compounds. Many photosynthetic bacteria utilize reduced S compounds or element S in anoxygenic photosynthesis.

Respiration

Respiration is a physiological process by which organic compounds are oxidized into CO_2 and water. Respiration is also an indicator of cell material synthesis. It is a complex process that consists of many interrelated biochemical reactions (Pearson, 2005). The bacteria involved in aerobic respiration are primarily responsible for degradation of waste products. In the presence of light, respiration and photosynthesis can occur simultaneously in algae. However, the respiration rate is low compared to the photosynthesis rate, which results in a net consumption of CO_2 and production of O_2. In the absence of light, on the other hand, algal respiration continues while photosynthesis stops, resulting in a net consumption of O_2 and production of CO_2 (USEPA, 2011).

Nitrogen Cycle

The N cycle occurring in a wastewater treatment pond consists of several biochemical reactions mediated by bacteria. Organic N and NH_3 enter with the influent wastewater. Organic N in fecal matter and other organic materials undergo conversion to NH_3 and ammonium ion NH_4^+ by microbial activity. The NH_3 may volatize into the atmosphere. The rate of gaseous NH_3 losses to the atmosphere is primarily a function of pH, surface to volume ratio, temperature, and the mixing conditions. An alkaline pH shifts the equilibrium of NH_3 gas and NH_4^+ towards gaseous NH_3 production, while the mixing conditions affect the magnitude of the mass transfer coefficient. Ammonium is nitri-fied to nitrite (NO_2) by the bacterium *Nitrosomonas* and then to NO_3^- by *Nitrobacter*. The overall nitrification reaction is:

$$NH_4^+ + 2O_2 \rightarrow NO_3^- + 2H^+ + H_2O$$

The NO_3^- produced in the nitrification process, as well as a portion of the NH_4^- produced from ammonification, can be assimilated by organisms to produce cell protein and other N-containing compounds. The NO_3^- may also be denitrified to form $NO_2^=$ and then N gas. Several species of bacteria may be involved in the denitrification process, including *Pseudomonas*, *Micrococcus*, *Achromobacter*, and *Bacillus*. The overall denitrification reaction is:

$$6NO_3^- + 5CH_3OH \rightarrow 3N_2 + 5CO_2 + 7H_2O + 6OH^-$$

Nitrogen gas may be fixed by certain species of cyanobacteria when N is limited. This may occur in N-poor industrial ponds, but rarely in municipal or agricultural ponds (USEPA, 1975, 1993).

Nitrogen removal in facultative wastewater ponds can occur through any of the following processes:

- Gaseous NH_3 stripping to the atmosphere;
- NH_4 assimilation in algal biomass;
- NO_3^- uptake by floating vascular plants and algae; and
- Biological nitrification-denitrification. Whether NH_4 is assimilated into algal biomass depends on the biological activity in the system and is affected by several factors such as temperature, organic load, detention time, and wastewater characteristics.

Dissolved Oxygen (DO)

Oxygen is a partially soluble gas. Its solubility varies in direct proportion to the atmospheric pres-sure at any given temperature. DO concentrations of approximately 8 mg/L are generally consid-ered to be maximum available under local ambient conditions. In mechanically aerated ponds, the limited solubility of O_2 determines its absorption rate (Sawyer et al., 1994).

The natural sources of DO in ponds are photosynthetic oxygenation and surface re-aeration. In areas of low wind activity, surface re-aeration may be relatively unimportant, depending on the water depth. Where surface turbulence is created by excessive wind activity, surface re-aeration can be significant. Experiments have shown that DO in wastewater ponds varies almost directly with the level of photosynthetic activity, which is low at night and early morning, and rises during daylight hours to a peak in the early afternoon. At increased depth, the effects of photosynthetic oxygenation and surface re-aeration decrease, as the distance from the water–atmosphere interface increases and light penetration decreases. This can result in the establishment of a vertical gradient. The microorganisms in the pond will segregate along the gradient.

pH and Alkalinity

In wastewater ponds, the H ion concentration, expressed as pH, is controlled through the carbonate buffering system. The equilibrium of this system is affected by the rate of algal photosynthesis. In photosynthetic metabolism, CO_2 is removed from the dissolved phase, forcing the equilibrium of the first expression (16.30) to the left. This tends to decrease the hydrogen ion (H^+) concentration and the bicarbonate (HCO_3) alkalinity. The decreased alkalinity associated with photosynthesis will simultaneously reduce the carbonate hardness present in the waste. Because of the close correlation between pH and photosynthetic activity, there is a diurnal fluctuation in pH when respiration is the dominant metabolic activity.

Physical Factors

Light

The intensity and spectral composition of light penetrating a pond surface significantly affect all resident microbial activity. In general, activity increases with increasing light intensity until the photosynthetic system becomes light saturated. The rate at which photosynthesis increases in proportion to an increase in light intensity, as well as the level at which an organism's photosynthetic system becomes light saturated, depends upon the biochemistry of the species (Lynch & Poole, 1979; Pearson, 2005). In ponds, photosynthetic O_2 production has been shown to be relatively constant with the range of 5,380–53,800 lumens/m^2 light intensity with a reduction occurring at higher and lower intensities (Pipes, 1961; Paterson & Curtis, 2005).

The spectral composition of available light is also crucial in determining photosynthetic activity. The ability of photosynthetic organisms to utilize available light energy depends primarily upon their ability to absorb the available wavelengths. This absorption ability is determined by the specific photosynthetic pigment of the organism. The main photosynthetic pigments are chlorophylls and phycobilins. Bacterial chlorophyll differs from algal chlorophyll in both chemical structure and absorption capacity. These differences allow the photosynthetic bacteria to live below dense algal layers where they can utilize light not absorbed by the algae (Spellman, 2020; Lynch & Poole, 1979; Pearson, 2005).

The quality and quantity of light penetrating the pond surface to any depth depend on the presence of dissolved and particulate matter as well as the water absorption characteristics. The organisms themselves contribute to water turbidity, further limiting the depth of light penetration. Given the light penetration interferences, photosynthesis is significant only in the upper pond layers. This region of net photosynthetic activity is called the euphotic zone (Lynch & Poole, 1979; Pearson, 2005). Light intensity from solar radiation varies with the time of day and difference in latitudes. In cold climates, light penetration can be reduced during the winter by ice and snow cover. Supplementing the treatment ponds with mechanical aeration may be necessary in these regions during that time of year.

Temperature

Temperature at or near the surface of the aerobic environment of a pond determines the succession of predominant species of algae, bacteria, and other aquatic organisms. Algae can survive at temperatures of 5–40°C. Green algae show most efficient growth and activity at temperatures of 30–35°C. Aerobic bacteria are viable within a temperature range of 10–40°C; 35–40°C is optimum for cyanobacteria (Anderson & Zweig, 1962; Gloyna 1976; Paterson & Curtis, 2005; Crites et al., 2006).

As the major source of heat for these systems in solar radiation, a temperature gradient can develop in a pond with depth. This will influence the rate of anaerobic decomposition of solids that have settled at the bottom of the pond. The bacteria responsible for anaerobic degradation are active in temperatures from 15°C to 65°C. When they are exposed to lower temperatures, their activity is reduced. The other major source of heat is the influent water. In sewerage systems with no major inflow or infiltration problems, the influent temperature is higher than that of the pond contents. Cooling influences are exerted by evaporation, contact with cooler groundwater, and wind action. The overall effect of temperature in combination with light intensity is reflected in the fact that nearly all investigators report improved performance during summer and autumn months when both temperature and light are at their maximum. The maximum practical temperature of wastewater ponds is likely less than 30°C, indicating that most ponds operate at less than optimum temperature for anaerobic activity (Oswald, 1996; Paterson & Curtis, 2005; Crites et al., 2006; USEPA, 2011).

During certain times of the year, cooler, denser water remains at depth, while the warmer water stays at the surface. Water temperature differences may cause ponds to stratify throughout their depth. As the temperature decreases during the fall and the surface water cools, stratification decreases, and the deeper water mixes with the cooling surface water. This phenomenon is call *mixis*, or pond or lake overturn. As the density of water decreases and the temperature falls below 4°C, winter stratification can develop. When the ice cover breaks up and the water warms, a spring overturn can also occur (Spellman, 1996).

Pond overturn, which releases odorous compounds into the atmosphere, can generate complaints from property owners living downwind of the pond. The potential for pond overturn during certain times of the year is the reason why regulations may specify that ponds be located downwind, based on prevailing winds during overturn periods, and away from dwellings.

DID YOU KNOW?

Common maintenance problems associated with pond systems include:

- *Weed control-cattails and other rooted aquatic plants.*
- *Algae control-blue-green and associated floating algae mats*
- *Burrowing animals-muskrats and turtles.*
- *Duckweed control and removal.*
- *Floating sludge mats.*
- *Dike vegetation mowing and removing woody plants.*
- *Dike erosion-rip rap and proper vegetation.*
- *Fence maintenance to restrict access.*
- *Mechanical equipment-pumps, blowers etc.*

Wind

Prevailing and storm-generated wind should be factored into pond design and siting as they influence performance and maintenance in several significant ways:

- Oxygen transfer and dispersal – By producing circulatory flows, winds provide the mixing needed for O_2 transfer and diffusion below the surface of facultative ponds. This mixing action also helps disperse microorganisms and augments the movement of algae, particularly green algae.
- Prevention of short circuiting and reduction of odor events – Care must be taken during design to position the pond inlet/outlet axis perpendicular to the direction of prevailing winds to reduce short circuiting, which is the most common cause of poor performance. Consideration must also be made for the transport and fate of odors generated by treatment by-products in anaerobic and facultative ponds.
- Disturbance of pond integrity – Waves generated by strong prevailing or storm winds are capable of eroding or overtopping embankments. Some protective material should extend one or more feet above and below the water level to stabilize earthen berms.
- A study indicates that wind effects can reduce hydraulic retention time.

Pond Nutritional Requirements

In order to function as designed, the wastewater pond must provide enough macro- and micronutrients for the microorganisms to grow and populate the system adequately. A treatment pond system should be neither overloaded nor underloaded with wastewater nutrients.

Nitrogen can be a limiting nutrient for primary productivity in a pond. The conversion of organic N to various other N forms results in a total net loss (Assenzo & Reid, 1966; Pano & Middlebrooks, 1982; Middlebrooks et al. 1982; Middlebrooks & Pano, 1983; Craggs, 2005). This N loss may be due to algal uptake or bacterial action. It is likely that both mechanisms contribute to the overall total N reduction. Another factor contributing to the reduction of total N is the removal of gaseous NH_3

under favorable environmental conditions. Regardless of the specific removal mechanism involved, NH_3 removal in facultative wastewater ponds has been observed at levels greater than 90%, with the major removal occurring in the primary cell of a multicell pond *system* (Middlebrooks et al., 1982; Shilton, 2005; Crites et al., 2006; USEPA, 2011).

Phosphorus

Phosphorus (P) is most often the growth-limiting nutrient in aquatic environments. Municipal wastewater in the United States is normally enriched in P even though restrictions on P-containing compounds in laundry detergents in some states have resulted in reduced concentrations since the 1970s. As of 1999, 27 states and the District of Columbia had passed laws prohibiting the manufacture and use of laundry detergents containing P. However, phosphate (PO_4^3) content limits in automatic dishwashing detergents and other household cleaning agents containing P remain unchanged in most states. With a contribution of approximately 15%, the concentration of P from wastewater treatment plants is still adequate to promote growth in aquatic organisms (Canadian Environmental Protection Act, 2009).

In aquatic environments, P occurs in three forms: particulate P, soluble organic P, and inorganic P. Inorganic P, primarily in the form of orthophosphate ($OP(OR)_3$), is readily utilized by aquatic organisms. Some organisms may store excess P as polyphosphate. At the same time, some PO_4^{-3} is continuously lost to sediments, where it is locked up in insoluble precipitates (Lynch & Poole, 1979; Craggs, 2005; Crites et al., 2006).

Phosphorus removal in ponds occurs via physical mechanisms such as adsorption, coagulation, and precipitation. The uptake of P by organisms in metabolic function as well as for storage can also contribute to its removal. Removal in wastewater ponds has been reported to range from 30% to 95% (Assenzo & Reid, 1966; Pearson, 2005; Crites et al., 2006).

Algae discharged in the final effluent may introduce organic P to receiving waters. Excessive algal "afterblooms" observed in waters receiving effluents have, in some cases, been attributed to N and P compounds remaining in the treated wastewater.

Sulfur

Sulfur (S) is a required nutrient for microorganisms, and it is usually present in enough concentration in natural waters. Because S is rarely limiting, its removal from wastewater is usually not considered necessary. Ecologically, S compounds such as hydrogen sulfide (H_2S) and sulfuric acid (H_2SO_4) are toxic, while the oxidation of certain S compounds is an important energy source for some aquatic bacteria (Lynch & Poole, 1979; Pearson, 2005).

Carbon

The decomposable organic C content of waste is traditionally measured in terms of its BOD_5, or the amount of O_2 required under standardized conditions for the aerobic biological stabilization of the organic matter over a certain period. Since complete treatment by biological oxidation can take several weeks, depending on the organic material and the organism present, standard practice is to use the BOD_5 as an index of the organic carbon content or organic strength of a waste. The removal of BOD_5 is a primary criterion by which treatment efficiency is evaluated.

BOD_5 reduction in wastewater ponds ranging from 50% to 95% has been reported in the literature. Various factors affect the rate of reduction of BOD_5. A very rapid reduction occurs in a wastewater pond during the first 5–7 days. Subsequent reductions take place at a sharply reduced rate. BOD_5 removals are generally much lower during winter and early spring than in summer and early fall. Many regulatory agencies recommend that pond operations do not include discharge during cold periods.

TRICKLING FILTERS

Trickling filters have been used to treat wastewater since the 1890s. It was found that if settled wastewater was passed over rock surfaces, slime grew on the rocks and the water became cleaner.

Today we still use this principle but, in many installations, instead of rocks we use plastic media. In most wastewater treatment systems, the *trickling filter* follows primary treatment and includes a secondary settling tank or clarifier as shown in Figure 14.4. Trickling filters are widely used for the treatment of domestic and industrial wastes. The process is a fixed film biological treatment method designed to remove BOD_5 and suspended solids. A trickling filter consists of a rotating distribution arm that sprays and evenly distributes liquid wastewater over a circular bed of fist-sized rocks, other coarse materials, or synthetic media (see Figure 14.5). The spaces between the media allow air to circulate easily so that aerobic conditions can be maintained. The spaces also allow wastewater to trickle down through, around, and over the media. A layer of biological slime that absorbs and consumes the waste trickling through the bed covers the media material.

The organisms aerobically decompose the solids producing more organisms and stable wastes, which either become part of the slime or are discharged back into the wastewater flowing over the media. This slime consists mainly of bacteria, but it may also include algae, protozoa, worms, snails, fungi, and insect larvae. The accumulating slime occasionally sloughs off (*sloughings*) individual media materials (see Figure 14.6) and is collected at the bottom of the filter, along with the treated wastewater, and passes on to the secondary settling tank where it is removed. The overall performance of the trickling filter is dependent on hydraulic and organic loading, temperature, and recirculation.

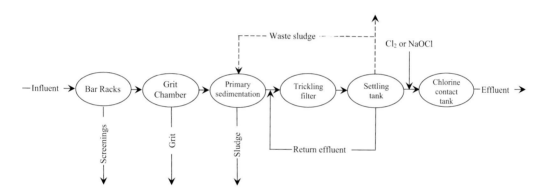

FIGURE 14.4 Simplified flow diagram of trickling filter used for wastewater treatment.

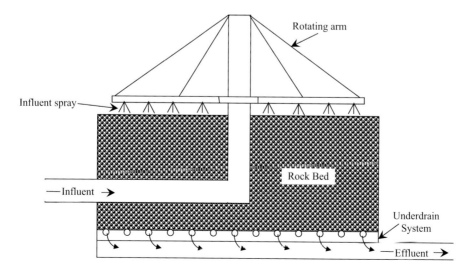

FIGURE 14.5 Schematic of cross section of a trickling filter

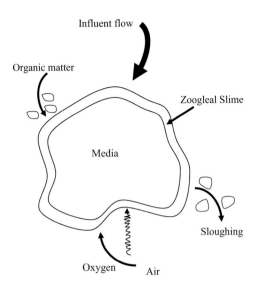

FIGURE 14.6   Filter media showing biological activities that take place on the surface area.

Trickling Filter Definitions

To clearly understand the correct operation of the trickling filter, the operator must be familiar with certain terms.

Note: The following list of terms applies to the trickling filter process. We assume that other terms related to other units within the treatment system (plant) are already familiar to operators.

- *Biological towers* – a type of trickling filter that is very deep (10–20 ft). Filled with a light-weight synthetic media, these towers are also known as oxidation or roughing towers or (because of their extremely high hydraulic loading) super-rate trickling filters.
- *Biomass* – the total mass of organisms attached to the media. Like solids inventory in the activated sludge process, it is sometimes referred to as the *zoogleal slime*.
- *Distribution arm* – the device most widely used to apply wastewater evenly over the entire surface of the media. In most cases, the force of the wastewater being sprayed through the orifices moves the arm.
- *Filter underdrain* – the open space provided under the media to collect the liquid (waste-water and sloughings) and to allow air to enter the filter. It has a sloped floor to collect the flow to a central channel for removal.
- *Hydraulic loading* – the amount of wastewater flow applied to the surface of the trickling filter media. It can be expressed in several ways: flow per square foot of surface per day (gpd/ft^2); flow per acre per day (MGAD); or flow per acre foot per day (MGAFD). The hydraulic loading includes all flow entering the filter.
- *High-rate trickling filters* – a classification in which the organic loading is in the range of 25–100 pounds of BOD$_5$ per 1,000 cubic feet of media per day. The standard rate filter may also produce a highly nitrified effluent.
- *Media* – an inert substance placed in the filter to provide a surface for the microorganism to grow on. The media can be field stone, crushed stone, slag, plastic, or redwood slats.
- *Organic loading* – the amount of BOD$_5$ or COD applied to a given volume of filter media. It does not include the BOD$_5$ or COD contributed to any recirculated flow and is commonly expressed as pounds of BOD$_5$ or COD per 1,000 cubic feet of media.
- *Recirculation* – the return of filter effluent back to the head of the trickling filter. It can level flow variations and assist in solving operational problems, such as ponding, filter flies, and odors.

- *Roughing filters* – a classification of trickling filters in which the organic is in excess of 200 pounds of BOD_5 per 1,000 cubic feet of media per day. A roughing filter is used to reduce the loading on other biological treatment processes to produce an industrial discharge that can be safely treated in a municipal treatment facility.
- *Sloughing* – the process in which the excess growth breaks away from the media and washes through the filter to the underdrains with the wastewater. These sloughings must be removed from the flow by settling.
- *Staging* – the practice of operating two or more trickling filters in series. The effluent of one filter is used as the influent of the next. This practice can produce a higher quality effluent by removing additional BOD_5 or COD.

Trickling Filter Equipment

The trickling filter distribution system is designed to spread wastewater evenly over the surface of the entire media. The most common system is the rotary distributor which moves above the surface of the media and sprays the wastewater on the surface. The force of the water leaving the orifices drives the rotary system. The distributor arms usually have small plates below each orifice to spread the wastewater into a fan-shaped distribution system. The second type of distributor is the fixed nozzle system. In this system, the nozzles are fixed in place above the media and are designed to spray the wastewater over a fixed portion of the media. This system is used frequently with deep-bed synthetic media filters.

Note: Trickling filters that use ordinary rock are normally only about 3 meters in depth because of structural problems caused by the weight of rocks – which also requires the construction of beds that are quite wide, in many applications, up to 60 feet in diameter. When synthetic media is used, the bed can be much deeper.

No matter which type of media is selected, the primary consideration is that it must be capable of providing the desired film location for the development of the biomass. Depending on the type of media used and the filter classification, the media may be 3–20 or more feet in depth. The underdrains are designed to support the media, collect the wastewater and sloughings and carry them out of the filter, and to provide ventilation to the filter.

Note: In order to ensure enough airflow to the filter, the underdrains should never be allowed to flow more than 50% full of wastewater.

The effluent channel is designed to carry the flow from the trickling filter to the secondary settling tank. The secondary settling tank provides 2–4 hours of detention time to separate the sloughing materials from the treated wastewater. Design, construction, and operation are like that of the primary settling tank. Longer detention times are provided because the sloughing materials are lighter and settle more slowly.

Recirculation pumps and piping are designed to recirculate (and thus improve the performance of the trickling filter or settling tank) a portion of the effluent back to be mixed with the filter influent. When recirculation is used, obviously, pumps and metering devices must be provided.

Filter Classifications

Trickling filters are classified by hydraulic and organic loading. Moreover, the expected performance and the construction of the trickling filter are determined by the filter classification. Filter classifications include standard rate, intermediate rate, high rate, super high rate (plastic media), and roughing rate types. Standard rate, high rate, and roughing rate are the filter types most used. The *standard rate filter* has a hydraulic loading (gpd/ft³) of from 25 to 90; has a seasonal sloughing frequency; does not employ recirculation; and typically has an 80–85% BOD_5 removal rate and 80–85% TSS removal rate. The *high rate filter* has a hydraulic loading (gpd/ft³) of 230–900; has a continuous sloughing frequency; always employs recirculation; and typically has a 65–80% BOD_5 removal rate and 65–80% TSS removal rate. The *roughing filter* has a hydraulic loading (gpd/ft³) of >900; has a continuous sloughing frequency; does not normally include recirculation; and typically has a 40–65% BOD removal rate and 40–65% TSS removal rate.

Standard Operating Procedures

Standard operating procedures for trickling filters include sampling and testing, observation, recirculation, maintenance, and expectations of performance. Collection of influent and processing effluent samples to determine performance and monitor process condition of trickling filters is required. Dissolved oxygen, pH, and settleable solids testing should be conducted daily. BOD_5 and suspended solids testing should be done as often as practical to determine the percent removal.

The operation and condition of the filter should be observed daily. Items to observe include the distributor movement, uniformity of distribution, evidence of operation or mechanical problems, and the presence of objectionable odors. In addition to the items above, the normal observation for a settling tank should also be performed. Recirculation is used to reduce organic loading, improve sloughing, reduce odors, and reduce or eliminate filter fly or ponding problems. The amount of recirculation is dependent on the design of the treatment plant and the operational requirements of the process. Recirculation flow may be expressed as a specific flow rate (i.e., 2.0 MGD). In most cases, it is expressed as a ratio (3:1, 0.5: 1.0, etc.). The recirculation is always listed as the first number and the influent flow listed as the second number. Because the second number in the ratio is always 1.0, the ratio is sometimes written as a single number (dropping the 1.0).

Flows can be recirculated from various points following the filter, to various points before the filter. The most common form of recirculation removes flow from the filter effluent or settling tank and returns it to the influent of the trickling filter as shown in Figure 14.7. Maintenance requirements include lubrication of mechanical equipment, removal of debris from the surface and orifices, as well as adjustment of flow patterns and maintenance associated with the settling tank.

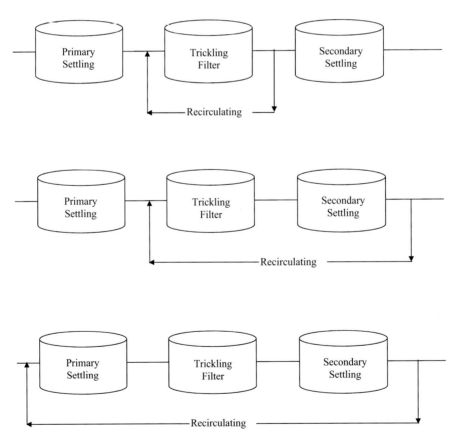

FIGURE 14.7   Common forms of recirculation.

General Process Description

The trickling filter process involves spraying wastewater over a solid media such as rock, plastic, or redwood slats (or laths). As the wastewater trickles over the surface of the media, a growth of microorganisms (bacteria, protozoa, fungi, algae, helminthes or worms, and larvae) develops. This growth is visible as a shiny slime very similar to the slime found on rocks in a stream. As the wastewater passes over this slime, the slime adsorbs the organic (food) matter. This organic matter is used for food by the microorganisms. At the same time, air moving through the open spaces in the filter transfers oxygen to the wastewater. This oxygen is then transferred to the slime to keep the outer layer aerobic. As the microorganisms use the food and oxygen, they produce more organisms, carbon dioxide, sulfates, nitrates, and other stable by-products; these materials are then discarded from the slime back into the wastewater flow and are carried out of the filter.

The growth of the microorganisms and the buildup of solid wastes in the slime make it thicker and heavier. When this slime becomes too thick, the wastewater flow breaks off parts of the slime. These must be removed in the final settling tank. In some trickling filters, a portion of the filter effluent is returned to the head of the trickling filter to level out variations in flow and improves operations (recirculation).

Overview and Brief Summary of Trickling Filter Process

A trickling filter consists of a bed of coarse media, usually rocks or plastic, covered with microorganisms.

Note: Trickling filters that use ordinary rock are normally only about 10 feet in depth because of structural problems caused by the weight of rocks, which also requires the construction of beds that are quite wide – in many applications, up to 60 feet in diameter. When synthetic media are used, the bed can be much deeper.

The wastewater is applied to the media at a controlled rate, using a rotating distributor arm or fixed nozzles. Organic material is removed by contact with the microorganisms as the wastewater trickles down through the media openings. The treated wastewater is collected by an underdrain system.

Note: To ensure enough air flow to the filter, the underdrains should never be allowed to flow more than 50% full of wastewater.

The trickling filter is usually built into a tank that contains the media. The filter may be square, rectangular, or circular.

The trickling filter does not provide any actual filtration. The filter media provides a large amount of surface area that the microorganisms can cling to and grow in a slime that forms on the media as they feed on the organic material in the wastewater.

The slime growth on the trickling filter media periodically sloughs off and is settled and removed in a secondary clarifier that follows the filter.

Trickling filter operation requires routine observation, meter readings, process control sampling and testing, and process control calculations. Comparison of daily results with expected "normal" ranges is the key to identifying problems and appropriate corrective actions.

Operator Observations

Slime – The operator checks the thickness of slime to ensure that it is thin and uniform (normal) or thick and heavy (indicates organic overload); the operator is also concerned with ensuring that excessive recirculation is not taking place and checks slime toxicity (if any). The operator is also concerned about the color of the slime: green slime is normal; dark green/black slime indicates organic overload; other colors may indicate industrial waste or chemical additive contamination. The operator should check the subsurface growth of the slime to ensure that it is normal (thin and translucent). If growth is thick and dark, organic overload conditions are indicated. Distribution arm operation is a system function important to slime formation; it must be checked regularly for proper operation. For example, the distribution of slime should be even and uniform. Striped conditions indicate clogged orifices or nozzles.

Flow – Flow distribution must be checked to ensure uniformity. If non-uniform, the arms are not level, or the orifices are plugged. Flow drainage is also important. Drainage should be uniform and rapid. If not, ponding may occur from media breakdown or debris on surface.

Distributor – Movement of the distributor is critical to proper operation of the trickling filter. Movement should be uniform and smooth. Chattering, noisy operation may indicate bearing failure. The distributor seal must be checked to ensure there is no leakage.

Recirculation – The operator must check the rate of recirculation to ensure that it is within design specifications. Rates above design specifications indicate hydraulic overloading; rates under design specifications indicate hydraulic underloading.

Note: *Recirculation* reduces the organic loading, improves sloughing, reduces odors, and reduces or eliminates filter fly or ponding problems. The amount of recirculation needed depends on the design of the treatment plant and the operational requirements of the process. Recirculation flow may be expressed as a specific flow rate (i.e., 2.0 MGD). In most cases, it is expressed as a ratio (3:1, 0.5:1.0, etc.). The recirculation is always listed as the first number, and the influent flow is listed as the second number.

Note: Because the second number in the ratio is always 1.0, the ratio is sometimes written as a single number (dropping the: 1.0).

Media – The operator should check to ensure that the medium is uniform.

Rotating Biological Contactors (RBCs)

The *rotating biological contactor* (RBC) is a biological treatment system (see Figure 14.9) and is a variation of the attached growth idea provided by the trickling filter. Still relying on microorganisms that grow on the surface of a medium, the RBC is instead a *fixed film* biological treatment device – the basic biological process, however, is like that occurring in the trickling filter. An RBC consists of a series of closely spaced (mounted side by side), circular, plastic (synthetic) disks, typically about 3.5 m in diameter and attached to a rotating horizontal shaft (see Figure 14.8). Approximately 40% of each disk is submersed in a tank containing the wastewater to be treated. As the RBC rotates, the attached biomass film (zoogleal slime) that grows on the surface of the disk moves into and out of the wastewater. While submerged in the wastewater, the microorganisms absorb organics; while

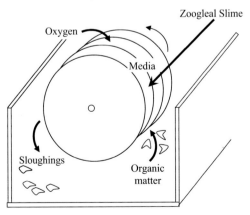

FIGURE 14.8   Rotating biological contactor (RBC) cross section and treatment system.

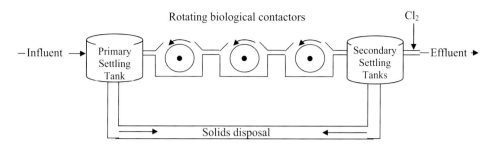

FIGURE 14.9 Rotating biological contactor (RBC) treatment system.

they are rotated out of the wastewater, they are supplied with needed oxygen for aerobic decomposition. As the zoogleal slime reenters the wastewater, excess solids and waste products are stripped off the media as sloughings. These sloughings are transported with the wastewater flow to a settling tank for removal.

Modular RBC units are placed in series (see Figure 14.9). Simply because a single contactor is not enough to achieve the desired level of treatment; the resulting treatment achieved exceeds conventional secondary treatment. Each individual contactor is called a stage and the group is known as a train. Most RBC systems consist of two or more trains with three or more stages in each. The key advantage in using RBCs instead of trickling filters is that RBCs are easier to operate under varying load conditions, since it is always easier to keep the solid medium wet. Moreover, the level of nitrification, which can be achieved by an RBC system, is significant – especially when multiple stages are employed.

RBC Equipment

The equipment that makes up an RBC includes the rotating biological contactor (the media: either standard or high density), a center shaft, drive system, tank, baffles, housing or cover, and a settling tank. The *rotating biological contactor* consists of circular sheets of synthetic material (usually plastic) which are mounted side by side on a shaft. The sheets (media) contain large amounts of surface area for growth of the biomass. The *center shaft* provides the support for the disks of media and must be strong enough to support the weight of the media and the biomass. Experience has indicated a major problem has been the collapse of the support shaft. The *drive system* provides the motive force to rotate the disks and shaft. The drive system may be mechanical, or air driven, or a combination of each. When the drive system does not provide uniform movement of the RBC, major operational problems can arise.

The *tank* holds the wastewater that the RBC rotates in. It should be large enough to permit variation of the liquid depth and detention time. *Baffles* are required to permit proper adjustment of the loading applied to each stage of the RBC process. Adjustment can be made to increase or decrease the submergence of the RBC. RBC stages are normally enclosed in some type of protective structure (*cover*) to prevent loss of biomass due to severe weather changes (snow, rain, temperature, wind, sunlight, etc.).

In many instances, this housing greatly restricts access to the RBC. The *settling tank* is provided to remove the sloughing material created by the biological activity and is similar in design to the primary settling tank. The settling tank provides 2–4 hours' detention time to permit settling of lighter biological solids.

RBC Operation

During normal operation, operator vigilance is required to observe the RBC movement, slime color, and appearance. However, if the unit is covered, observations may be limited to that portion of the media which can be viewed through the access door. Slime color and appearance can indicate process condition, for example:

- Gray, shaggy slime growth – indicates normal operation.
- Reddish brown, golden shaggy growth – indicates nitrification.
- White chalky appearance – indicates high sulfur concentrations.
- No slime – indicates severe temperature or pH changes.

Sampling and testing should be conducted daily for dissolved oxygen content and pH. BOD_5 and suspended solids testing should also be accomplished to aid in assessing performance.

RBC Expected Performance

The RBC normally produces a high-quality effluent with BOD_5 at 85–95% and suspended solids removal at 85–95%. The RBC treatment process may also significantly reduce (if designed for this purpose) the levels of organic nitrogen and ammonia nitrogen.

ACTIVATED SLUDGE

The biological treatment systems discussed to this point [ponds, trickling filters, and rotating biological contactors (RBCs)] have been around for years. The trickling filter, for example, has been around and successfully used since the late 1800s. The problem with ponds, trickling filters, and RBCs is that they are temperature sensitive, remove less BOD, and trickling filters, for example, cost more to build than the activated sludge systems that were later developed.

Note: Although trickling filters and other systems cost more to build than activated sludge systems, it is important to point out that activated sludge systems cost more to operate because of the need for energy to run pumps and blowers. As shown in Figure 14.10, the activated sludge process follows primary settling. The basic components of an activated sludge sewage treatment system include an aeration tank and a secondary basin, settling basin, or clarifier (see Figure 14.10). Primary effluent is mixed with settled solids recycled from the secondary clarifier and is then introduced into the aeration tank. Compressed air is injected continuously into the mixture through porous diffusers located at the bottom of the tank, usually along one side.

Wastewater is fed continuously into an aerated tank, where the microorganisms metabolize and biologically flocculate the organics. Microorganisms (activated sludge) are settled from the aerated mixed liquor under quiescent conditions in the final clarifier and are returned to the aeration tank. Left uncontrolled, the number of organisms would eventually become too great; therefore, some

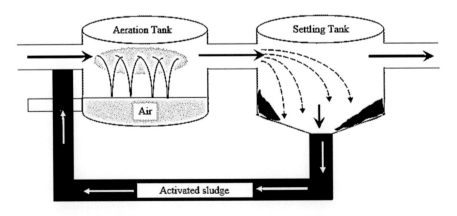

FIGURE 14.10   The activated sludge process.

must periodically be removed (wasted). A portion of the concentrated solids from the bottom of the settling tank must be removed from the process (waste activated sludge or WAS). Clear supernatant from the final settling tank is the plant effluent.

ACTIVATED SLUDGE TERMINOLOGY

To better understand the discussion of the activated sludge process presented in the following sections, you must understand the terms associated with the process. Some of these terms have been used and defined earlier in the text, but we list them here again to refresh your memory. Review these terms and remember them. They are used throughout the discussion.

Adsorption – taking in or reception of one substance into the body of another by molecular or chemical actions and distribution throughout the absorber.

Activated – to speed up reaction. When applied to sludge, it means that many aerobic bacteria and other microorganisms are in the sludge particles.

Activated sludge – a floc or solid formed by microorganisms. It includes organisms, accumulated food materials, and waste products from the aerobic decomposition process.

Activated sludge process – a biological wastewater treatment process in which a mixture or influent and activated sludge is agitated and aerated. The activated sludge is subsequently separated from the treated mixed liquor by sedimentation and is returned to the process as needed. The treated wastewater overflows the weir of the settling tank in which separation from the sludge takes place.

Adsorption – the adherence of dissolved, colloidal, or finely divided solids to the surface of solid bodies when they are brought into contact.

Aeration – mixing air and a liquid by one of the following methods: spraying the liquid in the air; diffusing air into the liquid; or agitating the liquid to promote surface adsorption of air.

Aerobic – a condition in which "free" or dissolved oxygen is present in the aquatic environment. Aerobic organisms must be in the presence of dissolved oxygen to be active.

Bacteria – single-cell plants that play a vital role in the stabilization of organic waste.

Biochemical oxygen demand – a measure of the amount of food available to the microorganisms in a waste. It is measured by the amount of dissolved oxygen used up during a specific time period (usually five days, expressed as BOD_5).

Biodegradable – from "degrade" (to wear away or break down chemically) and "bio" (by living organisms). Put it all together and you have a "substance, usually organic, which can be decomposed by biological action."

Bulking – a problem in activated sludge plants that results in poor settleability of sludge particles.

Coning – a condition that may be established in a sludge hopper during sludge withdrawal, when part of the sludge moves toward the outlet while the remainder tends to stay in place. Development of a cone or channel of moving liquids surrounded by relatively stationary sludge.

Decomposition – generally, in waste treatment, decomposition refers to the changing of waste matter into simpler, more stable forms that will not harm the receiving stream.

Diffuser – a porous plate or tube through which air is forced and divided into tiny bubbles for distribution in liquids. Commonly made of carborundum, aluminum, or silica sand.

Diffused air aeration – a diffused air-activated sludge plant takes air, compresses it, then discharges the air below the water surface to the aerator through some type of air diffusion device.

Dissolved oxygen – atmospheric oxygen dissolved in water or wastewater, usually abbreviated as DO.

Note: The typical required DO for a well-operated activated sludge plant is between 2.0-2.5 mg/L.

Facultative – facultative bacteria can use either molecular (dissolved) oxygen or oxygen obtained from food materials. In other words, facultative bacteria can live under aerobic or anaerobic conditions.

Filamentous bacteria – organisms that grow in thread or filamentous form.

Food-to-microorganisms ratio – a process control calculation used to evaluate the amount of food (BOD or COD) available per pound of mixed liquor volatile suspended solids.

Fungi – multicellular aerobic organisms.

Gould sludge age – a process control calculation used to evaluate the amount of influent suspended solids available per pound of mixed liquor suspended solids.

Mean cell residence time (MCRT) – the average length of time mixed liquor suspended solids particle remains in the activated sludge process. This is usually written as MCRT and may also be referred to as *solids retention time* (SRT).

Mixed liquor – the contribution of return activated sludge and wastewater (either influent or primary effluent) that flows into the aeration tank.

Mixed liquor suspended solids (MLSS) – the suspended solids concentration of the mixed liquor. Many references use this concentration to represent the amount of organisms in the liquor. Many references use this concentration to represent the amount of organisms in the activated sludge process. This is usually written MLSS.

Mixed liquor volatile suspended solids (MLVSS) – the organic matter in the mixed liquor suspended solids. This can also be used to represent the amount of organisms in the process. This is normally written as MLVSS.

Nematodes – microscopic worms that may appear in biological waste treatment systems.

Nutrients – substances required to support plant organisms. Major nutrients are carbon, hydrogen, oxygen, sulfur, nitrogen, and phosphorus.

Protozoa – single-cell animals that are easily observed under the microscope at a magnification of 100x. Bacteria and algae are prime sources of food for advanced forms of protozoa.

Return activated sludge – the solids returned from the settling tank to the head of the aeration tank. This is normally written as RAS.

Rising sludge – rising sludge occurs in the secondary clarifiers or activated sludge plant when the sludge settles to the bottom of the clarifier, is compacted, and then rises to the surface in a relatively short time.

Rotifers – multicellular animals with flexible bodies and cilia near their mouths used to attract food. Bacteria and algae are their major source of food.

Secondary treatment – a wastewater treatment process used to convert dissolved or suspended materials into a form that can be removed.

Settleability – a process control test used to evaluate the settling characteristics of the activated sludge. Readings taken at 30–60 minutes are used to calculate the settled sludge volume (SSV) and the sludge volume index (SVI).

Settled sludge volume – the volume of ml/L (or %) occupied by an activated sludge sample after 30 or 60 minutes of settling. Normally written as SSV with a subscript to indicate the time of the reading used for calculation (SSV_{30} or SSV_{60}).

Shock load – the arrival at a plant of a waste toxic to organisms, in enough quantity or strength to cause operating problems, such as odor or sloughing off the growth of slime on the trickling filter media. Organic overloads also can cause a shock load.

Sludge volume index – a process control calculation used to evaluate the settling quality of the activated sludge. Requires the SSV_{30} and mixed liquor suspended solids test results to calculate.

Solids – material in the solid state

Dissolved – solids present in solution. Solids that will pass through a glass fiber filter.

Fixed – also known as the inorganic solids. The solids that are left after a sample is ignited at 550°C for 15 minutes.

Floatable solids – solids that will float to the surface of still water, sewage, or other liquid. Usually composed of grease particles, oils, light plastic material, etc. Also called *scum*.

Non-settleable – finely divided suspended solids that will not sink to the bottom in still water, sewage, or other liquid in a reasonable period, usually 2 hours. Non-settleable solids are also known as colloidal solids.

Suspended – the solids that will not pass through a glass fiber filter.

Total – the solids in water, sewage, or other liquids; it includes the suspended solids and dissolved solids.

Volatile – the organic solids. Measured as the solids that are lost on ignition of the dry solids at 550°C.

Waste activated sludge – the solids being removed from the activated sludge process. This is normally written as WAS.

ACTIVATED SLUDGE PROCESS: EQUIPMENT

The equipment requirements for the activated sludge process are more complex than other processes discussed. Equipment includes an *aeration tank*, *aeration*, *system-settling tank*, *return sludge*, and *waste sludge system*. These are discussed in the following:

Aeration Tank

The *aeration tank* is designed to provide the required detention time (depends on the specific modification) and ensure that the activated sludge and the influent wastewater are thoroughly mixed. Tank design normally attempts to ensure no dead spots are created.

Aeration

Aeration can be mechanical or diffused. Mechanical aeration systems use agitators or mixers to mix air and mixed liquor. Some systems use sparge rings to release air directly into the mixer. Diffused aeration systems use pressurized air released through diffusers near the bottom of the tank. Efficiency is directly related to the size of the air bubbles produced. Fine bubble systems have a higher efficiency. The diffused air system has a blower to produce large volumes of low-pressure air (5–10 psi), air lines to carry the air to the aeration tank, and headers to distribute the air to the diffusers which release the air into the wastewater.

Settling Tank

Activated sludge systems are equipped with plain *settling tanks* designed to provide 2–4 hours' hydraulic detention time.

Return Sludge

The return sludge system includes pumps, a timer or variable speed drive to regulate pump delivery and a flow measurement device to determine actual flow rates.

Waste Sludge

In some cases, the *waste activated sludge* withdrawal is accomplished by adjusting valves on the return system. When a separate system is used it includes pump(s), timer or variable speed drive, and a flow measurement device.

OVERVIEW OF ACTIVATED SLUDGE PROCESS

The activated sludge process is a treatment technique in which wastewater and reused biological sludge full of living microorganisms are mixed and aerated. The biological solids are then separated from the treated wastewater in a clarifier and are returned to the aeration process or wasted. The microorganisms are mixed thoroughly with the incoming organic material, and they grow and reproduce by using the organic material as food. As they grow and are mixed with air, the individual organisms cling together (flocculate). Once flocculated, they more readily settle in the secondary clarifiers.

The wastewater being treated flows continuously into an aeration tank where the air is injected to mix the wastewater with the returned activated sludge and to supply the oxygen needed by the microbes to live and feed on the organics. Aeration can be supplied by injection through air diffusers in the bottom of the tank or by mechanical aerators located at the surface. The mixture of activated sludge and wastewater in the aeration tank is called the "mixed liquor." The mixed liquor flows to a secondary clarifier where the activated sludge can settle.

The activated sludge is constantly growing, and more is produced than can be returned for use in the aeration basin. Some of this sludge must, therefore, be wasted to a sludge handling system for treatment and disposal. The volume of sludge returned to the aeration basins is normally 40–60% of the wastewater flow. The rest is wasted.

FACTORS AFFECTING OPERATION OF THE ACTIVATED SLUDGE PROCESS

Several factors affect the performance of an activated sludge system. These include the following:

- Temperature
- Return rates
- Amount of oxygen available
- Amount of organic matter available
- pH
- Waste rates
- Aeration time
- Wastewater toxicity

To obtain the desired level of performance in an activated sludge system, a proper balance must be maintained between the amount of food (organic matter), organisms (activated sludge), and oxygen (dissolved oxygen). Most problems with the activated sludge process result from an imbalance between these three items.

To fully appreciate and understand the biological process taking place in a normally functioning activated sludge process, the operator must have knowledge of the key players in the process: the organisms. This makes a certain amount of sense when you consider that the heart of the activated sludge process is the mass of settleable solids formed by aerating wastewater containing biological degradable compounds in the presence of microorganisms. Activated sludge consists of organic solids plus bacteria, fungi, protozoa, rotifers, and nematodes.

GROWTH CURVE

To understand the microbiological population and its function in an activated sludge process, the operator must be familiar with the microorganism *growth curve*. In the presence of excess organic matter, the microorganisms multiply at a fast rate. The demand for food and oxygen is at its peak. Most of this is used for the production of new cells. This condition is known as the *log growth phase*. As time continues, the amount of food available for the organisms declines. Floc begins to form while the growth rate of bacteria and protozoa begins to decline. This is referred to as the *declining growth phase*.

The *endogenous respiration* phase occurs as the food available becomes extremely limited and the organism mass begins to decline. Some of the microorganisms may die and break apart, thus releasing organic matter that can be consumed by the remaining population.

The actual operation of an activated-sludge system is regulated by three factors:

1. The quantity of air supplied to the aeration tank;
2. The rate of activated-sludge recirculation; and
3. The amount of excess sludge withdrawn from the system. Sludge wasting is an important operational practice because it allows the operator to establish the desired concentration of MLSS, food/microorganism ratio, and sludge age.

Note: Air requirements in an activated sludge basin are governed by:

- biological oxygen demand loading and the desired removal effluent;
- volatile suspended solids concentration in the aerator; and
- suspended solids concentration of the primary effluent.

ACTIVATED SLUDGE FORMATION

The formation of activated sludge is dependent on three steps. The first step is the transfer of food from wastewater to organism. Second is the conversion of wastes to a usable form. Third is the flocculation step.

1. *Transfer* – Organic matter (food) is transferred from the water to the organisms. Soluble material is absorbed directly through the cell wall. Particulate and colloidal matter is absorbed into the cell wall, where it is broken down into simpler soluble forms, then absorbed through the cell wall.
2. *Conversion* – Food matter is converted to cell matter by synthesis and oxidation into end products such as CO_2, H_2O, NH_3, stable organic waste, and new cells.
3. *Flocculation* – Flocculation is the gathering of fine particles into larger particles. This process begins in the aeration tank and is the basic mechanism for the removal of suspended matter in the final clarifier. The concentrated *bio-floc* that settles and forms the sludge blanket in the secondary clarifier is known as activated sludge.

ACTIVATED SLUDGE: PERFORMANCE-CONTROLLING FACTORS

To maintain the working organisms in the activated sludge process, the operator must ensure that a suitable environment is maintained by being aware of the many factors influencing the process, and by monitoring them repeatedly. "Control" is defined as maintaining the proper solids (floc mass) concentration in the aerator for the incoming water (food) flow by adjusting the return and waste sludge pumping rate and regulating the oxygen supply to maintain a satisfactory level of dissolved oxygen in the process.

Aeration

The activated sludge process must receive enough aeration to keep the activated sludge in suspension and to satisfy the organism's oxygen requirements. Insufficient mixing results in dead spots, septic conditions, and/or loss of activated sludge.

Alkalinity

The activated sludge process requires enough alkalinity to ensure that pH remains in the acceptable range of 6.5–9.0. If organic nitrogen and ammonia are being converted to nitrate (nitrification), enough alkalinity must be available to support this process as well.

Nutrients

The microorganisms of the activated sludge process require nutrients (nitrogen, phosphorus, iron, and other trace metals) to function. If enough nutrients are not available, the process will not perform as expected. The accepted minimum ratio of carbon to nitrogen, phosphorus, and iron is 100 parts carbon to 5 parts nitrogen, 1 part phosphorus, and 0.5 parts iron.

pH

The pH of the mixed liquor should be maintained within the range of 6.5–9.0 (6.0–8.0 is ideal). Gradual fluctuations within this range will normally not upset the process. Rapid fluctuations or fluctuations outside this range can reduce organism activity.

Temperature

As temperature decreases, the activity of the organisms will also decrease. Cold temperatures also require a longer recovery time for systems that have been upset. Warm temperatures tend to favor denitrification and filamentous growth.

Note: The activity level of bacteria within the activated sludge process increases with a rise in temperature.

Toxicity

Enough concentrations of elements or compounds that enter a treatment plant that can kill the microorganisms (the activated sludge) are known as toxic waste (shock level). Common to this group are cyanides and heavy metals.

Note: A typical example of a toxic substance added by operators is the uninhabited use of chlorine for odor control or control of filamentous organisms (prechlorination).

Chlorination is for disinfection. Chlorine is a toxicant and should not be allowed to enter the activated sludge process; it is not selective with respect to the type of organisms damaged or killed. It may kill the organisms that should be retained in the process as workers. Chlorine is very effective in disinfecting the plant effluent after treatment by the activated sludge process, however.

Hydraulic Loading

Hydraulic loading is the amount of flow entering the treatment process. When compared with the design capacity of the system, it can be used to determine if the process is hydraulically overloaded or underloaded. If more flow is entering the system than it was designed to handle, the system is hydraulically overloadcd. If less flow is entering the system than it was designed for, the system is hydraulically underloaded.

Generally, the system is more affected by overloading than by underloading. Overloading can be caused by stormwater, infiltration of groundwater, excessive return rates, or many other causes. Underloading normally occurs during periods of drought or in the period following initial startup, when the plant has not reached its design capacity. Excess hydraulic flow rates through the treatment plant will reduce the efficiency of the clarifier by allowing activated sludge solids to rise in the clarifier and pass over the effluent weir. This loss of solids in the effluent degrades effluent quality and reduces the amount of activated sludge in the system, in turn, reducing process performance.

Organic Loading

Organic loading is the amount of organic matter entering the treatment plant. It is usually measured as biochemical oxygen demand. An organic overload occurs when the amount of BOD entering the system exceeds the design capacity of the system. An organic underload occurs when the amount of BOD entering the system is significantly less than the design capacity of the plant. Organic overloading may occur when the system receives more waste than it was designed to handle. It can also occur when an industry or other contributor discharges more wastes to the system than originally planned. Wastewater treatment plant processes can also cause organic overloads returning high-strength wastes from the sludge treatment processes.

Regardless of the source, an organic overloading of the plant results in increased demand for oxygen. This demand may exceed the air supply available from the blowers. When this occurs, the activated sludge process may become septic. Excessive wasting can also result in a type of organic overload. The food available exceeds the number of activated sludge organisms, resulting in increased oxygen demand and very rapid growth.

Organic underloading may occur when a new treatment plant is initially put into service. The facility may not receive enough waste to allow the plant to operate at its design level. Underloading can also occur when excessive amounts of activated sludge can remain in the system. When this occurs, the plant will have difficulty in developing and maintaining a good activated sludge.

Oxidation Ditches

An oxidation ditch is a modified extended aeration activated sludge biological treatment process that utilizes long solids retention times (SRTs) to remove biodegradable organics. Oxidation ditches are typically complete mix systems, but they can be modified to approach plug flow conditions.

Note: As conditions approach plug flow, diffused air must be used to provide enough mixing. The system will also no longer operate as an oxidation ditch.

Typical oxidation ditch treatment systems consist of a single- or multi-channel configuration within a ring, oval, or horseshoe-shaped basin. As a result, oxidation ditches are called "racetrack type" reactors. Horizontally or vertically mounted aerators provide circulation, oxygen transfer, and aeration in the ditch.

Preliminary treatment, such as bar screens and grit removal, normally precedes the oxidation ditch. Primary settling prior to an oxidation ditch is sometimes practiced but is not typical in this design. Tertiary filters may be required after clarification, depending on the effluent requirements. Disinfection is required and re-aeration may be necessary prior to final discharge. Flow to the oxidation ditch is aerated and mixed with return sludge from a secondary clarifier. Surface aerators, such as brush rotors, disc aerators, draft tube aerators, or fine bubble diffusers are used to circulate the mixed liquor. The mixing process entrains oxygen into the mixed liquor to foster microbial growth, and the motive velocity ensures contact of microorganisms with the incoming wastewater.

The aeration sharply increases the dissolved oxygen concentration but decreases as biomass uptakes oxygen as the mixed liquor travels through the ditch. Solids are maintained in suspension as the mixed liquor travels through the ditch. Solids are maintained in suspension as the mixed liquor circulated around the ditch. If design SRTs are selected for nitrification, a high degree of nitrification will occur. Oxidation ditch effluent is usually settled in a separate, secondary clarifier. An anaerobic tank may be added prior to the ditch to enhance biological phosphorus removal.

An oxidation ditch may also be operated to achieve partial denitrification. One of the most common design modifications for enhanced nitrogen removal is known as the Modified Ludzack-Ettinger (MLE) process. In this process an anoxic tank is added upstream of the ditch along with mixed liquor recirculation for the aerobic zone to the tank to achieve higher levels of denitrification. In the aerobic basin, autotrophic bacteria (nitrifiers) convert ammonia-nitrogen to nitrite-nitrogen and then to nitrate-nitrogen. In the anoxic zone, heterotrophic bacteria convert nitrate-nitrogen to nitrogen gas which is released into the atmosphere. Some mixed liquor from the aerobic basin is recirculated to the anoxic zone to provide mixed liquor with a high concentration of nitrate-nitrogen to the anoxic zone.

Several manufacturers have developed modifications to the oxidation ditch design to remove nutrients in conditions cycled or phased between the anoxic and aerobic states. While the mechanics of operation differ by manufacturer, in general, the process consists of two separate aeration basins, the first anoxic and the second aerobic. Wastewater and return activated sludge (RAS) are introduced into the first reactor, which operates under anoxic conditions.

Mixed liquor then flows into the second reactor operating under aerobic conditions. The process is then reversed, and the second reactor begins to operate under anoxic conditions. About applicability, the oxidation ditch process is a fully demonstrated secondary wastewater treatment technology, applicable in any situation where activated sludge treatment (conventional or extended aeration) is appropriate. Oxidation ditches are applicable in plants that require nitrification because the basins can be sized using an appropriate SRT to achieve nitrification at the mixed liquor minimum temperature.

This technology is very effective in small installations, small communities, and isolated institutions because it requires more land than conventional treatment plants (USEPA, 2000). There are currently more than 9,000 municipal oxidation ditch installations in the United States (Spellman, 2007). Nitrification to less than 1 mg/L ammonia nitrogen consistently occurs when ditches are designed and operated for nitrogen removal. An excellent example of an upgrade to the MLE process is provided in the following case. Keep in mind that the motivation for this upgrade was twofold: to increase optimal plant operation (DO optimization) and to conserve energy.

Advantages and Disadvantages

Advantages

The main advantage of the oxidation ditch is the ability to achieve removal performance objectives with low operational requirements and operation and maintenance costs.

Some specific advantages of oxidation ditches include:

- An added measure of reliability and performance over other biological processes owing to a constant water level and continuous discharge, which lowers the weir overflow rate and eliminates the periodic effluent surge common to other biological processes, such as SBRS.
- Long hydraulic retention time and complete mixing minimize the impact of a shock load or hydraulic surge.
- Produces less sludge than other biological treatment processes owing to extended biological activity during the activated sludge process.
- Energy-efficient operations result in reduced energy costs compared with other biological treatment processes.

Disadvantages

- Effluent suspended solids concentrations are relatively high compared to other modifications of the activated sludge process.
- Requires a larger land area than other activated sludge treatment options. This can prove costly, limiting the feasibility of oxidation ditches in urban, suburban, or other areas where land acquisition costs are relatively high (USEPA, 2000).

ACTIVATED SLUDGE PROCESS CONTROL PARAMETERS

When operating an activated sludge process, the operator must be familiar with the many important process control parameters, which must be monitored frequently and adjusted occasionally to maintain optimal performance.

Alkalinity

Monitoring alkalinity in the aeration tank is essential to control of the process. Insufficient alkalinity will reduce organism activity and may result in low effluent pH and, in some cases, extremely high chlorine demand in the disinfection process.

Dissolved Oxygen (DO)

The activated sludge process is an aerobic process that always requires some dissolved oxygen be present. The amount of oxygen required is dependent on the influent food (BOD), the activity of the activated sludge, and the degree of treatment desired.

pH

Activated sludge microorganisms can be injured or destroyed by wide variations in pH. The pH of the aeration basin will normally be in the range of 6.5–9.0. Gradual variations within this range will not cause any major problems; however, rapid changes of one or more pH units can have a significant impact on performance. Industrial waste discharges, septic wastes, or significant amounts of stormwater flows may produce wide variations in pH. pH should be monitored as part of the routine process control-testing schedule. Sudden changes or abnormal pH values may indicate an industrial discharge of strongly acidic or alkaline wastes. Because these wastes can upset the environmental balance of the activated sludge, the presence of wide pH variations can result in poor performance. Processes undergoing nitrification may show a significant decrease in effluent pH.

Mixed Liquor Suspended Solids, Mixed Liquor Volatile Suspended Solids, and Mixed Liquor Total Suspended Solids

The mixed liquor suspended solids (MLSS) or mixed liquor volatile suspended solids (MLVSS) can be used to represent the activated sludge or microorganisms present in the process. Process control calculations, such as sludge age and sludge volume index, cannot be calculated unless the MLSS is determined. To adjust the MLSS and MLVSS, the operator must increase or decrease the waste sludge rates. The mixed liquor total suspended solids or MLTSS is an important activated sludge control parameter. To increase the MLTSS, for example, the operator must decrease the waste rate and/or increase the MCRT. The MCRT must be decreased to prevent the MLTSS from changing when the number of aeration tanks in service is reduced.

Note: In performing the Gould Sludge Age Test, assume that the source of the MLTSS in the aeration tank is influent solids.

Return Activated Sludge Rate and Concentration

The sludge rate is a critical control variable. The operator must maintain a continuous return of activated sludge to the aeration tank, or the process will show a drastic decrease in performance. If the rate is too low, solids remain in the settling tank, resulting in solids loss and a septic return. If the rate is too high, the aeration tank can become hydraulically overloaded, causing reduced aeration time and poor performance. The return concentration is also important because it may be used to determine the return rate required to maintain the desired MLSS.

Waste Activated Sludge Flow Rate

Because the activated sludge contains living organisms that grow, reproduce, and produce waste matter, the amount of activated sludge is continuously increasing. If the activated sludge remains in the system too long, the performance of the process will decrease. If too much activated sludge is removed from the system, the solids become very light and will not settle quickly enough to be removed in the secondary clarifier.

Temperature

Because temperature directly affects the activity of the microorganisms, accurate monitoring of temperature can be helpful in identifying the causes of significant changes in organization populations or process performance.

Sludge Blanket Depth

The separation of solids and liquids in the secondary clarifier results in a blanket of solids. If solids are not removed from the clarifier at the same rate they enter, the blanket will increase in depth. If this occurs, the solids may carry over into the process effluent. The sludge blanket depth may be affected by other conditions, such as temperature variation, toxic wastes, or sludge bulking. The best sludge blanket depth is dependent upon such factors as hydraulic load, clarifier design, sludge characteristics, and many more. The best blanket depth must be determined on an individual basis by experimentation.

Note: In measuring sludge blanket depth, it is general practice to use a 15–20-feet-long clear plastic pipe marked at 6-inch intervals; the pipe is equipped with a ball valve at the bottom.

Activated Sludge Operational Control Levels

(Much of the information in this section is based on *Activated Sludge Process Control*, Part II, 2nd ed. Virginia Water Control Board, 1990.) The operator has two methods available to operate an activated sludge system. The operator can wait until the process performance deteriorates and make drastic changes, or the operator can establish normal operational levels and make minor adjustments to keep the process within the established operational levels.

Note: Control levels can be defined as the upper and lower values for a process control variable that can be expected to produce the desired effluent quality.

Although the first method will guarantee plant performance will always be maintained within effluent limitations, the second method has a much higher probability of achieving this objective. This section discusses methods used to establish *normal* control levels for the activated sludge process. Several major factors should be considered when establishing control levels for the activated sludge system.

These include the following:

- Influent characteristics
- Industrial contributions
- Process side streams
- Seasonal variations
- Required effluent quality

Influent Characteristics

Influent characteristics were discussed earlier. However, a major area to consider when evaluating influent characteristics is the nature and volume of industrial contributions to the system. Waste characteristics (BOD, solids, pH, metals, toxicity, and temperature), volume, and discharge pattern (continuous, slug, daily, weekly, etc.) should be evaluated when determining if a waste will require pretreatment by the industry or adjustments to operational control levels.

Industrial Contributions

One or more industrial contributors produce a significant portion of the plant loading (in many systems). Identifying and characterizing all industrial contributors is important. Remember that the volume of waste generated may not be as important as the characteristics of the waste. Extremely high-strength wastes can result in organic overloading and/or poor performance because of insufficient nutrient availability. A second consideration is the presence of materials that even in small quantities are toxic to the process microorganisms or that create a toxic condition in the plant effluent or plant sludge. Industrial contributions to a biological treatment system should be thoroughly characterized prior to acceptance, monitored frequently, and controlled by either local ordinance or by the implementation of a pretreatment program.

Process Side Streams

Process side streams are flows produced in other treatment processes that must be returned to the wastewater system for treatment prior to disposal.

Examples of process side streams include the following:

- Thickener supernatant
- Aerobic and anaerobic digester supernatant
- Liquids removed by sludge dewatering processes (filtrate, centrate, and substate)
- Supernatant from heat treatment and chlorine oxidation sludge treatment processes

Testing these flows periodically to determine both their quantity and strength is important. In many treatment systems, a significant part of the organic and/or hydraulic loading for the plant is generated by side stream flows. The contribution of the plant side stream flows can significantly change the operational control levels of the activated sludge system.

Seasonal Variations

Seasonal variations in temperature, oxygen solubility, organism activity, and waste characteristics may require several *normal* control levels for the activated sludge process. For example, during cold months of the year, aeration tank solids levels may have to be maintained at significantly higher levels than are required during warm weather. Likewise, the aeration rate may be controlled by the mixing requirements of the system during the colder months and by the oxygen demand of the system during the warm months.

Control Levels at Startup

Control levels for an activated sludge system during startup are usually based upon design engineer recommendations or information available from recognized reference sources. Although these levels provide a starting point, you should recognize that both the process control parameter sensitivity and control levels should be established on a plant-by-plant basis.

During the first 12 months of operation, you should evaluate all potential process control options to determine the following:

- Sensitivity to effluent quality changes
- Seasonal variability
- Potential problems

VISUAL INDICATORS FOR INFLUENT OR AERATION TANK

Wastewater operators are required to monitor or to make certain observations of treatment unit processes to ensure optimum performance – and to adjust when required. In monitoring the operation of an aeration tank, the operator should look for three physical parameters (turbulence, surface foam and scum, and sludge color and odor) that aid in determining how the process is operating and indicate if any operational adjustments should be made. This information should be recorded each time operational tests are performed. We summarize aeration tank and secondary settling tank observations in the following sections. Remember that many of these observations are very subjective and must be based upon experience. Plant personnel must be properly trained on the importance of ensuring that recorded information is consistent throughout the operating period.

Turbulence

Normal operation of an aeration basin includes a certain amount of turbulence. This turbulent action is, of course, required to ensure a consistent mixing pattern. However, whenever excessive, deficient, or nonuniform mixing occurs, adjustments may be necessary to airflow, or diffusers may need cleaning or replacement.

Surface Foam and Scum

The type, color, and amount of foam or scum present may indicate the required wasting strategy to be an employee.

Types of foam include the following:

- *Fresh, crisp, white foam* – moderate amounts of crisp white foam are usually associated with activated sludge processes producing an excellent final effluent. Adjustment: None, normal operation.
- *Thick, greasy, dark tan foam* – a thick greasy dark tan or brown foam or scum normally indicates an old sludge that is over-oxidized; high mixed liquor concentration; waste rate too high. Adjustment: Indicates old sludge, more wasting required.

- *White billowing foam* – large amounts of a white, soap suds-like foam, indicate a very young, under-oxidized sludge. Adjustment: Young sludge, less wasting required.

Sludge Color and Odor

Though not as reliable an indicator of process operations as foam, sludge colors and odor are also useful indicators. Colors and odors that are important include the following:

- *Chocolate brown/earthy odor* – indicates normal operation.
- *Light* tan *or brown/no odor* – indicates sand and clay from infiltration/inflow. Adjustment: extremely young sludge, decrease wasting.
- *Dark brown/earthy odor* – indicates old sludge, high solids. Adjustment: Increase wasting.
- *Black color/rotten egg odor* – indicates septic conditions; low dissolved oxygen concentration; airflow rate too low. Adjustment: Increase aeration.

Mixed Liquor Color

A light chocolate-brown mixed liquor color indicates a well-operated activated sludge process.

FINAL SETTLING TANK (CLARIFIER) OBSERVATIONS

Settling tank observations include flow pattern (normally uniform distribution), settling, amount and type of solids leaving with the process effluent (normally very low), and the clarity or turbidity of the process effluent (normally very clear).

Observations should include the following conditions:

- *Sludge bulking* – occurs when solids are evenly distributed throughout the tank and leaving over the weir in large quantities.
- *Sludge Solids Washout* – sludge blanket is down but solids are flowing over the effluent weir in large quantities. Control tests indicate good quality sludge.
- *Clumping* – large "clumps" or masses of sludge (several inches or more) rise to the top of the settling tank.
- *Ashing* – fine particles of gray to white material flowing over the effluent weir in large quantities.
- *Straggler Floc* – small, almost transparent, very fluffy, buoyant solids particles ($^1/8^1/4$ in. in diameter rising to the surface). Usually is accompanied by a very clean effluent. Usually new growth is most noted in the early morning hours. Sludge age is slightly below optimum.
- *Pin Floc* – very fine solids particles (usually less than $^1/32$ in. in diameter) suspended throughout the lightly turbid liquid. Usually the result of an over-oxidized sludge.

PROCESS CONTROL TESTING AND SAMPLING

The activated sludge process generally requires more sampling and testing to maintain adequate process control than any of the other unit processes in the wastewater treatment system. During periods of operational problems, both the parameters are tested, and the frequency of testing may increase substantially.

Process control testing may include settleability testing to determine the settled sludge volume; suspended solids testing to determine influent and mixed liquor suspended solids; return activated sludge solids and waste activated sludge concentrations; determination of the volatile content of the mixed liquor suspended solids; dissolved oxygen and pH of the aeration tank; BOD_5 and/or chemical oxygen demand (COD) of the aeration tank influent and process effluent; and microscopic evaluation of the activated sludge to determine the predominant organism.

SOLIDS CONCENTRATION: SECONDARY CLARIFIER

The solids concentration in the secondary clarifier can be assumed to be equal to the solids concentration in the aeration tank effluent. It may also be determined in the laboratory using a core sample taken from the secondary clarifier. The secondary clarifier solids concentration can be calculated as an average of the secondary effluent suspended solids and the return activated sludge suspended solids concentration.

ACTIVATED SLUDGE PROCESS RECORDKEEPING REQUIREMENTS

Wastewater operators soon learn that recordkeeping is a major requirement and responsibility of their jobs. Records are important (essential) for process control, for providing information on the cause of problems, for providing information for making seasonal changes, and for compliance with regulatory agencies. Records should include sampling and testing data, process control calculations, meter readings, process adjustments, operational problems and corrective action taken, and process observations.

DISINFECTION OF WASTEWATER

Like drinking water, liquid wastewater effluent is disinfected. Unlike drinking water, wastewater effluent is disinfected not to directly (direct end-of-pipe connection) protect a drinking water supply but instead is treated to protect public health in general. This is particularly important when the secondary effluent is discharged into a body of water used for swimming or for a downstream water supply. In the treatment of water for human consumption, treated water is typically chlorinated (although ozonation is also currently being applied in many cases). Chlorination is the preferred disinfection in potable water supplies because of chlorine's unique ability to provide a residual. This chlorine residual is important because when treated water leaves the waterworks facility and enters the distribution system, the possibility of contamination is increased. The residual works to continuously disinfect water right up to the consumer's tap.

In this section, we discuss basic chlorination and dechlorination. In addition, we describe ultraviolet (UV) irradiation, ozonation, bromine chlorine, and no disinfection. Keep in mind that much of the chlorination material presented in the following is like the chlorination information presented earlier.

CHLORINE DISINFECTION

Chlorination for disinfection, as shown in Figure 14.1, follows all other steps in conventional wastewater treatment. The purpose of chlorination is to reduce the population of organisms in the wastewater to levels low enough to ensure that pathogenic organisms will not be present in enough quantity to cause disease when discharged.

Note: Chlorine gas is heavier than air (vapor density of 2.5). Therefore, exhaust from a chlorinator room should be taken from floor level.

Note: The safest action to take in the event of a major chlorine container leak is to call the fire department.

Note: You might wonder why it is that chlorination of critical waters, such as natural trout streams, is not normal practice. This practice is strictly prohibited because chlorine and its by-products (i.e., chloramines) are extremely toxic to aquatic organisms.

Chlorination Terminology

Remember that there are several terms used in discussing disinfection by chlorination. Because it is important for the operator to be familiar with these terms, we repeat key terms again.

- *Chlorine* – a strong oxidizing agent which has a strong disinfecting capability. A yellow-green gas which is extremely corrosive and is toxic to humans in extremely low concentrations in air.
- *Contact Time* – the length of time the disinfecting agent and the wastewater remain in contact.
- *Demand* – the chemical reactions, which must be satisfied before a residual or excess chemical will appear.
- *Disinfection* – refers to the selective destruction of disease-causing organisms. All the organisms are not destroyed during the process. This differentiates disinfection from sterilization, which is the destruction of all organisms.
- *Dose* – the amount of chemical being added in milligrams/liter.
- *Feed Rate* – the amount of chemical being added in pounds/day.
- *Residual* – the amount of disinfecting chemical remaining after the demand has been satisfied.
- *Sterilization* – the removal of all living organisms.

Wastewater Chlorination: Facts and Process Description
Chlorine Facts

- Elemental chlorine (Cl_2 – gaseous) is a yellow-green gas, 2.5 times heavier than air.
- The most common use of chlorine in wastewater treatment is for disinfection. Other uses include odor control and activated sludge bulking control. Chlorination takes place prior to the discharge of the final effluent to the receiving waters (see Figure 14.1).
- Chlorine may also be used for nitrogen removal, through a process called breakpoint chlorination. For nitrogen removal, enough chlorine is added to the wastewater to convert all the ammonium nitrogen gas. To do this, approximately 10 mg/L of chlorine must be added for every 1 mg/L of ammonium nitrogen in the wastewater.
- For disinfection, chlorine is fed manually or automatically into a chlorine contact tank or basin, where it contacts flowing wastewater for at least 30 minutes to destroy disease-causing microorganisms (pathogens) found in treated wastewater.
- Chorine may be applied as a gas, a solid, or in liquid hypochlorite form.
- Chorine is a very reactive substance. It has the potential to react with many different chemicals (including ammonia), as well as with organic matter. When chlorine is added to wastewater, several reactions occur.
- Chlorine will react with any reducing agent (i.e., sulfide, nitrite, iron, and thiosulfate) present in wastewater. These reactions are known as chlorine demand The chlorine used for these reactions is not available for disinfection.
- Chlorine also reacts with organic compounds and ammonia compounds to form chlororganics and chloramines. Chloramines are part of the group of chlorine compounds that have disinfecting properties and show up as part of the chlorine residual test.
- After all the chlorine demands are met, the addition of more chlorine will produce free residual chlorine. Producing free residual chlorine in wastewater requires very large additions of chlorine.

Hypochlorite Facts
Hypochlorite is relatively safe to work with, though there are some minor hazards associated with its use (skin irritation, nose irritation, and burning eyes). It is normally available in dry form as a white powder, pellet, or tablet or in liquid form. It can be added directly using a dry chemical feeder or dissolved and fed as a solution.

Note: In most wastewater treatment systems, disinfection is accomplished by means of combined residual.

Wastewater Chlorination Process Description

Chlorine is a very reactive substance. Chlorine is added to wastewater to satisfy all chemical demands, that is, to react with certain chemicals (such as sulfide, sulfite, ferrous iron, etc.). When these initial chemical demands have been satisfied, chlorine will react with substances such as ammonia to produce chloramines and other substances which, although not as effective as chlorine, have disinfecting capability.

This produces a combined residual, which can be measured using residual chlorine test methods. If additional chlorine is added, free residual chlorine can be produced. Due to the chemicals normally found in wastewater, chlorine residuals are normally combined, rather than free residuals. Control of the disinfection process is normally based upon maintaining total residual chlorine of at least 1.0 mg/L for a contact time of at least 30 minutes at design flow.

Note: Residual level, contact time, and effluent quality affect disinfection. Failure to maintain the desired residual levels for the required contact time will result in lower efficiency and increased probability that disease organisms will be discharged.

Based on water quality standards, total residual limitations on chlorine are:

- *Fresh Water* – Less than 11 ppb total residual chlorine.
- *Estuaries* – Less than 7.5 ppb for halogen produced oxidants.
- *Endangered Species* – Use of chlorine is prohibited.

Chlorination Equipment

Hypochlorite Systems

Dependent on the form of hypochlorite selected for use, special equipment is required to control the addition of hypochlorite to the wastewater. Liquid forms require the use of metering pumps, which can deliver varying flows of hypochlorite solution. Dry chemicals require the use of a feed system designed to provide variable doses of the form used. The tablet form of hypochlorite requires the use of a tablet chlorinator designed specifically to provide the desired dose of chlorine. The hypochlorite solution or dry feed system dispenses the hypochlorite, which is then mixed with the flow. The treated wastewater then enters the contact tank to provide the required contact time.

Chlorine Systems

Because of the potential hazards associated with the use of chlorine, the equipment requirements are significantly greater than those associated with hypochlorite use. The system most widely used is a *solution feed system*. In this system, chlorine is removed from the container at a flow rate controlled by a variable orifice. Water moving through the chlorine injector creates a vacuum, which draws the chlorine gas to the injector and mixes it with the water. The chlorine gas reacts with the water to form hypochlorous and hydrochloric acid. The solution is then piped to the chlorine contact tank and dispersed into the wastewater through a diffuser. Larger facilities may withdraw the liquid form of chlorine and use evaporators (heaters) to convert it to the gas form. Small facilities will normally draw the gas form of chlorine from the cylinder. As gas is withdrawn, liquid will be converted to the gas form. This requires heat energy and may result in chlorine line freeze-up if the withdrawal rate exceeds the available energy levels.

In either type of system, normal operation requires adjustment of feed rates to ensure the required residual levels are maintained. This normally requires chlorine residual testing and adjustment based upon the results of the test. Other activities include removal of accumulated solids from the contact tank, collection of bacteriological samples to evaluate process performance, and maintenance of safety equipment (respirator-air pack, safety lines, etc.). Hypochlorite operation may also include making up of solution (solution feed systems), adding powder or pellets to the dry chemical feeder or tablets to the tablet chlorinator. Chlorine operations include adjustment of chlorinator feed rates, the inspection of mechanical equipment, testing for leaks using ammonia swab (white smoke means leaks), changing containers (requires more than one person for safety), and adjusting the injector water feed rate when required. Chlorination requires routine testing of plant effluent for total residual chlorine and may also require collection and analysis of samples to determine the fecal coliform concentration in the effluent.

Dechlorination

The purpose of *dechlorination* is to remove chlorine and reaction products (chloramines) before the treated waste stream is discharged into its receiving waters. Dechlorination follows chlorination usually at the end of the contact tank to the final effluent. Sulfur dioxide gas, sodium sulfate, sodium metabisulfate, or sodium bisulfates are the chemicals used to dechlorinate. No matter which chemical is used to dechlorinate, its reaction with chlorine is instantaneous.

CHLORINATION ENVIRONMENTAL HAZARDS AND SAFETY

Chlorine is an extremely toxic substance which can, when released to the environment, cause severe damage. For this reason, most state regulatory agencies have established a chlorine water quality standard (e.g., in Virginia, 0.011 mg/L in fresh waters for total residual chlorine and 0.0075 mg/L for chlorine produced oxidants in saline waters). Studies have indicated that above these levels chlorine can reduce shellfish growth and destroy sensitive aquatic organisms. This standard has resulted in many treatment facilities being required to add an additional process to remove the chlorine prior to discharge. As mentioned, the process, known as dechlorination, uses chemicals which react quickly with chlorine to convert it to a less harmful form. Elemental chlorine is a chemical with potentially fatal hazards associated with it. For this reason, many different state and federal agencies regulate the transport, storage and use of chlorine. All operators required to work with chlorine should be trained in proper handling techniques. They should also be trained to ensure that all procedures for storage, transport, handling, and use of chlorine follow appropriate state and federal regulations.

Note: Because chlorine is only shipped in full containers, unless asked specifically for chlorine required or used during a specified period, all decimal parts of a cylinder are rounded up to the next highest number of full cylinders.

ULTRAVIOLET RADIATION

Although ultraviolet (UV) disinfection was recognized as a method for achieving disinfection in the late nineteenth century, its application virtually disappeared with the evolution of chlorination technologies. However, in recent years, there has been a resurgence in its use in the wastewater field, largely because of concern for the discharge of toxic chlorine residual. Even more recently, UV has gained more attention because of the tough new regulations on chlorine use imposed by both OSHA and USEPA. Because of this relatively recent increased regulatory pressure, many facilities are actively engaged in substituting chlorine for other disinfection alternatives. Moreover, UV technology itself has made many improvements, which now makes UV attractive as a disinfection alternative. Ultraviolet light has very good germicidal qualities and is very effective in destroying microorganisms. It is used in hospitals, biological testing facilities, and many other similar locations. In wastewater treatment, the plant effluent is exposed to ultraviolet light of a specified wavelength and intensity for a specified contact period. The effectiveness of the process is dependent upon:

- UV light intensity;
- Contact time;
- Wastewater quality (turbidity)

For any one treatment plant, disinfection success is directly related to the concentration of colloidal and particulate constituents in the wastewater.

The Achilles' heel of UV for disinfecting wastewater is turbidity. If the wastewater quality is poor, the ultraviolet light will be unable to penetrate the solids, and the effectiveness of the process decreases dramatically. For this reason, many states limit the use of UV disinfection to facilities that can reasonably be expected to produce an effluent containing ≤30 mg/L, or less of BOD_5 and TSS. The main components of a UV disinfection system are mercury arc lamps, a reactor, and ballasts. The source of UV radiation is either the low-pressure or medium-pressure mercury arc lamp with low or high intensities.

Note: In the operation of UV systems, UV lamps must be readily available when replacements are required. The best lamps are those with a stated operating life of at least 7,500 hours and those that do not produce significant amounts of ozone or hydrogen peroxide. The lamps must also meet technical specifications for intensity, output, and arc length. If the UV light tubes are submerged in the waste stream, they must be protected inside quartz tubes, which not only protect the lights but also make cleaning and replacement easier. Contact tanks must be used with UV disinfection. They must be designed with the banks of UV lights in a horizontal position, either parallel or perpendicular to the flow or with banks of lights placed in a vertical position perpendicular to the flow.

Note: The contact tank must provide, at a minimum, 10-second exposure time.

We stated earlier that turbidity problems have been the problem with using UV in wastewater treatment – and this is the case. However, if turbidity is its Achilles' heel, then the need for increased maintenance (as compared to other disinfection alternatives) is the toe of the same foot. UV maintenance requires that the tubes be cleaned on a regular basis or as needed. In addition, periodic acid washing is also required to remove chemical buildup. Routine monitoring is required. Monitoring to check on bulb burnout, buildup of solids on quartz tubes, and UV light intensity is necessary.

Note: UV light is extremely hazardous to the eyes. Never enter an area where UV lights are in operation without proper eye protection. Never look directly into the ultraviolet light.

Advantages and Disadvantages

Advantages

- UV disinfection is effective at inactivating most viruses, spores, and cysts.
- UV disinfection is a physical process rather than a chemical disinfectant, which eliminates the need to generate, handle, transport, or store toxic/hazardous or corrosive chemicals.
- There is no residual effect that can be harmful to humans or aquatic life.
- UV disinfection is user-friendly for operators.
- UV disinfection has a shorter contact time when compared with other disinfectants (approximately 20–30 seconds with low-pressure lamps).
- UV disinfection equipment requires less space than other methods.

Disadvantages

- Low dosages may not effectively inactivate some viruses, spores, and cysts.
- Organisms can sometimes repair and reverse the destructive effects of UV through a "repair mechanism" known as photo reactivation, or in the absence of light known as "dark repairs."
- A preventive maintenance program is necessary to control fouling of tubes.
- Turbidity and total suspended solids in the wastewater can render UV disinfection ineffective. UV disinfection with low-pressure lamps is not as effective for secondary effluent with TSS levels above 30 mg/L.
- UV disinfection is not as cost competitive when chlorination dechlorination is used and fire codes are met (USEPA, 1999a).

Microbial Repair

Many microorganisms have enzyme systems that repair damage caused by UV light. Repair mechanisms are classified as either photorepair to dark repair (Knudson, 1985). Microbial repair can increase the UV dose needed to achieve a given degree of inactivation of a pathogen, but the process does not prevent activation. Even though microbial repair can occur, neither photorepair nor dark repair is anticipated to affect the performance of drinking water UV disinfection.

Photorepair

In photorepair (or photoreactivation), enzymes energized by exposure to light between 310 and 490 nm (near and in the visible range) break the covalent bonds that form the pyrimidine dimmers. Photoreceptor requires reactivating light and repairs only pyrimidine dimmers (Jagger, 1967). Knudson (1985) found that bacteria have the enzymes necessary for photorepair. Unlike bacteria, viruses lack the necessary enzymes for repair but can repair using the enzymes of a host cell (Rauth, 1965). Linden et al. (2002) did not observe photorepair of *Giardia* at UV does typical of UV disinfection applications (16 and 40 mJ/cm^2). However, unpublished data from the same study show *Giardia* reactivation in light conditions at very low UV doses (0.5 mJ/cm^2 (USEPA, 2006). Shin et al. (2001) reported that Cryptosporidium does not regain infectivity after inactivation by UV light. One study showed that Cryptosporidium can undergo some DNA photorepair (Oguma et al., 2001). Even though the DNA is repaired, however, infectivity is not restored.

Dark Repair

Dark repair is defined as any repair process that does not require the presence of light. The term is somewhat misleading because dark repair can also occur in the presence of light. Excision repair, a form of dark repair, is an enzyme-mediated process in which the damaged section of DNA is removed and regenerated using the existing complementary stand of DNA. As such, excision repair can occur only with double-stranded DNA and RNA. The extent of dark repair varies with the microorganism. With bacteria and protozoa, dark repair enzymes start to act immediately following exposure to UV light; therefore, reported dose-response data are assumed to account for dark repair. Knudson (1985) found that bacteria can undergo dark repair, but some lack the enzymes needed for dark repair. Viruses also lack the necessary enzymes for repair but can repair using the enzymes of a hot cell (Rauth, 1965). Oguma et al. (2001) used an assay that measures the number of dimmers formed in nucleic acid to show that dark repair occurs in Cryptosporidium, even though the microorganism did not regain infectivity. Linden et al. (2002) did not observe dark repair of *Giardia* at UV typical from UV disinfection applications (16 and 40 mJ/cm^2). Shin et al. (2001) reported Cryptosporidium does not regain infectivity after inactivation by UV light.

Applicability

When choosing a UV disinfection system, there are three critical areas to be considered. The first is primarily determined by the manufacturer; the second, by design and operations and maintenance (O&M); and the third must be controlled at the treatment facility.

Choosing a UV disinfection system depends on three critical factors listed below:

- *Hydraulic properties of the reactor* – Ideally, a UV disinfection system should have a uniform flow with enough axial motion (radial mixing) to maximize exposure to UV radiation. The path that an organism takes in the reactor determines the amount of UV radiation it will be exposed to before inactivation. A reactor must be designed to eliminate short-circuiting and/or dead zones, which can result in inefficient use of power and reduced contact time.
- *Intensity of the UV radiation* – Factors affecting the intensity are the age of the lamps, lamp fouling, and the configuration and placement of lamps in the reactor.

- *Wastewater characteristics* – These include the flow rate, suspended and colloidal solids, initial bacterial density, and other physical and chemical parameters. Both the concentration of TSS and the concentration of particle-associated microorganisms determine how much UV radiation ultimately reaches the target organism. The higher these concentrations, the lower the UV radiation absorbed by the organisms. UV disinfection can be used in plants of various sizes that provide secondary or advanced levels of treatment.

Ozonation

Ozone is a strong oxidizing gas that reacts with most organic and many inorganic molecules. It is produced when oxygen molecules separate, collide with other oxygen atoms, and form a molecule consisting of three oxygen atoms. For high-quality effluents, ozone is a very effective disinfectant. Current regulations for domestic treatment systems limit the use of ozonation to filtered effluents unless the system's effectiveness can be demonstrated prior to installation.

Note: Effluent quality is the key performance factor for ozonation.

For ozonation of wastewater, the facility must have the capability to generate pure oxygen along with an ozone generator. A contact tank with ≥10-minute contact time at design average daily flow is required. Off-gas monitoring for process control is also required. In addition, safety equipment capable of monitoring ozone in the atmosphere and a ventilation system to prevent ozone levels exceeding 0.1 ppm is required.

The actual operation of the ozonation process consists of monitoring and adjusting the ozone generator and monitoring the control system to maintain the required ozone concentration in the off-gas. The process must also be evaluated periodically using biological testing to assess its effectiveness.

Note: Ozone is an extremely toxic substance. Concentrations in air should not exceed 0.1 ppm. It also has the potential to create an explosive atmosphere. Enough ventilation and purging capabilities should be provided.

Note: Ozone has certain advantages over chlorine for disinfection of wastewater:

- Ozone increases DO in the effluent.
- Ozone has a briefer contact time.
- Ozone has no undesirable effects on marine organisms.
- Ozone decreases turbidity and odor.

Advantages and Disadvantages

Advantages

Ozone is more effective than chlorine in destroying viruses and bacteria.

- The ozonation process utilizes a short contact time (approximately 10–30 minutes).
- There are no harmful residuals that need to be removed after ozonation because ozone decomposes rapidly.
- After ozonation, there is no regrowth of microorganisms, except for those protected by the particulates in the wastewater stream.
- Ozone is generated on-site, and thus, there are fewer safety problems associated with shipping and handling.
- Ozonation elevates the dissolved oxygen concentration of the effluent. The increase in DO can eliminate the need for re-aeration and raise the level of DO in the receiving stream.

Disadvantages

- Low dosage may not effectively inactivate some viruses, spores, and cysts.

- Ozonation is a more complex technology than is chlorine or UV disinfection, requiring complicated equipment and efficient contacting systems.
- Ozone is very reactive and corrosive, thus requiring corrosion-resistant material such as stainless steel.
- Ozonation is not economical for wastewater with high levels of suspended soils (SS), BOD, COD, or total organic carbon.
- Ozone is extremely irritating and possibly toxic, so off-gases from the contactor must be destroyed to prevent worker exposure.
- The cost of treatment can be relatively high in capital and in power intensiveness.

Applicability

Ozone disinfection is generally used at medium to large-sized plants after at least secondary treatment. In addition to disinfection, another common use for ozone in wastewater treatment is odor control.

Ozone disinfection is the least used method in the US although this technology has been widely accepted in Europe for decades. Ozone treatment can achieve higher levels of disinfection than either chlorine or UV; however, the capital costs, as well as maintenance expenditures, are not competitive with available alternatives. Ozone is therefore used only sparingly, primarily in special cases where alternatives are not effective (USEPA, 1999b).

Operations and Maintenance

Ozone generation uses a significant amount of electrical power. Thus, constant attention must be given to the system to ensure that power is optimized for controlled disinfection performance.

There must be no leaking connections in or surrounding the ozone generator. The operator must on a regular basis monitor the appropriate subunits to ensure that they are not overheated. Therefore, the operator must check for leaks routinely, since a very small leak can cause unacceptable ambient ozone concentrations. The ozone monitoring equipment must be tested and calibrated as recommended by the equipment manufacturer.

Like oxygen, ozone has limited solubility and decomposes more rapidly in water than in air. This factor, along with ozone reactivity, requires that the ozone contactor be well covered and that the ozone diffuses into the wastewater as effectively as possible.

Ozone in gaseous form is explosive once it reaches a concentration of 240 g/m^3. Since most ozonation systems never exceed a gaseous ozone concentration of 50–200 g/m^3, this is generally not a problem. However, ozone in the gaseous form will remain hazardous for a significant amount of time; thus, extreme caution is needed when operating the ozone gas systems.

It is important that the ozone generator, distribution, contracting, off-gas, and ozone destructor inlet piping be purged before opening the various systems or subsystems. When entering the ozone contactor, personnel must recognize the potential for oxygen deficiencies or trapped ozone gas despite best efforts to purge the system. The operator should be aware of all emergency operating procedures required if a problem occurs. All safety equipment should be available for operators to use in case of an emergency.

BROMINE CHLORIDE

Bromine chloride is a mixture of bromine and chlorine. It forms hydrocarbons and hydrochloric acid when mixed with water. Bromine chloride is an excellent disinfectant that reacts quickly and normally does not produce any long-term residuals.

Note: Bromine chloride is an extremely corrosive compound in the presence of low concentrations of moisture.

The reactions occurring when bromine chloride is added to the wastewater are like those occurring when chlorine is added. The major difference is the production of bromamine compounds rather than chloramines. The bromamine compounds are excellent disinfectants but are less stable and dissipate quickly. In most cases, the bromamines decay into other, less toxic compounds rapidly and are undetectable in the plant effluent. The factors that affect performance are like those affecting the performance of the chlorine disinfection process. Effluent quality, contact time, etc., have a direct impact on the performance of the process.

No Disinfection

In a very limited number of cases, treated wastewater discharges without disinfection are permitted. These are approved on a case-by-case basis. Each request must be evaluated based upon the point of discharge, the quality of the discharge, the potential for human contact, and many other factors.

ADVANCED WASTEWATER TREATMENT

Advanced wastewater treatment is defined as the method(s) and/or process (es) that remove more contaminants (suspended and dissolved substances) from wastewater than are taken out by conventional biological treatment. Put another way, advanced wastewater treatment is the application of a process or system that follows secondary treatment or that includes phosphorus removal or nitrification in conventional secondary treatment.

Advanced wastewater treatment is used to augment conventional secondary treatment because secondary treatment typically removes only between 85% and 95% of the BOD and TSS in raw sanitary sewage. Generally, this leaves 30 mg/L or less of BOD and TSS in the secondary effluent. To meet stringent water-quality standards, this level of BOD and TSS in secondary effluent may not prevent violation of water-quality standards – the plant may not make a permit. Thus, advanced wastewater treatment is often used to remove additional pollutants from treated wastewater.

In addition to meeting or exceeding the requirements of water-quality standards, treatment facilities use advanced wastewater treatment for other reasons as well. For example, sometimes, conventional secondary wastewater treatment is not enough to protect the aquatic environment. In a stream, for example, when periodic flow events occur, the stream may not provide the amount of dilution of effluent needed to maintain the necessary DO levels for aquatic organism survival.

Secondary treatment has other limitations. It does not significantly reduce the effluent concentration of nitrogen and phosphorus (important plant nutrients) in sewage. An overabundance of these nutrients can over-stimulate plant and algae growth such that they create water quality problems.

For example, if discharged into lakes, these nutrients contribute to algal blooms and accelerated eutrophication (lake aging). Also, the nitrogen in the sewage effluent may be present mostly in the form of ammonia compounds. If in high enough concentration, ammonia compounds are toxic to aquatic organisms. Yet another problem with these compounds is that they exert a *nitrogenous* oxygen demand in the receiving water, as they convert to nitrates. This process is called nitrification.

Note: The term *tertiary treatment* is commonly used as a synonym for advanced wastewater treatment. However, these two terms do not have precisely the same meaning. Tertiary suggests a third step that is applied after primary and secondary treatment.

Advanced wastewater treatment can remove more than 99% of the pollutants from raw sewage and can produce an effluent of almost potable (drinking) water quality. However, obviously, advanced treatment is not cost-free. The cost of advanced treatment, for operations and maintenance as well as for retrofit of present conventional processes, is very high (sometimes doubling the cost of secondary treatment). Therefore, a plan to install advanced treatment technology calls for careful study – the benefit-to-cost ratio is not always big enough to justify the additional expense.

Even considering the expense, the application of some form of advanced treatment is not uncommon. These treatment processes can be physical, chemical, or biological. The specific process used is based upon the purpose of the treatment and the quality of the effluent desired.

CHEMICAL TREATMENT

The purpose of chemical treatment is to remove:

- Biochemical oxygen demand;
- Total suspended solids;
- Phosphorus;
- Heavy metals; and
- Other substances that can be chemically converted to a settleable solid.

Chemical treatment is often accomplished as an "add-on" to existing treatment systems or by means of separate facilities specifically designed for chemical addition.

In each case, the basic process necessary to achieve the desired results remains the same:

- Chemicals are thoroughly mixed with the wastewater.
- The chemical reactions that occur form solids (coagulation).
- The solids are mixed to increase particle size (flocculation).
- Settling and/or filtration (separation) then remove the solids.

The specific chemical used depends on the pollutant to be removed and the characteristics of the wastewater.

Chemicals may include the following:

- Lime
- Alum (aluminum sulfate)
- Aluminum salts
- Ferric or ferrous salts
- Polymers
- Bioadditives

MICROSCREENING

Microscreening (also called *microstraining*) is an advanced treatment process used to reduce suspended solids. The microscreens are composed of specially woven steel wire fabric mounted around the perimeter of a large revolving drum. The steel wire cloth acts as a fine screen, with openings as small as 20 micrometers (or millionths of a meter) – small enough to remove microscopic organisms and debris.

The rotating drum is partially submerged in the secondary effluent, which must flow into the drum then outward through the microscreen. As the drum rotates, captured solids are carried to the top where a high-velocity water spray flushes them into a hopper or backwash tray mounted on the hollow axle of the drum. Backwash solids are recycled to plant influent for treatment. These units have found the greatest application in the treatment of industrial waters and final polishing filtration of wastewater effluents. Expected performance for suspended solids removal is 95–99%, but the typical suspended-solids removal achieved with these units is about 55%. The normal range is from 10–80%.

According to Metcalf & Eddy (2003), the functional design of the microscreen unit involves the following considerations:

- The characterization of the suspended solids with respect to the concentration and degree of flocculation;
- The selection of unit design parameter values that will not only ensure capacity to meet maximum hydraulic loadings with critical solids characteristics but also provide desired design performance over the expected range of hydraulic and solids loadings; and
- The provision of backwash and cleaning facilities to maintain the capacity of the screen.

FILTRATION

The purpose of *filtration* processes used in advanced treatment is to remove suspended solids. The specific operations associated with a filtration system are dependent on the equipment used. A general description of the process follows.

Filtration Process Description

Wastewater flows to a filter (gravity or pressurized). The filter contains single, dual, or multimedia. Wastewater flows through the media, which removes solids. The solids remain in the filter. Backwashing the filter as needed removes trapped solids. Backwash solids are returned to the plant for treatment. Processes typically remove 95–99% of the suspended matter.

BIOLOGICAL DENITRIFICATION

Biological denitrification removes nitrogen from the wastewater. When bacteria meet a nitrified element in the absence of oxygen, they reduce the nitrates to nitrogen gas, which escapes the wastewater. The denitrification process can be done either in an anoxic activated sludge system (suspended growth) or in a column system (fixed growth). The denitrification process can remove up to 85% or more of nitrogen. After effective biological treatment, little oxygen demanding material is left in the wastewater when it reaches the denitrification process.

The denitrification reaction will only occur if an oxygen demand source exists when no dissolved oxygen is present in the wastewater. An oxygen demand source is usually added to reduce the nitrates quickly. The most common demand source added is soluble BOD or methanol. Approximately 3 mg/L of methanol is added for every 1 mg/L of nitrate-nitrogen. Suspended growth denitrification reactors are mixed mechanically, but only enough to keep the biomass from settling without adding unwanted oxygen. Submerged filters of different types of media may also be used to provide denitrification. A fine media downflow filter is sometimes used to provide both denitrification and effluent filtration. A fluidized sand bed where wastewater flows upward through a media of sand or activated carbon at a rate to fluidize the bed may also be used. Denitrification bacteria grow on the media.

CARBON ADSORPTION

The main purpose of *carbon adsorption* used in advanced treatment processes is the removal of refractory organic compounds (non-BOD$_5$) and soluble organic material that are difficult to eliminate by biological or physical/chemical treatment. In the carbon adsorption process, wastewater passes through a container filled either with carbon powder or carbon slurry. Organics adsorb onto the carbon (i.e., organic molecules are attracted to the activated carbon surface and are held there) with enough contact time. A carbon system usually has several columns or basins used as contactors. Most contact chambers are either open concrete gravity-type systems or steel pressure containers applicable to either up flow or downflow operation. With use, carbon loses its adsorptive capacity. The carbon must then be regenerated or replaced with fresh carbon. As head loss develops in carbon contactors, they are backwashed with clean effluent in much the same way the effluent filters are backwashed. Carbon used for adsorption may be in a granular form or in a powdered form.

Note: Powdered carbon is too fine for use in columns and is usually added to the wastewater and then later removed by coagulation and settling.

LAND APPLICATION

The application of secondary effluent onto a land surface can provide an effective alternative to the expensive and complicated advanced treatment methods discussed previously and the biological nutrient removal (BNR) system discussed later. A high-quality polished effluent (i.e., effluent with high levels of TSS, BOD, phosphorus, and nitrogen compounds as well as refractory organics are reduced) can be obtained by the natural processes that occur as the effluent flows over the vegetated ground surface and percolates through the soil. Limitations are involved with land application of wastewater effluent. For example, the process needs large land areas. Soil type and climate are also critical factors in controlling the design and feasibility of a land treatment process.

Type and Modes of Land Application

Three basic types or modes of land application or treatment are commonly used: irrigation (slow rate), overland flow, and infiltration-percolation (rapid rate). The basic objectives of these types of land applications and the conditions under which they can function vary. In *irrigation* (also called slow rate), wastewater is sprayed or applied (usually by ridge-and-furrow surface spreading or by sprinkler systems) to the surface of the land. Wastewater enters the soil. Crops growing in the irrigation area utilize available nutrients. Soil organisms stabilize the organic content of the flow. Water returns to the hydrologic (water) cycle through evaporation or by entering the surface water or groundwater.

The irrigation land application method provides the best results (compared with the other two types of land application systems) with respect to advanced treatment levels of pollutant removal. Not only are suspended solids and BOD significantly reduced by filtration of the wastewater, but also biological oxidation of the organics in the top few inches of soil occurs. Nitrogen is removed primarily by crop uptake, and phosphorus is removed by adsorption within the soil.

Expected performance levels for irrigation include:

- BOD_5 – 98%
- Suspended solids – 98%
- Nitrogen – 85%
- Phosphorus – 95%
- Metals – 95%

The overland flow application method utilizes physical, chemical, and biological processes as the wastewater flows in a thin film down the relatively impermeable surface. In the process, wastewater sprayed over sloped terraces flows slowly over the surface. Soil and vegetation remove suspended solids, nutrients, and organics. A small portion of the wastewater evaporates. The remainder flows to collection channels. Collected effluent is discharged to surface waters.

Expected performance levels for overflow include:

- BOD_5 – 92%
- Suspended solids – 92%
- Nitrogen – 70–90%
- Phosphorus – 40–80%
- Metals – 50%

In the infiltration-percolation (rapid rate) land application process, wastewater is sprayed/pumped to spreading basins (aka recharge basins or large ponds). Some wastewater evaporates. The remainder percolates/infiltrates into soil. Solids are removed by filtration. Water recharges the groundwater system. Most of the effluent percolates to the groundwater, very little of it is absorbed by vegetation. The filtering and adsorption action of the soil removes most of the BOD, TSS, and phosphorous from the effluent; however, nitrogen removal is relatively poor.

Expected performance levels for infiltration-percolation include:

- BOD_5 – 85–99%
- Suspended solids – 98%
- Nitrogen – 0–50%

- Phosphorus – 60–95%
- Metals – 50–95%

BIOLOGICAL NUTRIENT REMOVAL (BNR)

Nitrogen and phosphorus are the primary causes of cultural eutrophication (i.e., nutrient enrichment due to human activities) in surface waters. The most recognizable manifestations of this eutrophication are algal blooms that occur during the summer. Chronic symptoms of over-enrichment include low dissolved oxygen, fish kills, murky water, and depletion of desirable flora and fauna.

In addition, the increase in algae and turbidity increases the need to chlorinate drinking water, which, in turn, leads to higher levels of disinfection by-products that have been shown to increase the risk of cancer (USEPA, 2007c). Excessive amounts of nutrients can also stimulate the activity of microbes, such as *Pfisteria*, which may be harmful to human health (USEPA, 2001).

Approximately 25% of all water body impairments are due to nutrient-related causes (e.g., nutrients, oxygen depletion, algal growth, ammonia, harmful algal blooms, biological integrity, and turbidity) (USEPA, 2007d). In efforts to reduce the number of nutrient impairments, many point source discharges have received more stringent effluent limits for nitrogen and phosphorus. To achieve these new, lower effluent limits, facilities have begun to look beyond traditional treatment technologies.

Recent experience has reinforced the concept that biological nutrient removal (BNR) systems are reliable and effective in removing nitrogen and phosphorus. The process is based upon the principle that, under specific conditions, microorganisms will remove more phosphorus and nitrogen than is required for biological activity; thus, treatment can be accomplished without the use of chemicals. Not having to use and therefore having to purchase chemicals to remove nitrogen and phosphorus potentially has numerous cost-benefit implications. In addition, because chemicals are not required to be used, chemical waste products are not produced, reducing the need to handle and dispose of waste. Several patented processes are available for this purpose. Performance depends on the biological activity and the process employed.

Description

As mentioned, biological nutrient removal removes total nitrogen (TN) and total phosphorus (TP) from wastewater using microorganism under different environmental conditions in the treatment process (Metcalf & Eddy, 2013).

Nitrogen Removal

Total effluent nitrogen comprises ammonia, nitrate, particulate organic nitrogen, and soluble organic nitrogen. The biological processes that primarily remove nitrogen are nitrification and denitrification. During nitrification, ammonia is oxidized to nitrate by one group of autotrophic bacteria, most commonly *Nitrosomonas* (Metcalf & Eddy, 2013). Nitrite is then oxidized to nitrite by another autotrophic bacteria group, the most common being *Nitrobacter*.

Denitrification involves the biological reduction of nitrite to nitric oxide, nitrous oxide, and nitrogen gas (Metcalf & Eddy, 2013). Both heterotrophic and autotrophic bacteria are capable of denitrification. The most common and widely distributed denitrifying bacteria are Pseudomonas species, which can use hydrogen, methanol, carbohydrates, organic acids, alcohols, benzoates, and other aromatic compounds for denitrification (Metcalf & Eddy, 2013). In BNR systems, nitrification is the controlling reaction because ammonia-oxidizing bacteria lack functional diversity, have stringent growth requirements, and are sensitive to environmental considerations.

Note: Nitrification by itself does not actually remove nitrogen from wastewater. Rather, denitrification is needed to convert the oxidized form of nitrogen (nitrate) to nitrogen gas. Nitrification occurs in the presence of oxygen under aerobic conditions, and denitrification occurs in the absence of oxygen under anoxic conditions (USEPA, 2007c).

Note: Organic nitrogen is not removed biologically; rather only the particulate fraction can be removed through solids separated via sedimentation or filtration.

Phosphorus Removal

Total effluent phosphorus comprises soluble and particulate phosphorus. Particulate phosphorus can be removed from wastewater through solids removal. To achieve low effluent concentrations, the soluble fraction of phosphorus must also be targeted.

Biological phosphorus removal relies on phosphorus uptake by anaerobic heterotrophs capable of storing orthophosphate in excess of their biological growth requirements. The treatment process can be designed to promote the growth of these organisms, known as phosphate-accumulating organisms (PAOs) in mixed liquor. Under anaerobic conditions, PAOs convert readily available organic matter [e.g., volatile fatty acids (VFAs)] to carbon compounds called polyhydroxyalkanoates (PHAs). PAOs use energy generated throughout the breakdown of polyphosphate molecules to create PHAs. This breakdown results in the release of phosphorus (USEPA, 2007a).

Under subsequent aerobic conditions in the treatment process, PAOs use the stored PHAs as energy to take up the phosphorus that was released in the anaerobic zone, as well as any additional phosphate present in the wastewater. In addition to reducing the phosphate concentration, the process renews the polyphosphate pool in the return sludge so that the process can be repeated (USEPA, 2007b). Some PAOs use nitrate instead of free oxygen to oxidize stored PHAs and take up phosphorus. These denitrifying PAOs remove phosphorus in the anoxic zone, rather than the aerobic zone.

Process

There are several BNR process configurations available. Some BNR systems are designed to remove only TN or TP, while others remove both. The configuration most appropriate for any system depends on the target effluent quality, operator experience, influent quality, and exiting treatment processes, if retrofitting an existing facility. BNR configuration vary based on the sequencing of environmental conditions (i.e., aerobic, anaerobic, and anoxic) and timing (USEPA, 2007c).

Note: Anoxic is a condition in which oxygen is available only in the combined form (e.g., NO_2^- or NO_3. However, anaerobic is a condition in which neither free nor combined oxygen is available. Common BNR system configurations include:

- *Modified Ludzack-Ettinger (MLE) Process* – continuous-flow suspended-growth process with an initial anoxic stage followed by an aerobic stage; used to remove nitrogen.
- *A²/O Process* – MLE process preceded by an initial anaerobic stage; used to remove both TN and TP.
- *Step Feed Process* – alternating anoxic and aerobic stages; however, influent flow is split to several feed locations, and the recycle sludge stream is sent to the beginning of the process; used to remove TN.
- *Bardenpho Process (Four-Stage)* – continuous-flow suspended-growth process with alternating anoxic/aerobic/anoxic/aerobic stages; used to remove TN.
- *Modified Bardenpho Process* – Bardenpho process with the addition of an initial anaerobic zone; used to remove both TN and TP.
- *Sequencing Batch Reactor (SBR) Process* – suspended-growth batch process sequenced to simulate the four-stage process; used to remove TN (TP removal is inconsistent).
- *Modified University of Cape Town (UCT) Process* – A²/O Process with a second anoxic stage where the internal nitrate recycle is returned; used to remove both TN and TP.
- *Rotating Biological Contactor (RBC) Process* – continuous-flow process using RBCs with sequential anoxic/aerobic stages; used to remove TN.
- *Oxidation Ditch* – continuous-flow process using looped channels to create time-sequenced anoxic, aerobic, and anaerobic zones; used to remove both TN and TP.

Although the exact configurations of each system differ, BNR systems designed to remove TN must have an aerobic zone for nitrification and anoxic zone for denitrification, and BNR systems designed to remove TP must have an anaerobic zone free of dissolved oxygen and nitrate. Often, sand or other media filtration is used as a polishing step to remove particulate matter when low TN and TP effluent concentrations are required. Sand filtration can also be combined with attached growth denitrification filters to further reduce soluble nitrates and effluent TN levels (WEF, 1998).

Choosing which system is most appropriate for a facility primarily depends on the target effluent concentrations and whether the facility will be constructed as new or retrofit with BNR to achieve more stringent effluent limits. New plants have more flexibility and options when deciding which BNR configuration to implement because they are not constrained by existing treatment units and sludge handling procedures. Retrofitting an existing plant with BNR capabilities should involve consideration of the following factors (Park, 2012):

- Aeration basin size and configuration
- Clarifier capacity
- Type of aeration system
- Sludge processing units
- Operator skills

The aeration basin size and configuration dictates which BNR configurations are the most economical and feasible. Available excess capacity reduces the need for additional basins and may allow for a more complex configuration (e.g., five-stage Bardenpho versus four-state Bardenpho configuration). The need for additional basins could result in the need for more land if the space needed is not available. If the land is not available, another BNR process configuration may have to be considered.

The aeration system will most likely need to be modified to accommodate an anaerobic zone and to reduce the DO concentration in the return sludge. Such modifications could be as simple as removing aeration equipment from the zone designated for anaerobic conditions or changing the type of pump used for the recycled sludge stream (to avoid introduction oxygen).

The way sludge is processed at a facility is important in designing nutrient removal systems. Sludge is recycled within the process to provide the organisms necessary for the TN and TP removal mechanism to occur. The content and volume of sludge recycled directly impacts the system's performance. Thus, sludge handling processes may be modified to achieve optimal TN and TP removal efficiencies. For example, some polymers in sludge dewatering could inhibit nitrification when recycled.

Operators should be able to adjust the process to compensate for constantly varying conditions. BNR processes are very sensitive to influent conditions, which are influenced by weather events, sludge processing, and other treatment processes (e.g., recycling after filter backwashing). Therefore, operator skills and training are essential for achieving target TN and TP effluent concentrations (USEPA, 2007c).

ENHANCED BIOLOGICAL NUTRIENT REMOVAL (EBNR)

Removing phosphorus from wastewater in secondary treatment processes has evolved into innovative *enhanced biological nutrient removal* (EBNR) technologies. An ENBR treatment process promotes the production of phosphorus accumulating organisms that utilize more phosphorus in their metabolic processes than a conventional secondary biological treatment process (USEPA, 2007b). The average total phosphorus concentrations in raw domestic wastewater are usually between 6 and 8 mg/L, and the total phosphorus concentration in municipal wastewater after conventional secondary treatment is routinely reduced to 3 or 4 mg/L. Whereas, EBNR incorporated into the secondary treatment system can often reduce total phosphorus concentrations to 0.3 mg/L and less. Facilities using EBNR significantly reduced the amount of phosphorus to be removed through the subsequent chemical addition and tertiary filtration process. This improved the efficiency of the tertiary process and significantly reduced the costs of chemicals used to remove phosphorus. Facilities using EBNR reported that their chemical dosing was cut in half after EBNR was installed to remove phosphorus (USEPA, 2007b).

Treatment provided by these WWTPs also removes other pollutants that commonly affect water quality to very low levels (USEPA, 2007b). Biochemical oxygen demand and total suspended solids are routinely less than 2 mg/L and fecal coliform bacterial less than 10 fcu/100 mL. The turbidity of the final effluent is very low which allows for effective disinfection using ultraviolet light, rather than chlorination. Recent studies report that wastewater treatment plants using EBNR also significantly reduced the amount of pharmaceuticals and personal healthcare products from municipal wastewater, as compared to the removal accomplished by conventional secondary treatment.

SLUDGE: BACKGROUND INFORMATION

When we speak of *sludge* or *bio solids*, we are speaking of the same substance or material; each is defined as the suspended solids removed from wastewater during sedimentation and then concentrated for further treatment and disposal or reuse. The difference between the terms *sludge* and *biosolids* is determined by the way they are managed.

Note: The task of disposing, treating, or reusing wastewater solids is called *sludge* or *biosolids management*.

Sludge is typically seen as wastewater solids that are "disposed" of. Biosolids are the same substance managed for reuse – commonly called beneficial reuse (e.g., for land application as a soil amendment, such as biosolids compost).

Note: Even as wastewater treatment standards have become more stringent because of increasing environmental regulations, so has the volume of wastewater sludge increased.

Note: Before sludge can be disposed of or reused, it requires some form of treatment to reduce its volume, to stabilize it, and to inactivate pathogenic organisms.

Sludge forms initially as a 3–7% suspension of solids, and with each person typically generating about 4 gallons of sludge/week, the total quantity generated each day, week, month, and year is significant. Because of the volume and nature of the material, sludge management is a major factor in the design and operation of all water pollution control plants.

Note: Wastewater solids treatment, handling, and disposal account for more than half of the total costs in a typical secondary treatment plant.

SOURCES OF SLUDGE

Wastewater sludge is generated in primary, secondary, and chemical treatment processes. In primary treatment, the solids that float or settle are removed. The floatable material makes up a portion of the solid waste known as scum. Scum is not normally considered sludge; however, it should be disposed of in an environmentally sound way. The settleable material that collects on the bottom of the clarifier is known as *primary sludge*. Primary sludge can also be referred to as raw sludge because it has not undergone decomposition. Raw primary sludge from a typical domestic facility is quite objectionable and has a high percentage of water, two characteristics that make handling difficult.

Solids not removed in the primary clarifier are carried out of the primary unit. These solids are known as *colloidal suspended solids*. The secondary treatment system (i.e., trickling filter, activated sludge, etc.) is designed to change those colloidal solids into settleable solids that can be removed. Once in the settleable form, these solids are removed in the secondary clarifier. The sludge at the bottom of the secondary clarifier is called *secondary sludge*. Secondary sludges are light and fluffy and more difficult to process than primary sludges – in short, secondary sludges do not dewater well.

The addition of chemicals and various organic and inorganic substances prior to sedimentation and clarification may increase the solids capture and reduce the amount of solids lost in the effluent. This *chemical addition* results in the formation of heavier solids, which trap the colloidal solids or convert dissolved solids to settleable solids. The resultant solids are known as *chemical sludges*. As chemical usage increases, so does the quantity of sludge that must be handled and disposed of. Chemical sludges can be very difficult to process; they do not dewater well and contain lower percentages of solids.

SLUDGE CHARACTERISTICS

The composition and characteristics of sewage sludge vary widely and can change considerably with time. Notwithstanding these facts, the basic components of wastewater sludge remain the same. The only variations occur in the quantity of the various components as the type of sludge and the process from which it originated changes. The main component of all sludges is *water*. Prior to treatment, most sludge contains 95–99+ percentage of water. This high water content makes sludge handling and processing extremely costly in terms of both money and time. Sludge handling may represent up to 40% of the capital cost and 50% of the operation cost of a treatment plant. As a result, the importance of optimum design for handling and disposal of sludge cannot be overemphasized. The water content of the sludge is present in several different forms. Some forms can be removed by several sludge treatment processes, thus allowing the same flexibility in choosing the optimum sludge treatment and disposal method.

The forms of water associated with sludge include:

- *Free water* – water that is not attached to sludge solids in any way. This can be removed by simple gravitational settling.
- *Floc water* – water that is trapped within the floc and travels with them. Its removal is possible by mechanical de-watering.
- *Capillary water* – water that adheres to the individual particles and can be squeezed out of shape and compacted.
- *Particle water* – water that is chemically bound to the individual particles and can't be removed without inclination.

From a public health view, the second and probably more important component of sludge is the *solids matter*. Representing from 1% to 8% of the total mixture, these solids are extremely unstable. Wastewater solids can be classified into two categories based on their origin – organic and inorganic. *Organic solids* in wastewater, simply put, are materials that are or were at one time alive and that will burn or volatilize at 550°C after 15 minutes in a muffle furnace. The percentage of organic material within sludge will determine how unstable it is.

The inorganic material within sludge will determine how stable it is. The *inorganic solids* are those solids that were never alive and will not burn or volatilize at 550°C after 15 minutes in a muffle furnace. Inorganic solids are generally not subject to breakdown by biological action and are considered stable. Certain inorganic solids, however, can create problems when related to the environment, for example, heavy metals such as copper, lead, zinc, mercury, and others. These can be extremely harmful if discharged.

Organic solids may be subject to biological decomposition in either an aerobic or anaerobic environment. Decomposition of organic matter (with its production of objectionable by-products) and the possibility of toxic organic solids within the sludge compound the problems of sludge disposal.

The pathogens in domestic sewage are primarily associated with insoluble solids. Primary wastewater treatment processes concentrate these solids into sewage sludge, so untreated or raw primary sewage sludges have higher quantities of pathogens than the incoming wastewater. Biological wastewater treatment processes such as lagoons, trickling filters, and activated sludge treatment may substantially reduce the number of pathogens in wastewater (USEPA, 1989). These processes may also reduce the number of pathogens in sewage sludge by creating adverse conditions for pathogen survival.

Nevertheless, the resulting biological sewage sludges may still contain enough levels of pathogens to pose a public health and environmental concern. Moreover, insects, birds, rodents, and domestic animals may transport sewage sludge and pathogens from sewage sludge to humans and to animals. Vectors are attracted to sewage sludge as a food source, and the reduction of the attraction of vectors to sewage sludge to prevent the spread of pathogens is a focus of current regulations. Sludge-borne pathogens and vector attraction are discussed in the following section.

Sludge Pathogens and Vector Attraction

As discussed earlier, a pathogen is an organism capable of causing disease. Pathogens infect humans through several different pathways including ingestion, inhalation, and dermal contact. The infective dose, or the number of pathogenic organisms to which a human must be exposed to become infected, varies depending on the organism and on the health status of the exposed individual. Pathogens that propagate in the enteric or urinary system of humans and are discharged in feces or urine pose the greatest risk to public health regarding the use and disposal of sewage sludge. Pathogens are also found in the urinary and enteric systems of other animals and may propagate in non-enteric settings.

As mentioned earlier, the four major types of human pathogenic (disease-causing) organisms (bacteria, viruses, protozoa, and helminths) all may be present in domestic sewage. The actual species and quantity of pathogens present in the domestic sewage from a municipality (and the sewage sludge produced when treating the domestic sweater) depend on the health status of the local community and may vary substantially at different times. The level of pathogens present in treated sewage sludge (biosolids) also depends on the reductions achieved by the wastewater and sewage sludge treatment processes.

If improperly treated sewage sludge was illegally applied to land or placed on a surface disposal site, humans and animals could be exposed to pathogens directly by coming into contact with sewage sludge or indirectly by consuming drinking water or food contaminated by sewage sludge pathogens. Insects, birds, rodents, and even farm workers could contribute to these exposure routes by transporting sewage sludge and sewage sludge pathogens away from the site.

Potential routes of exposure include:

Direct Contact

Direct contact is the most common contact for exposure.

- Touching the sewage sludge;
- Walking through an area – such as a field, forest, or reclamation area – shortly after sewage sludge application;
- Handling soil from fields where sewage sludge has been applied; and
- Inhaling microbes that become airborne (via aerosols, dust, etc.) during sewage sludge spreading or by strong winds, or plowing or cultivating the soils after application.

Indirect Contact

Indirect contact is the least common contact for exposure.

- Consumption of pathogen-contaminated crops grown on sewage sludge-amended soil or of other food products that have been contaminated by contact with these crops or field workers, etc.;
- Consumption of pathogen-contaminated milk or other food products from animals contaminated by grazing in pastures or fed crops grown on sewage sludge-amended fields;

- Ingestion of drinking water or recreational waters contaminated by runoff from nearby land application sites or by organisms from sewage sludge migrating into groundwater aquifers;
- Consumption of inadequately cooked or uncooked pathogen-contaminated fish from water contaminated by runoff from a nearby sewage sludge application site; and
- Contact with sewage sludge or pathogens transported away from the land application or surface disposal site by rodents, insects, or other vectors, including grazing animals or pets.

DID YOU KNOW?

The purpose of USEPA's Part 503 regulation is to place barriers in the pathway of exposure either by reducing the number of pathogens in the treated sewage sludge (biosolids) to below detectable limits, in the case of Class A treatment, or, in the case of Class B treatment, by preventing direct or indict contact with any pathogens possibly present in the biosolids. Each potential pathway has been studied to determine how the potential for public health risk can be alleviated.

One of the lesser impacts to public health can be from inhalation of airborne pathogens. Pathogens may become airborne via the spray of liquid biosolids from a splash plate or high-pressure hose, or in fine particulate dissemination as dewatered biosolids are applied or incorporated. While high-pressure spray applications may result in some aerosolization of pathogens, this type of equipment is generally used on large, remote sites such as forests, where the impact on the public is minimal. Fine particulates created by the application of dewatered biosolids or the incorporation of biosolids into the soil may cause very localized fine particulate/dusty conditions, but particles in dewatered biosolids are too large to travel far, and the fine particulates do not spread beyond the immediate area. The activity of applying and incorporating biosolids may create dusty conditions. However, the biosolids are moist materials and do not add to the dusty condition, and by the time biosolids have dried sufficiently to create fine particulates, the pathogens have been reduced (Yeager & Ward, 1981).

Regarding vector attraction reduction, it can be accomplished in two ways: (1) by treating the sewage sludge to the point at which vectors will no longer be attracted to the sewage sludge; and (2) by placing a barrier between the sewage sludge and vectors.

Important Note: Release of wastewater solids without proper treatment could result in severe damage to the environment. Obviously, we must have a system to treat the volume of material removed as sludge throughout the system. Release without treatment would defeat the purpose of environmental protection. A design engineer can choose from many processes when developing sludge treatment systems. No matter what the system or combination of systems chosen, the ultimate purpose will be the same: the conversion of wastewater sludges into a form that can be handled economically and disposed of without damage to the environment or creating nuisance conditions. Leaving either condition unmet will require further treatment. The degree of treatment will generally depend on the proposed method of disposal. Sludge treatment processes can be classified into a few major categories. In *Part II*, we discuss the processes of thickening, digestion (or stabilization), dewatering, incineration, and land application. Each of these categories has then been further subdivided according to the specific processes that are used to accomplish sludge treatment. As mentioned, the importance of adequate, efficient sludge treatment cannot be overlooked when designing wastewater treatment facilities. The inadequacies of a sludge treatment system can severely affect a plant's overall performance capabilities. The inability to remove and process solids as fast as they accumulate in the process can lead to the discharge of large quantities of solids to receiving waters.

Even with proper design and capabilities in place, no system can be effective unless it is properly operated. Proper operation requires proper operator performance. Proper operator performance begins and ends with proper training.

Sludge Thickening

The solids content of primary, activated, trickling-filter, or even mixed sludge (i.e., primary plus activated sludge) varies considerably, depending on the characteristics of the sludge. The sludge removal and pumping facilities and the method of operation also affect the solids content. *Sludge thickening* (or *concentration*) is a unit process used to increase the solids content of the sludge by removing a portion of the liquid fraction. By increasing the solids content, more economical treatment of the sludge can be effected.

Sludge thickening processes include:

- Gravity thickeners;
- Flotation thickeners; and
- Solids concentrators.

Gravity Thickening

Gravity thickening is most effective on primary sludge. In operation, solids are withdrawn from primary treatment (and sometimes secondary treatment) and pumped to the thickener. The solids buildup in the thickener forms a solids blanket on the bottom. The weight of the blanket compresses the solids on the bottom and "squeezes" the water out. By adjusting the blanket thickness, the percent solids in the underflow (solids withdrawn from the bottom of the thickener) can be increased or decreased. The supernatant (clear water) which rises to the surface is returned to the wastewater flow for treatment. Daily operations of the thickening process include pumping, observation, sampling and testing, process control calculations, and maintenance and housekeeping.

Note: The equipment employed in thickening depends on the specific thickening processes used.

Equipment used for gravity thickening consists of a thickening tank, which is similar in design to the settling tank used in primary treatment. Generally, the tank is circular and provides equipment for continuous solids collection. The collector mechanism uses heavier construction than that in a settling tank because the solids being moved are more concentrated. The gravity thickener pumping facilities (i.e., pump and flow measurement) are used for the withdrawal of thickened solids. Solids concentrations achieved by gravity thickening are typically 8–10% solids from primary underflow, 2–4% solids from waste activated sludge, 7–9% solids from trickling filter residuals, and 4–9% from combined primary and secondary residuals.

The performance of gravity thickening processes depends on various factors, including:

- Type of sludge;
- Condition of influent sludge;
- Temperature;
- Blanket depth;
- Solids loading;
- Hydraulic loading;
- Solids retention time; and
- Hydraulic detention time.

Flotation Thickening

Flotation thickening is used most efficiently for waste sludges from suspended-growth biological treatment process, such as the activated sludge process. In operation, recycled water from the flotation thickener is aerated under pressure. During this time, the water absorbs more air than it would under normal pressure. The recycled flow together with chemical additives (if used) is mixed with the flow. When the mixture enters the flotation thickener, the excess air is released in the form of fine bubbles.

These bubbles become attached to the solids and lift them toward the surface. The accumulation of solids on the surface is called the *float cake*. As more solids are added to the bottom of the float cake it becomes thicker and water drains from the upper levels of the cake. The solids are then moved up an inclined plane by a scraper and discharged. The supernatant leaves the tank below the surface of the float solids and is recycled or returned to the waste stream for treatment. Typically, flotation thickener performance is 3–5% solids for waste activated sludge with polymer addition and 2–4% solids without polymer addition.

The flotation thickening process requires pressurized air, a vessel for mixing the air with all or part of the process residual flow, a tank for the flotation process to occur, and solids collector mechanisms to remove the float cake (solids) from the top of the tank and accumulated heavy solids from the bottom of the tank. Since the process normally requires chemicals be added to improve separation, chemical mixing equipment, storage tanks, and metering equipment to dispense the chemicals at the desired dose are required.

The performance of dissolved air-thickening process depends on various factors:

- Bubble size
- Solids loading
- Sludge characteristics
- Chemical selection
- Chemical dose

Solids Concentrators

Solids concentrators (belt thickeners) usually consist of a mixing tank, chemical storage and metering equipment, and a moving porous belt. In operation, the process residual flow is chemically treated and then spread evenly over the surface of the moving porous belt. As the flow is carried down the belt (like a conveyor belt), the solids are mechanically turned or agitated, and water drains through the belt. This process is primarily used in facilities where space is limited.

SLUDGE STABILIZATION

The purpose of sludge stabilization is to reduce volume, stabilize the organic matter, and eliminate pathogenic organisms to permit reuse or disposal. The equipment required for stabilization depends on the specific process used.

Sludge stabilization processes include:

- Aerobic digestion;
- Anaerobic digestion;
- Composting;
- Lime stabilization;
- Wet air oxidation (heat treatment);
- Chemical oxidation (chlorine oxidation); and
- Incineration.

Aerobic Digestion

Equipment used for *aerobic digestion* consists of an aeration tank (digester) which is similar in design to the aeration tank used for the activated sludge process. Either diffused or mechanical aeration equipment is necessary to maintain the aerobic conditions in the tank. Solids and supernatant removal equipment are also required. In operation, process residuals (sludge) are added to the digester and aerated to maintain a dissolved oxygen concentration of 1.0 mg/L. Aeration also ensures that the tank contents are well mixed. Generally, aeration continues for approximately 20 days' retention time. Periodically, aeration is stopped, and the solids can settle. Sludge and the clear liquid supernatant are withdrawn as needed to provide more room in the digester. When no additional volume is available, mixing is stopped for 12–24 hours before solids are withdrawn for disposal. Process control testing should include alkalinity, pH, percent Solids, percent Volatile solids for influent sludge, supernatant, digested sludge, and digester contents.

Anaerobic Digestion

Anaerobic digestion is the traditional method of sludge stabilization. It involves using bacteria that thrive in the absence of oxygen and is slower than aerobic digestion but has the advantage that only a small percentage of the wastes are converted into new bacterial cells. Instead, most of the organics are converted into carbon dioxide and methane gas.

Note: In an anaerobic digester, the entrance of air should be prevented because of the potential for air mixed with the gas produced in the digester which could create an explosive mixture.

Equipment used in anaerobic digestion includes a sealed digestion tank with either a fixed or a floating cover, heating and mixing equipment, gas storage tanks, solids and supernatant withdrawal equipment, and safety equipment (e.g., vacuum relief, pressure relief, flame traps, explosion proof electrical equipment). In operation, process residual (thickened or unthickened sludge) is pumped into the sealed digester. The organic matter digests anaerobically by a two-stage process. Sugars, starches, and carbohydrates are converted to volatile acids, carbon dioxide, and hydrogen sulfide.

The volatile acids are then converted to methane gas. This operation can occur in a single tank (single stage) or in two tanks (two stages). In a single-stage system, supernatant and/or digested solids must be removed whenever flow is added. In a two-stage operation, solids and liquids from the first stage flow into the second stage each time fresh solids are added. Supernatant is withdrawn from the second stage to provide additional treatment space. Periodically, solids are withdrawn for dewatering or disposal. The methane gas produced in the process may be used for many plant activities.

Note: The primary purpose of a secondary digester is to allow for solids separation.

Various performance factors affect the operation of the anaerobic digester. For example, percent Volatile Matter in raw sludge, digester temperature, mixing, volatile acids/alkalinity ratio, feed rate, percent solids in raw sludge and pH are all-important operational parameters that the operator must monitor. Along with being able to recognize normal/abnormal anaerobic digester performance parameters, wastewater operators must also know and understand normal operating procedures. Normal operating procedures include sludge additions, supernatant withdrawal, sludge withdrawal, pH control, temperature control, mixing, and safety requirements.

Other Sludge Stabilization Processes

In addition to aerobic and anaerobic digestion, other sludge stabilization processes include composting, lime stabilization, wet air oxidation, and chemical (chlorine) oxidation. These other stabilization processes are briefly described in this section.

Composting

The purpose of composting sludge is to stabilize the organic matter, reduce volume, and to eliminate pathogenic organisms. In a *composting operation,* dewatered solids are usually mixed with a bulking agent (i.e., hardwood chips) and stored until biological stabilization occurs. The composting mixture is ventilated during storage to provide enough oxygen for oxidation and to prevent odors. After the solids are stabilized, they are separated from the bulking agent. The composted solids are then stored for curing and applied to farmlands or other beneficial uses. Expected performance of the composting operation for both percent volatile matter reduction and percent moisture reduction ranges from 40% to 60%.

Definitions of Key Terms

Aerated Static Pile – Composting system using controlled aeration from a series of perforated pipes running underneath each pile and connected to a pump that draws or blows air through the piles.

Aeration (for composting) – Bringing about the contact of air and composted solid organic matter by means of turning or ventilating to allow microbial aerobic metabolism (bio-oxidation).

Aerobic – Composting environment characterized by bacteria active in the presence of oxygen (aerobes); generates more heat and is a faster process than anaerobic composting.

Anaerobic – Composting environment characterized by bacteria active in the absence of oxygen (anaerobes).

Bagged Biosolid – Biosolids that are sold or given away in a bag or other container (i.e., either an open or closed vessel containing 1 metric ton or less of biosolids).

Bioaerosol – Biological aerosols that can pose potential health risks during the composting and handling of organic materials. Bioaerosols are suspensions of particles in the air consisting partially or wholly of microorganisms. The bioaerosols of concern during composing include actinomycetes, bacteria, viruses, molds, and fungi.

Biosolids Composting – Is the process involving the aerobic biological degradation or bacterial conversion of dewatered biosolids, which works to produce compost that can be used as a soil amendment or conditioner.

Biosolids Quality Parameter – The EPA determined that three main parameters of concern should be used in gauging biosolids quality:

- The relevant presence or absence of pathogenic organisms;
- Pollutants; and
- The degree of attractiveness of the biosolids to vectors.

There can be several possible biosolids qualities. In order to express or describe those biosolids meeting the highest quality for all three of these biosolids quality parameters, the term *Exceptional Quality* or EQ has come into common use.

Bulking Agent – Materials, usually carbonaceous such as sawdust or woodchips, added to a compost system to maintain airflow by preventing settlement and compaction of the compost.

Bulk Biosolid – Biosolids that are not sold or given away in a bag or other container for application to the land.

Carbon to Nitrogen Ration (C:N Ratio) – Ratio representing the quantity of carbon (C) in relation to the quantity of nitrogen (N) in a soil or organic material; determines the composting potential of a material and serves to indicate product quality.

Compost – Is the product (innocuous humus) remaining after the composting process is completed.

Curing – Late stage of composting, after much of the readily metabolized material has been decomposed, which provides additional stabilization and allows further decomposition of cellulose and lignin (found in woody-like substances).

Curing Air – Curing piles are aerated primarily for moisture removal to meet final product moisture requirements and to keep odors from building up in the compost pile as biological activity is dissipating. Final product moisture requirements and summer ambient conditions are used to determine air requirements for moisture removal for the curing process.

Endotoxin – A toxin produced within a microorganism and released upon destruction of the cell in which it is produced. Endotoxins can be carried by airborne dust particles at composting facilities.

EPA's 503 Regulation – In order to ensure that sewage sludge (biosolids) is used or disposed of in a way that protects both human health and the environment, under the authority of the Clean Water Act as amended, the U.S. Environmental Protection Agency (EPA) promulgated, at 40 CFR Part 503, Phase I of the risk-based regulation that governs the final use or disposal of sewage sludge (biosolids).

Exceptional Quality (EQ) *Sludge* (Biosolids) – Although this term is not used in 40 CFR Part 503, it has become a shorthand term for biosolids that meet the pollutant concentrations in Table 3 of Part 503.13(b)(3); one of the six class A pathogen reduction alternatives in 503.32(a); and one of the vector attraction reduction options in 503.33(b)(1)-(8).

Feedstock – Decomposable organic material used for the manufacture of compost.

Heat Removal and Temperature Control – The biological oxidation process for composting bio-solids is an exothermic reaction. The heat given off by the composting process can raise the temperature of the compost pile high enough to destroy the organisms responsible for biodegradation. Therefore, the compost pile cells are aerated to control the temperature of the compost process by removing excess heat to maintain optimum temperature for organic solids degradation and pathogen reduction. Optimum temperatures are typically between 50°C and 60°C (122°F and 140°F). Using summer ambient air conditions, aeration requirements for heat removal can be calculated.

Metric Ton – One (1) metric ton, or 1,000 kg, equals about 2,205 lb, which is larger than the short ton (2,000 lb) usually referred to in the British system of units. The metric ton unit is used throughout this text.

Moisture Removal – When temperature increases, the quantity of moisture in saturated air increases. Air is required for the composting process to remove water that is present in the mix and produced by the oxidation of organic solids. The quantity of air required for moisture removal is calculated based on the desired moisture content for the compost product and the psychometric properties of the ambient air supply. Air requirements for moisture removal are calculated from summer ambient air conditions and required final compost characteristics.

Oxidation Air – The composting process requires oxygen to support aerobic biological oxidation of degradable organics in the biosolids and wood chips. Stoichiometric requirements for oxygen are related to the extent of organic solids degradation expected during the composting cycle time.

Pathogen Organism – specifically, salmonella and *E. coli* bacteria, enteric viruses, or visible helminth ova.

Peaking Air – The rate of organic oxidation, and therefore the heat release, can vary greatly for the composting process. If enough aeration capacity is not provided to meet peak requirements for heat or moisture removal, temperature limits for the process may be exceeded. Peaking air rates are typically 1.9 times the average aeration rate for heat removal.

Pollutant – An organic substance, an inorganic substance, a combination of organic and inorganic substances, or a pathogenic organism that, after discharge and upon exposure, ingestion, inhalation, or assimilation into an organism either directly from the environment or indirectly by ingestion through the food chain, could, on the basis of information available to the EPA, cause death, disease, behavioral abnormalities, cancer, genetic mutations, physiological malfunctions, or physical deformations in either organisms or offspring of the organisms.

Stability – State or condition in which the composted material can be stored without giving rise to nuisances or can be applied to the soil without causing problems there; the desired degree of stability for finished compost is one in which the readily decomposed compounds are broken down and only the decomposition of the more resistant biologically decomposable compounds remains to be accomplished.

Vector – Refers to the degree of attractiveness of biosolids to flies, rats, and mosquitoes that could meet pathogenic organisms and spread disease.

Aerated Static Pile (ASP)

Three methods of composting wastewater biosolids are common. Each method involves mixing dewatered wastewater solids with a bulking agent to provide carbon and increase porosity. The resulting mixture is piled or placed in a vessel where microbial activity causes the temperature of the mixture to rise during the "active composting" period. The specific temperatures that must be achieved and maintained for successful composing vary based on the method and use of the product. After active composting, the material is cured and distributed. Again, there are three commonly employed composting methods, but we only described the aerated static pile (ASP) method because it is commonly used. For an in-depth treatment of the other two methods, windrow and in-vessel, we refer you to F.R. Spellman's *Wastewater Biosolids to Compost* (1997).

The ASP Model Composting Facility uses the homogenized mixture of bulking agent (coarse hardwood wood chips) and dewatered biosolids piled by front-end loaders onto a large concrete composting pad where it is mechanically aerated via PVC plastic pipe embedded within the concrete slab. This ventilation procedure is part of the 26-day period of "active" composting when adequate air and oxygen is necessary to support aerobic biological activity in the compost mass and to reduce the heat and moisture content of the compost mixture. Keep in mind that a compost pile without a properly sized air distribution system can lead to the onset of anaerobic conditions and the appearance of putrefactive odors.

For illustration and discussion purposes, we assume a typical overall composting pad area is approximately 200 feet by 240 feet consisting of 11 blowers and 24 pipe troughs (troughs). Three blowers are 20 hp 2,400 cfm variable speed drive units capable of operating in either the positive or negative aeration mode. Blowers A, B, and C are each connected to two piping troughs that run the full length of the pad. The two troughs are connected at the opposite end of the composting pad to create an "aeration pipe loop." The other eight blowers are rated at 3 hp 1,200 cfm and are arranged 1 blower/6 troughs at half-length feeding 200 cfm/trough. These blowers can be operated in the positive or negative aeration mode. Aeration piping within the six pipe troughs is perforated PVC plastic pipe, 6 inches inside diameter and ¼ inch wall thickness. Perforation holes/orifices vary in size from 7/32 inch–1/2 inch, increasing in diameter as the distance from the blower increases.

DID YOU KNOW?

Aeration is an important process control parameter in the aerated static pile composting system. Air is required to supply oxygen for the biological degradation of organic solids in the biosolids and wood chips. Aeration is also needed for the removal of heat generated by the biological activity in the compost pile and excess moisture from the compost mix. Fans are used to ensure that sufficient quantities of air are supplied to meet composting process requirements and to provide the process control flexibility necessary for optimizing operations.

The variable-speed motor drives installed with blowers A, B, and C are controlled by five thermal probes mounted at various depths in the compost pile, and various parameters are fed back to the recorder, whereas the other eight blowers are constant speed, controlled by a timer that cycles them on and off. To ensure optimum composting operations, it is important to verify that these thermal probes are calibrated on a regular basis. In the constant speed system, thermal probes are installed, but all readings are taken and recorded manually.

For water and leachate drainage purposes, all aeration piping within the troughs slopes downward with the highest point at the center of the composting pad. Drain caps located at each end of the pipe length are manually removed on a regular basis so that any buildup of debris or moisture will not interfere with the airflow.

The actual construction process involved in building the compost pile will be covered in detail later, but for now a few key points should be made. For example, prior to the piling of the mixture on the composting pad, an 18-inch layer of wood chips is used as a base material. The primary purpose of the wood chips base is to keep the composting mixture clear of the aeration pipes, which reduces clogging of the air distribution openings in the pipes and allows free air circulation. A secondary benefit is that the wood chips insulate the composting mixture from the pad. The compost pad is like a heat sink, and this insulating barrier improves the uniformity of heat distribution within the composting mixture.

ADVANTAGES AND DISADVANTAGES

Biosolids composting has grown in popularity for the following reasons (WEF, 1995):

- Lack of availability of landfill space for solids disposal.
- Composting economics are more favorable when landfill tipping fees escalate.
- Emphasis on beneficial reuse at federal, state, and local levels.
- Ease of storage, handling, and use of the composted product.

The addition of biosolids compost to soil increases the soil's phosphorus, potassium, nitrogen, and organic carbon content. Composted biosolids can also be used in various land applications. Compost mixed with appropriate additives creates a material useful in wetland and mine land restoration. The high organic matter content and low nitrogen content common in compost provides a strong organic substrate that mimics wetland soils, prevents overloading of nitrogen, and absorbs ammonium to prevent transport to adjacent surface waters (Peot, 1998). Compost amended strip-mine spoils produce a sustainable cover of appropriate grasses, in contrast to inorganic-only amendments which seldom provide such a good or sustainable cover (Sopper, 1993).

Compost-enriched soil can also help suppress diseases and ward off pests. These beneficial uses of compost can help growers save money, reduce the use of pesticides, and conserve natural resources. Compost also plays a role in bioremediation of hazardous sites and pollution prevention. Compost has proven effective in degrading or altering many types of contaminants, such as wood-preservatives, solvents, heavy metals, pesticides, petroleum products, and explosives. Some municipalities are using compost to filter stormwater runoff before it is discharged to remove hazardous chemicals picked up when stormwater flows over surfaces such as roads, parking lots, and lawns. Additional uses for compost include soil mulch for erosion control, silviculture crop establishment, and production media (USEPA, 1997a).

Limitations of biosolids composting may include:

- Odor production at the composting site;
- Survival and presence of primary pathogens in the product;
- Dispersion of secondary pathogens such as *Aspergillus fumigatus*, particulate matter, other airborne allergens; and
- Lack of consistency in product quality with reference to metals, stability, and maturity.

Measuring Odors

In measuring odors, usually the panel or dilution methods are used. The dilution method is used in the water treatment process to detect odors in water and will not be discussed here. The panel method involves using ten or more people who make a judgment about the odor. These individual judgments are recorded and then analyzed. According to Vesilind (1980), the results of the panel method can be used to determine an "average opinion of the strength and nuisance value" of certain odors (p. 42).

When the panel method is used to measure odor, the parameter normally used for detecting odor is expressed as the number of effective dilutions-50 (ED-50). ED-50 is the number of fresh air dilutions required to reduce the odor level of a sample so that only 50% of the panel can smell it. Odor standards are based on odor control parameters such as ED 50, ED 10, ED 5, and others.

Malodorous Compounds in Biosolids-Derived Compost

To characterize composting as a smelly process is to correctly state the case. Whether or not this smelly process is offensive to the subject is another issue; it depends, almost entirely, on individual sensitivity. It is interesting to note that ingredients important to the composting process itself all smell. These smelly but important ingredients include the following: amines, aromatics, terpenes, organic and inorganic sulfur, and fatty acids.

Generally associated with fats and oils-based industrial operations, amines are more commonly known for their distinct fishy odor. In composting, amines are a by-product of microbial decomposition and generally form during anaerobic fermentation. Aromatics are usually present in biosolids and are volatilized during aeration. When woodchips are used as the bulking agent in the biosolids mix, aromatics are produced during the aerobic composting as the lignin's (in woodchips) breakdown. Likewise, terpenes (which are products of wood) are also present in compost piles that use woodchips as the bulking agent.

Probably, most wastewater specialists have been exposed to hydrogen sulfide and its characteristic rotten egg odor. Under normal circumstances, when biosolids are received at the composting site, any hydrogen sulfide emissions are quickly reduced when the biosolids and bulking agent are mixed and formed into aerobic piles. However, there can be a problem with hydrogen sulfide emissions, if the mix is incorrect or if the biosolids are too wet. When the biosolids is wet, it has the tendency to form into clumps. These clumps can become anaerobic and will form and release hydrogen sulfide.

Whether described as "stinking like a skunk" or smelling like "decayed cabbage," organic sulfurs are generally present in all biosolids-derived composting piles. Of the various organic sulfur compounds found in compost piles, probably the best known is the methyl mercaptans (smells like decayed cabbage). Fatty acids are generally produced under anaerobic conditions and do not add to odor generation problems unless the pile can go anaerobic.

Bottom Line on Composting Odor Control

At a biosolids composting facility, any sensible odor control management plan must consider all the areas and components of the composting process that might cause odors to be generated. While it is true that most odor problems are generated in the composting and curing process air systems, it is also true that at enclosed composting operations, odors generated from ancillary processes within the enclosure must be considered. Moreover, enclosed systems must have a way in which to control or scrub airflow within the structure prior to its release to the outside environment. Usually, not all process areas at a composting facility are enclosed. Keep in mind, these open areas, e.g., biosolids handling and mixing areas, can also cause odor control problems.

Lime Stabilization

Lime or alkaline stabilization can achieve the minimum requirements for both Class A (no detectable pathogens) and Class B (a reduced level of pathogens) biosolids with respect to pathogens, depending on the amount of alkaline material added and other processes employed. Generally, alkaline stabilization meets the Class B requirements when the pH of the mixture of wastewater solids and alkaline material is at 12 or above after 2 hours of contact.

Class A requirements can be achieved when the pH of the mixture is maintained at or above 12 for at least 72 hours, with a temperature of 52°C maintained for at least 12 hours during this time. In one process, the mixture is air dried to over 50% solids after the 72-hour period of elevated pH. Alternatively, the process may be manipulated to maintain temperatures at or above 70°F for 30 or more minutes, while maintaining the pH requirement of 12. This higher temperature can be achieved by overdosing with lime (that is, adding more than is needed to reach a pH of 12) by using a supplemental heat source, or by using a combination of the two. Monitoring for fecal coliforms or *Salmonella* is required prior to release by the generator for use.

Materials that may be used for alkaline stabilization include hydrated lime, quicklime (calcium oxide), fly ash, lime and cement kiln dust, and carbide lime. Quicklime is commonly used because it has a high heat of hydrolysis (491 British thermal units) and can significantly enhance pathogen destruction. Fly ash, lime kiln dust, or cement kiln dust are often used for alkaline stabilization because of their availability and relatively low cost.

The alkaline stabilized product is suitable for application in many situations, such as landscaping, agriculture, and mine reclamation. The product serves as a lime substitute, source of organic matter, and a specialty fertilizer. The addition of alkaline stabilized biosolids results in more favorable conditions for vegetative growth by improving soil properties such as pH, texture, and water holding capacity. Appropriate applications depend on the needs of the soil and crops that will be grown and the pathogen classification.

For example, a Class B material would not be suitable for blending in a topsoil mix intended for use in home landscaping but is suitable for agriculture, mine reclamation, and landfill cover where the potential for contact with the public is lower and access can be restricted. Class A alkaline stabilized biosolids are useful in agriculture and as a topsoil blend ingredient. Alkaline stabilized biosolids provide pH adjustment, nutrients, and organic matter, reducing reliance on other fertilizers.

Alkaline stabilized biosolids are also useful as daily landfill cover. They satisfy the federal requirement that landfills must be covered with soil or soil-like material at the end of each day (40 CFR 258). In most cases, lime stabilized biosolids are blended with other soil to achieve the proper consistency for daily cover.

As previously mentioned, alkaline stabilized biosolids are excellent for land reclamation in degraded areas, including acid mine spills or mine tailings. Soil conditions at such sites are very unfavorable for vegetative growth often due to acid content, lack of nutrients, elevated levels of heavy metals, and poor soil texture. Alkaline stabilized biosolids help to remedy these problems, making conditions more favorable for plant growth and reducing erosion potential. In addition, once a vegetative cover is established, the quality of mine drainage improves.

ADVANTAGES AND DISADVANTAGES

Alkaline stabilization offers several advantages, including:

- Consistency with the EPA's national beneficial reuse policy. Results in a product suitable for a variety of uses and is usually able to be sold.
- Simple technology requiring few special skills for reliable operation.
- Easy to construct from readily available parts.
- Small land area required.
- Flexible operation, easily started and stopped.

Several possible disadvantages should be considered in evaluating this technology:

- The resulting product is not suitable for use on all soils. For example, alkaline soils common in southwestern states will not benefit from the addition of a high pH material.

- The volume of material to be managed and moved off-site is increased by approximately 15–50% in comparison with other stabilization techniques, such as digestion. This increased volume results in higher transportation costs when material is moved off-site.
- There is potential for odor generation at both processing and end use site.
- There is a potential for dust production.
- There is a potential for pathogen regrowth if the pH drops below 9.5 while the material is stored prior to use.
- The nitrogen content in the final product is lower than that in several other biosolids products. During processing, nitrogen is converted to ammonia, which is lost to the atmosphere through volatilization. In addition, plant-available phosphorus can be reduced through the formation of calcium phosphate.
- There are fees associated with proprietary processes (Class A stabilization).

Thermal Treatment

Thermal treatment (or wet air oxidation) subjects sludge to high temperature and pressure in a closed reactor vessel. The high temperature and pressure rupture the cell walls of any microorganisms present in the solids and cause chemical oxidation of the organic matter. This process substantially improves dewatering and reduces the volume of material for disposal. It also produces a very high-strength waste, which must be returned to the wastewater treatment system for further treatment.

Chlorine Oxidation

Chlorine oxidation also occurs in a closed vessel. In this process, chlorine (100–1,000 mg/L) is mixed with a recycled solids flow. The recycled flow and process residual flow are mixed in the reactor. The solids and water are separated after leaving the reactor vessel. The water is returned to the wastewater treatment system, and the treated solids are dewatered for disposal. The main advantage of chlorine oxidation is that it can be operated intermittently. The main disadvantage is the production of extremely low pH and high chlorine content in the supernatant.

Sand Drying Beds

Sand beds have been used successfully for years to dewater sludge. Composed of a sand bed (consisting of a gravel base, underdrains, and 8–12 inches of filter grade sand), drying beds include an inlet pipe, splash pad containment walls, and a system to return filtrate (water) for treatment. In some cases, the sand beds are covered to provide drying solids protection from the elements.

In operation, solids are pumped to the sand bed and allowed to dry by first draining off excess water through the sand and then by evaporation. This is the simplest and cheapest method for dewatering sludge. Moreover, no special training or expertise is required. However, there is a downside, namely, drying beds require a great deal of manpower to clean; they can create odor and insect problems; and they can cause sludge buildup during inclement weather.

According to Metcalf and Eddy (2013), four types of drying are commonly used in dewatering biosolids:

1. Sand;
2. Paved;
3. Artificial media; and
4. Vacuum-assisted.

In addition to these commonly used dewatering methods, a few of the innovative methods of natural dewatering will also be discussed in this section. The innovative natural dewatering methods to be discussed include experimental work on biosolids dewatering via freezing. Moreover, dewatering biosolids with aquatic plants, which has been tested and installed in several sites throughout the US, is also discussed.

Drying beds are generally used for dewatering well-digested biosolids. Attempting to air dry raw biosolids is generally unsuccessful and may result in odor and vector control problems. Biosolids drying beds consist of perforated or open joint drainage system in a support media, usually gravel, covered with a filter media, usually sand but can consist of extruded plastic or wire mesh. Drying beds are usually separated into workable sections by wood, concrete, or other materials. Drying beds may be enclosed or open to the weather. They may rely entirely on natural drainage and evaporation processes or may use a vacuum to assist the operation (both types are discussed in the following sections).

Traditional Sand Drying Beds

This is the oldest biosolids dewatering technique and consists of 6–12 inches of coarse sand under-lain by layers of graded gravel ranging from $1/8$ to $1/4$ inches at the top and $3/4$–$1^1/2$ inches at the bottom. The total gravel thickness is typically about 1 foot. Graded natural earth (4–6 in.) usually makes up the bottom with a web of drain tile placed on 20–30 feet centers. Sidewalls and partitions between bed sections are usually of wooden planks or concrete and extend about 14 inches above the sand surface (McGhee, 1991).

Large open areas of land are required for sand drying biosolids. For example, it is not unusual to have drying beds that are up to 125+ feet long and from 20 to 35 feet in width. Even at the smallest wastewater treatment plants it is normal practice to provide at least two drying beds. The actual dewatering process occurs as a result of two different physical processes: evaporation and drainage. The liquor which drains off the biosolids goes to a central sump which pumps it back to the treatment process to undergo further treatment. The operation is very much affected by climate. In wet climates it may be necessary to cover the beds with a translucent material that will allow at least 85% of the sun's ultraviolet radiation to pass through. Typical loading rates for primary biosolids in dry climates range up to 200 kg/(sq m × year) and from 60 to 125 kg/(sq m × year) for mixtures of primary and waste activated biosolids.

When a drying bed is put into operation, it is generally filled with digested biosolids to a depth ranging from 8 to 12 inches. The actual drying time is climate-sensitive; that is, drying can take from a few weeks to a few months, depending on the climate and the season. After dewatering, the biosolids solids content will range from about 20% to 35%, and, more importantly, the volume will have been reduced up to 85%. Upon completion of the drying process, the dried biosolids are generally removed from the bed with handheld forks or front-end loaders. It is important to note that in the dried biosolids removal process a small amount of sand is lost and the bed must be refilled and graded periodically. Dried solids removed from a biosolids drying bed can be either incinerated or landfilled.

Drying Bed Operational Capacity

One of the primary considerations that must be taken into account when designing a biosolids drying bed is the determination of how many pounds of solids can be dried for every square foot of drying bed each year. In order to make this determination for the biosolids drying bed model described here, it is necessary to list certain parameters that will be needed to calculate the result. The following example lists these vital parameters and shows the calculation that is necessary to determine the answer.

Paved Drying Beds

The main reason for using paved drying beds is that they alleviate the problem of mechanical biosolids removal equipment damaging the underlain piping networks. The beds are paved with concrete or asphalt and are generally sloped toward center where a sump-like area with underlain pipes is arranged. These dewatering beds, like biosolids lagoons, depend on evaporation for the dewatering of the applied solids.

Paved drying beds are usually rectangular in shape with a center drainage strip. They can be heated via buried pipes in the paved section and generally are covered to prevent rain incursion. In this type of natural dewatering, solids contents of 45–50% can be achieved within 35 days in dry climates under normal conditions (McGhee, 1991). The operation of paved drying beds involves applying the biosolids to a depth of about 12 inches. The settled surface area is routinely mixed by a special vehicle-mounted machine which is driven through the bed. Mixing is important because it breaks up the crust and exposes wet surfaces to the environment. Supernatant is decanted in a manner like biosolids lagoons. Biosolids loadings in relatively dry climates range from about 120 to 260 kg/(sq m/year). High capital cost and larger land requirements than for sand beds are the two major disadvantages of paved drying beds.

Although the construction and operation methodologies for biosolids drying beds are well-known and widely accepted, this is not to say that the wastewater industry has not attempted to incorporate further advances into their construction and operation. For example, in an attempt to reduce the amount of dewatered biosolids that must be manually removed from drying beds, attempts have been made to construct drying beds in a specific manner whereby they can be planted with reeds, namely the *Phragmites communis* variety (to be covered in greater detail later). The intent of augmenting the biosolids drying bed with reeds is to effect further desiccation. Moreover, tests have shown that the plants extend their root systems into the biosolids mass.

This extended root system has the added benefit of helping to establish a rich microflora which eventually feeds upon the organic content of the biosolids. It is interesting to note that normal plant activity works to keep the system aerobic.

Vacuum-Assisted Drying Beds

For small plants which process small quantities of biosolids and have limited land area, vacuum-assisted drying beds may be the preferred method of dewatering biosolids. Vacuum-assisted drying beds normally employ the use of a small vacuum to accelerate the dewatering of biosolids applied to a porous medium plate. This porous medium is set above an aggregate-filled support underdrain which, as the name implies, drains to a sump. The small vacuum is applied to this underdrain, which works to extract free water from the biosolids; with biosolids loadings on the order of less than 10 kg/sq m/cycle, the time required to dewater conditioned biosolids is about 1 day (McGhee, 1991). Using this method of dewatering it is possible to achieve a solids content of >30%, although 20% solids is a more normal expectation. Removal of dewatered biosolids is usually accomplished with mechanized machinery such as front-end loaders.

Once the solids have been removed, it is important to wash the surface of the bed with high-pressure hoses to ensure residuals are removed. The main advantage cited for this dewatering method is the reduced amount of time that is needed for dewatering, which reduces the effects of inclement climatic conditions on biosolids drying. The main disadvantage of this type of dewatering may be its dependence on adequate chemical conditioning for successful operation (Metcalf & Eddy, 2013).

Natural Methods of Dewatering Biosolids

Two of the innovative methods of natural biosolids dewatering are discussed in this section:

- Dewatering via freezing
- Dewatering using aquatic plants

Note: Dewatering via freezing is primarily in the pilot study stage. Experimental work in this area has led to pilot plants and has yielded various mathematical models, but no full-scale operations (Outwater, 1994).

Rotary Vacuum Filtration

Rotary vacuum filters have also been used for many years to dewater sludge. The vacuum filter includes filter media (belt, cloth, or metal coils), media support (drum), vacuum system, chemical feed equipment, and conveyor belt(s) to transport the dewatered solids. In operation, chemically treated solids are pumped to a vat or tank in which a rotating drum is submerged. As the drum rotates, a vacuum is applied to the drum. Solids collect on the media and are held there by the vacuum as the drum rotates out of the tank. The vacuum removes additional water from the captured solids. When solids reach the discharge zone, the vacuum is released, and the dewatered solids are discharged onto a conveyor belt for disposal. The media is then washed prior to returning to the start of the next cycle.

Types of Rotary Vacuum Filters

The three principal types of rotary vacuum filters are rotary drum, coil, and belt. The *rotary drum* filter consists of a cylindrical drum rotating partially submerged in a vat or pan of conditioned sludge. The drum is divided length-wise into several sections that are connected through internal piping to ports in the valve body (plant) at the hub. This plate rotates in contact with a fixed valve plate with similar parts, which are connected to a vacuum supply, a compressed air supply, and an atmosphere vent. As the drum rotates, each section is thus connected to the appropriate service.

The *coil type* vacuum filter uses two layers of stainless-steel coils arranged in corduroy fashion around the drum. After a dewatering cycle, the two layers of springs leave the drum bed and are separated from each other so that the cake is lifted off the lower layer and is discharged from the upper layer. The coils are then washed and reapplied to the drum. The coil filter is used successfully for all types of sludges; however, sludges with extremely fine particles or ones that are resistant to flocculation dewater poorly with this system. The media on a *belt filter* leaves the drum surface at the end of the drying zone and passes over a small-diameter discharge roll to aid cake discharge. Washing of the media occurs next. Then the media are returned to the drum and to the vat for another cycle. This type of filter normally has a small-diameter curved bar between the point where the belt leaves the drum and the discharge roll. This bar primarily aids in maintaining belt dimensional stability.

Filter Media

Drum and belt vacuum filters use natural or synthetic fiber materials. On the drum filter, the cloth is stretched and secured to the surface of the drum. In the belt filter, the cloth is stretched over the drum and through the pulley system. The installation of a blanket requires several days. The cloth will (with proper care) last several hundred to several thousand hours. The life of the blanket depends on the cloth selected, the conditioning chemical, backwash frequency, and cleaning (i.e., acid bath) frequency.

Filter Drum

The filter drum is a maze of pipe work running from a metal screen and wooden skeleton and connecting to a rotating valve port at each end of the drum. The drum is equipped with a variable speed drive to turn the drum from 1/8 to 1 rpm. Normally, solids pickup is indirectly related to the drum speed. The drum is partially submerged in a vat containing the conditioned sludge. Normally, submergence is limited to 1/5 or less of the filter surface at a time.

Chemical Conditioning

Sludge dewatered using vacuum filtration is normally chemically conditioned just prior to filtration. Sludge conditioning increases the percentage of solids captured by the filter and improves the dewatering characteristics of the sludge. However, conditional sludge must be filtered as quickly as possible after chemical addition to obtain these desirable results.

Pressure Filtration

Pressure filtration differs from vacuum filtration in that the liquid is forced through the filter media by a positive pressure instead of a vacuum. Several types of presses are available, but the most used types are plate-and-frame presses and belt presses. *Filter presses* include the belt or plate-and-frame types. The belt filter includes two or more porous belts, rollers, and related handling systems for chemical makeup and feed, and supernatant and solids collection and transport.

The plate-and-frame filter consists of a support frame, filter plates covered with a porous material, hydraulic or mechanical mechanism for pressing plates together, and related handling systems for chemical makeup and feed, and supernatant and solids collection and transport. In the plate-and-frame filter, solids are pumped (sandwiched) between plates. Pressure (200–250 psi) is applied to the plates, and water is "squeezed" from the solids. At the end of the cycle the pressure is released, and as the plates separate, the solids drop out onto a conveyor belt for transport to storage or disposal. Performance factors for plate-and-frame presses include feed sludge characteristics, type and amount of chemical conditioning, operating pressures, and the type and amount of precoat.

The belt filter uses a coagulant (polymer) mixed with the influent solids. The chemically treated solids are discharged between two moving belts. First, water drains from the solids by gravity. Then, as the two belts move between a series of rollers, pressure "squeezes" additional water out of the solids. The solids are then discharged onto a conveyor belt for transport to storage/disposal. Performance factors for the belt press include sludge feed rate, belt speed, belt tension, belt permeability, chemical dosage, and chemical selection. Filter presses have lower operation and maintenance costs than vacuum filters or centrifuges. They typically produce a good quality cake and can be batch operated. However, construction and installation costs are high. Moreover, chemical addition is required, and the presses must be operated by skilled personnel.

CENTRIFUGATION

Centrifuges of various types have been used in dewatering operations for at least 30 years and appear to be gaining in popularity. Depending on the type of centrifuge used, in addition to centrifuge pumping equipment for solids feed and centrate removal, chemical makeup and feed equipment and support systems for removal of dewatered solids are required. Generally, in operation, the centrifuge spins at a very high speed. The centrifugal force it creates "throws" the solids out of the water. Chemically conditioned solids are pumped into the centrifuge. The spinning action "throws" the solids to the outer wall of the centrifuge. The centrate (water) flows inside the unit to a discharge point. The solids held against the outer wall are scraped to a discharge point by an internal scroll moving slightly faster or slower than the centrifuge speed of rotation. In the operation of the continuous feed, solids bowl, conveyor-type centrifuge (this is the most common type currently used), and other commonly used centrifuges, solid/liquid separation occurs as a result of rotating the liquid at high speeds to cause separation by gravity.

SLUDGE INCINERATION

Not surprisingly, incinerators produce the maximum solids and moisture reductions. The equipment required depends on whether the unit is a multiple hearth or fluid-bed incinerator. Generally, the system will require a source of heat to reach ignition temperature, solids feed system, and ash handling equipment. It is important to note that the system must also include all required equipment (e.g., scrubbers) to achieve compliance with air pollution control requirements. Solids are pumped to the incinerator. The solids are dried and then ignited (burned). As they burn, the organic matter is converted to carbon dioxide and water vapor, and the inorganic matter is left behind as ash or "fixed" solids. The ash is then collected for reuse or disposal.

Process Description

The incineration process first dries, then burns the sludge.

The process involves the following steps:

- The temperature of the sludge feed is raised to 212°F.
- Water evaporates from the sludge.
- The temperature of the water vapor and air mixture increases.
- The temperature of the dried sludge volatile solids rises to the ignition point.

Note: Incineration will achieve maximum reductions if enough fuel, air, time, temperature, and turbulence are provided.

Incineration Processes

Multiple Hearth Furnace

The *multiple hearth furnace* consists of a circular steel shell surrounding several hearths. Scrappers (rabble arms) are connected to a central rotating shaft. Units range from 4.5 to 21.5 feet in diameter and have from 4 to 11 hearths. In operation, dewatered sludge solids are placed on the outer edge of the top hearth. The rotating rabble arms move them slowly to the center of the hearth. At the center of the hearth, the solids fall through ports to the second level. The process is repeated in the opposite direction. Hot gases generated by burning on lower hearths dry solids. The dry solids pass to the lower hearths. The high temperature on the lower hearths ignites the solids. Burning continues to completion. Ash materials discharge to lower cooling hearths where they are discharged for disposal. Air flowing inside the center column and rabble arms continuously cools internal equipment.

Fluidized Bed Furnace

The *fluidized bed* incinerator consists of a vertical circular steel shell (reactor) with a grid to support a sand bed and an air system to provide warm air to the bottom of the sand bed. The evaporation and incineration process takes place within the super-heated sand bed layer. In operation, air is pumped to the bottom of the unit. The airflow expands (fluidize) the sand bed inside. The fluidized bed is heated to its operating temperature (1,200–1,500°F). Auxiliary fuel is added when needed to maintain operating temperature. The sludge solids are injected into the heated sand bed. Moisture immediately evaporates. Organic matter ignites and reduces to ash. Residues are ground to fine ash by the sand movement. Fine ash particles flow up and out of the unit with exhaust gases. Ash particles are removed using common air pollution control processes. Oxygen analyzers in the exhaust gas stack control the airflow rate.

Note: Because these systems retain a high amount of heat in the sand, the system can be operated for as little as 4 hours/day with little or no reheating.

Land Application of Biosolids

The purpose of land application of biosolids is to dispose of the treated biosolids in an environmentally sound manner by recycling nutrients and soil conditioners. In order to be land applied, wastewater biosolids must comply with state and federal biosolids management/disposal regulations. Biosolids must not contain materials that are dangerous to human health (i.e., toxicity, pathogenic organisms, etc.) or dangerous to the environment (i.e., toxicity, pesticides, heavy metals, etc.). Treated biosolids are land applied by either direct injection or application and plowing in (incorporation). Land application of biosolids requires precise control to avoid problems. The quantity and the quality of biosolids applied must be accurately determined. For this reason, the operator's process control activities include biosolids sampling/testing functions. Biosolids sampling and testing includes the determination of percent solids, heavy metals, organic pesticides and herbicide, alkalinity, total organic carbon (TOC), organic nitrogen, and ammonia nitrogen.

PERMITS, RECORDS, AND REPORTS

Permits, records, and reports play a significant role in wastewater treatment operations. In fact, regarding the "permit," one of the first things any new operator quickly learns is the importance of "making permit" each month. In this chapter, we briefly cover NPDES permits and other pertinent records and reports with which the wastewater operator must be familiar.

Note: The discussion that follows is general in nature; it does not necessarily apply to any state, but instead is an overview of permits, records, and reports that are an important part of wastewater treatment plant operations. For specific guidance on requirements for your locality, refer to your state water control board or other authorized state agency for information. In *Part II*, the term "Board" signifies the state-reporting agency.

DEFINITIONS

There are several definitions which should be discussed prior to discussing the permit requirements for records and reporting. These definitions are listed below.

Average Monthly Limitation – the highest allowable average over a calendar month, calculated by adding all the daily values measured during the month and dividing the sum by the number of daily values measured during the month.

Average Weekly Limitation – the highest allowable average over a calendar week, calculated by adding all the daily values measured during the calendar week and dividing the sum by the number of daily values determined during the week.

Average Daily Limitation – the highest allowable average over a 24-hour period, calculated by adding all the values measured during the period and dividing the sum by the number of values determined during the period.

Average Hourly Limitation – the highest allowable average for a 60-minute period, calculated by adding all the values measured during the period and dividing the sum by the number of values determined during the period.

Daily Discharge – means the discharge of a pollutant measured during a calendar day or any 24-hour period that reasonably represents the calendar for the purpose of sampling. For pollutants with limitations expressed in units of weight, the daily discharge is calculated as the total mass of the pollutant discharged over the day. For pollutants with limitations expressed in other units, the daily discharge is calculated as the average measurement of the pollutant over the day.

Maximum Daily Discharge – the highest allowable value for a daily discharge.

Effluent Limitation – any restriction by the State Board on quantities, discharge rates, or concentrations of pollutants which are discharged from point sources into state waters.

Maximum Discharge – the highest allowable value for any single measurement.

Minimum Discharge – the lowest allowable value for any single measurement.

Point Source – any discernible, defined, and discrete conveyance, including but not limited to any pipe, ditch, channel, tunnel, conduit, well, discrete fissure, container, rolling stock, vessel, or other floating craft, from which pollutants are or may be discharged. This definition does not include return flows from irrigated agricultural land.

Discharge Monitoring Report – forms for use in reporting of self-monitoring results of the permittee.

Discharge Permit – State Pollutant Discharge Elimination System permit which specifies the terms and conditions under which a point source discharge to state waters is permitted.

NPDES PERMITS

In the United States, all treatment facilities which discharge to state waters must have a discharge permit issued by the State Water Control Board or other appropriate state agency. This permit is known on the national level as the National Pollutant Discharge Elimination System (NPDES) permit and on the state level as the (State) Pollutant Discharge Elimination System (state-PDES) permit. The permit states the specific conditions which must be met to legally discharge treated wastewater to state waters.

The permit contains general requirements (applying to every discharger) and specific requirements (applying only to the point source specified in the permit). A general permit is a discharge permit, which covers a specified class of dischargers. It is developed to allow dischargers with the specified category to discharge under specified conditions. All discharge permits contain general conditions. These conditions are standard for all dischargers and cover a broad series of requirements. Read the general conditions of the treatment facility's permit carefully. Permittees must retain certain records.

Monitoring

These records include:

- Date, time, and exact place of sampling or measurements;
- Name(s) of the individual(s) performing sampling or measurement;
- Date(s) and time(s) analyses were performed;
- Name(s) of the individuals who performed the analyses;
- Analytical techniques or methods used;
- Observations, readings, calculations, bench data, and results;
- Instrument calibration and maintenance;
- Original strip chart recordings for continuous monitoring;
- Information used to develop reports required by the permit; and
- Data used to complete the permit application.

Note: All records must be kept at least three years (longer at the request of the State Board).

Reporting

Generally, reporting must be made under the following conditions/situations (requirements may vary depending on the state regulatory body with reporting authority):

Unusual or Extraordinary Discharge Report – must notify the Board by telephone within 24 hours of the occurrence and submit a written report within five (5) days.

Report must include:

- Description of the non-compliance and its cause;
- Non-compliance date(s), time(s), and duration;
- Steps planned/taken to reduce/eliminate; and
- Steps planned/taken to prevent reoccurrence.

Anticipated Non-compliance – Must notify the Board at least ten days in advance of any changes to the facility or activity which may result in non-compliance.

Compliance Schedule – Must report compliance or non-compliance with any requirements contained in compliance schedules no later than 14 days following the scheduled date for completion of the requirement.

24-Hour Reporting – Any non-compliance which may adversely affect state waters or may endanger public health must be reported orally with 24 hours of the time the permittee becomes aware of the condition. A written report must be submitted within five days.

Discharge Monitoring Reports (DMRs) – Reports self-monitoring data generated during a specified period (normally 1 month).

When completing the DMR, remember:

- More frequent monitoring must be reported;
- All results must be used to complete reported values;
- Pollutants monitored by an approved method but not required by the permit must be reported;
- No empty blocks on the form should be left blank;
- Averages are arithmetic unless noted otherwise;

- Appropriate significant figures should be used;
- All bypasses and overflows must be reported;
- The licensed operator must sign Report;
- Responsible official must sign Report; and
- Department must receive report by 10th of next month.

Sampling and Testing

The general requirements of the permit specify minimum sampling and testing which must be performed on the plant discharge. Moreover, the permit will specify the frequency of sampling, sample type, and length of time for composite samples. Unless a specific method is required by the permit, all sample preservation and analysis must follow the requirements set forth in the Federal Regulations' *Guidelines Establishing Test Procedures for the Analysis of Pollutants Under the Clean Water Act* (40 CFR 136).

Note: All samples and measurements must be representative of the nature and quantity of the discharge.

Effluent Limitations

The permit sets numerical limitations on specific parameters contained in the plant discharge.
Limits may be expressed as:

- Average monthly quantity (kg/day)
- Average monthly concentration (mg/L)
- Average weekly quantity (kg/day)
- Average weekly concentration (mg/L)
- Daily quantity (kg/day)
- Daily concentration (mg/L)
- Hourly average concentration (mg/L)
- Instantaneous minimum concentration (mg/L)
- Instantaneous maximum concentration (mg/L)

Compliance Schedule

If the facility requires additional construction or other modifications to fully comply with the final effluent limitations, the permit will contain a schedule of events to be completed to achieve full compliance.

Special Conditions

Any special requirements or conditions set for approval of the discharge will be contained in this section.
Special conditions may include:

- Monitoring required to determine effluent toxicity
- Pretreatment program requirements

Licensed Operator Requirements

The permit will specify, based on the treatment system complexity and the volume of flow treated, the minimum license classification required to be the designated responsible charge operator.

Chlorination/Dechlorination Reporting

Several reporting systems apply to chlorination or chlorination followed by dechlorination. It is best to review this section of the specific permit for guidance. If confused, contact the appropriate State Regulatory Agency.

CHAPTER SUMMARY

It has been said that wastewater treatment is more art than science. After completing this chapter and the previous chapters, it is more likely that the reader will understand that wastewater treatment is both art and science.

REVIEW QUESTIONS

1. Who must sign the DMR?
   a. maintenance operator
   b. lab supervisor
   c. general manager
   d. the licensed operator and the responsible person in charge

2. What does the COD test measure?
   a. amount of gold in the wastewater
   b. amount of salt in the wastewater
   c. amount of organic material that can be oxidized by strong oxidizing agent
   d. none of above

3. Three reasons for treating wastewater.
   a. remove gold ore, disease, and improve water quality
   b. prevent disease, protect aquatic organisms, protect water quality
   c. prevent disease, improve drinkability, remove salt
   d. none of above

4. Name two types of solids based on physical characteristics.
   a. gold and suspended
   b. silver and dissolved
   c. dissolved and suspended
   d. all above

5. What is the difference between organic and inorganic?
   a. organic contains carbon and inorganic does not decompose
   b. organics are plants and inorganic are animals
   c. organics are valuable and inorganic are worthless
   d. all above

6. Name four types of microorganisms, which may be present in wastewater.
   a. algae, rotifers, bacteria, virus
   b. corona virus, salt, virus, manure
   c. algae, salt, bacteria, virus
   d. none of above

7. When organic matter is decomposed aerobically, what materials are produced?
   a. silver, waste, manure, water
   b. water, carbon dioxide, stable solids, trash
   c. water, carbon dioxide, stable soils, more organics
   d. none of the above

8. Name three materials or pollutants, which are not removed by the natural purification process.
   a. toxic matter, inorganic dissolved solids, pathogenic organism
   b. pathogenic organism, organic dissolved solids, toxic matter
   c. pathogenic organism, inorganic suspended solids, toxic waste
   d. all above

9. What are the used water and solids from a community that flow to a treatment plant called?
    a. garbage
    b. oily waste
    c. raw effluent
    d. raw influent

10. Where do disease-causing bacteria in wastewater come from?
    a. from cattle
    b. from horses
    c. from body wastes of humans who have the disease
    d. all above

ANSWER KEY

1. Who must sign the DMR?
    a. maintenance operator
    b. lab supervisor
    c. general manager
    d. **the licensed operator and the responsible person in charge**

2. What does the COD test measure?
    a. amount of gold in the wastewater
    b. amount of salt in the wastewater
    c. **amount of organic material that can be oxidized by a strong oxidizing agent**
    d. none of above

3. Three reasons for treating wastewater.
    a. remove gold ore, disease, and improve water quality
    b. **prevent disease, protect aquatic organisms, and protect water quality**
    c. prevent disease, improve drinkability, and remove salt
    d. none of above

4. Name two types of solids based on physical characteristics.
    a. gold and suspended
    b. silver and dissolved
    c. **dissolved and suspended**
    d. all above

5. What is the difference between organic and inorganic?
    a. **organic contains carbon and inorganic does not decompose**
    b. organics are plants and inorganic are animals
    c. organics are valuable and inorganic are worthless
    d. all above

6. Name four types of microorganisms which may be present in wastewater.
    a. **algae, rotifers, bacteria, virus**
    b. corona virus, salt, virus, manure
    c. algae, salt, bacteria, virus
    d. none of above

7. When organic matter is decomposed aerobically, what materials are produced?
    a. silver, waste, manure, water
    b. water, carbon dioxide, stable solids, trash
    c. **water, carbon dioxide, stable soils, more organics**
    d. none of the above

8. Name three materials or pollutants which are not removed by the natural purification process.
   a. **toxic matter, inorganic dissolved solids, pathogenic organism**
   b. pathogenic organism, organic dissolved solids, toxic matter
   c. pathogenic organism, inorganic suspended solids, toxic waste
   d. all above

9. What are the used water and solids from a community that flow to a treatment plant called?
   a. garbage
   b. oily waste
   c. raw effluent
   d. **raw influent**

10. Where do disease-causing bacteria in wastewater come from?
   a. from cattle
   b. from horses
   c. **from body wastes of humans who have disease**
   d. all above

REFERENCES

Anderson, J.B., and Zwieg, H.P. (1962). Biology of Waste Stabilization Ponds. *Southwest Water Works Journal* 44(2): 15–18.

Assenzo, J.R., and Reid, G.W. (1966). Removing Nitrogen and Phosphorus by Bio-Oxidation Ponds in Central Oklahoma. *Water and Sewage Works* 13(8): 294–299.

Brockett, O.D. (1976). Microbial Reactions in Facultative Ponds-1. The Anaerobic Nature of Oxidation Pond Sediments. *Water Research* 10(1): 45–49.

Canadian Environmental Protection Act. (2009). Phosphorus. Accessed 12/6/20 @ https://law-lois.justice.gc. ca.eng/act.

Craggs, R. (2005). Nutrients. In *Pond Treatment Technology*, ed. A. Hilton. London, UK: IWA Publishing.

Crites, R.W., Middlebrooks, E.J., and Reed, S.C. (2006). *Natural Wastewater Treatment Systems*. Boca Raton, FL: CRC Press.

CWNS. (2000). *Treatment Ponds: Clean Water Needs Survey*. Washington, DC: USEPA.

Gallert, C., and Winter, J. (2005). *Bacterial Metabolism in Wastewater Treatment Systems. Environmental Biotechnology*, eds. H.J. Jordening and J. Winter. New York: Wiley VCH.

Gaudy, A.F., Jr., and Gaudy, E.T. (1980). *Microbiology for Environmental Scientists and Engineers*. New York: McGraw Hill.

Gloyna, E.F. (1976). Facultative Waste Stabilization Pond Design. In *Ponds as a Waste Treatment Alternative*, eds. Gloyna, E.F., Malina, J.F., Jr., and Davis, E.M. Water Resources Symposium No. 9, Austin, TX: University of Texas Press.

Grady, C.P.L, Jr., Daigger, G.T., Lover, N.G., and Filipe, C.D.M. (2011). *Biological Wastewater Treatment*, 3rd ed. Boca Raton, FL: CRC Press.

Grolund, E. (2002). *Microalgae at Wastewater Treatment in Cold Climates*. Department of Environmental Engineering. SE 971 87 LULEA Sweden, Lic Thesis 2002:35.

Jagger, J. (1997). *Introduction to Research in UV*. Englewood Cliffs, NJ: Prentice-Hall.

Knudson, G.B. (1985). Photoreactivation of UV-Irradiated *Legionella pneumoplila* and Other *Legionella* Species. *Applied and Environmental Microbiology* 49: 975–980.

Linden, K.G., Shin, G.A., Faubert, G., Cairns, W., and Sobsey, M.D. (2002). UV Disinfection of *Giardia lamblia* Cysts in Water. *Environmental Science and Technology* 36: 2519–2522.

Lynch, J.M., and Poole, N.J. (1979). *Microbial Ecology, A Conceptual Approach*. New York: John Wiley & Sons.

McGhee, T.J. (1991). *Water Supply and Sewerage*. New York: McGraw-Hill.

Metcalf & Eddy. (2003). *Wastewater Engineering: Treatment, Disposal and Reuse*. New York, NY: McGraw-Hill.

Metcalf & Eddy, Inc. (2013). *Wastewater Engineering: Treatment, Disposal, Reuse*, 5th ed. New York: McGraw-Hill.

Middlebrooks, E.J., Middlebrooks, C.H., Reynolds, J.H., Watters, G.Z., Reed, S.C., and George, D.B. (1982). *Wastewater Stabilization Lagoon Design, Performance and Upgrading*. Middlebrook, NY: Macmillan Publishing Co.

Middlebrooks, E.J., and Pano, A. (1983). Nitrogen Removal in Aerated Lagoons. *Water Research* 17(10): 1369–1378.

NEIWPCC. (1998). Guide for the Design of Wastewater Treatment Work TR-16. Washington, DC: New England Interstate Water Pollution Control Commission.

Oguma, K., Katayama, H., Mitani, H., Morita, S., Hirata, T., and Ohgaki, S. (2001). Determination of Pyrimidine Dimmers in *Escherichia coli* and *Cryptosporidium parvum* during UV Light Inactivation, Photoreactivation, and Dark Repair. *Applied and Environmental Microbiology* 67: 4630–4637.

Oswald, W.J. (1996). *A Syllabus on Advanced Integrated Pond Systems®*. Berkeley, CA: University of California.

Outwater, A.B. (1994). *Reuse of Sludge and Minor Wastewater Residuals*. Boca Raton, FL: Lewis Publishers.

Pano, A., and Middlebrooks, E.J. (1982). Ammonia Nitrogen Removal in Facultative Wastewater Stabilization Ponds. *Journal of Water Pollution Control Federation* 54(4): 2148.

Park, J. (2012). *Biological Nutrient Removal Theories and Design*. Online at http://www.dnr.state.wi,us/org/water/wm/ww/biophos/bnr_remvoal.htm.

Paterson, C., and Curtis, T. (2005). Physical and Chemical Environments. In *Pond Treatment Technology*, ed. A. Shilton. London, UK: IWA Publishing.

Pearson, H. (2005). Microbiology of Waste Stabilization Ponds. In *Pond Treatment Technology*, ed. A. Shilton. London, UK: IWA Publishing.

Peot, C. (1998). Compost Use in Wetland Restoration. Design for Success. In *Proceedings of the 12th Annual Residual and Biosolids Management Conference*. Alexandra, VA: Water Environment Federation.

Pipes, W.O., Jr. (1961). Basic Biology of Stabilization Ponds. *Water and Sewage Works* 108(4): 131–136.

Rauth, A.M. (1965). The Physical State of Viral Nucleic Acid and the Sensitivity of Viruses to Ultraviolet Light. *Biophysical Journal* 5: 257–273.

Sawyer, C.N., McCarty, P.L., and Parkin, G.F. (1994). *Chemistry for Environmental Engineering*. New York: McGraw Hill.

Shilton, A. (ed.). (2005). *Pond Treatment Technology*. London: IWA Publishing.

Shin, G.A., Linden, K.G., Arrowood, M.J., Faubert, G., and Sosbey, M.D. (2001). DNA Repair of UV-Irradiated *Cryptosporidium parvum* Oocysts and *Giardia lamblia* Cysts. In *Proceedings of the First International Ultraviolet Association Congress*, Washington, DC, June 14–16.

Sopper, W.E. (1993). *Municipal Sludge Use in Land Reclamation*. Boca Raton, FL: Lewis Publishers.

Spellman, F.R. (1996). *Stream Ecology and Self-Purification*. Boca Raton, FL: CRC Press.

Spellman, F.R. (1997). *Stream Ecology and Self-Purification*. Lancaster, PA: Technomic Publishing Company.

Spellman, F.R. (2000). *Microbiology for Water and Wastewater Operators*. Boca Raton, FL: CRC Press.

Spellman, F.R. (2007). *The Science of Water*, 2nd ed. Boca Raton, FL: CRC Press.

Spellman, F.R. (2020). *The Science of Water*, 4th ed. Boca Raton, FL: CRC Press.

Ullrich, A.H. (1967). Use of Wastewater Stabilization Ponds in Two Different systems. *JWPCF* 39(6): 965–977.

USEPA. (1975). *Process Design Manual for Nitrogen Control*, EPA-625/1-75-007. Cincinnati, OH: Center for Environmental Research Information.

USEPA. (1989). *Technical Support Document for Pathogen Reducing in Sewage Sludge. NTIS No. PB89-136618*. Springfield, VA: National Technical Information Service.

USEPA. (1993). *Manual: Nitrogen Control*, EPA-625/R-93/010. Cincinnati, OH.

USEPA. (2000). *Wastewater Technology Fact Sheet Package Plants*. Washington, DC: United States Environmental Protection Agency.

USEPA. (2001). *Memorandum: Development and Adoption of Nutrient Criteria into Water Quality Standards*. Online at http://oaspub.epa.gov/waters/national_rept.control#TOP_IMP.

USEPA. (2006). *UV Disinfection Guidance Manual*. Washington, DC: United States Environmental Protection Agency.

USEPA. (2007a). *Wastewater Management Fact Sheet: Membrane Bioreactors*. Washington, DC: United States Environmental Protection Agency.

USEPA. (2007b). *Advanced Wastewater Treatment to Achieve Low Concentration of Phosphorus*. Washington, DC: Environmental Protection Agency.

USEPA. (2007c). *Biological Nutrient Removal Processes and Costs*. Washington, DC: United States Environmental Protection Agency.

USEPA. (2007d) *National Section 303(d) List Fact Sheet*. Online at http://iaspub.epa.gov/waters/national_rept.control.

USEPA. (2011). *Principles of Design and Operations of Wastewater Treatment Pond Systems for Plant Operators, Engineers, and Managers*. Washington, DC: U.S. Environmental Protection Agency.

Vasconcelos, V.M., and Pereira, E. (April 2001). Cyanobacteria Diversity and Toxicity in a Wastewater Treatment Plant (Portugal). *Water Research* 35(5): 1354–1357.

Vesilind, P.A. (1980). *Treatment and Disposal of Wastewater Sludges*, 2nd ed. Ann Arbor, MI: Ann Arbor Science Publishers, Inc.

WEF (Water Environment Federation). (1995). *Wastewater Residuals Stabilization. Manual of Practice FD-9*. Alexandria, VA: Water Environment Federation.

WEF (Water Environment Federation). (1998). *Design of Municipal Wastewater Treatment Plants, Manual of Practice No. 8*, 4th ed. Vol. 2. Alexandria, VA: WEF.

Yeager, J.G., and Ward, R.I. (1981). Effects of Moisture Content on Long-Term Survival and Regrowth of Bacteria in Wastewater Sludge. *Applied and Environmental Microbiology* 41(5): 1117–1122.

SAMPLE FINAL EXAM QUESTIONS

1. What does the term pathogenic mean?
   a. disease-causing
   b. dangerous
   c. does not affect humans
   d. does not affect plant life

2. What is wastewater from a household called?
   a. industrial waste
   b. domestic waste
   c. farm waste
   d. none of above

3. What is wastewater called that comes from industrial complexes?
   a. farm waste
   b. factory waste
   c. domestic
   d. industrial waste

4. An example of grit _____.
   a. sand
   b. gravel
   c. coffee grounds
   d. all above

5. Who must sign the DMR?
   a. maintenance operator
   b. lab supervisor
   c. general manager
   d. the licensed operator and the responsible person in charge

6. What is the purpose of primary treatment?
   a. to remove bacteria
   b. to remove fungi
   c. to remove sand
   d. to remove settleable and floatable solids

7. What is the purpose of the settling tank in the secondary or biological treatment process?
   a. removes settleable solids formed by the biological activity
   b. removes sand
   c. removes eggshells
   d. kills bacteria

8. What does the COD test measure?
   a. amount of gold in the wastewater
   b. amount of salt in the wastewater

    c. amount of organic material that can be oxidized by a strong oxidizing agent
    d. none of above

9. Three reasons for treating wastewater.
    a. remove gold ore, disease, and improve water quality
    b. prevent disease, protect aquatic organisms, protect water quality
    c. prevent disease, improve drinkability, remove salt
    d. none of above

10. Three classifications of ponds based on location in the treatment system are _____.
    a. primary, secondary, tertiary
    b. polishing pond, deep pond, shallow pond
    c. polishing pond, oxidation pond, stabilization pond
    d. none of above

11. What is the advantage of using mechanical or diffused aeration equipment to provide oxygen?
    a. eliminates wide diurnal and seasonal variation in pond dissolved oxygen
    b. uses less energy
    c. increases efficiency
    d. no advantage

12. Lab results show a predominance of rotifers in the waste, what should be done?
    a. decrease waste rate
    b. increase waste rate
    c. do nothing
    d. shut down the plant

13. What does NPDES stand for?
    a. National Putting Day Extreme System
    b. Normal Procedures During Extreme Storms
    c. National Pollutant Discharge Elimination System
    d. National Pollution Discharge Elimination Specialty

14. A neutral solution has a pH of _____.
    a. 6.0
    b. 11.0
    c. 2.0
    d. 7.0

15. Name two types of solids based on physical characteristics.
    a. gold and suspended
    b. silver and dissolved
    c. dissolved and suspended
    d. all above

16. Why is the seeded BOD test required for some samples?
    a. microorganisms are dead or absent
    b. microorganisms need food
    c. microorganisms can grow
    d. none of above

17. What is the advantage of COD testing over the BOD test?
    a. COD is more expensive
    b. COD test takes less time than BOD test
    c. COD is easier to perform
    d. none of above

18. BOD measures the amount of _____ material in wastewater.
    a. organic
    b. inorganic
    c. solids
    d. mold

19. The activated sludge process requires _____ in the aeration tank to be successful.
    a. seeding
    b. oil
    c. grease
    d. living organisms

20. The activated sludge process cannot be successfully operated with a _____ clarifier.
    a. first
    b. dual system
    c. final
    d. none of above

21. The activated biosolids process can successfully remove _____ BOD.
    a. 100%
    b. colloidal
    c. 70%
    d. 5%

22. The bacteria in the activated biosolids process are either _____ or _____.
    a. solid, nonsolid
    b. present, or not
    c. anaerobic or facultative
    d. aerobic or facultative

23. An advantage of contact stabilization compared to complete mix is _____ aerated tank volume.
    a. reduced
    b. increased
    c. no effect
    d. none of above

24. Step feed activated biosolids processes have _____ mixed liquor concentrations in different parts of the tank.
    a. different
    b. same
    c. no
    d. 100%

25. Increasing the _____ of wastewater increases the BOD in the activated biosolids process.
    a. flow
    b. salt content
    c. volume
    d. temperature

26. Bacteria need phosphorus to successfully remove _____ in the activated biosolids process.
    a. salt
    b. BOD
    c. COD
    d. trash

27. The growth rate of microorganisms is controlled by the _____.
    a. BOD
    b. COD
    c. F/M
    d. salt

28. Adding chlorine just before the _____ can control alga growth.
    a. secondary clarifier weirs
    b. bar screens
    c. pumping station
    d. solids handling units

29. The purpose of the secondary clarifier in an activated biosolids process is _____.
    a. to remove chlorine
    b. to settle salt
    c. to avoid crossover
    d. to separate and return biosolids to the aeration tank

30. The _____ growth phase should occur in a complete mix activated biosolids process.
    a. increasing
    b. declining
    c. remains the same
    d. flat

31. In the activated biosolids process, what change must be made to increase the MLVSS?
    a. increase BOD
    b. increase waste rate
    c. decrease waste rate
    d. nothing

32. In the activated biosolids process, what change must be made to increase the F/M?
    a. decrease waste rate
    b. increase waste rate
    c. increase flow
    d. decrease volume

33. What is one advantage of complete mix over plug flow?
    a. loves shock loads
    b. more resistant to shock loads
    c. is inexpensive
    d. none of above

34. Exhaust air from a chlorine room should be taken from where?
    a. mid-room level
    b. at nearest window
    c. at ceiling level
    d. at floor level

35. What is the term that describes a normally aerobic system from which the oxygen has temporarily been depleted?
    a. anoxic
    b. might
    c. light
    d. none of above

36. The ratio that describes the minimum amount of nutrients theoretically required for an activated sludge system is 100:5:1. What are the elements that fit this ratio?

    a. PNP

    b. CNN

    c. C:N:P

    d. PLP

37. A flotation thickener is best used for what type of sludge.

    a. primary

    b. secondary

    c. activated

    d. all above

38. Drying beds are an example of a sludge stabilization process?

    a. true

    b. false

39. The minimum flow velocity in collection systems should be _____.

    a. 5 ft/s

    b. 10 ft/s

    c. 6 ft/s

    d. 2 ft/s

40. What effect will the addition of chlorine, acid, alum, carbon dioxide, or sulfuric acid have on the pH of wastewater?

    a. increase

    b. double

    c. lower

    d. no change

41. An amperometric titrator is used to measure _____.

    a. chlorine residual

    b. pH

    c. alkalinity

    d. acid level

42. The normal design detention time for primary clarifier is _____.

    a. 6 hours

    b. 12 hours

    c. 15 minutes

    d. 2 hours

43. An anaerobic digester is covered and kept under positive pressure to _____.

    a. decrease odor release

    b. maintain temperature

    c. collect gas

    d. all above

44. What is the advantage of using ponds to treat wastewater?

    a. low construction cost

    b. low operational cost

    c. very effective

    d. all above

45. What is the difference between organic and inorganic?

    a. organic contains carbon and inorganic does not decompose

    b. organics are plants and inorganic are animals

    c. organics are valuable and inorganic are worthless

    d. all above

SAMPLE FINAL EXAM ANSWER KEY

1. What does the term *pathogenic* mean?
   a. **disease-causing**
   b. dangerous
   c. does not affect humans
   d. does not affect plant life

2. What is wastewater from a household called?
   a. industrial waste
   b. **domestic waste**
   c. farm waste
   d. none of above

3. What is wastewater called that comes from industrial complexes?
   a. farm waste
   b. factory waste
   c. domestic
   d. **industrial waste**

4. An example of grit _____.
   a. sand
   b. gravel
   c. coffee grounds
   d. **all above**

5. Who must sign the DMR?
   a. maintenance operator
   b. lab supervisor
   c. general manager
   d. **the licensed operator and the responsible person in charge**

6. What is the purpose of primary treatment?
   a. to remove bacteria
   b. to remove fungi
   c. to remove sand
   d. **to remove settleable and floatable solids**

7. What is the purpose of the settling tank in the secondary or biological treatment process?
   a. **removes settleable solids formed by the biological activity**
   b. removes sand
   c. **removes eggshells**
   d. kills bacteria

8. What does the COD test measure?
   a. amount of gold in the wastewater
   b. amount of salt in the wastewater
   c. **amount of organic material that can be oxidized by a strong oxidizing agent**
   d. none of above

9. Three reasons for treating wastewater.
   a. remove gold ore, disease, and improve water quality
   b. **prevent disease, protect aquatic organisms, protect water quality**
   c. prevent disease, improve drinkability, remove salt
   d. none of above

10. Three classifications of ponds based on location in the treatment system are _____.
    a. primary, secondary, tertiary
    b. polishing pond, deep pond, shallow pond
    c. **polishing pond, oxidation pond, stabilization pond**
    d. none of above

11. What is the advantage of using mechanical or diffused aeration equipment to provide oxygen?
    a. **eliminates wide diurnal and seasonal variation in pond dissolved oxygen**
    b. uses less energy
    c. increases efficiency
    d. no advantage

12. Lab results show a predominance of rotifers in the waste, what should be done?
    a. decrease waste rate
    b. **increase waste rate**
    c. do nothing
    d. shut down the plant

13. What does NPDES stand for?
    a. National Putting Day Extreme System
    b. Normal Procedures During Extreme Storms
    c. **National Pollutant Discharge Elimination System**
    d. National Pollution Discharge Elimination Specialty

14. A neutral solution has a pH of _____.
    a. 6.0
    b. 11.0
    c. 2.0
    d. **7.0**

15. Name two types of solids based on physical characteristics.
    a. gold and suspended
    b. silver and dissolved
    c. **dissolved and suspended**
    d. all above

16. Why is the seeded BOD test required for some samples?
    a. **microorganisms are dead or absent**
    b. microorganisms need food
    c. microorganisms can grow
    d. none of above

17. What is the advantage of COD testing over the BOD test?
    a. COD is more expensive
    b. **COD test takes less time than BOD test**
    c. COD is easier to perform
    d. none of above

18. BOD measures the amount of _____ material in wastewater.
    a. **organic**
    b. inorganic
    c. solids
    d. mold

19. The activated sludge process requires _____ in the aeration tank to be successful.
    a. seeding
    b. oil

    c. grease

    d. **living organisms**

20. The activated sludge process cannot be successfully operated with a _____ clarifier.

    a. first

    b. dual system

    c. **final**

    d. none of above

21. The activated biosolids process can successfully remove _____ BOD.

    a. 100%

    b. **colloidal**

    c. 70%

    d. 5%

22. The bacteria in the activated biosolids process are either _____ or _____.

    a. solid, nonsolid

    b. present, or not

    c. anaerobic or facultative

    d. **aerobic or facultative**

23. An advantage of contact stabilization compared to complete mix is _____ aerated tank volume.

    a. **reduced**

    b. increased

    c. no effect

    d. none of above

24. Step feed activated biosolids processes have _____ mixed liquor concentrations in different parts of the tank.

    a. **different**

    b. same

    c. no

    d. 100%

25. Increasing the _____ of wastewater increases the BOD in the activated biosolids process.

    a. flow

    b. salt content

    c. volume

    d. **temperature**

26. Bacteria need phosphorus to successfully remove _____ in the activated biosolids process.

    a. salt

    b. **BOD**

    c. COD

    d. trash

27. The growth rate of microorganisms is controlled by the _____.

    a. BOD

    b. COD

    c. **F/M**

    d. salt

28. Adding chlorine just before the _____ can control alga growth.

    a. **secondary clarifier weirs**

    b. bar screens

c. pumping station

d. solids handling units

29. The purpose of the secondary clarifier in an activated biosolids process is _____.

    a. to remove chlorine

    b. to settle salt

    c. to avoid crossover

    d. **to separate and return biosolids to the aeration tank**

30. The _____ growth phase should occur in a complete mix activated biosolids process.

    a. increasing

    b. **declining**

    c. remains the same

    d. flat

31. In the activated biosolids process, what change must be made to increase the MLVSS?

    a. increase BOD

    b. increase waste rate

    c. **decrease waste rate**

    d. nothing

32. In the activated biosolids process, what change must be made to increase the F/M?

    a. **decrease waste rate**

    b. increase waste rate

    c. increase flow

    d. decrease volume

33. What is one advantage of complete mix over plug flow?

    a. loves shock loads

    b. **more resistant to shock loads**

    c. is inexpensive

    d. none of above

34. Exhaust air from a chlorine room should be taken from where?

    a. mid room level

    b. at nearest window

    c. at ceiling level

    d. **at floor level**

35. What is the term that describes a normally aerobic system from which the oxygen has temporarily been depleted?

    a. **anoxic**

    b. might

    c. light

    d. none of above

36. The ratio that describes the minimum amount of nutrients theoretically required for an activated sludge system is 100:5:1. What are the elements that fit this ratio?

    a. PNP

    b. CNN

    c. **C:N:P**

    d. PLP

37. A flotation thickener is best used for what type of sludge.

    a. primary

    b. **secondary**

    c. activated

    d. all above

38. Drying beds are an example of a sludge stabilization process?
    a. true
    b. **false**

39. The minimum flow velocity in collection systems should be _____.
    a. 5 ft/s
    b. 10 ft/s
    c. 6 ft/s
    d. **2 ft/s**

40. What effect will the addition of chlorine, acid, alum, carbon dioxide, or sulfuric acid have on the pH of wastewater?
    a. increase
    b. double
    c. **lower**
    d. no change

41. An amperometric titrator is used to measure _____.
    a. **chlorine residual**
    b. pH
    c. alkalinity
    d. acid level

42. The normal design detention time for primary clarifier is _____.
    a. 6 hours
    b. 12 hours
    c. 15 minutes
    d. **2 hours**

43. An anaerobic digester is covered and kept under positive pressure to _____.
    a. decrease odor release
    b. maintain temperature
    c. collect gas
    d. **all above**

44. What is the advantage of using ponds to treat wastewater?
    a. low construction cost
    b. low operational cost
    c. very effective
    d. **all above**

45. What is the difference between organic and inorganic?
    a. **organic contains carbon and inorganic does not decompose**
    b. organics are plants and inorganic are animals
    c. organics are valuable and inorganic are worthless
    d. all above

Part III

Practical Risk Management for Public Water/Wastewater Systems

Lorilee Medders

Photo Courtesy of David J. McNeil

DOI: 10.1201/9781003207146-20

Practical Risk Management for Public Water/Wastewater Systems: Introduction

2019 Photo Courtesy of San Diego County Water Authority.

INTRODUCTION

Parts I and II of this designation and learning series detailed the principles, purposes, and processes of public water/wastewater systems; in so doing, they also introduced many of the observable and imperceptible exposures facing public water/wastewater systems. *Part III – Practical Risk Management for Public Water/Wastewater Systems* shifts your attention to the management of these risks and encompasses the identification, analysis, and response to risk factors that form part of the life of a public water or wastewater system. Effective risk management means management of future outcomes proactively rather than reactively. Therefore, effective risk management offers the potential to reduce both the possibility and impact of an unfavorable event while capitalizing on opportunities that are deemed favorable. Ideally, for every material risk, a choice set is created between rejecting the risk or accepting it with a viable plan for coexistence with it.

DOI: 10.1201/9781003207146-21

Risk management is such an important process because it empowers a business with the necessary tools to adequately identify and mitigate potential losses. Of equal importance is that risk management provides an organization with a basis upon which it can undertake sound overall decision-making. Insurance has been and continues to be a powerful element of any best practice risk management program. For public water/wastewater systems, it is especially important given the regulatory environment (especially with regard to rate regulation) generally favors the purchase of insurance over most alternative risk financing options for large-impact events.

Part III – Practical Risk Management for Public Water/Wastewater Systems is divided into five (5) chapters, which together are intended to provide the reader with the fundamentals of risk management as it relates to public water/wastewater systems. This part will build on the operational knowledge provided by Parts I and II, and focus on the criticality of risk and risk management to water and wastewater systems as critical infrastructure. Part III also applies this critical infrastructure risk management framework to the challenges of emergency management, disaster recovery, continuity planning, and sustainability. Additionally, this part introduces risk management for the purpose of organizational value, the risk management process, and the risk treatment options available. Part IV of the course will focus almost entirely on the various insurance coverages available to a public water/wastewater system, with an emphasis on how each can be used to meet its business needs and requirements, so the focus in this part is on non-insurance options.

We focus in Part III on public water/wastewater systems risk management first with an emphasis on the role they play in the critical infrastructure of society, with a management focus on the sector as a prime target for those who would undermine infrastructure, and thus societal, security. We then turn to the challenges of emergency management, business continuity, and sustainability. Finally, we introduce enterprise risk management (ERM) and explore the risk management process as it relates to a water/wastewater system as a business organization. The ERM process is applied at a strategic level across the organization to identify potential events, assign likely outcomes, and communicate these to key stakeholders so that organizational objectives are not compromised by the exposures or by a failure to consider any exposure's full ramifications. The part is organized as follows: (1) Chapter 15: Risk, Risk Management, and Critical Infrastructure; (2) Chapter 16: Emergency Management, Business Continuity, and Sustainability; (3) Chapter 17: Risk Management for Value Creation and Preservation; (4) Chapter 18: Risk Management Process – Risk Identification and Risk Analysis; and (5) Chapter 19: Risk Management Process – Risk Treatment Selection, Implementation, and Monitoring.

AUTHOR'S COMMENTARY

It is the purpose of *Part III – Practical Risk Management for Public Water/Wastewater Systems* to provide an overview of the current knowledge and essential applications of risk management within a water/wastewater system context. Much of the basic financial and statistical foundation remains largely unchanged since the advent of risk management as a business discipline in the mid-20th century. What has changed and continues to change rapidly are the business environment and the practical availability of various risk treatment options. The business and market content included here is current as of the time of this writing.

15 Risk, Risk Management, and Critical Infrastructure

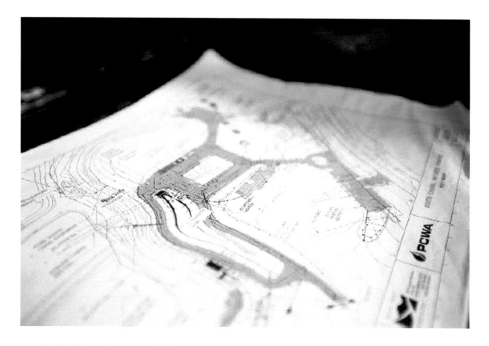

2019© Photo Courtesy of Placer County Water Agency, Placer County California 95604

LEARNING OBJECTIVES

- Define risk and its contributions – both positive and negative – to the financial health of organizations.
- Differentiate the concept of risk from the concept of uncertainty.
- Discover the importance of the International Standards Organization (ISO) to the development of risk and risk management.
- Recognize risk management, its importance as part of an organization's strategic planning process, and its evolution into enterprise risk management.
- Define critical infrastructure.
- Recognize the water and wastewater sector as critical infrastructure.
- State the interdependencies between the water sector and other infrastructure sectors.
- Apply the major threats to water and wastewater as critical infrastructure within a specific case illustration.
- Illustrate a "black sky" event.
- Compare and contrast the intents and purposes for each of the various laws, Presidential directives, programs, agencies, and other organizations that affect the management of water and wastewater risks as risks to critical infrastructure.
- Clarify ERM priorities as relates to water and wastewater as critical infrastructure.
- Describe SCADA and its contributions both to infrastructure security and to infrastructure security risks.

DOI: 10.1201/9781003207146-22

KEY DEFINITIONS

America's Water Infrastructure Act (AWIA) *of 2018* – a US federal law that provides for water infrastructure improvements throughout the country in the areas of: (1) flood control; (2) navigable waterways; (3) water resources development; (4) maintenance and repair of dams and reservoirs; (5) ecosystem restoration; (6) public water/wastewater systems; (7) financing of improvements; (8) hydropower development; and (9) technical assistance to small communities.

"Black Sky" Event – a catastrophic event that severely disrupts the normal functioning of our critical infrastructures in multiple regions for long durations (such as the entire power grid "going down" for an extended period).

Build America Investment Initiative – a 2014 Presidential initiative aimed at promoting private capital for building infrastructure.

Community Assistance for Resiliency and Excellence (WaterCARE) *Program* – an EPA strategy to provide assistance to communities addressing water infrastructure challenges.

Critical Infrastructure – a term used by governments to describe assets that are essential for the functioning of a society and economy.

Critical Infrastructure and Key Resource (CI/KR) – an umbrella term referring to the assets of the US essential to the nation's security, public health and safety, economic vitality, and way of life. It includes, among other items, power grids and water filtration plants; national monuments and government facilities; telecommunications and transportation systems; chemical facilities; and much more.

Critical Infrastructure Protection (CIP) – a concept that relates to the preparedness and response to serious incidents that involve the critical infrastructure of a region or nation.

Department of Homeland Security (DHS) – US federal agency established in 2002 that combines 22 federal departments and agencies into a unified, integrated Cabinet agency, with the mission to secure the nation from the many threats faced.

Emergency Support Function (ESF) – the grouping of governmental and certain private sector capabilities into an organizational structure to provide support, resources, program implementation, and services that are most likely needed to save lives, protect property and the environment, restore essential services and critical infrastructure, and help victims and communities return to normal following domestic incidents.

Enterprise Risk Management – an organizational model that represents any and all risks that could prevent value creation or erode existing organizational value.

Environmental Protection Agency (EPA) – an independent executive agency of the US federal government tasked with environmental protection matters.

Federal Water Pollution Control Act (aka Clean Water Act) – the primary US federal law governing water pollution.

Government Coordinating Council (GCC) – formed as the government counterpart for each Sector Coordinating Council (SCC) to enable interagency and cross-jurisdictional coordination. Each GCC comprises representatives from across various levels of government (federal, state, local, or tribal), as appropriate to the operating landscape of each individual sector.

Homeland Security Presidential Directive (HSPD) – a specific form of Executive Order that states the Executive Branch's national security policy, and carries the force and effect of law, stating requirements for the Executive Branch.

International Organization for Standardization (ISO) – an independent, nongovernmental international entity, having a membership of 165 national standard entities, which brings together experts to share knowledge and develop voluntary, market-relevant International Standards to support innovation and problem solving for international challenges.

National Infrastructure Protection Plan (NIPP) – a document called for by Homeland Security Presidential Directive 7, which aims to unify Critical Infrastructure and Key Resource protection efforts across the country for the purpose of strengthening and protecting CI/KR sectors.

National Institute of Standards and Technology (NIST) *Framework for Improving Critical Infrastructure Cybersecurity* – Required by Executive Order 13636 in 2013, and created through collaboration between government and the private sector, uses a common language to address and manage cybersecurity risk in a cost-effective way based on business needs without placing additional regulatory requirements on businesses.

National Response Framework (NRF) – a guide to how the nation responds to all types of disasters and emergencies. It is built on scalable, flexible, and adaptable concepts identified in the National Incident Management System to align key roles and responsibilities.

Precautionary Principle – an economic and legal principle by which the introduction of a new product or process whose ultimate effects are disputed or unknown should be resisted.

Public Health Security and Bioterrorism Preparedness and Response Act – Enacted in 2002, a federal US law that established procedures for preparation for bioterrorism and public health emergencies.

Public Wastewater System – system that provides sanitary sewer, stormwater, and other used water services to residences, businesses, and industries.

Public Water Systems – system that provides water for human consumption to 15 or more connections or regularly serves 25 or more people daily for at least 60 days out of the year.

Risk Assessment Methodology for Critical Asset Protection (RAMCAP) – a framework for analyzing and managing the risks associated with terrorist attacks against critical infrastructure assets.

Risk – uncertainty, the possibility of loss, and/or the effect of uncertainty on objectives.

Risk Management – practice of identifying and analyzing loss exposures and taking steps to minimize the financial impact of the risks they impose. Traditional risk management, sometimes called insurance risk management, has focused on pure risks (i.e., possible loss by fortuitous or accidental means) but not business risks (i.e., those that may present the possibility of loss or gain).

Safe Drinking Water Act (SDWA) – the principal federal law in the United States intended to ensure safe drinking water for the public.

Sector-Specific Agency (SSA) – a federal agency which serves as lead government body to a particular critical infrastructure sector, leveraging its particular knowledge, expertise, and network to support that sector's strength and maintain its security, function, and resiliency.

Sector-Specific Plan (SSP) – provide the means by which the National Infrastructure Protection Plan (NIPP) is implemented across all critical infrastructure and key resources (CI/KR) sectors, as well as a national framework for each sector to address its unique characteristics and risk landscape.

Supervisory Control and Data Acquisition (SCADA) – a centralized system that aims to monitor and control field devices at your remote sites. SCADA systems are used to monitor and control a plant or equipment in industries such as telecommunications, water and waste control, energy, oil and gas refining, and transportation.

U.S.A. Patriot Act of 2001 – an US federal act signed into law in 2001. USA PATRIOT is an acronym that stands for Uniting and Strengthening America by Providing Appropriate Tools Required to Intercept and Obstruct Terrorism.

Wastewater – water that has been used for various purposes around a community, including sewage, stormwater, and all other water used by residences, businesses, and industry. Wastewater requires treatment before it returns to waterways to protect the health of the waterbody and community.

Water Infrastructure and Resiliency Finance Center – an EPA-administered center that provides financing information to help local decision-makers make informed decisions for drinking water, wastewater, and stormwater infrastructure to protect human health and the environment.

Water Sector Coordinating Council (WSCC) – a policy, strategy, and coordination mechanism for the US water and wastewater systems sector in interactions with the government and other sectors on critical infrastructure security and resilience issues.

World Economic Forum (WEF) – an international nongovernmental organization, founded in 1971 and headquartered in Switzerland, having as its mission to improve the state of the world by engaging business, political, academic, and other leaders of society to shape global, regional, and industry agendas.

OVERVIEW

> The Frankensteined architectonic IoT microcosm of the prototypical critical infrastructure organization renders an infinite attack surface just begging to be exploited.

> **– James Scott, Senior Fellow, Institute for Critical Infrastructure Technology**

Generally, economic development requires risk taking, and development occurs at a greater rate than the risk needed to create it. Overall, business risk is known to present more opportunity than obstacles. The primary purpose of risk management then is to reduce risk, and in some cases virtually eliminate it via proper risk management without reducing the development rate. Moreover, risk management seeks to increase the rate of economic development over that which takes place without risk management.

For a public water/wastewater system, assessment and management of risks is the best way to prepare for eventualities that might otherwise stand in the way of progress and growth. When a water system evaluates potential threats and then develops structures to properly address them, it improves its odds of not only surviving adversity, but thriving in the face of it. Effective risk management also ensures risks of high priority are dealt with as assertively as possible and that management will have the necessary information to make informed decisions, ensuring the system remains financially sound. If an organization sets up risk management as a disciplined and continuous process for the purpose of identifying and resolving risks, then the risk management structures can be used to support other risk mitigation systems – planning, organization, cost control, and budgeting. In this way, the business will experience few surprises and will be well positioned to respond proactively in the event surprises do occur. Our starting point for building your risk management knowledge is to describe the integral role the risk management process plays in developing a best-practice risk management strategy.

DEFINING RISK

There are many risk definitions in the academic literature; the early risk management textbooks define risk as: "uncertainty" or "the possibility of loss." The standards most recognized at the international level – coming from the International Standards Organization (ISO) – denote risk similarly, albeit with nuances of difference from what is seen in the insurance literature; the standard ISO 31000:2009 defines risk as: "the effect of uncertainty on objectives," where "an effect is a deviation from what is expected (positive and/or negative), often expressed in terms of a combination of the consequences of an event (including changes in circumstances) and the associated likelihood of occurrence" and the uncertainty is "the lack of information about the understanding or knowledge of an event, its consequences and likelihood" [19,24].

Indeed, the concept of risk is more complex than the combination of likelihood and impact; it additionally comprises key risk concepts and issues considered by an organization's cognitive collective. This collective information-and-decision mindset determines an organization's risk *profile*, risk *perception*, risk *attitude*, risk *appetite*, risk-bearing *capacity*, and risk *tolerance*. All of these issues should be considered to assess the overall risk level of the organization. Each plays a role in determining the appropriate management measures to take in light of risks to which the organization is exposed. These terms will be developed and considered in Chapter 3.

Defining Risk Management

Risk management is an organizational model that supports decision-making processes, preparing the organization for challenges that could conflict with the mission or core values and/or hinder the achievement of strategic goals and objectives. Originally, the term *risk management* was used to denote a model that emphasized insurable risks, such as legal liability, property loss, or business succession planning in the early 1950s. Later, starting in the 1970s, banks and other financial institutions adopted the term to also apply to the handling of their operational and financial risks. Because neither of these meanings of risk management fully encompassed the management of risk holistically, across the entire organization, experts in the mid-1990s, propelled by increasing organizational and compliance complexity, developed the expression *enterprise risk management* (ERM) to denote an organizational model that represents any and all risks that could prevent value creation or erode existing organizational value. ERM explicitly cuts across all functional areas and operations of the organization, and is broader in concept than the sum of the insurable, operational, and financial risk management models previously defined. ERM encompasses the insurable and the uninsurable, the financial and the other-than-financial, and goes a step further to recognize connections between and among these [6,18,24].

The world around us is essentially a giant sphere of risks, with loss events inevitably going to occur. Since it is not a perfect world, public water/wastewater systems encounter risks that can adversely affect their survival and growth. As a result, it is important to understand the basic principle of risk management and how it can be used to help mitigate the effects of risks on business entities, and more specifically public water/wastewater systems. Risk management is the process of analyzing exposures to risk and determining how to best handle such exposure. It is a collaborative process where risk response strategies are developed in collaboration with the stakeholders who most understand the risks and are best able to manage them. For public water/wastewater systems, the added dimension of providing a public good/service introduces meaningful variations on the breadth and depth of the risks encountered as well as on the stakeholders involved.

Water as National Critical Infrastructure

Secure, resilient, and sustainable water infrastructure and services are essential to life, health, and livelihood. When water and wastewater services are lost, even for a short time, the consequences can be costly. If loss of service lasts for an extended period, the costs can be catastrophic. Nevertheless, loss of service occurs and can result from a failure within any element of the system. The three elements of critical infrastructure are physical, cyber, and human [12,15]. Whether partial or full, short-lived or extended, such losses not only result in direct problems, but also may create a cascade of problems because of the inherent connections and interdependent relationships between critical infrastructure sectors – including water – and extending outward from critical infrastructure to auxiliary sectors as well.

This chapter examines US drinking water and wastewater operators as purveyors of national critical infrastructure. The discussion is framed by the water sector's strategic ERM vision, goals, and objectives, and emphasizes the threats to water as critical infrastructure. It also explores the vulnerabilities that may exist within water and wastewater systems that can weaken a system's threat preparedness, response, and recovery as well as the potentially catastrophic consequences of a risk management failure. Finally, the chapter evaluates the strategies undertaken to ensure the safety of the water sector, as well as how and by whom these strategies are coordinated [26].

The term *critical infrastructure* is defined in §1016(e) of the Patriot Act of 2001 (42 U.S.C. 5195c(e)) as "systems and assets, whether physical or virtual, so vital to the United States that the incapacity or destruction of such systems and assets would have a debilitating impact on security, national economic security, national public health or safety, or any combination of those matters" [17]. This is the legal basis from which federal and state laws define critical infrastructure. Under this definition, water and wastewater systems are designated as critical infrastructure per Presidential Policy Directive 21: Critical Infrastructure Security and Resilience [15,29].

The full list of US Critical Infrastructure includes 16 sectors:

1. Chemical
2. Commercial facilities
3. Communications
4. Critical manufacturing
5. Dams
6. Defense industrial base
7. Emergency services
8. Energy
9. Financial services
10. Food and agriculture
11. Government facilities
12. Healthcare and public health
13. Information technology
14. Nuclear reactors, materials, and waste
15. Transportation systems
16. Water and wastewater systems

Four (4) of these critical sectors are designated lifeline functions – transportation, water, energy and communications, and reliable operations of these are so critical that disruption or loss will directly affect the security and resilience of critical infrastructure across sectors. The water sector is considered one of the lifeline sectors because its functions are essential to core operations in nearly every other critical sector.

Even when water services are lost for less than 8 hours, the functioning of multiple sectors is significantly degraded – emergency services, government facilities, hospitals, transportation command facilities, agriculture/food. As evidence, across all sites in all 16 critical infrastructure sectors that received Department of Homeland Security (DHS) assessments during 2011–2014, three-fourths depend on external water for their operations, and two-thirds depend on external wastewater services for their operations [23]. Figure 15.1 illustrates interdependencies between the water sector and other CI sectors.

Some key examples of interdependencies with the water sector include [16,23]:

- The energy sector relies on water services for multiple aspects of energy production and generation. In fact, 82% of electric generation plants report dependence upon water. In turn, the water sector relies on energy, specifically electricity, to operate its pumps, treatment facilities, delivery systems, and processing. Long-term power outages can overwhelm a water utility's backup energy supply or deplete fuel reserves. In the event of an electric

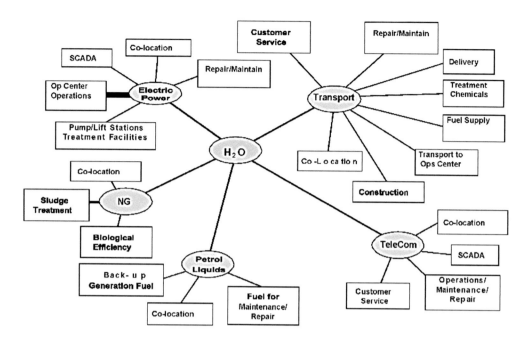

FIGURE 15.1 Water sector interdependencies with other critical infrastructure sectors. Source: Argonne National Laboratory [16].

power failure, water treatment and wastewater treatment facilities could lose 100% of operational capability if they do not have backup generation capability (e.g., emergency generators). This scenario is exacerbated if multiple energy utilities in a region are shut down or multiple public water/wastewater systems in a region have to compete for backup resources.

- The communications and information and technology (IT) sectors rely on water services for equipment cooling and facility operations, while the water sector relies on communications and IT for their operations and control systems, monitoring systems, internal communications, and communications with the public and emergency responders. Storing drinking water for short-term use to protect public health may seem almost routine – think of stocking up on water bottles and filling a bathtub before a major storm – yet it is impossible to store sufficient backup water or divert water resources to maintain water-intensive operations in places such as hospitals, office buildings, chemical plants, generators, and manufacturing facilities. Unlike electricity, water cannot easily be re-routed around disruptions, nor can facilities generate backup water on-site to maintain critical operations.

- The chemical sector has a strong dependency on water, with 100% of chemical facilities reporting water dependency. In turn, 100% of water treatment facilities and 90% of wastewater treatment facilities assessed by DHS are dependent upon chemicals for on site operations. One hundred percent of chemical facilities are dependent upon water.

THE RISK OF WATER CRISES AS A TOP GLOBAL RISK

Water and wastewater systems as the critical infrastructure on a global scale is unarguable, yet water-related infrastructure risk has not always been considered a top global risk. In years past, water was plentiful in most developed nations and the infrastructure for water management and distribution was in good repair. In recent years, however, the landscape for both water availability and the management of water infrastructure assets worldwide has been under growing pressure.

The World Economic Forum is an international nongovernmental organization committed to improving the world through international collaboration among businesses, policy-makers, and experts. It publishes an annual Global Risks Report, bringing light to the top risks as identified by an international group of business representatives, policy-makers, and subject matter experts. Global Risks Report 2020 highlights "Water Crises" as a top 10 global risk, both in terms of the likelihood and the potential magnitude of global impact. Figure 15.2 illustrates how water crises compare to other global risks that were identified.

Respondents to the 2020 Global Risks Survey collectively ranked water crises as eighth globally with respect to likelihood and fifth globally with respect to impact. The top 5 ranking for impact is especially disconcerting given that the global risks ranked first to fourth (1. climate action failure, 2. weapons of mass destruction, 3. biodiversity loss, and 4. extreme weather) are all considered to be interconnected with water crises. These rankings place water crises in the highest priority quadrant of the Global Risks Landscape 2020. Moreover, 69.3% of the respondents overall reported belief that the risk of water crises will rise in the near term. The youngest respondents (which the WEF dubs "global shapers") reported an even higher rate of belief, at 86%, that the risk of water crises is a growing concern in the short term, not just as a long-range risk [31].

RISK MANAGEMENT FOR WATER AS CRITICAL INFRASTRUCTURE

Homeland Security Presidential Directives (HSPDs) are issued by the US President on matters related to Homeland Security. Homeland Security Presidential Directive 7 (HSPD-7) identified the US critical infrastructure and key resources (CI/KR) sectors. It also designated federal government Sector-Specific Agencies (SSAs) for each of the sectors. Each sector is responsible for developing, implementing, and monitoring a Sector-Specific Plan (SSP) for sector risk management/security. As part of this charge, each sector provides sector-level performance review periodically to the Department of Homeland Security (DHS). SSAs are responsible for collaborating with public and private sector security partners and encouraging the development of appropriate information-sharing and analysis mechanisms within the sector.

HSPD-7 designated the Environmental Protection Agency (EPA) as the federal SSA for water and wastewater's critical infrastructure risk management/security. All EPA activities related to water security are carried out in cooperation with DHS and the EPA's water sector partners. In subsequent chapters, discussion of water sector risk management will broaden to emphasize a "value creation and preservation" approach to risky decision-making. Here, the focus is on the water sector as critical infrastructure to society, and thus the approach, at least in concept, employs the *precautionary principle*: an economic and legal principle by which the introduction of a new product or process whose ultimate effects are disputed or unknown should be resisted. Plainly stated, the precautionary principle dictates that the protection of the existing infrastructure (especially its reliability and security) take precedence over innovation within the sector (with the exception of innovations that clearly improve reliability and/or security of the sector).

Because of water – its delivery, storage, and treatment – as critical infrastructure, and because critical infrastructure is particularly vulnerable to degradation, natural disaster, and security threats, the focus in this chapter is on risk management in these areas. Thus, the discussion here bypasses some business risk management processes that will be explored in subsequent chapters and is almost solely devoted to (1) considering the likelihood that known threats will exploit vulnerabilities and the impact they have on the water sector's valuable assets; and (2) prioritizing the treatment of these risks.

WATER SECTOR SECURITY VISION AND GOALS

The water sector, in agreement with the government, recognizes its role as Critical Infrastructure and Key Resource (CI/KR). It also recognizes its interdependence with other critical sectors. In

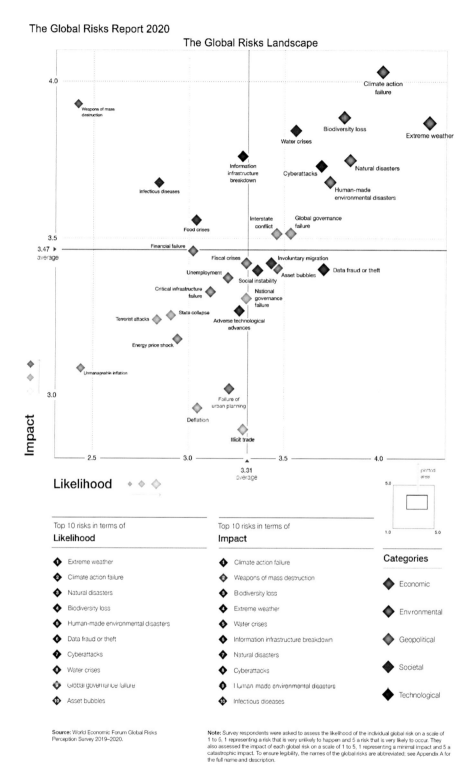

FIGURE 15.2 Global Risk Landscape 2020. Source: Global Risks Report 2020, World Economic Forum.

response to the risks the water sector faces as CI/KR, the EPA and joint working group of the Water Sector Coordinating Council (WSCC) and the Government Coordinating Council (GCC) have collaborated to develop a vision statement and security goals that provide clear direction for risk management efforts [27–29].

Security Vision Statement for the Water Sector

The water sector's Security Vision is a secure and resilient drinking water and wastewater infrastructure that provides clean and safe water as an integral part of daily life. This Vision assures the economic vitality of and public confidence in the Nation's drinking water and wastewater through a layered defense of effective preparedness and security practices in the sector.

The WSCC-GCC has developed four goals intended to drive the development of protective programs and performance measures: (1) sustain protection of public health and the environment; (2) recognize and reduce risks; (3) maintain a resilient infrastructure; and (4) increase communication, outreach, and public confidence. The joint group has developed goals and objectives related to this vision that can be found in Table 15.1.

TABLE 15.1
WSCC-GCC Working Group's Goals and Objectives for the Water Sector

Goals	Objectives
Sustain protection of public health and the environment	• Encourage integration of security concepts into daily business operations at utilities to foster a security culture. • Evaluate and develop security-related surveillance, monitoring, warning, and response capabilities to recognize risks introduced into water sector systems that affect public health and economic viability. • Develop a nationwide laboratory network for water quality security that integrates federal and state laboratory resources and uses standardized diagnostic protocols and procedures, or develop a supporting laboratory network capable of analyzing security threats to water quality.
Recognize and reduce risks	• Improve identification of vulnerabilities based on knowledge and best available information, with the intent of increasing the sector's overall security posture. • Improve identification of potential threats through sector partners' (public water/wastewater systems; national associations; and federal, state, and local governments) knowledge base and communications with the intent of increasing overall sector security posture. • Identify and refine public health and economic impact consequences of man-made or natural incidents to improve utility risk assessments and enhance the sector's overall security posture.
Maintain a resilient infrastructure	• Emphasize continuity of drinking water and wastewater services as it pertains to utility emergency preparedness, response, and recovery planning. • Explore and expand implementation of mutual aid agreements/compacts in the water sector. • Identify and implement key response and recovery strategies. • Increase understanding of how the sector is interdependent with other critical infrastructure sectors.
Increase communication, outreach, and public confidence	• Communicate with the public about the level of security and resilience in the water sector and provide outreach to ensure the public's ability to be prepared and respond to a natural disaster or man-made incident. • Enhance communication and coordination among utilities and federal, state, and local officials and agencies to provide information about threats. • Improve relationships among all water sector security partners through a strong public–private partnership characterized by trusted relationships.

Sources: Water and Wastewater Sector Strategic Roadmap Work Group [25]; Water Sector Government Coordinating Council [27].

A view of the objectives related to "Goal 1: Sustain protection of public health and the environment" reveal an intent to engage in ERM and internal risk reduction via knowledge of the risk. "Goal 2: Recognize and reduce risks" engages the enterprise risk management process. "Goal 3: Maintain a resilient infrastructure" objectives point out efforts to align ERM priorities with emergency risk management. And "Goal 4: Increase communication, outreach and public confidence" is a call to actively remember that the public is part of the water sector community (and vice versa), with a stake in its resiliency and sustainability.

THREATS TO WATER AS CRITICAL INFRASTRUCTURE

As discussed in previous chapters, organizations encounter myriad risks and must prioritize these in order to allocate resources prudently. Because of the diversity of assets in the water sector (e.g., size, treatment complexity, disinfection practices), a number of risk assessment methods were created and are used. These methods address the full range of utility components, including the physical plant (physical), employees (human), information technology (cyber), communications, and customers. Based on assessment via multiple methods, the risks faced by the water sector that most adversely impact its role as critical infrastructure currently are natural disasters, accidental contamination, intentional malicious attacks, degradation, and population changes [9,14,29].

Natural Disasters and Related Risks

Sudden natural disaster events, such as floods, hurricanes, and earthquakes, are among the most significant risks faced by the water sector, as well as chronic natural hazards such as drought and sea level rise. In addition to the direct damage they cause to physical infrastructure, natural disasters can also cause indirect damage by harming water quality and limiting service availability. Exacerbated by the increased intensity of severe weather linked to climate change, weather threatens water infrastructure [5,10,12,23,25,27,29]. Most water facilities are naturally located near natural sources of water. Sea level rise and higher storm surge result in greater risk of flooded facilities. Increasing precipitation and drought can also harm water quality.

High-severity events that are unpredictable (and also ambiguous in some cases) include various types of natural disasters. There are plausible scenarios that could cause a combination of severe, direct physical damage to infrastructure as well as widespread power outages (i.e., indirect damage) lasting multiple months. These threats are referred to as "black sky" events by experts, and include possibilities such as [23,27,29,30]:

- An earthquake in the New Madrid Fault Zone;
- High-magnitude earthquakes in sections of the San Andreas Fault; and
- An extreme geomagnetic storm.

The water sector has a track record of maintaining water and wastewater service during distressed conditions and minimizing the impact of disruptions that range from a few hours to a few weeks. The public is often unaware of the "near misses" that the sector has avoided. Disruptions are usually confined to local areas, but in rare cases can rise to a national-level event.

Accidental Contamination

High-profile contamination incidents have diminished public confidence in source water protection and the safety of public drinking water. Flint, Michigan, Elk River, West Virginia, Corpus Christi, Texas, Toledo, Ohio, and the Potomac River (DC area) have all suffered contaminations in recent years that played out in the national news. More disconcerting, they have exposed weaknesses in the structure for emergency preparedness, response, and recovery [14,23,29].

Intentional Malicious Attacks

Physical and/or cyberattacks, conducted either by employees, contractors, or external individuals/groups not associated with the facility, can pose a threat to the water and wastewater systems. Potential attack methods include: improvised explosive devices; vehicle-borne improvised explosive devices; hazardous material releases; explosive devices in wastewater collection systems; chemical, biological, or radiological contamination of drinking water distribution systems; assault; sabotage of water treatment systems; radiological dispersal devices; and cyberattacks on supervisory control and data acquisition (SCADA) systems. The use of and reliance on process control systems (e.g., SCADA), industrial Internet of Things, cloud services, and other connected technology have increased in recent years as have the cybersecurity threats and cyber vulnerabilities [3–5,8,11–13,22,23].

Water systems increasingly use SCADA industrial control systems to continuously control treatment processes and delivery, remotely monitor operations, and control the pressure and flows in pipelines. These automated systems allow small teams of operators to efficiently and remotely manage complex physical processes using digital controls. Growing reliance on industrial control systems over the last decade has resulted in increased connectivity, a proliferation of cyber access points, escalating system complexity, and wider use of common operating systems and platforms — factors that increase cyber risk and require sophisticated cyber protections. Attacks involving SCADA systems could risk customer health and erode public trust in the water system.

Industrial control systems monitor and control highly distributed physical processes, including remote control of often unmanned facilities. Public water/wastewater systems require the tools and expertise to rapidly detect and recognize cyberattacks. Cyber and physical security is intimately linked. A cyber intrusion could give a hacker the ability to manipulate physical processes (such as chemical treatment and water flows), while insufficient physical security (such as an unsecured control room door) could give an individual unauthorized access to critical cyber controls. Furthermore, water systems primarily rely on hardware and software vendors to develop secure control systems and patch vulnerabilities. They need a strong understanding of cybersecurity requirements to obtain secure technologies. Smaller water and wastewater operators often lack the resources and specialized personnel needed for cybersecurity improvements. For example, larger facilities may have the resources to maintain a separate, more secure system for operational systems, which is rare in smaller utilities.

US reports (coming from a combination of private sector and government intelligence) indicate that the nation's water and wastewater sector is under direct and serious threat. In March 2016, Verizon's monthly security breach publication pointed to a cyberattack on a major unnamed water utility [12,27]. The hack was suspected to have been conducted by a political activist hacker group, with motivations relating to the ongoing interstate conflict in Syria. The compromised system allowed the hackers to take control of the utility's water flow and chemical treatment system. Although the utility was able to identify and reverse the chemical and flow changes in time, this critical infrastructure failure could have resulted in a disaster to the local community via distribution of poorly (or maliciously) treated water, raising risks of infectious disease and water crises. The system in question ran on operating systems from over a decade prior, leaving it particularly vulnerable to such an attack. In 2018, a water utility based out of North Carolina self-reported that its internal computer systems – including servers and personal computers – sustained "a sophisticated ransomware attack" in October. Several of the Onslow Water and Sewer Authority's other databases were erased – forcing them to rebuild the systems from scratch [4].

Degradation of Infrastructure [21]

Water quality and operational reliability are increasingly suspect due to an overall aging and failing infrastructure. Old and worn water and wastewater treatment, distribution, and collection facilities require repair or replacement to maintain utility service. Meanwhile, the economics of water system infrastructure maintenance and repair in many areas is worsened by an aging population and a declining economic base. Aging infrastructure and limited resources for adequate response planning and resilience investments are inextricably linked, creating a complex risk. Much of the water infrastructure has or is approaching the age at which it needs to be replaced. For both drinking water and wastewater systems, the useful life of component parts ranges from 15 to 95 years, depending on the component and its materials. Addressing this risk requires a huge investment in infrastructure.

Physical infrastructure goes through a natural life process from creation to growth to maturation to decline to death. Physical systems, unlike natural systems, are not self-sustaining. They must be reinvigorated through maintenance, repair, renewal, and replacement on a periodic basis. These sustaining actions require the investment of capital, materials, labor, and other resources. The warning signs of infrastructure in distress may be subtle for a long period of time, and the relationship between infrastructure condition and its performance is nonlinear. Figure 15.3 is a conceptual depiction of the relationship between condition and performance, and can be applied to water infrastructure assets as well as to many other infrastructure assets and components.

When physical condition is new or excellent, performance is high; however, when the condition of infrastructure is poor, performance is low. Interestingly, the condition can deteriorate over a protracted period without noticeable performance loss such that investments made during the mid-life of infrastructure may not noticeably improve performance and could even prove counterproductive by leading decision-makers to believe that investment in routine inspection, maintenance, and repair is an unnecessary or imprudent expense. Unfortunately, although the time to failure for infrastructure may be lengthy, once failure begins, it proceeds rapidly and can be irreversible.

Estimates vary based on assumptions, time frames, and other factors; but it could cost several hundred billion dollars to as much as $1 trillion to address the nation's infrastructure needs and maintain current levels of service. The American Society of Civil Engineers (ASCE) has assessed the water and wastewater sectors' current and future investment needs to have huge gaps in funding. For 2040, the investment gaps are expected to grow to nearly $100 billion and $45 billion each for wastewater and drinking water, respectively [2].

Population Growth and Shifts

New plants are needed to meet the demands of growing populations in some areas, as existing water and wastewater facilities are already overburdened. In other areas, population growth has led to the use of decentralized systems. For example, septic systems currently serve about 40% of new developments in the US as populations have migrated from metropolitan areas into rural communities. Additionally, agricultural runoff and increasing urbanization not controlled by wastewater treatment lead to pollution [23,27,29].

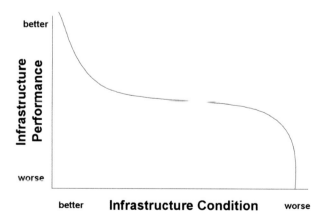

FIGURE 15.3   Relationship between infrastructure condition and performance [21].

SECURITY RISK ASSESSMENT TOOLS [13,15,22]

Given the increase in the use of SCADA and other cybertechnology, new security risks are created as cyber mediums reduce or eliminate others. The complexity of the technologies involved and the nesting of challenges that can result calls for special risk assessments devoted to security risk. Risk assessment tools have been developed specifically to support, support the DHS's risk management framework and take into consideration the criteria of DHS's Risk Assessment Methodology for Critical Asset Protection (RAMCAP) as outlined in the NIPP. All risk assessments are conducted at the local operator-utility level; identification of priority components at that level is dictated by local conditions. Threat information is provided by the federal government, and individual water systems apply that information to their risk assessments. EPA has supported development of risk assessment tools for drinking water and wastewater systems of all sizes. Consistent with DHS criteria, the tools aim to:

- *Deterrence capabilities* – The ability to deter an attack by affecting the adversary's perception of its capability or the level of effort required to execute a successful attack;
- *Detection capabilities* – The ability to identify or expose an attack before it takes place;
- *Devaluing capabilities* – The ability to reduce an attacker's incentive to attack by reducing the value of the target;
- *Defensive (delay and denial) capabilities* – The ability to prevent an attack or delay it long enough for security or law enforcement personnel to mount an effective response;
- *Response capabilities* – The ability to effectively respond to an attack underway;
- *Consequence reduction/mitigation capabilities* – The ability to limit consequences should an attack occur; and
- *Recovery capabilities* – The ability to return to an acceptable level of operations after an attempted or successful attack.

As part of the risk assessment, utilities develop an inventory of asset components including physical, cyber, IT, and personnel, and they identify which components are most critical to their mission. Common to all of these risk assessment methodologies are these six elements:

1. Characterization of the system, including its mission and objectives;
2. Identification and prioritization of adverse consequences to avoid;
3. Determination of critical assets that might be subject to malevolent acts that could result in undesired consequences;
4. Assessment of the likelihood (qualitative probability) of such malevolent acts by adversaries;
5. Evaluation of existing countermeasures; and
6. Analysis of current risk and development of a prioritized plan for risk reduction.

CONSEQUENCE-VULNERABILITY-THREAT ASSESSMENT [13,15,22]

An evaluation of these vulnerability assessments is initiated in the RAMCAP process to assess their compatibility with the DHS's criteria for risk assessments and to take into consideration the DHS's common definitions and analysis of the basic risk factors:

- Consequence analysis, which is the estimate of the potential public health and economic impacts that a successful attack could cause;
- Vulnerability assessment, which identifies weaknesses in a water or wastewater system's design, implementation, or operation that can be exploited by an adversary; and
- Threat analysis, which estimates the likelihood that a particular target, or type of target, will be selected for attack, and is based on intent and capability of an adversary.

When these three factors are combined, they describe the risk associated with a water asset.

Assessing Consequences

Among the factors to consider in assessing the consequences of any disruption of a water sector asset are the: (1) magnitude of service disruption; (2) number of illnesses or deaths resulting from an event; (3) impact on public confidence; (4) chronic problems arising from specific events; (5) economic impacts; and (6) other indicators of the impact of each event, as determined by the water system.

Assessing Vulnerabilities

Vulnerabilities are the characteristics of an asset's design, location, security posture, process, or operation that make it susceptible to destruction, incapacitation, or exploitation by mechanical failures, natural hazards, and terrorist attacks or other malicious acts. They are weaknesses that could result in consequences of concern, taking into account intrinsic structural weaknesses, protective measures, resiliency, and redundancies.

Assessing Threats

The water sector views threat analysis broadly, encompassing natural events, criminal acts, insider threats, and foreign and domestic terrorism. In the context of risk assessment, the threat component of risk analysis is based on the likelihood that an asset will be disrupted or attacked. To assist public water/wastewater systems in conducting risk assessments, baseline threat documents have been developed.

INTERDEPENDENCY AND RESILIENCE CONSIDERATIONS

Despite a great deal of attention and resources being brought to bear on CI risk assessment and modeling, the majority of these modeling techniques do not take into account interdependencies between and among CI sectors, or the resilience effects of events and decisions. Indeed, it is estimated that only 21% of modeling techniques directly consider sector interconnectedness and fewer than 15% consider resilience. Yet one can assume these considerations of interdependency and resilience may be connected to the core objectives for developing the assessment tools in the first place. Behavioral and cascading effects of both the CI operations and failures are at the core of the CI risk management objectives, so future consequence-vulnerability-threat modeling efforts can be expected to continue to converge on finding solutions to the challenges of sector interdependency and resilience [16,21].

TOP CRITICAL INFRASTRUCTURE RISK MANAGEMENT PRIORITIES

From an ERM perspective, critical infrastructure is subject to the greatest catastrophic loss potential and the strongest risk correlations of any economic sectors. It is infeasible to reduce the risk to zero (risk avoidance) or to fully cost share within the private sector (risk transfer-insurance). Therefore, risk and treatment prioritization is especially important when managing critical infrastructure, such as the water sector. The core US critical infrastructure risk management tenets include [7,23,27,29]:

- Risk should be identified and managed in a coordinated way within the critical infrastructure community to enable effective resource allocation.
- Critical infrastructure partnerships can greatly improve understanding of evolving risk to both cyber and physical systems and assets, and can offer data and perspectives from various stakeholders.
- Understanding and addressing risks from cross-sector dependencies and interdependencies are essential to enhancing overall critical infrastructure security and resilience.

- Gaining knowledge of and reducing infrastructure risk requires information sharing across all levels of the critical infrastructure community.
- A partnership approach, involving public and private stakeholders, recognizes the unique perspectives and comparative advantages of the diverse critical infrastructure community.
- Regional, state, and local partnerships are crucial to developing shared perspectives on gaps and improvement actions.
- Critical infrastructure transcends national boundaries, requiring bilateral, regional, and international collaboration; capacity building; mutual assistance; and other cooperative agreements.
- Security and resilience should be considered during the design of infrastructure elements.

Given the vision, goals, and objectives, and aligning them to the top threats yields a handful of high-priority goals and objectives for CI/KR risk management in the water sector [7,22,23,27,29]:

- Convert water-wastewater's definition as a lifeline sector into strong support for the sector's needs and capabilities.
- Improve detection, response, and recovery to contamination incidents.
- Enhance preparedness and improve capabilities of the sector for area-wide loss of water and power.
- Advance recognition of vulnerabilities and needed responses related to cybersecurity.

THE IMPORTANCE OF COOPERATION AND COLLABORATION TO ACHIEVING CI/KR GOALS AND OBJECTIVES [3,7,27,29]

The water sector operates within a complex and dynamic environment in which the consequences of water disruptions and potential cascading impacts are not fully understood. Thus, it is virtually impossible for the water operators, the operators of other critical infrastructure, public officials, or water customers to fully appreciate the risks. Because of the complexity and the widespread catastrophic potential of water risk, emergency management for water and wastepublic water/wastewater systems requires more focused efforts across economic sectors as well as public–private partnerships.

Primary Roles of Federal Government Agencies with Respect to Water

In addition to any disaster relief that may be available through the federal government, the primary national roles in ERM for water are related to risk identification, strategic objective setting, risk assessment, risk analysis, and risk management monitoring. Risk treatment options (realistically, retention, control, and/or transfer) fall to the state, regional and local government levels, and to the entities themselves in cases where they are private investor-owned.

The American Presidential Directive PDD-63 of May 1998 set up a national program of "Critical Infrastructure Protection" (CIP). The EPA was designated the federal lead agency for the water sector. The federal government's role in the water sector is primarily focused on water quality (EPA) and emergency response and recovery (FEMA). Resilience has not been substantially integrated into the actions of federal agencies and resilient outcomes are not typically part of federal programs and resources. In contrast to the energy and transportation sector, the water sector does not have a cabinet-level department and there is no dedicated Emergency Support Function (ESF) for water. Current authority for water is distributed across four ESFs and multiple federal agencies, leading to uncertainty, leadership challenges, information-sharing complications, and an overtaxing of water sector response resources – all of which can impede water service recovery during disasters. Additional authorities primarily consist of laws, entities, and individuals with limited authority for specific aspects of water as critical infrastructure.

Among its many other contributions to water sector risk management, the EPA has supported development of a suite of vulnerability (risk) assessment tools for drinking water and waste public water/wastewater systems of all sizes that address unique and fundamental security vulnerability concerns. All assessments were conducted at the utility level. Identification of priority components of a utility was dictated by local conditions, and determination of threats was made by the specific water sector utility conducting the assessment. As part of the vulnerability assessment, utilities develop inventories of system components, including physical, cyber, information technology, and personnel components, and identify which components are most critical to the system's mission.

The Secretary of Homeland Security is tasked with integrating and coordinating implementation efforts among federal departments and agencies, state and local governments, and the water sector. The Secretary is to establish uniform policies, approaches, guidelines, and methodologies for integrating CIP and risk management activities within and across sectors, and for developing metrics as part of a national CI/KR protection plan. The Secretary also maintains an organization to serve as a focal point for cybersecurity, and prepares an annual federal research and development plan to support this directive.

Related Homeland Security Presidential Directives

Three HSPDs directly expand the EPA's role in the national emergency response system, including that for the water sector. Several important HSPDs affecting EPA's role in water risk management are highlighted here.

Homeland Security Presidential Directive 5 (management of domestic incidents, 2003) – Establishes a single, comprehensive national incident management system (NIMS) and the National Response Framework (NRF), with EPA named to support and adopt both.

Homeland Security Presidential Directive 7 (critical infrastructure identification, prioritization, and protection, 2003) – Establishes a national policy for federal departments and agencies to identify and prioritize national CI/KR to protect them from terrorist attacks that could:

- Cause catastrophic health effects or mass casualties comparable to weapons of mass destruction (WMD);
- Impair the ability of federal departments and agencies to perform essential missions or ensure protection of public health and safety;
- Undermine state and local government capacities to maintain order and deliver minimum essential public services;
- Damage the water sector's capability to ensure the orderly functioning of the economy and delivery of essential services;
- Have a negative impact on the economy through the cascading disruption of other CI/KRs; and
- Undermine public morale and confidence in our national economic and political institutions.

Homeland Security Presidential Directive 8 (national preparedness, 2003) – Establishes policies to strengthen the US ability to prevent and respond to threatened or actual domestic terrorist attacks, major disasters, and other emergencies through development of a national, domestic all-hazards preparedness goal. It provides for state grants to build-through planning, training, and exercises – the capacity of first-responders to react to terrorist events. Funds also can be used to purchase equipment. States are required to develop state-specific plans. The directive also calls for development of quantifiable performance measures.

Homeland Security Presidential Directive 9 (defense of United States agriculture and food, 2004) – Establishes a national policy to defend the water, agriculture, and food system against terrorist attacks, major disasters, and other emergencies.

Homeland Security Presidential Directive 10 (biodefense for the 21st century, 2004) – Provides a comprehensive framework for the nation's biodefense. The directive focuses on threat awareness, prevention and protection, surveillance and detection, and response and recovery.

Related Federal Legislation

In addition to Presidential directives, statutes enacted by the legislative branch have affected water sector risk management substantially. Such law is briefly outlined here.

Safe Drinking Water Act (SDWA) (1974), 42 United States Code (U.S.C.) 300F-300J-26 – Provides a basis for drinking water security by protecting water quality and underground sources of drinking water. To protect the quality of public drinking water, EPA established regulations for national primary and secondary drinking water standards.

Public Health Security and Bioterrorism Preparedness and Response Act of 2002, Public Law 107-188 – Amends the SDWA such that: (1) EPA in 2002 provided the baseline probable threat information required to complete water system vulnerability assessments; (2) by the end of 2004, each community water system serving more than 3,300 persons conducted a vulnerability assessment, certified its completion, and submitted a copy to EPA; (3) each CWS serving more than 3,300 persons prepared or revised an ERP that incorporated the vulnerability assessment findings, and certified to EPA that the system had completed such a plan; (4) EPA developed a protocol to protect this information; (5) EPA developed vulnerability assessment guidance for systems serving 3,300 or fewer persons; and (6) EPA conducted research studies related to intentional contamination and other malicious attacks against water systems.

Federal Water Pollution Control Act (aka Clean Water Act), 33 U.S.C. 1251-1387 – The Clean Water Act governs the quality of discharges to surface and groundwater. It establishes national, technology-based standards for municipal waste treatment and numerous categories of industrial point-source discharges (discharges from such fixed sources as pipes and ditches); requires states, and in some cases tribes, to enact and implement water quality standards to attain designated waterbody uses; addresses water pollutants; and regulates dredge-and-fill activities and wetlands.

America's Water Infrastructure Act (AWIA) of 2018 – Requires community water systems that serve 3,300 or more persons to (1) conduct a risk and resilience assessment (RRA), and (2) prepare or revise an emergency response plan (ERP). The AWIA RRAs must evaluate risk to the water system from malevolent threats and natural hazards. The ERPs must include strategies to improve the resilience of the system, including the physical security and cybersecurity of the system.

Water Sector Collaborations with Other Stakeholders

AWWA Standard J100-10 (R13) – This was the first voluntary consensus standard encompassing an all-hazards risk and resilience management process for use specifically by water and wastepublic water/wastewater systems. It is a foundational, consensus-based standard that encompasses an all-hazards risk and resilience management process for use specifically by water and wastepublic water/wastewater systems.

CIPAC Water Sector Cybersecurity Strategy Workgroup: Final Report and Recommendations – The report recommends approaches to outreach and training to promote the use of the National Institute of Standards and Technology (NIST) Framework for Improving Critical Infrastructure Cybersecurity; identifies gaps in available guidance, tools, and resources for addressing this framework in the sector; and identifies measures of success that can be used by federal agencies to indicate the extent of use of the framework within water and wastepublic water/wastewater systems.

Roadmap to a Secure and Resilient Water Sector – Developed by the CIPAC Water Sector Strategic Priorities Working Group, it establishes a strategic framework that articulates the priorities of industry and government in the water sector to manage and reduce risk. It also produces a plan forward for the water sector GCC, SCC, and government and private sector security partners in the sector to improve the sector's security and resilience within the short term.

Water Infrastructure and Resiliency Finance Center – Established by the EPA in 2015, it supports the government-wide Build America Investment Initiative. The center provides communities, municipal utilities, and private entities with information and technical assistance on how to effectively use existing federal funding programs, access leading-edge financing solutions, and develop innovative procurement and partnership strategies. Although relatively new, the center has already undertaken several initiatives including establishing a network of university-based Environmental Finance Centers that correspond to the 10 EPA Regions; hosting Regional Finance Forums to bring together municipal officials and interested stakeholders to facilitate peer-to-peer interactions, share best practices, and build relationships; and providing technical assistance and tools through its Community Assistance for Resiliency and Excellence (WaterCARE) program. The center, which is advised by EPA's Environmental Financial Advisory Board, also works closely with other federal partners.

San Francisco Public Utilities Commission (SFPUC) Community Benefits Program – Engages neighborhoods that are directly affected by the operation of its water, wastewater, and power enterprises. The program includes education, workforce development, economic development, land use, neighborhood revitalization, support for climate change priorities, funding for the arts, localized professional services contracts, and philanthropic partnerships. This program serves as an example of local-regional collaborations with other sectors and stakeholders.

VARIATION IN PRACTICE BY WATER UTILITY SIZE AND RESOURCES

Despite the value of the collaborative resources, the adoption of leading-edge practices and resources has not been implemented across the water sector. Water and wastepublic water/wastewater systems vary widely in their individual ability and capacity to decide on and implement ERM best practices. Challenges that require new skill sets and training – such as cybersecurity – strain critical infrastructure in its efforts to adapt to a dynamic risk environment.

The largest water and wastewater systems serve the majority of the population – approximately 20% of water and wastewater systems serve more than 90% of the US population – and tend to adopt comparatively strong resiliency programs and practices. The many more numerous smaller systems meanwhile must rely on the transfer of knowledge and tools from outside the organization. Associations representing various areas and elements of the water sector, assisted by FEMA and EPA, have been quite active in developing and disseminating models, tools, and best practices, which thankfully are transferable to these smaller entities.

The affordability of systems—the ability of providers and their ratepayers to develop and maintain needed capabilities—is a cornerstone resilience issue for the sector. Utilities use a variety of rate structures to recover the costs of operating systems, including charging a flat fee regardless of the amount of water used, block rates based on usage, and seasonal rates. For utilities, there are several factors that come into play when setting rates: revenue, conservation, and affordability. The rates charged must bring in enough revenue to maintain the system; however, more and more customers are reducing the amount of water they use, decreasing revenue if rates are set based on usage. Finally, utilities have to ensure that rates are affordable for disadvantaged customers, but do not encourage wasting of water. In response, utilities are experimenting with different rate structures to try to balance these three factors.

In general, too many jurisdictions do not account for the full life-cycle cost of building, maintaining, upgrading, and replacing systems (whose life cycles can span decades). Moreover, it appears from our research and discussions that some utilities are diverting money collected as water fees for general revenue purposes. This was found to be true in Flint, when half of the collected fees were diverted in this manner. As a result, aging US water infrastructure has suffered from generations of underinvestment and is now prone to failure. In its 2021 Report Card for America's Infrastructure, the American Society of Civil Engineers gives drinking water and wastewater "C–" and "D+" ratings, respectively, on an A to F report card scale. State and local governments must increase investment into public water/wastewater systems to meet stricter federal water quality and drinking water safety standards – yet federal appropriations for water infrastructure have declined between 2008 and 2021 [1]. Often dominated by politics rather than engineering, decisions that set rates may simply reflect the least-cost path of patch and repair, ignoring resilience needs. This exacerbates longer-term problems and consequences, stretching the problems of a degrading infrastructure into future political cycles and generations of customers.

SUMMARY

This chapter has focused on risk and risk management as they apply to water/wastewater systems as critical infrastructure. Inherently, the discussion has centered on prioritized risks of degradation, natural disasters, and security. A water or wastewater system's risk management priorities are influenced largely by what is possible through coordination and collaboration with in-sector and external partners, such as the water sector working groups, WARNs, EPA, and FEMA.

The emphasis on degradation and security notwithstanding, there exist other risks that are a threat to water as infrastructure. Climate change in particular is a risk that over time may prove to be devastating to water as part of the societal fabric. In late 2019, UN Secretary-General António Guterres warned that a "point of no-return" on climate change is "in sight and hurtling toward us" [25]. Already, climate volatility has wreaked havoc in localized areas, and if overall climate change leads to greater weather disaster magnitude, water and wastewater systems must be prepared for these eventualities.

REVIEW QUESTIONS

1. T/F: Economic development, overall, is hindered by risk taking.
2. ISO defines risk as:
   a. Uncertainty
   b. The possibility of loss
   c. The effect of uncertainty on objectives
   d. All of the above
3. T/F: Critical infrastructure is primarily defined by physical (built) facilities necessary to the essential functions of the nation and economy.
4. T/F: The water and wastewater sector is defined as an area of critical infrastructure.
5. The water sector is interdependent with these other infrastructure sectors except for:
   a. Energy sector
   b. Transportation sector
   c. Communication sector
   d. Water is interdependent with all of these sectors
6. ABC Water System, LLC, is a rural water services provider that struggles to fund investment projects due to limited financial and other resources. Its water infrastructure is beyond the age at which it needs to be replaced. This threat to the system is an example of:
   a. Population growth-shift
   b. Natural disaster
   c. Degradation
   d. Contamination
7. A "black sky" event could be:
   a. An 18-hour disruption to the water services of a metropolitan area
   b. A two-hour service disruption of a rural water facility
   c. A one-year disruption to the entire power grid
   d. All of the above

8. The primary federal US law governing water pollution is the:
    a. Clean Water Act
    b. America's Water Infrastructure Act of 2018
    c. AWWA Standard J100-10 (R13)
    d. Homeland Security Presidential Directive 7

9. US ERM goals for critical infrastructure include all but:
    a. Sustain protection of public health and the environment
    b. Provide federal insurance for protection against losses to infrastructure
    c. Increase communication, outreach, and public confidence
    d. All of the above

10. SCADA contributions to:
    a. Infrastructure security
    b. Infrastructure risk
    c. Infrastructure complexity
    d. All of the above

11. T/F: Most critical infrastructure protection approaches directly account for sector interdependency and resilience.

ANSWER KEY

1. **T/F**: Economic development, overall, is hindered by risk taking.

2. ISO defines risk as:
    a. Uncertainty
    b. The possibility of loss
    c. **The effect of uncertainty on objectives**
    d. All of the above

3. T/**F**: Critical infrastructure is primarily defined by physical (built) facilities necessary to the essential functions of the nation and economy.

4. **T**/F: The water and wastewater sector is defined as an area of critical infrastructure.

5. The water sector is interdependent with these other infrastructure sectors except for:
    a. Energy sector
    b. Transportation sector
    c. Communication sector
    d. **Water is interdependent with all of these sectors**

6. ABC Water System, LLC, is a rural water services provider that struggles to fund investment projects due to limited financial and other resources. Its water infrastructure is beyond the age at which it needs to be replaced. This threat to the system is an example of:
    a. Population growth-shift
    b. Natural disaster
    c. **Degradation**
    d. Contamination

7. A "black sky" event could be:
    a. An 18-hour disruption to the water services of a metropolitan area
    b. A two-hour service disruption of a rural water facility
    c. **A one-year disruption to the entire power grid**
    d. All of the above

8. The primary federal US law governing water pollution is the:
    a. **Clean Water Act**
    b. America's Water Infrastructure Act of 2018

    c. AWWA Standard J100-10 (R13)

    d. Homeland Security Presidential Directive 7

9. US ERM goals for critical infrastructure include all but:

    a. Sustain protection of public health and the environment

    b. **Provide federal insurance for protection against losses to infrastructure**

    c. Increase communication, outreach, and public confidence

    d. All of the above

10. SCADA contributions to:

    a. Infrastructure security

    b. Infrastructure risk

    c. Infrastructure complexity

    d. **All of the above**

11. T/F: Most critical infrastructure protection approaches directly account for sector interdependency and resilience.

BIBLIOGRAPHY

1. ASCE Committee on America's Infrastructure. (2021). *2021 Report Card for America's Infrastructure.* American Society of Civil Engineers.
2. American Society of Civil Engineers and EBP. (2021). *Failure to Act: Economic Impacts of Status Quo Investment Across Infrastructure Systems.* American Society of Civil Engineers.
3. American Water Works Association. (2020). *Protecting the Water Sector's Critical Infrastructure Information – Analysis of State Laws.* Washington, DC: American Water Works Association.
4. AP News. (2018). Feds Investigate after Hackers Attack Water Utility. *AP News*, October 15.
5. Baylis, J., Edmonds, A., Grayson, M., Murren, J., McDonald, J., and Scott, B. (June, 2016). *Water Sector Resilience Final Report and Recommendations.* Washington, DC: National Infrastructure Advisory Council.
6. Beasley, M., Branson, B., and Hancock, B. (December, 2010). *Developing Key Risk Indicators to Strengthen Enterprise Risk Management.* Raleigh, NC: North Carolina State University and the Committee of Sponsoring Organizations of the Treadway Commission.
7. Black and Veatch. (2015). *Strategic Directions: Water Industry Report.* Overland Park, KS: Black and Veatch.
8. Burlinghame, G.A., and Chalker, R.T.C. (2017). Risk Management is Vital to Providing Safe Water. *Opflow Online* 2017: 20–23.
9. Chalker, R.T.C., Pollard, S.J.T., Leinster, P., and Jude, S. (2018). Appraising Longitudinal Trends in the Strategic Risks Cited by Risk Managers in the International Water Utility Sector, 2005–2015. *Science of the Total Environment* 618: 1486–1496.
10. Chalmers, B., and Basu, M. (February, 2020). *Global Risks for Infrastructure: The Climate Challenge.* Report by New York, NY: Marsh & McLennan Companies.
11. Clark, R., Panguluri, S., Nelson, T., and Wyman, R. (July, 2016). *Protecting Drinking Public Water/ Wastewater Systems from Cyber Threats.* Idaho Falls, ID: A report prepared for the U.S. Department of Energy Office of Nuclear Energy, Idaho National Laboratory.
12. Cybersecurity and Infrastructure Security Agency. (November, 2019). *A Guide to Critical Infrastructure Security and Resilience* Washington, DC: Cybersecurity and Infrastructure Security Agency.
13. Department of Homeland Security. (2020). *National Infrastructure Protection Plan – Water Sector.* Washington, DC: Department of Homeland Security.
14. EPA Office of Water. (March, 2020). *Building Security and Resilience in the Water Sector.*
15. GAO. (2003). *Critical Infrastructure Protection: Challenges in Securing Control System.* Washington, DC: United States General Accounting Office.
16. Gillette, J., Fisher, R., Peerenboom, J., and Whitfield, R. (2002). *Analyzing Water/Wastewater Interdependencies.* Infrastructure Assurance Center, Argonne National Laboratory.
17. H.R. 3162, USA Patriot Act. Version 1, Oct. 23, 2001.
18. The Institutes. (2013). The ERM Framework. In Elliott, M. (ed.), *Enterprise Risk and Insurance Management*, 1st ed. Malvern, PA: The Institutes.

19. International Standards Organization website: https://www.iso.org/about-us.html
20. International Water Association (2004). *The Bonn Charter for Safe Drinking Water International Water Association.* London, UK.
21. Little, R.G. (November, 2012). *Managing the Risk of Aging Infrastructure.* A report sponsored by International Risk Governance Council - Public Sector Governance of Emerging Risks.
22. National Institute of Standards and Technology. (2014). *Framework for Improving Critical Infrastructure Cybersecurity.* Gaithersburg, MD: National Institute of Standards and Technology.
23. Office of Cyber and Infrastructure Analysis (OCIA). (July, 2014). *Sector Resilience Report: Water and Wastewater Systems.* Washington, DC: Department of Homeland Security.
24. Rejda, G., McNamara, M., and Rabel, W. (2020). *Principles of Risk Management and Insurance*, 14th ed. New York: Pearson.
25. Rozsa, M. (December, 2019). U.N. Secretary-General Issues a Grave Warning about Climate Apocalypse. *Salon.com*, December 2, 2019.
26. United States. Pub.L. 93–523; 88 Stat. 1660; 42 U.S.C. § 300f *et seq.* 1974-12-16.
27. Water and Wastewater Sector Strategic Roadmap Work Group. (May, 2017). *Roadmap to a Secure and Resilient Water and Wastewater Sector.* WAshington, DC: Department of Homeland Security.
28. Water Research Foundation. (2017). *Preparedness and Response Practices to Support Water System Resilience: Fundamentals, Good Practices, and Innovations.* Denver, CO: Water Research Foundation.
29. Water Sector Government Coordinating Council. (May, 2007). *Water: Critical Infrastructure and Key Resources Sector-Specific Plan as input to the National Infrastructure Protection Plan.* Washington, DC: Water Sector Government Coordinating Council.
30. World Economic Forum. (January, 2012). Special Report: The Great East Japan Earthquake. *The Global Risks Report 2012*, 7th ed., a World Economic Forum Report, in collaboration with Marsh & McLennan Companies Swiss Reinsurance Company Wharton Center for Risk Management, University of Pennsylvania Zurich Financial Services, Geneva, Switzerland.
31. World Economic Forum. (January, 2020). Global Risks 2020 – An Unsettled World. *The Global Risks Report 2020*, 15th ed., a World Economic Forum Global Risks Initiative Insight Report, in partnership with Marsh & McLennan and Zurich Insurance Group, Geneva, Switzerland.

16 Emergency Management, Business Continuity, and Sustainability

Photo courtesy of California Water Service

LEARNING OBJECTIVES

- Differentiate between emergency management and business continuity.
- State the four phases of emergency management.
- Contrast the role of the Federal Emergency Management Agency (FEMA) with that of the Environmental Protection Agency (EPA) with respect to emergency management.
- Provide examples of why the precautionary principle may be an infeasible risk treatment option for most causes of catastrophes.
- Identify disaster vulnerabilities within an organization, and classify them according to whether they are shocks or stresses.
- Contrast resiliency with sustainability as an ERM goal.
- Describe the importance of collaborations and partnerships for effective emergency risk management.

DOI: 10.1201/9781003207146-23

- Highlight the ten core management areas that have been identified for sustainability within the water utility sector.
- Illustrate how a broad view of "community" is important for meeting ERM, resiliency and sustainability goals, and objectives.

KEY DEFINITIONS

America's Water Infrastructure Act (AWIA) – a federal act passed in 2018 requiring, among other things, community (drinking) water systems that serve more than 3,300 people develop or update risk assessments and emergency response plans (ERPs).

Business Continuity Planning – an evolving term that can mean preparing to continue operations during and after an emergency, or extend further to address longer-term enterprise resiliency or even enterprise-community sustainability.

Business Impact Analysis – a study that identifies which business processes would hurt the company the most to lose for various periods.

Centers for Disease Control and Prevention (CDC) – the primary national public health protection agency in the United States.

Consequence (or All-Consequence) Management Planning – emergency preparation that is driven by the fact that during a crisis or emergency it is more important to address the problem than to spend time and resources identifying the cause of the problem.

Disaster Mitigation – actions taken to help prevent disasters (loss prevention) or reduce or eliminate life and property loss (loss reduction) caused by hazards or disasters. Disaster mitigation equates to catastrophe risk control.

Disaster Recovery – strategies and tactics that enable an organization to maintain or quickly resume mission-critical functions following disruption due to an emergency.

Emergency Management – the organization and management of both resources and responsibilities for dealing with all humanitarian aspects of emergencies.

Emergency (or Disaster) Preparedness – taking necessary steps ahead of an emergency to ensure safety before, during, and after the event occurs.

Emergency (or Disaster) Response – the assessment of emergency situation as the event unfolds, and allocation of resources to meet needs as they evolve.

Emergency Response Plan (ERP) – the detailed preparation for emergency response. ERPs are required of water and wastepublic water/wastewater systems by federal law.

Environmental Protection Agency (EPA) – an independent executive agency of the US government tasked with environmental protection matters.

Federal Emergency Management Agency (FEMA) – an agency of the U.S. Department of Homeland Security having as its mission to lead the United States to prepare for, prevent, respond to, and recover from disasters.

Insurance Institute for Business and Home Safety (IBHS) – an independent, nonprofit, scientific research and communications organization supported solely by property and casualty insurers and reinsurers that conduct business in the US or reinsure risks located in the US.

Mutual Aid Agreement – allows entities to share resources in a more expedited way than would otherwise be possible without a formal disaster declaration.

National Incident Management System – a FEMA program that provides a consistent nationwide framework and approach to enable government at all levels, the private sector, and nongovernmental organizations to work together to prepare for, prevent, respond to, recover from, and mitigate the effects of incidents regardless of the cause or origin of the incident.

National Oceanic and Atmospheric Administration (NOAA) – a scientific agency within the US Department of Commerce that focuses on the conditions of the oceans, major waterways, and the atmosphere.

National Safety Council (NSC) – a nonprofit, public service organization promoting health and safety in the United States of America.

Nongovernmental Organization (NGO) – an organization which is independent of government involvement, founded by citizens to serve its members and others.

Preparedness, Emergency Response, and Recovery Critical Infrastructure Partnership Advisory Council (CIPAC) –a loosely organized group established by U.S. Homeland Security to facilitate interaction between governmental entities and representatives from the community of critical infrastructure owners and operators.

Resiliency – the capacity to bounce back, or recover, quickly from difficulties.

Shock – direct vulnerability, usually an intense, acute event that can disrupt an organization.

Small Business Administration (SBA) – a US government agency that provides support to entrepreneurs and small businesses.

Stress – indirect vulnerability, usually an underlying long-term economic, environmental, social, and governance condition that can adversely impact an organization and/or the stakeholders it serves.

Sustainability – ability to be maintained at rate or level that meets needs without compromising the ability of future generations to meet their own needs.

Vulnerability – a potential inability to withstand the effects of a hostile environment.

Wastewater Agency Response Network (WARN) – a cooperative based on a mutual aid agreement that allows utilities to share resources in a more expedited way than would otherwise be possible without a formal disaster declaration.

OVERVIEW

It wasn't raining when Noah built the ark

– Howard Ruff

The first chapter within *Part III – Practical Risk Management* covered the concept of risk, and water/wastewater risk management from the perspective of these systems as critical infrastructure. Chapter 2 expands on risk management within a water/wastewater context to focus on emergency planning and response, which is vital for survival in the immediate aftermath of a disaster event, and business impact and continuity planning for long-term organizational sustainability.

Chapter 1 closed with commentary on climate change, and the growing risk it represents for the utilities sector, and water/wastewater utilities in particular [13]. Climate change is but one factor among myriad factors that may contribute to organizational and/or sector-wide disasters. The job of emergency management and preparedness is to focus sights on shorter-term possibilities (e.g., extreme weather) while keeping longer-term event drivers (e.g., climate change) within peripheral vision. It is primarily with a shorter-range action orientation that Chapter 2 is concerned.

Disaster Vulnerability

Organizational vulnerabilities can be described in terms of shocks and stresses. Shocks are direct vulnerabilities; they are intense, acute events that can disrupt an organization. They include flash floods, wildfires, widespread loss of electrical power, dam failures, public health crises, and terrorist attacks. Shocks can lead to significant damage to infrastructure, as well as injuries and deaths. In contrast to shocks, stresses are underlying long-term economic, environmental, social, and governance conditions that can adversely impact an organization and/or the stakeholders it serves; these are indirect vulnerabilities. Stresses can also limit an organization's ability to address and recover from a shock. Stresses can include aging infrastructure, an economic downturn, and a lack of affordable housing. Stresses have the potential to intensify the impacts of shocks. For example, aging, weakened infrastructure may not be able to withstand a major flood event, which can impact the infrastructure directly (as a shock), but also have a detrimental effect on emergency services, community access to public services, and other needs. Multiple shocks can also lead to new stresses, for example, if multiple floods cause significant damage to a community's commercial sector, it could lead to an economic downturn.

Shocks and stresses are often interrelated. These interrelationships in vulnerability can be especially complex for public water/wastewater systems, as disruptions to the organization's operations and health are intermingled with disruptions to the community's operations and health, in both the short and long terms. A simple example is to imagine an event that disrupts employees' lives, and additionally may hinder some from being able to physically get to work. The consequences span the organization, the organization's direct stakeholders, and the community the organization serves. It is important to examine the impacts from shocks, identify stresses that can lead to shocks or weaken their ability to respond to shocks, and strategically plan to address these.

Effective Emergency Risk Management

Disasters and other emergency events are high-priority risks for any organization, but are especially important to utility entities because of the public good they provide. Emergencies may directly impact services, but utility services are also critical to much of the response and recovery effort for the wider community. Thus, emergency response and management must be given special attention within the water and wastewater risk management plan.

Emergency management is the organization and management of both resources and responsibilities for dealing with all *humanitarian* aspects of emergencies [9]. The aim is on life safety and property security to reduce the harmful effects of hazards, especially disasters. Emergency management consists of four elements – mitigation, preparedness, response, and recovery – each of which is necessary to effectively treat the risk. Figure 16.1 illustrates the cyclical relationship between these key components of emergency management.

Disaster mitigation describes actions taken to help prevent disasters (loss prevention) or reduce or eliminate life and property loss (loss reduction) caused by hazards or disasters. Disaster mitigation thereby equates to risk control for catastrophes. Emergency preparedness includes taking necessary steps ahead of an emergency to ensure safety before, during, and after the event occurs. Emergency response involves the assessment of the situation as the event unfolds and allocation of resources to meet needs as they evolve. Disaster recovery strategies and tactics enable an organization to maintain or quickly resume mission-critical functions following disruption due to an emergency. Each of these four phases of emergency management is explored here. First, a discussion of the Federal Emergency Management Agency (FEMA) and its National Incident Management System (NIMS) is in order.

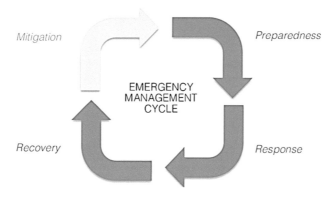

FIGURE 16.1   Emergency management cycle through its key components (FEMA) [9].

FEMA AND THE NIMS

FEMA's stated mission is simple and succinct: to help people before, during, and after disasters. FEMA's work and resources span across all phases of emergency management, and is a powerful place to begin for help when considering emergency risks. NIMS is a FEMA program that provides a consistent nationwide framework and approach to enable government at all levels, the private sector, and nongovernmental organizations (NGOs) to work together to prepare for, prevent, respond to, recover from, and mitigate the effects of incidents regardless of the cause or origin of the incident. Given the disaster mission of FEMA and more specifically NIMS, these programs are mentioned repeatedly throughout the balance of this chapter [9,12].

DISASTER MITIGATION

Disaster mitigation planning reduces the loss of life and property by minimizing the likelihood or impact of disasters. Mitigation aims to reduce vulnerability by reducing exposure (from a frequency or severity standpoint) to shocks. The primary FEMA emphasis is on natural disaster risks, including earthquake, fire, flood, windstorm, earthquake, and other perils. Unless and until scientific knowledge makes possible the human manipulation of weather and earth movement factors, the FEMA focus is on loss reduction rather than event prevention. FEMA-based disaster mitigation begins with the identification of natural disaster risks and vulnerabilities that are common in each local area.

After identifying these risks, local and state governments develop long-term strategies for protecting people and property from loss due to potential events. While government representatives at the federal, state, regional, and local levels take the lead in community disaster risk mitigation, private organizations can be actively involved. Private sector partners, such as utility systems, play a pivotal part in community disaster mitigation, both because they have direct organizational risks at stake and because they are stakeholders in the broader community. Mitigation strategies are key to breaking the historic cycle of disaster damage, rebuilding, and repeated damage, and water systems are participants as participants in this cycle are wise to be informed and active participants in breaking this cycle.

The inherently high-impact nature of catastrophes means that organizations may want to avoid or delay some innovations or activities until more is known about their risks. But for most operations and activities of public water/wastewater systems, risk avoidance is not an option. While it is important to consider the precautionary principle for new development plans, new infrastructure, and expanded or innovative services, the mainstay, core business activities must continue. For utilities and other organizations important to infrastructure, grants and special financing may be available to help ease the cost of disaster risk control, thus improving the affordability of taking measures to reduce likelihood and loss impacts of disaster events.

To the extent the organization is on its own to figure out cost-effective means for achieving risk mitigation, the rules of the game are increasingly complicated and may involve conflicting interests. A water system, whether government owned or investor owned, faces increasing hurdles to satisfy bondholders and other creditors that a project in need of financing is indeed worthy of financing. ESG (environmental, social, and governance) initiatives, for instance, are increasingly taken to account by analysts in the determination of creditworthiness. ESG initiatives themselves may be costly and require financing, thus resulting in a cycle of financing challenges.

EMERGENCY PREPAREDNESS

Emergency planning is effectively the planning for disasters, to include the response and recovery plans as well as portions of the loss reduction elements of mitigation plans. Emergency preparedness is complex, with myriad variables to consider, and many internal and external stakeholders to satisfy. It is a daunting process, especially with the dynamic nature of the external environment. Flexible emergency plan assessment and development solutions can suit the requirements of the organization's goals, objectives, and needs. Regardless, effective preparedness means a plan that is feasible, regulator-friendly, efficient, and effective, with a process that ensures the maximum number of planning assumptions are validated, and the maximum number of shocks, stresses, and improvement opportunities are identified.

Components of a Preparedness Plan

Even the most basic emergency preparedness plan must offer the following stated intentions [9,10,14]:

- Know what to do before, during, and after a disaster.
- Identify risks: Know what disasters are most likely to affect the organization.
- Develop a workplace emergency plan and ensure employees know it.
- Create a crisis communications plan to keep in contact with customers, suppliers, and employees during and after a disaster.
- Test and practice the preparedness plans.
- Have emergency supplies available at the workplace.
- Check insurance policies to ensure coverage is appropriate and adequate.
- Listen to and cooperate with government officials.

Achieving these elements is made easier by having an overall enterprise risk management (ERM) program in place, but the organization must pay special attention to disaster-related portions of the ERM plan. ERM is developed more fully in Chapter 17. Otherwise, all of the other efforts made to ensure the organization meets its strategic goals and objectives will be for naught should a catastrophic event or events occur.

Collaborations and Partnerships

The complexity of emergency preparedness and planning makes stakeholder collaborations and partnerships not only useful, but vital. Capitalize on the federal, state, regional, and local governments as collaborators. FEMA exists for this purpose, and holds high-minded goals and objectives for emergency planning. Indeed, its National Preparedness Goal is [9]:

> A secure and resilient nation with the capabilities required across the whole community to prevent, protect against, mitigate, respond to, and recover from the threats and hazards that pose the greatest risk.

Other federal agencies, like Ready.gov, the Environmental Protection Agency (EPA), National Oceanic and Atmospheric Administration (NOAA), and the Centers for Disease Control and

Prevention (CDC), are valuable resources for emergency preparedness as well. There are also regional and local government agencies that serve to help prepare for disasters. Additionally, private, nonprofit organizations that exist to educate and aid their constituents in planning for emergencies, of which the National Safety Council (NSC) and the Insurance Institute for Business & Home Safety (IBHS) are two examples [11,18]. Last, a growing number of for-profit businesses are available to advise and serve as project managers for emergency preparedness.

Preparedness Requirements for Water and Wastewater Systems

In addition to a water entity's internal concerns regarding catastrophic disruptions, external requirements have been placed on these organizations at the federal level. America's Water Infrastructure Act (AWIA) was signed into law in 2018, requiring community (drinking) water systems that serve more than 3,300 people develop or update risk assessments and emergency response plans (ERPs). The law specifies the elements that the risk assessments and ERPs must address, and establishes deadlines by which water systems must certify to EPA completion of the risk assessment and ERP. The most current template and instructions for the ERP can be accessed at: https://www.epa.gov/waterutilityresponse/develop-or-update-drinking-water-utility-emergency-response-plan [5–8].

AWIA mandates that within six months after certifying completion of its risk and resilience assessment, a water system must prepare or revise, where necessary, an ERP that incorporates the findings of the assessment. Overall, the Act requires:

- Strategies and resources to improve system resilience (including the physical security and cybersecurity);
- Plans and procedures that can be implemented, and identification of equipment that can be utilized, in case of an event that threatens the ability of the water system to deliver safe drinking water;
- Actions, processes, and equipment that can ameliorate the impact of a disruptive event on the public health and the safety and supply of drinking water provided to communities and individuals, including the development of (1) alternative source water options, (2) relocation of water intakes, and (3) construction of flood protection barriers;
- Strategies that can be used to assist in the detection of hazards – human induced or natural – that threaten the security or resilience of the system; and
- Coordination with local emergency planning committees when preparing or revising an assessment or ERP.

Training and Exercises

Education and practice are thought to play a key role in successful emergency preparedness. Most of these efforts are focused on ensuring smooth-working emergency response processes. FEMA and others provide numerous educational programs and exercises drills that can serve to increase emergency awareness among staff and build memory and confidence for handling emergencies if and when they arise.

The deliberateness of the training and exercises can contribute to quick mindedness when actual events occur, but can also create singlemindedness on the part of personnel who are involved. It is, therefore, worthwhile to provide multiple formats and incident examples as well as cross-train individuals to play differing roles within a drill exercise.

NIMS [12]

In the aftermath of the 2001 terrorist attacks, the need for an integrated nationwide incident management system with standard structures, terminology, processes, and resources became clear. The Department of Homeland Security (DHS) led a national effort to consolidate, expand, and enhance the previous work of several federal and state agencies to develop NIMS. FEMA published the first

NIMS document in 2004, and has revised it multiple times since. NIMS provides lessons learned, best practices, and changes in national policy, including updates to the National Preparedness System.

NIMS is applicable to all stakeholders with incident management and support responsibilities – emergency responders and other emergency management personnel, NGOs, the private sector, and elected and appointed officials responsible for making decisions regarding incidents. The scope of NIMS includes all incidents, regardless of size, complexity, or scope, even including planned events (e.g., special community events). Incident management priorities include saving lives, stabilizing the incident, and protecting property and the environment. Incident personnel apply and implement NIMS components in accordance with these principles:

- *Flexibility* – Scalable and applicable to wide variability of hazard, geography, demographics, climate, cultural and organizational authorities involved, including incidents that need multiagency, multijurisdictional, and/or multidisciplinary coordination.
- *Standardization* – Defines standard practices that allow incident personnel to work together effectively and foster cohesion among the various organizations involved; includes common terminology, which enables effective communication.
- *Unity of Effort* – Coordinating activities among various organizations to achieve common objectives. Unity of effort enables organizations with specific jurisdictional responsibilities to support each other while maintaining their own authorities.

EMERGENCY RESPONSE AND DISASTER RECOVERY

Emergency response centers on action – before, during, and after a disaster. The efficiency and effectiveness of organizational response can have a tremendous impact on the time it takes to fully recover from a catastrophic event. Effective response knows and acts on the priorities: first, to protect people (e.g., contact emergency authorities, evacuate the area, warn neighbors); second, to protect physical assets (e.g., guard the site, organize salvage operations); and third, to protect reputation (e.g., communicate with all economic stakeholders, maintain control of all media releases).

Just as proper emergency preparedness affects response, so does proper emergency response affect recovery. Emergency response keeps emergency recovery in mind. Survival is the short-term goal, but resiliency is the true end goal of effective emergency response. Take, for instance, internet outages. These are increasingly common as causes of emergency incidents. The US internet network has suffered a number of severe outages in recent years. In June 2019, major internet infrastructure and cybersecurity service Cloudfare suffered an outage that cost the company 15% of its global traffic. Earlier, in 2017, another internet infrastructure service provider experienced disruptions to its service resulting in a 90-minute internet outage [17,18]. These incidents have had significant repercussions for telecommunications companies. In both cases, several major internet service providers (ISPs) found their internet connections failing across the United States – including those of the nation's largest telecommunications providers. Telecommunications providers can indeed find themselves at the receiving end of both disruption and blame in the events of information infrastructure breakdowns, making it of paramount importance that they remain responsive to these incidents and work with authorities to address their root causes once they have occurred.

The Importance of Organizational Knowledge

Consistent with the ERM approach, knowledge must be both consolidated and shared. If an employee's knowledge is limited strictly to the work they do on a daily basis, this can lead to a myriad of problems. If there is little intra-organizational communication, what will happen when a disaster strikes? Response and recovery may fail due to lack of coordination; when it comes to maintaining a streamlined and secure organization, it is imperative all stakeholders have full knowledge of operations so that they can respond to any need across the organization on a well-informed basis. If all processes are known and accessible by every executive and employee within the entity, the likelihood of a healthy response and recovery is much greater. Everyone involved in operations should be able to easily view best practices and know how to implement them in the face of a threat. Only where stakeholders can locate all of the most up-to-date information necessary to protect and continue activities, this information can be converted into action. Yet it is nearly impossible to maintain organizational knowledge without a single repository where all of the organization's activities and processes can be viewed.

Emergency Command Center

Just as communities have Emergency Command Centers in the event of a disaster, so too should a business organization have centralized location and personnel setup for situational "command and control." Leadership never wants to lose control of a potentially damaging situation. Nothing can be accomplished in the face of a disaster if leadership does not have command over the events that are unfolding. Proactive planning and a firm grasp on every key piece of information create the best chance at a full recovery in the least amount of time possible.

It is important to possess and know how to effectively utilize visualization and decision support systems. An evolving virtual tool system can digest important information and update data in real time so that processes are always current and easily accessible. The ability to use that information to provide visual insights and deep analysis can materially change the effectiveness and efficiency of the outcomes you achieve. In the face of a threat, an enterprise needs to be able to immediately contact key decision-makers, review all assets, and determine which locations and stakeholders have been affected. People need to be able to trust this information and make decisions in real time. In other words, they can generate a reliable plan on the fly or be confident that existing plans are utilizing current information while filtering out information that is not relevant to the situation at hand. In case access to the command center or to electronic data is hindered or lost, a backup location, backup personnel, and backup data should be part of the response contingency plan.

Figure 16.2 provides an example of the organized information flow that may take place during an emergency incident. For the City and County of San Francisco, this describes the flow of information as organized within the Emergency Operation Center, Infrastructure Branch, and between the Water Utility Group Coordinator, Unit Leaders and various support agencies. The intent of such a flow is not only to optimize communication efforts directly, but also to indirectly optimize resource gathering and deployment.

Limited Resources

Emergency response is typically the most expensive part of emergency management. Shocks, stresses, needs, goals, and objectives are myriad. Resources are not. It is unlikely an organization can solve every challenge presented, so challenges must be prioritized both in advance (as previously discussed) and as they arise. Every public water and wastewater enterprise, no matter the size, has limited resources to dedicate to these efforts. Determine ahead of time (according to risk analysis) how to delegate time, money, and human resources. Then be adaptable as events unfold.

Business impact analysis (BIA) is a study that identifies which business processes would hurt the company the most to lose for various periods of time. It evaluates them based on a combination of mission criticality and time sensitivity. The BIA helps identify which processes most need to be protected. It provides a rational basis on which to allocate emergency resources. BIAs should be based on the primary mission as an organization and should be subjected to disciplined, rigorous analysis since limited resources limit what response measures are practicable.

Processes, People, and the Convergence of Emergency Response with ERM

Emergency response and recovery necessitates a strong understanding of the building blocks of the organization, including all the departmental processes involved and the interdependencies between them. This "bottom-up" approach to knowledge results in extensive process charting for understanding impacts and how to quickly become operational after a disruptive event. ERM, on the

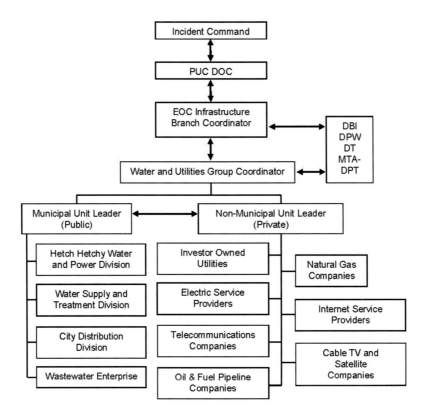

FIGURE 16.2 City and County of San Francisco emergency response information flow. Source: City and County of San Francisco [4].

other hand, because it is strategic risk management takes a "top-down" approach, and as such might not drill down into the innermost workings of the organization. Therefore, ERM, if not combined with resiliency planning, may provide less visibility into the dependencies that comprise the organization's ecosystem than one would assume. These dependencies are key to understanding shocks and especially stresses, so the convergence of emergency management and risk management yields direct benefits. Emergency response plans often contain the information that would benefit risk management in gaining a better perspective of operational risk.

Also thinking with respect to a "bottom-up" perspective, the frontline people in the organization are its first line of defense in the event of a catastrophe. A "top-down" approach may or may not engage operational managers to the point of giving them the ownership, responsibility, and accountability for directly assessing, controlling, and mitigating risks. While emergency preparedness planning does not necessarily provide operational managers with an understanding of the organization's risks in their entirety, it does enhance their awareness of the impact of an interruption and the potential losses that could result. Collaboratively, emergency management and risk management can align to develop a much broader, clearer, detailed picture of the enterprise. This, in turn, empowers risk management to identify and address risk to a finer degree to support organizational resiliency.

The Value of Qualitative Assessments

In terms of risk analysis, an enterprise's greatest inherent risks are those that pose the greatest probable maximum losses (PMLs) to the organization. Risks are categorically prioritized, controlled (mitigated), and transferred according to frequency and severity, where frequency times severity yields the overall quantitative loss potential. This emphasis on financial loss has its challenges, particularly as it relates to deployment of resources for emergency response (as opposed to pre-event risk control). The problem is that it fails to consider other nonfinancial impacts that can be devastating to the organization. Take a public water/wastewater system, for instance. The quantitative view works fine when it is applied to a process that is directly tied to water output, which can be tied to revenue. The impact of a disruption to the process can be measured in financial terms of reduced revenue due to a delay or shortage of quality drinking water. What the model cannot show is the community stress of stakeholders being without water; it cannot quantify the impact of such an event, and the purely quantitative view cannot prioritize such a risk. Emergency planning for response and recovery must develop qualitative ways of measuring impacts that, while not representing direct financial loss, deliver meaningful ways of rating and ranking risks.

Promoting the convergence of business continuity and risk management is key to strengthening organizational resiliency in a risk-filled world. Business continuity is able to support risk management through its expertise in process dependency mapping, its relationships with operational managers, and its ability to evaluate risk using qualitative rating methods. By converging these two disciplines, organizations can build a more effective and efficient risk and business continuity management program, and a more resilient enterprise.

Incident Action Checklists

Water and wastewater systems are potentially both "victim" and "helper" in an emergency. Because of the community responsibility these entities hold, they must respond (or at least appear to respond) quickly and confidently during times of disaster. The EPA has developed Incident Action Checklists specifically for water systems so that rapid, effective actions can be taken. At least 14 separate checklists are available to water systems, by peril – pandemic, power outage, harmful algal bloom, cybersecurity, extreme cold and winter storms, earthquake, drought, extreme heat, flooding, hurricane, tornado, tsunami, volcanic activity, and wildfire. These can be found at: https://www.epa.gov/waterutilityresponse/incident-action-checklists-water-utilities.

A Note on Service Disruption Incidents [3,17]

When considering risks, service disruption incidents are prioritized by drinking water and wastewater systems. Yet the individual risks may diverge from one another in a risk profile. Take, for instance, a drink water utility. The top frequency drivers of water supply interruption – burst and leaking pipes, contamination, flood, drought, planned and reactive maintenance – may not have similar frequency and severity characteristics. Even less likely to share risk characteristics are the top drivers of business impact – cyberattack/data breach, information technology (IT) and telecom outages, act of terrorism/security incident, interruption to utility supply, and supply chain disruption.

Although all of the risks mentioned above are likely prioritized by ERM (and with good reason), the emergency response may differ widely based on the details of the event. When a piped supply is disrupted, it is essential to restore the service to normal operation as quickly as possible. To ensure this happens and to reduce the interim impact, water systems are advised to engage the services of an experienced partner, through establishment of a framework agreement, who can provide alternative water supply during incidents. Bearing in mind the individuality of a disruption incident, these four immediate considerations are of utmost importance regardless of the precise event details [18]:

- *Quality assurance* – Water quality is of the utmost importance for any alternative supply provision.
- *Responsive capability* – Quick response and effective resource management are crucial both in and outside of working hours.
- *Appropriate solutions* – The response approach and the solutions provided need to be as variable as the event details, with multiple options available that can be adapted and used in unison with one another for diverse coverage. For example, visible drinking water stations so that customers can clearly see that the incident is being reactively managed, or innovative and less visible solutions where customers experience reduced interruption due to proactive mitigation methods.

- *Effective communications* – For the benefit of the water-consuming community, proactive, timely, and accurate communications between incident managers, responders, and the public through the lifecycle of an incident are crucial. From an incident support point of view, it is essential to ensure candor and accuracy in the communication of delays and to provide general regular updates.

A prime example of communications response efforts is offered by the City of Bozeman (Montana) Water Department. The department has Twitter and Facebook accounts, but reserves their use almost exclusively for emergencies. Thus, when a customer sees a message on his/her social media account, it is immediately recognizable as important. Only one or two people have access to the department's social media accounts, which provides for better control of messaging and information [18].

Mutual Aid Agreements [2]

There is special assistance available for emergency response within drinking water and wastepublic water/wastewater systems. Water and Wastewater Agency Response Networks (WARNs) are "utilities helping utilities," typically within a state. The idea is to respond to and recover from emergencies more efficiently and effectively by sharing resources with one another. WARNs are governed by a common mutual aid agreement. The WARN agreement allows utilities to share resources in a more expedited way than would otherwise be possible without a formal disaster declaration. The agreement outlines how liability, workers' compensation, insurance, and reimbursement will work. Other benefits include increased emergency preparedness and coordination, and enhanced access to specialized resources.

Cyclicality

Emergency response and recovery may appear on its surface to be a linear process – getting the organization from Point A (first moment of disruption) to Point B (full recovery from disruption). But it is in fact a circular process of adapting to events as they happen in real time and simultaneously monitoring the performance of the actions taken. Its cyclical nature involves repeated assessment, planning, action, and review, to respond appropriately.

SPECIAL SUPPLY CHAIN MANAGEMENT CONCERNS

A supply chain is a system of organizations, people, activities, information, and resources involved in supplying a product or service to a consumer. Thus, supply chain management is the handling of this system's flow. Supply chain risk management entails assessing and mitigating all the risks that might interrupt the normal, or expected, flow of goods and services from and to an organization's stakeholders. When applied to the production, distribution, and/or treatment of water, supply chain risk management encompasses managing the volatility related to producing, transporting, storing, and treating water.

Supply chain exposures are numerous, and can occur internally (within the organization) as well as externally (outside the organization). Internal exposures include: threats to facilities (e.g., natural disaster, man-made disaster, or terrorism); production bottlenecks (e.g., dependencies on key equipment and water supplies); dependence on information technology (e.g., information backup failures); strikes or other employment issues (e.g., stalled operations); machinery breakdown (e.g., critical backup in production may occur while new parts are ordered and installed); and so on.

External loss exposures to the supply chain include: sole source of supply (e.g., disruption at the source could undermine a water entity's ability to distribute water to satisfy demand); third party vendors (reliance on outside partners for water distribution could result in unsatisfied demand); change in demand level (e.g., incremental or substantial changes in demand due to changes in customer demographics can cause over- or underproduction); financial risks (e.g. increases in the cost of materials or transportation changes cause costs to rise); sociopolitical environments (e.g., inability to pass through increased costs in rates due to government regulations); natural and man-made catastrophes (e.g., storms, earthquakes, volcanic eruptions, and other natural disasters can interfere with transportation routes); and merger of a key vendor with a competitor or unfriendly partner (e.g., change in ownership of key vendor resulting in higher prices and/or reduced availability of materials). External exposures are as important to consider in the supply chain risk management process, especially since the organization has less control over the threats encountered. Increasingly, organizations are creating intentional redundancies (e.g., multiple vendor contracts) in their supply chain flows to curb these risks.

A compelling case for the need for redundancy in emergency planning is the December 2011 tsunami, and its indirect impacts on infrastructure sectors across the globe. In 2011, a nearly 45–foot-high tsunami, triggered by a magnitude 9.1 earthquake, struck three nuclear cores at the Fukushima Daiichi power plant in Japan. When inundated by the water, they melted. This event resulted not only in the involuntary evacuation of 100,000 people and many lost lives, but also contaminated food and water resources in the area – resulting in lawsuits against the company responsible and pressures on other surrounding infrastructure companies reliant on stable populations and water resources. An indirect impact was that several US car manufacturers could not obtain automotive microcontroller chips for a significant temporary period of time. It turns out, these manufacturers source parts from various companies from around the world, which in turn source microchip controllers from a company called Renesas. The plant where Renesas produces many of those chips, north of Tokyo, had been heavily damaged by the earthquake. With no alternative suppliers of automotive microcontroller chips, US car production temporarily shut down [19].

Utilities generally rely on outside contractors and suppliers for assistance, especially during emergencies. The East Bay Municipal Utility District (EBMUD) in Oakland, California, has multiple vendors for critical supplies and other contracted needs. EBMUD has established zero-dollar purchase orders with some of these vendors. In an emergency, EBMUD can allocate funds to the purchase orders, wherein the terms and specifications are already developed and pre-approved. Furthermore, EBMUD requires critical vendors to have a continuity plan in place for their own critical operations during a disaster [18].

ALL-CONSEQUENCE MANAGEMENT PLANNING

Most established emergency management plans – from disaster mitigation to preparedness to response to recovery – take a by-peril, or by-cause, approach. There is added value in taking a by-consequence approach. The concept of all-consequence management planning emanates from this mindset. A water sector working group in 2009 released consequence management planning guidelines for the water sector [7,16]. The Preparedness, Emergency Response, and Recovery Critical Infrastructure Partnership Advisory Council (CIPAC) Working Group prepared a document that complements a utility's overall emergency preparedness, recovery, and response planning. It is available at: https://www.waterisac.org/sites/default/files/2009_CIPAC_All-Hazard-CMP.pdf, and outlines hazards and consequences a utility should consider, guidelines for improving resiliency and consequence-specific actions to take.

Consequence management planning is driven by the fact that during a crisis or emergency it is more important to address the problem than to spend time and resources identifying the cause of the problem. For example, during a service outage, it is more important to restore power which returns service to the customers than to spend valuable resources looking for the underlying cause of the outage (at that moment). More importantly, effective planning allows a utility to mitigate the negative impacts associated with a service outage by taking steps to minimize the impacts and implement planning responses (actions) that result in a quick resolution to the problem. Planning to solve problems encourages a utility to take preparedness steps to decrease the vulnerability of its systems and improve its preparedness and response capabilities to reduce the impacts of any incident that might occur. Creating specific preparedness and action plans oriented to specific problems allows a utility to respond more quickly and effectively and makes a utility more resilient. For instance, by planning what would be needed to fix the problem of a power outage in advance, a utility reduces its vulnerability and is poised to quickly respond and recover if a power outage were to occur, regardless of the cause of the outage, reducing the severity of the impact.

Water utilities across the nation serve as examples of good emergency planning practice. Ohio's Tupper Plains-Chester Water District has managed its backup power and communications risk responsiveness in noteworthy ways. The District has contractually ensured that its fuel supplier has backup power and can continue to supply fuel to the District during an outage. To improve phone network reliability, the District invited its phone company to move their substation (which was previously located in a flood zone) to District property where it could also be provided with backup power supplied by the District [18].

RESILIENCY, BUSINESS CONTINUITY, AND SUSTAINABILITY

Overall risk management for drinking water and wastepublic water/wastewater systems must begin with resiliency and must expand to emphasize sustainability. Resiliency and sustainability are buzz terms most have heard repeatedly; they are THAT important.

Resiliency can be defined in multiple ways, but it is most simply the ability to survive in the midst of changing conditions or challenges. At the very least a community needs to be able to rebound from adverse events or situations. It is even better if a community can adapt to the possibility of future adverse events. And it is best if a community can find ways to thrive amid uncertainty and risk – to maintain quality of life, healthy growth, durable systems, economic vitality, and conservation of resources for present and future generations. Notice the use of the word "community." Importantly, a water utility is itself a community that serves a broader community. Its resiliency is interwoven with the resiliency of that larger community.

Resiliency is often conflated with other related goals, such as disaster preparedness, emergency management, and climate change mitigation and adaptation, but in fact each of these is just one facet of overall resiliency. Adaptable economic planning is another facet of resiliency that comes up less often than these other facets do, as people talk about resiliency, but it is equally critical. A community that has a diverse economic base will be more likely to withstand a shock to the local economy compared to a community that has only one or two major industries as economic drivers. The community is a system with interconnected pieces – economic, environmental, social, and governance – and is interconnected with other systems and other communities. Understanding this in abstract is critical to developing a truly successful concrete framework for resiliency. Long-term drought conditions can increase the risk of wildfires, threatening the lives and property of those in the wildland–urban interface and simultaneously threatening the agricultural and drinking water sectors. These impacts, both separately and together, can reduce long-term real estate demand and government revenues. Of course, these economic knock-on effects make it difficult to find affordable ways to combat the community's vulnerability to drought.

Emergency response and management are at the heart of the organizational community's resiliency. But business continuity planning typically is not. While almost every organization considers disaster risk a high priority for risk treatment, and likewise develops an emergency preparedness plan, most organizations do not place a high priority on business continuity planning. This is likely because the primary focus of emergency management is the protection against loss of life and direct property (humanitarian) while organizations more likely think of business continuity planning as the protection against indirect financial value loss to the organization (economic). Most enterprises behave as if they do not believe that disasters pose a substantial threat to the well-being of the enterprise. Yet statistics collected by the U.S. Small Business Administration (SBA) tell a different story; 40% of businesses fail to reopen following disaster and over 90% of businesses fail within two years after being struck by a disaster [15].

In its best and highest form, business continuity planning goes beyond resiliency to encompass sustainability. Developing sustainability requires reducing the vulnerability to adverse events as well as to underlying risks. Best-practice continuity planning looks beyond the hazard and emergency mitigation plans to address vulnerability to underlying risks, such as population growth, aging and deteriorating infrastructure, and changing climate conditions that if unaddressed can worsen impacts from loss events. Thus, best-practice continuity planning takes a holistic approach toward sustainability: protecting and improving, and leads to successful planning and decision-making for meeting long-term goals and visions, such as preserving and enhancing the overall quality of life for the broader community.

To make sustainability relatable, consider that water scarcity already affects a quarter of the global population. Business continuity for water and wastewater systems is critical for continued community health and global societal health. Water conservation is thus an integral part of water system risk management.

BUSINESS CONTINUITY AS BUSINESS IMPACT AND RESILIENCY PLANNING

A traditional approach to business continuity planning looks much like emergency management. It is broader in purpose, however, and is the advance planning and preparation undertaken to ensure that an organization will have the capability to operate its critical business functions during and after economically disruptive events. For public water/wastewater systems, economic disruptions are more likely to occur as a result of a natural disaster or other large-scale incident that disrupts economic activity in a state or region, but could also occur due to other sources, such as a disastrous liability loss or a catastrophic market failure. During these circumstances a utility may be unable to access its billing or banking systems, experience electrical service or communication outages, or be unable to access its facilities. It is critical that utilities prepare for the potential of economic disruption so they can continue to operate effectively during an incident. Business continuity and resiliency planning necessarily consists of strategies that [8,18]:

- Increase community cohesion and community identity;
- Improve comprehensive and more universal vulnerability identification;
- Decrease impacts from shock and stresses;
- Reduce costs associated with major events; and
- Provide a roadmap to receive mitigation and resiliency project funding.

The scope of business continuity planning and resiliency extends beyond emergency planning intents and purposes to also include governance, risk management, and compliance, and it is an approach that combines emergency management with these three programs for greater organizational coherence and efficiency. Business continuity planning is commonly seen as including the legal, audit, finance, and HR departments, as well as emergency management and IT/disaster recovery. Examples of specific resiliency performance indicators for a utility entity may include measures such as these [17,18]:

- *Power resiliency* – Period of time (e.g., hours or days) for which backup power is available for critical operations.
- *Treatment chemical resiliency* – Period of time (e.g., hours or days) minimum daily demand can be met with water treated to meet the Safe Drinking Water Act (SDWA) standards for acute contaminants.
- *Critical parts and equipment resiliency* – Current longest lead time (e.g., hours or days) for repair or replacement of operationally critical parts or equipment.
- *Critical staff resiliency* – Average number of response-capable backup staff for critical operation and maintenance positions.

- *Treatment operations resiliency (percent)* – Percentage of minimum daily demand met with the primary production or treatment plant offline for 24, 48, and 72 hours.
- *Source water resiliency* – Period of time (e.g., hours or days) minimum daily demand can be met with the primary raw water source unavailable.

Business Continuity as Sustainability Planning

Public water/wastewater systems have the ability to incorporate sustainability into their overall emergency management and business continuity plans. Sustainability is the ability to exist constantly. Developing a sustainability framework involves bringing together a number of community stakeholders, including representatives from government, community organizations, and the private sector to create a vision of what a sustainable community looks like, including but not limited to its water and wastewater systems. The framework can help [8]:

- Explore existing conditions;
- Understand the vulnerabilities – individually and collectively (i.e., shocks and stresses);
- Define a vision and goals;
- Define strategies and identify priority projects; and
- Establish an implementation roadmap, or action plan, for meeting goals and objectives.

Because an organization having sustainability as a goal is interested in the ability to always exist, it engages in a proactive process that is at once both opportunistic and risk aware, both present and futuristic. The vision, goals, vulnerabilities, projects, and the tactical roadmap can thus be viewed as multi-level. The EPA has loosely defined for utilities three levels of business to consider when planning for sustainability [8,18]:

Level 1 – Providing adequate, fundamental services: implementing practices that focus on meeting and maintaining compliance for all applicable regulations, ensuring adequate levels of operational resiliency, and implementing revenue and financing mechanisms that assure its mid- to long-term financial viability.

Level 2 – Optimizing operations and services: continual improvement and views optimizing its operations and services as central to mission success, such as actively engaging with its community to create operating conditions that are responsive to community needs and interests and actively seeking to ensure its operations support the community's economic and social well-being.

Level 3 – Transforming operations and services for the future: implementing practices consistent with many of the directions set forth in leading industry initiatives, such as employing practices that focus on managing treated wastewater and biosolids as valuable commodities, both to improve efficiency and as new revenue sources.

The EPA also has established ten core management areas on which public water/wastewater systems should focus sustainability efforts. These areas are based on the categories supported by EPA and major water sector associations, and a similar framework that EPA and the US Department of Agriculture (USDA) developed for small systems. The core management areas are [8]:

- Utility business planning
- Product quality and operational optimization
- Customer satisfaction and stakeholder understanding and support
- Employee and leadership development
- Financial viability
- Infrastructure stability

- Operational resiliency
- Water resource adequacy
- Community sustainability
- Performance measurement and continual improvement

Though each core management area addresses a specific set of goals and strategies to address shocks and stresses, integration of activities across sectors is key to developing a sustainable enterprise (especially as a key member of the broader community). These core areas are interdependent, and many strategies are likely to have a cross-sector impact. Integrated planning can also help reduce costs across core areas within the entity and across organizations. Thus, sustainability planning becomes a way to reduce indirect and future vulnerability by addressing and improving the underlying conditions that expose the utility and its interdependent community to hazards, and by developing a capacity to adapt to changing conditions.

SUMMARY

Organizations face the possibility of short-term emergencies and long-term-effect disasters in addition to everyday hazards. Regardless of how comprehensive the ERM plan is, a water utility must have separate and detailed plans in the event of disasters. Otherwise, the criticality of an event may not be fully managed until the event has occurred, resulting in nonoptimal outcomes at best and a cease to operations at worst. Water and wastepublic water/wastewater systems vary in purpose, size, environment, and resources. Nevertheless, all of these must proactively engage in emergency management (disaster mitigation, preparedness, response, and recovery) to enhance organizational resiliency and continuity. Governmental and other agencies not only serve as partners in emergency management but also as motivators. FEMA, the EPA, and working groups within the water utility sector each have set forth guidelines for appropriate emergency management and business continuity planning [1]. Business continuity planning in particular is taking on new and expanded meaning. Evolving from what was once equivalent to disaster recovery, it is now much more broadly about enterprise and community resiliency. It continues to expand, and is fast becoming a matter of sustainability planning.

REVIEW QUESTIONS

1. Underlying long-term economic, environmental, social and governance conditions that can adversely impact an organization and/or the stakeholders it serves are known as:
   a. Opportunities
   b. Responses
   c. Shocks
   d. Stresses
2. T/F: Emergency management includes mitigation, preparedness, response and recovery efforts.
3. Disaster risk typically is:
   a. Characterized by organizations as high priority and so emergency management and business continuity planning are also deemed as high priority
   b. Characterized by organizations as high priority, but while emergency management is also prioritized business continuity planning is not
   c. Characterized by organizations as low priority (due to low likelihood), so emergency management and business continuity planning are not prioritized
   d. Characterized by organizations as low priority and so emergency management and business continuity planning are also deemed as low priority
4. T/F: Disaster mitigation is equivalent to risk control for disasters.

5. Which of the following agencies requires an Emergency Response Plan (ERP) from all but the smallest public water/wastewater systems?
   a. Environmental Protection Agency
   b. Federal Emergency Management Agency
   c. National Security Council
   d. Small Business Administration

6. NIMS principles include(s):
   a. Flexibility
   b. Customization
   c. Individual effort
   d. All of the above

7. T/F: The concept of resiliency is interchangeable with the concept of sustainability.

8. The EPA's ten core management areas on which public water/wastewater systems should focus sustainability efforts include(s):
   a. Financial viability
   b. Infrastructure stability
   c. Operational resiliency
   d. Water resource adequacy
   e. All of the above

9. T/F: One quarter of the world's population is impacted by water scarcity.

10. T/F: The EPA considers that "optimal" water operations and services are part of sustainability.

ANSWER KEY

1. Underlying long-term economic, environmental, social, and governance conditions that can adversely impact an organization and/or the stakeholders it serves are known as:
   a. Opportunities
   b. Responses
   c. Shocks
   d. **Stresses**

2. **T/F**: Emergency management includes mitigation, preparedness, response, and recovery efforts.

3. Disaster risk typically is:
   a. Characterized by organizations as high priority and so emergency management and business continuity planning are also deemed as high priority
   b. **Characterized by organizations as high priority, but while emergency management is also prioritized business continuity planning is not**
   c. Characterized by organizations as low priority (due to low likelihood), so emergency management and business continuity planning are not prioritized
   d. Characterized by organizations as low priority and so emergency management and business continuity planning are also deemed as low priority

4. **T/F**: Disaster mitigation is equivalent to risk control for disasters.

5. Which of the following agencies requires an Emergency Response Plan (ERP) from all but the smallest public water/wastewater systems?
   a. **Environmental Protection Agency**
   b. Federal Emergency Management Agency
   c. National Security Council
   d. Small Business Administration

6. NIMS principles include(s):
    a. **Flexibility**
    b. Customization
    c. Individual effort
    d. All of the above

7. T/**F**: The concept of resiliency is interchangeable with the concept of sustainability.

8. The EPA's ten core management areas on which public water/wastewater systems should focus sustainability efforts include(s):
    a. Financial viability
    b. Infrastructure stability
    c. Operational resiliency
    d. Water resource adequacy
    e. **All of the above**

9. **T**/F: One quarter of the world's population is impacted by water scarcity.

10. **T**/F: The EPA considers that optimal water operations and services are part of sustainability.

REFERENCES

1. American Water Works Association (AWWA) website: https://www.awwa.org/.
2. AWWA. (September, 2008). *Economic Benefits of Forming and Participating in a Water*. Wastewater Agency Response Network (WARN).
3. Burlinghame, G.A., and Chalker, R.T.C. (2017). Risk Management is Vital to Providing Safe Water. *Opflow Online* 2017: 20–23.
4. City and County of San Francisco. *Emergency Support Function 12 – Water and Utilities Annex*.
5. Effective Utility Management (EUM) Utility Group. (January, 2017). *Effective Utility Management – A Primer for Water and Wastepublic Water/Wastewater Systems*. EPA Office of Wastewater Management.
6. Environmental Protection Agency (EPA) website: www.epa.gov.
7. EPA. (2008). *Water Security Initiative: Interim Guidance on Developing Consequence Management Plans for Drinking Water Utilities*.
8. EPA. (December, 2014). *Moving Toward Sustainability: Sustainable and Effective Practices for Creating Your Water Utility Roadmap*.
9. Federal Emergency Management Agency (FEMA) website: www.fema.gov.
10. Haddow, G., Bullock, J., and Coppola, D. (2013). *Introduction to Emergency Management*, 5th ed. Oxford, UK: Butterworth-Heinemann.
11. Institute for Business and Home Safety (IBHS) website: www.disastersafety.org.
12. National Incident Management System (NIMS) website: https://www.fema.gov/emergency-managers/nims.
13. Rozsa, M. (December, 2019). U.N. Secretary-General Issues a Grave Warning about Climate Apocalypse. *Salon.com*, December 2, 2019.
14. Rubin, C., and Cutter, S. (2019). *U.S. Emergency Management in the 21st Century from Disaster to Catastrophe*. Philadelphia, PA: Taylor & Francis.
15. Small Business Administration (SBA) website: www.sba.gov.
16. WaterISAC website: https://www.waterisac.org/.
17. Water Research Foundation. (2004). *Emergency Response Plan Guidance for Wastewater Systems*. Denver, CO: Water Research Foundation.
18. Water Research Foundation. (2017). *Preparedness and Response Practices to Support Water System Resilience: Fundamentals, Good Practices, and Innovations*. Denver, CO: Water Research Foundation.
19. World Economic Forum. (January, 2012). Special Report: The Great East Japan Earthquake. In *The Global Risks Report 2012*, 7th ed., a World Economic Forum Report, in collaboration with Marsh & McLennan Companies Swiss Reinsurance Company Wharton Center for Risk Management, University of Pennsylvania Zurich Financial Services, Geneva, Switzerland.

17 Risk Management for Value Creation and Preservation

LEARNING OBJECTIVES

- Define risk management according to the Committee of Sponsoring Organizations of the Treadway Commission (COSO).
- Contrast the enterprise risk management purposes and perspectives with those of traditional risk management.
- Recognize the advantages of an enterprise risk management approach over those of the traditional risk management approach.
- Differentiate between enterprise risk management and project management.
- Define the objective of enterprise risk management.
- Discover the importance of leadership and accountability for the enterprise risk management (ERM) program.
- Apply the concept of cascading responsibility to a public water or wastewater system.
- Contrast ERM with accounting.
- Examine the value of ERM practice variation and cross-functional communication.
- Differentiate between risk profile, risk perception, and risk attitude.
- Identify key risk indicators within an organization.
- Compare risk aversion with risk neutrality with risk seeking.
- Contrast risk appetite with risk tolerance.
- Articulate how risk-bearing capacity influences risk appetite.

DOI: 10.1201/9781003207146-24

- Describe an organizational value killer.
- Differentiate between physical and intangible assets.
- Outline the relationship between organizational ethics, organizational integrity, and reputation.
- Describe reputational risks and their drivers.
- Contrast triple bottom line (TBL) accounting with traditional accounting.
- Examine the challenges of tackling the reputation risk problem.
- Outline the innovation motivation – benefits – costs conflict within the utilities sector.

KEY DEFINITIONS

Aversion to Ambiguity – a preference for known risks over unknown risks.

Black Swan – an unpredictable event having severe consequences.

Cascading Responsibility – system of accountability wherein each manager is effectively a CEO for his or her sphere of responsibility.

Centralized Traditional Risk Management – an organizational risk management approach wherein one or a few individuals within the business hold the primary responsibility for all organizational risk management and insurance issues.

COSO – Committee of Sponsoring Organizations of the Treadway Commission formed in 1985 to sponsor the National Commission on Fraudulent Financial Reporting (later became popularly known as the Treadway Commission), an independent private-sector effort that studied the causal factors that can lead to fraudulent financial reporting. It was sponsored jointly by the American Accounting Association (AAA), the American Institute of Certified Public Accountants (AICPA), Financial Executives International (FEI), the Institute of Internal Auditors (IIA), and the Institute of Management Accountants (IMA). Representatives from industry, public accounting, investment firms, and the New York Stock Exchange sit on the Commission.

Decentralized Traditional Risk Management – an organizational risk management approach wherein responsibility for risk management is placed directly onto business unit leaders, making each of these functional leaders accountable for managing risks related to their key areas of responsibility.

Environmental, Social, Governance (ESG) – a combined category of organizational initiatives intended to reduce reputational and social responsibility risks of an organization.

Intangible Asset – an item of economic, commercial, or exchange value that is not physical in nature, such as customer goodwill, brand recognition, patents, trademarks, copyrights, and innovative ideas/processes.

Key Risk Indicator – metrics used by organizations to provide an early signal of increasing risk exposures in various areas of the enterprise.

Organizational Ethics – the principles and standards by which a business operates.

Organizational Integrity – the ethical adherence of the individual actors, the ethical quality of their interaction, as well as that of the overarching norms, activities, decision-making procedures, and results within a given organization.

Physical Assets – an item of economic, commercial, or exchange value that has a material existence, such as properties, equipment, and inventory (also known as tangible assets).

Reputational Risk – the potential loss to financial capital, social capital, and/or market share resulting from damage to a firm's reputation, often measured in lost revenue, increased operating, capital or regulatory costs, or destruction of shareholder value.

Risk Appetite – the total amount and type of risks that an organization decides to pursue, maintain, or adopt.

Risk Attitude – the decision response to a risk-return tradeoff.

Risk Aversion – an expectation, over the long term, to be compensated for taking risk, or a willingness to pay an extra over the long-run value of a risk to lay off (transfer) the risk to someone else.

Risk-bearing Capacity – the maximum potential impact of a risk event that an organization could financially withstand or the maximum level of risk that an organization can assume without violating regulations.

Risk Neutrality – neither a requirement of nor a willingness to pay a premium over and above the long-term average value of the risk-return tradeoff.

Risk Perception – the subjective view of potential loss likelihood and loss impact, and is informed not only by uncertainty but also according to personal values and interests.

Risk Profile – the set of risks that may affect all or part of an organization.

Risk Seeking – a willingness to pay a long-term expected premium to take on risk.

Risk Silo – a set of risks being considered (and managed) in isolation from consideration of other organizational risks.

Risk Tolerance – the level of variation that the entity is willing to accept around specific objectives.

Root Cause Analysis – reviewing a risk event that has affected the organization in the past (or present) and then working backward to pinpoint intermediate and root cause events that led to the ultimate loss or lost opportunity.

Sustainability – making, using, offering for sale or selling products and services that meet the needs of the present without compromising the ability of future generations to meet their own needs.

Triple Bottom Line Accounting – an accounting framework that incorporates environmental, social, and financial-governance performance of an organization in order to include the value of these initiatives in the overall business value.

Value Killer – a risk event that destroys 20% or more of organizational value in one month relative to the growth or decline of the MSCI (Morgan Stanley Capital International) All World Market Index in the same period (as defined by Deloitte).

OVERVIEW

The kinds of errors that cause plane crashes are invariably errors of teamwork and communication.

– Malcolm Gladwell

When disaster strikes or an organizational failure occurs, it is far easier to blame people than organizations. Top management may be asked to make public apologies, but the system design of process and assumptions within which they operate is less visible and because it is indeed not a person, also less accountable. Despite the individual responsibility that may be taken by those people with top decision authority, often system failures occur in an environment of intelligent people who are well intentioned. In any organization, risks must be calculated, then some must be taken. Otherwise, no activity – good or bad – occurs.

Chapters 1 and 2 viewed water sector (and utility) risk management primarily through the lens of water as critical infrastructure, subject to disasters. While this is an important place to begin learning about risk management in a water utilities context, it neglects a broader system of risk and relationships, value creation, and preservation. Chapter 3 thus turns to these concepts, and explores risk management of a water utility as a business venture, whether for profit, not-for-profit, or governmental. The discussion begins with the evolution of risk management practices to where they stand as of this writing – a system of dynamic decision processes referred to as enterprise risk management (ERM), which was referenced briefly in Chapter 1.

AN ENTERPRISE-WIDE PERSPECTIVE ON RISK MANAGEMENT

Traditionally, organizations managed risks through a specialist approach, either centralized or decentralized. The traditional, centralized approach was to find or hire one or a few individuals within the business who would hold the primary responsibility for risk management and insurance issues. These individuals would then serve in an advisory capacity to the business units to help them manage their individual unit risks, within a broader organizational framework. In the traditional, decentralized approach, the organization placed responsibility directly onto business unit leaders, making each of these functional leaders accountable for managing risks related to their key areas of responsibility. For example, the Chief Technology Officer (CTO) was responsible for managing risks related to the organization's information technology (IT) operations, the Treasurer was responsible for managing risks related to financing and cash flow, the Chief Operating Officer (COO) was responsible for managing production and distribution, and so on. Regardless of whether employing a centralized or decentralized approach, this traditional view of risk and its management suffered from significant limitations. These traditional approaches to risk management have in common that they result in risk silos, each of which contains a set of risks and risk management decisions that are considered in isolation from other organizational risks.

In the aftermath of the Enron and other financial market scandals of the early 2000s, the Committee of Sponsoring Organizations of the Treadway Commission (COSO) advanced a definition of risk management that effectively made its meaning interchangeable with ERM [12]:

> a process, effected by an entity's board of directors, management and other personnel, applied in a strategy setting and across the enterprise, designed to identify potential events that may affect the entity, and manage risk to be within its risk appetite, to provide reasonable assurance regarding the achievement of entity objectives.

The COSO definition reflects certain fundamental concepts; in particular, risk management is [19,21,23]:

- A process, ongoing and flowing through an entity;
- Effected by people at every level of an organization;
- Developed at a strategic level;
- Applied across the enterprise, at every level and unit, using a portfolio view of risk;
- Designed to identify potential events that, if they occur, will affect the entity and to manage the risk of these events within its risk appetite;
- Able to provide reasonable assurance to an entity's management and board of directors;
- Geared to achievement of objectives in one or more separate but overlapping categories.

Today, conversations and work effort in the arena of risk management have become so intertwined with the enterprise-wide perspective as to render any function-specific or insurable versus uninsurable view on risk management an incomplete organizational model at best. Best practice risk professionals now consider all categories of potential loss to which business value is exposed as part of an effective risk management program. The objective of enterprise risk management is to develop a holistic, portfolio view of the organization's significant risks to the achievement of the entity's important goals and objectives. ERM thus attempts to recognize and respond to the basket of risks that might have an impact – both positively and negatively – on the health and performance of the business. Several key advantages of an ERM approach over a traditional approach are offered below.

ADVANTAGES OF AN ERM APPROACH TO RISK MANAGEMENT [21,25,26]

Observe the Gaps

There may be risks that "fall between the silos" that none of the silo leaders can see. Risks do not follow management's organizational chart and, as a result, they can emerge anywhere in the business. As a result, a risk may be on the horizon that does not capture the attention of any of the silo leaders causing that risk to go unnoticed until it triggers a catastrophic risk event. For example, none of the silo leaders may be paying attention to demographic shifts occurring in the marketplace whereby population shifts toward large urban areas are happening at a faster pace than anticipated. Unfortunately, this oversight may drastically impact the strategy of a water system that underestimates and/or is ill prepared for the level of loss sustained by customers in areas that are thought to be outlying suburbs, rural areas, or small cities.

Recognize Overlaps and Interconnections

Some risks affect multiple silos in different ways. So, while a silo leader might recognize a potential risk, he or she may not realize the significance of that risk to other aspects of the business. A risk that seems relatively innocuous for one business unit, might actually have a significant cumulative effect on the organization if it were to occur and impact several business functions simultaneously. For example, the head of compliance may be aware of new proposed regulations that will apply to businesses operating in Brazil. Unfortunately, the head of compliance discounts these potential regulatory changes given the fact that the company currently only does business in North America and Europe. What the head of compliance doesn't understand is that a key element of the strategic plan involves entering into joint venture partnerships with entities doing business in Brazil and Argentina, and the heads of strategic planning and operations are not aware of these proposed compliance regulations.

Respond Insightfully

In a traditional approach to risk management, individual silo owners may not understand how an individual response to a particular risk might impact other aspects of a business. In that situation, a siloed manager might rationally decide to respond in a particular manner to a certain risk affecting his or her silo, but in doing so that response may trigger a significant risk in another part of the business. For example, in response to growing concerns about cyber risks, the IT function may tighten IT security protocols but in doing so, employees and customers find the new protocols confusing and frustrating, which may lead to costly "work-arounds" or even the loss of business.

Respond to External Threats and Opportunities

So often the focus of risk management focuses eye internally, for identifying and responding to risks. That is, management might emphasize risks related to internal operations at the expense of seeing the organization as a citizen of its physical, societal, competitive, and regulatory environment. For example, an entity may not be monitoring a competitor's move to develop a new technology that has the potential to significantly disrupt how products are used by consumers.

Recognize Emerging and Evolving Risks

Despite the fact that most business leaders understand the fundamental connection of "risk and return," business leaders sometimes struggle to connect their efforts in risk management to strategic planning. For example, the development and execution of the entity's strategic plan may not give adequate consideration to risks because the leaders of traditional risk management functions within the organization have not been involved in the strategic planning process. New strategies may lead to new risks not considered by traditional silos of risk management. Many organizations think they already have the means in place to manage their risks, when in reality, they are vulnerable to emerging, off-the-radar situations that could affect them.

ERM as a Proactive and Continuous Process

An effective ERM process should be an important strategic tool for leaders of the business. Insights about risks emerging from the ERM process should be an important input to the organization's strategic plan. As management and the board become more knowledgeable about potential risks on the horizon they can use that intelligence to design strategies to nimbly navigate risks that might emerge and derail their strategic success. Proactively thinking about risks should provide competitive advantage by reducing the likelihood that risks may emerge that might derail important strategic initiatives for the business and that kind of proactive thinking about risks should also increase the odds that the entity is better prepared to minimize the impact of a risk event should it occur.

Because risks constantly emerge and evolve, it is important to understand that ERM is an ongoing process. Unfortunately, some view ERM as a project that has a beginning and an end. While the initial launch of an ERM process might require aspects of project management, the benefits of ERM are only realized when management thinks of ERM as a process that must be active and alive, with ongoing updates and improvements. risk management is both a qualitative and quantitative problem-solving approach that uses various tools of assessment to work out and rank risks for the purpose of assessing and resolving them. When done well, the overall risk management process is circular and continuous, and necessarily requires the risk management professional to: (1) set risk management objectives; (2) identify risks; (3) analyze the risks; (4) evaluate and develop appropriate strategies for managing the risks; (5) implement the strategy and monitor strategic performance; and (6) repeat all of the above at regular, frequent intervals [19,21,25,26].

Assessment of Risks within an Opportunity Framework [8,22,23]

Because ERM seeks to provide information about risks affecting the organization's achievement of its core objectives, it is important to apply a strategic lens to the identification, assessment, and management of risks on the horizon. An effective starting point of an ERM process begins with gaining an understanding of what currently drives value for the business and what is in the strategic plan that represents new value drivers for the business. To ensure that the ERM process is helping management keep an eye on internal or external events that might trigger risk opportunities or threats to the business, a strategically integrated ERM process begins with a rich understanding of what's most important for the business's short-term and long-term success.

Consider a public water/wastewater system. A primary objective may be to provide clean drinking water at least cost. In that context, ERM should begin by considering what currently drives value for the business [e.g., what are the entity's key services, what differentiates the entity from substitute services (if any), what are the unique operations that allow the entity to deliver products and services]. In addition to thinking about the entity's primary value drivers, ERM also begins with an understanding of the organization's plans for growing value through new strategic initiatives outlined in the strategic plan (e.g., launch of a new product, pursuit of the acquisition of a competitor, or expansion of online offerings etc.). With this rich understanding of the current and future drivers of value for the enterprise, management is now in a position to move through the ERM process by next having management focus on identifying risks that might impact the continued success of each of the key value drivers. How might risks emerge that impact a "crown jewel" or how might risks emerge that impede the successful launch of a new strategic initiative?

Using this strategic lens as the foundation for identifying risks helps keep management's ERM focus on risks that are most important to the short-term and long-term viability of the enterprise.

Sometimes the emphasis on identifying risks to the core value drives and new strategic initiatives causes some to erroneously conclude that ERM is only focused on strategic risks and not concerned with operational, compliance, or reporting risks. That's not the case. Rather, when deploying a strategic lens as the point of focus to identify risks, the goal is to think about any kind of risk – strategic, operational, compliance, reporting, or whatever kind of risk – that might impact the strategic success of the enterprise. As a result, when ERM is focused on identifying, assessing, managing, and monitoring risks to the viability of the enterprise, the ERM process is positioned to be an important strategic tool where risk management and strategy leadership are integrated. It also helps remove management's "silo-blinders" from the risk management process by encouraging management to individually and collectively think of any and all types of risks that might impact the entity's strategic success.

Adapting is all about positioning companies to quickly recognize a unique opportunity or risk and use that knowledge to evaluate their options and seize the initiative either before anyone else or along with other organizations that likewise recognize the significance of what is developing in the marketplace. Failing to adapt can be fatal in today's complex and dynamic public water environment.

An ERM framework enables directors and senior management, as well as internal and external stakeholders to communicate more effectively. This enhanced risk-focused communication facilitates discussion about issues important to improving governance, assessing risk, designing risk responses and control activities, facilitating relevant information and communication flows, and monitoring ERM and internal control performance.

LEADERSHIP OF ERM

Given the goal of ERM is to create a top-down, enterprise view of risks to the entity, responsibility for setting the tone and leadership for ERM resides with executive management and the board of directors. They have the enterprise view of the organization and they are viewed as being ultimately responsible for understanding, managing, and monitoring the most significant risks impacting the business.

Top management is responsible for designing and implementing the enterprise risk management process for the organization. They determine what process should be in place and how it should function, and they are the ones tasked with keeping the process active and alive. The board of director's role is to provide risk oversight by (1) understanding and approving management's ERM process; and (2) overseeing the risks identified by the ERM process to ensure management's risk-taking actions are within the stakeholders' appetite for risk taking.

The overall purpose is to help the enterprise achieve its objectives and to optimize the enterprise's value. The board of directors and senior management have two very different roles. It is the duty of the board to approve strategic decisions, establish boundaries, and oversee execution. Senior management aligns strategy, processes, people, reporting, and technology to accomplish the organization's mission in accordance with its established values. The board sets boundaries in relation to risk. ERM instills within the organization a discipline around management and discusses opportunities and risks and how they are managed. The executive management and board members embody the commitment to integrity and ethical values. The board demonstrates oversight, independence, and objectivity. The management establishes reporting lines; the organization's commitment to competent individuals; and the organization's commitment to accountability. A business is only as good as its people, and these principles support that theory.

The tone at the top of the organization is critical to the effectiveness and long-term success of an ERM program. To achieve and maintain a tone that gives the ERM design life, often territorial barriers (of former operational and risk silos) must be overcome. Management must encourage and be inclusive of a diversity of thought and viewpoints, and must promote imagination rather than only rationality.

Responsibility for ERM [12,15,34]

Governance of the ERM process must ultimately reside at the top level of the organization, while accountability must reside at all levels. The COSO ERM concept defines risk management using a top-down approach that segments managerial responsibilities according to hierarchical positions. It states that each manager should be accountable to the next higher level for his or her portion of enterprise risk management, with the CEO ultimately accountable to the board. COSO ERM states that accountability develops from everyone in the organization knowing their responsibilities and contributing to the overall mission, vision, and objectives of the firm. To understand the COSO view of ERM accountability, visualize the hierarchy as a system of "cascading responsibilities," which meld together and thereby contribute to an organization's overall objectives; in other words, managerial alignment of organizational roles and tasks will ensure accountability. Following COSO's ERM model, in order to assemble risk management as an enterprise-wide activity, organizations must develop a code of compliance to measure whether everyone adheres to it and acts in accordance with their individual set of defined roles and responsibilities.

In addition to basic accountability, COSO stresses a clear separation of duties, checks, and balances. For example, it suggests measuring first the alignment of managerial achievement with a code of conduct and subsequently the contribution of the individual to the firm overall (i.e., the entire enterprise as one system). COSO ERM promotes the assumption that auditability is beneficial to the firm, and risk management work is represented in terms of formalized and disclosure-oriented types of work. COSO stresses the importance of alignment but does not address what alignment work entails, how it unfolds, and in relation to which particular agendas and/or objects it is produced. It depicts a cohesive entity, which builds its cohesiveness through a number of building blocks.

An organization in essence designs its own ERM puzzle, shaping the consistency of goals and objectives (and accountability for these) through how the managerial building blocks are put in place. ERM prescribes alignment of goals and objectives, while each entity must determine for itself the decomposition of enterprise-wide relationships into controllable building blocks, or areas of responsibility, to ensure that alignment (and accountability) operate as desired. ERM ultimately must address the management of uncertainties, and thus ERM in practice must erect a system of accountability that motivates this work.

The Value of Practice Variation and Cross-Functional Discussions

COSO ERM conceptualizes organizations as homogeneous and coherent organizations which define their risk management program consistently while being anchored with top management. Nevertheless, we know that divisions and professionals throughout firms may mobilize risk management in different ways. By design, the risk management process engages people at all levels with the identification and "problematization" of risks. The risk management process itself does not necessarily produce a common understanding of the organization's current status, its risks or risk characteristics (e.g., likelihood, impact, velocity of impact) because many different parties attempt to engage with these, if not daily then at least on a periodic basis.

A myopic perspective – although providing advantages when working through details – can yield inferior results in ERM planning. For instance, teams in different areas of the company, working in isolation, could easily spend most of their risk management work time on the problematization of the risks that have already been identified and documented – hashing and rehashing the risks of which they are aware and have some knowledge. Due to a natural human aversion to ambiguity [5], just about any absence of certainty – documented as either an identified risk with a greater-than-zero chance of occurring – could matter enough to warrant their attention [16,26]. Cross-functional team meetings and shared documentation can raise awareness of the lesser understood threats. Risk ambiguity can in many instances be reduced, but it is must first be brought to light. In so doing, teams also invariably identify, document, and debate risks of which they were otherwise not aware.

ERM AND ACCOUNTING

For accountants it may be appealing to think of ERM as possessing information hierarchies similar to those found in accounting. In the sense that the ERM structure can be (a) decomposed in stable, mobile, and combinable information units, and (b) assigned to individual responsibilities given hierarchical and functional characteristics of an organization, there is indeed a striking similarity to accounting. Nevertheless, ERM differs fundamentally from accounting. Unlike accounting, ERM documents uncertainty – the absence of certainty. Also unlike accounting, ERM "ledgers" do not add up and balance in a familiar manner. ERM systems instead reveal that uncertainty creates space for the unfamiliar, the possible, and the ambiguous that otherwise are difficult to observe in business operations.

In the face of the unfamiliar, however, accountants must play a significant role within the ERM risk process. Accountants inherently mitigate risk – with or without an ERM structure in place – via the operational information exposed within costing and preparation of financial statements. But accounting is needed also to promote and facilitate ethical and intelligent risk taking in support of long-term value creation and preservation. According to a 2019 report from the International Federation of Accountants, accounting professionals can contribute to organizational ERM through "a deep understanding of the business," "enhanced quantitative and statistics skills," and "the ability to lead and communicate across teams" [20].

ERM AND TRADITIONAL RISK MANAGEMENT [19,21,23]

Although effective ERM in concept involves a holistic and integrated view of organizational risks while traditional risk management sees risks with a more isolated perspective, this is not to say that ERM has nothing to learn from traditional risk management. Indeed, it can learn much from the mindset of the traditional risk manager. Indeed the "insurance manager" approach remains highly useful as a guide for how to enable the organization to simultaneously create and preserve value. Historically, the traditional insurance manager may have likely focused on one problem at a time (unlike today's risk manager, who is beset with a flurry of risks all at once), but nevertheless the best approach is the same regardless of whether the luxury of time and simplicity are available. The intention is to solve the problem(s) of how to enable the organization to take an action that holds value potential without taking more risk than that action is worth (if feasible, without taking any risk) in doing so. A traditional risk manager might not "make the call" as to whether the organization would manufacture and market a new product; he or she would most likely accept the decision if already approved by the heads of marketing and finance. The primary problem for the risk manager to solve would not be whether to act on the new product, but rather how best to plan for the new product's development and marketing in such a way as to mitigate any risks involved.

In a sense, ERM has always existed in the best operations; the responsibility for decision-making was shared across the organization and risk specialists contributed to value creation and protection by recommending (and typically implementing) risk management strategies. ERM and traditional risk management then are not far distant from one another. It is primarily the top-down approach of ERM, and fine alignment of the drill-down risk processes with the high-level strategic risk management plan that differs from the traditional approach. The assessment and treatment of a challenge, once identified, occurs much in the traditional way – with an eye toward enabling healthy and effective business operations. The identification and treatment of unhealthy and ineffective business operations occur as an inherent by-product of best practices risk management.

GUARDING AGAINST ERM AS THE MANAGEMENT OF EVERYTHING AND ALSO NOTHING

The basic concept of ERM – that risk processes should be clearly and explicitly aligned with organizational goals and objectives – seems simple enough. The design has been referred to as analogous to that of a thermostat which adjusts to changes in environment, subject to a predetermined target temperature. This overall rationale for ERM design has existed since the 1990s, and was developed further by COSO 2004, coming on the heels of renewed concerns about fraudulent financial reporting among US companies [9]. The COSO 2004 design details are traceable back to the accounting convention of internal control, or audit, which itself is traceable back to the use of control theory in an engineering and manufacturing context. So the ERM in practice today, if structured simply to be COSO compliant, is strongly influenced by internal audit norms, with an emphasis on process rather than on strategy. This is problematic, and arguably a horribly inferior outcome of the original ERM concept.

At least three problems present when process takes precedence over strategy. First, the organizational canopy view becomes diluted by its own documentation. The process-based organization can only see and manage on an enterprise-wide basis to the extent that the entire enterprise's opportunities and risks are not only documentable, but documented (presumably in some easy-to-audit way). Second, the essential risk management value of ERM becomes diluted by an insistence on process auditing. To the extent that detailed process-based rules for risk management are embedded in the organization's ERM design, the ERM program may strategically over-reach and risk becoming the risk management of everything. Before the reader jumps to the quick conclusion that the risk management of everything is a good thing, beware the third problem with the process-focused approach to ERM – that it risks becoming the risk management of nothing. If the responsibility (and blame) for compliance (and noncompliance) ultimately falls to an individual or small number of individuals, then the promise of ERM falls far short of being the full integration of risks and management, and indeed reduces to a sort of circular management of nothingness.

ERM has been compared with accounting in this chapter. While accounting is a critical component in an effective ERM program, an accounting-driven ERM blueprint which uses a controls-based approach, will have an inherent tendency to elevate detailed controls and their corresponding documents trails (i.e., processes) incorrectly to the strategic level. Since Sarbanes-Oxley (SOX) legislation was passed in 2002, demands under its section 404 for "evidence of effective controls over financial statements" have been amplified by many firms, emphasizing these over organizational strategies. Indeed, an "audit trail" and "box-ticking" mentality has permeated the ERM programs of most public and governmental entities despite later attempts by SOX authors to clarify and soften these demands. Excessive reliance on rules-based compliance increases the costs of doing business, overshadows the essence of good governance, creates an organizational culture that is rules dependent, and can result in legal absolutism. In the extreme, an organization may go to a great deal of expense to adhere to the "letter of the law," and wind up making decisions that erode rather than build organizational value [12].

STRATEGIC PERSPECTIVE AND OBJECTIVE SETTING

The overarching goal of an ERM program is to manage risks and uncertainties which have the potential to adversely impact the organization's ability to meet its business goals and objectives. Essentially, it is the entity's goals and objectives that inform what the ERM goals and objectives must be. In setting objectives, strategic and related objectives are established and the entity's risk position (risk appetite plus risk tolerances) is considered. A well-articulated risk appetite statement that is communicated effectively to operating units can provide clarity and focus to the business planning process. An important result of ERM objective setting, risk tolerances can be an effective tool in this regard if they are sufficiently granular and expressed in such a way that they can be: (a) mapped into the same metrics the organization uses to measure success in achieving an objective;

(b) applied to all categories of objectives (strategic, operations, reporting and compliance); and (c) implemented by operating personnel throughout the organization.

DEVELOPMENT OF ERM GOALS AND OBJECTIVES

The goal of an ERM process is to generate an understanding of the risks that management collectively believes are the current critical risks to the strategic success of the enterprise. One of the elements of the ERM internal environment is the risk management philosophy, which is the set of shared beliefs and attitudes characterizing how the entity considers risk in everything it does, from strategy development and implementation to its day-to-day activities. It communicates the importance of risk throughout the entire entity. Every organization has a risk management philosophy. Characteristics to consider of the philosophy is if it is developed, if it is implicit or explicit, and how its personnel understand and cooperate in its culture.

KEY DETERMINANTS OF RISK MANAGEMENT PHILOSOPHY [23,24]

The determination of an organization's risk management philosophy, as well as whether and how a particular risk fits into this context, depends on several factors and metrics – namely, the organization's risk profile, risk perception, risk attitude, risk appetite, risk-bearing capacity, and risk tolerance. Each of these issues should be considered to assess the organization's risk level as well as to evaluate the appropriate responses to take in light of its risks.

Risk Profile, Perception, and Attitude

A risk profile is the set of risks that may affect all or part of an organization. This profile includes as much objective risk information as possible, such as probability and potential financial impact. The risk profile, including individual risks as well as the whole portfolio of risks, has a subjective impact on the individuals considering them as well. This is because individuals within the organization each have personal risk perceptions that consider potential loss likelihood and loss impact in potentially differing ways. Collectively, these subjective risk perceptions create the entity's overall risk perception, and are informed not only by uncertainty but also according to values and interests.

Risk attitude is the response to a risk-return tradeoff, and so must be informed by both the objective risk profile and the subjective risk perception. According to the traditional economic theory of expected utility, three categories of risk attitudes exist – risk aversion, risk seeking, and risk neutrality. Risk-averse individuals or organizations must expect, over the long term, to be compensated for taking risk, or put differently, are willing to pay a "premium" (extra) over the long-run value of the risk to lay off (transfer) the risk to someone else. An example of risk aversion is the insurance purchase, which nearly always involves payment of a premium that is greater than the long-term average value of the risk because of the possibility of a high loss. Risk seekers, conversely, will pay a long-term expected premium to take on risk. An example is a gambler, who can expect over the long run, on average, to lose money, but prefers to take the risk because of the possibility of a high return. Risk neutral entities neither require nor are willing to pay a premium over and above the long-term average value of the risk-return tradeoff. An example is a stock purchase at par.

Risk aversion is the risk attitude most commonly and consistently attributed to organizations interested in effective risk management. Although some theories discuss the possibility that large, publicly traded organizations may be risk neutral in their decision-making, at least at times, the prevailing wisdom is that all organizations are prudent to behave in a risk-averse manner whenever the costs of a worst-case outcome could be financially or reputationally catastrophic.

Risk Appetite, Risk-bearing Capacity, and Risk Tolerance

Risk appetite is the total amount and type of risks that an organization decides to pursue, maintain or adopt. If an organization is particularly effective in managing certain types of risks, it may be willing to take on more risk in that category; conversely, it may not have any appetite in that area. Organizations will have different risk appetites depending on their sector, culture, and objectives. A range of appetites exist for different risks and these may change over time. While risk appetite will always mean different things to different people, a properly communicated, appropriate risk appetite statement can actively help organizations achieve goals and support continued viability.

Risk appetite reflects the enterprise's risk management philosophy, and in turn influences the entity's culture and operating style. Risk appetite is usually established in a risk appetite statement; which frames the risks the organization should accept, the risks it should avoid, and the strategic, financial, and operating parameters within which the organization should operate. Risk appetites are a fine balance. They must be flexible enough to respond to changes in the business environment, but not so flexible that the appetite is constantly changing. If an entity is altering the appetite frequently, it will lose value. Risk appetite is fundamental to any governance process that seeks to appropriately balance the organization around value creation and value protection.

Risk-bearing capacity represents the maximum potential impact of a risk event that an organization could financially withstand or the maximum level of risk that an organization can assume without violating regulations. Although in theory, an otherwise strong risk appetite is limited only by risk-bearing capacity, often in practice appetite will be well below capacity.

Risk tolerance is the level of variation that the entity is willing to accept around specific objectives. Objectives are usually quantitative, but may be qualitative. Risk appetite is about the pursuit of risk. Risk-bearing capacity is concerned with the financial and/or regulatory ability to pursue risk. Risk tolerance is at the intersection of these concepts since it is about the level of risk an organization is willing to accept, subject to the appetite and capacity to bear risk.

KEY RISK INDICATORS

Once the risk management philosophy is established, most organizations prioritize what management believes to be the top ten (or so) risks to the enterprise. Generally, the presentation of the top ten risks to the board focuses on key risk themes, with more granular details monitored by management. For example, a key risk theme for a business might be the attraction and retention of key employees. That risk issue may be discussed by the board of directors at a high level, while management focuses on the unique challenges of attracting and retaining talent in specific areas of the organization (e.g., IT, accounting, operations).

With knowledge of the most significant risks on the horizon for the entity, management then seeks to evaluate whether the current manner in which the entity is managing those risks is sufficient and effective. In some cases, management may determine that they and the board are willing to accept a risk while for other risks they seek to respond in ways to reduce or avoid the potential risk exposure. When thinking about responses to risks, it is important to think about both responses to prevent a risk from occurring and responses to minimize the impact should the risk event occur.

While the core output of an ERM process is the prioritization of an entity's most important risks and how the entity is managing those risks, an ERM process also emphasizes the importance of keeping a close eye on those risks through the use of key risk indicators (KRIs). Organizations are increasingly enhancing their management dashboard systems through the inclusion of KRIs linked to each of the entity's top risks identified through an ERM process. These KRI metrics help management and the board keep an eye on risk trends over time.

Key risk indicators are metrics used by organizations to provide an early signal of increasing risk exposures in various areas of the enterprise. In some instances, they may represent key ratios that management throughout the organization track as indicators of evolving risks (and potential opportunities) which signal the need for actions that need to be taken. Others may be more elaborate and involve the aggregation of several individual risk indicators into a multi-dimensional score about emerging events that may lead to new risks or opportunities.

Therefore, the selection and design of effective KRIs start with a firm grasp of organizational objectives and risk-related events that might affect the achievement of those objectives. Linkage of top risks to core strategies helps pinpoint the most relevant information that might serve as an effective leading indicator of an emerging risk. Mapping key risks to core strategic initiatives puts management in a position to begin identifying the most critical metrics that can serve as leading key risk indicators to help them oversee the execution of core strategic initiatives. KRIs related to legal and regulatory requirements, aging infrastructure and financial pressures have loomed large for utilities, and more specifically the water/wastewater sectors, in recent years [1,7,14].

KRI Factors to Consider

In addition to the consideration of risk attitudes, appetites, and tolerances, several factors should be considered when determining the KRIs. The nature and types of consequences that can occur impact how the organization will choose to describe the relevant indicators. Another set of important factors are related to how likelihood is defined and depicted. Is it a probability, rate of occurrence, or something qualitative, such as the designations of "rare," "sometimes," and "often." The immediacy of the risk and development of the consequences is also key, and may be as important as the consequences themselves. Once risk consequences, likelihood, and timing are under consideration, the impact of these needs to be measured in financial and/or operational terms. Although this may imply dollar values, the pertinent metrics for some criteria may instead be scientific, engineering, social, or compliance/governance in nature. Moreover, the impact can vary based on combinations of risks that could affect the organization – either on a related or unrelated basis – close together in time or situation. Finally, the impact should be thoroughly vetted, and potentially be redescribed, as to its effects on organizational reputation and/or sustainability.

Root Cause Analysis

An effective method for developing KRIs begins by analyzing a risk event that has affected the organization in the past (or present) and then working backward to pinpoint intermediate and root cause events that led to the ultimate loss or lost opportunity. The goal is to develop key risk indicators that provide valuable leading indications that risks may be emerging. The closer the KRI is to the ultimate root cause of the risk event, the more likely the KRI will allow for time to proactively take responsive action. Management can then use that analysis to identify information associated with the root cause event or intermediate event that might serve as a KRI related to either event. When KRIs for root cause events and intermediate events are monitored, management is in an enviable position to identify early mitigation strategies that can begin to reduce or eliminate the impact associated with an emerging risk event.

Underlying Strategic Risks

Strategic risks many times may lack historical precedent, and signals related to that risk are often weak, making them harder to uncover, and difficult to interpret. In addition, strategic risks can also be unique to one organization because of the strategy in place. Risks are also easier to overlook when they seem irrelevant, and can be difficult to address with customary risk methods. Some underlying strategic risks are knowable (and arguably should be known) because they are easily imaginable, even if they have not occurred yet (e.g., extreme weather events or catastrophic power failures). Others are less easily imaginable but are possible (even if only remotely). Such risks produce "black swan" events, which are extremely rare, unpredictable events having severe consequences when they occur. Black swans can cause catastrophic damage not only to an organization, but to an entire economy, and because they cannot be predicted, can only be prepared for by building robust systems [23,29].

As well as being hard to identify, strategic risks also tend to be difficult to quantify and track. If ignored, these risks can become "value killers" whereas if treated appropriately, they can become the drivers of value. Value killers can be categorized into five types: (1) High-impact/low-frequency risks; (2) Correlated or interdependent risks; (3) Liquidity risks; (4) Merger and acquisition risks; and (5) Culture and compensation risks.

OPEN-MINDED RISK PERSPECTIVE [4]

Recognizing the importance of strategic risks can be hard for an organization to convey to their employees and other stakeholders. A workplace that not only allows but encourages employees to shift their thinking away from the traditional, predictive view can benefit from a more insightful (and thereby also more successful) ERM program. Promote a broad view of risk and interconnectedness of risks over the traditional siloed view of financial and operational risks. Spend as much time thinking about (and responding to) unknown (and seemingly unknowable) risks as you do on known, fairly predictable risks faced by the organization. Encourage employees to be curious about risk rather than fear it. Scan for emerging risks, utilizing an outside-in view of the entity rather than the traditional, inside-out view.

Such an open-minded perspective on strategic risk and risk management objective setting results in the most proactive and creative approaches to the balance of the risk management process. First, employees can cohesively work together cross-functionally to identify, detect, monitor, and address risks that emerge. Second, risks are well understood, and well-informed strategic decisions can be made, including the less quantifiable and harder to predict. Third, risk infrastructure is aligned with the business strategy in place, which also enhances the lines of defense for risk governance discussed earlier in this module. Fourth, the entity's capital allocation decisions are aligned with your risk appetite, even as your risk profile changes. Finally, regulatory and compliance issues are considered, when deciding your risk management strategy, as cultural and governance resources rather than as obstacles or restrictions.

Think of the combination of these perspectives as having the "critical imagination of alternative futures." Rules-based compliance has its place for specific and necessary processes; it offers a regulated transparency to the risk management process that even a well-intentioned "imagination" does not. The attempt here is not to teach against rules and compliance within the rules, but rather not to focus on rules so exclusively that one risks "anchoring" risk management decisions to these rules. Such a mindset is expensive and potentially distracting from the intrinsic motivation to do better than the rules require. The drawbacks of an excessively heavy reliance on rules-based approaches – increased costs of doing business, overshadowing the essence of good governance, creating a culture of rules dependency and legal absolutism – are unintended consequences of some ERM designs. An open-minded perspective means that the way forward for effective organizational ERM is to strike an optimal balance between rules-based and principles-based approaches.

CURRENT STATE OF ERM AND RISK OVERSIGHT

It is incumbent upon the authors of a text that promotes best practices in the context of an evolving spectrum of practice to realistically assess for the reader how far evolved the world is with respect to best practice. In the ERM context, business leaders and other key stakeholders (e.g., governmental policy-makers, rating agencies, regulators, etc.) are recognizing the increasing complexities and challenges of risk oversight. A number of organizations have embraced the concept of ERM, but many still have not. And of those which have, most report significant challenges in achieving "best practice" ERM. The North Carolina State University Enterprise Risk Management Initiative, in partnership with the American Institute of Certified Public Accountants, periodically conducts a survey of US organizations to gain their perspectives on the state of risk oversight within their respective organizations [6]. A summary of ten key findings from the 2018 report are provided in Table 17.1.

TABLE 17.1

Summary of 2018 Key Risk Oversight Observations

1. Risk complexity	Most respondents (60%) believe the volume and complexity of risks is increasing extensively over time. And, 65% of organizations indicate they have recently experienced an operational surprise due to a risk they did not adequately anticipate.
2. Management pressure	Most boards of directors (68%) are putting pressure on senior executives to increase management involvement in risk oversight. Strong risk management practices are becoming an expected best practice. These pressures are getting harder and harder for senior executives to ignore.
3. ERM program status	22% of respondents describe their risk management as "mature" or "robust" with the perceived level of maturity declining over the past two years. 31% of organizations (48% of the largest organizations) have complete ERM processes in place.
4. Use of CROs	67% of large organizations and 63% of public companies report designating an individual to serve as chief risk officer (or equivalent) has increased over time. Most of those organizations (>80%) have management risk committees.
5. ERM alignment with strategic plan	Less than 20% of organizations view their risk management process as providing important strategic advantage. Only 29% of the organizations' board of directors substantively discuss top risk exposures in a formal manner when they discuss the organization's strategic plan.
6. Mixed reports on ERM processes	45% of the organizations have a risk management policy statement, with 43% maintaining risk inventories at an enterprise level. About 40% have guidelines for assessing risk probabilities and impact. Most (75%) update risk inventories at least annually.
7. Notable top management attention to top risks	Most boards of large organizations (82%) or public companies (89%) discuss written reports about top risks at least annually; however, just 60% of those describe the underlying risk management process as systematic or repeatable.
8. Mixed reports on satisfaction with KRIs	41% of the respondents admit they are "not at all" or only "minimally" satisfied with the nature and extent of internal reporting of key risk indicators that might be useful for monitoring emerging risks by senior executives.
9. Many ERM objectives are not incentivized	The lack of risk management maturity may be tied to the challenges of providing sufficient incentives for them to engage in risk management activities. Most (66%) have not included explicit components of risk management activities in compensation plans.
10. Many report lack of resources to implement ERM programs	Respondents of organizations that have not yet implemented an enterprise-wide risk management process indicate that one impediment is the belief that the benefits of risk management do not exceed the costs or there are too many other pressing needs.

Source: NC State Enterprise Risk Management Initiative [6].

Business accountants responding to the survey, in summary, view top management as under increasing pressure to optimally manage risks that are increasing in complexity. While most report having a designated chief risk officer (CRO) or equivalent, the vast majority do not perceive the over all ERM program as yet mature, robust, or providing strategic advantage. And reviews of organizational risk reporting, management attention devoted to risk, and the value of KRIs are mixed at best.

Another study, conducted by researchers through interviews with water utility risk managers over the periods 2005, 2011, and 2015, looked at risk oversight and governance differently. Focusing on the strategic risks themselves rather than on the state of the ERM process, the researchers looked for changes over time in the perspectives of the participating utility representatives. Figure 17.1 shows the top ten strategic risks cited in each of the three years (2005, 2011, and 2015) in these interviews. (There are not 30 separate cited risks because the lists of top 10 risks each year overlapped in risks cited.)

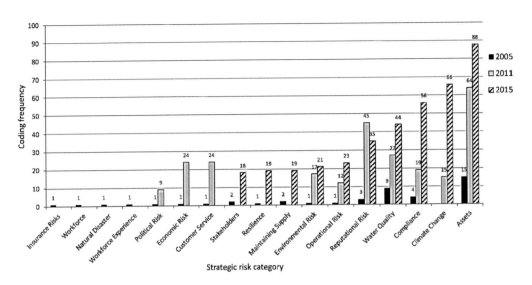

FIGURE 17.1 Histogram of strategic risks cited by water utility risk managers, 2005; 2011; and 2015 [11].

It is notable that "insurance risks" – those risks for which insurance is available and not prohibitively costly – rank lowest on this "top risks" list (in terms of frequency with which they came up in the interviews with risk managers). Thus, while insurable risks rank highly for these organizations as strategic risks (made the top ten in at least 2005, 2011, or 2015), they are outranked by largely uninsurable risks on this composite list. Rather than this ranking implies that insurable risks are not risks of the highest order for strategic planning and operations, this instead may imply that they rank lower than other top risks simply because insurance, in large part, provides a ready solution to the barriers that would otherwise exist to solving the risks.

The researchers additionally invited utility risk managers to share their insights on the barriers to managing these risks as well as how they might overcome the barriers. While issues of risk complexity/detail, lack of communication, change resistance, and lack of management buy-in all arose in these interviews, another issue also arose that may be related to, or even account for, some of the other challenges. The recurring theme that risk has a negative connotation. One participant even stated plainly, "the word risk itself suggests stress." Several interviewees indicated an awareness that the upside of risk must be demonstrated in order for many individuals outside of formal risk management to enthusiastically join in the discussion of it. These findings point again to the importance of risk as part of value creation, not just preservation.

REPUTATION RISK MANAGEMENT

Reputations result from perceptions that stakeholders have about an organization's intangible assets. Developing and implementing strategies to increase, protect, and recover the value of intangible assets, such as innovation, quality, security, safety, ethics and integrity, ethical sourcing, and environmental sustainability are essential to long-term sustainability. Customer and investor markets recognize and reward such actions. In our complex and interdependent global economy, public water/wastewater systems must go beyond protecting physical assets because cascading effects can create extreme consequences. Furthermore, small actors and individual events are stretching the boundaries of responsibility, increasing the risks faced in the course of doing business [18].

As a group, companies with higher reputation rankings have higher equity returns and lower credit costs. They have lower operating costs and higher net incomes. In rising markets, companies with superior reputations reward their investors with superior returns. In down markets, companies with superior reputations are more resilient and lose less value. The reason for this is that superior reputations pay off with pricing power, lower operating costs, greater earnings multiples, lower stock price volatility (i.e., a lower beta), and lower credit costs. When reputation is strong and resilient, untoward events are less damaging. When reputation is weak or impaired, these benefits are lost.

The "Values" of Reputation

Unfortunately, traditional accounting lacks the tools needed to measure and manage reputational assets. Traditional accounting methods are unable to measure the effects that intangible assets have on an organization's reputation and financial performance. Conventional thinking says that many vital assets are not quantifiable. If reputation assets are to be treated like physical assets, the financial community needs to make a market for them, much like markets for insurance and real estate. This asset class also needs transparency so investors and buyers know what they are getting. Triple bottom line accounting has developed during recent years as an expansion of traditional reporting to account for social, environmental, and financial-governance (ESG) performance that is difficult to quantify, such as the ecological impacts of water distribution and management or of wastewater treatment. Some organizations have adopted the TBL framework to evaluate performance in a broader perspective to enhance the business value of reputation and other ESG attributes of the organization. Managing the intangible assets of public water/wastewater systems through implementation of business processes for fostering reputational values is key to creating, preserving, and restoring reputation value. The reputational value of a public water or wastewater system is advanced by initiatives that meet the following criteria.

Create an Ethical Work Environment

Ethics are the moral principles by which a company operates; integrity is the act of adhering to those moral principles. Ethics are an integral part of governance that, when combined with integrity, affect the reputation value of all other intangible assets. Additionally, ethics are the keystone intangible asset because they form the basis for trust and confidence [4].

Drive Innovation

Innovation is the design, invention, development, and/or implementation of new or altered products, services, processes, systems, organizational structures, or business models for the purpose of creating new value for customers, and financial returns for the organization. Important keys to innovation include employee know-how and developing benefits that actually provide value to the organization.

Assure Quality

Quality is the extent to which a product is free from defects or deficiencies, the extent to which a service meets or exceeds the expectations of customers or clients, the extent to which products and services conform to measurable and verifiable criteria [13].

Uphold Safety

Safety is the state of being certain that a set of conditions will not accidentally cause adverse effects on the well-being of people or the environment.

Promote Sustainability

Sustainability means making, using, offering for sale, or selling products and services that meet the needs of the present without compromising the ability of future generations to meet their own needs.

Provide Security

Security is the degree of protection a company offers against events undertaken by actors intentionally, criminally, or maliciously for purposes that adversely affect the system or its stakeholders. It is intended to reduce the likelihood and/or magnitude of such events as well as minimize the system's residual uncertainty about the costs of security risks [9].

In consideration of initiatives that build and protect reputation value – whether the value proposition is intentional or inherent – it is important to see the difference between an organization's governance and its social responsibility. Governance is based on a set of parameters within which the organization operates – comprised of self-regulation (e.g. internal policies) and external regulation (e.g. laws and industry standards). Social responsibility views governance as a baseline level of acceptable organizational behavior, and is based on organizational beliefs and values. So an individual organization must provide self-governance to, say, control pollution, but may additionally engage in resisting employee layoffs or outsourcing simply to improve profitability, thus maintaining a commitment to the local community.

The Difficulty of Tackling the Reputation Risk Problem

The overarching solution to the challenges of reputation risk management encompasses two main elements: (1) improved visibility and control; and (2) fostering conformance with intangible asset management best practices through a system of process oversight and financial incentives. From a systems perspective, the operational and reputation benefits of improved visibility and control are not controversial. The solution is nonetheless challenging – to provide secure and reliable operational visibility and control over capital, physical assets, and the behavior of business network partners. Despite the challenge of developing and managing the ideal solution, adequate problem-solving approaches exist today that provide reasonable levels of visibility. What limits the effectiveness of these existing platforms are the reluctance of humans to conform and the difficulty of transferring any residual risk. For instance, if reputational assets are to become insurable like other classes of assets, businesses must remove doubt about the value of those assets. Indeed, recognition by the insurance industry and accounting profession that intangible assets deserve the same financial treatment as traditional assets will go a long way toward giving investors, traders, and other stakeholders an accurate picture of a company's health so that relevant information will become the basis on which markets reward (or punish) entities [17].

There is increasing pressure on US public utilities (water, electricity, gas, etc.) to provide infrastructure for continued development of the built environment while at the same time improving the natural and built environments' sustainability. The pace of technological change is increasing, while the cost of new technology overall is declining. The consumption of water and energy is flat or declining, while energy efficiency is increasing. Consumer preferences are evolving toward conservation, efficiency, and lower carbon sources of generation, while the majority of infrastructure of incumbent utility providers, including that of water systems, is aged or aging and built to produce and distribute their services along traditional, energy-inefficient lines. Thus, the economic pressures on incumbent providers are increasing. These financial pressures come at a time when the trend among regulators and the public is to place a greater burden on these entities to bear the costs of risk. Indeed, innovating into becoming the utility of the future is subject to both explicit and implicit design criteria. For public water/wastewater systems, innovation is driven by explicit criteria that are much as they have been traditionally: Provide safe, reliable and resilient water (or wastewater treatment) at an affordable and stable price. Implicit criteria for innovation are evolving quickly, and include well understood as well as ambiguous criteria [2,27,30–32].

Public water/wastewater systems understand that they are asked to promote economic development in environmentally friendly ways, subject to meeting social adequacy-responsibility objectives (even if systems still question how to measure and manage some of these values). What is less understood is the call to simultaneously satisfy vast numbers of individual customer preferences (some conflicting), utilizing a customer-centric framework, with quick responsiveness and adaptability. The business case is robust – a failure to innovate successfully is the greatest long-term reputational risk for public water/wastewater systems. (In the short-term, the biggest reputational risk is simple to understand – a safety breach that results in unsafe water.) Investments in infrastructure improvement and innovation are critical for reputation now, and for economic growth, and viability over time. Innovation (e.g., technological change) is a key element for economic growth and long-term prosperity – spawns new services, improvement of existing services, or higher efficiency of production processes. Investment in infrastructure and innovation is also critical for advancing long-term policy objectives, like safety, reliability, cheaper energy, and a cleaner environment.

Despite the importance of innovation, today there is a deficiency in research and development for public utilities, including water systems. Utilities are unlikely to innovate unless the payoff from successful innovation is large, and the investor market may also under-allocate resources to innovation because of public benefits, especially that benefits can be appropriated by others, such as competing utilities or the public at large ("free riders"). Innovation is expensive and for most systems is initiated only by technology-push or demand-pull incentives. Furthermore, it can take several years before the system recaps incurred costs via new revenues and/or other benefits. Innovation is inherently risky ("dry holes" are common) – costs and success are difficult to predict; benefits are often distant. Add to these factors the challenge that a technology can quickly become obsolete with the introduction of newer, more promising technologies. Exacerbating these traditional challenges are changes in the regulatory and liability landscapes, which impact a utility's risk attitude and appetite. Regulation affects: (1) the amount utilities spend to innovate; (2) the speed at which they innovate; (3) the nature of innovative activities; and (4) the management of R&D projects. History has shown that, depending on the operation of rate regulation, a utility could be either over-motivated or under-motivated to innovate. Utilities are both producers and consumers of innovation; research and history show two strong relationships prevail between regulation and innovation in this sector [2,27,32].

The Relationship to Innovation Motivation

When the emphasis is primarily on safety, reliability, and affordability (especially given the use of prudence tests), infrastructure improvements are given precedence over innovation investments. For instance, consider a public water/wastewater system with limited resources. If substantial repairs to equipment are needed to continue to provide safe drinking water to the community reliably and sustainably, the system may have to forego innovation opportunities in the short term as a safety and quality tradeoff [33].

The Relationship to Innovation Benefits

When benefits are allocated largely to consumers and risks are allocated largely to utilities, neither infrastructure improvements nor innovation investments may be pursued. Water system liability for wildfire damage is a prime example of such a nonoptimal allocation of risks and benefits [34]. The erosion of public service immunity and the expansion of utility liability (along with the nonrecoverable nature of liability judgments in many cases) discourage infrastructure improvements as well as innovation investments. For public water/wastewater systems in particular, these adverse impacts are exacerbated by the implicit (both the understood and ambiguous) expectations and design criteria that exist alongside these adverse developments in the regulatory and liability environment. Insurance of this expanded liability is typically helpful since insurance costs are generally recoverable. To the extent liability is deemed uninsurable or insurance costs are nonrecoverable, the negative impacts on infrastructure improvements and innovation investments can be severe. The ability to effectively manage the reputational risks of innovation (or failure to innovate) is contingent on the public water/wastewater systems's commitment to meeting the explicit demands of providing safe and reliable water as well to the rise to the implicit demands to provide water in environmentally friendly and sustainable ways, with a socially responsible focus on individual customer needs and preferences.

An Anthropological Perspective on Social Responsibility and Reputation

Because water and wastewater systems, like other public utilities, face often conflicting pressures to both preserve and innovate, limited resources dictate that at times decisions must be made to spend resources on one, and delay the other until more resources become available. Regardless of the choice, reputation can be enhanced or harmed (a "catch 22?"). There are, however, other ways in which to view these decisions. If decisions are made transparently and with the viewpoint of the system itself being a member of the community that is impacted (rather than simply viewing itself as the impacter), whatever decision is reached might be reached without harming the organization's reputation. Sometimes it is beneficial for businesses to see, and maybe even to borrow, from nonbusiness perspectives. In Figure 17.2, adapted from Schwarz and Thompson (1990), is one representation of a taxonomy of world views called cultural theory. It plots acceptance of social controls (indicated on the vertical axis) and levels of social commitment (indicated on the horizontal axis). In the quadrants, the worldviews or rationalities depicted are can be described as archetypes, or alternatively as ways of viewing decisions related to social responsibility.

Cultural theories classify individual and organizational ways of rational thinking and deciding into four types (quadrants as depicted) – the hierarchist who is a strict rules follower; the egalitarian who is torn between enjoying the world and protecting/saving it; the individualist for whom money or other self-interest is the primary motivator; and the fatalist who thinks decisions have little value or relationship to the outcomes. Most public utilities (and those who lead them) have been conditioned traditionally to be hierarchist in the context of decisions to be made. It is hopefully clear that moving to the individualist perspective may lead to more decisions that put the entity at risk via risk seeking; yet moving to the fatalist position is irresponsible, leading to decisions made by indecision. The egalitarian viewpoint, however, has promise. When viewed through the lens of risk appetite, the egalitarian perspective can be thought of as a natural tension that exists between value creation and value preservation. As such, taking the egalitarian view of risky decisions and risk management is akin to building an organizational risk appetite by weighing and balancing conflicts between the various strategic goals and objectives that lead to risk appetite statements and trickle down to risk

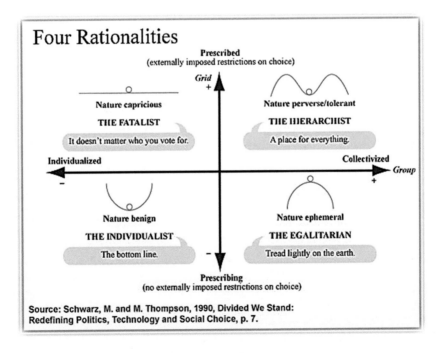

FIGURE 17.2 The four rationalities of cultural theory [28].

tolerances and guidelines. Reputation can be optimized via transparent discussions and risk-taking that acknowledge and value these tensions rather than ignore (and thus dismiss) them [3,22,28].

SUMMARY

Enterprise risk management is a holistic approach to the consideration and management of risk for organizations, placing a spotlight on individual risks as well as the interrelationships between and among risks. It holds multiple advantages over the traditional risk management, not the least of which is its ability to detect gaps and overlaps between risk planning that would otherwise be "siloed" and simultaneously see the external drivers of both existing and emerging risks.

The overall objective of an ERM program is to develop a holistic, portfolio view of the organization's significant risks to the achievement of the entity's important goals and objectives. In organizational objective setting, strategic objectives are prioritized, with risk appetite and risk tolerances considered. Risk appetite will differ by organization, not only due to differing strategic objectives but also due to differing risk attitudes and risk-bearing capacity. An effective risk appetite statement is well aligned with strategic objectives and key risk indicators, and risk tolerances should be adequately specific and articulated such that they can be aligned with key performance and risk indicators and applied throughout the organization. Strategic risks are appropriately and sustainably managed only if they are fully and consistently recognized and monitored. If identified, analyzed, and managed well, they can serve as value drivers rather than as value killers to the organization.

Despite the important role that attention to strategic goals and objectives has been shown to play in organizational value creation and protection, most organizations admit they still do not translate these into well-aligned risk management processes. Indeed, a recent survey of mostly US businesses, conducted jointly by North Carolina State University and the American Institute of CPAs [6], revealed that fewer than 20% of organizations see the organization's strategy or objectives evidenced in their risk processes, with another 69% of organizations admitting to not having a complete ERM process. The shortcomings in the practice of risk management are a major concern to these same organizations. There continue to be opportunities for improvement in how organizations identify, assess, manage, and monitor the risks that may significantly impact their ability to achieve strategic goals. Subsequent chapters highlight a number of practices that organizations may consider as they seek to strengthen their ability to proactively and strategically navigate risk at all levels.

Reputation value and risk are largely intangibles, but nevertheless impact an organization's overall performance and value. Defining, identifying, and quantifying reputation risk is difficult. Triple bottom line accounting has been created to enhance business value for organizations that incorporate ESG values, risks, and objectives into their planning and operations, but the relationships between such initiatives, reputation, and performance are indirect and not guaranteed, especially in the case of public utilities, such as public water/wastewater systems.

US public utilities face increasing economic pressure to provide infrastructure for continued development of the built environment while at the same time improving the natural and built environments' sustainability. Simultaneous with these financial pressures is a greater burden on these entities to shoulder the costs of risk. For public water/wastewater systems, this means promoting environmentally friendly development while also satisfying vast numbers of individual customer preferences quickly and efficiently.

Innovation is a must for public water/wastewater systems yet aging infrastructure and limited budgets necessitate financial tradeoffs. When the emphasis is primarily on safety, reliability, and affordability, infrastructure improvements are given precedence over innovation investments. As benefits are increasingly allocated largely to consumers and the risks are increasingly allocated to utilities, neither infrastructure improvements nor innovation investments may be affordable.

Reputation may best be protected by being realistic and transparent regarding the inherent tensions that exist in the public utility space. It need not be business versus society, but can instead be business as part of the society – openly interested both in benefitting and benefitting from its community and other stakeholders.

REVIEW QUESTIONS

1. T/F: Risk management according to the Committee of Sponsoring Organizations of the Treadway Commission (COSO) is participated in by people at every level of the organization even though it is developed at the strategic management level.

2. T/F: The focus of ERM is on strategic risks and not concerned with operational, compliance, or reporting risks.

3. The stated advantages of an enterprise risk management approach over the traditional risk management approach include:
   a. Better awareness of new and emerging risks
   b. Closing the gaps in risk management across functional areas
   c. Improved risk responsiveness
   d. All of the above

4. Key risk indicators (KRIs) within an organization:
   a. Are metrics that provide early warning of increasing risk exposures in various areas of the enterprise
   b. Help top management monitor risk trends over time
   c. May be identified through root cause analysis or other risk assessment methods
   d. All of the above

5. T/F: Risk aversion differs from risk neutrality in that the risk-averse decision-maker must expect to be compensated for taking risks while the risk neutral decision-maker does not require an expected risk compensation.

6. Risk appetite and risk tolerance are related in that:
   a. Risk tolerances determine the organization's risk appetite
   b. Risk appetite determines the organization's risk tolerances
   c. Risk tolerances are based on risk attitudes whereas risk appetite is based on risk profile
   d. None of the above are true

7. Organizational ethics and integrity contribute to organizational reputation:
   a. Directly, since triple bottom line (TBL) accounting recognizes ethics and integrity as having a quantified value
   b. Directly, as ethics and integrity guarantee the organization experiences fewer and less costly losses
   c. Indirectly, through a reduced likelihood of organizational value loss
   d. Indirectly, since ethics and integrity are virtually identical to ESG initiatives

8. Drivers of reputation risk include(s):
   a. Ethics
   b. Innovation
   c. Sustainability
   d. Safety and security
   e. All of the above

9. Triple bottom line (TBL) accounting differs from traditional accounting in that TBL accounting:
   a. Recognizes intangible assets and values related to ESG performance
   b. Does not quantify the values of reputation and other intangibles
   c. Does not use numbers at all, but instead uses qualitative measures to describe organizational value
   d. All of the above

10. T/F: The challenges of tackling the reputation risk problem within the utilities sector are largely due to tensions between the need for innovation, the need for reliability, and the need for affordability.

ANSWER KEY

1. **T/F**: Risk management according to the Committee of Sponsoring Organizations of the Treadway Commission (COSO) is participated in by people at every level of the organization even though it is developed at the strategic management level.

2. T/**F**: The focus of ERM is on strategic risks and not concerned with operational, compliance, or reporting risks.

3. The stated advantages of an enterprise risk management approach over the traditional risk management approach include:
   a. Better awareness of new and emerging risks
   b. Closing the gaps in risk management across functional areas
   c. Improved risk responsiveness
   d. **All of the above**

4. Key risk indicators (KRIs) within an organization:
   a. Are metrics that provide early warning of increasing risk exposures in various areas of the enterprise
   b. Help top management monitor risk trends over time
   c. May be identified through root cause analysis or other risk assessment methods
   d. **All of the above**

5. **T/F**: Risk aversion differs from risk neutrality in that the risk-averse decision-maker must expect to be compensated for taking risks while the risk neutral decision-maker does not require an expected risk compensation.

6. Risk appetite and risk tolerance are related in that:
   a. Risk tolerances determine the organization's risk appetite
   b. **Risk appetite determines the organization's risk tolerances**
   c. Risk tolerances are based on risk attitudes whereas risk appetite is based on risk profile
   d. None of the above are true

7. Organizational ethics and integrity contribute to organizational reputation:
   a. **Directly, since triple bottom line (TBL) accounting recognizes ethics and integrity as having a quantified value**
   b. Directly, as ethics and integrity guarantee the organization experiences fewer and less costly losses
   c. Indirectly, through a reduced likelihood of organizational value loss
   d. Indirectly, since ethics and integrity are virtually identical to ESG initiatives

8. Drivers of reputation risk include(s):
   a. Ethics
   b. Innovation
   c. Sustainability
   d. Safety and security
   e. **All of the above**

9. Triple bottom line (TBL) accounting differs from traditional accounting in that TBL accounting:
   a. **Recognizes intangible assets and values related to ESG performance**
   b. Does not quantify the values of reputation and other intangibles
   c. Does not use numbers at all, but instead uses qualitative measures to describe organizational value
   d. All of the above

10. **T/F**: The challenges of tackling the reputation risk problem within the utilities sector are largely due to tensions between the need for innovation, the need for reliability, and the need for affordability.

REFERENCES

1. Accenture and Oxford Economics. (2013). *Accenture 2013 Global Risk Management Study: Focus on Findings for the Energy and Utilities Sector.*
2. Akhmouch, A., and Clavreul, D. (2016). Stakeholder Engagement for Inclusive Water Governance: 'Practicing What We Preach' with the OECD Water Governance Initiative. *Water* 8(5): 204.
3. Allan, R., Jeffrey, P., Clarke, M., and Pollard, S. (2013). The Impact of Regulation, Ownership and Business Culture on Managing Corporate Risk within the Water Industry. *Water Policy* 15: 458–478.
4. Arjoon, S. (2006). Striking a Balance between Rules and Principles-Based Approaches for Effective Governance: A Risks-Based Approach. *Journal of Business Ethics* 68(1): 53–82.
5. Baron, J. (2019). *Thinking and Deciding.* Cambridge, UK: Cambridge University Press.
6. Beasley, M., Branson, B., and Hancock, B. (March, 2018). *The State of Risk Oversight: An Overview of Enterprise Risk Management Practices*, 9th ed. Raleigh, NC: NC State Enterprise Risk Management Initiative.
7. Black and Veatch. (2015). *Strategic Directions: Water Industry Report.* Overland Park, KS: Black and Veatch .
8. Bradford, R.W. (1999). *Simplified Strategic Planning.* Worcester, MA: Chandler House Press.
9. Briloff, A.J. (2001). Garbage In/Garbage Out: A Critique of Fraudulent Financial Reporting: 1987–1997 (the COSO Report) and the SEC Accounting Regulatory Process. *Critical Perspectives on Accounting* 12(2): 125–148.
10. Burlinghame, G.A., and Chalker, R.T.C. (2017). Risk Management is Vital to Providing Safe Water. *Opflow Online* 2017: 20–23.
11. Chalker, R.T.C., Pollard, S.J.T., , Leinster, P., and Jude, S. (2018). Appraising Longitudinal Trends in the Strategic Risks Cited by Risk Managers in the International Water Utility Sector, 2005–2015. *Science of the Total Environment* 618: 1486–1496.
12. COSO. (2004). *Enterprise Risk Management.* New York: Committee of the Sponsoring Organizations of the Treadway Commission.
13. Dale, B.G. (1999). *Managing Quality.* Malden, MA: Blackwell Publishers.
14. Deloitte. (2016). *Water Tight 2.0: The Top Trends in the Global Water Sector.* London, UK: Deloitte.
15. Doherty, N. (1985). *Corporate Risk Management: A Financial Exposition.* New York: McGraw-Hill.
16. Doherty, N. (2000). Risk Management Strategy: Duality and Globality. In *Integrated Risk Management: Techniques and Strategies for Managing Corporate Risk: Techniques and Strategies for Reducing Risk*, 1st ed, ed. N. Doherty. New York: McGraw-Hill.
17. Drori, G. (2006). Governed by Governance: The New Prism for Organizational Change. In *Globalization and Organization: World Society and Organizational Change* (pp. 91–118), eds. G. Drori, J. Meyer, and H. Hwang. Oxford, UK: Oxford University Press.
18. Eccles, R., Newquist, S., and Schatz, R. (February, 2007). Reputation and Its Risks. *Harvard Business Review.* https://hbr.org/2007/02/reputation-and-its-risks
19. The Institutes. (2013). The ERM Framework. In *Enterprise Risk and Insurance Management*, 1st ed. Malvern, PA: The Institutes.
20. International Federation of Accountants. (January, 2019). *Enabling the Accountant's Role in Effective Enterprise Risk Management.* New York: A report sponsored by the IFA.
21. Lam, J. (2014). *Enterprise Risk Management: From Incentives to Controls*, 2nd ed. Hoboken, NJ: Wiley.
22. Luís, A. (2014). *Strategic Risk Management in Water Utilities: Development of a Holistic Approach Linking Risks and Futures.* PhD Thesis, Cranfield University, Bedford, UK.
23. Martens, F., and Rittenberg, L. (2017). *Enterprise Risk Management: Integrating with Strategy and Performance.* A report sponsored by the Committee of Sponsoring Organizations of the Treadway Commission (COSO), New York.
24. Martens, F., and Rittenberg, L. (2020). *Risk Appetite – Critical to Success.* A report sponsored by Committee of Sponsoring Organizations of the Treadway Commission (COSO), New York.
25. Mccolis, J. (2003). Implementing Enterprise Risk Management: Getting the Fundamentals Right. *IRMI Expert Commentary.* https://www.irmi.com/articles/expert-commentary/enterprise-risk-management-whats-beyond-the-talk
26. Power, M. (2007). *Organized Uncertainty: Designing a World of Risk Management.* Oxford, UK: Oxford University Press.
27. Rouse, M. (2008). *Institutional Governance and Regulation of Water Services.* London, UK: The Essential Elements IWA Publishing.

28. Schwarz, M., and Thompson, M. (1990). *Divided We Stand: Redefining Politics, Technology, and Social Choice*. New York, NY: Harvester Wheatsheaf.

29. Taleb, N. (2010). *The Black Swan: The Impact of the Highly Improbable*, 2nd ed. New York: Random House.

30. Water Research Foundation. (2009). *Developing a Risk Management Culture — 'Mindfulness' in the International Water Utility Sector*. Denver, CO: Water Research Foundation.

31. Water Research Foundation. (2013). *Risk Governance: An Implementation Guide for Water Utilities*. Denver, CO: Water Research Foundation.

32. Water Research Foundation. (2015). *Risk Governance: Achieving Value by Aligning Risk Governance with Other Business Functions in Water Utilities*. Denver, CO: Water Research Foundation.

33. Westerhoff, G.P., Pomerance, H., and Sklar, D. (2005). Envisioning the Future Water Utility. *Journal of AWWA* 97: 67–74.

34. Wynne, B. (1996). May the Sheep Safely Graze? A Reflexive View of the Expert-Lay Knowledge Divide. In *Risk, Environment and Modernity: Towards a New Ecology* (pp. 44–83), eds. S. Lash, B. Szerszynski, & B. Wynne. London: Sage.

18 Risk Management Process – Risk Identification and Analysis

Photo courtesy of Las Virgenes Municipal Water District

LEARNING OBJECTIVES

- Articulate the meaning and importance of risk identification.
- Relate the risk register to the risk statement.
- Outline the components of an effective risk statement and evaluate the quality of a risk statement based on whether these elements are present.
- Recognize the various sources of risk information.
- Describe how sources of risk information may impact how risks are revealed.
- Select exposure revelation technique(s) that appropriately fits the situation and available sources of information.
- Differentiate between bowtie analysis and root cause analysis.
- Contrast scenario development with the case study approach to risk identification.
- Articulate the value and importance of utilizing the Delphi technique in risk identification.
- Contrast risk assessment with risk identification.
- Define risk analysis and its importance in the ERM process.
- Recognize the three primary characteristics of risks – frequency, severity, and velocity.
- Discriminate among statistical analysis, loss modeling analysis, and narrative-intuitive analysis.

DOI: 10.1201/9781003207146-25

- Decide whether a value represents a valid probability.
- Differentiate between discrete and continuous random variables.
- Describe how scenario analysis is expanded to conducted risk analysis rather than just risk identification.
- Classify the Poisson, gamma, and Tweedie distributions according to their characteristics and typical uses within risk analysis.
- Relate individual cognitive biases to the mind limitations and/or other circumstances that lead to them.
- Connect nonoptimal behaviors and thinking to the appropriate biases associated with them.
- Restate the correlation between the measures of cognitive reflection and measures of the need for cognitive closure.
- Identify the mode of thinking that, when practiced regularly, helps to reduce cognitive biases.

KEY DEFINITIONS

Ambiguity Effect – a cognitive bias based on the preference for options about which more is known than other available options.

Anchoring Effect – a cognitive bias marked by a heavy dependence on an initial piece of information offered to make subsequent judgments during decision-making.

Armchair Fallacy – a cognitive bias based on the adage that "actions speak louder than words," wherein an individual values action over the continuation of considering options.

Attribution Error – a cognitive bias wherein there is underemphasis on situational explanations for observed behaviors and overemphasis on dispositional and personality-based explanations for the behaviors.

Availability Heuristic – a cognitive bias based on reliance on examples that immediately come to mind.

Average Annual Loss (AAL) – the expected loss per year, averaged over many years.

Bandwagon Effect – a cognitive bias based on increased belief of information because increasingly others believe it.

Bizarreness Effect – a cognitive bias in which one recalls the details of the bizarre more or less accurately than the ordinary.

Blind Spot Bias – a cognitive bias based on the recognition of the impact of biases on the judgment of others, while failing to see the impact of biases on one's own judgment.

Bowtie Analysis – a risk evaluation method that can be used to analyze and demonstrate causal relationships, especially for medium- and high-risk scenarios.

Case Study – an approach to exposure identification that combines the experiential approach of the prepared checklist with the logical approach of scenario building.

Cognitive Bias – a systematic error in thinking that occurs when people are processing and interpreting information around them and impacts the decisions and judgments that they make.

Cognitive Reflection – an individual's tendency to think beyond initial or instinctive responses and engage in further thought to find correct answers.

Cognitive Reflection Test – a task-oriented test, designed to measure a person's tendency to override an incorrect "gut" response and engage in further reflection to find a correct answer.

Confabulation – a cognitive bias marked by confusion of what is imagined or dreamed with what is real.

Consensus Bias – a cognitive bias based on thinking of one's own behavioral choices and judgments as relatively common and appropriate to existing circumstances.

Conservatism Bias – a cognitive bias in which one perceives that preexisting beliefs only need a "slight course correction," if in need of correction at all.

Context Effect – a cognitive bias marked by the tendency to understand information according to a preconceived, or expected environment for it.

Curse of Knowledge Effect – a cognitive bias based on an incorrect perception that others have the background to understand the world's information to the same degree as we do, regardless of education of training differences.

Decoy Effect – a cognitive bias marked by the tendency to switch our preferences between two decisions when a third option presents.

Delphi Technique – an information-gathering method that uses remote interviewing with group feedback by asking a set of standardized questions to each participant, with the intention of obtaining a variety of perspectives in the responses.

Empathy Gap – a cognitive bias based on the underestimation of circumstances on a person's mental state and decisions.

Endowment Effect – a cognitive bias marked by the overvaluation of decision quality, especially purchases already made or assets already owned.

Exceedance probability – the likelihood that a particular value will be exceeded in a predefined future time period.

Framing Effect – a cognitive bias based on unconscious sensitivity to the manner in which a decision option is presented or articulated, giving the option negative or positive connotations.

Gambler's Fallacy – a cognitive bias marked by the belief that a random event is more or less likely because it has or has not occurred recently, such as a gambler's belief that the next hand must be a winner because "it is time for a win."

Gamma Distribution – any continuous probability distribution having both shape and mean parameters that are positive, real numbers. Probability distributions in this family of distributions are often used to describe loss amounts.

Halo Effect – a cognitive bias based on the belief that a person who has decided (or behaved) well (or poorly) in the past will do so again in the future, when the belief is otherwise unwarranted or there is evidence to the contrary.

Hindsight Bias – a cognitive bias wherein one erroneously thinks that past outcomes were more predictable than they actually were, thus implying that adverse outcomes were the result of poor decisions and favorable outcomes were the result of good decisions.

Hyperbolic Discounting – a cognitive bias wherein people choose smaller, immediate rewards over larger, later rewards, which is more likely the closer the delay is to the present.

Illusion of Control – a cognitive bias based on the belief that one has the ability to control events or outcomes that are in fact outside his or her control.

Illusory Superiority – a cognitive bias marked by the belief that one's own traits and abilities are superior to those of others.

In-group Bias – a cognitive bias wherein one believes friends and acquaintances are smarter than people who are not as well known.

Law of the Hammer – a cognitive bias marked by an overreliance on familiar tools.

Loss Exposure – a possibility of damage or loss that a particular entity or organization faces as a result of a particular peril striking a particular asset to which the organization has assigned value.

Loss Modeling Analysis – quantification of the financial impact of a range of potential future losses, which can be intended to estimate how many future events are likely to occur and how intense they are likely to be, usually based on probabilities, simulation, and expert judgment.

Mean – the average, or expected, value from a set of numbers, or if no set of numbers is given, the central tendency of the probability distribution of a random variable.

Median – the middle value that separates the upper half from the lower half of a data sample, a population, or a probability distribution.

Mode – the value that appears most often in a set of data values, or if no set of numbers, the value within a probability distribution's range that has the highest probability of occurring.

Murphy's Law – a cognitive bias that occurs when a person believes the facetious proposition, "If something can go wrong, it will go wrong." Although the origin of this adage being attributed to "Murphy" is unknown, the saying was first used during the 1940s and may have originated with members of the armed forces during the Second World War.

Narrative-intuitive Analysis – a framework for study whereby stories – factual and/or fictional – are told within the context of research and evaluation, typically for the purposes of discovering and examining how and why the stories unfold in the way they do.

Need for Closure Scale (NFCS) – a survey instrument that scores an individual's desire to finalize a decision or make a judgment.

Neglect of Probability – a cognitive bias that causes a person to disregard likelihood or chance when unsure about a decision, and can be marked by either a dismissiveness of probabilities or hugely overrated probabilities.

Ostrich Effect – a cognitive bias that causes people to avoid information that they perceive as potentially unpleasant. It is so named because of common legend that the ostrich sticks its head in the sand to avoid danger.

Overconfidence Effect – a cognitive bias in which an individual's subjective confidence in his or her judgments is consistently greater than the objective accuracy of those judgments.

Poisson Distribution – a discrete probability distribution that expresses the probability of a given number of events (x) occurring in a fixed interval of time or space, assuming these events occur with a known constant mean rate and independently of the time since the last event.

Probability – how likely an event is to occur, or how likely it is that a proposition is true. It is a number between 0 and 1, inclusive, where 0 indicates impossibility of the event and 1 indicates event certainty.

Probable Maximum Loss (PML) – a worst-case (or worst likely case) scenario loss amount, typically used in risk management and insurance to denote the dollar amount associated with a particular exceedance probability.

Random Variable – a factor with a value that is liable to change, and whose values depend on outcomes of an unknown or chance phenomenon.

Risk Analysis – the evaluation and measurement, if possible, of the qualitative and quantitative characteristics of loss exposures, where the key characteristics typically include likelihood/frequency, impact/severity, and/or velocity/timing.

Risk Assessment – the determination of possible losses, their likelihood and consequences, and the tolerances for such events.

Risk Frequency – a risk characteristic typically measured as a probability, likelihood, or rate with which an event occurs or is expected to occur.

Risk Identification – the process of determining risks that could potentially prevent the program, enterprise, or investment from achieving its objectives.

Risk Register – a tool for listing and documenting risks (and ultimately also actions to manage each risk).

Risk Severity – a risk characteristic typically measured as a dollar amount or other expression of the impact or magnitude that an event has, or is expected to have, when it occurs.

Risk Velocity – a risk characteristic typically measured by how fast an exposure can impact an organization and may be expressed either as the time within which an event is likely to occur or the time that passes between the occurrence of an event and when the organization feels its impacts.

Root Cause Analysis – a systematic process for identifying core (root) causes of problems or events and an approach for responding to them.

Scenario Development – identification of a specific set of uncertainties, and based on those imagine various alternative "realities" or states of the world that could occur in the future of the organization.

Skewness – a measure of the asymmetry of the probability distribution of a real-valued random variable around its mean, or expected value. Skewness can be positive (right skew), negative (left skew), zero (as in a normal or uniform distribution), or undefined.

Statistical Analysis – the science of collecting data and uncovering patterns and trends, typically for either summary and/or synthesis purposes.

Streetlight Effect – a cognitive bias wherein a person seeks answers or options in the easiest places to look rather than where the best answers or options may be. It is so named for the power of the streetlight in a dark place to draw the eye to where it casts its light.

Sunk Cost Fallacy – a cognitive bias in which a decision is influenced by a cost that has already been incurred and cannot be recovered, regardless of the decision. The sunk cost fallacy may result in valuing the past amounts already lost as greater than the prospective costs that could be avoided if a divergent path were chosen, thus serving as an emotional incentive to "stay the course" erroneously.

SWOT Analysis – a strategic planning technique used to identify an organization's strengths, weaknesses, opportunities, and threats.

System 1 Thinking – fast, instinctive, "gut" reaction cognition. Overall, it is regarded by scientists as inferior to that of System 2 thinking, unless time and circumstances substantially call for immediate, or quick decision-making.

System 2 Thinking – slow, reflective, analytical cognition. Overall, it is regarded by scientists as superior thinking to that of System 1, if and when time and circumstances allow.

Tail Value at Risk – a measure that quantifies the expected value (or mean) of the loss given that an event outside a given probability level has occurred.

Tweedie Distribution – a family of probability distributions, including the continuous normal, gamma, and inverse Gaussian distributions, the discrete Poisson distribution, and the combined Poisson–gamma distributions which have positive mass at zero, but are otherwise continuous. A Tweedie probability distribution is often used to describe aggregate loss costs, combining the frequency and severity of loss events.

Well-Traveled Road Effect – a cognitive bias in which individuals may overestimate the efficiency or underestimate the cost of an option because it is more familiar than other options. It is so named because it originated from studies of travelers who consistently estimate the time taken to traverse routes differently depending on their familiarity with the route.

OVERVIEW

> If I had an hour to solve a problem I'd spend 55 minutes thinking about the problem and 5 minutes thinking about solutions
>
> **Albert Einstein**

ERM has a process; this process necessarily begins with establishing the risk management context – the setting of strategic goals and objectives, as explored in Chapter 3. It is the purpose of Chapter 4 to discuss the next two steps – risk identification and risk analysis. Chapter 5 will home in on the selection, implementation, and continuous monitoring of risk treatment techniques. Figure 18.1 illustrates the flow of the ERM process, depicted more or less as a circular decision process.

Risk identification is the process of examining the mission, objectives, and operations of an entity to determine the multiple and various risks to which the organization is exposed. While it sounds like a simple step, it can be the most challenging of the entire risk management process. It is impossible to analyze or treat exposures that are unknown. Many failures in the risk management process can be linked to insufficient or improper risk identification. While it is impossible to identify every exposure to risk that an entity faces, a thoughtful and effective exposure identification

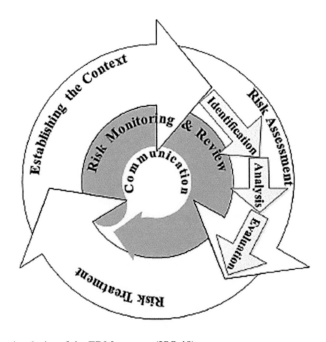

FIGURE 18.1   The circularity of the ERM process (ISO 10).

process, conducted periodically and repeatedly, can move the organization closer and closer to full awareness and appreciation of the risks faced [15].

Determining how much to spend to control the loss exposure or finance its potential consequences – or even the relativity of each exposure to another (i.e., is one "worse" than another?) – can only be assessed with some idea of the probability of an accident and its potential consequences. Senior management cannot rationally commit resources without this valuable information. Asking management to do so based strictly on exposure can lead to some seriously misguided decisions. Therefore, once exposures have been identified, their potential consequences and probability must be assessed and thoroughly analyzed. This is not simple, but it is an essential.

While risk identification is a crucial step in the risk assessment process, it is just a step. Risk identification can, of course, have some limited value in and of itself. Identifying the largest manufacturing facilities can provide a rough idea of where to focus attention. Only the most basic risk management decisions, however, can be made on the basis of mere exposure identification. Chapter 18 explores the types and sources of risks that public water/wastewater systems might face, the information sources and techniques they may apply for adequate risk identification, and also how the utilities may apply risk analysis to prioritize the risks that are uncovered.

For example, the potential for bioterrorism at a water treatment facility is worthy of attention. And that is important information. Often the first step in getting senior management's attention is showing them the potential for loss, in terms of assets and/or processes exposed, and perhaps their total value (or some other reasonable "value exposed" in the event of an accident).

THE RISK ENVIRONMENT AND TYPES OF EXPOSURES

The environment within which a water or wastewater system operates is rife with risks. Upon close study, this environment is in fact multiple environments, varied and occurring simultaneously, that individually and collectively present several sources of risk – physical, social, political, legal, operational, economic, and cognitive/behavioral environments. From these sources, various categories of risk can be evidenced [12].

Direct property exposures refer to situations that can cause immediate or direct losses to tangible and intangible organization assets. This can include losses to real or personal property, and the term typically refers to destruction, damage, or disappearance. *Indirect property exposures* refer to the loss of revenues, cash inflows, and/or income resulting from the damage or destruction of a person's property or a business's property.

Liability loss exposures stem from accidents, whether intentional or accidental, civil or criminal. Customer or visitor injuries on business premises, third-party injuries as a result of completed products or services, professional errors and omissions, employment practices liability, directors' and officers' liability, and automobile liability are all common types of liability loss exposures faced by businesses.

Human resource loss exposures, or people loss exposures, arise from injuries (or death) to employees, third parties, or volunteers. People loss exposures include the possibility of a loss to a small business from a disability, injury, resignation, death, or retirement of employees. Disability, workers compensation insurance, medical/health loss, and long-term care are just a few of the most common types of people loss exposures business policies available.

Economic/finance loss exposures are the risks that a company's cash flow, foreign investments, and earnings may suffer as a result of fluctuating foreign currency exchange rates, interest rates, unemployment levels, or other economic factors.

Environmental loss exposures involve the risk of loss due to climate change vulnerability; damage to natural resources; pollution, toxic emissions and waste; and environmental opportunity vulnerabilities. Water and wastewater systems are particularly susceptible to environmental risks as risks to reputation and financing opportunities as they are held to increasingly high standards for clean technology, renewable energy production, and zero waste guidelines while at the same time being responsible for providing or returning clean, safe water affordably.

Social loss exposures include human capital responsibilities, such as health and safety; employee development and supply chain labor standards; vulnerability to product and security liability, in the form of damage or loss to product/chemical safety or quality, privacy, or data security; stakeholder opposition due to controversial sourcing; and social opportunity vulnerabilities, such as lack of access to adequate financing for innovation.

Governance loss exposures can be categorized as issues of corporate, or organizational, governance and issues of corporate, or organizational behavior. Corporate governance challenges include board diversity, equity in compensation, ownership and control and accounting controls. Corporate behavior problems stem from lack of business ethics, anticompetitive practices, corruption and/or instability.

RISK STATEMENTS AND THE RISK REGISTER

One key characteristic of risk management and the risk management process is the progressive elaboration of the resulting risk management plan. The completeness of the risk list also needs to be elaborated progressively. An effective risk register is essential to the successful management of risk. As risks are identified they are logged on the register, a tool for documenting risks and their descriptions (and ultimately also the priority levels and actions to manage each risk). The register is typically provided as a scatterplot, a table (matrix), or both [6,13].

One key point to note for effective risk identification is the following: it is not only the list of risks that needs to be elaborated progressively, the same holds for the full description of each risk: a fully specified risk statement has to describe not only what may happen, but also why, when and to what effect.

People do not necessarily think in terms of fully specified risk statements: When you ask for potential risks, you will initially obtain a mixture of:

- effects (e.g., "the customer does not really know what they want");
- causes (e.g., the external influences on bidding successfully for a contract);
- impacts (e.g., "we could have to pay penalties");
- areas of risk (e.g., "I'm really worried about our delivery process");
- events (e.g., "we could fail the acceptance test").

This is not a problem once you understand how to use the various tools in order to elaborate, progressively, the complete list of fully specified risk statements.

One additional point to bear in mind is that risks are uncertain events which can have a positive or negative effect on the objectives of the organization. Positive risks are generally known as opportunities, negative risks as threats. In the identification process, you need to identify and specify all of these potential threats and opportunities.

An effective risk register consists of risk statements that fully describe and specify the risk and correspond to one or more elements in the organization's objectives related to risk appetite.

The components of the risk statements support all of the risk management processes that depend on risk identification:

- *Cause*: preemptive action can then be taken to prevent or reduce the probability or impact;
- *Event*: knowing the event allows you to determine symptoms and recognize when it occurs, so as to take any planned – or unplanned – action necessary;
- *Time Window*: this allows the project team to focus their efforts most effectively;
- *Impact*: this allows an evaluation of the effect and potential size;
- *Objective*: this shows which aspect of the project is impacted – and, therefore, who should take ownership.

Sources of Risk Information

Effective risk identification requires both a "big picture" overview of the business operations as well as a drill-down understanding of the detailed business processes and results. It is wise to involve people from all parts of the organization and from all levels of employment in the identification task. The identification efforts of various groups and individuals can be organized according to their responsibility areas and their typical access to information. As appropriate, individuals and groups can be asked to gather and share relevant information so that as many exposures as possible can be revealed. This risk information can come from many sources, but the most common sources are discussed here.

Financial Statements

Most risks ultimately impact an entity's assets. These assets can be in the form of real or personal property, or financial instruments, like cash and account balances. Perhaps the best guide to the organization's asset exposure is its financial statements. Financial statements are accounting documents that summarize the organization's books of account. These summaries are essentially "snapshots" of the firm's financial picture as of the day they are prepared.

Two types of financial statements can be quite useful for risk identification: balance sheets and income statements. Balance sheets list the tangible holding of firm assets, including buildings, machinery, equipment, and inventories, as well as financial holdings, such as cash and securities. When employing the creative exposure identification process to this list of assets, the question needs to be asked: "In what ways can the value of these assets be destroyed or reduced?"

Income statements depict the income and expenses incurred during the reporting period. They reflect the income generation process of the organization over time. Of course, profit, the life blood of the business, is determined by subtracting expenses from revenue. Both revenues and expenses therefore serve as variables that can affect the organization's survival. If expenses exceed revenues, or vice versa, and revenues fall below expenses, the firm incurs a loss.

Both hazard and financial risks can adversely affect either revenues or expenses. For example, an event causing the destruction of assets can also adversely affect revenues, as would be the case if a fire occurred at a key sales outlet. Alternatively, expenses can increase as a result of a sudden and unexpected rise in the cost of raw materials. By closely examining the structure of financial statements, a tremendous amount of information about the risks faced by an organization can be gained.

Flow Charts and Work Flows

As risk often arises from processes, charts or other information that summarize the processes within an organization can be a tremendous aid to exposure identification. This information often takes the form of production process flowcharts. For example, manufacturers frequently use a flowchart to show the process of moving from raw material input to final goods output as a means of identifying efficiency improvements. These charts can also be very helpful for risk assessment by showing vulnerabilities to accidental loss along the way.

The flow of many financial processes, such as the flow of receivables through the organization, is also charted in this way for a variety of uses. Such flowcharts can help identify a variety of perils, both hazard and financial, that can affect the flow of receivables. Key, for example, is the customer's ability to pay. This could be threatened by a downturn in the general economy. From a physical standpoint, the firm may rely on computers for processing the receivables. What if a natural peril threatens its computing capabilities?

Products and Services

Products and services are a key exposure area for many companies. Product and service literature is an excellent starting point for reviewing these exposures. Often, such literature offers quite detailed descriptions of the products. Further details can be gained from various engineering and service departments within the organization. A detailed product list is key to any analysis of product liability and related exposures. For entities that purely provide services, such as accounting, legal, and engineering firms, detailed descriptions of such services are also critical.

The assessment of product and service exposures can be especially challenging for businesses that provide many varied or customized products and services. A list of customers can serve as the starting point for a "backward" trail to such exposures. Often, accounting records on transactions record the item or service provided along with the customer. Many companies also keep an engineering records file. These can be reviewed periodically to identify new or unique product and service exposures.

Products and/or Services Distribution Network

Another key area of business operation is the distribution of goods and services. How do products get to the consumers? A variety of transit exposures can be identified in this way. These include both the exposures to these products along the distribution route as well as exposures arising from the methods of conveyance.

Many companies own fleets of trucks. This presents both an exposure to physical damage of the trucks as well as public liability for their operation. Additional exposures arise if the cargo is in some way volatile or dangerous. Even firms in pure service industries may have considerable transportation exposures. Consultants must get to their clients by plane, train, auto, or some other form of transportation. Service and repair people must be able to get to their customers as well.

Waste and By-products

Even the companies' wastes and byproducts must be considered. The pollution of land and water by production wastes is often a critical exposure that must be carefully assessed. All production has some sort of by-products, even service industries. The disposal of such byproducts presents exposures that must be reviewed.

Policies and Procedures

Clearly, a large variety of such internal documents are available to the risk manager to perform effective exposure assessments. The risk analyst should review as many of these documents as possible. Today, much of this "paper trail" has been replaced by electronic sources. This means that the Internet will become an ever more important source of company information.

Many of the company documents reviewed above are available on either the firm's internal intranet or as in the case of many public companies, available via the broader Internet. The organization's website is an excellent starting point for exposure awareness.

Physical Inspections

One of the most effective methods of systematic exposure identification is the process of physically touring the facilities and operations of the business. These tours allow the risk analyst to physically review operations with an eye toward possible exposures. A picture, as the saying goes, is worth a thousand words. In the case of exposure analysis, physical inspection can speak volumes.

Organizational Documentation

These documents can be as simple as a plot plan or building diagram. Most inspections based solely on these documents often prove inadequate, however, because written documentation cannot capture everything. For example, a plot plan of production facilities probably would not show the existence of a nearby stream that may present a significant flood exposure. While conducting physical inspections, the risk analyst may be accompanied by experts in physical exposure identification. For example, a tour of properties may include fire protection engineers from the firm's insurer. Such experts can help pinpoint exposures that might otherwise be missed by someone not as well versed in those particular hazards.

Employee and Business Partner Interaction and Interviews

Often, people who work in an area are able to perform insightful exposure assessments themselves. This rather direct approach can have the benefit of helping identify exposures that might otherwise remain hidden. In order to obtain the best results from an interview, it should be run as a project in its own right: define the objectives and desired outcome. Select the correct people and brief them (interviewers and interviewees). Allocate time and resources. Develop relevant questions.

Delphi Technique

The Delphi method can be seen as remote interviewing with group feedback [13]. A document is sent to each participant explaining the situation and asking a specific question or set of questions. The responses are then analyzed and assembled into a standardized form, intended to ensure a consistent and objective structure to each statement. The full list is then circulated to all participants with a follow-up question such as "what is the relative correctness and importance of each statement, and does this list suggest any additional points to you?" These responses are once again collated and a final list circulated for approval.

Contracts and Agreements

Of course, risks are routinely transferred from one party to another in business contracts and agreements. Any standard contracts used in the business should be carefully reviewed for risk management implications. Additionally, existing contracts dealing with particularly important aspects of the business or potentially high loss exposure activities should be examined. Examples include purchase orders, construction contracts, lease agreements, and service contracts.

Past Loss Experiences

Exposure to a particular type of loss, and thus the possibility of that sort of loss, does not require an organization to have experienced such a loss in the past. Nevertheless, to the extent that the business has experienced losses, the records related to this historical experience can be invaluable in the process of fostering awareness of future loss possibilities. Historical loss data may primarily reveal those areas where losses happen with some frequency but can also highlight areas of risk where even if not frequent, losses may be costly.

Insurance Policies

The contracts made with insurance companies can be extremely informative about what may be at risk and what causes of loss may be considered most likely and/or most costly. The insurance coverage provided within the insuring agreement spells out what the insurer thinks is in the "spirit" of the contract and what is of value to the insured. But the explicit noncovered items and causes of loss as spelled out in the policy conditions, exclusions, and other segments of the contract provide hints regarding what the insurer thinks is potentially too likely to occur or has the potential to be too costly to be insurable.

Insurer Loss Control Services

Aside from the exposure information that can be parsed from the insurance policies themselves, the insurance relationship can provide risk insights via loss control services. For example, in the area of property protection, insurers offer the services of professional engineers who may be experts in the property exposure identification process. Insurance attorneys can shed light on liability concerns and trends. And it is not uncommon for these services to be built into the insurance premium, and thus available at no additional charge to current policyholders.

Publicly Available Data and Loss Models

A variety of information sources on loss exposures are available. Geologic and meteorological factors are analyzed by a variety of governmental entities. Information on natural perils such as floods, windstorms, earthquakes, and fires is available from various governmental agencies. Among the most well known in the United States are the Federal Emergency Management Agency (FEMA) and the National Oceanic and Atmospheric Administration (NOAA), both of whom track natural events and event processes. Additionally, each state has divisions that contribute data to the federal databases and may keep detailed information on events occurring in their jurisdiction.

In addition to government and government-related entities, loss modeling firms track historic loss information as well as scientific assumptions that inform their expectations about future potential for losses. These firms periodically publish, or otherwise make available, the results of their efforts.

Exposure Surveys and Checklists

Among the aids to exposure identification are checklists or questionnaires, such as the IRMI Exposure Checklist (see end of chapter). These checklists and questionnaires can target enterprise risks in general or focus on a specific area of concern. The risk manager individually assesses the applicability of each potential exposure to his or her own organization by "checking it off" the list. Each check identifies a potential exposure for further study.

The questions are prepared by the experts based on their experience with particular types of losses, as well as their logical analysis for potential losses in those areas. One advantage of prepared checklists and questionnaires is that they apply the collective knowledge in the field. They often bring to mind obscure but significant exposures that may not be apparent to those with more limited experience. They also serve as guidelines to further analysis that may result in the discovery of other potentials for risk as well.

However, even the most comprehensive checklists can be only a partial representation of the reality facing the individual organization. While immensely helpful, they must be used with caution. They are a decision aid, and must not be used as a substitute for individualized investigation into the perils faced by the organization.

EXPOSURE REVELATION TECHNIQUES

Once exposure data sources have been thoroughly examined, it is important to imagine and develop the risk "stories" that are embedded in the information. Conscientious effort and a good imagination go a long way toward revealing the risks within the data. Below are several popular methods for pulling the stories from risk data sources.

SWOT Analysis

The four letters that make up the acronym SWOT stand for "Strengths," "Weaknesses," "Opportunities," and "Threats." The first two (SW) correspond to internal capabilities of the organization, whereas the last two (OT) refer to factors associated with the external environment in which the organization operates. Strengths and weaknesses can usually be related back to ERM strategic objectives easily and directly, as organizational factors that make the objectives easier or more difficult to achieve. Opportunities and threats can be described as physical, social, competitive, compliance – legal or economic in nature. Some external drivers (opportunities and threats) may be complex to tie back to ERM objectives as they may be ambiguous and/or difficult to predict. Social drivers as cultural norms and values, for instance, may place changing expectations onto an organization regarding its roles and values. Until such expectations are expressed through political actions, formation of nongovernment organizations (NGOs), or by other means, the entity may be largely unaware of the precise opportunities or threats represented. Economic drivers consist of both the macro- and microeconomic influences on an organization. While macroeconomic influences are generally systemic and nondiversifiable, their impacts on the organization's long-range ERM planning can be complex and diversifiable (e.g., the effects of business cycles, demographic migration, employment conditions, and levels of gross national product on water and wastewater demand).

Scenario Development

A valuable aid to the imagination in exposure identification is scenario building. Scenarios are stories or narratives that provide a descriptive summary of potential loss situations. Scenarios flesh out the identification of what can happen to an organization by suggesting how it may happen.

For example, consider the possibility that an organization might face a severe windstorm or hurricane as a possible natural exposure. Now expand on the hurricane event by imagining its effects. A hurricane of sufficient force would certainly shut down the business for a given period of time. This adversity is a consequence of the hurricane, in addition to the obvious physical damage the high winds would cause. By scenario building, another loss exposure was identified – loss of business due to a natural peril. While experienced risk managers may be able to identify the business interruption exposure independently from the hurricane scenario, the link to the hurricane exposure at least provides a double check of independent thought processes. Scenario building exercises can help in the identification of exposures not previously considered.

As an exposure identification device, the underlying logic of the scenario need not be explicit. It is more important to pick out exposures as they develop from the scenario. It is important that the scenario have an intuitive logic at this point. Is the story plausible? If so, the identified exposures are candidates for further analysis. The logical structure of scenarios can subsequently be used for risk analysis.

Case Study

Case studies of actual losses are an approach to exposure identification that combines the experiential approach of the prepared checklist with the logical approach of scenario building. The study of actual loss events is obviously rooted in reality. To be effective, however, cases should not merely constitute a catalog of losses (like the checklist), instead they should identify the complete spectrum of exposures resulting from any given loss event. The purpose of exposure identification is not to review the process involved to determine its logic or chain of causation, but to see what it may suggest with respect to the organization's own exposures.

Clearly, relevance is an important characteristic of case studies. From the tremendous number of losses that occur throughout the world, those pertinent to the organization's operations must be extracted. Sometimes the choice is easy; often it is difficult. There is the added dimension of risk here in that it is possible to rule out a risk as inapplicable when, in actuality, the organization does face it (or may face it in the future). The hallmark of a good exposure analysis is that it goes beyond the obvious. The decision between being overwhelmed with data, and possibly ignoring relevant data, is a tough one.

Root Cause Analysis

Root cause analysis (RCA) plays an important role in enterprise risk management (ERM) programs. RCA is part of the risk assessment process of an ERM program. It is a structured process designed to help an organization define causes to past or future risk events, understand the causes of those risk events, and, most importantly, prevent future incidents from happening. The practice of RCA is predicated on the belief that problems are best solved by attempting to address, correct, or eliminate root causes, as opposed to merely addressing the immediately obvious symptoms. As organizations begin the process of finding the true root cause of the problem, they will identify one of the three basic types of causes:

- *Physical* Causes – A tangible or material item failed in some way. For example, a car's brakes stopped working.
- *Human Causes* – People did something wrong or did not do something that was required. Human causes typically lead to physical causes. For example, no one filled the brake fluid, or the brake pads were not changed, which led to the brakes failing.

- *Organizational Causes* – A system, process, or policy that people use to make decisions in doing their work is faulty. For example, no one person was responsible for vehicle maintenance, and everyone assumed someone else had filled the brake fluid or changed the brake pads.

Bowtie Analysis

The bowtie method is a risk evaluation method that can be used to analyze and demonstrate causal relationships in medium- and high-risk scenarios. The method takes its name from the shape of the diagram that you create, which looks like a man's bowtie. A bowtie diagram does two things. First, the bowtie analyzes chains of events, or possible accident scenarios. Second, by identifying control measures the Bowtie reveals what an organization does, or can do, to control those scenarios.

Figure 18.2 illustrates a generic bowtie diagram. Notice the sources of risk as well as risk escalators (i.e., hazards) lead up to an event. Coming out of the event are the consequences, with mitigation and recovery interventions possible to reduce the consequences. (Note: When conducting forensic bowtie analysis it is not uncommon for anti-mitigation and recovery interventions to be discovered – factors that amplified the consequences.)

Once the control measures are identified, the bowtie method identifies the ways in which control measures fail. These factors or conditions are called escalation factors. There are possible control measures for escalation factors as well, which is why there is also a special type of control called an escalation factor control, which has an indirect but crucial effect on the main hazard. By visualizing the interaction between controls and their escalation factors one can see how the overall system weakens when controls have escalation factors.

A bowtie diagram visualizes the risk you are dealing with in one understandable picture. The diagram is shaped like a bowtie, creating a clear differentiation between the proactive (what happens or fails to happen, leading to the loss event) and reactive (what is, or can be, done to control or intervene post loss event) sides of risk management.

RISK ANALYSIS

The first and probably most important step in the risk management process is identifying the risks the organization faces. Once these exposures have been identified, the risk manager can proceed to an analysis of the risks' significance to the organization. Because it is not possible to mitigate all risks, prioritization ensures that those risks that can affect a business most significantly are dealt

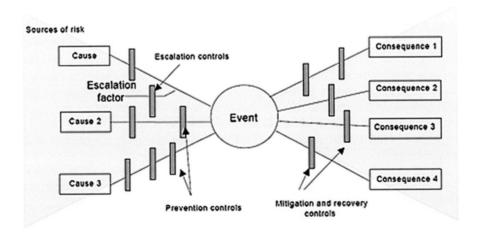

FIGURE 18.2 Sample Bowtie diagram.

with most urgently. A risk's significance may be based on its likelihood/frequency, materiality/severity of impact, velocity of impact, trending, or a combination of these characteristics. Analysis of risks makes it possible to prioritize them according to these characteristics [12].

Risk analysis includes the assessment and modeling of exposures so the business can decide how best to respond to them. Risk assessment and modeling answer questions, such as:

- Have we ever been sued? How often? How much did it cost us? (an extension of risk identification into risk analysis)
- How much property do we have exposed to loss? (significance of exposure)
- How much cash flow do we typically have available to pay for the unexpected? How profitable are we? (estimation of the organization's risk-bearing capacity)
- Which causes of loss are most frequent? Which locations experience the most losses? The worst losses? (implications of how best to treat risks)

CATEGORIES OF RISK ANALYSIS

Risk analysis techniques are divided into three categories:

- Statistical – based on statistical data and quantitative estimation.
- Loss modeling – based on logical simulation of the loss process and quantitative estimation.
- Narrative-intuitive – rely on logical models, actual historic cases of the loss process, and/or informed judgment or intuition.

Statistical Analysis

Statistical estimation requires loss data that can be adequately manipulated to provide indications of risk. As it attempts to extrapolate from actual observations, theoretically it should be a credible and reliable source of risk information. This implies one must begin with credible and reliable data regarding risky events; in fact, this may not always be available.

Loss Modeling Analysis

Loss modeling analysis simulates the loss process based on logical models and expert assumptions. Monte Carlo simulations mimic real loss situations, and can incorporate myriad input data to pinpoint current prevailing views on loss assumptions.

Narrative-Intuitive Analysis

Narrative analysis is based on scenario building. In risk analysis, narrative scenarios (or stories) about loss processes are reduced to a more formal structure that allows accurate analysis of risk using logical deductions.

MEASURING KEY RISK CHARACTERISTICS

Risk professionals look at key characteristics of risks for analysis. Frequency characteristics might be measured as total number of loss occurrences, probability, or incident rate. Severity characteristics could include total expected dollar impact, average expected dollar impact, or probable maximum dollar impact. Velocity of impact may also be of interest, answering the question of how quickly could the loss occur.

These key characteristics of a risk (especially frequency and severity) benefit from being measured, to the extent practicable. For instance, risk frequency is better understood if a specific likelihood of X happening in a given year (probability) or the number of events per unit of exposure can be estimated (rate).

STATISTICAL ANALYSIS

As stated, statistical analysis utilizes historic data to estimate future loss outcomes. Most often, these methods are used to arrive at measures of central tendency (typically mean, median, and mode), variability around the central tendency and Probable Maximum Loss (PML). Advanced statistical techniques can provide additional information, but are beyond the scope of this text [7]. Two critical concepts are the random variable and probability.

Random Variable

A random variable is a variable with a value that is unknown or a function that assigns values to each of an experiment's outcomes – whose value is a numerical outcome of a random phenomenon. A random variable can be either discrete (having specific values) or continuous (any value in a continuous range). A discrete random variable, X, has a finite number of possible values. A continuous random variable, Y, takes all values in an interval of numbers.

Probability

In the case of the discrete random variable, the probability distribution of X lists the values and their probabilities. In the case of the continuous random variable, the probability distribution is described by a density curve. The probability of any event is the area under the density curve and above the values of Y that make up the event.

The probability of an outcome is the proportion of times the outcome would occur if we repeated the procedure many times.

Examples include:

- Coin: What is the probability of obtaining heads when flipping a coin? (Discrete)
- A single die: What is the probability I will roll a four? (Discrete)
- Two dice: What is the probability I will roll a four? (Discrete)
- A jar of 30 red and 40 green jelly beans: What is the probability I will randomly select a red jelly bean? (Discrete)
- Computer: In the past 20 times I used my computer, it crashed 4 times and did not crash 16 times. What is the probability my computer will crash next time I use it? (Discrete)
- Computer: The average time between computer crashes is five days. What is the probability my computer will crash within the next 72 hours? (Continuous)
- Income: The average salary in my organization is $75,000. What is the probability that a randomly selected employee has a salary that exceeds $100,000? (Continuous)

Frequency and Severity Analysis

$$\text{Loss Frequency} = \text{Loss Count/Exposure}$$

$$\text{Loss Severity} = \text{Loss Dollars/Claim Count}$$

The addendum at the end of the chapter illustrates a normal probability distribution, with which most adults are familiar, at least in concept. It is a relatively easy distribution with which to do analysis since it conjures images that are nicely balanced and the mathematics is straightforward. Notwithstanding its appeal, the normal probability distribution has relatively little usefulness within a risk analysis context (unless loss frequency tends to be high, in which case losses could approximate a normal distribution).

It is a common assumption that in most situations, loss frequency, a discrete variable, is Poisson distributed, and loss severity, a continuous variable, is gamma distributed. The Poisson distribution is a probability distribution that shows how many times an event is likely to occur within a specified period of time. It is used for independent events which occur at a constant rate within a given

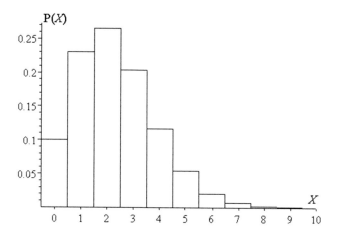

FIGURE 18.3   A sample Poisson probability distribution.

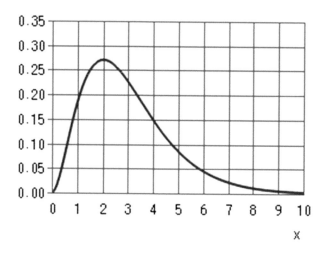

FIGURE 18.4   A sample gamma distribution.

interval of time. If a mean or average probability of an event happening per unit time/per page/per mile cycled etc., is given, and you are asked to calculate a probability of n events happening in a given time/number of pages/number of miles cycled, then the Poisson distribution is used [3,5].

Figure 18.3 illustrates how a Poisson probability distribution may appear. The number of times an event occurs is represented by X (along the horizontal axis), and the probability of X occurrences is represented by P(X) (along the vertical axis). The discreteness of the variable is depicted by the bars.

The gamma distribution is actually a family of right-skewed, continuous probability distributions. In right-skewed distributions the average (mean) is right of the peak of the distribution. These distributions are useful in real life where something has a natural minimum of zero (0). For example, it is commonly used in finance and insurance, for elapsed times, or during Poisson processes [3].

Figure 18.4 shows an illustrative gamma probability distribution. As with Figure 18.4, X represents the variable of interest (shown on the horizontal axis). For our purposes, it is the cost of a loss incident. The vertical axis indicates the probability of X occurring. In our case, the probability of certain levels of loss being reached as a result of a loss incident.

$$\text{Aggregate Loss Cost} = \text{Loss Frequency} \times \text{Loss Severity}$$

Since it is a common assumption that loss count is Poisson distributed and size of loss is gamma distributed, the aggregate loss cost (ALC) can be estimated using a gamma–Poisson compound distribution, more popularly called the Tweedie distribution.

$$LC = X1 + X2 + \cdots + XN - Xi \sim \text{Gamma for } i \in \{1, 2, \ldots, N\} - N \sim \text{Poission}$$

Although frequency and severity estimations can be much more complex than this, these three distributions fit most run-of-the-mill loss situations encountered in risk management [3,10].

In most cases, it is more insightful and thus more valuable to estimate frequency and severity separately, according to their appropriate distributions, rather than to estimate the ALC using the Tweedie distribution. The reason for this is that loss frequency and loss severity may show differing patterns or trends, and the optimal treatments for a loss exposure are better determined by seeing as many characteristics of the exposure as practicable.

Loss Modeling Analysis

Loss modeling is defined above as a simulation effort for estimating loss costs. Most loss models are utilized to determine Average Annual Loss (AAL), PML, and Tail Value at Risk (TVaR). Simulation imitates a real-life situation or system, and is useful for determining how sensitive a system is to changes in operating conditions or assumptions.

Simulation is different from statistical modeling in several important ways. The primary difference is that simulations model decision-making scenarios where at least one input parameter is a random variable. These random processes are typically too complex to be solved by analytical methods alone.

Simulations also do not require historic data, although they definitely benefit from incorporating such data if it is available. Thus, simulations can be helpful for risk analysis in situations where either historic loss data are not available or the future loss process is not expected to behave as it did in the past.

Once the simulation has been developed, it provides a convenient experimental laboratory to perform what if and sensitivity analysis. A nice benefit of simulations, and thus loss models, is they enable us to answer what-if questions without actually changing a system.

Despite their advantages, simulations are not necessarily better than other risk analysis techniques. Pitfalls of simulation include the large amount of time may be required to develop the simulation, the trial-and-error (assumptive) nature of simulation, and the sensitivity of loss models – small changes in inputs may result in large output changes.

In order to build a simulation model, one must:

- Determine the outcome that needs measuring (e.g., AAL from hurricanes in the State of North Carolina);
- Determine the input values that are needed to compute the output;
- Set the input values;
- Determine the proper probability distributions that best models each input value. Generate values for each input parameter;
- Establish the precise mathematical formulas and/or logical expressions that express how to compute output based upon input values;
- Run the simulation model, generating probabilistic inputs, and computing output based upon generated values.

Simulation can be run using actual past data. At least, if historic data are available, estimates from the simulation model should be compared with historical results. For each policy or recommendation under consideration by the decision maker, the simulation is run by considering a long sequence of potential input data values. Whenever possible, different policies/recommendations should be compared by using the same sequence of input data.

Catastrophe Loss Models

A special and widely used type of loss modeling simulation is the catastrophe model. Catastrophe loss models identify and quantify the likelihood of occurrence of specific natural disasters in a region and estimate the extent of future incurred losses. For catastrophic risks, the historical record is often limited or biased, and catastrophe models are used as the basis for estimating future incurred losses. In most areas of the United States that are catastrophe vulnerable, catastrophe loss modeling has become preferred by insurers, reinsurers, and regulators rather than relying purely on the historical record. The key elements of catastrophe models – hazard, building inventory (of the risk portfolio), vulnerability, and financial – each yield separate outputs, with final outputs producing information about loss estimates. Catastrophe modelers test the model outputs against real-life events that are in the historic record to verify their accuracy.

NARRATIVE-INTUITIVE ANALYSIS

Scenario analysis, also referred to as "what if" analysis, extends scenario development that is used for risk identification to build frequency and severity narratives for purposes of analysis. Scenario analysis does not try to show one exact picture of the future. Instead, it presents several alternative future developments. Consequently, a scope of possible future outcomes is observable. Not only are the outcomes observable, also the development paths leading to the outcomes. In contrast to prognoses, the scenario analysis is not based on the extrapolation of the past or the extension of past trends. It does not rely on historical data and does not expect past observations to remain valid in the future. Instead, it tries to consider possible developments and turning points, which may only be connected to the past. In short, several scenarios are fleshed out in a scenario analysis to show possible future outcomes. Each scenario normally combines optimistic, pessimistic, and more and less probable developments.

Scenario analysis can be used in conjunction with any of the three styles of risk analysis. It is particularly valuable in potentially high-severity situations as it can be used to suppose a particular loss of concern occurs (such that the probability of the event is assumed to be 1, or 100%). So the focus of estimation becomes solely about severity. This type of analysis makes it possible for organizations to consider various circumstances that could produce a "peak loss."

RISK ANALYSIS AND COGNITIVE CHALLENGES [1,4,9,11]

Risk analysis is subject to a variety of pitfalls, not the least of which is cognitive bias. Cognitive bias is a systematic error in thinking that occurs when people are processing and interpreting information around them and impacts the decisions and judgments that they make. If a person can avoid cognitive biases, intuitive analysis can be quite valuable, particularly as a method for recognizing patterns in information and experiences.

Data challenges have been mentioned previously. Statistical and loss modeling analysis both require heavy amounts of data. While the narrative-intuitive analysis style requires less data, it is also prone to greater cognitive bias. Cognitive bias results from information problems the brain must solve in order to arrive at conclusions and decisions. Several of these problems, and some of the biases to which they relate, are discussed here.

Problem 1 – Information Overwhelm

Incredible amounts of information exist in the world, and the brain needs to filter out most of it for decision-making purposes. The brain uses a few simple tricks to pick out the bits of information that are most likely going to be useful in some way. This trick can be useful, but can also cause distorted thinking and result in nonoptimal decisions. One example of a problem that can result is that we may notice more what is already "primed" in memory or repeated often. Associated biases include but are not limited to the: reliance on examples that immediately come to mind (availability heuristic); tendency to understand information according to a preconceived, or expected environment for it (context effect); and underestimation of circumstances on a person's mental state and decisions (empathy gap). While information overload may cause a greater propensity to notice the familiar over the unfamiliar, interestingly, it can also lead us to remember bizarre, funny, or visually striking experiences over familiar, ordinary experiences. We also may recall the details of the bizarre more or less accurately than the ordinary (bizarreness effect).

Furthermore, the mind notices when something has changed, and it may weigh the significance of the new data, or value by the direction the change happened (such as positive or negative) more than evaluating the new data, or value as if it had been presented alone. Known biases related to this effect include the: heavy dependence on an initial piece of information offered to make subsequent judgments during decision-making, such that all subsequent negotiation, argument, estimation, etc., are judged in relation to the initial data, or value (anchoring effect); and sensitivity to the connotations (positive or negative) with which options are presented, such that a decision may be opposite, or switch, depending on whether identical outcomes are presented as gains or losses (framing effect).

The mind is also drawn to details that confirm existing beliefs, maintains existing comfort zones, or allows us to think well of ourselves. This is a big problem, not only in immediate decision-making but also in overall ability to learn – from our own experiences, the experiences of others, as well as hard evidence. Several well-known selective perceptions can result, including the: tendency to search for, interpret, favor, and recall information in a way that confirms or supports one's prior beliefs or values (confirmation bias); and avoidance of information that is perceived as potentially unpleasant (ostrich effect). Even unwarranted self-confidence (overconfidence bias) and recognition of the impact of biases on the judgment of others, while failing to see the impact of biases on one's own judgment (blind spot bias) can be found within the mind's tendency to see flaws in others more easily than flaws in itself.

Problem 2 – Inadequate Meaning in the Data

Our minds may only get to observe tiny slivers of the world's information, yet we must make some sense of what we see in order to process information for decision-making. Once the reduced stream of information comes in, we connect the dots, fill in the gaps with what we already think we know, and update our mental models of the world. In so doing, we are at risk of finding patterns (or narratives) even in the sparsest data. Furthermore, the mind may connect dots and fill in data based on stereotypes, generalities, and history whenever there are information gaps or novel situations. Associated biases include the: disregard or mismeasurement of chance when making decisions under uncertainty (neglect of probability); confusion of what is imagined or dreamed with what is real (confabulation); belief that a random event is more or less likely because it has or has not occurred recently (gambler's fallacy); increase in belief of information because increasingly others believe it (bandwagon effect); and underemphasis on situational explanations for observed behaviors and overemphasis on dispositional and personality-based explanations for the behaviors (attribution error).

Exacerbating these direct responses to not finding enough meaning in the information we can readily observe, we also may superimpose our preferences – even if only indirectly – onto whatever information we do have. It is not necessarily biased to do so, but certainly can result in bias and nonoptimal decisions. Examples include the preference for the familiar over the unfamiliar, the simple over the complex, our understanding over that of others' understanding, and the present or the past over the future. Related biases that can arise include the unevidenced (or contrary to evidence) beliefs that: friends and acquaintances are smarter than the people we do not know or do not know as well (in-group, or peer, bias); a person who has decided (or behaved) well in the past will do so again in the future (halo effect); familiar decisions cost less/are more efficient/have better outcomes than unfamiliar or new decision options ("well-traveled road" effect); anything that can go wrong will go wrong ("Murphy's Law"); preexisting beliefs only need a "slight course correction" if in need of correction at all (conservatism bias); answers are best found in the easiest places to look (streetlight effect); others have the background to understand the world's information to the same degree we do (curse of knowledge effect); and/or past outcomes were more predictable than they actually were (hindsight bias).

Problem 3 – Inadequate Time to Fully Reflect

Decisions are constrained by time and information, and yet decisions must be made. With every piece of new information, the brain does its best to assess our ability to affect the situation, apply it to decisions, simulate the future to predict what might happen next, and otherwise act on our new insight.

In order to act, we need to be confident in our ability to make an impact and to feel like what we do is important. In reality, most of this confidence can be classified as overconfidence, but without it we might not act at all. Indeed, in order to stay focused, we might favor the immediate, relatable thing in front of us over the delayed and distant. Taken just a step farther, and in order to get anything done, we may be motivated to complete that in which we have already invested time and energy. This is inertia: an object in motion stays in motion. Inertia helps us finish things, even if we come across more and more reasons to give up. This need to act quickly can result in good decisions. It can also result in poor decision-making if erroneously seeing or thinking of: one's own behavioral choices and judgments as relatively common and appropriate to existing circumstances (consensus bias); an ability to control events or outcomes (illusion of control); their own traits and abilities as superior to those of others (illusory superiority); "actions speak louder than words" (armchair fallacy); money or effort already spent (or lost) as more valuable than money or effort that could be spent (sunk cost fallacy); and smaller-sooner reward as more valuable than a larger-later reward (hyperbolic discounting).

Since we are at least subconsciously aware of inertia and we rationally like to avoid mistakes, we also have an emotional/mental incentive to avoid irreversible decisions. If we must choose, we tend to choose the option that is perceived as the least risky or that preserves the status quo. ("Better the devil you know than the devil you do not.") Additionally, we may favor options that appear simple or that have more complete information over more complex, ambiguous options. Might we sometimes rather do the quick, simple thing than the important complicated thing, even if the important complicated thing is ultimately a better use of time and energy? Because of this, we may be subject to the: overreliance on familiar tools (law of the hammer); overvaluation of decisions already made (endowment effect); tendency to switch our preferences between two decisions when a third option presents (decoy effect); and preference for options about which more is known than other available options (ambiguity effect).

Problem 4 – Memory Limitations

The mind can only retain the information most likely to prove useful in the future. It makes constant bets and trade-offs around what it needs to remember and what it can afford to forget. For example, most people prefer generalizations over specifics because they take up less mind space. When there are lots of irreducible details, most individuals pick out a few standout items to save and discard the rest. What is saved is what is most likely to inform the filters related to problem 1's information overload, as well as inform what comes to mind during the processes mentioned in problem 2 around filling in incomplete information. Thus, we become susceptible to biases that are directly related to how we recall events and experiences as well as how we fill in details that we do not actually remember. We edit and reinforce memories, discard specifics to arrive at general observations, reduce events to their key elements, and store memories differently based on how they were experienced. These brain manipulations, which can serve us well, can also serve to our detriment when they result in implicit bias, prejudice, source confusion, suggestibility, false memory, and/or misinformation.

Cognitive Challenges Summarized

- We never observe everything. And some of the information we filter out is useful and important.
- Our search for meaning can conjure illusions. We sometimes imagine details that were filled in by our assumptions, and construct meaning and stories that are not really there.
- Quick decisions can be seriously flawed. Some of the quick reactions and decisions we jump to are unfair, self-serving, and/or counterproductive.
- Our memory reinforces errors. Some of the stuff we remember for later just makes all of the above systems more biased, and more damaging to our thought processes.

Correcting for Biases

Although we cannot eliminate our cognitive biases, we can accept that we are, by nature, biased, and that improvement in our thinking and deciding is always possible. Cognitive biases are useful in the right contexts, harmful in others. We might as well get familiar with them and even appreciate what they can do for us in helping us to process information. But we do well to attempt to control the biases rather than have them control us and our decisions.

Decision-making in risk management is particularly susceptible to unconscious use of these biases since such decisions are inherently made under situations of uncertainty. The subconscious mind generally gets all kinds of things wrong, especially if any data is missing. This happens regardless of intelligence or education, but certainly can be curbed by training and practice.

Rather than spend effort attempting to identify and correct every individual bias to which one might be susceptible, it may be more practical and effective to identify the extent to which one is in danger of bias more generally. Thankfully, a lot of rigorous research has been done in this area that not only provides helpful "food for thought," but also makes available to the public some methods for self-evaluation and correction. First, two modes of thinking have been identified that are used for decision-making – referred to simply as System 1 and System 2 thinking. System 1 is the fast, instinctive, "gut reaction" way of thinking and deciding. System 2 is the slow, methodical, "critical analysis" way of thinking and deciding. System 1 forms first impressions and often is the reason why we jump to conclusions. System 2 does reflective problem solving, and is the mode where we hopefully self-correct for any first impressions that were erroneous or have not been proven [8].

Given our human predilection for cognitive bias, individuals generally want to think we spend most of our time in System 2 thinking and/or whatever time we spend in System 1 thinking is valuable because we (individually) think we are particularly good at it. Scientists have developed and validated at least two instruments that can be used to measure a person's tendency toward one system of thinking over the other. The cognitive reflection test (CRT) is a task-oriented test, designed to measure a person's tendency to override an incorrect "gut" response and engage in further reflection to find a correct answer [4]. The Need for Closure Scale (NFCS) was designed to assess individuals' motivation to process information and make a judgment [16,17]. It is defined as a desire for an answer in order to end further information processing and judgment, even if that answer is not the correct or best answer. High cognitive reflection scores have been correlated with low NFCS scores, implying that individuals who reflect more on the available information (and thus arrive more often at correct answers) are more likely to have a lower need for closure (decision at any cost) than those who have low cognitive reflection scores.

The good news in thinking about cognitive reflection and bias is that every person can improve the quality of his or her thinking, and can make a continued improvement over time. To the extent that mental shortcuts remain, they may be necessary from a practical perspective to be an effective decision maker. The benefits of closure include the ability to act once a decision is made and the possibility of receiving action-related rewards. The costs of closure include the risk of making costly judgmental errors and the reduction of options and freedom resultant from decision-making.

SUMMARY

Risk analysis should be an integral part of the strategy-setting process. Strategic and other risks should be supported or rationalized by management [2]. Another reason why the risk assessment component is applicable to strategy setting and business planning is because strategic objectives are included within the scope of the ERM framework. Risk assessment and analysis have at their core the objectives to determine the frequency and severity characteristics of each identified risk. There are multiple methods for making conclusions about these characteristics – statistical, modeled, and narrative – and all of these may play a role in the risk management program. Cognitive bias is a hindrance to optimal decisions, and can even be disastrous to a public water or wastewater decision maker. Awareness and understanding one's individual propensity for decision bias is important to reducing it.

Identification of exposures to accidental loss is the critical first step in a comprehensive review of risks facing a public water or wastewater system. How, and how well, this initial exercise is performed will determine how effective the risk analysis can be. This in turn influences how effective any risk management plans built on this foundation will become. Sound risk management is, therefore, built on thorough and well-developed exposure identification.

Risk identification is important. Risk is in every phase of business. Risk is not always self-evident, and even when self-evident, it may be plain only to a few unless cross-communication of risks is incorporated within the process. More than one identification method should be used to maximize the quantity and quality of what is uncovered, even though often one method will uncover most risk.

Once exposures are identified, the risk management task is not over – it is just beginning. Decisions need to be made regarding how these exposures will be treated to minimize their impact on the organization. The next critical step is the assessment of the probability and consequence characteristics of the exposures identified. Once this information is ascertained, then and only then can a reasoned approach to the treatment of these risks be developed.

Enterprise risk management demands a systematic approach, from identification of loss exposures to their treatment. Risk managers need to understand and expand on these approaches, and apply them to the management of risk within his or her own organization, to be effective.

REVIEW QUESTIONS

1. Risk identification involves the development of:
   a. A risk list
   b. A risk register
   c. An overall risk statement
   d. A risk statement for each risk that is registered

2. A risk register outlines identified risks:
   a. In preparation for risk statement development
   b. According to category of impact
   c. In the form of a scatterplot or a table
   d. All of the above

3. The elements of an effective risk statement include all except:
   a. Objective and event
   b. Cause and impact
   c. How frequently it will occur
   d. Time window within which it could happen
   e. All of the above are included

4. Bowtie analysis differs from root cause analysis in that bowtie analysis:
   a. Focuses on cause and effect rather than on cause only
   b. Does not attempt to quantify the event

    c. Utilizes hypothetical loss scenarios rather than historical loss events

    d. All of the above are true

5. T/F: The Delphi technique is the best technique for identifying risks because it involves using multiple sources of information.

6. Risk analysis is important to the ERM process because it directly _____ risks, leading to the appropriate _____ of risks.

    a. Measures; treatment

    b. Discovers; measurement

    c. Treats; measurement

    d. Discovers; treatment

7. A risk's characteristic frequency may be expressed as its:

    a. Probability

    b. Rate of occurrence

    c. Qualitative likelihood

    d. All of the above

8. Statistical analysis may differ from loss modeling analysis in that statistical analysis relies on:

    a. Historic loss data

    b. A changing loss environment

    c. A simulated loss environment

    d. All of the above

9. Which of the following values is not a value that a probability can take?

    a. 0

    b. 0.5

    c. 1

    d. 1.5

10. _____ thinking, when practiced regularly, has been shown to increase CRT scores and reduce bias in decision-making.

    a. System 1

    b. System 2

    c. Confabulation

    d. Closure

ANSWER KEY

1. Risk identification involves the development of:

    a. A risk list

    b. A risk register

    c. An overall risk statement

    d. **A risk statement for each risk that is registered**

2. A risk register outlines identified risks:

    a. In preparation for risk statement development

    b. According to category of impact

    c. In the form of a scatterplot or a table

    d. **All of the above**

3. The elements of an effective risk statement include all except:

    a. Objective and event

    b. Cause and impact

    c. **How frequently it will occur**

    d. Time window within which it could happen

    e. All of the above are included

4. Bowtie analysis differs from root cause analysis in that bowtie analysis:

    a. **Focuses on cause and effect rather than on cause only**

    b. Does not attempt to quantify the event

    c. Utilizes hypothetical loss scenarios rather than historical loss events

    d. All of the above are true

5. T/F: The Delphi technique is the best technique for identifying risks because it involves using multiple sources of information.

6. Risk analysis is important to the ERM process because it directly _____ risks, leading to the appropriate _____ of risks.

    a. **Measures; treatment**

    b. Discovers; measurement

    c. Treats; measurement

    d. Discovers; treatment

7. A risk's characteristic frequency may be expressed as its:

    a. Probability

    b. Rate of occurrence

    c. Qualitative likelihood

    d. **All of the above**

8. Statistical analysis may differ from loss modeling analysis in that statistical analysis relies on:

    a. **Historic loss data**

    b. A changing loss environment

    c. A simulated loss environment

    d. All of the above

9. Which of the following values is not a value that a probability can take?

    a. 0

    b. 0.5

    c. 1

    d. **1.5**

10. _____ thinking, when practiced regularly, has been shown to increase CRT scores and reduce bias in decision-making.

    a. System 1

    b. **System 2**

    c. Confabulation

    d. Closure

REFERENCES

1. Baron, J. (2019). *Thinking and Deciding.* Cambridge, UK: Cambridge University Press.
2. Bradford, R.W. (1999). *Simplified Strategic Planning.* Worcester, MA: Chandler House Press.
3. Forbes, C., Evans, M., Hastings, N., and Peacock, B. (2011). *Statistical Distributions*, 4th ed. Hoboken, NJ: John Wiley & Sons.
4. Frederick, S. (2005). Cognitive Reflection and Decision Making. *Journal of Economic Perspectives* 19(4): 25–42. https://www.aeaweb.org/articles?id=10.1257/089533005775196732
5. Haight, F. (1967). *Handbook of the Poisson Distribution.* New York: John Wiley & Sons.
6. Hillson, D. (2002). The Risk Breakdown Structure (RBS) as an Aid to Effective Risk Management. In *PMI European Conference*, Cannes, France.
7. Hokimoto, T. (2017). *Advances in Statistical Methodologies and their Application to Real Problems.* London: IntechOpen.

8. Kahneman, D. (2011). *Thinking, Fast and Slow*. New York: Farrar, Straus and Giroux.
9. Kahneman, D., and Tversky, A. (2000). *Choices, Values, and Frames*. New York: Cambridge University Press.
10. Kleiber, C., and Kotz, S. (2003). *Statistical Size Distributions in Economics and Actuarial Sciences*. Hoboken, NJ: John Wiley & Sons.
11. Kruglanski, A.W., Webster, D.M., and Klem, A. (1993). Motivated Resistance and Openness to Persuasion in the Presence or Absence of Prior Information. *Journal of Personality and Social Psychology*, 65(5), 861–876.
12. The Institutes (2013). Risk Analysis. In *Enterprise Risk and Insurance Management*, 1st ed. Malvern, PA: The Institutes.
13. Meredith, J., and Mantel, S. (1995). *Project Management: A Managerial Approach*. New York: Wiley.
14. Pritchard, C.L. (1997). *Managing Risk – Concepts and Guidance*. Arlington, VA: ESI International.
15. Rejda, G., McNamara, M., and Rabel, W. (2020). *Principles of Risk Management and Insurance*, 14th ed. New York: Pearson.
16. Roets, A., and Van Hiel, A. (2011). Item Selection and Validation of a Brief, 15-Item Version of the Need for Closure Scale. *Personality and Individual Differences*, 50(1), 90–94.
17. Webster, D.M., and Kruglanski, A.W. (1994). Individual Differences in Need for Cognitive Closure. *Journal of Personality and Social Psychology*, 67(6), 1049–1062.

ADDENDUM: THE NORMAL PROBABILITY DISTRIBUTION

The mean for a normal distribution is called μ (pronounced "myu"), the Greek letter for m (for mean). The standard deviation for a normal distribution is called σ (pronounced "sigma"), the Greek letter for s (for standard deviation).

Standard deviation (σ) indicates the expected magnitude of the error from using the expected value as a predictor of the outcome (i.e., how far from the expected value an actual outcome might be).

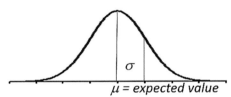

$$\text{Variance} = (\text{standard deviation})^2$$

Standard deviation (and variance) is higher when the outcomes have a greater deviation from the expected value and/or probabilities of the extreme outcomes increase. Every distribution has a variance and standard deviation, but the normal σ is easy to work with, and provides us with easy-to-interpret statistical rules that aid in decision-making.

The normal distribution enjoys an easy rule about interpreting the standard deviation:

67% of the time the outcome will fall within 1 standard deviation, so $X \leq +/- 1 \cdot \sigma$.
95% of the time X will fall within 2 σ, so $X \leq +/- 2 \cdot \sigma$.
99.75% of the time X will fall within 3 σ, so $X \leq +/- 3 \cdot \sigma$.

Because the normal distribution enjoys an easy rule about interpreting the standard deviation:

67% of the time the outcome will fall within 1 std deviation, so $X \leq +/- 1 \cdot \sigma$

But 67% confidence does not provide businesses with adequate confidence, so most businesses use 95% (or higher) confidence intervals to estimate PML:

95% of the time X will fall within $2s$, so $X \leq +/- 2 \bullet \sigma$
99.75% of the time X will fall within $3s$, so $X \leq +/- 3 \bullet \sigma$

Probable Maximum Loss at the 95% level is the number, PML, that satisfies the equation:

$$\text{Probability (Loss} < \text{PML)} < 0.95.$$

Losses will be less than PML 95% of the time.

19 Risk Management Process
Risk Treatment Selection, Implementation, and Monitoring

2019© Photo Courtesy of Placer County Water Agency, Placer County California 95604

LEARNING OBJECTIVES

- Prioritize risks based solely on risk characteristics.
- Enumerate the risk treatments that are available to manage various risks an organization may face.
- Illustrate risk characteristics on a risk map, or chart.
- Distinguish risk treatment categories from one another.
- Indicate which risk treatments may best fit a set of risk characteristics.
- Apply decision criteria to risk treatment options.
- Utilize guidelines to estimate an organization's risk-bearing capacity.
- Differentiate between pure and speculative risk.
- Recognize the characteristics of an (un)insurable risk.
- State the importance of good risk management plan implementation.
- Describe the importance of risk management plan monitoring.

DOI: 10.1201/9781003207146-26

KEY DEFINITIONS

Additional Insured – a person or entity that is covered formally under another party's insurance policy.

Discounted Cash Flow (DCF) *Analysis* – a valuation method used to estimate the value of an investment based on its future cash flows.

Diversification – allocating capital in a way that reduces the exposure to any one particular asset or risk.

Fortuitous Losses – losses happening by accident or chance rather than design.

Insurance – an arrangement by which an entity provides a guarantee of compensation for specified loss, damage, illness, or death in return for payment of a premium. True insurance includes risk pooling, risk transfer, and the indemnification of fortuitous losses only.

Internal Risk Reduction – increase in risk knowledge or decrease in risk ambiguity via diversification and/or investment in information.

Loss Prevention – measures taken to decrease the likelihood of loss or rate of loss occurrence.

Loss Reduction – measures taken to decrease the magnitude of loss, given a loss has occurred.

Precautionary Principle – an economic and legal principle by which the introduction of a new product or process whose ultimate effects are disputed or unknown should be resisted.

Pure Risk – a type of risk that cannot be controlled and has two outcomes: complete loss or no loss at all. There are no opportunities for gain or profit when pure risk is involved.

Redundancy – the inclusion of extra components, partners, or other elements which are not strictly necessary to functioning, in case of failure in other components, partners, or elements.

Risk Avoidance – the elimination of hazards, activities, and exposures that can negatively affect an organization's assets, with the intent to decrease likelihood of loss to zero.

Risk-Bearing Capacity – the ability to absorb additional risk-based volatility in its results without detrimental effects to key plans, strategies, operational and financial resources of the company.

Risk Control – a family of strategies that aim to reduce likelihood of loss and/or severity of loss.

Risk Correlation – a relationship between two or more risks, such that they are not independent of one another. Correlation implies that when one occurs another is more or less likely to also occur, but does not necessarily imply that one causes another to occur (or not to occur).

Risk Mapping – a visual depiction, usually as a chart or graph, of a select set of an organization's risks, designed to illustrate the impact or significance, the likelihood or frequency, and possibly other characteristics of the risks. Risk mapping is used to assist in prioritizing and managing risks.

Risk Retention – acceptance of losses, whether by deductibles, choosing against insurance, or other means.

Risk Transfer – the contractual shifting of a pure risk, or loss resulting from the risk, from one party to another.

Speculative Risk – a type of risk that, when undertaken, results in a possibility of gain, loss, or status quo. Speculative risks are made as conscious choices and are not just a result of uncontrollable circumstances.

Subrogation – the assumption by a third party of another party's legal right to collect a debt or damages; a legal doctrine whereby one person or entity is entitled to enforce the revived rights of another for one's own benefit.

OVERVIEW

> There are risks and costs to a program of action. But they are far less than the long-range risks and costs of comfortable inaction

John F. Kennedy

Several elements in the risk management process have been discussed in the preceding chapters. Only after the identified risks have been adequately analyzed and prioritized is it appropriate to consider the alternatives for how they can best be managed. Selection and implementation of risk treatment techniques in some cases are quite straightforward, but in other cases is highly complex [6]. The complexity of risk decision-making increases with the ambiguity around understanding the risk, the number of feasible treatment options as well as the uncertainty surrounding the organization's risk-bearing capacity. Complexity decreases as the understanding of the underlying risk increases, as fewer feasible options present, and as the financial outcomes become more certain – in the extremes of minimal risk (easily absorbed financially by the entity as the need arises) and maximum risk (where risk avoidance may be necessitated by the precautionary principle) the solutions are most clear. This chapter addresses decisions regarding measures that prevent risks from occurring/recurring, the best responses in the event the risks do occur/recur, and the cost–benefit trade-offs of various alternative strategies.

Regardless of how well organized and planned a risk management strategy may be, it must be executed properly for it to work as planned or hoped. Even with proper implementation, a risk management strategy may not prove as effective as the evaluation and decision process predicted, either because of randomness or because the risk environment changed. Thus, risk management performance must be monitored and strategies updated on a regular and recurring basis [6,8].

RISK MAPPING

Risk mapping is a process whereby you plot risks visually according to their likelihood and potential magnitude. (More than two risk characteristics are possible, however; for instance, velocity of impact is increasingly considered as well.) Building a risk map brings valuable benefits. You will have a thorough understanding of your risk environment and how individual risks compare to one another. It provides a qualitative determination of risks that are material—that is, that potentially can impact the organization's achievement of its financial and/or strategic objectives. You can use this to prioritize your risks – strategically and tactically – so you use your limited resources optimally.

This representation often takes the form of a two-dimensional matrix, or chart, with frequency on one axis and severity on the other axis. (The matrix takes on additional dimensions, of course, as we add additional risk characteristics, such as velocity.) The higher a risk ranks for these characteristics, the more threatening it is to the organization, and thus the higher priority or rank it may be given in decision-making. Figure 19.1 illustrates a simple risk map. This risk chart uses color gradients to denote the threat level to the organization (from low-threat green to high-threat red). Risk maps are a valuable tool as they allow for understanding the risk environment, prioritizing treatment strategies, allocating limited resources, and negotiating costs of treatment strategies (especially insurance).

Risk Environment/Profile Knowledge

Risk management begins with building a list of all risks the organization faces. Depending on the organization, this number could range from a handful to hundreds. Risk mapping requires an assessment of each risk, its causes and its consequences, individually. It also allows a holistic glimpse of the risk environment as a whole such that frequencies and severities can be compared across risks and potential correlations between risks can be observed. Finally, a risk map is a visual that anyone in your organization can use to see the big picture of risks most prominent in your industry or workplace.

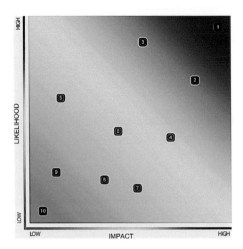

FIGURE 19.1   Risk map illustration.

Dependency Mapping

Consider the relationships between the fundamental building blocks of the organization. What are the dependencies associated with the business processes? This includes relationships between processes, sites, applications, suppliers, and people. In addition, business processes depend on one or more upstream business processes, and also feed into downstream business processes on which those depend. Articulating the relationships between organizational building blocks enables understanding how the organization works and how it can break, how disruptions can be prevented, and how the organization can recover when disruptions are unavoidable. By combining dependencies, criticalities, and impact profiles, a clearer understanding of the organization emerges – for example, whether an apparently noncritical process is actually highly critical because of critical processes depending on it. Bringing all this information together to conduct gap analysis reveals where risk treatments are adequate, lacking, or overlapping relative to the identified criticalities and impacts of the business processes. These discoveries represent opportunities to shift resources and reprioritize investments to where they are needed most. Dependency mapping defines these relationships within the context of the overall organization. It shows how a disruption to one fundamental building block can have impacts that extend beyond that single process, application, facility, etc [5,10].

Strategic Prioritization

With limited resources and risk-bearing capacity, it is important to be strategic about risk management techniques. Risk mapping allows you to determine how to prioritize risk management planning: Select and implement strategies and tactics for the most threatening risks before moving on to others. This prioritization method ensures that the risks presenting the most harmful potential are always addressed.

Resource Allocation

By determining which risks are the highest priority, the organization can spend the first and greatest efforts efficiently. By deciding which ones are the lowest priority, the organization knows what can be delayed or temporarily ignored if time and/or money are depleted. While no risk can be completely ignored, some can be low prioritized for later decision-making.

Treatment Cost Negotiation

Risk maps can also help an organization to negotiate competitive prices, especially insurance premiums. Insurers are looking for reliably good risks, and prefer to build relationships with buyers that they believe will have infrequent losses. Presenting a risk map for the insurer's inspection indicates the organization is self-aware and is actively managing its risks, which can result in more stable, if not lower, premiums.

THE BASIC RISK RESPONSE OPTIONS

Risk response is the most important component when applying the ERM framework to strategy setting and business planning. The way an entity responds to risk is just as important as the way an entity plans for risk. ERM focuses on strategic objectives while internal control provides an important risk response option in executing the strategy and business plan.

The basic methods for risk management – avoidance, retention, risk control, and transfer – collectively encompass virtually all opportunities for reducing the adverse impacts of risk on the organization, and increasing the possibilities for risks to be used to enhance the organization's value. These categories can be superimposed onto a risk map according to a decision heuristic as shown in Figure 19.2. Each of these categories is discussed below.

Note that, similar to the matrix in Figure 19.1, this matrix shows impact on the vertical axis and likelihood on the horizontal axis. Unlike Figure 19.2, however, this matrix is divided into four quadrants for decision-making purposes. The difficult element of this mapping process is to determine for the organization where precisely its dividing lines (both vertical and horizontal) are. Once that is decided, it becomes straightforward to utilize this simple structure in consideration of your categories of risk treatment options.

Risk Avoidance

The most severe and frequent risks are critical and would hinder your ability to conduct business. Some of these risks are too great – where the potential severity and/or frequency is deemed too high to go forward with the project or product. This treatment method of avoiding the possibility of loss is not feasible in most cases, especially if the risk is inherent to the organization's core business (e.g., "We provide water. We cannot choose to not provide water."). But there are risks so dire that the precautionary principle should at least be considered. The precautionary principle is an economic and legal foundation for making decisions to avoid risk (used more frequently in public policy than in business) when potentially catastrophic or irreversible events are possible, weakly understood (i.e., ambiguous), and where protective measures are costly and still may not solve the problem that they are designed to correct or reduce [10]. The precautionary principle certainly applies to water and wastewater risk management in the water sector's capacity as critical infrastructure (as discussed in Chapter 15 of this part).

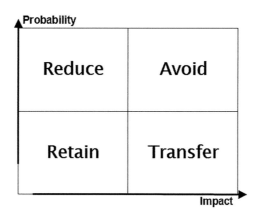

FIGURE 19.2 Classic risk treatment matrix.

ERM is intended to help the organization meet its strategic goals and objectives, within the framework of its appetite for risk. Any project or product that threatens the organization's survival and presents risks that are difficult to understand and protect against may warrant complete avoidance, or at least alterations that reduce the most catastrophic possibilities. Unfortunately, the risks that may best be avoided are likely to include innovations that public water/wastewater systems are under pressure to consider.

Risk Retention

On the other end of the frequency and severity spectrum are risks so minimal that they can be tolerated, and the resultant losses can be absorbed within the organization's risk-bearing capacity. Thus, the organization may make an informed decision to simply accept the risk. Even if a risk is not considered minimal, retention (or partial retention) may be selected if the cost of protective measures outweighs the benefit. Risk retention can be combined with other treatments in situations where a residual amount of risk remains after the preferred treatment options are implemented. Whenever a risk's frequency and severity are deemed as "low" or the residual risk after treatment is deemed "low," retention may be the best option.

Basic principles to be aware of include:

- Limitations of extrapolating prior loss record (retainable) into the future.
- Statistical risk analysis can have volatile and non-credible results, particularly in cases where there is little loss history from which to conduct the analysis.
- Statistical concepts like a "1-in-100-year" loss event can be misleading, and may be more accurately understood as critical probabilities, such that a "1-in-100-year" event actually has a 1% probability of occurring in any year, could occur early and frequently after a decision to retain.
- Risk appetite after a large loss tends to be smaller than before a large loss.

Because this treatment may tie up cash – necessarily set aside to pay for losses as they occur – the opportunity cost of cash should be considered in addition to considering whether there is sufficient cash to pay "out of pocket" for losses as they arise. Typically, public water and wastepublic water/wastewater systems (along with most other types of public utilities) do not have many high-expected-return projects competing for their attention, so this consideration is not as great as it might be for organizations within less-regulated and high-growth economic sectors [1–3].

Risk Control

Somewhere on the risk spectrum between the risks that can be retained and those that need to be avoided are those that best benefit from being controlled. Controls, which are protective measures that contribute to the reduction in the underlying risk, are worthy of consideration in almost every risk case. Risk control treatments include:

- Loss prevention (intended to reduce frequency);
- Loss reduction (aimed at reducing severity);
- Risk diversification and redundancy (intended to reduce the impact of losses from a single risk on the overall organizational value); and
- Internal risk reduction (aimed at reducing the uncertainty related to frequency and/or severity of a risk via improved knowledge of the risk).

Since this treatment category often requires substantial upfront expenditures with uncertain cost savings, any risk that contains ambiguity makes a cost–benefit analysis difficult. Thus, the very risks that might best benefit from risk control may be particularly disconcerting when various controls are considered. Nevertheless, if combined with retention and transfer, the control options can be quite effective in reducing future ambiguity and increasing predictability even when not as effective as predicted at reducing frequency and/or severity.

The most straightforward cases in which risk control is a best first option are the ones in which severity is clearly deemed low, but frequency is clearly deemed high by the organization. This is because for most risks the predictability of loss savings from prevention is greater than the predictability of loss savings from reduction (due to the organization often being better in control of whether an event occurs (loss frequency) than the magnitude of it once it occurs (loss severity)).

Diversification and redundancy add value by improving an organization's "anti-fragility" as well as its resilience to loss events. Strategically, not "putting all your eggs in one basket" is advised for any organization. For public utilities such as water systems, the importance is arguably greater than for other organizations since utilities provide critical infrastructure, serving the public good. A later chapter on critical infrastructure risk management delves into this topic more deeply.

Generally, when organizations think about risk controls they most commonly think of safety controls intended to directly reduce losses, such as loss to physical assets or liability settlements, but the indirect value of internal risk reduction cannot be overstated. Some ways companies might consider investing in internal risk reduction include:

- Learning from colleagues about their strategic risks – Brainstorm with key employees what undermines their overall strategic goals and objectives, and determine how best to treat these risks.
- Multiple models – Analyze risks from several perspectives, using more than one method, so as to improve the risk knowledge and decrease ambiguity.
- Engage with the Internet of Things (IoT) – Use sensor technology to monitor risky situations in real time that could cause a loss.
- Practice organizational response – Simulate loss events, including employee response, and develop plans to improve capabilities.

Risk Transfer

Also situated between the easy-to-retain risks and the avoid-at-all-costs risks on the risk spectrum are the risks that most benefit from being transferred to another entity. Insurance is the most well known of these transfer treatments, although not the only one. For instance, transfer can be accomplished through hold harmless clauses (transfer of the risk itself) and/or indemnity agreements (transfer of the loss consequences of a risk) within contracts. It can also be achieved through the issuance of securities that offer payments or that need not be repaid contingent upon a particular type of loss event occurring, or that fix the downside cost at a level that is acceptable to the organization.

Due to the combination of how insurance demand and the models used by insurance carriers work, insurers ideally prefer to take on risks that are deemed to have characteristically low frequency and high severity and are fortuitous in nature. Fortuitous (unexpected on the part of the insured) losses are an indication that losses occur randomly and thereby fit within the expectations of the probability distributions applied. A low probability or rate of loss occurrence makes a risk a non-nuisance for the carrier. That losses may have high magnitude when they do occur guarantees the insurer can successfully charge an adequate risk premium from a risk-averse organization to allow for reasonable expected profitability.

Thus, the cost–benefit trade-off for organizations facing low-frequency, high-severity risks usually calls for an insurance purchase for the largest mass of the loss potential. While the lowest amounts of loss may be retained (as a deductible or self-insured retention), and the highest amounts may require insurance policy layering for full coverage, insurance overall is commonly the most appealing option based on the premium–certainty trade-off it provides.

Insurance risk transfer is covered in depth in *Part IV - Insurance Essentials*. As part of the risk treatment selection process, however, it is worthwhile to note a couple of insurance items here. The value (benefits) of insurance coverage depends on the insurance policy limits (and sublimits), deductibles (i.e., de facto retentions), exclusions, and other terms of the insurance contract [7]. Careful consideration of these terms is important, and, at least in part, should be based on the risk analysis and mapping efforts.

Even in cases where full insurance (coverage for 100 percent of the known exposure) is deemed theoretically ideal, full insurance is not likely the best option in practice. Insurers generally prefer to do business with insureds (policyholders) who keep some "skin in the game," and thus have an incentive to reduce the underlying risk and/or control losses in the short term. It is common for partial insurance – not full insurance – to be available for exposures, so long as they are generally deemed insurable. Thus, insurance nearly always is accompanied by some retention and risk-control activities.

DECISION CRITERIA FOR TREATMENT

Decision criteria for risk treatment are both quantitative and qualitative. When risk management decisions affect cash flows over multiple periods, the expected effect on value is typically calculated using discounted cash flow (DCF) analysis as a decision starting point.

Discounted Cash Flow Analysis

Under DCF analysis, the Net Present Value (NPV) of the alternative decisions is utilized to determine the best option(s). NPV is an indicator of how much value an investment or project adds. Appropriately risked projects with a positive NPV could be accepted.

$$\mathbf{NPV} = \sum_{t=0}^{n} \frac{E(\mathrm{NCF}_t)}{(1+r)^t}$$

Where

E(NCFt) = expected net cash flow in time t for n periods; and r = opportunity cost of capital (reflects the riskiness of the cash flows and alternative uses of the capital).

Key factors in DCF analysis include time (the number of periods), interest rates, timing of investments/payments, number of payments, size of payments, taxes, and potential variability of cash flows from what is expected. Other related quantitative methods include the estimation of how long a treatment method takes before the entity expects to break even (i.e., the payback period) and the calculation of the discount rate that makes the NPV of all cash flows equal to zero in a DCF analysis (i.e., the internal rate of return).

Additional factors in the decision-making are often more qualitative than quantitative since making the final decision is not based solely on expected cost–benefit analysis [4]. Even within the quantitative analysis, qualitative assumptions must be made, such as cash flow certainty/uncertainty, interest rate/opportunity cost selection, and time horizons.

The Special Consideration of Non-insurance Risk Transfer

It is worthwhile at this juncture for a brief discussion of non-insurance risk transfer, particularly risks that are transferred to/from contractual counterparties and certain aspects of counterparty insurance requirements [9]. These may present in various forms, e.g., hold harmless and indemnity agreements, limitations on liability, and insurance requirements between the public water/wastewater companies and vendors, suppliers, contractors, and lenders. They are present in purchase orders, contracts, and other agreements of all sizes, up to and including contracts for development, design, construction, and operation of major capital and infrastructure projects. The risks assumed under contracts may be disproportionate to the size of the financial transaction. In identifying, selecting, and negotiating the optimal risk treatment mechanism, it is prudent to build a close partnership with procurement and contracts personnel, legal counsel (inside and outside the organization), and frontline operations personnel who may have a closer perspective on the goods and services to be provided and attendant risks.

Many counterparty transactions are fairly routine, and can be addressed with standard contract or purchase order templates. Occasionally, there may be occasion to decide (or recommend) whether to relax customary counterparty indemnity or insurance requirements, particularly for small vendors/contractors. Before relaxing those requirements, consider:

- Is there a competitive alternative that can meet your organization's requirements?
- Will it set a precedent that could create a problem later?
- Will the counterparty purchase insurance solely to meet your requirements and pass through cost, uneconomically?
- Does the counterparty have a good track record?
- Will your company bear the "deep pocket" risk? If so, how do the incremental contract/project risks compare to the operational risks your company would typically retain?
- Does the risk evaluation warrant it, and is there a compelling business reason to relax requirements?

The risk manager may want to periodically meet with supply chain and end users to discuss and jointly establish criteria for relaxing or strengthening counterparty insurance requirements. Efficiencies may be achieved by prequalifying certain vendor categories for reduced or modified insurance agreements.

Major contracts or new ventures often require custom contract language and rigorous negotiations, with the risk manager playing a key support role; in some instances even being called to the forefront to aid in negotiations around contractual risk transfer and insurance provisions. Ideally, the risk manager is brought in at the beginning rather than the end of the process. It is important to become familiar with the overall contract in the early draft stages and not to focus solely on counterparty indemnity and insurance requirements. It will be helpful to ascertain early:

- How much deal leverage, i.e., relative bargaining power between the parties? Are there competitive options? Sometimes, one party or another may have to bear more than its perceived "fair share" of the risk.
- It can be important to know where the company is likely to come to a landing on indemnity before advising on insurance requirements.
- Time constraints/urgency of getting a deal done.
- Are there important principal-to-principal relationships preexistent between the parties?
- What level of indemnification the company is willing to require of or offer the counterparty?
- Whether the terms of this contract set a precedent for future transactions.

Other high-level issues to consider in non-insurance transfer negotiations:

- Which party can/should best control the exposure
- Financial strength of the counterparty, i.e., wherewithal to back up an indemnity obligation
- Accepted norms on similar deals
- Both parties' existing insurance and leverage with insurance markets
- Relative efficiencies and economies in financing/insuring the risk
- Importance of having counterparty "skin in the game"
- How much risk your organization is willing/able to retain

In evaluating counterparty risks, consider:

- Nature of the goods or services to be provided under the contract;
- If this is a prototypical situation or if there is an established track record;
- Exposure to the company's personnel, property, or customers/third parties;
- Project duration and on-site/off-site activities;

- Counterparty personnel, autos or heavy equipment used on company premises or on the company's behalf, off premises; and
- Potential and impact of direct and indirect or consequential damages.

The Special Consideration of Risk Transfers Using Insurance as a Third Party Insured

Keeping in mind on which side of the contractual obligations your company resides, consider the advantages of being an additional insured:

- Insurer provided defense.
- Easier to access than contractual liability coverage.
- Insurer cannot subrogate against its additional insured.
- Depending on the additional insured contract language, it may offer more (or less) protection than the indemnity agreement.
- May avoid disputes with counterparty over relative fault and indemnity obligations.
- Perhaps greater confidence in insurer solvency versus contractor willingness or ability to indemnify.

Keep in mind that all additional insured commitments are not the same. Over time, the insurance industry has promulgated various contract language and policy endorsements that grant additional insured status and contractual liability coverage extensions with different scopes, some quite restrictive. For example, some additional insured endorsements may limit coverage for the additional insured's contributory negligence, or for completed operations (limiting coverage for ongoing work only).

Consider potential advantages of indemnification:

- Insurance does not cover everything (e.g., faulty workmanship, design errors, delays).
- Indemnity obligation may be broader than insurance.
- Insurer may be unwilling or unable to pay.
- You may have greater leverage with counterparty than with insurer.
- Coverage may be limited, expensive, and difficult to collect.
- Always consider other risk management techniques insurance for preventing or reducing the impact of a loss. These can operate apart from or in conjunction with insurance.

Finally, be especially vigilant around the contractual interplay between its status as an additional insured and indemnity obligations. These can operate in a mutually exclusive way or be contractually linked. While there may be some situations where it is desirable to have the scope of protection be broader in one area than the other, such may not be the case. Careful attention to the contract language is important. The contractual indemnity obligation may trigger an insurance response through the operation of contractual liability coverage. Factor in the overall limitations on liability that may be embedded in the contract.

The Special Consideration of Retention

Retention is the default treatment method should the organization not intentionally choose a particular treatment method, so risk retention should actually be considered carefully before the entity decides it is the best choice for a particular risk or set of risks. As stated previously, a basic guideline is to retain reasonably predictable and relatively small loss exposures that occur infrequently, control the risk of frequent loss exposures, and transfer potentially large, disruptive loss exposures.

Numerous factors impact the cost and benefits (and thus the NPV) of retention, and each should be considered before making the treatment decision.

- Ownership structure and regulation – Form of ownership and who the owners are impact the accountability and liability structure of the entity. Public utilities, for instance, typically are held to higher prudence and financial stewardship standards than are other enterprises because of the public good they provide.
- Firm size – Smaller entities typically experience more volatility in their losses, so predictability of losses is inherently difficult making large retentions riskier than for larger organizations.
- Correlation of losses – Relationships between risks and losses can result in a cascading of losses that could drive the cost of what initially appeared to be a single loss event substantially upward and/or could increase the likelihood of additional, similar events and thus drive aggregate losses substantially upward.
- Investment opportunities – Entities with multiple high-return investment opportunities surprisingly retain less, all else the same, than do entities with fewer or less attractive investment opportunities, at least when compared with the transfer/insurance option. This is because a risk transfer, especially via insurance, replaces financial uncertainty with a financial certainty effect, freeing up whatever cash is left over after the insurance purchase to be allocated to promising projects with relatively little fear of entity failure.
- Correlation of losses with other cash flows or investment opportunities – Relationships between potential losses and financial opportunities may make retention either more or less appealing. If investment opportunities are available that make losses less likely, or make the out-of-pocket payment of losses, such a relationship may enhance the appeal of retention as a treatment option. (The converse can also be true.)
- Financial leverage – Organizations having assets and projects that are highly leveraged must be more conservative than those with little debt, and thus are likely to retain less risk and loss, all else the same. This is a straightforward result of having less discretionary, free cash flow than would otherwise be available.
- Counterparty constraints – Counterparty contractual risk and insurance requirements may limit retention options. Many contracts, for example, require proof of adequate, relevant insurance coverage prior to commencement of an agreement or partnership.

Estimating Risk-Bearing Capacity

As previously discussed, the prudence of a treatment choice for risk depends on the characteristic frequency and severity of the risk (i.e., its profile). But what amounts are considered to constitute "high" severity? The threshold(s) vary by organization and time and are dependent on the organization's financial capacity to bear, or absorb, risk. The executive and financial officers and/or directors generally determine the extent to which the organization can/should retain risks and the extent to which the entity can comfortably absorb financial fluctuations in any given year. They can consider sales projections, cash flow requirements, profit expectations, loan covenants, legal and accounting tax position, etc. All of these factors influence the ability (and willingness) to assume given exposures to loss.

No precise formulas exist to determine a firm's proper risk retention level, but several guideline formulas have been developed. Risk-bearing capacity can be tested by looking at financial performance. Rules of thumb commonly used include the following:

- Total Assets – Threshold: 1% of assets.
- Net Income (Pretax) – Threshold: 5% of net income before taxes from continuing operations. Known as accountants' materiality test.
- Working Capital.
  - Net Working Capital (Current Assets – Current Liabilities) – Threshold: 1%–5% of net working capital. (The retention selected should not reduce a firm's current liability

ratio below 2:1.) Used to determine a company's ability to quickly fund an unexpected loss, rather than its long-term financial ability to absorb loss.

- • Quick Test of Working Capital (Net Working Capital – Inventory) – Threshold: 1%–5% of quick working capital. Measures a firm's ability to cover a sudden emergency using assets that can be quickly converted to cash, and may provide an indication of the appropriate "per occurrence" retained amount.
- • Earnings and Surplus – Threshold: 1%–5% of current earned surplus and an equal or lower percentage of the average pretax earnings for the past three to five years. (This approach logically assumes that retained losses are payable from either pretax or retained earnings.)
- • Earnings per Share – Threshold: $0.10–$0.20 per share is normally acceptable on an after-tax basis.
- • Annual Revenue, or Percentage of Sales – Threshold: 1/10 of 1% to 1% of annual sales. (The high range is normally associated with retention capacity for the sum of all retained occurrences in one 12-month period.)
- • Combination of these methods.

Insurability – Pure versus Speculative Risk

Organizations face exposure to loss from a variety of causes, or perils. Among these are natural perils, like fire, windstorm, flood, and earthquake. Other threats arise from a variety of man-made perils, including the safety of the workplace and liability for damages caused by use of certain products or services. Threats arising from physical events and activities such as these are often referred to as pure risks.

Risks can also arise from enterprise activities. Every business faces the threat of revenue downturns due to changes in customer tastes or due to a countrywide business recession. The financial operations of the enterprise can also be threatened by bad credit. These types of threats are known as speculative risks. Both types of risk – pure and speculative – are combined in the management of enterprise risk. Regardless of their differing characteristics, both types of exposures to risk – pure and speculative – share the characteristics of frequency and severity (and indeed may look quite similar if only compared on this basis).

The techniques for managing pure and speculative risk differ due to the different outcome characteristics that these risks possess. Pure risk offers only a downside or status quo outcome – losses – but no direct rewards. Appropriate risk management techniques for pure risk include the sharing of risk among participants of a group or pool of similar risks. This is the basis of insurance. Speculative risk can lead to gain in addition to the possibility of loss or status quo. Speculative risks are unique in that they often entail a degree of control by the impacted parties, making them potentially non-fortuitous. As such, pooling is not an effective mechanism since it introduces the possibility of intentional actions by pool members to manipulate the character of the risks that enter the pool. Other factors that can negatively impact a risk's insurability are uniqueness, ambiguity, unpredictability, catastrophe potential (even to insurers), and a prohibitively costly fair premium.

Notes on Selection of the Best Treatment(s)

Selecting the most appropriate risk treatment means balancing the costs of implementing each activity against the benefits derived. In general, the cost of managing the risks needs to be commensurate with the benefits obtained. When making cost versus benefit judgments the wider context should also be considered.

A range of treatments may be available for each risk and these options are not necessarily mutually exclusive or appropriate in all circumstances. Selection of the most appropriate risk treatment approach should be developed in consultation with relevant stakeholders and process owners.

Furthermore, management may wish to define expectations of the treatment strategies required for each risk level. While risks rated as high priority may be insured as a requirement, risks rated as low priority, even if they have affordable opportunities for improvement, may be left to the discretion of one employee.

Ultimately, the selection of a best treatment option for a given risk should not be made merely on an expected outcome basis, whether being addressed quantitatively or qualitatively. While expected outcomes are a good starting point, the organization's risk attitude and resulting risk appetite almost certainly will lead the prudent organization to consider its ability to survive a worst-case scenario that could occur in the short run rather than simply consider its expected case scenarios.

RISK MANAGEMENT PLAN IMPLEMENTATION

Effective risk treatment relies on securing commitment from key stakeholders and developing realistic objectives and timelines for implementation.

For each risk identified in the risk assessment, detail the following:

- Specify the treatment option agreed – avoid, reduce, share/transfer, or accept.
- Document the treatment plan – outline the approach to be used to treat the risk. Any relationships or interdependencies with other risks should also be highlighted.
- Assign an appropriate owner – who is accountable for monitoring and reporting on progress of the treatment plan implementation. Where the treatment plan owner and the risk owner are different, the risk owner has ultimate accountability for ensuring the agreed treatment plan is implemented.
- Specify a target resolution date – where risk treatments have long lead times, consider the development of interim measures. For example, it is unlikely to be acceptable for a residual risk to be rated "high" and to have a risk treatment with a resolution timeframe of two years.

THE IMPORTANCE OF RISK MONITORING

Risk treatment plans may involve the redesign of existing controls, introduction of new controls, or monitoring of existing controls. Low-impact risks may require periodic monitoring while major risks are likely to require more intense management oversight. The monitoring component of both frameworks plays an important role in an organization because it provides the discipline to improve risk management capabilities and internal control continuously in a changing business environment. Monitoring assesses progress toward attaining objectives and evaluates performance of processes, risk responses, and internal control. It identifies new issues, risks, and problems as well as deficiencies in ERM and/or internal control.

SUMMARY

Once adequately and appropriately analyzed, risks must be prioritized and treated. Risk mapping can aid these decisions. The choice of risk treatments – avoid, retain, control, and/or transfer – depend on careful analysis of cost–benefit trade-offs. Appropriate execution of a risk management plan, including a comprehensive strategy for treatment of all substantive risks, is important as well, followed by monitoring and regular "re-planning." Last, information and communication are vital to execution at all levels of the organization to identify, assess, and respond to risk on an ongoing basis and ensure the achievement of objectives.

REVIEW QUESTIONS

1. An unnamed risk is deemed to be characteristically "high frequency, low severity." What is likely the best starting point for the organization to consider managing the risk?
   a. Avoidance
   b. Retention
   c. Control
   d. Transfer

2. For which risks may the precautionary principle apply?
   a. Low frequency, low severity
   b. Low frequency, high severity
   c. High frequency, low severity
   d. High frequency, high severity

3. Which risks are likely to be the organization's lowest priority?
   a. Low frequency, low severity
   b. Low frequency, high severity
   c. High frequency, low severity
   d. High frequency, high severity

4. A public wastewater system implements a worker safety program to reduce the frequency of worker injuries. This risk treatment is an example of:
   a. Avoidance
   b. Retention
   c. Control
   d. Transfer

5. T/F: Insurance carriers generally prefer to insure risks that are characteristically low in frequency and high in severity.

6. An organization's risk-bearing capacity can be estimated based on
   a. Net income
   b. Working capital
   c. Revenues
   d. All of the above

7. T/F: Risk treatment decisions are ultimately made on the basis of expected outcomes, or expected (long-run average) values.

8. T/F: Risk treatments can be selected in the same manner regardless of whether the risk(s) being considered is(are) pure or speculative.

9. Non-insurance risk transfers can be achieved through
   a. Filing claims against other parties
   b. Informal (handshake) indemnity agreements
   c. Holding status as an additional insured on another party's insurance policy
   d. All of the above

10. The value of insurance, before considering the cost, is contingent upon
    a. Policy limits
    b. Policy exclusions
    c. Policy conditions
    d. All of the above

ANSWER KEY

1. An unnamed risk is deemed to be characteristically "high frequency, low severity." What is likely the best starting point for the organization to consider managing the risk?
    a. Avoidance
    b. Retention
    c. **Control**
    d. Transfer

2. For which risks may the precautionary principle apply?
    a. Low frequency, low severity
    b. Low frequency, high severity
    c. High frequency, low severity
    d. **High frequency, high severity**

3. Which risks are likely to be the organization's lowest priority?
    a. **Low frequency, low severity**
    b. Low frequency, high severity
    c. High frequency, low severity
    d. High frequency, high severity

4. A public wastewater system implements a worker safety program to reduce the frequency of worker injuries. This risk treatment is an example of:
    a. Avoidance
    b. Retention
    c. **Control**
    d. Transfer

5. T/F: Insurance carriers generally prefer to insure risks that are characteristically low in frequency and high in severity.

6. An organization's risk-bearing capacity can be estimated based on
    a. Net income
    b. Working capital
    c. Revenues
    d. **All of the above**

7. T/F: Risk treatment decisions are ultimately made on the basis of expected outcomes, or expected (long-run average) values.

8. T/F: Risk treatments can be selected in the same manner regardless of whether the risk(s) being considered is(are) pure or speculative.

9. Non-insurance risk transfers can be achieved through
    a. Filing claims against other parties
    b. Informal (handshake) indemnity agreements
    c. **Holding status as an additional insured on another party's insurance policy**
    d. All of the above

10. The value of insurance, before considering the cost, is contingent upon
    a. Policy limits
    b. Policy exclusions
    c. Policy conditions
    d. **All of the above**

REFERENCES

1. Accenture and Oxford Economics. (2013). *Accenture 2013 Global Risk Management Study: Focus on Findings for the Energy and Utilities Sector.*
2. Akhmouch, A., and Clavreul, D. (2016). Stakeholder Engagement for Inclusive Water Governance: 'Practicing What We Preach' with the OECD Water Governance Initiative. *Water* 8(5): 204.
3. Allan, R., Jeffrey, P., Clarke, M., and Pollard, S. (2013). The Impact of Regulation, Ownership and Business Culture on Managing Corporate Risk within the Water Industry. *Water Policy* 15: 458–478.
4. Baron, J. (2019). *Thinking and Deciding.* Cambridge, UK: Cambridge University Press.
5. Doherty, N. (2000). *Integrated Risk Management: Techniques and Strategies for Reducing Risk.* New York: McGraw-Hill.
6. The Institutes. (2013). Risk Analysis. In *Enterprise Risk and Insurance Management*, 1st ed. Malvern, PA: The Institutes.
7. Marsh. (October, 2020). *Managing Renewals in a Challenging Market.* New York: Marsh & McLennan Companies.
8. Rejda, G., McNamara, M., and Rabel, W. (2020). *Principles of Risk Management and Insurance*, 14th ed. New York: Pearson.
9. Rossman, V., and Moskin, M. (2012). *Commercial Contracts: Strategies for Drafting and Negotiating*, 2nd ed. New York: Wolters Kluwer Law International.
10. Yoe, C. (2019). *Principles of Risk Analysis: Decision Making under Uncertainty*, 2nd ed. Boca Raton: Taylor and Francis.

Practical Risk Management
for Public Water/Wastewater
Systems: Conclusion

Photo Courtesy of David J. McNeil

CONCLUSION

Part III – Practical Risk Management has taken an enterprise risk management (ERM) approach to organizational risk management, and homed in on the pertinent challenges and solutions for water and wastewater ERM in particular. ERM has brought forth an ability for enterprise organizations to manage risk more proactively, yet much work remains to make this a truly productive and value-added set of initiatives. The value of risk registers and risk assessment effectiveness have served to drive awareness of risk throughout the enterprise, resulting in more consistent risk management practices, and the recognition that over-controlling for some risks can be as much of a problem as under-controlling for others. Nonetheless, many organizations struggle to ensure that risks are addressed to appropriate levels both within the enterprise and among critical strategic partners. As organizations strive for higher levels of maturity in their risk management programs, more risk agendas are being established, more assessments are being conducted, more data is being generated, and more analysis is needed to make sense of it all.

DOI: 10.1201/9781003207146-27

Regulatory requirements, assessment methodologies, and industry best practices continue to expand and evolve. Tools and technologies continue to be acquired, configured, and reconfigured all in an effort to keep pace with the growing array of risks to be managed. Organizations continue to make significant investments, yet integrating disparate risk agendas into meaningful perspectives remains challenging. The concepts of risk appetite and risk tolerance remain elusive for many organizations, especially in the public sector, to embrace and leverage as decision-making tools. Resiliency is concerned with maintaining business operations within acceptable tolerances, irrespective of which risks may be the source of disruption. Consequence management planning, by definition, focuses on business processes, their dependencies on assets and resources inside and outside of the organization, and the impacts from any type of disruption that may occur. Moreover, resiliency strives to strike a balance between the investments required to withstand disruptive events versus those needed to recover quickly within levels of impact deemed acceptable to the business. A broad body of knowledge is on hand — principally from the water sector itself — on resilience strategy and practices for water and wastepublic water/wastewater systems. Converting this knowledge into sector-wide practice, however, is restricted by resource constraints and financial challenges that require innovative strategies and collaborative approaches.

Sustainability goes beyond resiliency rethinks how investment in resilient infrastructure can be leveraged to create new opportunities to reinvigorate communities, increase inclusion and stimulate local business investment. Yet chronic underinvestment, system failures, and service shortfalls are becoming increasingly common in the US infrastructure. Each water and wastewater owner and operator manages a unique set of assets and operates under a distinct risk map-profile. As such, each utility's risk-management priorities depend on many factors, including utility size, location, assets, distinct risks, and, perhaps most importantly, the resources and capabilities the utility can access. Some serve growing populations with increasing resources, while others serve shrinking populations with declining tax bases that must maintain systems, which are now oversized for the population they serve. While each utility is responsible for its own risk management, sector-wide collaboration and information sharing play a major role in enhancement of both the resiliency and sustainability of individual systems and the sector as a whole.

Part IV – Insurance Essentials explores the risk transfer options that are available to organizations, with emphasis on the insurance needs of the water and wastewater sector. As one reads *Part IV*, it is advisable to bear in mind the financial challenges of the water sector as a whole but also the variation in size and resources between and among systems within the water sector. For instance, the water sector is adept at maintaining water services during short-term disruptions, such as a power outage lasting less than 24 hours. Beyond that time frame, the ability to maintain services depends largely on the size of the system, its location, and access to resources. Large public water/wastewater systems may be able to "fail gracefully" compared to their smaller counterparts, with relatively easy access to mobile generators, access to nearby public water/wastewater systems in the region that can provide aid or relationships with other critical infrastructure partners that can share resources.

SAMPLE EXAM

1. Economic development typically:
   a. Occurs at a greater rate than the risk taken
   b. Occurs at a lesser rate than the risk taken
   c. Occurs at about the same rate as the risk taken
   d. Does not follow risk-taking, but instead follows risk avoidance

2. T/F: ISO defines risk as any uncertainty.

3. Critical infrastructure includes:
   a. Churches
   b. Telecommunications

c. Education
d. All of the above

4. T/F: The water and wastewater sector is defined as a lifeline sector in addition to being a sector of critical infrastructure.

5. T/F: The water sector is interdependent with all other sectors of critical infrastructure.

6. ABC Water System, LLC, is a rural water services provider that struggles to fund investment projects due to limited financial and other resources. Its water infrastructure is beyond the age at which it needs to be replaced. This threat to the system is an example of:
a. Population growth shift
b. Natural disaster
c. Degradation
d. Contamination

7. T/F: A "black sky" event is any utility disruption lasting longer than one hour.

8. T/F: The primary federal US law governing water pollution is Homeland Security Presidential Directive 7.

9. T/F: US ERM goals for critical infrastructure include communication, outreach, and public confidence.

10. T/F: SCADA contributes to both infrastructure security and infrastructure risk.

11. T/F: Most critical infrastructure protection approaches do not account for sector interdependency and resilience.

12. T/F: Underlying long-term economic, environmental, social, and governance conditions that can adversely impact an organization and/or the stakeholders it serves are known as stresses.

13. T/F: Emergency management includes mitigation, preparedness, response, and recovery efforts.

14. Business continuity planning typically _____ characterized by organizations as a high priority.
a. Is
b. Is not

15. T/F: Disaster mitigation is equivalent to emergency response.

16. T/F: The EPA requires an Emergency Response Plan (ERP) from all but the smallest public water/wastewater systems.

17. NIMS principles include:
a. Flexibility
b. Uniformity
c. Collective effort
d. All of the above

18. T/F: The concept of sustainability is broader and deeper than the concept of resilience.

19. The EPA's ten core management areas on which public water/wastewater systems should focus sustainability efforts include:
a. Financial viability
b. Infrastructure stability
c. Operational resiliency
d. Water resource adequacy
e. All of the above

20. T/F: Three quarters (3/4th) of the world's population is impacted by water scarcity.

21. T/F: The EPA considers that sustainability is important to water utility risk management.

22. T/F: According to the Committee of Sponsoring Organizations of the Treadway Commission (COSO), top management is solely responsible for ERM effectiveness.

23. T/F: The focus of ERM is on accounting and not concerned with strategic risks.

24. The stated advantages of an enterprise risk management approach over the traditional risk management approach include:
   a. Better awareness of new and emerging risks
   b. Closing the gaps in risk management across functional areas
   c. Improved risk responsiveness
   d. All of the above

25. T/F: Enterprise risk management was adopted from project management and is simply treated as an ongoing project.

26. T/F: The objective of enterprise risk management is to develop a holistic, portfolio view of the organization's significant risks to the achievement of the entity's important goals and objectives.

27. Who leads the ERM effort within an organization?
   a. Top management and Board of Directors
   b. Appropriate leadership depends on the organization
   c. Each functional area leads its own ERM program and process
   d. None of the above is true

28. T/F: Cascading responsibility for ERM within a public water or wastewater system means the most responsibility rests at the top of the organization and the least responsibility rests at the rank-and-file level of the organization.

29. T/F: ERM is essentially a legal function.

30. T/F: Risk attitude is related to how one makes decisions under uncertainty.

31. T/F: Key risk indicators (KRIs) within an organization are metrics that provide early warning of increasing risk exposures in various areas of the enterprise.

32. T/F: Risk aversion differs from risk neutrality in that the risk-averse decision maker must expect to be compensated for taking risks while the risk-neutral decision maker does not require an expected risk compensation.

33. T/F: Risk appetite and risk tolerance are related in that risk tolerances determine the organization's risk appetite.

34. T/F: Risk-bearing capacity influences risk appetite in that it impacts what is considered a high impact, or high severity, loss outcome.

35. The risks that result in organizational value killers can:
   a. Be organizational value drivers if they are well managed
   b. Involve events that erode at least 20% of the organization's relative value within one month
   c. Source from any category of risk, not just the strategic risk category
   d. All of the above are true

36. T/F: Both tangible and intangible assets contribute to reputation value and risks.

37. Organizational ethics and integrity contribute to organizational reputation:
   a. Directly, since triple bottom line (TBL) accounting recognizes ethics and integrity as having a quantified value
   b. Directly, as ethics and integrity guarantee the organization experiences fewer and less costly losses
   c. Indirectly, through a reduced likelihood of organizational value loss
   d. Indirectly, since ethics and integrity are virtually identical to ESG initiatives

38. T/F: One driver of reputation risk is innovation.

39. Triple bottom line (TBL) accounting differs from traditional accounting in that TBL accounting:
    a. Recognizes intangible assets and values related to ESG performance
    b. Does not quantify the values of reputation and other intangibles
    c. Does not use numbers at all but instead uses qualitative measures to describe organizational value
    d. All of the above

40. T/F: The challenges of tackling the reputation risk problem within the utilities sector are largely due to tensions between the need for innovation, the need for reliability, and the need for affordability.

41. Risk identification involves the development of:
    a. A risk list
    b. A risk register
    c. An overall risk statement
    d. A risk statement for each risk that is registered

42. T/F: A risk register outlines risks after a risk treatment is selected.

43. The elements of an effective risk statement include all except:
    a. Objective and event
    b. Cause and impact
    c. Time window within which it could happen
    d. All of the above are included

44. Suppose Water, Water Everywhere, Inc., an investor-owned water utility, identifies its risks and develops its risk statements according to the following template: **"X risk is a low/high frequency exposure having a low/high expected impact."** What is true about this template?
    a. It is a full description of the risk
    b. It is an incomplete description of the risk since it does not tie back to organizational objectives
    c. It is an incomplete description of the risk because it does not state when a loss is going to occur
    d. It is an overstatement of the risk because frequency and impact are not measurable until after a loss happens

45. Sources of risk information include:
    a. Flow charts
    b. Financial statements
    c. Interviews
    d. Insurance policies
    e. All of the above are sources

46. T/F: The sources of risk information that are utilized will likely impact which techniques can best be used to reveal exposures.

47. T/F: Bowtie analysis differs from root cause analysis in that bowtie analysis always utilizes hypothetical loss scenarios rather than historical loss events.

48. T/F: Scenario testing offers a potential advantage over the statistical approach to risk analysis because it engages creative thinking.

49. T/F: The Delphi technique is an inferior technique for identifying risks because it involves using multiple, and sometimes conflicting, sources of information.

50. T/F: Risk analysis is interchangeable with risk identification.

51. Risk analysis is important to the ERM process because it directly _____ risks, leading to the appropriate _____ of risks.
    a. Measures; treatment
    b. Discovers; measurement

    c. Treats; measurement
    d. Discovers; treatment

52. A risk's characteristic frequency may be expressed as its:
    a. Probability
    b. Impact
    c. Maximum
    d. All of the above

53. A risk's characteristic severity may be expressed as its:
    a. Mean loss amount
    b. Probable maximum loss
    c. Median loss amount
    d. Probability

54. T/F: The most likely outcome from any probability distribution is its mean.

55. Statistical analysis may differ from loss modeling analysis in that statistical analysis relies on:
    a. Historic loss data
    b. A changing loss environment
    c. A simulated loss environment
    d. All of the above

56. T/F: A probability cannot take a negative value.

57. Estimating the number of events that will occur involves the prediction of a:
    a. Continuous variable
    b. Discrete variable
    c. Constant
    d. All of the above

58. T/F: Scenario building is most often used in risk analysis to estimate risk frequency.

59. The Poisson distribution is commonly used to describe:
    a. Risk frequency alone
    b. Risk severity alone
    c. Risk frequency and severity
    d. Risk velocity

60. T/F: The Tweedie distribution is increasingly used to represent the compound frequency and severity of a risk.

61. T/F: The belief that a past outcome was 100% predictable when in reality it was not, is referred to as confabulation.

62. The armchair fallacy most likely results from:
    a. Missing information
    b. Inadequate time to reflect
    c. Inadequate meaning in the data
    d. Memory limitations

63. Which of the following is true regarding confirmation bias, in-group bias, and overconfidence bias?
    a. They are all three special cases of the same bias
    b. They may be found together in the same person, in the same decision moment
    c. They are mutually exclusive biases, meaning they are specifically never found together
    d. They are all resultant from the mind's memory limitations

64. T/F: The Cognitive Reflection Test (CRT) and the Need for Closure Scale (NFCS) scores are inversely correlated to each other: The higher the CRT score, the lower the NFCS score.

65. T/F: System 2 thinking, when practiced regularly, has been shown to increase CRT scores and reduce bias in decision-making.

66. An unnamed risk is deemed to be characteristically "low frequency, high severity." What is likely the best starting point for the organization to consider managing the risk?
    a. Avoidance
    b. Retention
    c. Control
    d. Transfer

67. For which risks may the precautionary principle apply?
    a. Low frequency, low severity
    b. Low, frequency, high severity
    c. High frequency, low severity
    d. High frequency, high severity

68. T/F: Risks having low-frequency, low-severity characteristics are likely to be the organization's lowest priority, all else the same.

69. A public wastewater system implements a worker safety program to reduce the severity of worker injuries. This risk treatment is an example of:
    a. Avoidance
    b. Retention
    c. Control
    d. Transfer

70. T/F: Insurance carriers generally prefer to insure risks that are characteristically high in frequency and low in severity.

71. T/F: An organization's risk-bearing capacity may be estimated for purposes of deciding how much risk to retain.

72. Risk treatment decisions are ultimately made on the basis of
    a. Expected outcomes, or expected (long-run average) values
    b. Risk-bearing capacity
    c. Probable maximum loss amounts
    d. All of the above

73. T/F: Exposures that are subject to rare (or zero) historic losses are less risky than predictable losses.

74. T/F: Noninsurance risk transfers are commonly achieved through hold harmless and/or indemnity agreements.

ANSWER KEY

1. Economic development typically:
    a. **Occurs at a greater rate than the risk taken**
    b. Occurs at a lesser rate than the risk taken
    c. Occurs at about the same rate as the risk taken
    d. Does not follow risk-taking, but instead follows risk avoidance

2. T/**F**: ISO defines risk as any uncertainty.

3. Critical infrastructure includes:
    a. Churches
    b. **Telecommunications**
    c. Education
    d. All of the above

4. **T/F**: The water and wastewater sector is defined as a lifeline sector in addition to being a sector of critical infrastructure.

5. **T/F**: The water sector is interdependent with all other sectors of critical infrastructure.

6. ABC Water System, LLC, is a rural water services provider that struggles to fund investment projects due to limited financial and other resources. Its water infrastructure is beyond the age at which it needs to be replaced. This threat to the system is an example of:
    a. Population growth shift
    b. Natural disaster
    c. **Degradation**
    d. Contamination

7. T/**F**: A "black sky" event is any utility disruption lasting longer than one hour.

8. T/**F**: The primary federal US law governing water pollution is Homeland Security Presidential Directive 7.

9. **T**/F: US ERM goals for critical infrastructure include communication, outreach, and public confidence.

10. **T**/F: SCADA contributes to both infrastructure security and infrastructure risk.

11. T/**F**: Most critical infrastructure protection approaches do not account for sector interdependency and resilience.

12. **T**/F: Underlying long-term economic, environmental, social, and governance conditions that can adversely impact an organization and/or the stakeholders it serves are known as stresses.

13. **T**/F: Emergency management includes mitigation, preparedness, response, and recovery efforts.

14. Business continuity planning typically _____ characterized by organizations as a high priority.
    a. Is
    b. **Is not**

15. T/**F**: Disaster mitigation is equivalent to emergency response.

16. **T**/F: The EPA requires an Emergency Response Plan (ERP) from all but the smallest public water/wastewater systems.

17. NIMS principles include:
    a. Flexibility
    b. Uniformity
    c. Collective effort
    d. **All of the above**

18. **T**/F: The concept of sustainability is broader and deeper than the concept of resilience.

19. The EPA's ten core management areas on which public water/wastewater systems should focus sustainability efforts include:
    a. Financial viability
    b. Infrastructure stability
    c. Operational resiliency
    d. Water resource adequacy
    e. **All of the above**

20. T/**F**: Three quarters (3/4th) of the world's population is impacted by water scarcity.

21. **T**/F: The EPA considers that sustainability is important to water utility risk management.

22. T/**F**: According to the Committee of Sponsoring Organizations of the Treadway Commission (COSO), top management is solely responsible for ERM effectiveness.

23. T/**F**: The focus of ERM is on accounting and not concerned with strategic risks.

24. The stated advantages of an enterprise risk management approach over the traditional risk management approach include:
   a. Better awareness of new and emerging risks
   b. Closing the gaps in risk management across functional areas
   c. Improved risk responsiveness
   d. **All of the above**

25. T/**F**: Enterprise risk management was adopted from project management and is simply treated as an ongoing project.

26. **T**/F: The objective of enterprise risk management is to develop a holistic, portfolio view of the organization's significant risks to the achievement of the entity's important goals and objectives.

27. Who leads the ERM effort within an organization?
   a. **Top management and Board of Directors**
   b. Appropriate leadership depends on the organization
   c. Each functional area leads its own ERM program and process
   d. None of the above is true

28. **T**/F: Cascading responsibility for ERM within a public water or wastewater system means the most responsibility rests at the top of the organization and the least responsibility rests at the rank-and-file level of the organization.

29. T/**F**: ERM is essentially a legal function.

30. **T**/F: Risk attitude is related to how one makes decisions under uncertainty.

31. **T**/F: Key risk indicators (KRIs) within an organization are metrics that provide early warning of increasing risk exposures in various areas of the enterprise

32. **T**/F: Risk aversion differs from risk neutrality in that the risk-averse decision maker must expect to be compensated for taking risks while the risk-neutral decision maker does not require an expected risk compensation.

33. T/**F**: Risk appetite and risk tolerance are related in that risk tolerances determine the organization's risk appetite.

34. **T**/F: Risk-bearing capacity influences risk appetite in that it impacts what is considered a high impact, or high severity, loss outcome.

35. The risks that result in organizational value killers can:
   a. Be organizational value drivers if they are well managed
   b. Involve events that erode at least 20% of the organization's relative value within one month
   c. Source from any category of risk, not just the strategic risk category
   d. **All of the above are true**

36. **T**/F: Both tangible and intangible assets contribute to reputation value and risks.

37. Organizational ethics and integrity contribute to organizational reputation:
   a. Directly, since triple bottom line (TBL) accounting recognizes ethics and integrity as having a quantified value
   b. Directly, as ethics and integrity guarantee the organization experiences fewer and less costly losses
   c. **Indirectly, through a reduced likelihood of organizational value loss**
   d. Indirectly, since ethics and integrity are virtually identical to ESG initiatives

38. **T**/F: One driver of reputation risk is innovation.

39. Triple bottom line (TBL) accounting differs from traditional accounting in that TBL accounting:
   a. **Recognizes intangible assets and values related to ESG performance**
   b. Does not quantify the values of reputation and other intangibles

    c. Does not use numbers at all, but instead uses qualitative measures to describe organizational value

    d. All of the above

40. **T/F**: The challenges of tackling the reputation risk problem within the utilities sector are largely due to tensions between the need for innovation, the need for reliability and the need for affordability.

41. Risk identification involves the development of:

    a. A risk list

    b. A risk register

    c. An overall risk statement

    d. **A risk statement for each risk that is registered**

42. T/F: A risk register outlines risks after a risk treatment is selected.

43. The elements of an effective risk statement include all except:

    a. Objective and event

    b. Cause and impact

    c. Time window within which it could happen

    d. **All of the above are included**

44. Suppose Water, Water Everywhere, Inc., an investor-owned water utility, identifies its risks and develops its risk statements according to the following template: **"X risk is a low/high frequency exposure having a low/high expected impact."** What is true about this template?

    a. It is a full description of the risk

    b. **It is an incomplete description of the risk since it does not tie back to organizational objectives**

    c. It is an incomplete description of the risk because it does not state when a loss is going to occur

    d. It is an overstatement of the risk because frequency and impact are not measurable until after a loss happens

45. Sources of risk information include:

    a. Flow charts

    b. Financial statements

    c. Interviews

    d. Insurance policies

    e. **All of the above are sources**

46. **T/F**: The sources of risk information that are utilized will likely impact which techniques can best be used to reveal exposures.

47. **T/F**: Bowtie analysis differs from root cause analysis in that bowtie analysis always utilizes hypothetical loss scenarios rather than historical loss events.

48. **T/F**: Scenario testing offers a potential advantage over the statistical approach to risk analysis because it engages creative thinking.

49. T/F: The Delphi technique is an inferior technique for identifying risks because it involves using multiple, and sometimes conflicting, sources of information.

50. T/F: Risk analysis is interchangeable with risk identification.

51. Risk analysis is important to the ERM process because it directly _____ risks, leading to the appropriate _____ of risks.

    a. **Measures; treatment**

    b. Discovers; measurement

    c. Treats; measurement

    d. Discovers; treatment

52. A risk's characteristic frequency may be expressed as its:
    a. **Probability**
    b. Impact
    c. Maximum
    d. All of the above

53. A risk's characteristic severity may be expressed as its:
    a. Mean loss amount
    b. Probable maximum loss
    c. Median loss amount
    d. **Probability**

54. T/**F**: The most likely outcome from any probability distribution is its mean.

55. Statistical analysis may differ from loss modeling analysis in that statistical analysis relies on:
    a. **Historic loss data**
    b. A changing loss environment
    c. A simulated loss environment
    d. All of the above

56. **T**/F: A probability cannot take a negative value.

57. Estimating the number of events that will occur involves the prediction of a:
    a. Continuous variable
    b. **Discrete variable**
    c. Constant
    d. All of the above

58. T/**F**: Scenario building is most often used in risk analysis to estimate risk frequency.

59. The Poisson distribution is commonly used to describe:
    a. **Risk frequency alone**
    b. Risk severity alone
    c. Risk frequency and severity
    d. Risk velocity

60. **T**/F: The Tweedie distribution is increasingly used to represent the compound frequency and severity of a risk.

61. T/**F**: The belief that a past outcome was 100% predictable when in reality it was not, is referred to as confabulation.

62. The armchair fallacy most likely results from:
    a. Missing information
    b. **Inadequate time to reflect**
    c. Inadequate meaning in the data
    d. Memory limitations

63. Which of the following is true regarding confirmation bias, in-group bias, and overconfidence bias?
    a. They are all three special cases of the same bias
    b. **They may be found together in the same person, in the same decision moment**
    c. They are mutually exclusive biases, meaning they are specifically never found together
    d. They are all resultant from the mind's memory limitations

64. **T**/F: The Cognitive Reflection Test (CRT) and the Need for Closure Scale (NFCS) scores are inversely correlated to each other: The higher the CRT score, the lower the NFCS score.

65. **T**/F: System 2 thinking, when practiced regularly, has been shown to increase CRT scores and reduce bias in decision-making.

66. An unnamed risk is deemed to be characteristically "low frequency, high severity." What is likely the best starting point for the organization to consider managing the risk?
    a. Avoidance
    b. Retention
    c. Control
    d. **Transfer**

67. For which risks may the precautionary principle apply?
    a. Low frequency, low severity
    b. Low, frequency, high severity
    c. High frequency, low severity
    d. **High frequency, high severity**

68. T/F: Risks having low-frequency, low-severity characteristics are likely to be the organization's lowest priority, all else the same.

69. A public wastewater system implements a worker safety program to reduce the severity of worker injuries. This risk treatment is an example of:
    a. Avoidance
    b. Retention
    c. **Control**
    d. Transfer

70. T/**F**: Insurance carriers generally prefer to insure risks that are characteristically high in frequency and low in severity.

71. **T**/F: An organization's risk-bearing capacity may be estimated for purposes of deciding how much risk to retain.

72. Risk treatment decisions are ultimately made on the basis of:
    a. Expected outcomes, or expected (long-run average) values
    b. Risk-bearing capacity
    c. Probable maximum loss amounts
    d. **All of the above**

73. T/**F**: Exposures that are subject to rare (or zero) historic losses are less risky than predictable losses.

74. T/F: Noninsurance risk transfers are commonly achieved through hold harmless and/or indemnity agreements.

DISCLAIMER

The information contained in *Part III – Practical Risk Management* is meant to provide the reader with a general understanding of certain aspects of risk management. The information isn't to be construed as risk management advice and isn't meant to be a substitute for risk management advice. AAWD&M, its stakeholders, and the authors expressly disclaim any and all liability with respect to any actions taken or not taken based upon the information contained in this section or with respect to any errors or omissions contained in such information. Readers are cautioned to thoroughly review their risk management policies. The recommendations contained in *Part III – Practical Risk Management* may not be appropriate for your organization and may be deemed inaccurate.

Part IV

Insurance Essentials for Public Water/Wastewater Systems

Paul Fuller

2019 Photo Courtesy of Las Virgenes Municipal Water District

DOI: 10.1201/9781003207146-28

Insurance Essentials for Public Water/Wastewater Systems: Introduction

LEARNING OBJECTIVES

- Explain the purpose and mechanics of insurance as an effective and necessary risk management technique.
- Learn how to properly structure and design your property and liability insurance policies to maximize the transfer of your most salient risks.
- Comprehend the role and benefit of a broker advocate when it comes to submitting your public water and wastewater system to the insurance marketplace, securing renewal terms, and ensuring you are paired with a best-in-class insurer.
- Understand the advantages and disadvantages of regulated insurers and unregulated pools.

DOI: 10.1201/9781003207146-29

KEY DEFINITIONS*

Admitted Insurer – an insurer licensed to do business in the state or country in which the insured exposure is located.

Application – form providing the insurer with certain information necessary to underwrite a given risk. The applicant completes it to receive insurance.

Assessments – the right to assess additional charges above an initial premium when those premiums are shown to be inadequate to cover the costs of actual and anticipated losses.

Audit – survey of the financial records of a person or organization conducted annually (in most cases) to determine exposures, limits, premiums, etc.

Broker – insurance intermediary who represents the insured rather than the insurer. Since they are not the legal representatives of insurers, brokers, unlike independent agents, often do not have the right to act on behalf of insurers, such as to bind coverage. While some brokers do have agency contracts with some insurers, they usually remain obligated to represent the interests of insureds rather than insurers (e.g., some state insurance codes impose a fiduciary responsibility to act on behalf of their customers or provide full disclosure of all their compensation from all sources).

Cumis Counsel – the 1984 California Court of Appeals decision in *San Diego Fed. Credit Union v. Cumis Ins. Soc'y, Inc.*, 162 Cal. App. 3d 358, 208 Cal. Rptr. 494 (Cal. App. 4th Dist. 1984), held that where an insurer is defending under a reservation of rights, a conflict of interest exists between the economic and litigation interest of the insurer and the insured. Under these circumstances, the traditional right of the insurer to select counsel and control the defense effort is lost. As a result, the insured selects counsel and controls the defense, but the insurer must pay the defense costs. Such counsel, selected and controlled by the insured at the expense of the insurer, is known as *Cumis Counsel.*

Department of Insurance – a governmental entity charged with the regulation and administration of insurance laws and other responsibilities associated with insurance.

Dividends – partial return of premium to the insured based on the insurer's financial performance or on the insured's own loss experience.

Governing Body – group of elected or appointed people who formulate the policy and direct the affairs of a public water and wastewater system.

Insurance – contractual relationship that exists when one party (the insurer) for a consideration (the premium) agrees to reimburse another party (the insured) for loss to a specified subject (the risk) caused by designated contingencies (hazards or perils).

Insurance Pools – public entity risk pools are defined as a cooperative group of governmental entities joining together to finance an exposure, liability, or risk (e.g., property and liability, workers compensation, or employee health care).

Insured – the person(s) protected under an insurance contract.

Insurer – the insurance company that undertakes to indemnify for losses and perform other insurance-related operations.

---

* Source material provided by International Risk Management Institute (IRMI) (December 20, 2020). In IRMI Insurance Glossary. Retrieved from www.irmi.com/glossary. Additional source material provided by policy forms and underwriting products promulgated by Insurance Services Office, (ISO) and their member companies, https://products.iso.com/productcenter/10. *Black's Law Dictionary*, 10th ed., 2014 was used for the definition of joint and several liability.

Joint and Several Liability – liability that may be apportioned either among two or more parties or to only one or a few select members of the group, at the adversary's discretion. Thus, each liable party is individually responsible for the entire obligation, but a paying party may have a right of contribution or indemnity from nonpaying parties.

Public Wastewater System – system that provides sanitary sewer, stormwater, and other used water services to residences, businesses, and industries.

Public Water System – system that provides water for human consumption to 15 or more connections or regularly serves 25 or more people daily for at least 60 days out of the year.

Reservation of Rights – an insurer's notification to an insured that coverage for a claim may not apply. Such notification allows an insurer to investigate (or even defend) a claim to determine whether coverage applies (in whole or in part) without waiving its right to later deny coverage based on information revealed by the investigation. Insurers use a reservation of rights letter because in many claim situations, all the insurer has at the inception of the claim are various unsubstantiated allegations and, at best, a few confirmed facts.

Risk Financing – business strategy of providing funds to cover the financial effect of unexpected losses experienced by an organization. Traditional forms of finance include risk transfer via insurance, self-insurance funding, and pooling of risks/funds with other organizations. The risk financing process consists of five steps: identifying and analyzing exposures, analyzing alternative risk financing techniques, selecting the best risk financing technique(s), implementing the selected technique(s), and monitoring the selected technique(s).

Risk Management – practice of identifying and analyzing loss exposures and taking steps to minimize the financial impact of the risks they impose. Traditional risk management, sometimes called insurance risk management, has focused on pure risks (i.e., possible loss by fortuitous or accidental means) but not business risks (i.e., those that may present the possibility of loss or gain).

Risk Transfer – use of contractual obligations such as indemnity and exculpatory agreements, waivers of recovery rights, and insurance requirements to pass along to others what would otherwise be one's own risks of loss.

Wastewater – water that has been used for various purposes around a community, including sewage, stormwater, and all other water used by residences, businesses, and industry. Wastewater requires treatment before it returns to waterways to protect the health of the waterbody and community.

INTRODUCTION

Insurance is an essential risk management technique for controlling many of the observable and imperceptible exposures facing public water/wastewater systems. Maximizing this technique requires a dual aptitude for risk identification and risk transfer. The former necessitates a thorough familiarity with operational and enterprise exposures, including emerging trends, to ensure all relevant risks are cataloged. The latter requires a capacity to properly design and structure insurance policies to augment the transfer of such risks. Other risk management techniques like avoidance, financing, and mitigation are necessary to address exposures that cannot be pragmatically transferred to insurers. *Part IV – Insurance Essentials* focuses on the proficient use of insurance as an effective risk transfer strategy. Its central purpose is to enhance the quality of your insurance placement by granularly examining those property and liability policies relevant to public water/wastewater systems. Achieving quality requires prudent amalgamation of policy terms, extra contractual services, and insurer resources. The ensuing outcome is value creation. Practical solutions will similarly be imparted to adeptly configure these policies and successfully transfer your most prominent exposures. Commentary from the author is bracketed to the core tenets of risk identification and risk transfer.

Property and liability insurance policies are not commodity purchases. Regrettably, buying decisions for these policies are commonly decided on the lowest common denominator: price. This focus is misguided, as the price of insurance should be subordinate to cost and compared against the likelihood of an uncovered or partially covered claim. In that context, the purchase of a lower priced, lower quality policy invariably costs more than the premium outlay of a higher priced, higher quality policy. Insurance decisions founded on value are always easier to defend than decisions rooted in price. That is not to say price is irrelevant; it is not. Pricing should be considered alongside other important factors such as coverages, limits, deductibles, and resources. The last intangible factor of resources is particularly important and embeds an insurer's subject matter expertise, seasoned claims professionals, knowledgeable risk engineers, and trial-tested defense attorneys. Ultimately, the quality of your purchase is a combination of policy terms and insurer resources. We will demonstrate the process to elevate your insurance policies where quality is prominent and premium is fair. The starting point of value begins with the integral role of your broker advocate in the insurance placement and risk management process.

BROKER ADVOCATE

A broker advocate serves as an intermediary, negotiator, and advisor who presents your public water and wastewater system to the insurance marketplace. A successful partnership involves timely communication, personal consultation, marketplace relationships, and organizational knowledge. Your insurance renewal process should commence a minimum of 120 days prior to policy expiration date and encompass a discussion on the state of the insurance marketplace, potential renewal challenges, and overall placement strategies. This meeting allows your broker advocate to share anticipated rate levels and projections on coverages, limits, and deductibles. It is similarly an opportunity to review and update your property, equipment, and automobile schedules as well as the operational and governance details of your organization. Insurance is a relationship business, and it behooves you to cogently and professionally present your public water and wastewater system to the marketplace. Examples of pertinent information to include in your renewal packet are as follows: operational summary; claims analysis; safety programs; board-and-management continuity; employee attrition; water operator licensures; pipeline-and-infrastructure replacement programs; financial stewardship; water quality conformance; technology updates; major projects; fire suppression/mitigation measures; and community relations.

You and your broker advocate should review incumbent renewal quotes and other alternatives under consideration. This review should comprise a detailed summary of the policy terms as well as a drilldown of pertinent coverages, exclusions, limits, and deductibles. Ambiguous or conflicting policy language should be codified. The objective is to reach consensus with your broker advocate as to the best renewal solution for your organization. Your broker advocate can then finalize your placement and reconvene with you to review the issued policies for accuracy, consistency, and thoroughness. Your post-renewal review should include confirmation that your property, equipment, and automobile schedules are correctly transcribed on the issued policies. Ad hoc meetings and telephone calls should take place throughout the policy period to proactively address emerging issues, open claims, coverage adequacy, and risk management services. That last point involves contract review, coverage options, and risk transfer solutions. An adept broker advocate will steer your renewal to a best-in-class insurer specializing in public water/wastewater systems. There are two policy options to seriously consider when evaluating your renewal placement: regulated or unregulated. Each option has its advantages and disadvantages, which we will discuss below.

Regulated vs. Unregulated*

Regulated insurance policies are frequently issued by admitted insurers and subject to insurance regulatory oversight at the state level. Such oversight is extensive and includes approval of rates and forms as well as strict capital and surplus requirements. States also have guaranty funds should an admitted insurer become insolvent. Regulated policies are governed by insurance law, which is based on the adhesion contract. That means any ambiguity inures to the benefit of the weaker party, which is the insured. There are no contractual requirements to renew with an admitted insurer for a specified number of years. Insureds can change their insurer with minimal penalties. Most regulated policies require access by a broker advocate who is paid a commission by the insurer and serves as your intermediary. Regulators require certain insured safeguards like reservation of rights letters when defense is extended without full commitment on indemnity and *Cumis* counsel when conflicts exist between insurer and insured. The levers of bad faith and punitive damages are similarly available for insureds to level the playing field. Regulated policies are written without joint and several liability, which means insureds are not financially liable should losses exceed premium. Policies are issued without assessments so costs are fixed. Regulated policies are typically more expensive than unregulated options. The regulation and legal protections associated with these policies, coupled with the utilization of a broker advocate, are integral reasons for this cost difference.

Unregulated insurance policies, referred to as Memorandums of Coverage (MOC), are issued by joint powers authorities (JPAs). These issuers, hereinafter referred to as pools, are groups of public entities that formally coalesce to share their risks and insurance costs. The basis of pooling is the inherent belief of strength in affiliation. When public entities share risk and premium, the combination should create lower costs than if the public entities acted alone. This unification is intended to augment safety, risk management, and loss control services as well as lessen insurance costs. Although all pools share the aforementioned values, each pool reflects their members' unique insurance and risk management priorities as well as their respective state's laws and regulations. An important distinction is its legal configuration: pools are considered self-insurance and operate outside of insurance regulatory oversight. Pools also require joint and several liability with its members unless they fully reinsure or cede 100% of their exposures to a regulated insurer. Any bearing of risk or assumption of future losses necessitates joint and several liability. The upside to such risks are opportunities for dividends, but the downside are possibilities for assessments. Members frequently pay lower insurance costs compared to regulated insurance policies due to these structural differences. Additional reasons include their not-for-profit nature, exemption of premium-and-income taxes, and infrequent utilization of broker advocates.

There are compelling reasons in support of regulated and unregulated policies. Consequently, the central focus of your selection criteria should be the quality, value, and reputation of the insurer. Your insurer should comprise best-in-class resources and demonstrable water and wastewater pedigree. This assessment should be conducted in parallel with the breadth of terms quoted and the depth of exposures transferred. Pricing and service should be included in the evaluation. Once the clinical review has been completed, it is prudent to examine the inherent advantages and disadvantages of regulated and unregulated policies. Your analysis should involve discussion with your governing body as to their comfort with joint and several liability, broker advocate utilization, multiyear contractual commitments, and any unique differences in policy terms. This diligence is necessary to select a best-in-class insurer that aligns with the cultural, financial, and risk levels of your public water and wastewater system. Our discussion throughout *Part IV – Insurance Essentials* applies equally to regulated and unregulated policies.

---

* Source material for unregulated policies provided by Association of Governmental Risk Pools (AGRIP), https://www.agrip.org.

Admitted Insurer vs. Nonadmitted Insurer*

Admitted insurers are insurance carriers that have been admitted or licensed to do business in a state and are subject to the control and regulation of the respective department of insurance or similar regulatory body. Such regulation is extensive and includes approval of coverage forms, rates, and underwriting rules. Each state has insurance statutes, and their department of insurance (or equivalent regulatory body) oversees admitted insurers to ensure compliance with all insurance laws. Admitted insurers pay a premium tax to the state, usually 1% to 1.5% of earned premium. An admitted insurer must participate in a state's Guaranty Fund (a fund created to pay the claims of an insolvent insurance company), FAIR Plan (distressed property), Assigned Risk Plans (workers compensation and automobile financial responsibility), and other pooling arrangements such as Wind Pools in those states with exposure to large windstorm losses.

Some insurers do not want to participate in a Guaranty Fund, FAIR Plan, Assigned Risk Plan, or other poling arrangements but would rather do business in the state on a nonadmitted basis. These insurers may choose this arrangement because they only write certain lines of coverage, are not involved in standard coverage, or specialize in certain lines of coverage. Therefore they chose to do business as a nonadmitted insurer. In order to do business as a nonadmitted insurer, the insurance carrier must apply to the state to do business on a nonadmitted basis, prove financial ability, and be of good character and reputation. If accepted, the carrier will become an approved or permitted nonadmitted insurer. A permitted nonadmitted insurer does not participate in any of the mandatory pooling arrangements, is not regulated by the respective state department of insurance, and pays a much higher earned premium tax. If an insured chooses to do business with a permitted nonadmitted insurer, the insured cannot seek help from the department of insurance in the event there is a claims problem or an insolvency. The insured's only recourse is to bring a lawsuit against the carrier.

Author's Commentary

The chapters that accompany *Part IV – Insurance Essentials* encapsulate a thorough review of property and liability policies deemed essential for public water/wastewater systems. Each of the policies necessitates a comprehension of their specific function and purpose. Their overall value centers on details, and these particulars will govern a policy grade standard of great, good, or average. The difference between grade standards is a matter of gradation, and the nuance among grade standards underscores the importance of identifying and grasping the minutia. *Part IV – Insurance Essentials* highlights these details and endeavors to serve as a practical reference manual. To that end, a checklist is inserted at the end of each chapter to compare your property and liability insurance policies with best-in-class benchmarks. We also included an article on an emerging legal theory that is problematic for public water/wastewater systems in states with broad constitutional protections on due process. In terms of stylistic consistency, we use the term insurer to embody regulated insurers and unregulated pools. Such reference is not intended to infer regulated policies are better than unregulated policies. They are not, as both have their advantages and disadvantages. The adoption is solely intended to improve readability.

We now begin our examination with the property policy. This policy encompasses coverage for your real property, business personal property, mobile equipment, business income, and extra expense as well as proprietary coverage extensions and equipment breakdown.

---

* Source: W. Kurt Fickling, CRM, CIC, CRIS, Instructor, Risk Management and Insurance, College of Business, Department of Finance and Insurance, East Carolina University.

POLICY CHECKLIST

REGULATED VS. UNREGULATED POLICIES

Structural Differences		
Description	**Regulated Insurers***	**Unregulated Pools**
Insurance or Self-Insurance	Insurance	Self-Insurance
Insurance or Contract Law	Insurance	Contract
Risk Bearer	Insurer	JPA
Time Commitment	None	Varies (typically 3 years)
Notification to Exit Requirement	None	Varies (up to1 year)
Common Anniversary Date	None	Yes
Mandatory Binding Arbitration	No	Yes
Joint and Several Liability	No	Yes (unless a fully reinsured pool)
Insurance Regulatory Oversight	Yes	No
Reservation of Rights Requirement	Yes	No
Cumis Counsel Requirement[†]	Yes	No
Unfair Claims Practices Act	Yes	No
Bad Faith Actions	Yes	No
Premium Taxes	Yes	No
Rate and Form Filings	Yes	No
Market Conduct Examination	Yes	No
Terrorism Risk Insurance Act (TRIA) Eligibility	Yes	No
State Guaranty Fund (Insurer Insolvency)	Yes	No

*Regulated insurer in this context refers to an admitter insurer.

[†]Cumis Counsel is an attorney employed by a defendant in a lawsuit when there is a liability insurance policy supposedly covering the claim, but there is a conflict of interest between the insurer and the insured defendant.

REVIEW QUESTIONS

1. T/F: When examining insurance policies, pricing should be considered alongside other important factors such as coverages, limits, deductibles, and insurer resources.

2. T/F: A broker advocate serves as an intermediary, negotiator, and advisor who presents your public water and wastewater system to the insurance marketplace.

3. Examples of information to include in your property and liability renewal submission include:
   a. Operational summary
   b. Loss runs including claims analysis
   c. Current property, equipment, and automobile schedules
   d. All of the above

4. T/F: Regulated insurance policies are subject to insurance regulatory oversight at the federal level.

5. T/F: Insurance pools are governed by insurance law, which is based on an adhesion contract.

6. Insurance regulators require certain insured safeguards with regulated policies such as:
   a. Reservation of rights letters
   b. Multiyear policy periods
   c. Joint and several liability
   d. Premium guarantees

7. T/F: The upside and downside of insurance pools with joint and several liability are the potential for members to receive dividends and assessments.

8. Insurance pools are commonly less expensive than regulated insurance policies because of:
    a. Joint and several liability
    b. Exemption of premium taxes
    c. Exemption of income taxes
    d. All of the above

9. T/F: Joint and several liability is legal doctrine applying in some states that allows an injured person to sue and recover from one or more of several wrongdoers at his or her option, regardless of that wrongdoer's degree of negligence. The injured party cannot receive double compensation but can choose to recover 100% of a damages award from any defendant who is found liable to any extent.

10. T/F: Insurance pools comprise a cooperative group of governmental entities joining together to finance an exposure, liability, or risk.

ANSWER KEY

1. **T/F:** When examining insurance policies, pricing should be considered alongside other important factors such as coverages, limits, deductibles, and insurer resources.

2. **T/F:** A broker advocate serves as an intermediary, negotiator, and advisor who presents your public water and wastewater system to the insurance marketplace.

3. Examples of information to include in your property and liability renewal submission include:
    a. Operational summary
    b. Loss runs including claims analysis
    c. Current property, equipment, and automobile schedules
    d. **All of the above**

4. **T/F:** Regulated insurance policies are subject to insurance regulatory oversight at the federal level.

5. **T/F:** Insurance pools are governed by insurance law, which is based on an adhesion contract.

6. Insurance regulators require certain insured safeguards with regulated policies such as:
    a. **Reservation of rights letters**
    b. Multiyear policy periods
    c. Joint and several liability
    d. Premium guarantees

7. **T/F:** The upside and downside of insurance pools with joint and several liability are the potential for members to receive dividends and assessments.

8. Insurance pools are commonly less expensive than regulated insurance policies because of:
    a. Joint and several liability
    b. Exemption of premium taxes
    c. Exemption of income taxes
    d. **All of the above**

9. **T/F:** Joint and several liability is legal doctrine applying in some states that allows an injured person to sue and recover from one or more of several wrongdoers at his or her option, regardless of that wrongdoer's degree of negligence. The injured party cannot receive double compensation but can choose to recover 100% of a damages award from any defendant who is found liable to any extent.

10. **T/F:** Insurance pools comprise a cooperative group of governmental entities joining together to finance an exposure, liability, or risk.

20 Property Policy

© Las Vegas Valley Water District and used with permission

LEARNING OBJECTIVES

- Explain the optimal structure for a property policy specifically designed for public water/wastewater systems.
- Identify the differences between a standardized and proprietary property policy.
- Articulate the necessity to integrate equipment breakdown within your property policy.
- Comprehend the deficiencies of a joint loss agreement, coinsurance penalty, and margin clause.
- Grasp how to properly integrate your property policy to include real property and business personal property, mobile equipment, loss of income, extra expense, equipment breakdown, and proprietary additional coverages into one policy.
- Identify the necessary modifications to the definition of covered real property and business personal property as well as property not covered.
- Explain the various exclusions and their exceptions as they relate to an integrated property policy.
- Describe the different types of additional coverages and their importance to a best-in-class property policy.
- Understand the various valuation methods for real property, business personal property, and mobile equipment.

DOI: 10.1201/9781003207146-30

- Comprehend the different deductibles for flood, windstorm, earthquake, and all other perils.
- Articulate the role of the Federal Emergency Management Agency (FEMA) and National Flood Insurance Program (NFIP) as it relates to insurance transfer and financial reimbursement for public water/wastewater systems.
- Describe the duties and obligations of both insurer and insurer after a loss or damage.

KEY DEFINITIONS*

Accounts Receivable Coverage – insures against loss of sums owed to the insured by its customers that are uncollectible because of damage by an insured peril to accounts receivable records.

Actual Cash Value (ACV) – in property and auto physical damage insurance, one of several possible methods of establishing the value of insured property to determine the amount the insurer will pay in the event of loss. ACV is calculated on the cost to repair or replace the damaged property, minus depreciation. In some states, it refers to fair market value of the dwelling, which is the amount a willing buyer would pay a willing seller for the dwelling (not including land) in its condition immediately before the loss.

Blanket Limit – a single limit of insurance that applies over more than one location or more than one category of property coverage, or both. A blanket limit is in contrast to specific or scheduled limits of insurance, which are separate limits that apply to each type of property at each location.

Business Income – loss of income suffered by an organization when damage to its premises by a covered cause of loss causes a slowdown or suspension of its operations. Coverage applies to loss suffered during the time required to repair or replace the damaged property. It may also be extended to apply to loss suffered after completion of repairs for a specified number of days.

Business Personal Property – non-real property owned by an organization. Similar to personal property coverage in a homeowner's policy, business personal property covers nearly all items of value that aren't considered a structure, fixture, automobile, watercraft, or aircraft.

Civil Authority – a property policy provision which outlines how the loss of business income coverage applies when a government entity denies access to the covered property.

Claim – used in reference to insurance, a claim may be a demand by an individual or organization to recover, under a policy of insurance, for loss that may be covered within that policy.

Climate Change – risk facing business and governmental entities that affect natural and human systems. A common approach in dealing with this loss exposure focuses on reducing the vulnerability associated with climate risk by incorporating climate-sensitive decision-making in the risk management process. The risk manager takes climate-related decisions or actions that make sense in overall business strategy terms, whether or not a specific climate threat actually materializes in the future. Three examples include greenhouse gas (GHG) emission reduction efforts, energy conservation, and the adoption of green building measures and approaches. Climate change risks include physical risks, litigation risks, reputational risks, stockholder risks, regulatory risks, and competition risks.

Coinsurance Clause – a property policy provision that penalizes the insured's loss recovery if the limit of insurance purchased by the insured is not at least equal to a specified percentage (commonly 80%) of the value of the insured property.

---

* Source material provided by International Risk Management Institute (IRMI), https://www.irmi.com/term/insurance-definitions/insurance, as well as policy forms and underwriting products promulgated by Insurance Services Office (ISO) and their member companies, https://products.iso.com/productcenter/10. Actual cash value includes definition from California FAIR Plan.

Coverage Territory – contractual provisions limiting coverage to geographical areas within which the insurance is affected.

Covered Causes of Loss – with respect to special causes of loss, direct physical loss or damage unless the loss or damage is specifically excluded or limited in the policy.

Debris Removal – coverage for the cost of removal of debris of covered property damaged by a covered cause of loss.

Deductible – amount the insurer will deduct from the loss before paying up to its policy limits. Most property policies contain a per-occurrence deductible provision that stipulates that the deductible amount specified in the supplemental policy declarations will be subtracted from each covered loss in determining the amount of the insured's loss recovery.

Deep Well Pumps – pumps designed for pumping water from wells with water levels more than 25 feet below the pump location. Such pumps are designed so that the pump cylinder is near the well water level and the water is forced to the surface rather than being sucked to it.

Described Premises – (1) In a property policy, the location where coverage applies. Usually described in the policy with a legal address. (2) Building or land occupied or owned by an insured.

Difference-in-Conditions (DIC) – an inland marine policy that is purchased in addition to a property policy to obtain coverage for perils not insured against in the commercial property policy (usually flood and earthquake).

Dividends – partial return of premium to the insured based on the insurer's financial performance or on the insured's own loss experience.

Earthquake – typically excluded (along with other earth movement) from most property policies, except ensuing fire. In most cases, earthquake coverage must be purchased by endorsement of a difference-in-conditions (DIC) policy or to an all risks policy. Normally, the coverage provided is subject to a per occurrence sublimit, an annual aggregate limit, and a separate deductible.

Electronic Data – exposure faced by individuals and organizations that may cause loss of, damage to, or inability to access or use electronically stored data.

Equipment Breakdown or Boiler and Machinery – coverage for loss due to mechanical or electrical breakdown of nearly any type of fixed equipment. Coverage applies to the cost to repair or replace the equipment and any other property damaged by the equipment breakdown. Resulting business income and extra expense loss is often covered as well.

Exclusion – policy provision that eliminates coverage for some type of risk. Exclusions narrow the scope of coverage provided by the insuring agreement. In many insurance policies, the insuring agreement is very broad. Insurers utilize exclusions to carve away coverage for risks they are unwilling to insure.

Extra Expense – coverage that pays for additional costs in excess of normal operating expenses that an organization incurs to continue operations while its property is being repaired or replaced after having been damaged by a covered cause of loss.

Federal Emergency Management Agency (FEMA) – federal agency that provides a single point of accountability for all federal emergency preparedness, mitigation, and response activities. FEMA's primary purpose is to coordinate the response to a disaster that overwhelms the resources of state and local governments. It works closely with these governmental bodies by funding emergency programs and offering technical guidance and training. FEMA administers the National Flood Insurance Program (NFIP) and advises communities on building codes, emergency response, and floodplain management.

Flood – coverage for damage to property caused by flood. Normally, this coverage is subject to a 'per occurrence' sublimit, an annual aggregate limit, and a separate deductible. Coverage may also be available from the National Flood Insurance Program (NFIP). The NFIP defines flood as a general and temporary condition of partial or complete inundation of two or more acres of normally dry land area or two or more properties (at least one of which is your property) from: overflow of inland waters, unusual and rapid accumulation or runoff of surface waters from any source, and mudflows.

Force Majeure – French term denoting a superior force. An unexpected or uncontrollable event that upsets one's plans and releases one from obligation.

Functional Replacement Cost – cost of acquiring another item of property that will perform the same function with equal efficiency, even if it is not identical to the property being replaced.

Insuring Agreement – portion of the insurance policy in which the insurer promises to make payment to or on behalf of the insured. The insuring agreement is usually contained in a coverage form from which a policy is constructed. Often, insuring agreements outline a broad scope of coverage, which is then narrowed by exclusions and definitions.

Loss – direct and accidental loss or damage, including reduction in value.

Margin Clause – nonstandard property policy provision, used with blanket insurance limits, stating the most an insured can collect for a loss at a given location is a specified percentage of the values reported for that location on the insured's statement of values. The maximum is normally stated as a percentage that is greater than 100%, such as 110% or 125%.

Market Value – price paid to purchase an asset in its particular condition immediately before the loss (i.e., willing buyer and willing seller).

Mobile Equipment – a term that refers to equipment such as earthmovers, tractors, diggers, farm machinery, forklifts, etc., that, even when self-propelled, are not considered automobiles for insurance purposes (unless they are subject to a compulsory or financial responsibility law or other motor vehicle insurance law).

National Flood Insurance Program (NFIP) – federally funded program established in 1968 to make flood insurance available at a reasonable cost for properties located in participating communities. NFIP flood insurance is available only for direct damage to buildings and contents; there is no time element coverage.

Ordinance or Law Provision – coverage for loss caused by enforcement of ordinances or laws regulating construction and repair of damaged buildings. Older structures that are damaged may need upgraded electrical, heating, ventilating, air-conditioning, and plumbing units based on building codes. Many communities have a building ordinance(s) requiring a building that has been damaged to a specified extent (typically 50%) must be demolished and rebuilt in accordance with current building codes rather than simply repaired.

Policy Conditions – section of an insurance policy that identifies general requirements of an insured and the insurer on matters such as loss reporting and settlement, property valuation, other insurance, subrogation rights, and cancellation and nonrenewal. The policy conditions are usually stipulated in the coverage form of the insurance policy.

Policy Period – term of duration of the policy. The policy period encompasses the time between the exact hour and date of policy inception and the hour and date of expiration.

Pollutants – any solid, liquid, gaseous, or thermal irritant or contaminant, including smoke, vapor, soot, fumes, acids, alkalis, chemicals, and waste. Waste includes materials to be recycled, reconditioned, or reclaimed.

Pollution Remediation Expenses – expenses incurred for or in connection with the investigation, monitoring, removal, disposal, treatment, or neutralization of pollution conditions to the extent required by: (a) federal, state, or local laws, regulations or statutes, or any subsequent amendments thereof, enacted to address pollution conditions; or (b) a legally executed state voluntary program governing the cleanup of pollution conditions.

Property Policy – first-party insurance that indemnifies the owner or user of property for its loss, or the loss of its income-producing ability, when the loss or damage is caused by a covered peril, such as fire or explosion.

Real Property – land and most things attached to the land, such as buildings and vegetation. The definition of land includes not only the surface of the earth, but also everything above and beneath it. Thus, the ownership of a tract of land theoretically includes both the airspace above it and the soil from its surface to the center of the earth.

Replacement Cost – property insurance term that refers to one of the two primary valuation methods for establishing the value of insured property for purposes of determining the amount the insurer will pay in the event of loss. The other primary valuation method is actual cash value (ACV). Replacement cost is defined as the cost to replace damaged property with materials of like kind and quality, without any deduction for depreciation.

Supervisory Control and Data Acquisition (SCADA) – acronym for supervisory control and data acquisition, a computer system for gathering and analyzing real time data. SCADA systems are used to monitor and control a plant or equipment in industries such as telecommunications, water and waste control, energy, oil and gas refining, and transportation.

Supplemental Declarations Page – the front page (or pages) of a coverage form that specifies the named insured, address, policy period, location of premises, policy limits, and other key information that varies from insured to insured. The supplemental declarations page is also known as the information page.

Tenants Leasehold Interest Loss – coverage that pays the loss suffered by a tenant due to termination of a favorable lease because of damage to the leased premises by a covered cause of loss. The principal coverage is the net leasehold interest, which is the present value of the difference between the total rent payable over the unexpired portion of the lease and the total estimated rental value of the property during the same period.

Vacant Buildings – vacancy comprises little or no furniture or other personal property. A building is frequently deemed vacant when it does not contain enough business personal property to conduct normal business operations. The wording in many property policies limits, reduces, or entirely eliminates coverage when a building has been vacant (or, in some forms, vacant or unoccupied) for a designated period such as 45 or 60 days.

Valuable Papers and Records – coverage that pays the cost to reconstruct damaged or destroyed valuable papers and records; usually is defined to include almost all forms of printed documents or records except money or securities; data processing programs, data, and media are usually excluded.

PROPERTY POLICY*

Overview

There are many ways to effectively structure your property policy. The composition we recommend is a fully integrated policy that embeds coverage for your insurable real property, business personal property, mobile equipment, business income, and extra expense. Your policy should also include proprietary coverage extensions tailored specifically for public water/wastewater systems. We will

---

* Source material provided by various online articles from International Risk Management Institute (IRMI), https://www.irmi.com/articles/expert-commentary/insurance, as well as policy forms and underwriting products promulgated by Insurance Services Office (ISO) and their member companies, https://products.iso.com/productcenter/10.

hereinafter refer to proprietary coverage extensions as additional coverages. This type of integrated policy should be written on a blanket policy limit for all covered property at described premises. There shouldn't be a margin clause or coinsurance clause, and the policy should be predicated on covering all causes of loss unless specifically excluded. The latter parlance is called a special causes of loss form or open perils unless specifically excluded form.

The peril of equipment breakdown should similarly be implanted as a specifically cited covered cause of loss. Such amalgamation eliminates the need for a separate boiler & machinery policy (i.e., equipment breakdown coverage) and prevents inevitable claim issues arising from conflicting policy language. These conflicts commonly lead to coverage disputes and processing delays. The preceding consolidation is especially significant for public water/wastewater systems as your fixed equipment is susceptible to property and equipment breakdown losses. To compound matters, many fixed equipment losses straddle the traditional coverage boundaries of property and boiler & machinery policies. A fully integrated property policy with equipment breakdown coverage removes the need for joint loss agreements, promotes coverage clarity, and reduces insured anguish on complex losses involving more than one peril.

Additional include a review and possible modification of the real property and business personal property definitions as well as those items specifically referenced as property not covered. These actions will ensure you and your insurer are aligned as to what comprises covered property. It will similarly allow you to articulate to your governing body those assets that fall outside the policy's scope. Such analysis is necessary to evaluate the prudence of purchasing supplemental coverage and additional policies. You should also have command of all exclusions and limitations within the policy as well as adept insight to properly assess and possibly amend the additional coverages. Final steps involve a sensible selection of deductibles and valuation methods as well as an understanding of your insured duties. This cognizance is necessary to certify the breadth, significance, and fitness of your property policy. We are now ready to begin our detailed analysis of designing a well-configured, fully integrated property policy.

STRUCTURAL DETAILS

Our journey commences with an examination of the policy limits. Your policy should be issued on a blanket policy basis and not constrained by the individual limits on your statement of values. A blanket policy limit offers coherence, clarity, and simplicity. Should there be a loss to covered property at described premises, then your loss will be adjusted up to the blanket policy limit shown on the supplemental declarations page. This structure ensures your loss will not be artificially capped at the limit shown on your statement of values for the respective property that is damaged. Such composition is particularly important should there be communitywide losses from a natural hazard like wildfire or wind. In catastrophic loss scenarios, the values contained in your statement of values may not adequately reflect the inflationary labor and material costs associated with the natural hazard. A prudent control measure for addressing this possibility is to attach an increase rebuilding expense following disaster endorsement to your blanket policy limit.

A blanket policy limit removes the foregoing concern and is particularly advantageous to public water/wastewater systems because of your wide-ranging, complex, and integrated assets. In addition to blanket policy limits, it is important to remove any margin clause, which caps the maximum amount your insurer will pay to a certain percentage, typically 125%, above the amount listed on your statement of values. The same applies to a coinsurance clause, which reduces your insurance settlement should the amount listed on your statement of values fall below a certain percentage, typically between 80% and 100%, of the actual replacement cost of your property. A margin clause and coinsurance clause should be avoided at all times, as they undermine the benefits of a blanket policy limit.

We have just explained the prudence of blanket policy limits without a margin clause and coinsurance clause. Now, we turn to the statement of values. Your blanket policy limit requires an accurate statement of values that identifies every location, otherwise known as described premises, where you have insurable real property and business personal property. An assiduous review of all described premises where you have insurable assets is critical, as blanket policy limits only trigger if there is direct physical loss or damage to covered property arising from a covered cause of loss at described premises. That means an otherwise covered loss will be excluded if your statement of values doesn't include a current listing of your described premises.

It is compulsory to list all described premises, even those with small amounts of insurable assets. Lightning, vandalism, windstorm, wildfire, and equipment breakdown are frequent causes of loss to insurable assets like lift stations, water storage tanks, and miscellaneous real property. These assets are often overlooked because they are commonly isolated from your larger, more prominent described premises. It is important to cover all your insurable assets at all described premises, as the cost of premium versus the benefit of protection is small for the value afforded. Equally important, additional coverages such as business income, extra expense, and debris removal are only triggered if there is direct physical loss or damage to covered property at described premises. The latter point underscores the necessity to schedule all described premises having any insurable real property and business personal property.

The proper description of real property and business personal property at described premises is also important. Such description doesn't require a granular breakdown of every structure, fixed equipment, and miscellaneous property at a described premise. It does, however, compel an accurate description of covered property. To illustrate, a water or wastewater treatment facility should reference the construction, year built, square footage, spacing, and protections (e.g., sprinklers, foam, firewalls, etc.) of all main buildings at the described premise. You do not need to list all the fixed equipment within the treatment facility or all ancillary structures, but you should provide an accurate encapsulation of the covered property at the described premise. The replacement cost should include conservative estimates to rebuild the above-referenced example should there be a total loss. Engineering costs, permit fees, and other soft costs like financing and legal fees are not typically considered insurable expenses and should not be incorporated in the replacement cost calculation.

Your ability to accurately convey replacement cost values and protection features of covered property, including sprinklers as well as noncombustible construction materials and firewalls - building separations, will maximize available credits and ensure your covered property receives the most preferred classification code. Those factors, coupled with proper valuation, will equate to a premium commensurate with your exposure. This level of detail should similarly apply to the identification of certain types of real property and business personal property that are not clearly codified in the definitions and, hence, possibly not subject to the blanket policy limit but instead relegated to an additional coverage sublimit. Worse yet, the lack of definitional codification may result in an uncovered loss should the asset be deemed property not covered.

COVERED PROPERTY

We are now ready to review and modify what is deemed covered property and what is considered property not covered in a best-in-class property policy. An example of verbiage that favorably codifies covered property is as follows:

"We will pay for direct physical loss of or damage to real property and business personal property caused by or resulting from a peril not otherwise excluded, not to exceed the applicable blanket limit of insurance for real property and business personal property described in the supplemental declarations, caused by or resulting from any covered cause of loss. The loss or damage to real property or business personal property must occur at, or within 100 feet of, the described premises shown in the supplemental declarations, unless otherwise stated. This coverage applies only at those premises shown in the supplemental declarations."

The above-referenced verbiage is broad and seemingly all-encompassing. However, the breadth of covered property is reliant on the definitions for real property and business personal property.

These definitions should be carefully evaluated and sufficiently expansive to ensure all your insurable assets qualify as covered property. Examples of water-wastewater centric definitions for real property and business personal property are as follows:

Real Property – items at described premises including: (1) aboveground piping; (2) aboveground and belowground penstock; (3) additions under construction; (4) all appurtenant buildings or structures, including restrooms; (5) alterations and repairs to the buildings or structures; (6) buildings; (7) completed additions; (8) exterior signs, meaning neon, automatic, mechanical, electric, or other signs either attached to the outside of a building or structure or standing free in the open; (9) fixtures, including outdoor fixtures; (10) glass which is part of a building or structure; (11) light standards; (12) paved surfaces such as sidewalks, patios or parking lots; (13) permanently installed machinery and equipment; (14) permanent storage tanks; (15) solar panels; (16) submersible pumps, pump motors, and engines; (17) underground piping located on or within 100 feet at a described premises; (18) underground vaults and machinery; (19) business personal property owned by you that is used to maintain or service the real property or structure or its premises, including fire-extinguishing equipment as well as outdoor furniture, floor coverings, and appliances used for refrigerating, ventilating, cooking, dishwashing, or laundering; and (20) if not covered by other insurance: additions under construction; alterations, and repairs to the real property or structure; material, equipment supplies, and temporary structures on or within 100 feet at a described premises if used for making additions, alterations, or repairs to the real property or structure.

Penstock – a conduit constructed of manmade materials built for the purpose of conveying water to a hydro turbine; it does not include tunnels, canals, aqueducts, or similar excavations or the cost of these excavations, which are excavated from or consist of natural materials.

Business Personal Property – (1) property you own that is used in your business; (2) furniture and fixtures; (3) machinery and equipment; (4) computer equipment; (5) communication equipment; (6) labor materials or services furnished or arranged by you on personal property of others; (7) stock; (8) your use interest as tenant in improvements and betterments - improvements and betterments are fixtures, alterations, installations, or additions made a part of the real property or structure you occupy but do not own, or you acquired or made at your expense, but cannot legally remove; and (9) leased personal property for which you have a contractual responsibility to insure, unless otherwise provided for under personal property of others. Stock means finished goods held in storage or for sale.

Special Commentary

The preceding definitions are wide-ranging and do not comprise sublimits, which is a further advantage. Nevertheless, even first-rate definitions require modification to remove the ambiguity of intent and gaps in coverage. Examples of necessary definitional additions for real property and business personal property to ensure proper codification of all your insurable assets are as follows: (1) fencing and retaining walls; (2) foundations; (3) uninstalled piping materials and related appurtenances; (4) telemetry-and-radio towers/antennas; (5) hydrants; and (6) water meters. Items 1-4 require scheduling to avoid coverage disputes. Fencing, retaining walls, foundations, and telemetry-and-radio towers/antennas are commonly assumed to be included within the definition of real property. These items, however, are not accurately codified and should be added to the real property definition to eliminate claims misunderstandings. Uninstalled piping materials and related appurtenances technically meet the business personal property definitional classification of "property you own that is used in your business," but the scheduling of such items on your statement of values as business personal property is a prudent action to certify insurer alignment.

The scheduling of hydrants and water meters (items 5-6) is not compulsory, but its contemplation serves as a valuable risk management exercise and perplexing business conundrum. Singular losses to these assets typically fall slightly above the selected deductible, but their aggregated values are substantial. They are similarly spread across your service territory, so their aggregated values are only exposed to natural disasters. You should be mindful that repairs to hydrants can be expensive in the rare event of damage to belowground piping connected to the barrel. Regarding water meters, the ownership of said assets is not always clear. They are, however, insurable because they meet the definition of critical infrastructure and must be functional to resume your operations after a loss. The determination to cover or not cover hydrants and water meters condenses down to premium. Insurers will rate for such exposures based on their aggregated values, which will be significant. One option to lower the premium is to secure a sublimit representing one-half of the aggregated replacement cost value and a restriction of the covered causes of loss to the perils of wildfire, windstorm, and other natural hazards. In such a compromise, the rate drops substantially as does the probability of loss. Coverage, however, is preserved with meaningful aggregate protection in the event of a catastrophic event.

PROPERTY NOT COVERED

The same level of careful review performed for covered property should also apply to property not covered. This latter category, even with best-in-class property policies, contains restrictions that should be understood and possibly removed to ensure your insurable assets are properly covered. The foregoing category represents potential for material constraints to insurable assets and is the impasse of many claims misunderstandings between insureds and insurers as to what was intended to be covered property and what was intended to be property not covered.

Common examples of property not covered in a best-in-class property policy include the following: (1) accounts, bills, currency, food stamps, or evidences of debt, money, notes, or securities (lottery tickets held for sale are not securities); (2) aircraft; (3) animals; (4) automobiles; (5) bridges greater than 50 feet in length; (6) canals, ditches, flumes, or aqueducts; (7) contraband, or property in the course of illegal transportation or trade; (8) cost of excavations, grading, backfilling, or filling; (9) cost to replace or restore the information on valuable papers & records, including those which exist as electronic data, except as provided as an additional coverage extension; (10) dams, locks, levees, or reservoirs; (11) electric utility power transmission & distribution lines, poles, and related equipment; (12) electronic data, except as provided as an additional coverage; and (13) land (including land on which the property is located). Additional examples include: (14) lawns; (15) property that is covered under another insurance policy in which it is more specifically described; (16) pump motors and engines exceeding 1,000 horsepower capacity; (17) retaining walls that are not part of a building except as provided as an additional coverage; (18) radio or television antennas (including satellite dishes) and their lead-in wiring, masts, or towers while outside of buildings except as provided as an additional coverage; (19) roadways; (20) saltwater piers, docks, and wharves; (21) trees, shrubs & plants except as provided as an additional coverage; (22) underground flues, drains, or well structures; (23) underground piping located more than 100 feet from described premises; (24) watercraft greater than 50 feet in length; and (25) water.

Special Commentary

Many of the preceding restrictions are acceptable and do not require modification. Items like money, aircraft, automobiles, and property more specifically described in another coverage form are all available through other insurance policies. Animals, illegal contraband, land, lawns, roadways, and saltwater structures are inapplicable to public water/wastewater systems. Assets like valuable papers & records, electronic data, and trees, shrubs & plants are properly addressed separately as additional coverages. It is compulsory to schedule your fencing and retaining walls as well as your foundations and telemetry-and-radio towers/antennas on the statement of values and specifically codify the items in the definition of real property. You must also delete said real property from the definition of property not covered. This approach removes claims misunderstandings, activates the blanket policy limit, and ensures the same covered causes of loss will apply to these assets as all other real property on described premises.

We now have 10 remaining restrictions to carefully examine. Some of the restrictions are acceptable and not overly onerous to public water/wastewater systems. They include canals, ditches, flumes, or aqueducts as well as dams, locks, levees, or reservoirs and the cost of excavations, grading, backfilling, or filling. The preceding restrictions share three factors: (1) high replacement cost values; (2) impervious to damage from normal covered causes of loss; and (3) made of or involve the movement of earthen materials. Removing said assets prevents prohibitively high premiums since insurance rates are based on aggregated replacement cost values. Such astute awareness is important as you balance protection benefits with protection costs. All underground piping infrastructure outside of 100 feet of described premises is similarly not covered for the aforesaid reasons. The main exposure for underground piping is earthquake and flood. A difference-in-conditions (DIC) policy will likely be required should you seek protection against these hazards. Underground piping can be damaged by negligent contractors, but subrogation minimizes the exposure to the variance between actual cash value and replacement cost.

Water, incidentally, is an acceptable restriction for the cost prohibitive reason that coverage of this asset would require a replacement cost value and corresponding premium charge on your entire water rights portfolio. The most common exposure to damage of water is contamination from pollutants, and this exposure is routinely excluded as a covered cause of loss. Notwithstanding, the cost of water you purchase to replenish lost water you had previously purchased would be covered under the policy's additional coverage for extra expense. Bridges and watercraft greater than 50 feet can be removed as property not covered and subsequently deemed covered property as long as you schedule the assets and incur a premium charge for their full replacement cost value. The same applies to pump motors and engines exceeding 1,000 horsepower capacity. In such instances, there will invariably be higher deductibles assigned to the equipment breakdown coverage for both property damage as well as business income and extra expense. Should the horsepower capacity be substantially above 1,000; then the asset may require a separate boiler & machinery policy.

The final restrictions are electric utility power transmission & distribution lines, poles, and related equipment as well as underground flues, drains, or well structures. In terms of the former, this exposure exists for public water/wastewater systems requiring an independent power connection to their utility provider's power source. It also exists with public water/wastewater systems generating power from their own operations. These assets can be removed as property not covered and subsequently deemed covered property as long as you schedule the assets, incur a premium charge for their full replacement cost value, and identify their location on described premises. A modification to the definition of real property is recommended to certify insurer alignment. Distribution lines situated over great distances will likely remain property not covered due to their propensity for loss. Underground flues, drains, and well structures are rarely added back as covered property because their exposure to loss is limited to traditionally excluded perils of earthquake and flood. Exposure to vandalism, which is a covered cause of loss involving rocks or debris being thrown into wells causing cracks in the casing and requiring extraction, is a legitimate concern warranting a cost-benefit analysis by you and your governing body.

It is important to thoroughly understand what is deemed covered property and what is deemed property not covered. This way, you can articulate the rationalization of placing certain assets as property not covered to your governing body. We will revisit property not covered when we review the additional coverages, as there are narrow coverage grants for some of these assets. As a reminder, the additional coverages do not qualify for full blanket policy limits and some have restricted covered causes of loss.

COVERED CAUSES OF LOSS

We have enunciated the importance of blanket policy limits, identified the clauses that can curb its breadth, and explained the process for maximizing the fullness of covered property. Now, we turn to the covered causes of loss. Your policy should cover all losses except as specifically excluded. The following verbiage achieves this requirement and should be compared against your property policy:

"Covered causes of loss means: (1) risks of direct physical loss unless the loss is excluded or limited by this coverage form; and (2) equipment breakdown is added as an additional covered cause of loss for direct physical loss to covered property."

Source: Allied World Assurance Company WaterPlus policy inclusive of material from Insurance Services Offices, Inc.

You will notice this verbiage is not restrictive, and, equally important, it specifically embeds the all-important equipment breakdown peril as an additional covered cause of loss. An example of a comprehensive equipment breakdown definition is as follows:

Equipment Breakdown: means direct damage to mechanical, electrical, or pressure systems as follows: (1) mechanical breakdown including rupture or bursting caused by centrifugal force; (2) artificially generated electrical current, including electrical arcing that disturbs electrical devices, appliances, or wires; (3) explosion of steam boilers, steam piping, steam engines, or steam turbines owned or leased by you or operated under your control; (4) loss or damage to steam boilers, steam pipes, steam engines, or steam turbines; (5) loss or damage to hot water boilers or other water heating equipment; or (6) if covered electrical equipment requires drying out as a result of a flood, then we will pay for the direct expenses for such drying out. (7) None of the following are covered objects as respects to equipment breakdown: (a) insulating or refractory material; (b) buried vessel or piping; (c) sewer piping, piping forming a part of a fire protection system or water piping other than: (i) feed water piping between any boiler and its feed pump or injector; (ii) boiler condensate return piping; or (iii) water piping forming a part of refrigerating and air conditioning vessels and piping used for cooling, humidifying, or space heating purposes; (8) structure, foundation, cabinet, or compartment containing the object; (9) power shovel, dragline, excavator, vehicle, aircraft, floating vessel or structure, penstock, draft tube, or well-casing; (10) conveyor, crane, elevator, escalator, or hoist but not excluding any electrical machine or electrical apparatus mounted on or used with this equipment; and (11) felt, wire, screen, die, lathes, extrusion, swing hammer, grinding disc, cutting blade, cable chain, belt, rope, clutch late, brake pad, non-metallic part, or any part or tool subject to frequent, periodic replacement.

EXCLUSIONS AND LIMITATIONS

Property policies are intended to protect against fortuitous events. They are not predicated to cover expected outcomes. Property policies are also not automatically structured to cover catastrophic perils of earthquake and flood. In some instances, wind and hail may be deemed catastrophic perils and excluded in coastal or high-risk regions. The same applies to high brush areas where wildfire may similarly be excluded. Catastrophic exposures can be secured for an additional premium via endorsement or through a separately purchased DIC policy. In either case, there will be restrictive terms and conditions as well as higher deductibles and lower limits. The other types of exclusions are more self-explanatory and logical. These common and acceptable exclusions underscore the fortuity doctrine, which requires a covered loss to be accidental or by chance and unexpected from the standpoint of the insured. Public policy would be violated if an insured was allowed to benefit from an insurance policy where losses were known or expected to occur. Such exclusions for public water/wastewater systems are frequently categorized into five groups, which we will detail below. The normal language introducing exclusions is as follows:

Unless otherwise stated, these exclusions and limitations apply to all coverages provided in this coverage form and to endorsements attached to this coverage form.

The five groups of exclusions commonly used by insurers are summarized as follows:

Group A: We will not pay for loss or damage caused directly or indirectly by any of the following exclusions, and such loss or damage is excluded regardless of any other cause or event that contributes concurrently or in any sequence to the loss: (1) earth movement including volcanic action and earthquake, but coverage will apply to any resulting loss arising from fire or explosion; (2) governmental action (i.e., confiscation), but coverage would apply to any resulting loss to prevent the spread of fire; (3) nuclear hazard, but coverage will apply to any resulting loss arising from fire; (4) war and military action; and (5) water including flood waters, mudslide, and seepage, but coverage will apply to any resulting fire, explosion, sprinkler leakage, or equipment breakdown.

Group B - Part I: We will not pay for loss or damage caused by or resulting from any of the following: (1) delay, loss of use, or loss of market; (2) smoke, vapor, or gas from agricultural smudging or industrial operations; (3) wear and tear; (4) rust, corrosion, fungus, decay, deterioration, hidden, or latent defect of any quality in property that causes it to damage or destroy itself; (5) smog; (6) settling, cracking, shrinking, or expansion; (7) nesting or infestation, or discharge or release of waste products or secretions by insects, birds, rodents, or other animals; and (8) the following causes of loss: (a) dampness or dryness of atmosphere; (b) changes in or extremes of temperature; or (c) marring or scratching. If an excluded cause of loss listed in B1 through B8 results from a specified cause of loss or building glass breakage, then coverage will apply for the loss or damage caused by that specified cause of loss or building glass breakage. Specified causes of loss include the following perils: fire; lightning; explosion; windstorm or hail; smoke; aircraft or vehicles; riot or civil commotion; vandalism; leakage from fire extinguishing equipment; sinkhole collapse; volcanic action; falling objects; weight of snow, ice, or sleet; water damage; and equipment breakdown. Water damage means accidental discharge or leakage of water or steam as the direct result of the breaking apart or cracking of any part of a system or appliance (other than a sump system including its related equipment and parts) containing water or steam.

Group B - Part II: We will not pay for loss or damage caused by or resulting from any of the following: (9) water, other liquids, powder, or molten material that leaks or flows from plumbing, heating, air conditioning, or other equipment (except fire protective systems) caused by or resulting from freezing, unless: (a) you do your best to maintain heat in the building or structure; or (b) you drain the equipment and shut off the supply if heat is not maintained; (10) voluntary parting with any property by you or anyone else to whom you have entrusted the property if induced to do so by any fraudulent scheme, trick, device, or false pretense; (11) rain, snow, ice, or sleet to business personal property in the open; and (12) collapse, except as provided as an additional coverage extension, but coverage will apply to any resulting loss arising from a covered cause of loss.

Group B - Part III: We will not pay for loss or damage caused by or resulting from any of the following: (13) pollution conditions, except as provided as an additional coverage extension, but coverage will apply to damage to glass caused by chemicals applied to the glass; (14) remediation expenses, except as provided as an additional coverage extension; (15) neglect to use all reasonable means to save and preserve property from further damage at and after the time of loss; (16) virus, bacterium, or other microorganism, other than fungus, wet rot, or dry rot, which is addressed in a separate exclusion, that induces or is capable of inducing physical distress, illness, or disease; and (17) the presence, growth, proliferation, spread, or any activity of fungus or wet rot or dry rot including loss, damage, or resulting remediation expenses, but coverage will apply to any resulting loss arising from a specified cause of loss, and this exclusion does not apply when fungus, wet rot, or dry rot results from fire or lightning or to the extent coverage is provided as an additional coverage extension.

Group C: We will not pay for loss or damage caused by or resulting from any of the following, but coverage will apply to any resulting loss or damage if the loss or damage results from a covered cause of loss: (1) weather conditions, but this exclusion only applies if weather conditions contribute in any way with a cause or event excluded above to produce the loss or damage; (2) acts or decisions, including the failure to act or decide, of any person, group, organization, or governmental body; (3) faulty, inadequate, or defective: (a) planning, zoning, development, surveying, siting; (b) design, specifications, workmanship, repair, construction, renovation, remodeling, grading, compaction; (c) materials used in repair, construction, renovation, or remodeling; or (d) maintenance of part or all of any covered property.

Group D: As it relates to business income and extra expense, we will not pay for: (1) any increase of loss caused by or resulting from: (a) delay in rebuilding, repairing, or replacing the property or resuming operations, due to interference at the location of the rebuilding, repair, or replacement by strikers or other persons; or (b) suspension, lapse, or cancellation of any license, lease, or contract, but coverage will apply if the suspension, lapse, or cancellation is directly caused by the suspension of operations as long as it affects your business income during the period of restoration in accordance with the additional coverage extension; (2) any extra expense caused by or resulting from suspension, lapse, or cancellation of any license, lease, or contract beyond the period of restoration; and (3) any other consequential loss.

Group E - Limitations: (1) We will not pay for loss of or damage to property, including any loss that is a consequence of loss or damage, as described and limited as follows: (a) property that is missing, where the only evidence of the loss or damage is a shortage disclosed on taking inventory, or other instances where there is no physical evidence to show what happened to the property; (b) property that has been transferred to a person or place outside the described premises on the basis of unauthorized instructions; (c) fragile articles such as statuary, marbles, chinaware, and porcelains, if broken, but coverage will apply to glass containers or property held for sale or photographic or scientific instrument lenses. (2) $2,500 for loss or damage due to theft of patterns, dies, and molds. (3) $500,000 for loss or damage caused directly or indirectly by water that backs up or overflows from a sewer, drain, or sump. (4) $250,000 in any one occurrence for loss or damage to real property or business personal property because of contamination as a result of a covered cause of loss. Contamination means direct damage to real property and business personal property caused by contact or mixture with any chemical used in the water treatment process.

Special Commentary

Let's revisit the catastrophic perils that are fortuitous and unexpected but commonly excluded under a property policy. The most common of these perils are earthquake and flood. The singular exception to the usual flood exclusion is coverage for expenses associated with the drying out of electrical equipment. This exposure is considered an equipment breakdown cause of loss. A public water and wastewater system should seriously evaluate the cost–benefit of buying back the perils of earthquake and flood via endorsement or through the purchase of a DIC policy, which is a policy designed to cover said perils and possibly other perils routinely excluded in a property policy. There are advantages and disadvantages to this approach.

The advantages include controlling the reimbursement of loss and removing budgetary pressures should earthquake or flood occur. The disadvantages comprise its costs, sublimits, and deductibles as well as restrictions regarding covered property and property not covered. It is especially important to understand the mechanics of the deductibles, as they are frequently designed to nullify any payment of loss except for moderate and severe events. In such instances, the coverage afforded has limited value and is only triggered when there is widespread damage to covered property.

Many public water/wastewater systems opt to rely on the Federal Emergency Management Agency (FEMA) for recovery in the event of earthquake and flood. FEMA views public water/wastewater systems as essential service providers and provides financial relief to quickly affect repairs. Frequently, financial relief is delivered in the form of pass-through grants or forgivable loans. Other times, it involves no-or-low interest loans. Reliance on FEMA is a legitimate risk management consideration, especially considering the costs, deductibles, and restrictions associated with earthquake and flood coverage in moderate- and high-hazard regions. Regarding the former, many public water/wastewater systems in moderate- and high-hazard earthquake areas wittingly forgo coverage for the abovementioned reasons and tie their fortunes to FEMA reimbursement.

One noteworthy policy restriction for earthquake coverage is underground piping infrastructure. It is frequently deemed property not covered and is one of the most exposed assets to this peril. A cost-benefit analysis is prudent to assess the payback period for earthquake coverage by examining the likely loss, premium payment, and deductible level. Should such coverage be purchased, it is compulsory to schedule your underground piping infrastructure as covered property.

The decision metrics for flood are different. Climate change necessitates the serious consideration of purchasing flood coverage, as damage from flood is increasingly unpredictable and no longer consigned to described premises in moderate- and high-hazard zones. Any purchase requires confirmation that your critical described premises are covered. Many insurers only cover described premises in 500-year flood plains, and they specifically exclude described premises in 100-year flood plains. Limits up to $5 million are commonly available at reasonable premiums for described premises in 500-year flood plains. Deductibles are customarily set at a minimum of $25,000. The onus is on you to secure accurate flood zone designations for each of your described premises. There are numerous flood zones, and the National Flood Insurance Program (NFIP) periodically changes zones after major flooding events. Examples of high-risk zones are A, AE, and V as well as variations of A, AE, and V.

Moderate-risk zones are X (Shaded) and B, whereas low-risk zones are X (Unshaded) and C. Assets in high- and moderate-risk zones are subject to higher deductibles than locations in low-risk zones. The deductibles in the former zones should be carefully reviewed, as they commonly have the greater of a per occurrence deductible or a percentage deductible based on the replacement cost value per damaged structure. Under no circumstances should you opt for a percentage deductible based on the blanket policy limit. The optimal approach is to negotiate a singular per occurrence flood deductible for all described premises. If a flooding event impacts multiple described premises and multiple flood zones, then you should seek confirmation that your deductibles do not stack.

Other exclusions requiring commentary include pollution remediation expenses as well as fungus, wet rot, or dry rot. These exclusions are slightly softened, as the policy's additional coverages provide sublimits for both exposures. The same applies to releases of water and wastewater treatment chemicals that damage real property and business personal property. Water and wastewater chemicals should apply to all types of chemicals as opposed to only chlorine and ammonia; a recommended minimum sublimit of $250,000 should apply. The foregoing pollution exposures necessitate a separate environmental pollution policy. A property policy, even a best-in-class version, is not equipped to adequately cover such exposures and frequently only provides a modest sublimit for specified hazards only.

Loss or damage caused directly or indirectly by water backing up or overflowing from a sewer, drain, or sump should also have a minimum sublimit of $500,000. The sublimit of $2,500 for theft of dies, patterns, and molds should similarly be carefully evaluated. Should you have irreplaceable pump impeller molds, dies, or patterns, then a higher limit should be afforded along with confirmation that the scheduled item is deemed covered property. The specific die, pattern, or mold should be codified in the definition of business personal property to certify insurer alignment. The remaining exclusions are deemed usual, customary, and reasonable and encompass non-fortuitous or predictable events. Incidentally, the exclusions for aircraft, automobile, and money are specifically addressed in other insurance policies catering specifically to those exposures. The exclusion pertaining to virus, bacteria, or another microorganism was historically deemed acceptable before the COVID-19 pandemic. It is the author's opinion that losses associated with COVID-19 do not trigger the property policy's direct physical loss or damage to covered property requirement. The reasoning is COVID-19 does not damage covered property but simply requires ongoing maintenance and cleaning. As such, this exclusion does not prejudice coverage for COVID-19 because the property policy trigger cannot be activated by a pandemic that does not cause direct physical loss or damage to covered property.

A final comment involves deep well pumps. These assets are subject to deferred maintenance problems and normal wear and tear, which are common exclusions. Coverage for said assets is compulsory considering their replacement cost value, but it does warrant a proactive maintenance program to ensure equipment breakdown losses are deemed covered accidents. Your property/equipment breakdown insurer will be obligated to extract deep well pumps for inspection and determination of a covered accident. Extractions are expensive, which shifts the cost from the public water and wastewater system to the insurer. However, that expense commonly is reverberated back in the form of higher deductibles, coverage restrictions, and rate increases. It also places the burden of repairs and reinstallation on your organization. As such, vertical line shafts may be a more practical and cost-effective long-term alternative to deep well pumps. This type of equipment is expensive and requires proper engineering and operational maintenance, but their service life is longer and equipment access is aboveground versus submerged hundreds of feet belowground.

ADDITIONAL COVERAGES

We have reviewed the mechanics of designing a best-in-class property policy and emphasized the importance of policy details. It is now time to address the necessity of rounding out your policy via proprietary coverage extensions tailored specifically for public water/wastewater systems. As previously mentioned, the parlance for this all-important policy burnishing is additional coverages. The additional coverages complete the fullness of the policy and provide overlapping tiers of protection. The tiers are broken down into three groups according to their importance:

- Tier 1 – Essential
- Tier 2 – Prudent
- Tier 3 – Recommended

The first tier is compulsory for public water/wastewater systems, whereas the second tier represents top-quartile enhancements that should be seriously considered. The third tier is strongly advised and predicated on convenience rather than necessity. A best-in-class property policy will contain the additional coverages and attendant limits listed in each of these tiers. The additional coverages deemed *essential* include: mobile equipment; business income; extended business income; extra expense; new locations or newly constructed property; collapse; debris removal; ordinance or law provision; electronic data; accounts receivable; valuable papers and records; and preservation of property. The additional coverages deemed *prudent* comprise: civil authority; commandeered property; utility services – direct damage, utility services – business income and extra expense; unintentional errors; property at other locations; tenants' leasehold interest loss; property in transit; pollution remediation expenses; SCADA upgrades; fungus, wet rot, or dry rot; water damage, other liquids, powder, or molten material damage; and indoor and outdoor signs. The additional coverages deemed recommended encompass: contract penalties; outdoor property; lock and key replacement; trees, shrubs, and plants; fine arts; fire protective devices; fire department service charge; arson reward; water contamination notification expenses; rental reimbursement – mobile equipment; tools and equipment owned by your employees; personal effects and property of others; cost of inventory or adjustment; non-owned detached trailers; and dependent business premises. We will now examine the three tiers and analyze each essential, prudent, and recommended additional coverage.

Tier 1 – Essential

Mobile Equipment

Denotation: Additional coverage that pays for direct physical loss or damage to owned, borrowed, rented, or leased mobile equipment while at any premises or in transit, caused by or resulting from a covered cause of loss. Mobile equipment means machinery or equipment, including accessories and spare parts for machinery or equipment, usual to your business. These items include but are not limited to: (1) forklifts; (2) tractors; (3) backhoes; (4) draglines; (5) excavators; (6) ATVs; (7) tools; and (8) watercraft less than 50 feet in length.

Commentary: This extension is not typically offered in a property policy, as it is commonly insured through an inland marine policy. Either approach is acceptable, but you should utilize the same insurer. Coverage should expand beyond owned mobile equipment and comprise borrowed, rented, and leased equipment. It is important to assess your legal liability for loss or damage to non-owned and borrowed mobile equipment. Owned mobile equipment that has a replacement cost value above $10,000 should be scheduled, whereas owned mobile equipment with cost new under $10,000 can be categorized as unscheduled mobile equipment with an appropriate aggregate limit. You should always schedule your unmanned aerial vehicles (aka drones) irrespective of value so your insurer is aware of the exposure. Newer and higher valued mobile equipment should be valued as replacement cost and not actual cash value. A recommended minimum limit for owned mobile equipment is subject to the cost new of your scheduled and unscheduled mobile equipment, and a recommended minimum limit for borrowed, rented, or leased mobile equipment is $1 million.

Business Income

Denotation: Additional coverage that pays for the actual loss of business income you sustain due to the necessary suspension of your operations during the period of restoration. The suspension must be caused by direct physical loss of or damage to covered property at described premises and caused by or resulting from any covered cause of loss. Business income means the net income (net profit or loss before income taxes) that would have been earned or incurred and continuing normal operating expenses incurred, including ordinary payroll. Period of restoration means the time period beginning with the date of direct physical loss or damage caused by or resulting from any covered cause of loss at the described premises and ending on the date when the property at the described premises should be repaired, rebuilt, or replaced with reasonable speed and similar quality. Suspension means the slowdown or cessation of your business activities or that part or all of the described premises rendered untenantable, if coverage for business income includes rental value applies.

Commentary: This extension is commonplace and necessary. The initial impression may be one of inapplicability for public water/wastewater systems. That assumption is incorrect, as all public water/wastewater systems (public, nonprofit, and investor owned) must have a source of income and reimbursement for normal operating expenses from the point of loss to full restoration. Business income coverage ensures continuity of your personnel and your operations once you have completed your restoration. A recommended time period is 24 months of actual loss sustained.

Extended Business Income

Denotation: Additional coverage that pays for the actual loss of business income you incur during the period beginning on the date property is actually repaired, rebuilt, or replaced and operations are resumed; and ending on the earlier of the date you could restore your operations with reasonable speed, to the level which would generate the business income amount that would have existed had no direct physical loss or damage occurred. To be triggered, coverage must result in direct physical loss or damage at the described premises to covered property and caused by and resulting from any covered cause of loss.

Commentary: This extension is not automatically offered by property insurers and represents an important resumption buffer after the expiration of the original business income time period. It is particularly relevant for catastrophic losses. A recommended time period is 12 months of actual loss sustained up.

Extra Expense

Denotation: Additional coverage that pays necessary extra expense you incur during the period of restoration that you would not have incurred had there been no direct physical loss or damage to covered property at described premises. Extra expense means expenses which are incurred to avoid or minimize the suspension of business and continue operations at the described premises or at replacement premises or at temporary locations, including relocation expenses and costs to equip and operate the replacement or temporary locations.

Commentary: This extension is commonplace and necessary; especially with public water/wastewater systems being designated essential services by federal and state governments. As such, your operations must not be suspended for long periods and necessitates expending resources to achieve this expectation. A recommended minimum period is 24 months of actual loss sustained.

New Locations or Newly Constructed Property

Denotation: Additional coverage that pays for direct physical loss or damage caused by a covered cause of loss to your new real property while being built on or off described premises and intended for use in your operations as well as real property you acquire, lease, or operate at locations other than the described premises intended for similar use. Coverage also applies to business personal property located at new premises as described above. A period of 180 days applies.

Commentary: This extension is not automatically offered by many property insurers. It ensures smaller projects under construction and meeting the real property definition as well as new locations meeting the real property and business personal property definitions are protected for covered causes of loss. Coverage should extend to theft of materials. A maximum period of 180 days applies to this automatic extension. A recommended minimum limit is $1 million.

Collapse

Denotation: Additional coverage that pays for direct physical loss or damage to covered property caused by abrupt collapse of real property. Abrupt collapse means an abrupt falling down or caving in of real property or any part of real property with the result that the real property or part of the real property cannot be occupied for its intended purpose. Coverage does not apply to well casings or to the perils of settling, cracking, shrinkage, bulging, or expansion.

Commentary: This extension is commonplace and necessary. It confirms abrupt collapse is a covered cause of loss for scenarios deemed insurable and serves as a reminder that well casings are commonly not deemed covered property on a property policy. There should be no minimum limit for this extension; it should qualify for the blanket policy limit.

Debris Removal

Denotation: Additional coverage that pays your expense to remove debris of covered property caused by or resulting from a covered cause of loss. Expenses will be paid only if they are reported within 180 days from the date of direct physical loss or damage.

Commentary: This extension is commonplace and necessary, but there are limitations. Many public water/wastewater systems have mature landscaping with larger trees. It is imperative to have windstorm as a covered cause of loss for damage to trees to trigger the additional coverage. Debris removal coverage also extends to removal of debris of real property and business personal property. It does not extend to any debris removal classified as a pollutant, which is a material gap for public water/wastewater systems that store chemicals, oil, petroleum, and water by-products on described premises. Such coverage is only available via an environmental pollution policy. A recommended minimum percentage is 25% of the blanket policy limit. It is important the percentage applies against the blanket policy limit and not the limit of real property shown on the statement of values. The limit should be in addition to the blanket policy limit.

Ordinance or Law Provision

Denotation: Additional coverage that applies to real property that has incurred direct physical loss or damage from a covered cause of loss. It pays for loss or damage to the undamaged portion of such real property caused by enforcement of any ordinance or law requiring the demolition of parts of the same property not damaged by the covered cause of loss, regulating the construction or repair of real property, or establishing zoning or land use requirements at the described premises in force at the time of the loss. When the real property is repaired or rebuilt, it must be intended for similar occupancy as the current property, unless otherwise required by zoning or land use law. Payment for the increased construction costs is deferred until the property is actually repaired or replaced, at the same premises or elsewhere. You must make the repairs within two years of the date of the covered loss.

Commentary: This extension is commonplace and necessary. Public water/wastewater systems with older facilities and subterranean reservoirs have a significant exposure to increased costs of construction due to updated ordinance, zoning, or land use laws. A recommended minimum percentage is 25% of the blanket policy limit. It is important the percentage applies against the blanket policy limit and not the limit of real property shown on the statement of values. The limit should be in addition to the blanket policy limit.

Electronic Data

Denotation: Additional coverage that pays for the cost to replace or restore electronic data which has been destroyed or corrupted by a covered cause of loss. A virus, harmful code, or similar instruction introduced into or enacted on a computer system (including electronic data) or a network to which it is connected, designed to damage or destroy any part of the system or disrupt its normal operation is included as a covered cause of loss unless caused by an employee or by an organization retained by you or for you to inspect, design, install, modify, maintain, repair, or replace that system. Electronic data means information, facts, or computer programs stored as or on, created or used on, or transmitted to or from computer software (including systems and applications software), on hard or floppy disks, CD-ROMs, tapes, drives, cells, data processing devices, or any other repositories of computer software which are used with electronically controlled equipment.

Commentary: This extension is commonplace and necessary but often overlooked. It represents a material exposure for public water/wastewater systems, especially from damage arising from outside virus infiltration. Electronic data comprises the data in your business and operational software. A cyber liability and network security policy is required to extend coverage for acts by employees or vendors. A recommended minimum limit is $500,000.

Accounts Receivable

Denotation: Additional coverage that applies to covered causes of loss to your accounts receivable if the records are at a described premise, at a safe place away from your described premises, or in or on a vehicle in transit between described premises.

Commentary: This extension is commonplace and necessary. It ensures your ability to receive reimbursement for an inability to collect outstanding amounts or reimbursement of reproduction costs should your accounts receivable ledger and associated records be damaged, destroyed, or stolen from a covered cause of loss. A recommended minimum limit is $500,000.

Valuable Papers and Records

Denotation: Additional coverage that pays for loss or damage to valuable papers and records whether on-site, off-site, or in transit. Valuable papers and records means inscribed, printed, or written documents as well as manuscripts, records, abstracts, books, deeds, drawings, films, maps, and mortgages. It does not mean money or securities, converted data, programs, or instructions used in your data processing operations, including the materials on which the data is recorded.

Commentary: This extension is commonplace and necessary, but it has limitations. Public water/wastewater systems are known to have old and important documents in their safes. These commonly include engineering plans, easements, deeds, shareholder stock, and water contracts. Valuable papers and records coverage, however, does not extend to your dies, patterns, or molds of original and essential equipment. Those items should be specifically scheduled on you statement of values and incorporated in the definition of business personal property. A recommended minimum limit is $500,000.

Preservation of Property

Denotation: Additional coverage that pays for loss or damage to covered property at described premises while it is being moved or temporarily stored at another location to preserve it from loss or damage by a covered cause of loss.

Commentary: This extension is commonplace and necessary. Your property policy has conditions requiring prudent protection of property that is subject to usual, customary, and reasonable standards of a public water and wastewater system. In return for this contractual obligation, your insurer agrees to extend coverage away from described premises. A recommended minimum limit is $500,000.

Tier 1 Special Commentary

Tier 1 essential additional coverages encompass an important outer layer of protection should there be direct physical loss or damage to your insurable real property and business personal property. This tier specifically complements and augments the protection for direct physical loss or damage to covered property arising from covered causes of loss. The assimilation of mobile equipment is another fundamental benefit, as that allows for the consolidation of real property, business personal property, and mobile equipment in one policy. This integration, along with the necessity of embedding equipment breakdown as a specifically cited covered cause of loss, improves pricing, streamlines deductibles, and removes conflicting language should one occurrence involve multiple assets of covered property.

Tier 2 – Prudent

Utility Services – Direct Damage

Denotation: Additional coverage that pays for loss of or damage to real property and business personal property at described premises caused by an interruption in water supply, communication supply, or power supply services. The interruption must result from direct physical loss or damage by a covered cause of loss to the utility service referenced above. Coverage does not apply to electronic data, including destruction or corruption of electronic data.

Commentary: This extension fills an important property coverage gap when you sustain damage because of an off-premises service interruption. A recommended minimum limit is $1 million.

Utility Services – Business Income and Extra Expense

Denotation: Additional coverage that applies to losses involving business income and/or extra expense from a suspension of operations at described premises caused by an interruption in water supply, communication supply, or power supply services. The interruption must result from direct physical loss or damage by a covered cause of loss to the utility service referenced above. Coverage does not apply to loss of or damage to electronic data, including destruction or corruption of electronic data, and requires a 12-hour waiting period.

Commentary: This extension fills an important financial reimbursement gap when you are unable to maintain operations after the waiting period because of a lack of power supply or other utility services to your described premises. A recommended minimum limit is $1 million.

Civil Authority

Denotation: Additional coverage that pays for the necessary extra expense caused by action of civil authority prohibiting access to the described premises due to direct physical loss of or damage to property, other than at the described premises, caused by or resulting from any covered cause of loss.

Commentary: This extension is rarely triggered since public water/wastewater systems are not normally precluded from accessing their described premises by action of a government body in charge. One exception is a natural disaster where access may be delayed until the event is contained. A recommended minimum limit is $500,000.

Commandeered Property

Denotation: Additional coverage that pays for direct physical loss or damage to commandeered property caused by or resulting from any covered cause of loss. Commandeered property means property belonging to someone else that you commandeer, seize, borrow, or take over for official use to manage an unexpected situation demanding an immediate official action by your entity during an emergency response.

Commentary: This extension protects you when an extraordinary emergency situation necessitates the appropriation of heavy equipment, materials, or miscellaneous property from others during an emergency response. A recommended minimum limit is $500,000.

Unintentional Errors

Denotation: Additional coverage that applies to any unintentional error or omission you make in determining or reporting values or in describing the covered property or covered location. Coverage does not apply to loss or damage caused directly or indirectly by flood, earth movement, or property which is otherwise insured.

Commentary: This extension primarily protects against mistakes you make on your statement of values. It is important to secure confirmation that the extension includes failing to accurately schedule real property and business personal property such as fences, radio towers, and uninstalled piping materials where you sought coverage and expected to pay a premium as well as omissions to list all your described premises. There is no protection for the failure to request certain specialized coverages like flood and earth movement. A recommended minimum limit is $250,000.

Property at Other Locations

Denotation: Additional coverage that applies to real property while it is off described premises to be cleaned, repaired, rebuilt, or restored as well as business personal property at any location you do not own, lease, or operate like exhibitions, fairs, trade shows, and conferences.

Commentary: This extension ensures any pump motors or other fixed equipment that are uninstalled and removed from described premises for maintenance or modifications are covered should there be damage from a covered cause of loss. A recommended minimum limit is $250,000.

Tenants Leasehold Interest Loss

Denotation: Additional coverage that pays for the cancellation of a favorable market rate lease resulting from direct physical loss of or damage to property at described premises caused by or resulting from any covered cause of loss.

Commentary: This extension is seldom triggered and only applicable should you lease any of your described premises. Examples of situations warranting the coverage include favorable leases negotiated through state or local public agencies or from a supportive business or fellow board member. A recommended minimum limit is $250,000.

Property in Transit

Denotation: Additional coverage that applies to the direct physical loss or damage arising from a covered cause of loss to business personal property, other than electronic data processing property or fine arts, while in transit.

Commentary: This extension ensures your machinery, equipment, computers, and communication equipment are covered should they require off-site servicing, repair, or upgrades. A recommended minimum limit is $100,000.

Pollution Remediation Expenses

Denotation: Additional coverage that pays for remediation expenses you incur as a result of the actual, alleged, or threatened presence of pollution conditions at described premises, but only if the pollution conditions result from a covered cause of loss occurring during the coverage period and reported within 180 days. Remediation expenses are expenses incurred for or in connection with the investigation, monitoring, removal, disposal, treatment, or neutralization of pollution conditions to the extent required by federal, state, or local laws and regulations or statutes as well as any legally executed state voluntary program governing the cleanup of pollution conditions. Coverage does not apply to the removal of fungus, wet rot, dry rot, virus, bacteria, asbestos, or underground petroleum tanks.

Commentary: This extension provides a baseline of protection should you incur a chemical or oil spill as well as a release of by-products or sludge at your described premises. Coverage also applies to aboveground petroleum tanks but not to underground petroleum tanks. The protection afforded has a time period for reporting claims that must be followed. A more comprehensive solution for pollution remediation expenses requires the purchase of an environmental pollution policy, which is critically important for the preceding exposures but also for others that will be addressed in a later chapter. A recommended minimum limit is $100,000.

SCADA Upgrades

Denotation: Additional coverage that pays additional costs for SCADA upgrades due to direct physical loss or damage from a covered cause of loss at a described premise. SCADA means the supervisory control and data acquisition system used in water and wastewater treatment and distribution to monitor leaks, water flow, water analysis, and other measurable items necessary to maintain operations.

Commentary: This extension safeguards against incompatibility and end of support issues for antiquated technology. Such upgrades are similarly necessary to ensure an operating platform is congruent with future system upgrades. A recommended minimum limit is $100,000. The protection afforded is intended to augment the replacement cost of SCADA that was damaged by a covered cause of loss.

Fungus, Wet Rot, or Dry Rot

Denotation: Additional coverage that pays for direct physical loss or damage to covered real property or business personal property caused by fungus, wet rot, or dry rot including the cost of its removal arising from specified causes of loss (other than fire or lightning) and flood (if flood coverage was purchased).

Commentary: This extension requires an awareness that only certain perils will trigger the coverage. Refer to Group B – Part I section for a review of what comprises specified causes of loss. Water damage is one of the specified causes of loss, but normal wear and tear as well as defective roofs will not trigger coverage. A more comprehensive solution for fungus, wet rot, or dry rot requires the purchase of an environmental pollution policy. A recommended minimum limit is $50,000.

Water Damage, Other Liquids, Powder, or Molten Material Damage

Denotation: Additional coverage that pays for loss or damage caused by or resulting from water, other liquids, powder, or molten material that leaks or flows from plumbing, heating, air conditioning, or other equipment (except fire protective systems) caused by or resulting from freezing if you do your best to maintain heat in the building or structure, or you drain the equipment and shut off the supply if heat is not maintained.

Commentary: This extension is especially important for vacant buildings or structures not regularly staffed as well as described premises where extreme low temperatures are possible. A recommended minimum limit is $50,000.

Indoor and Outdoor Signs

Denotation: Additional coverage that applies to loss or damage to signs, inside or outside a covered building or structure.

Commentary: This extension is a backstop should you fail to schedule your signs; especially signage that is detached and not located on a described premise. A recommended minimum limit is $50,000.

Tier 2 Special Commentary

Tier 2 prudent additional coverages encompass fundamental protection for public water/wastewater systems against losses that are not affirmed in the property policy. These coverages are commonly embedded in a best-in-class property policy without premium charge. Increases to sublimits are frequently available at an additional premium should there be a special reason or particular need. Certain coverages like fungus, wet rot, or dry rot as well as pollution remediation expenses represent serious exposures and warrant consideration of an environmental pollution policy to effectively encapsulate the hazard and provide more meaningful limits.

Tier 3 – Recommended

Contract Penalties

Denotation: Additional coverage that applies to penalties you are required to pay due to your failure to deliver your product according to contract terms solely as a result of direct physical loss or damage by a covered cause of loss to covered property.

Commentary: This extension should be viewed as a value-added perquisite and not a core coverage offered by your insurer. A force majeure clause within the contract can remove contract penalties by referencing direct physical loss or damage to your real property and business personal property. A recommended minimum limit is $100,000.

Outdoor Property

Denotation: Additional coverage that applies to your unscheduled outdoor property when the loss is caused by or results from any of the following covered causes of loss: (1) fire; (2) lightning; (3) explosion; (4) riot or civil commotion; (5) aircraft; (6) smoke; (7) vehicles; and (8) vandalism or malicious mischief. It is important you endorse the peril of wind to the covered causes of loss, as windstorm is a common trigger for such loss. Outdoor property commonly means fixed or permanent structures not scheduled on your statement of values including but not limited to: (a) historical markers or flagpoles; (b) sirens, antennas, towers, satellite dishes, or similar structures and their associated equipment or structures; (c) exterior signs not located at a premises; (d) fences or retaining walls; (e) storage sheds, garages, pavilions or other similar buildings or structures not located at a premises; (f) dumpsters, concrete trash containers, or permanent recycling bins; or (g) hydrants.

Commentary: This extension requires the addition of windstorm as a covered cause of loss to maximize its significance. Windstorm can create widespread damage to sirens, antennas, towers, satellite dishes, or similar structures as well as to fences and other types of unscheduled real property. The foregoing coverage is a backstop should you fail to list your antennas, towers, fences, and hydrants. It is best practice to modify the definition of real property to specifically reference these assets, as that approach will ensure blanket policy limits. A recommended minimum limit for this backstop coverage is $100,000.

Lock and Key Replacement

Denotation: Additional coverage that pays those expenses you incur to replace locks, lock cylinders, and keys in the event of covered theft of your covered property.

Commentary: This extension should be viewed as a value-added perquisite and not a core coverage offered by your insurer. A recommended minimum limit is $25,000.

Trees, Shrubs, and Plants

Denotation: Additional coverage that applies to your trees, shrubs, and plants when the loss is caused by or results from any of the following: (1) fire; (2) lightning; (3) explosion; (4) riot or civil commotion; (5) aircraft; (6) smoke; (7) vehicles; and (8) vandalism or malicious mischief.

Commentary: This extension requires the addition of windstorm as a covered cause of loss to maximize its significance. Windstorm can create widespread damage to trees and debris removal expenditures. A recommended minimum limit is $1,000 for any one tree, shrub, or plant and $25,000 in the aggregate. Higher limits on a 'per occurrence' and aggregate basis may be necessary should extensive landscaping or expensive plantings exist on your described premises.

Fine Arts

Denotation: Additional coverage that applies to irreplaceable fine arts in your headquarters; however, it is restrictive and limited to covered causes of loss only. There is no coverage for repairing, restoring, or retouching the fine art or for any defect that cause damage or destruction to the fine art. Voluntary parting is similarly excluded as are dishonest and criminal acts of employees and authorized agents as well as exhibitions where your fine art is on loan.

Commentary: This extension is more than a valued-added perquisite if you have irreplaceable fire arts in your headquarters. It requires an understanding of market value to assign accurate valuations. A recommended minimum limit is $25,000 for any one fine arts piece, with an aggregate limit reflecting the value of all your fine arts. The purchase of a specialty fine arts floater is required should you own a significant collection of irreplaceable pieces that requires specialized coverage.

Fire Protective Devices

Denotation: Additional coverage that applies to the reasonable cost to recharge or refill any fire protection devices discharged as a result of a fire or explosion.

Commentary: This extension should be viewed as a value-added perquisite and not a core coverage offered by your insurer. The protection afforded is not a warranty. It requires either the peril of fire or explosion to trigger coverage. A recommended minimum limit is $25,000.

Fire Department Service Charge

Denotation: Additional coverage that protects against liability for fire department service charges assumed by contract or agreement prior to loss or required by local ordinance.

Commentary: This extension should be viewed as a value-added perquisite and not a core coverage offered by your insurer. A recommended minimum limit is $25,000.

Arson Reward

Denotation: Additional coverage that provides a reward for information leading to the arrest and conviction of persons responsible for the crime of arson being committed against the insured. It only applies when a covered fire is deemed suspicious or to be arson by the fire department, and only when the person responsible is convicted of the crime.

Commentary: This extension should be viewed as a value-added perquisite and not a core coverage offered by your insurer. A recommended minimum limit is $10,000.

Water Contamination Notification Expenses

Denotation: Additional coverage that provides reimbursement for printing, mailing, and other expenses related to customer notification of a state-issued water contamination event.

Commentary: This extension should be viewed as a value-added perquisite and not a core coverage offered by your insurer. A recommended minimum limit is $5,000.

Rental Reimbursement – Mobile Equipment

Denotation: Additional coverage that applies to the cost of renting substitute mobile equipment that was damaged by a covered direct physical loss.

Commentary: This extension should be viewed as a value-added perquisite and not a core coverage offered by your insurer. A recommended minimum limit is $5,000.

Tools and Equipment Owned by Your Employees

Denotation: Additional coverage that pays for direct physical loss of or damage to tools and equipment owned by your employees or volunteers while at any premises or in transit, caused by or resulting from a covered cause of loss provided the loss or damage occurs during the course of your operations.

Commentary: This extension should be viewed as a value-added perquisite and not a core coverage offered by your insurer. It is an excellent source of employee goodwill. A recommended minimum limit is $5,000.

Personal Effects and Property of Others

Denotation: Additional coverage that applies to personal effects owned by your officers, your partners, members, managers, or employees as well as personal property of others in your care, custody, or control.

Commentary: This extension should be viewed as a value-added perquisite and not a core coverage offered by your insurer. It is an excellent source of employee and community goodwill. A recommend minimum limit is $5,000.

Cost of Inventory or Adjustment

Denotation: Additional coverage that applies to costs you incur to make an inventory or adjustment to prepare your proof of loss after a covered cause of loss occurs.

Commentary: This extension should be viewed as a value-added perquisite and not a core coverage offered by your insurer. A recommended minimum limit is $5,000.

Non-owned Detached Trailers

Denotation: Additional coverage that applies to loss of or damage to trailers you do not own, provided the trailer is not attached to any motorized conveyance, is used in your business, is in your care, custody, or control at the described premises, and you have a contractual responsibility to pay for loss of or damage to the trailer.

Commentary: This extension should be viewed as a value-added perquisite and not a core coverage offered by your insurer. A recommended minimum limit is $5,000.

Dependent Business Premises

Denotation: Additional coverage that applies to actual business income loss you incur due to the actual impairment of your operations; and the extra expense you incur due to the actual or potential impairment of your operations during the period of restoration. The actual or potential impairment of operations must be caused by or result from direct physical loss or damage from a covered cause of loss to real property or business personal property of a dependent business at a dependent business premises.

Commentary: This extension is quite narrow for public water/wastewater systems because the definition of a dependent business does not include premises from which you deliver your water or premises from which you receive any wholesale water. That constricts the exposure to premises like a waste disposal facility accepting your biosolids or sludge. A recommended minimum limit is $250,000.

Tier 3 Special Commentary

Tier 3 recommended additional coverages are intended to insulate public water/wastewater systems from budgetary constraints arising from smaller losses as well as promote employee and community goodwill. These coverages are not required to ensure continuity of operations or protection against capital events. They do, however, round-out an integrated and all-inclusive property policy. Awareness of the nuances encompassing fine arts, outdoor property, and trees, shrubs, and plants is especially important to ensure a well-balanced policy and certify insurer alignment.

DEDUCTIBLE

The selection of your deductible is an important step to maximize the value of your property policy. There is always a specific deductible level that is most advantageous to an insured. To achieve this optimal level, you should perform a cost-benefit analysis of the various deductible options. A level of diminishing returns exists when it comes to larger deductibles, and that necessitates a thorough review of the price points for the various options. Your per occurrence property deductible should only apply once when a claim involves real property and business personal property as well as equipment breakdown and mobile equipment. In such situations, the largest deductible should be used, and the stacking of multiple deductibles should be inapplicable. It is prudent to understand the pricing and application mechanics behind your deductible.

Another important consideration is the utilization of special deductibles for wind and hail as well as business income, extra expense, and utility interruption. Public water/wastewater systems with real property located in Tier 1 counties located in the Gulf and Atlantic coasts are exposed to named storm and non-named storm losses. Special deductibles can also apply to public water/wastewater systems with real property located in zones with known hail and tornadic activity. The preferred arrangement for windstorm and hail special deductibles is a flat per occurrence dollar amount for the event. Ideally, the deductible should apply to named storms only and not to any type of wind event. A less optimal approach is a percentage amount per damaged structure or business personal property. The least preferred method is a percentage amount per the blanket policy limit.

Public water/wastewater systems outside of designated windstorm and hail zones should not have special deductibles. Instead, their standard per occurrence property deductible should apply to these perils. The same methodology applies to earthquake and flood should you purchase these optional perils via a buyback endorsement or a difference-in-conditions policy. Regarding business income, extra expense, and utility interruption, there is commonly a waiting period deductible before coverage is triggered. A time period beyond 72 hours is ill advised, as most losses are resolved within that time frame.

VALUATION

Valuation is a critical piece to a properly configured property policy. It requires an understanding that different types of covered property and additional coverages are subject to dissimilar valuation methods. The preferred valuation method is replacement cost for your real property and business personal property as well as your newer and more expensive mobile equipment. Replacement cost, as the name applies, replaces damaged property without reduction for depreciation. It is important to secure replacement cost valuation for all real property and business personal property as well as building glass. The exceptions to the preceding rule comprise covered property deemed high hazard, older covered property without reasonable upgrades, and covered property with ornate fixtures where replacement is not practical. In such cases, functional replacement cost valuation is an acceptable alternative. This method provides replacement of the property but uses materials and designs deemed functional but not identical to the original construction.

Examples where functional replacement cost is prudent for a public water and wastewater system include compost facilities, subterranean reservoirs, repurposed facilities, and historical landmarks. Insurers will not pay on a replacement cost basis for any loss or damage until the lost or damaged property is actually repaired or replaced and unless the repairs or replacement are made within two years after the loss or damage. Moreover, insurers will not pay more for loss or damage on a replacement cost basis than the least of the following: (a) cost to replace covered property at the same premises; and (b) the cost of lost or damaged covered property with other comparable property that is used for the same purpose. Should any of the foregoing occur, then you will receive actual cash value.

Actual cash value is the least desirable valuation method since it incorporates a reduction for depreciation. This method is commonly used on mobile equipment as well as modular employee housing units. For mobile equipment, it is important you understand the limitation. Replacement cost is available on new mobile equipment, and that option should be exercised on high-valued equipment. There is a substantial premium difference between the two valuation methods, but the benefit of replacement cost should be seriously examined. Mobile equipment has a susceptibility to loss far exceeding real property and business personal property. The same applies to any rented or leased mobile equipment where you are contractually liable for the full replacement cost of the equipment. A review of these agreements is imperative to eliminate gaps in coverage. Community goodwill is also a consideration of replacement cost valuation for mobile equipment that is borrowed without an agreement.

There are special valuation and covered causes of loss rules pertaining to vacant buildings, which are commonly defined as being vacant for more than 60 consecutive days. Many insurers will apply actual cash value and not replacement cost for vacant buildings. As such, it is compulsory to understand the valuation method as well as restrictions to the covered causes of loss. Covered causes of loss are customarily restricted for said buildings to the following perils: (a) vandalism; (b) sprinkler leakage, unless you have protected the system against freezing; (c) building glass breakage; (d) water damage; (e) theft; or (f) attempted theft. Most insurers deem a building vacant when it does not contain enough business personal property to conduct customary operations. Buildings under construction are not considered vacant.

Fine arts is commonly valued at the least of its market value, the cost of reasonable restoration to its condition immediately before the loss, or the cost of replacement with substantially identical pieces. There is typically a sublimit limit for this additional coverage, which necessitates an evaluation of your most expensive piece of fine art as well as an aggregate value sufficient for your entire collection. Valuable papers and records, including those which exist on electronic or magnetic media, are normally valued at the cost of blank materials and labor to transcribe or copy the records. Accounts receivable are frequently valued by determining the total of the average monthly amounts of accounts receivable for the 12 months immediately preceding the month in which the loss or damage occurs and adjusting that total for any normal fluctuation.

Duties in the Event of Loss or Damage

Insurance policies are legally binding contracts requiring compliance by both insurer and insured. The insurer has specific contractual obligations that are codified in the form of coverages granted, policy conditions, and claims handling. The insured also has obligations, most notably your commitment to protect covered property from loss, provide accurate and truthful information, and assist the insurer when there is a claim submitted. It is important to take these obligations seriously, as any noncompliance could void your policy or prevent payment of what would have been a covered loss. Your compliance with duties involving claims are particularly important and require further discussion.

There are specific steps that must be completed to comply with your insured obligations in the event of loss or damage to covered property. The most common duties imposed by insurers on their insureds are as follows:

1. Notify the police if a law may have been broken.
2. Give prompt notice of the loss or damage and include a description of the property involved.
3. Provide a description as soon as practicable of how, when, and where the loss or damage occurred.
4. Take all reasonable steps to protect the covered property from further damage and keep a record of all such expenses.
5. Submit complete inventories of the damaged and undamaged property and list quantities, costs, values, and amount of loss claimed.
6. Allow permission for your insurer to inspect the property and take samples for further analysis as well as examine your books and records and make copies.
7. Promptly submit a signed, sworn proof of loss containing the necessary information to investigate the claim.
8. Cooperate in the investigation or settlement of the claim.
9. Allow examination of any insured under oath, while not in the presence of any other insured, relating to the insurance or the claim.

The preceding information is not difficult to follow, but it requires awareness and compliance of the specific rules-of-the-road. Honoring your contractual obligations will foster a strong rapport with your insurer and allow the cultivation of a long-term relationship based on mutual respect, fairness, and equity. This type of durable and committed adherence by both parties is preferred to incessant marketplace churning. Insurers will appreciate your level of conformance and partnership. The net result will be more competitive, predictable, and sustainable terms as well as augmented value from your insurer and property policy.

FINAL COMMENTS

The learning material in this chapter has been categorized to parallel the layout of a best-in-class property policy. The arrangement was purposeful to allow simultaneous review of both documents. Configuring a best-in-class property policy is not complicated, but it requires advanced knowledge and attention to details. Such fluency has been imparted through a streamlined process of problem identification and problem solution. A blanket policy limit without a margin clause or coinsurance clause is necessary, as are customized definitions of real property and business personal property. You similarly need to codify any real property or business personal property that is not incorporated in those definitions and modify the property not covered to remove those critical assets. When it comes to exclusions, the key is understanding intent and evaluating the prudence of supplemental coverage like flood and earthquake as well as an environmental pollution policy.

Your property policy should be tailored to encompass the additional coverages in the aforementioned three tiers, so losses are duly covered. Additional steps comprise selecting a deductible that represents the best blend of attachment, composition, and pricing as well as valuation methods that emphasize replacement cost. Understanding your insured obligations, especially as it relates to duties in the event of a loss, completes the policy encapsulation process. This level of knowledge will ensure maximum value and foster a meaningful rapport with your insurer. It will also provide a quiet confidence derived from knowing the status of your protection before you test it.

That concludes our review of the property policy. We will now segue our discussion to the Commercial General Liability (CGL) policy where policy mechanics as well as the importance of design and structure will be examined. Such knowledge will maximize the protective value of this important policy for your public water and wastewater system.

POLICY CHECKLIST

BEST-IN-CLASS PROPERTY POLICY: MINIMUM BENCHMARKS

Property

Description	Benchmark
Form	Proprietary
Covered Causes of Loss (special form)	Risk of direct physical loss unless the loss is excluded or limited by this coverage form; equipment breakdown is added as an additional covered cause of loss for direct physical loss to covered property.
Covered Property	Direct physical loss of or damage to real property and business personal property caused by or resulting from a peril not otherwise excluded, not to exceed the applicable blanket limit of insurance for real property and business personal property described in the supplemental declarations caused by or resulting from any covered cause of loss. The loss or damage to real property or business personal property must occur at, or within 100 feet of, the premises shown in the supplemental declarations.
Blanket Policy Limit	Yes (real property and& business personal property)

(Continued)

Description	Benchmark
Blanket Coverage Extensions Limit	Yes (24 months' actual loss sustained for business income, extended business income, and extra expense. $2,000,000 shared limit for commandeered property; civil authority; tenants leasehold interest; electronic data; and preservation of property)
Equipment Breakdown	Yes (embedded as a covered cause of loss and included in the blanket policy limit)
Replacement Cost	Yes (real property and business personal property as well as high valued and newer mobile equipment)
Mobile Equipment (owned and non-owned; actual cash value and replacement cost value options)	Scheduled Equipment @ actual cash value or replacement cost option Unscheduled Equipment @ actual cash value option Borrowed, Rented, and Leased Equipment @ $1,000,000 (actual cash value unless written agreement requires replacement cost option)
Windstorm and Hail Coverage	Yes (included in the blanket policy limit; no sublimit; confirm deductibles)
Coinsurance Clause	No
Margin Clause	No
Flood Coverage	Optional (requires clarification if coverage applies in all designated FEMA flood zones)
Earthquake Coverage	Optional (requires clarification if coverage applies in all zones)
New Locations or Newly Constructed Property	Yes – up to $1,000,000 for your new real property while being built on or off described premises as well as real property you acquire, lease, or operate at locations other than the described premises; and business personal property located at new premises.
Utility Services – Direct Damage; Utility Services – Business Income and Extra Expense	Yes – up to $1,000,000 for covered property damaged by an interruption in utility service to described premises. The interruption in utility service must result from direct physical loss or damage by a covered cause of loss and does not apply to loss or damage to electronic data, including destruction or corruption of electronic data. Separate limits apply to direct damage and business income and extra expense.
Contamination	Yes – up to $250,000 for loss or damage to covered property because of contamination as a result of a covered cause of loss. Contamination means direct damage to real property and business personal property caused by contact or mixture with any type of chemical used in the water treatment process.
Unintentional Errors	Yes – up to $250,000 for any unintentional error or omission you make in determining or reporting values or in describing the covered property or covered locations.
Pollution Remediation Expense	Yes – up to $100,000 or $250,000 for remediation expenses resulting from a covered cause of loss or specified cause of loss occurring during the policy period and reported within 180 days. Covered causes of loss means all risk of direct physical loss unless the loss is excluded or limited by the property policy. Specified cause of loss means: fire; lightning; explosion; windstorm or hail; smoke; aircraft or vehicles; riot or civil commotion; vandalism; leakage from fire extinguishing equipment; sinkhole collapse; volcanic action; falling objects; weight of snow; ice or sleet; water damage; and equipment breakdown.
SCADA Upgrades	Yes – up to $100,000 to upgrade your Supervisory Control and Data Acquisition (SCADA) system after direct physical loss from a covered cause of loss. The upgrade is in addition to its replacement cost
Contract Penalties	Yes – up to $100,000 for contract penalties you are required to pay due to your failure to deliver your product according to contract terms solely as a result of direct physical loss or damage by a covered cause of loss to covered property.
Property in Transit	Yes – up to $100,000 for direct physical loss or damage to covered property while in transit more than 1000 feet from the described premises. Shipments by mail must be registered for covered to apply. Electronic data processing property and fine arts are excluded.

(*Continued*)

Description	Benchmark
	Key Definitions
Pollution Conditions	The discharge, dispersal, release, seepage, migration, or escape of any solid, liquid, gaseous, or thermal irritant or contaminant, including smoke, vapor, soot, fumes, acids, alkalis, chemicals, minerals, chemical elements, and waste. Waste includes materials to be recycled, reconditioned, or reclaimed.
Real Property	The buildings, items, or structures described in the supplemental declarations that you own or that you have leased or rented from others in which you have an insurable interest. This includes: Aboveground and belowground penstock; Aboveground piping; Additions under construction; Alterations and repairs to buildings or structures; Buildings; Business personal property owned by you to maintain or service the described premises; Completed additions; Exterior signs (any type) attached to the outside of a building or standing free in the open; Fencing and retaining walls; Fixtures, including outdoor fixtures; Foundations; Glass which is part of a building or structure; Light standards; Materials, equipment, supplies, and temporary structures you own or for which you are responsible, on described premises or in the open (including property inside vehicles); Paved surfaces such as sidewalks, patios, or parking lots; Permanent storage tanks; Permanently installed machinery and equipment; Solar panels; Submersible pumps, pump motors, and engines; Telemetry-and-radio towers/antennas; Underground piping located on or within 100 feet of described premises; and Underground vaults and machinery.
Optional Real Property	Electric utility power transmission and distribution lines, poles, and related equipment (described premises only); Fire hydrants; and Water meters.
Business Personal Property	The property you own that is used in your business including: Communication equipment; Computer equipment; Furniture and fixtures; Irreplaceable dies, patterns, and molds for your pumps; Labor materials or services furnished or arranged by you on personal property of others; Leased personal property for which you have a contractual responsibility to insure; Machinery and equipment; Stock; Uninstalled piping materials and related appurtenances; and Your use interest as tenant in improvements and betterments.
Outdoor Property	Fixed or permanent structures that are outside covered real property including but not limited to: Dumpsters, concrete trash containers, or permanent recycling bins; or hydrants; Exterior signs not located at a premises; Fences or retaining walls; Historical markers or flagpoles; Sirens, antennas, towers, satellite dishes, or similar structures and associated equipment; and Storage sheds, garages, pavilions, or other similar structures not located at a premises.
	Additional Coverages
Above Ground Piping	Payable up to the blanket policy limit (described premises)
Below Ground Piping	Payable up to blanket policy limit (described premises)
Debris Removal	25% of blanket policy limit
Ordinance or Law Provision	25% of blanket policy limit
Business Personal Property at New Locations	$1,000,000
New Locations or Newly Constructed Property	$1,000,000
Utility Services – Direct Damage	$1,000,000

(*Continued*)

Description	Benchmark
Utility Services – Business Income and Extra Expense	$1,000,000
Accounts Receivable	$500,000
Valuable Papers and Records	$500,000
Overflow of Water from Sewer, Drain, Sump	$500,000
Dependent Business Premises	$250,000
Pollution Remediation Expense	$250,000 (specified causes of loss)
Property at Other Locations	$250,000
Tenants Leasehold Interest Loss	$250,000
Contamination	$250,000
Contract Penalties	$100,000
Outdoor Property	$100,000 (unscheduled and must include peril of windstorm)
Pollution Remediation Expense	$100,000 (covered causes of loss)
Property in Transit	$100,000
SCADA Upgrades	$100,000
Indoor and Outdoor Signs	$50,000 (unscheduled and described premises)
Limited Coverage for Fungus, Wet Rot, Dry Rot	$50,000
Water Damage, Other Liquids, Powder, or Molten Material Damage	$50,000
Fine Arts	$25,000
Fire Department Service Charge	$25,000
Fire Protection Devices	$25,000
Lock and Key Replacement	$25,000
Trees, Shrubs, and Plants	$25,000 ($1,000 maximum for any tree, plant, or shrub; mature landscaping requires higher sublimits)
Arson Reward	$10,000
Cost of Inventory or Adjustment	$5,000
Non-owned Detached Trailers	$5,000
Personal Effects of Others	$5,000
Personal Effects and Property of Others	$5,000
Rental Reimbursement – Mobile Equipment	$5,000
Tools and Equipment Owned by Your Employees	$5,000
Water Contamination Notification	$5,000
Patterns, Dies, and Molds	$2,500 (confirm you don't need higher limit for any irreplaceable pump impeller patterns, dies, and molds)

REVIEW QUESTIONS

1. T/F: A blanket policy limit in a property policy only applies to covered property on described premises.

2. A best-in-class property policy includes the following:
   a. Blanket policy limits
   b. Margin clause
   c. Coinsurance clause
   d. a and b only

3. T/F: A special causes of loss form in a property policy covers all causes of loss unless specifically excluded.

4. T/F: A property policy inclusive of equipment breakdown coverage requires a joint-loss agreement endorsement.

5. T/F: A blanket policy limit in a property policy is not constrained to the individual valuation limits of each covered property on an insured's statement of values.

6. The definition of real property in a property policy should be modified to include the following:
   a. Fencing
   b. Foundations
   c. Telemetry and radio towers
   d. All of the above

7. T/F: A prudent action to certify insurer alignment is to modify the definition of business personal property in a property policy to include uninstalled piping materials and related appurtenances.

8. T/F: You should schedule your underground piping infrastructure as covered property when purchasing earthquake coverage via endorsement to your property policy or through a difference-in-conditions policy.

9. T/F: Climate change does not necessitate the consideration of purchasing first-party flood coverage because damage from flood is increasingly predictable and consigned to high-hazard zones.

10. T/F: A property policy covers voluntary parting with any property by you or anyone else to whom you have entrusted the property if induced to do so by any fraudulent scheme, trick, device, or false pretense.

11. T/F: The additional coverages of business income, extra expense, and debris removal in a property policy are only triggered if there is direct physical loss or damage to covered property at described premises.

12. The following costs should be included when formulating replacement cost valuation for covered property on an insured's statement of values:
    a. Engineering fees
    b. Expected construction costs to rebuild the covered property
    c. Financing costs
    d. All of the above

13. The definition of real property in a property policy should include the following:
    a. Permanently installed machinery and equipment
    b. Money and securities
    c. Earthen dams and canals
    d. All of the above

14. The definition of business personal property in a property policy should include the following:
    a. Mobile equipment

   b. Submersible pumps
   c. Communications equipment
   d. All of the above

15. The definition of real property in a property policy should be expanded to include the following:
   a. Fencing and retaining walls
   b. Foundations
   c. Telemetry towers
   d. All of the above

16. Common examples of property not covered in a property policy include the following:
   a. Accounts, bills, currency, food stamps, or evidences of debt, money, notes, or securities
   b. Aircraft
   c. Animals
   d. All of the above

17. T/F: The additional coverage of extra expense in a property policy will pay the cost of replenishing previously purchased water after a covered cause of loss.

18. T/F: Electrical power distribution lines that are situated over great distances are considered a low loss probability to the perils of fire and windstorm.

19. T/F: The necessary drying out of covered electrical equipment as a result of a flood is included under equipment breakdown coverage.

20. T/F: Catastrophic exposures like earthquake and flood can be secured through the purchase of a DIC policy.

21. T/F: The fortuity doctrine requires a covered loss to be accidental or by chance and not based on exposures that are preventable and expected.

22. T/F: An earthquake exclusion in a best-in-class property policy affords coverage for any resulting loss arising from fire or explosion.

23. T/F: A property policy excludes losses arising from settling, cracking, shrinking, or expansion.

24. The contamination exclusion in a property policy should have exceptions for the following chemicals:
   a. Chlorine
   b. Ammonia
   c. a and b only
   d. All chemicals used in the water treatment process

25. T/F: Many public water/wastewater systems rely on FEMA and not insurance policies for first-party cost recovery in the event of earthquake and flood.

26. Examples of high-risk flood zones as designated by FEMA include the following:
   a. AE
   b. B
   c. Shaded X
   d. a and b only

27. T/F: The definition of valuable papers and records in a property policy includes irreplaceable pump dies and propeller molds.

28. T/F: Not-for-profit public water/wastewater systems do not need loss of income coverage.

29. T/F: The additional coverage of newly constructed property in a best-in-class property policy includes theft of building materials.

30. T/F: The additional coverage of debris removal in a property policy include materials classified as a pollutant.

31. T/F: The additional coverage of electronic data in a property policy includes employee or vendor-initiated viruses.

32. T/F: The additional coverage of unintentional errors in a property policy includes failure to request coverage for flood.

33. T/F: The additional coverage of property in transit in a property policy includes coverage for fine arts.

34. T/F: The additional coverage of dependent business premises in a property policy includes premises from which you deliver your water or premises from which you receive any wholesale water.

35. The following valuation methods may be used in a property policy:
    a. Replacement cost
    b. Functional replacement cost
    c. Actual cash value
    d. All of the above

36. T/F: Replacement cost valuation replaces damaged property without reduction for depreciation.

37. Examples where functional replacement cost valuation may be required for a public water and wastewater system include the following:
    a. Compost facilities
    b. Subterranean reservoirs
    c. Repurposed facilities
    d. All of the above

38. T/F: A property policy has special valuation and covered causes of loss rules pertaining to buildings vacant for more than 60 days.

39. T/F: Most property insurers deem a building vacant when it contains enough business personal property to conduct customary operations.

40. The most common duties imposed by insurers on their insureds in the event of loss or damage to covered property in a property policy include the following:
    a. Notify police if any laws were broken
    b. Take all reasonable measures to protect covered property from further damage
    c. Provide an inventory list of all damaged property within seven days
    d. a and b only

41. T/F: Accounts receivable are frequently valued by determining the total of the average monthly amounts of accounts receivable for the 12 months immediately preceding the month in which the loss or damage occurs and adjusting that total for any normal fluctuation.

42. Replacement cost valuation may not be offered in a property policy for the following:
    a. Older property without upgrades
    b. Older property with upgrades
    c. Older property with state-of-the-art technology
    d. a and c only

43. T/F: The additional coverages of business income, extra expense, and utility interruption in a property policy commonly include a waiting period deductible.

44. T/F: The preferred structure for windstorm and hail deductibles in a property policy is a flat per occurrence dollar amount for the event.

45. T/F: A deductible in a property policy that includes equipment breakdown and mobile equipment coverages should only apply once for a covered claim involving multiple losses from the same occurrence.

46. T/F: A force majeure clause within a contract can remove liability by exempting direct physical loss or damage to real property and business personal property from a natural and unavoidable catastrophe.

47. T/F: A property policy excludes fungus, wet rot, or dry rot arising from normal wear and tear.

48. T/F: The additional coverage of Supervisory Control and Data Acquisition (SCADA) upgrades in a property policy safeguards against incompatibility and end of support issues for antiquated technology.

49. T/F: Commandeered property means property belonging to someone else that you commandeer, seize, borrow, or take over for official use to manage an unexpected situation demanding an immediate official action by your public water and wastewater system during an emergency response.

50. T/F: Public water/wastewater systems with older facilities and subterranean reservoirs have a significant exposure to increased costs of construction due to updated ordinance, zoning, or land use laws.

51. The additional coverages of loss of income and extra expense in a property policy should include the following:
    a. Actual loss sustained verbiage up to a minimum of 12 months
    b. Actual loss sustained verbiage up to a maximum of 6 months
    c. No period of restoration
    d. No suspension of operations

52. Mobile equipment coverage should encompass the following:
    a. Owned mobile equipment
    b. Non-owned mobile equipment
    c. Borrowed mobile equipment
    d. All of the above

ANSWER KEY

1. **T**/F: A blanket policy limit in a property policy only applies to covered property on described premises.

2. A best-in-class property policy includes the following:
    a. **Blanket policy limits**
    b. Margin clause
    c. Coinsurance clause
    d. a and b only

3. **T**/F: A special causes of loss form in a property policy covers all causes of loss unless specifically excluded.

4. T/**F**: A property policy inclusive of equipment breakdown coverage requires a joint-loss agreement endorsement.

5. **T**/F: A blanket policy limit in a property policy is not constrained to the individual valuation limits of each covered property on an insured's statement of values.

6. The definition of real property in a property policy should be modified to include the following:
    a. Fencing
    b. Foundations
    c. Telemetry and radio towers
    d. **All of the above**

7. **T/F**: A prudent action to certify insurer alignment is to modify the definition of business personal property in a property policy to include uninstalled piping materials and related appurtenances.

8. **T/F**: You should schedule your underground piping infrastructure as covered property when purchasing earthquake coverage via endorsement to your property policy or through a difference-in-conditions policy.

9. T/F: Climate change does not necessitate the consideration of purchasing first-party flood coverage because damage from flood is increasingly predictable and consigned to high-hazard zones.

10. T/F: A property policy covers voluntary parting with any property by you or anyone else to whom you have entrusted the property if induced to do so by any fraudulent scheme, trick, device, or false pretense.

11. **T/F**: The additional coverages of business income, extra expense, and debris removal in a property policy are only triggered if there is direct physical loss or damage to covered property at described premises.

12. The following costs should be included when formulating replacement cost valuation for covered property on an insured's statement of values:
    a. Engineering fees
    b. **Expected construction costs to rebuild the covered property**
    c. Financing costs
    d. All of the above

13. The definition of real property in a property policy should include the following:
    a. **Permanently installed machinery and equipment**
    b. Money and securities
    c. Earthen dams and canals
    d. All of the above

14. The definition of business personal property in a property policy should include the following:
    a. Mobile equipment
    b. Submersible pumps
    c. **Communications equipment**
    d. All of the above

15. The definition of real property in a property policy should be expanded to include the following:
    a. Fencing and retaining walls
    b. Foundations
    c. Telemetry towers
    d. **All of the above**

16. Common examples of property not covered in a property policy include the following:
    a. Accounts, bills, currency, food stamps, or evidences of debt, money, notes, or securities
    b. Aircraft
    c. Animals
    d. **All of the above**

17. **T/F**: The additional coverage of extra expense in a property policy will pay the cost of replenishing previously purchased water after a covered cause of loss.

18. T/F: Electrical power distribution lines that are situated over great distances are considered a low loss probability to the perils of fire and windstorm.

19. T/F: The necessary drying out of covered electrical equipment as a result of a flood is included under equipment breakdown coverage.

20. **T/F**: Catastrophic exposures like earthquake and flood can be secured through the purchase of a DIC policy.

21. **T/F**: The fortuity doctrine requires a covered loss to be accidental or by chance and not based on exposures that are preventable and expected.

22. **T/F**: An earthquake exclusion in a best-in-class property policy affords coverage for any resulting loss arising from fire or explosion.

23. **T/F**: A property policy excludes losses arising from settling, cracking, shrinking, or expansion.

24. The contamination exclusion in a property policy should have exceptions for the following chemicals:
    a. Chlorine
    b. Ammonia
    c. a and b only
    d. **All chemicals used in the water treatment process**

25. **T/F**: Many public water/wastewater systems rely on FEMA and not insurance policies for first-party cost recovery in the event of earthquake and flood.

26. Examples of high-risk flood zones as designated by FEMA include the following:
    a. **AE**
    b. B
    c. Shaded X
    d. a and b only

27. **T/F**: The definition of valuable papers and records in a property policy includes irreplaceable pump dies and propeller molds.

28. **T/F**: Not-for-profit public water/wastewater systems do not need loss of income coverage.

29. **T/F**: The additional coverage of newly constructed property in a best-in-class property policy includes theft of building materials.

30. **T/F**: The additional coverage of debris removal in a property policy include materials classified as a pollutant.

31. **T/F**: The additional coverage of electronic data in a property policy includes employee or vendor-initiated viruses.

32. **T/F**: The additional coverage of unintentional errors in a property policy includes failure to request coverage for flood.

33. **T/F**: The additional coverage of property in transit in a property policy includes coverage for fine arts.

34. **T/F**: The additional coverage of dependent business premises in a property policy includes premises from which you deliver your water or premises from which you receive any wholesale water.

35. The following valuation methods may be used in a property policy:
    a. Replacement cost
    b. Functional replacement cost
    c. Actual cash value
    d. **All of the above**

36. **T/F**: Replacement cost valuation replaces damaged property without reduction for depreciation.

37. Examples where functional replacement cost valuation may be required for a public water and wastewater system include the following:
    a. Compost facilities
    b. Subterranean reservoirs
    c. Repurposed facilities
    d. **All of the above**

38. **T/F**: A property policy has special valuation and covered causes of loss rules pertaining to buildings vacant for more than 60 days.

39. **T/F**: Most property insurers deem a building vacant when it contains enough business personal property to conduct customary operations.

40. The most common duties imposed by insurers on their insureds in the event of loss or damage to covered property in a property policy include the following:
   a. Notify police if any laws were broken
   b. Take all reasonable measures to protect covered property from further damage
   c. Provide an inventory list of all damaged property within seven days
   d. **a and b only**

41. **T/F**: Accounts receivable are frequently valued by determining the total of the average monthly amounts of accounts receivable for the 12 months immediately preceding the month in which the loss or damage occurs and adjusting that total for any normal fluctuation.

42. Replacement cost valuation may not be offered in a property policy for the following:
   a. **Older property without upgrades**
   b. Older property with upgrades
   c. Older property with state-of-the-art technology
   d. a and c only

43. **T/F**: The additional coverages of business income, extra expense, and utility interruption in a property policy commonly include a waiting period deductible.

44. **T/F**: The preferred structure for windstorm and hail deductibles in a property policy is a flat per occurrence dollar amount for the event.

45. **T/F**: A deductible in a property policy that includes equipment breakdown and mobile equipment coverages should only apply once for a covered claim involving multiple losses from the same occurrence.

46. **T/F**: A force majeure clause within a contract can remove liability by exempting direct physical loss or damage to real property and business personal property from a natural and unavoidable catastrophe.

47. **T/F**: A property policy excludes fungus, wet rot, or dry rot arising from normal wear and tear.

48. **T/F**: The additional coverage of Supervisory Control and Data Acquisition (SCADA) upgrades in a property policy safeguards against incompatibility and end of support issues for antiquated technology.

49. **T/F**: Commandeered property means property belonging to someone else that you commandeer, seize, borrow, or take over for official use to manage an unexpected situation demanding an immediate official action by your public water and wastewater system during an emergency response.

50. **T/F**: Public water/wastewater systems with older facilities and subterranean reservoirs have a significant exposure to increased costs of construction due to updated ordinance, zoning, or land use laws.

51. The additional coverages of loss of income and extra expense in a property policy should include the following:
   a. **Actual loss sustained verbiage up to a minimum of 12 months**
   b. Actual loss sustained verbiage up to a maximum of 6 months
   c. No period of restoration
   d. No suspension of operations

52. Mobile equipment coverage should encompass the following:
   a. Owned mobile equipment
   b. Non-owned mobile equipment
   c. Borrowed mobile equipment
   d. **All of the above**

21 Commercial General Liability (CGL) Policy

2019 © Photo Courtesy of Placer County Water Agency, Placer County California 95604

LEARNING OBJECTIVES

- Understand how to structure your commercial general liability (CGL) policy to accommodate all your risks.
- Comprehend the differences between silence, exclusions, and affirmation in a CGL policy.
- Describe the necessary exceptions to your standardized CGL policy exclusions.
- Identify the optional coverages that are available within your CGL policy as well as those exclusions that must be avoided.
- Explain the individuals and organizations qualifying as named insureds in your CGL policy.
- Articulate the distinction between insureds and additional insureds in your CGL policy.
- Grasp the six limits of insurance afforded in your CGL policy.
- Comprehend the differences between Coverages A., B., and C. in your CGL policy.
- Explain the purpose and restrictions of no-fault medical payments in your CGL policy.
- Articulate the duties and obligations of insurer and insured after a loss in a CGL policy.
- Describe the difference between the insurer's obligation of defense and indemnity in a CGL policy.
- Understand the rationale of the separation of insureds condition in the CGL policy.

DOI: 10.1201/9781003207146-31

KEY DEFINITIONS*

Accident – includes continuous or repeated exposure to the same conditions resulting in bodily injury or property damage.

Additional Insured – person or organization not automatically included as an insured under an insurance policy who is included or added as an insured under the policy at the request of the named insured. A named insured's impetus for providing additional insured status to others may be a desire to protect the other party because of a close relationship with that party or to comply with a contractual agreement requiring the named insured to do so (e.g., project owners, customers, or owners of property leased by the named insured). In liability insurance, additional insured status is commonly used in conjunction with an indemnity agreement between the named insured (the indemnitor) and the party requesting additional insured status (the indemnitee).

Additional Insured Endorsement – policy endorsement used to add coverage for additional insureds by name (e.g., mortgage holders or lessors). There are a number of different forms intended to address various situations, some of which afford very restrictive coverage to additional insureds. Rather than naming each additional insured, a blanket additional insured endorsement is normally preferred whenever a written agreement requires it. Notwithstanding, some court cases have rejected additional insured status when a blanket endorsement is used.

Allocated Loss Adjustment Expense (ALAE) – loss adjustment expenses that are assignable or allocable to specific claims. Fees paid to outside attorneys, experts, and investigators used to defend claims are examples of ALAE.

Anti-stacking Provisions – provisions intended to avoid the application of multiple sets of deductibles or multiple sets of limits to a single loss event. Anti-stacking provisions stipulate that, in such an event, only one policy limit or one deductible (rather than the limit or deductible under each policy) applies to the occurrence.

Audit – survey of the financial and other records of a person or organization conducted annually (in most cases) to determine exposures, limits, premiums, etc.

Blanket Additional Insured Endorsement – an endorsement attached to liability insurance policies that automatically grants insured status to a person or organization that the named insured is required by contract to add as an insured. May apply only to specific types of contracts or entities.

Bodily Injury – liability policy term that includes bodily harm, sickness, or disease, including resulting death.

Claim – used in reference to insurance, a claim may be a demand by an individual or organization to recover, under a policy of insurance, for loss that may be covered within that policy.

Commercial General Liability (CGL) Policy – standard insurance policy issued to businesses and governmental entities to protect them against liability claims for bodily injury and property damage arising out of their premises, operations, products, and completed operations as well as personal and advertising injury offenses and no-fault medical expenses.

Completed Operations – under a commercial general liability (CGL) policy, work of the insured that has been completed as called for in a contract, or work completed at a single job site under a contract involving multiple job sites, or work that has been put to its intended use.

---

* Source material provided by International Risk Management Institute (IRMI), https://www.irmi.com/term/insurance-definitions/insurance, as well as policy forms and underwriting products promulgated by Insurance Services Office (ISO) and their member companies, https://products.iso.com/productcenter/10.

Contractual Liability – liability imposed on an entity by the terms of a contract. As used in insurance, the term refers not to all contractually imposed liability but to the assumption of the other contracting party's liability under specified conditions via an insured contract.

Coverage Territory – contractual provisions limiting coverage to geographical areas within which the insurance is affected.

Coverage Trigger – event that must occur before a particular liability policy applies to a given loss. Under an occurrence policy, the occurrence of injury or damage is the trigger; liability will be covered under that policy if the injury or damage occurred during the policy period. Under a claims-made policy, the making of a claim triggers coverage.

Damage to Premises Rented to You – one of the limits of liability prescribed by the standard commercial general liability policy (CGL). It applies to damage by fire to premises rented to the insured during the policy period and to damage regardless of cause to premises (including contents) occupied by the insured for seven days or less, although some insurers offer a great time period.

Damages – money whose payment a court orders as compensation to an injured plaintiff. Fines, penalties, or injunctive relief would not typically constitute damages.

Defamation – any written or oral communication about a person or thing that is both untrue and unfavorable. CGL policies provide coverage for claims alleging defamation.

Defense Clause – insurance provision in which the insurer agrees to defend, with respect to insurance afforded by the policy, all suits against the insured.

Defense Outside Limits – amounts paid by an insurer to defend claims or suits that are in addition to the policy limits and do not decrease the per occurrence claim limit.

Disclaimer – situation where an insurer refuses to accept your insurance claim on an existing policy.

Duty to Defend – term used to describe an insurer's obligation to provide an insured with defense to claims made under a liability insurance policy. As a general rule, an insured need only establish that there is potential for coverage under a policy to give rise to the insurer's duty to defend. Therefore, the duty to defend may exist even where coverage is in doubt and ultimately does not apply. Implicit in this rule is the principle that an insurer's duty to defend an insured is broader than its duty to indemnify. Moreover, an insurer may owe a duty to defend its insured against a claim in which ultimately no damages are awarded, and any doubt as to whether the facts support a duty to defend is usually resolved in the insured's favor.

Economic Damages – award to an injured person in an amount sufficient to compensate for his or her actual monetary loss.

Electronic Data – exposure faced by individuals and organizations that may cause loss of, damage to, or inability to access or use electronically stored data.

Employee – any natural person: (a) while in your service; (b) who you compensate directly by salary, wages, or commissions; and (c) who you have the right to direct and control while performing services for you. Employee includes a leased worker but does not include a temporary worker.

Employee Benefit Plans – benefits, such as health and life insurance, provided to employees at the workplace, usually paid for totally or in part by the employer. Liability of an employer for an error or omission in the administration of an employee benefit program, such as failure to advise employees of benefit programs.

Employers' Liability – part B of the workers compensation policy that provides coverage to the insured (employer) for liability to employees for work-related bodily injury or disease, other than liability imposed on the insured by a workers compensation law.

Exclusion – policy provision that eliminates coverage for some type of risk. Exclusions narrow the scope of coverage provided by the insuring agreement. In many insurance policies, the insuring agreement is very broad. Insurers utilize exclusions to carve away coverage for risks they are unwilling to insure.

Executive Officer – as defined in the standard commercial general liability (CGL) policy, an executive officer is a person holding any of the officer positions created by the named insured organization's governing document. Executive officers, like other employees, have insured status under the CGL policy of the organization; but unlike other employees, their insured status is not subject to the fellow employee exclusion.

Fellow Employee – endorsement that provides coverage for claims made by an injured employee against a fellow employee who caused or contributed to the injury.

Fiduciary Liability – responsibility on trustees, employers, fiduciaries, professional administrators, and the plan itself with respect to errors and omissions (E&O) in the administration of employee benefit programs as imposed by the Employee Retirement Income Security Act (ERISA).

Four Corners Test – legal principle that an insurer owes a duty of defense to its insured if the formal allegations of a claim or suit match the literal provisions of the insurance policy, regardless of any differing or contradictory facts known or knowable to the insurer.

General Aggregate Limit – the maximum limit of insurance payable during any given annual policy period for all losses other than those arising from specified exposures. Under the commercial general liability (CGL) policy, the general aggregate limit applies to all covered bodily injury and property damage (except for injury or damage arising out of the products-completed operations hazard) and all covered personal and advertising injury as well as no-fault medical expenses and damage to premises rented to you. When paid losses in these categories reach the specified aggregate limit, that limit is exhausted and no more losses in any of those categories will be paid under the policy. In other words, once the general aggregate limit is paid out, the only coverage remaining under the policy will be for products-completed operations claims which are paid out of a separate aggregate.

Governing Body – group of elected or appointed people who formulate the policy and direct the affairs of a public water and& wastewater system.

Impaired Property – under the commercial general liability (CGL) policy, property that has sustained loss of use because it incorporates the insured's defective product or work and can be restored to use by the repair, removal, or replacement of the defective product or work. Claims for loss of use of impaired property are excluded from the CGL policy.

Indemnity – compensation to a party for a loss or damage that has already occurred, or to guarantee through a contractual clause to repay another party for loss or damage that might occur in the future. The concept of indemnity is based on a contractual agreement made between two parties in which one party (the indemnitor) agrees to pay for potential losses or damages caused by the other party (the indemnitee).

Insured – any person or organization qualifying as an insured in the "who is an insured" provision of the applicable coverage. Except with respect to the limit of insurance, the coverage afforded applies separately to each insured who is seeking coverage or against whom a claim or suit is brought.

Insured Contract – a defined term common in liability policies that provides limited exceptions to the contractually assumed liability exclusion, by stating that the exclusion does not apply to liability assumed in an insured contract. The definition of the term varies, but in most cases, it will extend some coverage for liabilities assumed in an enforceable hold harmless provision of a commercial contract.

Insuring Agreement – portion of the insurance policy in which the insurer promises to make payment to or on behalf of the insured. The insuring agreement is usually contained in a coverage form from which a policy is constructed. Often, insuring agreements outline a broad scope of coverage, which is then narrowed by exclusions and definitions.

Insurance Services Office Inc. (ISO) – organization that collects statistical data, promulgates rating information, develops standard policy forms, and files information with state regulators on behalf of insurers that purchase its services.

Judgment – decision of a court regarding the rights and liabilities of parties in a legal action or proceeding. Judgments also generally provide the court's explanation of why it has chosen to make a particular court order.

Law Enforcement Wrongful Acts – provides errors and omissions (E&O) coverage for police departments. Unlike most professional liability coverage, such policies are often written on an occurrence (rather than on a claims-made) basis. Some of the more important covered acts include: false arrest, excessive force, and invasion of privacy. Common exclusions are: criminal/intentional acts, claims for injunctive relief, and motor vehicle operations.

Leased Worker – person leased to you by a labor leasing firm under an agreement between you and the labor leasing firm to perform duties related to the conduct of your business. Leased worker does not include a temporary worker.

Limits of Insurance – the most that will be paid by an insurer in the event of a covered loss under an insurance policy.

Loss – direct and accidental loss or damage, or a reduction in value.

Loss of Use – property damage is defined in the commercial general liability policy as physical injury to tangible property including resulting loss of use and loss of use of tangible property that has not been physically injured.

Manager – person serving in a directorial capacity for a limited liability company.

Member – owner of a limited liability company represented by its membership interest, who also may serve as a manager.

Montrose – type of known loss provision in a liability policy that restricts coverage for damage that occurs over multiple policy periods to those policies in which the insured was not aware of the occurrence at the inception of the policy period. The term "Montrose provision" is derived from the 1995 California Supreme Court ruling *Montrose Chem. Corp. v. Admiral Ins. Co.*, 10 Cal. 4th 645, 42 Cal. Rptr. 2d 324, 913 P.2d 878 (1995) that allowed recovery for damages even after the insured was aware of the loss, because the full extent of the loss was not yet known.

Non-owned Automobile – described in commercial auto policies as a private passenger-type auto owned by an employee that is used in connection with the named insured's business but that is not owned, leased, hired, rented, or borrowed by the named insured. *Hired Automobile* refers to autos the named insured leases, hires, rents, or borrows.

Occurrence – in a commercial general liability (CGL) policy, an accident, including continuous or repeated exposure to substantially the same general harmful conditions. CGL policies insure liability for bodily injury or property damage that is caused by an occurrence.

Premises-Operations – one of the categories of hazards ordinarily insured by a commercial general liability (CGL) policy. Composed of those exposures to loss that fall outside the defined products-completed operations hazard, it includes liability for injury or damage arising out of the insured's premises or out of the insured's business operations while such operations are in progress.

Products-Completed Operations – one of the hazards ordinarily insured by a CGL policy. It encompasses liability arising out of the insured's products or business operations conducted away from the insured's premises once those operations have been completed or abandoned.

Per Occurrence Limit – in liability insurance, the maximum amount the insurer will pay for all claims resulting from a single occurrence, no matter how many people are injured, how much property is damaged, or how many different claimants may make claims.

Personal and Advertising Injury – injury, including consequential bodily injury, arising out of one or more of the following offenses: (a) false arrest, detention, or imprisonment; (b) malicious prosecution; (c) the wrongful eviction from, wrongful entry into, or invasion of the right of private occupancy of a room, dwelling, or premises that a person occupies, committed by or on behalf of its owner, landlord, or lessor; (d) oral or written publication in any manner of material that slanders or libels a person or organization or disparages a person's or organization's goods, products, or services; (e) oral or written publication in any manner of material that violates a person's right of privacy; (f) the use of another's advertising idea in your advertisement; or (g) infringing upon another's copyright, trade dress, or slogan in your advertisement.

Physical Injury – losses caused by injuries to persons and legal liability imposed on the insured for such injury or for damage to property of others.

Policy Conditions – section of an insurance policy that identifies general requirements of an insured and the insurer on matters such as loss reporting and settlement, property valuation, other insurance, subrogation rights, and cancellation and nonrenewal. The policy conditions are usually stipulated in the coverage form of the insurance policy.

Policy Period – term of duration of the policy. The policy period encompasses the time between the exact hour and date of policy inception and the hour and date of expiration.

Pollutants – any solid, liquid, gaseous, or thermal irritant or contaminant, including smoke, vapor, soot, fumes, acids, alkalis, chemicals, and waste. Waste includes materials to be recycled, reconditioned, or reclaimed.

Primary and Noncontributory – term is commonly used in contract insurance requirements to stipulate the order in which multiple policies triggered by the same loss are to respond. [For example, a contractor may be required to provide liability insurance that is primary and noncontributory. This means that the contractor's policy must pay before other applicable policies (primary) and without seeking contribution from other policies that also claim to be primary (noncontributory)].

Products – defined in commercial general liability (CGL) policies to include property, other than real property manufactured, sold, handled, distributed, or disposed of by the named insured or others involved with the named insured in the stream of commerce. The definition of product includes containers, parts and equipment, product warranties, and provision of or failure to provide instructions and warnings.

Products-Completed Operations Aggregate Limit – one of the hazards ordinarily insured by a commercial general liability (CGL) policy. It encompasses liability arising out of the insured's products or business operations conducted away from the insured's premises once those operations have been completed or abandoned. This aggregate limit applies to all bodily injury and property damage included in the products and completed operations hazard and is the maximum amount that will be paid during the policy period.

Professional Liability – type of liability coverage designed to protect traditional professionals (e.g., accountants, attorneys) and quasi-professionals (e.g., real estate brokers, consultants) against liability incurred as a result of errors and omissions in performing their professional services. Most professional liability policies only cover economic or financial losses suffered by third parties, as opposed to bodily injury and property damage claims.

Property Damage – as defined in the commercial general liability (CGL) policy, physical injury to tangible property including resulting loss of use and loss of use of tangible property that has not been physically injured.

Railroad Protective Liability – insurance coverage protecting a railroad from liability it incurs because of the work of contractors on or near the railroad right-of-way.

Representation – statement made in an application for insurance that the prospective insured represents as being correct to the best of his or her knowledge. If the insurer relies on a representation in entering into the insurance contract and if it proves to be false at the time it was made, the insurer may have legal grounds to void the contract.

Reservation of Rights – an insurer's notification to an insured that coverage for a claim may not apply. Such notification allows an insurer to investigate (or even defend) a claim to determine whether coverage applies (in whole or in part) without waiving its right to later deny coverage based on information revealed by the investigation. Insurers use a reservation of rights letter because in many claim situations, all the insurer has at the inception of the claim are various unsubstantiated allegations and, at best, a few confirmed facts.

Risk Assumption – planned acceptance of losses by deductibles, deliberate noninsurance, and loss-sensitive plans where some, but not all, risk is consciously retained rather than transferred.

Settlement – resolution between disputing parties about a legal case, reached either before or after court action begins.

Silica – silicon dioxide, in the form of tiny, airborne crystals, which occurs naturally in soil and is present as an ingredient in many man-made substances such as brick, concrete, and asphalt. In sufficient concentrations, it can cause silicosis and other diseases. Silica and silica-related dust is commonly the subject of an exclusion in the commercial general liability (CGL) policy.

Stacking of Limits – the application of two or more policies' limits to a single occurrence or claim. This is common with products liability, construction defect, and pollution claims in which the occurrence has transpired over numerous years, and it is difficult to ascertain which policy provides coverage. It can also occur under auto liability or uninsured/underinsured motorists coverage when two or more vehicle limits can be stacked to apply to a single occurrence.

Stop-Gap Endorsement – an endorsement that is primarily used to provide employers' liability coverage for work-related injuries arising out of exposures in monopolistic fund states (fund workers compensation policies do not provide employers' liability coverage). If the employer has operations in non-monopolistic states, the endorsement is attached to the workers compensation policy providing coverage in those states. For employers operating exclusively in a monopolistic fund state, the endorsement is attached to the employer's commercial general liability (CGL) policy.

Suit – a civil proceeding in which: (1) damages because of bodily injury or property damage; or (2) a covered pollution cost or expense; to which this insurance applies, are alleged. Suit includes: (a) an arbitration proceeding in which such damages or covered pollution costs or expenses are claimed and to which the insured must submit or does submit with our consent; or (b) any other alternative dispute resolution proceeding in which such damages or covered pollution costs or expenses are claimed and to which the insured submits with our consent.

Supplementary Payments – term used in liability policies for the costs associated with the investigation and resolution of claims. Supplementary payments are normally defined to include such items as first aid expenses, premiums for appeal and bail bonds, pre- and post-judgment interest, and reasonable travel expenses incurred by the insured at the insurer's request when assisting in the defense of a claim. Actual settlements/judgments are considered damages rather than supplementary payments.

Tangible Property – property damage definition in a commercial general liability (CGL) policy includes physical injury to tangible property including resulting loss of use and loss of use of tangible property that has not been physically injured.

Temporary Worker – person who is furnished to you to substitute for a permanent employee on leave or to meet seasonal or short-term workload conditions.

Terrorism Risk Insurance Act (TRIA) of 2002 – federal legislation enacted in 2002 to guarantee the availability of insurance coverage against acts of international terrorism. Under the Act, commercial insurers are required to offer insurance coverage against such terrorist incidents and are reimbursed by the federal government for paid claims subject to deductible and retention amounts. This legislation was modified and extended by the Terrorism Risk Insurance Extension Act (TRIEA) in 2005.

Third-Party-Over Action – type of action in which an injured employee, after collecting workers compensation benefits from the employer, sues a third party for contributing to the employee's injury. Then, because of some type of contractual relationship between the third party and the employer, the liability is passed back to the employer by prior agreement.

Torts – a civil or private wrong giving rise to legal liability.

Vicarious Liability – the liability of a principal for the acts of its agents; can result from the acts of independent agents, partners, independent contractors, employees, and children.

Waiver of Subrogation – an agreement between two parties in which one party agrees to waive its right of recovery against another for legal liability in the event of a loss. The intent of the waiver is to prevent one party'-s insurer from pursuing subrogation against the other party. Insurance policies do not bar coverage if an insured waives subrogation against a third party before a loss. However, coverage is excluded from many policies if subrogation is waived after a loss because to do so would violate the principle of indemnity.

Water and Wastewater Professional Activities – your activities as a water or wastewater district, water utility, or as any other entity whose primary duty is the treatment and distribution of potable water, or the collection, treatment, and distribution of wastewater.

Wrongful Detention – the keeping in custody or confining a person in custody without any lawful reason.

Wrongful Entry – wrongful eviction from, wrongful entry into, or invasion of the right of private occupancy of a room, dwelling, or premises that a person occupies, committed by or on behalf of its owner, landlord, or lessor.

COMMERCIAL GENERAL LIABILITY (CGL) POLICY*

Overview

A commercial general liability (CGL) policy serves as the foundation for your public water and wastewater system's liability insurance protection. This policy safeguards your organization against claims involving third-party bodily injury and property damage arising from your premises,

---

* Source material provided by various online articles from IRMI, https://www.irmi.com/articles/expert-commentary/insurance as well as policy forms and underwriting products promulgated by ISO and their member companies, https://products.iso.com/productcenter/10.

operations, products, and completed operations. It similarly protects against offenses comprising personal and advertising injury. Coverage extends to property damage liability incurred to premises rented to you for a period of 30 days or for fire during the entire policy period as well as medical expenses for third parties injured, through no fault of your own, on or adjacent to your premises. CGL policies start with a broad promise, which is the insuring agreement.

The insurer pledges to pay indemnity (i.e., compensation) up to the respective limits of insurance if you are legally liable for damages because of bodily injury or property damage as well as personal and advertising injury subject to policy exclusions. The aforementioned pledge affords the insurer discretion to investigate and settle any claim or suit that is filed under your policy. This right, known as a duty to defend or pay on behalf policy, commonly requires the insurer to provide unlimited defense until an indemnification payment exhausts the limits of insurance.

Bodily injury and property damage are the most common liability claims filed against public water/wastewater systems. The former encompasses physical harm, including sickness or disease, whereas the latter comprises physical injury to tangible property as well as damages for the loss of use of tangible property that has been physically injured. Property damage also includes loss of use of tangible property that has not been physically injured. Electronic data, which is defined as information stored electronically, is not considered tangible property and must be specifically endorsed on the policy to qualify as property damage. Damages under a CGL policy do not include economic injury arising from your governing body's powers, policies, or decisions. Economic injury from the foregoing wrongful acts and other offenses is covered under a separate public officials and management liability (POML) policy, which we will discuss in a later chapter.

CGL policies are traditionally written on an occurrence form, which means the policy is triggered when the bodily injury or property damage transpired. The date of the actual occurrence has no bearing on the policy trigger, but the bodily injury or property damage must occur during the policy period and in the coverage territory. The latter is inapplicable to public water/wastewater systems since the coverage territory contains the United States. There simply needs to be an incident or event that qualifies as an occurrence. The occurrence date is irrelevant, but damages, which are defined as bodily injury or property damage, must occur during the policy period. There is no requirement for the claim or lawsuit seeking damages for bodily injury or property damage to be submitted or filed during the policy period. Such reporting can occur months or years after the actual damages. Conversely, the policy trigger or activation of personal and advertising injury necessitates an offense that takes place during the policy period. An offense is different than an occurrence because injury with the former takes place at the same time as the offense, whereas damages with the latter can transpire years after the actual occurrence.

The insuring agreement includes a separate promise by the insurer to defend suits against you where the damages demanded are *potentially* covered by the insuring agreement. The qualifier of *potentially* is needed because the broad insuring agreement is narrowed by the policy exclusions. These exclusions, which we will examine in detail, are foreshadowed by the following phrase in the insuring agreement "… to which this insurance applies." In other words, the policy does not apply to all damages but rather to damages to which this insurance applies. The parlance to provide defense for any allegations that could *potentially* trigger coverage is known as the duty to defend provision. The obligation of defense is much broader than the obligation of indemnity. The former simply needs to include the possibility of coverage to trigger the obligation, whereas the latter requires an affirmation that coverage exists.

The foregoing difference is logical, as an insurer cannot divine the outcome of litigation when there are various causes of action asserted. Insurers, therefore, must extend the obligation of defense and can withhold the obligation of indemnity until there is a rendering of judgment or settlement on the facts alleged. In other words, the ultimate determination of indemnity is based on a jury decision or a preemptive settlement by the insurer; and that takes place after the defense obligation. It is important to remember the indemnity obligation ends with the policy's limits of insurance. That is also when the defense obligation ends. It is common for the defense obligation to be unlimited, which means the costs associated with defense do not erode the limits of insurance. The word *unlimited* does has an ending point, and that is when the indemnity payments erode the limits of insurance. The parlance to provide unlimited defense is known as defense outside the limits.

Deductibles within a CGL policy also vary, as there are two options: inside and outside. The first is when allocated loss adjustment expenses (ALAE) are inside the deductible. With this option, your deductible will apply to expenses like field adjusters, expert witnesses, legal representation, and any other costs that correlate back to the respective claim. ALAE does not include salaries of the insurer's employees or its fixed costs. Your deductible will also, of course, apply to any indemnity payment on the claim. The second is when ALAE is outside the deductible. This option is preferred because you are not responsible for any expenses correlating back to the respective claim except for an indemnity payment.

It is common for larger deductibles to have ALAE inside and smaller deductibles to have ALAE outside. These mechanics are logical because larger deductibles equate to more premium credits and more risk assumption by the insured. The alignment of expectation between insurer and insured with larger deductibles warrant such an approach. Never select a deductible on a per claim basis. It should always be on a per occurrence basis. The former applies individually to each claimant of the occurrence, whereas the latter applies collectively to all claimants of the occurrence.

The selection of limits is equally important, as that defines the indemnity obligation of the insurer. Its exhaustion will also end the insurer's defense obligation. The customary limits within a CGL policy are $1 million per occurrence and $2 million aggregate. A per occurrence limit is the maximum amount for any one claim or multiple claims involving the same occurrence, whereas the aggregate limit allows the per occurrence limit to replenish and respond to other occurrences up to the aggregate limit shown on the supplemental declarations.

The preferred limit structure with a best-in-class CGL policy exceeds the customs presented above and offers the following: $10 million general aggregate (i.e., bodily injury, property damage, and medical expenses); $10 million products-completed operations aggregate; $1 million personal and advertising injury; $1 million each occurrence; $1 million damage to premises rented to you; and $10,000 medical expenses. It is important to secure a per location aggregate limit on your policy to prevent the exhaustion of your limits. A per location aggregate does exactly what the name implies: It provides separate general aggregate limits for each location.

There are two different types of CGL policies available to public water/wastewater systems. An industry standard policy is called Insurance Services Office, Inc. (ISO). This policy is uniform for all industries and provides consistency of language as well as standardized endorsements for various industry classifications. The alternative is proprietary forms and endorsements. Generally speaking, proprietary is preferred over standardized for the noticeable advantage of customization and germaneness. Such an approach enhances coverage breadth and reduces unintended coverage gaps. The outcome is a more appropriate, relevant, and comprehensive CGL policy. You also receive the benefit of a non-auditable policy, which means there is no additional premium should your ratable exposure basis increase during the policy period.

A proprietary policy, however, can be deficient if the insurer is not established and fluent within the water and wastewater segment. You cannot assume proprietary policies automatically confer the coverage you need. They do not, and not all insurers are created equal when it comes to public water/wastewater systems. Notwithstanding, proprietary policies are an excellent indicator of an insurer's subject matter expertise, resource depth, and segment commitment. It is important to thoroughly review these qualitative indicators alongside your policies to certify the prudence of your insurer selection and their best-in-class status. There are specific coverage grants, coverage extensions, and manuscript endorsements that should be codified to confirm insurer intent for many of the complex and overlapping exposures facing public water/wastewater systems. This method is called affirmation of coverage.

In some isolated circumstances, affirmation is not necessary because a particular exposure has a secondary or ancillary impact to bodily injury or property damage. The parlance to not affirm such coverage is known as silence. The preceding approach is acceptable for some exposures but not others. For silence to be appropriate, the exposure should be peripheral or tangential to the primary exposure causing the loss. We will provide details on the prudence or imprudence of affirmation and silence later in this chapter.

The best way to achieve a practical comprehension of your CGL policy is to granularly examine and dissect a best-in-class policy specifically designed for public water/wastewater systems. We will divide this exercise into seven sections:

1. Coverage A. Bodily Injury and Property Damage
2. Coverage B. Personal and Advertising Injury
3. Coverage C. Medical Expenses
4. Supplementary Payments
5. Who Is an Insured
6. Limits of Insurance
7. Key Policy Conditions

The three coverage parts have their own insuring agreements and exclusions. We will provide special commentary on these coverage parts, including their relationship with the insuring agreements and exclusions. An examination of the most salient exposures confronting public water/wastewater systems will similarly be imparted. The remaining sections will afford context and support for the above referenced coverages. These sections comprise the insurer's obligation for supplementary payments beyond the limits of insurance, the qualification for being an insured, the mechanics behind the limits of insurance, and the key policy conditions that require foreknowledge.

Since the majority of claims filed against public water/wastewater systems involve bodily injury and property damage, our review will be more detailed for Coverage A. versus Coverages B. and C. The assessment of Coverage A. will include a review of optional coverages including employee benefit plans, hired and non-owned auto liability, stop-gap liability, and law enforcement wrongful acts. We will similarly identify mandatory coverages that should be embedded in Coverage A. to ensure protection for the unique and specific exposures facing public water/wastewater systems.

Let's now proceed with our review of the above referenced sections. We begin with an assessment of Coverage A. Our learning method will encapsulate a reading of a best-in-class CGL policy followed by author commentary. The latter is intended to provide insight into the impact and meaning of the policy language as well as practical coverage solutions and removal strategies for onerous restrictions.

COVERAGE A: BODILY INJURY AND PROPERTY DAMAGE LIABILITY

Insuring Agreement

We will pay those sums that the insured becomes legally obligated to pay as damages because of bodily injury or property damage to which this insurance applies. We will have the right and duty to defend the insured against any suit seeking those damages. However, we will have no duty to defend the insured against any suit seeking damages for bodily injury or property damage to which this insurance does not apply. We may, at our discretion, investigate any occurrence and settle any claim or suit that may result. But: (1) the amount we will pay for damages is limited as described in the limits of insurance section; and (2) our right and duty to defend ends when we have used up the applicable limit of insurance in the payment of judgments or settlements under Coverages A., B., or medical expenses under Coverage C. No other obligation or liability to pay sums or perform acts or services is covered unless explicitly provided for under Supplementary Payments – Coverages A. and B.

This insurance applies to bodily injury or property damage only if: (1) the bodily injury or property damage is caused by an occurrence arising out of your operations that takes place in the coverage territory; (2) the bodily injury or property damage occurs during the coverage period; and (3) prior to the coverage period, no insured and no employee authorized by you to give or receive notice of an occurrence or claim, knew or had reason to know that the bodily injury or property damage had occurred in whole or in part. If such an insured or authorized employee knew or had reason to know, prior to the coverage period, that the bodily injury or property damage occurred in whole or part, then any continuation, change, or resumption of such bodily injury or property damage during or after the coverage period will be deemed to have been known to have occurred prior to the coverage period.

Bodily injury or property damage which occurs during the coverage period and was not, prior to the coverage period, known to have occurred by any insured or any employee authorized by you to give or receive notice of an occurrence, or claim, includes any continuation, change, or resumption of that bodily injury or property damage after the end of the coverage period. Bodily injury or property damage will be deemed to have been known to have occurred when any insured or any employee authorized by you to give or receive notice of an occurrence or claim: (1) reports all, or any part, of the bodily injury or property damage to us or any other insurer; (2) receives a written or verbal demand or claim for damages because of the bodily injury or property damage; or (3) becomes aware by any other means that bodily injury or property damage has occurred or has begun to occur.

Damages because of bodily injury include damages claimed by any person or organization for care, loss of services, or death resulting at any time from the bodily injury. The insurance provided under Coverage A. is extended to apply to bodily injury or property damage arising out of your water or wastewater professional activity. An act, error, or omission arising out of your water or wastewater professional activity shall be considered one occurrence.

Key Definitions

Bodily Injury – bodily injury, mental anguish or mental injury, sickness, or disease sustained by a person, including death resulting from any of these at any time.

Coverage Territory – (a) the United States of America (including its territories and possessions), Puerto Rico, and Canada; (b) international waters or airspace but only if the injury or damage occurs in the course of travel or transportation between any places included in (a) above; or (c) all other parts of the world if the injury or damage arises out of: (1) goods or products made or sold by you in the territory described in (a) above; (2) the activities of a person whose home is in the territory described in (a) but is away for a short time on your business; or (3) personal and advertising injury offenses that take place through the internet or similar electronic means of communication provided the insured's responsibility to pay damages is determined in a suit on the merits, in the territory described in (a) above or in a settlement we agree to.

Law Enforcement Wrongful Act – an actual or alleged act, error, omission, neglect, or breach of duty, including violation of any civil rights law, while performing law enforcement activity. All claims arising from a series of related acts, errors, omissions, neglects, or breaches of duty will constitute a single law enforcement wrongful act. Law enforcement activity means all operations authorized by you or a public law enforcement agency contracted by you but only with respect to those services or operations contracted by you for law enforcement.

Occurrence – an accident, including continuous or repeated exposure to substantially the same general harmful conditions.

Property Damage – (a) Physical injury to tangible property, including all resulting loss of use of that property. All such loss of use shall be deemed to occur at the time of the physical injury that caused it; or (b) Loss of use of tangible property that is not physically injured. All such loss of use shall be deemed to occur at the time of the occurrence that caused it. For the purposes of this insurance, electronic data is not tangible property.

Suit – a civil proceeding in which damages because of bodily injury, property damage, personal and advertising injury, law enforcement wrongful act, or a water or wastewater professional activity to which this insurance applies are alleged. Suit includes: (a) an arbitration proceeding in which such damages are claimed and to which the insured must submit or does submit with our consent; or (b) any other civil alternative dispute resolution proceeding in which such damages are claimed and to which the insured submits with our consent.

Water or Wastewater Professional Activity – an act, error, or omission which arises from your activities as a water or wastewater district, water utility, or any other entity whose primary duty is the treatment and distribution of potable water, or the collection, treatment, and distribution of wastewater.

Special Commentary

The insuring agreement offers a wide net of insurance protection for occurrences resulting in damages arising from bodily injury and property damage due to your products, completed operations, premises, and ongoing operations. It also extends coverage to water or wastewater professional activity (i.e., testing errors and omissions) and property damage liability incurred to premises rented to you for a period of 30 days or for fire during the entire policy period. This expansive coverage narrows with the exclusions pertaining to Coverage A., which we will review shortly.

As previously discussed, a best-in-class CGL policy is issued on an occurrence form with a duty to defend provision. Coverage extends to any claim involving damages from bodily injury and property damage that are not otherwise excluded. The damages must be the result of an occurrence, which isn't required to occur during the policy period, but the damages must be actualized during the policy period and within the coverage territory. The insurer's defense obligations are activated as long as the claim meets the foregoing criteria and is not explicitly excluded. The obligation of defense is exceedingly broad and simply requires a *possibility*, however remote, that the claim is covered. Indemnity obligations are much narrower and require confirmation of coverage to be triggered.

Best-in-class CGL policies offer broad coverage, but they also include restrictions. This reality is accentuated by the known or continuation of loss provision. The parlance to prevent insureds from activating successive CGL policies over a span of several years even though they were aware of continuation of bodily injury or property damage is known as the *Montrose* provision from the eponymous and seminal court case. Based on the *Montrose* decision, insurers modified their CGL policies to prevent activation of successive policy periods once the insured becomes aware of an occurrence that causes bodily injury or property damage.

The provision's purpose is to explicitly establish the stopping point for continuous trigger-type claims. Due to the broad nature of the duty to defend obligation, particularly in those states that follow the *Four Corners* test, insurers may still need to defend these allegations even if the facts ultimately show the insured knew, prior to the policy inception date, of the continuing bodily injury or property damage. Under the *Four Corners* test, an insurer has a duty to defend its insured if the complaint avers causes of action and fact patterns that are *potentially* covered by the policy. The *Montrose* provision allows insureds who are not aware of continuous bodily injury or property damage to activate multiple policy periods. This restrictive provision only applies to subsequent policies once knowledge has been obtained.

Similar to the *Montrose* provision, it is not unusual to have an intra-policy anti-stacking limit. This endorsement prevents the stacking of different liability policies from the same insurer in the same policy period should one claim possibly impact multiple lines of coverage. The exception is the excess liability policy, which is intended to offer additional limits of protection for an underlying loss. Intra-policy claims that impact multiple lines of coverage from the same occurrence in the same policy period are rare. An inter-policy anti-stacking endorsement, however, is not standard and should be removed. Unlike its intra-policy cousin, this endorsement would nullify the stacking of limits from previous policy periods should there be continuation of damages from a past occurrence that is unknown to the insured. Anti-stacking provisions can have a significant impact on claims, and that necessitates a thorough reading and comprehension of your policies.

The insuring agreement represents a broad coverage grant for occurrences resulting in damages from bodily injury and property damage due to your products, completed operations, premises, and ongoing operations. Such broadness is substantially narrowed by the policy's exclusions and limitations. We will now delve into these restrictions to better understand their impact on the scope of insured protection and insurer obligations.

EXCLUSIONS AND LIMITATIONS

Aircraft, Automobile, and Watercraft

Exclusion: Bodily injury or property damage arising out of the ownership, maintenance, use, or entrustment to others of any aircraft, auto, or watercraft owned or operated by or rented or loaned to any insured. Use includes operation and loading or unloading. This exclusion applies even if the claims against any insured allege negligence or other wrongdoing in the supervision, hiring, employment, training, or monitoring of others by that insured, if the occurrence which caused the bodily injury or property damage involved the ownership, maintenance, use, or entrustment to others of any aircraft, auto, or watercraft that is owned or operated by or rented or loaned to any insured.

This exclusion does not apply to: (1) an aircraft with a maximum passenger capacity of 20 persons (including crew) that you do not own and used solely for business travel of employees; (2) a watercraft while ashore on premises you own or rent; (3) a watercraft you do not own that is: (a) less than 26 feet long; (b) not being used to carry persons or property for a charge; (4) liability assumed under any insured contract for the ownership, maintenance, or use of aircraft or watercraft; (5) a watercraft you own that is: (a) powered by a motor or combination of motors of 250 horsepower or less; (b) not powered by a motor; or (c) a personal watercraft; (6) parking an auto on, or on the ways next to, premises you own or rent, provided the auto is not owned by or rented or loaned to you or the insured; or (7) bodily injury or property damage arising out of: (a) the operation of machinery or equipment that is attached to, or part of, a land vehicle that would qualify under the definition of mobile equipment if it were not subject to a compulsory or financial responsibility law or other motor vehicle insurance law in the state where it is licensed or principally garaged; or (b) the operation of any of the machinery or equipment listed in the definition of mobile equipment.

Mobile Equipment: any of the following types of land vehicles, including any attached machinery or equipment: (a) bulldozers, farm machinery, forklifts, and other vehicles designed for use principally off public roads; (b) vehicles maintained for use solely on or next to premises you own or rent; (c) vehicles that travel on crawler treads; (d) vehicles, whether self-propelled or not, maintained primarily to provide mobility to permanently mounted: (1) power cranes, shovels, loaders, diggers, or drills; or (2) road construction or resurfacing equipment such as graders, scrapers, or rollers; (e) vehicles not described in (a), (b), (c), or (d) above that are not self-propelled and are maintained primarily to provide mobility to permanently attached equipment of the following types: (1) air compressors, pumps, and generators, including spraying, welding, building cleaning, geophysical exploration, lighting, and well servicing equipment; or (2) cherry pickers and similar devices used to raise or lower workers; (f) vehicles not described in (a), (b), (c), or (d) above maintained primarily for purposes other than the transportation of persons or cargo.

Self-propelled vehicles with the following types of permanently attached equipment are not mobile equipment and will be considered autos: (1) equipment designed primarily for: (a) snow removal; (b) road maintenance, but not construction or resurfacing; or (c) street cleaning; (2) cherry pickers and similar devices mounted on automobile or truck chassis and used to raise or lower workers; and (3) air compressors, pumps, and generators, including spraying, welding, building cleaning, geophysical exploration, lighting, and well servicing equipment. Mobile equipment does not include any land vehicles that are subject to a compulsory or financial responsibility law or other motor vehicle insurance law in the state where it is licensed or principally garaged. Land vehicles subject to a compulsory or financial responsibility law or other motor vehicle insurance law are considered autos.

Commentary: This industry-standard exclusion necessitates the foregoing exceptions to prevent unintended gaps in coverage. It is imperative to codify these exceptions for non-owned aircraft as well as owned and non-owned watercraft and mobile equipment. Regarding mobile equipment, you must identify any equipment requiring Department of Motor Vehicles (DMV) placards. This type of mobile equipment must be scheduled on your business auto coverage (BAC) policy to ensure liability coverage, as there will be no coverage under a CGL policy. The same awareness is required of any owned watercraft that is powered by more than 250 horsepower or greater 26 feet in length. Any watercraft not conforming to the preceding exceptions will require scheduling on the policy via endorsement. Lastly, a clear review of all contracts involving your aircraft, watercraft, and mobile equipment is required to ensure these agreements meet the definition of insured contract. As a reminder, verbal agreements do not meet the definition of insured contract. Said definition is discussed under the contractual liability exclusion.

Asbestos

Exclusion: Any injury, damage, expense, cost, loss, liability, or legal obligation arising out of or in any way related to asbestos or asbestos-containing materials, or exposure thereto, or for the costs of abatement, mitigation, removal, elimination, or disposal of any of them. This exclusion does not apply to bodily injury or property damage arising out of potable water which you supply to others.

Commentary: This industry-standard exclusion requires an exception for liability arising from asbestos from potable water. Although asbestosis is an airborne disease, public water/wastewater systems with asbestos-lined piping and asbestos-containing fixed equipment and/or appurtenances have an exposure to asbestos litigation arising from allegations of bodily injury from water contamination. The cost of defending such litigation and the uncertainty of jury verdicts underscore the importance of this exception. The allegation of bodily injury arising from asbestosis from potable water will invariably involve large groups of plaintiffs and a substantial defense budget. It is an exposure you should not self-fund. The purchase of an environmental pollution policy is required to expand the asbestos coverage from potable water under a CGL policy to liability from gradual seepage and deterioration as well as cleanup costs, natural resources damages, and civil fines and penalties.

Contractual Liability

Exclusion: Bodily injury or property damage for which the insured is obligated to pay damages by reason of the assumption of liability in a contract or agreement. This exclusion does not apply to liability for damages: (1) that the insured would have in the absence of the contract or agreement; or (2) assumed in a contract or agreement that is an insured contract, provided the bodily injury or property damage occurs or takes place subsequent to the execution of the contract or agreement. Solely for the purposes of liability assumed in a insured contract, reasonable attorney fees and necessary litigation expenses incurred by or for a party other than an insured are deemed to be damages because of bodily injury or property damage provided: (a) liability to such party for, or for the cost of, that party's defense has also been assumed in the same insured contract; and (b) such attorney fees and litigation expenses are for defense of that party against a civil or alternative dispute resolution proceeding in which damages to which this insurance applies are alleged. Insured contract: (a) a contract for a lease of premises. However, that portion of the contract for a lease of premises that indemnifies any person or organization for damage by fire to premises while rented to you or temporarily occupied by you with permission of the owner is not an insured contract; (b) a sidetrack agreement; (c) any easement or license agreement, except in connection with construction or demolition operations on or within 50 feet of a railroad; (d) an obligation, as required by ordinance, to indemnify a municipality, except in connection with work for a municipality; (e) an elevator maintenance agreement; (f) that part of any other contract or agreement pertaining to your operations (including an indemnification of a municipality in connection with work performed for a municipality) under which you assume the tort liability of another party to pay for bodily injury or property damage to a third person or organization.

Tort liability means liability that would be imposed by law in the absence of any contract or agreement. Subpart (f) does not include that part of any contract or agreement: (1) that indemnifies a railroad for bodily injury or property damage arising out of construction or demolition operations, within 50 feet of any railroad property and affecting any railroad bridge or trestle, tracks, road-beds, tunnel, underpass, or crossing; (2) that indemnifies an architect, engineer, or surveyor for injury or damage arising out of: (a) preparing, approving, or failing to prepare or approve, maps, shop drawings, opinions, reports, surveys, field orders, change orders, or drawings and specifications; or (b) giving directions or instructions, or failing to give them, if that is the primary cause of the injury or damage; or (3) under which the insured, if an architect, engineer or surveyor, assumes liability for an injury or damage arising out of the insured's rendering or failure to render professional services, including those listed in (2) above and supervisory, inspection, architectural, or engineering activities.

Commentary: This industry standard exclusion is acceptable to public water/wastewater systems but requires modification of the insured contract definition to include operations within 50 feet of a railroad. Such modification is achieved via a Contract Liability – Railroad endorsement. The endorsement should be issued on a blanket basis to ensure compliance with all your railroad easement agreements. Some railroads require a separate railroad protective liability (RRPL) policy, which is fundamentally different than a CGL policy. Should a railroad require this policy, it is recommended you purchase said policy directly from the railroad to ensure compliance. Enrollment in the railroad's own RRPL master policy, albeit expensive, will remove potential delays and uncertainties. Several railroads, however, will accept the Contract Liability – Railroad endorsement in lieu of a RRPL policy. You should seek permission from your insurer to include the above referenced endorsement on a blanket basis for all railroads requiring status as an additional insured on your CGL policy for operations within 50 feet of their tracks.

A blanket approach should similarly apply to any organizations requiring additional insured, waiver of subrogation, and primary and noncontributory wording per written agreement. Verbal agreements are not afforded this protection. Only written agreements meet the definition of an insured contract. You must ensure your public water and wastewater system has no verbal agreements that were intended to apply as if they were written. A review of all written agreements is necessary to certify conformance with the insured contract definition. Written agreements that transfer liability to an insured do not always equate to a CGL policy accepting this liability. That is why a thorough review of each agreement is imperative. You should seek written confirmation from your insurer so that any incongruities between these agreements and your policy are clearly identified. Such a review should take place before contract execution. Caution: There can be potential problems when using a blanket additional insured endorsement; especially if the hold harmless and indemnity agreement requires any downstream subcontractor or other party to grant additional insured status. It is imperative to discuss such scenarios with your broker and insurer.

Dam, Reservoir, Levee Structural Failure or Collapse

Exclusion: Bodily injury or property damage, loss, costs, or expense arising directly or indirectly out of the structural failure, collapse, bursting, flooding, cracking, settling, seepage, under seepage, spillage, subsidence, landslide, or other earth movement of any dam, reservoir, levee, or dike owned, operated, maintained, constructed, or controlled by any insured. This exclusion does not apply to bodily injury or property damage, loss, costs, or expense arising directly or indirectly out of the structural failure, collapse, bursting, flooding, cracking, settling, spillage, subsidence, landslide, or other earth movement of any dam, reservoir, levee, or dike which is scheduled on the supplemental declarations.

Commentary: The foregoing exclusion is expansive and requires the scheduling of all your dams, reservoirs, levees, and dikes to avoid unintended gaps in coverage. Scheduling is necessary because many claims involving dams, reservoirs, levees, and dikes straddle structural failure or collapse as well as operational negligence and operator error. Examples of intersecting claims include damages from overtopping, overflow, velocity, and releases. You should also pair this scheduling with inverse condemnation coverage, which is a unique and overarching specie of liability encompassing the taking (or damaging in California) of private property without just compensation. A separate discussion on inverse condemnation will occur in the public officials and management liability policy chapter, and an expansive article is provided at the end of *Part IV – Insurance Essentials*. Seepage is deliberately excluded and deemed customary for impoundments and conveyances. Any third-party structures should not be allowed within proximity of normal seepage and under seepage from your water infrastructure.

To qualify for this exception, your water infrastructure must secure a favorable inspection from the respective state or federal regulatory agency. A minimum grade of conditional satisfactory is required. Many water impoundments and conveyances have volume or velocity restrictions as part of their inspection rating. Such restrictions do not foreclose eligibility for this exception, but the exception will correlate to the respective restriction. For levees that receive an unsatisfactory rating due to conflicting interests between state agencies (i.e., fish and game vs. water resources), a detailed letter from an outside engineer is useful in explaining the situation and requesting accommodation by your insurer for an exception. The letter should confirm the structural integrity of the levee, the maintenance processes, and the acceptability in all areas except those where state agencies diverge.

Conflicts can occur when the Department of Fish and Game do not approve a permit to clear brush, and the Department of Water Resources do not provide a passing inspection because of excessive brush. There should also be no restriction for irrigation canals or drainage ditches, as that water infrastructure falls outside the description of a dam, reservoir, levee, or dike. Reservoirs should similarly be defined as open and unenclosed water impoundments. Berms protecting wastewater treatment facilities necessitate a comparable level of coverage clarification. The attachment of any structural failure or collapse exclusion for a dam, reservoir, levee, and dike is problematic when determining the proximate cause of the loss. In many instances, the structural failure or collapse is based less on engineering or natural failure and more on operator error or maintenance deficiencies. Such an exclusion necessitates an affirmation by your insurer of its specific impact and intent on various claim scenarios.

Damage to Impaired Property or Property Not Physically Injured

Exclusion: Property damage to impaired property or property that has not been physically injured, arising out of: (1) a defect, deficiency, inadequacy, or dangerous condition in your product or your work; or (2) a delay or failure by you or anyone acting on your behalf to perform a contract or agreement in accordance with its terms. This exclusion does not apply to your potable water, nonpotable water, or wastewater as well as any loss of use of other property arising out of sudden and accidental physical injury to your product or your work after it has been put to its intended use.

Commentary: This industry-standard exclusion applies to property damage of property that has not suffered actual physical injury. It also applies to property that is considered impaired property, which generally means the property of another that cannot be used (or its use is limited) because of a problem with your product or work that was made a part of the impaired property. Failure to complete any work or project on time can similarly qualify as impaired property. The impetus behind this exclusion is the portion of the property damage definition that includes coverage for loss of use of tangible property that has not been physically injured. Its purpose is to restrict coverage for specific loss of use claims that may result even if no physical injury to the property of another has taken place. The specific loss of use claims excluded are either those caused by the assimilation of your defective product or faulty work into the property of another or those loss of use claims caused by your failure to finish any work or project on time.

There are two examples of impaired property impacting public water/wastewater systems. The first involves a defect in your potable water, whether the water is your own or purchased from a wholesaler. Quagga mussels from the Lower Colorado River is one such defect. This water can transport these mussels, which can then damage third-party property and require expensive eradication measures. The second example involves territory annexation of adjacent property owned by a developer whereby you agree to provide water or wastewater purveyance by a certain date, but you incur delays in providing this service. The exception to the foregoing exclusion provides coverage for both scenarios.

Damage to Property

Exclusion: Property damage to: (1) property you own, rent, or occupy, including any costs or expenses incurred by you, or any other person, organization, or entity, for repair, replacement, enhancement, restoration, or maintenance of such property for any reason, including prevention of injury to a person or damage to another's property; (2) premises you sell, give away, or abandon, if the property damage arises out of any part of those premises; (3) property loaned to you; (4) personal property in the care, custody, or control of the insured; (5) that particular part of real property on which you or any contractors or subcontractors working directly or indirectly on your behalf are performing operations, if the property damage arises out of those operations; or (6) that particular part of any property that must be restored, repaired, or replaced because your work was incorrectly performed on it.

Subparts (1), (3) and (4) of this exclusion do not apply to property damage (other than damage by fire) to premises, including the contents of such premises, rented or loaned to you for a period of 30 or fewer consecutive days. A separate limit of insurance applies to Damage to Premises Rented to You. Subpart (2) of this exclusion does not apply if the premises are your work and were never occupied, rented, or held for rental by you. Subparts (3), (4), (5), and (6) of this exclusion do not apply to liability assumed under a sidetrack agreement. Subpart (6) of this exclusion does not apply to property damage included in the products-completed operations hazard.

Commentary: This industry-standard exclusion prevents coverage for property damage to property occupied by a tenant, who is also an insured. There are important exceptions to this exclusion. The exclusion for property damage to premises occupied or rented to you does not apply to damage by fire. Further, if the premises are rented to you for 30 or fewer consecutive days (such as renting a hotel room), then the exclusion does not apply to the premises and the contents of the premises. In both cases, a sublimit (Damage to Premises Rented to You) applies. None of the exclusions in Coverage A. apply to damage by fire to premises while rented to you or temporarily occupied by you with permission of the owner. A separate limit applies to this coverage extension, with a recommended minimum limit of $1 million.

Damage to Your Product

Exclusion: Property damage to your product arising out of it or any part of it.

Commentary: This industry-standard exclusion is acceptable to public water/wastewater systems without any exceptions, as water or wastewater does not damage itself. Any resulting damage arising from your product would not be restricted by this exclusion.

Damage to Your Work

Exclusion: Property damage to your work arising out of it or any part of it and included in the products-completed operations hazard. This exclusion does not apply if the damaged work or the work out of which the damage arises was performed on your behalf by a subcontractor.

Commentary: This industry-standard exclusion is acceptable to public water/wastewater systems pending the exception to the foregoing exclusion that work performed on your behalf is covered. There is no intent for a CGL policy to pay damages arising from your faulty work. Practically speaking, this exclusion has little relevance as work performed by your employees is for the benefit of your public water and wastewater system. It is important that no work be performed by your employees to repair or replace lateral lines that are owned by third parties. Any such work should be contracted directly by the third party and a contractor. In these situations, you can provide reimbursement to the third party but not be a party to the contract. This approach will protect the public water and wastewater system from allegations of faulty work or the failure to coordinate work with a competent contractor. It also enhances community goodwill by facilitating work by others without direct involvement by the public water and wastewater system and clearly establishes the line of demarcation for lateral lines not owned by third parties.

Distribution of Material in Violation of Statutes

Exclusion: Bodily injury or property damage arising directly or indirectly out of any action or omission that violates or is alleged to violate: (1) The Telephone Consumer Protection Act (TCPA), including any amendment of or addition to such law; (2) The CAN-SPAM Act of 2003, including any amendment of or addition to such law; or (3) Any statute, ordinance, or regulation, other than the TCPA or CAN-SPAM Act of 2003, that prohibits or limits the sending, transmitting, communicating, or distribution of material or information.

Commentary: This industry-standard exclusion is acceptable to public water/wastewater systems because your operations do not involve unsolicited sales information but rather critical public service information.

Electronic Data

Exclusion: Damages arising out of the loss of, loss of use of, damage to, corruption of, inability to access, or inability to manipulate electronic data. As used in this exclusion, electronic data means information, facts, or programs stored as or on, created or used on, or transmitted to or from computer software, including systems and applications software, hard or floppy disks, CD-ROMS, tapes, drives, cells, data processing devices, or any other media which are used with electronically controlled equipment. This exclusion does not apply to property damage to electronic data that occurs in conjunction with damage to tangible property as a result of your product or operations.

Commentary: This industry-standard exclusion asserts that coverage does not apply for loss of electronic data because data is not deemed tangible property and thus not considered property damage. Damage to electronic data is a significant exposure and frequently occurs in tandem to tangible property that is damaged. As such, an exception to the electronic data exclusion is recommended to confirm damage to electronic data is covered as long as tangible property is also damaged from the same occurrence. Water and wastewater line breaks or emergency repairs can damage tangible property as well as electronic data. This exception is critical to remove the above referenced gap in coverage.

Employers' Liability

Exclusion: Bodily injury to: (1) an employee of the insured arising out of or in the course of: (a) employment by the insured; or (b) performing duties related to the conduct of the insured's operations; (2) a volunteer worker, if you provide or are required to provide benefits for such volunteer worker under any workers compensation, disability benefits, or unemployment compensation law, or any similar law; or (3) the spouse, child, parent, brother, or sister of that employee or volunteer worker as a consequence of (1) or (2) above. This exclusion applies whether the insured may be liable as an employer or in any other capacity and to any obligation to share damages with or repay someone else who must pay damages because of the injury. This exclusion does not apply to liability assumed by the insured under an insured contract.

Commentary: This industry-standard exclusion is acceptable to public water/wastewater systems, as the employers' liability coverage is traditionally found in your workers compensation policy. There are two scenarios, however, that require exceptions to the foregoing exclusion. The first involves an insured employer that is obligated to indemnify another for damages resulting from work-related injuries to the insured employer's own employees, provided the obligation of the insured employer to provide indemnity is due to liability assumed in an insured contract. This scenario should be covered under the definition of insurance contract. The second is an employee who is sued by another employee for negligently causing the other employee's injury. This scenario should be covered under a fellow employee endorsement. For public water/wastewater systems in monopolistic workers compensation states (i.e., North Dakota, Ohio, Washington, and Wyoming), you are able to purchase a stop-gap liability endorsement to your CGL policy to provide employers' liability. Monopolistic states only cover the legal benefits for work-related injuries to employees. Further details on stop-gap liability coverage is included later in this chapter.

Employment Practices and Employee Benefit Plans

Exclusion: Bodily injury or property damage arising out of your employment practices or administration of your employee benefit plans.

Commentary: This industry-standard exclusion is acceptable to public water/wastewater systems because said coverage is available under the POML policy.

Expected or Intended Injury

Exclusion: Bodily injury or property damage expected or intended from the standpoint of the insured. This exclusion does not apply to bodily injury or property damage resulting from reasonable force to protect persons or property.

Commentary: This industry-standard exclusion is acceptable to public water/wastewater systems, as it is against the fortuity doctrine to cover expected or intended injury from an insured unless such actions are to protect persons or property. Insurer clarification is required for scenarios involving the distribution of potable water that is known by you to violate water quality standards but where bodily injury or property damage was neither expected nor intended.

Fungi or Bacteria

Exclusion: (1) Any injury or damage which would not have occurred or taken place, in whole or in part, but for the actual, alleged, or threatened inhalation of, ingestion of, contact with, exposure to, existence of, or presence of, any fungi or bacteria on or within a building or structure, including its contents, regardless of whether any other cause, event, material, or product contributed concurrently or in any sequence to such injury or damage. (2) Any loss, cost, or expenses arising out of the abating, testing for, monitoring, cleaning up, removing, containing, treating, detoxifying, neutralizing, remediating, or disposing of, or in any way responding to, or assessing the effects of, fungi or bacteria, by any insured or by any other person or entity. (3) This exclusion does not apply: (a) to any fungi or bacteria that are, are on, or are contained in a good or product intended for consumption; or (b) to any injury or damage arising out of or caused by your water, irrigation, or wastewater intake, outtake, reclamation, treatment, and distribution processes.

Commentary: This industry-standard exclusion is acceptable to public water/wastewater systems with the foregoing exceptions. Coverage is not granted for liability arising from fungi or bacteria on your premises. However, the exceptions make clear that loss or damage arising from your operations, completed operations, and products are covered. Public water/wastewater systems have low customer traffic at their premises, and most of the interaction is brief in the form of paying bills or attending a board meeting or community event. The exceptions articulate that mold or bacteria remediation arising from your operations, completed operations, and products are covered. The purchase of an environmental pollution policy is required to expand fungi or bacteria coverage from potable water as well as water and wastewater operations under a CGL policy to liability arising from your premises as well as cleanup costs and civil fines and penalties.

Lead, Electromagnetic Radiation, Nuclear

Exclusion: (1) Any injury, damage, expense, cost, loss, liability, or legal obligation arising out of or in any way related to: (a) the toxic properties of lead, or any material or substance containing lead with the exception of potable water which you supply to others; or (b) electromagnetic radiation; or exposure thereto, or for the costs of abatement, mitigation, removal, elimination, or disposal of any of them. (2) Any loss, cost, or expense arising out of any actual, alleged, or threatened injury or damage to any person or property from any radioactive matter or nuclear material.

Commentary: This industry-standard exclusion is acceptable to public water/wastewater systems, but there should be a lead exception for bodily injury arising from potable water consumption. Lead in residential piping remains an exposure in some jurisdictions, and your public water and wastewater system can be subject to litigation associated with failure to properly mitigate the effects of lead poisoning. Electromagnetic radiation is an exposure that exists with public water/wastewater systems that have radio towers and larger generators. Sources of health risks from this radiation include radio frequencies, extremely low frequencies, and static magnetic fields. Mitigation measures to reduce the health risks from radiation involve monitoring radio frequencies, extremely low frequencies, and static magnetic fields. Risk transfer measures comprise proper indemnification and additional insured provisions for any tower leases with cellular phone companies. As of this writing, there is no widespread insurance coverage for electromagnetic radiation. Nuclear is not applicable to public water/wastewater systems unless you operate a nuclear desalination plant. It is prudent to ensure contractual releases are enacted with the engineering and construction firms as well as any third-party plant operators.

Mobile Equipment

Exclusion: Bodily injury or property damage arising out of: (1) the transportation of mobile equipment by an auto owned or operated by or rented or loaned to any insured; or (2) the use of mobile equipment in, or while in practice for, or while being prepared for, any prearranged racing, speed, demolition, or stunting activity.

Commentary: This industry-standard exclusion requires the purchase of a business auto coverage policy to cover injury or damage arising from the transportation of mobile equipment. The second component of the exclusion is inapplicable to public water/wastewater systems. Coverage for mobile equipment is referenced in the exclusion exception for aircraft, watercraft, and automobiles.

Personal and Advertising Injury

Exclusion: Bodily injury arising out of personal and advertising injury.

Commentary: This industry-standard exclusion is acceptable to public water/wastewater systems because said coverage is offered under Coverage B. Personal and Advertising Injury. Any subsequent bodily injury arising from a personal injury offense (i.e., heart attack from malicious prosecution, defamation, wrongful eviction, etc.) would not be covered under Coverage A. or Coverage B. of this policy.

Pollution

Exclusion: Any injury, damage, expense, cost, loss, liability, or legal obligation arising out of or in any way related to pollution, however caused. Pollution includes the actual, alleged, or potential presence in or introduction into the environment of any substance if such substance has, or is alleged to have, the effect of making the environment impure, harmful, or dangerous. Environment includes any air, land, structure (or the air within), watercourse, or other body of water, including underground water.

(1) This exclusion does not apply: (a) To bodily injury if sustained within a building which is or was at any time owned or occupied by, or rented or loaned to, any insured and is caused by smoke, fumes, vapor, or soot produced by or originating from equipment that is used to heat, cool, or dehumidify the building, or equipment that is used to heat water for personal use, by the building's occupants or their guests. (b) To bodily injury or property damage arising out of heat, smoke, or fumes from a hostile fire unless that hostile fire occurred or originated: (i) at any premises, site, or location which is or was at any time used by any insured or others for the handling, storage, disposal, processing, or treatment of waste; or (ii) at any premises, site, or location on which any insured or other contractors or subcontractors working directly or indirectly on any insured's behalf are performing operations to test for, monitor, clean up, remove, contain, treat, detoxify, neutralize, or in any way respond to, or assess the effects of, pollutants except to the extent coverage is provided in (c).

(c) To bodily injury or property damage which occurs or takes place as a result of your operations provided the bodily injury or property damage is not otherwise excluded in whole or in part and arises out of the following: (i) potable water which you supply to others; (ii) chemicals you use in your water or wastewater treatment process; (iii) natural gas or propane gas you use in your water or wastewater treatment process; (iv) urgent response for the protection of property, human life, health, or safety conducted away from premises owned by or rented to or regularly occupied by you; (v) your application of pesticide or herbicide if such application meets all standards of any statute, ordinance, regulation, or license requirement of any federal, state, or local government; (vi) smoke drift from controlled or prescribed burning that has been authorized and permitted by the respective regulatory agency; (vii) fuels, lubricants, or other operating fluids needed to perform the normal electrical, hydraulic, or mechanical functions necessary for the operation of mobile equipment or its parts, but only if: (a) the fuels, lubricants, or other operating fluids escape from a vehicle part designed to hold, store, or receive them; and (b) the fuels, lubricants, or other operating fluids are not: (i) intentionally discharged; or (ii) brought on or to a premises, site, or location with the intent to be discharged as part of the operations being performed by an insured, contractor, or subcontractor.

(d) Bodily injury and property damage arising out of the actual, alleged, or threatened discharge, dispersal, seepage, migration, release, or escape of pollutants if such bodily injury or property damage is sudden and accidental and neither expected nor intended by an insured. However, no coverage is provided under this exception for petroleum underground storage tanks; or (1)(c) and (1)(d) of this exclusion only apply if the discharge is accidental, unintended, and stopped as soon as possible. The entirety of any discharge or series of related discharges will be deemed a single discharge regardless of the length of time over which the pollutants are released. The entirety of any discharge or series of related discharges will be deemed to have only occurred at the date the earliest discharge commenced. (e) To bodily injury or to property damage if such bodily injury or property damage is caused by the escape or backup of sewage or wastewater from any sewage treatment facility or fixed conduit or piping that you own, operate, lease, control, or for which you have the right of way, but only if property damage occurs away from land you own or lease.

(2) This insurance does not apply to any loss, cost, or expense arising out of any: (a) request, demand, order, or statutory or regulatory requirement that any insured or others test for, monitor, clean up, remove, contain, treat, detoxify, or neutralize, or in any way respond to, or assess the effects of, pollutants; or (b) claim or suit by or on behalf of a governmental authority for damages because of testing for, monitoring, cleaning up, removing, containing, treating, detoxifying, neutralizing, or in any way responding to, or assessing the effects of, pollutants. However, this paragraph does not apply to liability for damages because of property damage that the insured would have in the absence of such request, demand, order, statutory or regulatory requirement, or such claim or suit by or on behalf of a governmental authority. Discharge as used in this exclusion includes dispersal, seepage, migration, release, or escape. Pollutants: any solid, liquid, gaseous, or thermal irritant or contaminant, including smoke, vapor, soot, fumes, acids, alkalis, chemicals, and waste. Waste includes materials to be recycled, reconditioned, or reclaimed.

Commentary: A best-in-class CGL policy should provide coverage on an accidental basis for the named pollution incidents referenced in the preceding exclusion as exceptions. It should similarly extend coverage as an exception for any other pollution incident (i.e., unnamed) that is sudden and accidental and neither expected nor intended from the standpoint of the named insured and stopped as soon as possible. The aforementioned exceptions reaffirm pollution coverage for bodily injury and property damage arising from your products, operations, and completed operations. These exceptions do not provide all-inclusive pollution liability, but they remove unintended coverage gaps for named pollution incidents that are accidental as well as unnamed pollution incidents that are sudden and accidental. The aforementioned affirmation mitigates coverage disputes and reservation of rights as it relates to specific pollution scenarios that overlap with your products and operations. Moreover, the word *sudden* cannot be included in named pollution incidents, as this limitation would prevent coverage for most bodily injury allegations involving water contamination. Such allegations involve a manifestation of consumption and do not occur suddenly.

The word *sudden* is appropriate for unnamed pollution incidents because the scope of said coverage does not encompass manifestation. Unnamed pollution incidents involve unplanned releases that are not gradual but rather precipitous. The preceding exceptions to the pollution exclusion represent material coverage for public water/wastewater systems, but there are coverage gaps that remain. The most notable include the following: midnight dumping; Comprehensive Environmental Response, Compensation & Liability Act (CERCLA) liability; natural resources damage; gradual seepage and deterioration; cleanup costs; civil fines; and by-product–biosolids–sludge–compost storage. These gaps can only be addressed through the purchase of an environmental pollution policy.

Professional Services

Exclusion: Any liability arising out of any act, error, omission, malpractice, or mistake of a professional nature committed by the insured or any person for whom the insured is legally responsible. It is understood this exclusion applies even if the claims against any insured allege negligence or other wrongdoing in the supervision, hiring, employment, training, or monitoring of others by that insured. However, this exclusion does not apply to bodily injury or property damage arising out of your water or wastewater professional activities. Water or wastewater professional activity means an act, error, or omission which arises from your activities as a water or wastewater district, water utility, or any other entity whose primary duty is the treatment and distribution of potable water, or the collection, treatment, and distribution of wastewater.

Commentary: This industry-standard exclusion should have an exception for bodily injury or property damage arising from water and wastewater testing errors and omissions. Your treatment and distribution operations involve licensed water and wastewater operators and may also involve laboratory analysts and chemists. It is imperative that any water contamination, aggressive water allegation, or wastewater activities not be disclaimed or reserved under rights because of a professional services exclusion. The exception to the foregoing exclusion prevents such a scenario. Should your public water and wastewater system engage in third-party watermaster services like billings, operations, design, and engineering, then the purchase of a separate professional liability policy should be considered. This type of policy would cover such operations as long these functions are codified in the definition of professional services. Coverage under a professional liability policy is limited to economic injury. A contingent property damage and bodily injury endorsement should be attached to reflect the overlapping nature of professional liability claims involving public water/wastewater systems. A separate professional liability policy is only needed if you perform third-party professional services.

Public Use of Property

Exclusion: Bodily injury or property damage arising out of the principles of eminent domain, condemnation, inverse condemnation, or adverse possession.

Commentary: This exclusion is acceptable to public water/wastewater systems for the preceding takings, but inverse condemnation requires additional commentary because of its unique specie of government liability. This legal theory comprises a violation of some state constitutions via the taking or damaging of private property without just compensation. Damages associated with inverse condemnation actions involve economic injury and not property damage. Coverage belongs under the POML policy, which encompasses economic injury arising from the wrongful acts of your governing body. The majority of inverse condemnation actions involve some semblance of property damage, so it is imperative your insurer codify their intent to interpret such POML claims as monetary damages involving the powers, policies, and decisions of your governing body.

Recall of Products, Work, or Impaired Property

Exclusion: Damages claimed for any loss, cost, or expense incurred by you or others for the loss of use, withdrawal, recall, inspection, repair, replacement, adjustment, removal, or disposal of: (1) your product; (2) your work; or (3) impaired property; if such product, work, or property is withdrawn or recalled from the market or from use by any person or organization because of a known or suspected defect, deficiency, inadequacy, or dangerous condition in it. This exclusion does not apply to any injury or damage arising out of or caused by your potable water, nonpotable water, or wastewater.

Commentary: This industry-standard exclusion should be modified to exempt injury or damage arising from your products. The foregoing exception will protect your public water and wastewater system in situations where distribution of your potable water is ceased by regulatory order due to unsafe contaminant or corrosive levels. Although there is no ability to recall potable water, there is an exposure for loss of use and repair claims from your customers based on your inability to deliver potable water that conforms to regulatory standards.

Riot, Civil Commotion, or Mob Action

Exclusion: Bodily injury or property damage arising out of: (1) riot, civil commotion, or mob action; or (2) any act or omission in connection with the prevention or suppression of said actions.

Commentary: This industry-standard exclusion is acceptable to public water/wastewater systems as you have no legal authority to suppress these activities. Notwithstanding, the exclusion underscores the importance of physical security and operational safeguards to ensure the public is not injured from your exposures should such activities occur in your service territory.

Sexual Abuse

Exclusion: Bodily injury arising out of the sexual abuse of any person. This exclusion shall not apply to the named insured if no elected or appointed official, executive officer, officer, director, or trustee of the insured knew or had reason to know of the sexual abuse. Also, we will defend an insured for covered civil action subject to the other terms of this coverage form until either a judgment or final adjudication establishes such an act, or the insured confirms such act.

Commentary: This industry-standard exclusion is acceptable to public water/wastewater systems, as it is against public policy to defend insureds against criminal acts. It is important to ensure vicarious liability is extended to your governing body should knowledge of such actions not be known. The same confirmation is required of defense for civil actions up to the point of adjudication or confession for the alleged employee.

Specific Operations

Exclusion: Bodily injury or property damage arising from the ownership, operation, maintenance, entrustment to others, or use of any: (1) gas or electric generation facility; or (2) sanitary landfill, dump, or other permanent waste disposal facility.

Commentary: This exclusion is acceptable to public water/wastewater systems as long as the foregoing operations are not applicable. Clarification should be secured regarding any gas or electric generation facilities. Your insurer should confirm gas distribution is permissible and distinct from gas generation. The same clarification is required for electric distribution versus electric generation. Moreover, public water/wastewater systems that produce electricity for their own purposes or sell such electricity to a local electric utility on a best-efforts contract should receive a carve-out of the exposure to avoid coverage misunderstandings.

War

Exclusion: Bodily injury or property damage, however caused, arising, directly or indirectly, out of: (a) war, including undeclared or civil war; (b) warlike action by a military force, including action in hindering or defending against an actual or expected attack, by any government, sovereign, or other authority using military personnel or other agents; or (c) insurrection, rebellion, revolution, usurped power, or action taken by governmental authority in hindering or defending against any of these.

Commentary: This industry-standard exclusion is acceptable to public water/wastewater systems but necessitates the purchase of optional terrorism coverage backstopped by the federal government. Terrorism coverage has specific trigger points and should include liability arising from nuclear, biological, and radioactive agents. You should evaluate the prudence of purchasing stand-alone terrorism coverage that covers actions not triggered by the federal government's reinsured backstop product. Such stand-alone coverage warrants consideration for those rarer public water/wastewater systems deemed high-valued targets by the federal government for reasons of size, location, or exposure.

Workers Compensation and Similar Laws

Exclusion: Any obligation of the insured under a workers compensation, disability benefits, or unemployment compensation law or any similar law.

Commentary: This exclusion is acceptable to public water/wastewater systems, as workers compensation coverage is statutorily required and offered through a separate policy.

Liquor Liability

Exclusion: Bodily injury or property damage for which any insured may be held liable by reason of: (1) causing or contributing to the intoxication of any person; (2) the furnishing of alcoholic beverages to a person under the legal drinking age or under the influence of alcohol; or (3) any statute, ordinance, or regulation relating to the sale, gift, distribution, or use of alcoholic beverages. This exclusion applies only if you are in the business of manufacturing, distributing, selling, serving, or furnishing alcoholic beverages.

Commentary: The foregoing exclusion is inapplicable to public water/wastewater systems since your operations comprise water purveyance. Some insurers expand the exclusion to include any organization that sells alcohol for a charge or where alcohol requires a license, even if no charge is made. Such an expansion is unacceptable, as it precludes holiday parties as well as fundraisers and charity events where you offer alcohol to guests. There should be no exclusion for liquor liability, host or no host, that extends to a public water and wastewater system.

Optional Coverages

There are optional coverages that can be attached to a CGL policy. The most prevalent involve employee benefit plans, hired and non-owned auto liability, stop-gap liability, and law enforcement liability wrongful acts. Employee benefit plans applies to ministerial acts and omissions associated with the administration of your public water and wastewater system's benefits plans. This coverage does not encompass fiduciary duties in operating your plan or managing its investments. Those exposures necessitate the purchase of a fiduciary liability policy. Hired and non-owned auto liability comprises the use of renting vehicles for your operations and exposure associated with employees driving their own vehicles for your operations. This coverage is essential when a public water and wastewater system does not have a BAC policy.

Stop-gap liability applies in monopolistic states where workers compensation is administered by a state fund, and employers' liability is offered in the commercial marketplace. Employers' liability pertains to litigation by your injured employees of unsafe working conditions. It does not include payment of benefits to injured employees. Law enforcement wrongful act means an actual or alleged act, error, omission, neglect, or breach of duty, including violation of any civil rights law, of your employees or contract employees while performing law enforcement activities authorized by you. Public water/wastewater systems that employ armed guards or park rangers as well as contract armed security should add this coverage to their CGL policy.

Special Commentary

The exclusions in Coverage A. substantially narrow the coverage grant offered in the insuring agreement. Securing the above referenced exceptions will curb the unintended consequences of these exclusions. There should also be no additional exclusions other than what was presented above. Examples of unacceptable exclusions include subsidence, failure to supply, and silica. Subsidence arising from your underground water and wastewater piping infrastructure is a major exposure, especially for organizations with undulating topography as well as a service territory with a history of earth movement. Underground pipes leak, which is why this exclusion is unacceptable. A prudent risk management measure is to monitor water pressure and leakage levels via a current system control and data acquisition (SCADA) system. Such records provide meaningful defense when litigation is initiated against your organization for property damage subsidence allegations arising from underground pipe leaks.

Failure to supply is another critical coverage. This coverage, along with subsidence, is best offered through silence. That means there should be no affirmation of coverage and absolutely no restriction of coverage. The former is not needed because failure to supply is a peripheral exposure resulting from bodily injury and property damage from your operations. The latter is frequently presented as a coverage enhancement. In actuality, it restricts failure to supply coverage to bodily injury and property damage from sudden and accidental injury to tangible property owned or used by any insured to procure, produce, process, or transmit the water.

This limitation removes failure to supply coverage arising from equipment malfunctions, emergency maintenance disruptions, as well as brownouts or blackouts from your electricity provider. Any substantial failure to supply claim will invariably involve allegations of improper maintenance scheduling, inadequate adequate equipment replacement, and insufficient infrastructure redundancy. Each of these scenarios does not injure tangible property owned or used by the public water and wastewater system.

Silica, often referred to as quartz, is a prevalent mineral within most public water and wastewater system jurisdictions. The respiration of silica dust can cause lung disease and lung cancer. It only takes a small amount of airborne silica dust to create a health hazard. The extraction or installation of underground piping infrastructure can be causally connected to the aforementioned lung disease and lung cancer. Employees of third-party contractors performing this type work can be exposed to this disease.

Most public water/wastewater systems are subjected to this exposure, which necessitates the coverage. The validity of such litigation is low, as it requires substantial and frequent exposure to be impacted with injury from silica dust. Notwithstanding, these suits can be expensive and time-consuming to defend and necessitate the obligation of defense and indemnity by your insurance. Similar to subsidence and failure to supply, this coverage is best offered through silence.

The inclusion of exceptions to the foregoing exclusions, coupled with no additional exclusions, will maximize the coverage grants within Coverage A. That leaves one final step, which is insurer affirmation as to the scope and impact of the exclusions to various claim scenarios. This affirmation will align claim expectations and mitigate interpretative overreach between you and your selected insurer. Your broker advocate should secure the aforementioned affirmation so that coverage and claims misunderstanding are avoided. We have now concluded our review of Coverage A. and will turn our attention to offenses involving personal and advertising injury, which are contained in Coverage B. of a CGL policy.

Coverage B: Personal and Advertising Injury

Insuring Agreement

We will pay those sums that the insured becomes legally obligated to pay as damages because of personal and advertising injury to which this insurance applies. We will have the right and duty to defend the insured against any suit seeking those damages. However, we will have no duty to defend the insured against any suit seeking damages for personal and advertising injury to which this insurance does not apply. We may, at our discretion, investigate any offense and settle any claim or suit that may result. But: (1) The amount we will pay for damages is limited as described in the section on Limits of Insurance; and (2) Our right and duty to defend end when we have used up the applicable limit of insurance in the payment of judgments or settlements under Coverages A. or B.

or medical expenses under Coverage C. No other obligation or liability to pay sums or perform acts or services is covered unless explicitly provided for under Supplementary Payments – Coverages A. and B. This coverage applies to personal and advertising injury only if: (1) the personal and advertising injury is caused by an offense arising out of your operations; and (2) the offense is committed in the coverage territory during the policy period.

Key Definitions

Personal and Advertising Injury – injury, including consequential bodily injury, arising out of one or more of the following offenses: (a) false arrest, detention, or imprisonment; (b) malicious prosecution; (c) the wrongful eviction from, wrongful entry into, or invasion of the right of private occupancy of a room, dwelling, or premises that a person occupies, committed by or on behalf of its owner, landlord, or lessor; (d) oral or written publication in any manner of material that slanders or libels a person or organization or disparages a person's or organization's goods, products, or services; (e) oral or written publication in any manner of material that violates a person's right of privacy; (f) the use of another's advertising idea in your advertisement; or (g) infringing upon another's copyright, trade dress, or slogan in your advertisement.

Advertisement – a notice that is broadcast or published to the general public or specific market segments about your goods, products, or services for the purpose of attracting customers or supporters. For the purposes of this definition: (a) notices that are published include material placed on the internet or on similar electronic means of communication; and (b) regarding websites, only that part of a website that is about your goods, products, or services for the purposes of attracting customers or supporters is considered an advertisement.

Special Commentary

Coverage B. is treated in the same manner as Coverage A. That means the duty to defend, who is an insured, supplementary payments, limits of insurance, and policy conditions impact both coverages equally. One important distinction between Coverage A. and Coverage B. is the latter is not triggered by physical harm. Instead, personal and advertising injury require an offense that is committed by a named insured for coverage to apply. An offense involves a violation or infringement of the rights of others. Further, an offense is often the result of an intentional act. The concept of an occurrence is irrelevant to Coverage B., as there is no requirement that personal or advertising injury be caused by an occurrence. That is not to suggest an intentional injury will be covered by Coverage B. The removal of coverage for intentional injury is accomplished through the exclusions as opposed to limitations found in the insuring agreement. Moreover, the actual offenses afforded under Coverage B. are specifically listed in the definition of personal and advertising injury. Personal and advertising injury is rarely triggered for public water/wastewater systems. When there is a claim, it is invariably based on personal injury as opposed to advertising injury. The offense will likely relate to defamation of character and involve a public dispute with a ratepayer or community leader. False arrest is rare but can involve an altercation involving a right-of-way easement between the property owner and the public water and wastewater system. The coverage for personal and advertising injury is undoubtedly complex and difficult to understand, in part because such claims are relatively uncommon. Comprehension is facilitated through the following deconstruction process: (1) coverage only applies to those offenses specifically listed in the definition of personal and advertising injury; and (2) the listed offenses may not always be covered because of the exclusions.

We will now delve into the exclusions and limitations and recommend you refer back to the definition of personal and advertising injury to maximize your understanding of this convoluted concept. The aim for most of these exclusions is to avoid unintended defense and indemnity obligations for bodily injury and property damage claims specifically excluded in Coverage A. The level of insurer precaution is spurred by the obligation to offer defense whenever there is a possibility of coverage. Although tangential, a personal injury offense of wrongful entry as it relates to conventional property damage claims can be packaged to trigger said obligation. The other exclusions are designed to narrow the coverage grant afforded in the insuring agreement, especially as it relates to intentional injury.

EXCLUSIONS AND LIMITATIONS

Asbestos

Exclusion: Any injury, damage, expense, cost, loss, liability, or legal obligation arising out of or in any way related to asbestos or asbestos-containing materials, or exposure thereto, or for the costs of abatement, mitigation, removal, elimination, or disposal of any of them.

Commentary: This exclusion is acceptable, as it is inapplicable to personal and advertising injury. Its inclusion removes unintended defense and indemnity obligations by the insurer for an asbestos claim alleging wrongful entry.

Breach of Contract

Exclusion: Personal and advertising injury arising out of a breach of contract, except an implied contract to use another's advertising idea in your advertisement.

Commentary: This exclusion is aimed at advertising injury and eliminates coverage (with one exception) for an insured not honoring the terms of a contract, even if the contract involves advertising activities. The singular exception is breach of an implied contract provided the alleged offense was from the use of a third party's advertising ideas in your advertisement. An example would be a logo, style, or tagline of advertisement that infringes on another idea.

Contractual Liability

Exclusion: Personal and advertising injury for which the insured has assumed liability in a contract or agreement. This exclusion does not apply to liability for damages that the insured would have in the absence of the contract or agreement.

Commentary: Coverage is not afforded for liability the insured assumes in a contract or agreement unless liability would have been imposed in absence of such a contract or agreement. This exclusion is problematic for hold harmless and indemnity agreements that include the term "personal injury" when the inference is bodily injury. It is not uncommon for these agreements to be overly broad and demand indemnity for any and all liability. You should review all agreements to ensure personal and advertising injury is limited to tort liability and not contractual liability.

Criminal Acts

Exclusion: Personal and advertising injury arising out of a criminal act committed by or at the direction of the insured.

Commentary: As some offenses can also constitute criminal acts, this exclusion expressly eliminates coverage for such criminal acts. The foregoing exclusion applies to criminal acts committed by the insured or committed at the direction of the insured. It does not apply to the vicarious liability of insureds who were not involved in the criminal act.

Distribution of Material in Violation of Statutes

Exclusion: Personal and advertising injury arising directly or indirectly out of any action or omission that violates or is alleged to violate: (1) The Telephone Consumer Protection Act, including any amendment of or addition to such law; (2) The CAN-SPAM Act of 2003, including any amendment of or addition to such law; or (3) Any statute, ordinance, or regulation, other than the TCPA or CAN-SPAM Act of 2003, that prohibits or limits the sending, transmitting, communication, or distribution of material or information.

Commentary: This exclusion is acceptable because public water/wastewater systems do not engage in solicitation.

Electronic Chatrooms or Bulletin Boards

Exclusion: Personal and advertising injury arising out of an electronic chatroom or bulletin board the insured hosts, owns, or over which the insured exercises control.

Commentary: This exclusion eliminates coverage for liability arising out of the electronic chatroom or bulletin board activities. Public water/wastewater systems are recommended not to host any chatrooms, as the exposure is not covered and would require a separate media and internet professional liability policy.

Employment Practices and Employee Benefit Plans

Exclusion: Personal and advertising injury arising out of your employment practices or administration of your employee benefit plans.

Commentary: Personal and advertising injury coverage is contemplated for oral or written publication of material that violates a person's right of privacy. This exclusion removes coverage for claims by employees alleging employment-related invasion of privacy as well as employment-related practices, policies, acts, omissions, and employee benefit plans. Coverage for employment practices and employee benefit plans is found under the POML policy. Employee benefit plans is also an optional coverage under the CGL policy.

Fungi or Bacteria

Exclusion: (1) Any injury which would not have occurred or taken place, in whole or in part, but for the actual, alleged, or threatened inhalation of, ingestion of, contact with, exposure to, existence of, or presence of, any fungi or bacteria on or within a building or structure, including its contents, regardless of whether any other cause, event, material, or product contributed concurrently or in any sequence to such injury or damage. (2) Any loss, cost, or expenses arising out of the abating, testing for, monitoring, cleaning up, removing, containing, treating, detoxifying, neutralizing, remediating, or disposing of, or in any way responding to, or assessing the effects of, fungi.

Commentary: Standard exclusion that does not impact the scope of coverage. Its inclusion removes unintended defense and indemnity obligations by the insurer for any fungi or bacteria claim alleging wrongful entry.

Infringement of Copyright, Patent, Trademark, or Trade Secret

Exclusion: Personal and advertising injury arising out of the infringement of copyright, patent, trademark, trade secret, or other intellectual property rights. Under this exclusion, such other intellectual property rights do not include the use of another's advertising idea in your advertisement. However, this exclusion does not apply to infringement, in your advertisement, of copyright, trade dress, or slogan.

Commentary: Infringement of the intellectual property rights of others is not covered unless those intellectual property rights are from your advertisement and then only for the infringement of the following intellectual property rights: copyright, trade dress, or slogan. This exclusion is acceptable for public water/wastewater systems, but it necessitates a service mark on your logo as well as confirmation your promotional collateral are not unintentionally coopted from another organization.

Insureds in Media and Internet Type Businesses

Exclusion: Personal and advertising injury committed by an insured whose business is: (1) advertising, broadcasting, publishing, or telecasting; (2) designing or determining content of websites for others; or (3) an internet search, access, content, or service provider. For the purposes of this exclusion, the placing of frames, borders, links, or advertising, for you or others anywhere on the internet, is not considered the business of advertising, broadcasting, publishing, or telecasting.

Commentary: This exclusion is acceptable to public water/wastewater systems because your operations fall outside these business activities. Public water/wastewater systems that develop their own website, including links to the websites of others, would not be considered an insured whose business is advertising, broadcasting, publishing, or telecasting.

Knowing Violation of the Rights of Another

Exclusion: Personal and advertising injury caused by or at the direction of the insured with knowledge that the act would violate the rights of another and would inflict personal and advertising injury.

Commentary: This restriction is one of the prime exclusions intended to eliminate coverage for claims involving intentional injury. For the exclusion to apply, you must know your actions violate the rights of another, and that violation would inflict personal or advertising injury. The exclusion applies whether the infliction was caused by the insured or caused by others under the direction of the insured. Wrongful detention of an individual with knowledge is an example of a claim that would not be covered.

Lead, Electromagnetic Radiation, Nuclear

Exclusion: (1) Any injury, expense, cost, loss, liability, or legal obligation arising out of or in any way related to: (a) toxic properties of lead, or any material or substance containing lead; or (b) electromagnetic radiation, or exposure thereto, or the costs of abatement, mitigation, removal, elimination, or disposal of any of them. (2) Any loss, cost, or expense arising out of any actual, alleged, or threatened injury to any person or property from radioactive matter or nuclear material.

Commentary. Standard exclusion that does not impact the scope of coverage. Its inclusion removes unintended defense and indemnity obligations by the insurer for any lead, electromagnetic radiation, or nuclear claim alleging wrongful entry.

Material Published Prior to Coverage Form Period

Exclusion: Personal and advertising injury arising out of the oral or written publication of material whose first publication took place before the beginning of the policy period.

Commentary: The intent of this exclusion is to limit coverage to the policy period in which the publication, and thus the offense, was first committed. Its inclusion excludes coverage for previously committed offenses.

Material Published with Knowledge of Falsity

Exclusion: Personal and advertising injury arising out of oral or written publication of material, if done by or at the direction of the insured with knowledge of its falsity.

Commentary: This exclusion is intended to eliminate coverage for specific types of intentional injury such as intentional libel or slander. As with the Knowing Violation of the Rights of Another exclusion, there is no coverage if you directly publish or direct others to publish information that is knowingly false. The exclusion only applies if you actually know the injurious information is false.

Pollution

Exclusion: Personal and advertising injury arising out of or in any way related to pollution, however caused. Pollution includes the actual, alleged, or potential presence in or introduction into the environment of any substance if such substance has, or is alleged to have, the effect of making the environment impure, harmful, or dangerous. Environment includes any air, land, structure (or the air therein), watercourse, or other body of water, including underground water.

Commentary: This exclusion is acceptable for public water/wastewater systems. Its inclusion removes unintended defense and indemnity obligations by the insurer for any pollution claim alleging wrongful entry.

Pollution Related

Exclusion: Any loss, cost, or expense arising out of any: (1) request, demand, order, or statutory or regulatory requirement that any insured or others test for, monitor, clean up, remove, contain, treat, detoxify, neutralize, or in any way respond to, or assess the effects of, pollutants; or (2) claim or suit by or on behalf of a governmental authority for damages because of testing for, monitoring, cleaning up, removing, containing, treating, detoxifying, neutralizing, or in any way responding to, or assessing the effects of, pollutants.

Commentary: This exclusion is acceptable to public water/wastewater systems. Its inclusion removes unintended defense and indemnity obligations by the insurer for any pollution-related claim alleging wrongful entry.

Professional Services

Exclusion: Any liability arising out of any act, error, omission, malpractice, or mistake of a professional nature committed by the insured or any person for whom the insured is legally responsible. It is understood this exclusion applies even if the claims against any insured allege negligence or other wrongdoing in the supervision, hiring, employment, training, or monitoring of others by that insured.

Commentary: Standard exclusion that does not impact the scope of coverage. Its inclusion removes unintended defense and indemnity obligations by the insurer for professional services claims alleging wrongful entry.

Public Use of Property

Exclusion: Personal and advertising injury arising out of the principles of eminent domain, condemnation, inverse condemnation, or adverse possession.

Commentary: Standard exclusion that does not impact the scope of coverage. This legal theory only applies to property damage and not third-party bodily injury or personal and advertising injury.

Quality or Performance of Goods – Failure to Conform to Statements.

Exclusion: Personal and advertising injury arising out of the failure of goods, products, or services to conform to any statement of quality or performance made in your advertisement.

Commentary: This exclusion eliminates coverage for claims alleging your goods, products, or services do not perform or are not of the quality advertised. An example of the foregoing would be corrosion to residential plumbing network arising from water hardness that violates state or federal standards. Such a scenario would not qualify as an advertising injury offense. Its inclusion also removes unintended defense and indemnity obligations by the insurer for any quality claim alleging wrongful entry.

Sexual Abuse

Exclusion: Personal and advertising injury arising out of the sexual abuse of any person.

Commentary: Standard exclusion that does not impact the scope of coverage. Its inclusion removes unintended defense and indemnity obligations by the insurer for any sexual abuse claim alleging wrongful entry.

Specific Operations

Exclusion: Personal and advertising injury arising from the ownership, operation, maintenance, entrustment to others, or use of any: (1) gas or electric generation facility; or (2) sanitary landfill, dump, or other permanent waste disposal facility.

Commentary: Standard exclusion that does not impact the scope of coverage. An endorsement is necessary should your public water and wastewater system have any exposure to these operations.

Unauthorized Use of Another's Name or Product

Exclusion: Personal and advertising injury arising out of the unauthorized use of another's name or product in your e-mail address, domain name, metatag, or any other similar tactics to mislead another's potential customers.

Commentary: This exclusion is acceptable to public water/wastewater systems because your business operations have preset service boundaries.

War

Exclusion: Personal and advertising injury, however caused, arising, directly or indirectly, out of: (1) war, including undeclared or civil war; (2) warlike action by a military force, including action in hindering or defending against an actual or expected attack, by any government, sovereign, or other authority using military personnel or other agents; or (3) insurrection, rebellion, revolution, usurped power, or any action taken by governmental authorities in hindering or defending against any of these scenarios.

Commentary: Standard exclusion that does not impact the scope and intent of coverage. Coverage for war is offered under the Terrorism Risk Insurance Act (TRIA) for specified events.

Wrong Description of Prices

Exclusion: Personal and advertising injury arising out of the wrong description of the price of goods, products, or services stated in your advertisement.

Commentary: This exclusion clarifies that coverage will not respond to claims alleging mistakes in prices. Coverage is not afforded for the erroneous publishing of water and wastewater rates.

Special Commentary

Coverage B. protects your governing body against offenses involving personal and advertising injury. Personal injury is considered the more pervasive of the two coverages, with the majority of Coverage B. offenses encompassing defamation of character, wrongful entry, wrongful eviction, and false arrest. Offenses involving advertising injury are less prevalent, but coverage is afforded for an infringement in one's advertising ideas. The exclusions within Coverage B. narrow the protection to areas of inadvertent offenses. Many of the exclusions are aimed to prevent unintended defense and indemnity obligations by the insurer for bodily injury or property damage claims that were specifically excluded under Coverage A. Lastly, it is important to thoroughly review all contracts to ensure any personal and advertising injury provisions are limited to tort liability only. There is no coverage for contractual liability as it relates to Coverage B.

We have now completed our review of Coverages A. and B. and will focus on Coverage C., which provides medical expenses to third parties injured on or adjacent to your premises without regard to fault or liability.

COVERAGE C: MEDICAL EXPENSE

Insuring Agreement

At your written request, we will pay medical expenses as described below for bodily injury caused by an accident: (1) on premises you own or rent; (2) on ways next to premises you own or rent; (3) on that portion of a right-of-way, easement, or similar interest in property, which you do not own or rent, upon which water, sewer, or other utility fixtures are installed as part of your operations, including any surface feature which directly results from the presence of such water, sewer, or other utility fixture, but does not include any other portion of a right-of-way, easement, or similar interest in property; or (4) because of your operations; provided that: (i) the accident takes place in the coverage territory and during the coverage period; (ii) the expenses are incurred and reported to us within one year of the date of the accident; and (iii) the injured person submits to examination, at our expense, by physicians of our choice as often as we reasonably require. We will make these payments regardless of fault. These payments will not exceed the applicable limit of insurance. We will pay reasonable expenses for: (1) first aid administered at the time of an accident; (2) necessary medical, surgical, X-ray, and dental services, including prosthetic devices; and (3) necessary ambulance, hospital, professional nursing, and funeral services.

Special Commentary

Coverage C. pays medical expenses incurred by a third party for an injury sustained in an accident that arises from your premises and operations and applies regardless of fault. This coverage reimburses the injured person for his or her expenses up to the respective limit of insurance and without the need for a lawsuit. Coverage C. differs substantially from Coverages A. and B. because these latter coverages apply only if you (or another insured) are legally liable for the injury. Coverage C. can serve as an effective hedge against lawsuits. An injured person may be less inclined to sue your public water and wastewater system for bodily injury if his or her medical expenses have been paid promptly. Coverage C. includes the following reasonable medical expenses: (a) first aid rendered at the time of an accident; (b) necessary medical, surgical, X-ray, and dental services including prosthetic devices; and (c) necessary ambulance, hospital, professional nursing, and funeral services.

For expenses to be covered under Coverage C., the accident must take place during the policy period. The expenses must be incurred and reported to the insurer within one year of the accident date. Coverage C. is subject to a limit of insurance that applies to each person, and this limit is typically set low (i.e., $10,000) because the coverage is intended for minor injuries. Medical expenses under Coverage C. are subject to each occurrence and general aggregate limits of insurance. It is subject to all the exclusions listed under Coverage A. and not intended for employees, temporary workers, or permanent residents.

EXCLUSIONS AND LIMITATIONS

Any Insured

Exclusion: To any insured, except volunteer workers not performing an emergency service activity or a law enforcement activity.

Commentary: No coverage applies for medical expenses incurred by anyone (including your employees) who qualifies as an insured under your CGL policy. An exception applies to medical expenses incurred by volunteer workers (these expenses are covered).

Athletic Activities

Exclusion: To a person injured while practicing, instructing, or participating in any physical exercises, sports, or athletic contests.

Commentary: No coverage applies to anyone injured while practicing, instructing, or participating in any physical exercises, sports, or athletic contests. A special accident and health policy is available to cover participant injuries on a no-fault basis.

Coverage A. Exclusions

Exclusion: Excluded under Coverage A. Bodily Injury and Property Damage Liability.

Commentary: No coverage applies to exclusions listed in the foregoing section. This exclusion reaffirms that coverage is limited to medical expenses incurred by third parties at your premises.

Hired Person

Exclusion: To a person hired to do work for or on behalf of any insured or a tenant of any insured.

Commentary: No coverage applies to injuries incurred by someone hired to do work for you or any other insured or any tenant. This exclusion reaffirms coverage is limited to third parties who are not your contractors, employees, or volunteers.

Injury on Normally Occupied Premises

Exclusion: To a person injured on that part of premises you own or rent that the person normally occupies.

Commentary: No coverage applies to injuries to a tenant or other persons who normally occupy some portion of your premises.

Products – Completed Operations Hazard

Exclusion: Included within the products-completed operations hazard.

Commentary: No coverage applies to injuries to third parties if the injuries are caused by accidents arising out of your products or completed work. Such injuries fall outside the scope of your premises and would not be eligible for no-fault medical expenses. Coverage does apply to liability under products-completed operations hazard, which is included under Coverage A.

Workers Compensation and Similar Laws

Exclusion: To a person, whether or not an employee of any insured, if benefits for the bodily injury are payable or must be provided under a workers compensation or disability benefits law or a similar law.

Commentary: No coverage applies to any employee or other eligible person entitled to workers compensation benefits. In most states, workers compensation benefits are intended to serve as an exclusive remedy for employment-related injuries. The purchase of a supplemental benefits policy that extends coverage to your employees for accidents occurring during and beyond the course of their employment duties will mitigate out-of-pocket expenses and remove any misunderstandings regarding the true intent of Coverage C.

Special Commentary

Coverage C. is a goodwill coverage grant intended to diffuse legal escalation from an accident of temporary inattention by an injured person on or adjacent to your premises. Public water/wastewater systems are not heavily trafficked by the public, but customers do visit the office for billing and miscellaneous questions. There are also public board meetings and other community events including student field trips. This coverage affords a small limit for accidents by guests who may not have private health insurance or personal funds to handle these expenses. It is important to remember that employees are not covered, including for sporting activities at a sanctioned work event. We will now discuss the additional obligations of Coverages A., B., and C. that comprise supplementary payments of a CGL policy.

SUPPLEMENTARY PAYMENTS

Coverages A and B

(1) We will pay, with respect to any claim we investigate or settle, or any suit against an insured we defend: (a) All expenses we incur; (b) Up to $1,000 for the cost of bail bonds required because of accidents or traffic law violations arising out of the use of any vehicle to which the bodily injury liability coverage applies. We do not have to furnish these bonds; (c) The cost of bonds to release attachments, but only for bond amounts within the applicable limit of insurance. We do not have to furnish these bonds; (d) All reasonable expenses incurred by the insured at our request to assist us in the investigation or defense of the claim or suit, including actual loss of earnings up to $500 a day because of time off from work; (e) All court costs taxed against the insured in the suit. However, these payments do not include attorneys' fees or attorneys' expenses taxed against the insured; (f) prejudgment interest awarded against the insured on that part of the judgment we pay. If we make an offer to pay the applicable limit of insurance, we will not pay any prejudgment interest based on that period of time after the offer; and (g) All interest on the full amount of any judgment that accrues after entry of the judgment and before we have paid, offered to pay, or deposited in court

the part of the judgment that is within the applicable limit of insurance. These payments will not reduce the limits of insurance.

(2) If we defend an insured against a suit and an indemnitee of the insured is also named as a party to the suit, we will defend that indemnitee if all of the following conditions are met: (a) The suit against the indemnitee seeks damages for which the insured has assumed the liability of the indemnitee in a contract or agreement that is an insured contract; (b) This insurance applies to such liability assumed by the insured; (c) The obligation to defend, or the cost of the defense of, that indemnitee, has also been assumed by the insured in the same insured contract; (d) The allegations in the suit and the information we know about the occurrence are such that no conflict appears to exist between the interests of the insured and the interests of the indemnitee; (e) The indemnitee and the insured ask us to conduct and control the defense of that indemnitee against such suit and agree that we can assign the same counsel to defend the insured and the indemnitee; and (f) The indemnitee: (1) agrees in writing to: (i) cooperate with us in the investigation, settlement, or defense of the suit; (ii) immediately send us copies of any demands, notices, summonses, or legal papers received in connection with the suit; (iii) notify any other insurer whose coverage is available to the indemnitee; and (iv) cooperate with us with respect to coordinating other applicable insurance available to the indemnitee; and (2) provides us with written authorization to: (i) obtain records and other information related to the suit; and (ii) conduct and control the defense of the indemnitee in such suit.

So long as the above conditions are met, attorneys' fees incurred by us in the defense of that indemnitee, necessary litigation expenses incurred by us, and necessary litigation expenses incurred by the indemnitee at our request will be paid as supplementary payments. Notwithstanding the provisions of Coverage A., such payments will not be deemed to be damages and will not reduce the limits of insurance. Our obligation to defend an insured's indemnitee and to pay for attorneys' fees and necessary litigation expenses as supplementary payments ends when: (1) we have used up the applicable limit of insurance in the payment of judgments or settlements; or (2) the conditions set forth above, or the terms of the agreement described in subpart (f) above, are no longer met.

Special Commentary

The first provision in the supplementary payments is the most important because it obligates your insurer to provide defense on an unlimited basis. This means all expenses associated with the claim, including legal expenses, are paid by the insurer without reduction of the limits of insurance. The duty of defense, however, ends when the limit of insurance is exhausted by an indemnity payment due to settlement or judgment. The payment of all defense expenses is reasonable because the insurer not only has the duty to defend you against covered claims but also the right to pay or settle any claim or lawsuit filed against the policy. This duty and right obviously involve expenses, and the insurer should be responsible for the payment of those expenses. Moreover, the insurer has the duty to pursue an appeal if there are reasonable grounds for such an action. An appeal is part of the duty to defend obligation so paying for appeal bonds is an integral part of the supplementary payments section.

Another important component is the duty to pay all reasonable expenses incurred by you if and when the insurer requests your assistance in the investigation or defense of the claim or lawsuit. Since you are required by the policy conditions to cooperate with the insurer in the investigation, settlement, or defense of any claim or lawsuit, it is appropriate for the insurer to pay the reasonable expenses incurred by you in such cooperation. By specifically using the phrase "at our request," insurers commit themselves to compensating you for only those expenses they are willing to incur or have incurred in connection with your assistance in defense or investigation of a suit or claim.

Incidentally, you are not at liberty to choose your defense counsel because supplementary payments fail to contemplate that choice. Interest on the full amount of any judgment against the insured that accrues after entry of the judgment is part of the supplementary payments. Normally, after a judgment is rendered against the insured, the court will allow interest to accrue on the amount due until the judgment is satisfied. Since the insurer may appeal the judgment, it is only fitting that the insurer pay for any interest that accrues on the judgment amount while the appeal process proceeds. The same obligation is imposed on the insurer for prejudgment interest. Supplementary payments also codify the insurer's duties of defense and indemnification as it relates to additional insureds, referred to as indemnitees, whereby an insured agreed to indemnify the indemnitee via written agreement. The indemnitee also has specific duties, which are codified and must be followed, to receive the aforementioned defense and indemnity protection by the insurer. Lastly, not all court costs taxed against the insured are automatically included as supplementary payments. Attorney fees and attorney expenses of opposing counsel taxed by the court against you are not covered as supplementary payments but rather as damages that you are obligated to pay because of the bodily injury or property damage claim. These fees and expenses are paid out of the limits of insurance, which underscores the prudence of an excess liability policy. We will now discuss the individuals and entities that qualify as a named insured under a CGL policy.

Who Is an Insured

(1) If you are designated in the supplemental declarations as: (a) An individual, you and your spouse are insureds, but only with respect to the conduct of a business of which you are the sole owner; (b) A partnership or joint venture, you are an insured. Your members, your partners, and their spouses are also insureds, but only with respect to the conduct of your business. However, if you are a public entity, you are insured as a partner in a partnership or as a joint venturer in a joint venture, but only if the partnership or joint venture is between you and another governmental organization or non-profit entity. Coverage does not extend to a partnership or joint venture that operates, controls, or funds a gas or electric generation facility; (c) A limited liability company, you are an insured. Your members are also insureds, but only with respect to the conduct of your business. Your managers are insureds, but only with respect to their duties as your managers; (d) An organization other than a partnership, joint venture, or limited liability company, you are an insured. Your executive officers and directors are insureds, but only with respect to their duties as your officers or directors. Your stockholders are also insureds, but only with respect to their liability as stockholders; (e) A public entity, you are an insured. Your operating authorities, boards, commissions, districts, or any other governmental units are insureds, provided that you operate, control, and fund the authority, board, commission, district, or other governmental unit. Coverage does not extend to an authority, board, commission, district or other governmental unit that operates, controls, or funds a gas or electric generation facility; (f) A trust, you are an insured. Your trustees are also insureds, but only with respect to their duties as trustees. (2) Each of the following is also an insured: (a) Elected or Appointed Officials: Your elected and appointed officials, including elected and appointed officials of your operating authorities, boards, commissions, districts, or other governmental units, but only for acts within the course and scope of their duties for the insured public entity or its operating authorities, boards, commissions, districts, or other governmental units; (b) Volunteer Workers or Employees: Your volunteer workers only while performing duties related to the conduct of your operations, or your employees, other than either your executive officers (if you are an organization other than a partnership, joint venture, or limited liability company) or your managers (if you are a limited liability company), but only for acts within the scope of their employment by you or while performing duties related to the conduct of your operations. However, none of these employees or volunteer workers are insureds for: (1) Bodily injury or personal and advertising injury: (i) to you, to your partners, or members (if you are a partnership or joint venture), to your members (if you are a limited liability company), to a co-employee while in the course of his or her employment or performing duties related to the conduct of your operations, or to your other volunteer workers while performing duties related to the conduct of your operations; (ii) to the spouse, child, parent, brother, or sister of that co-employee or volunteer worker as a consequence of (1)(i) above; or (iii) for which

there is any obligation to share damages with or repay someone else who must pay damages because of the injury described in (1)(i) or (1)(ii) above.

(2) Property damage to property: (i) owned, occupied, or used by, (ii) rented to, in the care, custody, or control of, or over which physical control is being exercised for any purpose by you, any of your employees, volunteer workers, any partner, or member (if you are a partnership or joint venture), or any member (if you are a limited liability company); (2) Each of the following is also an insured (continued): (c) Real Estate Managers: Any person (other than your employee or volunteer worker), or any organization while acting as your real estate manager; (d) Temporary Custodians: Any person or organization having proper temporary custody of your property if you die, but only: (i) with respect to liability arising out of the maintenance or use of that property; and (ii) until your legal representative has been appointed; (e) Legal Representatives: Your legal representative if you die, but only with respect to duties as such. That representative will have all your rights and duties under this coverage form; (f) Mutual Aid Agreements: Any persons or organizations providing service to you under any mutual aid or similar agreement, but only for acts within the scope of that mutual aid or similar agreement; (g) Good Samaritans: Employees and volunteer workers while acting as a Good Samaritan independently of his or her activities on your behalf, but only when he or she encounters the scene of an emergency requiring sudden action; (h) Owners of Commandeered Equipment: The owner of commandeered equipment other than an auto is an insured while the equipment is in your temporary care, custody, or control; (i) Lessors of Equipment: Persons or organizations from whom you lease equipment are insureds; but they are insureds only with respect to the maintenance or use by you of such equipment and only if you are contractually obligated to provide them with such insurance as is afforded by this contract. However, no such person or organization is an insured with respect to any: (i) damages resulting from their sole negligence; or (ii) occurrence that occurs, or offense that is committed, after the equipment lease ends; (j) Blanket Additional Insured: Any person or organization required to be an additional insured under an insured contract, if agreed to by you prior to the bodily injury, property damage, or personal and advertising injury, caused in whole or in part, by your acts or omissions or the acts or omissions of those acting on your behalf: (i) in the performance of your operations; or (ii) in connection with premises owned or rented by you;

(3) Any organization you newly acquire or form, other than a partnership, joint venture, or limited liability company and over which you maintain ownership or majority interest, will qualify as an insured if there is no other similar insurance available to that organization. However: (a) coverage under this provision is afforded only until the 90th day after you acquire or form the organization or the end of the coverage period, whichever is earlier; (b) Coverage A. does not apply to bodily injury or property damage that occurred before you acquired or formed the organization; and (c) Coverage B. does not apply to personal and advertising injury arising out of an offense committed before you acquired or formed the organization. No person or organization is an insured with respect to the conduct of any current or past partnership, joint venture, or limited liability company that is not shown as an insured in the supplemental declarations.

Special Commentary

This section codifies the individuals and organizations afforded protection under a CGL policy. Many of these descriptions are not relevant to public water/wastewater systems, but that inconvenience is acceptable as long as the insured codification includes all the pertinent organizations for your public water and wastewater system. It is imperative to list all organizations you seek to have insured status on the supplemental declarations. Comprehension of this section is critical to policy interpretation, and it begins with identifying who qualifies for named insured status and what protections are afforded. You must also understand the difference between insureds and additional insureds. The former has broad rights and coverage under your policy, whereas the latter has qualified or limited rights and coverage under your policy. It is also important to know the activities or events for which insureds and additional insureds are protected and when that protection applies. Two endorsements require specific commentary: (1) blanket additional insured; and (2) fellow employee. The former, which seamlessly adds individuals and organizations as required by written agreement, eliminates oversight and errors in securing singular additional insured endorsements. It is frequently embedded in a best-in-class CGL policy as illustrated above. The latter protects your public water and wastewater system and its employees for situations where a co-employee and your organization are sued based on the negligence of one employee that causes injury to another employee. Both endorsements should be provided by your insurer. We will now catalog the insureds into five classifications to better identify the individuals and organizations who receive insured status and their respective activities that qualify for coverage under your CGL policy.

Category 1 – Type of Organization

The first category of insured is determined by their form of business or organization. We will not discuss individuals because that classification is inapplicable to public water/wastewater systems. Partnerships and joint ventures, if designated as an insured on the supplemental declarations, receive protection as an insured organization. In addition, the partners of the insured partnership (and their spouses) or members of an insured joint venture (and their spouses as well) are also personally insured, but only with respect to the conduct of the business of the insured partnership or joint venture.

Limited liability companies, a form of organization that is a hybrid between a partnership and corporation, are protected if the limited liability company is designated as an insured on the supplemental declarations page. Members (owners) of an insured limited liability company are granted personal protection but only for conduct of business on behalf of the insured limited liability company. Managers (executives) are also personally insured, but only for their duties as managers of the insured limited liability company. No automatic coverage is granted for spouses of members or managers of a limited liability company unless the spouse is found to have acted as either an employee or volunteer worker of the insured. The importance of correct, complete, and timely insured codification of a partnership, joint venture, and limited liability company cannot be overemphasized. All such organizations must be designated on the supplemental declarations to qualify as an insured in a CGL policy.

No person or organization is an insured with respect to the conduct of any current or past partnership, joint venture, or limited liability company that is not shown as an insured on the supplemental declarations. For other organizations designated as an insured (often corporations), insured status applies to the organization itself. An insured corporation's executive officers (defined as a person holding any officer positions created by the named insured's charter, constitution, by-laws, or any other similar governing document) and directors are also personally insured by a CGL policy, but that protection is restricted to their duties as officers and directors of the insured. Stockholders are personally protected for their liability as stockholders of the insured. If a trust is designated on the supplemental declarations, then the trust is granted insured status. Any trustee is also an insured, but only with respect to their duty as a trustee for the insured trust.

Category 2 – Insured in Any Organization

Volunteer workers, employees, real estate managers (who are not employees), and a person with temporary custody of the insured's property if an insured dies as well as their legal representative all have limited insured status under this category. Volunteer workers are personally protected while performing duties related to the conduct of the insured's business. Employees are also personally protected but only for acts within the scope of their employment for the insured or while performing duties related to the conduct of the insured's business.

Protection is further restricted in that insured status does not apply to either a volunteer worker or an employee because of bodily injury or personal and advertising injury: (i) to the insured, partners, members, co-employees (while the other co-employee is in the course of the insured's employment or while performing duties related to the conduct of the insured's business) or other volunteer workers (while the other volunteer worker is performing duties related to the conduct of the insured's business); (ii) to the spouse, child, parent, or brother of the co-employee or volunteer worker that is a consequence of the injury to the co-employee or volunteer worker; (iii) to the obligation to share or repay damages that someone else must pay because of the injury to the co-employee or volunteer worker; or (iv) arising out of or failing to provide professional health care services. The fellow employee endorsement extends coverage for (i) above, whereas employers' liability under your workers compensation policy or a stop-gap liability endorsement in monopolistic states affords coverage for (ii) and (iii) above. Subcategory (iv) above is not applicable to public water/wastewater systems.

Similarly, volunteer workers and employees are not granted insured protection for property damage to property owned by, occupied by, used by, rented to, in the care, custody, or control of, or over which physical control is being exercised for any purposes by the insured or any of the insured's employees, volunteer workers, partners, or members. This coverage necessitates a property policy. Non-employees (independent contractors) who are real estate managers of the insured are automatically protected with insured status, but only while acting as the insured's real estate manager. Protection does not extend to the real estate manager's own activities if unrelated to the insured. If the insured dies, anyone with proper temporary custody of the insured's property, such as a relative maintaining a rental property while affairs of that business are being wound up, is an insured.

Category 3 – Mobile Equipment Registered under DMV Laws

Insured status extends under this category to any person (non-employee) the insured permits to drive mobile equipment on a public highway if that mobile equipment is registered to the insured under a motor vehicle registration law. Also included as an insured is anyone responsible for the permissive user (such as the employer of the permissive user) if the responsible person does not have any other insurance available. Insured status granted to such a permissive user does not apply to bodily injury to a co-employee of the permissive user or property damage to property owned by, rented to, or in the charge of or occupied by the insured or the employer of the permissive user.

Category 4 – Other Individuals and Organization

Insured status extends on a limited basis to other individuals and organizations entering into mutual aid agreements with the insured as well as individuals already listed as insureds for Good Samaritan acts, owners of commandeered equipment while the equipment is in the insured's possession, and lessors of equipment leased to you. Additional insured status, but not insured status, applies to counterparties of written agreements requiring such status. A best-in-class CGL policy automatically embeds additional insured status on a blanket basis whenever a written agreement requires it. If this coverage is not embedded, then it requires a blanket additional insured endorsement.

Category 5 – Newly Acquired or Formed Organizations

Insured status extends on a limited basis to newly acquired or newly formed organizations as long as those organizations do not have similar insurance available. This automatic coverage does not apply to partnerships, joint ventures, or limited liability companies. These types of organizations must immediately be listed on the policy as an insured to receive any protection. The insured status of newly acquired or newly formed organizations is also limited in time. It applies from the date of formation or acquisition and ends in 90 days or the end of the policy period, whichever occurs first. Further, no coverage applies to bodily injury or property damage as well as personal and advertising injury offenses that were committed before the insured formed or acquired the organization.

That concludes our examination of the "Who Is an Insured" section of a CGL policy. We will now review the mechanics behind the limits of insurance.

LIMITS OF INSURANCE

(1) The limits of insurance shown in the supplemental declarations and the rules below fix the most we will pay regardless of the number of: (a) insureds; (b) claims made or suits brought; or (c) persons or organizations making claims or bringing suits. (2) The general aggregate limit is the most we will pay for the sum of: (a) medical expenses under Coverage C.; (b) damages under Coverage A., except damages because of bodily injury or property damage included in the products-completed operations hazard; and (c) damages under Coverage B. (3) The products-completed operations aggregate limit is the most we will pay under Coverage A. for damages because of bodily injury and property damage included in the products-completed operations hazard. (4) Subject to subpart (2) above, the personal and advertising injury limit is the most we will pay under Coverage B. for the sum of all damages because of all personal and advertising injury sustained by any one person or organization. (5) Subject to subparts (2) or (3) above, whichever applies, the each occurrence limit is the most we will pay for the sum of: (a) damages under Coverage A.; and (b) medical expenses under Coverage C.; because of all bodily injury or property damage arising out of any one occurrence. (6) Subject to subpart (5) above, the damage to premises rented to you limit is the most we will pay under Coverage A. for damages because of property damage to any one premises while rented to you, or in the case of damage by fire while rented to you or temporarily occupied by you with permission of the owner. (7) Subject to subpart (5) above, the medical expense limit is the

most we will pay under Coverage C. for all medical expenses because of bodily injury sustained by any one person. The limits of insurance of this coverage form apply separately to each consecutive annual period and to any remaining period of less than 12 months, starting with the beginning of the coverage period shown in the supplemental declarations, unless the coverage period is extended after issuance for an additional period of less than 12 months. In that case, the additional period will be deemed part of the last preceding period for purposes of determining the limits of insurance.

Special Commentary

The limits of insurance codify the maximum amount your insurer will pay regardless of the number of insureds, claims made, suits brought, or persons or organizations making a claim or bringing suit. That means your policy limits do not increase if a claim or suit names multiple insureds, the incident involves multiple claims or suits, or numerous persons or organizations make a claim or file a suit. These mechanics must be understood to ensure your aggregate limits are adequate and to prevent noncompliance of minimum underlying limits with your excess liability policy. There are six different but interrelated limits codified in a CGL policy. This connectivity means a reduction of one limit by the payment of damages will also reduce another limit. The most important limits of a CGL policy are the aggregate limits, which comprise the general aggregate and the products-completed operations aggregate. An aggregate limit is the most the insurer will pay during the policy period. Once an insurer pays as damages the full amount of an aggregate limit (payment must be pursuant to a judgment or settlement), the insurer has no further obligation to any insured for any claims or suits that fall within the exhausted aggregate limit. Exhaustion of an aggregate limit also extinguishes an insurer's duty to defend any subsequent suits that fall within the fully depleted aggregate limit. Details regarding the six different limits are provided below:

Limit #1 – General Aggregate

The general aggregate limit is the most the insurer will pay for damages under Coverages A. and B. as well as expenses under Coverage C. The only circumstances in which the general aggregate limit does not apply is to damages because of bodily injury or property damage arising out of the products-completed operations hazard. These damages have their own products-completed operations aggregate limit. In other words, the general aggregate limit applies to all damages paid under insuring agreement Coverage A. (except damages arising out of products-completed operations), all damages paid under insuring agreement Coverage B., and all expenses paid under insuring agreement Coverage C.

Limit #2 – Products-Completed Operations Aggregate

The products-completed operations aggregate limit is the most the insurer will pay during the policy period for damages because of bodily injury or property damage and arising out of the products-completed operations hazard. The products-completed operations aggregate limit applies independently of the general aggregate limit, which means damages paid under the general aggregate limit do not reduce the products-completed operations aggregate limit and vice versa. An insurer's total liability or exposure in a policy period is the sum of the two aggregate limits.

Limit #3 – Personal and Advertising Injury

The personal and advertising injury limit is the most the insurer is required to pay for personal and advertising offenses. This limit is independent of the each occurrence limit applicable to Coverage A., which means the insurer may be required to pay the personal and advertising injury limit and the each occurrence limit. The personal and advertising injury limit applies not to each offense, but separately to each person or organization that sustains damages because of a covered offense or offenses. However, regardless of the number of persons or organizations claiming damages from a covered offense, or regardless of the number of offenses claimed during the policy period, the insurer is obligated to pay no more than the general aggregate limit for personal and advertising injury offenses.

Limit #4 – Each Occurrence

The each occurrence limit is the most the insurer will be obligated to pay for all damages paid within Coverage A. and all expenses paid within Coverage C. Although a separate products-completed operations aggregate limit applies to the payment of damages because of bodily injury and property damage arising out of the products-completed operations hazard, all damages paid under Coverage A. and all expenses paid under Coverage C. are nonetheless subject to the each occurrence limit.

Limit #5 - Damage to Premises Rented to You

The coverage for damage to premises rented to you is not provided by a specific coverage grant. Instead, it is offered by exceptions to certain exclusions found in Coverage A. The first exception provides coverage for property damage to premises, including the contents of such premises, rented to the insured for 30 or fewer consecutive days if an insured is legally obligated to pay for said property damage due to any cause except fire. The second exception, formerly known as fire damage legal liability, provides coverage for damage only to the premises (not to the contents of the premises) if an insured is legally obligated to pay for the property damage but only if fire caused the property damage. Neither exception provides coverage if the insured is obligated to pay solely because the insured has assumed liability for damage in a contract or agreement. It is imperative you review all contracts to confirm you are not contractually required to carry coverage for damage to premises rented to you. In such a situation, the appropriate solution will be coverage under your property policy.

The coverage granted by the exceptions noted above are subject to the damage to premises rented to you limit listed on the supplemental declarations. The limit applies to any one premises and is a sublimit of the each occurrence limit. Therefore, any damages paid because of property damage under the damage to premises rented to you limit will reduce the each occurrence limit for that same occurrence and will also reduce the general aggregate limit. The standard damage to premises rented to you limit is $1 million.

Limit #6 – Medical Expenses

Medical expenses is a separate insuring agreement that obligates the insurer to pay reasonable medical expenses, subject to the policy's terms and conditions, for bodily injury, caused by an accident, without regard to fault. Coverage C. is subject to the medical expense limit listed on the supplemental declarations. The medical expense limit applies separately to each person but is a sublimit of the each occurrence limit. As with the damage to premises rented to you limit, payments made as medical expenses will reduce the each occurrence limit for that same occurrence and will also reduce the general aggregate limit. The standard medical expense limit is $10,000 per person.

Per Location Endorsement

A per location endorsement is recommended for public water/wastewater systems, as it provides separate general aggregate limits for each of your described premises. This endorsement will not increase your products-completed operations aggregate limits, but it will bolster the general aggregate so that exhaustion of limits outside of the products-completed operations hazard is greatly mitigated.

Our review now concludes with an assessment of the key policy conditions that necessitate your understanding and conformance.

KEY POLICY CONDITIONS

Duties in the Event of an Occurrence, Offense, Claim, or Suit

Condition: (a) You must see to it that we are notified as soon as practicable of an occurrence or an offense which may result in a claim or suit. To the extent possible, notice should include: (i) how, when, and where the occurrence or offense took place; (ii) the names and addresses of any injured persons and witnesses; and (iii) the nature and location of any injury or damage arising out of the occurrence or offense. (b) If a claim is made or suit is brought against any insured, you must: (i) immediately record the specifics of the claim or suit and the date received; and (ii) notify us as soon as practicable. You must see to it that we receive written notice of the claim or suit as soon as practicable. (c) You and any other involved insured must: (i) immediately send us copies of any demands, notices, summonses, or legal papers received in connection with the claim or suit; (ii) authorize us to obtain records and other information; (iii) cooperate with us in the investigation or settlement of the claim or defense against the suit; and (iv) assist us, upon our request, in the enforcement of any right against any person or organization which may be liable to the insured because of injury or damage to which this insurance may also apply.

(d) No insured will, except at that insured's own cost, voluntarily make a payment, assume any obligation, or incur any expense, other than for first aid, without our consent. (e) If you report an occurrence or offense, to an insurer providing other than CGL insurance, which later develops into a CGL claim covered under this policy, failure to report such occurrence or offense to us at the time of the occurrence or offense shall not be deemed in violation of these conditions. However, you shall give notification to us, as soon as is reasonably possible, that the occurrence or offense is a CGL claim. (f) Knowledge of an occurrence or offense by any of your agents, volunteer workers, or employees shall not constitute knowledge by you unless one of your officers or anyone responsible for administering your insurance program has received a notification from the agent, volunteer worker, or employee.

Commentary: This provision codifies your compliance requirements for reporting a loss. The duties are not unreasonable but requires advanced knowledge to ensure compliance. Communication and cooperation with the insurer are critical to maintain your policy conformance. Equally important, it provides your insurer with the necessary information to defend your interests and adjudicate the loss.

Representations

Condition: By accepting this coverage form, you agree: (a) the information in the supplemental declarations is accurate and complete; (b) the information is based upon representations you made to us; and (c) we have issued this coverage form in reliance upon your representations. Your failure to disclose all hazards existing as of the inception date of the coverage form shall not prejudice you with respect to the coverage afforded, provided such failure or omission is not intentional. This coverage form is void if any material fact or circumstance relating to this insurance is intentionally omitted or misrepresented.

Commentary: This provision requires your thorough review, as it can possibly void the policy should material statements be intentionally misrepresented to your insurer. The key words in the provision are *material* and *intentional*. Although it varies by state jurisdiction, *material* usually means the insurer would have acted differently if the fact was known. Materiality is usually measured by how much, if any, the fact would have influenced the insurer. A representation provision is more advantageous than a warranty provision. The former requires the intentional withholding of material information for policy voidance. The latter can result in policy voidance even if the material information was withheld unintentionally. Warranty provisions should be avoided.

Separation of Insureds

Condition: Except with respect to the limits of insurance, and any rights or duties specifically assigned in this coverage form to the first named insured, this insurance applies: (a) as if each named insured were the only named insured; and (b) separately to each insured against whom claim is made or suit is brought.

Commentary: This provision affirms that each insured is viewed separately and distinct from all other insureds except for limits of insurance and specific duties assigned to the first named insured. Its intent is to restore coverage for several kinds of insured-vs.-insured claims that would otherwise fall within one of the policy's exclusions. The provision also clarifies exclusionary provisions that address acts or omissions of the insured. Such a feature is important, as it protects the rights of insureds that did not violate any policy provisions or exclusions. In certain situations, separate defense counsel is also imparted to each insured. This last point is an area of differential between regulated and unregulated policies. The former is mandated to offer separate counsel to different insureds in certain situations, whereas an unregulated policy is not under said obligation. These rights only apply to insureds and not additional insureds. The latter receive a subordinated status to insureds and do not have direct rights under the policy.

Transfer of Rights of Recovery against Others to Us

Condition: If the insured has rights to recover all or part of any payment, we have made under this coverage form, those rights are transferred to us. The insured must do nothing after loss to impair them. At our request, the insured will bring suit or transfer those rights to us and help us enforce them.

Commentary: This provision expresses your requirement to protect the insurer's rights of recovery after a loss. The integral phrase is *after the loss*. There is no requirement or obligation for you to protect the insurer's rights of recovery before a loss. As such, any insured contract with a waiver of subrogation or another type of release that is executed prior to the loss does not violate the provision. It is important to secure a blanket waiver of subrogation endorsement by your insurer for all written agreements with counterparties requiring this provision. A blanket endorsement will minimize conflicting verbiage and non-congruence between your policy and written agreement.

When We Do Not Renew

Condition: If we decide not to renew this coverage form, we will mail or deliver to the first named insured shown in the supplemental declarations written notice of the nonrenewal not less than 30 days before the expiration date. If notice is mailed, proof of mailing will be sufficient proof of notice.

Commentary: This provision ensures reasonable notification should your insurer choose to non-renew your policy. A 60-day minimum written notice of nonrenewal is recommended over the standard timeline of 30 days. It is also important to secure a 60-day written notice of material changes to policy terms and/or premium. Such requests are routinely granted as part of the renewal negotiation process.

Special Commentary

Details matter, and your policy conditions are a cornucopia of particulars. Some of the details inure to your benefit, while many are obligations imposed upon you by the insurer. The latter duties are frequently glossed over and not reviewed. These acts are imprudent and can lead to unintended consequences. The policy conditions are not complicated to understand, but they require a thorough reading to ensure proper comprehension. That deliverable falls to you, but your broker advocate should assist in providing written confirmation to ensure your accurate understanding of these provisions.

FINAL COMMENTS

CGL policies serve as the insurance foundation from which other liability policies are built upon. This policy protects your public water and wastewater system against third-party liability arising from your premises, operations, products, and completed operations. Coverage comprises bodily injury and property damage as well as personal and advertising injury. Additional coverages include damage to premises rented or occupied by you as well as no-fault medical expenses for third parties injured on or adjacent to your premises. An important feature of a CGL policy is its duty to defend provision, which requires a defense obligation for any potential claim that may trigger coverage. CGL policies should be written on an occurrence form. This type of form simply requires damages for bodily injury and property damage to occur during the policy period and in the coverage territory. The actual event or activity that triggers the damages is irrelevant, as is the date damages are demanded or litigation is filed. Conversely, an offense involving personal and advertising injury must occur during the policy period and in the coverage territory.

Defense costs should be unlimited and not erode the limits of insurance. Notwithstanding, the obligation of defense ends with the exhaustion of the limits of insurance from a settlement or judgment of indemnity (i.e., compensation). Since the insurer has an obligation of defense and indemnity, they have the right to settle any claim on your behalf. CGL policies also offer supplementary payments, which includes the payment of court costs and appeal bonds as well as prejudgment and post-judgment interest. These payments are in addition to your limits of insurance. Any court-ordered taxing of plaintiff attorney fees and attorney expenses, however, will erode the limits of insurance. Serious discussion is warranted with your governing body to ensure adequate limit protection through the purchase of an excess liability policy with your CGL policy as a scheduled underlying policy. Such a policy should be issued on a following form basis.

Details matter and your CGL policy should include tailored exceptions to many standardized exclusions. The exceptions include the following: asbestos, bacteria, fungi, and lead from your potable water; bacteria and fungi from your wastewater; damage to electronic data when there is concurrent damage to tangible property; pollution liability from your operations, potable water, and wastewater; blanket contractual liability for agreements with railroads; impaired property, inherent defects, and recall of your products from your operations as well as potable water and wastewater; fellow employee; and structural failure for your dams, reservoirs, levees, and dikes. It is similarly important to remove any limitations associated with subsidence, failure to supply, and silica. Coverage should be affirmed for water or wastewater professional activity, and, when necessary, law enforcement wrongful acts, hired and non-owned auto liability, stop-gap lability, and employee benefit plans.

The ability to issue blanket additional insured agreements inclusive of primary and noncontributory and waiver of subrogation verbiage where required by written agreement should be automatically embedded in the policy. These blanket endorsements mitigate incongruity between your CGL policy and written agreements. It also curbs scenarios where an otherwise covered claim becomes uncovered. Moreover, any indemnification agreement that includes personal and advertising injury should be limited to tort liability and not contractual liability, as CGL policies will not cover the latter.

The definition of insured should include the public water and wastewater system as well as past and present board members. Any partnerships or joint ventures should similarly be covered as long as the public water and wastewater system is the controlling party in these relationships. All organizations where you seek insured status should be scheduled on the supplemental declarations. The limits of insurance should have an aggregate limit not less than $10 million as well as a per location endorsement to mitigate scenarios involving exhaustion of your aggregate limit. Policy conditions should similarly be understood to ensure conformance with the duties imposed on you by the insurer. This knowledge will also safeguard your rights as an insured against your insurer. Policy conditions require thorough understanding and active discussion with your broker advocate. Such review should confirm no audit provision on the policy, which means any exposure changes during the policy period are reflective on the renewal premium only.

A properly designed CGL policy allows you to properly procure additional insurance policies like POML, BAC, environmental pollution, cyber liability and network risk, and excess liability to maximize your overall depth of liability protection. The aforementioned policies are designed to wraparound and augment your CGL policy. They are intended to cover exposures falling outside the scope and intent of a CGL policy. We will now begin our review of these other policies, beginning with POML, and examine their important role and value to your public water and wastewater system.

POLICY CHECKLIST: PART I

POLLUTION EXCLUSION EXEMPTIONS – DRILLDOWN

Commercial General Liability (CGL)	
Exemption	**Coverage**
Bodily Injury and Property Damage Arising Out of:	
Hostile Fire	Included
Potable Water	Included
Water and Wastewater Treatment Chemicals	Included
Propane and Natural Gas	Included
Pesticide and Herbicide Application	Included
Upset or Overturn of Equipment	Included
Wastewater Escape or Backup	Included
Smoke Drift from Controlled Burns	Included

Triggering Verbiage for Preceding Exemptions
Coverage for the preceding pollution exemptions must apply if the discharge is accidental, unintended, and stopped as soon as possible. The word *sudden* must not be included in this verbiage. The omission of the word *sudden* is necessary because you cannot cover water contamination litigation if you restrict your pollution coverage to sudden events. It must encompass accidental events that are unintended and stopped as soon as possible.

Additional Pollution Exemption
Coverage should also include bodily injury and property damage claims arising out of the actual, alleged, or threatened discharge, dispersal, seepage, migration, release, or escape of pollutants if such bodily injury or property damage is sudden and accidental and neither expected nor intended by an insured. This exception does not apply to petroleum underground storage tanks, which require a separate financial responsibility policy. The sudden restriction is acceptable for this exception because coverage extends to unnamed pollution events as opposed to named pollution events like water contamination. The latter necessitates the broader verbiage of accidental.

POLICY CHECKLIST: PART II

COMMERCIAL GENERAL LIABILITY (CGL) POLICY: MINIMUM BENCHMARKS

Commercial General Liability (CGL)	
Description	**Benchmark**
Form	
Occurrence	Yes
Proprietary	Yes
Duty to Defend	Yes
Defense Costs Outside Limits	Yes
Limits	
Per Occurrence	$1,000,000
General Aggregate	$10,000,000
Products and Completed Operations Aggregate	$10,000,000
Personal and Advertising Injury	$1,000,000
Damage to Premises Rented to You	$1,000,000
Medical Payments	$10,000
Special Coverages	
Contractual Liability – Railroads	Yes
Dams, Levees, Dikes, and Reservoirs (premises and operations)	Yes
Electronic Data	Yes
Failure to Supply (no ISO limitation)	Yes
Fellow Employee	Yes
Lead (potable water)	Yes
Pollution (named and unnamed events)	Yes
Silica	Yes
Subsidence	Yes
Water and Wastewater Testing Errors and Omissions	Yes
Waterborne Asbestos	Yes
Watercraft (owned and non-owned)	Yes
Policy Features	
Blanket Additional Insured, Waiver of Subrogation, and Primary and Noncontributory Verbiage	Yes
Broad Definition of Insured	Yes
Mental Anguish included as Bodily Injury	Yes
No Sublimits (singular exception for medical payments)	Yes
Non-auditable	Yes
Per Location Aggregate (inclusive of $10,000,000 general aggregate)	Yes
Pollution Coverage – bodily injury and property damage arising from the following pollution exclusion exemptions if the discharge is accidental, unintended, and stopped as soon as possible:	
Chemicals you use in your water or wastewater treatment process	Yes
Escape or backup of sewage or wastewater from any sewage treatment facility or fixed conduit or piping that you own, operate, lease, control, or for which you have the right of way, but only if property damage occurs away from land you own or lease	Yes
Fuels, lubricants, or other operating fluids needed to perform the normal electrical, hydraulic, or mechanical functions necessary for the operation of mobile equipment or its parts	Yes
Hostile fire at described premises	Yes
Natural gas or propane gas you use in your water or wastewater treatment process	Yes
Potable water which you supply to others	Yes

(Continued)

Commercial General Liability (CGL)

Description	Benchmark
Smoke drift from controlled or prescribed burning that has been authorized and permitted by an appropriate regulatory agency	Yes
Smoke, fumes, vapor, or soot produced by or originating from equipment that is used to heat, cool, or dehumidify buildings at described premises	Yes
Sudden and accidental events that are neither expected nor intended by an insured. However, no coverage is provided under this exception for petroleum underground storage tank	Yes
Urgent response for the protection of property, human life, health, or safety conducted away from premises owned by or rented to or regularly occupied by you	Yes
Your application of pesticide or herbicide chemicals if such application meets all standards of any statute, ordinance, regulation, or license requirement of any federal, state, or local government	Yes

Optional Coverages

Description	Benchmark
Dams, Levees, Dikes, and Reservoirs (structural collapse and failure) (mandatory if applicable)	Yes
Employee Benefit Plans	Yes
Hired and Non-owned Auto Liability	Yes
Law Enforcement Liability	Yes
Stop-Gap Liability (North Dakota, Ohio, Washington, and Wyoming)	Yes
Terrorism	Yes

Exceptions to Exclusions

Description	Benchmark
Damage to Impaired Property – This exclusion does not apply to your potable water, nonpotable water, or wastewater as well as any loss of use of other property arising out of sudden and accidental physical injury to your product or your work after it has been put to its intended use.	Yes
Damage to Your Work – This exclusion does not apply if the damaged work or the work out of which the damage arises was performed on your behalf by a subcontractor.	Yes
Electronic Data – This exclusion does not apply to property damage to electronic data that occurs in conjunction with damage to tangible property as a result of your product or operations.	Yes
Employers' Liability – This exclusion does not apply to liability assumed by the insured under an insured contract. Note: The addition of a fellow employee endorsement is necessary to extend coverage to your employees should they be sued by a co-employee for negligently causing their injury.	Yes
Fungi or Bacteria – This exclusion does not apply: (a) to any fungi or bacteria that is, is on, or is contained in a good or product intended for consumption; or (b) to any injury or damage arising out of or caused by your water, irrigation, or wastewater intake, outtake, reclamation, treatment, and distribution processes.	Yes
Lead, Electromagnetic Radiation, Nuclear – This exclusion does not apply to the toxic properties of lead contained in potable water which you supply to others.	Yes
Liquor Liability – This exclusion applies only if you are in the business of manufacturing, distributing, selling, serving, or furnishing alcoholic beverages.	Yes
Professional Services – This exclusion does not apply to bodily injury or property damage arising out of your water professional activities.	Yes
Recall of Products – This exclusion does not apply to any injury or damage arising out of or caused by: your potable water, nonpotable water, or wastewater for the loss of use, withdrawal, recall, inspection, repair, replacement, adjustment, removal or disposal of: your product, your work, or impaired property if such product, work, or impaired property is withdrawn or recalled from the market or from use by any person or organization because of a known or suspected defect, deficiency, inadequacy, or dangerous condition in it.	Yes
Sexual Abuse – This exclusion shall not apply to the named insured if no elected or appointed official, executive officer, officer, director, or trustee of the insured knew or had reason to know of the sexual abuse. Defense will similarly be afforded for the alleged insured for covered civil action subject to the other terms of this coverage form until either a judgment or final adjudication establishes such an act, or the insured confirms such act.	Yes

REVIEW QUESTIONS

1. A commercial general liability (CGL) policy safeguards your organization against the following claims:
   a. Bodily injury and property damage
   b. Personal and advertising injury
   c. No fault medical expenses
   d. All of the above

2. T/F: An insurer's obligation of defense and indemnity affords them the discretion to investigate and settle any claim or suit that is filed under a CGL policy.

3. T/F: Bodily injury and property damage are the most common claims filed against public water/wastewater systems CGL policy.

4. T/F: The definition of property damage in a CGL policy includes loss of use of tangible property that has not been physically injured.

5. T/F: Electronic data is not considered tangible property and must be specifically endorsed in a CGL policy to qualify as property damage.

6. T/F: Damages in a CGL policy do not include economic injury arising from a governing body's powers, policies, or decisions.

7. T/F: Economic injury from covered wrongful acts is included in a public officials' and management liability (POML) policy.

8. T/F: CGL policies are traditionally written on an occurrence form.

9. A CGL policy that is issued on an occurrence form is triggered when:
   a. The bodily injury or property damage transpired
   b. The date of the occurrence
   c. The date you report the claim
   d. All of the above

10. T/F: An offense must take place during the policy period to trigger personal and advertising injury coverage in a CGL policy.

11. T/F: An insurer's obligation of defense is greater than its obligation of indemnity in a CGL policy.

12. T/F: It is common for the defense obligation in a CGL policy to be unlimited, which means the costs associated with defense do not erode the limits of insurance.

13. T/F: Larger liability deductibles typically do not require reimbursement by an insured for allocated loss adjustment expenses (ALAE).

14. T/F: A per claim deductible is preferred over a per occurrence deductible.

15. T/F: The provision that prevents insureds from activating successive CGL policies over a span of several years even though they were aware of continuation of bodily injury or property damage is known as the Montrose provision.

16. T/F: An intra-policy anti-stacking limit is an endorsement that prevents the stacking of different liability policies from the same insurer during the same policy period should one claim potentially impact multiple lines of coverage.

17. T/F: You must schedule any mobile equipment requiring Department of Motor Vehicles (DMV) placards on your business auto coverage policy.

18. T/F: A careful review of all contracts involving aircraft, watercraft, and mobile equipment is required to ensure these agreements meet the definition of insured contract in a CGL policy.

19. T/F: Although asbestosis is an airborne disease, public water/wastewater systems with asbestos-lined piping and asbestos-containing fixed equipment and/or appurtenances have an exposure to asbestosis litigation.

20. T/F: The insured contract definition in a CGL policy should be broadened to include work within 50 feet of a railroad.

21. T/F: An offense in a CGL policy is different than an occurrence because injury with the former takes place at the same time as the offense, whereas damages with the latter can transpire years after the actual occurrence.

22. T/F: Defense outside the limits of liability in a CGL policy terminates once indemnity payments exhaust the limits of insurance.

23. ALAE includes the following:
    a. Expert witnesses
    b. Salaries of insurer's employees
    c. a and b only
    d. None of the above

24. The aircraft, watercraft, and automobile exclusion in a CGL policy includes the following exceptions:
    a. Watercraft greater than 26 feet
    b. Operation of machinery or equipment that is attached to, or part of, a land vehicle that would qualify under the definition of mobile equipment if it were not subject to a compulsory or financial responsibility law or other motor vehicle insurance law in the state where it is licensed or principally garaged
    c. Watercraft with 275 horsepower
    d. Parking a rented auto on, or on the ways next to, premises you own or rent

25. The contractual liability exclusion in a CGL policy includes the following exceptions:
    a. Damages the insured would not have had in the absence of the contract or agreement
    b. Assumed in a contract or agreement that is an insured contract, provided the bodily injury or property damage occurs or takes place before the execution of the contract or agreement
    c. a and b only
    d. None of the above

26. T/F: Failure to complete any work or project on time does not qualify as impaired property in a CGL policy.

27. The sexual abuse exclusion under Coverage A. in a CGL policy includes the following exceptions:
    a. Elected or appointed officials, executive officers, officers, directors, or trustees of the insured that did not know or have reason to know of the sexual abuse
    b. Defense of alleged insured for covered civil action subject to the other terms of this coverage form until either a judgment or final adjudication establishes such an act, or the insured confirms such act
    c. a and b only
    d. None of the above

28. The pollution exclusion in a best-in-class CGL policy will include the following exceptions:
    a. Potable water
    b. Smoke drift
    c. Seepage of underground petroleum tanks
    d. a and b only

29. The specific operations exclusion under Coverages A. and B. in a CGL policy include the following:
    a. Gas facility
    b. Sanitary landfill
    c. Electric generation facility
    d. All of the above

30. T/F: The recall of products, work, or impaired property exclusion in a best-in-class CGL policy does not apply to any injury or damage arising out of or caused by your potable water, nonpotable water, or wastewater.

31. T/F: Your elected and appointed officials, including elected and appointed officials of your operating authorities, boards, commissions, districts, or other governmental units qualify as an insured in a CGL policy but only for acts within the course and scope of their duties for the insured public entity or its operating authorities, boards, commissions, districts, or other governmental units.

32. T/F: Limited liability companies, a form of organization that is a hybrid between a partnership and corporation, are protected as a named insured if the limited liability company is designated on the supplemental declarations page in a CGL policy.

33. T/F: Insured status extends on a limited basis to newly acquired or formed organizations in the "Who Is an Insured" section in a CGL policy as long as those organizations do not have similar insurance available.

34. Newly acquired or formed organizations in the "Who Is an Insured" section in a CGL policy include the following:
    a. Partnerships
    b. Joint ventures
    c. a and b only
    d. None of the above

35. T/F: Never select a deductible in a CGL policy on a per claim basis.

36. T/F: The insured contract definition in a CGL policy does not apply to verbal agreements.

37. T/F: The parlance to not affirm coverage in a CGL policy is known as codification.

38. T/F: The impaired property exclusion in a CGL policy excludes claims asserting failure to complete any work or project on time.

39. T/F: An exception to the electronic data exclusion in a CGL policy is needed for damage to electronic data when there is concurrent damage to tangible property from the same occurrence.

40. T/F: A best-in-class CGL policy will provide coverage for named pollution incidents on an accidental basis and unnamed pollution incidents on a sudden and accidental basis.

41. T/F: Should your public water and wastewater system be engaged in third-party watermaster services such as billing, operations, design, and engineering, then the purchase of a separate professional liability policy should be considered.

42. T/F: Law enforcement wrongful act means an actual or alleged act, error, omission, neglect, or breach of duty, including violation of any civil rights law, of your employees or contract employees while performing law enforcement activities authorized by you.

43. Examples of unacceptable exclusions in a CGL policy include the following:
    a. Silica
    b. Subsidence
    c. Failure to supply
    d. All of the above

44. T/F: Personal and advertising injury coverage in a CGL policy is rarely triggered for public water/wastewater systems.

45. The most prevalent personal and advertising injury claims filed against public water/wastewater systems in a CGL policy include the following:
    a. Defamation of character
    b. Copyright infringement
    c. False arrest
    d. a and c only

46. T/F: The majority of personal and advertising injury exclusions in a CGL policy are designed to prevent unintended defense and indemnity obligations for bodily injury and property damage claims that were specifically excluded under Coverage A.

47. T/F: Coverage B. in a CGL policy protects your governing body against offenses involving personal and advertising injury.

48. Coverage for medical expenses in a CGL policy includes the following:
    a. First aid rendered at the time of an accident
    b. Necessary medical, surgical, X-ray, and dental services including prosthetic devices
    c. Necessary ambulance, hospital, professional nursing, and funeral services
    d. All of the above

49. For a claim to be covered under medical expenses in a CGL policy, the accident must:
    a. Take place before the policy period
    b. The expenses must be incurred and reported to the insurer within two years of the accident date
    c. Involve an employee
    d. None of the above

50. Supplementary payments in an occurrence CGL policy include the following:
    a. All expenses incurred by the insurer
    b. Up to $25,000 for the cost of bail bonds
    c. All expenses incurred by the insured
    d. None of the above

51. T/F: Attorney fees and attorney expenses of opposing counsel taxed by the court against an insured are not considered supplementary payments in a CGL policy.

52. T/F: All organizations that you seek to have named insured status should be listed on the supplemental declarations in a CGL policy.

53. T/F: A named insured has qualified or limited rights unlike an additional insured in a CGL policy.

54. T/F: A fellow employee endorsement in a CGL policy protects an insured's employees for situations where a co-employee is sued for a negligent act that causes injury to another employee.

55. The general aggregate limit in a CGL policy is the most the insurer will pay for the sum of:
    a. Medical expenses under Coverage C.
    b. Damages under Coverage A., except damages because of bodily injury or property damage included in the products-completed operations hazard
    c. Damages under Coverage B.
    d. All of the above

56. T/F: A per location endorsement in a CGL policy is recommended for public water/wastewater systems because it provides separate products-completed operations aggregate limits for each of your described premises.

57. T/F: Any partnerships or joint ventures are automatically included in the definition of an insured in a best-in-class CGL policy as long as the public water and wastewater system is the controlling party in these relationships.

58. T/F: Bodily injury or property damage must occur during the policy period and in the coverage territory to trigger coverage in a CGL policy.

59. T/F: A claim or lawsuit seeking damages for bodily injury or property damage under an occurrence CGL policy must be submitted or filed during the policy period for coverage to apply.

60. T/F: It is common for larger deductibles to require reimbursement by an insured for ALAE.

61. T/F: A per location aggregate in a CGL policy provides separate general aggregate limits for each location.

62. T/F: An insurer's obligation of indemnity in a CGL policy is exceedingly broad and simply requires a possibility, however remote, that the claim is covered.

63. T/F: Under the *Four Corners* test, an insurer does not have a duty to defend its insured under a CGL policy if the complaint has causes of action and fact patterns that are potentially covered by the policy.

64. T/F: The purchase of an environmental pollution policy is required if you seek third-party asbestos coverage from gradual seepage and deterioration as well as cleanup costs, natural resources damages, and civil fines and penalties.

65. T/F: The definition of insured contract in a CGL policy does not include tort liability that would be imposed by law in the absence of any contract or agreement.

66. T/F: The contractual liability – railroads endorsement in a CGL policy – offers the same coverage as a railroad protective liability (RRPL) policy.

67. T/F: The scheduling of dam failure coverage in a CGL policy should be paired with inverse condemnation coverage in a POML policy.

68. T/F: The scheduling of dam failure coverage in a CGL policy excludes seepage because seepage is deemed a normal and expected outcome of water impoundments and conveyances.

69. The damage to impaired property exclusion in a CGL policy prevents coverage for the following:
    a. Loss of use caused by the assimilation of your defective product
    b. Loss of use caused by your faulty work that is part of the work and done incorrectly
    c. Your failure to finish any work or project on time
    d. All of the above

70. The damage to property exclusion in a best-in-class CGL policy includes the following exceptions:
    a. Property damage from fire to premises occupied or rented to you for longer than 30 days
    b. Property damage from water damage to premises and contents rented to you for 30 or fewer consecutive days
    c. Property damage from fire to premises not occupied or rented to you for longer than 30 days
    d. a and b only

71. T/F: The damage to your work exclusion in a CGL policy necessitates best practices to avoid any work performed by your employees to repair or replace lateral lines that are owned by third parties.

72. The employers' liability exclusion in a CGL policy has an exception for the following:
    a. An insured employer that is obligated to indemnify another for damages resulting from work-related injuries to the insured employer's own employees, provided the obligation of the insured employer to provide indemnity is due to liability assumed in an insured contract
    b. Employee who is sued by another employee for negligently causing the other employee's injury
    c. a and b only
    d. None of the above

73. Examples of optional coverages in a CGL policy include the following:
    a. Hired and non-owned auto liability
    b. Employee benefit plans
    c. Stop-gap liability
    d. All of the above

74. T/F: It is against the fortuity doctrine to cover expected or intended injury by an insured in a CGL policy unless such actions were intended to protect persons or property.

75. T/F: Coverage is not provided for liability arising from fungi or bacteria on insured premises in a CGL policy.

76. T/F: Your public water and wastewater system can be subject to litigation associated with failure to properly mitigate the effects of lead poisoning.

77. Electromagnetic radiation can create the following health risks:
    a. Radio frequencies
    b. Extremely low frequencies
    c. Static magnetic fields
    d. All of the above

78. A best-in-class CGL policy includes the following pollution liability coverages:
    a. Potable water which you supply to others
    b. Chemicals you use in your water treatment process
    c. Natural gas or propane gas you use in your water treatment process
    d. All of the above

79. A best-in-class CGL policy includes bodily injury and property damage arising from the professional activities of:
    a. Licensed water operators
    b. Contracted chemists
    c. None of the above
    d. All of the above

80. T/F: Sexual abuse is uninsurable when a governing body was unaware of such actions.

81. T/F: Terrorism coverage is most commonly purchased through the Terrorism Risk Insurance Act (TRIA).

82. T/F: Coverage for employee benefit plans includes an insured's fiduciary duties in operating the plan and managing its investments.

83. Examples of coverage that is best addressed via silence and without an explicit exclusion in a CGL policy include the following:
    a. Subsidence
    b. Failure to supply
    c. Silica
    d. All of the above

84. Coverages A. and B. in a CGL policy share the following provisions:
    a. Duty to defend
    b. Who is an insured
    c. Supplementary payments
    d. All of the above

85. T/F: One important distinction between Coverages A. and B. in a CGL policy is Coverage B. is triggered by physical harm.

86. T/F: The definition of insured contract in a CGL policy does not afford liability the insured assumes for personal injury.

87. Infringement of the intellectual property rights of others from your advertisement in a CGL policy is covered for the following:
    a. Copyright
    b. Trade dress
    c. Slogan
    d. All of the above

88. T/F: Wrongful detention of an individual with knowledge by a named insured is an example of a claim that would be covered under Coverage B. in a CGL policy.

89. T/F: Under a duty to defend liability policy, an insurer has the right to pay or settle any claim or lawsuit filed against you.

90. T/F: Supplementary payments in a CGL policy allows the insured to choose their own defense counsel.

91. T/F: No person or organization is an insured in a CGL policy with respect to the conduct of any current or past partnership, joint venture, or limited liability company that is not listed as an insured on the supplemental declarations.

92. T/F: The limits of insurance in a CGL policy increases when a claim or suit names multiple insureds, the incident involves multiple claims or suits, or numerous persons or organizations make a claim or file a suit.

93. T/F: The most important limits in a CGL policy are the aggregate limits, which comprise the general aggregate and the products-completed operations aggregate.

94. T/F: Exhaustion of an aggregate limit in a CGL policy extinguishes an insurer's duty to defend any subsequent suits that fall within the fully depleted aggregate limit.

95. The separation of insureds provision in a CGL policy provides the following rights:
    a. Retains rights of a named insured that does not violate any policy provision
    b. Allows all named insureds the same rights as the first named insured
    c. Provides separate limits of insurance for each named insured
    d. All of the above

ANSWER KEY

1. A commercial general liability (CGL) policy safeguards your organization against the following claims:
   a. Bodily injury and property damage
   b. Personal and advertising injury
   c. No fault medical expenses
   d. **All of the above**

2. **T**/F: An insurer's obligation of defense and indemnity affords them the discretion to investigate and settle any claim or suit that is filed under a CGL policy.

3. **T**/F: Bodily injury and property damage are the most common claims filed against public water/ wastewater systems under a CGL policy.

4. **T**/F: The definition of property damage in a CGL policy includes loss of use of tangible property that has not been physically injured.

5. **T**/F: Electronic data is not considered tangible property and must be specifically endorsed in a CGL policy to qualify as property damage.

6. **T**/F: Damages in a CGL policy do not include economic injury arising from a governing body's powers, policies, or decisions.

7. **T**/F: Economic injury from covered wrongful acts is included in a public officials' and management liability (POML) policy.

8. **T**/F: CGL policies are traditionally written on an occurrence form.

9. A CGL policy that is issued on an occurrence form is triggered when:
   a. **The bodily injury or property damage transpired**
   b. The date of the occurrence
   c. The date you report the claim
   d. All of the above

10. **T**/F: An offense must take place during the policy period to trigger personal and advertising injury coverage in a CGL policy.

11. **T**/F: An insurer's obligation of defense is greater than its obligation of indemnity in a CGL policy.

12. **T**/F: It is common for the defense obligation in a CGL policy to be unlimited, which means the costs associated with defense do not erode the limits of insurance.

13. **T**/F: Larger liability deductibles typically do not require reimbursement by an insured for allocated loss adjustment expenses (ALAE).

14. T/**F**: A per claim deductible is preferred over a per occurrence deductible.

15. **T**/F: The provision that prevents insureds from activating successive CGL policies over a span of several years even though they were aware of continuation of bodily injury or property damage is known as the Montrose provision.

16. **T**/F: An intra-policy anti-stacking limit is an endorsement that prevents the stacking of different liability policies from the same insurer during the same policy period should one claim potentially impact multiple lines of coverage.

17. **T**/F: You must schedule any mobile equipment requiring Department of Motor Vehicles (DMV) placards on your business auto coverage policy.

18. **T**/F: A careful review of all contracts involving aircraft, watercraft, and mobile equipment is required to ensure these agreements meet the definition of insured contract in a CGL policy.

19. **T**/F: Although asbestosis is an airborne disease, public water/wastewater systems with asbestos-lined piping and asbestos-containing fixed equipment and/or appurtenances have an exposure to asbestosis litigation.

20. **T**/F: The insured contract definition in a CGL policy should be broadened to include work within 50 feet of a railroad.

21. **T**/F: An offense in a CGL policy is different than an occurrence because injury with the former takes place at the same time as the offense, whereas damages with the latter can transpire years after the actual occurrence.

22. **T**/F: Defense outside the limits of liability in a CGL policy terminates once indemnity payments exhaust the limits of insurance.

23. ALAE includes the following:
    a. **Expert witnesses**
    b. Salaries of insurer's employees
    c. a and b only
    d. None of the above

24. The aircraft, watercraft, and automobile exclusion in a CGL policy includes the following exceptions:
    a. Watercraft greater than 26 feet
    b. **Operation of machinery or equipment that is attached to, or part of, a land vehicle that would qualify under the definition of mobile equipment if it were not subject to a compulsory or financial responsibility law or other motor vehicle insurance law in the state where it is licensed or principally garaged**
    c. Watercraft with 275 horsepower
    d. Parking a rented auto on, or on the ways next to, premises you own or rent

25. The contractual liability exclusion in a CGL policy includes the following exceptions:
    a. Damages the insured would not have had in the absence of the contract or agreement
    b. Assumed in a contract or agreement that is an insured contract, provided the bodily injury or property damage occurs or takes place before the execution of the contract or agreement

    c. a and b only
    d. **None of the above**

26. **T/F**: Failure to complete any work or project on time does not qualify as impaired property in a CGL policy.

27. The sexual abuse exclusion under Coverage A. in a CGL policy includes the following exceptions:
    a. Elected or appointed officials, executive officers, officers, directors, or trustees of the insured that did not know or have reason to know of the sexual abuse
    b. Defense of alleged insured for covered civil action subject to the other terms of this coverage form until either a judgment or final adjudication establishes such an act, or the insured confirms such act
    c. **a and b only**
    d. None of the above

28. The pollution exclusion in a best-in-class CGL policy will include the following exceptions:
    a. Potable water
    b. Smoke drift
    c. Seepage of underground petroleum tanks
    d. **a and b only**

29. The specific operations exclusion under Coverages A. and B. in a CGL policy include the following:
    a. Gas facility
    b. Sanitary landfill
    c. Electric generation facility
    d. **All of the above**

30. **T/F**: The recall of products, work, or impaired property exclusion in a best-in-class CGL policy does not apply to any injury or damage arising out of or caused by your potable water, nonpotable water, or wastewater.

31. **T/F**: Your elected and appointed officials, including elected and appointed officials of your operating authorities, boards, commissions, districts, or other governmental units qualify as an insured in a CGL policy but only for acts within the course and scope of their duties for the insured public entity or its operating authorities, boards, commissions, districts, or other governmental units.

32. **T/F**: Limited liability companies, a form of organization that is a hybrid between a partnership and corporation, are protected as a named insured if the limited liability company is designated on the supplemental declarations page in a CGL policy.

33. **T/F**: Insured status extends on a limited basis to newly acquired or formed organizations in the "Who Is an Insured" section in a CGL policy as long as those organizations do not have similar insurance available.

34. Newly acquired or formed organizations in the "Who Is an Insured" section in a CGL policy include the following:
    a. Partnerships
    b. Joint ventures
    c. a and b only
    d. **None of the above**

35. **T/F**: Never select a deductible in a CGL policy on a per claim basis.

36. **T/F**: The insured contract definition in a CGL policy does not apply to verbal agreements.

37. **T/F**: The parlance to not affirm coverage in a CGL policy is known as codification.

38. **T/F**: The impaired property exclusion in a CGL policy excludes claims asserting failure to complete any work or project on time.

39. **T/F:** An exception to the electronic data exclusion in a CGL policy is needed for damage to electronic data when there is concurrent damage to tangible property from the same occurrence.

40. **T/F:** A best-in-class CGL policy will provide coverage for named pollution incidents on an accidental basis and unnamed pollution incidents on a sudden and accidental basis.

41. **T/F:** Should your public water & wastewater system be engaged in third-party watermaster services such as billing, operations, design, and engineering, then the purchase of a separate professional liability policy should be considered.

42. **T/F:** Law enforcement wrongful act means an actual or alleged act, error, omission, neglect, or breach of duty, including violation of any civil rights law, of your employees or contract employees while performing law enforcement activities authorized by you.

43. Examples of unacceptable exclusions in a CGL policy include the following:
    a. Silica
    b. Subsidence
    c. Failure to supply
    d. **All of the above**

44. **T/F:** Personal and advertising injury coverage in a CGL policy is rarely triggered for public water/wastewater systems.

45. The most prevalent personal and advertising injury claims filed against public water/wastewater systems in a CGL policy include the following:
    a. **Defamation of character**
    b. Copyright infringement
    c. False arrest
    d. a and c only

46. **T/F:** The majority of personal and advertising injury exclusions in a CGL policy are designed to prevent unintended defense and indemnity obligations for bodily injury and property damage claims that were specifically excluded under Coverage A.

47. **T/F:** Coverage B. in a CGL policy protects your governing body against offenses involving personal and advertising injury.

48. Coverage for medical expenses in a CGL policy includes the following:
    a. First aid rendered at the time of an accident
    b. Necessary medical, surgical, X-ray, and dental services including prosthetic devices
    c. Necessary ambulance, hospital, professional nursing, and funeral services
    d. **All of the above**

49. For a claim to be covered under medical expenses in a CGL policy, the accident must:
    a. Take place before the policy period
    b. The expenses must be incurred and reported to the insurer within two years of the accident date
    c. Involve an employee
    d. **None of the above**

50. Supplementary payments in an occurrence CGL policy include the following:
    a. **All expenses incurred by the insurer**
    b. Up to $25,000 for the cost of bail bonds
    c. All expenses incurred by the insured
    d. None of the above

51. **T/F:** Attorney fees and attorney expenses of opposing counsel taxed by the court against an insured are not considered supplementary payments in a CGL policy.

52. **T/F:** All organizations that you seek to have named insured status should be listed on the supplemental declarations in a CGL policy.

53. **T/F**: A named insured has qualified or limited rights unlike an additional insured in a CGL policy.

54. **T/F**: A fellow employee endorsement in a CGL policy protects an insured's employees for situations where a co-employee is sued for a negligent act that causes injury to another employee.

55. The general aggregate limit in a CGL policy is the most the insurer will pay for the sum of:
    a. Medical expenses under Coverage C.
    b. Damages under Coverage A., except damages because of bodily injury or property damage included in the products-completed operations hazard
    c. Damages under Coverage B.
    d. **All of the above**

56. **T/F**: A per location endorsement in a CGL policy is recommended for public water/wastewater systems because it provides separate products-completed operations aggregate limits for each of your described premises.

57. **T/F**: Any partnerships or joint ventures are automatically included in the definition of an insured in a best-in-class CGL policy as long as the public water and wastewater system is the controlling party in these relationships.

58. **T/F**: Bodily injury or property damage must occur during the policy period and in the coverage territory to trigger coverage in a CGL policy.

59. **T/F**: A claim or lawsuit seeking damages for bodily injury or property damage under an occurrence CGL policy must be submitted or filed during the policy period for coverage to apply.

60. **T/F**: It is common for larger deductibles to require reimbursement by an insured for allocated loss adjustment expenses.

61. **T/F**: A per location aggregate in a CGL policy provides separate general aggregate limits for each location.

62. **T/F**: An insurer's obligation of indemnity in a CGL policy is exceedingly broad and simply requires a possibility, however remote, that the claim is covered.

63. **T/F**: Under the *Four Corners* test, an insurer does not have a duty to defend its insured under a CGL policy if the complaint has causes of action and fact patterns that are potentially covered by the policy.

64. **T/F**: The purchase of an environmental pollution policy is required if you seek third party asbestos coverage from gradual seepage and deterioration as well as cleanup costs, natural resources damages, and civil fines and penalties.

65. **T/F**: The definition of insured contract in a CGL policy does not include tort liability that would be imposed by law in the absence of any contract or agreement.

66. **T/F**: The contractual liability – railroads endorsement in a CGL policy offers the same coverage as a railroad protective liability (RRPL) policy.

67. **T/F**: The scheduling of dam failure coverage in a CGL policy should be paired with inverse condemnation coverage in a POML policy.

68. **T/F**: The scheduling of dam failure coverage in a CGL policy excludes seepage because seepage is deemed a normal and expected outcome of water impoundments and conveyances.

69. The damage to impaired property exclusion in a CGL policy prevents coverage for the following:
    a. Loss of use caused by the assimilation of your defective product
    b. Loss of use caused by your faulty work that is part of the work and done incorrectly
    c. Your failure to finish any work or project on time
    d. **All of the above**

70. The damage to property exclusion in a best-in-class CGL policy includes the following exceptions:
    a. Property damage from fire to premises occupied or rented to you for longer than 30 days
    b. Property damage from water damage to premises and contents rented to you for 30 or fewer consecutive days
    c. Property damage from fire to premises not occupied or rented to you for longer than 30 days
    d. **a and b only**

71. **T/F**: The damage to your work exclusion in a CGL policy necessitates best practices to avoid any work performed by your employees to repair or replace lateral lines that are owned by third parties.

72. The employers' liability exclusion in a CGL policy has an exception for the following:
    a. **An insured employer that is obligated to indemnify another for damages resulting from work-related injuries to the insured employer's own employees, provided the obligation of the insured employer to provide indemnity is due to liability assumed in an insured contract**
    b. Employee who is sued by another employee for negligently causing the other employee's injury
    c. a and b only
    d. None of the above

73. Examples of optional coverages in a CGL policy include the following:
    a. Hired and non-owned auto liability
    b. Employee benefit plans
    c. Stop-gap liability
    d. **All of the above**

74. **T/F**: It is against the fortuity doctrine to cover expected or intended injury by an insured in a CGL policy unless such actions were intended to protect persons or property.

75. **T/F**: Coverage is not provided for liability arising from fungi or bacteria on insured premises in a CGL policy.

76. **T/F**: Your public water and wastewater system can be subject to litigation associated with failure to properly mitigate the effects of lead poisoning.

77. Electromagnetic radiation can create the following health risks:
    a. Radio frequencies
    b. Extremely low frequencies
    c. Static magnetic fields
    d. **All of the above**

78. A best-in-class CGL policy includes the following pollution liability coverages:
    a. Potable water which you supply to others
    b. Chemicals you use in your water treatment process
    c. Natural gas or propane gas you use in your water treatment process
    d. **All of the above**

79. A best-in-class CGL policy includes bodily injury and property damage arising from the professional activities of:
    a. Licensed water operators
    b. Contracted chemists
    c. None of the above
    d. **All of the above**

80. **T/F**: Sexual abuse is uninsurable when a governing body was unaware of such actions.

81. **T/F**: Terrorism coverage is most commonly purchased through the Terrorism Risk Insurance Act (TRIA).

82. **T/F**: Coverage for employee benefit plans includes an insured's fiduciary duties in operating the plan and managing its investments.

83. Examples of coverage that is best addressed via silence and without an explicit exclusion in a CGL policy include the following:
    a. Subsidence
    b. Failure to supply
    c. Silica
    d. **All of the above**

84. Coverages A. and B. in a CGL policy share the following provisions:
    a. Duty to defend
    b. Who is an insured
    c. Supplementary payments
    d. **All of the above**

85. T/F: One important distinction between Coverages A. and B. in a CGL policy is Coverage B. is triggered by physical harm.

86. **T/F**: The definition of insured contract in a CGL policy does not afford liability the insured assumes for personal injury.

87. Infringement of the intellectual property rights of others from your advertisement in a CGL policy is covered for the following:
    a. Copyright
    b. Trade dress
    c. Slogan
    d. **All of the above**

88. **T/F**: Wrongful detention of an individual with knowledge by a named insured is an example of a claim that would be covered under Coverage B. in a CGL policy.

89. **T/F**: Under a duty to defend liability policy, an insurer has the right to pay or settle any claim or lawsuit filed against you.

90. **T/F**: Supplementary payments in a CGL policy allows the insured to choose their own defense counsel.

91. **T/F**: No person or organization is an insured in a CGL policy with respect to the conduct of any current or past partnership, joint venture, or limited liability company that is not listed as an insured on the supplemental declarations.

92. **T/F**: The limits of insurance in a CGL policy increases when a claim or suit names multiple insureds, the incident involves multiple claims or suits, or numerous persons or organizations make a claim or file a suit.

93. **T/F**: The most important limits in a CGL policy are the aggregate limits, which comprise the general aggregate and the products-completed operations aggregate.

94. **T/F**: Exhaustion of an aggregate limit in a CGL policy extinguishes an insurer's duty to defend any subsequent suits that fall within the fully depleted aggregate limit.

95. The separation of insureds provision in a CGL policy provides the following rights:
    a. **Retains rights of a named insured that does not violate any policy provision**
    b. Allows all named insureds the same rights as the first named insured
    c. Provides separate limits of insurance for each named insured
    d. All of the above

22 Public Officials and Management Liability (POML) Policy

2019 Photo Courtesy of Las Virgenes Municipal Water District

LEARNING OBJECTIVES

- Discuss the interaction between your organization's indemnification agreements and its public officials and management liability (POML) policy.
- Understand how to properly structure your public officials and management liability (POML) policy.
- Explain the difference between wrongful acts and other offenses in a public officials and management liability (POML) policy.
- Comprehend the purpose and restrictions of the exclusions in a POML policy.
- Articulate the rational for integrating wrongful acts and employment practices in your POML policy.
- Grasp the importance of inverse condemnation in your public officials and management liability (POML) policy.
- Explain the difference between a POML policy and a commercial general liability (CGL) policy.
- Describe the individuals and organizations qualifying as named insureds in your POML policy.
- Identify the limits of insurance afforded in your POML policy.

- Comprehend the difference between Coverages A. and B. in your POML policy.
- Articulate the duties and obligations of insurer and insured after a loss in a POML policy.
- Discuss the difference between the insurer's obligation of defense and indemnity.

KEY DEFINITIONS*

Allocated Loss Adjustment Expense (ALAE) – loss adjustment expenses that are assignable or allocable to specific claims. Fees paid to outside attorneys, experts, and investigators used to defend claims are examples of ALAE.

Anti-Stacking Provisions – provisions intended to avoid the application of multiple sets of deductibles or multiple sets of limits to a single loss event. Anti-stacking provisions stipulate that, in such an event, only one policy limit or one deductible (rather than the limit or deductible under each policy) applies to the occurrence.

Bodily Injury – bodily injury, sickness, or disease sustained by a person, including death resulting from any of these.

Claim – used in reference to insurance, a claim may be a demand by an individual or organization to recover, under a policy of insurance, for loss that may be covered within that policy.

Claims-Made – policy providing coverage that is triggered when a claim is made against the insured during the policy period, regardless of when the wrongful act that gave rise to the claim took place. The one exception is when a retroactive date is applicable to a claims-made policy. In such instances, the wrongful act that gave rise to the claim must have taken place on or after the retroactive date. A claims-made policy is less preferred to an occurrence policy based on the stringent claims reporting requirements.

Coverage Territory – contractual provisions limiting coverage to geographical areas within which the insurance is affected.

Coverage Trigger – event that must occur before a particular liability policy applies to a given loss. Under an occurrence policy, the occurrence of injury or damage is the trigger; liability will be covered under that policy if the injury or damage occurred during the policy period. Under a claims-made policy, the making of a claim triggers coverage.

Damages – money whose payment a court orders as compensation to an injured plaintiff. Fines, penalties, or injunctive relief would not typically constitute damages.

Defense Clause – insurance provision in which the insurer agrees to defend, with respect to insurance afforded by the policy, all suits against the insured.

Defense Inside Limits – liability policy provision according to which amounts paid by the insurer to defend the insured against a claim or suit reduce the policy's applicable limit of insurance. Commercial general liability (CGL) policies are ordinarily not subject to such a provision. Defense within limits is more common in professional liability policies.

Defense Outside Limits – amounts paid by an insurer to defend claims or suits that are in addition to the policy limits and do not decrease the per occurrence claim limit.

Duty to Defend – term used to describe an insurer's obligation to provide an insured with defense to claims made under a liability insurance policy. As a general rule, an insured need only establish that there is potential for coverage under a policy to give rise to the insurer's duty to defend. Therefore,

---

* Source material provided by International Risk Management Institute, (IRMI), www.irmi.com/term/insurance-definit ions/insurance, as well as policy forms and underwriting products promulgated by Insurance Services Office (ISO) and their member companies, https://products.iso.com/productcenter/10.

the duty to defend may exist even where coverage is in doubt and ultimately does not apply. Implicit in this rule is the principle that an insurer's duty to defend an insured is broader than its duty to indemnify. Moreover, an insurer may owe a duty to defend its insured against a claim in which ultimately no damages are awarded, and any doubt as to whether the facts support a duty to defend is usually resolved in the insured's favor.

Economic Damages – award to an injured person in an amount sufficient to compensate for his or her actual monetary loss.

Eminent Domain – right of a government or its agent to expropriate private property for public use, with payment of compensation.

Employee – any natural person: (a) while in your service; (b) who you compensate directly by salary, wages, or commissions; and (c) who you have the right to direct and control while performing services for you. Employee includes a leased worker but does not include a temporary worker.

Employee Benefit Plans – benefits, such as health and life insurance, provided to employees at the workplace, usually paid for totally or in part by the employer. Liability of an employer for an error or omission in the administration of an employee benefit program, such as failure to advise employees of benefit programs.

Employment Practices – type of liability insurance covering wrongful acts arising from the employment process. The most frequent types of claims covered under such policies include: wrongful termination, discrimination, sexual harassment, and retaliation. These policies also cover claims from a variety of other types of inappropriate workplace conduct, including (but not limited to) employment-related: defamation, invasion of privacy, failure to promote, deprivation of a career opportunity, and negligent evaluation. The policies cover directors and officers, management personnel, and employees as insureds.

Exclusion – policy provision that eliminates coverage for some type of risk. Exclusions narrow the scope of coverage provided by the insuring agreement. In many insurance policies, the insuring agreement is very broad. Insurers utilize exclusions to carve away coverage for risks they are unwilling to insure.

Extended Reporting Period (ERP) – designated time period after a claims-made policy has expired during which a claim may be made and coverage triggered as if the claim had been made during the policy period.

Fiduciary Liability – responsibility on trustees, employers, fiduciaries, professional administrators, and the plan itself with respect to errors and omissions (E&O) in the administration of employee benefit programs as imposed by the Employee Retirement Income Security Act (ERISA).

Governing Body – group of elected or appointed people who formulate the policy and direct the affairs of a public water and wastewater system.

Hammer Clause – provision that allows an insurer to compel an insured to settle a claim. A hammer clause is also known as a blackmail clause, settlement cap provision, or consent to settlement provision.

Indemnity – compensation to a party for a loss or damage that has already occurred, or to guarantee through a contractual clause to repay another party for loss or damage that might occur in the future. The concept of indemnity is based on a contractual agreement made between two parties in which one party (the indemnitor) agrees to pay for potential losses or damages caused by the other party (the indemnitee).

Injunctive Relief – legal alternative to monetary damages in a civil suit involving a court ordering a party to take an affirmative action or restraining a party from taking a particular action. This measure is the appropriate remedy in any situation in which, if the defendant is not ordered to cease performing an action, the plaintiff will be unable to be properly compensated. Liability insurance policies typically provide coverage only for damages and not for injunctive relief.

Insured – any person or organization qualifying as an insured in the "who is an insured" provision of the applicable coverage. Except with respect to the limit of insurance, the coverage afforded applies separately to each insured who is seeking coverage or against whom a claim or suit is brought.

Insuring Agreement – portion of the insurance policy in which the insurer promises to make payment to or on behalf of the insured. The insuring agreement is usually contained in a coverage form from which a policy is constructed. Often, insuring agreements outline a broad scope of coverage, which is then narrowed by exclusions and definitions.

Inverse Condemnation – situation in which the government takes private property (or damages private property in California) but fails to pay just compensation as required by the federal and state constitutions.

Judgment – decision of a court regarding the rights and liabilities of parties in a legal action or proceeding. Judgments also generally provide the court's explanation of why it has chosen to make a particular court order.

Leased Worker – person leased to you by a labor leasing firm under an agreement between you and the labor leasing firm to perform duties related to the conduct of your business. Leased worker does not include a temporary worker.

Limits of Insurance – the most that will be paid by an insurer in the event of a covered loss under an insurance policy.

Loss – direct and accidental loss or damage, or a reduction in value.

Manager – person serving in a directorial capacity for a limited liability company.

Member – owner of a limited liability company represented by its membership interest, who also may serve as a manager.

Policy Conditions – section of an insurance policy that identifies general requirements of an insured and the insurer on matters such as loss reporting and settlement, property valuation, other insurance, subrogation rights, and cancellation and nonrenewal. The policy conditions are usually stipulated in the coverage form of the insurance policy.

Policy Period – term of duration of the policy. The policy period encompasses the time between the exact hour and date of policy inception and the hour and date of expiration.

Pollutants – any solid, liquid, gaseous, or thermal irritant or contaminant, including smoke, vapor, soot, fumes, acids, alkalis, chemicals, and waste. Waste includes materials to be recycled, reconditioned, or reclaimed.

Professional Liability – type of liability coverage designed to protect traditional professionals (e.g., accountants, attorneys) and quasi-professionals (e.g., real estate brokers, consultants) against liability incurred as a result of errors and omissions in performing their professional services. Most professional liability policies only cover economic or financial losses suffered by third parties, as opposed to bodily injury and property damage claims.

Property Damage – damage to or loss of use of tangible property.

Public Officials Liability – liability exposure faced by a public official from wrongful acts, usually defined under public officials and management liability (POML) policies as actual or alleged errors, omissions, misstatements, negligence, or breach of duty in his or her capacity as a public official or employee of the public entity.

Representation – statement made in an application for insurance that the prospective insured represents as being correct to the best of his or her knowledge. If the insurer relies on a representation in entering into the insurance contract and if it proves to be false at the time it was made, the insurer may have legal grounds to void the contract.

Reservation of Rights – an insurer's notification to an insured that coverage for a claim may not apply. Such notification allows an insurer to investigate (or even defend) a claim to determine whether coverage applies (in whole or in part) without waiving its right to later deny coverage based on information revealed by the investigation. Insurers use a reservation of rights letter because in many claim situations, all the insurer has at the inception of the claim are various unsubstantiated allegations and, at best, a few confirmed facts.

Retroactive Date – a provision found in many (although not all) claims-made policies that eliminates coverage for claims produced by wrongful acts that took place prior to a specified date, even if the claim is first made during the policy period.

Settlement – resolution between disputing parties about a legal case, reached either before or after court action begins.

Sexual Harassment – conduct involving unwelcome sexual advances, requests for sexual favors, and verbal, visual, or physical conduct of a sexual nature. There are two types of sexual harassment: quid pro quo sexual harassment, in which sexual contact is made a condition of employment, and hostile environment sexual harassment, in which such conduct creates an intimidating, hostile, or offensive working environment.

Stacking of Limits – the application of two or more policies' limits to a single occurrence or claim. This is common with product liability, construction defect, and pollution claims in which the occurrence has transpired over numerous years, and it is difficult to ascertain which policy provides coverage. It can also occur under auto liability or uninsured motorists/underinsured motorists coverage when two or more vehicle limits can be stacked to apply to a single occurrence.

Suit – a civil proceeding in which: (1) damages because of bodily injury or property damage; or (2) a covered pollution cost or expense; to which this insurance applies, are alleged. Suit includes: (a) an arbitration proceeding in which such damages or covered pollution costs or expenses are claimed and to which the insured must submit or does submit with our consent; or (b) any other alternative dispute resolution proceeding in which such damages or covered pollution costs or expenses are claimed and to which the insured submits with our consent.

Supplementary Payments – term used in liability policies for the costs associated with the investigation and resolution of claims. Supplementary payments are normally defined to include such items as first aid expenses, premiums for appeal and bail bonds, pre- and post-judgment interest, and reasonable travel expenses incurred by the insured at the insurer's request when assisting in the defense of a claim. Actual settlements/judgments are considered damages rather than supplementary payments.

Temporary Worker – person who is furnished to you to substitute for a permanent employee on leave or to meet seasonal or short-term workload conditions.

Third-Party Discrimination – coverage that applies to liability claims brought by nonemployees (typically, customers, clients, and vendors) against employees of the insured organization as it relates to sexual harassment and discrimination.

Tort or Governmental Immunities – describes the various doctrines or statutes that provide federal, state, or local governments immunity from tort-based claims.

Torts – a civil or private wrong giving rise to legal liability.

Vicarious Liability – the liability of a principal for the acts of its agents; can result from the acts of independent agents, partners, independent contractors, employees, and children.

Wage and Hour Indemnity – a type of insurance that covers indemnity costs (i.e., settlements and judgments) resulting from wage and hour claims. The two major categories of wage and hour claims are those alleging: (1) failure to pay overtime to nonexempt employees who are not exempt from, and thus eligible for, overtime pay; and (2) miscellaneous pay practices claims, which include but are not limited to misclassifying employees as independent contractors, failure to grant rest and meal breaks, and failure to pay wages when due.

Water and Wastewater Professional Activities – your activities as a water or wastewater district, water utility, or as any other entity whose primary duty is the treatment and distribution of potable water, or the collection, treatment, and distribution of wastewater.

Wrongful Act – any actual or alleged error, act, omission, neglect, misfeasance, nonfeasance, or breach of duty, including violation of any civil rights law, by any insured in the discharge of their duties for the named insured, individually or collectively, that results directly but unexpectedly and unintentionally in damages to others.

PUBLIC OFFICIALS AND MANAGEMENT LIABILITY POLICY*

OVERVIEW

Public officials and management liability policies are specifically designed to protect governing bodies from alleged wrongful acts and other offenses. Coverage also extends to the organization, as claims against governing bodies are frequently subject to indemnification agreements. POML policies are analogous to Directors & Officers Liability (D&O) policies in that both policies afford personal asset protection to governing bodies and extend coverage to their respective organizations. There is one fundamental distinction between the two policies, and that difference centers on organizational configuration. D&O policies are intended for investor-owned water and wastewater utilities that are publicly traded. This structure is notable, as publicly traded companies have unique exposures compared to public entities and nonprofit corporations. These exposures are specifically addressed within D&O policies and include shareholder derivative suits, Securities and Exchange (SEC) violations, and corporate indemnification. We will hereinafter refer to POML and D&O policies as simply POML policies since the vast majority of public water/wastewater systems are configured as public entities, nonprofit corporations, or non-publicly traded investor-owned water and wastewater utilities. Both policies are designed to protect governing bodies and their organizations, but our emphasis will not include the unique D&O exposures facing publicly traded companies.

The duties of your governing body are expansive and comprise significant liability for elected and appointed individuals. The governing body is responsible for overseeing and discharging the powers, policies, and decisions of your public water and wastewater system. As such, they are exposed to complex and overlapping federal and state torts as well as constitutional violations and an immeasurable ensemble of employment-related offenses. The resulting litigation is capacious, intricate, and pervasive. A best-in-class POML policy is like virtual personal armor that insulates your governing body from alleged wrongful acts or other offenses comprising malfeasance, misfeasance, breaches of fiduciary duty, and personnel-related matters. It is this combination of sweeping authority and unconstrained litigation that necessitates the purchase of a properly designed POML

---

* Source material provided by various online articles from International Risk Management Institute, (IRMI), www.irmi. com/articles/expert-commentary/insurance as well as policy forms and underwriting products promulgated by Insurance Services Office, (ISO) and their member companies, https://products.iso.com/productcenter/10.

policy to protect your governing body when acting in good faith and for the betterment of your public water and wastewater system.

The POML policy also supplements and monetizes the indemnification agreements enacted by your public water and wastewater system for the benefit of your governing body. Indemnification agreements are necessary because, despite partial tort immunities, your governing body can be held personally liable for alleged wrongful acts or other offenses in their official capacity as a director, officer, trustee, or executive. Indemnification agreements are intended to protect individuals who occupy these positions as long as their alleged wrongful acts or other offenses are not adjudicated to show civil fraud, personal enrichment, criminal acts, or intentional breaches of fiduciary duty. A best-in-class POML policy will transfer most of this financial liability from your public water and wastewater system to your insurer. The transfer is subject, of course, to policy terms and conditions, and it is the "most but not all" qualifier of one's financial liability that underscores the necessity of indemnification agreements. Litigation against governing bodies for their failure to procure or maintain insurance is one such example. This universal POML exclusion creates a material omission in protection, which subsequently imposes immeasurable personal liability for your governing body that can only be addressed through an indemnification agreement.

An expansive indemnification agreement, coupled with a best-in-class POML policy, encourages a diverse and adept grouping of civic, business, and government-minded individuals to serve on your governing body. These agreements act as a final personal layer of protective insulation for your governing body should litigation not be covered under the POML policy. That extra fortification is important because serving on your governing body is intensely time consuming and encompasses complex decisions and significant personal liability. The financial safeguards described above are exceedingly appropriate and will accentuate the quality and continuity of your governing body. Such precautions correlate back to a professionally managed public water and wastewater system.

Unlike a commercial general liability (CGL) policy, a POML policy is based solely on economic injury arising from the powers, policies, and decisions of your governing body in the oversight of your public water and wastewater system. The policy applies to wrongful acts as well as offenses involving employment practices and employee benefit plans. Claims comprising bodily injury and property damage fall outside the purview of the POML policy but are contemplated within a CGL policy. The POML policy starts with a broad promise, which is the insuring agreement. The insurer pledges to offer defense (i.e., legal representation) and pay indemnity (i.e., compensation) up to the respective limits of insurance if you are legally liable for damages because of economic injury arising from a wrongful act or other offense subject to policy exclusions.

POML policies for public water/wastewater systems should be written on an occurrence form. That means covered wrongful acts or other offenses simply need to occur during the policy period and in the coverage territory to trigger coverage. The latter point is inapplicable since the United States is considered part of the coverage territory. There is similarly no requirement for damages to be submitted or filed during the policy period. Such reporting can occur months or years after the wrongful act or offense and does not require an active POML policy or an extended reporting period (ERP) to comply with the policy's claims reporting requirements. That feature underscores an inherent advantage of an occurrence form compared to a claims-made form.

It is not advisable to secure a POML policy on a claims-made form. This type of form has strict claims reporting requirements, and it exposes insureds to marketplace vagaries by relying on the insurance industry to offer renewal terms consistent with past policies. The capriciousness and vulnerability never end because of the policy trigger. To trigger or activate a claims-made POML policy, the wrongful act or offense as well as the filing of a claim seeking damages, including its subsequent reporting to the insurer, must occur on or after the retroactive date and before policy expiration. An ERP is available at an additional premium to extend the time period to report claims that occurred within the aforementioned brackets.

The purchase of an ERP is recommended when you nonrenew your claims-made policy or receive renewal terms that are substantially more restrictive than past policies. As previously stated, there is no coverage should your claims-made policy be nonrenewed or your renewal terms no longer cover that particular claim scenario. The preceding drawback signifies the fundamental flaw of a claims-made form. It requires continual renewals to ensure coverage for your prior acts, and it requires marketplace reliance that coverage secured in the present will continue to be offered in the future. An occurrence form removes these material impediments and should be actively pursued over a claims-made form.

An equally important component within the POML policy is the duty to defend provision within the insuring agreement. Similar to a CGL policy, this provision obligates the insurer to provide defense for any allegations that could *potentially* trigger coverage. The qualifier of *potentially* is needed because the broad insuring agreement is narrowed by the policy's exclusions. These exclusions, which we will examine in detail, are foreshadowed by the following phrase in the insuring agreement "to which this insurance applies." In other words, the policy does not apply to all damages but rather only to damages to which this insurance applies. The obligation of defense is much broader than the obligation of indemnity. The former simply needs to include the possibility of coverage to trigger the obligation, whereas the latter requires an affirmation that coverage exists.

The difference is logical, as an insurer cannot divine the outcome of litigation from various causes of actions that are asserted. Insurers must offer the obligation of defense and can withhold the obligation of indemnity until there is a rendering of judgment or settlement on the facts or causes of action alleged. The ultimate determination of indemnity is either a jury decision or a preemptive settlement by the insurer, and that happens after the defense obligation. Incidentally, most wrongful acts and other offenses are concluded with a preemptive settlement and do not go to trial. This reality reaffirms the importance of the duty to defend obligation, as it frequently ensures a claim under a reservation of rights will trigger the indemnity obligation. It is also common for the defense obligation to be unlimited, which means the costs associated with defense do not erode the limits of insurance. The parlance to offer unlimited defense is known as defense costs outside the limits of insurance. The word "unlimited" does has an ending point, and that is when the indemnity payments exhaust the limits of insurance. Do not purchase a POML policy that has defense costs inside the limits of insurance and avoid a claims-made form if at all possible.

Deductibles within a POML policy vary, as there are two options. The first is when allocated loss adjustment expenses (ALAE) are within the deductible. This option means your deductible will apply to expenses like field adjusters, expert witnesses, legal representation, and other expenses that correlate back to the claim. Your deductible will also, of course, apply to any indemnity payment on the claim. The second is when ALAE is outside the deductible. This option is preferred because you are not responsible for any expenses correlating back to the claim except for indemnity payments. It is common for smaller deductibles to have ALAE outside and larger deductibles to have ALAE inside. The rationale is simple: larger deductibles equate to more credits because of more risk assumption by the insured. Hence, the cost of ALAE as well as any indemnity should be paid by the insured up to their deductible amount. The alignment of expectation between insurer and insured with larger deductibles warrant such an approach. It is also important to avoid a deductible that applies per claim versus one that applies per act or offense. The difference can be substantial if there are multiple claimants involved in the same action against you.

The selection of limits is equally important, as that will define the indemnity obligation of the insurer. Its exhaustion will also end the insurer's defense obligation. The preferred limit within a best-in-class POML policy is $1 million per act or offense and $10 million for the policy aggregate. A per occurrence limit is the maximum amount for any one claim or multiple claims involving the same act or offense, whereas the aggregate limit allows the per act or offense limit to replenish and respond to other acts or offenses up to the aggregate limit shown on the supplemental declarations. The preferred limit structure drill down is as follows: $10 million general aggregate (i.e., wrongful acts, employment-related practices, and employee benefit plans); and $1 million each act or offense. The $10 million aggregate is especially important since best-in-class POML policies typically include three separate coverage parts: wrongful acts; employment-related practices; and employee benefit plans. This approach ensures you will have ten separate $1 million limits for different acts or offenses that occur during the policy period.

Like all insurance policies, details matter. The quality of your POML policy is determined by these particulars, which means policy value is conditional on understanding policy nuances. The best way to achieve practical comprehension is to granularly examine and deconstruct a best-in-class POML policy specifically designed for public water/wastewater systems. We will divide this exercise into six sections:

1. Coverage A. Liability for Monetary Damages
2. Coverage B. Defense Expenses for Injunctive Relief
3. Coverage A. and B Exclusions
4. Who Is an Insured
5. Limits of Insurance
6. Key Policy Conditions

Coverages A. and B. have their own insuring agreement, which we will review. Coverage A also has its own supplementary payments, which we will similarly assess. There is an important distinction between the two coverage sections that must be recognized. Coverage A. has a duty to defend obligation as well as an indemnity obligation up to the policy limits, whereas Coverage B. is constrained to a reimbursement mechanism comprising an ancillary sublimit. Our review will be more detailed for Coverage A. because that coverage encapsulates the intent and rationale for purchasing the POML policy. It also signifies the majority of wrongful acts and other offenses filed against public water/wastewater systems.

We will extend the examination to comprise optional coverages that should be integrated within Coverage A. Examples include third-party discrimination, wage and hour, and inverse condemnation. Our review will also identify mandatory coverages that should be automatically embedded in Coverage A. The same scrutiny will apply to policy exclusions to ensure you are properly aware of which exclusions are acceptable and which require more analysis and discussion with your insurer. Important exceptions to certain exclusions will similarly be imparted. The rest of the sections provide context for Coverages A. and B. These include the qualification for being an insured, mechanics behind the limits of insurance, and required foreknowledge of key policy conditions. Our learning approach will include a reading from a best-in-class POML policy and commentary to properly structure this policy by examining the aforementioned sections. We will now begin our review with an examination of Coverage A. Liability for Monetary Damages.

COVERAGE A: LIABILITY FOR MONETARY DAMAGES

INSURING AGREEMENT

We will pay those sums that the insured becomes legally obligated to pay as damages arising out of a claim for: (a) a wrongful act; or (b) an employment practices offense; or (c) an offense in the administration of your employee benefit plans to which this insurance applies. We will have the right and duty to defend any claim seeking those damages. However, we will have no duty to defend any insured against any claim seeking damages for a wrongful act or an employment practices offense or an offense in the administration of your employee benefit plans to which this insurance does not apply.

We may, at our discretion, investigate any wrongful act, employment practices offense, or an offense in the administration of your employee benefit plans, and settle any claim that may result. However, the amount we will pay for damages is limited as described in the Limits of Insurance section of the policy. Our right and duty to defend ends when we have used up the applicable limit of insurance in the payment of damages or defense expenses under Coverages A and B. No other obligation or liability to pay sums or perform acts or services is covered under this insurance unless explicitly provided for below under Coverage A. Supplementary Payments. This insurance applies to claims for wrongful acts or offenses only if: (a) the wrongful act or offense takes place in the coverage territory and during the policy period; and (b) prior to the policy period, no insured listed in the "Who Is an Insured" section of the policy and no insured authorized by you to give or receive notice of a wrongful act, offense, or claim, knew, that the wrongful act or offense had taken place, in whole or in part. If such a listed insured knew prior to the policy period that the wrongful act or offense had taken place, then any continuation, change, or resumption of such wrongful act or offense during or after the policy period will be deemed to have been known prior to the policy period.

Wrongful acts or offenses will be deemed to have been known to have taken place at the earliest time when any insured listed in the "Who Is an Insured" section of the policy or any insured authorized by you to give or receive notice of a wrongful act, offense, or claim: (a) reports all, or any part, of the wrongful act or offense to any other insurer; (b) receives a written or verbal demand or claim for damages because of the wrongful act or offense; or (c) becomes aware, by any other means, that the wrongful act or offense has taken place or has begun to take place. All damages from all claims based on or arising out of the same or related wrongful acts or offenses, or the same act or interrelated acts of one or more insureds, regardless of the number of: (a) insureds; (b) plaintiffs; or (c) claims made; shall be subject to one each wrongful act or each offense limit of insurance, and one deductible. Related wrongful acts or offenses shall include wrongful acts or offenses which are the same, related, or continuous, or which arise from a common nucleus of facts, and shall be deemed to have taken place at the time of the first such wrongful act or offense.

Key Definitions

Administration – any of the following acts that you do or authorize a person to do: (a) counseling employees or volunteer workers, other than giving legal advice, on employee benefit plans; (b) interpreting your employee benefit plans; (c) handling records for your employee benefit plans; and (d) effecting enrollment, termination, or cancellation of employees or volunteer workers under your employee benefit plans.

Claim – (a) written notice, from any party, that it is their intention to hold the insured responsible for damages arising out of a wrongful act or offense by the insured; (b) a civil proceeding in which damages arising out of an offense or wrongful act to which this insurance applies are alleged; (c) an arbitration proceeding in which damages arising out of an offense or wrongful act to which this insurance applies are claimed and to which the insured must submit or does submit with our consent; (d) any other civil alternative dispute resolution proceeding in which damages arising out of an offense or wrongful act to which this insurance applies are claimed and to which the insured submits with our consent; or (e) a formal proceeding or investigation with the Equal Employment Opportunity Commission, or with an equivalent state or local agency. A claim does not mean any ethical conduct review or enforcement action or disciplinary review or enforcement action.

Coverage Territory – the United States of America (including its territories and possessions), Puerto Rico, and Canada.

Damages – monetary damages.

Employee Benefit Plans – group life insurance, group accident or health insurance, profit-sharing plans, pension plans, employee stock subscription plans, employee travel, vacation, or savings plans, workers compensation, unemployment insurance, social security and disability benefits insurance, and any other similar benefit program applying to employees or volunteer workers.

Employment Practices – the following, when alleged by a volunteer worker, employee, or applicant for employment, in connection with that person's actual or proposed employment relationship with the named insured: any actual or alleged improper employment-related act, error, or omission, including: (a) failing to hire or refusing to hire; (b) wrongful dismissal, discharge, or termination of employment or membership, whether actual or constructive; (c) wrongful deprivation of a career opportunity, or failure to promote; (d) wrongful discipline or demotion of volunteer workers or employees; (e) negligent evaluation of volunteer workers or employees; (f) retaliation against volunteer workers or employees for the exercise of any legally protected right or for engaging in any legally protected activity; (g) failure to adopt adequate workplace or employment-related policies and procedures, or the breach of any manual or employment-related policies of procedures; (h) harassment, including sexual harassment; (i) libel, slander, defamation, or invasion of privacy; or (j) violation of any federal, state, or local laws (whether common law or statutory) concerning discrimination in employment.

Policy Period – the term of duration of the policy shown in the supplemental declarations.

Wrongful Act – any actual or alleged error, act, omission, neglect, misfeasance, nonfeasance, or breach of duty, including violation of any civil rights law, by any named insured in the discharge of their duties for the named insured, individually or collectively, that results directly but unexpectedly and unintentionally in damages to others. The key phraseology in this definition involves damages that result directly but unexpectedly and unintentionally to others. The verbiage confirms a POML policy, even one that is best in class, will not respond to claims involving damages that are asserted to be expected and intentional by the governing body.

Sexual Harassment – any actual, attempted, or alleged unwelcome sexual advances, requests for sexual favors, or other conduct of a sexual nature by a person, or by persons acting in concert, which causes injury, but only when: (a) submission to or rejection of such conduct is made either explicitly or implicitly a condition of a person's employment, or a basis for employment decisions affecting a person; or (b) such conduct has the purpose or effect of interfering with a person's work performance or creating an intimidating, hostile or offensive work environment.

SPECIAL COMMENTARY

The insuring agreement starts with a broad promise that is substantially narrowed by the policy exclusions. It confirms the previously discussed defense and indemnity obligations and stipulates the ending of said obligations upon termination of the limits of insurance. The insurer specifies its obligations in the insuring agreement. These obligations also extend to the supplementary payments as well as to those individuals and organizations qualifying as an insured. Moreover, the insurer confirms it has the right to settle any claim or demand made against an insured to which this policy applies. We call this provision a hammer clause because the insurer has unilateral rights to conclude the matter with a preemptive settlement to the plaintiff.

Hammer clauses can be relaxed, but any softening transfers the defense and indemnity costs to the public water and wastewater system from the point where a settlement could have been reached by the insurer. Consequently, it is imprudent to activate its softening unless the alleged complaint is so egregious to the public water and wastewater system and its governing body's reputation that a forceful and absolute defense without settlement is warranted. The financial downside of such a decision is significant and should only be exercised for the most contemptible and malevolent of allegations.

Coverage A should include coverage for both wrongful acts and employment practice offenses. The inclusion for offenses involving the administration of your employee benefit plans is also important but said coverage can be included in your CGL policy. Irrespective of placement, coverage for employee benefit plans is essential and should be confirmed. It is also possible to secure your employment practices coverage via a stand-alone policy. An outside placement is acceptable, but the integration of wrongful acts and employment practices offenses within your POML policy is preferred because claims involving overlapping allegations can be consolidated into an integrated defense by one insurer. This approach removes finger pointing by different insurers as well as arguments on defense strategy and counsel appointments. The former prevents unnecessary and stressful delays in defense obligations as differing insurers attempt to resolve their obligation disputes. The latter removes disjointed legal representation where more emphasis is centered on causes of action impacting a particular line of coverage rather than a holistic strategy that evaluates the complaint, and all its causes of actions, in its entirety.

The remaining portion of the insuring agreement stipulates the obligations of the POML policy are limited to coverage grants codified in the insuring agreement as well as grants contained in the supplementary payments. It also articulates similar verbiage as a CGL policy that continuation of the same claim over multiple policy periods ceases the moment a qualified person of the named insured becomes aware of the wrongful act or employment practices offense or any offense in the administration of your employee benefit plans. This stipulation is critically important, as it can have an acute effect on claims-made forms where the act or offense and its reporting to the insurer must occur on or after the retroactive date and before the expiration of the policy. In contrast, occurrence forms only require the wrongful act or offense occur in the policy period. It can be reported to the insurer after the policy expires. The verbiage on continuation of a claim over multiple policy periods is similarly important for occurrence forms because it cuts off subsequent policy periods that will apply to the act or offense. The parlance to trigger multiple policy periods based on the continuation of the same claim from which the insured had no knowledge is called stacking of limits. Your knowledge of any continuation of the same claim over multiple policy periods will be the deciding factor if you can trigger your current POML or if you must tender the claim to a prior policy period. The insuring agreement encapsulates the essence of the coverage and should be thoroughly understood. It is now time to review the supplementary payments and their impact on Coverage A.

SUPPLEMENTARY PAYMENTS

We will pay, with respect to any claim we investigate, settle, or defend:

(1) All expenses we incur. (2) The cost of bonds to release attachments, but only for bond amounts within the applicable limits of insurance. We do not have to furnish these bonds. (3) All reasonable expenses incurred by the insured at our request to assist us in the investigation or defense of the claim, including actual loss of earnings up to $500 a day because of time off from work. (4) All court costs taxed against the insured in the claim. However, these payments do not include attorneys' fees or attorneys' expenses taxed against the insured. (5) Prejudgment interest awarded against the insured on that part of the judgment we pay. If we make an offer to pay the applicable limits of insurance, then we will not pay any prejudgment interest based on that period of time after the offer. (6) All interest earned on that part of any judgment within our limits of insurance after entry of the judgment and before we have paid, offered to pay, or deposited in court the part of the judgment that is within the applicable limits of insurance. These payments will not reduce the limits of insurance.

Special Commentary

There are three important points expressed in this section. The first point is confirmation that the insurer owes a duty to defend, and that duty does not impact the limits of insurance. In other words, all expenses incurred by the insurer to defend a claim are paid directly by the insurer and do not erode the limits of insurance. The duty to defend is not unlimited and ends when the limits of insurance have been exhausted. The second point is any attorney fees or expenses attached to an adverse judgment against your governing body are not covered under supplementary payments. Instead, these fees or expenses are applied against your limits of insurance. That restriction is standard within the industry and should facilitate discussion with your governing body to evaluate the cost of purchasing an excess liability policy that specifically includes your POML policy, and all its coverages, as a scheduled underlying policy. The third point is confirmation that interest paid for the appeal of an affirmed adverse judgment is covered under the supplementary payments and does not erode your limits of insurance. Standard interest rates imposed by courts can reach 10%, which accentuates the point's importance.

We now have concluded our review of Coverage A. Liability for Monetary Damages and will transition to the scope of Coverage. B. Defense Expenses for Injunctive Relief.

COVERAGE B: DEFENSE EXPENSES FOR INJUNCTIVE RELIEF

Insuring Agreement

(1) We will pay those reasonable sums the insured incurs as defense expenses to defend against an action for injunctive relief because of a wrongful act, an employment practices offense, or an offense in the administration of your employee benefit plans to which this insurance applies. However: (a) the amount we will pay for defense expenses is limited as described in the Limits of Insurance section of the policy; and (b) we have no obligation to arrange for or provide the defense for any action for injunctive relief. No other obligation or liability to pay sums or perform acts or services is covered. (2) This insurance applies only if: (a) the action seeking injunctive relief is brought in a legally authorized court or agency of the United States, any of its states or commonwealths, or any governmental subdivision of any of them; (b) such action is filed during the policy period; and (c) the insured: (1) first notifies us as soon as practicable after retaining counsel to respond to such action but in no case later than 60 days after the end of the policy period; and (2) is reasonably expedient in requesting us to reimburse any defense expenses incurred. (3) Related wrongful acts or offenses shall include wrongful acts or offenses which are the same, related, or continuous, or which arise from a common nucleus of facts, and shall be deemed to have taken place at the time of the first such wrongful act or offense. All defense expenses from all actions for injunctive relief based on or arising out of the same or related wrongful acts or offenses, or the same act or interrelated acts of one or more insureds, regardless of the number of: (a) insureds; (b) plaintiffs; (c) demands asserted or actions brought; or (d) injunctions, temporary restraining orders or prohibitive writs; shall be subject to one each action for injunctive relief limit of insurance.

Key Definitions

Defense Expenses – reasonable and necessary fees or expenses incurred by or on behalf of the insured for: (a) legal fees charged by the insured's attorney; (b) court costs; (c) expert witnesses; and (d) the cost of court bonds, but we do not have to furnish these bonds. Defense expenses do not include: (1) any salaries, charges, or fees for any insured, insured's volunteer workers or employees, or former volunteer workers or employees; or (2) any expenses other than (a), (b), (c), and (d) above.

Injunctive Relief – equitable relief sought through a demand for the issuance of a permanent, preliminary or temporary injunction, restraining order, or similar prohibitive writ against an insured, or order for specific performance by an insured.

Special Commentary

Coverage B. Defense Expenses for Injunctive Relief is ancillary to the protection offered under Coverage A. Liability for Monetary Damages. The reason is simple: Coverage B. is limited to the reimbursement of legal expenses associated with nonmonetary claims. This reimbursement is frequently set between $5,000 and $50,000. Its purpose is to provide some level of budgetary relief for nonmonetary claims filed against your public water and wastewater system where the impetus is to halt a project or redirect an initiative by your governing body. Public water/wastewater systems are frequently entangled in actions by environmental groups and concerned citizens who advocate for environmental preservation as well as habitat protection. Said actions can also be initiated by ratepayers opposing a public improvement in their neighborhood. The foregoing actions can be expensive but fall outside the true scope of coverage intended by a POML policy. The reason is simple: there is no monetary demand behind these actions but rather a demand to cease some project, operation, or initiative. As such, they are deemed business risks that cannot be settled with monetary payment. Notwithstanding, some level of reimbursement for legal fees expended on said litigation is recommended. Coverage B. also has tight reporting requirements of these expenses, which contraposes the normal flexibility of an occurrence form. All expenses associated with Coverage B. must be submitted within 60 days after the end of your policy period. This time limitation as it relates to claims reporting should be clarified to only apply if you select another insurer at renewal. In other words, there should be no 60-day restriction if you renew your policy with the same insurer.

Our review of Coverages A. and B. now transitions to the policy exclusions, which greatly narrows the scope of coverage afforded by the two insuring agreements. We will review each exclusion and provide commentary on their appropriateness. Recommended exceptions will be conveyed for any exclusion that materially impacts your rights under the policy. It is important to review your current policy to ensure there are no additional exclusions to what is presented below as well as confirmation your exclusions have the same exceptions as the policy we are reviewing.

COVERAGES A AND B: EXCLUSIONS

This insurance does not apply under either Coverage A or Coverage B to:

Other Applicable Coverage

Exclusion: Any wrongful act or offense which is insured by any other policy or policies except: (a) a policy purchased to apply in excess of this coverage form; or (b) that portion of damages otherwise covered by this coverage form which exceeds the limits of liability of such other policy or policies, subject to the Other Insurance condition in this policy.

Commentary: This industry standard exclusion is acceptable and doesn't impact the scope or intent of your policy. Its purpose is to tender an otherwise covered claim to another policy specifically designed to cover the alleged wrongful act or offense.

Asbestos

Exclusion: Any injury, damages, defense expenses, costs, loss, liability, or legal obligation arising out of or in any way related to asbestos or asbestos-containing materials or exposure thereto, or for the costs of abatement, mitigation, removal, elimination, or disposal of any of them.

Commentary: This industry standard exclusion is acceptable and doesn't impact the scope or intent of your policy. The asbestos exposure is appropriately addressed in your CGL policy and necessitates confirmation that coverage applies for bodily injury arising from asbestos in your potable water.

Attorney's Fees and Court Costs

Exclusion: Any award of court costs or attorney's fees which arises out of an action for injunctive relief.

Commentary: This industry standard exclusion is acceptable and doesn't impact the scope or intent of your policy because Coverage B. is capped at a smaller limit than Coverage A. and predicated on the reimbursement of your legal expenses. The reimbursement of nonmonetary legal defense fees is deemed an accommodation to what is otherwise considered a normal business risk. Coverage A. covers monetary damages and the attachment of legal fees for such claims.

Bodily Injury, Property Damage, or Personal and Advertising Injury

Exclusion: Damages, defense expenses, costs, or loss, based upon, attributed to, arising out of, in consequence of, or in any way related to bodily injury, property damage, or personal and advertising injury, other than mental anguish or mental injury resulting from a covered employment practices offense. Property Damage means: (a) physical injury to tangible property, including all resulting loss of use of that property; and (b) loss of use of tangible property that is not physically injured but results from (a) above.

Commentary: This industry standard exclusion is acceptable and doesn't impact the scope or intent of your policy. The aforementioned exposures are appropriately addressed and specifically covered in your CGL policy. Notwithstanding, it necessitates confirmation that inverse condemnation claims will be deemed economic injury and not subject to the property damage exclusion should physical injury to private property be alleged. Please refer to the article on inverse condemnation at the end of *Part IV – Insurance Essentials* for additional details on this legal theory.

Bonds

Exclusion: Any obligation related to a fidelity bond or a surety bond.

Commentary: This industry standard exclusion is acceptable and doesn't impact the scope or intent of your policy. Coverage for this exposure can be procured and appropriately addressed through one of the foregoing bonds.

Claims against Other Insureds

Exclusion: Any actions for injunctive relief or claims brought: (a) by a named insured against any other insured; or (b) by one named insured against another named insured.

Commentary: This industry standard exclusion is acceptable and doesn't impact the scope or intent of your policy. Coverage is not contemplated for litigation initiated by one insured against another insured. Such scenarios of governing body in fighting or disagreements are not deemed insurable. Coverage, however, should be affirmed if an insured sues the governing body in his or her capacity as a nonaffiliated third party or ratepayer. An example would be an inverse condemnation claim filed by a director or shareholder because his or her real property was damaged by an inherent risk in your water or wastewater delivery system. A simple confirmation by your insurer of this overt point is necessary to avoid misunderstandings or a misapplication of the exclusion.

Compliance with ADA Requirements

Exclusion: Damages, defense expenses, costs, or loss incurred as a result of physical modifications made to accommodate persons with disabilities as required by: (a) the Americans with Disabilities Act of 1990; or (b) any federal, state, or local disability discrimination or accommodation laws or regulations; including subsequent amendments or any regulations promulgated thereunder.

Commentary: This industry standard exclusion is acceptable and doesn't impact the scope or intent of your policy. Coverage is not contemplated to reimburse you for physical modifications of your facilities to comply with any law or regulation.

Sums Due under Contract

Exclusion: Any amount actually or allegedly due under the terms of any contract for the purchase of goods or services or any payment or performance contract.

Commentary: This industry standard exclusion is acceptable and doesn't impact the scope or intent of your policy. Coverage is not contemplated for noncompliance of any business contract. Such exposures are deemed a business risk and uninsurable. The definition of wrongful act excludes damages to others that are expected and intentional by the governing body.

Contractual Liability

Exclusion: Damages, defense expenses, costs, or loss based upon, attributed to, arising out of, in consequence of, or in any way related to any contract or agreement to which the insured is a party or a third-party beneficiary, including, but not limited to, any representations made in anticipation of a contract or any interference with the performance of a contract.

Commentary: This industry standard exclusion is acceptable and doesn't impact the scope or intent of your policy. Similar to the Sums Due Under Contract exclusion, coverage is not contemplated for noncompliance or interference of any business contract. Such exposures are deemed a business risk and uninsurable. Coverage is afforded under a CGL policy for bodily injury and property damage resulting from an insured contract. The aforementioned situation is distinct because it involves a contractual indemnification and assumption of risk involving bodily injury and property damage of a third party.

Debt Financing

Exclusion: Damages, defense expenses, costs, or loss arising out of or contributed to by any debt financing, including but not limited to bonds, notes, debentures, and guarantees of debt.

Commentary: This industry standard exclusion is acceptable and doesn't impact the scope or intent of your policy. Coverage is not contemplated for governing body decisions associated with assumption of debt or issuance of bonds. Such decisions are deemed a business risk and considered uninsurable.

Criminal Acts

Exclusion: Damages, defense expenses, costs, or loss arising out of or contributed to by any fraudulent, dishonest, criminal, or malicious act of the insured (except for sexual abuse which is excluded in the sexual abuse exclusion below), or the willful violation of any statute, ordinance, or regulation committed by or with the knowledge of the insured. However, we will defend the insured for covered civil action subject to the other terms of this coverage form until either a judgment or final adjudication establishes such an act, or the insured confirms such act.

Commentary: This industry standard exclusion is acceptable and doesn't impact the scope or intent of your policy. Criminal defense is not contemplated in best-in-class POML policies. Civil defense protection is afforded up to the point of adjudication or confession, which is an important and necessary coverage grant. The definition of wrongful act excludes damages to others that are expected and intentional by the governing body.

Employment Contracts

Exclusion: Any amount actually or allegedly due under the terms of any contract to commence or to continue employment, or as severance pay under any contract relating to the termination of employment.

Commentary: This industry standard exclusion is acceptable and doesn't impact the scope or intent of your policy. Coverage is not contemplated for noncompliance of any contract, including an employment agreement. Notwithstanding, damages arising from other employment-related offenses that fall outside the financial compliance of an employment contract are not subject to this exclusion.

ERISA, COBRA, and WARN Act Liability

Exclusion: Damages, defense expenses, costs, or loss arising out of or contributed to by any insured's obligations under: (a) the Employee Retirement Income Security Act of 1974 (ERISA); (b) the Comprehensive Omnibus Budget Reconciliation Act (COBRA); (c) the Worker Adjustment and Retraining Notification Act (WARN); or (d) Any similar federal, state, or local laws or regulations; including subsequent amendments or any regulations promulgated thereunder.

Commentary: This industry standard exclusion is acceptable and doesn't impact the scope or intent of your policy. Coverage is not contemplated for violations of the above-referenced laws. A fiduciary liability policy and ERISA bond are available for public water/wastewater systems that oversee investment decisions of any employee benefit plan or pension plan.

Failure to Maintain Insurance

Exclusion: Damages, defense expenses, costs, or loss arising out of or contributed to by the failure to maintain: (a) insurance of any kind, including adequate limits of insurance; or (b) suretyship or bonds. This exclusion does not apply to the extent coverage is provided under this coverage form for the administration of employee benefit plans.

Commentary: This industry standard exclusion is acceptable and doesn't impact the scope or intent of your policy. Coverage is not contemplated for governing body decisions to forego or modify insurance placements. Such decisions are deemed business risks and considered uninsurable. The policy is not designed to cover losses associated with a governing body's decision to forego insurance or maintain acceptable limits of insurance. As explained in the chapter's overview, the complexities and liabilities of this exclusion must be fully understood by the governing body. It necessitates an expansive indemnification agreement by the public water and wastewater system to the governing body to ensure maximum personal asset protection.

Fines

Exclusion: Fines, penalties, and taxes, including but not limited to those imposed by the Internal Revenue Service code or any similar state or local code.

Commentary: This industry standard exclusion is acceptable and doesn't impact the scope or intent of your policy. Coverage is not contemplated to pay monetary damages for violations of federal, state, and local tax obligations. Such decisions are deemed business risks and considered uninsurable.

Fungi or Bacteria

Exclusion: (a) Any liability, loss, injury, or damages which would not have occurred or taken place, in whole or in part, but for the actual, alleged, or threatened inhalation of, ingestion of, contact with, exposure to, existence of, or presence of, any fungi or bacteria on or within a building or structure, including its contents, regardless of whether any other cause, event, material, or product contributed concurrently or in any sequence to such injury or damage. (b) Any damages, defense expenses, costs, or loss arising out of the abating, testing for, monitoring, cleaning up, removing, containing, treating, detoxifying, neutralizing, remediating or disposing of, or in any way responding to, or assessing the effects of, fungi or bacteria, by any insured or by any other person or entity.

Commentary: This industry standard exclusion is acceptable and doesn't impact the scope or intent of your policy. The fungi or bacteria exposure is appropriately addressed in your CGL policy and necessitates confirmation that coverage applies for bodily injury and property damage claims arising from your products, completed operations, and operations. An environmental pollution policy is similarly needed to cover remediation expenses on described premises.

Violation of Law

Exclusion: Damages, defense expenses, costs, or loss due to an insured's willful violation of any local law, rule, or regulation.

Commentary: This industry standard exclusion is acceptable and doesn't impact the scope or intent of your policy. Coverage is not contemplated for the willful violation of law. Such decisions are considered uninsurable.

Law Enforcement

Exclusion: Damages, defense expenses, costs, or loss arising out of any law enforcement activity. This exclusion does not apply to an employment practices offense involving your law enforcement agency.

Commentary: This industry standard exclusion is acceptable and doesn't impact the scope or intent of your policy. Public water/wastewater systems rarely have peace officers except for situations involving park rangers or physical security of high-valued assets. Any ancillary law enforcement exposure should be specifically addressed in your CGL policy and necessitates confirmation that coverage applies to bodily injury and personal injury arising from law enforcement actions. A robust law enforcement exposure involving community policing requires a specialized law enforcement liability (LEL) policy.

Lead, Electromagnetic Radiation, Nuclear

Exclusion: (a) Any injury, damages, defense expenses, costs, loss, liability, or legal obligation arising out of or in any way related to: (1) the toxic properties of lead, or any material or substance containing lead; or (2) electromagnetic radiation; or exposure thereto, or for the costs of abatement, mitigation, removal, elimination, or disposal of any of them. (b) Any damages, defense expenses, costs, or loss arising out of any actual, alleged, or threatened injury or damage to any person or property from any radioactive matter or nuclear material.

Commentary: This industry standard exclusion is acceptable and doesn't impact the scope or intent of your policy. The lead exposure is better addressed in your CGL policy and necessitates confirmation that coverage applies to bodily injury for lead in potable water. Electromagnetic radiation is a standard exclusion in a CGL policy for bodily injury and property damage. Risk mitigation measures are required for any radio towers. Nuclear is an industry-wide exclusion under a CGL policy for bodily injury and property damage. These exposures are predicated on bodily injury and property damage. They are not contemplated to involve monetary damage arising from a wrongful act.

Performance of Employee Benefit Plans

Exclusion: Damages, defense expenses, costs, or loss arising from an employment practices offense or any offense in the administration of employee benefit plans arising out of: (a) the failure of any investment program, individual securities, or savings program to perform as held forth by or represented by an insured; (b) advice given by an insured in connection with participation or non-participation in any stock subscription plans, savings programs, or any other employee benefit plan; (c) errors in providing information or failing to provide information on past performance of investment vehicles; (d) failure of the insured, or any insurer, fiduciary, trustee , or fiscal agent, to perform any of their duties or obligations or to fulfill any of their guarantees with respect to the payment of benefits under employee benefit plans or the providing, handling, or investment of funds; (e) the liability of others which is assumed by the insured under a contract or agreement, except to the extent the insured would have been liable in the absence of the contract or agreement; (f) any claim for the return of compensation paid by the insured if a court determines that the payment was illegal; or (g) any claim for benefits that are lawfully paid or payable to a beneficiary from the funds of an employee benefit plan.

Commentary: This industry standard exclusion is acceptable and doesn't impact the scope or intent of your policy. Coverage is not contemplated for the performance guarantee of any investment vehicle or employee benefit plan. A fiduciary liability policy and ERISA bond are available for public water/wastewater systems that oversee investment decisions of any employee benefit or pension plan.

Pollution

Exclusion: (a) Any injury, damages, defense expenses, costs, loss, liability, or legal obligation arising out of or in any way related to pollution, however caused. Pollution includes the actual, alleged, or potential presence in or introduction into the environment of any substance if such substance has, or is alleged to have, the effect of making the environment impure, harmful, or dangerous. Environment includes any air, land, structure (or the air therein), watercourse, or other body of water, including underground water. (b) This insurance does not apply to any damages, defense expenses, costs, or loss arising out of any: (1) request, demand, order, or statutory or regulatory requirement that any insured or others test for, monitor, clean up, remove, contain, treat, detoxify, or neutralize, or in any way respond to, or assess the effects of, pollutants; or (2) claim by or on behalf of a governmental authority for damages because of testing for, monitoring, cleaning up, removing, containing, treating, detoxifying, or neutralizing, or in any way responding to, or assessing the effects of pollutants.

Commentary: This industry standard exclusion is acceptable and doesn't impact the scope or intent of your policy. The pollution exposure is appropriately addressed in your CGL policy and an environmental pollution policy. It necessitates confirmation that your CGL policy covers bodily injury and property damage claims arising from your potable water and wastewater as well as sudden and accidental releases of your water and wastewater chemicals and other specified operational perils. The same confirmation for all other salient pollution scenarios, subject to policy terms and conditions, is needed for your environmental pollution policy. These exposures are predicated on bodily injury and property damage. They are not contemplated to involve monetary damage arising from a wrongful act.

Preparation of Bid Specifications

Exclusion: Damages, defense expenses, costs, or loss arising out of estimates of probable costs or cost estimates being exceeded or faulty preparation of bid specifications or plans including architectural plans.

Commentary: This industry standard exclusion is acceptable and doesn't impact the scope or intent of your policy. Coverage is not contemplated for noncompliance of any contract or errors in preparing bids or architectural plans. These exposures are deemed business risks and considered uninsurable.

Professional Healthcare

Exclusion: Damages, defense expenses, costs, or loss arising out of providing or failing to provide professional healthcare services.

Commentary: This industry standard exclusion is acceptable and doesn't impact the scope or intent of your policy.

Professional Liability

Exclusion: Damages, defense expenses, costs, or loss arising out of the rendering of, or failure to render of professional services by a lawyer, engineer, architect, surveyor, or medical professional; except this exclusion will not apply to claims made against insureds while acting solely as public officials or employees on behalf of the named insured and in the conduct of its business, and not in their professional capacities as a lawyer, engineer, architect, surveyor, or medical professional.

Commentary: This industry standard exclusion is acceptable and doesn't impact the scope or intent of the policy. Coverage is not contemplated for the foregoing professional services, but coverage is provided for professionals in their official capacity as a public official or employee. Water and wastewater testing errors and omission is integrated in a best-in-class CGL policy for bodily injury and property damage. Any additional exposure associated with professional services like third-party billing services, design, construction management, or consulting necessitates a stand-alone professional liability policy. Public water/wastewater systems that employ general counsels and engineers should seriously contemplate the purchase of an employed counsel and employed engineer professional liability policy. Such policies will cover their moonlight activities as well as their in-house professional services.

Profit, Advantage, or Remuneration

Exclusion: Any damages, defense expenses, costs, or loss based upon or attributable to the insured gaining any profit, advantage, or remuneration to which the insured is not legally entitled.

Commentary: This industry standard exclusion is acceptable and doesn't impact the scope or intent of your policy. Coverage is not contemplated for illegal advantage. The exposure is considered uninsurable and against public policy to cover. The definition of wrongful act excludes damages to others that are expected and intentional by the governing body.

Public Use of Property

Exclusion: Damages, defense expenses, costs, or loss arising from any method or proceeding used to take control of private property for public use including condemnation, adverse possession, and dedication by adverse use or inverse condemnation.

Commentary: This industry standard exclusion should be modified to include an exception for unintended takings comprising inverse condemnation. A detailed description of the legal theory is contained in an article at the end of *Part IV – Insurance Essentials*. Although the damages may appear to be property damage, inverse condemnation actions comprise economic injury based on a constitutional violation of taking (or damaging in California) private property without just compensation. The taking arising from the powers, policies, and decisions of your governing body as it relates to an inherent risk in your water and wastewater delivery system.

Publications and Pronouncements

Exclusion: Damages, defense expenses, costs, or loss based upon, attributable to, or arising out of wrongful acts resulting from: (a) publications or pronouncements, including material placed on the internet or on similar electronic means of communication, concerning any organization or business enterprise or their products or services made by or at the direction of the insured with the knowledge of its falsity; or (b) printing of periodicals, advertising matter, or any or all jobs taken by any insured to be printed for a third party when the periodicals, advertising matter, or other printing is not within the scope of the organization's own activities.

Commentary: This industry standard exclusion is acceptable and doesn't impact the scope or intent of your policy.

Sexual Abuse

Exclusion: Damages, defense expenses, costs, or loss, based upon, attributed to, arising out of, in consequence of, or in any way related to: (a) sexual abuse of any person; or (b) the negligent: (1) employment; (2) investigation; (3) supervision; (4) reporting to the proper authorities, or failing to so report; or (5) retention; of a person for whom any insured is or ever was legally responsible and whose conduct would be excluded by paragraph (a) above. Sexual Abuse means any actual, attempted, or alleged sexual conduct by a person, or by persons acting in concert, which causes injury. Sexual abuse includes sexual molestation, sexual assault, sexual exploitation, or sexual injury, but not sexual harassment.

Commentary: This industry standard exclusion is acceptable and doesn't impact the scope or intent of your policy. Coverage for sexual abuse on a vicarious liability basis is contemplated within a CGL policy for bodily injury. Civil defense is similarly offered under a CGL policy up to the point of adjudication for the employee who is alleged to have caused the abuse.

Specific Operations

Exclusion: Damages, defense expenses, costs, or loss, arising out of or contributed to by the ownership, operation, maintenance, entrustment to others, or use of any: (a) gas or electric generation facility; or (b) sanitary landfill, dump, or other permanent waste disposal facility.

Commentary: This industry standard exclusion is acceptable and doesn't impact the scope or intent of your policy as long as you do not have these operations. Clarification should be secured regarding any gas or electric generation facilities. Your insurer should confirm gas distribution is permissible and distinct from gas generation. The same clarification is required for electric distribution versus electric generation. Moreover, public water/wastewater systems that produce electricity for their own purposes or sell such electricity to a local electric utility on a best-efforts contract should receive a carve-out provision of the exposure to avoid coverage misunderstandings.

Strikes, Riot, Civil Commotion, or Mob Action

Exclusion: Damages, defense expenses, costs, or loss arising out of or contributed to by any lockout, strike, picket line, replacement, or other similar actions resulting from labor disputes or labor negotiations or any act or omission in connection with the prevention or suppression of a riot, civil commotion or mob action.

Commentary: This industry standard exclusion is acceptable and doesn't impact the scope or intent of your policy. Coverage is not contemplated for collective bargaining disputes or civil commotion. These exposures are deemed business risks and considered uninsurable.

Tax Assessments

Exclusion: Damages, defense expenses, costs, or loss arising out of or contributed to by any tax assessments or adjustments, or the collection, refund, disbursement, or application of any taxes. This exclusion does not apply to the use or prioritization of your operating funds.

Commentary: This industry standard exclusion is acceptable and doesn't impact the scope or intent of your policy. Coverage is not contemplated for allegations you improperly derived your tax calculations or water rate methodology. These exposures are deemed business risks and considered uninsurable.

Wage and Hour Laws

Exclusion: Damages, defense expenses, costs, or loss arising out of or contributed to by any actual or alleged violation of the Fair Labor Standards Act of 1938, as amended, or any other federal, state, or local law related to wage and hour policies, improper payroll practices or the payment of overtime or vacation pay, including but not limited to back wages or other similar damages, or any monetary or nonmonetary compensation or benefits that may be owed to a past or present employee based upon misclassification of their job status, title, or duties.

Commentary: This industry standard exclusion should be modified to include a satisfactory sublimit for the misclassification of employees. The exposure is deemed preventable and a business risk, but a reasonable sublimit should be secured to avoid an uninsurable claim. The typical assertion for such clams involves the failure of the employer to pay overtime wages owed to the employee via their exempt versus nonexempt misclassification.

Water or Wastewater Professional Activity

Exclusion: Damages, defense expenses, costs, or loss arising out of an act, error, or omission in the performance of or failure to perform your water or wastewater professional activities, that also results in bodily injury or property damage.

Commentary: This industry standard exclusion is acceptable and doesn't impact the scope or intent of your policy. Coverage for these exposures is contemplated under a best-in-class CGL policy that specifically stipulates bodily injury for water and wastewater testing errors & omissions.

Workers Compensation and Similar Laws

Exclusion: Any obligation of the insured under a workers compensation, disability benefits, or unemployment compensation law, or any similar law.

Commentary: This industry standard exclusion is acceptable and doesn't impact the scope or intent of your policy. Coverage for these exposures is specifically contemplated under a workers compensation policy.

Private and Confidential Information

Exclusion: Damages, defense expenses, costs, or loss arising out of or contributed to by any misuse or improper release of confidential, private, or proprietary information.

Commentary: This industry standard exclusion is acceptable and doesn't impact the scope or intent of your policy as long as you secure an endorsement for cyber liability and network security or procure a stand-alone policy for said coverage.

Optional Coverages

There are optional coverages that should be attached to your POML policy. The most prevalent include inverse condemnation, wage and hour, and third-party discrimination. These optional coverages, especially inverse condemnation, are paramount and represent material exposures to your organization. The aforementioned coverages should be affirmed in writing and offered through endorsement. Limits for inverse condemnation should not be sublimited, and the coverage grant should not be restricted in any way. It is similarly important to affirm any excess liability policy includes inverse condemnation on a following form basis. Wage and hour, which comprise misclassifications of employees, should include a reasonable sublimit between $25,000 and 50,000. Third-party discrimination, which involves claims by third parties alleging employee misconduct, are more prevalent with service and retail business where interaction with the public is extensive. Notwithstanding, coverage is recommended up to a reasonable sublimit of $1 million.

Special Commentary

The foregoing exclusions include important commentary, which we will not recapitulate. Many of the exclusions remove unintended defense and indemnity obligations for claims better addressed under a CGL policy or other policies. As such, careful review and understanding of the exclusions is paramount. Should your POML have different exclusions, then it is incumbent to review these exclusions and secure written affirmation of their intent by your insurer. Any verbiage that appears overreaching should be curbed to reduce ambiguity or expansiveness. The same applies to any exclusionary phraseology that is more restrictive than what was described above. A best-in-class POML policy specifically designed for public water/wastewater systems will have readable exclusions that clearly articulate the intent of the insurer. We now turn our discussion to the named insureds afforded coverage and rights under the POML policy.

WHO IS AN INSURED

If you are designated in the supplemental declarations as: (a) An individual, you and your spouse are insureds, but only with respect to the conduct of a business of which you are the sole owner. (b) A partnership or joint venture, you are an insured. Your members, your partners, and their spouses are also insureds, but only with respect to the conduct of your business. However, if you are a public entity, you are insured as a partner in a partnership or as a joint venturer in a joint venture, but only if the partnership or joint venture is between you and another governmental organization or nonprofit entity. Coverage does not extend to a partnership or joint venture that operates, controls, or funds a school, hospital or medical clinic, nursing home, airport, port, public housing, gas, or electric generation facility. (c) A limited liability company, you are an insured. Your members are also insureds, but only with respect to the conduct of your business. Your managers are insureds, but only with respect to their duties as your managers. (d) An organization other than a partnership, joint venture, or limited liability company, you are an insured. Your executive officers and directors are insureds, but only with respect to their duties as your officers or directors. Your stockholders are also insureds, but only with respect to their liability as stockholders. (e) A public entity, you are an insured. Your operating authorities, boards, commissions, districts, or any other governmental units are insureds, provided that you operate, control, and fund the authority, board, commission, district, or other governmental units. Coverage does not extend to an authority, board, commission, district, or other governmental units that operates, controls, or funds a school, hospital or medical clinic, nursing home, airport, port, public housing, gas, or electric generation facility. (f) A trust, you are an insured. Your trustees are also insureds, but only with respect to their duties as trustees.

Each of the following is also an insured: (a) Elected or appointed officials: Your elected and appointed officials, including elected and appointed officials of your operating authorities, boards, commissions, districts, or other governmental units but only for acts within the course and scope of their duties for the insured public entity or its operating authorities, boards, commissions, districts, or other governmental units. (b) Volunteer workers or employees: Your volunteer workers only while performing duties related to the conduct of your operations, or your employees, other than either your executive officers (if you are an organization other than a partnership, joint venture, or limited liability company) or your managers (if you are a limited liability company), but only for acts within the scope of their employment by you or while performing duties related to the conduct of your operations.

However, none of these employees or volunteer workers are insureds for: (1) bodily injury or personal and advertising injury: (a) to you, to your partners, or members (if you are a partnership or joint venture), to your members (if you are a limited liability company), to a co-employee while in the course of his or her employment or performing duties related to the conduct of your operations, or to your other volunteer workers while performing duties related to the conduct of your operations; (b) to the spouse, child, parent, brother, or sister of that co-employee or volunteer worker as a consequence of paragraph (1)(a) above; or (c) for which there is any obligation to share damages with or repay someone else who must pay damages because of the injury described in paragraphs (1)(a) or (1)(b) above. (2) Property damage to property: (a) owned, occupied, or used by; (b) rented to, in the care, custody or control of, or over which physical control is being exercised for any purpose by, you, any of your employees, volunteer workers, any partner or member (if you are a partnership or joint venture), or any member (if you are a limited liability company). (c) Mutual Aid Agreements. Any persons or organizations providing service to you under any mutual aid or similar agreement.

Any organization you newly acquire or form, other than a partnership, joint venture, or limited liability company and over which you maintain ownership or majority interest, will qualify as a named insured if there is no other similar insurance available to that organization. However: (a) coverage under this provision is afforded only until the 90th day after you acquire or form the organization or the end of the policy period, whichever is earlier; and (b) coverage does not apply to a wrongful act, employment practices offense, or offense in the administration of employee benefit plans that took place before you acquired or formed the organization or of which you had notice or knowledge before you acquired the organization. No person or organization is an insured with respect to the conduct of any current or past partnership or joint venture that is not shown as a named insured in the supplemental declarations.

Your director, officer, employee, volunteer worker, or appointee, while serving on the board of directors of an organization that is a separate and distinct entity not subject to your direction and control, provided that the primary purpose of such organization is to support and further the efforts and welfare of individuals or organizations that provide water, sewer, or wastewater treatment.

KEY DEFINITIONS

Employee – natural persons who are past, present, or future, full-time and part-time employees, but only while acting with the scope of their employment for the named insured, including leased workers. Employee does not include a temporary worker.

Executive – a person holding any of the officer positions created by your charter, constitution, by-laws, or any other similar governing document.

Leased Worker – a person leased to you by a labor leasing firm under an agreement between you and the labor leasing firm, to perform duties related to your business. Leased worker does not include a temporary worker.

Temporary Worker – a person who is furnished to you to substitute for a permanent employee on leave or to meet seasonal or short-term workload conditions.

Volunteer Worker – a person who is not your employee, and who donates his or her work and acts at the direction of and within the scope of duties determined by you, and is not paid a fee, salary, or other compensation by you or anyone else for their work performed for you.

SPECIAL COMMENTARY

This section, which identifies the individuals and organizations qualifying as an insured, is nearly identical to that contained in a CGL policy. We redirect you to the CGL policy discussion for a refresher on the scope and depth of said section. There is, however, one item that warrants examination: coverage for your directors, officers, employees, volunteer workers, or appointees while they serve on a governing body of an organization that is unaffiliated with your public water and wastewater system but whose purpose is to support the overarching efforts of your public water and wastewater system. Such an extension is critically important because many nonprofit organizations have inadequate coverage or no coverage for their board of directors. Moreover, many committees formed by stakeholders are not legally structured and have no legal status. In situations like the foregoing, it is important for your governing body to port their POML policy so it extends to their involvement on boards whose principal purpose is to advance the goals of your public water and wastewater system. This portability has the indirect benefit of insulating your indemnification agreement from any resulting claims.

You should also ensure past and present directors, officers, and employees are included as insureds. It is unacceptable if the aforementioned definition only includes present directors, officers, and employees. The definition should include the present and past tense. Your governing body should also be reminded that coverage pertaining to their alleged wrongful acts and other offenses is limited to their official capacity as an elected or appointed official for your governing body. Coverage, for instance, would not apply to wrongful acts or other offenses that occur in venues where the individual is a private citizen and not representing the public water and wastewater system in his or her official capacity.

We can now transition to the limits of insurance section to understand the preferred mechanics for structuring your limits for wrongful acts, employment practices offenses, and offenses associated with the administration of employee benefit plans.

LIMITS OF INSURANCE

The limits of insurance shown in the supplemental declarations and the rules below establish the most we will pay regardless of the number of: (a) insureds; (b) claims made; or (c) persons or organizations making claims.

The aggregate limit of insurance set forth in the supplemental declarations for Coverage A. applies to all damages from all wrongful acts, employment practices offenses, and offenses in the administration of your employee benefit plans, and all defense expenses arising out of all actions for injunctive relief under Coverage B.

Subject to the aggregate limit of insurance, the each wrongful act or each offense limit of insurance set forth in the supplemental declarations for Coverage A., is the most we will pay under Coverage A. for all damages from all claims based upon or arising out of the same or related wrongful acts or offenses.

Subject to the aggregate limit of insurance, the each action for injunctive relief limit of insurance set forth in the supplemental declarations is the most we will pay under Coverage B. for all defense expenses from all actions for injunctive relief based upon or arising out of the same or related wrongful acts or offenses.

The aggregate limit of insurance applies separately to each policy period, starting with the effective date of the policy period shown in the supplemental declarations, unless the policy period is extended after issuance for an additional period of less than 12 months. In that case, the additional period will be deemed part of the last preceding policy period for purposes of determining the limits of insurance.

Our obligations to pay damages or defense expenses or to defend or continue to defend any claim under this coverage form ends when the aggregate limit of insurance is exhausted by the payment of damages or defense expenses. If we pay amounts for damages or defense expenses in excess of that limit of insurance, you agree to promptly reimburse us for such amounts upon our demand.

If the aggregate limit of insurance as set forth in the supplemental declarations is exhausted by the payment of damages or defense expenses, the entire premium for this coverage form will be deemed fully earned.

SPECIAL COMMENTARY

The limits of insurance are consistent in scope and intent as a CGL policy. As such, we refer you to the CGL policy discussion for a review of the mechanics. The aggregate limit of insurance, however, necessitates further discussion on the POML policy; especially if you integrate wrongful acts, employment practices offenses, and offenses in the administration of employee benefit plans into this policy. The advantage of integration is compelling. It streamlines the purchase process, removes insurer finger pointing of overlapping covered claims, and ensures a consistent legal defense strategy. The downside is the sharing of limits with three separate coverage parts.

To counter that exposure, it is imperative to secure an aggregate limit that is $10 million. This solution ensures ten separate $1 million limits for covered wrongful acts and other offenses that occur in the policy period. It also maximizes the protection afforded by an excess liability policy. Defense costs should always be outside or in addition to the limits of insurance. An aggregate limit of $10 million should be actively sought, as that will safeguard your public water and wastewater system from a scenario of multiple wrongful acts or offenses exhausting your policy limits. Equally important, your limits of insurance should replenish after every policy period. Any extension of an existing policy period will not afford replenishment. Consequently, extensions beyond 120 days are inadvisable.

Now, we discussed the key provisions that are most relevant to public water/wastewater systems within the policy conditions section of the POML policy.

KEY POLICY CONDITIONS

Duties in the Event of a Wrongful Act, Offense, or Claim

Condition: (a) You must see to it that we are notified as soon as practicable in writing of a wrongful act or offense which may result in a claim. To the extent possible, notice should include the following information: (1) how, when, and where the wrongful act or offense took place; and (2) the names and addresses of any persons seeking damages or of any witnesses. If a claim is made against any insured, you must: (1) immediately record the specifics of the claim and the date received; and (2) notify us in writing as soon as practicable. You must see to it that we receive written notice of the claim as soon as practicable. (b) You and any other insured named or identified in such claim must: (1) immediately send us copies of any demands, notices, summonses, or legal papers received in connection with the claim or suit; (2) authorize us to obtain records and other information; (3) cooperate with us in the investigation, settlement, or defense of the claim or suit; and (4) assist us, upon our request, in the enforcement of any right against any person or organization which may be liable to the insured because of damages to which this insurance may also apply. (c) No insureds will, voluntarily make a payment, assume any obligation, or incur any expense, without our written consent. Such payment, obligation, or expenses will be at their own cost. (d) Notice shall be deemed given as soon as practicable, if it is given to us by any person to whom you have delegated such responsibility as soon as practicable after they become aware of a wrongful act or offense.

Commentary: It is imperative to understand and comply with your policyholder duties so as not to compromise any policy rights afforded to you. These requirements are not overly onerous but noncompliance can result in voiding the policy or a declination or partial declination of an otherwise covered claim.

Legal Action against Us

Condition: No person or organization has a right under this coverage form: (a) To join us as a party or otherwise bring us into a claim seeking damages or defense expenses from an insured; or (b) To sue us on this coverage form unless all of the terms and conditions of this policy have been fully complied with. A person or organization may sue us to recover on an agreed settlement or on a final judgment against an insured; but we will not be liable for damages or defense expenses that are not payable under the terms of this coverage form or that are in excess of the applicable limit of insurance. Under Coverage A., an agreed settlement means a settlement and release of liability signed by us, the insured, and the claimant or the claimant's legal representative.

Commentary: To ensure you protect your legal rights against your insurer, you should understand your duties and comply with your obligations. Your right to sue the insurer is subject to your compliance of all policy conditions. Compliance is also required to inure all benefits afforded by the policy.

Representations

Condition: By accepting this policy, you agree: (a) The information in the application for this insurance is accurate and complete; (b) That the coverage form has been issued to you based upon the information provided by you and the representations you made to us in the application for this insurance. The application forms the basis of our obligations under this coverage form; and (c) This coverage form is void if any material fact or circumstance relating to this insurance is intentionally omitted or misrepresented in the application for this insurance.

Commentary: Representations are more favorable than warranties. Notwithstanding, you should understand the solemn and legal obligation to provide truthful and accurate information on material questions. This point is pronounced when an insurer predicates their policy terms on your answers.

Separation of Insureds

Condition: Except with respect to the limits of insurance as described in the policy, and any rights or duties specifically assigned to the first named insured, this insurance applies: (a) as if each named insured were the only named insured; and (b) separately to each insured against whom claim is made.

Commentary: This provision ensures each insured is treated independently for all provisions of the policy except the limits. The net result with a regulated insurer is separate legal representation, or *Cumis* counsel, for multiple named insureds named in the complaint as well as protection for those named insureds who comply with policy conditions when other named insured(s) do not.

Transfer of Rights of Recovery Against Others to Us

Condition: If the insured has rights to recover all or part of any payment, we have made under this coverage form, those rights are transferred to us. The insured must do nothing after loss to impair them. At our request, the insured will bring claim or transfer those rights to us and help us enforce them.

Commentary: This provision simply says the insured cannot impede subrogation after a loss, but it has no bearing on any contractual impediment that is initiated before a loss.

When We Do Not Renew

Condition: If we decide not to renew this coverage form, we will mail or deliver to the first named insured shown in the supplemental declarations written notice of such nonrenewal not less than 30 days before the expiration date. We will mail or deliver our notice to the first named insured's last mailing address known to us. If notice is mailed, proof of mailing will be sufficient proof of notice.

Commentary: It is important to negotiate a minimum 60 days, and preferably 90 days, of advanced notice by your insurer should they opt to nonrenew your policy. Thirty days is insufficient time for you and your broker to secure viable and competitive options.

Special Commentary

The policy conditions contain the legalese that summarizes the obligations of you and your insurer. These provisions are not onerous, but they require a thorough reading of all the provisions to ensure compliance with your obligations and do not jeopardize your rights under the policy. Although laborious, this section of the policy should be read and understood.

FINAL COMMENTS

POML policies provide personal asset protection for your governing body and financial transfer solutions for your indemnification agreements. These policies are important because your governing body is exposed to complex and extensive litigation involving alleged wrongful acts and other offenses in the oversight of your public water and wastewater system. It is this combination of sweeping authority and unconstrained litigation that necessitates the purchase of a POML policy to protect your governing body when acting in good faith and for the betterment of your public water and wastewater system. A best-in-class POML policy, coupled with an expansive indemnification agreement, will ensure an adept and diverse ensemble of individuals to serve on your governing body. The proper structure for your POML policy originates with an occurrence form with defense costs outside the limits of insurance. Coverage should encompass wrongful acts, employment practices offenses, and offenses involving the administration of your employee benefit plans. Recommended optional coverages include third-party discrimination, wage and hour, and inverse condemnation. The coverage of inverse condemnation is especially important for any public water and wastewater system whose state constitution requires just compensation for the damaging of private property substantially caused by an inherent risk in a public improvement without fault or foreseeability. The preceding coverage should not be restricted and should be affirmed by any excess liability policy on a following form basis.

The definition of named insured should include the public water and wastewater system as well as past and present board members. Any partnerships or joint ventures should similarly be covered as long as the public water and wastewater system is the controlling party in said relationships. POML policies should be portable and protect your governing body while serving on other organizations in their official capacity as a public water and wastewater system representative. Equally important, policy exclusions should not be onerous or ambiguous. The same level of clarity should apply to the insuring agreement for monetary liability as well as supplementary payments and policy conditions. Any softening of the hammer clause should not be implemented unless the litigation is egregious and malevolent against your governing body and public water and wastewater system. Lastly, your insurer should have subject matter expertise within the water and wastewater segment as well as trial-tested defense attorneys and claim examiners who have defended scores of wrongful acts and other offenses on behalf of public water/wastewater systems in your respective state. This knowledge is important as many wrongful acts are convoluted and involve overlapping causes of action of economic injury and property damage. In these situations, your insurer must have the aptitude to distill the essence of the complaint and the damages alleged. Such overlap underscores the importance of securing your CGL and POML policies from the same insurer. We now turn our attention to exposures pertaining to vehicular liability and physical damage and the importance of a properly structured business auto coverage (BAC) policy.

POLICY CHECKLIST

PUBLIC OFFICIALS AND MANAGEMENT LIABILITY (POML) POLICY: MINIMUM BENCHMARKS

Public Officials and Management Liability (POML)	
Description	**Benchmark**
Form	
Proprietary/Occurrence	Yes
Duty to Defend	Yes
Defense Costs Outside	Yes
Limits	
Wrongful Acts	$1,000,000
Employment Practices	$1,000,000
Employee Benefit Plans	$1,000,000
Third-Party Discrimination	$1,000,000
Wage and Hour	$25,000
Injunctive Relief	$25,000
Policy Aggregate	$10,000,000
Policy Highlights	
Broad Definition of Insured	Yes
Inverse Condemnation	Yes
Nonauditable	Yes
Outside Directorship (all types of entities)	Yes
Prior Acts (continuity)	Yes
Special Coverage: Inverse Condemnation	
Affirmation of Coverage	Yes
Defense Costs Outside	Yes
Strict Liability	Yes

(Continued)

Public Officials and Management Liability (POML)

Description	Benchmark
Physical and Nonphysical Injury	Yes
Unintended and Deliberate Takings (based on inverse condemnation assertion)	Yes
Defense Attorneys w/Subject Matter Expertise	Yes
Attachment of Prevailing Party's Legal Fees	Yes
Policy Sublimit	None
Property Damage Exclusion	None
Limitation for Negligence Only	None
Sharing of Occurrence Limits w/ Other Insureds	None

REVIEW QUESTIONS

1. T/F: Public officials and management liability (POML) policies are specifically designed to protect governing bodies from covered wrongful acts and other offenses.

2. T/F: POML policies also extend coverage to the public water and wastewater system, as claims against governing bodies are frequently subject to indemnification agreements.

3. T/F: Indemnification agreements between a public water and wastewater system and its governing body are necessary because, despite partial tort immunities, a governing body can be held personally liable for alleged wrongful acts and other offenses in their official capacity as a director, officer, trustee, or executive.

4. T/F: Unlike a commercial general liability (CGL) policy, Coverage A. in a POML policy is based on economic injury arising from the powers, policies, and decisions of the governing body in the oversight of the public water and wastewater system.

5. POML policies that are written on an occurrence form have the following advantages:
   a. Wrongful acts or other offenses must occur during the policy period
   b. Wrongful acts or other offenses must occur in the coverage territory
   c. Damages must be submitted or filed during the policy period
   d. a and b only

6. T/F: The purchase of an extended reporting period (ERP) is recommended when an insured non-renews and doesn't replace a claims-made policy or receives renewal terms that are substantially more restrictive than past claims-made policies.

7. Optional coverages in a POML policy that you should seriously consider include the following:
   a. Third-party discrimination
   b. Wage and hour
   c. Inverse condemnation
   d. All of the above

8. T/F: Wrongful act means any actual or alleged error, act, omission, neglect, misfeasance, non-feasance, or breach of duty, including violation of any civil rights law, by any insured in the discharge of their duties for the named insured, individually or collectively, that results directly but unexpectedly and unintentionally in damages to others.

9. T/F: A hammer clause in a POML policy affords the insurer unilateral rights to conclude any demand or claim filed against the policy with a preemptive settlement to the plaintiff.

10. T/F: A hammer clause in a POML policy can be relaxed, but any softening transfers the defense and indemnity costs to the insured from the point where a settlement could have been reached by the insurer.

11. T/F: The integration of wrongful acts and employment practices offenses in a POML policy is preferred because it allows for a claim involving overlapping allegations to be consolidated into an integrated defense by one insurer.

12. T/F: An expansive indemnification agreement act serves as a final personal layer of protective insulation for your governing body should litigation not be covered in a POML policy.

13. T/F: Bodily injury, property damage, and personal and advertising injury are included in a POML policy.

14. T/F: Any damages, defense expenses, costs, or loss based upon or attributable to the insured gaining any profit, advantage, or remuneration to which the insured is not legally entitled is commonly covered in a POML policy.

15. T/F: Damages, defense expenses, costs, or loss arising out of or contributed to by any debt financing, including but not limited to bonds, notes, debentures, and guarantees of debt is commonly excluded in a POML policy.

16. T/F: The definition of employee in a POML policy includes temporary workers.

17. T/F: The representation condition in a POML policy allows the insurer to void the policy if any material fact or circumstance relating to this insurance is unintentionally omitted or misrepresented in the application for this insurance.

18. T/F: A claims-made POML policy has strict claims reporting requirements and exposes insureds to marketplace vagaries by relying on the insurance industry to offer renewal terms consistent with past policies.

19. The public use of property exclusion includes the following types of takings:
    a. Adverse possession
    b. Dedicated use
    c. Eminent domain
    d. All of the above

20. The definition of injunctive relief in a POML policy includes the following:
    a. Equitable relief sought through a demand for the issuance of a permanent, preliminary, or temporary injunction, restraining order, or similar prohibitive writ against an insured, or order for specific performance by an insured
    b. Monetary damages less than $10,000
    c. Nonmonetary damages less than $10,000
    d. a and c only

21. The definition of claim in a POML policy includes the following:
    a. Written notice, from any party, that it is their intention to hold the insured responsible for damages arising out of a wrongful act or offense by the insured
    b. Formal proceeding or investigation with the Equal Employment Opportunity Commission, or with an equivalent state or local agency
    c. a and b only
    d. None of the above

22. T/F: Under a POML policy, reimbursement for injunctive relief defense expenses is ancillary to the obligation of defense and indemnity for monetary damages.

23. Examples of an injunctive relief claim include:
    a. Environment group suing public water and wastewater system to stop intake flows from reservoir
    b. Neighborhood community suing public water and wastewater system to stop construction of a water storage tank
    c. Regulatory agency suing public water and wastewater system for violation of their National Pollutant Discharge Elimination System (NPDES)
    d. a and b only

24. T/F: Coverage for wage and hour in a POML policy involves allegations of employee misclassification and the lack of rightful compensation based on their legal status as nonexempt employees.

25. T/F: Outside directorship coverage refers to a POML policy being portable and extending to directors and officers while they serve on a governing body of an organization that is unaffiliated with the public water and wastewater system but whose purpose is to support the overarching efforts of the public water and wastewater system.

ANSWER KEY

1. **T**/F: Public officials and management liability (POML) policies are specifically designed to protect governing bodies from covered wrongful acts and other offenses.

2. **T**/F: POML policies also extend coverage to the public water and wastewater system, as claims against governing bodies are frequently subject to indemnification agreements.

3. **T**/F: Indemnification agreements between a public water and wastewater system and its governing body are necessary because, despite partial tort immunities, a governing body can be held personally liable for alleged wrongful acts and other offenses in their official capacity as a director, officer, trustee, or executive.

4. **T**/F: Unlike a commercial general liability (CGL) policy, Coverage A. in a POML policy is based on economic injury arising from the powers, policies, and decisions of the governing body in the oversight of the public water and wastewater system.

5. POML policies that are written on an occurrence form have the following advantages:
   a. Wrongful acts or other offenses must occur during the policy period
   b. Wrongful acts or other offenses must occur in the coverage territory
   c. Damages must be submitted or filed during the policy period
   d. **a and b only**

6. **T**/F: The purchase of an extended reporting period (ERP) is recommended when an insured non-renews and doesn't replace a claims-made policy or receives renewal terms that are substantially more restrictive than past claims-made policies.

7. Optional coverages in a POML policy that you should seriously consider include the following:
   a. Third-party discrimination
   b. Wage and hour
   c. Inverse condemnation
   d. **All of the above**

8. **T**/F: Wrongful act means any actual or alleged error, act, omission, neglect, misfeasance, nonfeasance, or breach of duty, including violation of any civil rights law, by any insured in the discharge of their duties for the named insured, individually or collectively, that results directly but unexpectedly and unintentionally in damages to others.

9. **T**/F: A hammer clause in a POML policy affords the insurer unilateral rights to conclude any demand or claim filed against the policy with a preemptive settlement to the plaintiff.

10. **T**/F: A hammer clause in a POML policy can be relaxed, but any softening transfers the defense and indemnity costs to the insured from the point where a settlement could have been reached by the insurer.

11. **T**/F: The integration of wrongful acts and employment practices offenses in a POML policy is preferred because it allows for a claim involving overlapping allegations to be consolidated into an integrated defense by one insurer.

12. **T**/F: An expansive indemnification agreement act serves as a final personal layer of protective insulation for your governing body should litigation not be covered in a POML policy.

13. **T/F**: Bodily injury, property damage, and personal and advertising injury are included in a POML policy.

14. **T/F**: Any damages, defense expenses, costs, or loss based upon or attributable to the insured gaining any profit, advantage, or remuneration to which the insured is not legally entitled is commonly covered in a POML policy.

15. **T/F**: Damages, defense expenses, costs, or loss arising out of or contributed to by any debt financing, including but not limited to bonds, notes, debentures, and guarantees of debt is commonly excluded in a POML policy.

16. **T/F**: The definition of employee in a POML policy includes temporary workers.

17. **T/F**: The representation condition in a POML policy allows the insurer to void the policy if any material fact or circumstance relating to this insurance is unintentionally omitted or misrepresented in the application for this insurance.

18. **T/F**: A claims-made POML policy has strict claims reporting requirements and exposes insureds to marketplace vagaries by relying on the insurance industry to offer renewal terms consistent with past policies.

19. The public use of property exclusion includes the following types of takings:
    a. Adverse possession
    b. Dedicated use
    c. Eminent domain
    d. **All of the above**

20. The definition of injunctive relief in a POML policy includes the following:
    a. **Equitable relief sought through a demand for the issuance of a permanent, preliminary, or temporary injunction, restraining order, or similar prohibitive writ against an insured, or order for specific performance by an insured**
    b. Monetary damages less than $10,000
    c. Nonmonetary damages less than $10,000
    d. a and c only

21. The definition of claim in a POML policy includes the following:
    a. Written notice, from any party, that it is their intention to hold the insured responsible for damages arising out of a wrongful act or offense by the insured
    b. **Formal proceeding or investigation with the Equal Employment Opportunity Commission, or with an equivalent state or local agency**
    c. a and b only
    d. None of the above

22. **T/F**: Under a POML policy, reimbursement for injunctive relief defense expenses is ancillary to the obligation of defense and indemnity for monetary damages.

23. Examples of an injunctive relief claim include:
    a. Environment group suing public water and wastewater system to stop intake flows from reservoir
    b. Neighborhood community suing public water and wastewater system to stop construction of a water storage tank
    c. Regulatory agency suing public water and wastewater system for violation of their National Pollutant Discharge Elimination System (NPDES)
    d. **a and b only**

24. **T/F**: Coverage for wage and hour in a POML policy involves allegations of employee misclassification and the lack of rightful compensation based on their legal status as nonexempt employees.

25. **T/F**: Outside directorship coverage refers to a POML policy being portable and extending to directors and officers while they serve on a governing body of an organization that is unaffiliated with the public water and wastewater system but whose purpose is to support the overarching efforts of the public water and wastewater system.

23 Business Auto Coverage (BAC) Policy

2019 Photo Courtesy of Yorba Linda Water District

LEARNING OBJECTIVES

- Understand how to properly structure your business auto coverage (BAC) policy.
- Explain the difference between the liability and physical damage in a BAC policy.
- Identify the provisions in a fleet automatic endorsement in a BAC policy.
- Articulate the importance of the broadened pollution – covered auto endorsement in a BAC policy.
- Discuss the individuals and organizations qualifying as an insured in BAC policy.
- Comprehend the purpose and restrictions of the exclusions in a BAC policy.
- Grasp the importance of the auto symbols in the BAC policy.
- Describe the difference between a BAC policy and a personal auto policy.
- Comprehend the valuation differences among stated amount, actual cash value, agreed value, and replacement cost in a BAC policy.
- Articulate the coverage options of a BAC policy.
- Explain the difference between the insurer's obligation of defense and indemnity in a BAC policy.
- Identify the differences between auto, non-owned, and hired auto liability in a BAC policy.

DOI: 10.1201/9781003207146-33

KEY DEFINITIONS*

Accident – includes continuous or repeated exposure to the same conditions resulting in bodily injury or property damage.

Actual Cash Value (ACV) – in property and auto physical damage insurance, one of several possible methods of establishing the value of insured property to determine the amount the insurer will pay in the event of loss. ACV is calculated on the cost to repair or replace the damaged property, minus depreciation.

Additional Insured Endorsement – policy endorsement used to add coverage for additional insureds by name (e.g., mortgage holders or lessors). There are a number of different forms intended to address various situations, some of which afford very restrictive coverage to additional insureds. Rather than naming each additional insured, a blanket additional insured endorsement is normally preferred whenever a written agreement requires it. Notwithstanding, some court cases have rejected additional insured status when a blanket endorsement is used.

Audit – survey of the financial records of a person or organization conducted annually (in most cases) to determine exposures, limits, premiums, etc.

Auto – (1) a land motor vehicle, trailer, or semitrailer designed for travel on public roads; or (2) any other land vehicle that is subject to a compulsory or financial responsibility law or other motor vehicle insurance law where it is licensed or principally garaged. However, auto does not include mobile equipment.

Blanket Additional Insured Endorsement – an endorsement attached to liability insurance policies that automatically grants insured status to a person or organization that the named insured is required by contract to add as an insured. May apply only to specific types of contracts or entities.

Bodily Injury – bodily injury, sickness, or disease sustained by a person, including death resulting from any of these.

Claim – used in reference to insurance, a claim may be a demand by an individual or organization to recover, under a policy of insurance, for loss that may be covered within that policy.

Coverage Territory – contractual provisions limiting coverage to geographical areas within which the insurance is affected.

Covered Causes of Loss – direct physical loss or damage unless the loss or damage is specifically excluded or limited in the policy.

Covered Pollution Cost or Expense – any cost or expense arising out of: (1) any request, demand, order, or statutory or regulatory requirement that any insured or others test for, monitor, clean up, remove, contain, treat, detoxify, or neutralize, or in any way respond to, or assess the effects of, pollutants; or (2) any claim or suit by or on behalf of a governmental authority for damages because of testing for, monitoring, cleaning up, removing, containing, treating, detoxifying or neutralizing, or in any way responding to, or assessing the effects of, pollutants.

Covered pollution cost or expense does not include any cost or expense arising out of the actual, alleged, or threatened discharge, dispersal, seepage, migration, release, or escape of pollutants: (a) that are, or that are contained in any property that is: (1) being transported or towed by, handled or handled for movement into, onto or from the covered auto; (2) otherwise in the course of transit by or on behalf of the insured; or (3) being stored, disposed of, treated, or processed in or upon the covered

---

* Source material provided by International Risk Management Institute (IRMI), www.irmi.com/term/insurance-definit ions/insurance, as well as policy forms and underwriting products promulgated by Insurance Services Office (ISO) and their member companies, https://products.iso.com/productcenter/10.

auto; (b) before the pollutants or any property in which the pollutants are contained are moved from the place where they are accepted by the insured for movement into or onto the covered auto; or (c) after the pollutants or any property in which the pollutants are contained are moved from the covered auto to the place where they are finally delivered, disposed of, or abandoned by the insured.

Paragraph (a) above does not apply to fuels, lubricants, fluids, exhaust gases, or other similar pollutants that are needed for or result from the normal electrical, hydraulic, or mechanical functioning of the covered auto or its parts, if: (1) the pollutants escape, seep, migrate, or are discharged, dispersed, or released directly from an auto part designed by its manufacturer to hold, store, receive, or dispose of such pollutants; and (2) the bodily injury, property damage, or covered pollution cost or expense does not arise out of the operation of any equipment listed in paragraph (6)(b) or (6)(c) of the definition of mobile equipment. Paragraphs (b) and (c) above do not apply to accidents that occur away from premises owned by or rented to an insured with respect to pollutants not in or upon a covered auto if: (a) the pollutants or any property in which the pollutants are contained are upset, overturned, or damaged as a result of the maintenance or use of a covered auto; and (b) the discharge, dispersal, seepage, migration, release, or escape of the pollutants is caused directly by such upset, overturn or damage.

Damages – money whose payment a court orders as compensation to an injured plaintiff. Fines, penalties, or injunctive relief would not typically constitute damages.

Defense Clause – insurance provision in which the insurer agrees to defend, with respect to insurance afforded by the policy, all suits against the insured.

Diminution in Value – actual or perceived loss in market value or resale value which results from a direct and accidental loss.

Disclaimer – situation where an insurer refuses to accept your insurance claim on an existing policy.

Employee – any natural person: (a) while in your service; (b) who you compensate directly by salary, wages, or commissions; and (c) who you have the right to direct and control while performing services for you. Employee includes a leased worker but does not include a temporary worker.

Exclusion – policy provision that eliminates coverage for some type of risk. Exclusions narrow the scope of coverage provided by the insuring agreement. In many insurance policies, the insuring agreement is very broad. Insurers utilize exclusions to carve away coverage for risks they are unwilling to insure.

Fellow Employee – endorsement that provides coverage for claims made by an injured employee against a fellow employee who caused or contributed to the injury.

Financial Responsibility Laws – statutory provision requiring owners of automobiles to provide evidence of their ability to pay damages arising out of the ownership, maintenance, or use of an automobile.

Governing Body – group of elected or appointed people who formulate the policy and direct the affairs of a public water and wastewater system.

Indemnity – compensation to a party for a loss or damage that has already occurred, or to guarantee through a contractual clause to repay another party for loss or damage that might occur in the future. The concept of indemnity is based on a contractual agreement made between two parties in which one party (the indemnitor) agrees to pay for potential losses or damages caused by the other party (the indemnitee).

Insured – any person or organization qualifying as an insured in the "Who Is an Insured" section of the applicable coverage. Except with respect to the limit of insurance, the coverage afforded applies separately to each insured who is seeking coverage or against whom a claim or suit is brought.

Insurer – the insurance company that undertakes to indemnify for losses and perform other insurance-related operations.

Insuring Agreement – portion of the insurance policy in which the insurer promises to make payment to or on behalf of the insured. The insuring agreement is usually contained in a coverage form from which a policy is constructed. Often, insuring agreements outline a broad scope of coverage, which is then narrowed by exclusions and definitions.

Judgment – decision of a court regarding the rights and liabilities of parties in a legal action or proceeding. Judgments also generally provide the court's explanation of why it has chosen to make a particular court order.

Limits of Insurance – the most that will be paid by an insurer in the event of a covered loss under an insurance policy.

Loss – direct and accidental loss or damage, or a reduction in value.

Mobile Equipment – any of the following types of land vehicles, including any attached machinery or equipment: (1) bulldozers, farm machinery, forklifts, and other vehicles designed for use principally off public roads; (2) vehicles maintained for use solely on or next to premises you own or rent; (3) vehicles that travel on crawler treads; (4) vehicles, whether self-propelled or not, maintained primarily to provide mobility to permanently mounted: (a) power cranes, shovels, loaders, diggers, or drills; or (b) road construction or resurfacing equipment such as graders, scrapers, or rollers; (5) vehicles not described in paragraph (1), (2), (3), or (4) above that are not self-propelled and are maintained primarily to provide mobility to permanently attached equipment of the following types: (a) air compressors, pumps, and generators, including spraying, welding, building cleaning, geophysical exploration, lighting, and well-servicing equipment; or (b) cherry pickers and similar devices used to raise or lower workers; or (6) vehicles not described in paragraph (1), (2), (3), or (4) above maintained primarily for purposes other than the transportation of persons or cargo.

However, self-propelled vehicles with the following types of permanently attached equipment are not mobile equipment but will be considered autos: (a) equipment designed primarily for: (1) snow removal; (2) road maintenance, but not construction or resurfacing; or (3) street cleaning; (b) cherry pickers and similar devices mounted on automobile or truck chassis and used to raise or lower workers; and (c) air compressors, pumps, and generators, including spraying, welding, building cleaning, geophysical exploration, lighting, and well-servicing equipment. Mobile equipment does not include land vehicles that are subject to a compulsory or financial responsibility law or other motor vehicle insurance law where it is licensed or principally garaged. Land vehicles subject to a compulsory or financial responsibility law or other motor vehicle insurance law are considered autos.

Non-owned Automobile – described in commercial auto policies as a private passenger type auto owned by an employee that is used in connection with the named insured's business but that is not owned, leased, hired, rented, or borrowed by the named insured. Hired Automobile – refers to autos the named insured leases, hires, rents, or borrows.

Physical Injury – losses caused by injuries to persons and legal liability imposed on the insured for such injury or for damage to property of others.

Policy Conditions – section of an insurance policy that identifies general requirements of an insured and the insurer on matters such as loss reporting and settlement, property valuation, other insurance, subrogation rights, and cancellation and nonrenewal. The policy conditions are usually stipulated in the coverage form of the insurance policy.

Pollutants – any solid, liquid, gaseous, or thermal irritant or contaminant, including smoke, vapor, soot, fumes, acids, alkalis, chemicals, and waste. Waste includes materials to be recycled, reconditioned, or reclaimed.

Property Damage – damage to or loss of use of tangible property.

Replacement Cost – property insurance term that refers to one of the two primary valuation methods for establishing the value of insured property for purposes of determining the amount the insurer will pay in the event of loss. The other primary valuation method is actual cash value (ACV). Replacement cost is defined as the cost to replace damaged property with materials of like kind and quality, without any deduction for depreciation.

Representation – statement made in an application for insurance that the prospective insured represents as being correct to the best of his or her knowledge. If the insurer relies on a representation in entering into the insurance contract and if it proves to be false at the time it was made, the insurer may have legal grounds to void the contract.

Reservation of Rights – an insurer's notification to an insured that coverage for a claim may not apply. Such notification allows an insurer to investigate (or even defend) a claim to determine whether coverage applies (in whole or in part) without waiving its right to later deny coverage based on information revealed by the investigation. Insurers use a reservation of rights letter because in many claim situations, all the insurer has at the inception of the claim are various unsubstantiated allegations and, at best, a few confirmed facts.

Settlement – resolution between disputing parties about a legal case, reached either before or after court action begins.

Suit – a civil proceeding in which: (1) damages because of bodily injury or property damage; or (2) a covered pollution cost or expense; to which this insurance applies, are alleged. Suit includes: (a) an arbitration proceeding in which such damages or covered pollution costs or expenses are claimed and to which the insured must submit or does submit with our consent; or (b) any other alternative dispute resolution proceeding in which such damages or covered pollution costs or expenses are claimed and to which the insured submits with our consent.

Supplemental Declarations Page – the front page (or pages) of a policy that specifies the named insured, address, policy period, location of premises, policy limits, and other key information that varies from insured to insured. The supplemental declarations page is also known as the information page.

Supplementary Payments – term used in liability policies for the costs associated with the investigation and resolution of claims. Supplementary payments are normally defined to include such items as first aid expenses, premiums for appeal and bail bonds, pre- and post-judgment interest, and reasonable travel expenses incurred by the insured at the insurer's request when assisting in the defense of a claim. Actual settlements/judgments are considered damages rather than supplementary payments.

Temporary Worker – person who is furnished to you to substitute for a permanent employee on leave or to meet seasonal or short-term workload conditions.

Torts – a civil or private wrong giving rise to legal liability.

Vicarious Liability – the liability of a principal for the acts of its agents; can result from the acts of independent agents, partners, independent contractors, employees, and children.

Waiver of Subrogation – an agreement between two parties in which one party agrees to waive subrogation rights against another in the event of a loss. The intent of the waiver is to prevent one party's insurer from pursuing subrogation against the other party. Insurance policies do not bar coverage if an insured waives subrogation against a third party before a loss. However, coverage is excluded from many policies if subrogation is waived after a loss because to do so would violate the principle of indemnity.

BAC POLICY*

OVERVIEW

The operation of owned and non-owned automobiles represents a perceptible exposure for public water/wastewater systems. Your normal operations require the use of automobiles for disparate activities like equipment deliveries-to-employee transport, operational maintenance-to-emergency repairs, community meetings-to-ad hoc traveling, and meter readings-to-site development. The various scenarios are expansive, and the exposures range from first party to third party. Vehicular liability, in particular, represents one of the most financially significant and unpredictable risks facing public water/wastewater systems. Although less problematic and volatile than vehicular liability, first-party auto losses are similarly impactful and disruptive. Examples include damage or destruction of a vehicle, loss of a vehicle's value after an accident, and loss of use of a vehicle until it is repaired or replaced. A derivative of first-party exposures comprise injury to your drivers and passengers. This scenario necessitates coverage for no-fault medical payments or personal injury protection as well as coverage for bodily injury arising from at-fault accidents by other drivers who are either uninsured or underinsured. Third-party auto losses, in contrast, are significantly more complex and divergent than first-party losses. They are also exceedingly difficult to quantify. The three most common vehicular liability doctrines facing commercial and governmental entities include vicarious, contract, and common law liability.

Vicarious liability is the most substantial of the preceding legal doctrines. Owned or non-owned vehicles driven by your employees, volunteers, and governing body officials for public water and wastewater system purposes create a vicarious (principal-agent relationship) liability known as respondeat superior. This doctrine attaches liability to an employer (principal) for the acts of its employees, volunteers, and governing body officials (agents). The doctrine also applies to vehicles rented or borrowed by the foregoing individuals for public water and wastewater system purposes. Contractual liability is an equally material exposure and necessitates a thorough review of all agreements where vehicular liability is assumed. This review is important because the transfer of contractual exposure to your insurer is not automatic. You should secure written confirmation from your insurer to mitigate coverage misunderstanding as it relates to the assumption of vehicular contract liability. For those public water/wastewater systems that have garage operations and perform occasional mechanical work for other public water/wastewater systems or governmental entities, there is common law liability arising from your Bailee exposure. Such liability involves temporary possession of vehicles and related equipment from other entities as well as products-completed operations liability associated with your mechanical and garage work.

The core coverages encompassing commercial auto and personal auto policies are similar. The main difference is the type of liabilities they cover. While both policies are designed to pay for vehicle repairs, medical bills, and legal liabilities after an accident, commercial auto policies carry higher liability limits, cover a broader array of vehicles, and comprise more complex legal issues than personal auto policies. Coverage under your commercial auto policy is restricted to accidents occurring while you or your employees, volunteers, and governing body officials are using a vehicle for public water and wastewater system purposes. This usage can include owned and non-owned vehicles, but coverage for the latter is only intended to protect the public water and wastewater system. Your employees, volunteers, and governing body officials must rely on their personal auto policy for any individual liability and physical damage protection. Unlike owned vehicles, protection afforded under non-owned auto liability is excess of any personal auto policy.

Insurance Services Office (ISO) offers a widely accepted commercial auto policy, hereinafter referred to as a BAC form, which is extensively used by public water/wastewater systems and other commercial and governmental entities. This form is frequently augmented with customized features

---

* Source material provided by various online articles from International Risk Management Institute (IRMI), www.irmi.com/articles/expert-commentary/insurance as well as policy forms and underwriting products promulgated by Insurance Services Office (ISO) and their member companies, https://products.iso.com/productcenter/10.

from best-in-class insurers to expand its breadth, depth, and customization. BAC policies include coverage for liability and physical damage, with other coverages like medical payments/personal injury protection, uninsured/underinsured motorists, towing, and rental car reimbursement available by endorsement. Additional coverages and extensions, which are necessary to properly augment your BAC policy, will be examined along with the importance of auto symbols and valuation options later in this chapter.

We will sequentially dissect the BAC policy according to its core sections: (a) Supplemental Declarations; (b) Covered Autos; (c) Liability Coverage; (d) Physical Damage Coverage; (e) Key Conditions (loss and general); and (f) Available Endorsements and Extensions. Our discussion will now segue to the dynamics of the supplemental declarations, including their composition, and an examination of the breadth, depth, and customization of a properly structured BAC policy. The learning technique is similar to other chapters, with commentary intended to be paired with your current BAC policy. This approach will allow you to compare your BAC policy with a best-in-class benchmark.

SUPPLEMENTAL DECLARATIONS

Overview

The supplemental declarations function as the nucleus of any BAC policy. It states the named insured, policy number, form of business, mailing address of named insured, identity of the insurer and broker advocate, and policy period. Most importantly, the supplemental declarations list the vehicles insured and the coverages provided, along with rates, premiums, deductibles, endorsements, and other ancillary information. When numerous vehicles are insured, separate vehicle schedules may be attached. The foregoing data offers a notable and succinct encapsulation of the breadth, depth, and customization of the BAC policy.

There are six sections that comprise the supplemental declarations: (1) general information about the risk, including endorsements attached to the policy; (2) coverages, coverage symbols, limits of liability, and the premium for each coverage; (3) schedule of owned autos; (4) schedule of hired or borrowed autos; (5) schedule for non-ownership liability; and (6) schedule of gross receipts or mileage for liability coverage for public auto or leasing rental concerns. Various sections will be completed as appropriate for the coverages provided, but section six will always be deemed inapplicable for public water/wastewater systems.

Special Commentary

The BAC policy is extensively used for all types of commercial and governmental entities. Its core coverages include liability and physical damage. The policy is sufficiently flexible to include a wide array of essential optional coverages like no-fault medical payments/personal injury protection, uninsured/underinsured motorists, towing, and rental car reimbursement. More nuanced coverages such as garage liability and garage keepers liability, which are important for public water/wastewater systems with mechanics and repair facilities, can be added by endorsement or through the inclusion of a separate ISO policy form. Proprietary extensions are similarly available from best-in-class insurers to augment your BAC policy and include: broadened pollution liability; insured status for employees, volunteers, and governing body officials when using their personally owned vehicles for public water and wastewater system purposes; and other important but nuanced endorsements. The optional coverages and proprietary extensions are referenced by their respective form number in the supplemental declarations.

The scope of coverage in the BAC policy can be broadened or narrowed, depending on the coverage options chosen. The supplemental declarations codify these selections and provide a coherent outline of the various policy sections that follow. Most of the coverage options are indicated by the covered auto symbol numbers listed under item 2 of the supplemental declarations page. This section is particularly important to review, as it defines the specific protections afforded by the policy within each of the coverage grants. You can tailor coverage by selecting auto symbols that best represent your unique needs. Different symbols may be used for different coverages. If no symbol is shown for a particular coverage, then that coverage is not provided. More than one symbol may be shown for a given coverage if there is no conflict or overlap between the symbol descriptions. For example, the same coverage can apply to owned autos (symbol 2) and hired autos (symbol 8). With the supplemental declarations explained, we will now delve into the various sections of the coverage form that comprise the BAC policy.

SECTION I. COVERED AUTOS

OVERVIEW

Section I of the BAC form describes the coverage symbols listed in the supplemental declarations (as item two) and indicates the categories of autos that are cataloged as covered autos for the policy's various coverages. These symbols determine the policy's activation to different types of autos for each of its coverage grants. The following ten numerical symbols describe the vehicles that may be cataloged as covered autos:

Symbol 1 – Any Auto

Symbol 1 (any auto) is the broadest of all symbols because it catalogs any auto (whether owned, hired, borrowed, or used) as a covered auto for purposes of this policy. The symbol is intended for liability coverage only. Since the policy's definition of auto includes land vehicles that would otherwise qualify as mobile equipment except for being subject to a motor vehicle insurance law where licensed or principally garaged, it would also provide automatic liability coverage for the over-the-road exposure of any such equipment whether it is owned, leased, hired, borrowed, or used in connection with your operations. When this symbol applies, no other symbols are needed for the policy's liability coverage section because symbol 1 (any auto) applies to any auto.

Symbol 2 – Owned Autos Only, Symbol 3 – Owned Private Passenger Autos Only and Symbol 4 – Owned Autos Other Than Private Passenger Autos

Symbol 2 (owned autos only), symbol 3 (owned private passenger autos only), and symbol 4 (owned autos other than private passenger autos) refer to automobiles your public water and wastewater system owns. Use of any one of these symbols alone will not trigger coverage for hired, borrowed, or non-owned autos (except liability coverage for substitute autos and for trailers attached to one of your own power units). Coverage applies for newly acquired vehicles of the type described by the symbol. For example, with symbol 3 (owned private passenger autos only), you have automatic coverage for newly acquired private passenger autos but not for newly acquired trucks.

For purposes of liability coverage, either symbol 2 (owned autos only) or symbol 4 (owned autos other than private passenger autos) would include coverage for an owned land vehicle that would qualify as mobile equipment except for being subject to a compulsory or financial responsibility law or other motor vehicle insurance law. Symbol 2 (owned autos only), symbol 3 (owned private passenger autos only), and symbol 4 (owned autos other than private passenger autos) afford coverage for owned autos of the type described, regardless of whether such autos are specifically listed for coverage. Like symbol 1 (any auto), coverage triggered by the foregoing symbols would include an auto that was acquired during a previous policy period, even if it was inadvertently overlooked by you at renewal. It is not necessary for the policy wording to affirm coverage for such autos because the policy already includes these autos in the phrase "only those autos you own." Notwithstanding, you are advised to specifically list your insurable auto fleet to your insurer so as to avoid additional premium due when the policy is audited at the end of the policy period as well as unintended delays in claims processing. The auditable component of a BAC policy underscores the necessity of a fleet automatic endorsement. This feature will soon be examined in the special commentary.

With respect to physical damage (i.e., comprehensive and collision) coverage, the most common and recommended symbol is symbol 2 (owned autos only). It is inadvisable to purchase physical damage coverage for your private passenger type autos but not for your commercial autos, or vice-versa. Symbol 2 (owned autos only) is the appropriate covered auto designation for physical damage. In some cases, you may opt for a lower premium by selecting a higher collision deductible than a comprehensive deductible. It may similarly be prudent to evaluate the cost-benefit of securing a higher collision deductible for your larger commercial units than your private passenger autos and pickups. In such situations, symbol 2 (owned autos only) would still apply but item two in the supplemental declarations would not show the deductible amount and instead refer the reader to item three (where information is shown for each owned covered auto).

Symbol 2 (owned autos only) may also be used to provide automatic combined single limit and uninsured/underinsured motorists coverage for all your owned vehicles. You can opt for symbol 2 (owned autos only) or symbol 3 (owned private passenger autos only) when you choose to purchase coverage that is mandatory. Otherwise, you can select symbol 6 (owned autos subject to a compulsory uninsured motorists law) for owned autos required by law in the state where they are licensed or principally garaged to have uninsured motorists coverage. Symbol 2 (owned autos only) or symbol 3 (owned private passenger autos only) may also be used to provide certain no-fault coverages like medical payments and personal injury protection.

For public water/wastewater systems, symbol 1 (any auto) is recommended for liability, but symbol 2 (owned autos only) is a reasonable alternative if symbol 1 (any auto) is unavailable. The use of symbol 2 (owned autos only) requires the addition of symbol 8 (hired autos only) and symbol 9 (non-owned autos only) to protect against hired auto liability and non-owned auto liability. Symbol 3 (owned private passenger autos only) or symbol 4 (owned autos other than private passenger autos) is not advisable because it restricts coverage to either owned private passenger autos or owned autos other than private passenger autos. Most public water/wastewater systems have a combination of both, which makes those symbols inadvisable. Symbol 2 (owned autos only), as indicated above, is recommended for physical damage since it applies to all owned autos.

Symbol 5 – Owned Autos Subject to No-Fault

Some states mandate no-fault personal injury protection coverage. Symbol 5 (owned autos subject to no-fault) triggers such no-fault coverage but only for autos required by law to carry it. If some of your vehicles are garaged in states with mandatory personal injury protection requirements and some in states with optional personal injury protection provisions, then a careful analysis should be made of the no-fault laws in the applicable states to determine the appropriate coverage symbol(s) for your public water and wastewater system.

Personal injury protection, also known as no-fault insurance, pays the medical bills and rehabilitative costs for you and your passengers after a car accident involving injuries. It also extends if you are hit by a car while a pedestrian or cyclist while performing public water and wastewater system purposes. Personal injury protection is different than bodily injury liability because it pays for your own expenses, whereas bodily injury liability pays for medical expenses of drivers and passengers in other cars when you are at fault for the accident. This coverage may include some overlap with your individual or employer-based health policy, but personal injury protection is specifically written for car-related injuries. Such injuries are sometimes excluded from health policies. Personal injury protection also covers a number of additional expenses like lost wages, necessary medical procedures, rehabilitation expenses, and funeral costs.

Symbol 6 – Owned Autos Subject to a Compulsory Uninsured Motorists Law

Symbol 6 (owned autos subject to a compulsory uninsured motorists law), like symbol 5 (owned autos subject to no-fault), is reserved for a particular coverage: in this case, uninsured/underinsured motorists when said coverage is required by law in the state where the vehicles are licensed or principally garaged. This symbol triggers automatic coverage only in those states that do not allow you to reject the coverage. To obtain automatic coverage, you should use symbol 2 (owned autos only) or symbol 3 (owned private passenger autos only).

Symbol 7 – Specifically Described Autos

Symbol 7 (specifically described autos) may be used to activate liability, physical damage, medical payments, personal injury protection, and (in some states) uninsured/underinsured motorist coverage. Note that symbol 7 (specifically described autos) is not restricted to autos that are owned by your public water and wastewater system. It is technically possible for your organization to schedule autos it does not actually own but are used in your operations. Coverage also extends to non-owned attached trailers and provides restricted coverage for newly acquired vehicles. When this symbol is used, the onus is on you to ensure coverage is in place. Symbol 7 (specifically described autos) clearly and unambiguously restricts coverage to vehicles listed in the schedule of covered autos.

There are numerous court cases enforcing this coverage restriction by excluding coverage for non-listed vehicles. Symbol 7 (specifically described autos) is not recommended, as symbol 1 (any auto) or symbol 2 (owned autos only) are better choices for liability. Symbol 2 (owned autos only) is also a better choice for physical damage. As a reminder, the use of symbol 2 (owned autos only) for liability also requires symbol 8 (hired autos only) and symbol 9 (non-owned autos only).

Symbol 8 – Hired Autos Only

Symbol 8 (hired autos only) may be used to designate coverage for leased, hired, rented, or borrowed autos only. Autos that are owned by and/or hired from employees, partners, or members of their households are not included in this designation of covered autos. Symbol 8 (hired autos only) may be used for liability or physical damage coverage, either alone or jointly with symbol 2 (owned autos only), symbol 3 (owned private passenger autos only), symbol 4 (owned autos other than private passenger), symbol 7 (specifically described autos), or symbol 9 (non-owned autos only). Hired auto physical damage coverage does not apply to an auto leased, hired, rented, or borrowed with a driver.

Symbol 8 (hired autos only) restricts coverage for vehicles that are leased, hired, rented, or borrowed by you. If the vehicle is actually owned by the public water and wastewater system, then the insured cannot logically lease, hire, or borrow it from itself. Therefore, symbol 8 (hired autos only) provides no coverage for vehicles that are, in fact, owned by the public water and wastewater system. This can lead to coverage gaps if the use of symbol 8 (hired autos only) is not coordinated with your vehicle title and tax strategy. Although hired or borrowed is not defined in the BAC form, the term hired is customarily referred to as a rental auto. Borrowed would similarly refer to a vehicle that is temporarily used by the insured with permission by the owner.

Symbol 9 – Non-Owned Autos Only

Symbol 9 (non-owned autos only) is designed to trigger liability coverage for autos you do not own, hire, lease, rent, or borrow but are used in connection with your public water & wastewater system. Autos that are owned by your employees, volunteers, or governing body officials are included in this definition if such autos are used for public water and wastewater system purposes. Coverage afforded for these non-owned autos applies only for the benefit of the employer and not the foregoing individuals. Symbol 9 (non-owned autos only) is used to activate liability coverage and is not intended to provide physical damage for the non-owned vehicles.

Symbol 19 – Mobile Equipment Subject to Compulsory or Financial Responsibility or Other Motor Vehicle Insurance Law Only

Symbol 19 (mobile equipment subject to compulsory or financial responsibility or other motor vehicle insurance law only) is designed to trigger liability coverage for land vehicles that would ordinarily be classified as mobile equipment under a commercial general liability (CGL) policy except for being subject to a motor vehicle insurance law where licensed or principally garaged. The description of symbol 19 (mobile equipment subject to compulsory or financial responsibility or other motor vehicle insurance law only) makes no reference to ownership of the vehicle. The standard governing this symbol is the vehicle would ordinarily be considered mobile equipment except for being subject to a compulsory financial responsibility or other motor vehicle law where it is licensed or principally garaged.

Symbol 19 (mobile equipment subject to compulsory or financial responsibility or other motor vehicle insurance law only) is not the only way to activate liability coverage for vehicles subject to financial responsibility laws. The symbol may not be needed at all, depending on what other symbols are used in the policy to provide liability coverage for other autos (i.e., those that fit into the more traditional portion of the policy definition of auto). Symbol 1 (any auto), symbol 2 (owned autos only), symbol 4 (owned autos other than private passenger), symbol 8 (hired autos only), and symbol 9 (non-owned autos only) can all trigger liability coverage in some instances for this type of equipment.

Key Definitions

Auto – (1) a land motor vehicle, trailer, or semitrailer designed for travel on public roads; or (2) any other land vehicle that is subject to a compulsory or financial responsibility law or other motor vehicle insurance law where it is licensed or principally garaged. However, auto does not include mobile equipment.

Mobile Equipment – any of the following types of land vehicles, including any attached machinery or equipment: (1) bulldozers, farm machinery, forklifts, and other vehicles designed for use principally off public roads; (2) vehicles maintained for use solely on or next to premises you own or rent; (3) vehicles that travel on crawler treads; (4) vehicles, whether self-propelled or not, maintained primarily to provide mobility to permanently mounted: (a) power cranes, shovels, loaders, diggers, or drills; or (b) road construction or resurfacing equipment such as graders, scrapers, or rollers; (5) vehicles not described in paragraph (1), (2), (3), or (4) above that are not self-propelled and are maintained primarily to provide mobility to permanently attached equipment of the following types: (a) air compressors, pumps, and generators, including spraying, welding, building cleaning, geophysical exploration, lighting, and well-servicing equipment; or (b) cherry pickers and similar devices used to raise or lower workers; or (6) vehicles not described in paragraph (1), (2), (3), or (4) above maintained primarily for purposes other than the transportation of persons or cargo.

However, self-propelled vehicles with the following types of permanently attached equipment are not mobile equipment but will be considered autos: (a) equipment designed primarily for: (1) snow removal; (2) road maintenance, but not construction or resurfacing; or (3) street cleaning; (b) cherry pickers and similar devices mounted on automobile or truck chassis and used to raise or lower workers; and (c) air compressors, pumps, and generators, including spraying, welding, building cleaning, geophysical exploration, lighting, and well-servicing equipment. Mobile equipment does not include land vehicles that are subject to a compulsory or financial responsibility law or other motor vehicle insurance law where it is licensed or principally garaged. Land vehicles subject to a compulsory or financial responsibility law or other motor vehicle insurance law are considered autos.

Trailer – includes semitrailer.

Special Commentary

If symbol 1 (any auto), symbol 2 (owned autos only), symbol 3 (owned private passenger autos only), symbol 4 (owned autos other than private passenger autos), symbol 5 (owned autos subject to no-fault), symbol 6 (owned autos subject to a compulsory uninsured motorists law), or symbol 19 (mobile equipment subject to compulsory or financial responsibility or other motor vehicle insurance law only) is entered next to a coverage in item two of the supplemental declarations, then you have coverage for newly acquired autos of the type described for the remainder of the policy period. If symbol 7 (specifically described autos only) is entered next to a coverage in item two of the supplemental declarations, then an auto you acquire will be a covered auto for that coverage only if: (1) your insurer already covers all autos you own for that coverage, or it replaces an auto you previously owned that had that coverage; and (2) you tell the insurer within 30 days after acquisition that the coverage in question is desired.

With regard to liability coverage, symbol 2 (owned autos only), symbol 3 (owned private passenger autos only), or symbol 4 (owned autos other than private passenger autos) may be used in conjunction with symbol 8 (hired autos only) to provide hired auto liability coverage and symbol 9 (non-owned autos only) to provide non-ownership liability. The combination of symbol 2 (owned autos only), symbol 8 (hired autos only), and symbol 9 (non-owned autos only) affords coverage that is almost equivalent to symbol 1 (any auto). The preferred choice is unequivocally symbol 1 (any auto), but the combination of symbol 2 (owned autos only), symbol 8 (hired autos only), and symbol 9 (non-owned autos only) is a reasonable substitute. Small trailers (less than 2,000 pounds) and mobile equipment being carried or towed by a covered auto are automatically covered for liability. Autos used as a temporary substitute for covered autos because of breakdown, damage, or service are also covered for liability.

The utilization of fleet automatic is strongly recommended to ameliorate audit premiums and streamline mid-term additional/return premium invoices and mitigate unintended claims processing delays. This tool is only available for symbol 1 (any auto) or symbol 2 (owned autos only) under liability and symbol 2 (owned autos only) for physical damage. The insurer will not be inclined to offer this enhancement unless you cover all your owned vehicles for both liability and physical damage. In some situations, you can modify the fleet automatic for physical damage to apply to those vehicles seven years or newer or vehicles with cost new values not exceeding $100,000. This enhancement allows your fleet to be rated prospectively on renewal, which is preferred than a retrospective premium calculation of the expiring policy period. It is important to note any restrictions on fleet automatic as it pertains to physical damage. The BAC policy allows great flexibility in designating covered autos for the various coverages available under the policy. Coverage chosen by you does not need to apply to all covered autos but can be customized based on your unique needs and fleet characteristics. We will now turn our attention to the liability coverage within the BAC policy.

SECTION II. LIABILITY COVERAGE

INSURING AGREEMENT

We will pay all sums an insured legally must pay as damages because of bodily injury or property damage to which this insurance applies, caused by an accident and resulting from the ownership, maintenance, or use of a covered auto. We will also pay all sums an insured legally must pay as a covered pollution cost or expense to which this insurance applies, caused by an accident and resulting from the ownership, maintenance, or use of covered autos. However, we will only pay for the covered pollution cost or expense if there is either bodily injury or property damage to which this insurance applies that is caused by the same accident. We have the right and duty to defend any insured against a suit asking for such damages or a covered pollution cost or expense. However, we have no duty to defend any insured against a suit seeking damages for bodily injury or property damage or a covered pollution cost or expense to which this insurance does not apply. We may investigate and settle any claim or suit as we consider appropriate. Our duty to defend or settle ends when the liability coverage limit of insurance has been exhausted by payment of judgments or settlements.

Key Definitions

Accident – continuous or repeated exposure to the same conditions resulting in bodily injury or property damage.

Bodily Injury – bodily injury, sickness, or disease sustained by a person, including death resulting from any of these.

Covered Pollution Cost or Expense – any cost or expense arising out of: (1) any request, demand, order, or statutory or regulatory requirement that any insured or others test for, monitor, clean up, remove, contain, treat, detoxify, or neutralize, or in any way respond to, or assess the effects of, pollutants; or (2) any claim or suit by or on behalf of a governmental authority for damages because of testing for, monitoring, cleaning up, removing, containing, treating, detoxifying, or neutralizing, or in any way responding to, or assessing the effects of, pollutants.

Covered pollution cost or expense does not include any cost or expense arising out of the actual, alleged, or threatened discharge, dispersal, seepage, migration, release, or escape of pollutants: (a) that are, or that are contained, in any property that is: (1) being transported or towed by, handled, or handled for movement into, onto, or from the covered auto; (2) otherwise in the course of transit by or on behalf of the insured; or (3) being stored, disposed of, treated, or processed in or upon the covered auto; (b) before the pollutants or any property in which the pollutants are contained are moved from the place where they are accepted by the insured for movement into or onto the covered auto; or (c) after the pollutants or any property in which the pollutants are contained are moved from the covered auto to the place where they are finally delivered, disposed of, or abandoned by the insured.

Paragraph (a) above does not apply to fuels, lubricants, fluids, exhaust gases, or other similar pollutants that are needed for or result from the normal electrical, hydraulic, or mechanical functioning of the covered auto or its parts, if: (1) the pollutants escape, seep, migrate, or are discharged, dispersed, or released directly from an auto part designed by its manufacturer to hold, store, receive, or dispose of such pollutants; and (2) The bodily injury, property damage, or covered pollution cost or expense does not arise out of the operation of any cherry pickers and similar devices mounted on automobile or truck chassis and used to raise or lower workers as well as air compressors, pumps, and generators, including spraying, welding, building cleaning, geophysical exploration, lighting, and well-servicing equipment.

Paragraphs (b) and (c) above do not apply to accidents that occur away from premises owned by or rented to an insured with respect to pollutants not in or upon a covered auto if: (a) the pollutants or any property in which the pollutants are contained are upset, overturned, or damaged as a result of the maintenance or use of a covered auto; and (b) the discharge, dispersal, seepage, migration, release, or escape of the pollutants is caused directly by such upset, overturn, or damage.

Insured – any person or organization qualifying as an insured in the applicable coverage. Except with respect to the limit of insurance, the coverage afforded applies separately to each insured who is seeking coverage or against whom a claim or suit is brought.

Property Damage – damage to or loss of use of tangible property.

Suit – a civil proceeding in which: (1) damages because of bodily injury or property damage; or (2) a covered pollution cost or expense; to which this insurance applies, are alleged. Suit includes: (a) an arbitration proceeding in which such damages or covered pollution costs or expenses are claimed and to which the insured must submit or does submit with our consent; or (b) any other alternative dispute resolution proceeding in which such damages or covered pollution costs or expenses are claimed and to which the insured submits with our consent.

Special Commentary

The liability insuring agreement expresses the insurer's promise to pay all sums an insured legally must pay as damages because of bodily injury or property damage to which the insurance applies. The bodily injury or property damage must be caused by an accident and must result from the ownership, maintenance, or use of a covered auto. Coverage also extends on a limited basis to covered pollution cost or expense, but it only applies to a demand or order to clean up, remove, or neutralize pollution caused by the escape of substances (such as fuel or motor oil) from an automobile's normal operating system. There is no coverage for any pollution resulting from the escape of any substances being transported, stored, treated, or processed in or upon a covered auto. The insuring agreement makes no differentiation as to the damages the insurer is obligated to pay on behalf of the insured. Therefore, in the absence of a specific restriction, coverage will apply to punitive damages awarded against an insured where allowable by law.

Defense costs are in addition to policy limit, and the duty to defend ends when liability coverage limit has been exhausted through payment of judgments or settlements. The insurer must defend the insured defendant against false or fraudulent claims as long as the claims allege covered damages. The preceding is usually achieved through a reservation of rights letter. An insurer's reservation of rights is an important legal step, particularly in the context of liability insurance. A reservation of rights letter states that the insurer may deny coverage for some or all of the claim, even while the insurer is investigating the claim or beginning to treat the claim as if it were covered. The insurer may provide a defense to the insured, seemingly protecting the insured from the serious liabilities that may result from a civil suit. However, the insurer is alerting the insured defendant that insurance may ultimately not cover the resulting liability or a portion of the liability. We now turn our attention to the status of insured under the liability coverage section of the BAC policy.

WHO IS AN INSURED

The following are insureds: (a) You for any covered auto. (b) Anyone else while using with your permission a covered auto you own, hire, or borrow except: (1) the owner or anyone else from whom you hire or borrow a covered auto. This exception does not apply if the covered auto is a trailer connected to a covered auto you own; (2) your employee if the covered auto is owned by that employee or a member of his or her household; (3) someone using a covered auto while he or she is working in a business of selling, servicing, repairing, parking, or storing autos unless that business is yours; (4) anyone other than your employees, partners (if you are a partnership), members (if you are a limited liability company), or a lessee or borrower or any of their employees, while moving property to or from a covered auto; or (5) a partner (if you are a partnership) or a member (if you are a limited liability company) for a covered auto owned by him or her or a member of his or her household. (c) Anyone liable for the conduct of an insured described above but only to the extent of that liability.

Key Definitions

Employee – a leased worker. Employee does not include a temporary worker.

Leased Worker – a person leased to you by a labor leasing firm under an agreement between you and the labor leasing firm to perform duties related to the conduct of your business. Leased worker does not include a temporary worker.

Temporary Worker – a person who is furnished to you to substitute for a permanent employee on leave or to meet seasonal or short-term workload conditions.

Special Commentary

The definition of an insured is broken into three segments: (1) the named insured; (2) permissive users; and (3) anyone liable for the conduct of an insured. The broadest coverage, of course, is afforded to the named insured. A number of exceptions apply to permissive users, and the extent of liability coverage available for those responsible for the conduct of an insured is also limited. The named insured is insured for any covered auto. The symbol(s) indicated in the supplemental declarations for each particular coverage determines which autos are covered autos. Any permissive user is an insured for a covered auto owned, hired, or borrowed by the named insured, with four key exceptions: (1) anyone from whom the named insured borrows or hires a covered auto is not an insured; (2) an employee is not covered when using his or her or a family member's auto; (3) anyone working in the business of selling, servicing, repairing, or parking autos is not an insured; and (4) anyone other than named insured's employees is not covered while moving property to or from a covered auto.

Let's illustrate the foregoing restrictions in greater detail. A rental car company is not an insured under the BAC policy for your use of their rental car. An employee is not an insured under the BAC policy while using their own car for public water and wastewater system purposes. However, the public water and wastewater system is covered under the preceding scenario. A third-party mechanic is not an insured while test driving one of your covered autos. And, finally, only employees are insureds while loading and unloading your covered autos. It is critically important to expand the definition of employee to include your volunteer and governing body officials. This expansion will prevent claims misunderstandings with your BAC policy insurer. We will now delve into the coverage extensions within the liability coverage section of the BAC policy.

COVERAGE EXTENSIONS

Supplementary Payments – We will pay for the insured: (1) All expenses we incur. (2) Up to $2,000 for cost of bail bonds (including bonds for related traffic law violations) required because of an accident we cover. We do not have to furnish these bonds. (3) The cost of bonds to release attachments in any suit against the insured we defend, but only for bond amounts within our limit of insurance. (4) All reasonable expenses incurred by the insured at our request, including actual loss of earnings up to $250 a day because of time off from work. (5) All court costs taxed against the insured in any suit against the insured we defend. However, these payments do not include attorneys' fees or attorneys' expenses taxed against the insured. (6) All interest on the full amount of any judgment that accrues after entry of the judgment in any suit against the insured we defend, but our duty to pay interest ends when we have paid, offered to pay, or deposited in court the part of the judgment that is within our limit of insurance. These payments will not reduce the limit of insurance.

Out-of-State Coverage Extensions – While a covered auto is away from the state where it is licensed, we will: (1) Increase the limit of insurance for covered autos liability coverage to meet the limits specified by a compulsory or financial responsibility law of the jurisdiction where the covered auto is being used. This extension does not apply to the limit or limits specified by any law governing motor carriers of passengers or property. (2) Provide the minimum amounts and types of other coverages, such as no-fault, required of out-of-state vehicles by the jurisdiction where the covered auto is being used.

We will not pay anyone more than once for the same elements of loss because of these extensions.

Special Commentary

Supplementary liability payments under the BAC policy are similar to those found in the commercial general liability (CGL) policy. As the title implies, these supplementary payments are coverage extensions that will not reduce the liability coverage's limit of insurance. The insurer agrees to pay all expenses incurred by it for the insured including investigating and defending a claim. Notwithstanding, there is no duty to defend suits or claims for damages that are not covered under the liability coverage section or to defend after the limit of liability has been exhausted. This provision does not require the insurer to furnish bail bonds, but it does require the insurer to pay up to $2,000 for the cost of these bonds. The provision is not limited to bail bonds because of an accident. It also includes bonds for related traffic law violations. The bail bonds for which the insurer is responsible are those required because of an accident covered by the insurer.

The insurer also agrees to pay the cost of bonds for the release of attachments that have been made on the assets of the insured in any suit against the insured that the insurer is defending. For example, a claimant may have filed for a legal claim against the insured's equipment, pending the outcome of a lawsuit against the insured, to be assured that the insured is financially able to pay any judgment arising from the lawsuit. The attachment of the equipment would hamper the insured's business operations, and the insurer agrees to pay the cost of bonds to obtain the release of the attachment. The insurer is not obligated to pay that portion of the bond premium associated with a bond penalty (limit) in excess of the policy limit. Although this paragraph does not specifically deny the insurer's responsibility to furnish such bonds, the agreement applies only to the cost of the bonds.

The insurer also agrees to pay reasonable expenses incurred by the insured at the insurer's request, including actual loss of earnings up to $250 a day because of time off from work. Note that, for the insured's expenses to be paid by the insurer, the expenses must be incurred at the insurer's request. Expenses incurred solely at the discretion of the insured (e.g., voluntary appearance at a trial) are not covered expenses. Additionally, the insurer agrees to pay all court costs taxed against the insured in any suit the insurer defends as well as any interest that accrues between the time of a judgment and the time that judgment is paid.

The insurer agrees to pay interest on the full amount of the judgment even if the judgment is larger than the limit of insurance. Once the insurer has paid, offered to pay, or deposited in court the part of the judgment that is within its limits, however, the insurer's obligation to pay interest ends. This provision should encourage the insurer to pay its limits promptly, particularly if the judgment is significantly above the insured's limit of insurance. The taxing of attorney fees and related expenses are not in addition to the policy limit but applied against the policy limit. This point is important, as it can materially erode your indemnity protection and necessitates careful discussion with your governing body of higher limits via an excess liability policy where your BAC policy is scheduled as an underlying policy.

The second portion of the coverage extensions increases the limit of insurance or provides coverage to meet the minimum requirements of another jurisdiction while a covered auto is in that jurisdiction. The first paragraph of out-of-state coverage extensions deals with state compulsory or financial responsibility laws. When a covered auto is being used in a state other than the state where it is licensed, that auto's limit of liability is automatically increased to meet required limits.

Practically speaking, public water/wastewater systems will purchase liability limits well above the respective level mandated by the state.

The second out-of-state coverage extension deals with other coverages, such as no-fault. For example, under this paragraph, coverage is automatically provided for the minimum amount of no-fault coverage that is required by the state where the covered auto is being used. The extension is broad enough to apply to any type and amount of coverage required of out-of-state vehicles by the jurisdiction where the covered auto is being used. Finally, the insurer states that it will not pay more than once for the same elements of loss because of these out-of-state coverage extensions. That concludes our discussion on coverage extensions. We will now transition to exclusions found in the liability coverage section of the BAC policy.

Exclusions

Expected or Intended Injury: Bodily injury or property damage that was expected or intended by the insured has no liability coverage under the BAC policy. It is important to secure an exemption for situations where the insured is attempting to protect persons or property.

Contractual Liability: To the extent that the insured has assumed liability within a contract that is beyond that which would be imposed on the insured under general law, there is no liability coverage under the BAC policy. This exclusion does not apply (therefore coverage is provided) if liability would have existed in absence of the contract or damages are assumed in an insured contract, such as in the case of lease of premises, easement, auto rental agreement (except for damage to rented auto or with a driver), sidetrack agreement, or exculpatory agreement (i.e., a hold harmless contract, wherein the insured assumes tort liability of another).

Workers Compensation: If you have an employee who is injured due to an auto accident while working, then losses should be covered under the organization's workers compensation insurance policy.

Employee Indemnification and Employers' Liability: If you have an employee who is injured due to an auto accident while working, and as a result of a court case or legal settlement it is determined that the employer organization must compensate the injured employee or the injured employee's family, then these losses should be covered under the organization's workers compensation policy via its employers' liability coverage.

Fellow Employee: An illustrative example of this exclusion is when an injured passenger-employee sues a driver-employee. In such a scenario, there is no coverage for the driver. This exclusion can be removed by attaching a fellow employee endorsement. Such losses are intended to be covered by the driver's personal auto insurance policy, but liability limits of personal auto policies are frequently much lower than commercial auto policies and may be inadequate for the injuries sustained or alleged.

Care, Custody, or Control: There is no coverage for property owned by or in the care, custody, or control of the insured. Instead, an inland marine policy should be secured. Coverage for property in your care, custody, or control is available via garage keepers liability and should be sought if you have a garage facility that performs work for other public water/wastewater systems and public entities.

Handling of Property: CGL policies cover damages before property is moved and after it has been moved. The liability coverage section in your BAC policy does not apply to injury or damage that occurs before you begin loading property onto a covered auto or after the property has been unloaded. When property is deemed loaded or unloaded may vary from state to state. Accidents that fall within this exclusion are likely to be covered by your CGL policy. Thus, it is important to have both types of coverage.

Movement of Property by Mechanical Device: If the device is not attached to a covered auto, then any loss must be tendered to your CGL policy. Exceptions to the exclusion are if the device is a hand truck or the device is attached to a covered auto. Movement of property by hand, hand truck, or a device attached to the auto is covered until it reaches the place where it is finally delivered. Loading and unloading are covered unless using mobile equipment that is not attached to the auto. Example: Suppose an employee is using a forklift to unload bags when he accidentally drops the load, injuring a bystander. The injured party sues you for bodily injury. The suit is not covered by your auto policy since the forklift was not attached to your truck. The suit would instead require tendering to your CGL policy.

Operations: Operation of certain types of equipment attached to covered autos like cherry pickers, air compressors, pumps, generators, equipment used for spraying, and welding is excluded from coverage. Equipment attached to or part of a land vehicle that would qualify as mobile equipment if not subject to compulsory insurance law is also excluded. Such coverage is contemplated under your CGL policy.

Completed Operations: There is no coverage for completed operations performed with the insured's auto. Coverage for your completed operations is available via garage liability and should be sought if you have a garage facility that performs work for other public water/wastewater systems and public entities.

Pollution: This exclusion applies to bodily injury or property damage that arises out of the discharge, release, or escape of pollutants. It consists of the following five parts: (1) Pollutants Being Transported by You or on Your Behalf. The pollution exclusion eliminates coverage for claims arising from a release of pollutants being transported by you or on your behalf; (2) Pollutants Being Loaded onto, or Unloaded from, a Covered Auto. No coverage is provided for pollutants released while they are being loaded onto or unloaded from a covered auto; (3) Being Stored, Disposed of, Treated, or Processed in or on a Covered Auto. The exclusion applies to pollutants being stored, disposed of, treated, or processed in or on an insured vehicle; (4) Before They Are Loaded onto a Vehicle. Also excluded are claims stemming from pollutants released before they are loaded onto an insured auto; and (5) After They Have Been Unloaded from a Vehicle. This exclusion is similar to the previous one in that it applies to losses that occur on the customer's premises. While the pollution exclusion is broad, it does not exclude all claims involving pollution. This exclusion should be modified to include bodily injury and property damage arising from contents containing pollutants that are spilled or released after an accident. Public water/wastewater systems routinely transport substances deemed pollutants, and a broad form pollution liability endorsement is strongly recommended.

War: The war exclusion is common within insurance policies of all types. Such potentially widespread catastrophic losses are never intended to be covered by insurance.

Racing: Regardless of whether professional or organized racing, demolition contest, or stunting activity, losses due to racing are excluded.

Key Definitions

Insured Contract – (1) a lease of premises; (2) a sidetrack agreement; (3) any easement or license agreement, except in connection with construction or demolition operations on or within 50 feet of a railroad; (4) an obligation, as required by ordinance, to indemnify a municipality, except in connection with work for a municipality; (5) that part of any other contract or agreement pertaining to your business (including an indemnification of a municipality in connection with work performed for a municipality) under which you assume the tort liability of another to pay for bodily injury or property damage to a third party or organization.

Tort liability – means a liability that would be imposed by law in the absence of any contract or agreement; or (6) that part of any contract or agreement entered into, as part of your business, pertaining to the rental or lease, by you or any of your employees, of any auto. However, such contract or agreement shall not be considered an insured contract to the extent that it obligates you or any of your employees to pay for property damage to any auto rented or leased by you or any of your employees. An insured contract does not include that part of any contract or agreement: (a) that indemnifies a railroad for bodily injury or property damage arising out of construction or demolition operations, within 50 feet of any railroad property and affecting any railroad bridge or trestle, tracks, roadbeds, tunnel, underpass, or crossing; (b) that pertains to the loan, lease, or rental of an auto to you or any of your employees, if the auto is loaned, leased, or rented with a driver; or (c) that holds a person or organization engaged in the business of transporting property by auto for hire harmless for your use of a covered auto over a route or territory that person or organization is authorized to serve by public authority.

Pollutants – any solid, liquid, gaseous, or thermal irritant or contaminant, including smoke, vapor, soot, fumes, acids, alkalis, chemicals, and waste. Waste includes materials to be recycled, reconditioned, or reclaimed.

Special Commentary

The liability section of the BAC policy begins with an insuring agreement that presents a broad scope of afforded coverage. The liability exclusions substantially narrow its scope by restricting coverage for exposures that (a) may be more logically treated under other types of policies; (b) can be covered under the auto liability policy for an additional premium charge; or (c) are uninsurable. There are exclusions that can be modified, and we recommended you review the notes to ensure you secure exemptions for expected or intended injury, fellow employee, care, custody, or control (when applicable), completed operations (when applicable), and pollution. We now turn attention to the limit of insurance within the liability coverage section of the BAC policy.

Limit of Insurance

Regardless of the number of covered autos, insureds, premiums paid, claims made, or vehicles involved in the accident, the most we will pay for the total of all damages and covered pollution cost or expense combined resulting from any one accident is the limit of insurance for covered autos liability coverage shown in the supplemental declarations. All bodily injury, property damage, and covered pollution cost or expense resulting from continuous or repeated exposure to substantially the same conditions will be considered as resulting from one accident. No one will be entitled to receive duplicate payments for the same elements of loss under this coverage form and any medical payments, uninsured motorists, or underinsured motorists endorsements attached to this policy.

Key Definitions

Insured – any person or organization qualifying as an insured in the applicable coverage. Except with respect to the limit of insurance, the coverage afforded applies separately to each insured who is seeking coverage or against whom a claim or suit is brought.

Loss – direct and accidental loss or damage, or a reduction in value.

Special Commentary

This section stipulates the manner in which the limit of insurance of the liability coverage section is applied. It is designed to keep the policy's liability limit from stacking in the case of any one accident, regardless of the number of covered vehicles involved in the accident, or the number of insureds against whom claim is brought. Unlike the personal auto policy, the BAC policy has a combined single limit of insurance that is applicable to all bodily injury, property damage, and covered pollution cost from a single accident. No annual aggregate limit applies. The duty to defend and settle obligation by the insurer ends when the limit of insurance has been exhausted by payment of judgments or settlements. A minimum liability limit of $1 million combined single limit is recommended, and this limit should be increased with the purchase of a following form excess liability policy that includes your BAC policy as a scheduled underlying policy. Our review of the liability coverage section of the BAC policy is now completed. We will now delve into the physical damage section.

SECTION III. PHYSICAL DAMAGE COVERAGE

INSURING AGREEMENT

We will pay for loss to a covered auto or its equipment under: (a) Comprehensive coverage from any cause except: (1) the covered auto's collision with another object; or (2) the covered auto's overturn. (b) Specified causes of loss coverage caused by: (1) fire, lightning, or explosion; (2) theft; (3) windstorm, hail, or earthquake; (4) flood; (5) mischief or vandalism; or (6) the sinking, burning, collision, or derailment of any conveyance transporting the covered auto. (c) Collision coverage caused by: (1) the covered auto's collision with another object; or (2) the covered auto's overturn. *Towing* – We will pay up to the limit shown in the supplemental declarations for towing and labor costs incurred each time a covered auto of the private passenger type is disabled. However, the labor must be performed at the place of disablement. *Glass Breakage* – Hitting a bird or animal – Falling objects or missiles – If you carry comprehensive coverage for the damaged covered auto, then we will pay for the following: (a) glass breakage; (b) loss caused by hitting a bird or animal; and (c) loss caused by falling objects or missiles. However, you have the option of having glass breakage caused by a covered auto's collision or overturn considered a loss under collision coverage.

Coverage Extensions

(a) *Transportation Expenses* – We will pay up to $20 per day, to a maximum of $600, for temporary transportation expense incurred by you because of the total theft of a covered auto of the private passenger type. We will pay only for those covered autos for which you carry either comprehensive or specified causes of loss coverage. We will pay for temporary transportation expenses incurred during the period beginning 48 hours after the theft and ending, regardless of the policy's expiration, when the covered auto is returned to use or we pay for its loss. This extension is restricted to only theft and only private passenger type autos. A more complete solution requires rental car reimbursement coverage that applies to all covered autos. (b) *Loss of Use Expenses* – For hired auto physical damage, we will pay expenses for which an insured becomes legally responsible to pay for loss of use of a vehicle rented or hired without a driver under a written rental contract or agreement. We will pay for loss of use expenses if caused by: (1) other than collision only if the supplemental declarations indicates that comprehensive coverage is provided for any covered auto; (2) specified causes of loss only if the supplemental declarations indicate that specified causes of loss coverage are provided for any covered auto; or (3) collision only if the supplemental declarations indicate that collision coverage is provided for any covered auto. However, the most we will pay for any expenses for loss of use is $20 per day, to a maximum of $600. Note: This extension offers a low limit per day and maximum. A higher limit up to $150 per day up to 30 days for a maximum of $4,500 is recommended.

Special Commentary

Physical damage losses are one of the two primary risks covered by the BAC policy. Coverage applies to loss of a covered auto or its equipment from three sets of perils: (a) collision or overturn; (b) comprehensive coverage; and (c) specified causes of loss. Because comprehensive coverage and specified causes of loss are alternative approaches to insuring non-collision physical damage losses, one or the other may be selected by the insured to pair with collision coverage. Comprehensive is an open perils coverage approach that applies to any non-excluded cause of loss other than collision or overturn. Specified causes of loss coverage is a named perils coverage. It is recommended you opt for comprehensive over specified causes of loss. As with the policy's liability coverage, the supplemental declarations page governs what vehicles qualify as covered autos for each of the policy's coverages. Unlike property damage from the liability coverage section of the BAC policy, the physical damage coverage section does not apply automatically to loss of use claims. As previously noted in Section I, loss is defined in the BAC policy as direct and accidental loss or damage. Since loss of use (e.g., expenses incurred in temporarily replacing a damaged auto) is not a form of direct damage to a vehicle, but rather a consequence of direct damage, it does not trigger the BAC policy's physical damage coverage. The policy does, however, specify two sets of circumstances, total theft of a private passenger covered auto and legal liability for loss of use expenses caused by any covered peril associated with a hired auto, under which loss of use expenses, up to a per-day dollar maximum, will be covered.

The physical damage insuring agreement sets out the insurer's obligation to pay for physical damage loss to a covered auto or its equipment under comprehensive, specified causes of loss, or collision coverage. The insuring agreement also sets out provisions related to coverage for: (a) towing and labor costs; (b) optional provisions for glass breakage coverage; and (c) coverage extensions for temporary transportation expenses and for hired auto physical damage loss of use expenses an insured may become legally responsible to pay. The BAC policy extends limited amounts of coverage for substitute transportation if a covered auto (whether under comprehensive or specified causes) has been stolen, as well as for the insured's liability for loss of use of a rented vehicle through rental agreement.

It is recommended you expand the core physical damage coverage to include broader rental car reimbursement on all covered autos, auto loan/lease gap, airbag repair, towing on all covered autos, extra expense for stolen autos, glass repair waiver of deductible, and augmented repair expenses for customized vehicles. The foregoing enhancements will properly augment your BAC policy so that it provides the necessary breadth, depth, and customization required for public water/wastewater systems. We will now examine the exclusions of the physical damage coverage section of the BAC policy.

Exclusions

Catastrophic Perils – Nuclear hazards, war, and warlike action. The BAC policy does not exclude earthquake, flood, or other water damage.

Maintenance Losses – Wear and tear, freezing, mechanical breakdown, and road damage to tires. An exemption should be made for airbag repair.

Electronic Equipment – Mobile phones, laptops, tablets, iPods, cassette tapes, CDs, radar detectors, or any other equipment not permanently installed onto the vehicle.

Diminution in Value – Reduction in the value of a vehicle that has been damaged in an accident and then repaired. This exclusion may significantly limit recovery for a physical damage loss because a postaccident auto expertly repaired with high-quality parts will invariably have a lower resale value than a comparable vehicle without an accident.

Special Commentary

Physical damage exclusions set out circumstances in which there is no physical damage coverage available under the policy. Such exclusions are of special importance in connection with comprehensive coverage, under which loss is covered unless specifically excluded elsewhere in the policy. It is recommended you always select comprehensive coverage over specified causes of loss and pair it with collision coverage. The foregoing exclusions are reasonable, but you should secure an exemption under the mechanical breakdown exclusion for airbag repair. We are now ready to review the limit of insurance in the physical damage section of the BAC policy.

Limit of Insurance

The most we will pay for: (a) Loss to any one covered auto are the lesser of: (1) the actual cash value of the damaged or stolen property as of the time of the loss; or (2) the cost of repairing or replacing the damaged or stolen property with other property of like kind and quality. (b) All electronic equipment that reproduces, receives, or transmits audio, visual, or data signals in any one loss is $1,000, if, at the time of loss, such electronic equipment is: (1) permanently installed in or upon the covered auto in a housing, opening or other location that is not normally used by the auto manufacturer for the installation of such equipment; (2) removable from a permanently installed housing unit as described in paragraph (b)(1) above; or (3) an integral part of such equipment as described in paragraphs (b)(1) and (b)(2) above. An adjustment for depreciation and physical condition will be made in determining actual cash value in the event of a total loss. If a repair or replacement results in better than like kind or quality, we will not pay for the amount of the betterment.

Key Definitions

Actual Cash Value – the fair or reasonable cash price for which the property could be sold in the market in the ordinary course of business, and not at forced sale, or what property is worth in money, allowing for depreciation.

Agreed Value – you and your insurer agree upon the value of your vehicle when you take out the policy, and in the event of a covered loss, you will be reimbursed the lesser of the repair cost to fix the vehicle or the agreed value, regardless of any depreciation.

Replacement Cost Value – the cost to repair or replace the damaged or stolen property with other property of like kind and quality.

Special Commentary

Under the physical damage section of the BAC policy, the most that will be paid for covered damage or loss is the lesser of the actual cash value of the damaged property or replacement cost value, the cost to repair, or replace with like kind and quality.

The insurer is obligated only for the smallest amount that would be payable to the insured as determined by: (a) actual cash value; (b) cost to repair; or (c) cost to replace. In the event of a total loss, an adjustment for depreciation and physical condition is made to determine actual cash value and is determined as of the time of the loss. If a repair or replacement results in the insured's betterment, (restoration of property to a condition that is better than the property damaged or lost), the insurer will not pay for the amount of the betterment. There is a separate (lower) limit for loss to electronic equipment that is not excluded. This limit should be increased, as many public water/wastewater systems have expensive electronic equipment in their vehicles. The BAC policy preserves the concept of indemnity so that the insured does not profit from the loss when the circumstances are such that it is impossible for the insurer to repair or replace the property without bettering the insured's position.

You should avoid stated amount, which alters the physical damage loss settlement clause by limiting recovery to the lesser of: (a) actual cash value of damaged or stolen property; (b) replacement cost value (cost to repair or replace with like kind and quality); or (c) stated amount shown in endorsement's schedule. This valuation option is the least generous loss settlement method. For customized vehicles with specialized equipment or for high-valued commercial vehicles, an agreed value or replacement cost option is recommended. The former allows you a flat amount, the agreed value, should the vehicle be a total loss, whereas the latter provides cost new of a replacement vehicle. In some situations, it is acceptable to secure replacement cost for customized equipment installed on the chassis and actual cash value on all other vehicle components. That completes the limit of insurance within the physical damage coverage section of the BAC policy. We will now turn our focus on the deductible.

Deductible

For each covered auto, our obligation to pay for, repair, return, or replace damaged or stolen property will be reduced by the applicable deductible shown in the supplemental declarations. Any comprehensive deductible shown in the supplemental declarations does not apply to loss caused by fire or lightning.

Special Commentary

A deductible applies to most auto physical damage losses and is on a per-damaged vehicle basis. The amount of the deductible for a particular loss is stated in the supplemental declarations. The higher the deductible, the lower the cost of coverage for the insured. In the event of a loss, the amount of the applicable deductible is subtracted from the amount the insurer is obligated to pay as determined by the limit of insurance provision. It is common practice for collision and comprehensive deductibles in the same policy to be different, with collision deductibles usually being higher than comprehensive deductibles.

There is no deductible under non-collision coverage applying to loss caused by fire or lightning, regardless of whether the coverage is on a comprehensive or specified causes of loss basis. Two types of physical damage coverage that are not subject to a deductible are towing and transportation expense because of theft of a covered vehicle. In both cases, the insurer's obligation is limited to a certain specified amount, e.g., $50 for towing or $20 per day for temporary transportation expense, making a deductible unnecessary. We have now completed our review of the physical damage section of the BAC policy and will turn our attention to the key policy conditions.

SECTION IV. KEY POLICY CONDITIONS

Loss Conditions

Appraisal for Physical Damage Losses: This condition describes the process that will be followed if the insured or the insurer demands an appraisal of the damaged property.

Duties in the Event of Accident, Claim, Suit, or Loss: For liability coverage, key loss conditions include requirements that the insured cooperate with the insurer, provide copies of relevant legal papers, submit to physical examinations, and authorize the insurer to obtain relevant medical information. For physical damage coverage, these conditions include requirements that the insured promptly notify law enforcement if a covered auto is stolen, preserve property from further loss, permit the insurer to inspect and appraise damaged vehicle before it is repaired, and agree to examination under oath.

There is also a loss recovery condition that allows the insurer to choose to pay, repair, or replace property, return property, and repair damage caused by theft, or to keep property and pay the agreed or appraised value. Finally, among the loss conditions that apply to physical damage coverage, the insured agrees to a transfer of rights against others. This condition makes it possible for the insurer to subrogate against a third party on the insured's behalf to potentially recover some or all of the loss amounts paid should a third party be found liable for an accident after the insurer has compensated the insured for loss sustained. This clause gives the insurer the right to recover the amount it has paid for a loss from the party that caused it.

Applying to both the liability and physical damage coverages, there is an additional condition clause governing legal action against the insurer. This provision is often called the no-action clause because it limits your right to file an action (lawsuit) against your insurer. It typically bars you from suing unless you have fulfilled all of the requirements under the policy. Under the physical damage coverage, for example, you cannot sue your insurer with regard to a claim if you have failed to provide a description of the damaged property (a condition of coverage). Under the liability coverage, you are typically barred from suing your insurer to collect a settlement you made voluntarily (without your insurer's consent). Likewise, you may be prohibited from suing to collect damages until a final judgment has been made by a court.

General Conditions

Bankruptcy: The bankruptcy condition states that the insured's bankruptcy will not relieve the insurer of its duties under the policy.

Concealment, Misrepresentation, or Fraud: This clause allows the insurer to void the policy if the insured has committed a fraudulent act. An insured commits fraud when he or she intentionally deceives an insurer for the purpose of financial gain. Fraud may be committed when coverage is purchased, when a claim is filed, or at some other time. The clause also allows the insurer to deny coverage if an insured has intentionally misrepresented or concealed a material fact regarding the insurance coverage. The term "misrepresentation" means a misstatement of the truth. The misstatement is *material* if the insurer would have made a different decision had it known the true facts.

Liberalization: This clause automatically expands your policy to include any coverage your insurer has added to your coverage form. The clause typically applies to any extension that was made shortly before or during your policy period, if the extension is free of charge.

No Benefit to Bailee: The no benefit to Bailee clause applies to physical damage coverages. It states that no one, other than the insured, who has custody of the insured property will benefit from the policy. In other words, a Bailee is not entitled to a claim payment simply because he or she has possession of the insured property. Under the BAC policy, the Bailee may be a parking garage, towing company, repair shop, or anyone else who charges a fee to obtain control over your vehicle. The Bailee relationship is reversed should your public water and wastewater system operate a repair facility for other public water/wastewater systems and public entities. In such an instance, you will need to secure garage liability and garage keepers liability to protect against your Bailee exposures.

Other Insurance: This condition explains how the policy will respond when other coverage exists for a claim that is covered by your policy. As a rule of thumb, the BAC policy follows the vehicle, so the insurance covering the at-fault auto would generally be primary and other insurance would cover the excess primary policy. For instance, an employee driving his or her own vehicle for public water and wastewater system purposes would have primary coverage under a personal auto policy, and the BAC policy would serve as excess coverage. Coverage for a trailer follows the auto to which it is attached. If the other insurance situation involves two or more coverage forms or policies from same insurer, the clause is likely to dictate that whichever coverage pays the highest will pay the claim.

Premium Audit: The premium audit will examine your records to establish the actual exposure basis (e.g., vehicle fleet) and prior loss information and ensure that the correct codes and rates are used in determining your final premium. This condition should be removed via a fleet automatic endorsement, which removes mid-term premium changes for auto additions and deletions pending symbols 1 (any auto) or 2 (any owned auto) for liability, and symbol 2 (any owned auto) for physical damage. It also requires the purchase of physical damage (i.e., comprehensive and collision) on all or nearly all your vehicles.

Policy Period, Coverage Territory: The coverage trigger is occurrence based. Accidents and losses are covered only if they occur during the policy period shown in the supplemental declarations and in the coverage area, which is defined as: (a) the United States; (b) its territories and possessions; (c) Puerto Rico; (d) Canada; and (e) anywhere in the world for a private passenger vehicle (not a truck) that is hired, leased, rented, or borrowed without a driver for 30 days or less. To be covered, any claim or suit must be brought in the United States, its territories or possessions, Puerto Rico, or Canada. The BAC policy also covers physical damage to or an accident involving a covered auto that is being transported between places in the United States, its territories or possessions, Puerto Rico, or Canada.

Key Definitions

Subrogate – to substitute (something or someone, such as an insurer) for another with regard to a legal right or claim.

Bailee – someone who has been entrusted with another party's property for a particular purpose.

Material Misrepresentation – a misstatement of the truth that results in the insurer making a different decision that it would have made had it known the true facts.

Liberalization – an expansion of coverage during the policy effective period.

Special Commentary

The conditions section of a BAC policy lists the legal responsibilities of the insured and the insurer as well as the protocols for issues, such as loss reporting and settlement, property valuation, other insurance, subrogation rights, and cancellation and nonrenewal. These conditions are divided into loss conditions and general conditions. These conditions are not complicated or onerous, but they must be reviewed and understood to ensure you protect all your rights under the policy and avoid any unintentional voidance of policy terms and conditions. We are now ready to review the necessary coverages and enhancements to properly augment your BAC policy to achieve best-in-class status and provide the required protections for a public water and wastewater system.

SECTION V. ADDITIONAL COVERAGES AND ENHANCEMENTS

ADDITIONAL COVERAGES

Medical Payments – Coverage under medical payments, pays for the treatment of injuries you or your passengers suffer in a car accident, no matter who caused the crash. It also pays the medical bills if you are hit by a car while on foot or riding in someone else's vehicle for a public water and wastewater system purpose. Medical payments have many of the same benefits as your health insurance policy, but its purpose is to address gaps in coverage, low limits, and steep deductibles. This coverage is typically optional and does not involve deductible. Benefits commonly consist of the following: treatment of injuries you or your passengers sustain in a car accident, including medical, dental, surgical, and chiropractic care; treatment of injuries you or your passengers sustain while riding in someone else's car for a public water and wastewater system purpose; ambulance fees; X-rays, prostheses, and nursing; funeral costs following a fatal crash. It does not pay for health insurance deductibles or copays as well as wage reimbursement if injuries force you to miss work.

Personal Injury Protection – The purpose of the personal injury protection endorsement is to provide no-fault coverage in states where it is compulsory. Coverage extends to medical expenses, income loss, and other benefits stipulated. A separate personal injury protection endorsement exists for each no-fault state and pays first-party benefits regardless of which party was at fault. In some states, benefit levels can be increased above the minimum state-required levels with the added personal injury protection endorsement. It also includes childcare costs if you're limited by accident injuries and encompasses more extensive benefits than medical payments. Examples include wage reimbursement if injuries force you to miss work and funeral costs. Unlike medical payments, personal injury protection often includes a deductible, which is a predetermined amount that your insurer subtracts from claim payments.

Uninsured/Underinsured Motorists – This endorsement provides occupants of a covered auto and pedestrians a source of recovery when an at-fault motorist (or hit-and-run driver) has no insurance or inadequate insurance limits. In some states, uninsured/underinsured motorists also pays for property damage. This coverage is important, irrespective of your workers compensation policy because the latter may not pay all of an injured employee's damages. Moreover, an employee may be driving a covered auto for personal use, and uninsured/underinsured motorists coverage allows your policy to pick up liability for damages associated with loss incidents in these cases, given the at-fault third party either had no insurance (uninsured) or had insurance but with insufficient limits to cover the damages (underinsured).

Garage Liability and Garagekeepers Liability – The liability coverage provided within the garage coverage form pays third-party damages due to bodily injury or property damage resulting from your garage operations. There are two liability insuring agreements: (1) garage operations – other than covered autos; (2) and garage operations – covered autos. The first insuring agreement provides coverage for bodily injury and property damage that do not involve covered autos. The coverage is comparable to (but not as broad as) a CGL policy. It is subject to per occurrence and aggregate limits of insurance. The second insuring agreement covers bodily injury, property damage, and covered pollution liability arising out of ownership, maintenance, or use of covered autos. Products liability coverage is similar to the products liability coverage found under the CGL policy. Completed operations are also covered. Customers are insureds but only up to minimum limits required by state. Garage operations means your use of one or more locations for the service, repair, parking, or storage of autos other than your own, including all operations necessary or incidental thereto. Parking or storage of autos is a garage operation only when the autos are parked by you and are in your care, custody, or control. Customer's auto means a land motor vehicle, trailer, or semitrailer lawfully within your possession for service, repair, storage, or safekeeping, including a vehicle left in insured's care by employees and members of their households, who pay for services. Garage keepers coverage can be added to the garage form to protect the insured against liability for damage to an auto left in the insured's care while the insured is attending, servicing, repairing, parking, or storing the auto. This coverage is desirable because garage liability excludes damage to property in the insured's care, custody, or control. Covered causes of loss include collision and non-collision (comprehensive or specified causes of loss). When purchasing this coverage, there are two choices to consider. The first is the question of direct coverage, versus legal liability coverage. If you purchase direct coverage, then your policy will pay the claims for collision and comprehensive coverage for your client's vehicles regardless of whether you are at fault in the loss. If you choose legal liability coverage, then these claims will only be covered if your repair or body shop can be found to be legally liable for the loss. The second question to be considered is whether to purchase primary or excess coverage on your garage keepers coverage. Primary coverage means that your insurance policy will pay the first dollar on the covered loss, while excess coverage will only pay when there is no coverage on your client's insurance policy or if the limits of your client's policy are not high enough to cover the loss.

ADDITIONAL ENHANCEMENTS

Employees, Volunteers, and Governing Body Officials as Insureds – This endorsement extends coverage under the liability section of the BAC form to include insured status for employees, volunteers, and governing body officials while using their personally owned vehicles for public water and wastewater system purposes. The failure to include said coverage creates the possibility of your insurer subrogating against the foregoing individuals for any losses in excess of their personal auto policies. As a reminder, your BAC policy protects the public water and wastewater system for its vicarious liability arising from the use of a personally owned vehicle by your employees, volunteers, or governing body officials for organizational purposes.

The above-referenced endorsement extends protection to include the foregoing individuals. Even with this endorsement, the personal auto policy remains the primary protection. A sound risk management practice includes annual verification of personal auto policies with specified minimum liability limits for all individuals who may use their personally owned vehicles for public water and wastewater system purposes.

Fellow Employees – This endorsement extends coverage under the liability section of the BAC form by removing the fellow employee exclusion from the BAC form. Such removal allows the policy to respond on behalf of the at-fault employee following a vehicle-related injury to a fellow employee caused by a covered vehicle. Without this endorsement, employees who unintentionally cause an auto-related injury to a fellow employee may be left without any insurance protection. It is important to confirm the definition of employee includes volunteers and governing body officials.

Employee Hired Autos – This endorsement extends coverage under the liability section of the BAC form to include insured status for employees who rent autos in their own name for public water and wastewater system purposes. Symbol 1 (any auto) or symbol 8 (any hired auto) may cover the exposure, but this endorsement provides confirmation that your employees are covered. The endorsement is important because most employees rent vehicles in their own name for company business. Such transactions are common since auto rentals are often completed via personal credit cards. It is important to confirm the definition of employee includes volunteers and governing body officials.

Rental Reimbursement Coverage – This endorsement extends coverage under the physical damage section of the BAC form to reimburse the insured for the cost to rent a replacement vehicle while a covered vehicle is being repaired following a covered physical damage loss. The policy is subject to three maximums: a maximum per day limit; a maximum number of days; and a maximum total per loss, per vehicle. Further, the policy contains a 24-hour after the loss time deductible. Coverage limits should be based on the type of vehicles being replaced. Recommended minimum limits for private passenger autos are $50 per day up to 30 days for a total of $1,500 and all other autos are $150 per day up to 30 days for a total of $4,500. Specialized and high-valued commercial vehicles may require higher limits and necessitate a knowledge of the local rental market for such vehicles. Coverage is not extended if you have a spare or reserve auto available for use. The endorsement pays only when you need a replacement vehicle, not just because the covered vehicle is not available for use due to a covered cause of loss.

Optional Coverage for Pollutants Transported by Autos – This endorsement may be called pollution liability – broadened coverage for covered autos. It amends the first three parts of the pollution exclusion (described previously) under the liability section of the BAC form. When the endorsement is added, these three sections apply only to liability assumed under a contract. In other words, the three exclusions do not apply unless you have signed a contract in which you have assumed responsibility for pollutants released while being transported, loaded, or unloaded, stored, or treated on a covered auto. Two additional points should be mentioned regarding the endorsement: (1) it covers cleanup costs only if they result from an accident that also causes bodily injury or property damage covered by the policy – the endorsement does not cover cleanup costs in the absence of covered bodily injury or property damage; and (2) the endorsement has no effect on the last two sections of the pollution exclusion – there is no coverage for claims involving pollutants released before they are loaded onto a vehicle or after they have been unloaded.

Temporary Vehicle Substitute – This endorsement extends coverage under the liability section of the BAC form to include insured status for the owner or anyone else from whom you rent, lease, or borrow a substitute auto but only for such substitute auto. The substitute auto must be for a similar scheduled auto that is out of normal use because of its breakdown, repair, servicing, loss, or destruction. The substitute auto will be considered a covered auto you own and not a covered auto you rent, lease, or borrow.

Owner of Commandeered Auto – This endorsement extends coverage under the liability section of the BAC form to include insured status for the owner of a commandeered auto while the commandeered auto is in your temporary care, custody, or control and is being used as part of an emergency situation. A commandeered auto shall be deemed to be a covered auto you own but only while such commandeered auto is in your temporary care, custody, or control and is being used as part of an emergency situation.

Additional Insured – Automatic Status – This endorsement extends coverage under the liability section of the BAC form to include insured status for any person or organization with whom you have agreed in a written contract or written agreement to add as an additional insured under your policy but only with respect to liability arising out of your operations and to the extent such person or organization qualifies as an insured under the liability section of the BAC form.

Auto Loan/Lease Gap – This endorsement alters the amount paid under the physical damage section of the BAC form to include the difference between the actual cash value of the vehicle and the amount remaining on the loan or on the lease. It is intended to address situations where you are upside down on a vehicle loan or lease at time of loss. This first-party coverage is for your benefit only and is intended to satisfy your contract obligation with either the loss payee (lienholder) or lessor. Since vehicle values depreciate quickly and the difference between the actual cash value and the amount owed can be substantial, this endorsement should be considered if you lease your vehicles. The endorsement is only triggered when there is a total loss. Payment is limited to the value associated with the specific vehicle. Expenses such as overdue payments, high-mileage and usage penalties, security deposits, add-on costs (i.e., credit life), and balances from prior loans or leases carried over to the current financing agreement are excluded from coverage.

Airbag Repair – This endorsement confirms the mechanical breakdown exclusion under the physical damage section of the BAC form does not apply to the repair of an airbag because of an accidental discharge or deployment.

Towing – This endorsement extends coverage for towing under the physical damage section of the BAC form when comprehensive was included. It pays for reasonable labor costs incurred to make necessary repairs to a covered auto so it can be driven from the scene of disablement. The labor must be performed at a scene of disablement other than your normal garaging location for such auto. Coverage also extends for all reasonable towing costs incurred for towing the disabled covered auto from the scene of disablement to an appropriate repair facility. A recommended minimum limit is $2,500 for each disabled covered auto. Coverage should extend to both private passenger and any other autos.

Hired Auto Physical Damage – This endorsement extends hired auto coverage under the physical damage section of the BAC form. It covers any auto you lease, hire, rent, or borrow from anyone other than your employees or partners or members of their households for the broadest physical damage coverage that applies to at least one of your other covered autos. The recommended minimum limits are the lesser of the following: (1) the actual cash value of the damaged or stolen property as of the time of the loss; (2) The cost of repairing or replacing the damaged or stolen property with other property of like kind and quality; or (3) $500,000. A deductible representing an amount equal to your largest deductible shown in the supplemental declarations for any owned auto will apply. This deductible does not apply to loss caused by fire or lightning. Coverage is also deemed to be primary for any covered auto you hire without a driver and excess over any other collectible insurance for any covered auto that you hire with a driver.

Extra Expense for Stolen Autos – This endorsement extends coverage under the physical damage section of the BAC form by paying for the expense of returning a stolen covered auto to you.

Glass Repair Waiver of Deductible – This endorsement extends coverage under the physical damage section of the BAC form by waiving any deductible applicable to glass damage if the glass is repaired rather than replaced.

Customized Vehicles – This endorsement extends coverage under the physical damage section of the BAC form by paying the additional repair or replacement costs necessary to customize the damaged auto with permanently installed equipment of like kind and quality, without deduction for depreciation. Customization includes, but is not limited to, the following: (a) custom painting and gold leaf lettering; (b) light bars and sirens; (c) permanently installed communications equipment, including global positioning systems (GPS); and (d) computer or electronic equipment that receives or transmits audio, visual, or data signals. Coverage also extends to property owned by you that is permanently installed in an auto not owned by you.

Special Commentary

The BAC policy offers multiple endorsements, most of which serve to broaden coverage, although some narrow or otherwise alter the coverage. There are scores of endorsements available for use with the BAC policy, with the number varying by state. The foregoing endorsements represent the most salient coverage enhancements for public water/wastewater systems. You should compare your current BAC policy to the above-referenced best-in-class benchmark. It is important to assess the core coverages, optional coverages, policy endorsements, and respective limits so as to fully capture all elements for your examination process.

Final Comments

The autoexposure for public water/wastewater systems is expansive and involves most aspects of your daily operations. Its significance necessitates thoughtful risk mitigation measures inclusive of periodic Motor Vehicle Report (MVR) checks on new and existing employees, continuous driver training, active maintenance programs, driver assistance technologies, and a holistic safety culture of accountability. These measures, although necessary, are not sufficient on their own. They must be complemented by a deliberative risk transfer strategy, anchored by insurance, where your financial risk to first- and third-party auto losses is appropriately ported to a best-in-class insurer specializing in public water/wastewater systems. Your insurer should offer a comprehensive BAC policy that is augmented with customized endorsements and extensions unique to your specific needs.

Fundamental knowledge of your BAC policy is important, especially the supplemental declarations and auto symbols. The preferred auto symbol for liability is 1 (any auto). This symbol provides protection for any type of owned and non-owned auto. It does not necessitate the scheduling of autos to trigger coverage. Should symbol 1 (any auto) not be offered, then acceptable substitutes are symbol 2 (owned autos only), symbol 8 (hired autos only), and symbol 9 (non-owned autos only). Similar to symbol 1 (any auto), the preceding symbols provide extensive liability protection for owned and non-owned autos. These symbols also prevent the scheduling of autos to trigger coverage. The use of symbol 7 (specifically described autos) should be avoided, as this symbol triggers coverage only for the schedule of autos submitted to your insurer. Fleet automatic is also critical, as that will prevent mid-term premium changes as well as end-of-policy period audits. It also prevents unintended claims processing delays.

The preferred auto symbol for physical damage is symbol 2 (owned autos only) because it provides coverage for all owned autos. Should you have older vehicles in your fleet, then symbol 7 (specifically described autos) may be required for physical damage as a means of reducing premiums for fully or nearly depreciated vehicles. It is prudent to purchase comprehensive over specified causes of loss for physical damage losses not involving collision and overturn. Comprehensive covers all direct physical loss or damage to your autos unless specifically excluded. In contrast, coverage for specified causes of loss is limited to named perils only. The purchase of collision should be paired with comprehensive because collision represents the majority of physical damage claims. This combination provides maximum physical damage protection for your auto fleet. The cost of premium for the benefit of complete protection is warranted, as it provides financial protection to one of your most valuable capital assets. For the reasons referenced above for liability, physical damage fleet automatic should also be included.

Your BAC policy includes core coverage for liability and physical damage. It should also encompass optional coverages for uninsured/underinsured motorists as well as medical payments or personal injury protection (depending on your state). The addition of garage liability and garage keepers liability is equally important should you provide third-party automotive services to neighboring public water/wastewater systems and public entities. Recommended policy enhancements include the following: broad form pollution; rental car reimbursement (private passenger and all other autos); towing reimbursement (private passenger and all other autos); employees, volunteers, and governing body officials as insureds; fellow employee; glass repair waiver of deductible; customized vehicle decals and electronic equipment; airbag repair; hired autos by insureds; auto lease gap; commandeered property; and blanket additional insured and blanket loss payee where required by written agreement.

Liability limits should be set at $1 million combined single limit, $1 million hired and non-owned auto liability, $1 million uninsured/underinsured motorists, $500,000 hired physical damage, and $10,000 medical payments or personal injury protection. Garage liability and garage keepers liability should be $1 million and $500,000, respectively. Towing should be a flat $2,500 for all vehicles, whereas rental car reimbursement should be up to $50 per day up to 30 days for private passenger vehicles and up to $150 per day up to 30 days for all other vehicles. Physical damage should be based on an actual cash value basis except for high-valued and specialized apparatus. In such instances, the valuation should be agreed value or replacement cost. Stated amount should be avoided as a valuation option. The foregoing limits are minimum recommendations, and the purchase of higher limits via an excess liability policy where your BAC policy is scheduled as an underlying policy is strongly advised. Overall, your BAC policy is one of your most important risk transfer and insurance decisions. Vehicular liability across all industry segments is rising in frequency, severity, and complexity. The effectiveness of the foregoing risk transfer strategy requires a keen understanding of your policy structure and appropriate design dynamics. This knowledge will ensure your first- and third-party auto exposures are appropriately addressed in a customized and proprietary BAC policy that meets the benchmark of a best-in-class auto policy. We have now completed our review of the BAC policy and will begin our review of the commercial crime (crime) policy.

POLICY CHECKLIST

BAC Policy: Minimum Benchmarks

Business Auto Coverage (BAC)	
Description	**Benchmark**
Form	
Insurance Services Office (ISO)	Yes (with proprietary endorsements)
Symbol	
Symbol 1	Combined Single Limit (liability)
Symbol 2	Auto Physical Damage
Symbol 5	Medical Payments/Personal Injury Protection (PIP)
Symbol 8/9	Hired and Non-owned Auto Liability
Limits	
Combined Single Limit	$1,000,000
Hired and Non-owned Auto Liability	$1,000,000
Uninsured/Underinsured Motorists	$1,000,000
Medical Payments/PIP	$10,000
Hired Physical Damage	$500,000
Auto Physical Damage	Actual Cash Value (all vehicles); Replacement Cost (highly specialized equipment)
Garage Liability	$1,000,000
Garagekeepers Liability	$500,000
Rental Car Reimbursement	$50 per day / 30 days (private passenger auto) $150 per day / 30 days (all other vehicles)
Towing	$2,500 (any disablement of covered auto)
Coverage Enhancements and Extensions	
Employees, Volunteers, and Governing Body Officials as Insureds	Yes
Fellow Employee	Yes
Employee Hired Autos	Yes
Broad Form Pollution	Yes
Fleet Automatic	Yes (liability and physical damage)
Customized Vehicles	Yes (additional repair and replacement)
Temporary Vehicle Substitute	Yes
Blanket Additional Insured w/ Primary and Noncontributory and Waiver of Subrogation Verbiage	Yes (where required by written agreement)
Auto Loan/Lease Gap	Yes
Airbag Repair	Yes
Glass Repair Waiver of Deductible	Yes

REVIEW QUESTIONS

1. T/F: Vehicular liability represents one of the most financially significant and unpredictable risks facing public water/wastewater systems.

2. The most common vehicular liability doctrines facing commercial and governmental entities include the following:
   a. Vicarious
   b. Contract
   c. Common law
   d. All of the above

3. Owned or non-owned vehicles driven by your employees, volunteers, and governing body officials for public water and wastewater system purposes create the following liability:
   a. Contractual
   b. Common
   c. Vicarious
   d. None of the above

4. T/F: Respondeat superior is a legal doctrine that attaches liability to an employer (principal) for the acts of its employees, volunteers, and governing body officials (agents).

5. The following coverage is recommended for those public water/wastewater systems that have garage operations and perform occasional mechanical work for other public water/wastewater systems or governmental entities:
   a. Garage liability
   b. Garage keepers liability
   c. Fiduciary liability
   d. a and b only

6. T/F: Coverage under your commercial auto policy is restricted to accidents occurring while you or your employees, volunteers, and governing body officials are using a vehicle for public water and wastewater system purposes.

7. T/F: Unlike owned vehicles, protection afforded under non-owned auto liability within the BAC policy is primary over any valid and collectible personal auto policy.

8. T/F: Insurance Services Office (ISO) offers a widely accepted commercial auto policy referred to as a personal auto coverage (PAC) form, which is extensively used by public water/wastewater systems and other commercial and governmental entities.

9. Examples of coverages available in a BAC policy include the following:
   a. Uninsured motorists
   b. Medical payments
   c. Rental car reimbursement
   d. All of the above

10. T/F: There are five sections within the supplemental declarations page of the BAC policy.

11. T/F: Symbol 1 of a BAC policy refers to owned autos only.

12. T/F: Symbol 1 of a BAC policy is intended for liability coverage only.

13. T/F: Symbols 2, 3, or 4 of a business coverage (BAC) policy alone will not trigger coverage for hired, borrowed, or non-owned autos except for substitute autos and for trailers attaching to one of your own power units.

14. T/F: The recommended symbol for physical damage (i.e., comprehensive and collision) coverage in a BAC policy is symbol 2.

15. The symbols in a BAC policy that automatically include liability coverage for an owned land vehicle that would qualify as mobile equipment except for being subject to a compulsory or financial responsibility law or other motor vehicle insurance law are as follows:
    a. Symbol 1
    b. Symbol 2
    c. Symbol 3
    d. a and b only

16. T/F: Symbol 5 in a BAC policy triggers no-fault coverage but only for autos required by law to carry it.

17. T/F: Symbol 3 in a BAC policy triggers automatic uninsured motorists coverage only in those states that do not allow you to reject the coverage.

18. T/F: Symbol 7 in a BAC policy restricts coverage to vehicles listed in the schedule of covered autos.

19. The symbols that most closely duplicate liability coverage provided under symbol 1 in a BAC policy include the following:
    a. 2, 7, and 8
    b. 2, 7, and 9
    c. 2, 8, and 9
    d. None of the above

20. T/F: Hired auto physical damage coverage in a BAC policy does not apply to an auto leased, hired, rented, or borrowed without a driver.

21. T/F: Symbol 9 in a BAC policy includes autos that are owned by your employees, volunteers, or governing body officials if such autos are used for public water and wastewater system purposes.

22. T/F: Symbol 19 in a BAC policy includes vehicles that would ordinarily be considered mobile equipment except for being subject to a compulsory financial responsibility or other motor vehicle law where it is licensed or principally garaged.

23. The definition of auto in a BAC policy includes the following:
    a. Land motor vehicle, trailer, or semitrailer designed for travel on public roads
    b. Any other land vehicle that is subject to a compulsory or financial responsibility law or other motor vehicle insurance law where it is licensed or principally garaged
    c. a and b only
    d. None of the above

24. T/F: The utilization of fleet automatic in a BAC policy is recommended to ameliorate audit premiums as well as streamline mid-term additional/return premium invoices.

25. T/F: Defense costs are inside the limit of insurance in a BAC policy.

26. T/F: The insurer must defend the insured defendant against false or fraudulent claims as long as they allege covered damages in a BAC policy.

27. The definition of insured in a BAC policy comprise the following:
    a. Named insured
    b. Permissive users
    c. Anyone liable for the conduct of an insured
    d. All of the above

28. T/F: A rental car company is automatically classified as an insured under the BAC policy for your use of their rental car.

29. Standard liability exclusions in a BAC policy includes the following:
    a. Expected or intended injury
    b. Mechanical breakdown

    c. Glass repair

    d. a and b only

30. The pollution liability exclusion in a BAC policy encompasses the following:
    a. Pollutants being transported by you or on your behalf
    b. Pollutants being loaded onto, or unloaded from, a covered auto
    c. Pollutants being stored, disposed of, treated, or processed in or on a covered auto
    d. All of the above

31. T/F: The liability limit of insurance in a BAC policy automatically stacks when multiple covered autos, insureds, claims made, or vehicles are involved in the same accident.

32. T/F: No aggregate applies to the liability limit of insurance in a BAC policy.

33. T/F: Comprehensive coverage and specified causes of loss are alternative approaches to insuring non-collision physical damage losses in a BAC policy.

34. T/F: Specified causes of loss and collision coverage is the preferred pairing for physical damage coverage in a BAC policy.

35. Loss of use claims for physical damage in a BAC policy applies to the following scenarios:
    a. Total theft of a private passenger covered auto
    b. Legal liability for loss of use expenses caused by any covered peril associated with a hired auto
    c. a and b only
    d. None of the above

36. Standard physical damage exclusions in a BAC policy includes the following:
    a. Workers compensation
    b. Maintenance losses
    c. Diminution in value
    d. b and c only

37. T/F: Stated amount is the preferred valuation method for physical damage claims in a BAC policy.

38. T/F: Agreed value in a BAC policy means you and your insurer agree upon the value of your vehicle when you take out the policy, and in the event of a covered loss, you will be reimbursed the lesser of the repair cost to fix the vehicle or the agreed value, regardless of any depreciation.

39. There is no physical damage deductible in a BAC policy for the following covered autos:
    a. Fire
    b. Lightning
    c. a and b only
    d. None of the above

40. The liability and physical damage duties in the event of an accident, claim, suit, or loss in a BAC policy includes the following:
    a. Insured must cooperate with the insured
    b. Promptly notify law enforcement if a covered auto is stolen
    c. Insured must submit to physical examinations
    d. All of the above

41. T/F: Coverage under medical payments in a BAC policy pays for the treatment of injuries you or your passengers suffer in a car accident, no matter who caused the crash.

42. Medical payments in a BAC policy includes the following benefits:
    a. Medical, dental, surgical, and chiropractic care
    b. Health insurance deductibles and copays
    c. Wage reimbursement
    d. All of the above

43. Coverage under personal injury protection in a BAC policy includes the following benefits:
    a. Childcare costs if you are limited by accident injuries
    b. Wage reimbursement if injuries force you to miss work
    c. Funeral costs
    d. All of the above

44. T/F: Coverage for uninsured motorists in a BAC policy provides occupants of a covered auto and pedestrians a source of recovery when an at-fault motorist (or hit-and-run driver) has no insurance or inadequate insurance limits.

45. T/F: The garage liability coverage form pays first-party damages due to bodily injury or property damage resulting from your garage operations.

46. T/F: The garage liability coverage form pays for losses involving property in the insured's care, custody, or control.

47. T/F: Garage keepers coverage can be added to the garage liability coverage form to protect the insured against liability for damage to an auto left in the insured's care while the insured is attending, servicing, repairing, parking, or storing the auto.

48. Garage keepers coverage includes the following options:
    a. Direct coverage or legal liability coverage
    b. Primary coverage or excess coverage
    c. a and b only
    d. None of the above

49. T/F: The definition of employee in a BAC policy should be expanded to include volunteers and governing body officials.

50. T/F: Employees as insured endorsement in a BAC policy extends liability coverage for employees while using their personally owned vehicles for public water and wastewater system purposes.

51. T/F: Without the fellow employee endorsement in a BAC policy, employees who unintentionally cause an auto-related injury to a fellow employee may be left without any liability insurance protection.

52. T/F: Employees hired auto endorsement in a BAC policy extends liability coverage for employees who rent autos in their own name for public water and wastewater system purposes.

53. T/F: Rental reimbursement coverage endorsement in a BAC policy automatically includes both private passenger autos and other than private passenger autos.

54. T/F: Pollution liability – broadened coverage for covered autos endorsement in a BAC policy protects against spills of transported pollutant contents as long as there is no accident.

55. T/F: Auto loan/lease gap endorsement in a BAC policy alters the amount paid under a physical damage claim to include the difference between the actual cash value of the vehicle and the amount remaining on the loan or on the lease.

ANSWER KEY

1. **T**/F: Vehicular liability represents one of the most financially significant and unpredictable risks facing public water/wastewater systems.

2. The most common vehicular liability doctrines facing commercial and governmental entities include the following:
    a. Vicarious
    b. Contract
    c. Common law
    d. **All of the above**

3. Owned or non-owned vehicles driven by your employees, volunteers, and governing body officials for public water and wastewater system purposes create the following liability:
   a. Contractual
   b. Common
   c. **Vicarious**
   d. None of the above

4. **T/F**: Respondeat superior is a legal doctrine that attaches liability to an employer (principal) for the acts of its employees, volunteers, and governing body officials (agents).

5. The following coverage is recommended for those public water/wastewater systems that have garage operations and perform occasional mechanical work for other public water/wastewater systems or governmental entities:
   a. Garage liability
   b. **Garage keepers liability**
   c. Fiduciary liability
   d. a and b only

6. **T/F**: Coverage under your commercial auto policy is restricted to accidents occurring while you or your employees, volunteers, and governing body officials are using a vehicle for public water and wastewater system purposes.

7. **T/F**: Unlike owned vehicles, protection afforded under non-owned auto liability within the BAC policy is primary over any valid and collectible personal auto policy.

8. **T/F**: Insurance Services Office (ISO) offers a widely accepted commercial auto policy referred to as a personal auto coverage (PAC) form, which is extensively used by public water/wastewater systems and other commercial and governmental entities.

9. Examples of coverages available in a BAC policy include the following:
   a. Uninsured motorists
   b. Medical payments
   c. Rental car reimbursement
   d. **All of the above**

10. **T/F**: There are five sections within the supplemental declarations page of the BAC policy.

11. **T/F**: Symbol 1 of a BAC policy refers to owned autos only.

12. **T/F**: Symbol 1 of a BAC policy is intended for liability coverage only.

13. **T/F**: Symbols 2, 3, or 4 of a business coverage (BAC) policy alone will not trigger coverage for hired, borrowed, or non-owned autos except for substitute autos and for trailers attaching to one of your own power units.

14. **T/F**: The recommended symbol for physical damage (i.e., comprehensive and collision) coverage in a BAC policy is symbol 2.

15. The symbols in a BAC policy that automatically include liability coverage for an owned land vehicle that would qualify as mobile equipment except for being subject to a compulsory or financial responsibility law or other motor vehicle insurance law are as follows:
   a. Symbol 1
   b. Symbol 2
   c. Symbol 3
   d. **a and b only**

16. **T/F**: Symbol 5 in a BAC policy triggers no-fault coverage but only for autos required by law to carry it.

17. **T/F**: Symbol 3 in a BAC policy triggers automatic uninsured motorists coverage only in those states that do not allow you to reject the coverage.

18. **T**/F: Symbol 7 in a BAC policy restricts coverage to vehicles listed in the schedule of covered autos.

19. The symbols that most closely duplicate liability coverage provided under symbol 1 in a BAC policy include the following:
    a. 2, 7, and 8
    b. 2, 7, and 9
    c. **2, 8, and 9**
    d. None of the above

20. T/**F**: Hired auto physical damage coverage in a BAC policy does not apply to an auto leased, hired, rented, or borrowed without a driver.

21. **T**/F: Symbol 9 in a BAC policy includes autos that are owned by your employees, volunteers, or governing body officials if such autos are used for public water and wastewater system purposes.

22. **T**/F: Symbol 19 in a BAC policy includes vehicles that would ordinarily be considered mobile equipment except for being subject to a compulsory financial responsibility or other motor vehicle law where it is licensed or principally garaged.

23. The definition of auto in a BAC policy includes the following:
    a. Land motor vehicle, trailer, or semitrailer designed for travel on public roads
    b. Any other land vehicle that is subject to a compulsory or financial responsibility law or other motor vehicle insurance law where it is licensed or principally garaged
    c. **a and b only**
    d. None of the above

24. **T**/F: The utilization of fleet automatic in a BAC policy is recommended to ameliorate audit premiums as well as streamline mid-term additional/return premium invoices.

25. T/**F**: Defense costs are inside the limit of insurance in a BAC policy.

26. **T**/F: The insurer must defend the insured defendant against false or fraudulent claims as long as they allege covered damages in a BAC policy.

27. The definition of insured in a BAC policy comprise the following:
    a. Named insured
    b. Permissive users
    c. Anyone liable for the conduct of an insured
    d. **All of the above**

28. T/**F**: A rental car company is automatically classified as an insured under the BAC policy for your use of their rental car.

29. Standard liability exclusions in a BAC policy includes the following:
    a. **Expected or intended injury**
    b. Mechanical breakdown
    c. Glass repair
    d. a and b only

30. The pollution liability exclusion in a BAC policy encompasses the following:
    a. Pollutants being transported by you or on your behalf
    b. Pollutants being loaded onto, or unloaded from, a covered auto
    c. Pollutants being stored, disposed of, treated, or processed in or on a covered auto
    d. **All the above**

31. **T**/F: The liability limit of insurance in a BAC policy automatically stacks when multiple covered autos, insureds, claims made, or vehicles are involved in the same accident.

32. **T**/F: No aggregate applies to the liability limit of insurance in a BAC policy.

33. **T/F**: Comprehensive coverage and specified causes of loss are alternative approaches to insuring non-collision physical damage losses in a BAC policy.

34. **T/F**: Specified causes of loss and collision coverage is the preferred pairing for physical damage coverage in a BAC policy.

35. Loss of use claims for physical damage in a BAC policy applies to the following scenarios:
    a. Total theft of a private passenger covered auto
    b. Legal liability for loss of use expenses caused by any covered peril associated with a hired auto
    c. **a and b only**
    d. None of the above

36. Standard physical damage exclusions in a BAC policy includes the following:
    a. Workers compensation
    b. Maintenance losses
    c. Diminution in value
    d. **b and c only**

37. **T/F**: Stated amount is the preferred valuation method for physical damage claims in a BAC policy.

38. **T/F**: Agreed value in a BAC policy means you and your insurer agree upon the value of your vehicle when you take out the policy, and in the event of a covered loss, you will be reimbursed the lesser of the repair cost to fix the vehicle or the agreed value, regardless of any depreciation.

39. There is no physical damage deductible in a BAC policy for the following covered autos:
    a. Fire
    b. Lightning
    c. **a and b only**
    d. None of the above

40. The liability and physical damage duties in the event of an accident, claim, suit, or loss in a BAC policy includes the following:
    a. Insured must cooperate with the insured
    b. Promptly notify law enforcement if a covered auto is stolen
    c. Insured must submit to physical examinations
    d. **All of the above**

41. **T/F**: Coverage under medical payments in a BAC policy pays for the treatment of injuries you or your passengers suffer in a car accident, no matter who caused the crash.

42. Medical payments in a BAC policy includes the following benefits:
    a. **Medical, dental, surgical, and chiropractic care**
    b. Health insurance deductibles and copays
    c. Wage reimbursement
    d. All of the above

43. Coverage under personal injury protection in a BAC policy includes the following benefits:
    a. Childcare costs if you are limited by accident injuries
    b. Wage reimbursement if injuries force you to miss work
    c. Funeral costs
    d. **All of the above**

44. **T/F**: Coverage for uninsured motorists in a BAC policy provides occupants of a covered auto and pedestrians a source of recovery when an at-fault motorist (or hit-and-run driver) has no insurance or inadequate insurance limits.

45. **T/F**: The garage liability coverage form pays first-party damages due to bodily injury or property damage resulting from your garage operations.

46. T/**F**: The garage liability coverage form pays for losses involving property in the insured's care, custody, or control.

47. **T**/F: Garage keepers coverage can be added to the garage liability coverage form to protect the insured against liability for damage to an auto left in the insured's care while the insured is attending, servicing, repairing, parking, or storing the auto.

48. Garage keepers coverage includes the following options:
    a. Direct coverage or legal liability coverage
    b. Primary coverage or excess coverage
    c. **a and b only**
    d. None of the above

49. **T**/F: The definition of employee in a BAC policy should be expanded to include volunteers and governing body officials.

50. **T**/F: Employees as insured endorsement in a BAC policy extends liability coverage for employees while using their personally owned vehicles for public water and wastewater system purposes.

51. **T**/F: Without the fellow employee endorsement in a BAC policy, employees who unintentionally cause an auto-related injury to a fellow employee may be left without any liability insurance protection.

52. **T**/F: Employees hired auto endorsement in a BAC policy extends liability coverage for employees who rent autos in their own name for public water and wastewater system purposes.

53. T/**F**: Rental reimbursement coverage endorsement in a BAC policy automatically includes both private passenger autos and other than private passenger autos.

54. T/**F**: Pollution liability – broadened coverage for covered autos endorsement in a BAC policy protects against spills of transported pollutant contents as long as there is no accident.

55. **T**/F: Auto loan/lease gap endorsement in a BAC policy alters the amount paid under a physical damage claim to include the difference between the actual cash value of the vehicle and the amount remaining on the loan or on the lease.

24 Commercial Crime (Crime) Policy

2019 © Photo Courtesy of Placer County Water Agency, Placer County California 95604

LEARNING OBJECTIVES

- Understand how to properly structure your commercial crime (hereafter crime) policy.
- Explain the eight coverage parts that comprise a best-in-class crime policy.
- Discuss the difference between a crime policy and public officials bond.
- Articulate the coverage features of faithful performance with a crime policy.
- Comprehend the difference between a loss sustained and discovery form in a crime policy.
- Understand the definition of employee in a crime policy.
- Explain when a retroactive date is required in a crime policy.
- Identify the provisions where an employee is automatically disqualified as a covered employee in a crime policy.
- Grasp the insurer and insured duties in the event of a loss in a crime policy.
- Describe the scenarios of a crime policy applying to a loss that occurred before the policy period.
- Comprehend the criteria for selecting a limit in a crime policy.
- Discuss the process behind a social engineering attack.

DOI: 10.1201/9781003207146-34

KEY DEFINITIONS*

Application – form providing the insurer with certain information necessary to underwrite a given risk. The applicant completes it to receive insurance.

Audit – survey of the financial records of a person or organization conducted annually (in most cases) to determine exposures, limits, premiums, etc.

Banking Premises – interior of that portion of any building occupied by a banking institution or similar safe depository.

Claims Made – policy providing coverage that is triggered when a claim is made against the insured during the policy period, regardless of when the wrongful act that gave rise to the claim took place. The one exception is when a retroactive date is applicable to a claims-made policy. In such instances, the wrongful act that gave rise to the claim must have taken place on or after the retroactive date. Most professional, errors and omissions (E&O), directors and officers (D&O), and employment practices liability insurance (EPLI) are written as claims-made policies. A claims-made policy is less preferred to an occurrence policy based on the stringent claims reporting requirements.

Counterfeit Money – imitation of money that is intended to deceive and to be taken as genuine.

Coverage Territory – contractual provisions limiting coverage to geographical areas within which the insurance is affected.

Coverage Trigger – event that must occur before a particular liability policy applies to a given loss. Under an occurrence policy, the occurrence of injury or damage is the trigger; liability will be covered under that policy if the injury or damage occurred during the policy period. Under a claims-made policy, the making of a claim triggers coverage.

Discover or Discovered – time when you first become aware of facts which would cause a reasonable person to assume that a loss of a type covered by this insurance has been or will be incurred, regardless of when the act or acts causing or contributing to such loss occurred, even though the exact amount or details of loss may not then be known. Discover or discovered also means the time when you first receive notice of an actual or potential claim in which it is alleged that you are liable to a third party under circumstances which, if true, would constitute a loss under a crime policy.

Employee – (1) any natural person: (a) while in your service and for the first 30 days immediately after termination of service, unless such termination is due to theft or any dishonest act committed by the employee; (b) who you compensate directly by salary, wages or commissions; and (c) who you have the right to direct and control while performing services for you; (2) any natural person who is furnished temporarily to you: (a) to substitute for a permanent employee as defined in paragraph 1, who is on leave; or (b) to meet seasonal or short-term workload conditions; while that person is subject to your direction and control and performing services for you, excluding, however, any such person while having care and custody of property outside the premises; (3) any natural person who is leased to you under a written agreement between you and a labor leasing firm, to perform duties related to the conduct of your business, but does not mean a temporary employee as defined in paragraph (2); (4) any natural person who is: (a) a trustee, officer, employee, administrator or manager, except an administrator or manager who is an independent contractor, of any employee benefit plan; and (b) a director or trustee of yours while that person is engaged in handling funds or other property of any employee benefit plan; (5) any natural person who is a former employee, partner, member, manager, director, or trustee retained as a consultant while performing services

---

* Source material provided by International Risk Management Institute (IRMI), www.irmi.com/term/insurance-definit ions/insurance, as well as policy forms and underwriting products promulgated by Insurance Services Office, (ISO) and their member companies, https://products.iso.com/productcenter/10.

for you; (6) any natural person who is a guest student or intern pursuing studies or duties, excluding, however, any such person while having care and custody of property outside the premises; (7) any employee of an entity merged or consolidated with you prior to the effective date of this policy; or (8) any of your managers, directors, or trustees while: (a) performing acts within the scope of the usual duties of an employee; or (b) acting as a member of any committee duly elected or appointed by resolution of your board of directors or board of trustees to perform specific, as distinguished from general, directorial acts on your behalf; (9) any non-compensated natural person: (a) other than one who is a fund solicitor, while performing services for you that are usual to the duties of an employee; or (b) while acting as a fund solicitor during fund raising campaigns. Employee does not mean: any agent, broker, factor, commission merchant, consignee, independent contractor, or representative of the same general character not specified in paragraph (5). Employee Retirement Income Security Act (ERISA) Bond – federal requirement set at a limit of 10% of the funds handled, subject to a minimum limit of $1,000 and a maximum limit of $1 million. This type of surety bond is a form of insurance that protects employees against fraud or unethical actions.

Exclusion – policy provision that eliminates coverage for some type of risk. Exclusions narrow the scope of coverage provided by the insuring agreement. In many insurance policies, the insuring agreement is very broad. Insurers utilize exclusions to carve away coverage for risks they are unwilling to insure.

Forgery – the signing of the name of another person or organization with intent to deceive; it does not mean a signature which consists in whole or in part of one's own name signed with or without authority, in any capacity, for any purpose.

Fraudulent Instruction – (a) an electronic, telegraphic, cable, teletype, facsimile, or telephone instruction which purports to have been transmitted by you, but which was in fact fraudulently transmitted by someone else without your knowledge or consent; (b) a written instruction (other than those described in Coverage 2. Forgery or Alteration) issued by you, which was forged or altered by someone other than you without your knowledge or consent, or which purports to have been issued by you, but was in fact fraudulently issued without your knowledge or consent; or (c) an electronic, telegraphic, cable, teletype, facsimile, telephone, or written instruction initially received by you which purports to have been transmitted by an employee but which was in fact fraudulently transmitted by someone else without your or the employee's knowledge or consent.

Funds – money and securities.

Governing Body – group of elected or appointed people who formulate the policy and direct the affairs of a public water and wastewater system.

Insured – any person or organization qualifying as an insured in the "Who Is an Insured" section of the applicable coverage. Except with respect to the limit of insurance, the coverage afforded applies separately to each insured who is seeking coverage or against whom a claim or suit is brought.

Insuring Agreement – portion of the insurance policy in which the insurer promises to make payment to or on behalf of the insured. The insuring agreement is usually contained in a coverage form from which a policy is constructed. Often, insuring agreements outline a broad scope of coverage, which is then narrowed by exclusions and definitions.

Leased Worker – person leased to you by a labor leasing firm under an agreement between you and the labor leasing firm to perform duties related to the conduct of your business. Leased worker does not include a temporary worker.

Limits of Insurance – the most that will be paid by an insurer in the event of a covered loss under an insurance policy.

Loss – direct and accidental loss or damage.

Manager – person serving in a directorial capacity for a limited liability company.

Member – owner of a limited liability company represented by its membership interest, who also may serve as a manager.

Messenger – you, or a relative of yours, or any of your partners or members, or any employee while having care and custody of property outside the premises.

Money – (a) currency, coins, and bank notes in current use and having a face value; and (b) travelers' checks, register checks, and money orders held for sale to the public.

Occurrence – in a crime policy: (a) Employee Theft: An individual act, the combined total of all separate acts whether or not related, and a series of acts whether or not related committed by an employee acting alone or in collusion with other persons, during the policy period. (b) Forgery or Alteration: The definition is the same as employee theft except the forgery or alteration must be committed by a person other than an employee. This definition requires that the act or acts be committed by a person, other than an employee, and must involve one or more instruments. Therefore, if a ring of perpetrators forged multiple documents, the result would be considered one occurrence. (c) The definition of an occurrence for the other five insuring agreements (inside the premises – theft of money or securities, inside the premises – robbery or safe burglary, outside the premises, computer fraud, and money orders and counterfeit currency) is as follows: An individual act or event (same definition as employee theft, but the word or event is added), the combined total of all separate acts whether or not related, and a series of acts whether or not related committed by an employee acting alone or in collusion with other persons, during the policy period. Note: Event is added because some of the perils covered are events rather than acts committed by a person. Example: Theft of money and securities are covered for theft, disappearance, or destruction. Theft is an act by a person, but fire can destroy money, and that is an event.

Other Property – any tangible property other than money and securities that has intrinsic value. Other property does not include computer programs, electronic data, or any property specifically excluded under the crime policy.

Policy Conditions – section of an insurance policy that identifies general requirements of an insured and the insurer on matters such as loss reporting and settlement, property valuation, other insurance, subrogation rights, and cancellation and nonrenewal. The policy conditions are usually stipulated in the coverage form of the insurance policy.

Policy Period – term of duration of the policy. The policy period encompasses the time between the exact hour and date of policy inception and the hour and date of expiration.

Premises – the interior of that portion of any building you occupy in conducting your business.

Retroactive Date – a provision found in many (although not all) claims-made policies that eliminates coverage for claims produced by wrongful acts that took place prior to a specified date, even if the claim is first made during the policy period.

Robbery – unlawful taking of property from the care and custody of a person by one who has: (a) caused or threatened to cause that person bodily harm; or (b) committed an obviously unlawful act witnessed by that person.

Safe Burglary – unlawful taking of: (a) property from within a locked safe or vault by a person unlawfully entering the safe or vault as evidenced by marks of forcible entry upon its exterior; or (b) a safe or vault from inside the premises.

Securities – negotiable and nonnegotiable instruments or contracts representing either money or property and includes: (a) tokens, tickets, revenue, and other stamps (whether represented by actual stamps or unused value in a meter) in current use; and (b) evidence of debt issued in connection with credit or charge cards, which cards are not issued by you; but does not include money.

Settlement – resolution between disputing parties about a legal case, reached either before or after court action begins.

Social Engineering – the art of manipulating people in an online environment, encouraging them to divulge in good faith-sensitive, personal information, such as account numbers, passwords, or banking information. Social engineering can also take the form of the engineer requesting the wire transfer of monies to what the victim believes is a financial institution or person, with whom the victim has a business relationship, only to later learn that such monies have landed in the account of the engineer.

Suit – a civil proceeding in which: (1) damages because of bodily injury or property damage; or (2) a covered pollution cost or expense; to which this insurance applies, are alleged. Suit includes: (a) an arbitration proceeding in which such damages or covered pollution costs or expenses are claimed and to which the insured must submit or does submit with our consent; or (b) any other alternative dispute resolution proceeding in which such damages or covered pollution costs or expenses are claimed and to which the insured submits with our consent.

Supplemental Declarations Page – the front page (or pages) of a policy that specifies the named insured, address, policy period, location of premises, policy limits, and other key information that varies from insured to insured. The declarations page is also known as the information page.

Supplementary Payments – term used in liability policies for the costs associated with the investigation and resolution of claims. Supplementary payments are normally defined to include such items as first aid expenses, premiums for appeal and bail bonds, pre- and postjudgment interest, and reasonable travel expenses incurred by the insured at the insurer's request when assisting in the defense of a claim. Actual settlements/judgments are considered damages rather than supplementary payments.

Temporary Worker – person who is furnished to you to substitute for a permanent employee on leave or to meet seasonal or short-term workload conditions.

Theft – any act of stealing, including unlawful taking of property to the deprivation of the insured.

Transfer Account – an account maintained by you at a financial institution from which you can initiate the transfer, payment, or delivery of funds: (a) by means of electronic, telegraphic, cable, teletype, facsimile, or telephone instructions communicated directly through an electronic funds transfer system; or (b) by means of written instructions (other than those described in Coverage 2. Forgery or Alteration) establishing the conditions under which such transfers are to be initiated by such financial institution through an electronic funds transfer system.

Watchperson – any person you retain specifically to have care and custody of property inside the premises and who has no other duties.

CRIME POLICY*

Overview

The crime policy is an important auxiliary layer of first-party protection within your property and liability insurance portfolio. It safeguards your public water and wastewater system against employee theft and other first-party crimes, subject of course to notable exceptions. On the surface, crime policies appear overly complex due to their expansive coverage options and correlation to

---

* Source material provided by various online articles from International Risk Management Institute (IRMI) www.irmi. com/articles/expert-commentary/insurance as well as policy forms and underwriting products promulgated by Insurance Services Office, (ISO) and their member companies, https://products.iso.com/productcenter/10.

surety bonds. This perceived intricacy is intensified by its frequent inclusion of faithful performance of duty and routine exclusions for certain types of criminal losses specifically covered in other policies. Examples of the latter include theft of automobiles, mobile equipment, portable tools, and business personal property as well as physical injury to electronic data/software and monetary loss from social engineering/deception. The unique exposures of employee theft and other first-party crimes necessitate the proper syncing of your crime policy with your other policies to mitigate unintentional coverage gaps. We will demystify this byzantine policy by examining its structure and specifying proper design techniques to maximize its protection.

Similar to other policies, a crime policy is premised with an insuring agreement. The insurer pledges to compensate the insured up to the respective limits of insurance for certain types of employee theft and other first-party crimes. This promise affords the insurer discretion to investigate and settle any claim that is filed under the policy. It also grants the insurer rights to your recovered property after they have paid a loss. Although crime policies are quite similar for commercial and government entities, there are some noteworthy differences, especially as it relates to the treatment of bonded employees and the utilization of surety bonds versus insurance policies. For purposes of this chapter, we will focus on commercial entities but provide commentary on the necessary enhancements for governmental entities. The coverages afforded in a crime policy should satisfy the employee theft and other first-party crime exposures facing public water/wastewater systems, including those that are configured as governmental entities. In fact, most states allow these latter public water/wastewater systems to purchase crime policies instead of surety bonds because their public officials are not elected for the purpose of tax collection or treasury oversight. It is important to secure confirmation of this acceptable waiver from your respective state regulatory agency.

Crime policies are offered in two coverage forms: loss sustained and discovery of loss. The fundamental difference is their coverage trigger. The loss sustained form is akin to an occurrence trigger found in liability policies, whereas the discovery form is akin to a claims-made trigger. The discovery form covers losses discovered during the policy period regardless of when the loss may have occurred. There is an extended discovery period that allow for loss discovery up to 60 days after policy termination. The loss sustained form coverage applies to losses that occur during the policy period and discovered within one year after the policy period. The coverage grants are comparable, but there are material distinctions involving claim reporting requirements and prior policy period losses. Some crime policies state the loss sustained or discovery form as a subtitle on the first page, while others do not. You must read the policy to determine which form is used. The loss sustained and discovery forms have their respective advantages and disadvantages, which we will detail later in this chapter. Both can be properly structured to provide meaningful protection for your public water and wastewater system. Notwithstanding, you must be cognizant of their nuances to fully understand the depth of coverage afforded.

This chapter focuses on the eight standard coverages offered in a crime policy:

- Employee theft
- Forgery or alteration
- Inside the premises – theft of money and securities
- Inside the premises – robbery or safe burglary of other property
- Outside the premises – theft of money and securities
- Computer fraud
- Funds transfer fraud
- Money orders and counterfeit currency

A best-in-class crime policy will comprise all eight coverages as well as faithful performance of duty and designated agents operating on your behalf. Such a policy will specifically schedule all your entities on the supplemental declarations, as insured status only applies to scheduled entities. The balance of our discussion will address the complementary sections that support the foregoing coverages. These sections comprise limits of insurance, exclusions, key policy conditions, and coverage triggers. Since the majority of claims filed by public water/wastewater systems involve employee theft, our review will be more detailed for this peril.

We will not delve deeply into the unique coverages of kidnap, ransom, and extortion. These coverages are traditionally consolidated into what is known as a special crime policy. From a practical perspective, this type of policy only requires consideration for investor-owned utilities with operations outside the United States. It encompasses losses arising from the kidnap and holding for ransom of an employee or from the threat to do harm to a person or to certain property if ransom is not paid.

A special crime policy generally includes the following coverages: (1) kidnapping of an insured person; (2) bodily injury extortion (a threat to kidnap, injure, or kill an insured person); (3) property damage extortion (a threat to damage or pollute property, tamper with the insured's product, or reveal a trade secret or other proprietary information of the insured); (4) wrongful detention (involuntary confinement of an insured person); or (5) hijacking. Its inherent value rests not in the afforded coverages but rather in the post-policy services such as security and intelligence resources. These extra-judicial capabilities in the respective coverage territory provide critical extraction support for an insured's employees and/or property. We now begin with an examination of the eight standard coverages.

Coverage 1. Employee Theft

Insuring Agreement
We will pay for loss of or damage to money, securities, and other property resulting directly from theft committed by an employee, whether identified or not, acting alone or in collusion with other persons.

Key Definitions
Employee – (1) Any natural person: (a) while in your service and for the first 30 days immediately after termination of service, unless such termination is due to theft or any dishonest act committed by the employee; (b) who you compensate directly by salary, wages or commissions; and (c) who you have the right to direct and control while performing services for you. (2) Any natural person who is furnished temporarily to you: (a) to substitute for a permanent employee as defined in paragraph (1) above, who is on leave; or (b) to meet seasonal or short-term workload conditions; while that person is subject to your direction and control and performing services for you, excluding, however, any such person while having care and custody of property outside the premises. (3) Any natural person who is leased to you under a written agreement between you and a labor leasing firm, to perform duties related to the conduct of your business but does not mean a temporary employee as defined in paragraph (2) above. (4) Any natural person who is: (a) a trustee, officer, employee, administrator, or manager of any employee benefit plan (except an administrator or manager who is an independent contractor); and (b) a director or trustee of yours while that person is engaged in handling funds or other property of any employee benefit plan. (5) Any natural person who is a former employee, partner, member, manager, director, or trustee retained as a consultant while performing services for you. (6) Any natural person who is a guest student or intern pursuing studies or duties, excluding, however, any such person while having care and custody of property outside the premises. (7) Any employee of an entity merged or consolidated with you prior to the effective date of this policy. (8) Any of your managers, directors, or trustees while: (a) performing acts within the scope of the usual duties of an employee; or (b) acting as a member of any committee duly elected or appointed by resolution of your board of directors or board of trustees to perform specific, as distinguished from general, directorial acts on your behalf. (9) Any non-compensated natural person: (a) other than one who is a fund solicitor, while performing services for you that are usual to the duties of an employee; or (b) while acting as a fund solicitor during fundraising campaigns. Employee does not mean: any

agent, broker, factor, commission merchant, consignee, independent contractor, or representative of the same general character not specified in paragraph (5) above.

Money – currency, coins, and banknotes in current use, travelers' checks, and money orders.

Other Property – any tangible property other than money and securities that has intrinsic value. Other property does not include computer programs, electronic data, or any property specifically excluded under this policy.

Securities – negotiable and nonnegotiable instruments or contracts representing either money or other property – stocks, bonds, tokens, stamps, etc.

Theft – any act of stealing, including the *unlawful* taking of property to the deprivation of the insured.

Special Commentary

The purpose of this insuring agreement is to cover theft of money, securities, or other tangible property by an employee or employees, with or without outside assistance. Theft by an employee is purposefully excluded in the other insuring agreements (i.e., coverage grants 2–8). Coverage is commonly referred to as employee dishonesty and considered the policy's most important insuring agreement for two reasons: (1) the majority of crime claims arise from employee theft; and (2) these claims frequently span multiple policy periods and involve significant monetary loss. Proof of theft under this insuring agreement does not have to meet a criminal standard. It is intended to reimburse your public water and wastewater system for intentional theft losses from your own employees and their accomplices.

The term employee, which is defined above, is determined by whether the policy was issued on a blanket or schedule basis. If the former, then the foregoing definition applies and includes temporary and leased employees. If the latter, then a named schedule of employees and/or an employee position schedule is used to determine the employee status of a thief. A blanket basis is preferred, as it does not restrict the definition of employee like a scheduled basis. Irrespective of blanket or scheduled basis, you must endorse any natural person, partnership, or corporation that serves in the capacity as your agent. This endorsement only extends coverage to designated agents for the peril of theft and not faithful performance of duty.

Coverage 2. Forgery or Alteration

Insuring Agreement

(a) We will pay for loss resulting directly from forgery or alteration of checks, drafts, promissory notes, or similar written promises, orders, or directions to pay a sum certain in money that is: (1) made or drawn by or drawn upon you; or (2) made or drawn by one acting as your agent; or that is purported to have been so made or drawn. For the purposes of this insuring agreement, a substitute check as defined in the Check Clearing for the 21st Century Act shall be treated the same as the original it replaced. (b) If you are sued for refusing to pay any instrument covered in paragraph (a) above, on the basis that it has been forged or altered, and you have our written consent to defend against the suit, then we will pay for any reasonable legal expenses that you incur and pay in that defense. The amount that we will pay is in addition to the limits of insurance applicable to this insuring agreement.

Key Definitions

Alteration – something done to a written instrument that changes its meaning or terms without the consent of all parties to the instrument.

Forgery – signing the name of another person with the intent to deceive.

Special Commentary

The purpose of this insuring agreement is to compensate an insured for losses arising from fraudulent signatures or material modifications to a check, draft, promissory note, or similar instruments made or drawn by the insured or the insured's agent. Fraudulent signature (forgery) is straightforward, but material modifications (alteration) are less obvious. Any addition, deletion, or other modification to a written financial instrument that changes its meaning is considered an alteration for purposes of this insuring agreement. Adding a zero to a check amount is a common example of a covered alteration. Although coverage is extended worldwide, it only covers deceptive acts committed by outsiders. There is no coverage for acts by insureds, partners, members, directors, trustees, or employees. Coverage for those individuals and entities meeting the definition of employee is available under Coverage 1. Employee Theft.

Coverage 3. Inside the Premises – Theft of Money and Securities

Insuring Agreement

(a) We will pay for loss of money and securities inside the premises or banking premises: (1) resulting directly from theft committed by a person present inside such premises or banking premises; or (2) resulting directly from disappearance or destruction. (b) We will pay for loss from damage to the premises or its exterior resulting directly from an actual or attempted theft of money and securities, if you are the owner of the premises or are liable for damage to it. (c) We will pay for loss of or damage to a locked safe, vault, cash register, cash box, or cash drawer located inside the premises resulting directly from an actual or attempted theft of or unlawful entry into those containers.

Key Definitions

Banking Premises – the interior of that portion of any building occupied by a banking institution or similar safe depository.

Container – any locked safe, vault, cash register, cashbox, or cash drawer located inside the premises.

Premises – the inside of that portion of any building you occupy in conducting your business.

Special Commentary

The intent of this insuring agreement is to cover theft inside your premises or inside banking premises. Theft comprises the unlawful taking of property to the deprivation of the insured. It includes robbery, burglary, or any other unlawful taking as well as disappearance and destruction of said property. For disappearance to be covered, there must be some proof the property is missing. Inventory shortages, for instance, would not trigger coverage. The insuring agreement encompasses the loss of valuables and any damage to the premises and containers, regardless if the theft was successful. These extended coverages are not additional insurance amounts but fall within the stated limits in the supplemental declarations. Coverage is based on losses occurring inside and not outside of the premises. Theft of money and securities outside of premises are included within Coverage 5. Outside the Premises – Theft of Money and Securities. The singular exception would be losses occurring inside a bank, where the insured or agent of the insured has taken the money and/or securities for safekeeping.

Coverage 4. Inside the Premises – Robbery or Safe Burglary of Other Property

Insuring Agreement

(a) We will pay for loss of or damage to other property: (1) inside the premises resulting directly from an actual or attempted robbery of a custodian; or (2) inside the premises in a safe or vault resulting directly from an actual or attempted safe burglary. (b) We will pay for loss from damage to the premises or its exterior resulting directly from an actual or attempted robbery or safe burglary of other property, if you are the owner of the premises or are liable for damage to it. (c) We will pay for loss of or damage to a locked safe or vault located inside the premises resulting directly from an actual or attempted robbery or safe burglary.

Key Definitions

Custodian – the named insured, any of their employees etc. while having care and custody of the property inside the premises, excluding watch person or janitor.

Robbery – the unlawful taking of property from the care and custody of a person by one who has: (a) caused or threatened to cause that person bodily harm; or (b) committed an obviously unlawful act witnessed by that person.

Safe Burglary – the unlawful taking of: (a) property from within a locked safe or vault by a person unlawfully entering the safe or vault as evidenced by marks of forcible entry upon its exterior; or (b) a safe or vault from inside the premises.

Special Commentary

The intent of this insuring agreement is to compensate the insured for multiple types of losses that may result from the actual or attempted robbery or safe burglary of the insured. These perils, while possibly covered by your property policy, are included in this coverage grant. The custodian mentioned in the insuring agreement is much like the messenger mentioned in Coverage 5. Outside the Premises – Theft of Money and Securities except a custodian has care and control of property inside premises while a messenger has care and control of property outside premises. Coverage for the actual or attempted robbery of a custodian must involve the threat of bodily harm to the custodian and/or an unlawful act witnessed by the custodian. Coverage for safe burglary does not require evidence of forcible entry onto the premises (as someone could hide and wait for employees to leave), but it does require evidence of forcible entry to the safe (e.g., visible marks or scratches on the safe). An exception to the latter requirement involves taking of the whole safe or vault from inside the premises. Two coverage extensions include damage to premises and loss of or damage to locked safe or vault. Similar to outside premises coverage, these are not additional amounts of insurance but are instead subject to policy limits. Likewise, there is the same special limit of $5,000 per occurrence for loss of precious metals, stones, pearls, furs, etc., or manuscripts, drawings, or records of any kind.

Coverage 5. Outside the Premises – Theft of Money and Securities

Insuring Agreement

(a) We will pay for loss of money and securities outside the premises in the care and custody of a messenger or an armored motor vehicle company resulting directly from theft, disappearance, or destruction. (b) We will pay for loss of or damage to other property outside the premises in the care and custody of a messenger or an armored motor vehicle company resulting directly from an actual or attempted robbery.

Key Definitions

Messenger – the named insured and any of their employees/etc. while having care and custody of the property outside the premises.

Robbery – the unlawful taking of property from the care and custody of a person by one who has caused or threatened to cause that person bodily harm.

Special Commentary

The purpose of this insuring agreement is to cover the insured's loss of money, securities, or other property while in the care of a messenger or armored vehicle outside the insured's premises. Your property policy provides little coverage for property away from described premises. That reality underscores the importance of this coverage grant in protecting valuable assets away from your described premises. Money and securities are covered against theft, disappearance, or destruction. Other property is covered for actual or attempted robbery. There is a special limit of $5,000 per occurrence for loss of precious metals, stones, pearls, furs, etc. or manuscripts, drawings, or records of any kind.

Coverage 6. Computer Fraud

Insuring Agreement

We will pay for loss of or damage to money, securities, and other property resulting directly from the use of any computer to fraudulently cause a transfer of that property from inside the premises or banking premises: (a) to a person (other than a messenger) outside those premises; or (b) to a place outside those premises.

Key Definition

Computer Fraud – the use of a computer to deceptively cause the transfer of property.

Special Commentary

Nearly every insured relies on some form of electronic tool for its banking activities and cash management as well as core systems like order entry, billing, inventory controls, and accounts payable. Your computer system serves as the nucleus of all electronic tools, and the Internet creates endless opportunities for a thief to gain access to your money, securities, and other property by infiltrating said tool. The purpose of this insuring agreement is to protect the insured against loss of money, securities, and other property arising from computer fraud. The transfer of property from inside the premises or banking premises to a person or place outside those premises is the coverage trigger. Theft is the only covered peril; viruses are excluded. The insuring agreement does not require the computers used to commit the fraud be the property of the insured or located on an insured's premises. Not covered is computer fraud by owners, directors, or employees. Inventory shortages from computation are similarly not covered. There is a $5,000 sublimit for loss or damage to manuscripts, drawings, or records of any kind.

Coverage 7. Funds Transfer Fraud

Insuring Agreement

We will pay for loss of funds resulting directly from a fraudulent instruction directing a financial institution to transfer, pay, or deliver funds from your transfer account.

Key Definitions

Fraudulent Instructions – (1) an electronic, telegraphic, cable, teletype, facsimile, or telephone instruction which purports to have been transmitted by you (insured), but which was in fact fraudulently transmitted by someone else without your knowledge or consent; (2) a written instruction (other than those described in Coverage 2. Forgery or Alternation) issued by you, which was forged or altered by someone other than you without your knowledge or consent, or which purports to have been issued by you, but was in fact fraudulently issued without your knowledge or consent; or (3) an electronic, telegraphic, cable, teletype, facsimile, telephone, or written instruction initially received by you which purports to have been transmitted by an employee but which was in fact fraudulently transmitted by someone else without your or the employee's knowledge or consent.

Funds – money and securities.

Transfer Account – an account maintained by you at a financial institution from which you can initiate the transfer, payment, or delivery of funds: (a) by means of electronic, telegraphic, cable, teletype, facsimile, or telephone instructions communicated directly through an electronic funds transfer system; or (b) by means of written instructions (other than those described in Coverage 2. Forgery or Alteration) establishing the conditions under which such transfers are to be initiated by such financial institution through an electronic funds transfer system.

Special Commentary

The intent of this insuring agreement is to provide compensation for loss of funds resulting directly from a fraudulent instruction that directs a financial institution (bank) to transfer, pay, or deliver funds from the organization's transfer account. This coverage does not rely on a computer; it supplements Coverage 6. Computer Fraud. A common exposure for many organizations is a transfer of funds by electronic instruction (wire, facsimile) or voice-initiated transfer (telephone). This coverage is one of the most significant and emerging crime-related losses facing public water/wastewater systems and must be properly augmented via the removal of the voluntary party exclusion as it pertains to Coverage 7. Funds Transfer Fraud. The removal of said exclusion will ensure most forms of social engineering claim scenarios where the insured unwittingly transfers funds based on fraudulent actions of a third party is covered.

COVERAGE 8. MONEY ORDERS AND COUNTERFEIT MONEY

Insuring Agreement

We will pay for loss resulting directly from your having accepted in good faith, in exchange for merchandise, money, or services: (a) money orders issued by post office, express company, or bank that are not paid upon presentation; or (b) counterfeit money acquired during the regular course of business.

Key Definition

Counterfeit Money – an imitation of money that is intended to deceive and to be taken as genuine.

Special Commentary

The purpose of this insuring agreement is to cover good-faith acceptance of (1) money orders issued by post office, express company, or bank; and (2) counterfeit paper currency accepted during the regular course of business in exchange for merchandise, money, or services. It covers losses when you accept in good faith a non-valid money order supposedly issued by a post office or bank. The coverage territory includes the United States, its territories and possessions, Puerto Rico, and Canada.

LIMITS OF INSURANCE

The most we will pay for all loss resulting directly from an occurrence is the applicable limit of insurance shown in the supplemental declarations. If any loss is covered under more than one insuring agreement, then the most we will pay for such loss shall not exceed the largest limit of insurance available under any one of those insuring agreements.

Key Definitions

Loss Sustained Occurrence – (a) Under *Coverage 1. Employee Theft*: (1) an individual act; (2) the combined total of all separate acts whether or not related; or (3) a series of acts whether or not related; committed by an employee acting alone or in collusion with other persons, during the policy period shown in the supplemental declarations, except as provided under the policy conditions titled

Loss Sustained During Prior Insurance Issued by Us or Any Affiliate or *Loss Sustained During Prior Insurance Not Issued by Us or Any Affiliate*. (b) Under *Coverage 2. Forgery or Alteration*: (1) an individual act; (2) the combined total of all separate acts whether or not related; or (3) a series of acts whether or not related; committed by a person acting alone or in collusion with other persons, involving one or more instruments, during the policy period shown in the supplemental declarations, except as provided under the policy conditions titled *Loss Sustained During Prior Insurance Issued by Us or Any Affiliate* or *Loss Sustained During Prior Insurance Not Issued by Us or Any Affiliate*. (c) Under *Coverages 3–8*: (1) an individual act or event; (2) the combined total of all separate acts or events whether or not related; or (3) a series of acts or events whether or not related; committed by a person acting alone or in collusion with other persons, or not committed by any person, during the policy period shown in the supplemental declarations, except as provided under the policy conditions titled *Loss Sustained During Prior Insurance Issued by Us or Any Affiliate* or *Loss Sustained During Prior Insurance Not Issued by Us or Any Affiliate*.

Special Commentary

In selecting a limit, you should first contemplate the impact and severity of employee theft. This type of crime is pervasive and frequently remains undiscovered for multiple policy periods. The net result is a significant accumulation of loss from incremental events. Compounding the exposure is the inability of prudent risk management techniques like accounting controls and separation of duties to completely prevent such losses. Crime policies also do not stack their limits over multiple policy periods. As such, you must select a limit that properly encapsulates the maximum potential of employee theft should it span several years. A minimum limit of $500,000 to $2 million is recommended for public water/wastewater systems. This limit should be paired with active risk management processes that include annual audits from an outside accounting firm, prudent internal accounting protocol, and periodic background checks on existing personnel. The selection of your employee theft limit should also apply to your forgery or alteration, computer fraud, and funds transfer fraud coverages. Limits for burglary and robbery inside or outside of premises are easier to codify and should be based on the maximum amount of money and securities at these premises at any one time. The foregoing recommendations are minimum baselines that necessitate a thorough review of your organization's overall exposure to employee theft and other first-party crimes. Your crime policy must be paired with sound risk management strategies that blend risk mitigation, risk transfer, and risk avoidance. These strategies should evaluate the adequacy of your crime policy limits by encompassing peer group and industry information as well as the limit costs, loss history, and deductible options. Discussion should extend to your independent auditor and crime insurer as to limits of similarly sized organizations. Some states statutorily require the amount of your crime limits, but these limits are minimums and should be increased based on the propensity of loss from said exposures.

EXCLUSIONS

General Crime Policy Exclusions – All Coverages 1–8: (a) acts committed by you, your partners, or your members; (b) acts of employees learned of by you prior to the policy period (c) acts of employees, managers, directors, trustees, or representatives except as covered under Coverage 1. Employee Theft; (d) confidential information; (e) government action; (f) indirect loss; (g) legal fees, costs, and expenses except as covered under Coverage 2. Forgery or Alteration; (h) nuclear hazard; (i) pollution; and (j) war and military action.

Additional Exclusions Applying to Coverage 1: (a) inventory shortages; (b) trading; and (c) warehouse receipts.

Additional Exclusions Applying to Coverages 3–5: (a) accounting or arithmetical errors or omissions; (b) exchange or purchases; (c) fire; (d) money operated devices; (e) motor vehicles or equipment and accessories; (f) transfer or surrender of property; (g) vandalism; and (h) voluntary parting of title to or possession of property.

Additional Exclusions Applying to Coverage 6: (a) credit card transactions; (b) funds transfer fraud; and (c) inventory shortages.

Additional Exclusions Applying to Coverage 7: (a) computer fraud.

Special Commentary

The foregoing exclusions are divided into general exclusions that impact all coverage grants and specific exclusions that impact certain coverage grants. Many of the exclusions are self-explanatory, but some deserve additional commentary and clarification. Most significantly, there is no coverage for losses arising from an employee and any other individual classified in the definition of employee where you had knowledge of their previous dishonest act. There are no exceptions to this exclusion, and it necessitates proper action and controls should any individual affiliated with your public water and wastewater system have a history of theft or other dishonest acts. Inventory shortages are similarly not covered. The same applies to the voluntary parting of property as well as monetary loss from extortion and data breaches involving confidential material. Theft of automobiles, mobile equipment, personal tools, and business personal property is excluded in the crime policy but covered in other policies previously discussed.

KEY POLICY CONDITIONS

Ownership of Property: This condition confirms coverage only applies to property owned, leased, or held by the insured or for which the insured is legally liable.

Joint Insured: This condition addresses situations when more than one insured is named on the policy. Such situations include a parent company and its subsidiaries as well as partners in a partnership and various joint ventures. As it relates to this provision, knowledge of one insured is considered knowledge by every insured. That last point is particularly significant because it correlates back to knowledge of affiliated individuals who have committed a past theft or any other dishonest act. It also applies to any concealment, misrepresentation, and fraud against your insurer. It is important to confirm the employees of one insured are deemed employees of all insureds.

Consolidation/Merger: This condition applies when a covered organization consolidates or merges with another organization or purchases additional facilities. The policy automatically extends coverage to additional employees and premises for acts committed or events occurring within 90 days. You must notify your insurer within this 90-day period of any consolidation or merger.

Coverage Territory: The policy applies to losses within the United States, its territories, Puerto Rico, and Canada.

Duties in the Event of a Loss: The policy requires reporting a loss to the insurer as soon as possible upon discovering a loss or a situation that may result in loss. If that loss involves a crime except as covered under Coverage 1. Employee Theft and Coverage; 2. Forgery or Alteration, then you must also report the event to law enforcement. You must submit a proof of loss within 120 days of the event. If you fail to comply with reporting requirements, then you may void the policy.

Employee Benefit Plans: When an employee benefit plan is scheduled as an insured on the policy, the deductible is removed for the scheduled plan. The definition of an employee benefit plan includes any welfare or pension benefit plan shown in the supplemental declarations that you sponsor and which is subject to the Employee Retirement Income Security Act of 1974 (ERISA) and any amendments thereto. If your public water and wastewater system is configured as a governmental entity, then your employee benefit plans are not subject to ERISA. As such, you should secure confirmation that any of your employee benefit plans recognized by state and local law qualify for this coverage.

Concealment, Misrepresentation, or Fraud: If you or any other insured, at any time, intentionally conceal or misrepresent a material fact, then you risk voiding the policy. This condition applies to all insureds, including any employee benefit plan scheduled as an insured on the policy. Consequently, any act by any insured as well as any representative of a listed employee benefit plan that conceals or materially changes a fact in the application or in filing a claim, can void your policy. This restriction should be considered when adding an entity as an insured or any designated agent as an employee.

Valuation – Settlement: The value of loss of money is covered up to the face value. For loss of non-US currency, the insured can opt either for coverage at face value in the foreign currency or coverage in the US currency equivalent based on the rate of exchange posted in the Wall Street Journal on the day of loss (or the day the loss was discovered, if coverage is on a discovery form). Securities are covered at the market value as of close of business on the day the loss was discovered. Loss of other properties is settled on a replacement cost or cost to repair basis.

Specific to Employee Theft: Coverage is automatically canceled for any employee who commits (or has committed) a dishonest act known to the insured. The insurer can also cancel coverage for any employee with 30 days' notice. Employee benefit plans must be scheduled as an insured on the policy and satisfy the fidelity bonding requirement for ERISA.

Insurance Renewal Limitations: In discovery forms, an extended period to discover loss provision is included in the policy. This condition allows the insured to report a loss that is discovered no later than 60 days from date of cancellation but not after any other crime coverage has been obtained by the insured with the current insurer or other. The extended discovery period is one year from the date of cancellation for a loss discovered by an employee benefit plan. Typically, provision is made within the loss sustained form that replaces the discovery form to provide coverage for a loss discovered within the loss sustained policy but which occurred prior to the effective date of loss sustained coverage. So long as the prior coverage ended at the date that current coverage became effective, and the loss would have been covered under the loss sustained form had it been in effect at the time of occurrence, then the loss sustained form provides for such coverage. It is recommended that expiring and renewal policy conditions be reviewed carefully to ensure all discovery and loss sustained issues are understood and dealt with prior to changing policy forms on the renewal date.

Retroactive Dates: Retroactive dates may apply with discovery forms and limit coverage to losses occurring after the retroactive date shown in the endorsement. If a discovery form is proposed by the insurer, then you should ask about any retroactive date restrictions. The retroactive dates used by underwriters tend to range from policy inception to a point no further back than one year prior to the date the insurer first wrote your coverage. Insurers may reasonably impose a retroactive date either: (1) to limit coverage when an organization acquires or merges with another entity; or (2) when an entity adds a joint insured of which the insurer is wary, or when the insured asks for a large increase in its limit and the insurer suspects some preexisting knowledge of an employee theft may exist.

Reporting a Loss: The discovery form requires reporting suspected losses, even when loss conditions are not completely known. This condition requires the insured to report a loss as soon as possible after discovering a loss or a situation that may result in a loss. The discovery condition states the discovery of loss occurs when the insured receives notice of a claim against it or when the insured first becomes aware of facts that would cause a reasonable person to assume a loss has occurred, even though details may not be known. This description allows insurers wide latitude in designating the point at which you should be aware of the loss.

From a practical standpoint, the point at which an insured's suspicions turn into knowledge of a loss is difficult to pinpoint. Thus, it can be difficult for the insured to determine when it must report the loss. This is particularly true, as: (a) the insured cannot use inventory shortages to prove a loss, though the shortages may foster suspicions; and (b) changes in revenue may point to possible employee theft, but changes alone, even dramatic ones, may not provide evidence of a loss. This condition makes it clear that you must report a loss, or a situation that may lead to a loss, as soon as possible. If you wait, then you may violate the Duties in the Event of a Loss condition, and there may be no coverage due to late reporting. The loss sustained form has the same Duties in the Event of a Loss condition but has no discovery condition. Thus, the insured does not have the same reporting responsibility if facts simply point to a suspected loss.

Merger/Consolidation – Loss Sustained vs. Discovery Forms: In the event of a merger or consolidation, both forms automatically cover employee theft for the newly merged or consolidated entity for up to 90 days after the merger or consolidation. After that, there is no coverage unless the insurer has been notified in writing and agrees to accept the added exposure. Under both forms, the 90-day coverage extension applies only to occurrences that take place within that 90-day period. Since an occurrence is defined as an act or a series of acts by an individual, there is no coverage under the insured's policy for an ongoing theft by an employee of the acquired entity if it began before the acquisition date. Once the insurer has agreed to accept the risk of the acquired entity, the 90-day coverage extension does not apply. In that case, loss from acts of the acquired entity's employees is covered on the same basis as acts of the insured – on a discovery basis if the insured's policy is written on a discovery form or on a loss sustained basis if the insured's policy is written on a loss sustained form.

Additional Premises or Employees: This condition confirms any additional premises established or additional employees hired during the policy period, other than through consolidation or merger with, or purchase or acquisition of assets or liabilities of, another entity, are automatically covered under this policy without need to notice the insurer of the additions through the remainder of the policy period.

Cooperation: This condition requires the insured to cooperate with the insurer in all matters pertaining to the terms and conditions of the policy.

Transfer of Your Rights of Recovery Against Others to the Insurer: This condition compels the insured to transfer to the insurer all rights of recovery against any person or organization for any loss the insured sustained and for which the insurer has paid or settled. The insured must also do everything necessary to secure those rights and do nothing after loss to impair them.

Order of Recovery: This condition stipulates the following process of applying recoveries, net of the insurer's expenses, of the insured's lost or damaged assets after paying a claim for such loss or damage: (1) to the insured in satisfaction of the insured's covered loss, in excess of the amount paid under the insurance; (2) to the insurer in satisfaction of amounts paid in settlement of the insured's claim; and (3) to the insured in satisfaction of any deductible amount; and (4) to the insured in satisfaction of any loss not covered under the insurance.

Records: This condition necessitates the insured keep records of all property covered so the insurer can verify the amount of any loss.

Extended Period to Discover Loss and Policy Bridge – Discovery Replacing Loss Sustained: These two policy conditions effect coverage changes when the coverage trigger (discussed in detail below) needs to be modified for adequate coverage. Such policy changes may be effected either by condition or by endorsement.

Other Insurance: This condition states your policy does not apply to loss recoverable under other insurance unless the other insurance is insufficient to cover the whole loss. If so, then your policy will apply in excess of the coverage provided by the other insurance (which is considered primary).

COVERAGE TRIGGER

As previously discussed, the crime policy has two available coverage forms: loss sustained or discovery. The choice of form centers on its coverage trigger, and the type of trigger materially impacts your claims reporting requirements as well as the available limits for losses involving prior policy periods. The loss sustained trigger parallels the occurrence form of a liability policy, whereas the discovery trigger parallels a claims-made form. The key phraseology for a discovery form states: "loss ... discovered during the policy period ... or in the extended period." The loss sustained verbiage stipulates: "loss ... resulting directly from an occurrence taking place during the policy period"

The loss sustained form covers losses that occur during the policy period and are discovered during the policy period or during the one-year extended discovery period. There is no retroactive date with a loss sustained form. If a loss under a prior policy is not discovered during the prior policy, then the Loss Sustained During Prior Insurance condition affords coverage to such loss originating under the prior policy, but only if: (1) there has been no break in coverage between the prior policy and the current policy; and (2) both policies are worded such that they would apply to the loss. Thus, the loss sustained form can apply to a loss that originated years before, though coverage is restricted to the lowest limit found in either the current or prior policy. Losses will be paid by the current loss sustained form if each of the following applies: (1) loss occurred while prior insurance was in effect; (2) insured could have recovered except the discovery period has expired; (3) current insurance was effective as soon as prior insurance expired/was canceled; (4) loss would be covered under current policy if it had been in force at the time of occurrence; and (5) insurer will pay the lesser of amount recoverable under prior insurance or present insurance.

A discovery form covers losses discovered during the policy term regardless of when the loss may have occurred. Included in the form is an extended discovery period that allows for loss discovery for up to 60 days after policy termination. In the discovery form, the policy in force when the loss is discovered dictates coverage. This form can be a major advantage if: (1) the prior crime policy was not as broad as the current policy; (2) the prior policy's limit was inadequate; and/or (3) there was a gap during which no coverage applied. The major difference between the two triggers is the discovery option allows the insured's current policy to address losses that occur prior to the policy effective date without preconditions. Since coverage expansion (i.e., purchase of additional insuring and increased limits) usually occurs going forward, a current policy may have more coverage when losses are discovered than may have been in place at time of loss (i.e., act of embezzlement). This coverage enhancement should be considered whenever a loss sustained policy comes up for renewal.

Moving from loss sustained to discovery coverage may cause the insurer (incumbent or new) to underwrite the overall exposure more closely than that conducted for prior renewals. The insurer may pay special attention to controls and situations that existed prior to the prospective policy period as well as that which is current practice of the insured. If the insurer is not comfortable with certain exposure years prior to the one proposed for discovery, it may limit its discovery exposure by use of a retroactive date, similar in concept to the one used in a claims-made liability policy. You will need to review the potential coverage impact from use of the retroactive date to ensure any change from loss sustained to discovery is otherwise seamless in coverage and overall advantageous to the public water and wastewater system.

Final Comments

A best-in-class crime policy insulates public water/wastewater systems from losses arising from employee theft and other types of first-party crimes. Such a policy combines the following eight coverages: (1) employee theft; (2) forgery or alteration; (3) inside the premises – theft of money and securities; (4) inside the premises – robbery or safe burglary of other property; (5) outside the premises-theft of money and securities; (6) computer fraud; (7) funds transfer fraud; and (8) money orders and counterfeit currency. The most important of the foregoing coverages is employee theft, as that exposure is pervasive and weighty.

Noteworthy policy exclusions include losses arising from any individual classified in the definition of employee where you had knowledge of their previous dishonest act. Inventory shortages are similarly not covered. The same applies to the voluntary parting of property as well as monetary loss from extortion and data breaches involving confidential material. There should be an exception to the voluntary parting exclusion in Coverage 7. Funds Transfer Fraud so that social engineering claims are covered. Theft of automobiles, mobile equipment, personal tools, and business personal property are excluded in the crime policy but covered in other policies previously discussed.

Your crime policy should be paired with sound risk management strategies that combine risk mitigation, risk transfer, and risk avoidance. These strategies should assess the adequacy of your limits by encompassing peer group and industry information as well as limit costs, loss history, and deductible options. Discussion should extend to your independent auditor and crime insurer as to limit benchmarks for similarly sized organizations. As a general rule, the selection of your employee theft limit should extend to your forgery or alteration, computer fraud, and funds transfer fraud coverages. Limits for burglary and robbery inside or outside of premises are easier to properly codify and customize. Since crime policies do not stack their prior policy period limits, you must select a limit that contemplates a multiyear event. The scheduling of entities is equally important because only entities listed on the supplemental declarations, including employee benefit plans, qualify as an insured. Designated agents acting on your behalf must similarly be scheduled. There is a caveat to the foregoing scheduling recommendation: crime policies can be voided if any insured conceals, misrepresents, or frauds the insurer.

There are two available crime forms for your consideration: loss sustained and discovery. The loss sustained form is the more common of the two and covers losses that occur during the policy period and discovered during the policy period or the one-year extended discovery period. The loss sustained form also affords coverage to losses occurring in prior policy periods if certain conditions are met and does not have a retroactive date. A discovery form covers losses discovered during the policy term regardless of when the loss may have occurred. This form goes back to prior policy periods without restrictions, but that breadth is frequently narrowed with a retroactive date. Overall, your crime policy is an important sealant that addresses material gaps in your other first-party policies. Crime losses are increasing in frequency, severity, and creativity. It requires a keen understanding of your policy structure and the appropriate design dynamics to ensure coverage gaps are minimized. This knowledge will provide confidence to you and your governing body that employee theft and other first-party crimes are appropriately addressed in your crime policy and other policies comprising your proprietary property and liability insurance portfolio.

We have now completed our review of the crime policy and will turn our attention to miscellaneous policies requiring examination. These policies include cyber liability and network security, workers compensation, excess liability, and environmental pollution. We will begin with cyber liability and network security, which are emerging in depth and sophistication and accentuate the importance of risk transfer. Our focus will include exposure identification as well as the appropriate structure for an all-inclusive, best-in-class cyber liability and network security policy.

POLICY CHECKLIST

CRIME POLICY: MINIMUM BENCHMARKS

Crime	
Description	**Benchmark**
Form	
Insurance Services Office (ISO)	Yes (with proprietary endorsements)
Proprietary	Yes
Loss Trigger	Discovery or Loss Sustained
Coverages/Limits	
Employee Theft	$1,000,000
Forgery or Alteration	$1,000,000
Inside the Premises (theft)	$250,000
Inside the Premises (robbery or safe burglary)	$250,000
Outside the Premises	$250,000

(Continued)

Crime	
Description	**Benchmark**
Computer Fraud	$250,000
Funds Transfer Fraud	$250,000
Money Orders and Counterfeit Paper Currency	$250,000
Coverage Features	
Position versus Name	Yes
Designated Employee Benefit Plans	Yes
Faithful Performance	Yes
Legal Defense (forgery or alteration)	Yes
Social Engineering	Yes
(removal of voluntary parting exclusion for Coverage 7)	
Employees include Directors, Authorized Volunteers and Governing Body Officials	Yes
Coverage Extended for Damage to Premises and Loss of or Damage to Locked Safe or Vault (robbery or safe burglary).	Yes
Designated Agents (must be scheduled)	Yes
Retroactive Date (if purchasing discovery form)	Yes (secure as far back as possible)

REVIEW QUESTIONS

1. T/F: Banking premises means the exterior of that portion of any building occupied by a banking institution or similar safe depository.

2. T/F: Messenger means you, or any of your partners or members, or any employee while having care and custody of property inside the premises, excluding any person while acting as a watch person or janitor.

3. Discover or discovered in a standard crime policy means the following:
   a. Time when you first become aware of facts which would cause a reasonable person to assume that a loss of a type covered by this insurance has been or will be incurred, regardless of when the act or acts causing or contributing to such loss occurred, even though the exact amount or details of loss may not then be known
   b. Time when you first receive notice of an actual or potential claim in which it is alleged that you are liable to a third party under circumstances which, if true, would constitute a loss under a crime policy
   c. a and b only
   d. None of the above

4. The definition of employee in a standard crime policy includes the following:
   a. Any natural person who is: a trustee, officer, employee, administrator, or manager, except an administrator or manager who is an independent contractor, of any employee benefit plan
   b. Agent, broker, factor, commission merchant, consignee, independent contractor, or representative that is listed as a designated agent via endorsement
   c. a and b only
   d. None of the above

5. T/F: The definition of employee benefit plan in a standard crime policy means any welfare or pension benefit plan shown in the supplemental declarations that you sponsor and which is subject to the Employee Retirement Income Security Act of 1974 (ERISA) and any amendments thereto.

6. The definition of fraudulent instruction in a standard crime policy includes the following:
   a. An electronic, telegraphic, cable, teletype, facsimile, or telephone instruction which purports to have been transmitted by you, but which was in fact fraudulently transmitted by someone else without your knowledge or consent
   b. An unintentional error transmitted by an employee
   c. a and b only
   d. None of the above

7. T/F: Custodian means you, or a relative of yours, or any of your partners or members, or any employee while having care and custody of property outside the premises.

8. The definition of other property in a standard crime policy includes the following:
   a. Computer programs
   b. Any tangible property other than money and securities that has intrinsic value
   c. Electronic data
   d. All of the above

9. The definition of robbery in a standard crime policy includes the unlawful taking of property from the care and custody of a person by one who has:
   a. Caused or threatened to cause that person bodily harm
   b. Committed an obviously unlawful act witnessed by that person
   c. a and b only
   d. None of the above

10. The definition of transfer account in a standard crime policy means an account maintained by you at a financial institution from which you can initiate the transfer, payment, or delivery of funds:
    a. By means of electronic, telegraphic, cable, teletype, facsimile, or telephone instructions communicated directly through an electronic funds transfer system
    b. By means of telephone instructions
    c. a and b only
    d. None of the above

11. T/F: The definition of watch person in a standard crime policy means any person you retain specifically to have care and custody of property inside the premises and who has no other duties.

12. T/F: Some states statutorily require certain elected officials like tax collectors and treasurers to procure public officials and faithful performance surety bonds.

13. T/F: A standard crime policy has two available coverage forms: discovery of loss and loss sustained.

14. T/F: As it relates to a standard crime policy, the loss sustained form is akin to a claims-made trigger found in a liability policy.

15. T/F: As it relates to a standard crime policy, there are fundamental differences in claim reporting requirements between a loss sustained and discovery form.

16. T/F: Kidnap, extortion, and ransom are coverages automatically included in a standard crime policy.

17. T/F: Theft by an employee is purposefully excluded in a standard crime policy.

18. T/F: Employee dishonesty is the most common type of claim filed under a standard crime policy.

19. T/F: Many employee dishonesty claims span multiple policy periods and involve significant monetary loss.

20. T/F: The endorsement that lists designated agents in a standard crime policy extends coverage for the peril of theft and faithful performance of duty.

21. Forgery or alteration coverage in a standard crime policy includes losses arising from:
    a. Fraudulent signatures
    b. Material modifications to a check, draft, promissory note, or similar instruments
    c. Deceptive acts of employees
    d. a and b only

22. Coverage for inside the premises – theft of money and securities in a standard crime policy includes the following:
    a. Theft inside your premises
    b. Theft inside banking premises
    c. Inventory shortages
    d. a and b only

23. Coverage for inside the premises – robbery or safe burglary of other property in a standard crime policy requires the following:
    a. The threat of bodily harm to the custodian and/or an unlawful act witnessed by the custodian
    b. Evidence of forcible entry onto the premises
    c. a and b only
    d. None of the above

24. T/F: The purpose of coverage for outside the premises – theft of money and securities in a standard crime policy is to cover the insured's loss of money, securities, or other property while in the care of a messenger or armored vehicle outside the insured's premises.

25. Coverage for computer fraud in a standard crime policy includes the following:
    a. Coverage for viruses if caused by a third party
    b. Coverage for theft by a third party
    c. Coverage trigger involving the transfer of property from inside the premises or banking premises to a person or place outside those premises
    d. b and c only

26. T/F: Coverage for funds transfer fraud in a standard crime policy provides compensation for loss of funds resulting directly from a fraudulent instruction that directs a financial institution (bank) to transfer, pay, or deliver funds from the organization's transfer account.

27. T/F: There are eight coverage grants in a standard crime policy.

28. The following risk management steps should be considered when evaluating the limit adequacy of your crime policy:
    a. Peer group and industry information
    b. Discussion with an independent auditor as to the limits purchased by similarly sized entities
    c. Past loss history
    d. All of the above

29. T/F: There is no coverage in a standard crime policy for losses arising from an employee or any other individual classified in the definition of employee once an insured becomes aware of a previous dishonest act.

30. Examples of exclusions in a standard crime policy include the following:
    a. Inventory shortages
    b. Data breaches involving confidential material
    c. Voluntary parting of property
    d. All of the above

31. T/F: Retroactive dates only apply to discovery forms in a standard crime policy.

32. T/F: A loss sustained form in a standard crime policy covers losses that occur during the policy period and are discovered during the policy period or during the one-year extended discovery period.

33. The loss sustained during prior insurance condition in a loss sustained form of a standard crime policy affords coverage to losses originating under the prior policy if:
    a. There has been no break in coverage between the prior policy and the current policy
    b. Both policies are worded such that they would apply to the loss
    c. a and b only
    d. None of the above

34. T/F: A discovery form in a standard crime policy covers losses discovered during the policy term regardless of when the loss may have occurred.

35. A discovery form in a standard crime policy includes the following:
    a. An extended discovery period that allows for loss discovery up to 60 days after policy termination
    b. No penalty for gaps in coverage
    c. a and b only
    d. None of the above

ANSWER KEY

1. **T/F**: Banking premises means the exterior of that portion of any building occupied by a banking institution or similar safe depository.

2. **T/F**: Messenger means you, or any of your partners or members, or any employee while having care and custody of property inside the premises, excluding any person while acting as a watch person or janitor.

3. Discover or discovered in a standard crime policy means the following:
    a. Time when you first become aware of facts which would cause a reasonable person to assume that a loss of a type covered by this insurance has been or will be incurred, regardless of when the act or acts causing or contributing to such loss occurred, even though the exact amount or details of loss may not then be known
    b. Time when you first receive notice of an actual or potential claim in which it is alleged that you are liable to a third party under circumstances which, if true, would constitute a loss under a crime policy
    c. **a and b only**
    d. None of the above

4. The definition of employee in a standard crime policy includes the following:
    a. Any natural person who is: a trustee, officer, employee, administrator or manager, except an administrator or manager who is an independent contractor, of any employee benefit plan
    b. Agent, broker, factor, commission merchant, consignee, independent contractor, or representative that is listed as a designated agent via endorsement
    c. **a and b only**
    d. None of the above

5. **T/F**: The definition of employee benefit plan in a standard crime policy means any welfare or pension benefit plan shown in the supplemental declarations that you sponsor and which is subject to the Employee Retirement Income Security Act of 1974 (ERISA) and any amendments thereto.

6. The definition of fraudulent instruction in a standard crime policy includes the following:
    a. **An electronic, telegraphic, cable, teletype, facsimile, or telephone instruction which purports to have been transmitted by you, but which was in fact fraudulently transmitted by someone else without your knowledge or consent**
    b. An unintentional error transmitted by an employee

    c. a and b only

    d. None of the above

7. T/**F**: Custodian means you, or a relative of yours, or any of your partners or members, or any employee while having care and custody of property outside the premises.

8. The definition of other property in a standard crime policy includes the following:

    a. Computer programs

    b. **Any tangible property other than money and securities that has intrinsic value**

    c. Electronic data

    d. All of the above

9. The definition of robbery in a standard crime policy includes the unlawful taking of property from the care and custody of a person by one who has:

    a. Caused or threatened to cause that person bodily harm

    b. Committed an obviously unlawful act witnessed by that person

    c. **a and b only**

    d. None of the above

10. The definition of transfer account in a standard crime policy means an account maintained by you at a financial institution from which you can initiate the transfer, payment, or delivery of funds:

    a. **By means of electronic, telegraphic, cable, teletype, facsimile, or telephone instructions communicated directly through an electronic funds transfer system**

    b. By means of telephone instructions

    c. a and b only

    d. None of the above

11. **T**/F: The definition of watch person in a standard crime policy means any person you retain specifically to have care and custody of property inside the premises and who has no other duties.

12. **T**/F: Some states statutorily require certain elected officials like tax collectors and treasurers to procure public officials and faithful performance surety bonds.

13. **T**/F: A standard crime policy has two available coverage forms: discovery of loss and loss sustained.

14. T/**F**: As it relates to a standard crime policy, the loss sustained form is akin to a claims-made trigger found in a liability policy.

15. **T**/F: As it relates to a standard crime policy, there are fundamental differences in claim reporting requirements between a loss sustained and discovery form.

16. T/**F**: Kidnap, extortion, and ransom are coverages automatically included in a standard crime policy.

17. T/**F**: Theft by an employee is purposefully excluded in a standard crime policy.

18. **T**/F: Employee dishonesty is the most common type of claim filed under a standard crime policy.

19. **T**/F: Many employee dishonesty claims span multiple policy periods and involve significant monetary loss.

20. T/**F**: The endorsement that lists designated agents in a standard crime policy extends coverage for the peril of theft and faithful performance of duty.

21. Forgery or alteration coverage in a standard crime policy includes losses arising from:

    a. Fraudulent signatures

    b. Material modifications to a check, draft, promissory note, or similar instruments

    c. Deceptive acts of employees

    d. **a and b only**

22. Coverage for inside the premises – theft of money and securities in a standard crime policy includes the following:
   a. Theft inside your premises
   b. Theft inside banking premises
   c. Inventory shortages
   d. **a and b only**

23. Coverage for inside the premises – robbery or safe burglary of other property in a standard crime policy requires the following:
   a. **The threat of bodily harm to the custodian and/or an unlawful act witnessed by the custodian**
   b. Evidence of forcible entry onto the premises
   c. a and b only
   d. None of the above

24. T/F: The purpose of coverage for outside the premises – theft of money and securities in a standard crime policy is to cover the insured's loss of money, securities, or other property while in the care of a messenger or armored vehicle outside the insured's premises.

25. Coverage for computer fraud in a standard crime policy includes the following:
   a. Coverage for viruses if caused by a third party
   b. Coverage for theft by a third party
   c. Coverage trigger involving the transfer of property from inside the premises or banking premises to a person or place outside those premises
   d. **a and c only**

26. T/F: Coverage for funds transfer fraud in a standard crime policy provides compensation for loss of funds resulting directly from a fraudulent instruction that directs a financial institution (bank) to transfer, pay, or deliver funds from the organization's transfer account.

27. T/F: There are eight coverage grants in a standard crime policy.

28. The following risk management steps should be considered when evaluating the limit adequacy of your crime policy:
   a. Peer group and industry information
   b. Discussion with an independent auditor as to the limits purchased by similarly sized entities
   c. Past loss history
   d. **All of the above**

29. T/F: There is no coverage in a standard crime policy for losses arising from an employee or any other individual classified in the definition of employee once an insured becomes aware of a previous dishonest act.

30. Examples of exclusions in a standard crime policy include the following:
   a. Inventory shortages
   b. Data breaches involving confidential material
   c. Voluntary parting of property
   d. **All of the above**

31. T/F: Retroactive dates only apply to discovery forms in a standard crime policy.

32. T/F: A loss sustained form in a standard crime policy covers losses that occur during the policy period and are discovered during the policy period or during the one-year extended discovery period.

33. The loss sustained during prior insurance condition in a loss sustained form of a standard crime policy affords coverage to losses originating under the prior policy if:
   a. There has been no break in coverage between the prior policy and the current policy
   b. Both policies are worded such that they would apply to the loss

   c. **a and b only**
   d. None of the above

34. **T/F:** A discovery form in a standard crime policy covers losses discovered during the policy term regardless of when the loss may have occurred.

35. A discovery form in a standard crime policy includes the following:
   a. An extended discovery period that allows for loss discovery up to 60 days after policy termination
   b. No penalty for gaps in coverage
   c. **a and b only**
   d. None of the above

25 Miscellaneous Policies

Photo by David J. McNeil

LEARNING OBJECTIVES

- Understand how to properly structure your cyber liability and network risk policy.
- Articulate the first- and third-party coverages comprising a best-in-class cyber liability and network risk policy.
- Identify the regulatory framework associated with permissible pollutant releases into navigable waterways of the United States.
- Describe the reason environmental pollution policies are issued on claims-made basis.
- Explain the importance of closing the pollution exclusion gaps in your commercial general liability (CGL) and property policies with a properly structured and tailored environmental pollution policy.
- Comprehend how to properly structure your workers' compensation policy.
- Identify the best-in-class characteristics of workers' compensation insurer.
- Discuss the importance of insurer resources as it relates to improving employee safety and reducing workplace claims and workers' compensation premiums.
- Explain the best-in-class characteristics of a workers' compensation insurer.
- Describe the regulatory framework and policy mechanics of a workers' compensation policy.
- Understand how to properly structure your excess liability policy.
- Grasp the significance of following the scheduled underlying policies in an excess liability policy.

DOI: 10.1201/9781003207146-35

KEY DEFINITIONS*

Best Management Practices (BMPs) – methods that have been determined to be the most effective and practical means of preventing or reducing non-point source pollution to help achieve water quality goals. BMPs include both measures to prevent pollution and measures to mitigate pollution.

Blanket Additional Insured Endorsement – an endorsement attached to liability insurance policies that automatically grants insured status to a person or organization that the named insured is required by contract to add as an insured. May apply only to specific types of contracts or entities.

Bodily Injury – liability insurance term that includes bodily harm, sickness, or disease, including resulting death.

Bodily Injury by Accident-Policy Limit – the most the insurer will pay under Part Two, Employers' Liability, for all claims arising out of any one accident, regardless of how many employee claims or how many related claims (such as a loss of consortium suit brought by the injured worker's spouse) arise out of the accident.

Bodily Injury by Disease-Each Employee Limit – a policy limit within Part Two, Employers' Liability, establishing the most the insurer will pay for damages due to bodily injury by disease to any one employee.

Bodily Injury by Disease-Policy Limit – an aggregate limit of Part Two, Employers' Liability, stipulating the most the insurer will pay for employee bodily injury by disease claims during the policy period (normally a year) regardless of the number of employees who make such claims.

Bodily Injury – bodily injury, sickness, or disease sustained by a person, including death resulting from any of these.

Claim – used in reference to insurance, a claim may be a demand by an individual or organization to recover, under a policy of insurance, for loss that may be covered within that policy.

Claims-Made – policy providing coverage that is triggered when a claim is made against the insured during the policy period, regardless of when the wrongful act that gave rise to the claim took place. The one exception is when a retroactive date is applicable to a claims-made policy. In such instances, the wrongful act that gave rise to the claim must have taken place on or after the retroactive date. Most professional, errors and omissions (E&O), directors and officers (D&O), and employment practices liability insurance (EPLI) are written as claims-made policies. A claims-made policy is less preferred to an occurrence policy based on the stringent claims-reporting requirements.

Clean Water Act (CWA) of 1972 – federal act that requires monitoring of discharges into US waters. The CWA contains provisions for abatement actions, penalties, fines, and imprisonment of responsible parties.

Comprehensive Environmental Response, Compensation, and Liability Act (CERCLA) of 1980 – federal act establishing a system for reporting facilities where hazardous wastes are or have been disposed of, treated, or stored. It also encompasses trust funds financed by certain taxes to be used for cleanup costs. CERCLA establishes very broad liability standards for hazardous waste incidents that require liable parties to reimburse trust funds that finance cleanup operations.

---

* Source material provided by International Risk Management Institute (IRMI), https://www.irmi.com/term/insurance-definitions/insurance, as well as policy forms and underwriting products promulgated by Insurance Services Office (ISO) and their member companies, https://products.iso.com/productcenter/10. Additional material sourced through United States Environmental Protection Agency (USEPA), https://www.epa.gov/, and Occupational Safety & Health Administration (OSHA), https://www.osha.gov/. "Joint and several liability" definition provided by *Black's Law Dictionary*, 10th ed., 2014.

Consequential Bodily Injury – type of lawsuit covered by the employers' liability coverage of a workers' compensation policy. In this type of legal action, a member of the injured worker's family purports to have an injury that directly results from the injury to the employee. Often, mental injuries are alleged. Legislative action in many states has narrowed the applicability of this type of lawsuit.

Contractual Liability – liability imposed on an entity by the terms of a contract. As used in insurance, the term refers not to all contractually imposed liability but to the assumption of the other contracting party's liability under specified conditions via an insured contract.

Coverage Territory – contractual provisions limiting coverage to geographical areas within which the insurance is affected.

Coverage Trigger – event that must occur before a particular liability policy applies to a given loss. Under an occurrence policy, the occurrence of injury or damage is the trigger; liability will be covered under that policy if the injury or damage occurred during the policy period. Under a claims-made policy, the making of a claim triggers coverage.

Credit Monitoring – service provided within the privacy notification and crisis management insuring agreement of a cyber liability and network security policy. Credit monitoring is offered immediately following a data breach in which the personally identifiable information (PII) of various persons is exposed. Such monitoring includes but is not limited to the monitoring of one's credit history (typically for 1 year) to detect any suspicious activity or unauthorized charges. It also provides regular access to one's credit history as well as special alerts when there are significant changes to one's credit history.

Crisis Management Coverage – insuring agreement found within cyber liability and network security policies. Subject to a sublimit, crisis management coverage generally reimburses expenses incurred to restore confidence in the security of the insured's computer system.

Cyber and Privacy Liability – type of insurance designed to cover consumers of technology services or products. More specifically, the policies are intended to cover a variety of both liability and property losses that may result when a business engages in various electronic activities, such as selling on the internet or collecting data within its internal electronic network.

Cyber Extortion – act of cyber-criminals demanding payment through the use of or threat of some form of malicious activity against a victim, such as data compromise or denial of service attack.

Cyber Liability – term used to describe the liability exposures encountered when communicating or conducting business online. Potential liabilities include the internet and email. Online communication tools could result in claims alleging breaches of privacy rights, infringement, or misappropriation of intellectual property, employment discrimination, violations of obscenity laws, the spreading of computer viruses, and defamation.

Damages – money whose payment a court orders as compensation to an injured plaintiff. Fines, penalties, or injunctive relief would not typically constitute damages.

Data Forensics – the reasonable and necessary costs incurred by the insured to retain a qualified forensics firm to investigate, examine, and analyze the insured's network to determine the cause and source of the unauthorized misappropriation or disclosure of personally identifiable information and the extent to which such PII was accessed.

Defense Clause – insurance provision in which the insurer agrees to defend, with respect to insurance afforded by the policy, all suits against the insured.

Defense Inside Limits – liability policy provision according to which amounts paid by the insurer to defend the insured against a claim or suit reduce the policy's applicable limit of insurance.

Defense Outside Limits – amounts paid by an insurer to defend claims or suits that are in addition to the policy limits and do not decrease the per occurrence claim limit.

Described Premises – (1) In a property insurance policy, the location where coverage applies. Usually described in the policy with a legal address. (2) Building or land occupied or owned by an insured.

Disclaimer – situation where an insurer refuses to accept your insurance claim on an existing policy.

Dual Capacity Suits – principle, defined in a number of court cases, that a business may stand in relation to its employee not only as employer, but also as supplier of a product, provider of a service, owner of premises, etc. When a work-related injury arises out of one of these secondary relationships, the exclusivity of workers' compensation as a source of recovery to the injured worker may be challenged, and the employee may be allowed to sue the employer. Such actions are covered by the employer's liability insurance of the standard workers' compensation policy.

Duty to Defend – term used to describe an insurer's obligation to provide an insured with defense to claims made under a liability insurance policy. As a general rule, an insured need only establish that there is potential for coverage under a policy to give rise to the insurer's duty to defend. Therefore, the duty to defend may exist even where coverage is in doubt and ultimately does not apply. Implicit in this rule is the principle that an insurer's duty to defend an insured is broader than its duty to indemnify. Moreover, an insurer may owe a duty to defend its insured against a claim in which ultimately no damages are awarded, and any doubt as to whether the facts support a duty to defend is usually resolved in the insured's favor.

Economic Damages – award to an injured person in an amount sufficient to compensate for his or her actual monetary loss.

Employers' Liability – part B of the workers' compensation policy that provides coverage to the insured (employer) for liability to employees for work-related bodily injury or disease, other than liability imposed on the insured by a workers' compensation law.

The Endangered Species Act (ESA) – federal and state laws designed to protect critically imperiled species from extinction as a consequence of economic growth and development.

Environmental Protection Agency (EPA) – independent federal agency, created in 1970 that sets and enforces rules and standards that protect the environment and control pollution.

Excess Liability – excess insurance that is subject to all of the terms and conditions of the policy beneath it. In the event of a conflict, it is the underlying policy provisions that take precedence. Many excess liability policies state that they are follow form except with respect to certain terms and conditions. When this is the case, the excess liability policy is not truly on a follow form basis.

Exclusion – policy provision that eliminates coverage for some type of risk. Exclusions narrow the scope of coverage provided by the insuring agreement. In many insurance policies, the insuring agreement is very broad. Insurers utilize exclusions to carve away coverage for risks they are unwilling to insure.

Experience Modification – an adjustment of an employer's premium for workers' compensation coverage based on the losses the insurer has experienced from that employer.

Extended Reporting Period (ERP) – designated time period after a claims-made policy has expired during which a claim may be made and coverage triggered as if the claim had been made during the policy period.

Following Form – excess liability policy provision that agrees to follow the underlying policies as to how the provision applies. An excess liability policy may stand alone for certain exclusions, conditions, etc., while relating back to the underlying coverage for most provisions.

Governing Body – group of elected or appointed people who formulate the policy and direct the affairs of a public water and wastewater system.

Indemnity – compensation to a party for a loss or damage that has already occurred, or to guarantee through a contractual clause to repay another party for loss or damage that might occur in the future. The concept of indemnity is based on a contractual agreement made between two parties in which one party (the indemnitor) agrees to pay for potential losses or damages caused by the other party (the indemnitee).

Insured – any person or organization qualifying as an insured in the "Who Is an Insured" section of the applicable coverage. Except with respect to the limit of insurance, the coverage afforded applies separately to each insured who is seeking coverage or against whom a claim or suit is brought.

Insuring Agreement – portion of the insurance policy in which the insurer promises to make payment to or on behalf of the insured. The insuring agreement is usually contained in a coverage form from which a policy is constructed. Often, insuring agreements outline a broad scope of coverage, which is then narrowed by exclusions and definitions.

Joint and Several Liability – liability that may be apportioned either among two or more parties or to only one or a few select members of the group, at the adversary's discretion. Thus, each liable party is individually responsible for the entire obligation, but a paying party may have a right of contribution or indemnity from nonpaying parties.

Leased Worker – person leased to you by a labor leasing firm under an agreement between you and the labor leasing firm to perform duties related to the conduct of your business. Leased worker does not include a temporary worker.

Limits of Insurance – the most that will be paid by an insurer in the event of a covered loss under an insurance policy.

Loss – direct and accidental loss or damage.

Malware – software that is designed to infiltrate and damage a computer system without the computer owner's knowledge or consent. The term is a shortened combination of the words malicious and software and is a generic term used to refer to a variety of forms of hostile or intrusive software, such as computer viruses, worms, Trojan horses, and spyware.

Manager – person serving in a directorial capacity for a limited liability company.

Member – owner of a limited liability company represented by its membership interest, who also may serve as a manager.

Messenger – you, or a relative of yours, or any of your partners or members, or any employee while having care and custody of property outside the premises.

Midnight Dumping – general term for the illegal disposal of hazardous wastes in remote locations, often at night.

Montrose – type of known loss provision in a liability policy that restricts coverage for damage that occurs over multiple policy periods to those policies in which the insured was not aware of the occurrence at the inception of the policy period. The term "Montrose provision" is derived from the 1995 California Supreme Court ruling *Montrose Chem. Corp. v. Admiral Ins. Co.,* 10 Cal. 4th 645, 42 Cal. Rptr. 2d 324, 913 P.2d 878 (1995) that allowed recovery for damages even after the insured was aware of the loss, because the full extent of the loss was not yet known.

National Council on Compensation Insurance (NCCI) – organization that collects workers' compensation statistical data (such as premiums, exposure units, and losses), computes advisory workers' compensation rating information, develops workers' compensation policy forms, and files workers' compensation information with regulators on behalf of insurance companies that purchase its services.

National Pollutant Discharge Elimination System (NPDES) – provision of the Clean Water Act that prohibits discharge of pollutants into waters of the US unless a special permit is issued by the EPA, a state, or a tribal government.

Network Security – use of hardware, software, and firmware, including, without limitation, firewalls, filters, routers, intrusion detection software, antivirus software, automated password management applications, and other authentication mechanisms, which are designed to control or restrict the access to a network, or parts thereof; also includes the use of third-party service providers which provide, or assist in the provisioning, of such hardware, software, and firmware.

Notification Costs – the cost of notifying parties affected by a data breach as required by government statutes or regulations.

Policy Conditions – section of an insurance policy that identifies general requirements of an insured and the insurer on matters such as loss reporting and settlement, property valuation, other insurance, subrogation rights, and cancellation and nonrenewal. The policy conditions are usually stipulated in the coverage form of the insurance policy.

Policy Period – term of duration of the policy. The policy period encompasses the time between the exact hour and date of policy inception and the hour and date of expiration.

Pollutants – any solid, liquid, gaseous, or thermal irritant or contaminant, including smoke, vapor, soot, fumes, acids, alkalis, chemicals, and waste. Waste includes materials to be recycled, reconditioned, or reclaimed.

Pollution Remediation Expenses – expenses incurred for or in connection with the investigation, monitoring, removal, disposal, treatment, or neutralization of pollution conditions to the extent required by: (a) federal, state, or local laws, regulations or statutes, or any subsequent amendments thereof, enacted to address pollution conditions; or (b) a legally executed state voluntary program governing the cleanup of pollution conditions.

Primary and Noncontributory – the term is commonly used in contract insurance requirements to stipulate the order in which multiple policies triggered by the same loss are to respond [e.g., a contractor may be required to provide liability insurance that is primary and noncontributory; if so, then the contractor's policy must pay before other applicable policies (primary) and without seeking contribution from other policies that also claim to be primary (noncontributory)].

Privacy Wrongful Act – any actual or alleged act, error, misstatement, misleading statement, omission, neglect, or breach of duty committed by any insured or third-party contractor, which results in: (1) the misappropriation or disclosure of personally identifiable information; (2) a breach or violation of US federal or state law or regulations associated with the control and use of personally identifiable information, or any similar or related laws or regulations of any foreign jurisdiction; (3) identity theft; or (4) the unauthorized release of a third party's confidential and proprietary information.

Property Damage – damage to or loss of use of tangible property.

Regulatory Actions – any substantive action by an agency (normally published in the Federal Register) that promulgates or is expected to lead to the promulgation of a final rule or regulation, including notices of inquiry, advance notices of proposed rulemaking, and notices of proposed rulemaking.

Replacement Cost – property insurance term that refers to one of the two primary valuation methods for establishing the value of insured property for purposes of determining the amount the insurer will pay in the event of loss. The other primary valuation method is actual cash value (ACV). Replacement cost is defined as the cost to replace damaged property with materials of like kind and quality, without any deduction for depreciation.

Reservation of Rights – an insurer's notification to an insured that coverage for a claim may not apply. Such notification allows an insurer to investigate (or even defend) a claim to determine whether coverage applies (in whole or in part) without waiving its right to later deny coverage based on information revealed by the investigation. Insurers use a reservation of rights letter because in many claim situations, all the insurer has at the inception of the claim are various unsubstantiated allegations and, at best, a few confirmed facts.

Retroactive Date – a provision found in many (although not all) claims-made policies that eliminates coverage for claims produced by wrongful acts that took place prior to a specified date, even if the claim is first made during the policy period.

Return to Work (RTW) – post-injury program that returns injured employees to some type of work as soon as medically possible. Even if the injured workers are impaired, temporary or modified duties can be assigned that take the impairments into consideration. The end result is the reduction of indemnity costs associated with the claims.

Settlement – resolution between disputing parties about a legal case, reached either before or after court action begins.

Social Engineering – the art of manipulating people in an online environment, encouraging them to divulge in good faith sensitive, personal information, such as account numbers, passwords, or banking information. Social engineering can also take the form of the engineer requesting the wire transfer of monies to what the victim believes is a financial institution or person, with whom the victim has a business relationship, only to later learn that such monies have landed in the account of the engineer.

Special Investigative Unit (SIU) – unit within an insurer involved in detecting and pursuing action against fraudulent activities on the part of insureds or claimants.

Stop-Gap Endorsement – an endorsement that is primarily used to provide employers' liability coverage for work-related injuries arising out of exposures in monopolistic fund states (fund workers' compensation policies do not provide employers' liability coverage). If the employer has operations in non-monopolistic states, the endorsement is attached to the workers' compensation policy providing coverage in those states. For employers operating exclusively in a monopolistic fund state, the endorsement is attached to the employer's commercial general liability (CGL) policy.

Suit – a civil proceeding in which: (1) damages because of bodily injury or property damage; or (2) a covered pollution cost or expense; to which this insurance applies, are alleged. Suit includes: (a) an arbitration proceeding in which such damages or covered pollution costs or expenses are claimed and to which the insured must submit or does submit with our consent; or (b) any other alternative dispute resolution proceeding in which such damages or covered pollution costs or expenses are claimed and to which the insured submits with our consent.

Supplementary Payments – term used in liability policies for the costs associated with the investigation and resolution of claims. Supplementary payments are normally defined to include such items as first aid expenses, premiums for appeal and bail bonds, pre- and post-judgment interest, and reasonable travel expenses incurred by the insured at the insurer's request when assisting in the defense of a claim. Actual settlements/judgments are considered damages rather than supplementary payments.

Temporary Worker – person who is furnished to you to substitute for a permanent employee on leave or to meet seasonal or short-term workload conditions.

Third-Party-Over Action – type of action in which an injured employee, after collecting workers' compensation benefits from the employer, sues a third party for contributing to the employee's injury. Then, because of some type of contractual relationship between the third party and the employer, the liability is passed back to the employer by prior agreement.

Torts – a civil or private wrong giving rise to legal liability.

Waiver of Subrogation – an agreement between two parties in which one party agrees to waive subrogation rights against another in the event of a loss. The intent of the waiver is to prevent one party's insurer from pursuing subrogation against the other party. Insurance policies do not bar coverage if an insured waives subrogation against a third party before a loss. However, coverage is excluded from many policies if subrogation is waived after a loss because to do so would violate the principle of indemnity.

Workers' Compensation – system by which no-fault statutory benefits prescribed in state law are provided by an employer to an employee (or the employee's family) due to a job-related injury (including death) resulting from an accident or occupational disease.

CYBER LIABILITY AND NETWORK SECURITY POLICY*

OVERVIEW

Public water/wastewater systems are susceptible to unauthorized access of their confidential ratepayer and employee information. These releases can lead to significant liability from those individuals and entities impacted by the breach. Liability is frequently exacerbated by civil fines and penalties from regulatory authorities for your failure to maintain satisfactory security protocol. Cyber and network security attacks can also generate property losses associated with forensic programming expenses to patch coding errors and restoration costs to recover lost or corrupted data. Extortion is equally prevalent and can immobilize your computer systems where forced payments or massive reengineering of software and data are required.

Moreover, there is reputational exposure to your public water and wastewater system should you encounter adverse local media coverage arising from a data breach involving your ratepayers and employees. This type of adversity requires the utilization of a public relations firm to properly manage and dissipate the negative reporting. These costs, although expensive, are necessary to assuage public sentiment and restore confidence in your governing body.

The preceding list is not exhaustive and underscores the importance of a sound risk management strategy to ensure your network is properly secured, and your employees are appropriately trained and vetted to mitigate these physical and electronic exposures. A best-in-class cyber liability and network security policy, inclusive of first- and third-party coverage as well as specific post-loss services, plays an integral role in this strategy. Such a policy is important because commercial general liability (CGL) policies do not cover economic injury, and public officials & management liability (POML) policies, which do cover economic injury, routinely exclude cyber liability and network security unless specifically provided via proprietary endorsement.

Your property policy, incidentally, may provide some ancillary first-party coverage for damage arising from viruses initiated by third parties that infect your electronic data. Coverage would not apply for any malevolent acts by your employees or vendors. We will now detail the form structure, coverage grants, and key provisions that comprise a best-in-class cyber liability and network security policy.

---

* Source material provided by various online articles from IRMI, https://www.irmi.com/articles/expert-commentary/in surance as well as policy forms and underwriting products promulgated by ISO and their member companies, https://pr oducts.iso.com/productcenter/10.

STRUCTURE

Cyber liability and network security policies vary greatly among insurers. This incongruence is exacerbated by the absence of an Insurance Services Office, Inc. (ISO) policy standard. The lack of a baseline necessitates a thorough review of policy terms and conditions, including definitions and exclusions, to ensure your organization is properly safeguarded. These policies are mostly written on claims-made forms, which, as previously discussed, encompass unique policy triggers and onerous claims reporting requirements. Limits should also be carefully examined, with a recommended minimum policy limit of $1 million, since most policies include defense costs inside the limit of liability. Specialty enhancements like data forensics, credit monitoring, and crisis management should not have sublimits less than $250,000. Coverage should encompass third-party liability for impacted ratepayers and employees as well as first-party protection for losses suffered by your public water and wastewater system. Important policy features are summarized as follows:

Third-Party (Liability) Coverages

Privacy Liability: This coverage includes liability arising from a data breach that discloses confidential information of your ratepayers and employees. The coverage trigger should be your failure to protect confidential information, regardless of the cause (e.g., any failure to protect). Additionally, it should not be limited to intentional breaches and include your failure to disclose breaches in accordance with privacy laws.

Regulatory Actions: Defense obligations should be triggered on regulatory and other governmental actions as opposed to a formal lawsuit. Any such limitation is unacceptable, as it would invariably preclude defense during the investigative stage of government actions. These actions, incidentally, often comprise the most expensive phase of the investigation. The coverage trigger should also include a civil investigative demand or similar request for information as a pairing to the defense obligation for preliminary investigative actions by regulatory agencies. Civil fines and penalties are equally important and should be provided.

Notification Costs: This coverage includes the costs of notifying third parties possibly affected by a data breach. Its inclusion is necessary, as there is an ever-increasing and constantly evolving landscape of breach notification laws on a state-by-state basis. Such coverage should be triggered by a voluntary notification versus the normal standard of legally liable. This flexibility is necessary to afford your organization more control in addressing notification timing, including when potentially breached parties are notified.

Crisis Management: This coverage includes the costs of managing the public relations outfall from a significant data breach scenario. You will need such a resource should your public water and wastewater system experience a breach that is adversely reported by the local media and causes community concern.

Call Centers: This coverage represents one of the higher costs associated with data breaches and should be provided within the notification and crisis management coverages. It is important to confirm no unreasonable restrictions as it relates to the number of affected persons who are eligible to receive call center services, the hours and locations of the call center, and the specific services the call center staff provides.

Credit/Identity Monitoring: This coverage should not be limited by the number of affected individuals that can receive these services. A minimum of three years of monitoring services should be available to any party to a data breach.

Transmission of Viruses/Malicious Code: This coverage protects against liability claims alleging damages from transmission of viruses and other malicious code or data.

First-Party Coverages

Theft and Fraud: This coverage applies to theft or destruction of your data as well as theft of your funds via fraudulent instruction. The latter coverage is commonly referred to as social engineering and can be covered under your cyber liability and network security policy or crime policy. The frequency of funds transfer fraud via social engineering has increased exponentially and mandates the purchase of said coverage. Adding to your crime policy simply requires the removal of the *voluntary separation of funds exclusion* from the funds transfer fraud coverage.

The process is more detailed with your cyber liability and network security policy because this policy is structured to deal with indirect loss, which is akin to economic damages, arising from the failure of network security or privacy controls. Since social engineering involves direct loss, your cyber liability and network security policy must be modified to cover loss from the fraudulent instruction of a financial institution to wire funds from your transfer account to the duplicitous recipient's banking account.

Forensic Investigation: This coverage applies to the costs of determining the cause of a data breach. Should there be a breach, it is imperative to find the hole and patch the error.

Network/Business Interruption: This coverage applies to the costs of revenue lost and additional expense due to an interruption of your computer systems.

Extortion: This coverage applies to the costs of ransom if a third party demands payment to refrain from publicly disclosing or causing damage to your confidential electronic data.

Data Loss and Restoration: This coverage applies to the costs of restoring data if it is lost, and in some cases, diagnosing and repairing the cause of the loss.

S.I.T.E: These four endorsements should always be bundled to afford maximum protection: (1) *S*upply chain interruption from cyberattacks against your organization or suppliers; (2) *I*nvoice manipulation resulting in payments being misdirected or fraudulently directed; (3) *T*echnology disruption affecting operational industrial control hardware and/or software; and (4) *E*-crime losses from payment or delivery of money or securities as a result of fraud.

Key Provisions

Trigger – Loss or Claim: Cyber liability and network security policies are triggered either by an event that results in the loss of data during the policy period, or a claim that is made against the public water and wastewater system arising from such an event during the policy period. The preferred claims trigger should be the actual loss or event. This way, it is easy to surmise when your policy has been triggered and when you can report the claim in conformance with the policy's claims reporting requirements. That preference is critical because claims-made forms can limit and even preclude available coverage due to their onerous coverage triggers and claims reporting requirements.

Retroactive Coverage: Cyber liability and network security policies often contain a retroactive date, which means losses arising from events prior to the retroactive date will not be covered. Insurers often endeavor to fix the retroactive date at the initial date of coverage by the public water and wastewater system. If securing a new policy, it is important to negotiate at least a one-year, but preferably three-year, retroactive date. This time cushion is necessary because bad actors often infiltrate an organization's computer system several months before triggering an attack. A one-to-three-year retroactive date will increase the likelihood of triggering coverage for an attack that commenced after policy inception but where forensic investigation determined it was initiated several months prior to the attack date.

Acts and Omissions of Third Parties: Vendor negligence is a serious exposure, especially if you use cloud services to maintain confidential ratepayer or employee information. In these situations, it is important for coverage to apply to breaches of data maintained by third parties as long as there is a written agreement between the public water and wastewater system and vendor to provide such services.

Coverage for Unencrypted Devices: Many cyber and network security policies exclude coverage for data lost from unencrypted devices, which is a substantial restriction that should be avoided.

Coverage for Corporations and Other Entities: Your policy should include natural persons as well as corporations and other business entities, as these classifications of customers are also affected by data breaches.

Breaches Not Related to Electronic Records: Coverage should encompass loss or theft of both electronic data and of nonelectronic data (i.e., paper records), as both can be used in a breach.

Location of Security Failure: Coverage should not be limited to physical theft of data from company premises, as theft of laptops, personal digit assistants (PDAs), and external drives are portable and frequently away from such premises. There should similarly be no restriction for password theft of nonelectronic means.

Exclusions for Generalized Acts or Omissions: Your policy should not include any ambiguous and overarching exclusions including the following: (a) shortcomings in security of which you were aware prior to the inception of coverage; (b) your failure to take reasonable steps to design, maintain, and upgrade its security; and (c) certain failures of security software. The preceding exclusions are overly broad, lack adequate definitions, and are subjective in their interpretation.

Exclusions for Acts of Terrorism or War: Your policy will include this standardized exclusion but exceptions should apply to state-sponsored infiltration or data breaches.

Risk Management

As a part of your overall cyber liability and network security hygiene program, the following areas should be verified and addressed: (1) confirm your cyber liability and network security policy properly aligns with your CGL, property, and crime policies to avoid unintended coverage gaps; (2) develop a proactive software and patch management protocol that is continually evaluated; (3) apply a robust log management protocol to review current network usage conditions and assist in post-breach forensic investigations; (4) install network segmentation to isolate access of the network and compartmentalize bad actors who gain entry to one segment; (5) block suspicious activity through strong filters; (6) install credential management to augment security practices (i.e., cyber hygiene) and reduce intruder exploitation [e.g., a just-in-time (JIT) approach to privileged access management (PAM) allows for access/authorization on an as-needed basis while reducing attack vectors for bad actors]; (7) encrypt devices on the network; (8) establish a baseline for host-and-network activity; (9) implement organization-wide IT guidance and policies; and (10) deploy notice and consent banners for computer systems.

Final Comments

Cyber liability and network security are emerging issues of material concern for all public water/wastewater systems. The frequency and sophistication of these attacks underscore the need for a concerted risk mitigation and risk transfer strategy that includes governing body support as well as meaningful intellectual and financial resources. This risk impacts both your first-party data and software systems as well as third-party liability from ratepayers and employees. There is also serious exposure to civil fines and penalties from regulatory authorities should there be a data breach based on your purported failure to maintain adequate security protocol.

A proactive risk management strategy includes steps to mitigate loss. Examples include a thorough evaluation of your internal processes, hiring practices, employee training, software protection, and augmented network security. It similarly comprises consultancy services to probe your network for vulnerabilities as well as utilization of county–state–federal resources to ensure your public water and wastewater system is maximizing its cyber and network protections. Such a strategy necessitates risk transfer via the purchase of a best-in-class cyber liability and network security policy.

Details matter with these policies, and that requires comprehension of key policy features and policy provisions as well as selection of the proper limits of insurance and policy deductibles. It similarly requires a review of the post-loss services available from your insurer to ensure you can accommodate regulatory mandates for notification and monitoring as well as crisis management should the breach generate adverse public reaction. The combination of risk mitigation and risk transfer should seamlessly align with the overlapping objectives of minimizing, resolving, communicating, and curbing breaches should they occur.

We now turn our attention to workers' compensation and the importance of maintaining a safe work environment for your employees. This central tenet is anchored by the prudent selection of a best-in-class workers' compensation insurer. The right insurer will complement your mission of instilling a holistic culture and philosophy of safety throughout the organization.

WORKERS' COMPENSATION POLICY*

OVERVIEW

Employee safety is paramount to the successful operation and performance of all public water/wastewater systems. Achieving this overarching objective requires the identification and mitigation of workplace risks so employee injuries are diminished. The minimization of employee injuries is possible, but its complete avoidance is unfeasible. Notwithstanding, the design of deliberative human and operational processes as well as an organizational culture of safety will mitigate the frequency and severity of such injuries. Federal and state laws also play a role, as they were enacted to assist with this key precept and to establish a social compact between employer and employee. The abovementioned laws establish safety baseline requirements for employers. The Occupational Safety & Health Administration (OSHA) is empowered to enforce these minimum requirements. States similarly mandate the compulsory purchase by employers of workers' compensation policies as a means to facilitate safe workplaces and to provide a financial safety net for injured employees.

Workers' compensation policies can be expensive depending on the classification of work performed by your employees, your state's prescribed workers' compensation benefits, your state's legal environment, and the overall loss experience of your public water and wastewater system. The last component is referred to as your experience modification (x mod). The x mod applies a credit or debit based on your actual loss experience compared to the average for your industry and can be a source of reward or punishment. Even though this coverage is compulsory and the prescribed benefits are commoditized, there are material differences between insurers when it comes to the delivery of policy services. These variances should be carefully evaluated and are segmented as follows: safety and health; claims administration; underwriting stability; and financial security. We will delve into each of the differentials as well as the origins of workers' compensation and its key features.

ORIGINS

Workers' compensation laws originated to establish compensation for injured employees while simultaneously reducing the amount of litigation against employers when employees are injured on the job. These laws are administered at the state level, with the federal government administering its own separate program for federal workers and federal contract employees. Workers' compensation is an insurance program for employers, whereby the employer purchases a workers' compensation policy, and that policy pays the medical bills and lost wages for an injured employee. In return, the employee agrees to accept this form of compensation and not file a legal claim against the employer for further compensation after an injury. The laws establish a system in which employees need not fear having to pay medical bills in the case of a workplace injury, and they need not sue an

---

* Source material provided by Davidson Pattiz, Chief Operating Officer at The Zenith, and his American Association of Water Distribution & Management 2018 video Becoming Better Purchasers of Workers' Compensation Insurance Claims Services, www.aawdm.org. Additional material provided by: various online articles from International Risk Management Institute, https://www.irmi.com/articles/expert-commentary/insurance; policy forms and underwriting products promulgated by Insurance Services Office and their member companies, https://products.iso.com/productcenter/10; and Occupational Safety & Health Administration (OSHA), https://www.osha.gov/.

employer (and win) to get compensation for their injuries. Workers' compensation laws were also created to help employers avoid bankruptcy over a workplace accident since, before the enactment of such laws, there were no limits on the payout for cases of negligence on the part of the employer. In practice, this system eliminates the question of negligence or causation. If an injury occurs on the job, then it will usually be covered by workers' compensation regardless of the root cause of the accident or injury. Payment will apply, irrespective if the employer or employee was at fault.

STRUCTURE

A workers' compensation policy typically consists of the following sections: (1) declarations page; (2) policy form; and (3) various endorsements. It may include one or more schedules, such as a list of described premises. Under a workers' compensation policy, your public water and wastewater system is assigned one or more classifications based on the type of work performed by your employees. The most common classifications comprise waterworks operations (7520), salespersons (8742), and clerical (8810). The premium you pay for workers' compensation coverage depends on the classifications used, the rate charged for each classification, and the remuneration you pay your workers (i.e., payroll). Your premium may also be affected by an x mod, which reflects your prior claims experience.

Workers' compensation rates, classifications, policy forms, and other issues related to workers' compensation are administered by a state workers' compensation bureau. Many states delegate functions like rate making, statistical analysis, and form development to an organization called the National Council on Compensation Insurance (NCCI). The states are referred to as NCCI states. Four states (North Dakota, Ohio, Washington, and Wyoming) are unique in that they do not allow private insurance. In these states, which are called monopolistic states, insurance policies are issued by a state insurance fund. The NCCI has developed a standard workers' compensation policy that is utilized in all NCCI states. This policy is also used in many independent states as well. It provides two basic coverages: Coverage A. Workers' Compensation; and Coverage B. Employers' Liability.

Coverage A. Workers' Compensation Coverage: Workers' compensation coverage affords benefits to employees who have been injured in the course of employment. The coverage is extended regardless of fault. That is, an injured employee need not sue you for negligence to obtain benefits. Moreover, an injured worker is generally eligible for benefits even if his or her negligence contributed to the injury.

State workers' compensation laws typically provide the following types of benefits:

Medical Coverage: this benefit includes doctor visits, hospital care, prescription medications, physical therapy, and other medical treatments.

Disability: this benefit provides a partial replacement of income lost when employees are unable to work due to an on-the-job injury. The disability may be temporary or permanent, and partial or total.

Vocational Rehabilitation: this benefit enables employees who cannot return to their prior occupation to learn a new skill based on their current capabilities.

Death Benefits: this benefit supports the spouse and minor children of an employee killed on the job.

While the types of benefits employees receive for work-related injuries are fairly consistent from one state to another, the amount of prescribed benefits they receive vary widely from state to state. Thus, the applicable workers' compensation law of your state is incorporated into the policy. Its inclusion means the provisions of this law become part of your policy, and that equates to not having express limits but rather a reference to statutory benefits. Workers' compensation covers bodily injury by accident or bodily injury by disease (i.e., occupational disease). State law determines which occupational diseases are covered. An example of an occupation disease is asbestosis. The policy covers injuries caused by accidents that occur during the policy period. For a disease to be covered, it must be caused or aggravated by the conditions of employment. An injured employee typically has one year to file a workers' compensation claim. Using California as an example, regulators can extend that timeline under the following circumstances: (a) if the employee is under 18 at the time of injury, then the one-year statute of limitations begins when the person becomes a legal adult; (b) if there is a repetitive stress injury, then the employee may file a claim up to one year from the date that he or she became aware of the injury; or (c) an employee has up to five years from the date of injury to file a claim if the original injury caused additional or further injury.

Your insurer also has the right to obtain subrogation from the responsible party should they pay benefits to an employee who was injured because of someone else's negligence. That is, your insurer may recoup the amount of its loss payment from the party that caused the injury. You are obligated to protect your insurer's right to recover its payments from the person or organization responsible for your employee's injury.

Coverage B. Employers' Liability: Employers' liability protects an employer from legal liability arising out of employee injuries that are not covered by Coverage A. of your workers' compensation policy. Although Coverage B. applies to all employers' liability claims not specifically excluded, the policy lists the four most common types of claims:

(1) *Third-Party-Over Actions*: This type of claim involves a lawsuit filed by a third party seeking indemnity because it was held liable for an employee's injury. An example would be an employee injured using a piece of machinery that the employer had not properly maintained. In addition to the benefits received under the workers' compensation program, the employee also sues the manufacturer of the equipment. The manufacturer of the equipment, in turn, sues the employer for contributory negligence due to poor maintenance.

(2) *Loss of Consortium*: This type of claim involves a lawsuit typically filed by an injured employee's spouse for loss of the services of his or her spouse who was injured in the course of employment.

(3) *Dual Capacity Suits*: This type of claim involves a lawsuit brought by an injured employee against his or her employer when the injury arises from a product the employer manufacturers. In such a case, the employer is liable not only as an employer but also as a manufacturer.

(4) *Consequential Bodily Injury*: This type of claim involves a lawsuit filed by a family member for injuries suffered as a consequence of the employee's injury. An example would be the spouse of a severely injured employee who suffers a heart attack or a nervous breakdown upon learning of the injury.

Coverage B. has specific limits of insurance for covered claims. These limits should be set at an amount no less than $1 million per accident, $1 million per policy, and $1 million per employee. Employers' liability applies to damages that the insured must pay, much like a CGL policy, and hence is subject to the limits shown on the policy's declarations page. The coverage is written on a per occurrence basis, which is straightforward when dealing with bodily injury claims. The bodily injury must simply occur during the policy period. For occupational disease, coverage would only apply if the employee was exposed to the conditions causing or aggravating the disease during the policy period. Coverage B. does not apply in those states or jurisdictions that have monopolistic state funds (i.e., North Dakota, Ohio, Washington, and Wyoming). Defense costs are paid outside the limit, and, hence, will not reduce the limits of insurance. An excess liability policy should extend over Coverage B. to provide additional limits of insurance. Coverage B. does contain exclusions that should be read and understood.

These exclusions are straightforward and summarized as follows: (1) Employers' liability only covers liability imposed by law, which means liability assumed under a contract is not covered. However, claims for employee injuries assumed under contract are covered under a best-in-class

CGL policy via an exemption to the employers' liability exclusion for contractually assumed liability. Coverage B. covers third-party-over actions unless such claim is the result of assumed liability from a contract. (2) Punitive or exemplary damages because of bodily injury to an employee employed in violation of the law. (3) Bodily injury to an employee while employed in violation of the law with the employer's knowledge. (4) Any obligation imposed by a workers' compensation, occupational disease, unemployment compensation, or disability benefits law, or any similar law.

These types of losses would be covered under the specific policy designed for the exposures, such as Coverage A. of your workers' compensation policy. (5) Bodily injury intentionally caused or aggravated by the employer. (6) Bodily injury occurring outside the United States of America, its territories, possessions, and Canada. This exclusion does not apply to bodily injury if a citizen or resident of the United States of America or Canada is temporarily outside of the country. (7) Damages arising out of wrongful termination, discrimination, harassment, and other workplace-related torts. Coverage for this exposure should be provided under your POML policy or a stand-alone employment practices liability policy. (8) Exclusions 8–10. These three exclusions carve out coverage for various exposures covered by federal law. (11) Fines or penalties imposed for violation of federal or state law. (12) Damages payable under the Migrant and Seasonal Agricultural Worker Protection Act.

Insurer Selection

As previously discussed, the primary responsibility of your public water and wastewater system is to provide a safe work environment for your employees. That obligation correlates with your ability to deliver safe, reliable, and affordable drinking water and wastewater services. Not only is the preceding responsibility good for your employees, it's good for your public water and wastewater system and ratepayers. A best-in-class workers' compensation insurer will help you focus on workplace safety in a manner that prioritizes your employees and your operations. Although the policy and prescribed benefits are commoditized, the delivery of services and resources by insurer is distinct and should be carefully evaluated. Rate comparisons by insurer is certainly important, but the value of their services and resources should take precedence. Simply stated, low rates do not mitigate workplace injuries or prevent an uptick in your x mod. Long-term value is derived by an insurer specializing in your industry and offering expansive services and resources.

Central to the fundamental mission of a safe work environment are cogent safety and health measures. These measures, when effectively designed, reduce the costs of workers' compensation claims and protect your public water and wastewater system from extra expense, customer dissatisfaction, and lost goodwill. A best-in-class workers' compensation insurer will partner with you to create an organizational culture of safety. This all-encompassing way of thinking will prompt your employees to take actions and make decisions that decrease their risk of injury and illness. To achieve an organizational culture of safety, your insurer should understand your unique operations and the various ways your employees can get hurt on the job. Your insurer should similarly offer meaningful services, tools, and resources to improve workplace safety and track your progress toward fewer injuries and claims. And when injuries occur, your insurer should be proactive via injury-management and return-to-work assistance. These resources will help your employees heal while also controlling your insurance costs.

Your insurer's safety engineers should possess a blend of technical expertise, insurance knowl-edge, and segment experience. Their core focus should be facilitating and promoting an organi-zational culture of safety so that workplace injuries are reduced. An integral part of this focus is customized risk assessments unique to your public water and wastewater system. Examples include the following: *Manual Material Handling*: handling heavy, bulky, or awkwardly shaped items. *Repetitive Motion*: repetitive tasks with potential for muscle fatigue, awkward motion, sus-tained position, forceful exertion, overexertion, or overuse. *Machinery and Equipment*: machin-ery and equipment use, maintenance, and repair with emphasis on guarding and lockout/tagout.

Falls: elevated areas, walking/work surfaces, parking lots, ramps, stairs, elevators, escalators, ladders, and scaffolds. *Occupational Disease*: hazardous substances that can be inhaled, absorbed, or ingested as well as high noise levels, temperature extremes, biological hazards, radiation, lasers, and vibration. *Motor Vehicle*: operation of company or personal vehicles for business use. *Catastrophic Potential*: group transportation, explosion, chemical release, and other events with potential for multiple employees to suffer serious injury, illness, or death. *Injury Management*: immediate claim reporting, return to work, and litigation control.

Traditional inspections, safety program audits, and general training should be supplemented with an evidence-based safety and health curriculum that follows process improvement models of resource prioritization and problem definition. This approach is empirically proven to reduce workers' compensation claims, productivity disruptions, and overall expenses. An evidence-based safety and health curriculum comprises four well-defined steps: (1) reviewing your operations and analyzing past injuries to determine your critical hazards and activities that have, or might, lead to employee injuries; (2) identifying problems so they can be solved; (3) defining injury cause-and-effect relationships through research, observation, interviews, and other problem-solving tools that include implementing safety measures to counter such relationships; and (4) reviewing and tracking your claims costs and exposure reduction to quantify the impact. The foregoing steps underscore the necessity of selecting an insurer that utilizes comprehensive and segment-specific risk management and safety resources as well as proprietary training programs, compliance tools, HR management practices, and other support tools for your organization.

A common theme for best-in-class workers' compensation insurers is their commitment to in-house personnel as well as comprehensive claims administration resources. The use of subcontracted personnel and services should be avoided, as the quality and control of such outsourcing is unreliable. Your insurer should have a dedicated team of in-house claims specialists who will guide you through the claims process and keep you and your injured employees fully informed. A proactive approach to claims management leads to lower premiums by improving your x mod. It is prudent to ask your prospective insurer for a five-year average x mod percentage drop of insureds with x mods greater than 115%. A marked decrease of more than 20% will indicate an insurer whose practices are achieving actionable results.

It is important for an insurer to invest in top-tier, in-house personnel who prioritize employee recovery and their return to work as quickly and safely as possible. That equates to using a collaborative team of in-house claims specialists, nurses, medical bill review experts, and fraud fighters. There should also be in-house doctors who consult with claims specialists and communicate with treating physicians to optimize care and outcomes. If an injury has legal implications, then your insurer's in-house team of claims experts, attorneys, and former judges should work directly with claims adjusters to advance a fair outcome.

Proactive communication between all stakeholders is necessary to ensure proper decisions are made on behalf of your injured employees. Because public water/wastewater systems have a diverse workforce, it is important to select an insurer with multilingual specialists who can speak to your employees in their native language, and field examiners who can meet with them at their home, doctor's office, or job site. There should also be services to help injured employees get the medical care they need. When injuries are serious, your insurer should deploy a specialized catastrophic injury response team so they can engage in real time with doctors and nurses. This service includes spending time with employees and their families so their needs are better understood and recover can be facilitated.

Your insurer should similarly have a robust network of qualified medical providers whose utilization is convenient to your employees. These networks should encompass medical providers who regularly treat patients with workplace injuries and illnesses and who understand the importance of collaboration with you, your employee, and your insurer's specialists throughout the course of treatment. Equally important, your insurer should support your efforts to build relationships with specific medical providers before injuries occur. The more your medical providers know about your public water and wastewater system, the clearer the path to returning injured employees to work. That is especially true when your provider is aware of your return-to-work (RTW) options. This level of assiduousness should extend to medical provider's financial and treatment assessments. Such diagnostics should be conducted by a specialized audit team who review treatment procedures and medical bills to ensure payments are reasonable and treatments are appropriate. A well-resourced audit team will aggressively dispute any liens and payments that are not deemed usual and customary.

When your employees are injured, it is essential they receive the necessary medical treatment to promptly recover and return to work. A best-in-class workers' compensation insurer will work closely with you to achieve the foregoing objective and to develop transitional work assignments within your employee's abilities. A robust RTW program positively accentuates employee morale and reinforces your commitment to an organizational culture of safety. Statistics show that employees who return to work generally recover more quickly because they remain active, productive, and connected to their workplace.

It is also proven to boost employee morale as well as reduce operational disruption and expenses from lost product, replacement workers, and administrative time. The net result is a lower x mod and better control of your insurance costs. Your insurer should take a proactive approach and develop light duty, transitional work assignments for your public water and wastewater system. There should similarly be strategies to ease back injured employees into the workplace and optimize recovery. These strategies should be formulated by your insurer in consultation with the injured employee, their doctors, and you. A best-in-class workers' compensation insurer will guide you in implementing or refining an effective RTW program and will provide you the resources to make your efforts a success.

Another important component when selecting an insurer is their commitment to insurance fraud. Your insurer should have an in-house Special Investigation Unit (SIU) whose purpose is to expose and fight those seeking to take advantage of you. The spectrum of fraud is widespread and can include an employee, medical provider, lawyer, medical billing firm, or others who feed off the workers' compensation system. The SIU should also investigate providers for excessive treatment and exaggerated billing. When an investigation indicates fraud, your insurer's SIU should team up with their claims and legal specialists to prepare and submit a detailed report to law enforcement agencies. A fraudulent claim that goes unchecked is costly, disruptive, and can put your public water and wastewater system at risk. Not only does it impact your operations, it can increase claim costs with unnecessary medical and disability payments. This often results in a higher x mod, which can raise your premiums for years to come.

The final components in selecting a best-in-class workers' compensation insurer are their reputation for underwriting stability and financial security. Underwriting stability includes a track record of rate consistency and a commitment to avoid disruptive rate variances. It also comprises meaningful partnerships with their insureds whereby shock losses are not met with overcorrection and employee classifications are continually evaluated to ensure your organization is charged an appropriate rate for your exposure. Financial security encompasses a strong balance sheet and a history of prudent underwriting results that avoid the vagaries of cash-flow underwriting. A best-in-class insurer will generate a substantial amount of their overall premium writings from workers' compensation. This level of commitment is necessary to warrant the necessary investment of in-house personnel and proprietary resources to tangibly reduce workers' compensation costs and workplace injuries for their insureds.

FINAL COMMENTS

Employee safety is paramount to the operations of all public water/wastewater systems. One mechanism to ensure employee safety is regulatory oversight and rules established by the OSHA. Another is the required purchase of workers' compensation policies to ensure employees receive prescribed benefits should they be injured on the job. Although these policies are commoditized and dictated by the laws of your respective state, there are key differentials that should be evaluated before selecting your workers' compensation insurer. The differences include: safety and health; claims administration; underwriting stability; and financial stability.

Safety and health resources are designed to mitigate accidents by evaluating human and operational processes of your public water and wastewater system as well as formulating a culture of safety throughout your organization. Claims administration involves handling employee injuries from start to finish in a way that facilitates their return to work while also ensuring their benefits are rightfully discharged and any fraud or abuse is actively avoided. Underwriting stability correlates to an insurer that evaluates an insured from a longer lens of time than one policy period. Such evaluation should comprise the insured's commitment to reduce losses and enhance workplace safety. Financial security is the means from which an insurer pays claims and provides coverage over a span of decades and not years. The utilization of in-house personnel for all facets of the workers' compensation process is preferred, as this approach underscores your insurer's commitment to this line of coverage. It reaffirms their dedication to the consistent delivery of services to their insureds.

Selecting the right workers' compensation insurer will reduce your costs by lowering your losses. That combination safeguards your ratepayers and, most importantly, your employees. The net result is a productive, safe, and motivated workforce whose focus is doing the right job, at the right pace, with the right process, and through a continual emphasis of safety throughout the organization. A holistic culture of safety will ensure a professionally managed public water and wastewater system that has high employee morale and strong community support. Achieving these objectives correlate to a governing body philosophy of prioritizing insurance value over insurance price.

Your broker advocate also plays an integral role as communicator, facilitator, and validator throughout the claims process. A broker advocate will ensure communication flows quickly, consistently, and accurately to all parties; but especially to your injured employee. Equally important, he or she will ensure your claims are correctly reported to the respective workers' compensation rating bureau while concurrently authenticating your x mod calculation. These actions buttress the efforts of your insurer and align all parties for the prompt, fair, and proper resolution of all claims. Equally important, your broker advocate facilitates the timely return to work of your injured employees. We now turn our attention to excess liability and the importance of purchasing higher limits of liability while ensuring the same continuity of coverage afforded in the scheduled underlying policies apply to this policy.

EXCESS LIABILITY POLICY*

Overview

The purchase of higher liability limits via an umbrella or excess liability policy (hereinafter referred to as excess liability) is a prudent risk management strategy for all public water/wastewater systems. Limits, however, are subordinate to coverage. It is imperative these higher limits are written over your primary policies, which are traditionally the CGL, POML, and business auto coverage (BAC) policies as well as employers' liability of your workers' compensation policy. Equally important, you need confirmation the higher liability limits are written on an excess liability policy that follows the depth and breadth of the coverages offered in the scheduled underlying policies. We call that configuration a following form policy whereby your higher limits seamlessly attach and follow the primary policies underneath it.

---

* Source material provided by various online articles from International Risk Management Institute (IRMI) https://www. irmi.com/articles/expert-commentary/insurance as well as policy forms and underwriting products promulgated by Insurance Services Office (ISO) and their member companies, https://products.iso.com/productcenter/10.

There once was a difference between umbrella and excess liability policies. Both were structured to provide higher limits to the aforementioned primary policies, but the umbrella policy could also drop down and possibly cover exposures that were excluded or not contemplated in the underlying policies. That material advantage is no longer available, as most umbrella policies have a host of exclusions that mitigate the possibility of dropping down to cover an uninsured loss. And more concerning, many of their exclusions may actually prevent the triggering of coverage for a loss that is covered under a scheduled underlying policy. The same type of restrictive exclusions is also prevalent with traditional excess liability policies. That means coverages automatically embedded in best-in-class CGL and POML policies may not be included for higher limits unless your excess liability policy is written as a following form over all the coverages granted in the schedule of underlying policies.

One remedy for the foregoing scenario is to secure a following form excess liability policy from the same insurer offering your CGL, POML, and BAC policies. This situation should ensure continuity of coverage and prevent situations where your higher liability limits are not activated because of exclusions in your excess liability policy that are not included in the scheduled underlying policies. The focus should be on seamless coverage continuity and not simply limits and premium. Inattention to your excess liability placement can create material misunderstandings and misalignment between you, your governing body, and your insurer.

STRUCTURE

Excess liability policies are independent policies subject to their own insuring agreements, definitions, conditions, and exclusions. That statement applies in all instances, including excess liability policies that expressly state the policy follows the terms, conditions, exclusions, limitations, and definitions of the scheduled underlying insurance policies. The preceding promise is invariably qualified with a phrase "except to the extent that the terms, conditions, definitions, and exclusions of this policy differ … In the event of any conflict, the terms, conditions, definitions, and exclusions of this policy shall control." This verbiage is not trickery but a usual and customary practice of excess liability insurers because their policy is independent from the scheduled underlying policies. A following form excess liability policy issued by the same insurer as the underlying CGL, POML, and BAC policies avoids coverage gaps, prevents duty to defend misunderstandings, and ensures your excess liability policy will provide higher limits on a following form basis over the scheduled underlying policies.

This notion of excess liability independence is evident in the Other Insurance clause of the scheduled underlying policies. An excess liability policy does not automatically follow amendments to these policies. One pertinent example is a contractual requirement to list a counterparty as an additional insured on your CGL policy and offer coverage on a primary and noncontributory basis. That illustration is important, as an additional insured does not automatically receive these contractual rights in the excess liability policy unless it is specifically amended to include such rights. The best way to ensure continuity is to secure the following verbiage in your excess liability policy:

> If the written contract in which you have agreed to provide insurance for that person or organization [as an insured] specifically requires that this insurance apply on a primary and noncontributory basis, this insurance will apply as if other insurance available to that person or organization under which that person or organization qualifies as a named insured does not exist, and we will not share with that other insurance.

The larger point is the Other Insurance clause of an excess liability policy should be amended to provide primary and noncontributory coverage to an additional insured. The same applies to other contractual requirements like waiver of subrogation.

Excess liability policies, including those written on a following form basis, will have qualifiers regarding any coverage with a sublimit below the standardized $1 million limit as well as well as any coverage where there is no legal liability imposed on the insureds. Examples of the former include damage to premises rented to you (i.e., fire legal liability should the limit be below $1 million) as well as property in your care, custody, and control. Examples of the latter comprise medical payments, uninsured motorists, nonmonetary damages, and wage and hour. Those limitations are acceptable and are customary. Notwithstanding, it does require a keen review of any coverages that have a sublimit so that you can articulate this limitation to your governing body. The lack of coverage where there is no legal liability is less important and does not necessitate governing body disclosure. Excess liability policies do not always expand their scheduled underlying policies beyond CGL, POML, BAC, and employers' liability of the workers' compensation policy. That means stand-alone policies like cyber liability and network security, environmental pollution liability, professional liability, and peripheral liability policies are routinely not eligible for the higher limits of your excess liability policy. In such situations, it is incumbent you select appropriate primary limits for these stand-alone liability policies so that your public water and wastewater system is adequately protected and your governing body is apprised of which primary policies qualify as an underlying scheduled policy for an excess liability policy and which ones do not.

FINAL COMMENTS

The sole purpose of excess liability policies is to offer higher limits should a covered claim, or the demand for damages from a covered claim, exceed the primary limit in a scheduled underlying policy. An excess liability policy requires the same level of assiduous review as your other policies. This review is not complex and necessitates confirmation that the excess liability policy will honor any non-sublimitted liability claim that is paid by the scheduled underlying carrier should said claim exceed the primary limit. It is also advisable to use the same insurer as the scheduled underlying policies for CGL, POML, and BAC. The approach safeguards continuity of coverage as well as consistency of defense attorneys and legal strategies. Equally important, it ensures seamless alignment between your primary and excess liability insurers. That point is especially important when you are dealing with a catastrophic loss or a loss involving a large demand.

You should also remain cognizant that a following form excess liability policy is independent of your scheduled underlying policies. That means any contractual liability requirements like additional insured, primary and noncontributory verbiage, or waiver of subrogation that is included in your CGL policy should also be endorsed on your excess liability policy. Lastly, you must remain cognizant of any stand-alone liability policies that are not eligible as a scheduled underlying policy. In these situations, you should carefully review and select an appropriate primary limit for such stand-alone liability policies. The aforementioned roadmap will ensure a properly designed excess liability policy where higher limits properly integrate with the coverages afforded in the underlying scheduled policies.

We now turn our attention to pollution exposures that transcend best-in-class CGL and property policies. These exposures should not be assumed but rather transferred to an environmental pollution insurer providing comprehensive first- and third-party solutions specifically designed for public water/wastewater systems.

ENVIRONMENTAL POLLUTION POLICY*

OVERVIEW

Public water/wastewater systems have significant but often overlooked pollution exposures. The conventional wisdom has been to focus mostly on pollutants within potable water that could cause bodily

* Source material provided by various online articles from International Risk Management Institute (IRMI), https://www. irmi.com/articles/expert-commentary/insurance as well as policy forms and underwriting products promulgated by Insurance Services Office (ISO) and their member companies, https://products.iso.com/productcenter/10. Additional material provided through United States Environmental Protection Agency (USEPA), https://www.epa.gov/.

injury to your consumers as well as property damage arising from wastewater line breaks and back-ups. These exposures are noteworthy and a best-in-class CGL policy tailored specifically for public water/wastewater systems should afford coverage. There is similarly a rightful focus on the release of water and wastewater chemicals as well as other pollutants into the air or ground during the course of your operations, which could cause bodily injury or property damage to third parties. Again, a best-in-class CGL policy should provide coverage for these types of claims as long as the release is sudden and accidental and the loss involves bodily injury or property damage. Some public water/wastewater systems apply pesticide and herbicide chemicals around their intake structures and distribution infrastructure to mitigate pestilence and overgrowth. An overspray or wind drift claim can easily occur during the course of said operations, and, once more, a best-in-class CGL policy should pick up this exposure for losses involving bodily injury or property damage as long as you complied with all regulatory requirements and standards in the use and application of such chemicals.

Regarding first-party losses, a best-in-class property policy tailored specifically for public water/wastewater systems should have a sublimit, not less than $100,000, for any pollution remediation expenses associated with pollutant releases on described premises that do not involve an underground petroleum tank and are reported within a set time period, usually six months. This coverage should apply to all pollutant releases from the following sources: raw untreated, treated, and potable water and wastewater effluent; water and wastewater treatment chemicals; process tanks; aboveground petroleum tanks; biosolids, by-products or brine storage; and piping infrastructure. Covered causes of loss should be all risk for pollution remediation expenses except for the standardized exclusions in the property policy. Underground petroleum tanks, which require a special financial responsibility pollution policy or enrollment in a state tank fund insurance program to comply with state and federal requirements, are specifically excluded as a covered pollution remediation expense.

Based on probable coverage of the foregoing exposures within best-in-class CGL and property policies, there is a common misperception that a stand-alone environmental pollution policy is not necessary for public water/wastewater systems. That understanding is fundamentally incorrect and can have devastating financial consequences. There are several salient exposures facing public water/wastewater systems that are not addressed or inadequately addressed in best-in-class CGL and property policies. For instance, civil fines and penalties as well as natural resources damage and off-premises cleanup costs are all commonly excluded in a best-in-class CGL policy. Similarly, a best-in-class property policy will routinely exclude civil fines and penalties and only cover the cleanup at the described premise. This cleanup coverage, however, is frequently capped at a sublimit intended to cover the smallest of spills. Equally important, POML policies universally exclude claims involving the failure to maintain or secure insurance. Your governing body's decision to not purchase an environmental pollution policy could subsequently result in two uncovered claims from one incident: (1) damages associated with the pollution condition; and (2) damages associated with a wrongful act for not securing such a policy via the failure to maintain or procure insurance exclusion.

The combination of these first- and third-party restrictions necessitate serious contemplation of a proprietary environmental pollution policy. The purchase of such a policy, however, is not a guarantee that you are addressing the coverage gaps within your CGL and property policies. You must buy the right environmental pollution policy, and that requires a policy tailored specifically for public water/wastewater systems. To do so, you should thoroughly review the pollution exclusions and the respective exceptions in your CGL and property policies. A coverage encapsulation provides the necessary platform to begin your evaluation of environmental pollution policies. The discussion that follows will identify your most pronounced pollution exposures and provide a roadmap to ensure your environmental pollution policy envelops these risks. A checklist will be included at the end of the chapter to ensure this information is streamlined and easy to apply.

Exposure

Public water/wastewater systems are subject to strict pollution laws and requirements. The Federal Water Pollution Control Act, hereinafter referred to as the Clean Water Act (CWA), mandates strict legal conformance whenever public water/wastewater systems discharge any pollutant or combination of pollutants to surface waters that are deemed waters of the United States. This conformance, with certain exceptions, is regulated by a National Pollutant Discharge Elimination System (NPDES) permit. The CWA is administered by the Environmental Protection Agency (EPA), which frequently grants their state regulatory counterparts authority to issue such permits. As previously mentioned, these permits are required for any public water and wastewater system that discharge pollutants from their water delivery system or wastewater treatment facility into waters of the United States. Discharges are normal and customary, and most discharges ultimately flow into waters under the jurisdiction of the CWA. The regulatory definition of pollutant is broad enough to include any chlorine residual or influent in a discharge.

Water discharges are a core part of providing safe and reliable drinking water to the public. Drinking water system releases are classified as planned and unplanned. Planned releases typically result from routine operation and maintenance activities like disinfection of mains, testing of hydrants, storage tank maintenance, cleaning and lining a section of pipe, and routine flushing of distribution systems for maintenance. The volume, flow, duration, and potential pollutants of concern vary with each type of activity and the source of the release. Planned releases may involve drinking water, raw water, groundwater, or low volume drinking water system releases. In general, planned releases are easier to control because Best Management Practices (BMPs) can be prepared and implemented in advance. Unplanned releases are the result of accidents or incidents that cannot be scheduled or planned for in advance. Examples include water main breaks, leaks, overflows, fire hydrant shearing, and emergency flushing activities. Unplanned releases associated with wastewater can lead to even more catastrophic damage, especially if the release involves untreated influent.

The CWA has four enforcement mechanisms that can impose significant fines and penalties on public water/wastewater systems whose releases result in natural resources damage and environmental degradation as well as aquatic and nonaquatic habitat loss: (1) federal-enforcement by EPA through civil and criminal actions; (2) state-enforcement by the EPA's state counterpart through criminal and civil actions; (3) local-enforcement and further regulation by local municipalities; and (4) citizen-enforcement by the private citizen and groups pursuant to enforcement section of the CWA. To confuse matters, the Endangered Species Act (ESA) and Natural Resource Damage Act (NRD) are distinct from the CWA and provide separate jurisdictional and enforcement powers to federal and state fish and wildlife agencies as well as the federal department of interior. ESA and NRD are empowered to assess fees and assessments for resource damages such as fish and habitat kills, destruction of wetlands, and native habitat destruction.

NRD is acutely focused on compliance, protection, and cleanup rather than impacted resource replacement, whether by monetary assessment or replacement of resource in kind. Once such NRD claim involved a 5-gallon container of motor oil accidentally displaced into a river. The resulting oil sheen was quickly cleaned up and reported by the responsible party. The respective state regulatory agency imposed no fine, but the NRD leveed a $3 million fine two years after the incident alleging one dozen fish may have been killed by the motor oil. This example underscores the significance and complexity of overlapping and often conflicting enforcement criteria from three distinct regulatory agencies; each of which have the ability to levy criminal and civil fines and penalties against your public water and wastewater system.

Drinking water system releases, planned and unplanned, are a potential concern for enforcement actions because these releases frequently contain constituents or cause conditions that pose a threat to land habitat as well as freshwater and saltwater aquatic life. Chlorine at or above certain concentrations in the receiving water is known to be toxic to land and aquatic life. In a similar way, discharges of sediment and debris as part of a drinking water system release have the potential to impact downstream water quality. There are many additional exposures that could be accentuated beyond planned and unplanned releases and the resulting financial impact of regulatory enforcement, but this particular exposure aptly highlights a material risk that is not covered by your CGL and property policies. Its importance is further underscored by fish kills being deemed a class "A" violation and subject to significant fines and penalties. Releases associated with untreated influent are even more problematic. Such releases invariably lead to large civil fines and penalties as well as costly cleanup orders from the respective regulatory enforcement agency. We will now delve into the structure of a best-in-class environmental pollution policy tailored specifically for public water/wastewater systems.

STRUCTURE

The central purpose of a properly designed environmental pollution policy is to fill the insurance coverage gaps created by pollution exclusions in your CGL and property policies. Because pollution exclusions vary a great deal in the aforementioned policies, environmental insurance policies vary substantially as well. Therefore, when evaluating environmental pollution policies, it is necessary to perform a coverage review of the actual policy forms and endorsements while also referring back to the pollution exclusions and exceptions in your CGL and property policies. You should gravitate toward an environmental pollution policy that is specifically designed for public water/wastewater systems. These policies should not be evaluated solely on premium, deductible, and limits. The proper measurement should be coverage breadth and coverage customization, as many such policies are not structured to accommodate your unique pollution exposures.

Environmental pollution policies have specific mechanics that should be understood. These policies are issued on a claims-made basis, which means coverage applies to pollution incidents that commence and are reported on or after the retroactive date and during the policy period. The reporting provision can be augmented through the purchase of an extended reporting period (ERP) should your policy be cancelled, nonrenewed, or willingly lapsed. An ERP should be pre-negotiated and comprise a minimum time period of three years at a cost not to exceed 200% of the policy's annualized premium. Defense costs are frequently inside the limits, which necessitates a minimum limit of $1 million and a minimum provision that adds back at least 25% of the policy limits for defense costs should your limit be exhausted.

Pollution incidents should be defined as broadly as possible and include some likeness to the following verbiage:

> discharge, dispersal, release, seepage, or escape of any solid, liquid, gaseous or thermal irritant or contaminant, including but not limited to, smoke, vapors, soot, fumes, acids, alkalis, toxic chemicals, hazardous substances, petroleum hydrocarbons, low level radioactive materials, medical waste & waste materials.

Some definitions may also cite pathological materials, methamphetamines, or red-bagged materials. The inclusion of fungus and legionella is also required and may be included in the definition of pollutant or separately defined. An expansive definition is needed to encompass unintended dumping of pollutants on your described premises or unanticipated pollutants within your products and by-products, including sludge and biosolids. Products liability is particularly problematic and requires a clear understanding of the corresponding level of coverage afforded by both your CGL and environmental pollution policies.

Your discovery trigger for first-party losses should include cleanup costs arising from pollution incidents, whereas your demand trigger for third-party claims should include alleged bodily injury, property damage, off-site cleanup costs, and natural resources damage arising from pollution incidents. First- and third-party coverage triggers should encompass gradual pollution incidents as well as sudden and accidental and time-limited releases. Civil fines and penalties should also be embedded in the policy. To avoid unintended coverage gaps, the definitions for bodily injury and property damage should dovetail with your CGL and property definitions. Cleanup costs should be activated on environmental laws, regulatory edict, or advice from an independent engineer. This approach mitigates uncertainty and provides maximum flexibility as to the activation of first- and third-party cleanup costs immediately after a pollution incident.

A properly structured environmental pollution policy will include coverage for all described premises, with a blanket description encompassing your infrastructure piping and distribution-treatment fixed equipment. Coverage should include first-party cleanup costs, business interruption, loss of rents, and extra expenses for any pollution incidents occurring on described premises. Any third-party liability associated with the migration of pollution from the confines of described premises to other real property should similarly be covered. First- and third-party coverage should similarly include contracting activities on described and off-site premises. Such coverage should extend to emergency pollution incidents while performing repairs on piping infrastructure and distribution-treatment fixed equipment. Contracting operations should also include operations and maintenance (O&M) performed for third parties and defined as any activity that is usual, customary, and expected from a public water and wastewater system. The codification of infrastructure piping and distribution-treatment fixed equipment as described premises, coupled with a broad definition of contracting operations, will maximize your first- and third-party coverage triggers.

Coverage must extend beyond described premises and include third-party liability for non-owned locations as well as transportation of pollutants from a scheduled location to a non-owned location. Both coverages are essential, as there is strict liability under the Comprehensive Environmental Response, Compensation, and Liability Act (CERCLA) should a public water and wastewater system dispose of a by-product to an authorized waste disposal facility and that facility subsequently becomes bankrupt. Even if you did everything right, you can be sued, and you will be subject to strict liability by the federal government. The transportation of by-products is also important, as coverage for the loading and unloading of said by-products is not covered under a best-in-class auto liability policy with the *Pollution Liability – Broadened Coverage* endorsement. This endorsement will only cover upset and overturn of pollutants from its container on a covered auto that is in route to its final destination. It is also important that the definition of mold and fungus include all bacteria and not just legionella bacteria. Larger deductibles, incidentally, are commonly applied to environmental pollution policies due to the high severity nature of these claims. A deductible above $25,000 should be avoided unless it is being used as a means to reduce premium.

Blanket additional insured and waiver of subrogation endorsements, if required by written agreement, should be stipulated in the definition of insured contract. The same applies to easements and right of way agreements. First party cleanup costs should include natural resources damage, civil fines and penalties, and actual cash value damage to your real property and business personal property. First party business interruption is also available under a separate insuring agreement. Third-party liability should comprise actual cash value for damage of real property and business personal property owned by others, including loss of use and business interruption, as well as natural resources damage, cleanup costs, and civil fines and penalties. Important policy enhancements comprise the following: (a) 24/7 emergency spill response hotline answered by experts who provide oversight and assistance with environmental remediation contractor selection; (b) expert negotiation and representation on your behalf with federal, state, and local regulators, including fine and penalty negotiation as well as legal and engineering services; (c) crisis management services including public and media relations; and (d) green standards to upgrade damaged real property to conform with leadership in energy and environmental design guidelines.

An environmental pollution policy is not a panacea for all pollution claims. There are naturally exclusions that apply even with a best-in-class policy. The most common exclusions include the following: (a) operational noncompliance of NPDES or MS4 permits resulting from inadequate upgrades and improvements; (b) violations of deed restrictions, institutional, and engineered controls; (c) criminal fines and penalties; (d) odors originating from described premises that are not sudden and accidental; (e) landfill materials permanently stored at scheduled insured sites; (f) pollutants contained within infrastructure or treatment apparatus (there must be a discharge or release); (g) pollution incidents that commenced before the policy's retroactive date; (h) pollution incidents reported after the policy's expiration date or extended reporting period (ERP); (i) violation of an activity or use limitation (AUL); (j) communicable disease transmission; and (k) intentional, willful, or deliberate noncompliance of a permit or order. Each of the foregoing exclusions are acceptable and reasonable but require cognizance so that your public water and wastewater system can implement countermeasures to mitigate such claims.

Prior to purchasing an environmental pollution policy, it is important to confirm your most pronounced pollution exposures are covered. Written confirmation should include the following loss scenarios and comprise bodily injury, property damage, first- and third-party cleanup costs, natural resources damage, and civil fines and penalties arising from pollution incidents encompassing: (a) rupture or leaking of wastewater piping; (b) discharge of chlorinated water from a broken main or scheduled maintenance; (c) seepage of biosolids or sludge from an insured site into groundwater; (d) seepage of pollutants from an insured site that contaminates adjacent property, waterways, or aquifers; (e) illicit dumping or abandonment of hazardous substances onto described premises; (f) CERCLA liability from a previously authorized waste disposal; (g) overflow, seepage, or escape of pollutants from an agricultural or brine line; (h) contamination from agricultural smudging, fertilizer, or compost of converted by-product, including from perfluorooctanoic acid (PFOA); (i) inadvertent disturbance of asbestos from lined piping, gaskets, and insulation as well as lead-based paint; (j) civil fines and penalties associated with fish kills from state wildlife and/or state or federal regulatory agency; (k) runoff from construction activities; and (l) violations of NPDES or MS4 permits from fortuitous events. The issue of product liability requires additional commentary, as many environmental pollution policies exclude products due to emerging and expanding litigation associated with PFOAs. This exposure must be clearly codified and properly correlated with the products liability coverage in your CGL policy. A best-in-class CGL policy will cover products liability involving potable water as well as influent and effluent that causes third party bodily injury or property damage. Your environmental pollution policy should augment this coverage by including property damage arising from unintended pollutants within your products. The singular exception may be property damage or cleanup costs arising from PFOAs. That exposure is commonly excluded, thus making risk transfer via insurance not viable for many organizations.

FINAL COMMENTS

There is a substantive need for all public water/wastewater systems to contemplate the purchase of an environmental pollution policy. However, such a purchase should incorporate the necessary coverages addressing the salient exposures facing your organization. The reality is public water/wastewater systems face significant regulation associated with pollutants that move into and out of their infrastructure and operations. Civil fines and penalties are a near certainty should a planned or unplanned release of chlorinated water or influent result in fish or aquatic life kill. Moreover, public water/wastewater systems face substantial exposure to midnight dumping, CERCLA liability from defunct waste disposal facilities, transportation of by-products, fungus and mold bacteria, seepage and gradual deterioration of underground aquifers, and disruption of asbestos-containing material.

The vast majority of best-in-class CGL and property policies will not cover the foregoing exposures with the possible exception of an ancillary sublimit for fungus and mold cleanup on described premises. The need for a combination first- and third-party environmental pollution policy is critical to effectively manage this risk and transfer the exposure to a specialty insurer. The claim examples highlight the exposures, and pollution losses are increasing in both frequency and severity. The civil fines and penalties are similarly expanding and becoming more disruptive. The fact that enforcement involves a compilation of federal, state, and local regulatory agencies, including conflicting interpretation from differing regulatory agencies as well as concerned citizens and activist groups, necessitates the need to transfer this risk and not assume it. Such totality of knowledge will allow you to properly design a meaningful environmental pollution policy while also communicating the starting and stopping points of coverage so that your governing body can prudently authorize appropriate policies and procedures.

POLICY CHECKLIST: PART I

Cyber Liability and Network Security Policy: Minimum Benchmarks

Cyber Liability and Network Security	
Description	**Benchmark**
Form	
Claims-Made	Yes
Proprietary	Yes
Duty to Defend	Yes
Defense Inside the Limits	Yes
Retroactive Date	Yes (secure 1–3 years prior to policy inception)
Limits	
Coverage A.	$1,000,000
Coverage B.	$1,000,000
Coverage C.	$250,000
Coverage D.	$250,000
Policy Aggregate	$2,000,000
Coverage A.: Privacy Liability	Protects against any actual or alleged act, error, misstatement, misleading statement, omission, neglect, or breach of duty committed by an insured or third party contractor, which results in a breach of the insured's network security, the consequences of which are: Unauthorized access to, use of or tampering with a third party's network; The inability of an authorized third party to gain access to the insured's services; Denial or disruption of internet service to an authorized third party; Identity theft or credit/debit card fraud; The transmission of malicious code; or The unauthorized release of a third party's confidential business.
Coverage B.: Breach Consultation Services	Protects against any reasonable and necessary costs incurred by or on behalf of the insured to: Determine the applicability of, and the insured's obligation to comply with, any breach notification law; Draft a notification letter to be sent to any affected individual required to be notified by the insured; Retain a qualified forensics firm from the attached schedule of services to investigate, examine, and analyze the insured's network to determine the cause and source of the unauthorized misappropriation or disclosure of personally identifiable information and the extent to which such personally identifiable information was accessed; Retain a qualified public relations firm, crisis management firm, or law firm from the attached schedule of services to minimize potential harm arising from a public relations event; and Retain a qualified service provider from the attached schedule of services to provide breach response services.
Coverage C.: Breach Response Services	Provides reimbursement to the named insured for the costs incurred by the named insured for notification to, and for credit monitoring of, any persons residing in the United States, including employees, arising from a privacy wrongful act, which takes place during the policy period.
Coverage D.: Public Relations and Data-Forensic Expense	Covers the reasonable and necessary costs incurred by the named insured to retain a qualified public relations firm to mitigate brand impairment and ill will as well as a forensics firm to investigate, examine, and analyze the named insured's network, to find the cause, source, and extent of a data breach.
Policy Features	
Defense Obligation for Regulatory Actions	Yes

(*Continued*)

Cyber Liability and Network Security

Description	Benchmark
Acts and Omissions of Third Parties	Yes
Call Centers (no minimum)	Yes
Civil Investigative Demands	Yes
Corporations and Others Entities	Yes
Cyber Extortion	Yes
Electronic and Nonelectronic Data	Yes
Failure to Disclose Breach	Yes
Network/Business Interruption	Yes
No Generalized Exclusions	Yes
Policy Trigger – Loss	Yes
Unencrypted Devices	Yes
S.I.T.E. Endorsement	Yes
Deductible	$2,500 (recommended but higher deductibles prudent for larger public water/ wastewater systems)

POLICY CHECKLIST: PART II

Excess Liability Policy: Minimum Benchmarks

Excess Liability

Description	Benchmark
Form	
Proprietary	Yes
Following Form	Yes
Occurrence and Claims Made	Yes (follows underlying)
Defense Costs Outside	Yes (follows underlying)
Limits	
Per Occurrence	$10,000,000
Annual Aggregate	$10,000,000
Underlying Coverages	
Commercial General Liability	Yes
Public Officials and Management Liability	Yes
Business Auto Coverage	Yes
Employers' Liability	Yes
Policy Highlights	
Following Form	Yes (exceptions apply to first-party coverages, including medical payments and uninsured/underinsured motorists, as well as any coverages with sublimits)
Nonauditable	Yes
Waiver of Subrogation including Primary and Noncontributory Verbiage	Yes

(*Continued*)

Excess Liability	
Description	**Benchmark**
Following Form	
Asbestos (potable water)	Yes
Dam, Levee and Dike Collapse and Failure	Yes
Employee Benefit Plans	Yes
Employment Practices	Yes
Failure to Supply	Yes
Inverse Condemnation	Yes
Lead (potable water)	Yes
Outside Directorship	Yes
Pollution Liability	Yes
Third-Party Discrimination	Yes
Waterborne Asbestos	Yes
Watercraft	Yes
Wrongful Acts	Yes

POLICY CHECKLIST: PART III

Environmental Pollution Policy: Minimum Benchmarks

Environmental Pollution Policy	
Description	**Benchmark**
Form	
Claims Made	Yes
Tailored to Water-Related Entities	Yes (proprietary)
Duty to Defend	Yes
Defense Costs	Inside (plus additional 25% of limits)
Coverages	
Pollution Liability for Your Insured Site(s)	Yes
Pollution Liability for Your Off-site Activities	Yes
Environmental Pollution Policy	
Inclusive of First-Party Cleanup Costs	Yes
Limits	
Per Occurrence	$1,000,000
Annual Aggregate	$1,000,000
Sublimits	
Business Interruption and Extra Expense	$1,000,000
Fungus/Legionella	$500,000
Emergency Remediation Expense/Job Site	$250,000
Crisis Management Event	$250,000
Green Standards	$250,000

(Continued)

Environmental Pollution Policy

Description	Benchmark
Deductible	$25,000 (higher deductibles prudent for larger public water/ wastewater systems)
Retroactive Date	Continuity

Covered Causes of Loss	
Sudden and Accidental	Yes
Gradual Seepage and Deterioration	Yes
Time Release	Yes

Coverage Extensions	
Cleanup Costs (first- and third-party)	Yes
Civil Fines and Penalties	Yes
Natural Resources Damage	Yes
Bodily Injury including Medical Monitoring Costs	Yes
Midnight Dumping	Yes
Unintended Asbestos Disturbance	Yes
Unintended Lead Paint Disturbance	Yes
Transportation and Hauling of Cargo	Yes
Contracting Pollution Liability	Yes
Emergency Cleanup Costs	Yes
Non-owned Waste Disposal Sites	Yes
Temporary On-site Storage of Biosolids and Sludge	Yes
Off-site and On-site Spreading of Biosolids	Yes
CERCLA Liability	Yes
Contractual Liability	Yes
Unburied Storage Tanks and Vessels	Yes
Terrorism	Yes

Policy Features	
Broad Definition of Insured	Yes
Blanket Additional Insured	Yes
Blanket Waiver of Subrogation	Yes
Blanket Non-owned Waste Disposal Sites	Yes
Definition of Cleanup Costs includes Advice by Environmental Professionals	Yes
24/7 Emergency Spill Response Support Hotline including Response Oversight	Yes
90-Day Automatic Extended Reporting Period	Yes
3 Year Supplemental Extended Reporting Period	Yes
No Policy Scheduling of Contracting Operations, Transportation Activities, or Waste Disposal Facilities	Yes
Insured Sites to include All Piping Infrastructure and All Real Property Locations Referenced on Application	Yes
Non-auditable Premium	Yes
First- and Third-Party Coverage Triggers	Combination first-party discovery coverage trigger for cleanup costs along with a third-party demand trigger for claims alleging bodily injury, property damage, cleanup costs, and natural resources damage arising from new pollution incidents from policy inception or from the retroactive date if coverage has been continuous through time.

(Continued)

Environmental Pollution Policy

Description	Benchmark
Pollutant Incident Definition	Definition to encompass the discharge, dispersal, release, seepage, or escape of any solid, liquid, gaseous or thermal irritant or contaminant, including but not limited to, smoke, vapors, soot, fumes, acids, alkalis, toxic chemicals, hazardous substances, petroleum hydrocarbons, low level radioactive materials, medical waste, and waste materials. Definition also includes a sublimit for fungus and legionella.

Acceptable Exclusions

Non-disclosed Locations	Yes
Noncompliance Exclusion	Yes
Products Liability	Yes (requires correlation w/ CGL policy)
Criminal Fines, Penalties, and Assessments	Yes
Communicable Diseases	Yes
Material Change in Use or Operation	Yes
Insured vs. Insured	Yes

Policy Conditions

Claims Reporting	As soon as practicable
Notice of Cancellation	90 days
Other Insurance – Primary	Yes (if no valid and collectible insurance)
Declarations and Representations Condition	Yes
Jurisdiction and Venue and Choice of Law	Removed

REVIEW QUESTIONS

1. T/F: Public water/wastewater systems are susceptible to unauthorized access of their confidential ratepayer and employee information.

2. Privacy liability coverage in a cyber liability and network security policy should include the following coverages:
   a. Liability arising from a data breach that discloses confidential information of your ratepayers and employees
   b. Defense obligations for regulatory and other governmental actions as well as formal suits
   c. Civil fines and penalties
   d. All of the above

3. T/F: Credit and identity monitoring coverage in a cyber liability and network security policy should include a minimum of three years of monitoring services to any party impacted by a data breach.

4. T/F: Transmission of viruses and malicious code coverage in a cyber liability and network security policy protects against liability claims alleging damages from transmission of viruses and other malicious code or data.

5. T/F: An insured is better protected with a claim policy trigger and not a loss policy trigger when it comes to a cyber liability and network security policy.

6. Exclusions to avoid in a cyber liability and network security policy include the following:
   a. Acts and omissions of third parties
   b. Unencrypted devices

    c. a and b only

    d. None of the above

7. Coverage in a cyber liability and network security policy should include the following:

    a. Breaches not related to electronic records

    b. Any location of security failure

    c. a and b

    d. None above

8. A cyber liability and network security policy should not include any ambiguous and overarching exclusions including the following:

    a. Shortcomings in software security of which you were aware prior to the inception of coverage

    b. Your failure to take reasonable steps to design, maintain, and upgrade software security

    c. Certain failures of security software

    d. All of the above

9. T/F: The Occupational Safety & Health Administration (OSHA) is empowered to enforce minimum workplace safety standards.

10. T/F: The federal government and not the states regulate the compulsory purchase by public water/wastewater systems of workers' compensation policies.

11. Workers' compensation policies are priced on the following methodology:

    a. Classification of work performed by your employees

    b. Your total payroll

    c. Your overall loss experience

    d. All of the above

12. T/F: Workers' compensation laws originated to establish compensation for injured employees while simultaneously reducing the amount of litigation against employers when employees are injured on the job.

13. The most common workers' compensation classifications for public water/wastewater systems are the following:

    a. Waterworks operations (7520)

    b. Salespersons (8742)

    c. Clerical (8810)

    d. All of the above

14. The premium for workers' compensation coverage includes many factors including the following:

    a. Classifications used

    b. Rates charged

    c. Total payroll

    d. All of the above

15. Examples of monopolistic workers' compensation states include the following:

    a. North Dakota

    b. Ohio

    c. West Virginia

    d. a and b only

16. State workers' compensation laws typically provide the following types of benefits:

    a. Medical

    b. Disability

    c. Vocational rehabilitation

    d. All of the above

17. T/F: While the types of benefits employees receive for work-related injuries are fairly consistent from one state to another, the amount of prescribed workers' compensation benefits vary widely from state to state.

18. T/F: The applicable workers' compensation law of your state is incorporated into the policy.

19. An injured employee in California typically has one year to file a workers' compensation claim, but regulators can extend that timeline under the following circumstances:
   a. If the employee is under 18 at the time of injury, then the one-year statute of limitations begins when the person becomes a legal adult
   b. If there is a repetitive stress injury, then the employee may file a claim up to one year from the date that he or she became aware of the injury
   c. a and b only
   d. None of the above

20. The four most common types of employers' liability claims are as follows:
   a. Third-party-over actions
   b. Loss of consortium
   c. Dual capacity suits
   d. All of the above

21. Examples of workers' compensation risk assessments include the following:
   a. Manual material handling
   b. Machinery and equipment operations
   c. Falls
   d. All of the above

22. T/F: Statistics show that employees who return to work generally recover less quickly because they remain inactive, unproductive, and not connected to their workplace.

23. T/F: Limits are more important than exclusions in an excess liability policy.

24. T/F: Excess liability policies are independent policies subject to their own insuring agreements, definitions, conditions, and exclusions.

25. T/F: The central purpose of excess liability policies is to offer higher limits should a covered claim, or the demand for damages from a covered claim, exceed the primary limit in a scheduled underlying policy.

26. Examples of salient pollution exposures not addressed or inadequately addressed in best-in-class commercial general liability (CGL) and property policies include the following:
   a. Civil fines and penalties
   b. Natural resources damage
   c. None of the above
   d. All of the above

27. T/F: The Clean Water Act (CWA) mandates strict legal conformance whenever public water/wastewater systems discharge any pollutant or combination of pollutants to surface waters that are deemed waters of the United States.

28. T/F: The Clean Water Act (CWA) is administered by the Environmental Protection Agency (EPA), which frequently grants its state regulatory counterparts authority to issue National Pollutant Discharge Elimination System (NPDES) permits.

29. T/F: Water discharges from a water delivery system are a core part of providing safe and reliable drinking water to the public.

30. T/F: Water delivery system releases are classified as planned and unplanned.

31. Example of planned water releases include the following:
   a. Disinfection of mains
   b. Testing of hydrants

c. Storage tank maintenance

d. All of the above

32. T/F: Best Management Practices (BMPs) can be prepared and implemented in advance for planned water releases.

33. Examples of unplanned water releases include the following:

a. Water main breaks, leaks, and overflows

b. Fire hydrant shearing

c. Scheduled flushing activities

d. a and b only

34. The Clean Water Act (CWA) enforcement mechanisms include the following:

a. Federal enforcement

b. Citizen enforcement

c. State enforcement

d. All of the above

35. T/F: Endangered Species Act (ESA) is distinct from the Clean Water Act (CWA) and provides separate jurisdiction and enforcement powers to federal and state fish and wildlife agencies.

36. T/F: Endangered Species Act (ESA) is part of the Clean Water Act (CWA).

37. Components of a best-in-class environmental pollution policy include the following:

a. Blanket description of all infrastructure and distribution piping as described premises

b. Coverage for first-party cleanup costs, business interruption, loss of rents, and extra expense

c. Coverage for third-party liability associated with the migration of a pollutant from the confines of described premises

d. All of the above

38. Pollution exposures extending beyond described premises include the following:

a. Transportation of pollutants from described premises to non-owned locations

b. Authorized waste disposal facilities not situated on described premises

c. Agricultural smudging and spreading of biosolids on described premises

d. a and b only

39. Special enhancements in a best-in-class environmental pollution policy include the following:

a. 24/7 emergency spill response hotline answered by experts who provide oversight and assistance with environmental remediation contractor selection

b. Expert negotiation and representation on your behalf with federal, state, and local regulators, including fine and penalty negotiation as well as legal and engineering services

c. Crisis management services including public and media relations

d. All of the above

40. T/F: Environmental pollution policies are traditionally written on an occurrence form with defense outside the limit of insurance.

ANSWER KEY

1. **T**/F: Public water/wastewater systems are susceptible to unauthorized access of their confidential ratepayer and employee information.

2. Privacy liability coverage in a cyber liability and network security policy should include the following coverages:

a. Liability arising from a data breach that discloses confidential information of your ratepayers and employees

b. Defense obligations for regulatory and other governmental actions as well as formal suits

c. Civil fines and penalties

d. **All of the above**

3. **T/F:** Credit and identity monitoring coverage in a cyber liability and network security policy should include a minimum of three years of monitoring services to any party impacted by a data breach.

4. **T/F:** Transmission of viruses and malicious code coverage in a cyber liability and network security policy protects against liability claims alleging damages from transmission of viruses and other malicious code or data.

5. T/**F:** An insured is better protected with a claim policy trigger and not a loss policy trigger when it comes to a cyber liability and network security policy.

6. Exclusions to avoid in a cyber liability and network security policy include the following:
   a. Acts and omissions of third parties
   b. Unencrypted devices
   c. **a and b only**
   d. None of the above

7. Coverage in a cyber liability and network security policy should include the following:
   a. Breaches not related to electronic records
   b. Any location of security failure
   c. **a and b**
   d. None above

8. A cyber liability and network security policy should not include any ambiguous and overarching exclusions including the following:
   a. Shortcomings in software security of which you were aware prior to the inception of coverage
   b. Your failure to take reasonable steps to design, maintain, and upgrade software security
   c. Certain failures of security software
   d. **All of the above**

9. **T/F:** The Occupational Safety & Health Administration (OSHA) is empowered to enforce minimum workplace safety standards.

10. T/**F:** The federal government and not the states regulate the compulsory purchase by public water/wastewater systems of workers' compensation policies.

11. Workers' compensation policies are priced on the following methodology:
    a. Classification of work performed by your employees
    b. Your total payroll
    c. Your overall loss experience
    d. **All of the above**

12. **T/F:** Workers' compensation laws originated to establish compensation for injured employees while simultaneously reducing the amount of litigation against employers when employees are injured on the job.

13. The most common workers' compensation classifications for public water/wastewater systems are the following:
    a. Waterworks operations (7520)
    b. Salespersons (8742)
    c. Clerical (8810)
    d. **All of the above**

14. The premium for workers' compensation coverage includes many factors including the following:
    a. Classifications used
    b. Rates charged
    c. Total payroll
    d. **All of the above**

15. Examples of monopolistic workers' compensation states include the following:
    a. North Dakota
    b. Ohio
    c. West Virginia
    d. **a and b only**

16. State workers' compensation laws typically provide the following types of benefits:
    a. Medical
    b. Disability
    c. Vocational rehabilitation
    d. **All of the above**

17. **T**/F: While the types of benefits employees receive for work-related injuries are fairly consistent from one state to another, the amount of prescribed workers' compensation benefits vary widely from state to state.

18. **T**/F: The applicable workers' compensation law of your state is incorporated into the policy.

19. An injured employee in California typically has one year to file a workers' compensation claim, but regulators can extend that timeline under the following circumstances:
    a. If the employee is under 18 at the time of injury, then the one-year statute of limitations begins when the person becomes a legal adult
    b. If there is a repetitive stress injury, then the employee may file a claim up to one year from the date that he or she became aware of the injury
    c. **a and b only**
    d. None of the above

20. The four most common types of employers' liability claims are as follows:
    a. Third-party-over actions
    b. Loss of consortium
    c. Dual capacity suits
    d. **All of the above**

21. Examples of workers' compensation risk assessments include the following:
    a. Manual material handling
    b. Machinery and equipment operations
    c. Falls
    d. **All of the above**

22. T/**F**: Statistics show that employees who return to work generally recover less quickly because they remain inactive, unproductive, and not connected to their workplace.

23. T/**F**: Limits are more important than exclusions in an excess liability policy.

24. **T**/F: Excess liability policies are independent policies subject to their own insuring agreements, definitions, conditions, and exclusions.

25. **T**/F: The central purpose of excess liability policies is to offer higher limits should a covered claim, or the demand for damages from a covered claim, exceed the primary limit in a scheduled underlying policy.

26. Examples of salient pollution exposures not addressed or inadequately addressed in best-in-class CGL and property policies include the following:
    a. Civil fines and penalties
    b. Natural resources damage
    c. None of the above
    d. **All of the above**

27. **T**/F: The Clean Water Act (CWA) mandates strict legal conformance whenever public water/wastewater systems discharge any pollutant or combination of pollutants to surface waters that are deemed waters of the United States.

28. **T/F**: The Clean Water Act (CWA) is administered by the Environmental Protection Agency (EPA), which frequently grants its state regulatory counterparts authority to issue National Pollutant Discharge Elimination System (NPDES) permits.

29. **T/F**: Water discharges from a water delivery system are a core part of providing safe and reliable drinking water to the public.

30. **T/F**: Water delivery system releases are classified as planned and unplanned.

31. Example of planned water releases include the following:
    a. Disinfection of mains
    b. Testing of hydrants
    c. Storage tank maintenance
    d. **All of the above**

32. **T/F**: Best Management Practices (BMPs) can be prepared and implemented in advance for planned water releases.

33. Examples of unplanned water releases include the following:
    a. Water main breaks, leaks, and overflows
    b. Fire hydrant shearing
    c. Scheduled flushing activities
    d. **a and b only**

34. The Clean Water Act (CWA) enforcement mechanisms include the following:
    a. Federal enforcement
    b. Citizen enforcement
    c. State enforcement
    d. **All of the above**

35. **T/F**: Endangered Species Act (ESA) is distinct from the Clean Water Act (CWA) and provides separate jurisdiction and enforcement powers to federal and state fish and wildlife agencies.

36. **T/F**: Endangered Species Act (ESA) is part of the Clean Water Act (CWA).

37. Components of a best-in-class environmental pollution policy include the following:
    a. Blanket description of all infrastructure and distribution piping as described premises
    b. Coverage for first-party cleanup costs, business interruption, loss of rents, and extra expense
    c. Coverage for third-party liability associated with the migration of a pollutant from the confines of described premises
    d. **All of the above**

38. Pollution exposures extending beyond described premises include the following:
    a. Transportation of pollutants from described premises to non-owned locations
    b. Authorized waste disposal facilities not situated on described premises
    c. Agricultural smudging and spreading of biosolids on described premises
    d. **a and b only**

39. Special enhancements in a best-in-class environmental pollution policy include the following:
    a. 24/7 emergency spill response hotline answered by experts who provide oversight and assistance with environmental remediation contractor selection
    b. Expert negotiation and representation on your behalf with federal, state, and local regulators, including fine and penalty negotiation as well as legal and engineering services
    c. Crisis management services including public and media relations
    d. **All of the above**

40. **T/F**: Environmental pollution policies are traditionally written on an occurrence form with defense outside the limit of insurance.

Insurance Essentials for Public Water/Wastewater Systems: Conclusion

2019 Photo Courtesy of San Diego County Water Authority

CONCLUSION

Part IV – Insurance Essentials endeavors to provide meaningful and practical knowledge of insurance as it relates to public water and wastewater system operations and oversight. Although insurance is an indispensable and effective transfer tool in managing and mitigating risk, it is not a panacea. Value lies in the quality of policies you procure. We reviewed the salient property and liability policies for public water/wastewater systems and presented a granular review of their pertinent sections. The information imparted provides a methodical roadmap to properly transfer the distinct exposures facing your public water and wastewater system. The policy checklists are similarly intended to streamline this information so you can effectively design and structure your property and liability renewal policies. The addendum on inverse condemnation offers additional perspective

DOI: 10.1201/9781003207146-36

on an emerging legal theory that poses a contagion risk to public water/wastewater systems in states where their constitutional takings clause encompasses damage to private property. Both documents are meant to be reviewed with your governing body so that meaningful discussion can commence on procuring value over price when examining your insurance renewals.

In addition to a drill down on policy structure and exposure analysis, *Part IV – Insurance Essentials* reaffirms the importance of selecting the right insurer. This selection is critical because a long-term relationship with your insurer facilitates coverage continuity, premium stability, and claims fairness. Selection should be based on the insurer's subject matter expertise, proprietary coverages, length of service, premium competitiveness, claims handling philosophy, and post-policy services. Critical public water and wastewater system metrics to consider when evaluating your insurer include customer retention ratios, premium size, insured count, coverage fluency, defense panel experience, and trade group involvement. The utilization of an admitted insurer or a pool presents advantages and disadvantages that should be carefully evaluated. They both offer compelling attributes and necessitate serious contemplation, especially as it relates to joint and several liability and broker advocate utilization.

The last element, and arguably the most important, is the knowledge and support of your broker advocate. There is no substitute for a value-based broker advocate. This individual will be your guide when it comes to navigating the circuitous and complex path that represents the insurance industry. A broker advocate represents your public water and wastewater system to the marketplace and provides critical activism as it relates to negotiation of terms and coordination of post-policy services as well as timely and fair resolution of claims. The American Association of Water Distribution & Management (AAWD&M), its stakeholders, and the authors are available to assist you with questions you may have after reading *Part IV – Insurance Essentials*. Contact information has been provided and communication is welcomed.

Lorilee Medders, PhD, CPRO-W[2]
Joseph F. Freeman Distinguished Professor of Insurance
Walker College of Business
Appalachian State University
meddersla@appstate.edu
(828) 262-6234

Paul Fuller, CPCU, CPRO-W[2]
CEO, Allied Public Risk, LLC
Insurance Administrator, CalMutuals Joint Powers Risk & Insurance Management Authority
pfuller@alliedpublicrisk.com
(415) 205-8648

DISCLAIMER

The information contained in *Part IV – Insurance Essentials* is meant to provide the reader with a general understanding of certain aspects of insurance. The information isn't to be construed as insurance advice and isn't meant to be a substitute for insurance advice. AAWD&M, its stakeholders, and the authors expressly disclaim any and all liability with respect to any actions taken or not taken based upon the information contained in this section or with respect to any errors or omissions contained in such information. Readers are cautioned to thoroughly review their insurance policies with their broker advocates and risk bearers. The recommendations on coverage, limits, and various other policy terms may not be appropriate for your organization. Interpretation of policy language is the ultimate right of the risk bearer. The interpretations contained in *Part IV – Insurance Essentials* may be rejected or deemed inaccurate. All actual and hypothetical claims must be discussed and answered by your risk bearer and broker advocate.

ADDENDUM: INSURANCE COVERAGE RULES FOR INVERSE CONDEMNATION ACTIONS INVOLVING PUBLIC WATER SYSTEMS (*DEFENSE COUNSEL JOURNAL*/JULY 2019)

2019 Photo Courtesy of California Water Service Group

REVIEW QUESTIONS

1. T/F: Inverse condemnation is covered in a commercial general liability (CGL) policy.

2. T/F: Inverse condemnation is based on a constitutional violation of the takings clause.

3. T/F: An inverse condemnation action is based on property damage arising from the operations of a water delivery system.

4. Inverse condemnation actions involve the following types of inherent risks in a deliberately designed, constructed, or maintained public improvement:
   a. Design
   b. Construction
   c. a and b only
   d. None of the above

5. T/F. Proximate causation and not substantial cause and effect is the causation standard for inverse condemnation actions.

6. Inverse condemnation involves the following injuries:
   a. Physical injury to real property and personal property
   b. Nonphysical injury to real property and personal property
   c. Bodily injury to minors
   d. a and b only

7. Damages under inverse condemnation comprise the following:
   a. Economic injury
   b. Prevailing party's legal fees
   c. Prevailing party's appraisal fees
   d. All of the above

ANSWER KEY

1. **T/F**: Inverse condemnation is covered in a commercial general liability (CGL) policy.

2. **T/F**: Inverse condemnation is based on a constitutional violation of the takings clause.

3. **T/F**: An inverse condemnation action is based on property damage arising from the operations of a water delivery system.

4. Inverse condemnation actions involve the following types of inherent risks in a deliberately designed, constructed, or maintained public improvement:
   a. Design
   b. Construction
   c. a and b only
   d. **None of the above**

5. **T/F**: Proximate causation and not substantial cause and effect is the causation standard for inverse condemnation actions.

6. Inverse condemnation involves the following injuries:
   a. Physical injury to real property and personal property
   b. Nonphysical injury to real property and personal property
   c. Bodily injury to minors
   d. **a and b only**

7. Damages under inverse condemnation comprise the following:
   a. Economic injury
   b. Prevailing party's legal fees
   c. Prevailing party's appraisal fees
   d. **All of the above**

Index